AF615909

Biothiols in Health and Disease

ANITOXIDANTS IN HEALTH AND DISEASE

Series Editors

LESTER PACKER, PH.D.
University of California
Berkeley, California

JÜRGEN FUCHS, PH.D, M.D.
Johann Wolfgang Goethe University
Frankfurt, Germany

1. Vitamin A in Health and Disease, *edited by Rune Blomhoff*
2. Biothols in Health and Disease, *edited by Lester Packer and Enrique Cadenas*

Additional Volumes in Preparation

Handbook of Antioxidants, *edited by Enrique Cadenas and Lester Packer*

Related Volumes

Vitamin E in Health and Disease, *edited by Lester Packer and Jürgen Fuchs*

Biothiols in Health and Disease

edited by

Lester Packer

University of California
Berkeley, California

Enrique Cadenas

University of Southern California
School of Pharmacy
Los Angeles, California

Marcel Dekker, Inc. New York • Basel • Hong Kong

ISBN: 0-8247-9654-3

The publisher offers discounts on this book when ordered in bulk quantities. For more information, write to Special Sales/Professional Marketing at the address below.

This book is printed on acid-free paper.

Marcel Dekker, Inc.
270 Madison Avenue, New York, New York 10016

Current printing (last digit):
10 9 8 7 6 5 4 3 2 1

PRINTED IN THE UNITED STATES OF AMERICA

Series Introduction

In June of 1992, 17 international researchers in the field of free radical and antioxidant biology and preventive medicine met at the village of Saas Fee, Switzerland, and drew up the Saas Fee Declaration to recognize the importance of prevention in medicine and health. Since then, hundreds of researchers from around the world have signed the declaration:

Saas Fee Declaration

On the significance of antioxidants in preventive medicine.

1. The intensive research on free radicals of the past 15 years by scientists worldwide has led to the statement in 1992 that antioxidant nutrients may have major significance in the prevention of a number of diseases. These include cardiovascular and cerebrovascular disease, some forms of cancer and several other disorders, many of which may be age-related.
2. There is now general agreement that there is a need for further work at the fundamental scientific level, as well as in large-scale randomized trials and in clinical medicine, which can be expected to lead to more precise information being made available.
3. The major objective of this work is the prevention of disease. This may be achieved by use of antioxidants which are natural physiological substances. The strategy should be to achieve optimal intakes of these antioxidant nutrients as part of preventive medicine.
4. It is quite clear that many environmental sources of free radicals exist, such as ozone, sunlight, and other forms of radiation, smog, dust, and other atmospheric pol-

lutants. The optimal intake of antioxidants provides a preventive measure against these hazards.

5. There is a great need for improvement in public awareness of the potential preventive benefits of antioxidant nutrient intake. There is overwhelming evidence that the antioxidant nutrients such as vitamin E, vitamin C, carotenoids, alpha-lipoic acid and others are safe even at very high levels of intake.

6. Moreover, there is now substantial agreement that governmental agencies, health professionals and the media should promote information transfer to the general public, particularly when evidence exists that benefits for human health and public expenditure are overwhelming.

This declaration arose from the overwhelming evidence now available indicating that antioxidants play a critical role in wellness, health maintenance, and the prevention of chronic and degenerative diseases. Antioxidants neutralize free radicals that are generated during normal metabolism and during exposure to environmental insult. Free radicals play a role in most major health problems of the industrialized world, including cardiovascular disease, cancer, and disorders of aging.

Some antioxidants are quite familiar as vitamins or vitamin-forming compounds: vitamin E, vitamin C, and the carotenoids, including beta-carotene. These antioxidants must be constantly replenished through the diet. Others, such as ubiquinols and the thiol antioxidants, including glutathione and lipoic acid, are manufactured by the body, but the levels of many of these can be bolstered through dietary supplementation. Until recently, it was thought that each antioxidant played its role in isolation from the others. But work in several laboratories indicates that there is a dynamic interplay among the systems. For example, when vitamin E neutralizes a free radical in a membrane, it becomes itself a relatively harmless free radical, which decomposes. However, vitamin C can regenerate vitamin E from the vitamin E radical, in effect "recycling" vitamin E. Vitamin C becomes a radical in the process, but it, too, can be recycled by interacting with other antioxidant systems. It has been shown that these interactions occur in the test tube, and nutritional supplementation studies support this idea for the whole organism. Thus, a picture is emerging of a complex interplay among the defense systems, with the various antioxidant cycles acting to prevent cell damage and disease. Our knowledge is far from complete but these findings already have implications in terms of recommendations for supplementation.

Hence, it seems particularly appropriate to offer this series at the present time. Never has the demand for knowledge about antioxidants been greater, and never has their potential for treating disease and improving health been clearer. The series highlights natural antioxidants and artificial antioxidants that mimic natural systems.

In the past few years unprecedented progress has been made in the recognition and understanding of the role of thiol antioxidants in protection against disease and cell regulation, and in mechanisms where thiols and thiol enzymes protect against environmental factors that are sources of damage to living systems.

Thiols play a central role in the structure and activity of many vital systems. These include receptor activity, hormone action, oxidation and reduction of key cellular reductants, cell signaling, gene expression, and cell proliferation and differentiation. Thiols are also involved in enzyme activity, not only structurally but also catalytically as, for example, in the glutathione peroxidase gene superfamily. Further, thiols such as glutathione, thioredoxin, mono- and dithiols like cysteine or *N*-acetyl-cysteine and lipoic acid are important in antioxidant protection. Because of these numerous functions, nutritional approaches to modulate thiol status constitute a new and emerging field of research, in which the potential beneficial role of thiols in prevention and protection against acute and chronic diseases is being sought. This volume will therefore be of interest to investigators in numerous disciplines in which new knowledge in the field of thiols is pertinent, including biochemists and molecular biologists, nutritional scientists, environmental toxicologists, plant scientists, clinicians, and others in the biomedical community.

Lester Packer
Jürgen Fuchs

In memory of Alton Meister
1922–1995

Preface

Thiols participate in numerous cellular functions, such as biosynthetic pathways, detoxification by conjugation, and cell division. Studies on oxidative stress in recent years have amply documented the key role of thiols—more specifically, the thiol-disulfide status of the cell—in a wide array of biochemical and biological responses. Awareness of the great importance of biothiols in cellular oxidative injury has grown along with the recognition of free radicals in biological processes. The reactions of thiols with free radicals are of interest not only in free radical chemistry: the most abundant nonprotein thiol in the cell, glutathione, is essential for the detoxification of peroxides as cofactors of various selenium-dependent peroxidases. The high concentration of glutathione in cells clearly indicates its general importance in metabolic and oxidative detoxification processes. In many ways, glutathione may be considered the master antioxidant molecule, a phrase that Alton Meister, one of the pioneers in glutathione research and a contributor to this volume, has used. Bolstering of glutathione by other thiols, both natural (such as α-lipoic acid) and synthetic (such as Ebselen and several other drugs), has been investigated as a therapeutic approach to the oxidative component of various pathologies. Moreover, the redox changes of several thiol-containing proteins may be involved in key regulatory steps of the enzyme as well as cell proliferation.

The chapters in this book provide a comprehensive and in-depth account of the molecular mechanisms underlying the multiple functions of biothiols, with emphasis on their interaction with oxidants and the biological and clinical implications of this process. The topics covered range from the reactions of thiyl radicals in in vitro models to complex processes in clinical medicine and diverse therapeutic approaches involving thiols. Contributors have been selected, based on their expertise in various specialized areas, to provide a unique view on the subject of biothiol/oxidant interactions.

Lester Packer
Enrique Cadenas

Contents

Contributors

Silvia Alvarez, Ph.D. Institute of Biochemistry and Biophysics, School of Pharmacy and Biochemistry, University of Buenos Aires, Buenos Aires, Argentina

M. W. Anders, D.V.M., Ph.D. Paul Stark Professor of Pharmacology and Chairman, Department of Pharmacology, University of Rochester, Rochester, New York

Ohara Augusto, Ph.D. Professor, Department of Biochemistry, Instituto de Química, Universidade de São Paulo, São Paulo, Brazil

Aalt Bast, Ph.D. Professor, Department of Pharmacochemistry, Leiden/Amsterdam Center for Drug Research, Vrije Universiteit, Amsterdam, The Netherlands

Alberto Boveris, Ph.D. Department of Biochemistry and Biophysics, School of Pharmacy and Biochemistry, University of Buenos Aires, Buenos Aires, Argentina

Carroll E. Cross, M.D. Department of Internal Medicine, University of California, Davis, California

Janice A. DeGray, Ph.D. Laboratory of Molecular Biophysics, National Institute of Environmental Health Sciences, National Institutes of Health, Research Triangle Park, North Carolina

Wolfgang Dekant, Ph.D., M.D. Associate Professor, Institut für Toxikologie, Universität Würzburg, Würzburg, Germany

Jason P. Eiserich, M.S. Department of Internal Medicine, University of California, Davis, California

Steven A. Everett, B.Sc., D.Phil. Gray Laboratory Cancer Research Trust, Mount Vernon Hospital, Northwood, Middlesex, England

Henry Jay Forman, Ph.D. Professor, Department of Molecular Pharmacology and Toxicology, University of Southern California, Los Angeles, California

Kazuko Fujiwara, Ph.D. Assistant Professor, The Institute for Enzyme Research, The University of Tokushima, Tokushima, Japan

Stephen A. Gravina, Ph.D. Howard Hughes Medical Institute, Rockefeller University, New York, New York

Carlo Gregolin, M.D. Dipartimento di Chimica Biologica, Universitá di Padova, Padova, Italy

F. Arnold Gries, M.D. Professor of Internal Medicine and Cardiology, Diabetes Research Institute, Heinrich Heine University, Düsseldorf, Germany

Guido R.M.M. Haenen, Ph.D. Associate Professor, Department of Pharmacochemistry, Leiden/Amsterdam Center for Drug Research, Vrije Universiteit, Amsterdam, The Netherlands

Barry Halliwell, Ph.D., D.Sc. Pharmacology Group, University of London King's College, London, England

Keiko Inoue Department of Biochemistry, Osaka City University Medical School, Osaka, Japan

Masayasu Inoue, M.D., Ph.D. Professor and Chairman, Department of Biochemistry, Osaka City University Medical School, Osaka, Japan

Neil Kaplowitz, M.D. Professor of Medicine, Physiology and Molecular Pharmacology/Toxicology and Director, USC Center for Liver Diseases, University of Southern California School of Medicine, Los Angeles, California

Rui-Ming Liu, Ph.D. Department of Molecular Pharmacology and Toxicology, University of Southern California, Los Angeles, California

Shelly Chi-Loo Lu, M.D. Assistant Professor, Department of Medicine, University of Southern California School of Medicine, Los Angeles, California

Matilde Maiorino Dipartimento di Chimica Biologica, Universitá di Padova, Padova, Italy

Ronald P. Mason, Ph.D. Laboratory of Molecular Biophysics, National Institute of Environmental Health Sciences, National Institutes of Health, Research Triangle Park, North Carolina

Alton Meister, M.D., Ph.D. Professor, Department of Biochemistry, Cornell University Medical College, New York, New York

John J. Mieyal, Ph.D. Professor and Acting Chairman, Department of Pharmacology, Case Western Reserve University School of Medicine, Cleveland, Ohio

Paul A. Mieyal Department of Pharmacology, New York Medical College, Valhalla, New York

Yukiko Minamiyama Department of Biochemistry, Osaka City University Medical School, Osaka, Japan

Masashi Mizuno, Ph.D. Assistant Professor, Faculty of Agriculture, Kobe University, Kobe, Japan

Yutaro Motokawa, M.D. Professor, The Institute for Enzyme Research, The University of Tokushima, Tokushima, Japan

Kazuko Okamura-Ikeda, Ph.D. The Institute for Enzyme Research, The University of Tokushima, Tokushima, Japan

Charles A. O'Neill, Ph.D. Department of Internal Medicine, Division of Pulmonary/Critical Care Medicine, University of California, Davis, California

Murad Ookhtens, M.S., Ph.D. Associate Professor, Department of Medicine, University of Southern California School of Medicine, Los Angeles, California

Lester Packer, Ph.D. Professor, Department of Molecular and Cell Biology, University of California, Berkeley, California

Mulchand S. Patel Professor and Chairman, Department of Biochemistry, School of Medicine and Biomedical Sciences, State University of New York at Buffalo, Buffalo, New York

Rafael Radi, M.D., Ph.D. Department of Biochemistry, Facultad de Medicina, Universidad de la República, Montevideo, Uruguay

Donald J. Reed, Ph.D. Distinguished Professor of Biochemistry and Director, Department of Biochemistry and Biophysics, Environmental Health Sciences Center, Oregon State University, Corvallis, Oregon

Abraham Z. Reznick, Ph.D. Department of Morphological Sciences, The Bruce Rappaport Faculty of Medicine, Technion, Haifa, Israel

Antonella Roveri Dipartimento di Chimica Biologica, Università di Padova, Padova, Italy

Eisuke F. Sato Assistant Professor, Department of Biochemistry, Osaka City University Medical School, Osaka, Japan

Christian Schöneich, Ph.D. Assistant Professor, Department of Pharmaceutical Chemistry, University of Kansas, Lawrence, Kansas

Michael Ming Shi, M.D., Ph.D. Department of Environmental Health, Harvard School of Public Health, Boston, Massachusetts

Usha Srinivasan, M.Sc. Department of Chemistry, Case Western Reserve University School of Medicine, Cleveland, Ohio

David W. Starke, M.S. Department of Chemistry, Case Western Reserve University School of Medicine, Cleveland, Ohio

Yuichiro J. Suzuki Ph.D. Assistant Research Biochemist, Department of Molecular and Cell Biology, University of California, Berkeley, California

Hans Jürgen Tritschler, Ph.D. Medical Department, ASTA Medica AG, Frankfurt, Germany

Heinz Ulrich, M.D. Head, Medical Department, ASTA Medica AG, Frankfurt, Germany

Fulvio Ursini, M.D. Professor of Biochemistry, Department of Chemical Sciences and Technology, Università di Udine, Udine, Italy

Spyridon Vamvakas, M.D. Institut für Toxikologie, Universität Würzburg, Würzburg, Germany

Albert van der Vliet, Ph.D. Department of Internal Medicine, University of California, Davis, California

Nataraj N. Vettakkorumakankav Department of Biochemistry, School of Medicine and Biomedical Sciences, State University of New York at Buffalo, Buffalo, New York

Peter Wardman, Ph.D., D.Sc., C.Chem., F.R.S.C. Gray Laboratory Cancer Research Trust, Mount Vernon Hospital, Northwood, Middlesex, England

Klaus Wessel, Ph.D. ASTA Medica AG, Frankfurt, Germany

Christine C. Winterbourn, Ph.D. Professor, Department of Pathology, Christchurch School of Medicine, Christchurch, New Zealand

Eric H. Witt Research Biochemist, Department of Molecular and Cell Biology, University of California, Berkeley, California

Seiichi Yamamasu Department of Biochemistry, Osaka City University Medical School, Osaka, Japan

Hidenori Yu Department of Biochemistry, Osaka City University Medical School, Osaka, Japan

Dan Ziegler, M.D. Consultant Physician, Diabetes Research Institute, Heinrich Heine University, Düsseldorf, Germany

1

Reactions of Thiyl Radicals

Peter Wardman
Gray Laboratory Cancer Research Trust, Mount Vernon Hospital, Northwood, Middlesex, England

I. INTRODUCTION

Biothiols are generally regarded as protective agents, since they can "repair" (but frequently not restitute) cellular free radicals:

$$\text{radical} + \text{thiol} \rightarrow \text{“repaired” radical} + \text{thiyl radical} \tag{1}$$

or

$$R'^{\bullet} + RSH \rightleftharpoons R'H + RS^{\bullet} \tag{2}$$

Spin conservation requires the species left when the thiol has donated a hydrogen atom or electron in the "repair" reaction [Eqs. (1) and (2)] to be a free radical: the thiyl radical. Denoting a thiol as RSH, the thiyl radical can be represented by $RS^{\bullet}$.

The distinction between radical repair and restitution can be seen by considering a radical produced by hydrogen abstraction from a carbon atom with three other bonds in a tetrahedral configuration. As soon as the radical is formed, a planar configuration will be adopted and donation of a hydrogen atom to repair the radical may produce a species with a molecular formula nominally identical to the original species but with different stereochemical identity (1).

Because of the phenomenological importance of thiols as protective agents, the reactivity of the thiyl radicals necessarily produced in the repair reaction has received relatively little attention. This is inappropriate since many repair reactions involve an equilibrium [e.g., Eq. (2)] and the overall repair efficiency may reflect other reactions that remove thiyl radicals from the system, pulling the repair equilibrium in favor of net repair.

This chapter summarizes the present knowledge of the reactions of thiyl radicals, particularly of the most important biothiols, such as glutathione (GSH) and cysteine (CySH). The kinetic factors that control the fate of thiyl radicals in cells have been discussed recently (2), and there have been several reviews of the chemistry of sulfur-centered radicals, including the chemistry of thiyl radicals (see especially Refs. 3–6). The importance of conjugation of thiyl radicals with thiol/thiolate or oxygen has been emphasized previously (7) and illustrated in recent studies (8–10). This chapter therefore presents only an overview of the chemistry of thiyl radicals. Areas of present uncertainty are highlighted and the factors which must be considered in studying the kinetics of thiyl radical reactions are stressed.

II. CONJUGATION (ADDITION) REACTIONS OF THIYL RADICALS

A. Conjugation with Thiol or Thiolate

The absorption of the disulfide radical anion, $(RSSR)^{\bullet-}$, at ~420 nm ($\varepsilon \sim 800$ $m^2\ mol^{-1}$ or ~ 8000 $dm^3\ mol^{-1}\ cm^{-1}$) has long formed (11) the basis of detecting thiyl radical formation through the conjugation equilibria:

$$RS^{\bullet} + RS^{-} \rightleftharpoons (RSSR)^{\bullet-} \tag{3}$$

$$(RSSR)^{\bullet-} + H^{+} \rightleftharpoons (RSSR)H^{\bullet} \rightleftharpoons RS^{\bullet} + RSH \tag{4}$$

which will be pH dependent because of thiol ionization ($pK_5 \sim 9$):

$$RSH \rightleftharpoons RS^{-} + H^{+} \tag{5}$$

Except for dithiols (e.g., dithiothreitol), Eq. (4) can probably be disregarded for our purposes (12). The fraction f_6 of thiyl radicals present at equilibrium in the free (thiyl, $RS^{\bullet}$) form, i.e., unconjugated as $(RSSR)^{\bullet-}$, is given by:

$$f_6 = \left(\frac{1 + K_3[RSH]_{tot}}{1 + 10^{pK-pH}} \right)^{-1} \tag{6}$$

where $[RSH]_{tot} = ([RSH] + [RS^{-}])$ and $pK = pK_5$. For GSH a value of $K_3 \sim 3500\ dm^3\ mol^{-1}$ has recently been confirmed (8), and using $pK_5 \sim 9.2$ it is easily calculated that in many cellular systems at pH ~ 7, most of the glutathione thiyl radicals remain in the unconjugated form, considering only Eq. (3) (7).

Equation (3) linked to Eq. (5) is of **exceptional importance** in thiyl radical chemistry because it serves to "switch" the moderately *oxidizing* thiyl radical to the powerfully *reducing* disulfide radical anion.

B. Conjugation with Oxygen

Coincidentally, thiyl radicals conjugate with oxygen about as efficiently (and with similar kinetics) as the conjugation [Eq. (3)] with thiolate:

$$RS^{\bullet} + O_2 \rightleftharpoons RSOO^{\bullet} \tag{7}$$

Combining the equilibria in Eqs. (3), (5), and (7), the fraction of $RS^{\bullet}$ not conjugated with RS^- or O_2 is given by:

$$f_8 = \left(\frac{1 + K_3[\mathrm{RSH}]_{\mathrm{tot}}}{(1 + 10^{\mathrm{p}K - \mathrm{pH}})} + K_7[\mathrm{O_2}] \right)^{-1} \tag{8}$$

where pK = pK_5. This function is plotted in Figure 1 for a range of physiological conditions: total thiol up to 3 mmol dm^{-3} (typical intracellular GSH levels are ~2 mmol dm^{-3}) and pH 6.5–8.5, using equilibrium and ionization constants reported for GSH. The upper surface is for anoxic conditions, and the lower for 40 μmol dm^{-3} oxygen (representative of well-oxygenated tissue). The presence of oxygen up to physiological levels makes only a few percent difference to the fraction of $GS^{\bullet}$ in the free (thiyl) form. At pH 7.4 and with

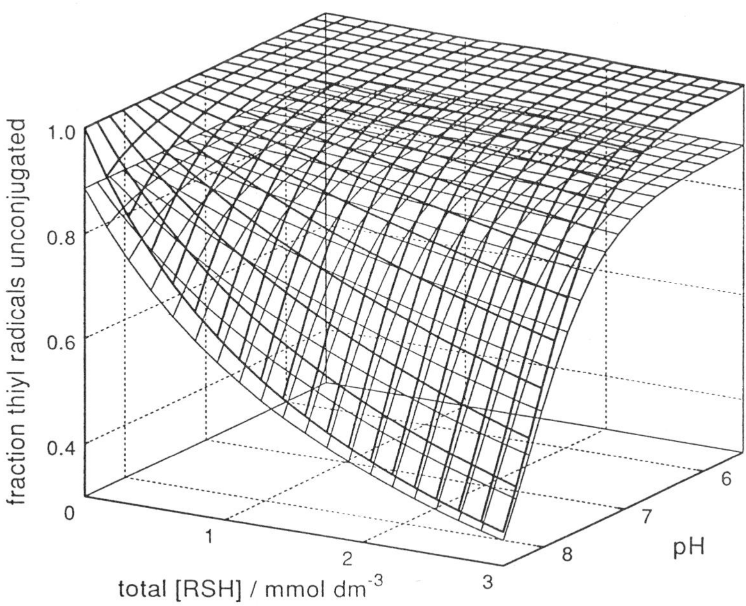

Figure 1 The fraction of thiyl radicals not conjugated with thiolate or oxygen, for a thiol with ionization and equilibrium constants similar to those of GSH, as a function of total thiol concentration and pH. Upper surface: anoxic; lower surface: 40 μmol dm^{-3} oxygen.

2 mmol dm^{-3} GSH, ~80% of thiyl radicals are unconjugated even in well-oxygenated tissue, considering Eqs. (3) and (7).

C. Adding to Unsaturated (Double) Bonds

Radical addition of thiyl radicals to double bonds can be represented as follows:

$$RS^{\bullet} + \text{-CH=CH-} \rightleftharpoons \text{-CH(SR)-}\dot{\text{C}}\text{H-} \quad (9)$$

Rapid reactions of this type (k_9 ~ 10^8–10^9 dm^3 mol^{-1} s^{-1}) have been reported for styrene muconate (*trans, trans*-1,3-butadiene-1,4-dicarboxylate) and the conjugated polyene retinol (vitamin A) (13). A different type of reaction (hydrogen abstraction) is likely to arise if the double bonds are not conjugated, as in the bis-allylic bond system in polyunsaturated fatty acids (see below). The C-S bond energy in the radical product of Eq. (9) is low, so that the reaction can be viewed as an equilibrium, and with some olefins and thiyl radicals *cis-trans* isomerization results (14).

D. Dimerization of Thiyl Radicals

Thiyl radicals dimerize to form the disulfide [Eq. (10), $2k_{10}$ ~ 1.5×10^9 dm^3 mol^{-1} s^{-1} for GSH (15)]:

$$RS^{\bullet} + RS^{\bullet} \rightarrow RSSR \quad (10)$$

The first half-life (= $1/2k_{10}[RS^{\bullet}]_0$) depends on the steady-state concentration of thiyl radicals. In cells this seems likely to be much less than micromolar. If GS$^{\bullet}$ radicals achieved a steady-state concentration as high as 10 nmol dm^{-3}, then the first half-life of Eq. (10) would be ~70 ms. This reaction seems most unlikely to play a role in the physiological reactions of thiols, since the relaxation times of the conjugation equilibria [Eqs. (3), (4), and (7) with thiolate and O_2] are of the order of a few microseconds—at least three orders of magnitude faster than thiyl/thiyl addition—under typical physiological conditions (2). Electron transfer from the disulfide radical anion to oxygen (see below) occurs over about a millisecond even in severely hypoxic cells. Measurements involving enzyme-catalyzed generation of GS$^{\bullet}$ in the presence or absence of oxygen support this conclusion (16).

III. HYDROGEN ABSTRACTION REACTIONS OF THIYL RADICALS

A. Abstraction at Carbon Activated by Alcohol, Ether, or Carboxylate Substitution

As noted above, the hydrogen-donation reaction [Eq. (2)] is now recognized to be an equilibrium (1,17). Pulse radiolysis has been used to characterize the rates of the reverse, or hydrogen-abstraction, pathway. One approach is to add an indicator that forms a convenient chromophore on reaction with the radical produced on H-abstraction. 4-Nitroacetophenone (17) or benzyl viologen (18) have proved convenient. The approach is exemplified by the data shown in Figure 2, which shows the absorption at 400 nm of the benzyl viologen radical cation ($BV^{\bullet +}$) produced via $CO_2^{\bullet -}$ via H-abstraction from formate/formic acid by $GS^{\bullet}$:

$$GS^{\bullet} + HCO_2^- \ (HCO_2H) \rightleftharpoons GSH + CO_2^{\bullet -} \ (+ H^+) \tag{11}$$

In these experiments, bromide present in high concentration scavenged $^{\bullet}OH$ radicals, and the resulting $Br_2^{\bullet -}$ radical oxidized GSH to yield $GS^{\bullet}$ within a few microseconds after the radiation pulse. $GS^{\bullet}$ then abstracted H from formate/formic acid over tens/hundreds of microseconds. The stepwise pathway shown in Figure 2 is annotated with the approximate half-lives of each step. It is important that the chromophore-producing final step is not rate-limiting, but that the desired H-abstraction step is. Appropriate experimental conditions are easily found, and by varying the concentrations in a manner similar to that described in an earlier study (17) it was shown that H abstraction occurs with $k_{11} \sim 8 \times 10^5$ dm^3 mol^{-1} s^{-1} at pH 4.2.

The $\sim$0.9 Gy pulse used in these experiments generated only $\sim$0.6 μmol dm^{-3} $GS^{\bullet}$, so competing reactions such as Eq. (10) would have a first half-life of $\sim$1 ms (10% loss of $GS^{\bullet}$ through Eq. (10) in $\sim$120 μs, or over two "divisions" in Fig. 2). It is therefore difficult to exclude *completely* unwanted, competing reactions such as Eq. (10) in these experiments, although the errors introduced are frequently negligible, provided concentrations of thiyl radicals well below 1 μmol dm^{-3} are used.

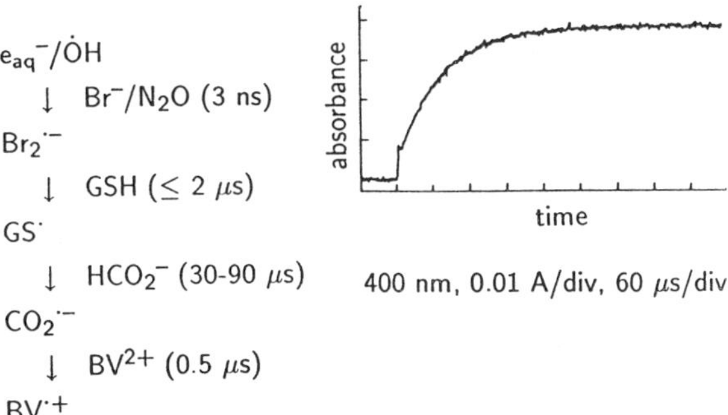

Figure 2 Formation of the absorption at 400 nm of the benzyl viologen radical cation, $BV^{\bullet +}$, produced via $CO_2^{\bullet -}$ arising from H abstraction from formate/formic acid by $GS^{\bullet}$. The solution contained 0.18 mol dm^{-3} KBr, 20 mmol dm^{-3} HCO_2Na, 5 mmol dm^{-3} HCO_2H, 0.25 mmol dm^{-3} BV^{2+}, and $\sim$25 mmol dm^{-3} N_2O (pH 4.2). The absorbance trace covers 600 μs full sweep.

Analogous experiments by the author using BV^{2+} to monitor the formation of the $(CH_3)\dot{C}OH$ radical produced in Eq. (12) using a dose of ~0.09 Gy (~60 nmol dm^{-3} $GS^{\bullet}$) yielded an estimate of $k_{12} \sim 8 \times 10^3$ dm^3 mol^{-1} s^{-1}:

$$GS^{\bullet} + (CH_3)_2CHOH \rightleftharpoons GSH + (CH_3)_2\dot{C}OH \tag{12}$$

This value is consistent with that obtained [1.2×10^4 dm^3 mol^{-1} s^{-1} (19)] using 1:1 water:2-propanol as solvent and monitoring the formation of the vitamin E phenoxyl radical obtained on one-electron oxidation of the vitamin by $GS^{\bullet}$.

B. Abstraction from Carbon with Bis-Allylic Character (PUFAs)

While abstraction by thiyl radicals of hydrogen from carbon activated only by adjacent alcohol or ether functions is characterized by rate constants typically ~ 10^4 dm^3 mol^{-1} s^{-1}, the much weaker C-H bond energies associated with bis-allylic centers lead to much higher rate constants for abstraction of H by $GS^{\bullet}$ from polyunsaturated fatty acids (PUFAs) (19):

$$GS^{\bullet} + R^1CH{=}CH{-}CH_2{-}CH{=}CHR^2 \rightarrow$$
$$GSH + R^1CH{=}CH{-}\dot{C}H{-}CH{=}CHR^2 \tag{13}$$

Thus, estimates of k_{13} are in the range ~1–3 $\times$ 10^7 dm^3 mol^{-1} s^{-1} for common PUFAs.

IV. ELECTRON TRANSFER REACTIONS OF THIYL RADICALS

A. Thermodynamics

A thiyl radical $RS^{\bullet}$ may oxidize a substrate (which acts as an electron donor or reductant) in a single-electron transfer reaction:

$$RS^{\bullet} + \textit{reductant} \rightleftharpoons RS^- + \textit{radical} \tag{14}$$

where *radical* is that obtained on one-electron oxidation of the *reductant*. The equilibrium [Eq. (14)] will be thermodynamically favorable ($K_{14} > 1$, $\Delta G_{14} < 0$, $\Delta E_{14} > 0$) if the reduction potentials for the two single-electron couples reflect the inequality:

$$E_{mi}(RS^{\bullet},H^+/RSH) > E_{mi}(\textit{radical/reductant}) \tag{15}$$

where the potentials are midpoint potentials of the couples at pH_i (20). As illustrated in Figure 3, the thiyl radical couple is pH-dependent, with E_{mi} being typically about 0.8–0.9 V vs. NHE at high pH (pH > pK_5), where $E_{mi} \cong$

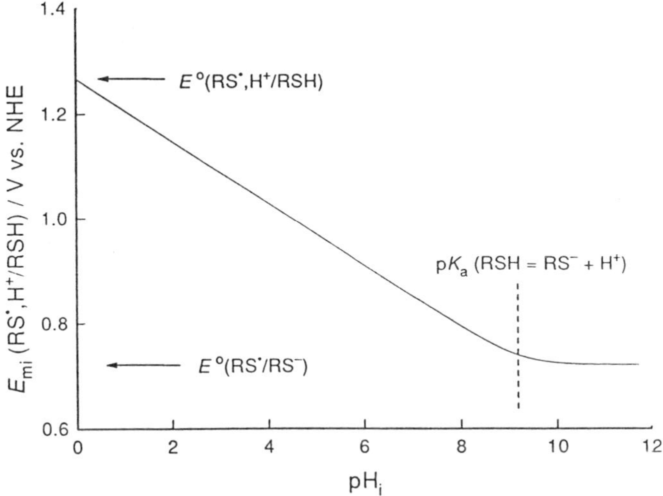

Figure 3 Typical variation of the midpoint reduction potential of the thiyl radical/thiol one-electron couple with pH. Values are based on glutathione (see Sec. IV.D).

$E^0(RS^•/RS^-)$, and increasing by ~0.06 V/pH unit as the pH is decreased below pK_5. The principles of this pH dependence of the couples have been discussed previously (20). Thus the standard potential of the couple $E^0(GS^•,H^+/GSH)$ is ~0.55 V higher (i.e., ~$pK_5 \times 0.06$ V) than the standard potential $E^0(GS^•/GS^-)$.

Practically, the most important implications of the reduction potentials of the reactant/product couples in Eq. (14) are the magnitude of the equilibrium constant K_{14} and the influence of this on the rate of electron transfer. The equilibrium constant is related to the potentials by:

$$(\mathbf{R}T/F) \ln K_{14} = E(RS^•/RS^-) - E(\textit{reductant/radical}) \quad (16)$$

where the subscripts have been omitted for simplicity, but it must be understood that K_{14} may be pH-dependent not only because of thiol ionization [Eq. (5)], but also because of prototropic equilibria involving the other reactants and products in Eq. (14). **R** is the gas constant, T the absolute temperature, and F the Faraday constant.

From Eq. (16) it can easily be seen that thiyl radicals may oxidize ascorbate (AH^-):

$$RS^• + AH^- \rightleftharpoons RSH + AH^• \quad (17)$$

with the reaction being thermodynamically very favorable since $E(AH^•/AH^-)$ ~ 0.3 V (20). Thus at pH ~ 7, if $E_{m7}(RS^•,H^+/RSH) = 0.9$ V, ΔE_{17} ~ 0.6 V. This corresponds to ΔG_{17} ~ −60 kJ mol^{-1} and K_{17} ~ 10^{10}.

B. Kinetics

Although a high rate constant for Eq. (17) has been reported [$k_{17} \sim 6 \times 10^8$ $dm^3\ mol^{-1}\ s^{-1}$ for GSH (21,22)], it must not be assumed that a reaction that is favorable thermodynamically will be kinetically fast or that a reaction with much lower "driving energy" will be necessarily much slower. Thus oxidation of phenothiazines (PZ) has been used to estimate the reduction potential of the glutathionyl radical (see below) by measuring the position of equilibrium:

$$GS^{\bullet} + PZ + H^+ \rightleftharpoons GSH + PZ^{\bullet +} \tag{18}$$

$E_m(PZ^{\bullet +}/PZ) \sim 0.9$ V for PZ = promethazine (23), and at pH $\sim$ 4.3, $K_{18}[H^+]$ $\sim$ 80 (24), so the equilibrium position of Eq. (18) is $\sim 10^8$ less favorable than that of Eq. (17). Nonetheless, $k_{18} \sim 1 \times 10^8$ $dm^3\ mol^{-1}\ s^{-1}$ (ignoring H^+ in the rate expression).

The problem of predicting which electron transfer reactions are likely to be kinetically "sluggish" is outside the scope of this article. It is sufficient to note that deprotonation frequently facilitates reactions, so that phenolate anions are oxidized much faster than phenols, for example. Conversely, protonation of an electron donor may slow down electron transfer. Thus thiyl radicals easily oxidize aromatic diamines such as *N,N,N′N′*-tetramethyl-*p*-phenylenediamine (TMPD) except at low pH when the amine substituent(s) may be protonated (25).

Within a given series of structurally similar reactants, however, the effect on kinetics of variations in energetics can be predicted with more confidence. The Marcus theory (26) has proven valuable in relating rates to energetics. Marcus developed an Arrhenius-like expression:

$$k = A \exp(-\Delta G^*/\mathbf{R}T) \tag{19}$$

where the free energy of activation, ΔG^*, is related to ΔG^0 by a solvent reorganization parameter, λ:

$$\Delta G^* = (\lambda/4)(1 + \Delta G^0/\lambda)^2 \tag{20}$$

The relationship between log k and ΔG^0 (or $-\Delta E^0$) is parabolic, and k is subject to the diffusion-controlled limit. However, if λ is $\sim$80–120 kJ mol^{-1} and ΔE is around 0.2 to −0.4 V, then the trend of the increase of log k with ΔE is expected to be around 7–11 V^{-1}. Roughly speaking, k will often increase with ΔE by about an order of magnitude for each $\approx$0.1 V increase in ΔE. This broad generalization only applies within a single chemical reaction series involving structurally related analogs (the "self-exchange" rates are important) and for reactions well below the diffusion-controlled limit. Fuller details may be found elsewhere (27–29).

C. Examples

A reaction that does not strictly involve thiyl radicals directly is perhaps best mentioned first in view of its overall importance in thiyl radical reactions in biology. Thiyl radicals conjugated with thiolate, as in Eq. (3) to form disulfide radical-anions, react with oxygen by electron transfer at rates approaching the diffusion-controlled limit. Thus an estimate of $k_{21} \sim 5.1 \times 10^8$ dm^3 mol^{-1} s^{-1} for RSH = GSH has been reported (8):

$$(RSSR)^{\bullet -} + O_2 \rightarrow RSSR + O_2^{\bullet -} \tag{21}$$

Since, as noted above and discussed elsewhere (2), the relaxation time for Eq. (3) is a few microseconds or less under typical physiological conditions, the conjugation equilibrium will usually not be rate-limiting. The half-life for Eq. (21) (= (ln 2)/$k_{21}[O_2]$) is about 30–40 μs in well-oxygenated tissue and about an order of magnitude longer in hypoxic cells.

Oxidation of ascorbate [Eq. (17)], phenothiazines [Eq. (18)], or aromatic amines such as TMPD [Eq. (22)] has been noted above:

$$RS^{\bullet} + TMPD \rightarrow RS^{-} + (TMPD)^{\bullet +} \tag{22}$$

ABTS [2,2′-azinobis(3-ethylbenzothiazoline-6-sulfonic acid)] is another convenient oxidizable substrate for thiyl radicals yielding an intense chromophore on one-electron oxidation (30). The similar reaction with aminopyrine (AP, 4-(dimethylamino)-1,2-dihydro-1,5-dimethyl-2-phenyl-3*H*-pyrazol-3-one) has provided an instructive illustration of the importance of kinetics driving thermodynamically unfavorable reaction sequences:

$$RS^{\bullet} + AP \rightleftharpoons RS^{-} + AP^{\bullet +} \tag{23}$$

Equation (23) can be observed directly from either direction: the forward reaction after generating $RS^{\bullet}$ by pulse radiolysis and the reverse by producing $AP^{\bullet +}$ by γ-radiolysis and mixing the solution containing the relatively stable $AP^{\bullet +}$ radical with thiol in a stopped-flow system (31). At pH ~5–6 the forward reaction (RSH=GSH) was complete in ~100 μs with 100 μmol dm^{-3} AP, whereas the reverse reaction occurred over ~0.5 s with 200 μmol dm^{-3} GSH. Evidently Eq. (23) is thermodynamically very favorable, but $AP^{\bullet +}$ does disappear rapidly on adding GSH (32), i.e., the equilibrium is driven to the left, because thiyl radicals are removed from the equilibrium. The removal of thiyl radicals can occur by conjugation with thiolate [Eq. (3)] followed by reaction of $(RSSR)^{\bullet -}$ with O_2 (or with $AP^{\bullet +}$ in anoxic systems).

Thiyl radicals oxidize a variety of transition metal complexes (25,33). Rate constants for oxidation of some complexes by $C_2H_5S^{\bullet}$ at pH 1 (25) are plotted in Figure 4 vs. the reduction potential of the couples. Although there are insufficient data to analyze the relationship, the data appear to follow a not unfa-

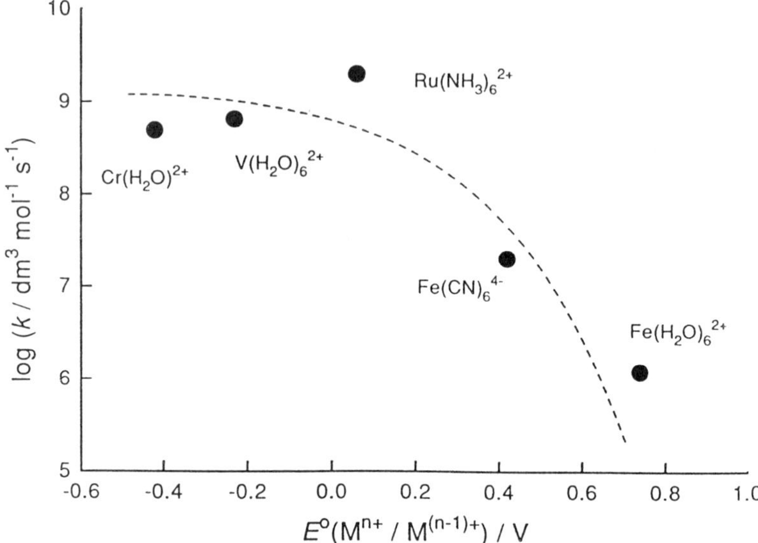

Figure 4 Rate constants for oxidation of metal complexes by $C_2H_5S^{\bullet}$ at pH 1 (25) plotted against reduction potentials of the metal complexes (29).

miliar pattern for electron-transfer reactions, with rate constants "plateauing" near the diffusion-controlled limit, eventually decreasing with a slope of ~0.05–0.1 V^{-1} (see above). Similar behavior is seen for reaction of semiquinone radicals with oxygen (34,35). However, this behavior should not be taken as an indication that these reactions of thiyl radicals proceed via an "outer-sphere" mechanism to which the Marcus theory should apply. Many reactions that formally involved electron transfer proceed through an intermediate or "inner-sphere" mechanism. Indeed, some reactants such as $^{\bullet}OH$ seldom, if ever, react with organic compounds via an outer-sphere mechanism, and in view of the electronic similarity between $RS^{\bullet}$ and $RO^{\bullet}$ it seems likely that many electron-transfer reactions of thiyl radicals involve intermediate complex formation.

D. Problems

Redox equilibration between phenothiazines and thiyl radicals has been used to estimate reduction potentials of thiyl radicals (24,36) using well-established techniques (20) and to investigate other aspects of thiyl radical chemistry (18,22). There are, however, three problems presently unresolved.

First, it has been observed that the absorption produced on oxidation of the phenothiazine chlorpromazine by the glutathionyl radical [see Eq. (18)], which appears to be largely that from the phenothiazine radical cation, is different at

$\lambda > 560$ nm from the spectrum of the radical cation. Hence some adduct formation cannot be ruled out.

The second problem is uncertainty in the value of the "reference" reduction potential, $E_{mi}(PZ^{\bullet+}/PZ)$, where PZ = a phenothiazine. Although values for common phenothiazines such as chlorpromazine and promethazine have been reported (20) and the latter compound has been studied recently in some detail (23), there is still some uncertainty. If the reduction potential of the promethazine radical cation is 0.90 V (23), then from the measured $K_{18}[H^+] = 79$ at pH 4.33 (24) for Eq. (18) (promethazine/glutathione), it is estimated that $E_m(GS^{\bullet},H^+/GSH) = 1.01$ V at pH 4.33, corresponding to 0.85 V at pH 7 and 0.83 V at pH 7.4 ($E^0(GS^{\bullet},H^+/GSH) = 1.27$ V) assuming the reaction involved simple electron transfer without adduct formation. The absorption produced on oxidation of promethazine by $Br_2^{\bullet-}$ was found to be identical with that produced from the thiyl radical from cysteine (21). These values are lower than those previously deduced from the data (24) because of the revised value for the reference potential.

The third area where conflicting results have been reported involves the use of phenothiazines to compare the oxidizing properties of thiyl radicals with those of thiyl peroxyl radicals, $RSOO^{\bullet}$ [see Eq. (7)]. The rate of formation of the chlorpromazine cation radical following generation of $GS^{\bullet}$ in solutions containing varying amounts of oxygen was measured (18). It was shown that the data could be fitted satisfactorily without assuming significant (relative) reactivity of $GSOO^{\bullet}$ toward chlorpromazine (CPZ), i.e., $k_{24} \geq 10\ k_{25}$:

$$GS^{\bullet} + CPZ\ (+\ H^+) \rightarrow GSH + CPZ^{\bullet+} \tag{24}$$

$$GSOO^{\bullet} + CPZ \rightarrow GSOO^- + CPZ^{\bullet+} \tag{25}$$

However, another study concluded that $k_{24} = 1.8\ k_{25}$ (22).

V. INTRAMOLECULAR REARRANGEMENTS OF THIYL RADICALS

One-electron oxidation of thiols may lead to carbon-centered as well as sulfur-centered radicals. Intramolecular rearrangements of the thiyl radicals from GSH (37–40) and 2-mercaptoethanol (9) have been studied in detail.

In the case of GSH, base-catalyzed deprotonation of the protonated amino function in the glutamyl moiety leads to transfer of H from the adjacent carbon atom to the sulfur center:

$$^+H_3NCH(CO_2^-) \cdots CH_2S^{\bullet} \rightarrow {}^+H_3N\dot{C}(CO_2^-) \cdots CH_2SH \tag{26}$$

Carbon-centered radicals were observed by EPR on warming irradiated glasses containing GSH above 77 K (10,38), and the higher resolution associated with

EPR observations in aqueous solutions permitted detailed study of the identity of the carbon-centered radical (40). Schemes illustrating in more detail the possible mechanism of the rearrangement have been presented (2,40).

This rearrangement is more likely to be important in severely hypoxic cells or cells/organelles with lower-than-average thiol concentrations. If the rate-limiting step is deprotonation by base (of the protonated amino function) with a rate constant of $5 \times 10^9\ dm^3\ mol^{-1}\ s^{-1}$ (40), then rearrangement would have a half-life of ~0.6 ms at $[OH^-] = 2.5 \times 10^{-7}\ mol\ dm^{-3}$ (pH 7.4). The competing, conjugative reactions of thiyl radicals [Eqs. (3) and (7)] occur on much faster time scales at pH 7.4 except at very low oxygen or thiol concentrations (2).

The tautomerism of the thiyl radical from 2-mercaptoethanol [Eq. (27)] has been studied in detail (9):

$$HOCH_2CH_2S^{\bullet} \rightleftharpoons HOCH_2\dot{C}HSH \tag{27}$$

In this case the kinetics are such that physiological oxygen levels could irreversibly "trap" the carbon-centered radical via a peroxyl radical.

VI. IMPORTANCE OF THIYL RADICAL REARRANGEMENTS AND THIYL PEROXYL RADICALS IN BIOLOGY

The time scales of radical rearrangements and of the main competing, conjugative reactions of thiyl radicals have been discussed above. The same comparison may be made for the involvement of thiyl peroxyl radicals in reaction systems representative of many biological environments.

A full discussion of the consequences of formation of thiyl peroxyl radicals via the conjugation equilibrium [Eq. (7)] is outside the scope of this chapter. A recent paper (10) has summarized the rather complex chemistry. Key reactions include arrangement of the peroxyl radical to a sulfonyl radical ($RSO_2^{\bullet}$, both oxygens bonded to sulfur), itself adding oxygen to form a sulfonyl peroxyl radical:

$$RS^{\bullet} + O_2 \rightleftharpoons RSOO^{\bullet} \rightarrow RSO_2^{\bullet}\ (+\ O_2) \rightarrow RS(O_2)OO^{\bullet} \tag{28}$$

and formation of a sulfinyl radical, $RSO^{\bullet}$:

$$RSOO^{\bullet} + RSH \rightarrow RSO^{\bullet} + RSOH \tag{29}$$

The isomerization of $RSOO^{\bullet}$ is probably the rate-limiting step in the sequence of reactions shown in Eq. (28). In the case of 2-mercaptoethanol, this step has been reported to have a half-life of ~350 μs (9). With this thiol, $k_{29} = 2 \times 10^6\ dm^3\ mol^{-1}\ s^{-1}$ (9), and so the half-life of Eq. (29) will be at least ~100 μs in most biological systems if the corresponding reaction with GSH has a rate constant of the same order.

As noted above, "bleeding off" thiyl radicals to form superoxide through Eq. (21) coupled to Eq. (3) has a half-life of 30–40 μs in well-oxygenated tissue. Thus, even considering this "sink" (41) alone for radicals [ignoring other routes, such as reaction with ascorbate (see below)], it seems unlikely that Eqs. (28) and (29) play a major role in the chemistry of thiyl radicals in many biological environments. Of course, this conclusion is subject to the proviso that all the key rate and equilibrium constants under biological conditions are of the same order as the data discussed here; in some organelles (low pH, low thiol concentration) there is quite likely to be involvement of Eq. (28) and/or Eq. (29).

VII. PROPERTIES OF ARENETHIYL RADICALS AND OF THIYL RADICALS FROM DITHIOLS OR THIOLS WITH A LOW PK_A FOR IONIZATION

The ground states and radicals obtained on one-electron oxidation of thiol functions substituted in aromatic systems [e.g., thiophenols (42) or methimazole (43)] differ in their properties as much as those from aliphatic alcohols and phenols (44). The chemistry of arenethiyl radicals has much in common with that of phenoxyl radicals, although there are quite large differences in the redox properties associated with substituting sulfur for oxygen. Thus $E^0(C_6H_5S^\bullet/C_6H_5S^-)$ = 0.70 V, ~0.4 V *lower* than the phenoxyl analog. In contrast, for the 4-mercaptophenylthiyl radical, $E^0(4\text{-}^-SC_6H_4S^\bullet/4\text{-}^-SC_6H_4S^-)$ = 0.33 V, ~0.3 V *higher* than the 4-benzosemiquinone analog (42). These differences reflect different spin density distributions and resonance stabilization.

Dithiols such a lipoic acid (1,2-dithiolane-3-pentanoic acid) and dithiothreitol (1,4-dimercapto-2,3-butanediol) similarly represent special cases. Formation of a cyclic disulfide radical anion occurs in equilibrium with the open-chain form. Some examples have been reviewed (12). Enhanced formation of disulfide radical anions compared to free thiyl radicals at physiological pH is also a feature of those thiols with a higher propensity to dissociation [Eq. (5)] (i.e., pK_5 reduced from the 8.5–9.5 region typical of many thiols). This is simply because Eq. (3) is shifted to the right by the increased fraction of thiol present at physiological pH as thiolate [Eq. (5)]. An example is the radioprotector WR 1065, where pK_{30} = 7.7 (45):

$$H_2N(CH_2)_3NH(CH_2)_2SH \rightleftharpoons H_2N(CH_2)_3NH(CH_2)_2S^- + H^+ \quad (30)$$

The structure above omits protonation of the two amine functions, important at physiological pH. The strong inductive effect resulting is transmitted through the carbon chain to the S-H bond. Ionization also effectively removes hydrogen for H-donation reactions, so this form of radical "repair" reaction is pH-dependent over the physiological range (46). Other equilibria will be similarly influenced, so the quantitative conclusions above relating to glutathione or simpler thiols may be not immediately applicable to all thiols. Nevertheless, the logic of comparing the various competing reactions applies equally.

Substitution of R in a thiol RSH by RS to form a perthiol, RSSH, is a case where ionization of the S-H bond is enhanced [by about 1.5 units in the perthiol ionization analogous to (46)]. However, the chemistry of perthiyl radicals ($RSS^{\bullet}$) differs in several important respects from that of thiyl radicals (see Chap. 3). The properties of sulfur-centered radicals in proteins may also differ quantitatively from those in corresponding free thiols (see Chap. 2 for a discussion of the behavior of radicals in proteins).

VIII. SUPEROXIDE OR ASCORBYL RADICAL AS THE "RADICAL SINK" FROM THIYL RADICALS?

The role of superoxide as a radical "sink" for radical "repair" involving thiols [via Eqs. (1), (2), (3), and (21)] has been the subject of considerable discussion (41). The likely dominance of Eq. (21) as the main reaction pathway for thiyl radicals, compared to routes involving thiylperoxyl radicals, has been noted above. However, many tissues contain ascorbate, highly reactive towards thiyl radicals [Eq. (17)]. It is instructive to calculate the approximate fraction of thiyl radicals that ultimately lead to superoxide in the presence of physiological levels of ascorbate. This fraction f_{31}, can be estimated to be the rate of Eq. (21) divided by the sum of the rates of Eqs. (17) and (21):

$$f_{31} = \frac{k_{21}K_3[O_2][RSH]_{tot}/(1 + 10^{pK-pH})}{k_{21}K_3[O_2][RSH]_{tot}/(1 + 10^{pK-pH}) + k_{17}[AH^-]} \tag{31}$$

where $pK = pK_5$. This function is plotted in Figure 5 as a function of ascorbate and oxygen concentrations over the physiological range, using rate and equilibrium data for GSH. The upper surface is for 5 mmol dm^{-3} GSH, the lower for 1 mmol dm^{-3} GSH. It is seen that at the lower limit of ascorbate plotted [50 μmol dm^{-3}, appropriate for blood serum (47)], even in well-oxygenated tissue only ~3% of thiyl radicals go down the superoxide "sink" at low GSH concentrations. At high GSH concentrations (5 mmol dm^{-3}), this rises to ~15%. However, even with 5 mmol dm^{-3} GSH, at the ascorbate levels found in many tissues [around 0.5 mmol dm^{-3} (48)], only ~2% of thiyl radicals follow the superoxide pathway: most oxidize ascorbate.

From this simple calculation, the importance of ascorbate rather than superoxide as a radical "sink" is evident. Of course, the reactivity of ascorbate (AH^-) toward superoxide has been ignored:

$$O_2^{\bullet -} + AH^- (+2H^+) \rightarrow H_2O_2 + AH^{\bullet} \tag{32}$$

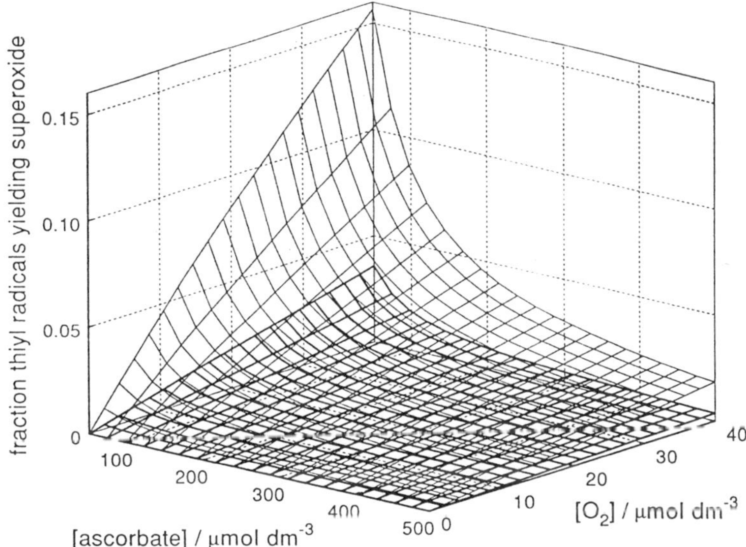

Figure 5 The fraction of thiyl radicals generating superoxide radicals, calculated according to Eq. (31), as a function of ascorbate and oxygen concentrations. Upper surface: 5 mmol dm^{-3} total thiol; lower surface; 1 mmol dm^{-3} total thiol. Calculations are for GSH at pH 7.4.

Since both reactants are involved in prototropic equilibria, the overall reaction is quite complex (49), but the effective rate constant k_{32} at pH 7.4 is $\sim 1.5 \times 10^5$ dm^3 mol^{-1} s^{-1}. Hence if the ascorbate concentration is ~ 0.5 mmol dm^{-3}, superoxide reacts with ascorbate with a half-life of ~ 10 ms. Disproportionation of $O_2^{\bullet-}$ to form hydrogen peroxide, catalyzed by superoxide dismutase (SOD), would compete with Eq. (32) at this ascorbate concentration if the SOD level was as low as ~ 30 nmol dm^{-3}. Ascorbate radicals are finally "sunk" through disproportionation:

$$2\ AH^{\bullet} \rightleftharpoons \text{ascorbate} + \text{dehydroascorbate} \qquad (33)$$

IX. CONCLUSIONS

Three equilibria define the key features of thiyl radical chemistry: ionization [Eq. (5)], conjugation with thiolate [Eq. (3)], and conjugation with oxygen [Eq. (7)]. The subsequent fate of thiyl radicals is dominated by reaction with oxygen through Eq. (21) [linked to Eqs. (3) and (5)] or reaction with ascorbate [Eq. (17)]. The latter is likely to provide the main "sink" for radicals that have been "repaired" by thiols under many physiological conditions. Thiyl radicals are potential prooxidants because of a high reactivity toward polyunsaturated fatty acids. For a complete understanding, it is necessary to model the chemical kinetics by including spatial information and diffusion to take into account microscopic concentration gradients (2). Further, Eq. (31) will be increasingly inaccurate at low oxygen/thiol and high ascorbate concentrations, since reaction with ascorbate, Eq. (17), may occur on a faster time scale than the conjugation equilibria. These factors will be the subject of a subsequent paper.

ACKNOWLEDGMENTS

This work was supported by the Cancer Research Campaign. I thank K.-D. Asmus, L. P. Candeias, S. A. Everett, and C. von Sonntag for helpful discussions.

REFERENCES

1. Akhlaq, M. S., Schuchmann, H.-P., and von Sonntag, C. (1987) The reverse of the 'repair' reaction of thiols: H-abstraction at carbon by thiyl radicals. Int. J. Radiat. Biol. 51:91–102.
2. Wardman, P., and von Sonntag, C. (1995) Evaluation of the kinetic factors which control the fate of thiyl radicals in cells. Methods Enzymol. 251:31–45.
3. Chatgilialoglu, C., and Asmus, K.-D., ed. (1990) Sulfur-Centered Reactive Intermediates in Chemistry and Biology. Plenum Press, New York.
4. Asmus, K.-D. (1990) Sulfur-centered free radicals. Methods Enzymol. 186:168–180.
5. Asmus, K.-D. (1993) Recent aspects of thiyl and perthiyl free radical chemistry. In: Active Oxygen, Lipid Peroxides, and Antioxidants (Yagi, K., ed.), pp. 57–67, Japan Sci. Soc. Press/CRC Press, Tokyo/Boca Raton, FL.
6. Lal, M. (1994) Radiation induced oxidation of sulphydryl molecules in aqueous solutions. A comprehensive review. Radiat. Phys. Chem. 43:595–611.
7. Wardman, P. (1988) Conjugation and oxidation of glutathione via thiyl free radicals. In: Glutathione Conjugation. Mechanisms and Biological Significance (Sies, H., and Ketterer, B., eds.), pp. 43–72, Academic Press, London.
8. Prütz, W. A., Butler, J., and Land, E. J. (1994) The glutathione free radical equilibrium, $GS^{\bullet} + GS^{-} = GSSG^{\bullet -}$, mediating electron transfer to Fe(III)-cytochrome *c*. Biophys. Chem. 49:101–111.
9. Zhang, X., Zhang, N., Schuchmann, H.-P., and von Sonntag, C. (1994) Pulse radiolysis of 2-mercaptoethanol in oxygenated aqueous solution. Generation and reactions of the thiylperoxyl radical. J. Phys. Chem. 98:6541–6547.
10. Becker, D., Summerfield, S., Gillich, S., and Sevilla, M. D. (1994) Influence of oxygen on the repair of direct radiation damage to DNA by thiols in model systems. Int. J. Radiat. Biol. 65:537–548.
11. Adams, G. E., McNaughton, G. S., and Michael, B. D. (1967) The pulse radiolysis

of sulphur compounds. Part I. Cysteamine and cystamine. In: The Chemistry of Ionization and Excitation (Johnson, G. R. A., and Scholes, G., eds.), pp. 281–293. Taylor & Francis, London.
12. von Sonntag, C. (1990) Free-radical reactions involving thiols and disulphides. In: Sulfur-Centered Reactive Intermediates in Chemistry and Biology (Chatgilialoglu, C., and Asmus, K.-D., eds.), pp. 359-366. Plenum Press, New York.
13. Dunster, C., and Willson, R. L. (1990) Thiyl free radicals: electron transfer, addition or hydrogen abstraction reactions in chemistry and biology, and the catalytic role of sulphur compounds. In: Sulfur-Centered Reactive Intermediates in Chemistry and Biology (Chatgilialoglu, C., and Asmus, K.-D., eds.), pp. 377–387. Plenum Press, New York.
14. von Sonntag, C. (1987). The Chemical Basis of Radiation Biology. Taylor & Francis, London, p. 360.
15. Hoffman, M. Z., and Hayon, E. (1973) Pulse radiolysis study of sulfhydryl compounds in aqueous solution. J. Phys. Chem. 77:990–996.
16. Ross, D., Norbeck, K., and Moldéus, P. (1985) The generation and subsequent fate of glutathionyl radicals in biological systems. J. Biol. Chem. 260:15028-15032.
17. Schöneich, C., Bonifacic, M., and Asmus, K.-D. (1989) Reversible H-atom abstraction from alcohols by thiyl radicals: determination of absolute rate constants by pulse radiolysis. Free Radical Res. Commun. 6:393–405.
18. Wardman, P. (1990) Thiol reactivity towards drugs and radicals: some implications in the radiotherapy and chemotherapy of cancer. In: Sulfur-Centered Reactive Intermediates in Chemistry and Biology (Chatgilialoglu, C., and Asmus, K.-D., eds.), pp. 415–427. Plenum Press, New York.
19. Schöneich, C., Bonifacic, M., Dillinger, U., and Asmus, K.-D. (1990) Hydrogen abstraction by thiyl radicals from activated C-H bonds of alcohols, ethers and polyunsaturated fatty acids. In: Sulfur-Centered Reactive Intermediates in Chemistry and Biology (Chatgilialoglu, C., and Asmus, K.-D., ed.), pp. 367–376. Plenum Press, New York.
20. Wardman, P. (1989) Reduction potentials of one-electron couples involving free radicals in aqueous solution. J. Phys. Chem. Ref. Data 18:1637–1755.
21. Forni, L. G., Mönig, J., Mora-Arellano, V. O., and Willson, R. L. (1983) Thiyl free radicals: direct observations of electron transfer reactions with phenothiazines and ascorbate. J. Chem. Soc., Perkin Trans. 2:961–965.
22. Tamba, M., and O'Neill, P. (1991) Redox reactions of thiol free radicals with the antioxidants ascorbate and chlorpromazine: role in radioprotection. J. Chem. Soc., Perkin Trans. 2:1681–1685.
23. Jonsson, M., Lind, J., Eriksen, T. E., and Merényi, G. (1994) Redox and acidity properties of 4-substituted aniline radical cations in water. J. Am. Chem. Soc. 116:1423–1427.
24. Wardman, P. (1990) The fate of glutathione thiyl radicals in biological systems. Free Radical Biol. Med. 9 (suppl. 1):62.
25. Huston, P., Espenson, J. H., and Bakac, A. (1992) Reactions of thiyl radicals with transition-metal complexes. J. Am. Chem. Soc. 114:9510–9516.
26. Marcus, R. A. (1993) Electron transfer reactions in chemistry: theory and experiment (Nobel lecture). Angew. Chem. Int. Ed. Engl. 32:1111–1121.

27. Cannon, R. D. (1980) Electron Transfer Reactions. Butterworths, London.
28. Eberson, L. (1987) Electron Transfer Reactions in Organic Chemistry. Springer-Verlag, Berlin.
29. Lappin, G. (1994) Redox Mechanisms in Inorganic Chemistry. Ellis Horwood, Chichester.
30. Wolfenden, B. S., and Willson, R. L. (1982) Radical cations as reference chromogens in kinetic studies of one-electron transfer reactions: pulse radiolysis studies of 2,2-azinobis-(3-ethylbenzthiazoline-6-sulphonate). J. Chem. Soc., Perkin Trans. 2:805–812.
31. Wilson, I., Wardman, P., Cohen, G. M., and d'Arcy Doherty, M. (1986) Reductive role of glutathione in the redox cycling of oxidizable drugs. Biochem. Pharmacol. 35:21–22.
32. Eling, T. E., Mason, R. P., and Sivarajah, K. (1985) The formation of aminopyrine cation radical by the peroxidase activity of prostaglandin H synthase and subsequent reactions of the radical. J. Biol. Chem. 260:1601–1607.
33. Armstrong, D. A. (1990) Redox systems with sulphur-centered species. In: Sulfur-Centered Reactive Intermediates in Chemistry and Biology (Chatgilialoglu, C., and Asmus, K.-D., eds.), pp. 341–357. Plenum Press, New York.
34. Meisel, D., and Czapski, G. (1975) One-electron transfer equilibria and redox potentials of radicals studies by pulse radiolysis. J. Phys. Chem. 79:1503–1509.
35. Wardman, P. (1990) Bioreductive activation of quinones: redox properties and thiol reactivity. Free Radical Res. Commun. 8:219–229.
36. Surdhar, P. S., and Armstrong, D. A. (1987) Reduction potentials and exchange reactions of thiyl radicals and disulfide anion radicals. J. Phys. Chem. 91:6532–6537.
37. Sjorberg, L., Eriksen, T. E., and Revesz, L. (1982) The reaction of the hydroxyl radical with glutathione in neutral and alkaline aqueous solution. Radiat. Res. 89:255–263.
38. Becker, D., Swarts, S., Champagne, M., and Sevilla, M. D. (1988) An ESR investigation of the reactions of glutathione, cysteine and penicillamine thiyl radicals: competitive formation of $RSO^{\bullet}$, $R^{\bullet}$, $RSSR^{\bullet -}$, and RSS. Int. J. Radiat. Biol. 53:767–786.
39. Eriksen, T. E., and Fransson, G. (1988) Formation of reducing radicals on radiolysis of glutathione and some related compounds in aqueous solution. J. Chem. Soc., Perkin Trans. 2:1117–1122.
40. Grierson, L., Hildenbrand, K., and Bothe, E. (1992) Intramolecular transformation reaction of the glutathione thiyl radical into a non-sulphur-centred radical: a pulse radiolysis and EPR study. Int. J. Radiat. Biol. 62:265–277.
41. Winterbourn, C. C. (1993) Superoxide as an intracellular radical sink. Free Radical Biol. Med. 14:85–90.
42. Armstrong, D. A., Sun, Q., Tripathi, G. N. R., Schuler, R. H., and McKinnon, D. (1993) Spectra, ionization constants, and rates of oxidation of 1,4-dimercaptobenzene and properties of the *p*-mercatophenylthiyl and *p*-benzodithiyl radicals. J. Phys. Chem. 97:5611–5617.
43. Taylor, J. J., Willson, R. L., and Kendall-Taylor, P. (1984) Evidence for direct interactions between methimazole and free radicals. FEBS Lett. 176:337–340.

44. Simic, M. G., and Hunter, E. P. L. (1986) Reaction mechanisms of peroxyl and C-centered radicals with sulfhydryls. J. Free Radicals Biol. Med. 2:227–230.
45. Newton, G. L., Dwyer, T. J., Kim, T., Ward, J. F., and Fahey, R. C. (1992) Determination of the acid dissociation constants for WR-1065 by proton NMR spectroscopy. Radiat. Res. 131:143–151.
46. Everett, S. A., Folkes, L. K., Wardman, P., and Asmus, K.-D. (1994) Free-radical repair by a novel perthiol: reversible hydrogen transfer and perthiyl radical formation. Free Radical Res. 20:387–400.
47. Lunec, J., and Blake, D. R. (1985) The determination of dehydrascorbic acid and ascorbic acid in the serum and synovial fluid of patients with rheumatoid arthritis (RA). Free Radical Res. Commun. 1:31–39.
48. Hornig, D. (1975) Distribution of ascorbic acid, metabolites and analogues in man and animals. Ann. NY Acad. Sci. 258:103–117.
49. Cabelli, D. E., and Bielski, B. H. J. (1983) Kinetics and mechanism for the oxidation of ascorbic acid/ascorbate by HO_2/O_2^- radicals. A pulse radiolysis and stopped-flow photolysis study. J. Phys. Chem. 87:1809–1812.

2

Thiyl Radicals, Perthiyl Radicals, and Oxidative Reactions

Christian Schöneich
University of Kansas, Lawrence, Kansas

I. INTRODUCTION

Thiols are highly abundant in biological systems and serve key roles in many biochemical processes such as (1) the detoxification of reactive oxygen species and xenobiotics [1], (2) the binding of transition metals in redox active complexes [2], (3) the folding of proteins and stabilization of protein structure through the formation of disulfide bonds [3], and (4) the activation/inactivation of enzymes via rapid chemical transformation of the mercapto group, for example, by nitrosylation with NO [4]. In fact, the importance of many proteins, oligopeptides, and polypeptides, e.g., glutathione and the metallothionines, relies on the presence of the amino acid cysteine, which constitutes the biologically most significant thiol.

Chemically important for the reactivity of thiols toward various reagents is the protonation state of the mercapto group:

$$RSH \rightleftharpoons RS^- + H^+ \qquad (1)$$

For the tripeptide glutathione the pK_a of equilibrium (1) is located at 9.2 [5], whereas the single amino acid cysteine exhibits a pK_a of 8.36 [5,6]. Thus, under physiological conditions thiols will exist predominantly in their protonated states. However, these pK_a values may be lowered through the complexation of thiols by various transition metals.

As an important biological feature, thiols readily involve in redox processes. Among the most prominent oxidation products of thiols are the corresponding disulfides, RSSR, sulfinic acids, RSO_2H, and sulfonic acids, RSO_3H, respectively [7]. However, frequently the formation of these products is preceded by the initial generation of thiyl free radicals, $RS^\bullet$, through either electron or hydrogen transfer from the thiol to an appropriate oxidant (Ox) or free radical ($R'^\bullet$):

$$RS^- + Ox \rightarrow RS^\bullet + Ox^{\bullet -} \quad (2)$$

$$RSH + R'^\bullet \rightarrow RS^\bullet + R'H \quad (3)$$

The latter processes represent, in fact, an important task of thiols, namely, the interception of reactive oxygen species or free radicals in biological systems, present at high levels, for example, under conditions of oxidative stress [8]. In this regard, thiols have traditionally be employed as radioprotectors against the damaging effects of ionizing radiation. The protection by thiols against reactive oxygen species requires, however, that the resulting thiyl radicals be removed by subsequent processes yielding less reactive radicals or molecular products. From the redox potentials of the two following half-reactions [9], it appears that thiyl radicals are relatively potent oxidants under physiological conditions:

$$RS^\bullet + e^- \rightarrow RS^- \qquad E_0 = 0.75\ V \quad (4)$$

$$RS^\bullet + e^- + H+ \rightarrow RSH \qquad E_0 = 1.33\ V \quad (5)$$

Thus, thiyl radicals will react with many physiological electron (D^-) or hydrogen (DH) donors according to the following general reactions:

$$RS^\bullet + D^- \rightleftharpoons RS^- + D^\bullet \quad (6)$$

$$RS^\bullet + DH \rightleftharpoons RSH + D^\bullet \quad (7)$$

The extent of biological protection by thiols will, therefore, depend on the extent to which such reactions occur randomly with surrounding biomolecules or controlled with antioxidants particularly devoted to the further interception of thiyl and/or related radicals. It may, consequently, be assumed that under certain conditions of biological dysfunctions, e.g., antioxidant deficiency, thiols may well intercept initially generated reactive oxygen species, but that the resulting thiyl radicals subsequently attack biological targets.

The present chapter discusses the kinetics and mechanisms of oxidation of biomolecules or representative model compounds by thiyl radicals and perthiyl radicals, $RSS^\bullet$. The latter class of compounds was included due to a recent proposal [10] that catalytic trace amounts of trisulfides or higher polysulfides, potent precursors for perthiyl radicals, may play an important role in the protection by thiols and disulfides against ionizing radiation.

The structures of the thiols and perthiols most frequently discussed in this chapter are shown in Figure 1.

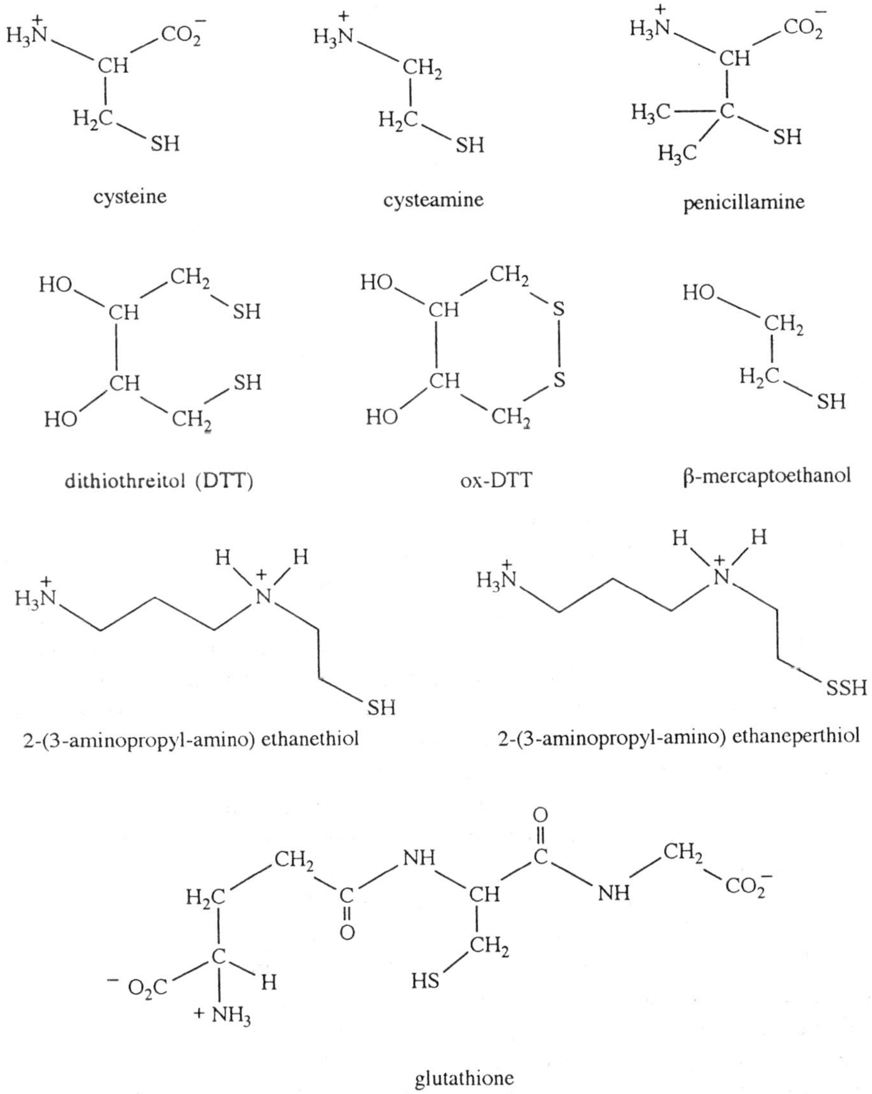

cysteine

cysteamine

penicillamine

dithiothreitol (DTT)

ox-DTT

β-mercaptoethanol

2-(3-aminopropyl-amino) ethanethiol

2-(3-aminopropyl-amino) ethaneperthiol

glutathione

Figure 1 Structures of some thiols and perthiols.

II. FORMATION OF THIYL RADICALS

A. Reaction of Thiols with Carbon-Centered Radicals

Thiyl radicals are readily formed by the reaction of carbon-centered radicals with thiols according to the general hydrogen transfer reaction (3). In radiation biology this reaction is generally referred to as the "repair reaction," as it serves to repair carbohydrate-based radicals of DNA strands [8,11]. However, such hydrogen transfer only reconstitutes the broken C-H bond of the carbohydrate without completely restoring the respective stereochemistry, representing a radical interception rather than a true "repair reaction." Rate constants for Eq. (3) have been measured by means of pulse radiolysis for a variety of carbon-centered radicals from aliphatic alcohols, ethers, and various carbohydrates [12–18]. For unsubstituted alkyl radicals as well as for α-hydroxyl- or α-alkoxylalkyl radicals, they are generally on the order of 10^6–10^8 M^{-1} s^{-1}, depending on the structure of the thiol and the attacking radical, respectively. Interestingly, the more negative the redox potentials of the attacking alkyl radicals, the faster they react with thiols, contrary to any expectation for this formal one-electron oxidation of the thiol. Not surprisingly then, oxidizing carbon-centered radicals such as the formylmethyl radical, $^{\bullet}CH_2\text{-}CH{=}O$, do not measurably (i.e., $k_3 \ll 10^5$ $M^{-1}s^{-1}$) react under hydrogen transfer with protonated thiols but rather under electron transfer with the deprotonated mercapto group [15]. The hydrogen transfer process from protonated thiols onto carbon-centered radicals appears, therefore, to include yet uncharacterized intermediates and/or transition states. For example, it may be assumed that carbon-centered radicals associate with the mercapto group prior to the hydrogen transfer, as proposed by Henglein and coworkers for the reaction of carbon-centered radicals with H_2S [19]. The considerations above are of particular importance for carbon-centered radicals derived from the reaction of hydroxyl radicals with DNA. Hydrogen abstraction from the DNA carbohydrates predominantly yields derivatives of 1,2-dihydroxyethyl radicals. The latter undergo hydrogen transfer with thiols with $k_3 \approx 10^8$ $M^{-1}s^{-1}$ [17]. On the other hand they competitively eliminate water in a hydroxide ion and buffer-catalyzed process, yielding derivatives of α-keto alkyl radicals that cannot be repaired by protonated thiols but rather undergo electron transfer with the deprotonated mercapto group. Furthermore, the addition of hydroxyl radicals to DNA bases yields a variety of reducing and oxidizing radical intermediates of which generally only the reducing species abstract hydrogen atoms from thiols [20].

B. Reaction of Thiols with Reactive Oxygen Species

Among the most prominent reactive oxygen species potentially formed under conditions of oxidative stress are hydroxyl radicals ($HO^{\bullet}$), alkoxyl radicals

($RO^\bullet$), peroxyl radicals ($ROO^\bullet$), superoxide anion ($O_2^{\bullet-}$), singlet oxygen ($^1O_2$), hydrogen peroxide (H_2O_2), organic peroxides (ROOH), nitrogen dioxide (NO_2), and oxoperoxonitrate ($ONOO^-$). By means of pulse radiolysis and ESR spectroscopy it has been demonstrated that $HO^\bullet$ [12,13,18,21,22], NO_2 [23], and the $ONOO^-$/ONOOH couple [23] generate thiyl radicals upon reaction with thiols. The reaction of superoxide anion with dithiothreitol, HS-(DTT)-SH, was shown not to yield thiyl radicals via hydrogen transfer but suggested rather to proceed via formation of an adduct, which decomposes into hydroxide ion and a sulfinyl radical [24]:

$$\text{HS-(DTT)-S}^- + H^+ + O_2^{\bullet-} \rightleftharpoons [\text{HS-(DTT)-S} \cdots \text{OOH}]^- \tag{8}$$

$$[\text{HS-(DTT)]S} \cdots \text{OOH}]^- \rightarrow \text{HS-(DTT)-SO}^\bullet + OH^- \tag{9}$$

The formation of thiyl radicals by the rather slow reaction of unsubstituted peroxyl radicals with protonated thiols has yet not been unambiguously demonstrated [25,26]. In addition to the possible hydrogen transfer [Eq. (10)], peroxyl radicals may directly transfer an oxygen atom to yield alcohols and sulfinyl radicals according to Eq. (11):

$$ROO^\bullet + RSH \rightarrow ROOH + RS^\bullet \tag{10}$$

$$ROO^\bullet + RSH \rightarrow ROH + RSO^\bullet \tag{11}$$

On the other hand, it has been argued that thiyl radicals were obtained during the reaction of the halothane-derived peroxyl radical ($CF_3CH(Cl)OO^\bullet$) with thiols [Eq. (12)] $k_{12} = (0.49 - 2.9) \times 10^7\ M^{-1}s^{-1}$ [27]. It should be noted, however, that under the employed experimental conditions, thiyl radicals could neither be observed directly nor indirectly by monitoring their strongly absorbing radical anion complex [Eq. (13)], since both species react rapidly with molecular oxygen (see below).

$$CF_3CH(Cl)OO^\bullet + RSH \rightarrow CF_3CH(Cl)OO^- + RS^\bullet + H^+ \tag{12}$$

$$RS^\bullet + RS^- \rightleftharpoons [RSSR]^{\bullet-} \tag{13}$$

Surprisingly, aliphatic alkoxyl radicals, although powerful oxidants, react rather slowly with thiols, and thiyl radicals were not detected [28]. On the other hand, one-electron oxidation reactions both of thiol and thiolate anion have been reported for phenoxyl radicals, e.g., of the water-soluble vitamin E analog trolox C [29] and of the amino acid tyrosine [30]. Such redox reactions between tyrosyl and cysteinyl radicals may play an important role in intramolecular electron-transfer reactions in proteins. In a recent study, Adam et al. [31] provided evidence for the one-electron reduction of 1,2-dioxetanes by thiols affording the corresponding thiyl radicals, and, thus, such mechanisms may also operate for hydrogen peroxide and organic hydroperoxides.

C. Enzymatic Oxidation of Thiols

The oxidation of various thiols including cysteine, *N*-acetyl cysteine, penicillamine, and glutathione by peroxidases, mainly horseradish peroxidase and lactoperoxidase, has been shown by means of ESR to yield thiyl free radicals, characterized as their adducts to spin traps such as 5,5′-dimethyl-1-pyrroline-*N*-oxide (DMPO) [32–34]. These reactions proceed aerobically as well as anaerobically in the presence of hydrogen peroxide and may represent an important source for thiyl free radicals in biological systems and/or for the synthesis of thiol metabolites such as cysteine sulfinic acid.

D. Oxidation of Thiols by Transition Metals

The autoxidation of thiols and its catalysis by transition metals has long been known [35,36]. By the spin-trapping technique it was demonstrated that Cu- and Fe-catalyzed thiol oxidation involves the intermediary formation of thiyl free radicals [37]. However, these processes simultaneously generate hydroxyl radicals or hydroxyl radical–like species [37], and it remains to be demonstrated whether the observed thiyl radicals directly originate from a one-electron transfer between the thiol and the transition metal [Eq. (14)] or from the subsequent oxidation of thiols by oxygen-centered free radicals (see above).

$$RSH + M^{(n+1)+} \rightarrow RS^{\bullet} + H^{+} + M^{n+} \tag{14}$$

In this respect, the observation that the addition of metal chelators such as EDTA to the reaction mixtures inhibited the metal-catalyzed formation of thiyl radical adducts but not of the hydroxyl radical adducts seems to indicate that it may be a direct one-electron transfer from thiol to the metal that produces the thiyl radical.

E. Reduction, Oxidation, and Photolysis of Disulfides

The one-electron reduction of a disulfide yields a disulfide radical anion that exists in equilibrium [Eq. (13)] with free thiyl radicals and thiolate:

$$RSSR + e^{-} \rightarrow [RSSR]^{\bullet -} \tag{15}$$

The actual yield of thiyl radicals from Eq. (15) at a given pH then depends on the respective equilibrium constant (K_{13}), the presence or absence of excess thiolate, the structure and pK_a of the thiol, and the pK_a of the initial disulfide radical anion [Eq. (16)]. Protonation of acyclic disulfide radical anions results in the instantaneous decomposition of the latter into thiyl radical and thiol [38]. On the other hand, protonated cyclic disulfide radical anions, e.g., from 4,5-dihydroxy-1,2-dithiane (ox-DTT), are relatively stable [39]. For ox-DTT, the reaction is characterized by $pK_{a,16} = 5.2$ [39]:

$$[RSSR]^{\bullet -} + H^+ \rightleftharpoons [R\text{-}\overset{H}{\overset{\wedge}{S\text{-}S}}\text{-}R]^{\bullet} \tag{16}$$

In the absence of excess thiolate, acyclic disulfide radical anions will rapidly decompose stoichiometrically into thiolate and thiyl radicals with $t_{1/2} \approx 1\text{–}2\ \mu s$, whereas cyclic disulfide radical anions such as from ox-DTT persist [39]. Relatively long-lived disulfide radical anions are also observed upon one-electron reduction of protein disulfide bonds where the decomposition would require the conformational change of a large biomolecule [40]:

$$\text{HO-HC(CH}_2\text{S)(HO-HC)(CH}_2\text{S)} + {}^{\bullet}C(CH_3)_2\text{-OH} \rightleftharpoons \text{HO-HC(CH}_2\text{S)(HO-HC)(CH}_2\text{S)}\cdot{}^{\bullet}\cdot C(CH_3)_2\text{-OH} \tag{17}$$

$$\text{HO-HC(CH}_2\text{S)(HO-HC)(CH}_2\text{S)}\cdot{}^{\bullet}\cdot C(CH_3)_2\text{-OH} \longrightarrow \text{HO-HC(CH}_2\text{S)(HO-HC)(CH}_2\text{S)}^{\bullet -} + H^+ + (CH_3)_2C{=}O \tag{18}$$

For experimental purposes the reduction of disulfides is easily achieved by the reaction with the hydrated electron formed via radiation chemical techniques [22]. However, employing ox-DTT, it has been shown that the formate radical, α-hydroxyalkyl radicals, and, even better, their respective anions (ketyl radicals) reduce disulfides, most probably via formation of a short-lived adduct between the carbon-centered radical and the disulfide [Eqs. (17) and (18)] [41].

Thiyl radicals are also formed upon the reaction of hydroxyl radicals with disulfides. This mechanism proceeds via the initial addition of a hydroxyl radical to the disulfide function with subsequent cleavage into $RS^{\bullet}$ and RSOH [42]:

$$HO^{\bullet} + RSSR \rightarrow R\text{-}{}^{\bullet}S(OH)\text{-}S\text{-}R \tag{19}$$

$$R\text{-}{}^{\bullet}S(OH)\text{-}S\text{-}R \rightarrow R\text{-}SOH + RS^{\bullet} \tag{20}$$

However, this pathway accounts for only approximately 50% of the decay of the hydroxyl radical adduct at the disulfide, with the residual fraction decomposing via formation of a disulfide radical cation and hydroxide [42].

The photolysis of α-unsubstituted disulfides at $\lambda \approx 254$ nm predominantly yields thiyl radicals [Eq. (21)] and only a minor fraction of perthiyl radicals via cleavage of the C-S bond [Eq. (22)] [43–45]:

$$RSSR + h\nu \rightarrow 2\ RS^{\bullet} \tag{21}$$

$$RSSR + h\nu \rightarrow R^{\bullet} + RSS^{\bullet} \quad (22)$$

However, the latter pathway gains importance for α-substituted disulfides such as ditertbutyldisulfide and penicillamine disulfide, where perthiyl radical formation constitutes the major pathway [44–46]. The dihedral angle along the C-S-S-C unit appears to be one important factor by influencing the energy levels of the nonbonding sulfur orbitals from which photolytic excitation occurs [47].

III. FORMATION OF PERTHIYL RADICALS

Besides the direct photolytic cleavage of disulfides, perthiyl radicals may be produced by triplet photosensitization [Eq. (23)] [48]:

$$Ph_2CO^* + t\text{-Bu-SS-Bu-}t \rightarrow Ph_2CO + t\text{-BuSS}^{\bullet} + t\text{-Bu}^{\bullet} \quad (23)$$

Several other pathways lead to the formation of perthiyl radicals from perthiols and from trisulfides. Hydrogen abstraction by alkyl radicals from perthiols occurs with rate constants higher by an order of magnitude than the analogous "repair reaction" for thiols, i.e., $k_{24} = 2.4 \times 10^9\ M^{-1}s^{-1}$ for the reaction of $(CH_3)_2C^{\bullet}(OH)$ with 2-(3-aminopropylamino) ethaneperthiol [49]:

$$R'^{\bullet} + RSSH \rightarrow R'H + RSS^{\bullet} \quad (24)$$

One-electron reduction of a trisulfide by the hydrated electron, hydrogen atom, formate radical anion, and by α-hydroxyalkyl radicals yields either perthiyl radical and thiolate or thiyl radical and perthiolate [45,50]:

$$RSSSR + e^- \rightarrow [RSSSR]^{\bullet -} \quad (25)$$

$$[RSSSR]^{\bullet -} \rightleftharpoons RSS^{\bullet} + RS^- \quad (26)$$

$$[RSSSR]^{\bullet -} \rightleftharpoons RS^{\bullet} + RSS^- \quad (27)$$

Whereas Eq. (26) is the predominant pathway for the reduction of penicillamine trisulfide, it accounts for only approximately 20% of the products obtained from cysteine trisulfide, with the residual fraction of 80% being formation of thiyl radicals according to Eq. (27) [45]. Clearly, the structure of the substituents adjacent to the trisulfide function plays an important, although as yet unidentified, role. This is indicated also by the fact that a relatively stable trisulfide anion intermediate ($t_{1/2} = 3.9\ \mu s$) could be observed for cysteine trisulfide, whereas its lifetime was much shorter for penicillamine trisulfide. The structural differences become important again for the reaction of hydroxyl radicals with both trisulfides, affording perthiyl radicals with 63% yield from penicillamine trisulfide and only 10% from cysteine trisulfide [45].

In biological systems, glutathione is the most prominent natural thiol involved in radioprotection, and, consequently, any trace amounts of perthiols or trisulfides generated by the interaction of ionizing radiation with GSH containing systems would originate from GSH. Although it has still to be demonstrated experimentally, we may speculate that the chemistry of GSH trisulfide should resemble the chemistry of cysteine trisulfide, i.e., the reactions outlined above should yield mainly thiyl rather than perthiyl radicals for the GSH trisulfide as well.

The reaction of hydroxyl radicals and $Br_2^{\bullet -}$ with the disulfides from cysteine, penicillamine, and cysteamine, but not from glutathione and homocysteine, afforded appreciable yields of perthiyl radicals. These reactions are proposed to proceed via initial abstraction of a C_α-H atom followed by β-cleavage yielding perthiyl radical and a dehydroalanine derivative [51]:

$$^{-}O_2C-CH(NH_2)-CH_2-S-S-CH_2-CH(NH_2)-CO_2^{-} \xrightarrow[-H_2O]{+HO^{\bullet}} {}^{-}O_2C-{}^{\bullet}C(NH_2)-CH_2-S-S-CH_2-CH(NH_2)-CO_2^{-} \quad (28)$$

$$^{-}O_2C-{}^{\bullet}C(NH_2)-CH_2-S-S-CH_2-CH(NH_2)-CO_2^{-} \longrightarrow {}^{-}O_2C-C(NH_2){=}CH_2 + {}^{\bullet}S-S-CH_2-CH(NH_2)-CO_2^{-} \quad (29)$$

IV. ONE-ELECTRON OXIDATIONS BY THIYL AND PERTHIYL RADICALS

Thiyl radicals undergo one-electron transfer reactions with many biologically relevant electron donors [Eq. (6)] such as NADH [Eq. (30)]. Rate constants for the forward reactions with various electron donors are summarized in Table 1. A remarkably high rate constant is obtained for the reaction of $GS^{\bullet}$ with vitamin A (retinol) [60]. This reaction occurs most probably via thiyl radical adduct formation [Eq. (31)] rather than direct electron transfer [Eq. (32)].

$$RS^{\bullet} + D^{-} \rightleftharpoons RS^{-} + D^{\bullet} \quad (6)$$

$$RS^{\bullet} + NADH \rightleftharpoons RS^{-} + H^{+} + NAD^{\bullet} \quad (30)$$

$$GS^{\bullet} + \text{retinol} \rightarrow [\text{GS-retinol}]^{\bullet} \quad (31)$$

$$GS^{\bullet} + \text{retinol} \rightarrow GS^{-} + [\text{retinol}]^{\bullet +} \quad (32)$$

In general, the rate constants for the oxidation of the electron donors by $RS^{\bullet}$ are at least two orders of magnitude higher than for the comparable oxidations by carbon-centered radicals, e.g., $k_{33} < 10^5\ M^{-1}s^{-1}$ (53):

$$(CH_3)_2C^{\bullet}(OH) + NADH \rightarrow (CH_3)_2CH(OH) + NAD^{\bullet} \quad (33)$$

Table 1 Rate Constants for One-Electron Oxidation of Various Electron Donors by Thiyl Radicals

Electron donor	k	$(M^{-1}s^{-1})$	Radical	Solvent	Ref.
Trolox C	pH 6.2	1.0×10^8	$CyaS^{\bullet}$	H_2O, 1.0 M ^tBuOH	29
	pH 12.2	8.0×10^8	$CyaS^{\bullet}$	H_2O, 1.0 M ^tBuOH	29
Vitamin E	pH 5.0	4.4×10^5	$PenS^{\bullet}$	H_2O, 6.7 M ^iPrOH	52
NADH	pH 6.0	2.3×10^9	$CyaS^{\bullet}$	H_2O, 1.0 M ^iPrOH/acetone	53
		5.8×10^8	$CysS^{\bullet}$	H_2O, 1.0 M ^iPrOH/acetone	53
		3.1×10^8	$PenS^{\bullet}$	H_2O, 1.0 M ^iPrOH/acetone	53
		2.3×10^8	$GS^{\bullet}$	H_2O, 1.0 M ^iPrOH/acetone	53
Ferrocyto-chrome *c*	pH 6.0	2.5×10^8	$GS^{\bullet}$	H_2O, 1.0 M ^iPrOH/acetone	54
		1.1×10^8	$CysS^{\bullet}$	H_2O, 1.0 M ^iPrOH/acetone	54
		4.7×10^7	$PenS^{\bullet}$	H_2O, 1.0 M ^iPrOH/acetone	54
Ascorbic acid	pH 6.5	1.3×10^9	$CyaS^{\bullet}$	H_2O, 1.0 M ^iPrOH/acetone	55
		1.2×10^9	$CysS^{\bullet}$	H_2O, 1.0 M ^iPrOH/acetone	55
		4.9×10^8	$PenS^{\bullet}$	H_2O, 1.0 M ^iPrOH/acetone	55
		6.0×10^8	$GS^{\bullet}$	H_2O, 1.0 M ^iPrOH/acetone	55
	pH 5.2	6.0×10^8	$GS^{\bullet}$	H_2O	56
Chloropromazine	pH 3.0	3.0×10^8	$CysS^{\bullet}$	H_2O, 1.0 M ^iPrOH/acetone	55
		2.7×10^8	$CyaS^{\bullet}$	H_2O, 1.0 M ^iPrOH/acetone	55
		1.2×10^8	$PenS^{\bullet}$	H_2O, 1.0 M ^iPrOH/acetone	55
		1.4×10^8	$GS^{\bullet}$	H_2O, 1.0 M ^iPrOH/acetone	55
	pH 5.5	9.9×10^8	$GS^{\bullet}$	H_2O	56
	pH 6.0	1.75×10^8	β-$MeS^{\bullet}$	H_2O	9
ABTS	pH 7.0	5.0×10^8	$CysS^{\bullet}$	H_2O	57
		3.3×10^8	$GS^{\bullet}$	H_2O	57
		1.05×10^9	$CyaS^{\bullet}$	H_2O, 1.0 M ^iPrOH/acetone	55
Aminopyrine	pH 6.0	3.9×10^8	$CyaS^{\bullet}$	H_2O, 1.0 M ^iPrOH/acetone	58
		2.6×10^8	$CysS^{\bullet}$	H_2O, 1.0 M ^iPrOH/acetone	58
		2.5×10^8	$GS^{\bullet}$	H_2O, 1.0 M ^iPrOH/acetone	58
	pH 4.9	3.0×10^8	$GS^{\bullet}$	H_2O	59
Vitamin A	pH 6.0	1.4×10^9	$GS^{\bullet}$	H_2O, MeOH	60
TMPD	pH 7.0	2.6×10^9	$C_2H_5S^{\bullet}$	H_2O	61

$CyaS^{\bullet}$ = cysteamine thiyl radical; $PenS^{\bullet}$ = penicillamine thiyl radical; $GS^{\bullet}$ = glutathione thiyl radical; $CysS^{\bullet}$ = cysteine thiyl radical; β-$MeS^{\bullet}$ = mercaptoethanol thiyl radical; ABTS = 2,2′-azinobis-(3-ethyl-benzthiazoline-6-sulfonate); TMPD = N,N,N′,N′-tetramethyl-1,4-phenylenediamine.

On the other hand, most carbon-centered radicals react with thiols with $k_3 = 10^6–10^8\ M^{-1}s^{-1}$. As pointed out by Willson and coworkers, it appears, therefore, reasonable to consider thiols as efficient "catalysts" for the repair of carbon-centered radicals by antioxidants through a sequence of hydrogen and electron transfer [53]. The oxidation of NADH and its deuterated analog, NAD^2H, did not reveal any primary kinetic isotope effect, indicating that Eq. (30) is a true electron transfer [53]. There is, at present, no information available as to whether thiyl radicals react with the electron donors via outer-sphere or inner-sphere, i.e., addition-elimination, mechanisms. Many examples have demonstrated the propensity of thiyl radicals to form intermediary thiyl radical adducts, e.g., in the displacement reactions with disulfides [62]. Therefore, it appears reasonable to assume, although it has still to be shown, that one-electron oxidations carried out by thiyl radicals involve intermediary adducts.

The back reaction of Eq. (6), i.e., the oxidation of thiolate by the oxidized electron donor, is generally slow as compared to the forward reaction. For example, the oxidation of aminopyrine by glutathione thiyl radicals, $GS^\bullet$, occurs with $k_{34} = (2.5–3.0) \times 10^8\ M^{-1}s^{-1}$, whereas the reverse exhibits a rate constant of $k_{-34} = (2\text{-}3) \times 10^4\ M^{-1}s^{-1}$, i.e., four orders of magnitude lower.

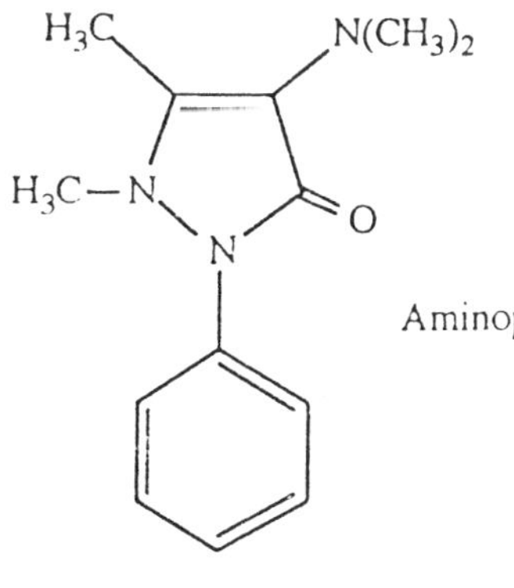

Aminopyrine

$$GS^\bullet + AP \rightleftharpoons GS^- + AP^{\bullet +} \tag{34}$$

Nevertheless, the addition of reduced glutathione to a solution containing aminopyrine radical cations resulted in a rapid overall reduction of the latter [59]. This observation can be explained by the fact that thiyl radicals are removed from Eq. (34) by additional pathways, namely, diffusion-controlled recombination [Eq. (35)] and/or the addition of oxygen [see Eq. (37)].

$$2\ RS^\bullet \rightarrow RSSR \tag{35}$$

On the other hand, the aminopyrine radical cations are relatively stable, decomposing with $2k = 52\ M^{-1}s^{-1}$ [59].

There are as yet few rate constants available for one-electron oxidation reactions carried out by perthiyl radicals. Penicillamine perthiyl radicals, $PenSS^{\bullet}$, oxidize ascorbate with $k_{36} = 4.1 \times 10^6 \ M^{-1}s^{-1}$ [45], i.e., with a rate constant two orders of magnitude smaller than that for the corresponding reaction of the penicillamine thiyl radical.

$$PenSS^{\bullet} + AH^{-} \rightarrow PenSS^{-} + H^{+} + A^{\bullet -} \quad (36)$$

This lower reactivity can be explained by the resonance stabilization of perthiyl radicals through delocalization of the electron spin over the two sulfur atoms. The resonance stabilization energy amounts to approximately 8.8 kJ mol^{-1} [49].

As displayed in Eq. (37), the addition of oxygen to a thiyl radical yields a thiylperoxyl radical [63–67]. This oxygen addition is reversible [65,68], and the corresponding rate constants have been subject to much controversy.

$$RS^{\bullet} + O_2 \rightleftharpoons RSOO^{\bullet} \quad (37)$$

Recent values of $k_{37} = 2.2 \times 10^9 \ M^{-1}s^{-1}$ and $k_{-37} = 6.2 \times 10^5 \ s^{-1}$ have been obtained for thiyl radicals from β-mercaptoethanol [68] and confirm earlier rate constants for glutathione thiyl radicals of $k_{37} = 2.0 \times 10^9 \ M^{-1}s^{-1}$ and $k_{37} = 6.5 \times 10^5 \ s^{-1}$, respectively [65]. A value of $k_{37} = 3.75 \times 10^9 \ M^{-1}s^{-1}$ for thiyl radicals from 2-mercaptoethanol was also obtained via competition with hydrogen transfer from polyunsaturated fatty acids [69] (see below).

Thiylperoxyl radicals are weaker oxidants than their thiyl radical precursors. Table 2 displays a comparison of rate constants for the oxidation of ascorbate and chloropromazine by $GS^{\bullet}$ and $GSOO^{\bullet}$, obtained at pH 5.2 by Tamba and O'Neill [56]. The respective values for $GSOO^{\bullet}$ are approximately half of those for $GS^{\bullet}$. On the other hand, also on the basis of experiments with $GS^{\bullet}$ and chloropromazine, Wardman suggested that the rate constant k_{38} should be at least an order of magnitude smaller than k_{39} [70].

Table 2 Rate Constants for Oxidation of Ascorbate and Chloropromazine by Glutathione Thiyl and Thiylperoxyl Radicals at pH 5.2

Radical	Antioxidant	k ($M^{-1}s^{-1}$)
$GS^{\bullet}$	Ascorbate	6.0×10^8
$GS^{\bullet}$	Chloropromazine	9.9×10^8
$GSOO^{\bullet}$	Ascorbate	2.1×10^8
$GSOO^{\bullet}$	Chloropromazine	5.3×10^8

$GS^{\bullet}$ = glutathione thiyl radical; $GSOO^{\bullet}$ = glutathione thiylperoxyl radical.
Source: Ref. 56.

$$GSOO^{\bullet} + CPZ \rightarrow CPZ^{\bullet +} + \text{products} \quad (38)$$

$$GS^{\bullet} + CPZ \rightarrow CPZ^{\bullet +} + GS^{-} \quad (39)$$

One reason for such discrepancy might be the fact that thiylperoxyl radicals have been shown to undergo additional unimolecular reactions. Promoted by elevated temperatures and light, thiylperoxyl radicals rearrange to sulfonyl radicals [Eq. (40)] [71]. A corresponding rate constant of $k_{40} = 2 \times 10^3\ s^{-1}$ has been derived for thiyl radicals from 2-mercaptoethanol [68].

$$RSOO^{\bullet} \rightarrow RS(O)O^{\bullet} \quad (40)$$

In contrast to thiylperoxyl radicals, the sulfonyl radicals appear to be very powerful one-electron oxidants. In one example, the rate constant for ABTS oxidation by the model sulfonyl radical $CH_3S(O)O^{\bullet}$ was determined to be $k_{41} = 1.9 \times 10^9\ M^{-1}s^{-1}$ [72], i.e., approximately four times higher than rate constants for ABTS oxidation by thiyl radicals.

$$CH_3S(O)O^{\bullet} + ABTS \rightarrow CH_3S(O)O^{-} + ABTS^{\bullet +} \quad (41)$$

Under aerobic conditions sulfonyl radicals would also add molecular oxygen to yield sulfonyl peroxyl radicals [Eq. (42)], which also potentially oxidize ABTS. A preliminary rate constant of $k_{42} \approx 3.0 \times 10^8\ M^{-1}s^{-1}$ has been obtained without further examining whether such a process is reversible or not [72].

$$CH_3S(O)O^{\bullet} + O_2 \rightarrow CH_3S(O_2)OO^{\bullet} \quad (42)$$

An overall rate constant for the addition of oxygen to penicillamine perthiyl radicals was reported to be $k_{43} = 5.1 \times 10^6\ M^{-1}s^{-1}$, again, without further examining the potential reversibility of such a reaction [45].

$$PenSS^{\bullet} + O_2 \rightarrow PenSSOO^{\bullet} \quad (43)$$

Steady-state radiolysis experiments revealed the efficient formation of sulfate from perthiylperoxyl radicals, a process involving unimolecular rearrangement yielding perthiylsulfonyl radicals, which add oxygen, recombine, and eliminate SO_3 [45].

V. HYDROGEN ABSTRACTION REACTIONS BY THIYL AND PERTHIYL RADICALS

Taking the bond dissociation energy of the mercapto group of ~90 kcal/mol [73], it is expected that thiyl radicals abstract hydrogen atoms from hydrogen donors with C-H bond dissociation energies of comparable and/or lower magnitude. Experimentally, the potential involvement of thiyl radicals in hydrogen abstraction reactions has been documented, for example, by their reaction with "activated" C-H bonds of 2-mercaptoethanol, inducing a chain process [74], formate [75], amines [76], various ethylbenzene, cumene, and toluene derivatives [77], and with phosphite, HPO_3^{2-} [78]. The homolytic cleavage of such "activated" C-H bonds yields carbon-centered radicals, which derive stabilization from conjugation of the radical center with π-systems or with heteroatom electron lone pairs (such as from oxygen, sulfur, and nitrogen). Such stabilization results in a considerable reduction of the C-H bond dissociation energies to values close or below the S-H bond dissociation energy of thiols [see Ref. 79, e.g., 105 kcal/mol for H_3C-H vs. 91 kcal/mol for $(CH_3)_2(OH)C$-H and 82 kcal/mol for bisallylic methylene groups, $(R_2C{=}CH)_2CH$-H]. In recent years, it has been shown that the repair reaction [Eq. (3)] is, in fact, reversible and should be more exactly formulated as Eq. (3a) [16,52,80,81]:

$$R'^{\bullet} + RSH \rightleftharpoons R'H + RS^{\bullet} \quad (3a)$$

As described above, the rate constants for the forward reactions [Eq. (3a)] are, in most cases, on the order of $k_{3a} = 10^6$–$10^8\ M^{-1}s^{-1}$. In contrast, absolute rate constants for the back reactions, measured for a number of model alcohols and ethers, are on the order of $k_{-3a} = 10^3$–$10^4\ M^{-1}s^{-1}$ [16,52,80,81], i.e., lower by two to five orders of magnitude. However, in order to assess the potential biological importance of the reverse reaction, we have to take into account not only these individual rate constants but also (1) the respective concentrations of the reactants and (2) the presence of potential scavengers of either the thiyl or the carbon-centered radical on both sides of Eq. (3a), respectively. In a model system consisting of 2×10^{-2} M PenSH and 1.0–6.7 M 2-propanol, it was shown, for example, that an added concentration of approximately 5×10^{-4} M carbon tetrachloride, CCl_4, was sufficient to shift equilibrium 44 completely onto the left-hand side, i.e., to induce a thiyl radical–induced oxidation of 2-propanol [52]. This occurred because of the efficient removal of (2-hydroxy)propan-2-yl radicals from Eq. (44) by electron transfer to CCl_4 according to Eq. (45) (in the presence of 6.7 M 2-propanol: $k_{45} = 1.2 \times 10^8\ M^{-1}s^{-1}$ [52]).

$$(CH_3)_2C^{\bullet}OH + PenSH \rightleftharpoons (CH_3)_2CHOH + PenS^{\bullet} \quad (44)$$

$$(CH_3)_2C^{\bullet}OH + CCl_4 \rightarrow (CH_3)_2C{=}O + {}^{\bullet}CCl_3 + H^+ + Cl^- \quad (45)$$

Molecular oxygen adds diffusion controlled with $k_{46} \approx 2 \times 10^9\ M^{-1}s^{-1}$ to most of the carbon-centered radicals [82] (with the exception of carbon-centered radicals that are part of a conjugated system, e.g., pentadienyl radicals; see below):

$$R'^{\bullet} + O_2 \rightarrow R'\text{-}OO^{\bullet} \quad (46)$$

Therefore, molecular oxygen constitutes an ideal scavenger of carbon-centered radicals in biological systems even at physiological oxygen concentrations of approximately 5×10^{-5} M in tissue and should theoretically be able to shift reactions such as Eq. (3a) onto the left-hand side. It is well established in radiation biology that molecular oxygen participates in the "fixation" of DNA damage by nonreversible reaction with carbon-centered radicals in competition to their repair by thiols [11]. The considerations above offer an additional aspect by which oxygen may enhance radiation-induced DNA damage, namely, by reversing the "repair reaction" through Eq. (3a). Molecular oxygen will, of course, also remove thiyl radicals from Eq. (3a), but only in a reversible manner [Eq. (37)]. On the other hand, electron donors that efficiently reduce thiyl radicals such as vitamins E and ABTS have been shown to inhibit the oxidation of C-H bonds by thiyl radicals [52,72]. It should be noted that an efficient attack on C-H bonds by thiyl radicals requires high concentrations of the target molecules because of the low rate constants. Such conditions may exist locally when thiyl radicals are produced in the close vicinity of DNA macromolecules through "repair processes," i.e., the forward reaction of Eq. (3a).

Recently, the first rate constants for the analogous Eq. (47) for 2-(3-aminopropylamino) ethaneperthiyl radicals have been determined to be $k_{47} = 2.4 \times 10^9\ M^{-1}s^{-1}$ and $k_{-47} = 3.8 \times 10^3\ M^{-1}s^{-1}$ [49].

$$(CH_3)_2C^{\bullet}OH + RSSH \rightleftharpoons (CH_3)_2CHOH + PenS^{\bullet} \tag{47}$$

From the corresponding equilibrium constant it appears, therefore, that perthiols should be better "repair" agents than thiols. However, the perthiol group deprotonates with $pK_a = 6.2 \pm 0.1$ [49]:

$$RSSH \rightleftharpoons RSS^- + H^+ \tag{48}$$

This means that at physiological pH approximately 95% of the perthiol exists in the deprotonated form, which, like its deprotonated thiol analog, 2-(3-aminopropylamino) ethanethiolate, is unreactive toward carbon-centered radicals from alcohols [49]. On the other hand, the thiol analog exhibits a $pK_a = 7.6 \pm 0.1$, corresponding to a fraction of 20% being deprotonated at physiological pH, resulting in a considerably larger "effective" concentration of protonated thiol available for the repair reaction. It has therefore been argued [49] that the benefit of higher "repair" efficiency of the perthiol through hydrogen atom donation is most probably compensated by the disadvantageous pK_a. However, the deprotonated perthiol may still serve as a good electron donor for the interception of oxidizing radicals such as peroxyl radicals.

Since thiyl radicals participate in *inter*molecular abstractions of hydrogen atoms from "activated" C-H bonds, it is not surprising that they also undergo *intra*molecular hydrogen abstraction processes. The thiyl radical from dithiothreitol was shown to react via a 1,4-H shift to produce a β-mercaptoalkyl radical, which eliminates either H_2S or $HS^{\bullet}$, with the latter oxidizing a second dithiothreitol molecule, propagating a chain reaction, and producing the final product, H_2S, which was experimentally detected [83]:

HO-HC(CH_2-S$^\bullet$)-C(OH)(H)-CH_2-SH $\xrightarrow{(48)}$ HO-HC(CH_2-SH)-C$^\bullet$(OH)-CH_2-SH

(50): HO-HC(CH_2-SH)-C(=O)-CH_3 + $HS^\bullet$

(49): HO-HC(CH_2-SH)-C(=O)-$^\bullet CH_2$ + H_2S

Thiyl radicals from 2-mercaptoethanol participate in a 1,2-H shift [Eq. (51)] [68], analogous to the processes established for alkoxyl radicals [84]. The corresponding rate constants have been measured to be $k_{51} \approx 10^3\ s^{-1}$ and $k_{-51} \approx 3 \times 10^4\ s^{-1}$ [68].

$$HO\text{-}CH_2\text{-}CH_2\text{-}S^\bullet \rightleftharpoons HO\text{-}CH_2C^\bullet H\text{-}SH \qquad (51)$$

Finally, glutathione thiyl radicals abstract hydrogen atoms from the C_α atom of the γ-glutamyl moiety according to Eq. (52) [85]. This pathway constitutes an interesting conversion of an oxidizing radical, $RS^\bullet$, into a reducing species, an α-amino–substituted carbon-centered radical, $H_2N\text{-}C^\bullet H(CO_2^-)$-*peptide*. The latter process is catalyzed by base, promoting deprotonation of the N-terminal amino group, which enables the nitrogen electron lone pair to stabilize the developing carbon-centered radical. So far glutathione is the only peptide for which such an intramolecular hydrogen transfer has been demonstrated. It remains to be shown whether such processes can also lead to the degradation of proteins, e.g., by way of hydrogen transfer from C_α-H bonds located in the protein backbone.

Polyunsaturated fatty acids (PUFAs) constitute an important class of biomolecules containing particularly labile C-H bonds. Thiyl radicals abstract hydrogen atoms from bisallylic methylene groups of PUFAs, forming pentadienyl radicals, with rate constants on the order of $k_{53} = 10^6\text{–}10^8\ M^{-1}s^{-1}$, depending on the structure of both the thiyl radicals and the PUFAs [69,80,81,86]. Absolute rate constants for individual thiyl radicals and PUFAs are summarized in Table 3. Although, theoretically, Eq. (53) should be formulated as an equi-

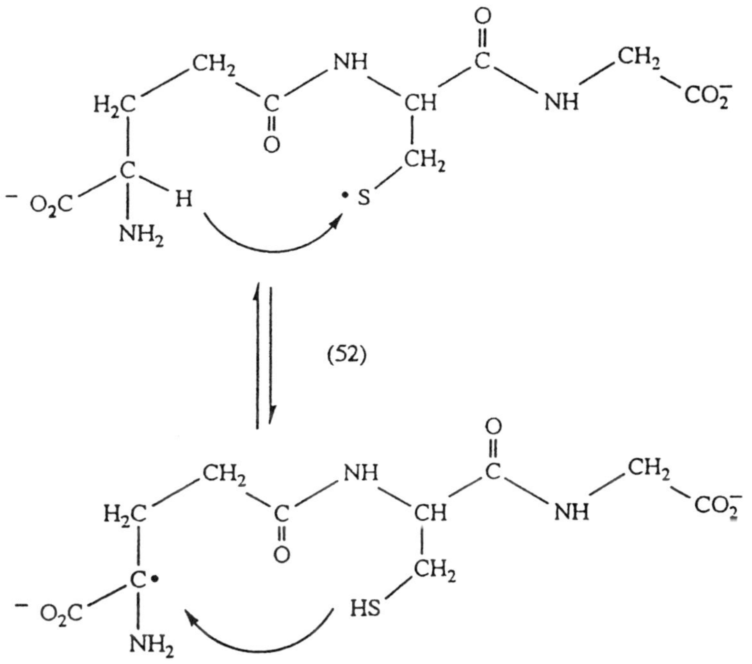

(52)

Table 3 Absolute Rate Constants for Hydrogen Abstraction by Thiyl Radicals from Polyunsaturated Fatty Acids

	k (10^7 $M^{-1}s^{-1}$)			
Fatty acid	$GS^{\bullet}$	$CysS^{\bullet}$	$PenS^{\bullet}$	β-$MeS^{\bullet}$
t-18:2	0.7	—	—	—
c-18:2	0.8	0.6	0.3	3.1
c-18:3	1.9	0.9	0.4	4.5
c-22:3	1.9	—	—	—
c-20:4	3.1	1.6	0.5	6.8
c-22:6	3.9	2.3	1.1	7.5

c = *cis*; *t* = *trans*; $GS^{\bullet}$ = glutathione thiyl radical; $CysS^{\bullet}$ = cysteine thiyl radical; $PenS^{\bullet}$ = penicillamine thiyl radical; β-$MeS^{\bullet}$ = β-mercaptoethanol thiyl radical.
Source: Ref. 69.

librium process, the "repair reaction," i.e., the hydrogen transfer from thiol to the resulting pentadienyl radicals, is extremely slow, and so far rate constants have not been measured. The pentadienyl radicals constitute precursors for various lipid peroxidation products, and, thus, thiyl radicals have the potential to initiate lipid peroxidation:

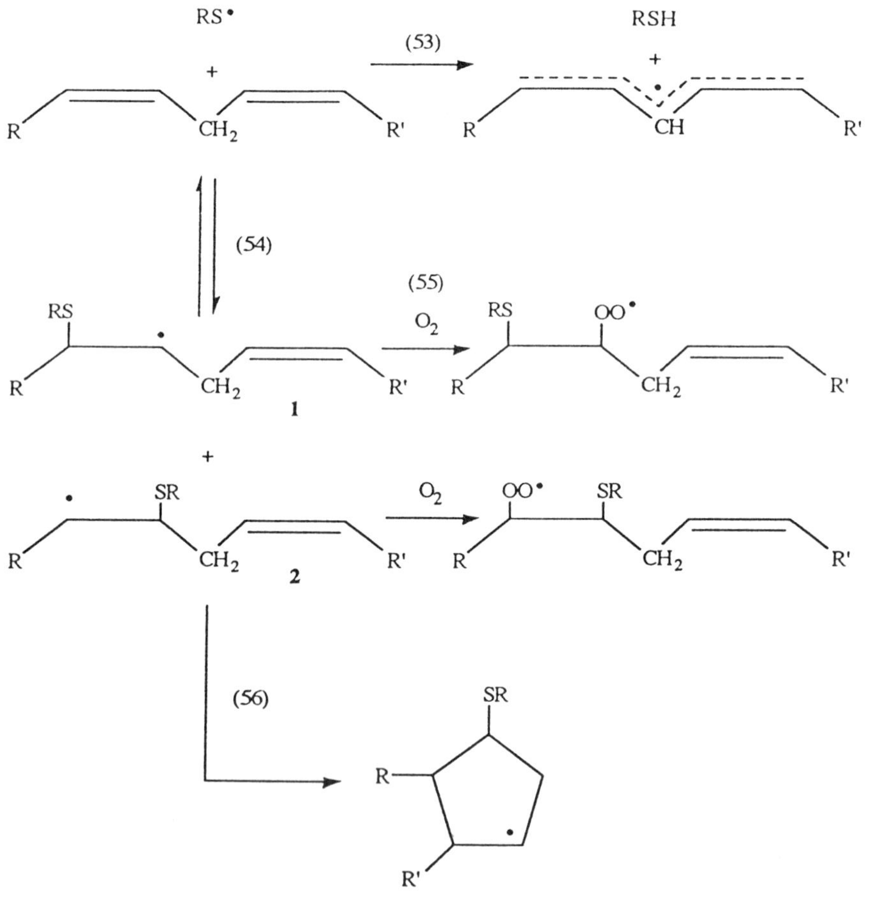

RS• + R–CH=CH–CH$_2$–CH=CH–R' —(53)→ RSH + R–CH=CH–•CH–CH=CH–R'

(54) ⇌

R–CH(SR)–•CH–CH$_2$–CH=CH–R' (**1**) —(55) O_2→ R–CH(SR)–CH(OO•)–CH$_2$–CH=CH–R'

+

R–•CH–CH(SR)–CH$_2$–CH=CH–R' (**2**) —O_2→ R–CH(OO•)–CH(SR)–CH$_2$–CH=CH–R'

(56) → cyclopentyl radical bearing SR, R and R'

Alternatively, thiyl radicals may add to the double bonds yielding β-(alkylthio) alkyl radicals of type **1** or **2**, respectively [Eq. (54)]. The latter processes are reversible, and rate constants for Eq. (54) may be on the order of $k_{54} \approx 1 \times 10^8$ $M^{-1}s^{-1}$ [86,87] and $k_{-54} = (2\text{–}3) \times 10^5$ s^{-1} [88]. In biological systems β-(alkylthio)alkyl radicals may subsequently be trapped by addition of molecular oxygen [Eq. (55)], yielding peroxyl radicals that may initiate further lipid peroxidation processes. In fact, Ito and Matsuda [86] have employed molecular oxygen to scavenge β-(alkylthio)alkyl radicals, formed by the addition of various substituted thiyl radicals to alkenes, for measuring the absolute rate constants for these addition reactions. Thus, also by addition reactions, thiyl radicals may initiate lipid peroxidation. In addition, the β-(alkylthio)alkyl radicals of type **2** have a theoretical propensity for undergoing 1,5-cyclization reaction [Eq. (56)], processes that occur with rate constants on the order of $k \approx 10^5\ s^{-1}$ [89], and may further contribute to trapping thiyl radical addition products of PUFAs. The subsequent addition of molecular oxygen to the resulting cyclopentane radicals will most probably be diffusion controlled [82].

Biologically, it seems rather improbable that thiyl radicals from water soluble thiols such as glutathione or cysteine would attack PUFAs which are incorporated in the lipid bilayer of membranes. However, thiyl radicals from protein bound thiols, particularly of membrane spanning proteins, or from more lipophilic thiols such as dithiothreitol or 2-mercaptoethanol may do so.

VI. REACTIONS WITH METAL COMPLEXES

Thiyl radicals have been found to react via one-electron transfer, hydrogen abstraction, addition, and homolytic displacement with various organometallic complexes. For example, the oxidation of $[Co(sep)]^{2+}$ [=(S)-(1,3,6,8,10,13,16,19-octaazabicyclo[6.6.6]eicosane)cobalt (II)], $[Fe(CN)_6]^{4-}$, and $[Ru(NH_3)_6]^{2+}$ by ethylthiyl radicals is believed to proceed via outer-sphere electron transfer, whereas an inner-sphere mechanism appears to prevail for the oxidation of $[Cr(H_2O)_6]^{2+}$ and $[Fe(H_2O)_6]^{2+}$ [Eqs. (57)–(59)] [61]. The relatively stable intermediate ($t_{1/2} \approx 30$ min), $[(H_2O)_5Cr\text{-}S\text{-}C_2H_5]^{2+}$, formed according to Eq. (57), was isolated, but no thiol, which would be expected upon its further decomposition. On the other hand, the analogous intermediate of Eq. (58), $[C_2H_5\text{-}S\text{-}Fe(H_2O)_5]^{2+}$, could not be isolated, but the resulting thiolate formed according to Eq. (59):

$$C_2H_5S^\bullet + [Cr(H_2O)_6]^{2+} \rightarrow [C_2H_5\text{-}S\text{-}Cr(H_2O)_5]^{2+} + H_2O \qquad (57)$$

$$C_2H_5S^\bullet + [Fe(H_2O)_6]^{2+} \rightarrow [C_2H_5\text{-}S\text{-}Fe(H_2O)_5]^{2+} + H_2O \qquad (58)$$

$$[C_2H_5S\text{-}Fe(H_2O)_5]^{2+} + H_2O \rightarrow C_2H_5S^- + [Cr(H_2O)_6]^{3+} \qquad (59)$$

The reaction of ethylthiyl radicals with complexes of the general structure $C_2H_5M^{2+}$ (M = Co, Cr) produced the corresponding thiol and ethylene and was characterized by a significant primary kinetic isotope effect [61], facts that are accommodated by a hydrogen transfer mechanism:

$$C_2H_5S^\bullet + C_2H_5M^{2+} \rightarrow C_2H_5SH + C_2H_4 + M^{2+} \qquad (60)$$

Rate constants of $k_{60} \approx 10^8\ M^{-1}s^{-1}$ [61] strongly suggest that the C-H bonds in these complexes are highly activated, as a comparably high rate constant for the hydrogen abstraction by a thiyl radical, $k_{61} = 1.6 \times 10^9\ M^{-1}s^{-1}$, was only reported for the highly activated C-H bond of the 1,6-diazabicyclo[4.4.4]dodecane radical cation, $[4.4.4]^{\bullet +}$ [91]:

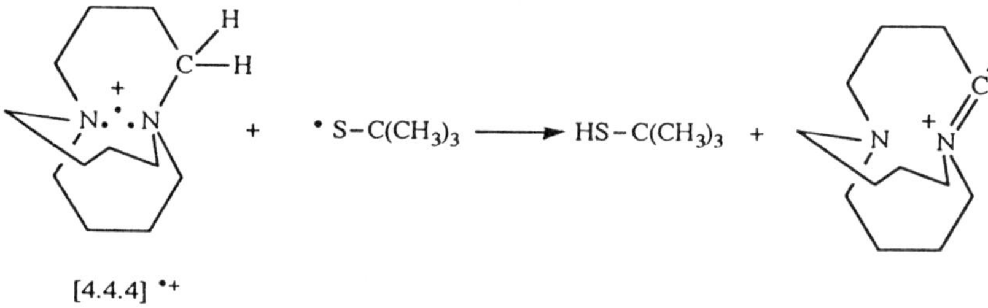

$$[4.4.4]^{\bullet +} + {}^{\bullet}S{-}C(CH_3)_3 \longrightarrow HS{-}C(CH_3)_3 + [\text{iminium cation}] \quad (61)$$

$[4.4.4]^{\bullet +}$

A homolytic displacement process instead of hydrogen transfer was observed upon the reaction of ethylthiyl radicals with $CH_3Co([14]aneN_4)H_2O)^{2+}$, yielding ethylmethylsulfide ($[14]aneN_4$ = 1,4,8,11-tetraazacyclotetradecane).

Detailed investigation of the reaction of thiyl radicals from cysteine and penicillamine with RSH/Cu(I) complexes has revealed a rather complex reaction scheme [Eqs. (62)–(68)] [92–94]. Depending on the Cu(I)/thiol ratio, these compounds exist as mono- or binuclear complexes $Cu(I)L_2$ or $Cu(I)_2L_3$, respectively [Eq. (62)]. Both species are oxidized by cysteinyl radicals, $CysS^{\bullet}$, with $k_{64} = 1.8 \times 10^9\ M^{-1}s^{-1}$ and $k_{66} = 1.0 \times 10^9\ M^{-1}s^{-1}$ (in this case L = CysSH). Moreover, the $Cu(I)L_2$ complex is efficiently oxidized also by the $[CysSSCys]^{\bullet -}$ species [Eq. (65)] ($k_{65} = 2.7 \times 10^8\ M^{-1}s^{-1}$), formed upon complexation of $CysS^{\bullet}$ with excess $CysS^-$ [Eq. (63)]. Whereas the oxidation of $Cu(I)L_2$ directly yields $Cu(II)L_2$, the oxidation of $Cu(I)_2L_3$ yields $Cu(I)Cu(II)L_3$, which converts into $Cu(I)L_2$ by complexation with an additional ligand L [Eqs. (67) and (68)]:

$$Cu(I)L_2 + Cu(I)L_2 \rightleftharpoons Cu(I)_2L_3 + L \quad (62)$$

$$CysS^{\bullet} + CysS^- \rightleftharpoons [CysSSCys]^{\bullet -} \quad (63)$$

$$CysS^{\bullet} + Cu(I)L_2 \rightarrow CysS^- + Cu(II)L_2 \quad (64)$$

$$[CysSSCys]^{\bullet -} + Cu(I)L_2 \rightarrow 2\ CysS^- + Cu(II)L_2 \quad (65)$$

$$CysS^{\bullet} + Cu(I)_2L_3 \rightarrow CysS^- + Cu(I)Cu(II)L_3 \quad (66)$$

$$Cu(I)Cu(II)L_3 + L \rightarrow Cu(I)Cu(II)L_4 \quad (67)$$

$$Cu(I)Cu(II)L_4 \rightarrow Cu(I)L_2 + Cu(II)L_2 \quad (68)$$

A similar sequence was also established for penicillamine thiyl radicals, PenS$^\bullet$, although the corresponding rate constants were generally lower due to the steric hindrance imposed by the additional methyl substituents. Interestingly, PenS$^\bullet$ radicals were also found to reduce Cu(II)(PenS$^-$)$_2$ according to Eqs. (69) and (70) ($k_{69} = 3.4 \times 10^8$ M^{-1}s^{-1} and an estimate of $k_{70} \approx 10^8$ s^{-1}) [94]:

$$\mathrm{PenS^\bullet + Cu(II)(PenS^-)_2 \rightarrow (PenS^-)Cu(II)(PenSSPen)^{\bullet -}} \quad (69)$$

$$\mathrm{(PenS^-)Cu(II)(PenSSPen)^{\bullet -} \rightarrow (PenS^-)Cu(I)(PenSSPen)} \quad (70)$$

VII. CONCLUSION

This chapter discusses the variety of oxidation reactions of potential biological significance carried out by thiyl and perthiyl radicals. Many more possible pathways would have to be discussed to include other reactions of potential synthetic use, such as addition reactions or the involvement of thiols and thiyl radicals in organic chain reactions like the reduction of carbon-halogen bonds in the presence of silanes [95]. Thiyl radicals not only participate in oxidation reactions, but may initiate reduction processes by complexation to thiolate. For example, the disulfide radical anion has been shown to reduce molecular oxygen to the superoxide anion [18,96]:

$$\mathrm{RSSR^{\bullet -} + O_2 \rightleftharpoons RSSR + O_2^{\bullet -}} \quad (71)$$

The reactions discussed here have mostly been investigated for small molecular weight thiols and thiyl radicals. In the future, it will be necessary to extend such studies to biological macromolecules, such as proteins. The amino acid cysteine is extremely sensitive to oxidation, and, in fact, loss of protein thiols is one prominent pathway of protein oxidation and/or inactivation. It has been shown that cysteine is not only converted to the respective disulfide [97,98], and the characterization of protein thiyl radical reactivity will be helpful for a detailed understanding of the stability of proteins.

REFERENCES

1. Ketterer, B., Meyer, D. J., and Tank, K. H. (1988) The role of glutathione transferase in the detoxication and repair of lipid and DNA hydroperoxides. In: Oxygen Radicals in Biology and Medicine (Simic, M. G., Taylor, K. A., Ward, J. F., and von Sonntag, C., eds.), pp. 669–674. Plenum Press, New York.
2. Que, Jr., L., ed. (1988) Metal Clusters in Proteins. American Chemical Society Symposium Series 372, ACS, Washington, DC.
3. Creighton, T. E. (1992) Folding pathways determined using disulfide bonds. In: Protein Folding (Creighton, T. E., ed.), pp. 301–351. W.H. Freeman and Company, New York.

4. Stamler, J. S., Singel, D. J., and Loscalzo, J. (1992) Biochemistry of nitric oxide and its redox-activated forms. Science 258:1898–1902.
5. Benesch, R. E., and Benesch R. (1955) The acid strength of the -SH group in cysteine and related compounds. J. Am. Chem. Soc. 77:5877–5881.
6. Long, C., ed. (1968) Biochemist's Handbook, p. 46, E.&F.N. SPON Ltd., London.
7. Packer, J. E. (1974) The radiation chemistry of thiols. In: The Chemistry of the Thiol Group, Part 2 (Patai, S., ed.), pp. 481–517. John Wiley & Sons, London.
8. Von Sonntag, C. (1987) The Chemical Basis of Radiation Biology. Taylor and Francis, London.
9. Surdhar, P. S., and Armstrong, D. A. (1987) Reduction potentials and exchange reactions of thiyl radicals and disulfide anion radicals. J. Phys. Chem. 91:6532–6537.
10. Prütz, W. A. (1992) Catalytic reduction of Fe(III)-cytochrome-c involving stable radiolysis products derived from disulfides, proteins and thiols. Int. J. Radiat. Biol. 61:593–602.
11. Von Sonntag, C., and Schuchmann, H.-P. (1990) Sulphur compounds and "chemical repair" in radiation biology. In: Sulfur-Centered Reactive Intermediates in Chemistry and Biology (Chatgilialoglu, C., and Asmus, K.-D., eds.), pp. 409–414. NATO ASI Series A, Vol. 197, Plenum Press, New York.
12. Adams, G. E., McNaughton, G. S., and Michael, B. D. (1968) Pulse radiolysis of sulphur compounds. Part 2. Free radical "repair" by hydrogen transfer from sulfydryl compounds. Trans. Faraday Soc. 64:902–910.
13. Adams, G. E., Armstrong, R. C., Charlesby, A., Michael, B. D., and Willson, R. L. (1969) Pulse radiolysis of sulphur compounds. Part 3. Repair by hydrogen transfer of a macromolecule irradiated in aqueous solution. Trans. Faraday Soc. 65:732–742.
14. Baker, M. Z., Badiello, R., Tamba, M., Quintiliani, M., and Gorin, G. (1982) Pulse radiolytic study of hydrogen transfer from glutathione to organic radicals. Int. J. Radiat. Biol. 41:595–602.
15. Von Sonntag, C. (1990) Free-radical reactions involving thiols and disulfides. In: Sulfur-Centered Reactive Intermediates in Chemistry and Biology (Chatgilialoglu, C., and Asmus, K.-D., eds.), pp. 359–366. NATO ASI Series A, Vol. 197, Plenum Press, New York.
16. Akhlaq, M. S., Schuchmann, H.-P., and von Sonntag, C. (1987) The reverse of the "repair" reaction of thiols: H-abstraction at carbon by thiyl radicals. Int. J. Radiat. Biol. 51:91–102.
17. Akhlaq, M. S., Al-Baghdadi, S., and von Sonntag, C. (1987) On the attack of hydroxyl radicals on polyhydric alcohols and sugars and the reduction of the so-formed radicals by 1,4-dithiothreitol. Carbohydr. Res. 164:71–83.
18. Redpath, J. L. (1973) Pulse radiolysis of dithiothreitol. Radiat. Res. 54:364–374.
19. Karmann, W., and Henglein, A. (1967) Pulse radiolytic study of the reactions of organic radicals with hydrogen sulfide. Ber. Bunsenges. Phys. Chem. 71:421–424.
20. O'Neill, P. (1983) Pulse radiolytic study of the interaction of thiols and ascorbate with OH adducts of dGMP and dG: Implications for DNA repair processes. Radiat. Res. 96:198–210.

21. Purdie, J. W., Gillis, H. A., and Klassen, N. V. (1971) Pulse radiolysis of penicillamine in aqueous solution: The thiyl radical and the disulfide radical anion. J. Chem. Soc. Chem. Comm.: 1163–1165.
22. Hoffman, M. Z., and Hayon, E. (1973) Pulse radiolysis study of sulfhydryl compounds in aqueous solution. J. Phys. Chem. 77:990–996.
23. Augusto, O., Gatti, R. M., and Radi, R. (1994) Spin-trapping studies of peroxynitrite decomposition and of 3-morpholinosydnonimine N-ethylcarbamide autooxidation: Direct evidence for metal-independent formation of free radical intermediates. Arch. Biochem. Biophys. 310:118–125.
24. Zhang, N., Schuchmann, H.-P., and von Sonntag, C. (1991) The reaction of superoxide radical anion with dithiothreitol: A chain process. J. Phys. Chem. 95:4718–4722.
25. Schulte-Frohlinde, D., Behrens, G., and Önal, A. (1986) Lifetime of peroxyl radicals of poly(U), poly(A) and single- and double-stranded DNA and their rate of reaction with thiols. Int. J. Radiat. Biol. 50:103–110.
26. Swarts, S. G., Becker, D., DeBolt, S., and Sevilla, M. D. (1989) An electron spin resonance investigation of the structure and formation of sulfinyl radicals: reaction of peroxyl radicals with thiols. J. Phys. Chem. 93:155–161.
27. Mönig, J. (1984) Ph.D. thesis, Technical University Berlin.
28. Erben-Russ, M., Michel, C., Bors, W., and Saran, M. (1987) Absolute rate constants of alkoxyl radical reactions in aqueous solution. J. Phys. Chem. 91:2362–2365.
29. Davies, M. J., Forni, L. G., and Willson, R. L. (1988) Vitamin E analogue trolox C. ESR and pulse radiolysis studies of free radical reactions. Biochem. J. 255:513–522.
30. Prütz, W. A. (1990) Free radical transfer involving sulphur peptide functions. In: Sulfur-Centered Reactive Intermediates in Chemistry and Biology (Chatgilialoglu, C., and Asmus, K.-D., eds.), pp. 389–399. NATO ASI Series A, Vol. 197, Plenum Press, New York.
31. Adam, W., Epe, B., Schiffmann, D., Vargas, F., and Wild, D. (1988) Facile reduction of 1,2-dioxetanes by thiols as potential protective measure against photochemical damage of cellular DNA. Angew. Chem. Int. Ed. Engl. 27:429–431.
32. Harman, L. S., Mottley, C., and Mason, R. P. (1984) Free radical metabolites of L-cysteine oxidation. J. Biol. Chem. 259:5606–5611.
33. Mason, R. P., and Ramakrishna Rao, D. N. (1990) Electron spin resonance investigation of the thiyl free radical metabolites of cysteine, glutathione, and drugs. In: Sulfur-Centered Reactive Intermediates in Chemistry and Biology (Chatgilialoglu, C., and Asmus, K.-D., eds.), pp. 401–408, NATO ASI Series A, Vol. 197, Plenum Press, New York.
34. Ross, D., Norbeck, K., and Moldeus, P. (1985) The generation and subsequent fate of glutathionyl radicals in biological systems. J. Biol. Chem. 260:15028–15032.
35. Lamfrom, H., and Nielsen, S. O. (1957) The iron catalysis of thioglycolate oxidation by oxygen. J. Am. Chem. Soc. 79:1966–1970.
36. Lambeth, D. O., Ericson, G. R., Yorek, M. A., and Ray, P. D. (1982) Implications for in vitro studies of the autoxidation of ferrous ion and the iron-catalyzed autoxidation of dithiothreitol. Biochim. Biophys. Acta 719:501–508.

37. Saez, G., Thornalley, P. J., Hill, H. A. O., Rems, R., and Bannister, J. V. (1982) The production of free radicals during the autoxidation of cysteine and their effect on isolated rat hepatocytes. Biochim. Biophys. Acta 719:24–31.
38. Hoffman, M. Z., and Hayon, E. (1972) One-electron reduction of the disulfide linkage in aqueous solution. Formation, protonation, and decay kinetics of the $RSSR^-$ radical. J. Am. Chem. Soc. 94:7950–7957.
39. Akhlaq, M. S., and von Sonntag, C. (1987) Intermolecular H-abstraction of thiyl radicals from thiols and the intramolecular complexing of the thiyl radical with the thiol group in 1,4-dithiothreitol. A pulse radiolysis study. Z. Naturforsch. 42c:134–140.
40. Clement, J. R., Armstrong, D. A., Klassen, N. V., and Gillis, H. A. (1972) Pulse radiolysis of aqueous papain. Can. J. Chem. 50:2833–2840.
41. Akhlaq, M. S., Murthy, C. P., Steenken, S., and von Sonntag, C. (1989) Reaction of α-hydroxyalkyl radicals and their anions with oxidized dithiothreitol. A pulse radiolysis and product analysis study. J. Phys. Chem. 93:4331–4334.
42. Bonifacic, M., and Asmus, K.-D. (1976) Free radical oxidation of organic disulfides. J. Phys. Chem. 80:2426–2430.
43. Dixon, C. J., and Grant, D. W. (1973) The photolysis of cystine in acidic aqueous solution. Photochem. Photobiol. 18:387–391.
44. Morine, G. H., and Kuntz, R. R. (1981) Observations of C-S and S-S bond cleavage in the photolysis of disulfides in solution. Photochem. Photobiol. 33:1–5.
45. Everett, S. A., Schöneich, C., Stewart, J. H., and Asmus, K.-D. (1992) Perthiyl radicals, trisulfide radical ions, and sulfate formation. A combined photolysis and radiolysis study on redox processes with organic di- and trisulfides. J. Phys. Chem. 96:306–314.
46. Grant, D. W., and Stewart, J. H. (1984) The photolysis of penicillamine disulphide in acidic aqueous solution. Photochem. Photobiol. 40:285–290.
47. Boyd, D. B. (1972) Conformational dependence of the electronic energy levels in disulfides. J. Am. Chem. Soc. 94:8799–8804.
48. Burkey, T. J., Hawari, J. A., Lossing, F. P., Lusztyk, J., Sutcliffe, R., and Griller, D. (1985) The tert-butylperthiyl radical. J. Org. Chem. 50:4966–4967.
49. Everett, S. A., Folkes, L. K., Wardman, P., and Asmus, K.-D. (1994) Free radical repair by a novel perthiol: Reversible hydrogen transfer and perthiyl radical formation. Free Rad. Res. 20:387–400.
50. Wu, Z., Back, T. G., Ahmad, R., Yamdagni, R., and Armstrong, D. A. (1982) Mechanism of reduction of bis(2-hydroxyethyl)trisulfide by e^-_{aq} and $^{\bullet}CO_2^-$. Spectrum and scavenging of $RSS^{\bullet}$ radicals. J. Phys. Chem. 86:4417–4422.
51. Elliot, A. J., McEachern, R. J., and Armstrong, D. A. (1981) Oxidation of amino-containing disulfides by $Br_2^{\bullet-}$ and $^{\bullet}OH$. A pulse radiolysis study. J. Phys. Chem. 85:68–75.
52. Schöneich, C., Bonifačić, M., and Asmus, K.-D. (1989) Reversible H-atom abstraction from alcohols by thiyl radicals: Determination of absolute rate constants by pulse radiolysis. Free Rad. Res. Comms. 6:393–405.
53. Forni, L. G., and Willson, R. L. (1986) Thiyl and phenoxyl free radicals and NADH. Direct observation of one-electron oxidation. Biochem. J. 240:897–903.

54. Forni, L. G., and Willson, R. L. (1986) Thiyl free radicals and the oxidation of ferrocytochrome c. Direct observation of coupled hydrogen-atom- and electron transfer reactions. Biochem. J. 240:905–907.
55. Forni, L. G., Mönig, J., Mora-Arellano, V. O., and Willson, R. L. (1983) Thiyl free radicals: Direct observations of electron transfer reactions with phenothiazines and ascorbate. J. Chem. Soc. Perkin Trans. II:961–965.
56. Tamba, M., and O'Neill, P. (1991) Redox reactions of thiol free radicals with the antioxidants ascorbate and chloropromazine: Role in radioprotection. J. Chem. Soc. Perkin Trans. II:1681–1685.
57. Wolfenden, B. S., and Willson, R. L. (1982) Radical-cations as reference chromogens in kinetic studies of one-electron transfer reactions: Pulse radiolysis studies of 2,2′-azinobis-(3-ethylbenzthiazoline-6-sulphonate). J. Chem. Soc. Perkin Trans. II:805–812.
58. Forni, L. G., Mora-Arellano, V. O., Packer, J. E., and Willson, R. L. (1988) Aminopyrine and antipyrine free radical-cations: Pulse radiolysis studies of one-electron transfer reactions. J. Chem. Soc. Perkin Trans. II:1579–1584.
59. Wilson, I., Wardman, P., Cohen, G. M., and D'Arcy Doherty, M. (1986) Reductive role of glutathione in the redox cycling of oxidizable drugs. Biochem. Pharmacol. 35:21–22.
60. D'Aquino, M., Dunster, C., and Willson, R. L. (1989) Vitamin A and glutathione-mediated free radical damage: Competing reactions with polyunsaturated fatty acids and vitamin C. Biochem. Biophys. Res. Commun. 161:1199–1203.
61. Huston, P., Espenson, J. H. and Bakac, A. (1992) Reactions of thiyl radicals with transition-metal complexes. J. Am. Chem. Soc. 114:9510–9516.
62. Bonifačić, M., and Asmus, K.-D. (1984) Adduct formation and absolute rate constants in the displacement reaction of thiyl radicals with disulfides. J. Phys. Chem. 88:6286–6290.
63. Quintiliani, M., Badiello, R., Tamba, M., Esfandi, A., and Gorin, G. (1977) Radiolysis of glutathione in oxygen-containing solutions of pH 7. Int. J. Radiat. Biol. 32:195–202.
64. Schäfer, K., and Asmus, K.-D. (1978) Addition of oxygen to organic sulfur radicals. J. Phys. Chem. 82:2777–2780.
65. Tamba, M., Simone, G., and Quintiliani, M. (1986) Interactions of thiyl free radicals with oxygen: A pulse radiolysis study. Int. J. Radiat. Biol. 50:595–600.
66. Mönig, J., Asmus, K.-D., Forni, L. G., and Willson, R. L. (1987) On the reaction of molecular oxygen with thiyl radicals: A re-examination. Int. J. Radiat. Biol. 52:589–602.
67. Sevilla, M. D., Yan, M., and Becker, D. (1988) Thiol peroxyl radical formation from the reaction of cysteine thiyl radical with molecular oxygen: An ESR investigation. Biochem. Biophys. Res. Commun. 155:405–410.
68. Zhang, X., Zhang, N., Schuchmann, H.-P., and von Sonntag, C. (1994) Pulse radiolysis of 2-mercaptoethanol in oxygenated solution. Generation and reactions of the thiylperoxyl radical. J. Phys. Chem. 98:6541–6547.
69. Schöneich, Ch., Dillinger, U., von Bruchhausen, F., and Asmus, K.-D. (1992) Oxidation of polyunsaturated fatty acids and lipids through thiyl and sulfonyl radi-

cals: Reaction kinetics, and influence of oxygen and structure of thiyl radicals. Arch. Biochem. Biophys. 292:456–467.
70. Wardman, P. (1990) Thiol reactivity towards drugs and radicals: Some implications in the radiotherapy and chemotherapy of cancer. In: Sulfur-Centered Reactive Intermediates in Chemistry and Biology (Chatgilialoglu, C., and Asmus, K.-D., eds.), pp. 415–427. NATO ASI Series A, Vol. 197, Plenum Press, New York.
71. Sevilla, M. D., Becker, D., and Yan, M. (1990) The formation and structure of the sulfoxyl radicals $RSO^{\bullet}$, $RSOO^{\bullet}$, $RSO_2^{\bullet}$, and $RSO_2OO^{\bullet}$ from the reaction of cysteine, glutathione, and penicillamine thiyl radicals with molecular oxygen. Int. J. Radiat. Biol. 57:65–81.
72. Schöneich, C. (1990) Quantitative kinetische Untersuchungen zum reversiblen H-Atomtransfer durch Radikale von Thiolen und biologisch relevanten Verbindungen. Ph.D. thesis, Technical University Berlin.
73. Benson, S. W. (1978) Thermochemistry and kinetics of sulfur-containing molecules and radicals. Chem. Rev. 78:23–35.
74. Huyser, E. S., and Kellog, R. M. (1966) Free-radical cleavage of β-hydroxythioethers to ketones and mercaptanes. J. Org. Chem. 31:3366–3369.
75. Morita, M., Sasai, K., Tajima, M., and Fujimaki, M. (1971) The roles of thiyl radicals in the radiolysis of a mixed aqueous solution of cysteine and formate-hydrogen abstraction from formate and the formation of carbon dioxide. Bull. Chem. Soc. Jpn. 44:2257–2258.
76. Stone, P. G., and Cohen, S. G. (1982) Effects of structure on rates and quantum yields in photoreduction of fluorenone by amines. Catalysis and inhibition by thiols. J. Am. Chem. Soc. 104:3435–3440.
77. Pryor, W. A., Gojon, G., and Church, D. F. (1978) Relative rate constants for hydrogen atom abstraction by the cyclohexanethiyl and benzenethiyl radicals. J. Org. Chem. 43:793–800.
78. Schäfer, K., and Asmus, K.-D. (1981) Reaction of thiols and disulfides with phosphite radicals. A chain mechanism and $RS^{\bullet}/PO_3^{2-\bullet}$ equilibrium. J. Phys. Chem. 85:852–855.
79. McMillen, D. F., and Golden, D. M. (1982) Thermochemistry of the C-H bond. Ann. Rev. Phys. Chem. 33:493–532.
80. Schöneich, C., Bonifačić, M., Dillinger, U., and Asmus, K.-D. (1990) Hydrogen abstraction by thiyl radicals from activated C-H bonds of alcohols, ethers, and polyunsaturated fatty acids. In: Sulfur-Centered Reactive Intermediates in Chemistry and Biology (Chatgilialoglu, C., and Asmus, K.-D., eds.), pp. 367–376. NATO ASI Series A, Vol. 197, Plenum Press, New York.
81. Schöneich, C., and Asmus, K.-D. (1990) Reaction of thiyl radicals with alcohols, ethers and polyunsaturated fatty acids. Radiat. Environ. Biophys. 29:263–271.
82. Von Sonntag, C., and Schuchmann, H.-P. (1991) The elucidation of peroxyl radical reactions in aqueous solution with the help of radiation-chemical methods. Angew. Chem. Int. Ed. Engl. 30:1229–1253.
83. Akhlaq, M. S., and von Sonntag, C. (1986) Free-radical-induced elimination of H_2S from dithiothreitol. A chain reaction. J. Am. Chem. Soc. 108:3542–3544.
84. Schuchmann, H.-P., and von Sonntag, C. (1981) Photolysis at 185 nm of dimethyl

ether in aqueous solution: Involvement of the hydroxymethyl radical. J. Photochem. 16:289–295.
85. Grierson, L., Hildenbrand, K., and Bothe, E. (1992) Intramolecular transformation reaction of the glutathione thiyl radical into a non-sulphur-centred radical: A pulse radiolysis and EPR study. Int. J. Radiat. Biol. 62:265–277.
86. Ito, O., and Matsuda, M. (1988). Chemical kinetics of reactions of sulfur-containing organoradicals. In: Chemical Kinetics of Small Organic Radicals, Vol. III (Alfassi, Z. B., ed.), pp. 133–164. CRC Press, Boca Raton, FL.
87. McPhee, D. J., Campredon, M., Lesage, M., and Griller, D. (1989) Reactions of the tert-butylthiyl radical with organometallic compounds and alkenes. J. Am. Chem. Soc. 111:7563–7567.
88. Wagner, P. J., Sedon, J. H., and Lindstrom, M. J. (1978) Rates of radical β-cleavage in photogenerated diradicals. J. Am. Chem. Soc. 100:2579–2580.
89. Griller, D., and Ingold, K. U. (1980) Free-radical clocks. Acc. Chem. Res. 13:317–323.
90. Lunazzi, L., Placucci, G., and Grossi, L. (1979) Temperature-controlled addition vs. abstraction by methylthiyl radical with cyclopentene. J. Chem. Soc. Chem. Comm. 533–534.
91. Alder, R. W., Bonifačić, M., and Asmus, K.-D. (1986) Reaction of a stable N∴N bonded radical cation with free radicals generated by pulse radiolysis: Exceedingly rapid hydrogen abstraction from C-H bonds. J. Chem. Soc. Perkin Trans. II:277–284.
92. Leu, A-D., and Armstrong, D. A. (1986) Thiyl radical oxidation of Cu^I: Formation and spectra of oxidized copper thiolates. J. Phys. Chem. 90:1449–1454.
93. Mezyk, S. P., and Armstrong, D. A. (1988) Kinetics of oxidation of Cu(I) complexes of cysteine and penicillamine: Nature of intermediates and reactants at pH 10.0. Can. J. Chem. 67:736–745.
94. Mezyk, S. P., and Armstrong, D. A. (1991) Radical reactions of Cu(II)-penicillamine complexes: Evidence for a Cu(II)-disulfide anion intermediate. Can. J. Chem. 69:533–539.
95. Chatgilialoglu, C., and Guerra, M. (1993) Thiyl radicals. In: Supplement S: The Chemistry of Sulfur-Containing Functional Groups (Patai, S., and Rappoport, Z., eds.), pp. 363–394. John Wiley & Sons, Chichester, United Kingdom.
96. Willson, R. L. (1970) Pulse radiolysis studies of electron transfer in aqueous disulfide solutions. J. Chem. Soc. Chem. Comm. 1425–1426.
97. Armstrong, D. A. (1990) Redox systems with sulphur-centered species. In: Sulfur-Centered Reactive Intermediates in Chemistry and Biology (Chatgilialoglu, C., and Asmus, K.-D., eds.), pp. 341–351. NATO ASI Series A, Vol. 197, Plenum Press, New York.
98. Radi, R., Beckman, J. S., Bush, K. M., and Freeman, B. A. (1991) Peroxynitrite oxidation of sulfhydryls. The cytotoxic potential of superoxide and nitric oxide. J. Biol. Chem. 266:4244–4250.

3

Antioxidant Drug Design: A Comparison of Thiol and Perthiol Antiradical and Prooxidant Reaction Mechanisms

Steven A. Everett
Gray Laboratory Cancer Research Trust, Mount Vernon Hospital, Northwood, Middlesex, England

I. INTRODUCTION

Thiols have been employed as potential therapeutic agents in the prevention of free radical–mediated pathological disorders (1) and as radioprotective adjuncts to clinical cancer radiotherapy (2–4). As the molecular mechanisms associated with free radical generation and cellular damage unfold (5–9), drug design strategies have developed to enhance thiol efficacy. These have been based on the modification of side groups to enhance targeting of cellular constituents and/or modification of the parent drug to a pro-drug, thereby targeting the thiol to specific sites in the body. Both of these targeting strategies are best exemplified by the polyaminothiophosphate ester ethiofos or WR-2721 (Fig. 1, **1**), a pro-drug that undergoes biotransformation to the active free thiol form WR-1065 (Fig. 1, **2**) when dephosphorylated by alkaline phosphatase in endothelial cells (10). The preferential uptake of WR-1065 into nontumor tissues has been exploited in the application of WR-2721 in clinical cancer radiotherapy (2) and chemotherapy (11,12). The ability of WR-1065 to protect DNA is enhanced by the doubly protonated polyamino side chain, which targets the negatively charged DNA backbone via an electrostatic counter-ion condensation phenomenon (13,14). However, variations in side group structure tend to have a modest effect on rates of free radical scavenging by aliphatic thiols.

Structural modification of thiols to their di-sulfur perthiol analogs (RSSH) represents a novel strategy whereby relative differences in S-H bond energies

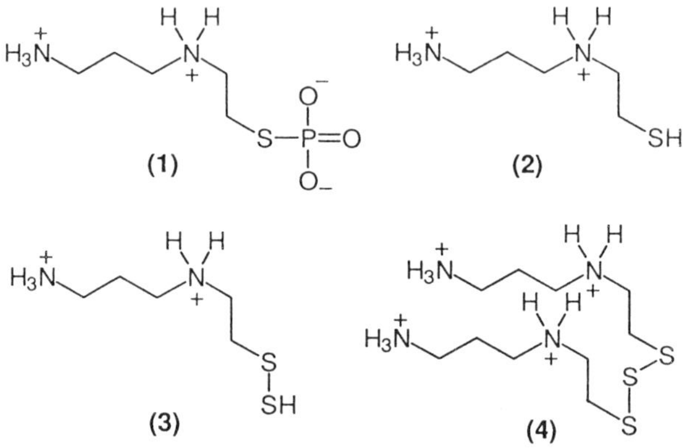

Figure 1 (**1**), The pro-drug ethiofos, WR-2721; (**2**), WR-1065; (**3**) perthiol analog of WR-1065; and (**4**) the symmetrical trisulfide of WR-1065.

form the basis for manipulation of thiol antiradical properties (15,16). Radiation chemistry and pulse radiolysis can contribute to rational antioxidant drug design by characterizing both drug/free radical interactions and potential pro-oxidative effects of drug-derived radicals (17). Mechanistic studies of this type are an important first step in assessing the potential antioxidant properties of perthiols.

II. STRUCTURAL MODIFICATION OF THIOLS TO PERTHIOLS

A. S-H Bond Energies in Thiols and Perthiols

The design of an effective free radical scavenger to interact with a known structural site or the search for structurally related antioxidant scavengers after the discovery of drugs with promising clinical activity generally relies on an understanding of the factors that control structural preference and drug/free radical interactions. Isosteric modification to molecular structure involves the replacement of an atom, or group of atoms, in a molecule by another group with similar electronic and steric configurations (18). The replacement of the sulfur atom in a thiol by a disulfide bond can be loosely classified as an isosteric modification since the electronic and steric properties of the S-H bond remains unchanged. Disulfide bonds are capable of exerting a bond-weakening effect on adjacent bonds, and theoretical thermodynamic calculations have indicated that the disulfide bond in perthiols weakens the RSS-H bond by ≈90 kJ/mol relative to RS-H (19,20).

B. Synthesis of Perthiols

The structural modification of thiols to their perthiol analogs requires the preparation of thiol-methoxycarbonyl mixed disulfide ($RSSCO_2CH_3$) by the condensation of the thiol with methoxycarbonyl-sulfonyl chloride ($ClSCO_2CH_3$) (21,22):

$$RSH + ClSCO_2CH_3 \rightarrow RSSCO_2CH_3 + HCl \quad (1)$$

The second step is an alkoxide-induced nucleophilic displacement of the perthiolate anion (RSS^-) from $RSSCO_2CH_3$ (15):

$$RO^- + RSSCO_2CH_3 \rightarrow ROCO_2CH_3 + RSS^- \quad (2)$$

Equations (1) and (2) are performed under an inert atmosphere to prevent oxidation of both reactants and products. The ^{1}H-NMR spectra of $RSSCO_2CH_3$ are characterized by a singlet peak with a chemical shift $\delta = 3.95$ ppm relative to 2,2 dimethyl-2-silapentane-5-sulfonate, which corresponds to the three uncoupled hydrogens of the methoxycarbonyl grouping (CH_3O_2C-) (21).

The versatility of this method facilitates the synthesis of perthiols with specific targeting abilities. The perthiol analog of the radioprotective drug WR-1065 (Fig. 1, **3**) is expected to exhibit a DNA-targeting capability similar to that of the parent thiol. Unequivocal identification of the perthiols was obtained by combined gas chromatography/mass spectroscopy (GC-MS) with chemical ionization (CI). Prior to GC the perthiols were converted to their corresponding trimethylsilyl (TMS) derivatives using *bis*-(TMS)-triflouroacetamide (BSTFA) as the derivatizing reagent (23). The CI mass spectra of totally derivatized perthiol contain the molecular ions $(M + H)^+$, which are 32 m/z (where m/z = charge/mass ratio) higher than the corresponding thiols, as expected from the introduction of an extra sulfur atom. The fragmentation pattern of the totally derivatized perthiol analog of WR-1065 was also characterized by the fragment ions $(TMS)_2\text{-N-}(CH_2)_3N^{\oplus}{=}CH_2$ (231 m/z) and $(TMS)_2\text{-}N^{\oplus}{=}CH_2$ (174 m/z) generated as a consequence of cleavage β to the secondary and primary amino groups in the polyamino side chain, respectively.

III. PERTHIOL IONIZATION

Structural modification of thiols to perthiols introduces important changes in their relative acid-base chemistry. At physiological pH, the position of the perthiol acid-base equilibrium in Eq. (3) shifts toward the deprotonated perthiolate anion compared to the corresponding ionization of a thiol.

$$RSSH \rightleftharpoons RSS^- + H^+ \quad (3)$$

For example, the acid dissociation constant or pK_a for the SH ionization in WR-1065 is 7.7 (24) but is reduced to 6.2 in the corresponding perthiol analog (15).

Consequently, at physiological pH, 22% of WR-1065 exists as the thiolate anion, whereas 85% of the perthiol analog exists in the form of the perthiolate anion. Much of the radioprotective effects of WR-1065 have been attributed to repair of DNA radicals by hydrogen atom transfer (25,26). However, the RS^- anion undergoes rapid electron-transfer reactions with, for example, peroxyl radicals ($ROO^\bullet$) (27) and some DNA base radicals (28), suggesting that neutralization of DNA radicals by electron transfer is also a distinct possibility. Both hydrogen- and electron-transfer processes are therefore expected to contribute to the free radical–scavenging properties of perthiols.

IV. HYDROGEN ATOM TRANSFER REACTIONS OF PERTHIOLS

A. Repair Reactions of Perthiols

The classical chemical repair reaction refers to the ability of the thiols to restore free radical–damaged sites on biomolecules by hydrogen atom transfer from the S-H moiety (3,29,30). Typically the S-H bond strength is lower than many C-H bonds, and as a consequence many carbon-centered radicals are repaired by thiols. Theoretically a widening of the differential between S-H and C-H bond energies (as exemplified by perthiols) should facilitate increased rates of hydrogen atom transfer and potentiate repair capability. Alcohols and ethers constitute simple models for the sugar moieties of DNA nucleotides and have been utilized to study thiol repair [Eq. (4)] and the reverse repair reaction [Eq. (4a)] (i.e., hydrogen abstraction) by thiyl radicals ($RS^\bullet$) (31,32):

$$RSH + (CH_3)_2CHOH \rightarrow (CH_3)_2C^\bullet OH + RS^\bullet \quad (4)$$

$$RS^\bullet + (CH_3)_2C^\bullet OH \rightarrow (CH_3)_2CHOH + RSH \quad (4a)$$

Pulse radiolysis of perthiols in water/alcohol mixtures at pH 4.5 generate a transient absorption spectrum with $\lambda_{max} = 374$ nm (15,16). By comparison with similar absorption spectra observed for the one-electron reduction of symmetrical trisulfides, penicillamine trisulfide and *bis*-(2-hydroxyethyl) trisulfide (33–35), the one-electron oxidation of amino acid disulfides (36), and the photolysis of disulfides (37), the observed absorption spectrum was assigned to the perthiyl radical ($RSS^\bullet$) generated according to the following reactions:

$$H_2O \rightarrow e^-_{aq}, {}^\bullet OH, H^\bullet \cdots \quad (5)$$

$$e^-_{aq} + N_2O + H_2O \rightarrow {}^\bullet OH + OH^- + N_2 \quad (6)$$

$$^\bullet OH/H^\bullet + (CH_3)_2CHOH \rightarrow H_2O/H_2 + (CH_3)_2C^\bullet OH + {}^\bullet CH_2(CH_3)CHOH \quad (7)$$

$$(CH_3)_2C^\bullet OH + RSSH \rightarrow (CH_3)_2CHOH + RSS^\bullet \quad (8)$$

Spectrum a in Figure 2 is the absorption spectrum obtained for the perthiol analog of WR-1065. In this case $RSS^\bullet$ radicals were formed with a total radiation chemical yield of $G\varepsilon_{374} = 1.09 \times 10^{-4}$ m²/J. In this system the radiation chemical yield of carbon-centered radicals $G((CH_3)_2C^\bullet OH) = G(e^-_{aq} + H^\bullet + {}^\bullet OH) \approx 0.65$ μmol/J, yielding an extinction coefficient of $\varepsilon_{374} = 1680 \pm 20$ $mol^{-1}\ dm^3\ cm^{-1}$, which is in good agreement with those previously reported for other perthiyl radicals (34–38). No evidence was obtained for the existence of thiyl radicals (see Fig. 2, spectrum c), clearly ruling out possible reduction of the perthiol disulfide bond by the alcohol radicals according to the following equations:

$$(CH_3)_2C^\bullet OH + RSSH \rightleftharpoons [RSSH(CH_3)_2COH]^\bullet \quad (9)$$

$$[RSSH(CH_3)_2COH]^\bullet \rightarrow RS{\therefore}SH_2 + (CH_3)_2C{=}O \quad (10)$$

$$RS{\therefore}SH_2 \rightarrow RS^\bullet + H_2S \quad (11)$$

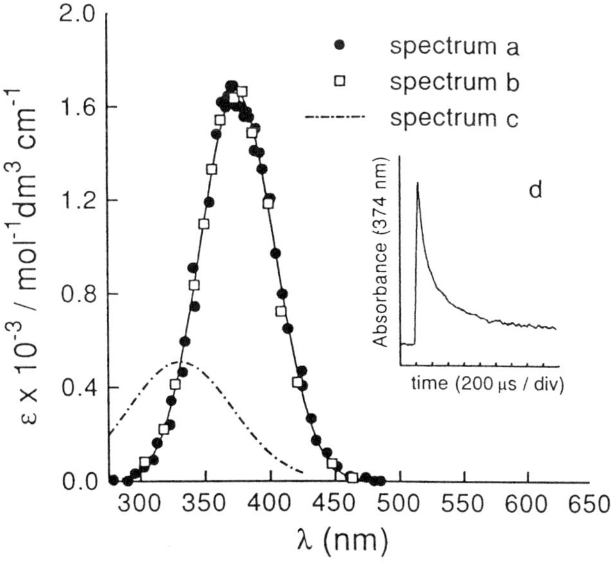

Figure 2 Perthiyl radical spectra obtained on pulse radiolysis (10Gy) of the WR-1065 perthiol analog 0.5 mmol dm^{-3} in (a) an N_2O-saturated $(CH_3)_2CHOH/H_2O$ mixture at pH 4.5, (b) an air-saturated $(CH_3)_2CHOH/CCl_4/H_2O$ mixture at pH 8. Spectrum (c) is that of the corresponding thiyl radical derived from WR-1065 in an N_2O-saturated $(CH_3)_2CHOH/H_2O$ mixture at pH 4.5. Inset (d) is a typical trace recorded at 374 nm showing fast formation and subsequent bimolecular decay of perthiyl radicals generated by the experimental conditions described in (a) above.

For comparison and support of the assigment of the perthiyl radical, a one-electron reduction of the corresponding symmetrical trisulfide (Fig. 1, **4**) was performed by both e^-_{aq} and $CO_2^{\bullet -}$ according to Eqs. (5) and (12)–(14), which generated a transient absorption spectrum virtually identical to that of the perthiol/alcohol system.

$$e^-_{aq} + RSSSR \rightarrow (RSSSR)^{\bullet -} \rightleftharpoons RS^- + RSS^\bullet \quad (12)$$

$$^\bullet OH/H^\bullet + HCO_2^- \rightarrow H_2O/H_2 + CO_2^{\bullet -} \quad (13)$$

$$CO_2^{\bullet -} + RSSSR \rightarrow RS^- + RSS^\bullet + CO_2 \quad (14)$$

The characterization of perthiyl radicals represents an unequivocal identification of Eq. (8) and therefore of perthiol hydrogen transfer to carbon-centered radicals. In this respect Eq. (8) clearly parallels the classical repair reaction of thiols in Eq. (4) (29,32,39). The kinetics of perthiyl radical formation follow an exponential rate law with the half-life proportional to the perthiol concentration. The rate constant $k_8 = 2.5 \times 10^9$ mol^{-1} dm^3 s^{-1} obtained for the perthiol analog of WR-1065 was found to be an order of magnitude faster than the analogous reaction with the parent thiol (26), where RSH = WR-1065, $k_4 = 2 \times 10^8$ mol^{-1} dm^3 s^{-1}.

B. The Reverse Repair Reaction: Hydrogen Abstraction by Perthiyl Radicals

In low alcohol concentrations (≈0.1 mol dm^{-3}) the perthiyl radicals generated in Eq. (8) undergo radical-radical recombination via Eq. (15), presumably to the tetrasulfide RS_4R (shown in Fig. 2, inset) by analogy to the bimolecular decay of thiyl radicals to the disulfide (30).

$$2RSS^\bullet \rightarrow RS_4R \quad (15)$$

However, at high alcohol concentrations of >5 mol dm^{-3}, perthiyl radicals were capable of abstracting a hydrogen atom from the activated C-H bond of the alcohol in what is essentially a reverse repair reaction [Eq. (8a)], indicating that the reaction is in fact a reversible equilibrium reaction (15).

$$RSS^\bullet + (CH_3)_2CHOH \rightarrow RSSH + (CH_3)_2C^\bullet OH \quad (8a)$$

Rate constants for both the forward and reverse reactions were obtained by an indirect competition method based on irreversible scavenging of alcohol radicals by methyl viologen (40) according to the following:

$$MV^{2+} + (CH_3)_2C^\bullet OH \rightarrow MV^{\bullet +} + (CH_3)_2C{=}O + H^+ \quad (16)$$

The absorption at 600 nm indicated a fast and then significantly slower build-up of the $MV^{\bullet +}$ radical cation absorption, consistent with fast perthiol hydrogen transfer [Eq. (8)] and the slower hydrogen abstraction reaction [Eq. (8a)] in analogy to Eqs. (4) and (4a) (31,32). From the first-order RSS$^•$-mediated build-up of the $MV^{•+}$ it was possible to obtain a rate constant $k_{8a} = 3.8 \times 10^3$ $dm^3\ mol^{-1}\ s^{-1}$, which was almost an order of magnitude slower than that of the corresponding WR-1065 thiyl radical (15), where $k_{4a} = 1.4 \times 10^4\ mol^{-1}\ dm^3\ s^{-1}$.

For perthiols, the equilibrium clearly lies on the side of perthiol repair, since $k_8 >> k_{8a}$. A comparison of the equilibrium constants $K_8 = (k_8/k_{8a}) = (6.3 \pm 0.1) \times 10^5$ and $K_1 = (k_4/k_{4a}) = (1.4 \pm 0.1) \times 10^4$ indicate that for the perthiol analog, the reversible repair reaction [Eqs. (8) and (8a)] is shifted further toward repair than the corresponding reversible repair reaction of WR-1065 [Eqs. (4) and (4a)]. Clearly the presence of the disulfide bond has introduced a bond-weakening effect in the adjacent perthiol S-H as predicted from thermodynamic calculations, thereby enhancing rates of hydrogen-transfer processes.

C. The Thermodynamic Basis for Perthiol Hydrogen Atom Transfer

Thermodynamic calculations have predicted that the S-H bond in higher sulfanes, i.e., H_2S_{n+1}, might be as much as 90 kJ/mol lower than in S-H bonds in aliphatic thiols (19,20). This bond-weakening effect has previously been attributed to the inherent stability of perthiyl radicals. Molecular orbital calculations have predicted that polysulfide radicals of the type $RS_n^•$ ($n > 1$) are highly stabilized in comparison to RS$^•$ radicals. While the sulfur nonbonding $3p^2$ atomic orbitals in a thiyl radical remain degenerate, the overlapping $3p^2$ atomic orbitals in the sulfur-sulfur bond of perthiyl radicals establish π-bonding molecular orbitals, which accommodate the RSS$^•$ radical electron spin (41). The electronic nature of the perthiyl radical can be interpreted as a resonance hybrid of the normal sulfur-sulfur sigma-bonded state (2σ) and a five-electron bonded state ($2\sigma/2\pi/1\pi^*$) (42). The resonance stabilization energy of the RSS$^•$ radical over the unstabilized RS$^•$ radical is 8.8 kJ/mol, calculated from the differences in the Gibbs free energies for the reversible hydrogen transfer reactions [Eqs. (8a) and (8), (4a) and (4)], i.e., $(\Delta G_{(4)/(4a)} - \Delta G_{(8)/(8a)}) = (-RT \ln K_4 + RT \ln K_8)$ (15). The thermodynamic driving force for perthiol repair is therefore derived from the stability of the perthiyl radical. Moreover, the generation of perthiyl radicals has important implications for the potential prooxidative effects derived from the free radical–scavenging activity of perthiols (see Sec. VIII).

V. FREE RADICAL SCAVENGING BY PERTHIOLS INVOLVING ELECTRON TRANSFER

The radioprotective effect of WR-1065 is considered to depend largely upon chemical repair of DNA radicals by hydrogen atom transfer from the S-H group, but electron transfer from the thiolate anion may also be important. The rate of perthiol hydrogen atom transfer to alcohol radicals was found to decrease with increasing pH, as would be expected if the S-H hydrogen was involved in Eq. (8). The pH dependence of k_8 was analyzed using the appropriate function assuming no reactivity of RSS^- anion and gave a pK_a (RSS-H) $= 6.2 \pm 0.1$. At physiological pH a greater fraction of the perthiol is ionized to the RSS^- anion in relation to the amount of thiol present in the deprotonated RS^- anion form. Indirect evidence has been obtained for perthiyl radical formation from the reduction of Fe(III) cytochrome *c* by the glutathione perthiolate anion GSS^- in Eq. (17), suggesting that free radical scavenging may occur by perthiolate electron transfer (43).

$$GSS^- + Fe(III)cytc \rightarrow GSS^{\bullet} + Fe(II)cytc \quad (17)$$

The halogenated peroxyl radical ($CCl_3OO^{\bullet}$) can be generated by pulse radiolysis of air-saturated $CCl_4/(CH_3)_2CHOH/H_2O$ mixtures via Eqs. (5), (7), and (18)–(20):

$$e^-_{aq} + CCl_4 \rightarrow CCl_3^{\bullet} + Cl^- \quad (18)$$

$$(CH_3)_2C^{\bullet}OH + CCl_4 \rightarrow CCl_3^{\bullet} + (CH_3)_2C{=}O + H^+ + Cl^- \quad (19)$$

$$CCl_3^{\bullet} + O_2 \rightarrow CCl_3OO^{\bullet} \quad (20)$$

$$CCl_3OO^{\bullet} + RSS^- \rightarrow CCl_3OO^- + RSS^{\bullet} \quad (21)$$

At pH 8, where the perthiol analog of WR-1065 exists totally in the RSS^- form, there is no RSSH available to form perthiyl radicals via hydrogen transfer processes as exemplified by Eq. (7) or alternatively via Eq. (22):

$$CCl_3OO^{\bullet} + RSSH \rightarrow RSS^{\bullet} + CCl_3OOH \quad (22)$$

The introduction of perthiol generates a perthiyl radical spectrum similar to that obtained by perthiol hydrogen transfer to alcohol radicals (Fig. 2, spectrum b), but in this case as a consequence of the electron transfer reaction in Eq. (21). The rate constant $k_{21} = 4.2 \times 10^9\ mol^{-1}\ dm^3\ s^{-1}$ was obtained from the slope of the linear plot of the first-order buildup of the perthiyl radical absorption, k_{obs} [RSS^-].

Perthiols can therefore scavenge free radicals by either hydrogen- or electron-transfer processes, depending on the nature of the free radical species. The fact that perthiyl radicals are a common radical product in both cases would indicate that the thermodynamic driving force for both hydrogen and electron transfer is due to $RSS^{\bullet}$ radical resonance stabilization.

VI. HYDROXYL RADICAL SCAVENGING BY PERTHIOLS

Highly damaging hydroxyl radicals ($^{\bullet}OH$) generated in vivo as a consequence of cellular oxidative stress or ionizing radiation are capable of reacting with almost all biomolecules found in living cells with rate constants of 10^9–10^{10} mol^{-1} dm^3 s^{-1}. Thiols scavenge hydroxyl radicals by hydrogen atom transfer with the subsequent formation of $RS^•$ radicals (30,39):

$$RSH + {}^•OH \rightarrow RS^• + H_2O \tag{23}$$

Not surprisingly, rate constants for Eq. (23) for aliphatic thiols approach diffusion-controlled rates, e.g., when RSH = WR-1065, $k_{23} = 9 \times 10^9$ mol^{-1} dm^3 s^{-1} (26). Pulse radiolysis of N_2O-saturated aqueous solutions containing the perthiol analog of WR-1065 at pH 4.5 and 8 (pHs at which the perthiols exist exclusively in the protonated, RSSH, and deprotonated, RSS^-, form, respectively) generated spectra characteristic of perthiyl radicals according to Eqs. (5), (6), (24a), and (24b):

$$RSSH + {}^•OH \rightarrow RSS^• + H_2O \tag{24a}$$

$$RSS^- + {}^•OH \rightarrow RSS^• + OH^- \tag{24b}$$

However, for both of the above pathways, perthiyl radicals account for 70% of the $^•OH$, indicating the existence of alternative reaction pathways derived from the dissociation of possible intermediate perthiol hydroxyl radical adduct, i.e., $[RSSH \cdots OH]^•$ in analogy to the one-electron oxidation of disulfides (34,36) and trisulfides (33) by $^•OH$ radicals. Considering the indiscriminate reactivity of $^•OH$ radicals, it is reasonable to assume that a fraction of the radicals are lost via hydrogen abstraction from the perthiol-polyamino side chain. The rate constants for the reaction of the perthiol analog of WR-1065 were determined by the thiocyanate competition method (44) with $k_{24a} = 1.5 \times 10^{10}$ mol dm^3 s^{-1} and $k_{24b} = 1.8 \times 10^{10}$ mol^{-1} dm^3 s^{-1}.

VII. POTENTIAL PROOXIDATIVE THREAT OF PERTHIYL RADICAL GENERATION FROM PERTHIOLS

An important consideration in antioxidant drug design is whether the antioxidant-derived radical exhibits prooxidative effects (45). The perthiyl radical is a common free-radical product following the free radical–scavenging activity of perthiols. In general $RSS^•$ radicals exhibit properties similar to those of $RS^•$ radicals, i.e., in their reactions with cellular antioxidants such as ascorbate, molecular oxygen, and hydrogen abstraction from activated C-H bonds of alcohols. However, delocalization of the $RSS^•$ radical electron spin leads to perthiyl radicals exhibiting a lower reactivity than thiyl radicals. Perthiyl radicals are moderately good oxidants (33,46), as indicated by the reactivity of the penicillamine perthiyl radical, $PenSS^•$, toward ascorbate ($k_{25} = 5.1 \times 10^6$ mol^{-1} dm^3 s^{-1}):

$$PenSS^• + \text{ascorbate } (AH^-) \rightarrow PenSSH + A^{•-} \tag{25}$$

A. Reaction of Perthiyl Radicals with Molecular Oxygen

Perthiyl radicals exhibit a similar reactivity toward molecular oxygen as with ascorbate to generate perthiyl peroxyl radicals ($RSSOO^{\bullet}$) in Eq. (22) with rate constants in the order of $k_{26} \approx 5\text{–}8 \times 10^{6}$ mol^{-1} dm^{3} s^{-1} (33,46).

$$RSS^{\bullet} + O_2 \rightarrow RSSOO^{\bullet} \tag{26}$$

The detection by high-performance ion chromatography of inorganic sulfate ions (SO_4^{2-}) following Eq. (26) has provided some indication as to the fate of perthiyl radicals in the presence of mo!-cular oxygen (33). The mechanism of SO_4^{2-} anion formation (33) has been suggested to proceed in analogy to the well-characterized reactions of thiyl peroxyl radicals (46–48) and peroxyl radicals in general (49):

$$RSSOO^{\bullet} \rightarrow RSS(O)(O)^{\bullet} \tag{27}$$

$$RSSO_2^{\bullet} + O_2 \rightarrow RSSO_2OO^{\bullet} \tag{28}$$

$$RSSO_2OO^{\bullet} + H_2O \rightarrow RSSO^{\bullet} + 2H^{+} + SO_4^{2-} \tag{29}$$

Most of the oxygenated intermediates formed as a consequence of Eqs. (26)–(29) must be considered good oxidants, as in the case of the corresponding radical intermediates believed formed in the corresponding $RS^{\bullet}/O_2$ systems (47,50,51), with the potential to inflict biological damage. The importance of molecular oxygen in the fate of thiyl radicals in cells remains a matter for debate (52) (see Chap. 1).

B. Conjugation of Perthiyl Radicals with Perthiolate and Thiolate Anions

By analogy to the conjugative reaction of thiyl radicals with thiolate anion (53,54) Eq. (30) may be of importance because of (as demonstrated by the perthiol analog of WR-1065) the existence of the perthiolate anion at physiological pH

$$RSS^{\bullet} + RSS^{-} \rightleftharpoons RS_4R^{\bullet-} \tag{30}$$

However, no evidence has been obtained for a stable tetrasulfide radical anion ($(RS_4R)^{\bullet-}$), which must have a half-life < 1 μs. The central S-S bond in tetrasulfides is very weak, where RSS–SSR ≈ 148 kJ/mol (41). Thus, an extra electron in the antibonding orbital of this already weak bond is expected to have a highly destabilizing effect, assuming that this radical species has a localized three-electron bond that is ($2\sigma/1\sigma^{*}$) in electronic nature, i.e., $(RSS \therefore SSR)^{\bullet-}$ (55). The interaction of $RSS^{\bullet}$ radicals with endogenous glutathione thiolate anions (GS^{-}) is expected to generate the corresponding trisulfide radical anion $(RSSSG)^{\bullet-}$. Despite the fact that the RS–SSR bond is significantly stronger than

RSS–SSR by ≈69 kJ/mol, no evidence has been obtained for the transient stabilization of trisulfide radical anions from the one-electron reduction of trisulfides (15,33,35,55). Clearly, Eq. (30) is thermodynamically unfavorable, with the equilibrium lying well to the side of the reactants. However, perthiyl radicals may be driven to react more by Eq. (30) by the removal of the $RS_4R^{\bullet -}$ from the equilibrium by molecular oxygen in the following equation:

$$RS_4R^{\bullet -} + O_2 \rightarrow RS_4R + O_2^{\bullet -} \quad (31)$$

The relative importance of Eqs. (22)–(25), (26), and (27) in a cellular system will depend on perthiol pK_a, perthiol concentration, and oxygen concentration. In the absence of any significant conjugation of perthiyl radicals with RSS^- or RS^- anions, Eqs. (27)–(29) may have important implications for the prooxidative effects of perthiyl radicals in cells.

C. Reactions of Perthiyl Radicals with Polyunsaturated Fatty Acids

Free radical attack on polyunsaturated fatty acids (PUFAs) is known to initiate the chain of lipid peroxidation in cellular biomembranes (9,56). Any radical with sufficient oxidizing power to abstract a hydrogen atom from the weak C-H bonds of PUFA biallylic groups can therefore be classified as a potential initiator of lipid peroxidation. Since $RSS^\bullet$ radicals are also capable of abstracting a hydrogen atom from alcohols (15), it follows that they may also be potential initiators of lipid peroxidation by abstracting a hydrogen atom from PUFAs:

$$\text{PUFA-H} + RSS^\bullet \rightarrow (\text{PUFA})^\bullet + RSSH \quad (32)$$

One-electron reduction of a symmetrical trisulfide in Eq. (12) as a source of $RSS^\bullet$ radicals provides a suitable pathway by which to study Eq. (32) in isolation without contributions from the reverse repair reaction [Eq. (33)] or hydrogen transfer to a perthiyl radical–PUFA adduct radical [Eqs. (34) and (35)]:

$$(\text{PUFA})^\bullet + RSSH \rightarrow \text{PUFA} + RSS^\bullet \quad (33)$$

$$RSS^\bullet + \text{PUFA} \rightarrow [RSS \cdots \text{PUFA}]^\bullet \quad (34)$$

$$[RSS \cdots \text{PUFA}]^\bullet + RSSH \rightarrow \text{RSS-PUFA-H} + RSS^\bullet \quad (35)$$

Perthiyl radicals were found to react almost exclusively by hydrogen abstraction via Eq. (33), as indicated by a quantitative conversion of perthiyl radicals to pentadienyl radicals (16). Figure 3 shows the pentadienyl radical absorption spectrum obtained after the reaction of the polyamino-perthiyl radical $H_2N(CH_2)_3NH(CH_2)_2SS^\bullet$ [derived from the one-electron reduction of the symmetrical trisulfide of WR-1065, (Fig. 1, **4**)] with linolenic acid. The rate constant $k_{32} = 1.1 \times 10^6$ mol^{-1} dm^3 s^{-1} was obtained by direct measurements of

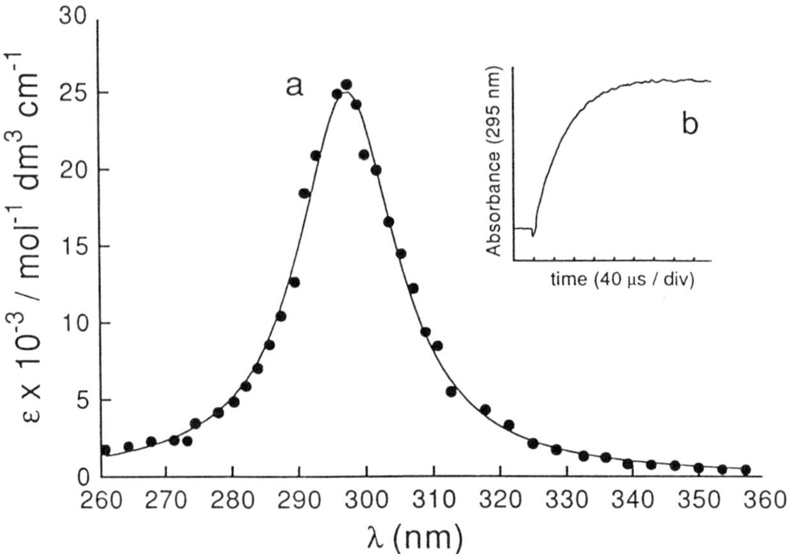

Figure 3 The pentadienyl radical absorption spectrum obtained on pulse radiolysis (10Gy) of (a) an N_2-saturated $(CH_3)_3COH/H_2O$ (1:1, v/v) mixture containing 1 mmol dm^{-3} linolenic acid and 5 mmol dm^{-3} symmetrical trisulfide of WR-1065. Inset (b): Optical trace recorded at 295 nm showing first-order buildup of the pentadienyl radical following perthiyl radical hydrogen abstraction of a bisallylic hydrogen from linolenic acid.

the first-order build-up of the pentadienyl radical absorption at 295 nm (Fig. 3, inset). Thiyl radicals have also been shown to abstract hydrogen from the activated C-H bonds of alcohols and ethers with rate constants in the order of 10^4 mol^{-1} dm^3 s^{-1} (32,57). Rate constants for $RS^{\bullet}$ radical hydrogen abstraction from the weaker C-H bonds in PUFAs to form pentadienyl radicals are significantly faster—by three orders of magnitude (31,57,58). For example, the rate constants for the reaction of GS^- with linoleic, linolenic, and arachadonic acids are in the order of $1\text{–}3 \times 10^7$ mol^{-1} dm^3 s^{-1}. Although perthiyl radicals must also be considered potential initiators of lipid peroxidation, they are less likely to do so than are thiyl radicals.

VIII. CONCLUSIONS

In general, the free radical–scavenging reactions of perthiols are qualitatively similar to but quantitatively quite different from those of the corresponding thiol antioxidants. Perthiols are not only efficient hydrogen atom donors, but in the form of perthiolate anions they can also behave as effective electron donors.

Moreover, the antioxidant-derived radicals, in this case perthiyl radicals, are significantly less reactive than their thiol counterparts and, consequently, are less likely to pose a threat to cellular integrity. In view of the essential role that thiols play in limiting free radical–induced cellular oxidative stress, the encouraging performance of perthiols as free radical scavengers suggests a possible role as antioxidants. Of course for the desired therapeutic effect, a potential therapeutic agent must be capable of reaching its site or sites of action in a concentration sufficient to initiate a therapeutic response. Drug absorption, distribution, metabolism, and excretion are equally important factors in perthiol antiradical activity. However, an understanding of the molecular mechanisms associated with perthiol antiradical activity is the first step in rational antioxidant drug design.

ACKNOWLEDGMENTS

This work is supported by the Cancer Research Campaign. I thank P. Wardman and K.-D. Asmus for helpful discussions.

REFERENCES

1. Rice-Evans, C., and Diplock, A. (1993) Current status of antioxidant therapy. Free Rad. Biol. Med. 15: 77–96.
2. Denekamp, J., and Rojas, A. (1987) Radioprotection in vivo: Cellular heterogeneity and fractionation. In: Anticarcinogenesis and Radiation Protection (Cerutti, P. A., Nygaard, O. F., and Simic, M. G., eds.), pp. 421–431. Plenum Press, New York.
3. Wardman, P. (1991) Chemical properties of radiation modifiers of radiation damage and their radiobiological effects. In: The Early Effects of Radiation on DNA (Fielden, E. M., and O'Neill, P., eds.), pp. 249–264. Springer-Verlag, Berlin.
4. Wardman, P. (1993) Sensitization and protection of oxidative damage caused by high energy radiation. In: Atmospheric Oxidation and Antioxidants (Scott, G., ed.), pp. 101–127. Elsevier.
5. Poli, G., Albano, E., and Dianzani, M. U., ed. (1993) Free Radicals: From Basic Science to Medicine. Birkhauser Verlag, Berlin.
6. Sies, H., ed. (1991) Oxidative Stress: Oxidants and Antioxidants, 2nd ed. Academic Press, London.
7. Halliwell, B., and Gutteridge, J. M. C. (1989) Free Radicals in Biology and Medicine. Clarendon Press, Oxford.
8. Halliwell, B., and Arouma, O. I., ed. (1993) DNA and Free Radicals. Ellis Horwood, London.
9. Halliwell, B., and Chirico, S. (1993) Lipid peroxidation: its mechanism, measurement, and significance. Am. J. Clin. Nutr. 57: 15S–25S.
10. Calabro-Jones, P. M., and Aguilera, J. A. (1988) Uptake of WR-2721 derivatives by cells in culture: Identification of the transported form of the drug. Cancer Res. 48: 3634–3640.

11. Glover, D., Glick, J. H., Weiler, C., Hurowitz, S., and Kigerman, M. M. (1986) Clinical trials of WR-2721 prior to alkylating agent chemotherapy and radiotherapy. Pharmacol. Ther. 39: 3–7.
12. Glover, D., Grabelsky, S., Fox, K., Weiler, C., Cannon, L., and Glick, J. (1989) Clinical trials of WR-2721 and cis-platinum. Int. J. Radiat. Oncol. Biol. Phys. 16: 1201–1204.
13. Fahey, R. C., Voynovic, B., and Michael, B. D. (1991) The effects of counter-ion condensation and co-ion depletion upon the rates of chemical repair of poly(U) radicals by thiols. Int. J. Radiat. Biol. 59: 885–899.
14. Fahey, R. C., Prise, K. M., Stratford, M. R. L., Wafta, R. R., and Michael, B. D. (1991) Rates of repair of pBR322 radicals by thiols as measured by the gas explosion technique: Evidence that counter-ion condensation and co-ion depletion are significant at physiological ionic strength. Int. J. Radiat. Biol. 59: 901–917.
15. Everett, S. A., Folkes, L. K., Asmus, K.-D., and Wardman, P. (1994) Free-radical repair by a novel perthiol: Reversible hydrogen transfer and perthiyl radical formation. Free Rad. Res. 20: 387–400.
16. Everett, S. A., and Wardman, P. Perthiols as antioxidants: Radical-scavenging and pro-oxidative mechanisms. Methods Enzymol. 251: 55–69.
17. Wardman, P., Candeias, L. P., Everett, S. A., and Tracey, M. (1994) Pulse radiolysis applied to drug design. Int. J. Radiat. Biol. 65: 35–41.
18. Smith, H. J. (1988) Introduction to the Principles of Drug Design. Butterworth & Co, London.
19. Benson, S. D. (1978) Thermochemistry and kinetics of sulfur-containing molecules and radicals. Chem. Rev. 78: 23–35.
20. Griller, D., Martinho Simoes, J. A., and Wayner, D. D. M. (1990) Thermochemistry of sulfur-centered intermediates. In: Sulfur-Centered Reactive Intermediates in Chemistry and Biology (Chatgilialoglu, C., and Asmus, K.-D., ed.), pp. 37–52, Plenum Press, New York.
21. Mott, A., and Barany, G. (1984) A new method for the synthesis of unsymmetrical trisulfanes. Synthesis 51: 657–660.
22. Barany, G., Schroll, A. L., Mott, A. W., and Halsrud, D. A. (1983) A general strategy for elaboration of the dithiocarbonyl functionality, -(C=O)SS-: Application to the synthesis of *bis*-(chlorocarbonyl) disulfane and related derivatives of thiocarbonic acids. J. Org. Chem. 48: 4750–4761.
23. Leimer, K. R., Rice, R. H., and Gehrke, C. W. (1977) Complete mass spectra of the per-trimethysilylated amino acids. J. Chromatogr. 141: 355–375.
24. Newton, G. L., Dwyer, T. J., Kim, T., Ward, J., and Fahay, R. C. (1992) Determination of the acid dissociation constants of WR-1065 by proton NMR spectroscopy. Rad. Res. 131: 143–151.
25. Zheng, S., Newton, G. L., Ward, J. F., and Fahay, R. C. (1992) Aerobic radioprotection of pBR322 by thiols: Effect of thiol net charge upon scavenging of hydroxyl radicals and repair of DNA radicals. Rad. Res. 130: 183–193.
26. Ward, J. F., and Mora-Arrellano, V. O. (1984) Pulse radiolysis studies of WR-1065. Int. J. Radiat. Oncol. Biol. Phys. 10: 1533–1536.
27. Simic, M. G., and Hunter, E. P. L. (1986) Reactions of peroxyl and C-centred radicals with sulfhydryls. Free Radical Biol. Med. 2: 227–230.

28. O'Neill, P. (1983) Pulse radiolytic study of the interaction of thiols and ascorbate with OH adducts of dGMP and dG: Implications for DNA repair processes. Rad. Res. 96: 198–210.
29. von Sonntag, C. (1987) The Chemical Basis of Radiation Biology. Taylor & Francis, London.
30. Asmus, K.-D. (1990) Sulfur-centered free radicals. Meth. Enzymol. 186: 168–180.
31. Schöneich, C., Asmus, K.-D., Dillinger, U., and Bruchhausen, F. V. (1989) Thiyl radical attack on polyunsaturated fatty acids: A possible route to lipid peroxidation. Biochem. Biophys. Res. Comms. 161: 113–120.
32. Schöneich, C., Bonifačić, M., and Asmus, K.-D. (1989) Reversible H-atom abstraction from alcohols by thiyl radicals: Determination of absolute rate constants by pulse radiolysis. Free Rad. Res. Comms. 6: 393–405.
33. Everett, S. A., Schöneich, C., Stewart, J. H., and Asmus, K.-D. (1992) Perthiyl radicals, trisulfide radical ions and sulfate formation. A combined photolysis and radiolysis study on redox processes with organic di- and trisulfides. J. Phys. Chem. 96: 306–314.
34. Armstrong, D. A. (1990) Applications of pulse radiolysis for the study of short-lived species. In: Sulfur-Centered Reactive Intermediates in Chemistry and Biology (Chatgilialoglu, C., and Asmus, K.-D., ed.), pp. 121–134. Plenum Press, New York.
35. Wu, Z., Back, T. G., Ahmad, R., Yamdagni, R., and Armstrong, D. A. (1982) Mechanisn of reduction of *bis*-(2-hydroxyethyl) trisulphide by e^-_{aq} and $^{\bullet}CO_2^-$. Spectrum and scavenging of $RSS^{\bullet}$ radicals. J. Phys. Chem.: 4417–4422.
36. Elliot, A. J., McEachern, R. J., and Armstrong, D. A. (1981) Oxidation of Amino-containing disulfides by $Br_2^{\bullet-}$ and $^{\bullet}OH$. A pulse radiolysis study. J. Phys. Chem. 85: 68–75.
37. Morine, G. H., and Kuntz, R. R. (1981) Observations of C-S and S-S bond cleavage in the photolysis of disulfides in solution. Photochem. Photobiol. 33: 1–5.
38. Burkey, T. J., Hawari, J. A., Lossing, F. P., Lusztyk, J., Sutcliffe, R., and Griller, D. (1985) The *tert*-butylperthiyl radical. J. Org. Chem. 50: 4966–4967.
39. Chatgilialoglu, C., and Asmus, K.-D., ed. (1990) Sulfur-Centered Reactive Intermediates in Chemistry and Biology. NATO ASI, Vol. 197, Plenum Press, New York.
40. Wardman, P. (1990) Thiol reactivity towards drugs and radicals: some implications in the radiotherapy and chemotherapy of cancer. In: Sulfur-Centered Reactive Intermediates in Chemistry and Biology (Chatgilialoglu, C., and Asmus, K. D., eds.), pp. 415–427. Plenum Press, New York.
41. Guerra, M. (1990) Electronic transitions in sulfur-centred radicals by means of MSXα calculations. In: Sulfur-Centered Reactive Intermediates in Chemistry and Biology (Chatgilialoglu, C., and Asmus, K.-D., ed.), pp. 7–12. Plenum Press, New York.
42. Kice, J. L. (1971) The sulfur-sulfur bond. In: Sulfur in Organic and Inorganic Chemistry (Senning, A., ed.), pp. 153–207. Marcel Dekker, New York.
43. Prütz, W. A. (1992) Catalytic reduction of Fe(III)-cytochrome-*c* involving stable radiolysis products derived from disulphides, proteins and thiols. Int. J. Radiat. Biol. 61: 593–602.

44. Scholes, G., Shaw, P., Willson, R. L., and Ebert, M. (1965) Pulse radiolysis studies of nucleic acid and related substances. In: Pulse Radiolysis (Ebert, M., Keene, J. P., Swallow, A. J., and Baxendale, J. H., ed.), pp. 151–164. Academic Press, London.
45. Halliwell, B. (1990) How to characterize a biological antioxidant. Free Rad. Res. Commun. 9: 1–32.
46. Asmus, K.-D. (1993) Recent aspects of thiyl and perthiyl free radical chemistry. In Active Oxygens, Lipid Peroxides, and Antioxidants (Yagi, K., ed.), pp. 57–67. CRC Press, Tokyo.
47. Sevilla, M. D., Becker, D., and Yan, M. (1990) The formation and structure of the sulfoxyl radicals $RSO^{\bullet}$, $RSOO^{\bullet}$, $RSO_2^{\bullet}$, and $RSO_2OO^{\bullet}$ from the reaction of cysteine, glutathione and penicillamine thiyl radicals with molecular oxygen. Int. J. Radiat. Biol. 57: 65–81.
48. Tamba, M., Simone, G., and Quintiliani, M. (1986) Interactions of thiyl radicals with oxygen: a pulse radiolysis study. Int. J. Radiat. Biol. 50: 595–600.
49. von Sonntag, C., and Schuchmann, H.-P. (1991) The elucidation of peroxyl radical reactions in aqueous solution with the help of radiation-chemical methods. Angew. Chem. 30: 1229–1253.
50. Becker, D., Swarts, S., Champagne, M., and Sevilla, D. (1988) An ESR investigation of the reactions of glutathione, cysteine and penicillamine thiyl radicals: Competitive formation of $RSO^{\bullet}$, $R^{\bullet}$, $RSSR^{\bullet-}$ and $RSS^{\bullet}$. Int. J. Radiat. Biol. 53: 767–786.
51. Becker, D., Summerfield, S., Gillich, S., and Sevilla, M. D. (1994) Influence of oxygen on the repair of direct radiation damage to DNA by thiols in model systems. Int. J. Radiat. Biol. 65: 537–548.
52. Wardman, P., and von Sonntag, C. Evaluation of the kinetic factors which control the fate of thiyl radicals in cells. Methods Enzymol. (in press).
53. Hoffman, M. Z., and Hayon, E. (1972) Reactions of $RSSR^{\bullet-}$ radical anions. J. Am. Chem. Soc. 94: 7950–7958.
54. Hoffman, M. Z., and Hayon, E. (1972) One-electron reduction of the disulfide linkage in aqueous solution. Formation, protonation, and decay kinetics of the $RSSR^{\bullet-}$ radical. J. Am. Chem. Soc. 94: 7950–7957.
55. Asmus, K.-D. (1990) Sulfur-centred three electron bonded radical species. In: Sulfur-Centered Reactive Intermediates in Biology and Medicine (Chatgilialoglu, C., and Asmus, K.-D., ed.), pp. 155–172. Plenum Press, New York.
56. Dix, T. A., and Aikens, J. (1992) Mechanisms and biological relevance of lipid peroxidation initiation. Chem. Res. Toxicol. 6: 2–18.
57. Schöneich, C., and Asmus, K.-D. (1990) Reaction of thiyl radicals with alcohols, ethers and polyunsaturated acids: A possible role of thiyl free radicals in thiol mutagenesis? Radiat. Env. Biophys. 29: 263–271.
58. Schöneich, C., Dillinger, U., Bruchhausen, F. V., and Asmus, K.-D. (1992) Oxidation of polyunsaturated fatty acids and lipids through thiyl and sulfonyl radicals: Reaction kinetics, and influence of structure of thiyl radicals. Arch. Biochem. Biophys. 292: 456–467.

4

Biothiyls: Free Radical Chemistry and Biological Significance

Janice A. DeGray and Ronald P. Mason
National Institute of Environmental Health Sciences, National Institutes of Health, Research Triangle Park, North Carolina

Thiol-containing compounds are important in many biochemical and pharmacological reactions: disulfide bonds are important in determining the overall structure of proteins, and in many drugs the thiol group is an important reactive center that determines both their effects and their side effects. Thiols are among the most easily metabolized compounds. They are easily oxidized, forming thiyl and thiyl-derived radicals, leaving the oxidizing agent in a reduced state. This reductive activity of thiols is generally considered to be beneficial. Reduced glutathione (GSH) has a radioprotective action (1), cysteine is a bactericidal agent (2), and *N*-acetylcysteine is used in the treatment of acetaminophen overdose (3). This reductive ability of glutathione and its cellular presence makes it an effective bioactive antioxidant in vivo and in vitro. However, thiyl radicals formed upon oxidation of thiols may lead to the formation of other sulfur-centered free radicals and related reactive species and thus may cause cellular damage (4). The glutathionyl radical generated from glutathione thionitrite has been shown to cause mutagenicity in *Salmonella typhimurium* TA100 (5), a number of thiyl radicals have been implicated in lipid peroxidation (6,7), and penicillamine thiyl radical may be involved in the oxidation of alcohols (8).

I. THIYL RADICAL CHEMISTRY

Thiols and disulfides are related to each other by redox chemistry. Thiols (RSH) are easily oxidized to form thiyl radicals ($RS^{\bullet}$):

$$RSH \rightarrow RS^{\bullet} + H^{+} + e^{-}$$

and two thiyl radicals can combine to form the disulfide.

$$2\ RS^{\bullet} \rightarrow RSSR$$

In addition to direct enzymatic oxidation of thiols, the thiyl radical can be formed by hydrogen abstraction in the reaction with hydroxyl radical,

$$RSH + {}^{\bullet}OH \rightarrow RS^{\bullet} + H_2O$$

in the reaction with the superoxide radical anion,

$$RSH + O_2^{\bullet -} + H^{+} \rightarrow RS^{\bullet} + H_2O_2$$

by photolysis of the disulfide,

$$RSSR \xrightarrow{h\nu} 2\ RS^{\bullet}$$

and by direct one-electron oxidation by metals.

$$RSH + M^{n} \rightarrow RS^{\bullet} + M^{n-1} + H^{+}$$

Once formed, thiyl radicals can participate in a number of different reactions. They can react with the thiolate or parent thiol to form the disulfide radical anion.

$$RS^{\bullet} + RS^{-}(RSH) \rightarrow (RS\overset{\bullet}{-}SR)^{-} + (H^{+})$$

The thiyl radical and the disulfide radical anion both react with molecular oxygen. The thiyl radical reacts with oxygen to form the thiol peroxyl radical (9–11), which may be a damaging species involved in ionizing radiation therapy (12) and protein deactivation (13).

$$RS^{\bullet} + O_2 \rightarrow RSOO^{\bullet}$$

The disulfide radical anion also reacts with oxygen to form the superoxide radical anion, itself a potentially damaging species.

$$(RS\overset{\bullet}{-}SR)^{-} + O_2 \rightarrow RSSR + O_2^{\bullet -}$$

In addition to these species, further oxidation products of thiols include thioperoxide (RSOH), the sulfinyl radical ($RSO^{\bullet}$), sulfinic acid (RSO_2H), the sulfonyl radical ($RSO_2^{\bullet}$), and sulfonic acid (RSO_3H).

$$RS^{\bullet} \rightarrow RSOH \rightarrow RSO^{\bullet} \rightarrow R\overset{O}{\overset{\|}{S}}OH \rightarrow R\underset{O}{\underset{\|}{\overset{O}{\overset{\|}{S}}}}{}^{\bullet} \rightarrow R\underset{O}{\underset{\|}{\overset{O}{\overset{\|}{S}}}}-OH$$

II. DETECTION OF THIYL RADICALS

A. Electron Paramagnetic Resonance Detection

Alkyl thiyl radicals exhibit large anisotropy in their **g** tensors, which broadens the electron paramagnetic resonance (EPR) signal beyond detection at room temperature (14). This anisotropy is due to the near degeneracy of the π-type orbitals at the sulfur center. Therefore, direct detection of thiyl radicals by EPR is usually impossible. Thiyl radicals and thiyl-derived radicals can, however, be detected at low temperature in aqueous glasses (10,11). On the other hand, the glutathione and cystine disulfide anion radicals have been detected at room temperature in a fast-flow system (15). The only other definitive method of detecting the presence of thiyl radicals is spin trapping. In the spin-trapping experiment, a radical reacts with a nitroso or nitrone spin trap to form a nitroxide adduct, which is detectable with EPR.

The spin trap usually employed for detection of thiyl radicals is 5,5-dimethyl-1-pyrroline *N* oxide (DMPO), a nitrone. The EPR spectra of DMPO radical adducts are more distinct for the radical trapped than the spectra obtained using *C*-phenyl-*N*-*t*-butyl nitrone (PBN). The DMPO/thiyl radical adducts are, in general, more intense than PBN adducts due to higher formation rates and slower decay rates (16,17). There are a few cases where the use of PBN is preferred. The DMPO thiyl radical adduct rapidly oxidizes in the presence of strong oxidizing agents such as $K_3Fe(CN)_6$ or Ce^{4+} complexed to nitrilotriacetate. Under such conditions PBN may be more useful than DMPO due to the higher stability of the PBN radical adducts under these conditions (18).

Information about the free radical species can be elicited by analyzing the hyperfine coupling constants and the number of hyperfine lines observed from the radical adduct. Usually one observes hyperfine coupling due to the nitrogen and the β-hydrogen from the spin trap. It has recently been shown that the extra hyperfine coupling constant seen in most DMPO thiyl radical adducts is due to the γ-protons from the thiyl group. In the same study it was also found that the thiyl radical can add, depending on the steric hindrance, at either the N or O atoms of nitroso spin traps (19).

This chapter will concentrate on thiyl radical formation and detection by EPR. The reader may wish to consult a review article by Ross, which discusses the role of glutathione in protecting against free radical–induced toxicity from a more biochemical standpoint (20).

B. Other Methods of Detection

Other less definitive methods for the detection of thiyl radicals include oxygen-consumption studies and optical detection. In oxygen-consumption studies, the rate of oxygen disappearance due to its reactivity with thiyl and disulfide anion radicals is monitored. Optical methods for the detection of thiyl-derived radicals have been reviewed by Asmus (21).

III. ENZYMATIC FORMATION OF THIYL RADICALS

A. Peroxidase Enzymes

1. *Horseradish Peroxidase*

The plant enzyme horseradish peroxidase (HRP) is a model system for mammalian peroxidase-catalyzed oxidation of thiols.

$$HRP + H_2O_2 \rightarrow \text{Compound I} + H_2O$$

$$\text{Compound I} + RSH \rightarrow \text{Compound II} + RS^{\bullet}$$

$$\text{Compound II} + RSH \rightarrow HRP + RS^{\bullet}$$

The first detection of thiyl radical adducts was in a horseradish peroxidase/hydrogen peroxide system (22). Until this time the formation of compound II was usually reported as inferential evidence of a one-electron oxidation of thiols (23).

Glutathione was oxidized in the presence of horseradish peroxidase/hydrogen peroxide and DMPO, and the glutathionyl radical adduct was detected ($a^N = 15.4$ G and $a^H_\beta = 16.2$ G) (24). The reaction was totally dependent on the presence of exogenous hydrogen peroxide. This system is frequently used as a control in many spin-trapping experiments involving horseradish peroxidase and glutathione and, as such, the glutathionyl radical adduct in this system has been confirmed by a number of researchers.

Some thiols such as cysteine are known to autooxidize readily and, thus, under an oxygen atmosphere, do not require an outside source of hydrogen peroxide. The formation of the cysteinyl radical during autooxidation of cysteine was proven when the radical was trapped with DMPO and the adduct positively identified by mass spectrometry and NMR (25).

The thiyl radical adduct of dithiothreitol was observed when dithiothreitol was incubated in a system of horseradish peroxidase/hydrogen peroxide and DMPO ($a^N = 15.2$ G and $a^H_\beta = 16.3$ G) (13). This same radical adduct was formed upon photolysis (26).

2. *Lactoperoxidase*

Lactoperoxidase is a mammalian peroxidase found in saliva, tears, and milk and is a prototype for other mammalian hematoporphyrin peroxidases. Lactoperoxidase in the presence of hydrogen peroxide was found to catalyze the oxidation of a variety of thiols to their respective thiyl radicals. These include glutathione ($a^N = 15.1$ G, $a^H_\beta = 16.0$ G, and $a^H_\gamma = 0.64$ G), cysteine ($a^N = 15.1$ G and

a_{β}^{H} = 17.4 G), *N*-acetylcysteine (a^{N} = 15.3 G, a_{β}^{H} = 16.7 G, and a_{γ}^{H} = 0.48 G), captopril (a^{N} = 15.4 G and a_{β}^{H} = 16.0 G), penicillamine (a^{N} = 15.3 G and a_{β}^{H} = 20.0 G) (27) and cysteamine (a^{N} = 15.2 G, a_{β}^{H} = 17.0 G, and a_{γ}^{H} = 0.72 G) (27,28). Penicillamine and cysteine did not require exogenous hydrogen peroxide, presumably due to their ability to autooxidize. Svensson et al. did not find micromolar concentrations of glutathione or cysteine to be substrates of lactoperoxidase (28).

3. *Myeloperoxidase*

Myeloperoxidase is an abundant neutrophil protein involved in microbial killing and inflammatory tissue damage. Myeloperoxidase in the presence of hydrogen peroxide was found to catalyze the oxidation of cysteamine, cysteine methyl ester, cysteine ethyl ester, and, to a lesser extent, mercaptoethanol and thioglycollic acid (28). The corresponding thiyl radical adducts were detected in the presence of DMPO. Using oxygen-consumption evidence, myeloperoxidase has been proposed to generate the thiyl radical with cysteamine and the methyl and ethyl esters of cysteine (29,30).

4. *Eosinophil Peroxidase*

Eosinophil peroxidase, found in the eosinophilic leukocytes, catalyzes the one-electron oxidation of cysteamine and cysteine methyl ester to their respective thiyl radical adducts in the presence of DMPO (28).

5. *Prostaglandin H Synthase*

Ram seminal vesicle microsomes (a source of prostaglandin H synthase), in the presence of arachidonic acid or 15-hydroperoxy-5,8,11,13-eicosatetraenoic acid, catalyze the oxidation of glutathione. The purified prostaglandin H synthase also forms the glutathionyl radical adduct in the presence of DMPO. A small signal was detectable in the absence of added hydroperoxide, presumably due to the formation of hydrogen peroxide during the autooxidation of glutathione (31).

B. Xanthine/Xanthine Oxidase

Superoxide radicals are generated during normal physiological and biological functions. Xanthine/xanthine oxidase is a common superoxide-generating system, and the superoxide anion generated by this enzyme oxidizes glutathione to the glutathionyl radical. The EPR spectrum observed is indicative of a mixture of DMPO adducts, probably the superoxide radical adduct and the thiyl radical adduct (32).

C. Hemoglobin

When rats are exposed to the methemoglobin-forming agent phenylhydrazine, a hemoglobin thiyl free radical forms that reacts with DMPO to form a very stable radical adduct (33–35). This adduct was also detected in vitro with purified rat hemoglobin (33). In addition to the hemoglobin thiyl free radical formation after phenylhydrazine exposure, iproniazid, phenelzine, and hydrazine were shown to generate this radical adduct in vivo (34). The DMPO thiyl hemoglobin free radical adduct formed after phenylhydrazine exposure was also detected in the blood of rats treated with DMPO and a variety of hydroperoxides (36). Pretreatment of animals to lower the concentration of nonprotein sulfhydryl groups in red blood cells enhanced free radical production upon exposure to *t*-butyl hydroperoxide. Aniline, nitrobenzene, and their metabolites, phenylhydroxylamine and nitrosobenzene, also produce the hemoglobin thiyl radical in vivo. In vitro studies indicated that the oxidizing species was the phenylhydronitroxide radical, which was produced by the reaction of phenylhydroxylamine with oxyhemoglobin (37). The hemoglobin thiyl radical adduct was assigned based on thiol-blocking experiments. Rat hemoglobin contains one cysteine, which is not reactive with iodoacetamide (38). This cysteine appears to be the one that forms the stable, highly immobilized hemoglobin thiyl radical adduct, because although formation of this radical was inhibited by most thiol-blocking agents, iodoacetamide was ineffective.

D. Leghemoglobin

Leghemoglobin is a plant hemoprotein found in N_2-fixing root nodules and is involved in the oxygen transport to bacteroids. In incubations of leghemoglobin, hydrogen peroxide, DMPO, and either GSH or cysteine, the respective thiyl radical adducts are generated (39–41). The same glutathionyl adduct was also detected upon incubation of the deoxyferroleghemoglobin with hydrogen peroxide, GSH, and DMPO (39). This thiyl radical appears to be generated by ferryl leghemoglobin and the protein-derived radical formed from the reaction of leghemoglobin with hydrogen peroxide. Thus, glutathione may protect the enzyme from inactivation by regenerating the native form of the enzyme.

E. Myoglobin

Myocytes have low amounts of glutathione peroxidase and catalase. During reperfusion injury, reactive oxygen species are formed. Under these circumstances, further injury may occur due to the low amounts of inherent antioxidants. The addition of hydrogen peroxide to ruptured cardiac myocytes leads to the formation of ferryl myoglobin, which is capable of initiating peroxidative damage to membranes. The thiol compounds *N*-(2-mercaptopropionyl)glycine and *N*-acetylcysteine reduced the ferryl myoglobin to metmyoglobin and, in the process, generated their respective thiyl radicals which were trapped with DMPO. The same thiyl radicals could be detected in a system using purified metmyoglobin and hydrogen peroxide. In the presence of Desferal, an iron chelator, thiol-containing drugs slightly inhibited the peroxidative damage caused by either the model system or ruptured myocytes, whereas lipid peroxidation was stimulated in the absence of Desferal. Thus *N*-(2-mercaptopropionyl)glycine or *N*-acetylcysteine can be considered antioxidants except in the presence of Fe^{3+}, where they may act as prooxidants (42).

The ability of a metmyoglobin and the H_2O_2 system to oxidize *N*-acetylcysteine was confirmed by Romero et al. (43). In addition to *N*-acetylcysteine, the thiyl radical adducts of glutathione and cysteine were formed in a system of metmyoglobin, H_2O_2, and DMPO.

F. Bovine Serum Albumin

The anaerobic reaction of Ce^{4+} (a thiol-selective oxidant) with bovine serum albumin in the presence of DMPO generates an anisotropic EPR spectrum assigned as a DMPO/thiyl radical adduct from a cysteine residue (44). Unlike the thiyl radical in myoglobin, this protein thiyl radical transfers its unpaired electron to carbon sites in the protein, causing irreparable damage.

IV. DIRECT OXIDATION OF THIOLS BY NONENZYMATIC SYSTEMS

A. Oxidation by Bioactive Compounds

N-Methyl-*N′*-nitro-*N*-nitrosoguanidine (MNNG) is a gastric carcinogen, and the addition of an antioxidant decreases its gastric carcinogenesis. In vitro experiments indicate that MNNG is susceptible to nucleophilic attack by L-cysteine, producing the thionitrite derivative of cysteine, and that the cysteine thiyl free radical might be produced by homolytic cleavage of this thionitrite (45).

$$R—S—N{=}O \rightarrow RS^{\bullet} + {}^{\bullet}NO$$

This was confirmed by trapping the cysteinyl radical adduct of DMPO (a^N = 15.3 G and a_β^H = 17.0 G) in the presence of MNNG and L-cysteine. It is postulated that this thiyl radical may be prooxidant and account for the toxicity of MNNG (46).

Nitroprusside is a potent hypotensive agent whose use in the treatment of high blood pressure and heart attack has been limited due to accidental deaths, apparently as a result of cyanide poisoning (47,48). Nitroprusside has been found to be directly reduced by cysteine and glutathione, and the formation of cysteinyl and glutathionyl radical adducts of DMPO was verified by EPR. Although the biological importance of this direct reduction of nitroprusside by thiols is uncertain, the ability of thiols to reduce nitroprusside correlates well with cyanide liberation from nitroprusside in the presence of thiol compounds and the rat liver microsome/NADPH system (49).

Cysteine is thought to reduce peroxynitrite by a one-electron process (50). It was confirmed in a spin-trapping study that glutathione and cysteine directly reduce the peroxynitrite anion to form thiyl radical (51). In a similar mechanism the glutathionyl radical adduct of DMPO was detected during the autooxidation of 3-morpholinosydnonimine. This autooxidation generates a flux of nitric oxide and superoxide anion, which combine to form peroxynitrite. This peroxynitrite is then reduced by glutathione.

$$GSH + ONOO^- \xrightarrow{H+} GS^\bullet + H_2O + {}^\bullet NO_2$$

Presumably, the nitrogen dioxide formed in this reaction also oxidizes thiols to thiyl radicals (51,52).

$$GSH + {}^\bullet NO_2 \rightarrow GS^\bullet + NO_2^- + H^+$$

The reductive addition of glutathione to 2- and 6-hydroxymethyl-1,4-naphthoquinone forms hydroquinone, which autooxidizes to form the semiquinone radical. This semiquinone radical can react with oxygen to form the corresponding quinone and superoxide anion. In the presence of the spin trap, the glutathionyl radical adduct and the hydroxyl radical adduct of DMPO were detected. The mechanism of formation of the glutathionyl radical in this system is uncertain. It may be formed via the oxidation of glutathione by the superoxide anion radical, the hydroxyl radical, or even the semiquinone radical (53). The reduction of 2-methylmethoxynaphthoquinone by NADPH–cytochrome P-450 reductase, a one-electron transfer enzyme, or DT-diaphorase, a two-electron transfer enzyme, was studied. In the presence of glutathione, the formation of DMPO/thiyl radical adducts was observed, and the addition of superoxide dismutase suppressed the thiyl radical adduct. Superoxide dismutase enhanced GSSG accumulation during NADPH–cytochrome P-450 reductase catalysis of this quinone, whereas it inhibited GSSG formation during reduction by DT-diaphorase. The production of thiyl radicals was mainly coupled to semiquinone reduction during NADPH–cytochrome P-450 catalysis and to superoxide radical reduction during DT-diaphorase catalysis (54).

Hematoporphyrin derivative, a mixture of many porphyrins used in photodynamic therapy of cancer, has been shown upon irradiation with filtered white light to generate the DMPO/cysteinyl radical adduct. The oxidation of cysteine is presumably by reaction with singlet oxygen (a product of irradiation of hematoporphyrin derivative in the presence of oxygen). Cysteine may act as an antioxidant by scavenging singlet oxygen (55).

Cysteine and glutathione may also react directly with hematoporphyrin triplet. Illumination of a deaerated solution of hematoporphyrin IX in the presence of glutathione or cysteine leads to photoreduction of hematoporphyrin to its radical form and is concomitant with photooxidation of these thiols to their corresponding thiyl radicals. The latter have been detected as adducts to the spin trap 2-methyl-2-nitrosopropane (MNP) (56).

N-Hydroxypyridine-2-thione, known as pyrithione or Omadine, possesses microbicidal activity when complexed with Fe^{3+}. The pyrithione-zinc complex is frequently used in antifungal and antibacterial formulations and is the active ingredient of several antidandruff shampoos. Upon exposure of pyrithione solutions to ultraviolet radiation, $^{\bullet}OH$ radicals are released and the pyridine-2-thiyl radical is formed. The latter species has been identified in aqueous and toluene solutions using DMPO, MNP, and *aci*-nitromethane as trapping agents (57,58). 6-Mercaptopurine (6MP), the antitumor and immunosuppressant drug, appears to be phototoxic to many patients. Using EPR and the spin traps DMPO, MNP, and *aci*-nitromethane, it has been shown that the ultraviolet radiation of 6MP results in photogeneration of the corresponding thiyl radical via electron transfer from the photoexcited 6MP to either O_2 (in aerated solutions) or the trapping agents (in air-free samples) (59). Thus, both pyrithione and 6MP possess photoreducing capacity. These are also the first examples of aromatic thiyl radical being spin-trapped in aqueous solutions.

B. One-Electron Oxidation by Metals

Thiols can directly reduce electron acceptors such as transition metal ions.

1. Thiol-Driven Fenton Reactions

Hydroxyl radicals are thought to be one of the most reactive intermediates involved in oxidative stress. Captopril, a thiol-containing angiotensin-converting enzyme inhibitor, and its stereoisomer epicaptopril have been shown to generate the hydroxyl radical in a thiol-driven Fenton reaction.

$$Fe^{3+} + RSH \rightarrow Fe^{2+} + RS^{\bullet} + H^+$$

$$Fe^{2+} + H_2O_2 \rightarrow Fe^{3+} + {}^{\bullet}OH + HO^-$$

Thiols also react with the $^{\bullet}OH$ at a diffusion-controlled rate.

$$RSH + {}^{\bullet}OH \rightarrow RS^{\bullet} + H_2O$$

In the absence of iron, captopril can exert a protective effect against lipid peroxidation and act as an antioxidant. In the presence of iron, its prooxidant action may dominate (60).

2. Manganese

MnO_2 and Mn^{3+} are known atmospheric pollutants and can cause neurotoxic effects with exposure. Prolonged exposure also reduces the pool of nonprotein sulfhydryls. MnO_2 and Mn^{3+} catalyzed GSH oxidation, generating the glutathionyl radical, which was trapped by DMPO and the PBN analog, α-4-pyridyl-1-oxide *N*-*t*-butyl nitrone (4-POBN). The neurotoxic effects may be due to the generation of glutathionyl radical, which decreases the pool of antioxidants (61).

3. *Vanadium*

The in vivo toxicity of V^{5+} is known to correlate with the depletion of cellular glutathione (62). V^{5+} has been found to catalyze the oxidation of thiols, generating the thiyl radicals from glutathione, cysteine, *N*-acetylcysteine, and penicillamine as detected by trapping with DMPO (63). The results suggest that free radical reactions play a significant role in the depletion of cellular thiols and hence, V^{5+}, toxicity.

4. *Mercury*

Hg^{2+}, a neurotoxic metal, is known to induce porphyrinuria, presumably due to the oxidation of porphyrins by superoxide and hydroxyl radicals. Hg^{2+} may increase the concentration of these reactive oxygen species by decreasing cellular levels of glutathione via complexation with Hg^{2+}. In the presence of glutathione, Hg^{2+}, hydrogen peroxide, and DMPO, a complex spectrum consisting of both DMPO/$^{\bullet}$OH and DMPO/$^{\bullet}$SG was observed. Thus, in addition to compromising the glutathione concentration via complexation, formation of glutathionyl radical can occur (64).

V. THIOL PUMPING

Glutathione and related thiols can exert a protective effect by reducing potentially reactive compounds such as free radicals. This reduction of free radicals ($A^{\bullet}$) by GSH leading to the regeneration of the parent molecule (AH) and the corresponding thiyl radical is known as "thiol pumping" or "futile metabolism."

$$AH \rightarrow A^{\bullet} + H^{+} + e^{-}$$

$$GSH + A^{\bullet} \rightarrow GS^{\bullet} + AH$$

In the first report of thiol pumping, Yamazaki et al. (65) monitored the formation of the chlorpromazine radical cation by EPR in the presence of glutathione. This rate of formation was suppressed by the reaction of glutathione with the chlorpromazine radical cation, presumably forming the glutathionyl radical.

The glutathionyl radical is also shown to be formed by a reaction with the aminopyrine cation radical. Aminopyrine is oxidized by prostaglandin H synthase to the aminopyrine radical cation, which then disproportionates to the iminium cation. The iminium cation is further hydrolyzed to the demethylated amine and formaldehyde. In the presence of GSH, the aminopyrine radical cation is reduced, resulting in the formation of aminopyrine and the glutathionyl radical (66). Ram seminal vesicles, a source of prostaglandin H synthase, oxidizes both aminopyrine and phenol to the aminopyrine radical cation and the phenoxyl radical, respectively. The glutathionyl radical was trapped when either substrate was incubated in the presence of ram seminal vesicles, GSH, arachidonic acid, and DMPO. Mouse keratinocytes also contain the enzyme prostaglandin H synthase. When phenol was incubated in the presence of whole mouse keratinocytes, GSH, arachidonic acid, and DMPO, the DMPO/glutathionyl radical adduct was detected. No stimulation of the glutathionyl radical was detected in the presence of aminopyrine and whole keratinocytes, presumably due to the fact that aminopyrine does not penetrate the cell membrane as well as hpenol. This is the first detection of thiol pumping within a cell (67). The phenoxyl radical of the synthetic estrogen diethylstilbestrol, which is proposed as a possible determinant of the genotoxicity of this compound, was also shown to react with GSH to generate the glutathionyl radical (68). 17β-estradiol, a natural estrogen, has been shown to generate the glutathionyl radical in the presence of lactoperoxidase, GSH, hydrogen peroxide, and DMPO. In the absence of spin trap, thiyl radical chemistry results in hydrogen peroxide generation and this could explain the hydroxyl radical-induced DNA base lesions recently reported for female breast cancer tissue (69).

Ross et al. have done extensive studies on the fate of GSH during the horseradish peroxidase- and prostaglandin H synthase–catalyzed oxidation of *p*-phenetidine (70) and acetaminophen (71). They have shown that the free radical metabolites of both compounds react readily with glutathione to form the parent compounds and the GSSG. Glutathione, cysteine, and *N*-acetylcysteine all react with the acetaminophen phenoxyl radical to form their respective thiyl radicals. In the presence of DMPO, the glutathionyl, cysteinyl, and *N*-acetylcysteinyl adducts were detected by EPR.

Pulse-radiolysis experiments failed to detect any reaction between the acetaminophen phenoxyl radical and cysteine due to the high oxidation potential of the thiol in comparison to the acetaminophen phenoxyl radical (72). Under physiological conditions, the reaction between cysteine and the acetaminophen phenoxyl radical may still proceed if the cysteinyl radical is removed from the system by other reactions. In a similar system involving the reaction between the aminopyrine radical cation and GSH, glutathionyl radical reacts with oxygen and/or thiolate, pushing the equilibrium to the right due to the stability of the GSSG product (73). At a high concentration of GSH, the acetaminophen phenoxyl radical oxidizes glutathione to the thiyl radical, which reacts with the thiolate anion to form the disulfide radical anion. The disulfide radical anion was detected by direct EPR in a flow system (15).

Etoposide, an antitumor drug, when activated by the horseradish peroxidase/hydrogen peroxide system, forms the etoposide phenoxyl radical. In the presence of glutathione or cysteine, the parent etoposide compound and the respective DMPO/thiyl radical adducts were formed (74). In a similar experiment, eugenol was oxidized by the horseradish peroxidase/hydrogen peroxide system to generate its phenoxyl radical metabolite. Addition of glutathione and DMPO resulted in the trapping of the glutathionyl radical adduct (75).

In a model system consisting of horseradish peroxidase, hydrogen peroxide, and GSH, the glutathionyl radical adduct formation was found to be augmented by the addition of clozapine, an antipsychotic drug, the reactive clozapine radical being reduced by GSH (76). Similar results were obtained with a system consisting of gentian violet (an antimicrobial carcinogen), horseradish peroxidase, hydrogen peroxide, and GSH (77).

VI. THIYL-SPECIFIC ANTIOXIDANT ENZYME

Although thiols themselves may have beneficial properties, their thiyl metabolites have been shown to exert deleterious effects. Kim et al. have isolated a new antioxidant protein, the thiyl-specific antioxidant enzyme TSA, which provides protection against this damage (78). Thiyl radicals from dithiothreitol were produced by horseradish peroxidase/hydrogen peroxide under aerobic and anaerobic conditions and by the Fe(III)/oxygen system. The formation of DMPO/thiyl radical adducts was inhibited by TSA regardless of the thiyl radical–generating system. Only active mutants that maintained their ability to protect target enzymes against inactivation catalyzed the removal of thiyl radical. Comparison of active with inactive mutants showed that cysteine 47 is required for this activity. In addition, thiyl radicals react with oxygen to generate an unidentified thiol peroxyl species. Fe(III)-ethylenediaminetetraacetic acid is proposed to react with this species to generate a more reactive radical that can abstract a hydrogen atom from ethanol to produce the α-hydroxyethyl radical. This reactive thiyl-oxygen radical is believed to be responsible for causing deleterious effects on biomolecules (13).

REFERENCES

1. Baker, M. Z., Badiello, R., Tamba, M., Quintiliani, M., and Gorin, G. (1982) Pulse radiolytic study of hydrogen transfer from glutathione to organic radicals. Int. J. Radiat. Biol. 41:595–602.
2. Nyberg, G. K., Granberg, G. P. D., and Carlsson, J. (1979) Bovine superoxide dismutase and copper ions potentiate the bactericidal effect of autoxidizing cysteine. Appl. Environ. Microbiol. 38:29–34.
3. Prescott, L. F., Park, J., Ballantyne, A., Adriaenssens, P., and Proudfoot, A. T. (1977) Treatment of paracetamol (acetaminophen) poisoning with N-acetylcysteine. Lancet 432–434.
4. Mason, R. P. (1984) Spin trapping free radical metabolites of toxic chemicals. In: Spin Labeling in Pharmacology (Holtzman, J. L., ed.), pp. 87–129. Academic Press, New York.
5. Carter, M. H., and Josephy, P. D. (1986) Mutagenicity of thionitrites in the Ames test. The biological activity of thiyl free radicals. Biochem. Pharmacol. 35:3847–3851.

6. Schöneich, C., Asmus, K.-D., Dillinger, U., and v. Bruchhausen, F. (1989) Thiyl radical attack on polyunsaturated fatty acids: A possible route to lipid peroxidation. Biochem. Biophys. Res. Commun. 161:113–120.
7. Schöneich, C., Dillinger, U., v. Bruchhausen, F., and Asmus, K.-D. (1992) Oxidation of polyunsaturated fatty acids and lipids through thiyl and sulfonyl radicals: Reaction kinetics, and influence of oxygen and structure of thiyl radicals. Arch. Biochem. Biophys. 292:456–467.
8. Schöneich, C., Bonifačić, M., and Asmus, K.-D. (1989) Reversible H-atom abstraction from alcohols by thiyl radicals: Determination of absolute rate constants by pulse radiolysis. Free Rad. Res. Commun. 6:393–405.
9. Tamba, M., Simone, G., and Quintiliani, M. (1986) Interactions of thiyl free radicals with oxygen: A pulse radiolysis study. Int. J. Radiat. Biol. 50:595–600.
10. Sevilla, M. D., Yan, M., and Becker, D. (1988) Thiol peroxyl radical formation from the reaction of cysteine thiyl radical with molecular oxygen: An ESR investigation. Biochem. Biophys. Res. Commun. 155:405–410.
11. Sevilla, M. D., Becker, D., and Yan, M. (1990). The formation and structure of the sulfoxyl radicals $RSO^{\bullet}$, $RSOO^{\bullet}$, $RSO_2^{\bullet}$ and $RSO_2OO^{\bullet}$ from the reaction of cysteine, glutathione and penicillamine thiyl radicals with molecular oxygen. Int. J. Radiat. Biol. 57:65–81.
12. Quintiliani, M. (1986) The oxygen effect in radiation inactivation of DNA and enzymes. Int. J. Radiat. Biol. 50:573–594.
13. Yim, M. B., Chae, H. Z., Rhee, S. G., Chock, P. B., and Stadtman, E. R. (1994) On the protective mechanism of the thiol-specific antioxidant enzyme against the oxidative damage of biomacromolecules. J. Biol. Chem. 269:1621–1626.
14. Symons, M. C. R. (1974) On the electron spin resonance detection of $RS^{\bullet}$ radicals in irradiated solids: Radicals of type $RSSR^-$, RS-SR_2, and $R_2SSR_2^+$. J. Chem. Soc. Perkin Trans. 2:1618–1620.
15. Rao, D. N. R., Fischer, V., and Mason, R. P. (1990) Glutathione and ascorbate reduction of the acetaminophen radical formed by peroxidase. Detection of the glutathione disulfide radical anion and the ascorbyl radical. J. Biol. Chem. 265:844–847.
16. Josephy, P. D., Rehorek, D., and Janzen, E. G. (1984) Electron spin resonance spin trapping of thiyl radicals from the decomposition of thionitrites. Tetrahedron Lett. 25:1685–1688.
17. Ito, O., and Matsuda, M. (1984) Flash photolysis study for substituent and solvent effects on spin-trapping rates of phenylthiyl radicals with nitrones. Bull. Chem. Soc. Jpn. 57:1745–1749.
18. Graceffa, P. (1988) Spin trapping the cysteine thiyl radical with phenyl-*N*-*t*-butylnitrone. Biochim. Biophys. Acta 954:227–230.
19. Mile, B., Rowlands, C. C., Sillman, P. D., and Fildes, M. (1992) The EPR spectra of thiyl radical spin adducts produced by photolysis of disulfides in the presence of 2,4,6-tri-tert-butylnitrosobenzene and 5,5-dimethyl-1-pyrroline *N*-oxide. J. Chem. Soc. Perkin Trans. 2:1431–1437.
20. Ross, D. (1988) Glutathione, free radicals and chemotherapeutic agents. Mechanisms of free-radical induced toxicity and glutathione-dependent protection. Pharmacol. Ther. 37:231–249.

21. Asmus, K.-D. (1984) Pulse radiolysis methodology. Meth. Enzymol. 105:167–178.
22. Harman, L. S., Mottley, C., and Mason, R. P. (1984) Free radical metabolites of L-cysteine oxidation. J. Biol. Chem. 259:5606–5611.
23. Olsen, J., and Davis, L. (1976) The oxidation of dithiothreitol by peroxidases and oxygen. Biochim. Biophys. Acta 445:324–329.
24. Harman, L. S., Carver, D. K., Schreiber, J., and Mason, R. P. (1986) One- and two-electron oxidation of reduced glutathione by peroxidases. J. Biol. Chem. 261: 1642–1648.
25. Saez, G., Thornalley, P. J., Hill, H. A. O., Hems, R., and Bannister, J. V. (1982) The production of free radicals during the autoxidation of cysteine and their effect on isolated rat hepatocytes. Biochim. Biophys. Acta 719:24–31.
26. Davies, M. J., Forni, L. G., and Shuter, S. L. (1987) Electron spin resonance and pulse radiolysis studies on the spin trapping of sulphur-centered radicals. Chem.-Biol. Interact. 61:177–188.
27. Mottley, C., Toy, K., and Mason, R. P. (1987) Oxidation of thiol drugs and biochemicals by the lactoperoxidase/hydrogen peroxide system. Mol. Pharmacol. 31:417–421.
28. Svensson, B. E., Gräslund, A., Ström, G., and Moldeus, P. (1993) Thiols as peroxidase substrates. Free Rad. Biol. Med. 14:167–175.
29. Svensson, B. E. (1988) Thiols as myeloperoxidase-oxidase substrates. Biochem. J. 253:441–449.
30. Svensson, B. E., and Lindvall, S. (1988) Myeloperoxidase-oxidase oxidation of cysteamine. Biochem. J. 249:521–530.
31. Eling, T. E., Curtis, J. F., Harman, L. S., and Mason, R. P. (1986) Oxidation of glutathione to its thiyl free radical metabolite by prostaglandin H synthase. A potential endogenous substrate for the hydroperoxidase. J. Biol. Chem. 261:5023–5028.
32. Ross, D., and Moldeus, P. (1986) Thiyl radicals—their generation and further reactions. In: Biological Reactive Intermediates III. Mechanisms of Action in Animal Models and Human Disease (Kocsis, J. J., Jollow, D. J., Witmer, C. M., Nelson, J. O., and Snyder, R., eds.), pp. 329–335. Plenum Press, New York.
33. Maples, K. R., Jordan, S. J., and Mason, R. P. (1988) In vivo rat hemoglobin thiyl free radical formation following phenylhydrazine administration. Mol. Pharmacol. 33:344–350.
34. Mason, R. P., and Maples, K. R. (1989) In vivo rat hemoglobin thiyl free radical formation following phenylhydrazine or hydrazine-based drug administration. In: Medical, Biochemical and Chemical Aspects of Free Radicals (Hayaishi, O., Niki, E., Kondo, M., and Yoshikawa, T., eds.), pp. 775–779. Elsevier Science Publishers, B. V., Amsterdam.
35. Maples, K. R., Jordan, S. J., and Mason, R. P. (1988) In vivo rat hemoglobin thiyl free radical formation following administration of phenylhydrazine and hydrazine-based drugs. Drug Metab. Disposit. 16:799–803.
36. Maples, K. R., Kennedy, C. H., Jordan, S. J., and Mason, R. P. (1990) In vivo thiyl free radical formation from hemoglobin following administration of hydroperoxides. Arch. Biochem. Biophys. 277:402–409.
37. Maples, K. R., Eyer, P., and Mason, R. P. (1990) Aniline-, phenylhydroxyl-

amine-, nitrosobenzene-, and nitrobenzene-induced hemoglobin thiyl free radical formation in vivo and in vitro. Mol. Pharmacol. 37:311–318.
38. Kelman, D. J., and Mason, R. P. (1993) Characterization of the rat hemoglobin thiyl free radical formed upon reaction with phenylhydrazine. Arch. Biochem. Biophys. 306:439–442.
39. Puppo, A., Monny, C., and Davies, M. J. (1993) Glutathione-dependent conversion of ferryl leghaemoglobin into the ferric form: A potential protective process in soybean (*Glycine max*) root nodules. Biochem. J. 289:435–438.
40. Davies, M. J., and Puppo, A. (1992) Direct detection of a globin-derived radical in leghaemoglobin treated with peroxides. Biochem. J. 281:197–201.
41. Davies, M. J., and Puppo, A. (1993) Identification of the site of the globin-derived radical in leghaemoglobins. Biochim. Biophys. Acta 1202:182–188.
42. Turner, J. J. O., Rice-Evans, C. A., Davies, M. J., and Newman, E. S. R. (1991) The formation of free radicals by cardiac myocytes under oxidative stress and the effects of electron-donating drugs. Biochem. J. 277:833–837.
43. Romero, F. J., Ordoñez, I., Arduini, A., and Cadenas, E. (1992) The reactivity of thiols and disulfides with different redox states of myoglobin. Redox and addition reactions and formation of thiyl radical intermediates. J. Biol. Chem. 267:1680–1688.
44. Davies, M. J., Gilbert, B. C., and Haywood, R. M. (1993) Radical induced damage to bovine serum albumin: Role of the cysteine residue. Free Rad. Res. Commun. 18:353–367.
45. Schulz, U., and McCalla, D. R. (1969) Reactions of cysteine with N-methyl-N-nitroso-*p*-toluenesulfonamide and *N*-methyl-*N′*-nitro-*N*-nitrosoguanidine. Can. J. Chem. 47:2021–2027.
46. Mikuni, T., and Tatsuta, M. (1991) Production of the thiyl free radical by the reaction of *N*-methyl-*N′*-nitro-*N*-nitrosoguanidine with L-cysteine. Biochem. Int. 24:585–591.
47. Vesey, C. J., Cole, P. V., Linnell, J. C., and Wilson, J. (1974) Some metabolic effects of sodium nitroprusside in man. Br. Med. J. 2:140–142.
48. Vesey, C. J., Cole, P. V., and Simpson, P. J. (1976) Cyanide and thiocyanate concentrations following sodium nitroprusside infusion in man. Br. J. Anaesth. 48:651–660.
49. Rao, D. N. R., Elguindi, S., and O'Brien, P. J. (1991) Reductive metabolism of nitroprusside in rat hepatocytes and human erythrocytes. Arch. Biochem. Biophys. 286:30–37.
50. Radi, R., Beckman, J. S., Bush, K. M., and Freeman, B. A. (1991) Peroxynitrite oxidation of sulfhydryls. The cytotoxic potential of superoxide and nitric oxide. J. Biol. Chem. 266:4244–4250.
51. Augusto, O., Gatti, R. M., and Radi, R. (1994) Spin-trapping studies of peroxynitrite decomposition and of 3-morpholinosydnonimine *N*-ethylcarbamide autooxidation: Direct evidence for metal-independent formation of free radical intermediates. Arch. Biochem. Biophys. 310:118–125.
52. Prütz, W. A., Mönig, H., Butler, J., and Land, E. J. (1985) Reactions of nitrogen dioxide in aqueous model systems: Oxidation of tyrosine units in peptides and proteins. Arch. Biochem. Biophys. 243:125–134.

53. Goin, J., Gibson, D. D., McCay, P. B., and Cadenas, E. (1991) Glutathionyl- and hydroxyl radical formation coupled to the redox transitions of 1,4-naphthoquinone bioreductive alkylating agents during glutathione two-electron reductive addition. Arch. Biochem. Biophys. 288:386–396.
54. Giulivi, C., and Cadenas E. (1994) One- and two-electron reduction of 2-methyl-1,4-naphthoquinone bioreductive alkylating agents: Kinetic studies, free-radical production, thiol oxidation and DNA-strand-break formation. Biochem. J. 301:21–30.
55. Buettner, G. R. (1984) Thiyl free radical production with hematoporphyrin derivative, cysteine and light: a spin-trapping study. FEBS Lett. 177:295–299.
56. Felix, C. C., Reszka, K., and Sealy, R. C. (1983) Free radicals from photoreduction of hematoporphyrin in aqueous solution. Photochem. Photobiol. 37:141–147.
57. Reszka, K. J., and Chignell, C. F. (1994) Photochemistry of 2-mercaptopyridines. Part 1. An EPR and spin trapping investigation using 5,5-dimethyl-1-pyrroline N-oxide in aqueous and toluene solutions. Photochem. Photobiol. 60: 442–449.
58. Reszka, K. J., and Chignell, C. F. (1994) Photochemistry of 2-mercaptopyridines. Part 2. An EPR and the spin trapping investigation using 2-methyl-2-nitrosopropane and *aci*-nitromethane as spin traps in aqueous solutions. Photochem. Photobiol. 60: 450–454.
59. Moore, D. E., Sik, R. H., Bilski, P., Chignell, C. F., and Reszka, K. J. (1994) Photochemical sensitization by azathioprine and its metabolites. III. A direct EPR and spin-trapping study of light induced free radicals from 6-mercaptopurine and its oxidation products. Photochem. Photobiol. 60: 574–581.
60. Mišík, V., Mak, I. T., Stafford, R. E., and Weglicki, W. B. (1993) Reactions of captopril and epicaptopril with transition metal ions and hydroxyl radicals: An EPR spectroscopy study. Free Rad. Biol. Med. 15:611–619.
61. Shi, X., and Dalal, N. S. (1990) The glutathionyl radical formation in the reaction between manganese and glutathione and its neurotoxic implications. Med. Hypotheses 33:83–87.
62. Bruech, M., Quintanilla, M. E., Legrum, W., Koch, J., Netter, K. J., and Fuhrmann, G. F. (1984) Effects of vanadate on intracellular reduction equivalents in mouse liver and the fate of vanadium in plasma, erythrocytes and liver. Toxicology 31:283–295.
63. Shi, X., Sun, X., and Dalal, N. S. (1990) Reaction of vanadium(V) with thiols generates vanadium (IV) and thiyl radicals. FEBS Lett. 271:185–188.
64. Woods, J. S., Calas, C. A., Aicher, L. D., Robinson, B. H., and Mailer, C. (1990) Stimulation of porphyrinogen oxidation by mercuric ion. I. Evidence of free radical formation in the presence of thiols and hydrogen peroxide. Mol. Pharmacol. 38:253–260.
65. Ohnishi, T., Yamazaki, H., Iyanagi, T., Nakamura, T., and Yamazaki, I. (1969) One-electron-transfer reactions in biochemical systems. II. The reaction of free radicals formed in the enzymic oxidation. Biochim. Biophys. Acta 172:357–369.
66. Eling, T. E., Mason, R. P., and Sivarajah, K. (1985) The formation of aminopyrine cation radical by the peroxidase activity of prostaglandin H synthase and subsequent reactions of the radical. J. Biol. Chem. 260:1601–1607.
67. Schreiber, J., Foureman, G. L., Hughes, M. F., Mason, R. P., and Eling, T. E.

(1989) Detection of glutathione thiyl free radical catalyzed by prostaglandin H synthase present in keratinocytes. Study of cooxidation in a cellular system. J. Biol. Chem. 264: 7936–7943.
68. Ross, D., Mehlhorn, R. J., Moldeus, P., and Smith, M. T. (1985) Metabolism of diethylstilbestrol by horseradish peroxidase and prostaglandin-H synthase. Generation of a free radical intermediate and its interaction with glutathione. J. Biol. Chem. 260:16210–16214.
69. Sipe, H. J., Jr., Jordan, S. J., Hanna, P. M., and Mason, R. P. (1994) The metabolism of 17β-estradiol by lactoperoxidase: A possible source of oxidative stress in breast cancer. Carcinogenesis 15: 2637–2643.
70. Ross, D., Larsson, R., Andersson, B., Nilsson, U., Lindquist, T., Lindeke, B., and Moldéus, P. (1985) The oxidation of *p*-phenetidine by horseradish peroxidase and prostaglandin synthase and the fate of glutathione during such oxidations. Biochem. Pharmacol. 34:343-351.
71. Ross, D., Albano, E., Nilsson, U., and Moldéus, P. (1984) Thiyl radicals—formation during peroxidase-catalyzed metabolism of acetaminophen in the presence of thiols. Biochem. Biophys. Res. Commun. 125:109–115.
72. Bisby, R. H., and Tabassum, N. (1988) Properties of the radicals formed by one-electron oxidation of acetaminophen—a pulse radiolysis study. Biochem. Pharmacol. 37:2731–2738.
73. Wilson, I., Wardman, P., Cohen, G. M., and Doherty, M. D. (1986) Reductive role of glutathione in the redox cycling of oxidizable drugs. Biochem. Pharmacol. 35:21–22.
74. Katki, A. G., Kalyanaraman, B., and Sinha, B. K. (1987) Interactions of the antitumor drug, etoposide, with reduced thiols in vitro and in vivo. Chem.-Biol. Interact. 62:237–247.
75. Thompson, D., Norbeck, K., Olsson, L.-I., Constantin-Teodosiu, D., Van der Zee, J., and Moldéus, P. (1989) Peroxidase-catalyzed oxidation of eugenol: Formation of a cytotoxic metabolite(s). J. Biol. Chem. 264:1016–1021.
76. Fischer, V., Haar, J. A., Greiner, L., Lloyd, R. V., and Mason, R. P. (1991) Possible role of free radical formation in clozapine (clozaril)-induced agranulocytosis. Mol. Pharmacol. 40:846–853.
77. Gadelha, F. R., Hanna, P. M., Mason, R. P., and Docampo, R. (1992) Evidence for free radical formation during horseradish peroxidase-catalyzed *N*-demethylation of crystal violet. Chem.-Biol. Interact. 85:35–48.
78. Kim, K., Kim, I. H., Lee, K.-Y., Rhee, S. G., and Stadtman, E. R. (1988) The isolation and purification of a specific "protector" protein which inhibits enzyme inactivation by a thiol/Fe(III)/O_2 mixed-function oxidation system. J. Biol. Chem. 263:4704–4711.

5

Peroxynitrite Reactivity: Free Radical Generation, Thiol Oxidation, and Biological Significance

Ohara Augusto
Instituto de Química, Universidade de São Paulo, São Paulo, Brazil

Rafael Radi
Facultad de Medicina, Universidad de la República, Montevideo, Uruguay

I. INTRODUCTION

The discovery that the free radical nitric oxide is synthesized and used as a major transducer molecule by mammalian cells has vastly expanded the horizon of free radical reactions in physiology and pathology. Physiological roles of nitric oxide include blood pressure control, inhibition of platelet aggregation, neurotransmission, and the cytotoxic activity of stimulated macrophages, among many others, which are just beginning to be disclosed. The discovery that nitric oxide is used as widespread biological signal initially came as a surprise for several reasons, particularly because the compound was long known to be an environmental toxin. For the same reason, it was easier to conceive of nitric oxide as a cytotoxic effector molecule directed towards invading microorganisms and cancer cells (1–5).

The chemical properties of nitric oxide that make it an adequate biological messenger include its relatively low reactivity as a free radical species, which results in a biological half-life in the range of seconds, its small size and hydrophobicity, which permit easy transmembrane diffusion, and its selective reactivity with heme, iron-sulfur, and thiol-containing proteins, which influence signal transduction mechanisms (5). In addition, the ability of nitric oxide to react with molecular oxygen (6) and oxygen-derived free radicals (7–9) provides tissues with a nonenzymatic pathway for modulating its local concentration.

The reaction between superoxide anion and nitric oxide, in particular, may represent a critical control point in cells producing both species, leading either to downregulation of the biological effects of superoxide and nitric oxide or to potentially toxic oxidation of nearby biomolecules secondary to the formation of peroxynitrite. Indeed, the toxicity of nitric oxide involves oxidation reactions, but the compound itself is only a weak oxidant. However, it reacts with the relatively mild reductant, superoxide anion, to produce peroxynitrite (7–9) and its conjugate acid, peroxynitrous acid, both of which are potent oxidants:

$$NO^{\bullet} + O_2^{\bullet-} \xrightarrow{k = 6.7 \times 10^9\ M^{-1}\ s^{-1}} ONOO^- \tag{1}$$

$$ONOO^- + H^+ \underset{pK_a = 6.8}{\rightleftharpoons} ONOOH \tag{2}$$

Thus, two relatively unreactive free radicals give rise to strong oxidizing species, which are able to oxidize several biomolecules such as thiols (10), deoxyribose (8), methionine (11), and lipids (12). Consequently, peroxynitrite and peroxynitrous acid are becoming increasingly recognized as reactive intermediates that may participate in many injury processes associated with oxidative damage.

Thiols represent a key defense mechanism against free radicals and reactive oxygen species; glutathione and albumin, in particular, account for most of the antioxidant function of thiols in the intra- and extracellular space, respectively (13–15). In addition, oxidation of protein sulfhydryls and thiol-containing cofactors such as coenzyme A, lipoic acid, and thioredoxin can influence essential metabolic processes. Consequently, it is timely to discuss the formation and reactivity of peroxynitrite in biological systems focusing on free radical generation and thiol oxidation processes as they constitute key events during biological oxidative damage.

II. PEROXYNITRITE AS A NOVEL BIOLOGICAL INTERMEDIATE

A. Cellular Production of Superoxide and Nitric Oxide

Superoxide formation is a continuous phenomenon for aerobic cells. About 1% of oxygen consumption evolves to superoxide mostly due to electron leakage from the electron transport chains, such as those of the mitochondria and the endoplasmic reticulum. Rates of superoxide production can increase severalfold depending on disruption of normal electron flow, enhanced oxygen tension, or enhanced activity of superoxide-producing enzymes such the membrane-bound NADPH oxidase of activated macrophages and neutrophils (17–20).

Nitric oxide is an endogenously synthesized free radical first characterized as part of the endothelial-derived relaxation factor (21). Nitric oxide is now known to be produced by a variety of mammalian cells including neurons, smooth muscle cells, macrophages, neutrophils, platelets, type II neumocytes, and others (for a review, see Ref. 22). Nitric oxide is synthesized by the oxidation of L-arginine to L-citrulline in the presence of molecular oxygen and NADPH. The enzyme responsible for the synthesis of nitric oxide, nitric oxide synthase, is analogous to cytochrome P-450 reductase, and both constitutive and inducible forms of nitric oxide synthase have been characterized (22,23). The constitutive enzyme is activated by mediators that increase intracellular calcium concentrations such as ATP, bradykinin, and calcium ionophores. The inducible nitric oxide synthase is induced by different cytokines as well as by bacterial and parasite antigens; in this case, nitric oxide synthesis starts after a lag phase of about 3 h and can be maintained for 48 h (4,22,23).

B. Peroxynitrite Formation Under Physiological Conditions

The reaction of nitric oxide and superoxide anion is known to occur in vitro at an almost diffusion-controlled rate. Thus, the rate constant for the combination reaction [Eq. (1)] ($k = 6.7 \times 10^9\ M^{-1}\ s^{-1}$) (9) is larger than those for the reactions of superoxide with superoxide dismutase ($k = 2 \times 10^9\ M^{-1}\ s^{-1}$) (24) or of nitric oxide with heme compounds ($k = 10^5$–$10^7\ M^{-1}\ s^{-1}$) (25,26), suggesting that it can occur in vivo.

Indeed, several lines of evidence indicate that peroxynitrite is formed under physiological conditions. Historically, when nitric oxide was identified as a major component of the endothelial-derived relaxation factor, it was observed that addition of superoxide dismutase doubled its half-life, suggesting a fast reaction between superoxide and nitric oxide (21). The formation of peroxynitrite by cells under metabolic conditions that favor the simultaneous enhancement of both superoxide and nitric oxide has been recently demonstrated. Secondary to macrophage activation, quantitative generation of peroxynitrite with respect to the limiting reagent, nitric oxide, has been shown by Beckman and coworkers (27). Also, formation of peroxynitrite has been demonstrated in the case of Kuppfer cells (28), endothelial cells (29), and neutrophils (30). The formation of peroxynitrite in human tissues is strongly suggested by the immunohistochemical detection of nitrotyrosine residues in atherosclerotic lesions of fixed human arteries (31). Indeed, nitrotyrosine has been proposed to be used as a marker for peroxynitrite formation, since the latter nitrates tyrosine efficiently, whereas both nitric oxide and nitrogen dioxide are weak nitrating agents (31–33) (see also Sec. III.B).

III. PEROXYNITRITE REACTIVITY

A. Peroxynitrite Stability and Half-Life

The anionic form of peroxynitrite can be conveniently synthesized and stored at highly alkaline pH, but peroxynitrite is unstable at pH 7.4 and 37°C, with a

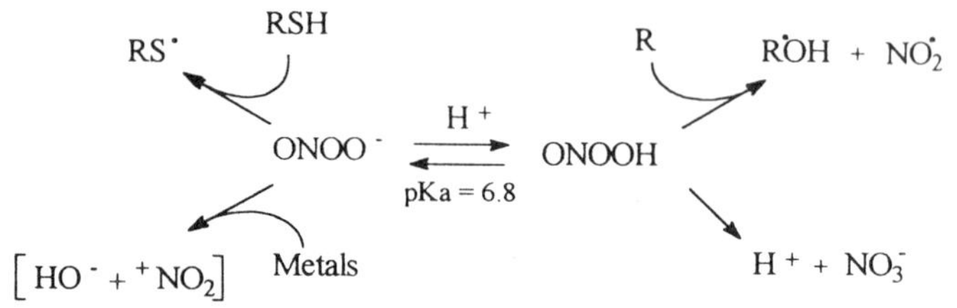

Figure 1 Schematic representation of possible routes for peroxynitrite decomposition.

half-life of less than 1 s (8). Peroxynitrite has a pK_a of 6.8 (10,34,35), and therefore about 20% will be in the conjugate acid form, peroxynitrous acid, at pH 7.4; the rapid decomposition of peroxynitrous acid displaces the equilibrium of Eq. (2) to the right, ultimately leading to the complete disappearance of the anion. Since peroxynitrite has a broad absorption band in the 300-nm region (E_{302} = 1.67 M^{-1} cm^{-1}), the proton-catalyzed decomposition of peroxynitrite can be followed using stopped-flow spectrophotometric techniques (10). The pseudo–first-order rate constant of peroxynitrite decomposition at pH 7.4 and 37°C is 0.6 s^{-1} (8,10).

Early work from the chemical literature demonstrated that peroxynitrite decomposition and reactivity was a rather complex process, involving free radical chemistry (36–39). The redox state of the nitrogen atom in peroxynitrite is +5, and its decomposition in phosphate buffer leads mostly to the formation of the stable isomer nitrate. The precise decomposition pathway, however, is influenced by what types of target molecules come in contact with peroxynitrite and is strongly pH dependent in the region of physiological interest (Fig. 1).

B. Direct Reactions

Peroxynitrite anion is a strong oxidant (Table 1) (40–46), which directly oxidizes sulfhydryl groups (10) (see Sec. V) and reacts with transition metals, including copper of superoxide dismutase, to form a strong nitrating species resembling nitronium ion (NO_2^+) (27,32,33,47). Indeed, peroxynitrite has been shown to nitrate several phenolics, including tyrosines, salicylic acid, and *p*-hydroxyphenylacetic acid. Nitration reactions are catalyzed by low molecular weight metallic complexes as well as by metals bound to proteins, suggesting a metal-catalyzed heterolytic cleavage to a nitroniumlike species [Eqs. (3) and (4)] (27,32,33,47). In the absence of transition metal ions, peroxynitrite can nitrate through an alternative mechanism, which involves the oxidation of the phenol group to a phenoxyl radical, which then reacts with nitrogen dioxide in

Table 1 Reduction Potentials of Relevant Pairs at pH 7.0

Redox pair	E°′ (V)	Ref.
$^{\bullet}OH/H_2O$	2.31	40
$ONOOH^{*}/N^{\bullet}O_2$[a]	2.10	41
$NO_2^+/N^{\bullet}O_2$	1.60	41
$ONOOH/N^{\bullet}O_2$	1.40	41
$N^{\bullet}O/NO^+$	1.21	41
$ONOOH/NO_2^-$	0.99	41
Fe(IV)=O/Fe(III)[b]	0.97	42
$O_2^{\bullet -}/H_2O_2$	0.94	40
$CysteineS^{\bullet}/CysteineS^-$	0.92	40
$GS^{\bullet}/GS^-$	0.90	43
$N^{\bullet}O_2/NO_2^-$	0.87	44
$PUFA^{\bullet},H^+/PUFA\text{-}H$[c]	0.60	45
$H_2O_2/^{\bullet}OH$	0.32	40
Fe(III)EDTA/FE(II)EDTA	0.12	40
Cysteine/Cystine	−0.22	46
GSH/GSSG	−0.33	44
Lipoate(red)/Lipoate(ox)	−0.29	46
$GSSG/GSSG^{\bullet -}$	−1.50	44

[a]Activated (vibrationally excited) *trans*-peroxynitrous acid.
[b]Horseradish peroxidase compound II/horseradish peroxidase.
[c]Polyunsaturated fatty acid, bis-allylic-H.

a free radical termination reaction (see Sec. III.C). This mechanism, however, renders low nitration yields (32,47).

$$\text{SOD-Cu}^{2+} \cdots {}^{-}\text{O-O-N=O} \rightarrow \text{SOD-Cu}^{+}\text{O}^{-} \cdots \text{O=N}^{+}\text{=O} \tag{3}$$

$$\text{SOD-Cu}^{+}\text{O}^{-} \cdots \text{O=N}^{+}\text{=O} + \text{phenol} \rightarrow \text{SOD-Cu}^{2+} + {}^{-}\text{OH} + \text{phenol-NO}_2 \tag{4}$$

Nitration of tyrosine residues by tetranitromethane has been shown to alter the function of several proteins such as cytochrome P-450 (48), α-thrombin (49), and mitochondrial ATPase (50). In addition, phosphorylation of tyrosines, a key event in the regulation of cell functions, will be impaired by nitration. Consequently, endogenous nitration of cellular proteins could be a major pathological mechanism of peroxynitrite-induced cellular injury.

It is also worth pointing out that, in addition to being a good nitrating agent, nitronium ion is a good one-electron oxidant (Table 1). Also, tetranitromethane has been used to oxidize critical protein thiols (51,52).

C. The Hydroxyl Radical–like Pathway

Peroxynitrite is formed from the combination reaction of two free radicals, superoxide and nitric oxide [Eq. (1)], and therefore is not itself a free radical. However, peroxynitrite is capable of initiating many of the reactions commonly attributed to hydroxyl radicals, particularly under acidic conditions, such as polymerization of methyl acrylate (37), oxygen release from hydrogen peroxide (39), oxidation of dimethylsulfoxide and deoxyribose to products indicative of hydroxyl radical attack (8), and oxidation of 5,5-dimethyl-1-pyrroline-*N*-oxide (DMPO) to the DMPO–hydroxyl radical adduct (53). The maximum yield of hydroxyl radical–like oxidant formed from peroxynitrite was estimated to be about 30%, a value comparable to the 20% nitrogen dioxide yield obtained under the same experimental conditions (8,35). Initially, these data supported the view that peroxynitrous acid undergoes a homolytic scission to hydroxyl radical and nitrogen dioxide in about 30% yields, with the remaining compound rearranging to nitrate without the release of free radical intermediates (Fig. 1). However, the assumption of complete homolytic cleavage of peroxynitrous acid was recently considered to be an oversimplification based on kinetic and thermodynamic arguments (41). Homolytic cleavage generally involves a loose and floppy transition state, whereas the low activation entropy for peroxynitrous acid (3 kcal mol^{-1} T^{-1}) suggests a rigid and relatively symmetric structure for the transition state. From the activation energy and the forward rate of reaction for the hydroxyl radical-like pathway, it was calculated that the rate of hydroxyl radical recombination with nitrogen dioxide would be 10^{14} M^{-1} s^{-1}, a value at least 10^2–10^3 times higher than the diffusion limit. Consequently, this impossibly rapid reverse reaction and the low activation entropy led to the suggestion that the hydroxyl radical-like reactivity of peroxynitrous acid is mediated by a vibrationally excited intermediate that does not separate into free hydroxyl radical and nitrogen dioxide (41). These arguments, considered together with studies on the pH-dependent yield of hydroxyl radical products from peroxynitrite (35), were recently summarized by the mechanism shown in Figure 2 and described below.

The energetically more stable *cis* form of peroxynitrite predominates at high pH, but its protonation decreases charge repulsion and thereby promotes isomerization to *trans*-peroxynitrous acid, which can then dissociate to *trans*-peroxynitrite anion. Both *trans*-peroxynitrite anion and *trans*-peroxynitrous acid can readily undergo transition to a vibrationally excited state, which involves bending of the N-O-O bond and lengthening of the O-O bond as demonstrated by quantum mechanical calculations (54). The activated intermediates can directly rearrange to nitrate or nitric acid, respectively, via attack of the terminal oxygen on the nitrogen. Activated *trans*-peroxynitrous acid is also capable of oxidizing a target molecule exhibiting a hydroxyl radical–like reactivity but without the release of free hydroxyl radical (Fig. 2). The estimated redox potential of

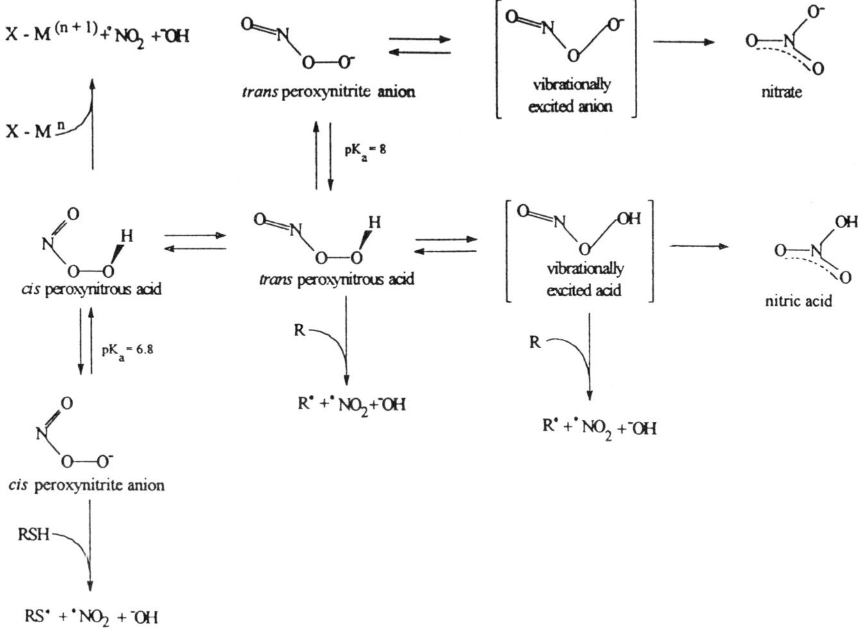

Figure 2 Mechanistic view of peroxynitrite isomerization and reactivity.

the activated *trans*-peroxynitrous acid is 2.1 V, a value comparable to the 2.3 V value of the hydroxyl radical/H_2O pair (Table 1). Direct oxidation of a target molecule by activated *trans*-peroxynitrite anion, however, is energetically unfavorable, explaining the loss of the hydroxyl radical-like reactivity of peroxynitrite at high pH (35).

D. Peroxynitrite-Mediated One-Electron Oxidation

In the context of oxidative biological damage, it is important to emphasize that peroxynitrous acid can act as a one- or two-electron oxidant [Eqs. (5) and (6)] since both reduction potentials, $E^{\circ\prime}$ (ONOOH, $H^+/N^{\bullet}O_2$, H_2O) and $E^{\circ\prime}$(ONOOH, H^+/NO_2^-, H_2O) (1.4 V and 0.99 V, respectively), are high (Table 1). The possibility of the two processes competing with each other has been considered in the case of methionine, which is oxidized by peroxynitrite to both methionone sulfoxide and ethylene (11,55), depending on the initial concentration of the amino acid.

$$ONOOH + RH \rightarrow R^{\bullet} + H_2O + N^{\bullet}O_2 \quad (5)$$

$$ONOOH + R \rightarrow R{=}O + H^+ + NO_2^- \quad (6)$$

The experimental data currently available, however, are still insufficient to establish the thermodynamic and kinetic factors that control the multiple reaction pathways available for peroxynitrite (Figs. 1 and 2). Still, free radical reactions are expected to occur due to the favorable one-electron redox potential of both peroxynitrous acid and activated *trans*-peroxynitrous acid (Table 1). Also, activated *trans*-peroxynitrous acid, when triggering free radical production from nearby molecules, also yields nitrogen dioxide (Fig. 2). This species is also a free radical intermediate, which can participate in secondary reactions, further propagating peroxynitrite-mediated one-electron oxidations. Most important, peroxynitrite-mediated oxidations will occur in the absence of transition metal ions either from metalloenzymes or low molecular weight complexes.

The biological formation of free radicals through enzymatic processes, including the biosynthesis of nitric oxide, is dependent on metalloproteins, flavoproteins, or metallo-flavoproteins (16–20,22). On the other hand, much of the damage resulting from oxidative stress has been attributed to the highly reactive hydroxyl radical (Table 1) formed in the Haber-Weiss or superoxide-driven Fenton reaction, which requires a suitable metal ion promoter such as copper or iron:

$$2\,O_2^{-} + 2H^+ \rightarrow H_2O_2 + O_2 \tag{7}$$

$$O_2^{-} + Fe^{3+}(\text{ligand}) \rightarrow O_2 + Fe^{2+}(\text{ligand}) \tag{8}$$

$$H_2O_2 + Fe^{2+}(\text{ligand}) \rightarrow {}^{\bullet}OH + {}^{-}OH + Fe^{3+}(\text{ligand}) \tag{9}$$

Superoxide-mediated toxicity has been largely explained through this mechanism, in view of the ubiquitous occurrence of superoxide dismutase in aerobic organisms and the restricted direct reactivity of superoxide toward biomolecules. There are, however, some limitations to the occurrence of the superoxide-driven Fenton reaction in vivo: in particular the low steady-state concentrations of superoxide in tissues ranging from pM to nM levels, the slow rate constant for Eq. (8) ($k = 10^6\ M^{-1}\ s^{-1}$) (56), and the low availability of catalytic transition metal ions in biological fluids. In addition, cells maintain an internal reducing environment, keeping GSH, ascorbate, and nicotinamide adenine nucleotides largely in a reduced state, suggesting that any of these reductants, existing at much higher levels than superoxide, will easily outcompete the latter for the reduction of ferric iron. Thus, other mechanisms have been proposed to explain superoxide toxicity (57–59), including formation of peroxynitrite (8,10–12,33,60), which could promote free radical reactions by a transition metal ion–independent pathway.

1. Peroxynitrite-Mediated Lipid Peroxidation

The possibility of peroxynitrite mediating one-electron oxidation processes in biological systems was tested by studying the interaction between peroxynitrite and egg and soybean phosphatidylcholine liposomes (12), both of which are good models to understand the chemical interactions of an oxidant with biomembranes. In these studies we found that peroxynitrite initiated lipid peroxidation of phosphatidycholine liposomes as evidenced by formation of thiobarbituric acid reactive substances (TBARS) and of conjugated dienes and by induced oxygen consumption (Fig. 3). The peroxidation process was shown to be metal-independent, since neither iron depletion of the buffer systems nor the metal chelator DTPA significantly inhibited the oxidation process. Similarly, addition of exogenous iron did not potentiate peroxynitrite-mediated lipid peroxidation. The yields of this process were greater at acidic to neutral pH, suggesting that proton-catalyzed decomposition of peroxynitrite to the hydroxyl radical-like oxidant (Fig. 2) was the main mechanism leading to lipid oxidation. Indeed, lipid peroxidation is a free radical process that involves an initial hydrogen abstraction from a polyunsaturated fatty acid molecule to form an allylic radical, a process that is not trivial and that requires oxidizing intermediates with redox potential higher than 0.6 V (Table 1) (61,62). Consequently, lipid peroxidation

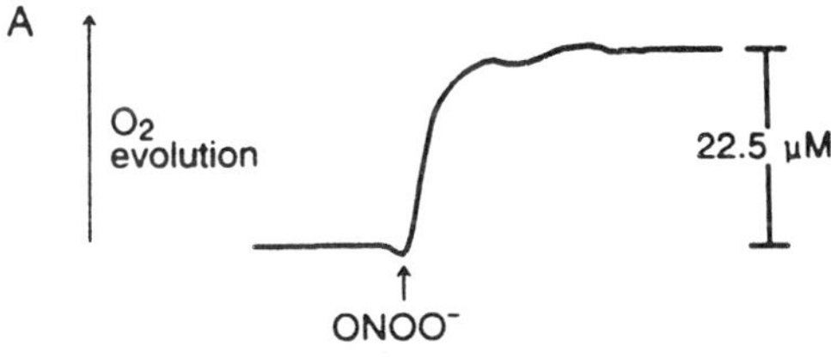

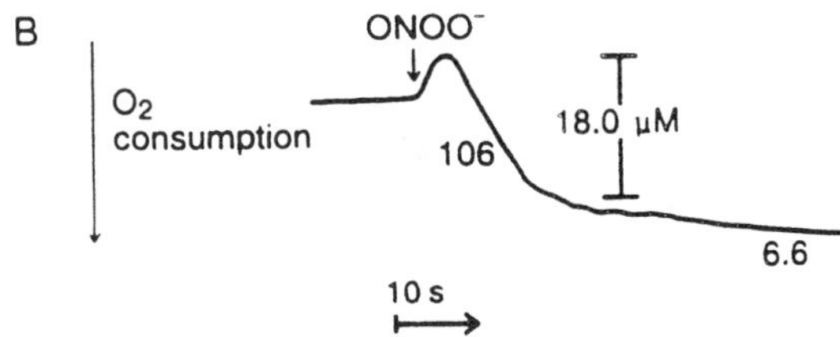

Figure 3 Polarographic assessment of peroxynitrite decomposition and stimulation of lipid peroxidation. Peroxynitrite (1 mM) was added in the absence (A) or in the presence (B) of liposomes (5.6 μmol liposome phospholipid/ml) in potassium phosphate (100 μM), DTPA (50 μM), pH 6.0, 37°C. The slope represents rates of oxygen consumption in mM/min. (From Ref. 12.)

can only be initiated by strong oxidizing intermediates such as the hydroxyl radical (63,64) and peroxynitrous acid, as shown by our studies (12). Overall, the results obtained can be the explained by following reactions:

$$RH + ONOOH \rightarrow R^{\bullet} + N^{\bullet}O_2 + H_2O \quad (10)$$

$$R_1H + N^{\bullet}O_2 \rightarrow R_1^{\bullet} + NO_2^- + H^+ \quad (11)$$

$$R^{\bullet} + O_2 \rightarrow ROO^{\bullet} \quad (12)$$

$$ROO^{\bullet} + RH \rightarrow ROOH + R^{\bullet} \quad (13)$$

In the above mechanism, the hydroxyl radical–like oxidant would react with a second-order rate constant of about $10^9\ M^{-1}s^{-1}$ (8,12,41,53), while the by-product of peroxynitrous reaction with fatty acids, nitrogen dioxide, is a lipophilic oxidizing free radical, which is capable of abstracting an allylic hydrogen in a relatively slow reaction ($k = 10^5\ M^{-1}\ s^{-1}$) (65). In Figure 3, a fast phase of oxygen consumption that lasts for about 10 s is evident and corresponds to the total life of peroxynitrous acid under these experimental conditions. Next, a slower oxygen consumption occurs, which can be attributed to propagation reactions, including those mediated by nitrogen dioxide. As nitrogen dioxide can also participate in addition reactions, formation of acyl-nitro derivatives are expected to occur, as has just been demonstrated (66). In summary, these studies showed that peroxynitrite can cause membrane lipid oxidation through the formation of a hydroxyl radical–like oxidant without the participation of transition metal chemistry. Later work demonstrated that simultaneous superoxide and nitric oxide release by the sydnonimine SIN-1 resulted in LDL oxidation (67).

2. *EPR Evidence for Peroxynitrite-Mediated One-Electron Oxidations*

EPR is the only direct method for detecting free radicals, since it is based on the absorption of energy by free radicals in the presence of a magnetic field. Measurement of the absorbed energy in an EPR spectrometer permits scanning of the characteristic EPR spectrum of the radical. Under physiological conditions, however, most free radicals do not attain steady-state concentrations higher than the detection limit of the EPR spectrometer (10^{-6}–10^{-9} M, depending on the radical structure), and consequently their EPR detection and identification have been performed by freeze-trapping (68,69) or spin-trapping techniques (70–75). In the latter, a diamagnetic molecule (spin trap) reacts with a free radical to produce a more stable radical (spin adduct), which then accumulates, attaining concentrations high enough to be detected by the EPR spectrometer (Fig. 4). Commonly utilized spin traps are 5,5-dimethyl-1-pyrroline *N*-oxide (DMPO), which is extensively used for the detection of oxygen and thiyl radicals, α-phenyl-*N*-*tert*-butyl nitrone (PBN), and *tert*-nitrosobutane (t-NB), which are usually employed for detection of carbon-centered radicals. The spin-trapping

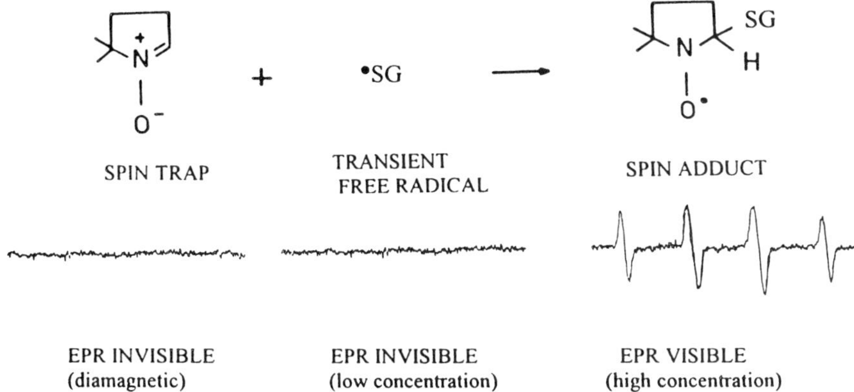

Figure 4 Schematic representation of the spin-trapping reaction beween the spin trap DMPO and the glutathionyl radical formed during peroxynitrite-mediated oxidation of glutathione. The EPR spectrum of the spin adduct was reproduced from Ref. 53.

technique is considered to be a less ambiguous method for detecting transient free radicals, such as the hydroxyl radical, in biological systems. Although it has limitations, the technique has been successfully applied in several experimental models, including animals, and has contributed to our understanding of several aspects of free radical reactions in biological systems (70–76). Consequently, in order to obtain more direct evidence for the formation of free radical intermediates during the decomposition of peroxynitrite, we decided to use electron paramagnetic resonance (EPR) spectroscopy both in the presence and in the absence of spin traps (53,77–79).

For instance, we have studied the decomposition of peroxynitrite by spin-trapping experiments with DMPO (53). Proton-catalyzed decomposition of micromolar concentrations (10–60 μM) of peroxynitrite at pH 7.5 resulted in the formation of the DMPO–hydroxyl radical adduct (a_N = 1.49 mT; a_H = 1.49 mT). Yields, however, were low, and the adduct became undetectable with concentrations of peroxynitrite higher than 60 μM. This behavior was attributed to DMPO–hydroxyl radical decomposition to EPR silent products by direct reactions with peroxynitrite and/or nitrogen dioxide (53). Indeed, 100 mM DMPO did not change the rate of peroxynitrite decomposition as followed by stopped flow experiments, indicating that the trap does not react in a bimolecular fashion with the anion. By contrast, low concentrations (5–10 μM) of peroxynitrite completely decompose preformed DMPO-hydroxyl adduct, whereas high concentrations lead to the formation of oxidized DMPO (DMPO-X) (53,80). Considering that both peroxynitrite anion (10) and nitrogen dioxide (65,81) rapidly react with sulfhydryls, we reasoned that addition of thiols could increase the yields of the DMPO-hydroxyl radical adduct obtained during peroxynitrite decomposition. Indeed, the yield of the adduct greatly increased in the presence of cysteine or glutathione (Fig. 5). In the presence of glutathione and scavengers such as ethanol, formate, and dimethyl sulfoxide, the corresponding carbon-centered radical adduct of the scavenger was detected, whereas the concentration of the DMPO–hydroxyl radical adduct decreased (see, for instance, Fig. 5). In spite of the complexity of the system under study due to the presence of GSH and to the fast decomposition of peroxynitrite, we decided to perform static

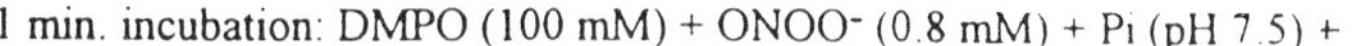

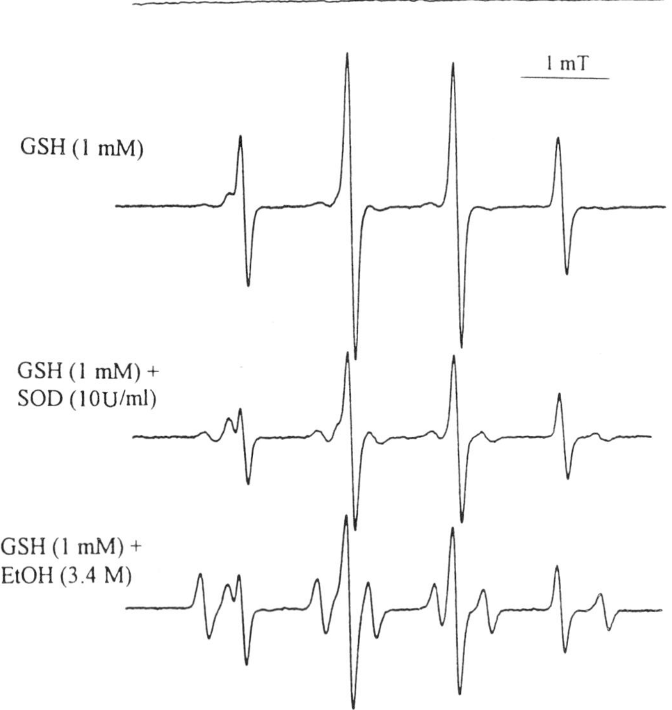

Figure 5 EPR spectra of DMPO radical adducts obtained during peroxynitrite decomposition. The spectra were obtained after 1.0 min of incubation at room temperature of DMPO (80 mM) and peroxynitrite (0.8 mM) in phosphate buffer (100 mM), pH 7.5, in the absence (first spectrum) or in the presence of the reagents indicated in the other spectra. The experimental conditions are as described in Ref. 53. (Adapted from Ref. 53.)

competition experiments with various concentrations of ethanol and formate to calculate the relative rate constants of the peroxynitrite-derived oxidant with DMPO and scavengers ($k_{scavenger}/k_{DMPO}$) according to the following:

$$\frac{k_{scavenger}}{k_{DMPO}} = \frac{[DMPO]([DMPO\text{-}OH]_0 - [DMPO\text{-}OH])}{[scavenger][DMPO\text{-}OH]} \tag{14}$$

The mean values of the ratio $k_{scavenger}/k_{DMPO}$ obtained with formate and ethanol, 0.74 and 0.065, respectively, were compared with the values obtained in pulse radiolysis experiments (0.85 and 0.55), which are attributed to free hydroxyl radical (53). This comparison suggested that the oxidant formed during peroxynitrite decomposition, probably activated *trans*-peroxynitrous acid (Fig. 2), reacts with nearby molecules with rate constants similar to those expected for free hydroxyl radical and forming the same free radical products, including the DMPO–hydroxyl radical adduct. Actually, these spin-trapping studies were the first to provide direct evidence for peroxynitrite-mediated one-electron oxidations, since the radicals glutathionyl (a_N = 1.54 mT; a_H = 1.62 mT), α-ethylhydroxy (a_N = 1.58 mT; a_H = 2.29 mT), methyl (a_N = 1.62 mT; a_H = 2.33 mT), and carbon dioxide anion (a_N = 1.58 mT; a_H = 1.88 mT) were unambiguously identified by the EPR parameters of the corresponding DMPO adducts obtained during oxidation of glutathione, ethanol, dimethyl sulfoxide, and formate, respectively. The same radicals were detected by substitution of peroxynitrite by the nitric oxide and superoxide anion donor, SIN-1, which mimics physiological situations by generating small and continuous fluxes of peroxynitrite (53). Also, decomposition of peroxynitrite in the presence of GSH, dimethyl sulfoxide, and another spin trap, PBN, also led to the detection of the PBN-glutathionyl and PBN–methyl radical adducts (Fig. 6); in this case the PBN–hydroxyl radical adduct was not detected, as expected from its known instability (72–75).

We have also used direct EPR for detecting peroxynitrite-mediated one-electron oxidations (77,78). For instance, we used the vasodilatory molsidomine, SIN-1, to induce oxidative stress in human plasma and followed it by measurements of the ascorbyl radical level by direct EPR (77). Addition of 1 mM SIN-1 to human plasma at room temperature led to a quick increase in the concentration of the ascorbyl radical, which doubled in 10 min (Fig. 7) and slowly decayed, probably due to depletion of the dissolved oxygen in the flat cell (77). These results suggested that peroxynitrite is able to mediate, directly or indirectly, the one-electron oxidation of ascorbate present in human plasma. Indeed, recent studies by Koppenol and coworkers have demonstrated that peroxynitrite is able to oxidize ascorbate, leading to the formation of ascorbyl radical (82). More recently, we have demonstrated peroxynitrite-mediated oxidation of desferrioxamine to the corresponding nitroxide radical, which was detected by

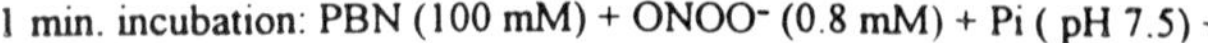

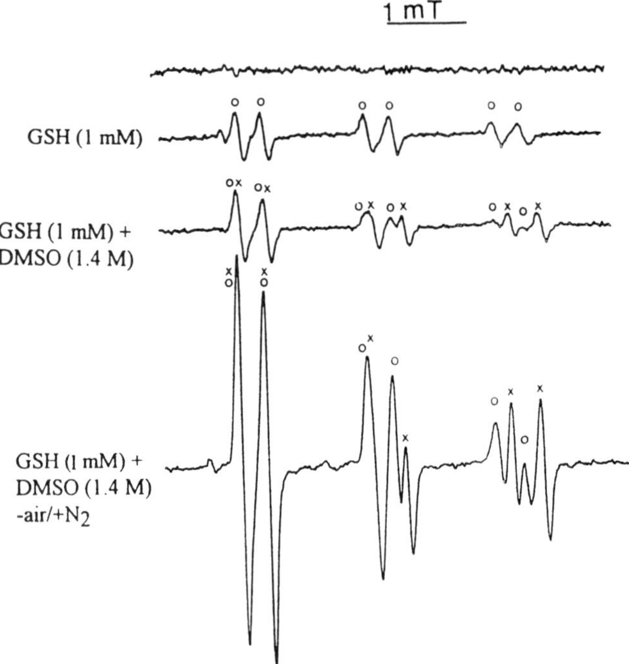

Figure 6 EPR spectra of PBN radical adducts obtained during peroxynitrite decomposition. The spectra were obtained after 1.0 min of incubation at room temperature of PBN (100 mM) and peroxynitrite (0.8 mM) in phosphate buffer (100 mM), pH 7.5, in the absence (first spectrum) or in the presence of the reagents indicated in the other spectra. The EPR spectra are labeled to characterize the formed adducts: (o) PBN-glutathionyl (a_N = 1.55 mT; a_H = 0.31 mT) and (x) PBN-methyl (a_N = 1.64 mT; a_H = 0.36 mT). Experimental conditions as described in Refs. 53, 77–79.

direct EPR. The yield of the desferroxamine nitroxide was higher at low pHs, indicating *trans*-peroxynitrous acid as the main oxidizing species (Fig. 2) (78).

Also, we have recently been able to trap with both DMPO and PBN the protein-thiyl free radical resulting from the oxidation of bovine serum albumin by peroxynitrite (79). The yield of albumin radical adducts increased with pH, indicating peroxynitrite anion as their main forming agent (see also Sec. IV.B). The important point to emphasize here is that our studies employing EPR spectroscopy have directly demonstrated that both peroxynitrite anion and peroxynitrous acid can oxidize target molecules, including biomolecules, to free radical intermediates (53,77–79). Consequently, the use of EPR and EPR–spin-

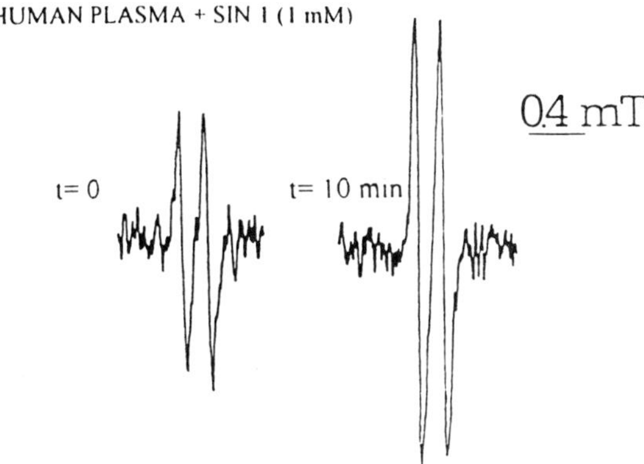

Figure 7 EPR spectrum of human plasma obtained at room temperature before and after 10 minutes of addition of 1 mM SIN-1. The observed EPR spectrum corresponds to the ascorbyl radical (a_H = 0.18 mT) formed from one-electron oxidation of endogenous ascorbate. (Modified from Ref. 77.)

trapping techniques can greatly contribute to the understanding of peroxynitrite toxicity.

IV. THIOL REACTIVITY

A. Thiols as Redox-Active Nucleophiles

Thiol-containing compounds are important in many biochemical and pharmacological processes, such as maintenance of protein structure by disulfide bonds, the beneficial and side effects of thiol-containing drugs, and the antioxidant effects of the reduced cellular tripeptide glutathione (13,14,83,84). Thiols are nucleophiles that have a negative redox potential (Table 1) and therefore may form conjugates with electrophiles or act as reducing agents. They participate in most reactions as the thiolate anion, and consequently these processes are favored at high pH. Indeed, the pK_a of the sulfhydryl groups of cysteine and glutathione are 8.37 and 9.2, respectively. The pK_a of the thiol group in proteins, however, can greatly differ from the pK_a of thiols in solution due to electrostatic influences of neighboring charged residues. In general, the proximity of a positively charged group tends to decrease, whereas a negative group tends to increase the pK_a value of the thiol group. The reactivity of a protein thiol can also be greatly diminished when the group is hindered in a protein region poorly accessible to different reactants.

Among the several reactions undergone by sulfhydryl groups such as acylation, alkylation, arylation, and reaction with metal ions and carbonyl compounds, we will emphasize the oxidation reactions due to their importance in oxidative damage. Indeed, thiols are considered to be key biological antioxidants due to their redox properties, ubiquitous presence, and high intra- and extracellular concentrations such as the 1–10 mM level of glutathione in cells and the 0.3–0.5 mM albumin level in plasma (13–16). Glutathione can directly react with oxidants, including free radicals, and it has been proposed that the tripeptide acts in conjunction with superoxide dismutase as an integral component of the cellular antioxidant defense (44,59). In extracellular fluids where the levels of glutathione are minimal (e.g., <2 μM in human plasma) (85), the single thiol group of albumin appears to be a very key antioxidant as indicated by several studies (15,16,85).

In addition to reacting directly with free radicals and other oxidants and being able to complex transition metal ions (86), glutathione participates in a very efficient enzymatic pathway to detoxify peroxides composed by glutathione peroxidase and glutathione reductase with the ancillary contribution of enzymatic systems that provide NADPH, including the first enzyme of the pentose phosphate shunt pathway, glucose-6-phosphate dehydrogenase [Eqs. (15)–(17)] (19,87). The enzyme glutathione peroxidase is able to metabolize not only excess hydrogen peroxide but also organic peroxides. Moreover, a new variety of this enzyme that is able to decompose phospholipid hydroperoxides in membranes at the expense of GSH has been characterized (88).

$$H_2O_2 + 2\ GSH \xrightarrow{\text{glutathione peroxidase}} 2\ H_2O + GSSG \tag{15}$$

$$GSSH + NADPH + H^+ \xrightarrow{\text{glutathione reductase}} 2\ GSH + NADP^+ \tag{16}$$

$$\text{glucose-6-Pi} + NADP^+ \xrightarrow{\text{g-6-Pi-dehydrogenase}} \text{6-phosphogluconate} + NADP^+ + H^+ \tag{17}$$

B. Oxidation States of Thiols

The oxidation of low molecular weight thiol compounds, including cysteine, by metal ions has been studied for a number of years (13,89). Among the stable products of oxidation are disulfide, sulfinic acid, and sulfonic acid (Fig. 8). The two-electron oxidation product sulfenic acid has not been isolated in the case of low molecular weight thiols due to its instability (see below). The diamagnetic products of oxidation form via free radical intermediates with the thiyl radical, either dimerizing to the disulfide or reacting with molecular oxygen or hydrogen peroxide to form products containing oxygen with sulfur in a higher oxidation state (Fig. 8) (90–94). Under mild conditions of oxidation, formation of the disulfide predominates (13,89). More recently, it became established that

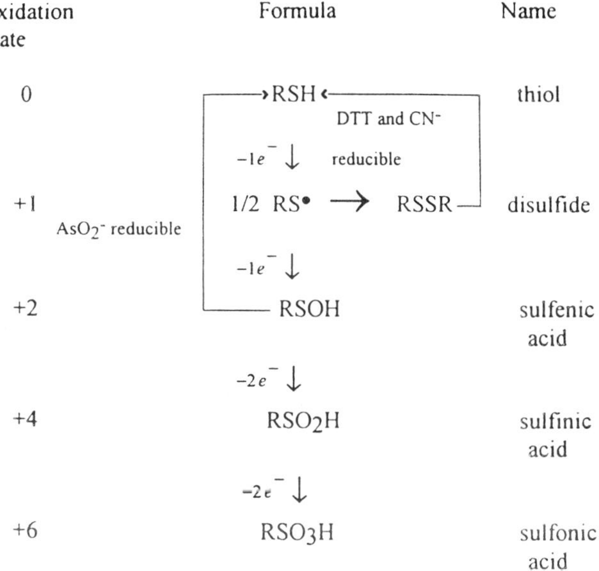

Figure 8 Products of thiol oxidation. (Modified from Ref. 10.)

the thiyl radical of glutathione can also be produced enzymatically through reactions catalyzed by peroxidases or by direct reactions of glutathione with free radical metabolites of drugs and toxic chemicals [Eq. (18)] (93). The thiyl radical can also undergo further reactions to produce superoxide anion [Eqs. (19) and (20)] or oxysulfur radicals, which ultimately decay to diamagnetic products [Eq. (21)] (44,59,62,93).

$$GSH + R^{\bullet} \rightarrow RH + GS^{\bullet} \tag{18}$$

$$GS^{\bullet} + GS^{-} \xrightleftharpoons{k = 3.3 \times 10^{3}\ M^{-1}} (GSSG)^{\bullet -} \tag{19}$$

$$(GSSG)^{\bullet -} + O_2 \xrightarrow{k = 1.6 \times 10^{8}\ M^{-1}\ s^{-1}} GSSG + O_2^{\bullet -} \tag{20}$$

$$GS^{\bullet} + O_2 \xrightleftharpoons{K = 3.2 \times 10^{3}\ M^{-1}} GSOO^{\bullet} \tag{21}$$

Disulfide formation, which is the most common product of the interaction of thiols with oxidants, does not necessarily involve the formation of thiyl radicals, as has been demonstrated in the case of the oxidation of glutathione by diamide and by alkyl nitrates catalyzed by glutathione-S-transferase (95), or by hydrogen peroxide catalyzed by glutathione peroxidase (93). The oxidation of xenobiotics bearing thiol or thione functional groups also does not involve free radical intermediates. These organic sulfur drugs are oxidized by microsomal monooxygenases to sulfenic acids that are reduced nonenzymatically to the parent drug by glutathione (95). As mentioned above, sulfenic acid forms of low molecular weight thiols usually cannot be isolated, and their formation is inferred from indirect experimental data. Indeed, simple sulfenic acids are either oxidized to sulfinic or sulfonic acid (Fig. 8) or react with a molecule of thiol to yield the disulfide [Eq. (22)] or undergo a self-condensation to thiosulfinate [Eq. (23)] (96). However, sulfenic acid derivatives can be stabilized within the protein structure in some sterically isolated thiol groups, which has been demonstrated in the case of the sulfhydryl groups of glyceraldehyde-3-phosphate dehydrogenase and papain (97,98). Recently, cysteine–sulfenic acid derivatives have been suggested to play important roles in redox regulation of DNA-binding activities of transcription such as Fos and Jun, OxyR, and bovine papillomavirus type 1 E2 protein (for a review, see Ref. 99).

$$RSOH + RSH \rightarrow RSSR + H_2O \tag{22}$$

$$RSOH + RSOH \rightarrow RSOSR + H_2O \tag{23}$$

V. PEROXYNITRITE-MEDIATED OXIDATION OF THIOLS

A. Kinetics and Mechanism

The biological importance of sulfhydryl groups led us to study the oxidation of low and high molecular weight thiols by peroxynitrite (10). We found that peroxynitrite oxidized cysteine to cystine and the single thiol group of bovine serum albumin to sulfinic or sulfonic acid derivatives. Sulfhydryl oxidation yields were much greater at alkaline pH than at acidic pH (Fig. 9), indicating that peroxynitrite anion rather than peroxynitrous acid was the primary oxidizing species. It should be noted, however, that in the acid pH range there was still a significant fraction of cysteine (34% yield) (Fig. 9) that was oxidized, probably by peroxynitrous acid and/or its decomposition products (Fig. 2). The role of peroxynitrite as the main thiol-oxidizing species was confirmed by rapid kinetic, stopped-flow experiments, which showed a bimolecular reaction between the cysteine sulfhydryl group and the anion (10). Indeed, the pH dependence of the apparent bimolecular rate constant showed a bell-shaped curve with two apparent pK_a values and a maximal apparent rate constant at pH 7.4:

$$k_2' = k_2 \frac{k_{a1}}{(K_{a1} + [H^+])} \frac{[H^+]}{(K_{a2} + [H^+])} \tag{24}$$

where k_2' is the apparent rate constant at a given pH, k_2 is the second-order rate constant of cysteine reaction with peroxynitrite anion, K_{a1} is the dissociation constant of peroxynitrite, and K_{a2} is the dissociation constant of cysteine thiol.

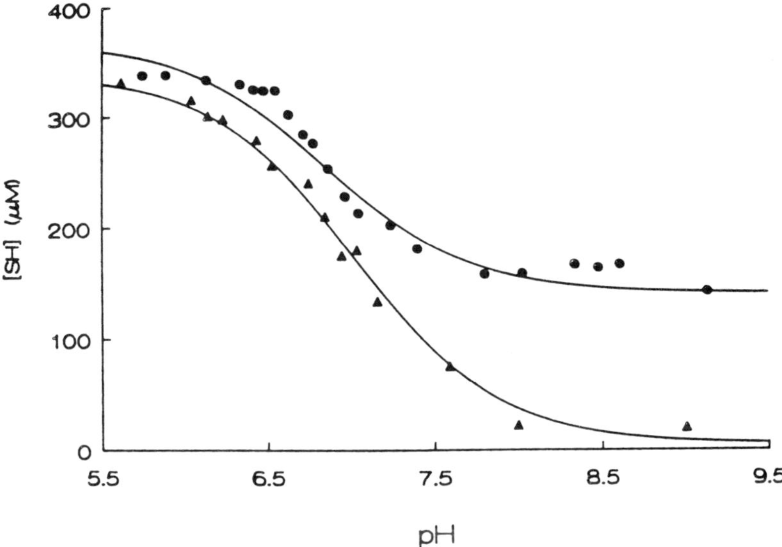

Figure 9 Influence of pH on peroxynitrite-mediated sulfhydryl oxidation. Upper curve, reactions were started by addition of peroxynitrite (0.38 mM) to bovine serum albumin (0.55 mM) (●) and incubated for 5 min at 37°C in potassium phosphate (50 mM), pH 5.0–9.0. Final pH was determined after peroxynitrite addition. Lower curve, reactions were started upon addition of peroxynitrite (0.25 mM) to cysteine (0.5 mM) (▲) and were performed as described above. The solid lines represent nonlinear regression using the Henderson-Hasselbach equation. (From Ref. 10.)

The stopped-flow experiments indicated that peroxynitrite at pH 7.4 oxidizes cysteine at a rate of 5000 M^{-1} s^{-1} (Fig. 10) and the sulfhydryl group of bovine serum albumin at 2500 M^{-1} s^{-1} (10). The experiments with albumin indicate that its single thiol group accounts for about 50% of the total reactivity of peroxynitrite with the protein. This is a remarkable observation, considering that albumin is a protein composed of 582 amino acids, only one of which accounts for half of the reactivity of the oxidant. Thus, sulfhydryl groups appear to be the major target of peroxynitrite attack on proteins.

The pH dependence of the second-order rate constant strongly suggested that the protonated form of the thiol (and not the thiolate anion) was the reacting species with peroxynitrite anion (10). Although thiolate anion is frequently more susceptible to oxidation, the reaction of peroxynitrite anion with the mercaptide would be unexpected considering the electrostatic repulsion between two negatively charged species. Kinetic data are not sufficient to unambiguously distinguish between the reaction of *cis*-peroxynitrite anion with thiol and the reaction

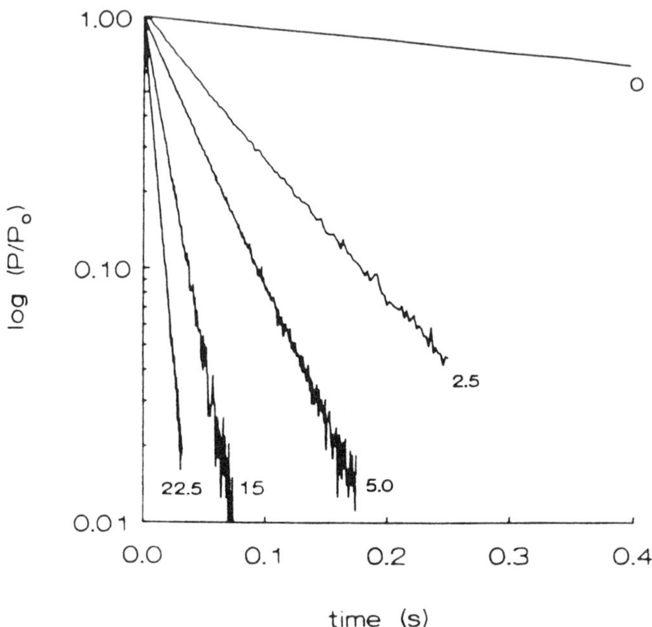

Figure 10 Time course of peroxynitrite decomposition in the presence of cysteine. Peroxynitrite (0.60 mM) was added to a reaction medium containing 0, 2.5, 5.0, 15, or 22.5 mM cysteine, in potassium phosphate (50 mM), pH 7.5, at 37°C. P_0 is the initial peroxynitrite concentration, and P is its concentration at a given time. (Modified from Ref. 10.)

of *cis*-peroxynitrous acid with thiolate anion. However, other evidence indicates that the direct reaction with peroxynitrite anion is the main mechanism accounting for sulfhydryl oxidation. First, oxidation yields were much higher under alkaline pH conditions, where very little peroxynitrite will protonate and decompose (Figs. 2 and 9). One could argue that at alkaline pH not much thiol will be in the protonated form, but the pK_a values of thiols are higher than the pK_a of peroxynitrite, and once a thiol molecule is consumed another dissociated molecule will protonate to maintain the equilibrium. Second, desferrioxamine, a scavenger by peroxynitrous acid (8,12), was unable to inhibit the oxidation of thiols mediated by peroxynitrite.

It is important to stress that the rate constants of peroxynitrite reactions with thiols are three orders of magnitude higher than the corresponding rate constants for the reactions of hydrogen peroxide with the sulfhydryls at pH 7.4. Also, in contrast to peroxynitrite, hydrogen peroxide oxidizes the thiolate anion rather than its protonated form and yields different oxidation products (10).

B. Oxidation Products Including Thiyl Radical Formation

Depending on the nature and relative concentration of an oxidant, thiols can be oxidized to different states, which can be determined by the use of compounds such as arsenite, cyanide, dithiothreitol, and borohydride (Fig. 8). Under mild oxidative conditions, peroxynitrite- and hydrogen peroxide–dependent cysteine oxidation was more than 90% cyanide and dithiothreitol reducible, indicating predominant oxidation to the disulfide form (10). The single thiol group of bovine serum albumin was oxidized to the sulfenic acid level by hydrogen peroxide and to higher oxidation states (either sulfinic or sulfonic acid levels) by peroxynitrite (10).

The oxidation of thiols by peroxynitrite is mediated by the formation of thiyl free radicals, as we demonstrated by the use of EPR spin-trapping techniques (53,79) (see also Sec. III.D.2). We have detected DMPO radical adducts of glutathionyl, cysteinyl, and albumin-cysteinyl during peroxynitrite-mediated oxidation of glutathione, cysteine, and bovine serum albumin, respectively. The corresponding adducts could also be detected with another spin trap, PBN (see Fig. 6) (53,79).

Taking into account the optimal thiol oxidation stoichiometry (two molecules of oxidized cysteine/molecule of peroxynitrite), the EPR observations, and the known oxidizing potential of nitrogen dioxide, which reacts with thiols at 10^8 M^{-1} s^{-1} (65), we propose that peroxynitrite-mediated oxidation of low molecular weight thiols is mediated by two separate one-electron transfer steps yielding thiyl radical intermediates [Eqs. (25) and (26)]. In the case of glutathione, the latter can initiate a free radical chain reaction with the participation of molecular oxygen [Eq. (19) and (20)] (53).

$$GSH + ONOO^- \rightarrow GS^\bullet + N^\bullet O_2 + OH^- \quad (25)$$

$$GSH + N^\bullet O_2 \rightarrow GS^\bullet + NO_2^- + H^+ \quad (26)$$

Although the above mechanism remains to be proved, our spin-trapping studies of peroxynitrite decomposition in the presence of GSH presented some evidence for the formation of superoxide anion (53). Indeed, the fact that the DMPO–hydroxyl radical adduct did not disappear even in the presence of 3.4 M ethanol (Fig. 5) suggested that part of it is formed from DMPO–superoxide radical adduct decay. In agreement, nitrogen-saturated buffers or 10 μ/ml superoxide dismutase inhibited the yield of the DMPO–hydroxyl radical adduct in about 40% (Fig. 5) (53).

In the case of bovine serum albumin, a single thiol group is present in each protein molecule, making an attack of a second thiol by the formed nitrogen dioxide [Eq. (26)] more difficult. Albumin sulfhydryl oxidation to high thiol oxidation states should, most likely, involve oxygen addition as follows (10,79,93,94):

$$BSA\text{-}SH \rightarrow BSA\text{-}S^\bullet \rightarrow BSA\text{-}SOO^\bullet \rightarrow BSA\text{-}SO_2H$$

C. Biological Significance

The demonstration of the oxidation of sulfhydryls by peroxynitrite constituted the first evidence of a direct reaction between peroxynitrite anion and a biological target without the involvement of the hydroxyl radical–like species and experimentally confirmed that peroxynitrite is a potent oxidant in its own right (10). This led us to postulate that direct reactions of peroxynitrite anion with critical cellular targets such as thiols may be quantitatively more relevant in its toxicology than the slower route leading to the formation of the hydroxyl radical–like oxidant *trans*-peroxynitrous acid (Figs. 1 and 2) (60). Indeed, if in a reaction mixture a thiol such as glutathione and a hydroxyl radical scavenger such as deoxyribose are competing for the reactivity of peroxynitrite, the first for peroxynitrite anion and the second for the hydroxyl radical-like oxidant arising from its proton-catalyzed decomposition, we will find that with thiol levels in the millimolar range, most of the peroxynitrite will be consumed by them. While the reactivity of peroxynitrite toward thiols, metal centers, and ascorbate is first order in regard to all of them, the reactivity of peroxynitrite is zero order in regard to hydroxyl radical scavengers because the formation of the activated *trans*-peroxynitrous acid is the rate-limiting step after protonation (Fig. 2). Thus, taking the bimolecular rate constant of the reaction between peroxynitrite and glutathione as 1350 $M^{-1}\ s^{-1}$ (k_{GSH}) (41), the average value of the cellular concentration of glutathione as 5 mM (actual concentration = 1–10 mM) (13), and the apparent rate constant of peroxynitrite decay at pH 7.4 as 0.6 s^{-1} (k_{app}) (8), we can calculate the rate of peroxynitrite disappearance as:

$$-d[ONOO\text{-}]/dt = k_{GSH}\ [GSH][ONOO^-] + k_{ap}\ [ONOO^-] \quad (27)$$

$$= (1.35\ mM^{-1}\ s^{-1} \times 5\ mM + 0.6\ s^{-1})\ [ONOO^-] \quad (28)$$

$$= (6.75\ s^{-1} + 0.6\ s^{-1})\ [ONOO^-] \quad (29)$$

These calculations show that in the presence of 5 mM glutathione, more than 90% of the peroxynitrite will directly react with the thiol and less than 10% will react by the hydroxyl radical–like pathway. Since other direct targets of peroxynitrite exist in the cell, such as other thiols, methionine (11,55), and metal centers (32,33,47), we can predict that a larger percentage of peroxynitrite will decompose through bimolecular attack to biological targets. In agreement with the kinetic predictions, we found that peroxynitrite anion was the main intermediate responsible for the inactivation of enzymes, inhibition of mitochondrial electron transport (100), and cytotoxicity to *Trypanosoma cruzi* (101,102) (see below). However, since the apparent rate of protonation increases at acidic pH, the hydroxyl radical–like pathway may be relevant and preferentially operate in low pH compartments such as the phagosomes and the Gouy-Stern-Chapman boundary layer of the plasma membrane. Also, the hydroxyl radical-like pathway may be favored in extracellular compartments where the concentration of thiols and metals is low.

The fact that thiyl radicals are formed during peroxynitrite-mediated oxidation of both low and high molecular weight thiols (53,79) is particularly important, since these radicals are reactive species that can initiate free radical chain reactions (93,94), further propagating peroxynitrite-induced biological damage. Although little is still known about the toxicology of thiyl radicals, a thiol-specific antioxidant enzyme has been recently characterized (103), and its function has been suggested to be the removal of thiyl radicals ($RS^{\bullet}$) and/or the corresponding disulfide radical anions [$(RSSR)^{\bullet -}$)] before they generate more reactive species (104). This suggestion remains to be proved. In the case of the ubiquitous glutathione, its high cellular concentration and the high rate constants for the reaction of the glutathionyl radical with the thiolate anion [Eq. (19)] and of the formed disulfide radical anion with oxygen [Eq. (20)] make it likely that, as proposed by Winterbourn (59), the thiyl radical can be detoxified, to some extent, by superoxide dismutase. In this case, the concerted action of thiols and superoxide dismutase would contribute to the cellular protection against peroxynitrite damage. This is an intriguing possibility that remains to be explored, since during peroxynitrite formation [Eq. (1)] the enzyme can be outcompeted by nitric oxide in its reaction with superoxide and also because superoxide dismutase can react with peroxynitrite although at slower rates (32,47). The biological fates of formed protein-thiyl radicals need to be more thoroughly studied (74,105–107). Under low oxygen tensions, these radicals could lead to protein aggregation through the formation of inter- and/or intramolecular disulfide bonds (108,109). On the other hand, further oxidation of thiyl radicals to sulfenic, sulfinic, or sulfonic acid can lead to partial or total loss of the biological activity of proteins (110,111). Particularly important is the possibility of the formation of stable protein-sulfenic acid products, which have been implicated in the regulation of the genetic response to oxidative stress (99). Undoubtedly, the oxidation of biothiols to thiyl radicals by peroxynitrite may have several biological consequences and represents a fertile area for further investigations.

1. Inhibition of Electron Transport in Mammalian Mitochondria

We have investigated the effects of biologically relevant concentrations of peroxynitrite toward isolated intact rat heart mitochondria to determine whether peroxynitrite can alter mitochondrial functions and, if so, to study the damaging mechanism (100). Oxygen-consumption studies (Fig. 11) indicated that peroxynitrite caused a profound inhibition of glutamate-malate (complex I)- and succinate (complex III)–dependent respiration, while N,N,N′,N′-tetramethyl-*p*-phenylenediamine/ascorbate (complex IV)–dependent respiration remained unchanged (100). Studying the different enzymic activities involved in mitochon-

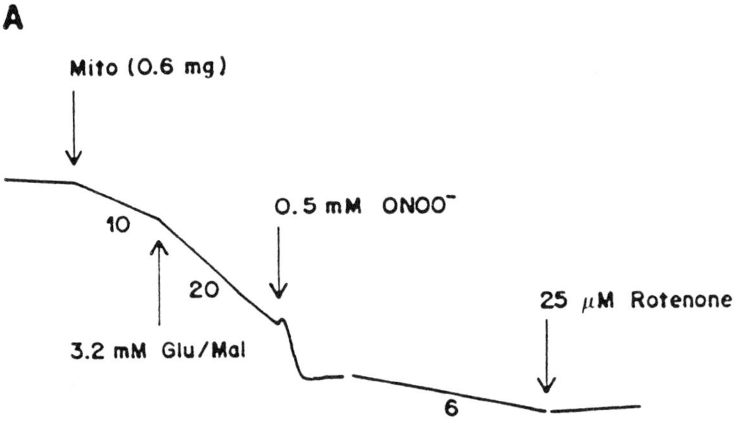

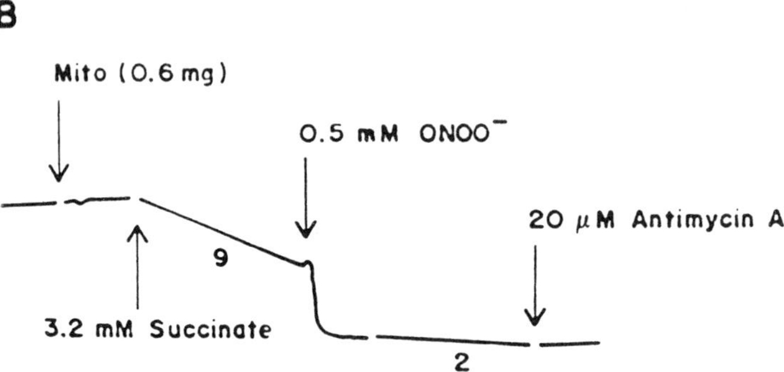

$[O_2]$ = 25 μM

1 min

Figure 11 Effect of peroxynitrite on NADH-linked (A) and succinate-dependent (B) mitochondrial oxygen consumption. State 4 oxygen consumption in the presence of respiratory substrates and inhibitors was studied using a polarographic Clark-type electrode. The reaction vessel contained 1.0 ml final reaction volume. Numbers are rates of oxygen consumption in $nmol^{-1}\ min^{-1}\ mg^{-1}$. (From Ref. 100.)

drial respiration, we found a preferential inactivation of enzymes that contain critical sulfhydryl groups in their active sites such as succinate dehydrogenase and ATPase. NADH dehydrogenase and cytochrome *c* oxidase were more resistant to the oxidative inactivation mediated by peroxynitrite. Both succinate dehydrogenase and ATPase were protected by compounds that react directly with peroxynitrite, such as cysteine and methionine, whereas typical hydroxyl radical scavengers such as dimethyl sulfoxide and mannitol were ineffective (100). These studies support peroxynitrite anion rather than peroxynitrous acid or its derived oxidants as the main damaging species to mitochondria.

2. *Toxicity to Trypanosoma cruzi*

The parasite *Trypanosoma cruzi* is the etiological agent of Chagas' disease. Macrophage activation has been shown to be a key mechanism against acute *T. cruzi* invasion of humans, and both reactive oxygen species and nitric oxide are considered to be critical mediators of the killing of the parasite (112,113). Peroxynitrite, at concentrations estimated to be formed in activated macrophages, was shown to be cytotoxic to *T. cruzi* by inhibiting parasite growth (101). Part of the cytotoxic effects of peroxynitrite can be attributed to inhibition of the energetic metabolism of the parasite since the compound inhibited NADH- and succinate-linked oxygen consumption (102). The NADH-dependent respiration of the parasite greatly depends on an enzymic activity not present in mammalian cells, that of NADH–fumarate reductase. This enzyme as well as *T. cruzi* succinate dehydrogenase have critical sulfhydryls at their active sites. We found that both NADH–fumarate reductase and succinate dehydrogenase were inactivated by peroxynitrite and that postincubation with dithiothreitol recovered fumarate reductase activity by 60–70% and succinate dehydrogenase activity by 100%, indicating that most enzyme inactivation was due to peroxynitrite-mediated oxidation of critical sulfhydryls. Similar to the observation made with intact mitochondria, protection against the peroxynitrite-mediated process was extensive in the case of cysteine and methionine but much less effective in the case of hydroxyl radical scavengers (102). These results provide further support for our hypothesis that at pH 7.4 a direct attack, in particular on sulfhydryl groups, is a major mechanism accounting for the biological damage caused by peroxynitrite (60). Finally, our results on mitochondria and *T. cruzi* also indicate that peroxynitrite is able to cross cell membranes.

VI. CONCLUSION AND PERSPECTIVES

The discovery of the endogenous production of nitric oxide by mammalian cells and its role as a major signal-transducing and effector molecule vastly expanded the horizon of free radical reactions in biology. Nitric oxide is well known as an environmental pollutant, and nearly the same properties that make it toxic may enable it to be used as a rapid, locally acting messenger. For instance, reaction with oxygen and oxygen-derived radicals may be an important route to modulate the local concentration of nitric oxide. The fast reaction between superoxide anion and nitric oxide in particular may represent a critical control point in cells producing both species, leading either to downregulation of the biological effects of superoxide and nitric oxide or to potentially toxic oxidation of nearby biomolecules through the formation of the peroxynitrite anion.

The biochemical observations reviewed herein emphasize the mechanisms by which peroxynitrite can initiate free radical processes in target molecules. The most studied route is the proton-catalyzed decomposition of peroxynitrite anion at physiological pH. This process leads to a species with the reactivity of the hydroxyl radical, probably activated *trans*-peroxynitrous acid, which can trigger free radical production from nearby molecules and yields, in addition, nitrogen dioxide, another free radical species that can further propagate peroxynitrite-mediated one-electron oxidations. Most important, these free radicals will be formed in the absence of transition metal ions either from metalloenzymes or from low molecular weight complexes. Peroxynitrite anion, however, is a strong oxidant by itself and can directly oxidize low and high molecular weight thiols with the intermediate formation of thiyl radicals, as demonstrated by the use of EPR spin-trapping techniques (53,79). As is the case for nitrogen dioxide, thiyl radicals are reactive species, which can further propagate oxidation reactions. Indeed, the high intra- and extracellular concentrations of thiols led us to propose that direct reactions of peroxynitrite anion with critical cellular targets may be quantitatively more relevant to its cytotoxicity than the route leading to the hydroxyl radical-like oxidant activated *trans*-peroxynitrous acid (60). Accordingly, peroxynitrite anion was shown to be the main intermediate responsible for the inhibition of mammalian mitochondrial respiration (100) and for cytotoxicity to the parasite *T. cruzi* (101,102). In addition to thiyl radicals, several other radicals could be detected by EPR techniques during the oxidation of target molecules by both peroxynitrite anion and peroxynitrous acid (53,77–79). Consequently, peroxynitrite-mediated one-electron oxidation of biomolecules may be an important event in its cytotoxic mechanism, and EPR techniques can greatly contribute to a better understanding of peroxynitrite reactivity under physiological conditions.

The cytotoxic potential of peroxynitrite will be modulated by the environment in which the oxidant is formed, since peroxynitrite decomposition is influenced by pH and by the presence of different target molecules. Thus, better knowledge of the different possible reaction pathways under physiologically relevant conditions is critical for the ultimate goal of therapeutic intervention in those clinical conditions where excess peroxynitrite formation contributes to tissue damage. Indeed, an important series of papers has been published supporting the concept that peroxynitrite is formed in vivo and may be a critical mediator in the development of several pathological conditions, including immunocomplex-stimulated pulmonary edema (114), heart ischemia-reperfusion (115), neurotoxicity (116,119), pulmonary emphysema secondary to smoke inhalation (11), and atherogenesis (31,118). Although much work is needed to determine the precise contribution of peroxynitrite to diverse clinical conditions, it is now clear that the interactions between reactive oxygen- and nitrogen-derived species to yield secondary oxidizing intermediates are of key importance for the understanding of free radical–dependent damage to biological systems.

ACKNOWLEDGMENTS

We warmly acknowledge the continuous support and contributions of Drs. B. A. Freeman and J. S. Beckman. We also thank the Fundação de Amparo à Pesquisa do Estado de São Paulo (FAPESP), the Financiadora de Estudos e Projetos (FINEP), the Conselho Nacional de Desenvolvimento Científico e Tecnológico (CNPq), the Consejo Nacional de Investigaciones Científicas y Técnicas (CONICYT), and the Swedish Agency for Research and Cooperation with Developing Countries (SAREC) for financial support.

REFERENCES

1. Gally, J. A., Montague, P. R., Reeke, J., G. N., and Edelman, G. M. (1990) The NO hypothesis: Possible effects of a short-lived, rapidly diffusible signal in the development and function of the nervous system. Proc. Natl. Acad. Sci. USA 87:3547–3551.
2. Marletta, M. (1989) Nitric oxide: Biosynthesis and biological significance. Trends Biochem. Sci. 14:488–492.
3. Ignarro, L. J. (1991) Signal transduction mechanisms involving nitric oxide. Biochem. Pharmacol. 41:485–490.
4. Moncada, S., Palmer, R. M. J., and Higgs, E. A. (1991) Nitric oxide: Physiology, pathophysiology, and pharmacology. Pharmacol. Rev. 43:109–142.
5. Moncada, S. (1992) Nitric oxide: Mediator, modulator and pathophysiologic entity. J. Lab. Clin. Med. 120:187–191.
6. Ford, P. C., Wink, D. A., and Stanbury, D. M. (1993) Autoxidation kinetics of aqueous nitric oxide. FEBS Lett. 326:1–3.
7. Blough, N. V., and Zafiriou, O. C. (1985) Reaction of superoxide with nitric oxide to form peroxynitrite in alkaline aqueous solution. Inorg. Chem. 24:3504–3505.
8. Beckman, J. S., Beckman, T. W., Chen, J., Marshall, P. M., and Freeman, B. A. (1990) Apparent hydroxyl radical production by peroxynitrite: Implications for endothelial injury from nitric oxide and superoxide. Proc. Natl. Acad. Sci. USA 87:1620–1624.
9. Huie, R. E., and Padmaja, S. (1993) The reaction of NO with superoxide. Free Rad. Res. Commun. 18:195–199.

10. Radi, R., Beckman, J. S., Bush, K. M., and Freeman, B. A. (1991) Peroxynitrite oxidation of sulfhydryls. The cytotoxic potential of superoxide and nitric oxide. J. Biol. Chem. 266:4244–4250.
11. Moreno, J., and Pryor, W. (1992) Inactivation of α1-proteinase inhibitor by peroxynitrite. Chem. Res. Toxicol. 5:425–431.
12. Radi, R., Beckman, J. S., Bush, K. M., and Freeman, B. A. (1991) Peroxynitrite-induced membrane lipid peroxidation: The cytotoxic potential of superoxide and nitric oxide. Arch. Biochem. Biophys. 288:481–487.
13. Jocelyn, P. C. (1972) Biochemistry of the SH Group. Academic Press, New York.
14. Meister, A. (1984) New aspects of glutathione biochemistry and transport. Fed. Proc. 43:3031–3041.
15. Halliwell, B. (1988) Albumin-an important extracellular antioxidant? Biochem. Pharmacol. 37:569–571.
16. Radi, R., Bush, K.M., Cosgrove, T. P., and Freeman, B. A. (1991) Reaction of xanthine oxidase-derived oxidants with lipid and protein of human plasma. Arch. Biochem. Biophys. 286:117–125.
17. McCord, J. M., Keele, B. B., Jr., and Fridovich, I. (1971) An enzyme-based theory of obligate anaerobiosis: The physiological function of superoxide dismutase. Proc. Natl. Acad. Sci. USA 68:1024–1027.
18. Naqui, A., Chance, B., and Cadenas, E. (1986) Reactive oxygen intermediates in biochemistry. Ann. Rev. Biochem. 55:137–166.
19. Halliwell, B., and Gutteridge, J. M. C. (1989) Free Radicals in Biology and Medicine. Clarendon Press, Oxford.
20. Jones, O. T. G., and Cross, A. R. (1992). Composition of NADPH oxidase of phagocytes and other cell types. In: Biological Oxidants: Generation and Injurious Consequences (Cochrane, C. G., and Gimbrone, M. A., Jr., eds.), pp. 119–131. Academic Press, New York.
21. Gryglewsky, R. J., Palmer, R. M. J., and Moncada, S. (1986) Superoxide anion is involved in the breakdown of the endothelium-derived vascular relaxing factor. Nature 320:454–456.
22. Nathan, C. (1992) Nitric oxide as a secretory product of mammalian cells. FASEB J. 6:3051–3064.
23. Leone, A. M., Palmer, R. M. J., Knowles, R. G., Francis, P. L., Ashton, D. S., and Moncada, S. (1991) Constitutive and inducible nitric oxide synthases incorporate molecular oxygen into both nitric oxide and citrulline. J. Biol. Chem. 266:23790–23795.
24. Fridovich, I. (1989) Superoxide dismutases. An adaptation to a paramagnetic gas. J. Biol. Chem. 264:7761–7764.
25. Sharma, V. S., Traylor, T. G., and Gardiner, R. (1987) Reaction of nitric oxide with heme proteins and model compounds of hemoglobin. Biochemistry 26:3837–3843.
26. Traylor, T. G., and Sharma, V. S. (1992) Why NO? Biochemistry 31:2847–2849.
27. Ischiropoulos, H., Zhu, L., and Beckman, J. S. (1992) Peroxynitrite formation from macrophage-derived nitric oxide. Arch. Biochem. Biophys. 298:446–451.
28. Wang, J. F., Komarov, P., Sies, H., and de Groot, H. (1991) Contribution of

nitric oxide synthase to luminol-dependent chemiluminescence generated by phorbol-ester-activated Kuppfer cells. Biochem. J. 279:311–314.
29. Kooy, N. W., and Royal, J. A. (1994) Agonist-induced peroxynitrite production from endothelial cells. Arch. Biochem. Biophys. 310:352–359.
30. Carreras, M. C., Pargament, G. A., Catz, S., Poderoso, J. J., and Boveris, A. (1994) Nitric oxide production during the respiratory burst of human neutrophils. FEBS Lett. 341:65–68.
31. Beckman, J. S., Ye, Y. Z., Anderson, P., Chen, J., Accavitti, M. A., Tarpey, M. M., and White, C. R. (1994) Extensive nitration of protein tyrosine in human atherosclerosis detected by immunohistochemistry. Biol. Chem. Hoppe-Seyler 357: 81–88.
32. Ischiropoulos, H., Zhu, L., Chen, J., Tsai, M., Martin, J. C., Smith, C., and Beckman, J. S. (1992) Peroxynitrite-mediated tyrosine nitration catalyzed by superoxide dismutase. Arch. Biochem. Biophys. 298:431–437.
33. Beckman, J. S., and Crow, J. P. (1993) Pathological implications of nitric oxide, superoxide and peroxynitrite formation. Biochem. Soc. Trans. 21:330–334.
34. Keith, W. G., and Powell, P. E. (1969) Kinetics of decomposition of peroxynitrous acid. J. Chem. Soc. A 453:90.
35. Crow, J. P., Spruell, C., Chen, J., Gunn, C., Schiropoulos, H., Tsai, M., Smith, C. D., Radi, R., Koppenol, W. H., and Beckman, J. S. (1994) On the pH dependent yield of hydroxyl radical products from peroxynitrite. Free Rad. Biol. Med. 16:331–338.
36. Halfpenny, E., and Robinson, P. L. (1952) The nitration and hydroxylation of aromatic compounds by pernitrous acid. J. Chem. Soc. 939–946.
37. Halfpenny, E., and Robinson, P. L. (1952) Pernitrous acid. The reaction between hydrogen peroxide and nitrous acid, and the properties of an intermediate product. J. Chem. Soc. 928–938.
38. Hughes, M. N., and Nicklin, H. G. (1968) The chemistry of pernitrites. Part I. Kinetics of decomposition of pernitrous acid. J. Chem. Soc. 450–452.
39. Mahoney, L. R. (1979) Evidence for the formation of hydroxyl radicals in the isomerization of pernitrous acid to nitric acid in aqueous solution. J. Am. Chem. Soc. 92:5262–5263.
40. Koppenol, W. H., and Butler, J. (1985) Energetics of interconversion reactions of oxyradicals. Adv. Free Rad. Biol. Med. 1:91–131.
41. Koppenol, W. H., Pryor, W. A., Moreno, J. J., Ischiropoulos, H., and Beckman, J. S. (1992) Peroxynitrite, a cloaked oxidant formed by nitric oxide and superoxide. Chem. Res. Toxicol. 5:834–842.
42. Hayashi, Y., and Yamazaki, I. (1979) The oxidation reduction potentials of compound I/compound II and compound II/ferric couples of horseradish peroxidase A_2 and C. J. Biol. Chem. 254:9101–9106.
43. Surdhar, P. S., and Armstrong, D. A. (1987) Reduction potentials and exchange reactions of thiyl radicals and disulfide anion radicals. J. Phys. Chem. 91:6532–6537.
44. Koppenol, W. H. (1993) A thermodynamic appraisal of the radical sink hypothesis. Free Rad. Biol. Med. 14:91–94.

45. Koppenol, W. H. (1990). Oxyradical reactions: From bond-dissociation energies to reduction potentials. FEBS Lett. 264:165–167.
46. Segel, I. W. (1976) Biochemical Calculations. John Wiley and Sons, Inc., New York.
47. Beckman, J. S., Ischiropoulos, H., Zhu, I., van der Woerd, M., Chen, J., Harrison, J., Martin, J. C., and Tsai, M. (1992) Kinetics of superoxide dismutase and iron catalzyed nitration of phenolics. Arch. Biochem. Biophys. 298:438–445.
48. Janing, G. R., Kraft, R., Blanck, J., Rabe, H., and Ruckpaul, K. (1987) Chemical modification of cytochrome P-450 LM4. Identification of functionally linked tyrosine residues. Biochim. Biophys. Acta 916:512–523.
49. Lundblad, R. L., Noyes, C. M., Featherstone, G. L., Harrison, J. H., and Jenzano, J. W. (1988) The reaction of α-thrombin with tetranitromethane. J. Biol. Chem. 263:3729–3734.
50. Guerrieri, F., Yagi, T., and Papa, S. (1984) On the mechanism of H^+ translocation by mitochondrial H^+-ATPase. Studies with chemical modifier of tyrosine residues. J. Bioenerg. Biomembr. 16:251–262.
51. Sokolovsky, M., Harrel, D., and Riordan, J. F. (1969) Reaction of tetranitromethane with sulfhydryl groups in proteins. Biochemistry 8:4740–4745.
52. Riordan J. F., and Vallee, B. L. (1972) Nitration with tetranitromethane. In: Methods in Enzymology, Vol. XXV, Part B (Hirs, C. H. W., and Timasheff, S. N., eds.), pp. 515–521. Academic Press, New York.
53. Augusto, O., Gatti, R. M., and Radi, R. (1994) Spin-trapping studies of peroxynitrite decomposition and of 3-morpholinosydnonimime N-ethylcarbamide autooxidation. Arch. Biochem. Biophys. 310:118–125.
54. Tsai, J. H. M., Harrison, J. G., Martin, J. C., Hamilton, T. P., Vanderwoerd, M., Jablonski, M. J., and Beckman, J. S. (1994) Role of conformation of peroxynitrite anion ($ONOO^-$) and its stability and toxicity. J. Am. Chem. Soc. 116:4115–4116.
55. Pryor, W. A., Jin, X., and Squadrito, G. L. (1994) One- and two-electron oxidations of methionine by peroxynitrite. Proc. Natl. Acad. Sci. USA 91:11173–11177.
56. Aust, S. D., Morehouse, L. A., and Thomas, C. E. (1985) Role of metals in oxygen radical reactions. J. Free Rad. Biol. Med. 1:3–25.
57. Fridovich, I. (1986) Biological effects of the superoxide radical. Arch. Biochem. Biophys. 247:1–11.
58. Liochev, S. I., and Fridovich, I. (1994) The role of $O_2^{\bullet-}$ in the production of $HO^\bullet$: In vitro and in vivo. Free Rad. Biol. Med. 16:29–33.
59. Winterbourn, C. C. (1993) Superoxide as an intracellular radical sink. Free Rad. Biol. Med. 14:85–90.
60. Radi, R. (1995) Oxidative reactions of peroxynitrite in biological systems: direct attack versus the hydroxyl radical-like pathway. In: The Oxygen Paradox in Biology and Medicine (Davies, K. J. A., ed.). CLEUP Press, Padova (in press).
61. Gardner, H. W. (1989) Oxygen radical chemistry of polyunsaturated fatty acids. Free Rad. Biol. Med. 7:65–86.
62. Buettner G. R. (1993) The pecking order of free radicals and antioxidants: Lipid

peroxidation, α-tocopherol, and ascorbate. Arch. Biochem. Biophys. 300:535–543.
63. Kellogg, E. W., and Fridovich, I. (1977) Liposome oxidation and erythrocyte lysis by enzymically generated superoxide and hydrogen peroxide. J. Biol. Chem. 252:6721–6728.
64. Schaich, K. M., and Borg, D. C. (1988) Fenton reactions in lipid phases. Lipids 23:570–579.
65. Prütz, W. A., Mönig, H., Butler, J., and Land, E. J. (1985). Reactions of nitrogen dioxide in aqueous model systems: Oxidation of tyrosine units in peptides and proteins. Arch. Biochem. Biophys. 243:125–134.
66. Rubbo, H., Radi, R., Trujillo, M., Telleri, R., Kalyanaraman, B., Barnes, S., Kick, M., and Freeman, B. A. (1994) Nitric oxide regulation of superoxide and peroxynitrite-dependent lipid peroxidation: Formation of novel-nitrogen containing oxidized lipid derivatives. J. Biol. Chem. 269:26066–26075.
67. Darley-Usmar, V. M., Hogg, N., O'Leary, V., Wilson, M. T., and Moncada, S. (1992) The simultaneous generation of superoxide and nitric oxide can initiated lipid peroxidation in human low density lipoprotein. Free Rad. Res. Commun. 17:9–20.
68. Schaich, K., and Borg, D. C. (1980) EPR studies in autoxidation: In: Autoxidation in Food and Biological Systems (Simic, M. G., and Karel, M., eds.), pp. 45–69. Plenum Press, New York.
69. Baker, J. E., Felix, C. C., and Kalyanaraman, B. (1988) Myocardial ischemia and reperfusion: Direct evidence for free radical generation by electron spin resonance spectroscopy. Proc. Natl. Acad. Sci. 85:2786–2789.
70. Perkins, M. J. (1980) Spin trapping. Adv. Phys. Org. Chem. 17:1–64.
71. Janzen, E. G. (1980) A critical review of spin trapping in biological systems. In: Free Radicals in Biology, Vol. IV (Pryor, W. A., ed.), pp. 115–154. Academic Press, New York.
72. Finkelstein, E., Rosen, G. M., and Rauckman, E. J. (1980) Spin trapping of superoxide and hydroxyl radical: practical aspects. Arch. Biochem. Biophys. 200:1–16.
73. Augusto, O. (1989) Spin trapping studies of xenobiotic-mediated toxicity. In: CRC Handbook of Free Radicals and Antioxidants, Vol. 3 (Miquel, L., Quintanilha, A. T., and Weber, H., eds.), pp. 193–208. CRC Press, Inc., Boca Raton, FL.
74. Motley, C., and Mason, R. P. (1990) Nitroxide radical adducts in biology: Chemistry, applications, and pitfalls. In: Biological Magnetic Resonance, Vol. 8 (Berliner, L. J., and Reuben, J., eds.), pp. 489–546. Plenum, New York.
75. Knecht, K. T., and Mason, R. P. (1993) In vivo spin trapping of xenobiotic free radical metabolites. Arch. Biochem. Biophys. 303:185–194.
76. Laurindo, F. R. M., Pedro, M. A., Barbeiro, H. V., Pileggi, F., Carvalho, M. H. C., Augusto, O., and da Luz, P. (1994) Vascular free radical release. Ex-vivo and in vivo evidence for a flow-dependent endothelial mechanism. Circ. Res. 74:700–709.
77. Pedro, M. A., Gatti, R. M., and Augusto, O. (1993) In vivo free radical formation monitored by the ascorbyl radical. Quím. Nova 16:370–372.

78. Denicola, A., Souza, J., Gatti, R. M., Augusto, O., and Radi, R. (1995) Desferrioxamine inhibition of the hydroxyl radical-like reactivity of peroxynitrite: Role of hydroxamic groups. Free Rad. Biol. Med. (in press).
79. Gatti, R. M., Radi, R., and Augusto, O. (1994) Peroxynitrite-mediated oxidation of albumin to the protein thiyl free radical. FEBS Lett. (in press).
80. Gatti, R. M., Radi, R., and Augusto, O. (1993) Unpublished work.
81. Pryor, W. A., Church, D. F., Govidan, C. K., and Crank, G. (1982). Oxidation of thiols by nitric oxide and nitrogen dioxide: synthetic utility and toxicological implications. J. Org. Chem 47:156–159.
82. Bartlett, D., Church, D. F., Bounds, P. L., and Koppenol, W. H. (1994) The kinetics of the oxidation of L-ascorbic acid by peroxynitrite. Free Rad. Biol. Med. (in press).
83. Prescott, L. F., Park, J., Sutherland, G. R., Smith, I. J., and Proudfoot, A. T. (1976) Cysteamine, methionine and penicillamine in the treatment of paracetamol poisoning. Lancet II:109–113.
84. Bird, R. P. (1980) Cysteamine as a protective agent with high-LET radiations. Radiat. Res. 82:290–296.
85. Stocker, R., and Frei, B. (1991) Endogenous antioxidant defenses in human blood plasma. In: Oxidative Stress: Oxidants and Antioxidants (Sies, H. ed.), pp. 213–243. Academic Press, New York.
86. Rotilio, G., Paci, M., Sette, M., and Ciriolo, M. R. (1991) GSH as an antioxidant: new functions in the physiological activation of Cu, Zn superoxide dismutase, copper transport and trans-membrane transduction of reducing power. In: Oxidative Damage and Repair (Davies, K. J. A., ed.), pp. 71–76. Pergamon Press, Oxford.
87. Flohé, L. (1982) Glutathione peroxidase brought into focus. In: Free Radical in Biology, Vol. V (Pryor, W. A., ed.), pp. 223–254. Academic Press, New York.
88. Maiorino, M., Gregolin, C., and Ursini, F. (1990) Phospholipid hydroperoxide glutathione peroxidase. In: Methods in Enzymology, Vol. 186, Part B (Packer L., and Glazer, A. N., eds.), pp. 448–457. Academic Press, New York.
89. Saez, G., Thornally, P. J., Hill, H. A. O., Hems, R., and Bannister, J. V. (1982) The production of free radicals during the autoxidation of cysteine and their effect on isolated rat hepatocytes. Biochim. Biophys. Acta 719:24–31.
90. Barton, J. P., Packer, J. E., and Sims, R. J. (1973) Kinetics of the reaction of hydrogen peroxide with cysteine and cysteamine. J. Chem. Soc. Perkin Trans. II:1547–1549.
91. Harman, L. S., Mottley, C., and Mason, R. P. (1984) Free radical metabolites of L-cysteine oxidation. J. Biol. Chem. 259:5606–5611.
92. Eling, T. E., Curtis, J. F., Harman, L. S., and Mason, R. P. (1986) Oxidation of glutathione to its thiyl free radical metabolite by prostaglandin H synthase. J. Biol. Chem. 261:5023–5028.
93. Mason, R. P., and Rao, D. N. R. (1990) Thiyl free radical metabolites of thiol drugs, glutathione, and proteins. In: Methods in Enzymology, Vol. 186, Part B (Packer, L., and Glazer, A. N., eds.), pp. 319–329. Academic Press, New York.
94. Asmus, K.-D. (1990) Sulfur-centered free radicals. In: Methods in Enzymology, Vol. 186, Part B (Packer, L., and Glazer, A. N., eds.), pp. 169–180. Academic Press, New York.

95. Ziegler, D. M. (1991) Non-radical mechanisms for the oxidation of glutathione. In: Oxidative Stress: Oxidants and Antioxidants (Sies, H., ed.), pp. 85–97. Academic Press, New York.
96. Kice, J. L. (1980) Mechanisms and reactivity in reactions of organic oxyacids of sulfur and their anhydrides. Adv. Phys. Org. Chem. 17:65–181.
97. Allison, W. S. (1976) Formation and reactions of sulfenic acids in proteins. Acc. Chem. Res. 9:293–299.
98. Liu, T.-Y. (1977) The role of sulfur in proteins. In: The Proteins, 3rd ed., Vol. 3 (Neurath, H., and Hill, R. L., eds.), pp. 239–402. Academic Press, New York.
99. Clairborne, A., Miller, H., Parsonage, D., and Ross, R. P. (1993) Protein-sulfenic acid stabilization and function in enzyme catalysis and gene regulation. FASEB J. 7:1483–1490.
100. Radi, R., Rodriguez, M., Castro, L., and Telleri, R. (1994) Inhibition of mitochondrial electron transport by peroxynitrite. Arch. Biochem. Biophys. 308:89–95.
101. Denicola, A., Rubbo, H., Rodriguez, D., and Radi, R. (1993) Peroxynitrite-mediated cytotoxicity to *T. cruzi*. Arch. Biochem. Biophys. 304:279–287.
102. Rubbo, H., Denicola, A., and Radi, R. (1994) Peroxynitrite inactivates thiol-containing enzymes of *Trypanosoma cruzi* energetic metabolism and inhibits cell respiration. Arch. Biochem. Biophys. 308:96–102.
103. Chae, H. Z., Kim, I.-H., Kim, K., and Rhee, S. G. (1993) Cloning, sequencing, and mutation of thiol-specific antioxidant gene of Saccharomyces cerevisae. J. Biol. Chem. 268:16815–16821.
104. Yim, M. B., Chae, H. Z., Rhee, S. G., Chock, P. B., and Stadtman, E. R. (1994) On the protective mechanism of the thiol-specific antioxidant enzyme against the oxidative damage to biomacromolecules. J. Biol. Chem. 269:1621–1626.
105. Graceffa, P. (1983) Spin labelling of protein sulfhydryl groups by spin trapping of a sulfur radical: Application to bovine serum albumin and myosin. Arch. Biochem. Biophys. 225:802–808.
106. Maples, K. R., Kennedy, C. H., Jordan, S. J., and Mason, R. P. (1990) In vivo thiyl free radical formation from haemoglobin following administration of hydroperoxydes. Arch. Biochem. Biophys. 277:402–409.
107. Davies, M. J., Gilbert, B. C., and Haywood, R. M. (1994) Radical-induced damage to bovine serum albumin: role of cysteine residue. Free Rad. Res. Commun. 18:353-367.
108. Schuessler, H., and Freundle, K. (1983) Reactions of formate and ethanol radicals with bovine serum albumin studied by electrophoresis. Int. J. Radiat. Biol. 44:17–29.
109. Schuessler, H., and Schilling, K. (1984) Oxygen effect in the radiolysis of proteins. Part 2. Bovine serum albumin. Int. J. Radiat. Biol. 45:267–281.
110. Clement, J. R., Armstrong, D. A., Klassen, N. V., and Gillis, H. A. (1972) Pulse radiolysis of aqueous papain. Can. J. Chem. 50:2833–2840.
111. Buchanan, J. D., and Armstrong, D. A. (1978) The radiolysis of glyceraldehyde-3-phosphate dehydrogenase. Int. J. Radiat. Biol. 33:409–418.
112. Tanaka, Y., Kiyotaki, C., Tanowitz, H., and Bloom, B. R. (1982) Reconstitution of a variant macrophage cell line defective in oxygen metabolism with a H_2O_2-generating system. Proc. Natl. Acad. Sci. USA 79:2584–2588.

113. Muñoz-Fernández, M. A., Fernández, M. A., and Fresno, M. (1992) Activation of human macrophages for the killing of intracellular *Trypanosoma cruzi* by TNF and IFN-γ through a nitric oxide-dependent mechanism. Eur. J. Immunol. 22:301–307.
114. Mulligan, M. S., Hevel, J. M., Marletta, M. A., and Ward, P. A. (1991) Tissue injury caused by deposition of immune complexes is L-arginine dependent. Proc. Natl. Acad. Sci. USA 88:6338–6342.
115. Mahteis, G., Sherman, M. P., Buckberg, G. D., Haybron, D. M., Young, H. H., and Ignarro, L. (1992) Role of L-arginine-nitric oxide pathway in myocardial reoxygenation injury. Am. J. Physiol. 262:H616–H620.
116. Cazevielle, C., Muller, A., Meynier, F., and Bonne, C. (1993) Superoxide and nitric oxide cooperation in hypoxia/reoxygenation-induced neuron injury. Free Radical Biol. Med. 14:389–395.
117. Lipton, S. A., Chol, Y.-B., Pan, Z.-H., Lei, S. Z., Chen, H.-S. V., Sucher, N. J., Loscalzo, J., Singel, D. J., and Stamler, J. S. (1993) A redox-based mechanism for the neuroprotective and neurodestructive effects of nitric oxide and related nitroso compounds. Nature 364:626–632.
118. White, R. C., Brock, T. A., Chang, L. Y., Crapo, J. D., Briscoe, P., Ku, D., Bradley, W. A., Gianturco, S. H., Gore, J., Freeman, B. A., and Tarpey, M. M. (1994) Superoxide and peroxynitrite in atherosclerosis. Proc. Natl. Acad. Sci. USA 91:1044–1048.

6

Concerted Antioxidant Activity of Glutathione and Superoxide Dismutase

Christine C. Winterbourn
Christchurch School of Medicine, Christchurch, New Zealand

I. INTRODUCTION

Reduced glutathione (GSH) has a multifaceted role in antioxidant defense. It is a direct scavenger of free radicals as well as a cosubstrate for peroxide detoxification by glutathione peroxidases and for conjugation by glutathione-S-transferases. In addition, through disulfide exchange reactions, GSH is able to reduce protein disulfides and regulate the thiol/disulfide status of the cell. As a direct scavenger, GSH is considered to be a major contributor to radioprotection and radical repair (1,2) and to protect against xenobiotics that are metabolized to free radicals by one-electron oxidation or reduction (3,4). GSH is also essential for controlling endogenous oxidants. Animals administered buthionine sulfoximine, an inhibitor of GSH synthesis, die within a few days (5). The main injury is to mitochondria, where the greatest amounts of oxidants are generated (6). Ascorbate is able to protect against some of the effects of GSH deficiency, illustrating the importance of the interaction between the GSH and ascorbate antioxidant pathways (7,8).

The emphasis of this chapter is on the radical reactions of GSH, and in particular the interrelationship between different radical species and scavengers. It describes how GSH can channel radicals to superoxide, so that radical stress can be controlled by removal of this one terminal radical by superoxide dismutase (SOD) (9,10). Other thiols exhibit similar free radical chemistry, with variations due to differences in properties such as pK and redox potential (11).

In a biological situation, physical properties such as membrane permeability will also influence behavior as will ability to act as enzyme substrates. For example, cysteamine but not GSH is a substrate for myeloperoxidase (12), and cysteine but not GSH promotes superoxide-dependent modification of low-density lipoprotein (13). Thus, while many of the reactions that will be discussed for GSH can be applied to other physiological and pharmacological thiols, differences in their chemical and physical properties will affect how they interact in radical pathways in biological systems.

II. RADICAL REACTIONS OF GSH

A. Scavenging

GSH has been shown to react with a diverse range of radicals (14–16). This has been detected using a variety of methods. The most direct is ESR using a spin trap. 5,5-Dimethyl-1-pyrrolidone-*N*-oxide, which forms a characteristic adduct with the glutathionyl radical ($GS^{\bullet}$), has been the most widely used (17). A large amount of information has been obtained from pulse radiolysis studies in which rapid reactions of radicals are followed spectroscopically and reaction rates determined. Other studies have measured inhibitory effects of GSH on product yields from radical reactions and used SOD to probe involvement of superoxide. It should be noted that all of these methods have limitations, not the least being that the detection method can distort the path of the reaction. For example, trapping one radical in a system can influence equilibria and prevent alternative reactions, so it cannot be concluded that the trapped radical is necessarily the most reactive. Thus, spin trapping $GS^{\bullet}$ prevents formation of the disulfide radical ($GSSG^{\bullet -}$) and its reaction with oxygen to give superoxide. Alternatively, many pulse radiolysis studies rely on this reaction because of the good spectral characteristics of $GSSG^{\bullet -}$ (18). Oxygen therefore needs to be excluded and the useful pH range restricted. Product analysis studies of course do not detect radical species directly.

A representative list of compounds whose radicals react with GSH is given in Table 1. These include phenoxyl, alkoxyl, arylamino, peroxyl, semiquinone, and carbon-centered radicals. The parent compounds are physiological low molecular weight compounds and macromolecules, xenobiotics encountered as drugs or environmentally, and other compounds such as vitamin E that are considered themselves to be radical scavengers. It is apparent, therefore, that a large proportion of the radicals generated in a cell could theoretically be scavenged by GSH.

There are many potential sources of such radicals (15,16). These include generation either radiolytically or by oxidations catalyzed by a peroxidase or another heme protein or reduction by an enzyme such as xanthine oxidase or

Table 1 Compounds Whose Radicals Are Reported to Oxidize GSH to Its Thiyl Radical

Compound or class of compound	Radical-generating system	Ref.
Tyrosine	Peroxidase	45, 75
Sugars	Radiolysis	18
DNA bases	Radiolysis	8, 98
α-Tocopherol	Radiolysis	84
	Flash photolysis	85
Aliphatic alcohols	Radiolysis	18
Phenols	Peroxidase	14, 42, 45, 49, 99
	Radiolysis	100
	Autoxidation	44
Aromatic amines	Peroxidase	14, 42, 49, 101, 102
	Autoxidation	43, 103
Phenothiazines	Peroxidase	14
Dialuric acid, divicine	Autoxidation	28, 46

cytochrome P-450 reductase. Intracellular formation of $GS^•$ derived from a peroxidase-catalyzed reaction has been demonstrated (9). Glutathionyl radicals are also produced in the reaction of GSH with hydroxyl radicals formed in Fenton reactions (20) and by one-electron oxidation by transition metal complexes or heme proteins (21–23). The reaction of GSH with peroxynitrite gives $GS^•$(24), as does the decomposition of nitrosoglutathione (GSNO) [Eq. (1)], which can be formed from either $NO^•$ or $NO_2^•$ and GSH (25,26). Equation (1) is frequently considered as a source of $NO^•$ without considering that it gives another radical product:

$$GSNO \rightarrow GS^• + NO^• \tag{1}$$

The reaction of GSH with semiquinones can occur, but it is frequently in competition with conjugation (17,27). Many quinones form conjugates with GSH via a nonenzymatic reductive Michael addition. However, it is possible to achieve conditions, e.g., with alloxan (28), where scavenging of the semiquinone is highly favored over conjugation.

B. Reactions of Glutathionyl Radicals

As is characteristic of radical reactions, scavenging by GSH generates another radical. Therefore, depending on the properties of the secondary radical, it is not necessarily a defense or protective process. In almost all cases, radical scavenging by GSH generates $GS^•$. (Superoxide appears to be an exception, as discussed below.) Until recently, the prevailing view was that $GS^•$ radicals were

Table 2 Compounds Reported to Be Oxidized by Glutathionyl Radicals

Compound or class of compound	Ref.
NADH	104
Ascorbic acid	105
Polyunsaturated fatty acids	106, 107
Phenothiazines	105
Ferrocytochrome *c*	104
Retinol	108

biologically benign, undergoing few reactions other than dimerization to oxidized glutathione (GSSG). This is not the case. Although the rate constant for dimerization is high ($2k = 9 \times 10^8\ M^{-1}\ s^{-1}$), it competes unfavorably with other reactions of $GS^•$ (29,30) and is likely to be significant only if radical concentrations build up to the micromolar range. $GS^•$ is an oxidizing radical that is in fact quite reactive (16). Some of its reactions (Table 2) include the reverse of reactions in Table 1, oxidation of compounds such as NADH and ascorbate, and hydrogen abstraction from polyunsaturated fatty acids to initiate lipid peroxidation. $GS^•$ formation is therefore not necessarily protective.

$GS^•$ radicals can also undergo addition reactions, for example, with styrene (31). An interesting variation of this has been observed with dihydrodiols of polycyclic hydrocarbons (32,33) and the carcinogen 2-amino-4-(5-nitro-2-furyl)-thiazole (34). The products are the same as formed by the traditional conjugation reaction and provide an alternative radical-mediated route to conjugate formation.

C. Superoxide Generation and Reactions of Glutathione Radicals with Oxygen

The above discussion highlights the apparently paradoxical situation of $GS^•$ having prooxidant properties, yet GSH acting as an efficient radical-scavenging antioxidant. To explain these observations, it is necessary to consider alternative reactions of $GS^•$, and in particular how they are influenced by oxygen. Although a wide range of radicals (represented as $R^•$) react with GSH, for many the reaction is not thermodynamically favorable and Eq. (2) lies well to the left.

$$R^• + GSH \leftrightarrow RH + GS^• \quad (2)$$

The reason why radical scavenging and radical repair by GSH can occur is that the forward reaction can be kinetically driven by efficient removal of the thiyl radical product (14). Dimerization to GSSG would not produce this driving force under most physiological conditions and is unlikely to be a major route for $GS^{\bullet}$ decay. A key feature of thiyl radical chemistry is the equilibrium between $GS^{\bullet}$, an oxidizing radical, and the strongly reducing disulfide radical, $GSSG^{\bullet -}$.

$$GS^{\bullet} + GS^{-} \leftrightarrow GSSG^{\bullet -} \quad (3)$$

The position of the equilibrium ($K_3 = 3 \times 10^3\ M^{-1}$) depends on the thiolate anion concentration, which is determined by the pH (pK of GSH 8.8) and the GSH concentration. At pH 7.4 and 5 mM GSH, the ratio of $GS^{\bullet}$ to $GSSG^{\bullet -}$ is approx 2:1 (30). However, because the equilibrium is established rapidly, reactions of $GSSG^{\bullet -}$ may dominate even though its relative concentration is low.

Both radicals react with oxygen [Eqs. (4) and (5)]. The reaction in Eq. (5) is very fast ($k = 2 \times 10^8\ M^{-1}\ s^{-1}$) and irreversible. It is able to provide the driving force to displace Eq. (2) (35), and the combination of Eqs. (2), (3), and (5) accounts for the good scavenging ability of GSH (10).

$$GS^{\bullet} + O_2 \leftrightarrow GSOO^{\bullet} \quad (4)$$

$$GSSG^{\bullet -} + O_2 \rightarrow GSSG + O_2^{\bullet -} \quad (5)$$

There is a somewhat confusing picture as to the significance of Eq. (4). The peroxylsulfenyl radical ($GSOO^{\bullet}$) is highly unstable, and can undergo reactions leading to sulfinyl ($GSO^{\bullet}$) and sulfonyl radicals. These have been observed in frozen solution (36,37), but they are difficult to detect (38), and more basic understanding of the solution chemistry of sulfur radicals is required before their relevance to biological situations can be truly assessed. However, the consensus is that Eq. (4) is either not very fast or reversible (29,39–41), and because the peroxysulfenyl radical is relatively unreactive, at physiological pH and GSH concentrations it is a minor decay route compared with Eq. (5). This is consistent with a number of studies in which $GS^{\bullet}$ was produced in the presence of oxygen. The disulfide was found to account for >85% of the product (28,42–45), and superoxide production was detected (14,28,43,45,46). Small amounts of glutathione sulfonic acid have been reported in some cases (47,48), while no higher oxidation products were found in others (14,45,49).

The reaction of the disulfide radical with oxygen to give superoxide appears to be a fundamental aspect of radical scavenging by GSH. The reaction of the thiyl radical with oxygen, however, should become more significant at lower pH where $GSSG^{\bullet -}$ formation by Eq. (3) is less favorable. $GSSG^{\bullet -}$ is a strong reducing agent that can reduce compounds such as quinones and heme proteins, and it reacts rapidly with phenoxyl radicals (29,30). These reactions could become alternatives to Eq. (5) at low oxygen concentrations.

From the above considerations it would be expected that the majority of the oxygen consumed in GSH-scavenging reactions yields superoxide and GSSG, rather than higher oxidation states of the thiol. It also follows that GSH would be a poor scavenger in the absence of oxygen and that superoxide would always be produced as a consequence of its scavenging activity. Furthermore, it is feasible that under the appropriate conditions, all the radicals listed in Table 1 could be scavenged by GSH with consequent generation of superoxide. Such conditions could be met by the millimolar GSH concentrations and typical pH and pO_2 that exist intracellularly. Superoxide, therefore, would act as a radical sink (10). With SOD present, the superoxide would be removed enzymatically. This sequence, therefore, provides an elegant mechanism for a single enzyme to control the effects of radical generation.

III. REACTIONS OF SUPEROXIDE GENERATED FROM THIYL RADICALS

Radical scavenging by GSH, in the absence of SOD, could lead to injurious reactions mediated by superoxide. Since the discovery of SOD, there has been immense interest in the nature of such reactions. There is now convincing evidence that superoxide itself impairs cell function through oxidative inactivation of iron-sulfur dehydratases such as aconitase (50). It may also participate in the generation of secondary more reactive oxidants, by releasing iron from iron-sulfur proteins (51) and possibly ferritin (52) and participation in hydroxyl radical production or lipid peroxidation through reduction of transition metal complexes (53). It also reacts with $NO^{\bullet}$ to give peroxynitrite (54). When superoxide is generated in a GSH-mediated process, it could cause oxidative stress by the following additional mechanisms.

A. Reaction with GSH

Suggestive evidence for a reaction between superoxide and thiols was obtained initially by Packer and coworkers in their study of the radiolysis of cysteine (55,56). Several investigations of the reaction with GSH have since been performed (57–59), but they were inconsistent in their conclusions. Most agree that a reaction occurs, but whereas Asada and Kanematsu (57) reported a rate constant of $7 \times 10^5\ M^{-1}\ s^{-1}$, the estimate from radiolytic studies by Bielski et al. (60) was $<15\ M^{-1}\ s^{-1}$. However, the former study examined epinephrine oxidation without considering possible complications due to the thiyl radicals participating in its chain oxidation, and the latter may have been complicated by superoxide regeneration in the reaction. We have recently reinvestigated this reaction, using hypoxanthine and xanthine oxidase as a source of superoxide (48). As shown in Figure 1 and in agreement with Wefers and Sies (47), adding GSH causes a superoxide-dependent increase in oxygen consumption and GSSG formation. This is due to a short-chain reaction that regenerates super-

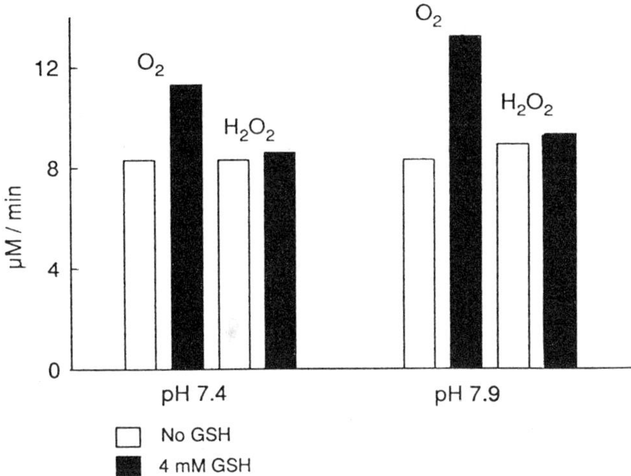

Figure 1 Oxygen uptake and hydrogen peroxide production by a hypoxanthine/xanthine oxidase system in air in the presence and absence of glutathione. Rates of oxygen uptake and hydrogen peroxide production were measured with an oxygen electrode and peroxide electrode, respectively. The additional oxygen uptake in the presence of GSH was fully inhibited by SOD. (From Ref. 48.)

oxide, with a rate constant, determined by competitive inhibition by SOD, of about 500–1000 $M^{-1} s^{-1}$. Hydrogen peroxide is not produced in the reaction (Fig. 1), indicating that it does not occur by simple hydrogen abstraction to give the thiyl radical [Eq. (6)] as is usually assumed.

$$O_2^{\bullet -} + GSH + H^+ \rightarrow H_2O_2 + GS^{\bullet} \quad (6)$$

The reaction mechanism appears similar to that proposed by von Sonntag and coworkers for the reaction of superoxide with dithiothreitol (61), although in neither case have the intermediates been identified directly. The kinetics and product analysis are consistent with the initial step being the formation of a sulfinyl radical [Eq. (7)] followed by Eqs. (8) and (9) leading to $GS^{\bullet}$ formation. This would then regenerate superoxide by Eqs. (3) and (5).

$$O_2^{\bullet -} + GSH \rightarrow [GSO_2H]^{\bullet -} \rightarrow GSO^{\bullet} + OH^- \quad (7)$$

$$GSO^{\bullet} + GSH \rightarrow GS^{\bullet} + GSOH \quad (8)$$

$$GSOH + GSH \rightarrow GSSG + H_2O \quad (9)$$

$$GS^{\bullet} + GS^- \leftrightarrow GSSG^{\bullet -} \quad (3)$$

$$GSSG^{\bullet -} + O_2 \rightarrow GSSG + O_2^{\bullet -} \quad (5)$$

About 10% of the GSH is converted to the sulfonic acid, and singlet oxygen formation attributed to peroxyl radical decay has been detected (47,62). This implies that Eq. (4) and subsequent reactions occur as a minor pathway. The combination of these reactions gives a chain that is terminated primarily by superoxide dismutation, and for every cycle consumes GSH with the stoichiometry of Eq. (10). Thus GSH is consumed in excess of the original superoxide generated.

$$1.4O_2 + 3.6GSH \rightarrow 1.6GSSG + 0.4GSO_3H + 1.4H_2O \quad (10)$$

B. Superoxide-Dependent Chain Reactions

A number of physiological and xenobiotic molecules (RH_2) undergo autoxidation, in many cases by a radical chain involving superoxide [Eqs. (11) and (12)]. Compounds that autoxidize by this mechanism include hydroquinones (63), hydroxypyrimidines (28,64), aromatic amines (65,66), catecholamines (67–69), thiols (70), hydrazines (71), enzyme-bound NADH (72), and sugars that can form enediols (73). A kinetic study of the autoxidation of dialuric acid and other hydroxypyrimidines (64) showed that while SOD inhibits the chain, this is transitory because buildup of product allows a second chain, consisting of Eqs. (12) and (13), to proceed.

$$O_2^{\bullet -} + RH_2 + H^+ \rightarrow RH^\bullet + H_2O_2 \quad (11)$$

$$RH^\bullet + O_2 \leftrightarrow R + O_2^{\bullet -} + H^+ \quad (12)$$

$$R + RH_2 \leftrightarrow 2RH^\bullet \quad (13)$$

$$RH_2 + O_2 \rightarrow R + H_2O_2 \quad (14)$$

The stoichiometry is given by Eq. (14), and because chain lengths can be long, many molecules of hydrogen peroxide can be formed per initiating event. If GSH is present (46), it scavenges the semiquinone radical intermediate ($RH^\bullet$) and regenerates the hydroquinone (R) by Eq. (2) but does not inhibit the chain because superoxide is regenerated via Eqs. (3) and (5).

$$RH^\bullet + GSH \leftrightarrow RH_2 + GS^\bullet \quad (2)$$

$$GS^\bullet + GS^- \leftrightarrow GSSG^{\bullet -} \quad (3)$$

$$GSSG^{\bullet -} + O_2 \rightarrow GSSG + O_2^{\bullet -} \quad (5)$$

The overall result is oxidation of GSH and production of hydrogen peroxide until all the GSH is consumed. Thus, GSH increases oxidative stress caused by superoxide-dependent chain autoxidations, and only the combination of GSH and SOD provides an effective defense.

C. Radical-Superoxide Interactions

When a primary radical is generated and reacts with GSH to give thiyl, disulfide, and ultimately superoxide radicals, there is the potential for mixed radical reactions to occur. Even though radical-radical reactions are fast [e.g., for dityrosine formation from tyrosyl radicals, $2k = 4 \times 10^8\ M^{-1}\ s^{-1}$ (74)], they generally compete poorly with radicals reacting with nonradicals that are present at much higher concentrations. For example, GSH at $< 100\ \mu M$ is able to prevent measurable dimerization of tyrosyl radicals (75). Superoxide, however, is relatively unreactive with nonradicals so its reactions with other radicals are more favorable. For example, the tyrosyl radical reacts with superoxide with a rate constant $2k = 1.5 \times 10^9\ M^{-1}\ s^{-1}$ (76). It has generally been considered that this is an electron-transfer reaction to regenerate tyrosine (74), but a recent radiolytic study (76) has shown that a major product is a peroxide addition product, which is stable for some hours before undergoing cyclization and peroxide loss. We have evidence that a similar product is formed when tyrosyl radicals are generated by a peroxidase in the presence of GSH. A product containing a peroxide group, detected with the FOX assay (77), was formed in a superoxide-dependent reaction at GSH concentrations low enough to scavenge some but not all of the phenoxyl radicals (75). These findings suggest that radical scavenging by GSH can lead to superoxide-dependent formation of organic hydroperoxides with possible undesirable biological reactivities.

IV. BIOLOGICAL IMPLICATIONS OF CONCERTED ANTIOXIDANT ACTION OF GSH AND SUPEROXIDE DISMUTASE

Biological systems have an integrated pattern of antioxidant defense (78–80). It can be considered to be directed against two categories of oxidant stress, namely, free radicals and longer-lived oxidants, such as peroxides, each with a separate mechanism for control. Peroxides are detoxified primarily by catalase or the glutathione peroxidases. The latter all require GSH, but different enzymes in the family exhibit specificities for different organic hydroperoxides and hydrogen peroxide. With respect to radical defense, the free radical species that can be produced in the cell are varied, and it would seem important to have an enzymic mechanism for preventing their undesirable reactions. Yet SOD is the only known enzyme that occurs ubiquitously whose substrate is a radical and purpose is the disposal of radicals. However, the mechanism whereby GSH enables superoxide to act as a radical sink enables this one enzyme, acting on a single substrate, to suppress such reactions. The concerted action of GSH and SOD, therefore, can be regarded as a general radical defense.

Figure 2 represents schematically how this pathway interacts with other radical reactions. It shows how superoxide, if not removed, could participate in chain oxidations and includes alternative routes to superoxide through direct reaction of radicals such as semiquinones with oxygen. These are also equilibria, and the presence of SOD is often required to pull them in the direction of superoxide production (81). It also shows the link between radical and nonradical pathways through the common involvement of GSH.

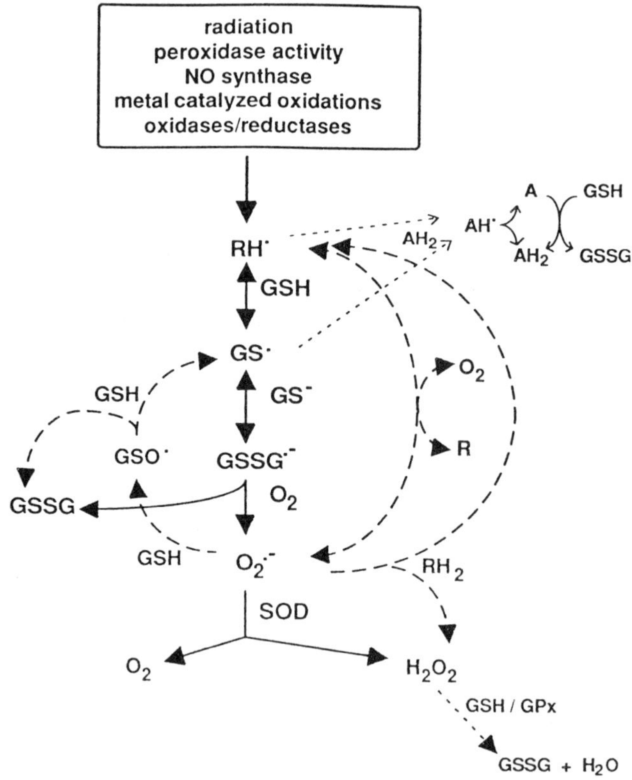

Figure 2 Scheme showing production of radical species and removal by a pathway involving reactions with GSH and SOD. Also shown are other pathways that can integrate into the scheme: regeneration of radicals by $O_2^{\bullet -}$-dependent reactions if $O_2^{\bullet -}$ is not removed, reaction of either the primary or glutathionyl radical with ascorbate (AH_2) to give the ascorbyl radical ($AH^{\bullet}$), and removal of hydrogen peroxide by the glutathione peroxidase pathway. $RH^{\bullet}$ represents a generic radical that could be generated in the cell (e.g., phenoxyl, C-centered, protein thiyl, semiquinone, hydroxyl). Of the parent compounds R and RH_2, some but not all would be capable of the one-electron reductions or oxidations shown.

While GSH and SOD alone may provide adequate protection against radical injury, ascorbate is an alternative physiological scavenger that could function as a free radical trap. The ascorbyl radical does not react with oxygen. It is relatively stable, and dismutation is considered to be its major route of decay (82). It is a substrate for monodehydroascorbate reductase, but this enzyme does not have the wide distribution of SOD (7,83) and therefore it is unlikely to be crucial for antioxidant activity for ascorbate. Ascorbate is known to act synergistically with other antioxidants. It can regenerate vitamin E from its radical (84,85), although the extent to which this occurs physiologically is debatable (79). In vivo studies have demonstrated that it can protect against the effects of GSH deficiency and conversely that GSH can spare ascorbate (5,86). Reactions that could contribute to this synergism (see Fig. 2) include the reduction of dehydroascorbate by GSH either through a direct or enzyme-catalyzed reaction and reduction of $GS^{\bullet}$ by ascorbate (7). Scavenging by ascorbate and subsequent dismutation of the ascorbyl radicals produced could bypass superoxide production from GSH. However, more information on the interaction between these two scavengers is needed before their relative input into cellular antioxidant defense can be assessed.

While this discussion has stressed the protective role of GSH and SOD against oxidative damage, it is increasingly apparent that free radicals and oxidants generally have an important regulatory function. The expression of genes controlled by the transcription factors NFκB (87–89) and AP-1 (90,91) and processes such as lymphocyte differentiation (92) and apoptosis (93,94) can all be activated by oxidants. A variety of oxidative stresses, including peroxides, exogenous radical-generating systems, and redox cycling drugs, as well as antioxidants such as butylated hydroxytoluene, GSH, *N*-acetylcysteine, and increased expression of SOD can influence these processes. It is well known that redox state plays an important role in cell metabolism, to a large extent through alterations in thiol/disulfide status (95,96). This can regulate enzymes that require a free thiol or disulfide for activity, with specific enzymes being switched on or off at widely different GSH:GSSG ratios (97). It appears that oxidation and reduction of thiol groups also is important in activation or suppression of gene expression. A mechanism such as proposed in Figure 2 enables different radicals and nonradical oxidants and antioxidants to interact, with GSH and thiol status being a common factor. It may be through this common pathway that diverse oxidants and antioxidants modulate thiol groups on regulatory proteins and thus control their activity.

ACKNOWLEDGMENT

This work was supported by the Health Research Council of New Zealand.

REFERENCES

1. Bigalow, J. E., and Varnes, M. E. (1983) The role of thiols in cellular response to radiation and drugs. Radiat. Res. 95:437–455.
2. von Sonntag, C., and Schuchmann, H. (1990) Sulphur compounds and "chemical repair" in radiation biology. In: Sulfur-Centered Reactive Intermediates in Chemistry and Biology (Chatgilialoglu, C., and Asmus, K. D., eds.), pp. 409–414. Plenum Press, New York.
3. Ross, D. (1988) Glutathione, free radicals and chemotherapeutic agents. Mechanisms of free-radical induced toxicity and glutathione-dependent protection. Pharmacol. Ther. 37:231–249.
4. Shan, X., Aw, T. Y., and Jones, D. P. (1990) Glutathione-dependent protection against oxidative injury. Pharmacol. Ther. 47:61–71.
5. Mårtensson, J., and Meister, A. (1991) Glutathione deficiency decreases tissue ascorbate levels in newborn rats: Ascorbate spares glutathione and protects. Proc. Natl. Acad. Sci. USA 88:4656–4660.
6. Jain, A., Mårtensson, J., Stole, E., Auld, P. A. M., and Meister, A. (1991) Glutathione deficiency leads to mitochondrial damage in brain. Proc. Natl. Acad. Sci. USA 88:1913–1917.
7. Meister, A. (1994) Glutathione-ascorbic acid antioxidant system in animals. J. Biol. Chem. 269:9397–9400.
8. Willson, R. L. (1983) Free radical repair mechanisms and the interactions of glutathione and vitamins C and E. In: Radioprotectors and Anticarcinogens (Nygaard, O. F., and Simic, M. G., eds.), pp. 1–22. Academic Press, New York.
9. Munday, R., and Winterbourn, C. C. (1989) Reduced glutathione in combination with superoxide dismutase as an important biological antioxidant defence mechanism. Biochem. Pharmacol. 38:4349–4352.
10. Winterbourn, C. C. (1993) Superoxide as an intracellular radical sink. Free Radical Biol. Med. 14:85–90.
11. Munday, R. (1994) Bioactivation of thiols by one-electron oxidation. Adv. Pharmacol. 27:237–270.
12. Svensson, B. E., Graslund, A., Strom, G., and Moldeus, P. (1993) Thiols as peroxidase substrates. Free Radical Biol. Med. 14:167–175.
13. Heinecke, J. W., Kawamura, M., Suzuki, L., and Chait, A. (1993) Oxidation of low density lipoprotein by thiols: Superoxide-dependent and -independent mechanisms. J. Lipid Res. 34:2051–2061.
14. Subrahmanyam, V. V., McGirr, L. G., and O'Brien, P. J. (1987) Glutathione oxidation during peroxidase catalysed drug metabolism. Chem. Biol. Interact. 61:45–59.
15. O'Brien, P. J. (1988) Radical formation during the peroxidase catalyzed metabolism of carcinogens and xenobiotics: The reactivity of these radicals with GSH, DNA, and unsaturated lipid. Free Radical Biol. Med. 4:169–183.
16. D'Aquino, M., Bullion, C., Chopra, M., Devi, D., Devi, S., Dunster, C., James, G., Komuro, E., Kundu, S., Niki, E., Raza, F., Robertson, F., Sharma, J., and Willson, R. L. (1994) Sulfhydryl free radical formation enzymatically by sonolysis, by radiolysis, and thermally: Vitamin A, curcumin, muconic acid, and related conjugated olefins as references. Methods Enzymol. 233:34–46.

17. Goin, J., Gibson, D. D., McCay, P. B., and Cadenas, E. (1991) Glutathionyl- and hydroxyl radical formation coupled to the redox transitions of 1,4-naphthoquinone bioreductive alkylating agents during glutathione two-electron reductive addition. Arch. Biochem. Biophys. 288:386–396.
18. Baker, M. Z., Badiello, R., Tamba, M., Quintiliani, M., and Gorin, G. (1982) Pulse radiolytic study of hydrogen transfer from glutathione to organic radicals. Int. J. Radiat. Biol. 41:595–602.
19. Schreiber, J., Foureman, G. L., Hughes, M. F., Mason, R. P., and Eling, T. E. (1989) Detection of glutathione thiyl free radical catalyzed by prostaglandin H synthase present in keratinocytes. J. Biol. Chem. 264:7936–7943.
20. Sjöberg, L., Eriksen, T. E., and Révész, L. (1982) The reaction of the hydroxyl radical with glutathione in neutral and alkaline aqueous solution. Radiat. Res. 89:255–263.
21. Harman, L. S., Carver, K. D., Schreiber, J., and Mason, R. P. (1986) One- and two-electron oxidation of reduced glutathione by peroxidases. J. Biol. Chem. 261:1642–1648.
22. Mottley, C., Toy, K., and Mason, R. P. (1987) Oxidation of thiol drugs and biochemicals by the lactoperoxidase/hydrogen peroxide system. Mol. Pharmacol. 31:417–421.
23. Romero, F. J., Ordonez, I., Arduini, A., and Cadenas, E. (1992) The reactivity of thiols and disulfides with different redox states of myoglobin. Redox and addition reactions and formation of thiyl radical intermediates. J. Biol. Chem. 267:1680–1688.
24. Radi, R., Beckman, J. S., Bush, K. M., and Freeman, B. A. (1991) Peroxynitrite oxidation of sulfhydryls: The cytotoxic potential of superoxide and nitric oxide. J. Biol. Chem. 266:4244–4250.
25. Stamler, J. S., Singel, D. J., and Loscalzo, J. (1992) Biochemistry of nitric oxide and its redox-active forms. Science 258:1898–1902.
26. Augusto, O., Gatti, R. M., and Radi, R. (1994) Spin-trapping studies of peroxynitrite decomposition and of 3-morpholinosydnonimine N-ethylcarbamide autooxidation: Direct evidence for metal-independent formation of free radical intermediates. Arch. Biochem. Biophys. 310:118–125.
27. Brunmark, A., and Cadenas, E. (1989) Redox and addition chemistry of quinoid compounds and its biological implications. Free Radical Biol. Med. 7:435–477.
28. Winterbourn, C. C., and Munday, R. (1989) Glutathione-mediated redox cycling of alloxan. Mechanisms of superoxide dismutase inhibition and of metal-catalyzed OH formation. Biochem. Pharmacol. 38:271–277.
29. Asmus, K. D. (1990) Sulfur-centered radicals. Methods Enzymol. 186:168–180.
30. Wardman, P. (1988) Conjugation and oxidation of glutathione via thiyl radicals. In: Glutathione Conjugation: Mechanisms and Biological Significance (Sies, H., and Ketterer, B., eds.). Academic Press, New York.
31. Stock, B. H., Schreiber, J., Guenat, C., Mason, R. P., Bend, J. R., and Eling, T. E. (1986) Evidence for a free radical mechanism of styrene-glutathione conjugate formation catalysed by prostaglandin H synthase and horseradish peroxidase. J. Biol. Chem. 261:15915–15922.

32. Foureman, G. L., and Eling, T. E. (1989) Peroxidase-mediated formation of glutathione conjugates from polycyclic aromatic dihydrodiols and insecticides. Arch. Biochem. Biophys. 269:55–68.
33. Foureman, G. L., Knecht, K. T., and Eling, T. E. (1992) Peroxidase-medited glutathione conjugation of benzo[*a*]pyrene-7,8-dihydrodiol is enhanced by benzo[*a*] pyrene phenols in vitro. Carcinogenesis 13:515–518.
34. Lakshmi, V. M., Zenser, T. V., Sohani, S., and Davis, B. B. (1992) Mechanism of formation of the thioether conjugate of the bladder carcinogen 2-amino-4-(5-nitro-furyl)-thiazole (ANFT). Carcinogenesis 13:2087–2093.
35. Koppenol, W. H. (1993) A thermodynamic appraisal of the radical sink hypothesis. Free Radical Biol. Med. 14:91–94.
36. Sevilla, M. D., Becker, D., and Yan, M. (1990) The formation and structure of the sulfoxyl radicals $RSO^{\bullet}$, $RSOO^{\bullet}$, $RSO_2^{\bullet}$, and $RSO_2OO^{\bullet}$ from the reaction of cysteine, glutathione and penicillamine thiyl radicals with molecular oxygen. Int. J. Radiat. Biol. 57:65–81.
37. Swarts, S. G., Becker, D., DeBolt, S., and Sevilla, M. D. (1989) An electron spin resonance investigation of the structure and formation of sulfinyl radicals: Reaction of peroxyl radicals with thiols. J. Phys. Chem. 93:155–161.
38. von Sonntag, C., and Schuchmann, H. P. (1991) The elucidation of peroxyl radical reactions in aqueous solution with the help of radiation-chemical methods. Angew. Chem. Int. Ed. Engl. 30:1229–1253.
39. Mönig, J., Asmus, K. D., Forni, L. G., and Willson, R. L. (1987) On the reaction of molecular oxygen with thiyl radicals: A re-examination. Int. J. Radiat. Biol. 52:589–602.
40. Tamba, M., Simone, G., and Quintiliani, M. (1986) Interactions of thiyl free radicals with oxygen: a pulse radiolysis study. Int. J. Radiat. Biol. 50:595–600.
41. Mason, R. P., and Ramakrishna Rao, D. N. (1990) Electron spin resonance investigation of the thiyl free radical metabolites of cysteine, glutathione and drugs. In: Sulfur-Centered Reactive Intermediates in Chemistry and Biology (Chatgilialoglu, C., and Asmus, K. D., eds.), pp. 401–408. Plenum Press, New York.
42. Subrahmanyam, V., and O'Brien, P. J. (1985) Peroxidase catalysed oxygen activation by arylamine carcinogens and phenol. Chem. Biol. Interact. 56:185–199.
43. Munday, R. (1987) Oxidation of glutathione and reduced pyridine nucleotides by the myotoxic and mutagenic aromatic amine, 1,2,4-triaminobenzene. Chem. Biol. Interact. 62:131–141.
44. Eyer, P., and Lengfelder, E. (1984) Radical formation during autoxidation of 4-dimethylaminophenol and some properties of the reaction products. Biochem. Pharmacol. 33:1005–1013.
45. Nakamura, M., Yamazaki, I., Ohtaki, S., and Nakamura, S. (1986) Characterization of one- and two-electron oxidations of glutathione coupled with lactoperoxidase and thyroid peroxidase reactions. J. Biol. Chem. 261:13923–13927.
46. Winterbourn, C. C. (1989) Inhibition of autoxidation of divicine and isouramil by the combination of superoxide dismutase and reduced glutathione. Arch. Biochem. Biophys. 271:447–455.
47. Wefers, H., and Sies, H. (1983) Oxidation of glutathione by the superoxide radical

to the disulfide and the sulfonate yielding singlet oxygen. Eur. J. Biochem. 137:29–36.
48. Winterbourn, C. C., and Metodiewa, D. (1994) The reaction of superoxide with reduced glutathione. Arch. Biochem. Biophys. 314:284–290.
49. Ross, D., Norbeck, K., and Moldeus, P. (1985) The generation and subsequent fate of glutathionyl radicals in biological systems. J. Biol. Chem. 260:15028–15032.
50. Flint, D. H., Tuminello, J. F., and Emptage, M. H. (1993) The inactivation of Fe-S cluster containing hydro-lyases by superoxide. J. Biol. Chem. 268:22369–22376.
51. Liochev, S. I., and Fridovich, I. (1994) The role of $O_2^{\bullet-}$ in the production of $OH^{\bullet}$: In vitro and in vivo. Free Radical Biol. Med. 16:29–33.
52. Biemond, P., van Eijk, H. G., Swaak, A. J. G., and Koster, J. F. (1984) Iron mobilization from ferritin by superoxide derived from phagocytosing polymorphonuclear leucocytes. Possible mechanism in inflammation diseases. J. Clin. Invest. 73:1576–1579.
53. Halliwell, B., and Gutteridge, J. M. C. (1989) Free Radicals in Biology and Medicine. Clarendon Press, Oxford.
54. Beckman, J. S., Beckman, T. W., Chen, J., Marshall, P. A., and Freeman, B. A. (1990) Apparent hydroxyl radical production by peroxynitrite: Implications for endothelial injury from nitric oxide and superoxide. Proc. Natl. Acad. Sci. USA 87:1620–1624.
55. Al-Thannon, A. A., Barton, J. P., Packer, J. E., Sims, R. J., Trumbore, C. N., and Winchester, R. V. (1974) The radiolysis of aqueous solutions of cysteine in the presence of oxygen. Int. J. Radiat. Phys. Chem. 6:233–248.
56. Barton, J. P., and Packer, J. E. (1970) The radiolysis of oxygenated cysteine solutions of neutral pH: The role of RSSR and superoxide. Int. J. Radiat. Phys. Chem. 2:159–166.
57. Asada, K., and Kanematsu, S. (1976) Reactivity of thiols with superoxide radicals. Agric. Biol. Chem. 40:1891–1892.
58. Ross, D., Cotgreave, I., and Moldeus, P. (1985) The interaction of reduced glutathione with active oxygen species generated by xanthine oxidase-catalysed metabolism of xanthine. Biochim. Biophys. Acta 841:278–282.
59. Crank, G., and Makin, M. I. H. (1984) Oxidation of thiols by superoxide ion. Aust. J. Chem. 37:2331–2337.
60. Bielski, B. H. J., and Shiue, G. G. (1979) In: Oxygen Free Radicals and Tissue Damage, pp. 43–48. Excerpta Medica, Amsterdam.
61. Zhang, N., Schuchmann, H., and von Sonntag, C. (1991) The reaction of superoxide radical anion with dithiothreitol: A chain process. J. Phys. Chem. 95:4718–4722.
62. Wefers, H., Riechmann, E., and Sies, H. (1985) Excited species generation in horseradish peroxidase-mediated oxidation of glutathione. J. Free Rad. Biol. Med. 1:311–318.
63. Ollinger, K., Buffington, G. D., Ernster, L., and Cadenas, E. (1990) Effect of superoxide dismutase on the autoxidation of substituted hydro- and semi-naphthoquinones. Chem. Biol. Interact. 73:53–76.

64. Winterbourn, C. C., Cowden, W. B., and Sutton, H. C. (1989) Auto-oxidation of dialuric acid, divicine and isouramil. Superoxide dependent and independent mechanisms. Biochem. Pharmacol. 38:611–618.
65. Munday, R. (1986) Generation of superoxide radical and hydrogen peroxide by 1,2,4-triaminobenzene, a mutagenic and myotoxic aromatic amine. Chem. Biol. Interact. 60:171–181.
66. Munday, R. (1988) Generation of superoxide radical, hydrogen peroxide and hydroxyl radical during the autoxidation of N,N,N′,N′-tetramethyl-p-phenylenediamine. Chem. Biol. Interact. 65:133–143.
67. Cohen, G., and Heikkila, R. E. (1974) The generation of hydrogen peroxide, superoxide radical and hydroxyl radical by 6-hydroxydopamine, dialuric acid, and related cytotoxic agents. J. Biol. Chem. 249:2447–2452.
68. Sullivan, S. G., and Stern, A. (1981) Effects of superoxide dismutase and catalase on catalysis of 6-hydroxydopamine and 6-aminodopamine autoxidation by iron and ascorbate. Biochem. Pharmacol. 30:2279–2285.
69. Misra, H. P., and Fridovich, I. (1972) The role of superoxide anions in the autoxidation of epinephrine and a simple assay for superoxide dismutase. J. Biol. Chem. 217:3170–3175.
70. Munday, R. (1985) Toxicity of aromatic disulphides. I. Generation of superoxide radical and hydrogen peroxide by aromatic disulphides in vitro. J. Appl. Toxicol. 5:402–408.
71. Misra, H. P., and Fridovich, I. (1976) The oxidation of phenylhydrazine: Superoxide and mechanism. Biochemistry 15:681–687.
72. Bielski, B. H. J., and Chan, P. C. (1980) Studies on free and enzyme-bound nicotinamide adenine dinucleotide free radicals. J. Am. Chem. Soc. 102:1713–1716.
73. Mashino, T., and Fridovich, I. (1987) Superoxide radical initiates the autoxidation of dihydroxyacetone. Arch. Biochem. Biophys. 254:547–551.
74. Hunter, E. P. L., Desrosiers, M. F., and Simic, M. G. (1989) The effect of oxygen, antioxidants, and superoxide radical on tyrosine phenoxyl radical dimerization. Free Rad. Biol. Med. 6:581–585.
75. Winterbourn, C. C., Metodiewa, D., and Pichorner, H. Generation and fate of superoxide generated from peroxidase-catalysed tyrosine oxidation in the presence of glutathione (submitted).
76. Jin, F., Leitich, J., and von Sonntag, C. (1993) The superoxide radical reacts with tyrosine-derived phenoxyl radicals by addition rather than by electron transfer. J. Chem. Soc. Perkin Trans. II:1583–1588.
77. Wolff, S. P. (1994) Ferrous ion oxidation in presence of ferric ion indicator xylenol orange for measurement of hydroperoxides. Methods Enzymol. 233:182–189.
78. Freisleben, H. J., and Packer, L. (1993) Free radical scavenging activities, interactions and recycling of antioxidants. Biochem. Soc. Trans. 21:325–330.
79. Liebler, D. C. (1993) The role of metabolism in the antioxidant function of vitamin E. Crit. Rev. Toxicol. 23:147–169.
80. Buettner, G. R. (1993) The pecking order of free radicals and antioxidants: Lipid peroxidation, α-tocopherol, and ascorbate. Arch. Biochem. Biophys. 300:535–543.

81. Winterbourn, C. C. (1981) Cytochrome c reduction by semiquinone radicals can be indirectly inhibited by superoxide dismutase. Arch. Biochem. Biophys. 209:159–167.
82. Bielski, B. H. J. (1982) Chemistry of ascorbic acid radicals. In: Ascorbic Acid: Chemistry, Metabolism and Uses (Seib, P. A., and Tolbert, B. M., eds.), pp. 81–100. American Chemical Society, Washington, DC.
83. Villalba, J. M., Canalejo, A., Burón, M. I., Córdoba, F., and Navas, P. (1993) Thiol groups are involved in NADH-ascorbate free radical reductase activity of rat liver plasma membrane. Biochem. Biophys. Res. Commun. 192:707–713.
84. Niki, E., Tsuchiya, J., Tanimura, R., and Kamiya, Y. (1982) Regeneration of vitamin E from α-chromanoxyl radical by glutathione and vitamin C. Chem. Lett. 789–792.
85. Bisby, R. H., and Parker, A. W. (1991) Reactions of the alpha-tocopherol radical in micellar solutions studied by nanosecond laser flash photolysis. FEBS Lett. 290:205–208.
86. Mårtensson, J., Jain, A., Stole, E., Frayer, W., Auld, P. A. M., and Meister, A. (1991) Inhibition of glutathione synthesis in the newborn rat: A model for endogenously produced oxidative stress. Proc. Natl. Acad. Sci. USA 88:9360–9364.
87. Staal, F. J. T., Roederer, M., and Herzenberg, L. A. (1991) Intracellular thiols regulate activation of nuclear factor κB and transcription of human immunodeficiency virus. Proc. Natl. Acad. Sci. USA 87:9943–9947.
88. Schreck, R., Meier, B., Männel, D. N., Dröge, W., and Baeuerle, P. A. (1992) Dithiocarbamates as potent inhibitors of nuclear factor κB activation in intact cells. J. Exp. Med. 175:1181–1194.
89. Schreck, R., Rieber, P., and Baeuerle, P. A. (1991) Reactive oxygen intermediates as apparently widely used messengers in the activation of the NF-κB transcription factor and HIV-1. EMBO J. 10:2247–2258.
90. Schenk, H., Klein, M., Erdbrügger, W., Dröge, W., and Schulze-Osthoff, K. (1994) Distinct effects of thioredoxin and antioxidants on the activation of transcription factors NF-κB and AP-1. Proc. Natl. Acad. Sci. USA 91:1672–1676.
91. Abate, C., Patel, L., Rauscher III, F. J., and Curran, T. (1990) Redox regulation of Fos and Jun DNA-binding activity in vitro. Science 249:1157–1161.
92. Ivanov, V., Merkenschlager, M., and Ceredig, R. (1993) Antioxidant treatment of thymic organ cultures decreases NF-κB and TCF1(α) Transcription factor activities and inhibits αβ T Cell development. J. Immunol. 151:4694–4704.
93. Buttke, T. M., and Sandstrom, P. A. (1994) Oxidative stress as a mediator of apoptosis. Immunol. Today 15:7–10.
94. Kane, D. J., Sarafian, T. A., Anton, R., Hahn, H., Gralla, E. B., Valentine, J. S., Ord, T., and Bredesen, D. E. (1993) Bcl-2 inhibition of neural death: Decreased generation of reactive oxygen species. Science 262:1274–1277.
95. Brigelius, R. (1985) Mixed disulfides: Biological functions and increase in oxidative stress. In: Oxidative Stress (Sies, H., ed.), pp. 243–272. Academic Press, London.
96. Claiborne, A., Miller, H., Parsonage, D., and Ross, R. P. (1993) Protein-sulfenic acid stabilization and function in enzyme catalysis and gene regulation. FASEB J. 7:1483–1490.

97. Cappel, R. E., and Gilbert, H. F. (1984) Thiol/disulfide exchange between 3-hydroxy-3-methyglutaryl-CoA reductase and glutathione. J. Biol. Chem. 263:12204–12212.
98. Held, K. D., Harrop, H. A., and Michael, B. D. (1984) Effect of oxygen and sulphydryl-containing compounds on irradiated transforming DNA. II. Glutathione, cysteine and cysteamine. Int. J. Radiat. Biol. 45:615–626.
99. Ramakrishna Rao, D. N., Fischer, V., and Mason, R. P. (1990) Glutathione and ascorbate reduction of the acetaminophen radical formed by peroxidase. J. Biol. Chem. 265:844–847.
100. D'Arcy Doherty, M., Wilson, I., Wardman, P., Basra, J., Patterson, L. H., and Cohen, G. M. (1986) Peroxidase activation of 1-naphthol to naphthoxy or naphthoxy-derived radicals and their reaction with glutathione. Chem. Biol. Interact. 58:199–215.
101. Moldeus, P., O'Brien, P. J., Thor, H., Berggren, M., and Orrenius, S. (1983) Oxidation of glutathione by free radical intermediates formed during peroxidase-catalyzed *N*-demethylation reactions. FEBS Lett. 162:411–415.
102. Wilson, I., Wardman, P., Cohen, G. M., and D'Arcy Doherty, M. (1986) Reductive role of glutathione in the redox cycling of oxidizable drugs. Biochem. Pharmacol. 35:21–22.
103. Munday, R., Manns, E., Fowke, E. A. and Hoggard, G. K. (1990) Structure-activity relationships in the myotoxicity of ring-methylated *p*-phenylenediamines in rats and correlation with autoxidation rates in vitro. Chem. Biol. Interact. 76:31–45.
104. Forni, L. G., and Willson, R. L. (1986) Thiyl and phenoxyl free radicals and NADH. Biochem. J. 240:897–903.
105. Forni, L. G., Mönig, J., Mora-Arellano, V. O., and Willson, R. L. (1983) Thiyl free radicals: Direct observations of electron transfer reactions with phenothiazines and ascorbate. J. Chem. Soc. Perkin Trans. II:961–965.
106. Schöneich, C., Asmus, K.-D., Dillinger, U., and Bruchhausen, F. (1989) Thiyl radical attack on polyunsaturated fatty acids: A possible route to lipid peroxidation. Biochem. Biophys. Res. Commun. 161:113–120.
107. Schoneich, C., Dillinger, U., von Bruchhausen, F., and Asmus, K. D. (1992) Oxidation of polyunsaturated fatty acids and lipids through thiyl and sulfonyl radicals: Reaction, kinetics, and influence of oxygen and structure of thiyl radicals. Arch. Biochem. Biophys. 292:456–467.
108. D'Aquino, M., Dunster, C., and Willson, R. L. (1989) Vitamin A and glutathione-mediated free radical damage: Competing reactions with polyunsaturated fatty acids and vitamin C. Biochem. Biophys. Res. Commun. 161:1199–1203.

7

Glutathione-Dependent Bioactivation of Xenobiotics

M. W. Anders
University of Rochester, Rochester, New York

Wolfgang Dekant and Spyridon Vamvakas
Institut für Toxikologie, Universität Würzburg, Würzburg, Germany

I. INTRODUCTION

The glutathione-dependent biotransformation of xenobiotics is historically associated with the detoxication of xenobiotics. The mercapturic acid pathway is an important mechanism for the detoxification of a range of electrophilic compounds or metabolites. Mercapturic acid formation involves glutathione *S*-transferase–catalyzed glutathione S-conjugate formation, hydrolysis of the glutathione *S*-conjugates to the corresponding cysteine *S*-conjugates, and *N*-acetyltransferase–catalyzed *N*-acetylation to give *S*-substituted *N*-acetyl-L-cysteine conjugates (1). The mercapturic acids thus formed are highly polar and are excreted in the urine. Although mercapturic acid formation usually leads to the formation of products of low toxicity, aminoacylase-catalyzed deacetylation of mercapturic acids may afford cysteine *S*-conjugates that undergo cysteine conjugate β-lyase–dependent bioactivation (2). *N*-Acetyl-*S*-(3-oxopropyl)-L-cysteine, the mercapturic acid derived from acrolein, and *N*-acetyl-*S*-(3-chloro-2-propenyl)-L-cysteine, the mercapturic acid formed by the glutathione-dependent biotransformation of 1,3-dichloropropene, apparently undergo flavoprotein monooxygenase-catalyzed sulfoxidation to reactive, cytotoxic intermediates (3,4).

In addition to its role in xenobiotic detoxification, studies conducted over the past decade have provided evidence that glutathione conjugate formation is also an important bioactivation mechanism for several groups of compounds, including haloalkanes and haloalkenes. This review will focus on the glutathione-dependent bioactivation of haloalkanes and haloalkenes. It should be noted that a role for thiols, including glutathione, in the bioactivation of a range of drugs and chemicals, including organic isothiocyanates (5,6), quinones (7), and *N*-methyl-*N′*-nitro-*N*-nitrosoguanidine (8), has been identified. Reviews about the glutathione-dependent bioactivation of xenobiotics have appeared (7,9–13).

II. GLUTATHIONE-DEPENDENT BIOACTIVATION OF HALOALKANES

Glutathione *S*-transferase–catalyzed glutathione *S*-conjugate formation is the initial step in the glutathione-dependent bioactivation of haloalkanes, but the conjugates formed are unstable and give rise to toxic metabolites. With dihalomethanes, formaldehyde is formed as a product, whereas with vicinal dihaloalkanes, half-sulfur mustards, which give rise to episulfonium ions, are formed.

A. Dihalomethanes

Dichloromethane is the most widely used dihalomethane and enjoys wide use as a solvent, degreaser, and paint remover and in the manufacture of photographic film. Carbon monoxide is formed by the cytochrome P-450–dependent biotransformation of dihalomethanes (14–16), and formaldehyde is formed by the glutathione *S*-transferase–catalyzed biotransformation of dihalomethanes (17,18). In addition, the bacterial mutagenicity of dihalomethanes is associated with glutathione-dependent bioactivation (19–21). Pharmacokinetic investigations demonstrated that dichloromethane is metabolized by both oxidative and conjugative pathways: a high-affinity, low-capacity oxidative pathway is associated with cytochrome P-450–dependent metabolism, and a low-affinity, high-capacity conjugative pathway is associated with glutathione *S*-transferase–dependent metabolism (22). The tumorigenicity of dichloromethane in mice is associated with glutathione-dependent bioactivation (23,24).

1. Bioactivation Mechanism of Dihalomethanes

The first step in the bioactivation of dihalomethanes (**1**, Fig. 1) is a glutathione *S*-transferase–catalyzed S_N2 attack of glutathione to give *S*-(halomethyl)glutathione (**2**, Fig. 1) (18). Halomethyl sulfides are subject to hydrolysis (25), and *S*-(halomethyl)glutathione affords *S*-(hydroxymethyl)glutathione (**3**, Fig. 1), which is the hemithioacetal formed by reaction of formaldehyde (**4**, Fig. 1) and glutathione (26,27). *S*-(Hydroxymethyl)glutathione is a substrate for the NAD^+-dependent formaldehyde dehydrogenase and gives *S*-formylglutathione (**5**, Fig. 1) as a product. *S*-Formylglutathione is a substrate for *S*-formylglutathione hydrolase, and formic acid (**6**, Fig. 1) and glutathione are formed as products (18,28). Hence the glutathione-dependent bioactivation of dihalomethanes requires glutathione, but glutathione is not irreversibly consumed in the reaction.

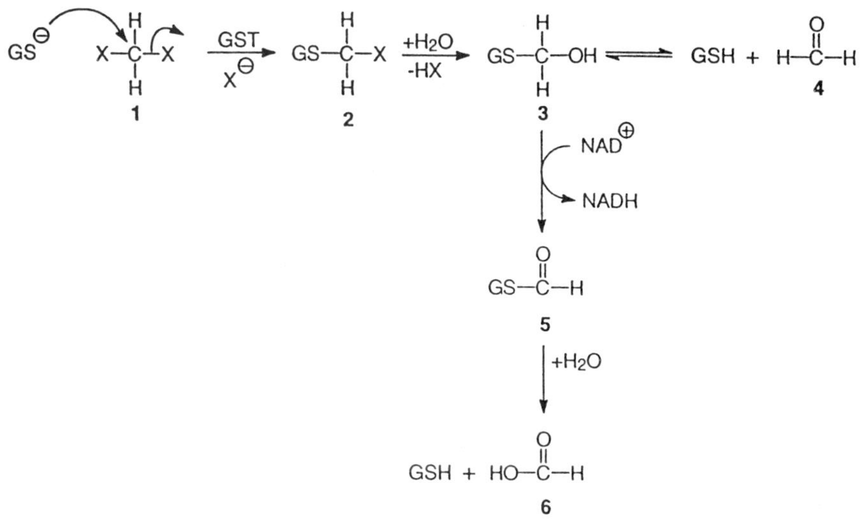

Figure 1 Glutathione-dependent bioactivation of dihalomethanes **1**, Dihalomethane (X = Cl, Br); **2**, *S*-(halomethyl)glutathione; **3**, *S*-(hydroxymethyl)glutathione; **4**, formaldehyde; **5**, *S*-formylglutathione; **6**, formic acid. GSH, Glutathione; GST, glutathione *S*-transferase; FD, formaldehyde dehydrogenase; FGH, *S*-formylglutathione hydrolase.

The biotransformation of [^{13}C]dichloromethane has recently been studied with ^{13}C-NMR (29). These studies identified *S*-(hydroxymethyl)glutathione and formaldehyde as products, but no adducts derived from the attack of nucleophiles on *S*-(chloromethyl)glutathione were detected. Methanol was detected as a minor metabolite and apparently arises from the attack of glutathione on *S*-(hydroxymethyl)glutathione.

2. *Biological Effects of Dihalomethanes*

As mentioned above, dichloromethane is tumorigenic in mice, but not in rats and hamsters, and its tumorigenicity is associated with glutathione-dependent bioactivation (23,24,30). The mechanism by which dichloromethane induces tumors in mice has not been elucidated. The frequency and profile of H-*ras* codon 61 mutations and K-*ras* mutations were the same in spontaneous and dichloromethane-induced liver and lung tumors, respectively (31). In addition, inactivation of the *p53* and retinoblastoma tumor suppressor genes is an infrequent event in lung and liver tumors from dichloromethane-exposed mice (32). Finally, sustained, enhanced cell proliferation is not detectable in lung and liver tissues of dichloromethane-exposed mice (33).

The rate of metabolism of dichloromethane by glutathione conjugation is much higher in mice than in other species. Evidence for the formation of DNA adducts in dichloromethane-treated animals is lacking, although formaldehyde, formed as a metabolite of dichloromethane, may participate in the production of DNA-protein crosslinks (34). The differences in the mutagenicity of dichloromethane in several strains of bacteria support a role for formaldehyde in the genotoxicity of dichloromethane (35).

Other studies, however, also suggest a role for a reactive glutathione conjugate. When dichloromethane-induced DNA-damage was measured in freshly isolated rat and mouse hepatocytes, mouse hepatocytes were much more sensitive to dichloromethane than were rat hepatocytes. Moreover, the concentration dependency of induction of DNA single-strand breaks in mouse hepatocytes indicates a role for S-(chloromethyl)glutathione as a DNA-damaging metabolite in mice (36), and the DNA adduct expected to be formed from *S*-(chloromethyl)glutathione, *S*-[1-(N^2-deoxyguanosinyl)methyl] glutathione, is stable (37).

The dose-dependent incorporation of [^{14}C]dichloromethane-derived radioactivity into lung and liver DNA in mice, but not in hamsters, indicates that species differences in the bioactivation of dichloromethane may be the basis for species differences in dichloromethane-induced tumorigenicity (23,38–41).

B. Vicinal Dihaloalkanes

Vicinal dihaloalkanes, such as 1,2-dichloroethane, 1,2-dibromoethane, and 1,2-dibromo-3-chloropropane, have enjoyed wide use. 1,2-Dibromoethane and 1,2-dibromo-3-chloropropane were formerly used extensively as soil fumigants and nematocides. 1,2-Dichloroethane and 1,2-dibromoethane are mutagenic and tumorigenic, and their use has been curtailed. 1,2-Dibromo-3-chloropropane produces testicular atrophy, oligospermia, and infertility in exposed males and is no longer used commercially.

The cytochrome P-450–dependent biotransformation of 1,2-dichloroethane and 1,2-dibromoethane gives chloroacetaldehyde and bromoacetaldehyde as products (42–44), but the mutagenicity of vicinal dihaloalkanes is associated with glutathione-dependent biotransformation (45,46), as will be discussed below. The glutathione-dependent metabolism of 1,2-dibromoethane yields *S*-(2-hydroxyethyl)glutathione, *S*-(2-hydroxyethyl)-L-cysteine, and *S*-(2-hydroxyethyl)-*N*-acetyl-L-cysteine, and their sulfoxides, as excretory products (47,48). In addition, the glutathione-dependent biotransformation of 1,2-dichloroethane also affords ethene and glutathione disulfide as products (49).

The bacterial mutagenicity of vicinal dihaloethanes is associated with their glutathione-dependent bioactivation (see Sec. II.B.2).

1. *Bioactivation Mechanism of Vicinal Dihaloalkanes*

The bioactivation of vicinal dihaloalkanes involves a glutathione *S*-transferase–catalyzed S_N2 attack of glutathione to give the half-sulfur mustard *S*-(2-haloalkyl) glutathione (**1**, Fig. 2). The half-sulfur mustards thus formed may undergo an

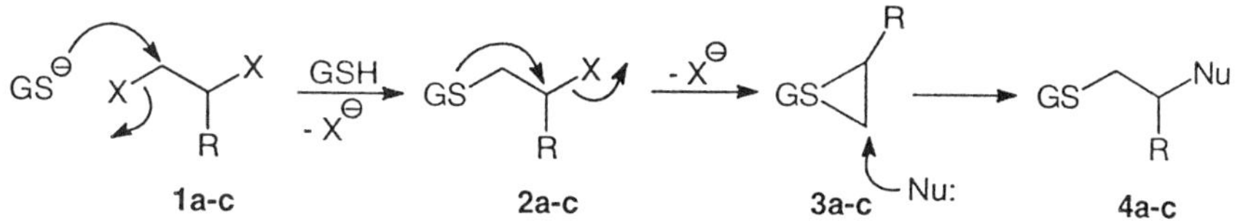

Figure 2 Glutathione-dependent bioactivation of vicinal dihaloalkanes. **1a**, 1,2-Dichloroethane (X = Cl, R = H); **1b**, 1,2-dibromoethane (X = Br, R = H); **1c**, 1,2-dibromo-3-chloropropane (X = Br, R = —CH_2Cl); **2a**, *S*-(2-chloroethyl)glutathione (X = Cl, R = H); **2b**, *S*-(2-bromoethyl)glutathione (X = Br, R = H); **2c**, *S*-(2-bromo-3-chloropropyl)glutathione (X = Br, R = —CH_2Cl); **3a–c**, episulfonium ions derived from **2a–c; 4a–c**, adducts of **3a–c** formed by reaction with DNA, protein nucleophiles, or water.

intramolecular displacement reaction to give an episulfonium (thiiranium) ion (**3**, Fig. 2), which may react with cellular nucleophiles to produce covalently bound adducts (**4**, Fig. 2).

There is considerable evidence for this reaction pathway in the bioactivation and toxicity of vicinal dihaloalkanes. For example, half-sulfur mustards are intermediates in the biotransformation of vicinal dihaloalkenes to ethene (50). Also, *S*-(2-chloroethyl)-DL-cysteine is nephrotoxic in rats, but the analogs *S*-(3-chloropropyl)-DL-cysteine and *S*-(2-hydroxyethyl)-DL-cysteine, which do not readily form episulfonium ions, are not nephrotoxic (51). NMR spectroscopic studies have provided direct evidence for episulfonium ion formation from *S*-(2-haloethyl)-L-cysteines (52).

In addition to vicinal dihaloethanes, analogous vicinal dihaloalkanes also undergo glutathione-dependent bioactivation. 1,2-Dibromo-3-chloropropane (**1c**, Fig. 2) is mutagenic and carcinogenic and is selectively toxic to the kidney and testes (53). The testicular genotoxicity and the nephrotoxicity of 1,2-dibromo-3-chloropropane are associated with glutathione-dependent reactions (54,55). Although the DNA-damaging effects of 1,2-dibromo-3-chloropropane in rats are due to glutathione-dependent bioactivation, the bacterial mutagenicity of 1,2-dibromo-3-chloropropane is associated with its cytochrome P-450–dependent bioactivation (56). Studies on the glutathione-dependent bioactivation of 1,2-dibromo-3-chloropropane show that the first step in the reaction is the formation of *S*-(2-bromo-3-chloropropyl)glutathione (**2c**, Fig. 2) with the subsequent formation of an episulfonium ion (**3c**, Fig. 2). The episulfonium ion may react with DNA, protein nucleophiles, or water (57).

2. *Biological Effects of Vicinal Dihaloalkanes*

1,2-Dibromoethane is carcinogenic in several species, producing tumors at multiple sites including liver, lung, stomach, breast, skin, and kidney. Although 1,2-dibromoethane is metabolized extensively by oxidative pathways, it is now established that the binding of 1,2-dibromoethane metabolites to DNA as well as its mutagenicity and carcinogenicity are associated with glutathione-dependent bioactivation. The conjugation of 1,2-dibromoethane with glutathione yields the half-mustard *S*-(2-bromoethyl)glutathione, which cyclizes to form a reactive episulfonium ion intermediate. The identification of *S*-[2-(N^7-guanyl)ethyl]glutathione both in vivo and in vitro indicates a role for the glutathione-dependent bioactivation pathway in the toxicity and carcinogenicity of 1,2-dibromoethane. In a recent study (58), a forward mutation assay utilizing the bacteriophage M13 lac Z gene and regulatory region was carried out to assess the importance of the *S*-[2-(N^7-guanyl)ethyl]glutathione adduct in the bacterial mutagenicity of the glutathione-conjugate of 1,2-dibromoethane and to gain insight into the mechanisms of carcinogenicity of the parent compound. The results strongly implicated *S*-[2-(N^7-guanyl)ethyl]glutathione as a mutagenic lesion that gives rise to a G:C $\rightarrow$ A:T transition.

1,2-Dichloroethane is both hepatotoxic and nephrotoxic and produces tumors at multiple sites in long-term bioassays. The bioactivation of 1,2-dichloroethane also involves conjugation with glutathione to form *S*-(2-chloroethyl)glutathione. There is, however, a subtle difference in the bioactivation mechanisms of 1,2-dibromoethane and 1,2-dichloroethane. With 1,2-dichloroethane, both *S*-(2-chloroethyl)glutathione and the cysteine analog *S*-(2-chlorethyl)cysteine appear to be involved in the induction of toxicity and mutagenicity. The cysteine conjugate of 1,2-dichloroethane has more alkylating power towards the N^7-position of deoxyguanosine than the corresponding glutathione conjugate, although the latter is a more potent mutagen in *Salmonella typhimurium* TA 100 (59,60). Both the cysteine and glutathione conjugates of 1,2-dichloroethane are cytotoxic and genotoxic in hepatocytes and in renal proximal tubule cells (51,61).

Glutathione-dependent bioactivation also appears to be responsible for the renal and testicular toxicity of 1,2-dibromo-3-chloropropane. The lack of significant isotope effects with perdeutero-1,2-dibromo-3-chloropropane indicates that carbon-hydrogen bond breakage is not the rate-limiting step in 1,2-dibromo-3-chloropropane–induced renal and testicular damage. Moreover, renal and testicular necrosis and the ability of 1,2-dibromo-3-chloropropane to induce DNA damage are not dependent on cytochrome P-450, but rather display a requirement for glutathione (56). These observations indicate that some adverse effects of 1,2-dibromo-3-chloropropane are associated with a glutathione-dependent pathway and the formation of a reactive episulfonium ion, which may react with DNA to form *S*-[1-(hydroxymethyl)-2-(N^7-guanyl)ethyl]glutathione and *S*-[bis(N^7-guanyl)methyl]glutathione (62).

Finally, tris(2,3-dibromopropyl)phosphate is nephrotoxic in animals after acute dosage and is a selective renal carcinogen in long-term bioassays. The mechanisms underlying tris(2,3-dibromopropyl)phosphate have been partially elucidated. Incubation of tris(2,3-dibromopropyl)phosphate with cytosol, [^{35}*S*]glutathione, and DNA results in [^{35}*S*]glutathione-binding to DNA (63). The structures of the glutathione conjugates found in rats given tris(2,3-dibromopropyl)phosphate also support the involvement of episulfonium ion formation. Alternatively (or additionally) oxidative biotransformation by cytochrome P-450 to 2-bromoacrolein may play a role in the mutagenicity of tris(2,3-dibromopropyl)phosphate.

III. GLUTATHIONE-DEPENDENT BIOACTIVATION OF HALOALKENES: CYSTEINE CONJUGATE β-LYASE PATHWAY

Studies conducted over the past decade have elaborated a multiorgan, multistep pathway for the bioactivation of haloalkenes that explains their selective nephrotoxicity and, perhaps, nephrocarcinogenicity. This pathway involves the hepatic formation of glutathione *S*-conjugates of haloalkenes, hydrolysis of glutathione *S*-conjugates to give cysteine *S*-conjugates, translocation of cysteine *S*-conjugates to the kidney, uptake of cysteine *S*-conjugates by amino acid transport systems, and bioactivation by cysteine conjugate β-lyase. Reviews about the cysteine conjugate β-lyase pathway have appeared (9,10,12,64–67).

A. Bioactivation Mechanism of Haloalkenes

1. *Biosynthesis of Glutathione S-Conjugates*

The liver is the primary site of glutathione *S*-conjugate formation, and glutathione *S*-transferase activity is present in hepatic cytosolic, microsomal, and mitochondrial fractions. The reaction of haloalkenes with glutathione is preferentially catalyzed by the microsomal glutathione *S*-transferase (EC 2.5.1.18), whose activities are highest in the liver (68). The microsomal transferase is unrelated in protein structure to the cytosolic or mitochondrial transferases (69).

The microsomal glutathione *S*-transferase catalyzes the reaction of haloalkenes with glutathione, and two reaction mechanisms are observed: 1,1-dichloroalkenes (**1**, Fig. 3) undergo addition-elimination reactions to give *S*-(1-chloro-1-alkenyl) glutathione conjugates (**2**, Fig. 3). 1,1-Difluoroalkenes (**3**, Fig. 3) undergo addition reactions to afford *S*-(1,1-difluoroalkylglutathione conjugates (**4**, Fig. 3) (70). Dichloroethyne (**5**, Fig. 3) is also a substrate for the microsomal glutathione *S*-transferases and undergoes an addition reaction to give *S*-(1,2-dichlorovinyl)glutathione (**6**, Fig. 3) as a product (71).

1,1-Difluoroalkenes undergo regiospecific, microsomal glutathione *S*-transferase–catalyzed biotransformation to *S*-(1,1-difluoroalkyl)glutathione conjugates (7). Tetrafluoroethene is biotransformed to *S*-(1,1,2,2-tetrafluoroethyl)glutathione (72), 2,2-dichloro-1,1-difluoroethene is biotransformed to *S*-(2,2-dichloro-1,1-difluoroethyl)glutathione (73), and 2-bromo-2-chloro-1,1-difluoroethene is metabolized to *S*-(2-bromo-2-chloro-1,1-difluoroethyl)glutathione (74,75).

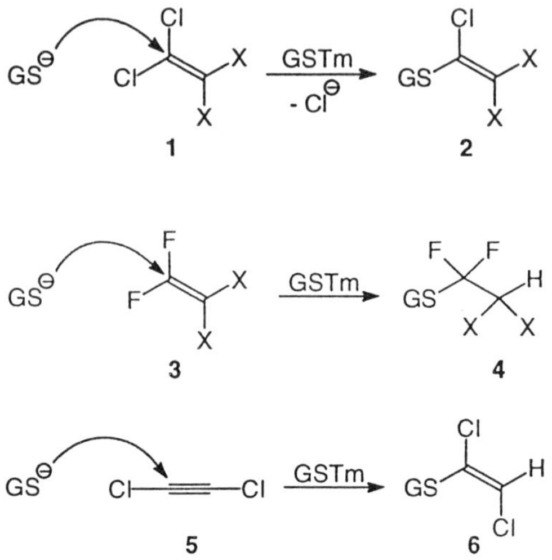

Figure 3 Glutathione *S*-transferase–catalyzed reaction of haloalkenes with glutathione. **1**, 1,1-Dichloroalkene; **2**, *S*-(1-chloro-1-alkenyl)glutathione; **3**, 1,1-difluoroalkene; **4**, *S*-(1,1-difluoroalkyl)glutathione; **5**, dichloroethyne; **6**, *S*-(1,2-dichlorovinyl)glutathione. GSTm, Microsomal glutathione *S*-transferase.

Chlorotrifluoroethene undergoes regiospecific and stereoselective biotransformation to *S*-(2-chloro-1,1,2-trifluoroethyl)glutathione (76). The conjugate formed with the microsomal glutatione *S*-transferase as the catalyst contains a new chiral center that is enriched in the 2*S*-diastereoisomer, whereas the product formed by the cytosolic glutathione *S*-transferase contains equal amounts of the 2*S*- and 2*R*-diastereoisomers (77); this observation was exploited to demonstrate that conjugate formation in isolated rat hepatocytes is catalyzed preferentially by the microsomal glutathione *S*-transferase. Finally, hexafluoropropene differs from other 1,1-difluoroalkenes studied in that it undergoes a microsomal glutathione *S*-transferase–catalyzed addition-elimination reaction to give *S*-(1,2,3,3,3-pentafluoro-1-propenyl)glutathione (78).

2. *Fate of Glutathione S-Conjugates*

Glutathione *S*-conjugates are primarily eliminated from the liver in the bile (see Ref. 67 for a recent review about the physiological disposition of toxic glutathione *S*-conjugates). The canalicular, ATP-dependent transport system accepts a range of polar, high molecular–weight compounds, including glutathione *S*-conjugates, as substrates (79–81). Although sinusoidal transport of glutathione *S*-conjugates occurs, this system has a low affinity for glutathione *S*-conjugates (80). Studies with hexachlorobutadiene in the isolated perfused rat liver show that *S*-(1,2,3,4,4-pentachlorobutadienyl)glutathione biosynthesized in the liver is excreted in the bile and that release into the caval effluent occurs only after perfusion with toxic concentrations of hexachlorobutadiene (82). Moreover, cannulation of the bile duct protects rats from hexachlorobutadiene-induced nephrotoxicity (83), illustrating the importance of biliary excretion of glutathione *S*-conjugates.

After elimination from the liver in the bile, glutathione *S*-conjugates are either absorbed intact from the small intestine or are absorbed as cysteine *S*-conjugates after sequential hydrolysis by γ-glutamyltransferase (EC 2.3.3.2.2) and aminopeptidase M (EC 3.4.11.2) or cysteinylglycine dipeptidase (EC 3.4.13.6). γ-Glutamyltransferase is a heterodimeric serine hydrolase that catalyzes the transfer of the glutamyl group from γ-glutamyl di- and tripeptides to acceptor amino acids or dipeptides (84). With glutathione *S*-conjugates, the products are L-cysteinylglycine *S*-conjugates and L-γ-glutamyl amino acids or dipeptides (Fig. 4). γ-Glutamyltransferase also catalyzes the hydrolysis of glutathione *S*-conjugates to give L-glutamate and L-cysteinylglycine *S*-conjugates as products (85). γ-Glutamyltransferase is an ectoenzyme present in many epithelial tissues, including the bile duct, intestinal microvilli, brush-border membranes of the renal proximal tubules, and in the vasculature and basal-lateral membranes of the kidney. Selective inhibitors of γ-glutamyltransferase include L-serine/borate, which is a competitive, transition-state inhibitor (86), and L-(αS,5S)-α-amino-3-chloro-4,5-dihydro-5-isoxazoleacetic acid (acivicin, AT-125), which is an irreversible inhibitor (87). Aminopeptidase M is a homodimeric, zinc-dependent hydrolase that catalyzes the hydrolysis of N-terminal amino acids from di- and oligopeptides (88). As with γ-glutamyltransferase, aminopeptidase M activity is high in epithelial tissues, including the kidney. Aminopeptidase M is inhibited by 1,10-phenanthroline and bestatin (89). Cysteinylglycine dipeptidase is a homodimeric, zinc-dependent hydrolase that catalyzes the hydrolysis of a range of dipeptides, including *S*-substituted cysteinylglycines (90,91). Cysteinylglycine dipeptidase is inhibited by 1,10-phenanthroline, but not by bestatin. Both 1,10-phenanthroline and phenylalanylglycine block the cytotoxicity of *S*-(1,2-dichlorovinyl)-L-cysteinylglycine, indicating that hydrolysis to the cysteine conjugate is required for expression of toxicity (92).

When *S*-(1,2,3,4,4-pentachlorobutadienyl)glutathione was infused into the lumen of the rat intestine via the bile duct, both the intact glutathione conjugate and *S*-(1,2,3,4,4-pentachlorobutadienyl)-L-cysteine were present in portal blood, although the concentration of the cysteine conjugate was higher than that of the glutathione conjugate (93). The fate of circulating glutathione and cysteine *S*-conjugates is not well understood, although interorgan biotransformation, transport, and elimination are likely involved (94). Cysteine *S*-conjugates may be taken up by the liver and converted to the corresponding mercapturates (*S*-substituted *N*-acetyl-L-cysteines) by hepatic *N*-acetyltransferase (95). *S*-(2,4-Dinitro-

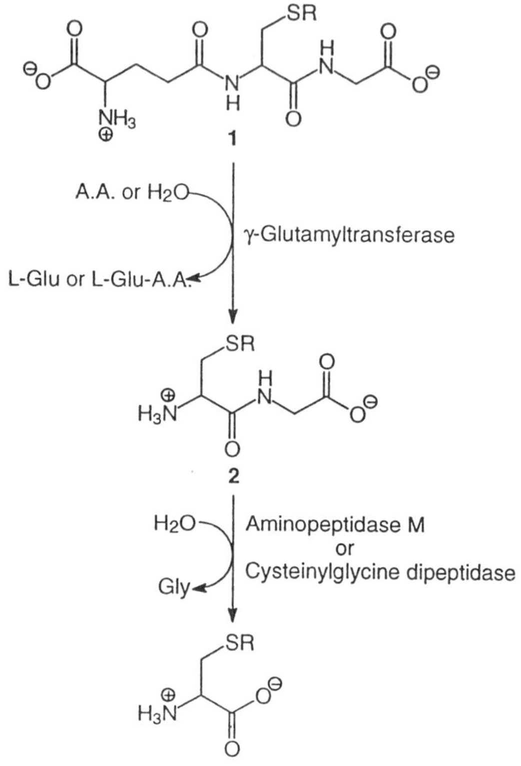

Figure 4 Enzymatic processing of glutathione *S*-conjugates. **1**, Glutathione *S*-conjugate; **2**, L-cysteinylglycine *S*-conjugate; **3**, cysteine *S*-conjugate. A.A., Amino acid or dipeptide.

phenyl)glutathione may be cleared from the circulation by a γ-glutamyltransferase–dependent process or by uptake of the intact conjugate followed by transport into bile, but these mechanisms may not be applicable to all glutathione conjugates (96).

Circulating glutathione *S*-conjugates that are delivered to the kidney may be removed from the plasma by filtering or nonfiltering mechanisms. Filtered glutathione *S*-conjugates may be biotransformed to the corresponding cysteine *S*-conjugates by γ-glutamyltransferase and aminopeptidase M or cysteinylglycine dipeptidase and the cysteine *S*-conjugates taken up by amino acid transport systems (97). Nonfiltering mechanisms may be quantitatively more important than filtering mechanisms for uptake of *S*-conjugates in the kidney. Glutathione *S*-conjugates present in the peritubular circulation may be metabolized to cysteine *S*-conjugates by γ-glutamyltransferase present in the renal vasculature or basolateral membranes (98). The probenecid-sensitive organic-anion transport system, which is located on the basolateral side of renal proximal tubular cells, appears to play a role in the accumulation of *S*-conjugates in the kidney (99). Probenecid partially blocks the nephrotoxicity of some cysteine *S*-conjugates (100,101), indicating a role for uptake by a probenecid-sensitive transporter.

3. *Bioactivation of Cysteine S-Conjugates by Cysteine Conjugate β-Lyase*

Several pyridoxal phosphate–dependent enzymes catalyze the elimination of leaving groups from the β-carbon of amino acids. The enzymatic metabolism of cysteine *S*-conjugates to pyruvate, ammonia, and thiols has been observed in several strains of bacteria and in rat, rabbit, and human liver and kidney. The C-S lyase was purified from rat liver and kidney and termed cysteine conjugate β-lyase (EC 4.4.1.13) (102–113). Subsequent studies demonstrated that hepatic β-lyase is attributable to kynurenine aminotransferase and that the renal cytosolic β-lyase is identical with glutamine transaminase K. Hepatic β-lyase does not cross-react with an antibody to the renal enzyme, and only a small fraction of renal β-lyase activity can be accounted for as kynureninase. β-Lyase activity is also present in the mitochondrial outer membrane and matrix, and the mitochondrial β-lyase is similar to the renal cytosolic glutamine transaminase K.

The role of metabolism of cysteine *S*-conjugates in *S*-conjugate-induced toxicity has been established by the use of aminooxyacetic acid, an inhibitor of pyridoxal phosphate–dependent enzymes, and by α-keto acids, such as α-keto-γ-methiolbutyrate, which enhance both renal and cytosolic β-lyase activities (114,115) and increase the toxicity of *S*-haloalkenyl and *S*-haloalkyl cysteine *S*-conjugates. Moreover, α-methyl analogs of nephrotoxic cysteine *S*-conjugates, which cannot be cleaved by β-lyase, are not toxic (75,101,116).

These results suggest that the thiol metabolites formed by enzymatic cleavage of cysteine *S*-conjugates and their interaction with cellular macromolecules are responsible for *S*-conjugate–induced toxicity. Thus, the chemical structure and, hence, the chemical reactivity of the thiols formed may control the expression of cysteine *S*-conjugate toxicity. Several different types of unstable and potentially toxic thiols are formed by cysteine conjugate β-lyase–dependent metabolism.

Cysteine *S*-conjugates biosynthesized from haloalkenes are cleaved by cysteine conjugate β-lyase to give unstable α-halothiolates (Fig. 5). α-Fluoroethanethiolates are products of the cysteine conjugate β-lyase–catalyzed biotransformation of the fluoroalkene-derived conjugates *S*-(2-chloro-1,1,2-trifluoroethyl)-L-cysteine and of *S*-(1,1,2,2-tetrafluoroethyl)-L-cysteine (117,118) (**2**, Fig. 5). α-Fluoroethanethiolates can be trapped in model systems mimicking β-lyase

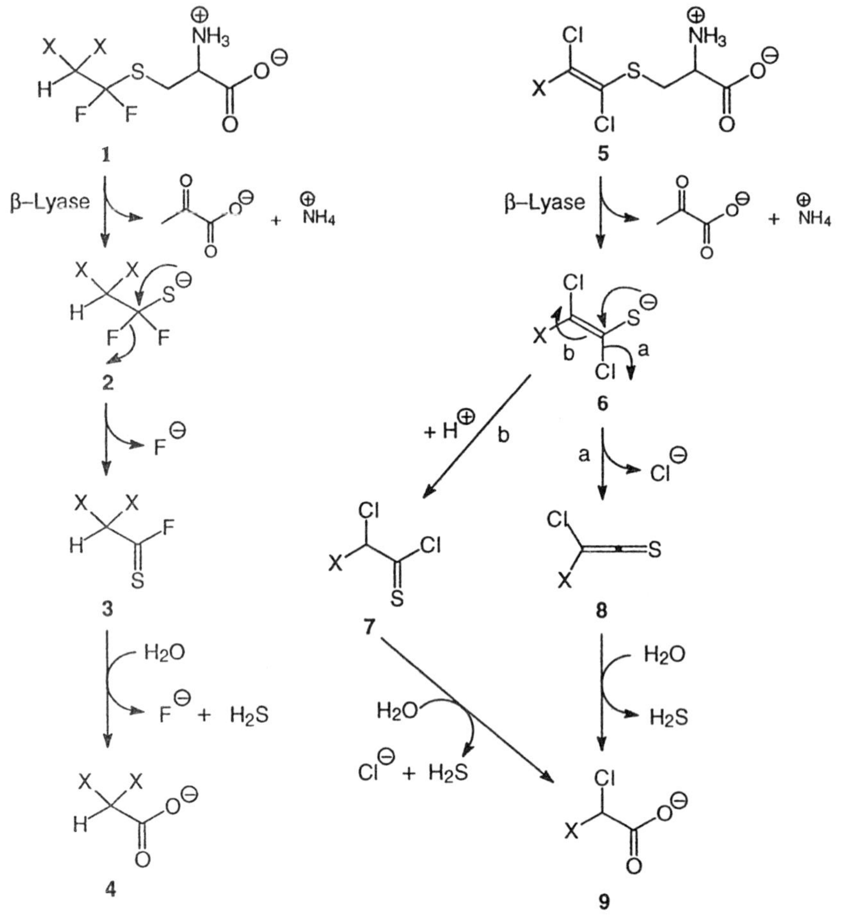

Figure 5 Cysteine conjugate β-lyase–dependent bioactivation of cysteine *S*-conjugates. **1**, Fluoroalkene-derived *S*-(2,2-dihalo-1,1-difluoroalkyl)-L-cysteine (X = Br, Cl, F); **2**, 2,2-dihalo-1,1-difluoroethanethiolate; **3**, dihalothionoacetyl fluoride; **4**, dihaloacetate; **5**, chloroalkene-derived *S*-(1-chloro-1-alkenyl)-L-cysteine (X = Cl, $-CF_3$, $-CCl{=}CCl_2$); **6**, 1-chloroethenethiolate; **7**, chlorohalothioacyl chloride; **8**, chlorohalothioketene; **9**, chlorohaloacetate.

activity, but rapidly eliminate fluoride to yield thioacyl fluorides (**3**, Fig. 5), which yield dihaloacetates (**4**, Fig. 5), fluoride, and hydrogen sulfide as terminal products (118,119).

The concept of intermediate thioacyl fluoride formation is supported by several observations:

1. Incubation of *S*-(2-chloro-1,1,2-trifluoroethyl)-L-cysteine with a pyridoxal model system or with purified bovine kidney β-lyase yields chlorofluoroacetate and inorganic fluoride as terminal products. Incubation with the pyridoxal model system in the presence of the model nucleophile diethylamine results in formation of *N,N*-diethylchlorofluorothioacetamide (118).
2. Studies on the chemical reactivity of synthetic 2-chloro-1,1,2-trifluoroethanethiol and 1,1,2,2-tetrafluoroethanethiol show the formation of thioamides and thiono acids in the presence of amines and water, respectively (120).
3. 1,1-Difluoroalkyl 2-nitrophenyl disulfides have been developed as proreactive intermediates to characterize the reactivity of α-haloalkyl- and α-haloalkenythiolates. Such disulfides can be cleaved in inert solvents to yield the corresponding thiolates and their derived products, which can then be trapped can characterized. When 2-chloro-1,1,2-trifluoroethyl 2-nitrophenyl disulfide was cleaved in the presence of cyclopentadiene, which undergoes a cycloaddition reaction with carbon-sulfur double bonds, thianorbornenes were identified as stable products formed from the thiolate.

These observations strongly indicate that thioacyl fluorides are formed as reactive intermediates from α-fluoroalkyl-L-cysteine *S*-conjugates (121).

The chloroalkene-derived *S*-conjugates *S*-(1,2-dichlorovinyl)-L-cysteine, *S*-(1,1,2-trichlorovinyl)-L-cysteine, and *S*-(1,2,3,4,4-pentachlorobutadienyl)-L-cysteine (**5**, Fig. 5) are biotransformed by cysteine conjugate β-lyase to yield α-chloroalkenylthiolates (**6**, Fig. 5), which give rise to electrophilic species whose interaction with cellular macromolecules is responsible for the cytotoxicity and mutagenicity of *S*-(1,2-dichlorovinyl)-L-cysteine, *S*-(1,1,2-trichlorovinyl)-L-cysteine, and *S*-(1,2,3,4,4-pentachlorobutadienyl)-L-cysteine. Incubation of these *S*-conjugates with a β-lyase model system or with partially purified bacterial β-lyase results in the formation of chloroacetate and chlorothionoacetate from *S*-(1,2-dichlorovinyl)-L-cysteine, dichloroacetate from *S*-(1,1,2-trichlorovinyl)-L-cysteine, and 2,3,4,4-tetrachlorobutenoate and 2,3,4,4-tetrachlorothionobutenoate from *S*-(1,2,3,4,4-pentachlorobutadienyl)-L-cysteine (**9**, Fig. 5) (122,123). Incubation of cysteine *S*-conjugates with the β-lyase model system in the presence of diethylamine gave the corresponding thioamides, indicating that thioacylating agents are formed as reactive intermediates. Two different types of reactive intermediates may be formed:

1. Thioacyl chlorides (**7**, Fig. 5) may be formed by tautomerization of α-chloroalkenylthiolates (**6**, Fig. 5).
2. Alternatively, α-chloroalkenylthiolates (**6**, Fig. 5) may eliminate chloride to afford thioketenes (**8**, Fig. 5). Thioketenes are known thio-

acylating agents and are highly reactive with nitrogen nucleophiles (124,125).

To distinguish between thioacyl chlorides and thioketenes, α-chloroalkenylthiolates were generated from cysteine *S*-conjugates and from the corresponding α-chloroalkenyl 2-nitrophenyl disulfides in the presence of cyclopentadiene (121). Cyclopentadiene undergoes a [4 + 2]cycloaddition reaction with the carbon-sulfur double bond of thioketenes and gives products that differ from thioacyl chlorides and thioketenes in their mass spectra and chromatographic properties (121). In these experiments, only thianorbornenes indicative of the formation of thioketenes as reactive intermediates were observed, indicating that thioketenes are the exclusive reactive intermediates formed from *S*-(α-haloalkenyl)-L-cysteine *S*-conjugates.

Both thioketenes are thioacyl fluorides are reactive acylating agents and may react with amino and hydroxyl groups in proteins and lipids. N^{ε}-(Difluorothioacetyl)-L-lysine and N^{ε}-(chlorofluorothioacetyl)-L-lysine have been identified in proteins from subcellular fractions incubated with *S*-(2-chloro-1,1,2-trifluoroethyl)-L-cysteine or *S*-(1,1,2,2–tetrafluoroethyl)-L-cysteine (126–130). Difluorothioacetamido-lipid adducts have also been identified in mitochondria. Moreover, N^{ε}-(chlorofluoroacetyl)-L-lysine is formed in renal proteins of rats given *S*-(2-chloro-1,1,2-trifluoroethyl)-L-cysteine (131). Covalent binding of thioketenes or thioacyl fluorides to cellular macromolecules is associated with *S*-conjugate-induced toxicity and mutagenicity.

B. Biological Effects of Cysteine *S*-Conjugates

The haloalkenes hexachlorobutadiene, hexafluoropropene, and chlorotrifluoroethene and the alkyne dichloroethyne are selectively nephrotoxic in rats and induce renal proximal tubular damage (132–134). Moreover, the widely used solvents tetrachloroethene and trichloroethene (135,136) as well as dichloroethyne and hexachlorobutadiene (137) induce carcinomas of the proximal tubules in rats. Glutathione-dependent pathways have been implicated in the renal toxicity of these compounds.

S-(1,2-Dichlorovinyl)-L-cysteine and *S*-(2-chloro-1,2,2-trifluoroethyl)-L-cysteine cause dose-dependent increases in blood urea-nitrogen concentrations and in urinary glucose excretion rates indicative of proximal tubular damage (101,116). Aminooxyacetic acid, an inhibitor of pyridoxal phosphate–dependent enzymes and of β-lyase (101), blocks *S*-conjugate–induced proximal tubular damage in vivo. Also, *S*-(1,2-dichlorovinyl)-DL-α-methylcysteine, *S*-(1-chloro-1,2,2-trifluoroethyl)-DL-α-methylcysteine, and *S*-(2-bromo-2-chloro-1,1-difluoroethyl)-DL-α-methylcysteine are not nephrotoxic (75,101,116). Such α-methyl cysteine *S*-conjugates cannot undergo β-lyase–catalyzed β-elimination reactions, and reactive intermediates cannot, therefore, be formed.

Both *S*-(1,1,2,2-tetrafluoroethyl)-L-cysteine and *S*-(1,2,3,4,4-pentachlorobutadienyl)-L-cysteine (138), metabolites of the nephrotoxic alkenes tetrafluoroethene and hexachlorobutadiene, respectively, produce damage to the pars recta of the proximal tubules. Similar lesions were observed after inhalation of tetrafluoroethene (72,139) and after giving hexachlorobutadiene (83,137,140–142).

The glutathione *S*-conjugates *S*-(2-chloro-1,2,2-trifluoroethyl)glutathione (116), *S*-(1,2-dichlorovinyl)glutathione (101), and *S*-(1,2,3,4,4-pentachlorobutadienyl)glutathione (83) are also nephrotoxic in rats. The metabolism of these glutathione *S*-conjugates to the cysteine *S*-conjugates via the intermediate cysteinylglycine *S*-conjugates appears to be essential for the expression of toxicity. Inhibition of γ-glutamyltransferase with acivicin blocks *S*-(1,2-dichlorovinyl)glutathione- and *S*-(2-chloro-1,2,2-trifluoroethyl)glutathione–induced nephrotoxicity in vivo. Aminooxyacetic acid also blocks *S*-(1,2-dichlorovinyl)glutathione–induced nephrotoxicity. These observations support the proposed mechanism. The β-lyase–dependent cytotoxicity of α-haloalkyl and α-haloalkenyl cysteine *S*-conjugates was confirmed in freshly isolated rat proximal tubule cells and in cultured renal cells in vitro (for reviews, Refs. 67 and 143).

The demonstration of high cysteine conjugate β-lyase activities in *Salmonella typhimurium* strains was decisive in elucidating the mechanism of the bacterial mutagenicity and metabolism of α-haloalkenyl cysteine and glutathione *S*-conjugates. *S*-(1,2,2-Trichlorovinyl)-L-cysteine, *S*-(1,2-dichorovinyl)-L-cysteine, and *S*-(1,2,3,4,4-pentachlorobutadienyl)-L-cysteine are mutagenic in the Ames preincubation assay without the addition of an exogenous activating system (144). Incubation of these *S*-conjugates with homogenates of *Salmonella typhimurium* results in the time- and concentration-dependent production of pyruvate, which is formed by the β-lyase–catalyzed biotransformation of the *S*-conjugates in equimolar amounts with the presumed mutagenic intermediates. Both mutagenicity and pyruvate production are blocked by aminooxyacetic acid, an inhibitor of the β-lyase. Moreover, *S*-(1,2-dichlorovinyl)-DL-α-methylcysteine, which cannot be cleaved by β-lyase, is not mutagenic and does not produce pyruvate (145). The corresponding mercapturic acids are also potent bacterial mutagens in the presence of soluble aminoacylases from rat kidney cytosol, and ^{14}C-labeled *N*-acetyl-*S*-(1,2,3,4,4-pentachlorobutadienyl)-L-cysteine is metabolized to covalently bound metabolites in the presence of rat aminoacylases. (146).

In structure-mutagenicity studies with α-haloalkyl cysteine *S*-conjugates, only the bromine-containing cysteine *S*-conjugates *S*-(2-bromo-2-chloro-1,1-difluoroethyl)-L-cysteine, *S*-(2-bromo-1,1,2-trifluoroethyl)-L-cysteine, and *S*-(2-2-dibromo-1,1-difluoroethyl)-L-cysteine were mutagenic in *Salmonella typhimurium*, whereas the bromine-lacking conjugates *S*-(2-chloro-1,1,2-trifluoroethyl)-L-cysteine, *S*-(2,2-chloro-1,1-difluoroethyl)-L-cysteine, and *S*-(1,1,2,2-tetrafluoroethyl)-L-cysteine were not mutagenic (147). In contrast to these clear differences in mutagenicity, both bromine-lacking and bromine-containing *S*-conjugates were cytotoxic and nephrotoxic (see below).

Rat kidney microsomes contain high γ-glutamyltransferase and dipeptidase activities. Accordingly, *S*-(1,2-dichlorovinyl)glutathione, *S*-(1,2,2-trichlorovinyl)glutathione, and *S*-(1,2,3,4,4-pentachlorobutadienyl)glutathione are biotransformed by rat kidney particulate fractions to the corresponding cysteine *S*-conjugates, which are potent mutagens. Inhibition of γ-glutamyltransferasc by L-serine-borate blocks the mutagenicity of the glutathione *S*-conjugates (145).

The mutagenicity of the haloalkenes hexachlorobutadiene, trichloroethene, and tetrachloroethene in the Ames test has been the subject of debate over the last 10 years. The results of studies conducted in absence or presence of rat liver subcellular fractions that catalyze oxidative metabolism and do not mimic the renal processing of *S*-conjugates were largely negative or confounded by the presence of contaminants (148–150) (for a review, see Ref. 151). A clear mutagenic effect was, however, obtained with pure hexachlorobutadiene and tetrachloroethene when the mercapturic acid pathway was simulated in the Ames preincubation assay by sequential addition of rat liver microsomes and glutathione followed by rat kidney microsomes (152,153).

Cysteine and glutathione *S*-conjugates of hexachlorobutadiene, tetrachloroethene, trichloroethene, and dichloroethyne are also genotoxic in mammalian cells and induce γ-glutamyltransferase– and β-lyase–dependent DNA repair in LLC-PK1 cells, a porcine kidney cell line. LLC-PK1 cells exhibit many characteristics of renal proximal tubule cells, including the presence of the enzymes of mercapturic acid formation and β-lyase activity (153–155). Induction of DNA repair occurs at concentrations that do not cause severe cytotoxicity. The observed extent of DNA repair is low compared with the effects of known alkylating agents, such as streptozotocin or 4-nitroquinoline-1-oxide; more importantly, the concentration range over which DNA repair occurs in absence of cell death is also narrow.

S-(1,2,3,4,4-Pentachlorobutadienyl)-L-cysteine failed to induce single-strand breaks in isolated rabbit renal tubules (138); in the same study, evidence was presented showing that DNA-DNA cross-links were formed. In contrast, *S*-(1,2-dichlorovinyl)-L-cysteine induces DNA single-strand breaks in kidney tubules both in vivo and in vitro and DNA double-strand breaks in LLC-PK1 cells (156,157). Finally, covalent binding of ^{35}S to DNA was demonstrated with ^{35}S-labeled cysteine conjugates, both in bacteria and in renal cells (158–161).

Genotoxicity studies in renal cells confirmed the bacterial mutagenicity of α-chloroalkenyl *S*-conjugates. The *S*-conjugates are, however, weakly mutagenic in renal cells in culture, and cytotoxicity and induction of cell death are the predominant effects (92,155,162,163). Moreover, in in vivo long-term studies, renal cell tumors are formed only with dose regimens that cause marked nephrotoxicity.

In an attempt to establish a link between the extranuclear effects of these *S*-conjugates and the formation of kidney tumors, the effects of *S*-(1,2-dichlorovinyl)-L-cysteine on Ca^{2+} homeostasis in renal cells were investigated. Studies with fluorescence digital imaging microscopy showed that *S*-(1,2-dichlorovinyl)-L-cysteine increases cytosolic Ca^{2+} concentrations prior to the onset of cell death. The increases in cytosolic Ca^{2+} concentrations are associated with the impaired ability of the mitochondria to sequester cytosolic Ca^{2+} and precede the collapse of the mitochondrial membrane potential (164,165). Similar effects are seen with some tumor promoters, such as organic hydroperoxides, that induce oxidative stress. *S*-(1,2-Dichlorovinyl)-L-cysteine also induces oxidation of pyridine nucleotides in kidney mitochondria and promotes dimethylnitrosoamine-induced renal adenocarcinomas in mice (166,167).

Among the Ca^{2+}-dependent degradative pathways, induction of DNA double-strand breaks by Ca^{2+}- and Mg^{2+}-dependent endonucleases is of particular interest in terms of tumor formation. *S*-(1,2-Dichlorovinyl)-L cysteine induces DNA fragmentation at concentrations that produce little or no impairment of cell growth. This Ca^{2+}-dependent DNA damage is followed by increased polyADP-ribosylation of nuclear proteins (167). A nonlethal increase in the amount of nuclear polyADP-ribosyl conjugates may change the structure and function of nuclear proteins and thereby alter the expression of genes involved in cell proliferation and differentiation and possibly in tumor formation. Furthermore, *S*-(1,2-dichlorovinyl)-L-cysteine induces the expression of the protooncogenes *c-fos* and *c-myc* in LLC-PK1 cells (168).

Both the bromine-containing and bromine-lacking α-fluoroalkyl cysteine *S*-conjugates induce β-lyase–dependent toxicity in LLC-PK1 cells, induce Ca^{2+} release from pig kidney mitochondria, and produce DNA double-strand breaks in LLC-PK1 cells (147). The effects of these conjugates on gene expression have not been investigated; hence it is not possible to assess the role of mutagenicity versus indirect genotoxic effects on the biological endpoint gene expression.

ACKNOWLEDGMENTS

The authors thank Sandra E. Morgan and Hannelore Popa-Henning for their assistance in the preparation of the manuscript. Research conducted in the authors' laboratories was supported by National Institute of Environmental Health Sciences grant ES03127 (M.W.A.), by the Deutsche Forschungsgemeinschaft, Bonn (W.D., S.V.), by the Bundesministerium für Forschung und Technologie, Bonn (W.D.), the Doktor-Robert-Pfeger Stiftung, Bamberg (W.D.), by the Hauptverband der Berufsgenossenschaften, St. Augustin (W.D.), and by NATO grant 901032 (M.W.A., W.D.).

REFERENCES

1. Stevens, J. L., and Jones, D. P. (1989) The mercapturic acid pathway: Biosynthesis, intermediary metabolism, and physiological disposition. In: Glutathione: Chemical, Biochemical, and Medical Aspects (Dolphin, D., Poulson, R., and Avramovic, O., eds.), Part B, pp. 46–84. Wiley, New York.
2. Anders, M. W., and Dekant, W. (1994) Aminoacylases. Adv. Pharmacol. 27:433–450.
3. Hashmi, M., Vamvakas, S., and Anders, M. W. (1992) Bioactivation mechanism of S-(3-oxopropyl)-*N*-acetyl-L-cysteine, the mercapturic acid of acrolein. Chem. Res. Toxicol. 5:360–365.
4. Park, S. B., Osterloh, J. D., Vamvakas, S., Hashmi, M., Anders, M. W., and Cashman, J. R. (1992) Flavin-containing monooxygenase-dependent stereoselective S-oxygenation and cytotoxicity of cysteine S-conjugates and mercapturates. Chem. Res. Toxicol. 5:193–201.
5. Baillie, T. A., and Slatter, J. G. (1991) Glutathione: A vehicle for the transport of chemically reactive metabolites in vivo. Accts. Chem. Res. 24:264–270.
6. Baillie, T. A., and Kassahun, K. (1994) Reversibility in glutathione-conjugate formation. Adv. Pharmacol. 27:165–183.
7. Dekant, W., Anders, M. W., and Monks, T. J. (1993) Bioactivation of halogenated xenobiotics by S-conjugate formation. In: Renal Disposition and Nephrotoxicity of Xenobiotics (Anders, M. W., Dekant, W., Henschler, D., Oberleithner, H., and Silbernagl, S., eds.), pp. 187–215. Academic Press, San Diego.
8. Lawley, P. D., and Thatcher, C. J. (1970) Methylation of deoxyribonucleic acid in cultured mammalian cells by N-methyl-N′-nitro-N-nitrosoguanidine. The influence of cellular thiol concentrations on the extent of methylation and the 6-oxygen atom of guanine as a site of methylation. Biochem. J. 116:693–707.
9. Anders, M. W., Lash, L., Dekant, W., Elfarra, A. A., and Dohn, D. R. (1988) Biosynthesis and biotransformation of glutathione S-conjugates to toxic metabolites. CRC Crit. Rev. Toxicol. 18:311–341.
10. Koob, M., and Dekant, W. (1991) Bioactivation of xenobiotics by formation of toxic glutathione conjugates. Chem.-Biol. Interact. 77:107–136.
11. Anders, M. W. (1991) Glutathione-dependent bioactivation of xenobiotics: Implications for mutagenicity and carcinogenicity. In: Xenobiotics and Cancer. Implications for Chemical Carcinogenesis and Cancer Chemotherapy (Ernster, L., Esumi, H., Fujii, Y., Gelboin, H. V., Kato, R., and Sugimura, T., eds.), pp. 89–99. Japan Sci. Soc. Press, Tokyo.
12. Anders, M. W., Dekant, W., and Vamvakas, S. (1992) Glutathione-dependent toxicity. Xenobiotica 22:1135–1145.
13. Dekant, W., Vamvakas, S., and Anders, M. W. (1992) The kidney as a target organ for xenobiotics bioactivated by glutathione conjugation. In: Tissue-Specific Toxicity: Biochemical Mechanisms (Dekant, W., and Neumann, H.-G., eds.), pp. 163–194. Academic Press, London.
14. Kubic, V. L., Anders, M. W., Engel, R. R., Barlow, C. H., and Caughey, W. S. (1974) Metabolism of dihalomethanes to carbon monoxide. I. In vivo studies. Drug Metab. Dispos. 2:53–57.

15. Kubic, V. L, and Anders, M. W. (1975) Metabolism of dihalomethanes to carbon monoxide. II. In vitro studies. Drug Metab. Dispos. 3:104–112.
16. Kubic, V. L., and Anders, M. W. (1978) Metabolism of dihalomethanes to carbon monoxide. III. Studies on the mechanisms of the reaction. Biochem. Pharmacol. 27:2349–2355.
17. Ahmed, A. E., and Anders, M. W. (1976) Metabolism of dihalomethanes to formaldehyde and inorganic halide. I. In vitro studies. Drug Metab. Dispos. 4:357–361.
18. Ahmed, A. E., and Anders, M. W. (1978) Metabolism of dihalomethanes to formaldehyde and inorganic halide. II. Studies on the mechanism of the reaction. Biochem. Pharmacol. 27:2021–2025.
19. Jongen, W. M. F., Harmsen, E. G. M., Alink, G. M., and Koeman, J. H. (1982) The effect of glutathione conjugation and microsomal oxidation on the mutagenicity of dichloromethane in *S. typhimurium*. Mutat. Res. 95:183–189.
20. Green, T. (1983) The metabolic activation of dichloromethane and chlorofluoromethane in a bacterial mutation assay using *Salmonella typhimurium*. Mutat. Res. 118:277–288.
21. Dillon, D., Edwards, I., Combes, R., McConville, M., and Ziegler, E. (1992) The role of glutathione in the bacterial mutagenicity of vapour phase dichloromethane. Environ. Mol. Mutagen. 20:211–217.
22. Gargas, M. L., Clewell III, H. J., and Andersen, M. E. (1986) Metabolism of inhaled dihalomethanes in vivo: Differentiation of kinetic constants for two independent pathways. Toxicol. Appl. Pharmacol. 82:211–223.
23. Reitz, R. H., Mendrala, A. L., and Guengerich, F. P. (1989) In vitro metabolism of methylene chloride in human and animal tissues: Use in physiologically based pharmacokinetic models. Toxicol. Appl. Pharmacol. 97:230–246.
24. Green, T. (1989) A biological data base for methylene chloride risk assessment. In: Biologically Based Methods for Cancer Risk Assessment (Travis, C. C., ed.), pp. 289–300. Plenum Press, New York.
25. Böhme, H., Fischer, H., and Frank, R. (1949) Darstellung and Eigenschaften der α-halogenierten Thioäther. Justus Liebigs Chem. Ann. 563:54–72.
26. Uotila, L., and Koivusalo, M. (1974) Formaldehyde dehydrogenase from human liver. Purification, properties, and evidence for the formation of gluathione thiol esters by the enzyme. J. Biol. Chem. 249:7653–7663.
27. Naylor, S., Mason, R. P., Sanders, J. K. M., and Williams, D. H. (1988) Formaldehyde adducts of glutathione: Structure elucidation by two-dimensional N.M.R. spectroscopy and fast-atom-bombardment tandem mass spectrometry. Biochem. J. 249:573–579.
28. Uotila, L., and Koivusalo, M. (1974) Purification and properties of *S*-formylglutathione hydrolase from human liver. J. Biol. Chem. 249:7664–7672.
29. Hashmi, M., Dechert, S., Dekant, W., and Anders, M. W. (1994) Bioactivation of [^{13}C]dichloromethane in mouse, rat, and human liver cytosol: ^{13}C Nuclear magnetic resonance spectroscopic studies. Chem. Res. Toxicol. 7:291–296.
30. Kari, F., Foley, J. F., Seilkop, S. K., Maronpot, R. R., and Anderson, M. W. (1993) Effect of varying exposure regimens on methylene chloride-induced lung and liver tumors in female B6C3F1 mice. Carcinogenesis 14:819–826.

31. Devereux, T. R., Foley, J. F., Maronpot, R. R., Kari, F., and Anderson, M. W. (1993) *Ras* proto-oncogene activation in liver and lung tumors from B6C3F1 mice exposed chronically to methylene chloride. Carcinogenesis 14:795–801.
32. Hegi, M. E., Söderkvist, P., Foley, J. F., Schoonhoven, R., Swenberg, J. A., Kari, F., Maronpot, R., Anderson, M. W., and Wiseman, R. W. (1993) Characterization of *p53* mutations in methylene chloride-induced lung tumors from B6C3F1 mice. Carcinogenesis 14:803–810.
33. Foley, J. F., Tuck, P. D., Ton, T.-V. T., Frost, M., Kari, F., Anderson, M. W., and Maronpot, R. R. (1993) Inhalation exposure to a hepatocarcinogenic concentration of methylene chloride does not induce sustained replicative DNA synthesis in hepatocytes of female B6C3F1 mice. Carcinogenesis 14:811–817.
34. Casanova, M., Deyo, D. F., and Heck, H. D. (1989) Covalent binding of inhaled formaldehyde to DNA in the nasal mucosa of Fischer 344 rats: Analysis of formaldehyde and DNA by high-performance liquid chromatography and provisional pharmacokinetic interpretation. Fundam. Appl. Toxicol. 12:397–417.
35. Graves, R. J., Callender, R. D., and Green, T. (1994) The role of formaldehyde and S-chloromethylglutathione in the bacterial mutagenicity of methylene chloride. Mutat. Res. 320:235–243.
36. Graves, R. J., Coutts, C., Eyton-Jones, H., and Green, T. (1994) Relationship between DNA damage and methylene chloride-induced hepatocarcinogenicity in B6C3F1 mice. Carcinogenesis 15:991–996.
37. Thier, R., Taylor, J. B., Pemble, S. E., Humphreys, W. G., Persmark, M., Ketterer, B., and Guengerich, F. P. (1993) Expression of mammalian glutathione S-transferase 5-5 in *Salmonella typhimurium* TA1535 leads to base-pair mutations upon exposure to dihalomethanes. Proc. Natl. Acad. Sci. USA 90:8576–8580.
38. Green, T., Provan, W. M., Collinge, D. C., and Guest, A. E. (1988) Macromolecular interactions of inhaled methylene chloride in rats and mice. Toxicol. Appl. Pharmacol. 93:1–10.
39. Green, T., Provan, W. M. Dugard, P. H., and Cook, S. K. (1988) Methylene chloride (dichloromethane): Human risk assessment using experimental animal data. ECETOC Technical Report 32, pp. 1–62, Brussels.
40. Kermani, H. R. S., Sloane, R. A., Moorman, M. P., Yang, R. S. H., Ray, C., and Reitz, R. H. (1990) Enzyme kinetics of methylene chloride: MFO and GST activities in female B6C3F1 mice liver and lung in relation to aging and chronic dosing. Toxicologist 10:186.
41. Andersen, M. E., Clewell III, H. J., Gargas, M. L., MacNaughton, M. G., Reitz, R. H., Nolan, R. J., and McKenna, M. J. (1991) Physiologically based pharmacokinetic modeling with dichloromethane, its metabolite, carbon monoxide, and blood carboxyhemoglobin in rats and humans. Toxicol. Appl. Pharmacol. 108:14–27.
42. Guengerich, F. P., Crawford, J. W. M., Domoradzki, J. Y., MacDonald, T. L., and Watanabe, P. G. (1980) In vitro activation of 1,2-dichloroethane by microsomal and cytosolic enzymes. Toxicol. Appl. Pharmacol. 55:303–317.
43. Shih, T.-W., and Hill, D. L. (1981) Metabolic activation of 1,2-dibromoethane by glutathione transferase and by microsomal mixed function oxidase: Further evi-

dence for formation of two reactive metabolites. Res. Commun. Chem. Pathol. Pharmacol. 33:449–461.

44. McCall, S. N., Jurgens, P., and Ivanetich, K. M. (1983) Hepatic microsomal metabolism of the dichloroethanes. Biochem. Pharmacol. 32:207–213.
45. White, R. D., Petry, T. W., and Sipes, I. G. (1984) The bioactivation of 1,2-dibromoethane in rat hepatocytes: Deuterium isotope effect. Chem.-Biol. Interact. 49:225–233.
46. Storer, R. D., and Conolly, R. B. (1985) An investigation of the role of microsomal oxidative metabolism in the in vivo genotoxicity of 1,2-dichloroethane. Toxicol. Appl. Pharmacol. 77:36–46.
47. Edwards, K., Jackson, H., and Jones, A. R. (1970) Studies with alkylating esters-II. A chemical interpretation through metabolic studies of the antifertility effects of ethylene dimethanesulphonate and ethylene dibromide. Biochem. Pharmacol. 19:1783–1789.
48. Nachtomi, E. (1970) The metabolism of ethylene dibromide in the rat. The enzymic reaction with glutathione in vitro and in vivo. Biochem. Pharmacol. 19:2853–2860.
49. Livesey, J. C., and Anders, M. W. (1979) In vitro metabolism of 1,2-dihaloethanes to ethylene. Drug Metab. Dispos. 7:199–203.
50. Livesey, J. C., Anders, M. W., Langvardt, P. W., Putzig, C. L., and Reitz, R. H. (1982) Stereochemistry of the glutathione-dependent biotransformation of vicinal-dihaloalkanes to alkenes. Drug Metab. Dispos. 10:201–204.
51. Elfarra, A. A., Baggs, R. B., and Anders, M. W. (1985) Structure-nephrotoxicity relationships of *S*-(2-chloroethyl)-DL-cysteine and analogs. Role for an episulfonium ion. J. Pharmacol. Exp. Ther. 233:512–516.
52. Dohn, D. R., and Casida, J. E. (1987) Thiiranium ion intermediates in the formation and reactions of *S*-(2-haloethyl)-L-cysteines. Bioorg. Chem. 15:115–124.
53. Whorton, M. D., and Foliart, D. E. (1983) Mutagenicity, carcinogenicity and reproductive effects of dibromochloropropane (DBCP). Mutat. Res. 123:13–30.
54. Omichinski, J. G., Brunborg, G., Søderlund, E. J., Dahl, J. E., Bausano, J. A., Holme, J. A., Nelson, S. D., and Dybing, E. (1987) Renal necrosis and DNA damage caused by selectively deuterated and methylated analogs of 1,2-dibromo-3-chloropropane in the rat. Toxicol. Appl. Pharmacol. 91:358–370.
55. Omichinski, J. G., Brunborg, G., Holme, J. A., Søderlund, E. J., Nelson, S. D., and Dybing, E. (1988) The role of oxidative and conjugative pathways in the activation of 1,2-dibromo-3-chloropropane to DNA-damaging products in rat testicular cells. Mol. Pharmacol. 34:74–79.
56. Holme, J. A., Søderlund, E. J., Brunborg, G., Omichinski, J. G., Bekkedal, K., Trygg, B., Nelson, S. D., and Dybing, E. (1989) Different mechanisms are involved in DNA damage, bacterial mutagenicity and cytotoxicity induced by 1,2-dibromo-3-chloropropane in suspensions of rat liver cells. Carcinogenesis 10:49–54.
57. Pearson, P. G., Omichinski, J. G., Myers, T. G., Soderlund, E. J., Dybing, E., and Nelson, S. D. (1990) Metabolic activation of 1,2-dibromo-3-chloropropane to mutagenic metabolites: Detection and mechanism of formation of (Z)- and (E)-2-chloro-3-(bromomethyl)oxirane. Chem. Res. Toxicol. 3:458–466.

58. Cmarik, J. L., Humphreys, W. G., Bruner, K. L., Lloyd, R. S., Tibbetts, C., and Guengerich, F. P. (1992) Mutation spectrum and sequence alkylation selectivity resulting from modification of bacteriophage M13mp18 DNA with *S*-(2-chloroethyl)gluathione. J. Biol. Chem. 267:6672–6679.
59. Foureman, G. L., and Reed, D. J. (1987) Formation of S-[2-(N^7-guanyl)ethyl] adducts by the postulated S-(2-chloroethyl)cysteinyl and S-(2-chloroethyl)glutathionyl conjugates of 1,2-dichloroethane. Biochemistry 26:2028–2033.
60. Humphreys, W. G., Kim, D.-H., Cmarik, J. L., Shimada, T., and Guengerich, F. P. (1990) Comparison of the DNA-alkylating properties and mutagenic response of a series of *S*-(2-haloethyl)-substituted cysteine and glutathione derivatives. Biochemistry 29:10342–10350.
61. Vamvakas, S., Dekant, W., and Henschler, D. (1989) Assessment of unscheduled DNA synthesis in a cultured line of renal epithelial cells exposed to cysteine *S*-conjugates of haloalkenes and haloalkanes. Mutat. Res. 222:329–335.
62. Humphreys, W. G., Kim, D. H., and Guengerich, F. P. (1991) Isolation and characterization of N^7-guanyl adducts derived from 1,2-dibromo-3-chloropropane. Chem. Res. Toxicol. 4:445–453.
63. Inskeep, P. B., and Guengerich, F. P. (1984) Glutathione-mediated binding of dibromoalkanes to DNA: Specificity of rat glutathione S-transferases and dibromoalkane structure. Carcinogenesis 5:805–808.
64. Elfarra, A. A., and Anders, M. W. (1984) Renal processing of glutathione conjugates: Role in nephrotoxicity. Biochem. Pharmacol. 33:3729–3732.
65. Dekant, W., Vamvakas, S., and Anders, M. W. (1989) Bioactivation of nephrotoxic haloalkenes by glutathione conjugation; Formation of toxic and mutagenic intermediates by cysteine conjugate β-lyase. Drug Metab. Rev. 20:43–83.
66. Dekant, W., Vamvakas, S., and Anders, M. W. (1992) The kidney as a target organ for xenobiotics bioactivated by glutathione conjugation. In: Tissue-Specific Toxicity. Biochemical Mechanisms (Dekant, W., and Neumann, H. G., eds.), pp. 163–194. Academic Press, London.
67. Dekant, W., Vamvakas, S., and Anders, M. W. (1994) Formation and fate of nephrotoxic and cytotoxic glutathione S-conjugates: Cysteine conjugate β-lyase pathway. Adv. Pharmacol. 27:117–164.
68. Morgenstern, R., Lundqvist, G., Andersson, G., Balk, L., and DePierre, J. W. (1984) The distribution of microsomal glutathione transferase among different organelles, different organs, and different organisms. Biochem. Pharmacol. 33:3609–3614.
69. Morgenstern, R., Guthenberg, C., and DePierre, J. W. (1982) Microsomal glutathione *S*-transferase. Purification, initial characterization and demonstration that it is not identical to the cytosolic gluathione S-transferases A, B and C. Eur. J. Biochem. 128:243–248.
70. Dekant, W., Vamvakas, S., and Anders, M. W. (1990) Bioactivation of hexachlorobutadiene by glutathione conjugation. Food Chem. Toxicol. 28:285–293.
71. Kanhai, W., Dekant, W., and Henschler, D. (1989) Metabolism of the nephrotoxin dichloracetylene by glutathione conjugation. Chem. Res. Toxicol. 2:51–56.
72. Odum, J., and Green, T. (1984) The metabolism and nephrotoxicity of tetrafluoroethylene in the rat. Toxicol. Appl. Pharmacol. 76:306–318.

73. Commandeur, J. N. M., Oostendorp, R. A. J., Schoofs, P. R., Xu, B., and Vermeulen, N. P. E. (1987) Nephrotoxicity and hepatotoxicity of 1,1-dichloro-2,2-difluoroethylene in the rat. Biochem. Pharmacol. 36:4229–4237.
74. Wark, H., Earl, J., Chau, D. D., Overton, J., and Cheung, H. T. A. (1990) A urinary cysteine-halothane metabolite: Validation and measurement in children. Br. J. Anaesth. 64:469–473.
75. Finkelstein, M. B., Baggs, R. B., and Anders, M. W. (1992) Nephrotoxicity of the glutathione and cysteine conjugates for 2-bromo-2-chloro-1,1-difluoroethene. J. Pharmacol. Exp. Ther. 261:1248–1252.
76. Dohn, D. R., Quebbemann, A. J., Borch, R. F., and Anders, M. W. (1985) Enzymatic reaction of chlorotrifluoroethene with glutathione: ^{19}F NMR evidence for stereochemical control of the reaction. Biochemistry 24:5137–5143.
77. Hargus, S. J., Fitzsimmons, M. E., Aniya, Y. and Anders, M. W. (1991) Stereochemistry of the microsomal glutathione *S*-transferase catalyzed addition of glutathione to chlorotrifluoroethene. Biochemistry 30:717–721.
78. Koob, M., and Dekant, W. (1990) Metabolism of hexafluoropropene. Evidence for bioactivation by glutathione conjugate formation in the kidney. Drug Metab. Dispos. 18:911–916.
79. Inoue, M., Akerboom, T. P. M., Sies, H., Kinne, R., Thao, T., and Arias, I. M. (1984) Biliary transport of glutathione *S*-conjugate by rat liver canalicular membrane vesicles. J. Biol. Chem. 259:4998-5002.
80. Inoue, M., Kinne, R., Tran, T., and Arias, I. M. (1984) Glutathione transport across hepatocyte plasma membranes. Analysis using isolated rat-liver sinusoidal-membranes vesicles. Eur. J. Biochem. 138:491–495.
81. Akerboom, T. P. M., Narayanaswami, V., Kunst, M., and Sies, H. (1991) ATP-Dependent *S*-(2,4-dinitrophenyl)glutathione transport in canalicular plasma membrane vesicles from rat liver. J. Biol. Chem. 266:13147–13152.
82. Gietl, Y., and Anders, M. W. (1991) Biosynthesis and biliary excretion of *S*-conjugates of hexachlorobuta-1,3-diene in the perfused rat liver. Drug Metab. Dispos. 19:274–277.
83. Nash, J. A., King, L. H., Lock, E. A., and Green, T. (1984) The metabolism and disposition of hexachloro-1:3-butadiene in the rat and its relevance to nephrotoxicity. Toxicol. Appl. Pharmacol. 73:124–137.
84. Tate, S. S., and Meister, A. (1985) γ-Glutamyl transpeptidase from kidney. Methods Enzymol. 113:400–419.
85. Meister, A. (1988) Glutathione metabolism and its selective modification. J. Biol. Chem. 263:17205–17208.
86. Tate, S. S., and Meister, A. (1978) Serine-borate complex as a transition-state inhibitor of γ-glutamyl transpeptidase. Proc. Natl. Acad. Sci. USA 75:4806–4809.
87. Reed, D. J., Ellis, W. W., and Meck, R. A. (1980) The inhibition of γ-glutamyl transpeptidase and glutathione metabolism of isolated rat kidney cells by L-(αS,5S)-α-amino-3-chloro-4,5-dihydro-5-isoxazoleacetic acid (AT-125; NSC-163501). Biochem. Biophys. Res. Commun. 94:1273–1277.
88. McDonald, J. K., and Barrett, A. J. (1986) Mammalian Proteases: A Glossary and Bibliography, Vol. 2, Exopeptidases, pp. 59–71. Academic Press, London.

89. Tate, S. S. (1985) Microvillus membrane peptidases that catalyze hydrolysis of cysteinylglycine and its derivatives. Methods Enzymol. 113:471–484.
90. Kozak, E. M., and Tate, S. S. (1982) Glutathione-degrading enzymes of microvillus membranes. J. Biol. Chem. 257:6322–6327.
91. McIntyre, T., and Curthoys, N. P. (1982) Renal catabolism of glutathione. Characterization of a particulate rat renal dipeptidase that catalyzes the hydrolysis of cysteinylglycine. J. Biol. Chem. 257:11915–11921.
92. Lash, L. H., and Anders, M. W. (1986) Cytotoxicity of *S*-(1,2-dichlorovinyl)-glutathione and *S*-(1,2-dichlorovinyl)-L-cysteine in isolated rat kidney cells. J. Biol. Chem. 261:13076–13081.
93. Gietl, Y., Vamvakas, S. and Anders, M. W. (1991) Intestinal absorption of *S*-(pentachlorobutadienyl)glutathione and *S*-(pentachlorobutadienyl)-L-cysteine, the glutathione and cysteine *S*-conjugates of hexachlorobuta-1,3-diene. Drug Metab. Dispos. 19:703–707.
94. Inoue, M., Okajima, K., and Morino, Y. (1984) Hepato-renal cooperation in biotransformation, membrane transport, and elimination of cysteine *S*-conjugates of xenobiotics. J. Biochem. 95:247–254.
95. Duffel, M. W., and Jakoby, W. B. (1982) Cysteine *S*-conjugate *N*-acetyltransferase from rat kidney microsomes. Mol. Pharmacol. 21:444–448.
96. Hinchman, C. A., Truong, A. T., and Ballatori, N. (1993) Hepatic uptake of intact glutathione *S*-conjugate, inhibition by organic anions, and sinusoidal catabolism. Am. J. Physiol. 265:G547–G554.
97. Lash, L. H., and Anders, M. W. (1989) Uptake of nephrotoxic *S*-conjugates by isolated rat renal proximal tubular cells. J. Pharmacol. Exp. Ther. 248:531–537.
98. Spater, H. W., Poruchynsky, M. S., Quintana, N., Inoue, M., and Novikoff, A. B. (1982) Immunocytochemical localization of γ-glutamyltransferase in rat kidney with protein A-horseradish peroxidase. Proc. Natl. Acad. Sci. USA 79:3547–3550.
99. Ullrich, K. J., Rumrich, G., Wieland, T., and Dekant, W. (1989) Contraluminal para-aminohippurate (PAH) transport in the proximal tubule of the rat kidney. VI. Specificity: Amino acids, their *N*-methyl-, *N*-acetyl- and *N*-benzoyl derivatives glutathione- and cysteine conjugates, di- and oligopeptides. Pflügers Arch. 415:342–350.
100. Lock, E. A., and Ishmael, J. (1985) Effect of the organic acid transport inhibitor probenecid on renal cortical uptake and proximal tubular toxicity of hexachloro-1,3-butadiene and its conjugates. Toxicol. Appl. Pharmacol. 81:32–42.
101. Elfarra, A. A., Jakobson, I., and Anders, M. W. (1986) Mechanism of *S*-(1,2-dichlorovinyl)glutathione-induced nephrotoxicity. Biochem. Pharmacol. 35:283–288.
102. Tateishi, M., Suzuki, S., and Shimizu, H. (1978) Cysteine conjugate β-lyase in rat liver. A novel enzyme catalyzing formation of thiol-containing metabolites of drugs. J. Biol. Chem. 253:8854–8859.
103. Suzuki, S., Tomisawa, H., Ichihara, S., Fukazawa, H., and Tateishi, M. (1982) A C-S bond cleavage enzyme of cysteine conjugates in intestinal microorganisms. Biochem. Pharmacol. 31:2137–2140.
104. Stevens, J., and Jakoby, W. B. (1983) Cysteine conjugate β-lyase. Mol. Pharmacol. 23:761–765.

105. Tomisawa, H., Suzuki, S., Ichihara, S., Fukazawa, H., and Tateishi, M. (1984) Purification and characterization of C-S lyase from *Fusobacterium varium*. J. Biol. Chem. 259:2588–2593.
106. Larsen, G. L., and Stevens, J. L. (1986) Cysteine conjugate β-lyase in the gastrointestinal bacterium *Eubacterium limosum*. Mol. Pharmacol. 29:97–103.
107. Stevens, J. L. (1985) Cysteine conjugate β-lyase activities in rat kidney cortex: Subcellular localization and relationship to the hepatic enzyme. Biochem. Biophys. Res. Commun. 129:499–504.
108. Stevens, J. L. (1985) Isolation and characterization of a rat liver enzyme with both cysteine conjugate β-lyase and kynureninase activity. J. Biol. Chem. 260:7945–7950.
109. Stevens, J. L., Robbins, J. D., and Byrd, R. A. (1986) A purified cysteine conjugate β-lyase from rat kidney cytosol. J. Biol. Chem. 261:15529–15537.
110. Tomisawa, H., Ichihara, S., Fukazawa, H., Ichimoto, N., Tateishi, M., and Yamamoto, I. (1986) Purification and characterization of human hepatic cysteine-conjugate β-lyase. Biochem. J. 235:569–575.
111. Nelson, R., Van Dyke, R., Powis, G., Lash, L. H., and Anders, M. W. (1988) Characterization of human liver and kidney cysteine conjugate β-lyase activity. FASEB J. 2:A1355.
112. MacFarlane, M., Foster, J. R., Gibson, G. G., King, L. J., and Lock, E. A. (1989) Cysteine conjugate β-lyase of rat kidney cytosol: Characterization, immunocytochemical localization, and correlation with hexachlorobutadiene nephrotoxicity. Toxicol. Appl. Pharmacol. 98:185–197.
113. Lash, L. H., Nelson, R. M., Van Dyke, R. A., and Anders, M. W. (1990) Purification and characterization of human kidney cytosolic conjugate β-lyase activity. Drug Metab. Dispos. 18:50–54.
114. Lash, L. H., Elfarra, A. A., and Anders, M. W. (1986) Renal cysteine conjugate β-lyase: Bioactivation of nephrotoxic cysteine *S*-conjugates in mitochondrial outer membrane. J. Biol. Chem. 261:5930–5935.
115. Elfarra, A. A., Lash, L. H., and Anders, M. W. (1987) α-Ketoacids stimulate renal cysteine conjugate β-lyase activity and potentiate the cytotoxicity of *S*-(1,2-dichlorovinyl)-L-cysteine. Mol. Pharmacol. 31:208–212.
116. Dohn, D. R., Leininger, J. R., Lash, L. H., Quebbemann, A. J., and Anders, M. W. (1985) Nephrotoxicity of *S*-(2-chloro-1,1,2-trifluoroethyl)glutathione and *S*-(2-chloro-1,1,2-trifluoroethyl)-L-cysteine, the glutathione and cysteine conjugates of chlorotrifluoroethene. J. Pharmacol. Exp. Ther. 235:851–857.
117. Green, T., and Odum, J. (1985) Structure/activity studies of the nephrotoxic and mutagenic action of cysteine conjugates of chloro- and fluoroalkenes. Chem.-Biol. Interact. 54:15–31.
118. Dekant, W., Lash, L. H., and Anders, M. W. (1987) Bioactivation mechanism of the cytotoxic and nephrotoxic *S*-conjugate *S*-(2-chloro-1,1,2-trifluoroethyl)-L-cysteine. Proc. Natl. Acad. Sci. USA 84:7443–7447.
119. Banki, K., Elfarra, A. A., Lash, L. H., and Anders, M. W. (1986) Metabolism of *S*-(2-chloro-1,1,2-trifluoroethyl)-L-cysteine to hydrogen sulfide and the role of hydrogen sulfide in *S*-(2-chloro-1,1,2-trifluoroethyl)-L-cysteine-induced mitochondrial toxicity. Biochem. Biophys. Res. Commun. 138:707–713.

120. Fokin, A. V., Skladnev, A. A., and Knunyants, I. L. (1961) Reaction of fluoroolefins with hydrogen sulfide. Doklady Akad. Nauk SSR 138:1132–1135.
121. Dekant, W., Berthold, K., Vamvakas, S., and Henschler, D. (1988) Thioacylating agents as ultimate intermediates in the β-lyase catalyzed metabolism of *S*-(pentachlorobutadienyl)-L-cysteine. Chem.-Biol. Interact. 67:139–148.
122. Dekant, W., Berthold, K., Vamvakas, S., Henschler D., and Anders, M. W. (1988) Thioacylating intermediates as metabolites of *S*-(1,2-dichlorovinyl)-L-cysteine and *S*-(1,2,2-trichlorovinyl)-L-cysteine formed by cysteine conjugate β-lyase. Chem. Res. Toxicol. 1:175–178.
123. Raasch, M. S. (1970) Bis(trifluoromethyl)thioketene. I. Synthesis and cycloaddition reactions. J. Org. Chem. 35:3470–3483.
124. Raasch, M. S. (1972) Bis(trifluoromethyl)thioketene. II. Acyclic derivatives. J. Org. Chem. 37:1347–1356.
125. Dekant, W., Urban, G., Görsmann, C., and Anders, M. W. (1991) Thioketene formation from α-haloalkenyl 2-nitrophenyl disulfides: Models for biological reactive intermediates of cytotoxic S-conjugates. J. Am. Chem. Soc. 113:5120-5122.
126. Hargus, S. J., and Anders, M. W. (1991) Immunochemical detection of covalently modified kidney proteins in *S*-(1,1,2,2-tetrafluoroethyl)-L-cysteine-treated rats. Biochem. Pharmacol. 42:R17–R20.
127. Hayden, P. J., Ichimura, T., McCann, D. J., Pohl, L. R., and Stevens, J. L. (1991) Detection of cysteine conjugate metabolite adduct formation with specific mitochondrial proteins using antibodies raised against halothane metabolite adducts. J. Biol. Chem. 266:18415–18418.
128. Hayden, P. J., Welsh, C. J., Yang, Y., Schaefer, W. H., Ward, A. J. I., and Stevens, J. L. (1992) Formation of mitochondrial phospholipid adducts by nephrotoxic cysteine conjugate metabolites. Chem. Res. Toxicol. 5:231–237.
129. Hayden, P. J., Yang, Y., Ward, A. J. I., Dulik, D. M., McCann, D. J., and Stevens, J. L. (1991) Formation of difluorothionoacetyl-protein adducts by *S*-(1,1,2,2-tetrafluoroethyl)-L-cysteine metabolites: Nucleophilic catalysis of stable lysyl adduct formation by histidine and tyrosine. Biochemistry 30:5935–5943.
130. Fisher, M. B., Hayden, P. J., Bruschi, S. A., Dulik, D. M., Yang, Y., Ward, A. J. I., and Stevens, J. L. (1993) Formation, characterization, and immunoreactivity of lysine thioamide adducts from fluorinated nephrotoxic cysteine conjugates in vitro and in vivo. Chem. Res. Toxicol. 6:223–230.
131. Harris, J. W., Dekant, W., and Anders, M. W. (1992) In vivo detection and characterization of protein adducts resulting from bioactivation of haloethene cysteine S-conjugates by ^{19}F NMR: Chlorotrifluoroethene and tetrafluoroethene. Chem. Res. Toxicol. 5:34–41.
132. Reichert, D., Ewald, D., and Henschler, D. (1975) Generation and inhalation toxicity of dichloroacetylene. Food Cosmet. Toxicol. 13:511–515.
133. Potter, C. L., Gandolfi, A. J., Nagle, R., and Clayton, J. W. (1981) Effects of inhaled chlorotrifluoroethylene and hexafluoropropene on the rat kidney. Toxicol. Appl. Pharmacol. 59:431–440.
134. Ishmael, J., Pratt, I., and Lock, E. A. (1982) Necrosis of the pars recta (S_3 seg-

ment) of the rat kidney produced by hexachloro 1:3 butadiene. J. Pathol. 138:99–113.
135. National Cancer Institute (1986). Carcinogenesis Bioassay of Tetrachloroethylene. National Toxicology Program Technical Report 232. U.S. Department of Health, Education, and Welfare, Public Health Service, Washington, D.C.
136. National Cancer Institute (1986). Carcinogenesis Bioassay of Trichloroethylene. National Toxicology Program Technical Report 311. U.S. Department of Health, Education, and Welfare, Public Health Service, Washington, D.C.
137. Kociba, R. J., Keyes, D. G., Jersey, G. C., Ballard, J. J., Dittenber, D. A., Quast, J. F., Wade, L. E., Humiston, C. G., and Schwetz, B. A. (1977) Results of a two year chronic toxicity study with hexachlorobutadiene in rats. Am. Ind. Hyg. Assoc. J. 38:589–602.
138. Jaffe, D. R., Hassall, C. D., Brendel, K., and Gandolfi, A. J. (1983) In vivo and in vitro nephrotoxicity of the cysteine conjugate of hexachlorobutadiene. J. Toxicol. Environ. Health 11:857–867.
139. Dilley, J. V., Carter, J., V. L., and Harris, E. S. (1974) Fluoride ion excretion by male rats after inhalation of one of several fluoroethylenes or hexafluoropropene. Toxicol. Appl. Pharmacol. 27:582–590.
140. Gradiski, D., Duprat, P., Magadur, J.-L., and Fayein, E. (1975) Etude toxicologique experimentale de l'hexachlorobutadiene. Eur. J. Toxicol. 8:180–187.
141. Harleman, J. H., and Seinen, W. (1979) Short-term toxicity and reproduction studies in rats with hexachloro-(1,3)-butadiene. Toxicol. Appl. Pharmacol. 47:1–14.
142. Lock, E. A., and Ishmael, J. (1979) The acute toxic effects of hexachloro-1:3-butadiene on the rat kidney. Arch. Toxicol. 43:47–57.
143. Monks, T. J., Anders, M. W., Dekant, W., Stevens, J. L., Lau, S. S., and van Bladeren, P. J. (1990) Glutathione conjugate mediated toxicities. Toxicol. Appl. Pharmacol. 106:1–19.
144. Dekant, W., Vamvakas, S., Berthold, K., Schmidt, S., Wild, D., and Henschler, D. (1986) Bacterial β-lyase mediated cleavage and mutagenicity of cysteine conjugates derived from the nephrocarcinogenic alkenes trichlorethylene, tetrachloroethylene and hexachlorobutadiene. Chem.-Biol. Interact. 60:31–45.
145. Vamvakas, S., Elfarra, A. A., Dekant, W., Henschler, D., and Anders, M. W. (1988) Mutagenicity of amino acid and glutathione *S*-conjugates in the Ames test. Mutat. Res. 206:83–90.
146. Vamvakas, S., Dekant, W., Berthold, K., Schmidt, S., Wild, D., and Henschler, D. (1987) Enzymatic transformation of mercapturic acids derived from halogenated alkenes to reactive and mutagenic intermediates. Biochem. Pharmacol. 36:2741–2748.
147. Finkelstein, M. B., Vamvakas, S., Bittner, D., and Anders, M. W. (1994) Structure-mutagenicity and structure-cytotoxicity studies on bromine-containing cysteine-*S*-conjugates and related compounds. Chem. Res. Toxicol. 7:157–163.
148. Reichert, D., Neudecker, T., Spengler, U., and Henschler, D. (1983) Mutagenicity of dichloroacetylene and its degradation products trichloroacetyl chloride, trichloroacryloyl chloride, and hexachlorobutadiene. Mutat. Res. 117:21–29.
149. Reichert, D., Neudecker, T., and Schütz, S. (1984) Mutagenicity of hexachloro-

butadiene, perchlorobutenoic acid and perchlorobutenoic acid chloride. Mutat. Res. 137:89–93.
150. Reichert, D., and Schütz, S. (1986) Mercapturic acid formation is an activation and intermediary step in the metabolism of hexachlorobutadiene. Biochem. Pharmacol. 35:1271–1275.
151. Henschler, D. (1987) Mechanisms of genotoxicity of chlorinated aliphatic hydrocarbons. In: Selectivity and Molecular Mechanisms of Toxicity (DeMatteis, F., and Lock, E. A., eds.), pp. 153–181. Macmillan, New York.
152. Vamvakas, S., Kordowich, F. J., Dekant, W., Neudecker, T., and Henschler, D. (1988) Mutagenicity of hexachloro-1,3-butadiene and its S-conjugates in the Ames test—role of activation by the mercapturic acid pathway in its nephrocarcinogenicity. Carcinogenesis 9:907–910.
153. Vamvakas, S., Herkenhoff, M., Dekant, W., and Henschler, D. (1989) Mutagenicity of tetrachloroethylene in the Ames test—metabolite activation by conjugation with glutathione. J. Biochem. Toxicol. 4:21–27.
154. Vamvakas, S., Dekant, W., Schiffmann, D., and Henschler, D. (1988) Induction of unscheduled DNA synthesis and micronucleus formation in Syrian hamster embryo fibroblasts treated with cysteine S-conjugates of chlorinated hydrocarbons. Cell Biol. Toxicol. 4:393–403.
155. Vamvakas, S., Köchling, A., Berthold, K., and Dekant, W. (1989) Cytotoxicity of cysteine *S*-conjugates: Structure-activity relationships. Chem.-Biol. Interact. 71:79–90.
156. Jaffe, D. R., Hassall, C. D., Gandolfi, A. J., and Brendel, K. (1985) Production of DNA single strand breaks in renal tissue after exposure to 1,2-dichlorovinylcysteine. Toxicology 35:25–33.
157. McLaren, J., Boulikas, T., and Vamvakas, S. (1984) Induction of poly(ADP-ribosyl)ation in the kidney after in vivo application of renal carcinogens. Toxicology 88:101–104.
158. Bhattacharya, R. K., and Schultze, M. O. (1972) Properties of DNA treated with S-(1,2-dichlorovinyl)-L-cysteine and a lyase. Arch. Biochem. Biophys. 153:105–115.
159. Bhattacharya, R. K., and Schultze, M. O. (1973) Hybridization of DNA modified by interaction with a metabolic fragment from S-(1,2-dichlorovinyl)-L-cysteine. Biochem. Biophys. Res. Commun. 54:538–543.
160. Bhattacharya, R. K., and Schultze, M. O. (1973) Modification of polynucleotides by a fragment produced by enzymatic cleavage of S-(1,2-dichlorovinyl)-L-cysteine. Biochem. Biophys. Res. Commun. 53:172–181.
161. Vamvakas, S., Müller, D. A., Dekant, W., and Henschler, D. (1988) DNA-binding of sulfur-containing metabolites from ^{35}S-(pentachlorobutadienyl)-L-cysteine in bacteria and isolated renal tubular cells. Drug. Metab. Drug Interact. 6:349–358.
162. Stevens, J., Hayden, P., and Taylor, G. (1986) The role of glutathione conjugate metabolism and cysteine conjugate β-lyase in the mechanism of *S*-cysteine conjugate toxicity in LLC-PK1 cells. J. Biol. Chem. 261:3325–3332.
163. Stevens, J. L., Ayoubi, N., and Robbins, J. D. (1988) The role of mitochondrial

matrix enzymes in the metabolism and toxicity of cysteine conjugates. J. Biol. Chem. 263:3395–3401.

164. Vamvakas, S., and Anders, M. W. (1990) Perturbation of Ca^{2+} homeostasis in kidney cells induced by nephrotoxic cysteine *S*-conjugates: New insights with fluorescence digital imaging microscopy. Toxicologist 10:130.
165. Vamvakas, S., and Anders, M. W. (1990) Perturbation of calcium homeostasis as a link between acute cell injury and carcinogenesis in the kidney. Toxicol. Lett. 53:115–120.
166. Meadows, S. D., Gandolfi, A. J., Nagle, R. B., and Shively, J. W. (1988) Enhancement of DMN-induced kidney tumors by 1,2-dichlorovinylcysteine in Swiss-Webster mice. Drug Chem. Toxicol. 11:307–318.
167. Vamvakas, S., Bittner, D., Dekant, W., and Anders, M. W. (1992) Events that precede and that follow *S*-(1,2-dichlorovinyl)-L-cysteine-induced release of mitochondrial Ca^{2+} and their association with cytotoxicity to renal cells. Biochem. Pharmacol. 44:1131–1138.
168. Vamvakas, S., and Köster, U. (1993) The nephrotoxin dichlorovinylcysteine induces expression of the protooncogenes *c-fos* and *c-myc* in LLC-PK_1 cells-A comparative investigation with growth factors and 12-*O*-tetradecanoylphorbolacetate. Cell Biol. Toxicol. 9:1–13.

8

Strategies for Increasing Cellular Glutathione

Alton Meister
Cornell University Medical College, New York, New York

I. INTRODUCTION

This chapter, which considers the means currently available for increasing the levels and supply of cellular glutathione, is based on the premise that such increase would be of value in experiments and in therapy. There is now much evidence that supports this premise and the conclusion that glutathione is the preeminent cellular antioxidant substance. Present knowledge of glutathione metabolism suggests several approaches that can effectively increase cellular glutathione.

Glutathione is a component of the essential antioxidant system that provides animal cells with their reducing milieu (1–4). Glutathione (GSH) is synthesized in the thiol form and is maintained as a thiol by the action of glutathione disulfide (GSSG) reductase, which uses NADPH:

$$GSSG + H^+ + NADPH \rightarrow 2GSH + NADP^+$$

Glutathione functions in the maintenance of the thiol groups of proteins and also of the reduced forms of various other cellular constituents such as ascorbate and α-tocopherol; these reactions may occur nonenzymatically and are also catalyzed by the activity of transhydrogenases. Glutathione not only protects cells against oxidative damage, but also against a variety of toxic compounds of endogenous and exogenous origin. Mitochondria normally produce a substantial quantity of reactive oxygen species (e.g., hydrogen peroxide and superoxide), which are normally destroyed by glutathione-dependent reactions; for example, glutathione peroxidase catalyzes the reduction of hydrogen peroxide:

$$H_2O_2 + 2GSH \rightarrow GSSG + 2H_2O$$

The thiol of glutathione interacts with compounds involved in metabolism, and also with a large variety of drugs and drug metabolites to form S-conjugates of glutathione. For example, S-conjugation of glutathione is involved in the metabolism and functions of compounds of the leukotriene series (5), steroid metabolism (6), estradiol metabolism (7), and in the formation of melanin (8). There is much interest in glutathione S-transferases, especially their interactions with various drugs (9).

The crucially important role of cellular glutathione is dramatically illustrated by observations of animals that have been treated with buthionine sulfoximine, a transition-state inhibitor of γ-glutamylcysteine synthetase, the enzyme that catalyzes the first and rate-limiting reaction in glutathione synthesis (10). Such animals exhibit very low cellular levels of glutathione because cellular export of glutathione continues in the absence of significant intracellular synthesis of the tripeptide. Glutathione deficiency in newborn rats and guinea pigs leads to multiorgan damage, e.g., in lung, liver, kidney, and brain (newborn rats), that is lethal within a few days. Cellular damage is especially evident in the mitochondria, which do not synthesize glutathione, but transport it from the cytosol; cell damage usually occurs when the mitochondrial glutathione levels are decreased to less than about 40% of the normal levels. Glutathione deficiency in adult rats and mice, animals that can synthesize ascorbate (and whose synthesis is enhanced in these animals by glutathione deficiency), leads to morbidity associated with lung and intestinal damage, but this is less severe than that seen in newborn rats and in guinea pigs, animals that do not synthesize ascorbate.

Administration of oxidizing agents or of certain toxic compounds to animals may lead to decrease of glutathione levels. Glutathione deficiency has been observed in several human diseases and may also occur as a result of exposure to certain toxic compounds. Treatment of glutathione deficiency by administration of glutathione might be thought to be a logical and useful approach. However, glutathione, given orally or parenterally, is generally ineffective because it is very poorly transported into most cells. The available data indicate that effective treatments require that glutathione be synthesized intracellularly or that it be delivered to the cell, and the strategies for increasing cellular glutathione discussed here include these two general types. It should also be noted that cellular protection is likely to depend not only on the cellular level of glutathione at the time of challenge, but often on the capacity of the cell to synthesize glutathione. Protection is also afforded by other antioxidants (e.g., ascorbate and α-tocopherol), which are maintained in reduced form by reactions involving glutathione.

II. BIOSYNTHESIS OF GLUTATHIONE

The synthesis of glutathione requires the presence of two synthetases, ATP, Mg^{2+}, and the amino acid substrates (glycine, cysteine, glutamate) (11). The tripeptide is formed by the consecutive actions of γ-glutamylcysteine synthetase and glutathione synthetase:

$$\text{L-glutamate} + \text{L-cysteine} + \text{ATP} \rightleftharpoons \text{L-}\gamma\text{-glutamyl-L-cysteine} + \text{ADP} + \text{Pi}$$

$$\text{L-}\gamma\text{-glutamyl-L-cysteine} + \text{glycine} + \text{ATP} \rightleftharpoons \text{glutathione} + \text{ADP} + \text{Pi}$$

In general, the rate of synthesis of γ-glutamylcysteine determines the rate of glutathione synthesis. Cysteine is usually the rate-limiting substrate; however, under certain conditions, the level of glutamate may also affect the rate of glutathione synthesis (12,13). The reaction catalyzed by γ-glutamylcysteine synthetase is feedback-inhibited by glutathione. This control is nonallosteric; inhibition by glutathione is competitive with respect to glutamate (14). The findings indicate that the γ-glutamyl moiety of glutathione binds to the enzyme site that also binds glutamate.

It is notable that ophthalmic acid (γ-glutamyl-α-aminobutyrylglycine), a glutathione analog in which the thiol is replaced by a methyl group, is only slightly inhibitory as compared to glutathione (13,14). This finding was elucidated by studies on the subunit structure of the enzyme. The holoenzyme (Mr 103,162; rat kidney) consists of a heavy subunit (Mr 72,614), which exhibits synthetase activity feedback-inhibited by glutathione, and a light subunit (Mr 30,548), which by itself has no catalytic activity (13,15–17). Although the heavy subunit of the enzyme has the structural requirements for synthetase activity and for feedback inhibition, both the separated heavy subunit isolated from the holoenzyme and the corresponding recombinant heavy subunit exhibit very low affinity for glutamate ($K_m \approx 18$ mM) as compared to the holoenzyme ($K_m \approx 1.4$ mM). Furthermore, the affinity of the separated heavy subunit for glutathione is greatly increased as compared to the holoenzyme. These and related considerations have led to the conclusion that the light subunit plays a significant regulatory role under physiological conditions. Thus, the heavy subunit alone would not be expected to function effectively under physiological conditions (in which the level of glutamate is ≈3 mM and that of glutathione is ≈4 mM). The in vitro studies show that heavy subunit preparations have much lower affinity for glutamate and much higher affinity for glutathione as compared to the isolated holoenzyme and that binding of the light subunit to the heavy subunit promotes formation of a holoenzyme that has properties that closely match those of the isolated enzyme (17).

Aspects of the feedback inhibition of the enzyme by glutathione were further elucidated by the finding that ophthalmic acid, which inhibits the holoenzyme only slightly, inhibits to a much greater extent after the holoenzyme is treated with dithiothreitol (13). This suggests that a reductive step is involved in inhibition by glutathione and that this significantly affects interactions of glutamate and tripeptides at the active site. A tentative scheme for the oxidized and reduced forms of the enzyme has been suggested (13). According to this idea interaction of the subunits is associated with intersubunit disulfide bond formation, which promotes an enzyme form that is highly active; reduction of this disulfide bond formation, which promotes an enzyme form that is highly active; reduction of this disulfide seems to decrease affinity for glutamate and to increase it for glutathione. This would provide a way in which the cellular level of glutathione could provide a signal that influences γ-glutamylcysteine synthetase activity.

III. CELLULAR DELIVERY OF CYSTEINE

Cellular cysteine is derived from dietary protein and glutathione as well as from dietary methionine by transsulfuration. The availability of cysteine for glutathione synthesis is affected by various factors, for example, the rate of cysteine (or cystine) transport into cells, its metabolic breakdown, and the extent of its use for protein synthesis (18). The observed diurnal variation of cellular glutathione levels, which is correlated with feeding times in rodents (19–22), indicates that the cellular levels of glutathione are not always maximal and suggests that increased intake of cysteine (or of its precursors) might promote higher glutathione levels. Although administration of cysteine may lead to higher cellular glutathione levels, this approach is limited by the fact that cysteine exhibits a relatively high degree of toxicity. For example, addition of cysteine to a basal essential amino acid diet for mice led to weight loss and death (23). Orally administered cysteine was reported to produce necrosis of neurons in mice (24). A single injection of cysteine (1.2 mg/g of body weight) into 4-day-old mice led to brain atrophy (25). Addition of cysteine to culture media was found to be toxic to cells (26).

Cellular levels of cysteine seem to be in the range of 10–100 μM, which is lower than that estimated for the apparent K_m value for γ-glutamylcysteine synthetase (0.35 mM) (18). In contrast, the K_m value for cysteine for human cysteine–tRNA synthetase is about 3 μM. On the basis of the data now available, it would seem that even a moderate increase in cellular cysteine level would promote increased synthesis of glutathione. Fasting or decreased feeding of protein leads to 20–50% lower levels of cellular glutathione. Although administered glutathione is not effectively transported into most cells (see Ref. 27), it can be degraded extracellularly and thus serve as a source of cysteine moieties.

Delivery of cysteine to cells may be achieved by administration of cysteine derivatives that are effectively transported and converted to cysteine intracellularly. Various derivatives of cysteine (peptides, amides, esters, etc.) have been considered (18). It is probable that most of the toxicity exhibited by cysteine itself is exerted extracellularly and is related to the free thiol of this amino acid. Therefore, it seems desirable to seek compounds that are not converted to cysteine extracellularly and that do not have a thiol moiety. Despite these considerations, much attention has been given to *N*-acetylcysteine, which is reported to have low toxicity in humans. This compound is now used as a therapy for acetaminophen toxicity (28–33). Its administration leads to increased glutathione levels, indicating that *N*-acetylcysteine is transported and deacylated. The site or sites of deacetylation have not as yet been established, nor has the nature of the enzyme (or enzymes) involved been determined.

Several thiazolidines have been examined as potentially useful delivery compounds for cysteine. L-Thiazolidine-4-carboxylic acid, which is formed by reaction of L-cysteine and formaldehyde (34,35) can replace dietary cysteine in the growth of rats (36,37). This thiazolidine, a substrate of mitochondrial proline oxidase, is converted to *N*-formylcysteine, which is hydrolyzed to cysteine. 2-Substituted thiazolidine-4-carboxylic acids may also be converted to cysteine by nonenzymatic ring opening followed by hydrolysis (38,39).

L-2-Oxothiazolidine-4-carboxylate is well transported into cells and split by 5-oxoprolinase to CO_2 and L-cysteine (40–42). 5-Oxoprolinase is found in all rodent tissues except for erythrocytes and lens. After administration of L-2-oxothiazolidine-4-carboxylate, cysteine (and glutathione) is formed in many tissues; the experimental findings have been reviewed (18,43). It is notable that L-2-oxothiazolidine-4-carboxylate is more effective than *N*-acetyl-L-cysteine in protecting liver glutathione in acetaminophen-treated mice (41). A substantial increase in brain cysteine levels was found after mice were given L-2-oxothiazolidine-4-carboxylate (44).

Another approach to the increase of tissue glutathione is administration of γ-glutamylcystine and related compounds (45). Such compounds are effectively transported, especially into the kidney. After transport of γ-glutamylcystine, this compound is reduced to form cysteine and γ-glutamylcysteine, which serve as substrates, respectively, of γ-glutamylcysteine synthetase and glutathione synthetase. Such transport, reduction, and utilization are steps in an alternative or "salvage" pathway of glutathione formation in which the reaction catalyzed by γ-glutamylcysteine synthetase is bypassed.

IV. CELLULAR DELIVERY OF GLUTATHIONE

Many animal cells normally export glutathione, but there is little, if any, evidence for the reverse process, i.e., uptake of intact glutathione. [Oral administration of glutathione to mice led to an increase of mucosal glutathione in colon, jejunum, and stomach (46), but this was not demonstrated to be due to direct uptake of intact glutathione. Vina et al. (47) found that oral administration of glutathione to fasted rats led to a significant increase in the level of cysteine in the portal blood plasma.] Although it cannot be ruled out that some (or certain) cells can transport the intact tripeptide, there are considerable data indicating that there is no appreciable (i.e., quantitatively significant) transport of glutathione into most cells (see Ref. 27). These observations led to the search for derivatives of glutathione that might be transported into cells and would be converted into glutathione after transport.

Of the several derivatives thus far examined, esters of glutathione appear to be most suitable. In the initial work, fasted mice were injected intraperitoneally with glutathione or with glutathione mono (glycyl) methyl or ethyl ester, and the tissue levels of glutathione were examined at various times after injection (48). After giving the monoester, there was a significant increase in the tissue glutathione levels. An increase was also found after pretreating the mice with buthionine sulfoximine, thus excluding the possibility that the observed increase is due to cellular synthesis of glutathione. In subsequent studies these findings were confirmed and extended. Experiments with [^{35}S]-labeled glutathione mono ethyl ester showed that the ester is taken up by many tissues of the mouse and that the uptake is greater than found in control studies with [^{35}S]glutathione (18,43). Other work showed preferential uptake of the ester as compared to glutathione by cultured cells and cell suspensions (49). The effectiveness of such esters appears to be related to the absence of a negatively charged group on the glycine residue and to the greater hydrophobicity of the molecule as compared to glutathione itself. Evidence from a number of studies has shown that glutathione monoesters are transported into many types of cells (50,51), that there is absorption of the ester from the gastrointestinal tract after oral administration to mice (50), and that the monoesters of glutathione can protect cells in suspension against radiation (49,52,53) and animals against the effects of a variety of toxic compounds, including heavy metals (54–56), anticancer agents (54,56), and acetaminophen (48).

Glutathione monoesters have been found to protect many tissues against the effects of glutathione deficiency induced by administration of buthionine sulfoximine (27). Treatment of adult mice with buthionine sulfoximine led to skeletal muscle lesions associated with myofiber necrosis; electron microscopy revealed mitochondrial swelling and vacuolization with rupture of cristae and membrane distintegration (12). These effects, which were not found in the heart, were prevented by simultaneous treatment of the mice with glutathione monoester, but not by treatment with glutathione. Treatment with the ester led to maintenance of normal skeletal muscle mitochondrial glutathione levels; the buthionine sulfoximine–treated mice had skeletal muscle mitochondrial levels that were decreased to about 20% of the controls, whereas treatment with the ester led to control levels of glutathione. Interestingly, the mitochondrial glutathione levels in the heart, in the same study, decreased on treatment with buthionine sulfoximine to about 50% of the controls and increased to about 150% of the controls after simultaneous treatment with the ester. The heart is apparently protected against the effects of glutathione deficiency because it contains significant amounts of catalase (57).

Treatment of adult mice with buthionine sulfoximine led to a substantial decrease of the glutathione levels of liver, lung, lymphocytes, and plasma (Table 1) (58). The level of glutathione in the liver mitochondria of the treated animals was about 60% of the controls, whereas that of the lung was about 18% of the controls. Electron microscopy showed no damage in the liver, but there was a significant decline in the number of mitochondria in the lung; there was marked mitochondrial degeneration and also major disruption of the lamellar bodies of the type 2 cells with swelling and loss of their characteristic lattice structure. Simultaneous administration of glutathione monoester (but not of glutathione) prevented the buthionine sulfoximine–induced decrease of mitochondrial glutathione in the liver and lung (Table 1), in fact, the ester-treated animals exhibited mitochondrial glutathione levels higher than those of untreated controls. The ester-treated animals showed no evidence of mitochondrial or lamellar body damage in the lung. These studies show that administration of glutathione ester (but not of glutathione itself) not only prevents decrease of the cellular and mitochondrial levels of glutathione in several tissues of buthionine sulfoximine–treated mice, but also protects against cellular damage. It is notable that administration of glutathione intraperitoneally led to a great increase in plasma glutathione (Table 1), but that even such high levels did not lead to significantly increased levels of glutathione in the tissues or mitochondria, nor did they protect against cellular damage.

Treatment of newborn rats with buthionine sulfoximine leads to much greater morbidity than found in the studies described above on adult animals. Thus, electron microscopy showed decreased numbers of mitochondria in brain, liver, lung, and kidney. In all tissues examined, except for the heart, there was marked mitochondrial degeneration (59,60). In contrast to the results observed in adult animals, there was focal necrosis in liver, proximal tubular degeneration in the kidney, significant cerebral damage, and severe lung damage. The affected animals died within a few days. Tissue damage and mortality were significantly decreased by co-administration of glutathione monoesters [or of ascorbate (see below)]. Similar effects have been found in buthionine sulfoximine–treated guinea pigs (61). In another experimental model, newborn rats were treated with two doses of buthionine sulfoximine, given at 36 and 60 h of age, and their lenses were examined when the rats opened their eyes at 14–15 days of age. Virtually all of the treated animals exhibited cataracts, which were not prevented by co-administration of glutathione, but which were significantly prevented by co-administration of glutathione monoester (62) or of ascorbate (59). In these studies it was shown that administration of the ester increased lens glutathione levels.

Table 1 Effect of Glutathione Monoester on GSH Levels

Experiment no.	Treatment[a]	GSH levels					
		Liver		Lung			
		Total (μmol/g)	Mitochondrial (nmol/mg of protein)	Total (μmol/g)	Mitochondrial (nmol/mg of protein)	Lymphocyte Total (nmol per 10^6 cells)	Plasma Total (μM)
1	None	8.26	6.4	1.69	4.7	1.39	58.3
2	BSO	0.98	3.8	0.36	0.86	0.20	0.90
3	BSO + GSH monoester	8.29	10.8	0.71	10.0	0.45	43.0
4	BSO + GSH	1.10	3.9	0.42	0.87	0.24	5080[b]

[a]Mice were treated with saline (experiment 1) and with BSO (experiments 2–4) for 9 days. The BSO was given intraperitoneally (4 mmol/kg) at 10 a.m. and 6 p.m. In experiment 3, GSH monoisopropyl ester (5 mmol/kg; 1–1.3 ml) was injected intraperitoneally as an isosmolar solution (pH 6.5–6.8) twice daily (8 a.m. and 4 p.m.) for 9 days. In experiment 4, mice were injected with GSH, isopropanol, and Na_2SO_4 in amounts equimolar to the ester given in experiment 3. Mice (4–5 per experiment) were sacrificed at 11 a.m. on day 9, and GSH determinations were done as described. Data are given as means; SDs were ± 0.04–0.08 for tissues and 0.2–480 for plasma.
[b]The GSH levels in plasma (from right ventricle) were 25,100 and 12,080 mM at 50 and 120 min, respectively, after GSH administration.
Source: Ref. 58.

Glutathione monoesters were initially used because of their effectiveness and because they can be made in good yield without much difficulty. Several studies suggested that glutathione diesters might be toxic (27). Furthermore, the diesters are more difficult to prepare than the monoesters; the yields are lower and the possibilities for formation of impurities are increased. Glutathione diethyl ester was found to be rapidly converted to the corresponding mono (glycyl) ester by plasma from rats and mice. The effects of injecting mono- and diethyl esters of glutathione into mice on the tissue levels of glutathione were found to be similar, consistent with such rapid conversion in the plasma. However, when it was found that human plasma lacks glutathione diester α-esterase, it became of interest to compare the transport properties of glutathione mono- and diesters in human cells (63). It was found that human cells (erythrocytes, peripheral blood mononuclear cells, fibroblasts, ovarian tumor cells, and purified T cells) transport glutathione diethyl ester much more effectively than the corresponding monoethyl (glycyl) ester (Fig. 1). The more rapid transport of the diester as compared to the monoester may be ascribed to its greater lipophilicity. Human cells rapidly convert the transported glutathione diethyl ester to the monoester. The monoester, which is converted to glutathione intra-

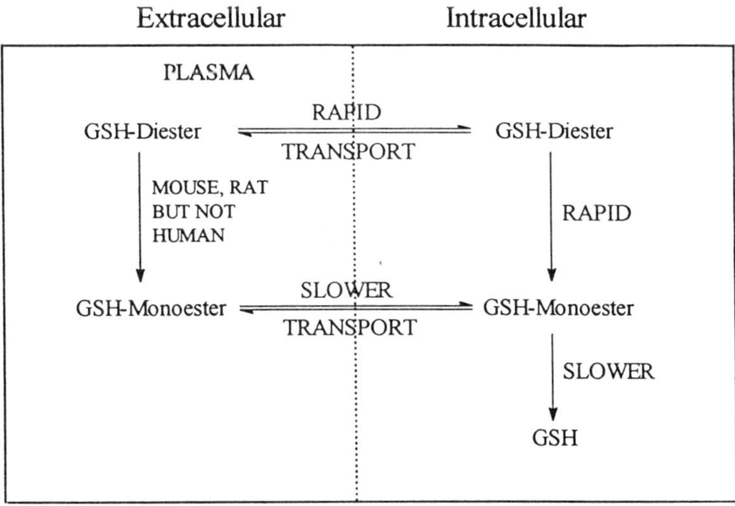

Figure 1 Scheme for the transport and hydrolysis of glutathione monoethyl ester and glutathione diethyl ester. After injection into animals, the diester is split to the mono-(glycyl)ester in the plasma of mice and rats (but not human, hamster, and several others). The diester is transported rapidly into cells, which convert it rapidly to the monoester, which is more slowly converted to glutathione. So the diester is a more effective transport vehicle than the monoester. (From Ref. 63.)

cellularly, is only slowly exported; thus, the diester serves as a monoester delivery compound, which is "trapped" as monoester within the cells. The intracellular levels of the monoester increase to levels that are much higher than the levels of monoester found after application of the monoester to the cells. High levels of monoester provide the cells with a means of producing glutathione over a period of time. The observation that plasma from a number of animals (e.g., rabbit, guinea pig, hamster), like human plasma, does not split the diester to the monoester, provides several possible animal models for further work. Initial studies on the hamster showed that injection of the diethyl ester led to a threefold increase in the level of glutathione in the liver; at this dose level the increase found after giving the monoester is about 0.7-fold.

V. SPARING OF GLUTATHIONE

The availability of cellular glutathione might be increased by giving compounds that would perform functions normally carried out by glutathione. Thus far, ascorbate is the most extensively studied molecule of this type. The important relationships between glutathione and ascorbate in animals became evident from studies on newborn rats that were rendered glutathione deficient by administration of buthionine sulfoximine. It was found that the lethal and other effects of glutathione deficiency can be prevented by giving ascorbate (59). Later work on guinea pigs (61), which, like newborn rats, do not synthesize ascorbate, and on adult mice, which do, have shown that ascorbate and glutathione have certain physiological properties in common and that they can spare each other under appropriate conditions (64–66).

Glutathione-deficient newborn rats develop multiorgan damage and die within a few days. These effects are prevented by giving ascorbate (1–2 mmol/kg of body weight/day), but not by giving dehydroascorbate. Glutathione deficiency in newborn rats leads to marked decline of the ascorbate levels of the tissues. This decrease in tissue ascorbate can be prevented by giving ascorbate, as expected, but interestingly, giving ascorbate also leads to increased levels of glutathione in the tissues and in the mitochondria isolated from them. As discussed above, treatment of newborn rats with buthionine sulfoximine at 36 and 60 h of age leads to cataract formation at a 97% incidence, but the incidence of cataract formation was only 9% in newborn rats also given ascorbate (2 mmol/kg of body weight/day). The mitochondrial glutathione levels of tissues (liver, lung, kidney, brain, lens) of buthionine sulfoximine–treated newborn rats were increased 2.7- to 60-fold when ascorbate was also given (59). Similar findings were made in guinea pigs treated with buthionine sulfoximine. These animals exhibited decreased tissue levels of ascorbate and died within a few days. These effects were largely prevented by giving ascorbate (61).

Evidence that glutathione is spared by ascorbate was also obtained in studies on adult mice given buthionine sulfoximine. Such treatment does not produce damage in liver and kidney, as detected by electron microscopy, because these tissues are protected by the presence of ascorbate. Notably the level of ascorbate in the liver of adult mice increases substantially when the animals are treated with buthionine sulfoximine. Induction of ascorbate synthesis in the liver is apparently triggered by glutathione deficiency (66); the mechanism involved in this phenomenon, which occurs also in the livers of rats treated with certain drugs (67–69), is not yet known. Such induction of ascorbate synthesis does not occur in newborn rats (1–5 days of age). In contrast to the liver and kidney, which are evidently protected by endogenous synthesis of ascorbate, the lungs of adult mice treated with buthionine sulfoximine show structural damage characterized by type 2 cell lamellar body degeneration and decreased levels of phosphatidylcholine in lung and in bronchoalveolar fluid. These changes are prevented in coadministration of ascorbate (70).

Evidence that ascorbate may be spared by glutathione was obtained in studies on guinea pigs given an ascorbate-deficient diet. Such animals develop scurvy and die within 21–24 days. When these guinea pigs were given glutathione monoester, the onset of scurvy was significantly delayed; there were no signs of scurvy after 40 days (65). Treatment with glutathione monoester of guinea pigs receiving a scorbutic diet with glutathione monoester led to increased tissue levels of both ascorbate and glutathione as compared to those of saline-treated controls. This provides evidence that glutathione, given in ester form, can spare ascorbate. The reactions involved in the relationships between glutathione and ascorbate are summarized in Figure 2.

VI. INCREASE OF CELLULAR CAPACITY FOR GLUTATHIONE SYNTHESIS BY GENE TRANSFER

Considerations relating to drug- and radiation-resistant tumor cells that have increased levels of glutathione (27,71–76) and studies on a model bacterial system (77) support the hypothesis that cellular *capacity* for glutathione synthesis—as opposed to the cellular *level* of glutathione—is likely to be a major factor in the protection of cells. In the model system, a strain of *Escherichia coli*, enriched in its content of γ-glutamylcysteine synthetase and glutathione synthetase by recombinant DNA techniques, was examined for radiation resistance (77). The gene-enriched strain was found to be much more radioresistant than the wild strain of origin. Although the gene-enriched strain had higher glutathione levels than the wild type, the observed radioresistance was found to be associated with the capacity of the gene-enriched strain to synthesize glutathione when irradiated rather than to the cellular levels of glutathione per se. This conclu-

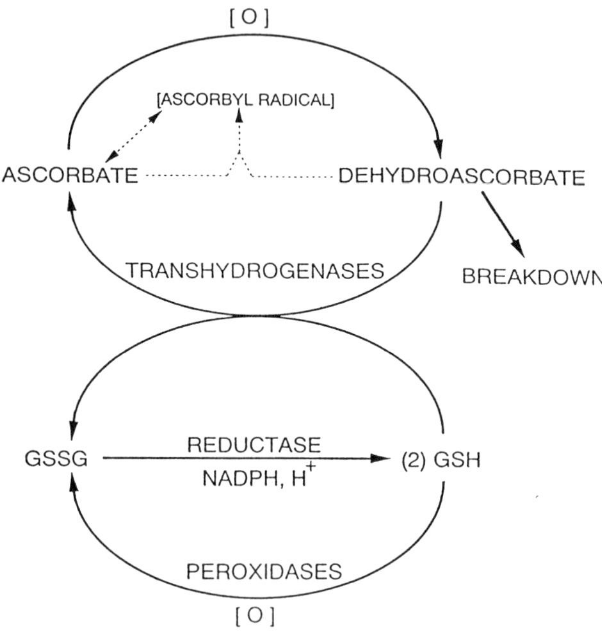

Figure 2 Relationships between ascorbate and glutathione in animals. Reactive oxygen species ([O]) can interact with ascorbate and with glutathione. Dehydroascorbate may be reduced nonenzymatically and enzymatically by reaction with glutathione and also by the semidehydroascorbate reductase pathway (see Ref. 64).

sion was indicated by the finding that radioresistance was abolished by adding buthionine sulfoximine, which does not act directly to decrease glutathione levels, but which decreases the ability of the cells to synthesize glutathione in response to a challenge.

This concept appears applicable to findings on certain mammalian cell systems. Although cells with high capacity for glutathione synthesis may have high levels of glutathione, the cellular capacity for glutathione synthesis may be the underlying factor in cellular resistance. The relative importance of such capacity as compared to glutathione levels may depend to some extent on the intensity and type of challenge. These considerations may explain some of the data in the literature that do not show a close correlation between cellular glutathione levels and resistance (to drugs and radiation) and cell protection (27). A recent study of adriamycin-sensitive and adriamycin-resistant human ovarian cells obtained by biopsy led to findings that support the view that the cellular capacity to synthesize glutathione is a more important indicator of resistance than the cellular level of glutathione (78). Thus, although there was a poor correlation between drug resistance or sensitivity and cellular glutathione content, the resistant cells maintained a high level of glutathione through synthesis during exposure to adriamycin, whereas the sensitive cell lines did not turn on synthesis.

The studies reviewed above indicate that cellular glutathione may be increased by increasing the quantities of the two synthetases needed for glutathione synthesis. This approach has been achieved in a bacterial system, which may serve as a model for studies involving gene transfer in animals. In contrast to *E. coli*, whose γ-glutamylcysteine synthetase consists of a single polypeptide chain (Mr ≈ 56,000) (79), mammalian γ-glutamylcysteine has, as noted above, two subunits, which are separately coded for (16,17). The preparation of transgenic mice that have increased capacity to catalyze glutathione synthesis is now feasible and is in progress. Since γ-glutamylcysteine synthetase catalyzes the rate-limiting reaction in glutathione synthesis, it may be possible to obtain animals with such increased capacity by increasing only the level of this enzyme. Alternatively, the levels of both synthetases may need to be increased. The possibility that transgenic animals of this type may in addition require increased dietary levels of cysteine (or of its precursors) to maintain high cellular levels of glutathione also needs to be considered.

VII. DISCUSSION

There is now much interest in the idea that a wide range of human conditions such as aging, cancer, atherosclerosis, arthritis, viral infections (including AIDS), stroke, myocardial infarction, pulmonary conditions, bowel disease, neurodegenerative disease, and others may be produced (or made worse) by reactive oxygen species [e.g., oxygen free radicals, hydrogen peroxide (80)]. There is related interest in the treatment or prevention of such conditions by administration of antioxidants including ascorbate, α-tocopherol, and β-carotene. Some of the studies that underlie this "free radical hypothesis of disease" and the proposed preventatives or therapies are based on epidemiological approaches that involve data collected on nutritional intake. In some instances administration of vitamin supplements is suggested; experience has shown that these are essentially nontoxic. At this time, the hypothesis, though highly attractive, is unproven (see, for example, Refs. 81 and 82).

Studies in our laboratory on the functions of glutathione appear relevant to this hypothesis. Thus, there is clear evidence for an essential glutathione-ascorbate antioxidant pathway in animals (64). The dramatic effects of glutathione deficiency emphasize the importance of maintaining the intracellular reducing milieu and the cellular reducing power. Normal cells produce substantial quantities of reactive oxygen species, which are normally destroyed by the glutathione system. A deficiency of this system leads to significant morbidity. It has been suggested (27,60) that glutathione deficiency, induced by giving an inhibitor of glutathione synthesis, would serve as a useful model of disease. These considerations do not exclude the possibility that antioxidants other than glutathione may play roles in cellular protection. Indeed, the data thus far obtained indicate that ascorbate, α-tocopherol, probably other substances, and certain enzymes are of importance. If the actions of various "antioxidants" are indeed *antioxidant* in nature as postulated, it would be expected that after oxidation, they would need to be regenerated by reduction. Or if they are not reduced, they would need to be replaced. For example, dehydroascorbate, if not reduced, is irreversibly degraded. There is now substantial evidence that glutathione functions in the regeneration of ascorbate from dehydroascorbate, and also that it functions directly, or via ascorbate, in reducing the tocopheroxy radical (83–88). These considerations suggest that the cellular reducing power may be of major significance in combatting various types of oxidative injury associated with formation of reactive oxygen substances. Thus, an increase in tissue glutathione levels and/or glutathione synthesis capacity might be advantageous, especially when tissue or plasma glutathione levels are decreased by disease or when there is excessive oxidative stress and chemical toxicity.

It seems evident that individuals who have genetically induced deficiencies of either of the synthetases needed for the formation of glutathione would be expected to benefit from therapy that supplies the tripeptide or the missing enzyme. An inborn error involving the cystathionine pathway may lead to a deficiency of cysteine and thus of glutathione. Such a deficiency may occur in premature infants (89); decreased levels of glutathione have been found in preterm as compared to full-term infants (70). Cysteine deficiency may, in principle, be remedied by administration of compounds such as L-2-oxothiazolidine-4-carboxylate and *N*-acetyl-L-cysteine. Treatment of patients with inborn deficiency to glutathione synthetase with glutathione esters might be useful, but it is not yet certain as to how effectively the esters are transported across the blood-brain barrier. The possibility that treatment with ascorbate might be of value in patients with deficiency of the synthetases has been suggested (59) and has been examined in one patient with apparent favorable result (90). Other aspects of the treatment of inborn errors of glutathione metabolism have been recently reviewed (91).

As discussed elsewhere (64,92), there is now good reason to think that cellular glutathione may function to decrease the formation of oxidized low-density lipoprotein (LDL), which is currently thought to play an important role in the development of atherosclerosis (93). Many years ago, Willis (94) came to the conclusion that acute and chronic scurvy produced in guinea pigs leads to atherosclerosis. The basic observation was confirmed by Ginter (95). Willis found that administration of ascorbic acid inhibited atherosclerosis induced by feeding cholesterol, and he suggested the use of ascorbate as a therapy for atherosclerosis. It is of interest to see that there is recent evidence that ascorbic acid can prevent oxidation of LDL (96). Although human trials with α-tocopherol, ascorbate, and β-carotene are apparently in progress (97), it would also be of interest to try another relatively nontoxic therapy, such as L-2-oxothiazolidine-4-carboxylate, that would increase cellular glutathione.

Mention should also be made of the reports indicating that a deficiency of glutathione is associated with AIDS. Thus, decreased levels of glutathione were found in the venous plasma and lung epithelial lining fluid of symptom-free HIV-seropositive individuals (98); earlier studies led to similar findings on peripheral blood monocytes, lymphocytes, and plasma (99). Clinical trials of L-2-oxothiazolidine-4-carboxylate and *N*-acetyl-L-cysteine are now in progress.

An important role of glutathione in the lectin-activated proliferation of lymphocytes was indicated by the finding that this process is inhibited by treatment of the cells with buthionine sulfoximine (100). Later work (101–105) has confirmed and extended these findings and supports the suggestion (98,101) that treatment of AIDS patients with compounds that increase cellular glutathione might be beneficial. Additional interest in this approach is stimulated by findings that suggest that cellular thiols play a role in regulation of genes affected by the human immunodeficiency virus (HIV-LTR) (106–110).

Another condition that may be favorably affected by therapy that leads to increased cellular glutathione is the adult respiratory distress syndrome (ARDS). In this disease there is diffuse lung infiltration, apparently by activated granulocytes and perhaps other cells, which leads to progressive pulmonary capillary endothelial damage and development of complications involving failure of other organs (111). Plasma and erythrocyte levels of glutathione are reported to be decreased. In a recent study, 46 patients with ARDS were treated at random with placebo, *N*-acetyl-L-cysteine, and L-2-oxothiazolidine-4-carboxylate. Significant protection was observed in the treated patients, who had fewer organ failures (112). Further trials are needed to verify this apparently protective effect.

Currently understood relationships between carcinogenesis and oxygen free radicals, which are complex and depend upon the carcinogen, the target tissue, and perhaps other factors, have not as yet led to a global answer (113). Nevertheless, increased cellular glutathione in normal cells, relative to the tumor cells, as has been discussed previously (27), might improve the usefulness of certain anticancer agents. [Increased tumor cell glutathione is the basis for resistance of some tumors to anticancer agents and radiation (see Ref. 27).] Mirvish (114) has recently reviewed the "strong circumstantial evidence" that nitrosation caused by reaction of nitrites with secondary amines and *N*-substituted amides under the acidic conditions of the stomach are carcinogenic. Ascorbic acid protects against this reaction and would thus be expected to decrease carcinogenesis in the stomach and several other organs. The apparent protection of the stomach by ascorbic acid, which is actively secreted into human stomach, may be relevant to the observation that, as compared to jejunum and colon, the stomach mucosa was found to be relatively more resistant to the effects of experimental generalized glutathione deficiency (115).

It has long been known that glutathione protects animals against acetaminophen toxicity (see, for example, Refs. 28–33,41). Glutathione ester was found to protect cells against radiation in vitro (49,53,116). There is evidence that glutathione ester can protect normal cells against the toxic effects of a variety of anticancer agents without significantly affecting the antitumor activity. It is postulated that inflammatory bowel disease (117) and certain degenerative diseases of the nervous system (118), possibly including amyotrophic lateral sclerosis (119), may involve formation of oxidants. The suggested connections between glutathione and the immune system (101,120) and the process of aging (121) may reflect still other potentially significant effects of increasing cellular glutathione on human welfare. Of the several strategies described above, administration of cysteine-delivery or glutathione-delivery compounds is the most available at this time. Future developments based on gene-transfer systems may ultimately turn out to be more effective.

The studies reviewed here indicate that cellular glutathione in animals can be increased by several approaches based on current knowledge of the biochemistry of this molecule, which is the major cellular antioxidant compound. It seems reasonable to think that the glutathione levels of cells fluctuate between a high, perhaps maximal level (controlled by feedback inhibition of γ-glutamylcysteine synthetase and by cellular export), and a significantly lower level (dependent upon the availability of cysteine and other factors). We have at this time only rough estimates of what these values might actually be and only general ideas about the temporal variations in glutathione levels of different cells and about the capacity of cells for glutathione synthesis. It seems clear that cellular protection depends also on other molecules, such as ascorbate. The availability of strategies that can increase cellular glutathione should facilitate experimental and clinical approaches that will illuminate relationships postulated to occur between the formation of reactive oxygen substances and the development of human disease.

ACKNOWLEDGMENTS

Research carried out in the author's laboratory was supported in part by National Institutes of Health Grant 2R37 DK12034 from the U.S. Public Health Service (National Institute of Diabetes and Digestive and Kidney Diseases). The author acknowledges the skillful assistance of Mrs. Selma Haschemeyer in the preparation of this manuscript.

REFERENCES

1. Larsson, A., Orrenius, A., Holmgren, A., and Mannervik, B., eds. (1983) Function of Glutathione. Biochemical, Physiological, Toxicological, and Clinical Aspects. Raven Press, New York.
2. Dolphin, D., Poulson, R., and Avramovic, O., eds. (1989). Glutathione: Chemical, Biochemical, and Medical Aspects, Parts A and B, Coenzyme and Cofactors, Vol. III. John Wiley, New York.
3. Taniguchi, N., Higashi, T., Sakamoto, Y., and Meister, A., eds. (1989). Glutathione Centennial. Molecular Perspectives and Clinical Implications. Academic Press, New York.
4. Meister, A., and Anderson, M. E. (1983). Glutathione. Annu Rev. Biochem. 52:711–760.
5. Orning, L., Hammarstrom, S., and Samuelsson, B. (1980) Leukotriene D: A slow reacting substance from rat basophilic leukemic cells. Proc. Natl. Acad. Sci. USA 77:2014–2017.
6. Benson, A. M., Talalay, P., Keen, J. H., and Jakoby, W. B. (1977) Relationship between the soluble glutathione-dependent Δ^5-3-ketosteroid isomerase and the glutathione S-transferases of the liver. Proc. Natl. Acad. Sci. USA 74:158–162.
7. Kuss, E. (1971) Microsomale Oxidation des Ostradiols-17β. Hoppe-Seyler's Z. Physiol. Chem. 352:817–836.
8. Prota, G. (1980) Cysteine and glutathione in mammalian pigmentation. In: Natural Sulfur Compounds; Novel Biochemical and Structural Aspects (Cavallini, D., Gaull, G. E., and Zappia, V., eds.). Plenum Press, New York.
9. Tew, K. D., and Clapper, M. L. (1988). Glutathione S-transferases and anticancer drug resistance. In: Mechanisms of Drug Resistance in Neoplastic Cells (Woolley, P. V., III, and Tews, K. D., eds.), pp. 141–159. Academic Press, New York.
10. Meister, A. (1988) Glutathione metabolism and its selective modification. J. Biol. Chem. 263:17205–17208.
11. Meister, A. (1974) Glutathione synthesis. Enzymes 10:671–697.
12. Martensson, J., and Meister, A. (1989). Mitochondrial damage in muscle occurs after marked depletion of glutathione and is prevented by giving glutathione mono ester. Proc. Natl. Acad. Sci. USA 86:471–475.
13. Huang, C.-S., Chang, L-S., Anderson, M. E., and Meister, A. (1993) Catalytic and regulatory properties of the heavy subunit of rat kidney γ-glutamylcysteine synthetase. J. Biol. Chem. 268:19675–19678.
14. Richman, P. G., and Meister, A. (1975) Regulation of γ-glutamyl cysteine synthetase by nonallosteric feedback inhibition by glutathione. J. Biol. Chem. 250:1422–1426.
15. Seelig, G. F., and Meister, A. (1985) γ-Glutamylcysteine synthetase from erythrocytes, Methods Enzymol. 113:390–392.
16. Yan, N., and Meister, A. (1990) Amino acid sequence of rat kidney γ-glutamylcysteine synthetase. J. Biol. Chem. 265:1588–1593.
17. Huang, C.-S., Anderson, M. E., and Meister, A. (1993) Amino acid sequence and

function of the light subunit of rat kidney γ-glutamylcysteine synthetase. J. Biol. Chem. 268:20578–20583.
18. Anderson, M. E., and Meister, A. (1987) Intracellular delivery of cysteine. Methods Enzymol. 143:313–325.
19. Edwards, S., and Westerfield, W. W. (1952) Blood and liver glutathione during protein deprivation. Proc. Soc. Exp. Biol. Med. 79:57–59.
20. Beck, L. V., Rieck, V. D., and Duncan, B. (1958) Diurnal variation in mouse and rat liver sulfhydryl. Proc. Soc. Exp. Biol. Med. 97:229–231.
21. Tateishi, M., Higashi, T., Shinya, S., Naruse, A., and Sakamoto, Y. (1974) Studies on the regulation of glutathione level in rat liver. J. Biochem. Tokyo 75:93–103.
22. Tateishi, M., Higashi, T., Naruse, A., Nakashima, K., Shiozaki, H., and Sakamoto, Y. (1977) Rat liver glutathione: Possible role as a reservoir of cysteine. J. Nutr. 107:51–60.
23. Birnbaum, S. M., Winitz, M., and Greenstein, J. P. (1957) Quantitative nutritional studies with water-soluble, chemically defined diets. III. Individual amino acids as sources of "non-essential" nitrogen. Arch. Biochem. Biophys. 72:428–436.
24. Olney, J. W., Ho, O.-L, and Rhee, V. (1971). Cytotoxic effects of acidic and sulphur containing amino acids on the infant mouse central nervous system. Brain Res. 14:61–76.
25. Karlsen, R. L., Grofava, I., Malthe-Sorensen, D., and Fornum, E. (1981) Morphological changes in rat brain induced by L-cysteine injection in newborn animals. Exp. Brain Res. 208:167–180.
26. Nishiuch, Y., Sasaki, M., Nakayasu, M., and Oikawa, A. (1976) Cytotoxicity of cysteine in culture media. In Vitro 12:635.
27. Meister, A. (1991). Glutathione deficiency produced by inhibition of its synthesis and its reversal: Applications in research and therapy. Pharmacol. Ther. 51:155–194.
28. Mitchell, J. R., Jollow, D. L., Potter, W. Z., Davis, D. C., Gillette, J. R., and Brodie, B. B. (1973) Acetaminophen-induced hepatic necrosis. I. Role of drug metabolism. J. Pharmac. Exp. Ther. 187:185.
29. Mitchell, J. R., Jollow, D. J., Potter, W. Z., Gillette, J. R., and Brodie, B. B. (1973) Acetaminophen-induced hepatic necrosis. IV. Protective role of glutathione. J. Pharmac. Exp. Ther. 187:211.
30. Potter, W. Z., Davis, D. C., Mitchell, J. R., Jollow, D. J., Gillette, J. R., and Brodie, B. B. (1973) Acetaminophen-induced hepatic necrosis. III. Cytochrome P-450-mediated covalent binding in vitro. J. Pharmac. Exp. Ther. 187:203.
31. Jollow, D. J., Mitchell, J. R., Potter, W. Z., David, D. C., Gillette, J. R., and Brodie, B. B. (1973) Acetaminophen-induced hepatic necrosis. II. Role of covalent binding in vivo. J. Pharmac. Exp. Ther. 187:195.
32. Piperno, E., and Bersennbruegge, D. A. (1976) Reversal of experimental paracetamol toxicosis with N-acetylcysteine. Lancet (Oct. 2):738–739.
33. Smilkstein, M. J., Knapp, G. I., Kulig, K. W., and Rumack, B. H. (1988) Efficacy of oral N-acetylcysteine in the treatment of acetaminophen overdose. Analysis of the National multicenter study (1976 to 1985). N. Engl. J. Med. 319:1557–1562.

34. Schubert, M. P. (1936) Compounds of thiol acids with aldehydes. J. Biol. Chem. 114:341.
35. Ratner, S., and Clarke, H. T. (1937) The action of formaldehyde upon cysteine, J. Am. Chem. Soc. 59:200.
36. MacKenzie, C. G., and Harris, J. (1975) N-Formylcysteine synthesis in mitochondria from formaldehyde and L-cysteine via thiazolidinecarboxylic acid. J. Biol. Chem. 227:393.
37. Debey, H. H., MacKenzie, J. B., and MacKenzie, C. G. (1958) The replacement by thiazolidinecarboxylic acid of exogenous cysteine and cysteine. J. Nutr. 66:607–619.
38. Nagasawa, H. T., Goon, D. J. W., Zera, R. T., and Yuzon, D. L. (1982) Prodrugs of L-cysteine as liver protective agents. 2(RS)-methylthiazolidine-4(R)-carboxylic acid, a latent cysteine. J. Med. Chem. 25:489–491.
39. Nagasawa, H. T., Goon, D. J. W., Muldoon, W. P., and Zera, R. T. (1984) 2-Substituted thiazolidine-4(R)-carboxylic acids as prodrugs of L-cysteine. Protection of mice against acetaminophen hepatotoxicity. J. Med. Chem. 27:591–596.
40. Williamson, J. M., and Meister, A. (1981) Stimulation of hepatic glutathione formation by administration of L-2-oxothiazolidine-4-carboxylate, A 5-oxo-L-prolinase substrate. Proc. Natl. Acad. Sci. USA 78:936–939.
41. Williamson, J. M., Boettcher, B., and Meister, A. (1982) An intracellular cysteine delivery system that protects against toxicity by promoting glutathione synthesis. Proc. Natl. Acad. Sci. USA 79:6246–6249.
42. Williamson, J. M., and Meister, A. (1982) New substrates of 5-oxo-L-prolinase, J. Biol. Chem. 257:12,039–12,042.
43. Meister, A., Anderson, M. E., and Hwang, O. (1986) Intracellular cysteine and glutathione delivery systems. J. Am. Coll. Nutr. 5:137–151.
44. Anderson, M. E., and Meister, A. (1989) Marked increase of cysteine levels in many regions of brain after administration of 2-oxothiazolidine-4-carboxylate. FASEB J. 3:1632–1636.
45. Anderson, M. E., and Meister, A. (1983) Transport and direct utilization of γ-glutamylcyst(e)ine for glutathione synthesis. Proc. Natl. Acad. Sci. USA 80:707–711.
46. Martensson, J., Jain, A., and Meister, A. (1990) Glutathione is required for intestinal function. Proc. Natl. Acad. Sci. USA 87:1715–1719.
47. Vina, J., Perez, C., Furukawa, T., Palacin, M., and Vina, J. R. (1989) Effect of oral glutathione on hepatic glutathione levels in rats and mice. Br. J. Nutr. 62:683–691.
48. Puri, R. N., and Meister, A. (1983) Transport of glutathione as γ-glutamylcysteinylglycyl ester, into liver and kidney. Proc. Natl. Acad. Sci. USA 80:5258–5260.
49. Wellner, V. P., Anderson, M. E., Puri, R. N., Jensen, G. L., and Meister, A. (1984) Radioprotection by glutathione ester; transport of glutathione ester into human lymphoid cells and fibroblasts. Proc. Natl. Acad. Sci. USA 81:4732–4735.
50. Anderson, M. E., Powrie, F., Puri, R. N., and Meister, A. (1985) Glutathione monoethyl ester; preparation, uptake by tissues, and conversion to glutathione. Archs. Biochem. Biophys. 239:538–548.

51. Anderson, M. E., and Meister, A. (1989) Glutathione monoesters. Analyt. Biochem. 183:16–20.
52. Jensen, G. L., and Meister, A. (1983) Radioprotection of human lymphoid cells by exogenously-supplied glutathione is mediated by γ-glutamyl transpeptidase. Proc. Natl. Acad. Sci. USA 80:4714–4717.
53. Astor, M. B., Anderson, M. E., and Meister, A. (1988) Relationship between intracellular glutathione levels and hypoxic cell radiosensitivity. Pharmacol. Ther. 39:115–121.
54. Anderson, M. E., Naganuma, A., and Meister, A. (1990) Protection against cisplatin toxicity by administration of glutathione ester. FASEB J. 4:3251–3255.
55. Naganuma, A., Anderson, M. E., and Meister, A. (1990) Cellular glutathione is a determinant of sensitivity to mercuric chloride toxicity: Prevention of toxicity by giving glutathione monoester. Biochem. Pharmacol. 40:693–697.
56. Teicher, B. A., Crawford, J. M., Holden, S. A., Lin, Y., Cathcart, K. N. S., Luchette, C. A., and Flatow, J. (1988) Glutathione monoethyl ester can selectively protect liver from high dose BCNU or cyclophosphamide. Cancer 62:1275–1281.
57. Radi, R., Turrens, J. F., Change, L. Y., Bush, K. M., Crapo, J. D., and Freeman, B. A. (1991) Detection of catalase in rat heart mitochondria. J. Biol. Chem. 266:22028–22034.
58. Martensson, J., Jain, A., Frayer, W., and Meister, A. (1989) Glutathione metabolism in the lung: Inhibition of its synthesis leads to lamellar body and mitochondrial defects. Proc. Natl. Acad. Sci. USA 86:5926–5300.
59. Martensson, J., and Meister, A. (1991) Glutathione deficiency decreases tissue ascorbate levels in newborn rats; ascorbate spares glutathione and protects. Proc. Natl. Acad. Sci. USA 89:4656–5660.
60. Martensson, J., Jain, A., Stole, E., Frayer, W., Auld, P. A. M., and Meister, A. (1991) Inhibition of glutathione synthesis in the newborn rat; a model for endogenously-produced oxidative stress. Proc. Natl. Acad. Sci. USA 88:9360–9364.
61. Griffith, O. W., Han, J., and Martenson, J. (1991) Vitamin C protects adult guinea pigs against tissue damage and lethality caused by buthionine sulfoximine-mediated glutathione depletion. FASEB J. 5:4708 (abstract).
62. Martensson, J., Steinherz, R., Jain, A., and Meister, A. (1989) Glutathione ester prevents buthionine sulfoximine-induced cataracts and lens epithelial cell damage. Proc. Natl. Acad. Sci. USA 86:8727–8731.
63. Levy, E. J., Anderson, M. E., and Meister, A. (1993) Transport of glutathione diethyl ester into human cells. Proc. Natl. Acad. Sci. USA 90:9171–9175.
64. Meister, A. (1994) Glutathione-ascorbic acid antioxidant system in animals. J. Biol. Chem. 269:9397–9400.
65. Martensson, J., Han, J., Griffith, O. W., and Meister, A. (1993). Glutathione ester delays the onset of scurvy in ascorbate-deficient guinea pigs. Proc. Natl. Acad. Sci. USA 90:317–321.
66. Martensson, J., and Meister, A. (1992) Glutathione deficiency increases hepatic ascorbic acid synthesis in adult mice. Proc. Natl. Acad. Sci. USA 89:11566–11568.
67. Holloway, D. E., and Peterson, F. J. (1984) Ascorbic acid in drug metabolism. Drugs Pharm. Sci. 21:225–295.

68. Burns, J. J., Mosbach, E. H., and Schulenberg, S. (1954), Ascorbic acid synthesis in normal and drug-treated rats, studied with L-ascorbic-1-C^{14} acid. J. Biol. Chem. 207:679–687.
69. Zannonis, V. G., Brodfuehrer, J. I., Smart, R. C., and Susick, R. L., Jr. (1987) Ascorbic acid, alcohol, and environmental chemicals. Ann. NY Acad. Sci. 458:364–388.
70. Jain, A., Martensson, J., Mehta, T., Krauss, A. N., Auld, P. A. M., and Meister, A. (1992) Ascorbic acid prevents oxidative stress in glutathione-deficient mice; effects on lung type-2-cell lamellar bodies, lung surfactant, and skeletal muscle, Proc. Natl. Acad. Sci. USA 89:5093–5097.
71. Meister, A., and Griffith, O. W. (1979) Effects of methionine sulfoximine analogs on the synthesis of glutamine and glutathione; possible chemotherapeutic implications. Cancer Treat. Rep. 63:1151–1121.
72. Clement, J., Griffith, O. W., and Meister, A., cited in Ref. 27.
73. Vistica, D. T., and Ahmad, S. (1989) Acquired resistance of tumor cells to L-phenylalanine mustard: Implications for the design of a clinical trial involving glutathione depletion. In: Glutathione Centennial Molecular Perspectives and Clinical Implications (Taniguchi, N., Higashi, T., Sakamoto, Y., and Meister, A., eds.), pp. 301–315. Academic Press, New York.
74. Hamilton, T. C., Winker, M. A., Louie, K. G., Batist, G., Behrens, B. C., Tsuro, T., Grotzinger, K. R., McKoy, W. M., Young, R. C., and Ozols, R. F. (1985) Augmentation of adriamycin, melphalan, cisplatin cytotoxicity in drug-resistant and -sensitive human ovarian carcinoma cell lines by buthionine sulfoximing mediated glutathione depletion. Biochem. Pharmacol. 34:2583–2586.
75. Ozols, R. F., Hamilton, T. C., Louie, K. G., Behrens, B. C., and Young, R. C. (1986) Glutathione depletion with buthionine sulfoximine: Potential clinical applications. In: Thioredoxin and Glutaredoxin Systems: Structure and Function (Holmgren, A., Branden, C. I., Jornvall, H., and Sjoberg, B.-M., eds.), pp. 277–293. Raven Press, New York.
76. Ozols, R. F., Louie, K. G., Plowman, J., Behrens, B. C., Fine, R. L., Dykes, D., and Hamuilton, T. C. (1987) Enhanced melphalan cytotoxicity in human ovarian cancer in vitro and in tumor-bearing nude mice by buthionine sulfoximine depletion of glutathione. Biochem. Pharmacol. 36:147–153.
77. Moore, W. B., Anderson, M. E., Meister, A., Murata, K., and Kimura, A. (1989) Increased capacity for glutathione synthesis enhances resistance to radiation in *Escherichia coli*: A possible model for mammalian cell protection. Proc. Natl. Acad. Sci. USA 86:1461–1464.
78. Lee, F. Y. F., Siemann, D. W., and Sutherland, R. M. (1989) Changes in cellular glutathione content during adriamycin treatment in human ovarian cancer—a possible indicator of chemosensitivity. Br. J. Cancer 60:291–298.
79. Huang, C.-S., Moore, W., and Meister, A. (1988) On the active site thiol of γ-glutamylcysteine synthetase: relationships to catalysis, inhibition, and regulation, Proc. Natl. Acad. Sci. USA 85:2464–2468.
80. Halliwell, B., and Gutteridge, J. M. C. (1989). Free Radicals in Biology and Medicine, 2nd ed. Clarendon Press, Oxford.

81. Herbert, V. (1994) The antioxidant supplement myth. Am. J. Clin Nutr. 60:157–158.
82. Hennekens, C. H., Buring, J. E., and Peto, R. (1994) Antioxidant vitamins—benefits not yet proved. N. Engl. J. Med. 330:1080–1081.
83. Bendich, A., Machlin, L. J., and Scandurra, O. (1986) Antioxidant role of vitamin C. Adv. Free Radical Biol. Med. 2:419–444.
84. Reddy, C. C., Scholz, R. W., Thomas, C. E., and Massaro, E. J. (1982) Vitamin E dependent reduced glutathione inhibition of rat liver microsomal lipid peroxidation. Life Sci. 31:571–576.
85. Hill, K. E., and Burk, R. F. (1984) Influence of vitamin E and selenium on glutathione-dependent protection against microsomal lipid peroxidation. Biochem. Pharmacol. 33:1065–1068.
86. Haenen, G. R. M. M., Tai Tin Tsoi, J. N. L., Vermeulen, N. P. E., Timmerman, H., and Bast, A. (1987) 4-Hydroxy-2,3-trans-nonenal stimulates microsomal lipid peroxidation by reducing the glutathione-dependent protection. Arch. Biochem. Biophys. 259:449–456.
87. McCay, P. B., and Powell, S. R. (1989) Relationship between glutathione and chemically-induced lipid peroxidation. In: Glutathione: Chemical, Biochemical and Medical Aspects, Part B (Dolphin D., et al., eds.), pp. 111–151. John Wiley and Sons, New York.
88. McCay, P. B., Brueggemann, G., Lai, E. K., and Powell, S. R. (1989) Evidence that alpha-tocopherol functions cyclically to quench free radicals in hepatic microsomes. Requirement for glutathione and a heat-labile factor. Ann. NY Acad. Sci. 570:32–45.
89. Sturman, J. A., Gaul, G., and Raiha, N. C. R. (1970) Absence of cystathionase in human fetal liver: Is cystine essential? Science 169:74–76.
90. Jain, A., Buist, N. R. M., Kennaway, N. G., Powell, B. R., Auld, P. A. M., and Martensson, J. (1994) Effect of ascorbate or N-acetylcysteine treatment in a patient with herditary glutathione synthetase deficiency. J. Ped. 124:229–233.
91. Meister, A., and Larsson, A. (1994) Glutathione synthetase deficiency and other disorders of the γ-glutamyl cycle. In: The Metabolic Basis of Inherited Disease, 7th ed. (Scriver, C. R., Beaudet, A. L., Sly, W. S., and Valle, D., eds.), pp. 1461–1477.
92. Meister, A. (1992) On the antioxidant effects of ascorbic acid and glutathione. Biochem. Pharmacol. 44:1905–1915.
93. Yla-Herttuala, S., Palinski, W., Rosenfeld, M. E., Parthasarathy, S., Carew, T. E., Butler, S., Witztum, J. L., and Steinberg, D. (1989) Evidence for the presence of oxidatively modified low density lipoprotein in atherosclerotic lesions of rabbit and man. J. Clin. Invest 84:1086–1095.
94. Willis, G. C. (1953) An experimental study of the intimal ground substance in atherosclerosis. Can. Med. Assoc. J. 69:17–22.
95. Ginter, E. (1978) Marginal vitamin C deficiency, lipid metabolism, and atherogenesis. Adv. Lipid Res. 16:167–220.
96. Jialal, I., and Grundy, S. M. (1991) Preservation of the endogenous antioxidants in low density lipoprotein by ascorbate but not probucol during oxidative modification. J. Clin. Invest. 87:597–601.

97. Witztum, J. L. (1994) The oxidation hypothesis of atherosclerosis. Lancet 344:793–795.
98. Buhl, R., Holroyd, K. J., Mastrangeli, A., Cantin, A. M., Jaffe, H. A., Wells, F. B., Saltini, C., and Crystal, R. G. (1989) Systemic glutathione deficiency in symptom-free HIV-seropositive individuals. Lancet 8765:1294–1298.
99. Eck, H.-P., Gmuender, H., Hartmann, M., Petzoldt, D., Daniel, V., and Droege, W. (1989) Low concentrations of acid-soluble thiol (cysteine) in the blood plasma of HIV-1 infected patients. Biol. Chem. Hoppe Seyler 370:101–108.
100. Meister, A. (1983) Selective modification of glutathione metabolism. Science 220:471–477.
101. Suthanthiran, M., Anderson, M. E., Sharma, V. K., and Meister, A. (1990) Glutathione regulates activation-dependent DNA synthesis in highly purified normal human T lymphocytes stimulated via the CD2 and CD3 antigens. Proc. Natl. Acad. Sci. USA 87:3343–3347.
102. Hamilos, D. L., and Wedner, H. J. (1985) The role of glutathione in lymphocyte activation. Comparison of inhibitory effects of buthionine sulfoximine and 2-cyclohexene-1-one by nuclear size transformation. J. Immunol. 135:2740–2747.
103. Hamilos, D. L., Zelarney, P., and Mascali, J. J. (1989) Lymphocyte proliferation in glutathione-depleted lymphocytes: Direct relationship between glutathione availability and the proliferative response. Immunopharmacology 18:223–235.
104. Messina, J. P., and Lawrence, P. P. (1989) Cell cycle progression of glutathione-depleted human peripheral blood mononuclear cells is inhibited at S phase. J. Immunol. 143:1974–1981.
105. Fidelus, R. K., Ginouves, P., Lawrence, D., and Tsan, M. (1987). Modulation of intracellular glutathione concentrations alters lymphocyte activation and proliferation. Exp. Cell Res. 170:269–275.
106. Duh, E. J., Maury, W. J., Folks, T. M., Fauci, A. S., and Rabson, A. B. (1989): Tumor necrosis factor α activates human immunodeficiency virus type 1 through induction of nuclear factor binding to the NF-κB sites in the long terminal repeat. Proc. Natl. Acad. Sci. USA 86:5974–5978.
107. Staal, F. J. T., Roederer, M., Herzenberg, L. A., and Herzenberg, L. A. (1990) Intracellular thiols regulate activation of nuclear factor κB and transcription of human immunodeficiency virus. Proc. Natl. Acad. Sci. USA 87:9943–9947.
108. Roederer, M., Staal, F. J. T., Raju, P. A., Ela, S. W., Herzenberg, L. A., and Herzenberg, L. A. (1990) Cytokine-stimulated human immunodeficiency virus replication is inhibited by N-acetyl-L-cysteine. Proc. Natl. Acad. Sci. USA 87:4884–4888.
109. Toledano, M. B., and Leonard, W. J. (1991) Modulation of transcription factor NF-κB binding activity by oxidation-reduction in vitro. Proc. Natl. Acad. Sci. USA 88:4328–4332.
110. Kalebic, T., Inter, A., Poli, G., Anderson, M. E., Meister, A., and Fauci, A. (1991) Suppression of HIV expression in chronically infected monocytic cells by glutathione, glutathione ester, and N-acetyl cysteine. Proc. Natl. Acad Sci. USA 88:986–990.
111. Cross, C. E., van der Vliet, A., O'Neill, C. A., and Eiserich, J. P. (1994) Reactive oxygen species and the lung. Lancet 344:930–932.

112. Bernard, G. R., Dupont, W., Edens, T., Higgins, S., Jiang, K., Morris, P., Paz, H., Russell, J., Steinberg, K., Stroud, M., Swindell, B., Wheeler, A. P., and Wright, P. (1994) Antioxidants in the acute respiratory distress syndrome (ARDS). Am. J. Resp. Crit. Care Med. 149:A241.
113. Cerutti, P. A. (1994) Oxy-radicals and cancer. Lancet 344:862–863.
114. Mirvish, S. S. (1994) Experimental evidence for inhibition of N-nitroso compound formation as a factor in the negative correlation between vitamin C consumption and the incidence of certain cancers. Cancer Res. (suppl.):19485–19515.
115. Martensson, J., Jain, A., and Meister, A. (1990) Glutathione is required for intestinal function. Proc. Natl. Acad. Sci. USA 87:1715–1719.
116. Vos, O., and Roos-Verhey, W. S. D. (1988) Radioprotection by glutathione esters and cysteamine in normal and glutathione-depleted mammalian cells. Int. J. Radiat. Biol. 53:273–281.
117. Grisham, M. B. (1994) Oxidants and free radicals in inflammatory bowel disease. Lancet 344:859–861.
118. Jenner, P. (1994) Oxidative damage in neurodegenerative disease. Lancet 344:796–798.
119. Deng, H.-X., Hentati, A., Tainer, J. A., Iqbal, Z., Cayabyab, A., Hung, W.-Y., Getzoff, E. D., Hu, P, Herzfeldt, B., Roos, R. P., Warner, C., Deng, G., Soriano, E., Smyth, C., Parge, H. E., Ahmed, A., Roses, A. D., Hallewell, R. A., Pericak-Vance, M. A., and Siddique, T. (1993) Amyotrophic lateral sclerosis and structural defects in Cu, Zn superoxide dismutase. Science 261:1047–1052.
120. Furukawa, T., Meydani, S. N., and Blumberg, J. B. (1987) Reversal of age-associated decline in immune responsiveness by dietary glutathione supplementation in mice. Mech. Ageing Dev. 38:107–117.
121. Lang, C., Naryshkin, S., Schneider, D., Mills, B., and Lindeman, R. (1992) Low blood glutathione levels in healthy aging adults. J. Lab. Clin. Med. 120:720–725.

9

Glutathione Synthesis in Oxidative Stress

Henry Jay Forman and Rui-Ming Liu
University of Southern California, Los Angeles, California

Michael Ming Shi
Harvard School of Public Health, Boston, Massachusetts

I. INTRODUCTION

Glutathione [γ-L-glutamyl-L-cysteinyl-glycine (GSH)] is the most abundant intracellular nonprotein thiol and is implicated in many cellular functions, including detoxification of xenobiotics, synthesis of cellular compounds, cell cycle regulation, and the regulation of gene expression (1–5). The most studied function of GSH is as an antioxidant. GSH can directly scavenge free radicals or act as a cosubstrate in the reduction of H_2O_2 and lipid hydroperoxides by the glutathione peroxidases (GSHPx). GSH also forms conjugates with a variety of compounds either nonenzymatically or through the action of glutathione S-transferases (see Chap. 7). These various activities of GSH protect cells against oxidative damage produced by radiation, ultraviolet light, photodynamic effects, and xenobiotic agents. After reviewing the pathways responsible for altering GSH concentrations in the cell, we will describe the role of GSH in protection against oxidative damage and the effects oxidative stress has on the intracellular GSH concentration. Finally, we will focus on recent evidence demonstrating the induction by oxidative stress of enzymes responsible for GSH synthesis.

II. PATHWAYS DETERMINING GLUTATHIONE CONCENTRATION

Intracellular GSH concentration is determined principally by four pathways: the γ-glutamyl cycle, conversion of other amino acids, the glutathione redox cycle,

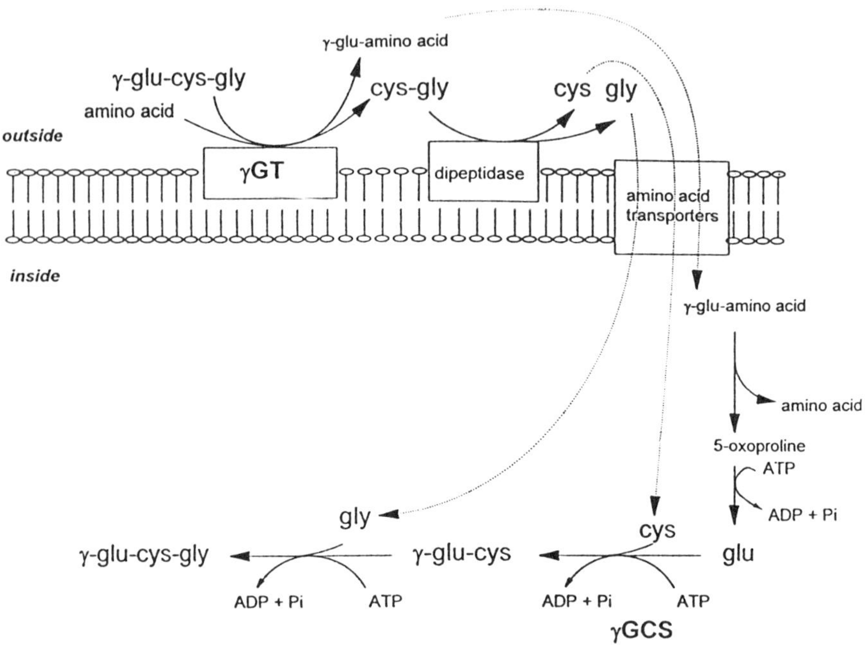

Figure 1 Breakdown of extracellular GSH (γ-glu-cys-gly) and de novo GSH synthesis in the γ-glutamyl cycle. The steps composing the cycle (except for transport of GSH from inside to outside) are shown. The enzymes γ-glutamyl transpeptidase (γGT) and γ-glutamylcysteine synthetase that are discussed in the chapter are shown in bold.

and the mercapturate pathway. In a few cells, GSH is directly taken up from the surrounding extracellular fluid (6–10).

A. The γ-Glutamyl Cycle

The γ-glutamyl cycle described by Meister (11) (Fig. 1) is responsible for most de novo synthesis of GSH, utilization of extracellular GSH for this purpose, and may also function as an amino acid transport system. De novo synthesis of GSH is a two-step process: the formation of γ-glutamylcysteine from the constituent amino acid glutamic acid and cysteine catalyzed by γ-glutamylcysteine synthetase (γ-GCS), and the formation of the final product GSH from γ-glutamylcysteine and glycine catalyzed by glutathione synthetase. The first step, the synthesis of γ-glutamylcysteine, is the rate-limiting step in GSH synthesis when sufficient cysteine is present. The activity of γGCS is feedback-inhibited by the final product GSH, and it has been calculated that 80% of cytosolic γGCS is inactive due to binding of GSH under physiological conditions (12,13). A substantial depletion in cellular GSH concentrations would therefore be expected to cause a release of the GSH bound to γGCS, which would then result in an enhanced synthesis of GSH. This is one way in which cells control their GSH level when challenged by agents that lead to an initial depletion of intracellular GSH (14,15).

The rate of GSH synthesis is also controlled by the intracellular availability of constituent amino acids, especially intracellular cysteine concentration. Because the apparent K_M value for cysteine for γGCS (0.35 mM) is not far from intracellular cysteine concentration, the intracellular availability of cysteine is considered as another rate-limiting factor in the synthesis of GSH (12,13). For most cells, the cysteine required for GSH synthesis is obtained by degradation of circulating GSH catalyzed by a membrane bound enzyme, γ-glutamyl transpeptidase (γGT) (16,17). γGT transfers the γ-glutamyl moiety of extracellular GSH to an amino acid (cystine is the most active amino acid acceptor of the γ-glutamyl group), forming γ-glutamyl amino acid and cysteinylglycine. The γ-glutamyl amino acids are transported into the cell and metabolized to release the amino acid and 5-oxoproline, which can be converted to glutamate. Cysteinylglycine is degraded by dipeptidases on the surface of cells, and the two resulting amino acids are then transported into the cell. The uptake of the precursor amino acids, cysteine, glutamic acid, and glycine is mediated by several transport systems (18). Many cell types in culture respond to GSH depletion or oxidative stress by increasing the activities of these transport systems and thereby increase their supplies for de novo GSH synthesis (19–26). Nevertheless, the crucial element of this part of the γ-glutamyl cycle is to provide a source of cysteine that, unlike the other two constituent amino acids of GSH, is not found in circulation.

Although most cell types cannot directly transport GSH into the cytosol from the plasma, a few epithelial cells have been reported to have this capacity (9,27). It appears, however, that GSH uptake is by the same transporter as is used for uptake of γ-glutamyl amino acids so that under physiological conditions GSH uptake will compete with products of the γ-GT-catalyzed reaction. The physiological role of the γ-glutamyl amino acid transporter remains an unsettled question; is its function to transport GSH, to act as an auxiliary amino acid transport system, or both?

B. Conversion of Other Amino Acids to Cysteine

The supply of cysteine is crucial to GSH synthesis. Cystine can be taken up by cells and reduced to cysteine; however, due to its solubility (the lowest of all common amino acids) and the tendency to undergo disulfide exchange with proteins, its concentration is very low in plasma. Methionine, the other sulfur-containing amino acid, can also serve as a source of cysteine in some tissues.

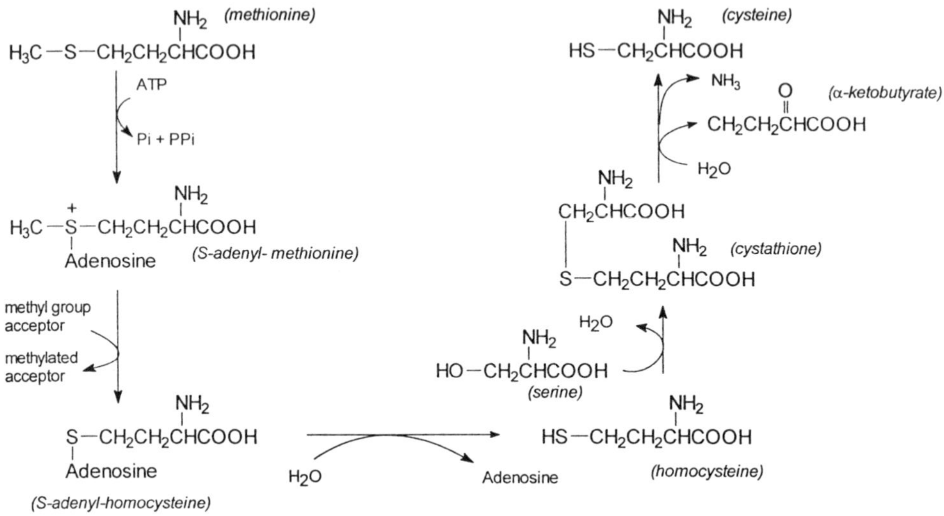

Figure 2 The cystathionine pathway. Methionine and serine are used to produce cysteine.

In the liver, which is a major exporter of GSH, the cystathionine pathway (Fig. 2) provides a significant source of cysteine for GSH synthesis. Recently, it has been found that the lung has the capacity for synthesis of cysteine from methionine and that this increases significantly during hyperoxic exposure (28); however, which of the 40 cell types in the lung are involved and whether this provides a significant source of cysteine for GSH synthesis remains to be further studied.

C. The GSH Redox Cycle

Probably the best characterized aspect of GSH metabolism is its oxidation and reduction during oxidative stress (Fig. 3). During the exposure to exogenous or endogenous oxidants, H_2O_2 and lipid peroxides can be produced. The major enzymes in degradation of these compounds are the glutathione peroxidases (GSHPx), which are selenoproteins that catalyze the reduction of these hydropeptides by GSH (29–32). The oxidized glutathione (GSSG) produced is then reduced to GSH by NADPH through the glutathione reductase reaction. Under normal conditions, most of intracellular GSH is maintained in its reduced state. If the production of H_2O_2 or lipid hydroperoxides is rapidly increased in oxidative stress, the depletion of intracellular GSH can occur due to the oxidation of GSH into GSSG. If the rate of oxidation of GSH is greater than the reduction of GSSG, GSSG can accumulate. GSSG can either be excreted from the cell or react with a protein sulfhydryl leading to formation of a mixed disulfide.

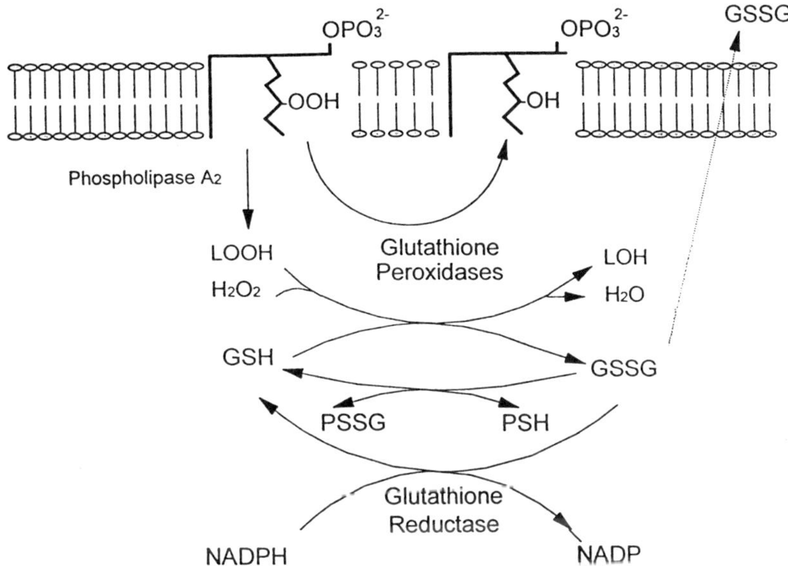

Figure 3 Oxidation and reduction of GSH by the enzymatic pathways. The glutathione peroxidases (GSHPx) reduce either phospholipid hydroperoxides, or lipid hydroperoxides released from membrane by phospholipase A_2 and H_2O_2 with GSH. The GSSG produced may be transported from the cell, reduced by glutathione reductase to GSH, or react with protein sulfhydryls (PSH) to produce mixed disulfides (PSSG). The PSSG may then undergo further exchange to produce protein disulfides (PSSP).

D. Mercapturate Pathway

Xenobiotics, of which most are electrophiles, their metabolites, and various endogenous compounds can form conjugates with reduced glutathione (see Chap. 7). These can be formed either nonenzymatically or enzymatically by reactions catalyzed by glutathione S-transferases. The conjugates formed are usually excreted from the cell. Although this pathway is a very important detoxification pathway, the net result of GSH conjugate formation is a loss of intracellular GSH.

III. EFFECT OF INTRACELLULAR GSH CONCENTRATION ON OXIDATIVE STRESS

A. Depletion of Intracellular GSH Augments Oxidative Stress

GSH is one of the most important reducing agents in the cell. In conjunction with ascorbate and vitamin E, GSH provides a renewable supply of antioxidant capacity to cells. A decrease in the intracellular GSH content of cells increases susceptibility to the toxicity of oxidant exposure (33–37). Radiotherapy is a major modality for the treatment of malignant tumors. Although the mechanism of radiation damage is not completely understood, clearly the free radicals produced by radiation are likely to be responsible for much of its biological effect. As GSH is a strong free radical scavenger, the depletion of GSH would be expected to enhance the lifetime of oxygen-reactive radicals (38) and to increase the sensitivity of tumor cells to the radiation treatment. This hypothesis is supported by the results of many studies (33,37,39–44). For example, Miller and Henderson found that cells treated with the γGCS inhibitor, buthionine sulfoximine (BSO), and then exposed to photodynamic therapy and x-irradiation decreased survival directly correlated with the extent of GSH depletion (35). These authors also found that genetically GSH-deficient human fibroblasts, in which the GSH level is only 6% that of the parent cells, were much more sensitive to photodynamic and x-irradiation treatment than the parent cells. Similarly, the radiosensitivity of retinoblastoma cells is increased when GSH is depleted by BSO exposure (37). Several mechanisms of radioprotection by GSH have been suggested including radical scavenging, restoration of damaged molecules by reduction, reduction of peroxides, and maintenance of protein thiols in the reduced state (35,45).

Low GSH level in cells, induced by chemical treatment, low-protein diets, or genetic mutation, also can make cells more sensitive to the toxicity of oxidant chemicals and oxygen (34,36,44,46–50). Animals that received *N,N*-bis(2-chloroethyl)-*N*-nitrosourea (BCNU), an inhibitor of glutathione reductase, or diethylmaleate (DEM), an agent that depletes GSH through conjugation, show enhanced toxicity in exposure to hyperbaric O_2 (34,48). Depletion of intracellular GSH by BSO or BCNU increased the sensitivity of tumor cells to oxidative cytolysis and radiosensitization by oxygen (44,47,49,50). Deneke et al. (36) found that feeding rats with low-protein diets increases sensitivity to hyperoxia toxicity apparently by inhibiting the increase in GSH synthesis in the lung that occurs during adaptation to elevated oxygen exposure. A diet specifically deficient in sulfur-containing amino acids has the same effects on lung GSH content as general protein deficiency and exacerbates the effect of selenium deficiency (which causes a marked decrease in GSHPx) on oxygen toxicity of (51). In two cell lines with marked differences in GSH, the one with lower GSH was significantly more sensitive to the toxicity of menadione (MQ), an agent that both redox cycles to generate superoxide and H_2O_2 and directly conjugates with GSH (46). Thus, the relationship between either decreased GSH content and/or inhibition of GSH synthesis and increased sensitivity to a variety of agents that induce oxidative stress is well established.

It has been demonstrated that when the content of GSH in liver is decreased approximately 80% by conjugation, lipid peroxidation occurs spontaneously (52). Some authors have suggested that this result implies that an 80% decrease in

GSH is required for oxidative damage. Nonetheless, one needs to keep in mind that hepatocytes have almost five times as much GSH as other cells and that lipid peroxidation was observed without an exogenous oxidative stress. Alterations in cell function and metabolism occur in oxidative stress with much less GSH depletion. For example, loss of alveolar macrophage respiratory burst activity following either in vivo exposure to hyperoxia or in vitro exposure to hydroperoxides occurs with little or at most transient depletion of GSH (53,54).

B. Protection by Increased Intracellular GSH

Several studies have demonstrated that increasing GSH content provides protection against oxidative stress (7,8,47,48,55–66). The methods used for increasing GSH include incubating cells in medium containing GSH, GSH monoethyl or diethyl esters, *N*-acetyl-cysteine, or other precursors of cysteine or by pretreating cells with low doses of oxidants or GSH-depleting agents. The protection results from either the uptake of intact GSH or increased GSH resynthesis in the cells.

The lung is a primary target for oxidant injury because its large epithelial surface area is in direct contact with the atmosphere. The lung therefore is the target of atmospheric pollutants, such as ozone and nitrogen dioxide, and is at risk of oxidant injury during oxygen therapy. Lungs also have the largest endothelial surface area of all tissues, which makes them a major target of circulating oxidants and xenobiotics. Thus, many investigations into the role of GSH in protection against oxidative injury have focused on the lung. Several studies have shown that extracellular GSH can protect alveolar macrophages, pulmonary epithelial cells, and pulmonary endothelial cells from oxidative damage (8,48,55–59). In most of these studies, it was found that the principal mechanism for use of extracellular GSH was not direct uptake but the breakdown by γGT, followed by de novo GSH synthesis within the cell. Extracellular GSH can protect bovine pulmonary artery endothelial cells from MQ or H_2O_2 (48,58,67). The protection is partially inhibited by acivicin, an inhibitor γGT, or BSO, suggesting that the protection is due to the extracellular breakdown and subsequent intracellular resynthesis of GSH. Part of the protection by extracellular GSH, however, was due to nonenzymatic reactions outside the cell (67). A stably γGT-cDNA–transfected NIH-3T3 fibroblast cell line has been established that uses extracellular GSH much more efficiently than their parent cells when challenged with DEM (68). This cell line could also more efficiently use extracellular GSH to resist the toxicity of 2,3-dimethoxy-1,4-naphthoquinone (DMNQ), a redox-cycling quinone that cannot conjugate with GSH (57). In this case, inhibition of γGT more markedly prevented protection by extracellular GSH.

Neonatal alveolar epithelial type II cells exposed to oxidative stress from paraquat, a redox-cycling herbicide, or 80% O_2 increased in intracellular GSH content due to both an increase of de novo synthesis and GSH uptake (56). Similar results were reported for adult type II lung epithelial cells (8). Protection against the toxicity of *tert*-butylhydroperoxide (*t*BOOH) in type II cells by extracellular GSH was blocked by γ-glutamyl-L-glutamate, an agent that blocks GSH transport, suggesting that the protection resulted from direct uptake of extracellular GSH (59).

Nevertheless, as most cells cannot efficiently take up intact GSH directly from the plasma, another method for increasing intracellular GSH without de novo synthesis is the treatment of cells with GSH monoethylester or GSH diethylester (65,69–71). The esters are permeable to cells and are cleaved within the cytosol to release GSH. Treating cells with cysteine or derivatives of cysteine can also increase the intracellular GSH level; however, cysteine autooxidizes in media generating H_2O_2, which limits its use (47,64,72,73). The protection of cells or animals against oxidant damage by treating cells or animals with GSH esters or cysteine precursors has been reported by several authors (47,64,65).

Exposure to sublethal oxidative stress also increases GSH content (see Sec. IV.A). We have recently investigated the effect of pretreatment with sublethal quinone doses on the tolerance of rat epithelial L2 cells to the toxicity of subsequently administered high doses of quinones. The pretreated L2 cells, which had higher GSH levels, were more resistant than carrier-pretreated cells, which had low GSH content. Protection afforded by this pretreatment was diminished by acivicin and BSO, suggesting that increased GSH synthesis was responsible for the protection (Liu, R.-M., Hu, H., and Forman, H. J., in preparation).

IV. EFFECT OF OXIDATIVE STRESS ON INTRACELLULAR GSH CONTENT

A. Depletion of Intracellular GSH in Oxidative Stress

Depletion of intracellular GSH has been reported with various sources of oxidative stress (1,15,74–80). During the exposure to many xenobiotics, ozone, nitrogen dioxide, or hyperoxia, H_2O_2 is produced. When transition metals such as iron are present, hydroxyl radicals can be formed from H_2O_2 (81), which then initiates peroxidation of fatty acid, producing lipid hydroperoxides. The principal mechanism for depletion of intracellular GSH in oxidative stress begins with its oxidation by hydroperoxides, as shown in Figure 3. Detoxification of H_2O_2 is catalyzed by either catalase or GSHPx, but the detoxification of lipid hydroperoxides only can be catalyzed by GSHPx. Detoxification of H_2O_2 and lipid hydroperoxides by GSHPx leads to the consumption of GSH and formation of GSSG. If the rate of oxidation of GSH exceeds the rate of reduction of GSSG or the activity of glutathione reductase is inhibited, GSSG can accumulate in the cell (74,82–84). GSSG can be excreted from cells (85,86) or can form a mixed disulfide with protein (74,75,82). While the latter reaction partially restores GSH, making protein sulfhydryls a pool of added reducing power, the production of GSSG results in a net loss of intracellular GSH and potential inactivation of sulfhydryl-containing enzymes (87,88). Perfusing lungs with paraquat causes an increase in GSSG release; however, lungs that are deficient in GSHPx released very little GSSG when perfused with paraquat, showing the dependence of the enzymatic pathway for oxidation of GSH (89). An immediate drop in GSH and elevation in GSSG occurs in rat blood treated with *t*BOOH, which is accompanied by a *t*BOOH dose–dependent decrease in total glutathione and a corresponding increase of glutathione-protein mixed disulfides (74). In hydroperoxide-treated rat liver mitochondria, concomitant oxidation of GSH and formation of mixed protein sulfhydryl-glutathione disulfides is not accompanied by GSSG efflux, as mitochondria apparently cannot export GSSG (82).

B. Conjugation of GSH with Oxidant Compounds

Several compounds that can redox cycle to produce H_2O_2, such as MQ, can also form conjugates with GSH. Such conjugates can also redox cycle and may actually be better substrates for reductases than their parent compound (90). For some compounds the nonenzymatic conjugation with GSH proceeds rapidly, while for other compounds conjugate formation is catalyzed by glutathione S-transferases. (Formation of GSH conjugates is reviewed in detail in Chap. 7.) Formation of these conjugates results in depletion of GSH. In oxidative stress the formation of conjugates with GSH (besides GSSG, which is a conjugate itself) occurs in several circumstances. For example, perfusion of isolated rat liver with H_2O_2 or reperfusion of the liver after 90 min of ischemia results in increased activity of microsomal glutathione S-transferase with a concomitant decrease in GSH content (76). Whether their metabolism generates H_2O_2 or not, many electrophiles deplete intracellular GSH by forming a conjugate with GSH that is then transported out of the cell (19,20,46,91).

C. Increased GSH Content in Response to Oxidative Stress

Some oxidants, such as *t*BOOH, H_2O_2, benzoylperoxide, and diamide cause a decrease in GSH content followed by an increase to a content higher than the initial intracellular GSH content (15). To a certain extent this may be due to relief of feedback inhibition of γGCS. Nevertheless, some oxidants can increase GSH content without first producing GSH depletion (28,92–96). The significance of this increase in GSH content in oxidant-treated cells is considered as an adaptive reaction. The mechanisms by which oxidants increase intracellular GSH content include an increase in the rate of transport of GSH or its precursors, an increase in the activities of GSH synthesis–related enzymes (especially γGCS), and increases in the activities of other enzymes such as γGT and glutathione reductase.

D. Increased Transport of GSH and Amino Acids

As described above, normal maintenance of intracellular GSH involves either uptake of GSH or the amino acids used for de novo synthesis. There is evidence suggesting that uptake of both intact GSH and precursor amino acids occurs during oxidative stress. When the intestinal mucosa is exposed to oxidants, GSSG increases but so does the content of total glutathione (22). As preincubation with BSO did not alter this elevation, the results suggest that the mucosa can take up GSH in response to oxidative stress (22). The transport systems for the GSH precursor amino acids cysteine, glutamate, and glycine are common features of cells. The activities of these amino acid transport systems is stimulated when the cellular GSH content is decreased by an oxidant or GSH conjugation (19–26). Bovine pulmonary artery endothelial cells depleted of GSH by DEM, BCNU, arsenite, or hyperoxia respond with enhanced rates of transport of cysteine (23,24,26) and glutamic acid (25,26). The increased rate of uptake of these GSH precursors is associated with the elevation of intracellular GSH above the initial steady-state content. In human fibroblasts and rat hepatocytes, the electrophilic agents DEM, sulfobromophthalein, cyclohex-2-en-1-one, ethacrynic acid, and 1,2-epoxy-3(*p*-nitrophenox)propane can increase the uptake of cysteine (19,20). This increase in cysteine transport is completely blocked by either actinomycin D or cycloheximide, suggesting that new transport protein(s) are induced by exposure to these compounds (19,20). Transport of another GSH precursor, glycine, is increased in rat kidneys after GSH depletion by DEM (21). It has been suggested that intracellular GSH content regulates cystine and glutamate transport so that the depletion of intracellular GSH increases uptake of these amino acids (25,26).

E. Increase of γGCS Activity by Oxidative Stress

γGCS, which catalyzes the first and rate-limiting step in GSH synthesis, is inducible in a variety of circumstances, such as in drug-resistant tumor cells (97–99), heat-shocked erythroid cells (100), antioxidant-treated mice (101), and oxidant-treated cells or animals (102–104). The increased activity of γGCS is accompanied by an elevation of intracellular GSH content. A two- to threefold increase in GSH content is accompanied by a dose- and time-dependent elevation of γGCS-mRNA content in methyl mercury–treated rat kidneys (102). In cultured Chinese hamster V79 cells, the rate of incorporation of cysteine into GSH and the total glutathione content increase after initial depletion by *t*BOOH (15). MQ, which can produce oxidative stress both by depleting GSH through formation of a conjugate and by producing H_2O_2 through redox cycling, increases GSH content and γGCS activity after causing initial depletion of intracellular GSH in bovine pulmonary artery endothelial cells (103). Nevertheless,

DMNQ, which cannot form a conjugate with GSH but can generate H_2O_2 through redox cycling (Fig. 4), increases both GSH content and γGCS activity in the bovine pulmonary artery endothelial cells without causing a significant initial decrease in GSH content (103). DMNQ also can increase the GSH content and γGCS activity in the rat lung epithelial L2 cell line (104).

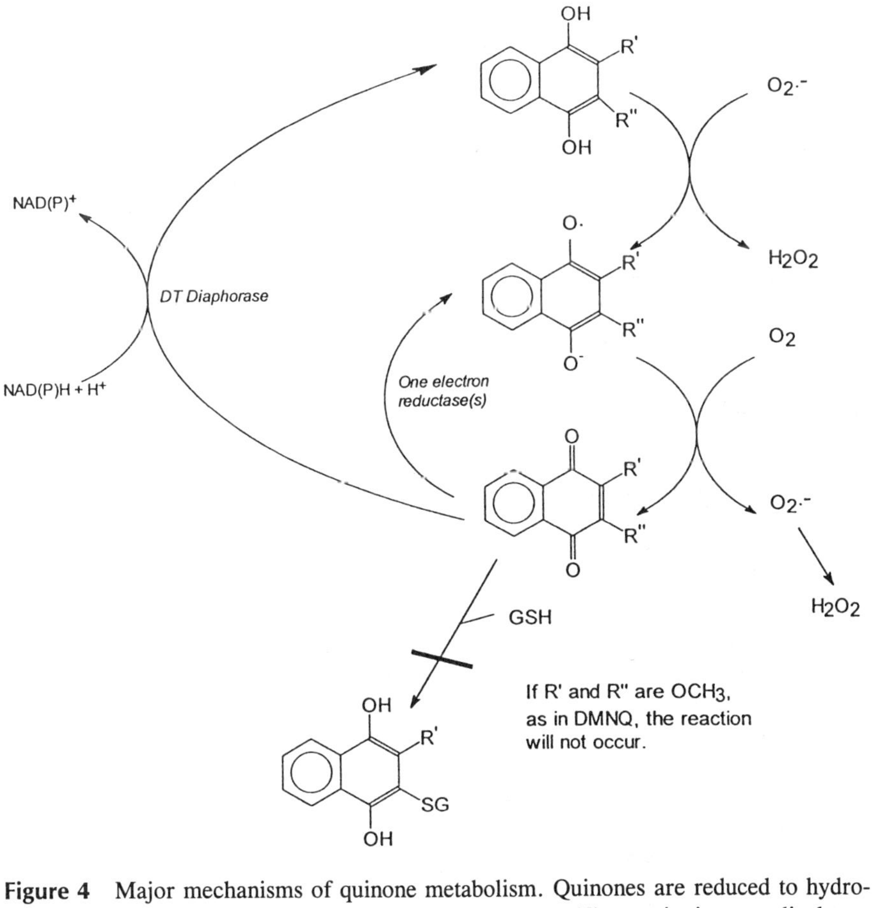

Figure 4 Major mechanisms of quinone metabolism. Quinones are reduced to hydroquinones or semiquinone radicals by cellular reductases. The semiquinone radical rapidly autoxidizes with the regeneration of the parent quinone and formation of superoxide, which can dismute to form H_2O_2. The hydroquinone can react with superoxide to form H_2O_2 and the semiquinone. Most quinones also react with cellular nucleophiles, such as GSH. DMNQ cannot conjugate with GSH.

Wallig et al. (14) suggested several mechanisms for an increase of γGCS activity caused by cyanohydroxybutene: release of feedback inhibition by GSH depletion, an increase in γGCS synthesis, an increase in substrate transport, and stimulation of γGCS activity by conjugation to glutathione. Similarly, oxidative stress may act to increase γGCS activity through all four mechanisms. Presumably, the increase γGCS activity is then responsible for at least part of the increase in intracellular GSH content.

Rat γGCS is composed of two subunits (73 and 30 kDa) that can be dissociated by native gel electrophoresis after dithiothreitol treatment (105). The heavy subunit has independent catalytic activity and demonstrates feedback inhibition by GSH (12). Nonetheless, the light subunit, which is not essential for catalysis, has recently been shown to have a physiologically important regulatory function—the lowering of the K_M for glutamate into the physiological range of concentrations (106). Although Meister and coworkers have suggested that the active form of the enzyme involves formation of a disulfide bridge between the subunits (107), the reducing environment of the cytosol would seem to make this difficult to achieve. Nevertheless, this does offer a potential mechanism for activation of the enzyme by oxidative stress that would promote disulfide exchange. Formation of protein disulfides in oxidative stress probably relies upon the action of GSHPx to produce GSSG, which provides a possible link between hydroperoxide-induced stress and increased γGCS activity (108).

F. Increased Activities of γGT and Other GSH-Related Enzymes

The activities of other enzymes, which directly or indirectly relate to GSH metabolism, have been shown to increase in response to oxidative stress. For example, the activities of glutathione reductase and GSHPx are increased in the lungs of hyperoxia-adapted rats along with other antioxidant enzymes (94,107, 109–112). Interestingly, in an early study correlating an increase in antioxidant enzyme activities in the lung with adaptation to hyperoxia, Cross and coworkers reported that an increase in nonprotein sulfhydryls also occurs (107).

The activity of γGT is increased in MQ-resistant human cancer cells (113). We have recently shown that the activity of γGT increases sevenfold in lung epithelial L2 cells after MQ treatment (114). An accompanying increase in intracellular GSH content caused by MQ is blocked by acivicin, a specific inhibitor of γGT, suggesting an important role of this enzyme in maintaining cellular GSH content in oxidative stress (114). In recent studies, we have found that other oxidants, such as H_2O_2 and *tert*-butylhydroquinone, also induce γGT in L2 cells (unpublished data).

V. γ-GLUTAMYL CYCLE ENZYME GENE REGULATION BY OXIDATIVE STRESS

A. Regulation of γGCS Gene Expression by Oxidative Stress

Although the mechanism of regulation of γGCS at the gene level has not yet been clarified, several studies have shown that the heavy catalytic subunit of γGCS (γGCS-HS-mRNA) increases in oxidant-treated animals or cell lines (101–104,115). Recent evidence from our laboratory has shown that the level of γGCS-HS-mRNA is increased in MQ- and DMNQ-treated bovine pulmonary artery endothelial cells and in DMNQ-treated rat lung epithelial L2 cells (103,104). These increases in mRNA are inhibited by actinomycin D, an inhibitor of transcription, and are only slightly affected by cycloheximide, an inhibitor of translation (103,104). Nuclear run-on experiments, which measure relative transcription rates, showed that the transcription rate of γGCS-HS-mRNA in L2 cells was increased by DMNQ exposure (103). In both the endothelial and epithelial cells, the increase in γGCS-HS-mRNA was accompanied by increases in both γGCS enzyme activity and GSH content.

Potential signaling mechanisms for induction of γGCS include direct or indirect effects of decreased GSH and increased GSSG or mixed disulfides, formation of a glutathione conjugate with a xenobiotic agent, and production of $O_2^{\bullet -}$ and H_2O_2. Talalay et al. (116) demonstrated that several inducers of phase II enzymes are Michael reaction acceptors, i.e., an electrophilic olefin or related electron-deficient center. As 2(3)-*tert*-butyl-4-hydroxyanisole, diethylmaleate, or phorone-treatment increases both hepatic GSH and γGCS activity and γGCS-HS-mRNA in mice, it is postulated that the increase in γGCS-HS-mRNA content could be transcriptionally regulated through the Michael addition of those compounds to GSH to form conjugates (117,118). Michael addition to 1,4-naphthoquinones is limited to the C_2 and/or C_3 position in the quinoid ring (Fig. 4). In DMNQ, however, the electron-donating properties of the methoxy groups at the C_2 and C_3 positions markedly diminish the potential of the compound to act as a Michael acceptor, such that DMNQ cannot conjugate with GSH. Thus, the induction of γGCS by DMNQ is most likely mediated through generation of $O_2^{\bullet -}$ and H_2O_2 via redox cycling and dismutation (Fig. 4). Nevertheless, it is entirely possible that more than one mechanism is responsible for γGCS induction, as has been observed by Pickett and coworkers for the glutathione S-transferase Ya subunit, which has a *cis*-acting element that responds to xenobiotics and a second element that responds to oxidative stress (119–122).

It has been demonstrated that the intracellular redox state regulates a broad array of genes in response to oxidative stress through either induction and/or stabilization of mRNAs (119,123,124). In mammalian systems, oxidative stress influences transcription by activation of transcription factors, such as NFκB, which can be activated by H_2O_2 (125). The *cis*-acting regulatory element antioxidant response element (ARE), which Pickett and coworkers found in the 5'-flanking region of the rat glutathione transferase Ya subunit gene and rat NAD(P)H:quinone reductase, has also been identified to be responsive to H_2O_2 (119,126). Whether the transcriptional induction of γGCS-HS gene by oxidative stress is through an ARE-like element is under further investigation.

Galter (5) suggested that GSSG production may be required for NFκB activation. Bergelson (4) suggested that diverse chemicals that induce the AP-1(activator protein-1) complex, which also activates of glutathione S-transferase Ya gene expression, may act through production of reactive oxygen species and depletion of reduced GSH. Yao et al. found, following GSH depletion in peripheral mononuclear cells of cancer patients by BSO treatment, that the level of γGCS-HS-mRNA increased, suggesting that the cellular GSH content may be involved in the regulation of γGCS gene expression (115). While we observed that MQ caused a marked decrease in GSH and increased GSSG before the increase in γGCS-HS-mRNA, we did not observe a significant decrease in GSH with DMNQ at concentrations that induced γGCS expression (103,104). Nevertheless, as the rate of H_2O_2 production increases with DMNQ, as we observed (104), the rate of GSSG production should also increase even if the steady state is unaffected. Thus, both the chemical signals and elements of molecular regulation involved in the increased transcription of γGCS remain to be identified.

B. Regulation of γGT Gene Expression by Oxidative Stress

γGT gene expression is elevated in some rat and human liver cancers, *ras*-transformed or retinoic acid–treated liver epithelial cells, and in liver treated with alcohol, ethoxyquin, or aflatoxin B1 (127–131). We have recently found that γGT-mRNA level increases in lung epithelial L2 cells in response to MQ exposure (114). Five γGT genes have been found in humans, but only one is present in rats and mice (128,131,132). Nevertheless, the regulation of transcription of γGT in the rodents may not be any less complicated as multiple mRNAs are transcribed and/or spliced from the one gene (128,131). The mechanism of regulation of this gene by oxidants or any other agent is not clear. Kurauchi et al. (133) and Rajagopalan et al. (134) have shown that the rat promoter lacks a TATA box but has two CCAAT boxes, making it unusual for an apparently inducible gene; however, intriguingly the 5'-flanking region of the mouse γGT gene contains a consensus sequence that matches the antioxidant response element found in the promoter region of the glutathione S-transferase Ya subunit gene and NADPH:quinone reductase gene, which is discussed in the preceding section (121,135).

VI. SUMMARY

Glutathione is a major component of antioxidant defense. Mechanisms for its maintenance in the reduced state and increased synthesis have evolved such that an increase in intracellular GSH appears to be an adaptive response in oxidative stress. The response to oxidative stress includes increased transport activities for intact GSH as well as the precursor amino acids for de novo synthesis and increased activities of the γ-glutamyl cycle enzymes γGT and γGCS. Oxidative stress causes an increase the mRNA levels of both γGT and the catalytic subunit of γGCS. Regulation of the regulatory subunit of γGCS, translational and posttranslational modifications and stabilization of the mRNAs and enzyme activities are also potentially affected by oxidative stress. Together, these elements constitute the mechanisms for increased intracellular GSH that occurs in oxidative stress.

ACKNOWLEDGMENTS

The studies from our lab cited here were supported by grants ES05511 from the National Institute of Environmental Health Sciences and HL37556 from the National Institutes of Heart, Lung, and Blood. We thank our past and present colleagues, Dr. Timothy W. Robison, Dr. Ewa Rajpert-De Meyts, Dr. John Groffen, Dr. Nora Heisterkamp, Dr. Minyuen Chang, Dr. Amir Kugelman, Dr. Mark W. Sutherland, Dr. George A. Loeb, Dr. Raymond J. Dorio, Dr. Henry A. Choy, Dr. Aron B. Fisher, Dr. Mitchell Glass, Dr. Thomas Aldrich, Evelyne Gozal, Li Tian, Hepeng Hu, Gerald Harrison, Dianne Skelton, June Nelson, and Eric Rotman, for their contributions to this work.

REFERENCES

1. Katoh, T., Ohmori, H., Murakami, T., Karasaki, Y., Higashi, K., and Muramatsu, M. (1991) Induction of glutathione-S-transferase and heat-shock proteins in rat after ethylene oxide exposure. Chem. Biol. Interact. 42:1247–1254.
2. Meyer, M., Pahl, H. L., and Baeuerle, P. A. (1994) Regulation of the transcription factors NF-κB and AP-1 by redox changes. Chem. Biol. Interact. 91:91–100.
3. Freeman, M. L., Sierra-rivera, E., Voorhees, G. J., Eisert, D. R., and Meredith, M. J. (1993) Synthesis of hsp-70 is enhanced in glutathione-depleted Hep G2 cells. Radiat. Res. 135:387–393.
4. Bergelson, S., Pinkus, R., and Daniel, V. (1994) Intracellular glutathione levels regulate fos/jun induction and activation of glutathione S-transferase gene expression. Cancer Res. 54: 36–40.
5. Galter, D., Mihm, S., and Droge, W. (1994) Distinct effects of glutathione disulphide on the nuclear transcription factors κB and the activator protein-1. Eur. J. Biochem. 221: 639–648.

6. Hagen, T. M., Aw, T. Y., and Jones, D. P. (1988) Glutathione uptake and protection against oxidative injury in isolated kidney cells. Kidney Int. 34:74–81.
7. Lash, L. H., Hagen, T. M., and Jones, D. P. (1986) Exogenous glutathione protects intestinal epithelial cells from oxidative injury. Proc. Natl. Acad. Sci. USA 83:4641–4645.
8. Hagen, T. M., Brown, L. A., and Jones, D. P. (1986) Protection against paraquat-induced injury by exogenous GSH in pulmonary alveolar type II cells. Biochem. Pharmacol. 35: 4537–4542.
9. Sze, G., Kaplowitz, N., Ookhtens, M., and Lu, S. C. (1993) Bidirectional membrane transport of intact glutathione in Hep G2 cells. Am. J. Physiol. 265: G1128–G1134.
10. Zlokovic, B. V., Mackic, J. B., McComb, J. G., Weiss, M. H., Kaplowitz, N., and Kannan, R. (1994) Evidence for transcapillary transport of reduced glutathione in vascular perfused guinea-pig brain. Biochem. Biophys. Res. Commun. 201: 402–408.
11. Meister, A. (1973) On the enzymology of amino acid transport. Science 180: 33–39.
12. Richman, P. G., and Meister, A. (1975) Regulation of gamma-glutamyl-cysteine synthetase by nonallosteric feedback inhibition by glutathione. J. Biol. Chem. 250:1422–1426.
13. Morris, P. E., and Bernard, G. R. (1994) Significance of glutathione in lung disease and implications for therapy. Am. J. Med. Sci. 307:119–127.
14. Wallig, M. A., Kore, A. M., Crawshaw, J., and Jeffery, E. H. (1992) Separation of the toxic and glutathione-enhancing effects of the naturally occurring nitrile, cyanohydroxybutene. Fund. Appl. Toxicol. 19:598–606.
15. Ochi, T. (1993) Mechanism for the changes in levels of glutathione upon exposure of cultured mammalian cells to tertiary-butylhydroperoxide and diamide. Arch. Toxicol. 67:401–410.
16. Meister, A. (1994) Glutathione, ascorbate, and cellular protection. Cancer Res. 54:1969s–1975s.
17. Suzuki, H., Hashimoto, W., and Kumagai, H. (1993) Escherichia coli K-12 can utilize an exogenous gamma-glutamyl peptide as an amino acid source, for which gamma-glutamyltranspeptidase is essential. J. Bacteriol. 175: 6038–6040.
18. Deneke, S. M., and Fanburg, B. L. (1989) Regulation of cellular glutathione. Am. J. Physiol. 257: L163–L173.
19. Bannai, S. (1984) Induction of cystine and glutamate transport activity in human fibroblasts by diethylmaleate and other electrophilic agents. J. Biol. Chem. 259: 2435–2440.
20. Bannai, S., Takada, A., Kasuga, H., and Tateishi, N. (1986) Induction of cystine transport activity in isolated rat hepatocytes by sulfobromophthalein and other electrophilic agents. Hepatology 6: 1361–1368.
21. Torres, A. M., Ochoa, E. J., Guibert, E., Rodriguez, J. V., and Elias, M. M. (1993) Renal transport of glycine during glutathione replenishment in rats. Biochem. Med. Metab. Biol. 50:159–168.
22. Benard, O., and Balasubramanian, K. A. (1993) Effect of oxidant exposure on thiol status in the intestinal mucosa. Biochem. Pharmacol. 45: 2011–2015.

23. Deneke, S. M. (1992) Induction of cystine transport in bovine pulmonary artery endothelial cells by sodium arsenite. Biochim. Biophys. Acta 1109: 127–131.
24. Deneke, S. M., Lawrence, R. A., and Jenkinson, S. G. (1992) Endothelial cell cystine uptake and glutathione increase with N,N-bis(2-chloroethyl)-N-nitrosourea exposure. Am. J. Physiol. 262: L301–L304.
25. Deneke, S. M., Steiger, V., and Fanburg, B. L. (1987) Effect of hyperoxia on glutathione levels and glutamic acid uptake in endothelial cells. J. Appl. Physiol. 63:1966–1971.
26. Deneke, S. M., Baxter, D. F., Phelps, D. T., and Fanburg, B. L. (1989) Increase in endothelial cell glutathione and precursor amino acid uptake by diethyl maleate and hyperoxia. Am. J. Physiol. 257: L265–L271.
27. Yi, J. R., Lu, S., Fernandez-Checa, J., and Kaplowitz, N. (1994) Expression cloning of a rat hepatic reduced glutathione transporter with canalicular characteristics. J. Clin. Invest. 93:1841–1845.
28. Rusakow, L. S., White, C. W., and Stabler, S. P. (1993) O2-induced changes in lung and storage pool thiols in mice: effect of superoxide dismutase. J. Appl. Physiol. 74: 989–997.
29. Cohen, G., and Hochstein, P. (1963) Glutathione peroxidase; the primary agent for the elimination of hydrogen peroxide in erythrocytes. Biochemistry 2: 1420–1428.
30. Rotruck, J. T., Pope, A. L., Ganther, H. E., Swanson, A. B., Gafeman, D. G., and Hoekstra, W. G. (1973) Selenium: Biochemical role as a component of glutathione peroxidase. Science 179:588–590.
31. Maiorino, M., Ursini, F., Leonelli, M., Finato, N., and Gregolin, C. (1982) A pig heart peroxidation inhibiting protein with glutathione peroxidase activity on phospholipid hydroperoxides. Biochem. Int. 5:575–583.
32. Maiorino, M., Thomas, J. P., Girotti, A. W., and Ursini, F. (1991) Reactivity of phospholipid hydroperoxide glutathione peroxidase with membrane and lipoprotein lipid hydroperoxides. Free Rad. Res. Comm. 12-13:131–135.
33. Koch, C. J., and Skov, K. A. (1994) Enhanced radiation-sensitivity by preincubation with nitroimidazoles: Effect of glutathione depletion. Int. J. Radiat. Oncol. Biol. Phys. 29: 345–349.
34. Deneke, S. M., Lynch, B. A., and Fanburg, B. L. (1985) Transient depletion of lung glutathione by diethylmaleate enhances oxygen toxicity. J. Appl. Physiol. 58: 571–574.
35. Miller, A. C., and Henderson, B. W. (1986) The influence of cellular glutathione content on cell survival following photodynamic treatment in vitro. Radiat. Res. 107: 83–94.
36. Deneke, S. M., Lynch, B. A., and Fanburg, B. L. (1985) Effects of low protein diets or feed restriction on rat lung glutathione and oxygen toxicity. J. Nutr. 115: 726–732.
37. Yi, X., Ding, L., Jin, Y., Ni, C., and Wang, W. (1994) The toxic effects of GSH depletion and radiosensitivity by BSO on retinoblastoma. Int. J. Radiat. Oncol. Biol. Phys. 29: 393–396.
38. Hodgkiss, R. J., Jones, N. R., Watts, M. E., and Woodcock, M. (1984) Glutathione depletion enhances the lifetime of oxygen-reactive radicals in mammalian cells. Int. J. Radiat. Biol. 46:673–674.

39. Morse, M. L., and Dahl, R. H. (1978) Cellular glutathione is a key to the oxygen effect in radiation damage. Nature 271:660–662.
40. Deschavanne, P. J., Malaise, E. P., and Revesz, L. (1981) Radiation survival of glutathione-deficient human fibroblasts in culture. Br. J. Radiol. 54: 361–362.
41. Bump, E. A., Yu, N. Y., and Brown, J. M. (1982) Radiosensitization of hypoxic tumor cells by depletion of intracellular glutathione. Science 217: 544–545.
42. Vos, O., Van der Schans, G. P., and Roos-Verhey, W. S. D. (1984) Effects of BSO and DEM on thiol-level and radiosensitivity in hela cells. Int. J. Radiat. Oncol. Biol. Phys. 10: 1249–1253.
43. Clark, E. P., Epp, E. R., Biaglow, J. E., Morse-Gaudio, M., and Zachgo, E. (1984) Glutathione depletion, radiosensitization, and isonidazole potentiation in hypoxic Chinese hamster ovary cells by buthionine sulfoximine. Radiat. Res. 98: 370–380.
44. Shrieve, D. C., Denekamp, J., and Minchinton, A. I. (1985) Effects of glutathione depletion by buthionine sulfoximine on radiosensitization by oxygen and misonidazole in vitro. Radiat. Res. 102: 283–294.
45. Bump, E. A., and Brown, J. M. (1990) Role of glutathione in the radiation response of mammalian cells in vitro and in vivo. Pharmacol. Ther. 47: 117–136.
46. Liu, R. M., Nebert, D. W., and Shertzer, H. G. (1993) Menadione toxicity in two mouse liver established cell lines having striking genetic differences in quinone reductase activity and glutathione concentrations. Toxicol. Appl. Pharmacol. 122: 101–107.
47. Tsan, M. F., Danis, E. H., Del Vecchio, P. J., and Rosano, C. L. (1985) Enhancement of intracellular glutathione protects endothelial cells against oxidant damage. Biochem. Biophys. Res. Commun. 127: 270–276.
48. Jenkinson, S. G., Jordan, J. M., and Lawrence, R. A. (1988) BCNU-induced protection from hyperbaric hyperoxia: Role of glutathione metabolism. J. Appl. Physiol. 65: 2531–2536.
49. Astor, M. B., Hall, E. J., Biaglow, J. E., and Hartog, B. (1984) Effects of D,L-Buthionine-S-R-sulfoximine on cellular thiol level and the oxygen effect in Chinese hamster V79 cells. Int. J. Radiat. Oncol. Biol. Phys. 10: 1239–1242.
50. Arrick, B. A., Nathan, C. F., Griffith, O. W., and Cohn, Z. A. (1982) Glutathione depletion sensitizes tumor cells to oxidative cytolysis. J. Biol. Chem. 257: 1231–1237.
51. Forman, H. J., Rotman, E. I., and Fisher, A. B. (1983) The roles of selenium and sulfur-containing amino acids in protection against oxygen toxicity. Lab. Invest. 49: 148–153.
52. Younes, M., and Siegers, C.-P. (1980) Lipid peroxidation as a consequence of glutathione depletion in rat and mouse liver. Res. Commun. Chem. Pathol. Pharmacol. 27: 119–128.
53. Sutherland, M. W., Nelson, J., Harrison, G., and Forman, H. J. (1985) Effects of t-butyl hydroperoxide on NADPH, glutathione, and repiratory burst of rat alveolar macrophages. Arch. Biochem. Biophys. 243: 325–331.
54. Sutherland, M. W., Glass, M., Nelson, J., Lyen, Y., and Forman, H. J. (1985) Oxygen toxicity: Loss of lung macrophage function without metabolite depletion. J. Free. Rad. Biol. Med. 1: 209–214.

55. Forman, H. J., and Skelton, D. C. (1990) Protection of alveolar macrophages from hyperoxia by gamma-glutamyl transpeptidase. Am. J. Physiol. 259: L102–L107.
56. Brown, L. A. S., Bai, C., and Jones, D. P. (1992) Glutathione protection in alveolar type II cells from fetal and neonatal rabbits. Am. J. Physiol. 262: L305–L312.
57. Shi, M., Gozal, E., Choy, H. A., and Forman, H. J. (1993) Extracellular glutathione and gamma-glutamyl transpeptidase prevent H_2O_2-induced injury by 2,3-dimethoxy-1,4-naphthoquinone. Free Rad. Biol. Med. 15: 57–67.
58. Tsan, M. F., White, J. E., and Rosano, C. L. (1989) Modulation of endothelial GSH concentrations: Effect of exogenous GSH and GSH monoethyl ester. J. Appl. Physiol. 66: 1029–1034.
59. Brown, L. A. S. (1994) Glutathione protects signal transduction in type II cells under oxidant stress. Am. J. Physiol. 266: 1172–1177.
60. Hagen, T. M., Aw, T. Y., and Jones, D. P. (1988) Glutathione uptake and protection against oxidative injury in isolated kidney cells. Kidney Int. 34:74–81.
61. Burk, R. F., Patel, K., and Lane, J. M. (1983) Reduced glutathione protection against rat liver microsomal injury by carbon tetrachloride. Biochem. J. 215: 441–445.
62. Jensen, G. L., and Meister, A. (1983) Radioprotection of human lymphoid cells by exogenously supplied glutathione is mediated by gamma-glutamyl transpeptidase. Proc. Natl. Acad. Sci. USA 80: 4714–4717.
63. Hiraishi, H., Terano, A., Ota, S., Mutoh, H., Sugimoto, T., Harada, T., Razandi, M., and Ivey, K. J. (1994) Protection of cultured rat gastric cells against oxidant-induced damage by exogenous glutathione. Gastroenterology 106: 1199–1207.
64. Leff, J. A., Wilke, C. P., Hybertson, B. M., Shanley, P. F., Beehler, C. J., and Repine, J. E. (1993) Postinsult treatment with N-acetyl-L-cysteine decreases IL-1-induced neutrophil influx and lung leak in rats. Am. J. Physiol. 265: L501–L506.
65. Wellner, V. P., Anderson, M. E., Puri, R. N., Jensen, G. L., and Meister, A. (1984) Radioprotection by glutathione ester: transport of glutathione ester into human lymphoid cells and fibroblasts. Proc. Natl. Acad. Sci. USA 81: 4732–4735.
66. Patterson, C. E., and Rhoades, R. A. (1988) Protective role of sulfhydryl reagents in oxidant lung injury. Exp. Lung Res. 14: 1005–1019.
67. Chang, M., Shi, M., and Forman, H. J. (1992) Exogenous glutathione protects endothelial cells from menadione toxicity. Am. J. Physiol. 6: L637–L634.
68. Rajpert-De Meyts, E., Shi, M., Chang, M., Robison, T. W., Groffen, J., Heisterkamp, N., and Forman, H. J. (1992) Transfection with gamma-glutamyl transpeptidase enhances recovery from glutathione depletion using extracellular glutathione. Toxicol. Appl. Pharmacol. 114:56–62.
69. Levy, E. J., Anderson, M. E., and Meister, A. (1993) Transport of glutathione diethyl ester into human cells. Proc. Natl. Acad. Sci. USA 90: 9171–9175.
70. Campbell, E. B., and Griffith, O. W. (1989) Glutathione monoethyl ester: High-performance liquid chromatographic analysis and direct preparation of the free base form. Anal. Biochem. 183: 21–25.
71. Anderson, M. E., Powrie, F., Puri, R. N., and Meister, A. (1985) Glutathione

monoethyl ester: Preparation, uptake by tissues, and conversion to glutathione. Arch. Biochem. Biophys. 239:538–548.
72. Roberts, J. C., and Francetic, D. J. (1991) Time course for the elevation of glutathione in numerous organs of L1210-bearing CDF1 mice given the L-cysteine prodrug, Ribcys. Toxicol. Lett. 59: 245–251.
73. Takada, A., and Bannai, S. (1984) Transport of cystine in isolated rat hepatocytes in primary culture. J. Biol. Chem. 259: 2441–2445.
74. Simplicio, P. D., and Rossi, R. (1994) The time-course of mixed disulfide formation between GSH and proteins in rat blood after oxidative stress with tert-butyl hydroperoxide. Biochim. Biophys. Acta 1199: 245–252.
75. DeLucia, A. J., Mustafa, M. G., Hussain, M. Z., and Cross, C. E. (1975) Ozone interaction with rodent lung. J. Clin. Invest. 55: 794–802.
76. Aniya, Y., and Naito, A. (1993) Oxidative stress-induced activation of microsomal glutathione *S*-transferase in isolated rat liver. Biochem. Pharmacol. 45: 37–42.
77. Comporti, M., Maellaro, E., Bello, B. D., and Casini, A. F. (1991) Glutathione depletione: Its effects on other antioxidant systems and hepatocellular damage. Xenobiotica 21: 1067–1076.
78. Jenkinson, S. G., Black, R. D., and Lawrence, R. A. (1988) Glutathione concentrations in rat lung bronchoalveolar lavage fluid: Effects of hyperoxia. J. Lab. Clin. Med. 112: 345–351.
79. Adams, J. D., Jr., Lauterburg, B. H., and Mitchell, J. R. (1983) Plasma glutathione and glutathione disulfide in the rat: Regulation and response to oxidative stress. J. Pharmacol. Exp. Ther. 227: 749–754.
80. Gustafson, D. L., Swanson, J. D., and Pritsos, C. A. (1993) Modulation of glutathione and glutathione dependent antioxidant enzymes in mouse heart following doxorubicin therapy. Free Rad. Res. Comm. 19: 111–120.
81. Hochstein, P., and Atallah, A. S. (1988) The nature of oxidants and antioxidant systems in the inhibition of mutation and cancer. Mutation Res. 202: 363–375.
82. Olafsdottir, K., and Reed, D. J. (1988) Retention of oxidized glutathione by isolated rat liver mitochondria during hydroperoxide treatment. Biochim. Biophys. Acta 964: 377–382.
83. Bunnell, E., and Pacht, E. R. (1993) Oxidized glutathione is increased in the alveolar fluid of patients with the adult respiratory distress syndrome. Am. Rev. Respir. Dis. 148: 1174–1178.
84. Nishiki, K., Jamieson, D., Oshino, N., and Chance, B. (1976) Oxygen toxicity in the perfused rat liver and lung under hyperbaric conditions. Biochem. J. 160: 343–355.
85. Masuda, Y., Ozaki, M., and Aoki, S. (1993) K^+-driven sinusoidal efflux of glutathione disulfide under oxidative stress in the perfused rat liver. FEBS Lett. 334: 109–113.
86. Jenkinson, S. G., Spence, T. H., Jr., Lawrence, R. A., Hill, K. E., Duncan, C. A., and Johnson, K. H. (1987) Rat lung glutathione release: Response to oxidative stress and selenium deficiency. J. Appl. Physiol. 62: 55–60.
87. Brigelius, R., Muckel, C., Akerboom, T. P. M., and Sies, H. (1983) Identification and quantitation of glutathione and its relationship to glutathione disulfide. Biochem. Pharmacol. 32:2529–2534.

88. Loeb, G. A., Skelton, D. C., Coates, T. D., and Forman, H. J. (1988) Role of the selenium dependent glutathione peroxidase in antioxidant defenses in rat alveolar macrophages. Exp. Lung Res. 14: 921–936.
89. Glass, M., Sutherland, M. W., Forman, H. J., and Fisher, A. B. (1985) Selenium deficiency potentiates paraquat-induced lipid peroxidation in isolated perfused rat lung. J. Appl. Physiol. 59: 619–622.
90. Buffinton, G. D., Ollinger, K., Brunmark, A., and Cadenas, E. (1989) DT-diaphorase-catalyzed reduction of 1,4-naphthoquinone derivatives and glutathionyl-quinone conjugates: Effect of substituents on autoxidation rates. Biochem. J. 257: 561–571.
91. Ong, F. B., Wan Ngah, W. Z., Top, A. G. M., Khalid, B. A. K., and Shamaan, N. A. (1994) Vitamin E, glutathione S-transferase and gamma-glutamyl transpeptidase activities in cultured hepatocytes of rats treated with carcinogens. Int. J. Biochem. 26: 397–402.
92. Rister, M., and Wustrow, C. (1985) Effect of hyperoxia on reduced glutathione in alveolar macrophages and polymorphonuclear leukocytes. Res. Exp. Med. 185: 445–450.
93. Kennedy, K. A., and Lane, N. L. (1994) Effect of in vivo hyperoxia on the glutathione system in neonatal rat lung. Exp. Lung Res. 20: 73–83
94. Kimball, R. E., Reddy, K., Peirce, T. H., Schwartz, L. W., Mustafa, M. G., and Cross, C. E. (1976) Oxygen toxicity: Augmentation of antioxidant defense mechanisms in rat lung. Am. J. Physiol. 230: 1425–1431.
95. Siems, W., and Brenke, R. (1992) Changes in the glutathione system of erythrocytes due to enhanced formation of oxygen free radicals during short-term whole body cold stimulus. Arctic. Med. Res. 51: 3–9.
96. Purucker, E., and Lutz, J. (1992) Effect of hyperbaric oxygen treatment and perfluorochemical administration of glutathione status of the lung. Adv. Exp. Med. Biol. 317: 131–136.
97. Bailey, H. H., Gipp, J. J., Ripple, M., Wilding, G., and Mulcahy, T. (1992) Increase of gamma-glutamylcysteine synthetase activity and steady-state messenger RNA levels in melphalan-resistant DU-145 human prostate carcinoma cells expressing elevated glutathione levels. Cancer Res. 52: 5115–5118.
98. Ruszczewski, P., Truskolaski, P., Dabrowski, Z., Rap, Z. M., and Herbaczyn'ska-Cedro, K. (1979) Effect of prostaglandin E_2 and of indomethacin upon cerebral and pulmonary consequences of exposure to hyperbaric oxygen in rats. Acta Neurol. Scand. 59: 188–199.
99. Misra, H. P., and Gorsky, L. D. (1981) Paraquat and NADPH-dependent lipid peroxidation in lung microsomes. J. Biol. Chem. 256: 9994–9998.
100. Kondo, T., Yoshida, K., Urata, Y., Goto, S., Gasa, S., and Taniguchi, N. (1993) Gamma-glutamylcysteine synthetase and active transport of glutathione S-conjugate are responsive to heat shock in K562 erythroid cells. J. Biol. Chem. 268: 20366–20372.
101. Borroz, K. I., and Eaton, D. L. (1993) Development and utilization of a cDNA probe for gamma-glutamylcysteine synthetase to study transcriptional regulation of glutathione biosynthesis. Toxicologist 13: 408.

102. Woods, J. S., Davis, H. A., and Baer, R. P. (1992) Enhancement of gamma-glutamylcysteine synthetase mRNA in rat kidney by methyl mercury. Arch. Biochem. Biophys. 296: 350–353.
103. Shi, M. M., Iwamoto, T., and Forman, H. J. (1994) Gamma-glutamylcysteine synthetase and glutathione increase in quinone-induced oxidative stress in bovine pulmonary artery endothelial cells. Am. J. Physiol. 267: 1414–1421.
104. Shi, M. M., Kugelman, A., Iwamoto, T., Tian, L., and Forman, H. J. (1994) Quinone-induced oxidative stress elevates glutathione and induces gamma-glutamylcysteine synthetase activity in rat lung epithelial L2 cells. J. Biol. Chem. 269: 26512–26517.
105. Seelig, G. F., Simondsen, R. P., and Meister, A. (1984) Reversible dissociation of gamma-glutamylcysteine synthetase into two subunits. J. Biol. Chem. 259: 9345–9347.
106. Huang, C. S., Chang, L. S., Anderson, M. E., and Meister, A. (1993) Catalytic and regulatory properties of the heavy subunit of rat kidney gamma-glutamylcysteine synthetase. J. Biol. Chem. 268: 19675–19680.
107. Huang, C.-S., Anderson, M. E., and Meister, A. (1993) The function of the light subunit of gamma-glutamylcysteine synthetase (rat kidney). FASEB J. 7: A1102.
108. Loeb, G. A., Skelton, D. C., and Forman, H. J. (1989) Dependence of mixed disulfide formation in alveolar macrophages upon production of oxidized glutathione: Effect of selenium depletion. Biochem. Pharmacol. 38: 3119–3121.
109. Crapo, J. D., and Tierney, D. F. (1974) Superoxide dismutase and pulmonary oxygen toxicity. Am. J. Physiol. 226: 1401–1407.
110. Frank, L., Bucher, J. R., and Roberts, R. J. (1978) Oxygen toxicity in neonatal and adult animals of various species. J. Appl. Physiol. 45: 699–704.
111. Forman, H. J., and Fisher, A. B. (1981) Antioxidant enzymes in rat granular pneumocytes: Constitutive levels and effect of hyperoxia. Lab. Invest. 45: 1–6.
112. Tierney, D., Ayres, L., Herzog, S., and Yang, J. (1973) Pentose pathway and production of reduced nicotinamide adenine dinucleotide phosphates: A mechanism that may protect the lung from oxidants. Am. Rev. Respir. Dis. 108: 1348–1351.
113. Ngo, E. O., and Nutter, L. M. (1994) Status of glutathione and glutathione-metabolizing enzymes in menadione-resistant human cancer cells. Biochem. Pharmacol. 47: 421–424.
114. Kugelman, A., Choy, H. A., Liu, R., Shi, M. M., Gozal, E., and Forman, H. J. (1994) Gamma-glutamyl transpeptidase is increased by oxidative stress in rat alveolar L2 epithelial cells. Am. J. Respir. Cell Mol. Biol. 11: 586–592.
115. Yao, K. S., Godwin, A. K., Ozols, R. F., Hamilton, T. C., and O'Dwyer, P. J. (1993) Variable baseline gamma-glutamylcysteine synthetase messenger RNA expression in peripheral mononuclear cells of cancer patients, and its induction by buthionine sulfoximine treatment. Cancer Res. 53: 3662–3666.
116. Talalay, P., De-Long, M. J., and Prochaska, H. J. (1988) Identification of a common chemical signal regulating the induction of enzymes that protect against chemical carcinogenesis. Proc. Natl. Acad. Sci. USA 85: 8261–8265.
117. Eaton, D. L., and Hamel, D. M. (1994) Increase in gamma-glutamylcysteine synthetase activity as a mechanism for butylated hydroxyanisole-mediated elevation of hepatic glutathione. Toxicol. Appl. Pharmacol. 126: 145–149.

118. Borroz, K. I., Buetler, T. M., and Eaton, D. L. (1994) Modulation of gamma-glutamylcysteine synthetase large subunit mRNA expression by butylated hydroxyanisole. Toxicol. Appl. Pharmacol. 126: 150–155.
119. Rushmore, T. H., Morton, M. R., and Pickett, C. B. (1991) The antioxidant responsive element. Activation by oxidative stress and identification of the DNA consensus sequence required for functional activity. J. Biol. Chem. 266: 11632–11639.
120. Rushmore, T. H., King, R. G., Paulson, K. E., and Pickett, C. B. (1990) Regulation of glutathione S-transferase Ya subunit gene expression: Identification of a unique xenobiotic-responsive element controlling inducible expression by planar aromatic compounds. Proc. Natl. Acad. Sci. USA 87: 3826–3830.
121. Nguyen, T., and Pickett, C. B. (1992) Regulation of rat glutathione S-transferase Ya subunit gene expression. DNA-protein interaction at the antioxidant responsive element. J. Biol. Chem. 267: 13535–13539.
122. Nguyen, T., Rushmore, T. H., and Pickett, C. B. (1994) Transcriptional regulation of a rat liver glutathione *S*-transferase Ya subunit gene. Analysis of the antioxidant response element and its activation by the phorbol ester 12-*O*-tetradecanoylphorbol-13-acetate. J. Biol. Chem. 269:13656–13662.
123. Storz, G., Tartaglia, L. A., and Ames, B. N. (1990) Transcriptional regulator of oxidative stress-inducible genes: direct activation by oxidation. Science 248: 189–194.
124. Clerch, L. B., and Massaro, D. (1992) Oxidation-reduction-sensitive binding of lung protein to rat catalase mRNA. J. Biol. Chem. 267: 2853–2855.
125. Schreck, R., Rieber, P., and Baeuerle, P. A. (1991) Reactive oxygen intermediates as apparently widely used messengers in the activation of the NFκB transcription factor and HIV-1. EMBO J. 10: 2247–2258.
126. Favreau, L. V., and Pickett, C. B. (1993) Transcriptional regulation of the rat NAD(P)H:quinone reductase gene. Characterization of a DNA-protein interaction at the antioxidant responsive element and induction by 12-O-tetradecanoylphorbol 13-acetate. J. Biol. Chem. 268: 19875–19881.
127. Power, C. A., Griffiths, S. A., Simpson, J. L., Laperche, Y., Guellaen, G., and Manson, M. M. (1987) Induction of gamma-glutamyl transpeptidase mRNA by aflatoxin B1 and ethoxyquin in rat liver. Carcinogenesis 8: 737–740.
128. Rajagopalan, S., Wan, D. F., Habib, G. M., Sepulveda, A. R., Mcleod, M. R., Lebovitz, R. M., and Lieberman, M. W. (1993) Six mRNAs with different 5′ ends are encoded by a single gamma-glutamyltransferase gene in mouse. Proc. Natl. Acad. Sci. USA 90: 6179–6183.
129. Habib, G. M., Rajagopalan, S., Godwin, A., Lebovitz, R. M., and Lieberman, M. W. (1992) The same gamma-glutamyl transpeptidase RNA species is expressed in fetal liver, hepatic carcinomas, and ras T24-transformed rat liver epithelial cells. Mol. Carcinog. 5: 75–80.
130. Tsao, M. S., and Batist, G. (1988) Induction of gamma-glutamyl transpeptidase activity by all-trans retinoic acid in cultured rat liver epithelial cells. Biochem. Biophys. Res. Commun. 157: 1039–1045.
131. Chobert, M. N., Lahuna, O., Lebargy, F., Kurauchi, O., Darbouy, M., Bernaudin, J. F., Guellaen, G., Barouki, R., and Laperche, Y. (1990) Tissue-

specific expression of two gamma-glutamyl transpeptidase mRNAs with alternative 5′ ends encoded by a single copy gene in the rat. J. Biol. Chem. 265: 2352–2357.
132. Courtay, C., Heisterkamp, N., Siest, G., and Groffen, J. (1994) Expression of multiple gamma-glutamyltransferase genes in man. Biochem. J. 297: 503–508.
133. Kurauchi, O., Lahuna, O., Darbouy, M., Aggerbeck, M., Chobert, M.-N., and Laperche, Y. (1991) Organization of the 5′ end of the rat gamma-glutamyl transpeptidase gene: Structure of a promoter active in the kidney. Biochemistry 30: 1618–1623.
134. Rajagopalan, S., Park, J., Patel, P. D., Lebovitz, R. M., and Lieberman, M. W. (1990) Cloning and analysis of the rat gamma-glutamyltransferase gene. J. Biol. Chem. 265: 11721–11725.
135. Friling, R. S., Bensimon, A., Tichauer, Y., and Daniel, V. (1990) Xenobiotic-induced expression of murine glutathione S-transferase Ya subunit gene is controlled by an electrophile-responsive element. Proc. Natl. Acad. Sci. USA 87: 6258–6262.

10

Sinusoidal Efflux of Hepatic GSH: Significance, Physiology, and Regulation

Neil Kaplowitz, Murad Ookhtens, and Shelly Chi-Loo Lu
University of Southern California School of Medicine, Los Angeles, California

The plasma contains a finite glutathione (GSH) concentration, which turns over very rapidly. The major source of plasma GSH is the liver, and the release of hepatic GSH into plasma is carrier-mediated. In this chapter we will discuss the significance, physiology, and regulation of sinusoidal GSH efflux. It is important to recognize that the hepatocyte releases GSH at both poles of the cell. Although the sinusoidal (basolateral) release is rather liver specific, the canalicular (apical) release is seen widely in tissues, including kidney and lung. The recently cloned putative canalicular transporter, indeed, is expressed widely (1). Thus, the apical secretion of GSH by various organs is probably a general mechanism to provide GSH to the lumen of these organs to protect against oxidants and to provide the substrate for γ-glutamyltransferase. We have recently reviewed the current state of knowledge of the canalicular GSH transport system and will not discuss this further (2). Two distinct polypeptides representing two distinct genes are responsible for the secretion of GSH at the two poles of the hepatocyte. Our focus in this chapter will be the basolateral transport of hepatic GSH.

I. SIGNIFICANCE OF SINUSOIDAL GSH EFFLUX

The liver secretes GSH into plasma, which is most readily detected as a higher GSH concentration in hepatic venous plasma compared to portal (3,4). In the rat, nearly all the released GSH is available systemically (no breakdown in liver sinusoids) and, as noted above, disappears very rapidly (5). In other species, such as guinea pig and human, γ-glutamyltransferase in the sinusoidal liver plasma membrane alters much, if not all, the released GSH (6) and probably provides the systemic circulation with the transpeptidation reaction products, γ-glutamylcystine and cysteinylglycine, as well as cysteine (Fig. 1). In adult rat much of the plasma GSH derived from the liver is broken down rapidly in extrahepatic organs leading to production of cysteine and other products (Fig. 2). Thus, depending on the species, the GSH released from the liver cells is broken down mainly in the periphery or in the liver sinusoids. The net result is to make GSH precursors, cysteine and perhaps γ-glutamylcystine, available systemically.

Cysteine is a very toxic and unstable molecule, which autooxidizes rapidly. Tissue and extracellular cysteine is kept low teleologically as a protective mechanism through reincorporation into GSH. GSH is stable and nontoxic and is therefore an ideal vehicle to store and transfer cysteine. Thus, to maintain steady-state cysteine concentrations, an intracellular-extracellular cycle is needed: cysteine is incorporated into GSH in cells and is recovered from GSH outside cells. The recovery requires the action of γ-glutamyltransferase, which is an integral membrane protein with an ecto-active site. Therefore, integral to this cycle is the need to release GSH from cells, a process accomplished by carrier-mediated transport.

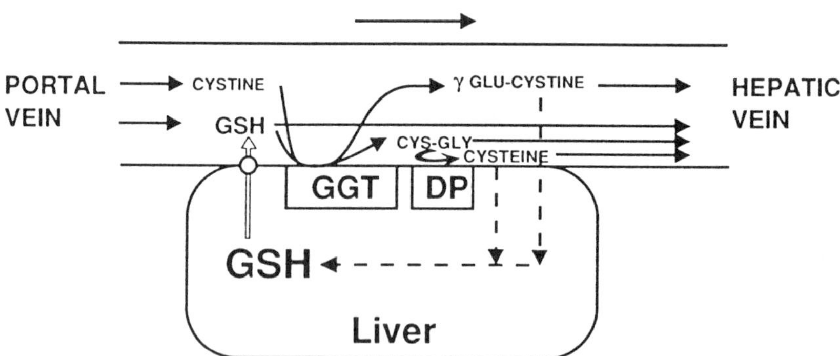

Figure 1 Handling of GSH in the hepatic sinusoid of guinea pig and presumably humans. The GSH that enters the liver via portal vein along with that contributed to from hepatic efflux can exit via the hepatic vein unchanged or undergo metabolism via γ-glutamyltransferase (GGT) and dipeptidase (DP). Cystine can serve as a γ-glutamyl acceptor, generating γ-glutamylcystine and cysteinylglycine. (Not shown, GSH can be broken down to its three constituents by these same enzymes.) The products might be available to more distally perfused hepatocytes (dashed lines) or exit for systemic availability.

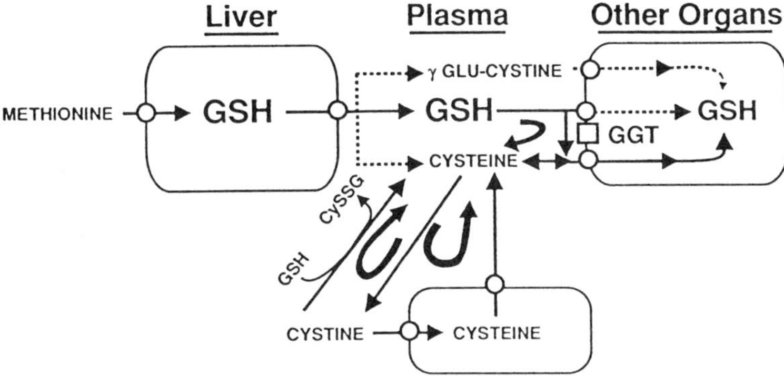

Figure 2 Interorgan homeostasis of GSH in relation to cysteine/cystine homeostasis. Three major mechanisms appear to maintain plasma cysteine (curved plasma arrows), as discussed in the text.

The synthesis of GSH can be accomplished from several plasma precursors. Least well characterized is γ-glutamylcysteine; it can be taken up by some cells, e.g., renal epithelium (7), although its availability to other cells is not well characterized. Once inside the cell, it is reduced to cysteine and γ-glutamylcysteine. The latter is an obligate precursor of GSH, which bypasses the feedback inhibition by GSH on the rate-limiting enzyme in GSH synthesis, namely, γ-glutamylcysteine synthetase. Therefore, the extracellular production of γ-glutamylcystine represents a novel mechanism for determining intracellular GSH synthesis. Cystine, the cosubstrate with GSH for this γ-glutamyltransferase reaction, is the best physiological acceptor substrate for γ-glutamyl transfer. It should be noted that further work is needed to define the contribution of this pathway to GSH homeostasis, and more information is needed on the transport system(s) for γ-glutamylcystine. We have previously shown that γ-glutamyl amino acids can trans-stimulate GSH transport in liver plasma membrane vesicles (8). This finding would predict that they share the same transporter, and, therefore, the outwardly directed GSH gradient would serve as the driving force for uptake (through countertransport) of γ-glutamyl amino acids, at the same time providing more extracellular GSH for transpeptidation. However, other mechanisms and transport systems for γ-glutamyl amino acids have not been excluded.

The sulfur amino acid precursors of GSH, methionine, cystine, and cysteine are differentially available for GSH synthesis in various cell types. Methionine undergoes significant transsulfuration to cysteine only in hepatocytes and lens epithelium. Cystine, the predominant plasma sulfur amino acid, is widely transported into cells with some notable exceptions, which therefore depend mainly on availability and uptake of plasma cysteine to produce GSH. These latter include hepatocytes (which alternatively utilize methionine), erythrocytes, brain capillary endothelial cells, and lymphocytes (9). Therefore, maintenance of steady-state concentrations of plasma cysteine is critical for these cells. This concept has recently received wide attention in relation to HIV infection (10).

Three principal mechanisms exist to maintain plasma cysteine (which is rapidly lost through oxidation and/or tissue uptake):

1. Uptake of cystine, intracellular reduction, and release of cysteine: this mechanism, proposed by Bannai, is demonstrated in cell culture (11).
2. Breakdown of plasma GSH to cysteine via the action of γ-glutamyltransferase and dipeptidases.
3. Thiol disulfide exchange between GSH released from the liver and plasma cystine generating a mixed disulfide and free cysteine.

The relative contribution of these mechanisms in vivo is uncertain.

Recent in vivo tracer-kinetic studies using i.v. bolus injections of labeled GSH, cysteine, and cystine combined with multicompartmental analysis and modeling (Fig. 3) have delineated the following in the adult male rat (5,12): the irreversible disposal (i.e., nonrecycling component) of plasma GSH is maintained by hepatic (sinusoidal) GSH efflux (18 nmol/min/ml of plasma ≅ 18 nmol/min/g liver). The irreversible disposal of plasma GSH is completely accounted for

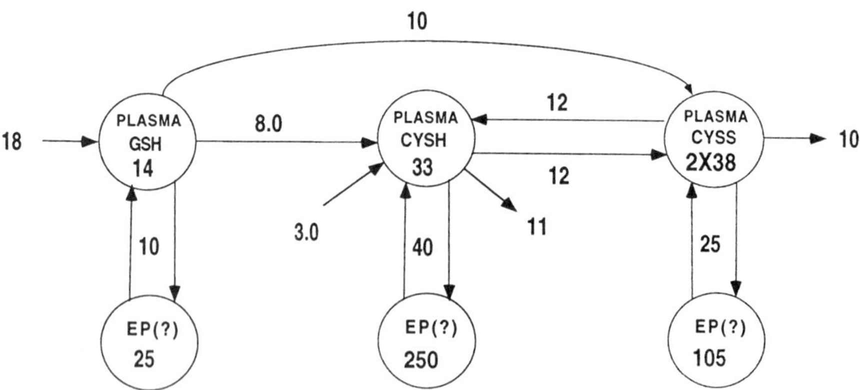

Figure 3 Multicompartmental model of plasma GSH, cysteine (CYSH), and cystine (CYSS) turnover-disposal in the adult male rat. The magnitude of the rates (nmol/min/ml plasma) are shown next to the arrows and the measured plasma concentrations (μM) inside the respective compartments. Concentrations of and transport rates from the CYSS compartment are presented in CYSH equivalents. EPs indicate unidentified extraplasma pools through which recycling of label take place. The respective EP equivalents (μM) are computed based on steady-state criteria.

by breakdown to and appearance of cysteine + cystine. Plasma GSH appears to be the major *net* source, i.e., 86%, of plasma cysteine + cystine (18 out of 21 cysteine equivalent units in the model). The remaining 14% is supplied through sources untraceable by plasma radioactivities. This fraction could possibly represent the release of unlabeled cysteine from tissues, subsequent to reduction of cellular cystine. There is also considerable interconversion of labeled cysteine and cystine in plasma (12 cysteine equivalent units), presumably due to autoxidation of cysteine to cystine and reduction of cystine to cysteine by circulating GSH. The irreversible disposal of plasma cystine is completely supplied through plasma GSH and cysteine pools. Thus, it is clear that the release of hepatic GSH into plasma is a major determinant of steady-state plasma cysteine and cystine.

An alternative mechanism for handling of plasma GSH has been suggested, i.e., sodium-coupled uptake of intact GSH. This has been demonstrated in kidney, lung, and intestine by Jones and coworkers (13–15). We have demonstrated a similar phenomenon in brain capillaries (16–18). This mechanism involves coupling the uphill transport of GSH into cells against its electrochemical gradient to the downhill transport of sodium. Considerable controversy currently exists as to the existence and importance of such a mechanism. Further progress must await cloning of such a transporter(s).

In addition to sinusoidal GSH efflux providing GSH itself or precursors to other tissues, extracellular GSH may serve a protective function in maintaining thiol redox balance in plasma and on cell surfaces and serve as an important extracellular scavenger of radicals such as superoxide anion.

II. PHYSIOLOGICAL CHARACTERISTICS OF SINUSOIDAL GSH TRANSPORT

The efflux of GSH into sinusoidal plasma was first described by Bartoli and Sies (19). The carrier-mediated nature of this efflux was subsequently demonstrated by our laboratory in the perfused liver (20), isolated (21) or cultured hepatocytes (22), and basolateral membrane vesicles (8,23). In all models the transport exhibited capacity limitation (i.e., saturation kinetics) with K_m in the 3–7 mM range, competitive *cis*-inhibition by BSP-GSH (8,24) and *trans*-inhibition by methionine and cystathionine (25–27). The *trans*-inhibition is consistent with an allosteric effect on the sinusoidal transporter. The physiological significance of this phenomenon is uncertain but could represent a mechanism to accelerate GSH repletion when hepatic GSH falls; as repletion occurs the higher cell GSH overcomes the inhibition (i.e., competitive kinetics). These inhibitors did not affect the canalicular GSH transport. In isolated cells and cultured hepatocytes as well as basolateral liver plasma membrane vesicles, trans-stimulation of GSH transport was observed (8). Thus, an inwardly directed concentration gradient would permit net uptake of GSH by hepatocytes, a feature one would predict for a facilitative transport system. Normally, however, the physiological outwardly directed concentration gradient of GSH results in a net efflux from the liver.

The sinusoidal transport exhibits sigmoidal kinetics in the perfused liver (20) and intact cells (21,22). However, the basis for the prediction of multiple interactive transport sites remains to be elucidated. Recently, we have cloned the sinusoidal GSH transporter from a rat liver cDNA library utilizing the Xenopus oocyte expression system and demonstrated nearly exclusive expression of its transcript in liver (27a). The distribution of this transporter, namely, in liver only, is consistent with the liver supplying nearly all the GSH in plasma.

III. REGULATION OF SINUSOIDAL GSH EFFLUX

In view of the importance of sinusoidal GSH efflux to the interorgan homeostasis of cysteine and to extracellular defense, it is critical to recognize the circumstances that regulate this system. Increased sinusoidal efflux has been observed in chronic alcohol-fed rats (28), in exhaustive exercise (29), in fasting (30), and in response to endotoxin administration (31). We will review briefly the issue of driving forces for GSH transport and normal regulation and then discuss more extensively an area of recent interest to us, namely, the influence of thiols and disulfides on the sinusoidal transporter.

A. Driving Forces

The sinusoidal efflux of hepatic GSH occurs down its concentration gradient, 5–10 mM in hepatocytes and 10–20 μM in portal venous plasma. GSH has a net negative charge at physiological pH, and its transport is driven by membrane potential, i.e., it is electrogenic (8,23,32). In hepatocytes, efflux is markedly inhibited by high extracellular potassium (33). In basolateral liver plasma membrane vesicles, uptake of GSH is enhanced by an inside positive potential created with an inwardly directed potassium gradient plus valinomycin and inhibited by an outwardly directed potassium gradient plus valinomycin, both in comparison to voltage clamped conditions (potassium equilibrated plus valinomycin).

Recently, we have discovered that GSH transport in basolateral liver plasma vesicles exhibits an alternative driving force, namely, the proton gradient. Thus, an inwardly directed proton gradient enhanced GSH uptake by vesicles. Since this was inhibited by an inside positive potential, the system operated as movement of net positive charge, indicating coupling of more than one proton per GSH molecule (K. Tsukahara and N. Kaplowitz, unpublished observations). Under physiological conditions, an outward proton gradient exists in hepatocytes, which is countered by outside positive potential. Thus, the importance of this newly identified driving force is uncertain. However, despite the unfavorable effect of membrane potential on this mode of transport, the possibility exists that changes in state of phosphorylation or thiol redox status may modulate the relative influence of membrane potential and proton gradient on this transport system. Also, it remains possible (although in our view less likely) that the response to these two driving forces represent two distinct transport systems with the same or different subcellular distributions.

B. Hormonal Regulation

Cyclic AMP (cAMP), as the proximate transduction signal of glucagon, is known to hyperpolarize the plasma membrane potential of hepatocytes. We observed in cultured hepatocytes, perfused rat liver, and in vivo that the sinusoidal efflux of GSH increased very rapidly in response to glucagon and to dibutyryl cAMP (22). Sies and coworkers observed that both the cAMP-dependent hormone, glucagon, as well as Ca^{+2}/C-kinase–dependent hormones, vasopressin and phenylephrine, increased sinusoidal GSH release in the perfused liver (34). cAMP is known to increase Na + K + ATPase activity as its mechanism for hyperpolarization; consistent with this, we found that ouabain blocked the stimulation of GSH efflux by dibutyryl cAMP (23). Glucagon (cAMP)–induced stimulation of GSH efflux is therefore mediated by an enhancement in the driving force for GSH efflux, namely, membrane potential. Thus, the enhanced GSH efflux with fasting and perhaps stressful conditions may be at least partially explained by this phenomenon.

We were unable to confirm an effect of vasopressin on GSH efflux (23), but others subsequently did so in hepatocytes (35). It appears that at the supraphysiological concentrations we employed, no effect is seen. However, at physiological concentrations of vasopressin, a modest stimulation of GSH efflux is observed in cultured hepatocytes, which presumably represents an effect on the sinusoidal transport. This effect is at least partially dependent on protein kinase C. In perfused livers, a more marked enhancement of efflux was observed (34), but interpretation is complicated by the possibility of paracellular leak of canalicular GSH back into the perfusate in response to hormone-induced increase in paracellular permeability (36). Now that the sinusoidal and canalicular transporters have been cloned, it should be possible to generate antisera to be utilized in immunoprecipitation studies to directly examine the influence of various hormones and signal-transduction pathways on the phosphorylation of the transporter polypeptides and correlate these findings with functional effects.

C. Thiol-Disulfide Effects

In examining the effects of various precursor sulfur amino acids on GSH in primary cultured rat hepatocytes, we observed a marked difference in GSH efflux when cells were cultured overnight in standard (cystine-containing) DME/

F12 supplemented with methionine versus cystine-free DME-F12 supplemented with methionine. After exposure to cystine-containing medium, the efflux of GSH was less than after exposure to cystine-free medium (Fig. 4). This effect could not be explained by a difference in cell GSH levels or the composition of the extracellular medium during efflux measurements (simple Krebs-Hepes buffer). This observation led us to explore the influence of thiols and disulfides on GSH transport (37,38).

The influence of cystine concentration and duration of exposure on GSH efflux were examined in cultured hepatocytes (37). The effect was no greater after 1 h of exposure than after overnight exposure and persisted as above after removal of the cystine. Shorter exposure, i.e., 15 min, was associated with reversible inhibition. Cystine concentrations in the ≥0.1 mM range gave near maximal inhibition, and inhibition of efflux was seen with as little as 20 μM cystine (Fig. 5). The effect was also observed in the perfused liver (37), indicating a sinusoidal effect (Fig. 6). Since cystine is not taken up by the intact liver, it could also be concluded that the effect was extracellular.

A number of mechanisms that might explain inhibition by cystine were excluded, including ATP depletion and depolarization of the plasma membrane potential. However, DTT reversed the inhibition and in fact markedly enhanced efflux. The effect of DTT was time and concentration dependent, being maximal at 30 min with 1 and 2 mM DTT and requiring more time to achieve the same maximal effect at 0.5 mM DTT. Submaximal effects were seen with lower DTT concentrations. As seen in Figure 7, when cells were treated with 2 mM DTT for 30 min and then followed by 0.5 mM cystine for 1 h, cystine reversed the stimulation, suggesting that both DTT and cystine act at the same site.

The effect of DTT was also observed in the perfused rat liver (Fig. 8), indicating that sinusoidal transport is stimulated (37). We examined a wide range of mono- and dithiols (38). The following compounds, studied at 1 mM dithiol or 2 mM monothiol, exerted a stimulatory effect in the perfused liver in the single pass mode: dimercaprol, 1,2-ethane dithiol, 1,3-propane dithiol, 1,4-butane dithiol, monothioglycerol, 2-mercaptoethanol, and cysteamine; the dithiols tended to be more potent (≈ four- to fivefold increase) than monothiols (≈ twofold increase). Interestingly, the following exerted no effect under the same experimental conditions: dihydrothioctic acid, penicillamine, *N*-acetylcysteine, mercaptopropionylglycine, mercaptoethane sulfonic acid, mercaptoacetic acid, mercaptopropionic acid, and captopril as well as various alcohols (glycerol, 1-propanol, 1-2-propane-diol, and ethanol). All of the thiol compounds that were ineffective share the feature of having a negative-charged substituent.

The thiol stimulatory effect on sinusoidal GSH transport could not be explained on the basis of an effect on cell ATP or membrane potential, nor was

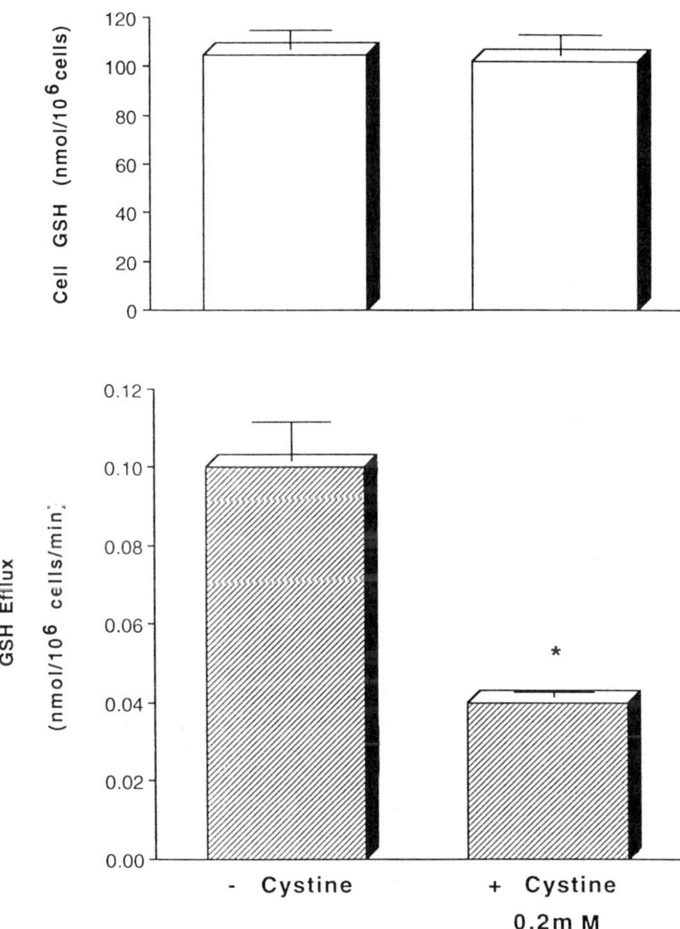

Figure 4 Comparison of GSH efflux in primary cultured rat hepatocytes grown in the presence or absence of cystine. DME/F12 media either missing or containing 0.2 mM cystine was employed. Media in both conditions contained methionine. Following overnight exposure, media were replaced by Krebs-Hepes buffer and efflux determined over 1 h. The upper panel shows that cell GSH was not different in the two conditions at the time of instituting efflux determinations. The lower panel shows a marked decrease in GSH efflux after exposure to medium containing cystine. $^{*}p < 0.01$ vs. -Cystine. (From Ref. 37.)

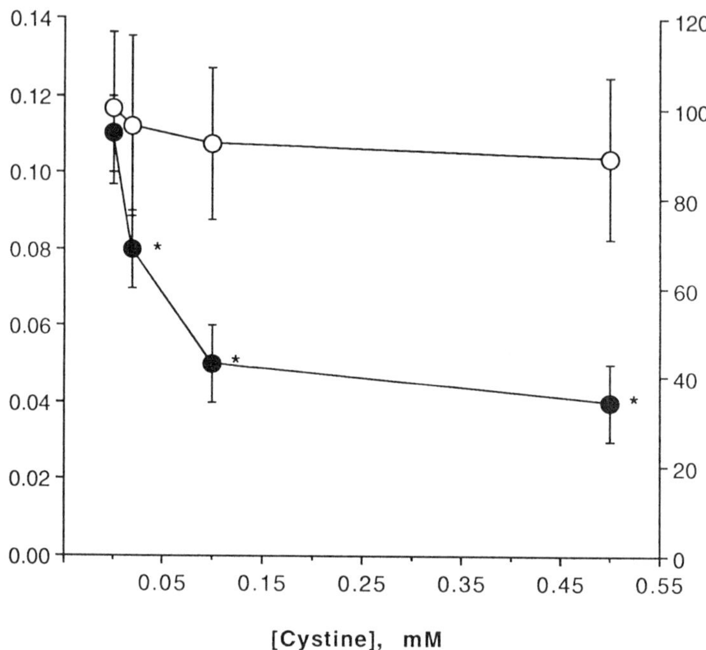

Figure 5 Dose-dependent inhibition of GSH efflux after exposure to cystine in primary rat hepatocyte culture. Medium consisted of DME/F12 minus cystine and supplemented with methionine (1 mM) to which 0–0.5 mM cystine was added. Cells were exposed overnight and efflux and cell GSH determined as in Figure 3. $^{*}p < 0.05$ vs. control by ANOVA; (—●—) GSH efflux rate (nmol/10^6 cells/min); (—○—) Cell GSH (nmol/10^6 cells). (Adapted from Ref. 37.)

the stimulation altered by antagonists of C-kinase, calmodulin, or microtubules. However, it remains theoretically possible that DTT mobilizes a pool of transporters in submembrane vesicles not dependent on microtubules.

As noted above, the sinusoidal GSH transporter exhibits features of a facilitative carrier, thus having the capacity to operate bidirectionally. In cultured hepatocytes, we have examined the influence of cystine or DTT on the kinetics of GSH transport at varying cell GSH. Cystine and DTT both exerted a V_{max} effect on GSH efflux; cystine decreased V_{max} and DTT increased V_{max}; neither affected the K_m for GSH efflux (37). In contrast, DTT decreased V_{max} for GSH uptake in cell culture (38). This effect indicates that thiol-reducing agents change the transport function to favor outward transport of GSH, namely, increased V_{max} for efflux and decreased V_{max} for uptake. This phenomenon is illustrated by the

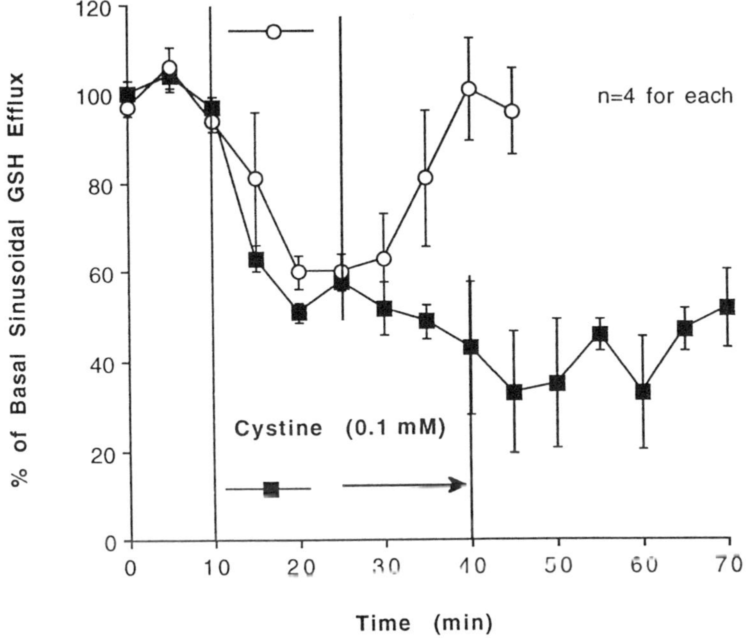

Figure 6 Effect of cystine on GSH efflux in the single-pass perfused rat liver. Sinusoidal GSH efflux was determined by HPLC; all the GSH during the cystine perfusion interval was recovered as mixed disulfide with cysteine. Biliary total GSH and bile flow were unaffected (not shown). After 15-min infusion of cystine, GSH efflux recovered, whereas inhibition persisted after 30-min infusion.

effect of high extracellular GSH on cell GSH levels in the presence and absence of DTT. Without extracellular DTT, cell GSH increases in the presence of 10 mM GSH. DTT alone leads to a decrease in cell GSH (studied in culture without GSH precursor, i.e., no synthesis). High extracellular GSH does not abrogate the fall in cell GSH (Fig. 9), indicating that in the presence of DTT, extracellular GSH does not enter cells at a rate sufficient to compensate for enhanced efflux (38).

Of interest, the DTT-stimulated transport is no longer responsive to inhibition by depolarization of the plasma membrane potential (S. Lu and N. Kaplowitz, unpublished observations). Thus, some structural or conformational change in the transporter alters its properties and the major driving force. Thus, while no longer responsive to membrane potential, it is conceivable that the proton gradient becomes the driving force. This possibility remains to be tested. Alterna-

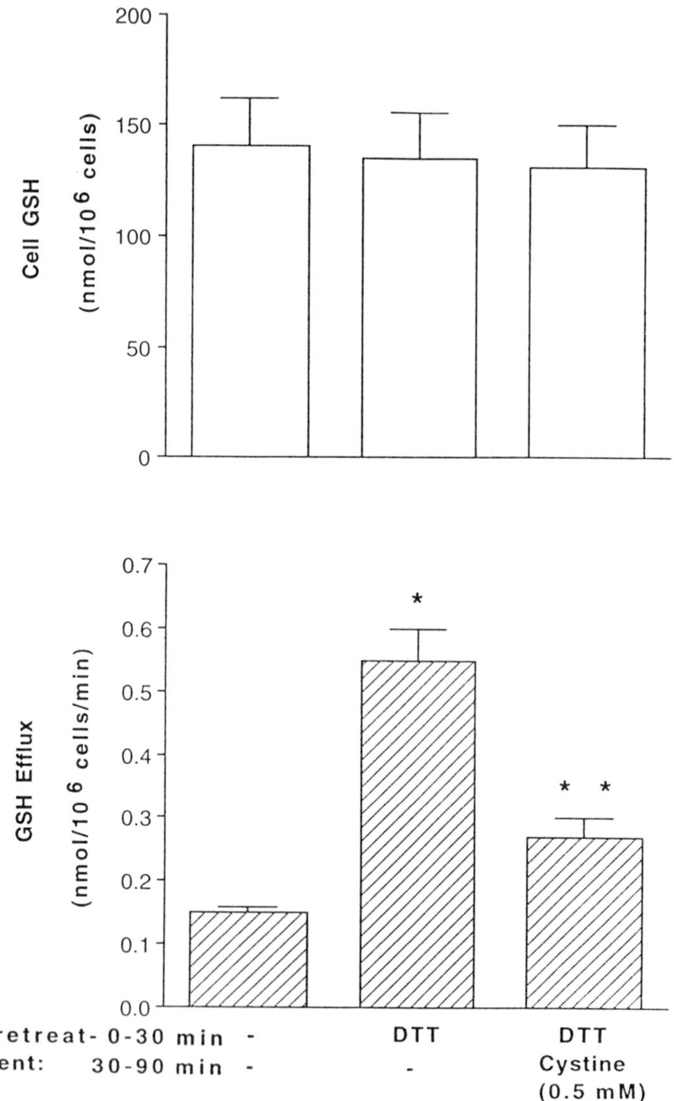

Figure 7 Effect of cystine on DTT-stimulated GSH efflux in cultured rat hepatocytes. Cells were treated for 30 min with DTT (2 mM). Cells were then washed and exposed for 60 min to cystine (0.5 mM) or vehicle. Subsequently cell GSH (upper panel) and efflux (lower panel) were determined in the absence of DTT or cystine. $^{*}p < 0.05$ vs. control; $^{**}p < 0.05$ vs. control + DTT (by ANOVA). (Adapted from Ref. 37.)

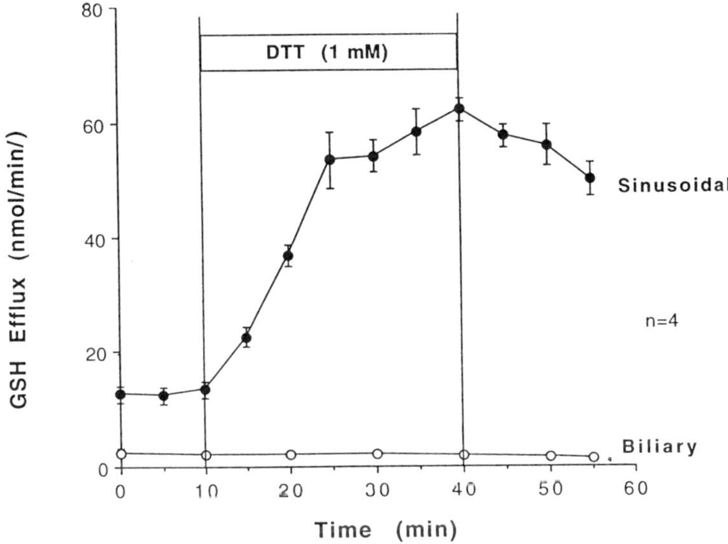

Figure 8 Effect of DTT on GSH efflux in the perfused rat liver. Thirty-minute infusion of DTT (1 mM) markedly enhanced GSH efflux, which persisted following cessation of DTT. No change in biliary GSH secretion was observed. (From Ref. 37.)

tively, opening a pore or channel in the transporter with diffusion of GSH down its concentration gradient is conceivable. However, high extracellular GSH does not slow the DTT-stimulated efflux of GSH and the efflux remains temperature sensitive with little or no efflux at 4°C. At present, we suspect that DTT and membrane potential act at the same step in the carrier's kinetic cycle, most likely accelerating the rate of return of the unloaded carrier from outside to inside orientation.

At present, we are left with the likely possibility that thiol groups intrinsic to the sinusoidal transporter polypeptide are involved and that modulation of their redox state influences transporter function. Aside from the possibility that further exploration of this interesting phenomenon may provide new insights into the mechanism of the sinusoidal GSH transporter function, pathophysiological insights and therapeutic applications may evolve from these studies. Thus, changes in disulfide status in plasma (e.g., with oxidant stress or aging) may have an adverse effect on sinusoidal GSH efflux leading to impaired plasma defense and cysteine/GSH availability in extrahepatic locations. Conversely, pharmacological exploitation of thiol stimulation may improve sinusoidal antioxidant defense (e.g., in reperfusion after liver graft storage) as well as systemic defense against inflammation-induced oxidant stress.

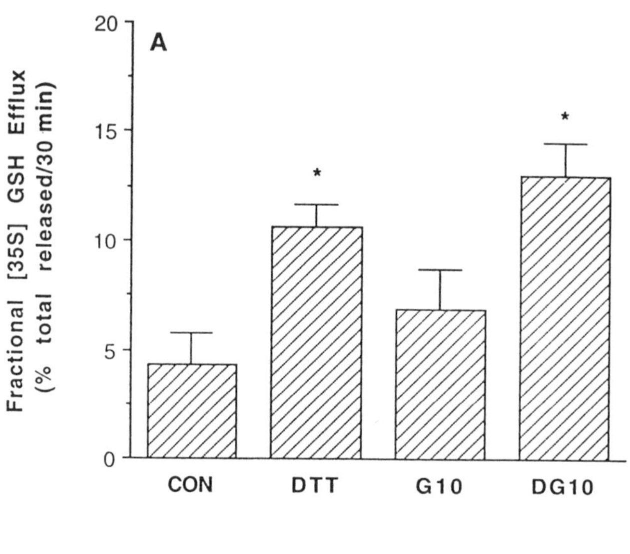

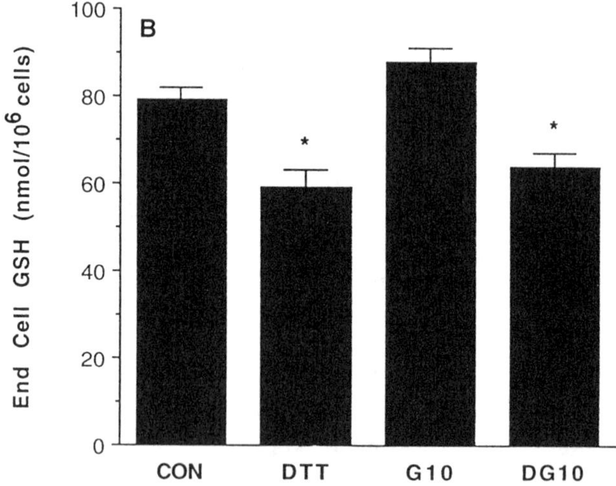

Figure 9 Effect of extracellular GSH and/or DTT on cell GSH and efflux in cultured rat hepatocytes. Cell GSH was first depleted with diethylmaleate (0.5 mM, 20 min) and then incubated with ^{35}S-methionine (1 mM) and serine (1 mM) for 2 h to replete labeled GSH. (A) Efflux was then performed in Krebs buffer alone (CON), with DTT (2 mM), GSH (G10 = 10 mM), or DTT (2 mM) plus GSH (10 mM) (DG10) in the buffer. GSH efflux is expressed as % total ^{35}S-GSH released in 30 min. (B) At the end of efflux, cell GSH was measured. (Modified from Ref. 38.)

IV. CONCLUSIONS

Sinusoidal efflux of hepatic GSH represents a key step in the interorgan homeostasis of cysteine and GSH and the major source of plasma GSH for antioxidant protection. GSH release is governed by a polypeptide carrier, which is regulated by hormone-mediated changes in the driving force for transport or phosphorylation of the carrier. In addition, the transporter function is altered by plasma disulfides and nonphysiological thiol compounds, which may have pathophysiological and pharmacological significance.

ACKNOWLEDGMENTS

This work was supported by NIH grant DK30312, DK45334, AG07467, and Department of Veterans Affairs Medical Research Funds.

REFERENCES

1. Yi, J., Fernandez-Checa, J. C., and Kaplowitz, N. (1994) Expression cloning of a rat hepatic reduced glutathione transporter with canalicular characteristics. J. Clin. Invest. 93:1841–1845.
2. Kaplowitz, N., Fernandez-Checa, J. C., and Yi, J. R. (1994) Mechanisms of canalicular GSH secretion. In: Transport and the Liver (Keppler, D., and Jungermann, K., eds.). Kluwer Academic Publishers, Lancaster, United Kingdom, 166–177.
3. Anderson, M. E., Bridges, R. J., and Meister, A. (1980) Direct evidence for interorgan transport of glutathione. Biochem. Biophys. Res. Commun. 96:848–853.
4. Kaplowitz, N., Eberle, D. E., Petrini, J., Touloukian, J., Corvasce, M. C., and Kuhlenkamp, F. (1983) Factors influencing the efflux of hepatic glutathione into bile in rats. J. Pharm. Exp. Ther. 224:141–147.
5. Ookhtens, M., Mittur, A. V., and Erhart, N. A. (1994) Changes in plasma glutathione concentrations, turnover, and disposal in developing rats. Am. J. Physiol. (Reg. Integ. Comp-Physiol) 226:R979–R988.
6. Speisky, H., Shackel, N., Varghese, A., Wade, D., and Israel, Y. (1990) Role of hepatic γ-glutamyltransferase in the degradation of circulating glutathione. Hepatology 11:843–849.
7. Anderson, M., and Meister, A. (1983) Transport and direct utilization of γ-glutamylcysteine for glutathione synthesis. Proc. Natl. Acad. Sci. USA 80:707–711.
8. Garcia-Ruiz, C., Fernandez-Checa, J., and Kaplowitz, N. (1992) Bidirectional mechanism of plasma membrane transport of reduced GSH in hepatocytes and membrane vesicles. J. Biol. Chem. 267:22256–22264.
9. Dröge, W., Eck, H. P., Naher, H., Pekar, U., and Daniel, V. (1988) Abnormal amino acid concentrations in the blood of patients with acquired immunodeficiency syndrome (AIDS) may contribute to the immunologic defect. Biol. Chem. Hoppe-Seyler 369:143–148.
10. Staal, F. J., Ela, S. W., Roederer, M., Anderson, M. T., Herzenberg, L. A., and

Herzenberg, L. A. (1992) Glutathione deficiency and human immunodeficiency versus infection. Lancet 339:909–912.

11. Takada A., and Bannai, S. (1984) Transport of cystine in isolated rat hepatocytes in primary culture. J. Biol. Chem. 259:2441–2445.
12. Ookhtens, M., and Mittur, A. V. (1994) Developmental changes in plasma thiol/disulfide turnover in rats: A multicompartmental approach. Am. J. Physiol. (Reg. Integ. Comp. Physiol.) 267:R415–425.
13. Lash, L. H., and Jones, D. P. (1984) Renal glutathione transport: Characteristics of the sodium-dependent system in the basal-lateral membrane. J. Biol. Chem. 259:14508–14514.
14. Lash, L. H., Hagen, T. M., and Jones, D. P. (1986) Exogenous glutathione protects intestinal epithelial cells from oxidative injury. Proc. Natl. Acad. Sci. USA 83:4641–4645.
15. Hagen, T. M., Brown, L. A., and Jones, D. P. (1986) Protection against paraquat-induced injury by exogenous GSH in pulmonary alveolar type II cells. Biochem. Pharm. 353:4537–4542.
16. Kannan, R., Kuhlenkamp, J., Jeandidier, E., Trinh, H., Ookhtens, M., and Kaplowitz, N. (1990) Evidence for carrier-mediated transport of glutathione across the blood-brain barrier in the rat. J Clin. Invest. 85:2009–2013.
17. Kannan, R., Kuhlenkamp, J., Ookhtens, M., and Kaplowitz, N. (1992) Transport of GSH at the blood brain barrier of the rat: Inhibition and age-dependence. J. Pharm. Exp. Ther. 263:964–970.
18. Zlokvic, B., Mackic, J., McComb, J., Weiss, M., Kaplowitz, N., and Kannan, R. (1994) Evidence for transcapillary transport of reduced glutathione in vascular perfused guinea-pig brain. Biochem. Biophys. Res. Commun. 201:402–408.
19. Bartoli, G. M., and Sies, H. (1978) Reduced and oxidized glutathione efflux from liver. FEBS Lett. 86:89–91.
20. Ookhtens, M., Corvasce, C., Hobdy, K., Aw, T. Y., and Kaplowitz, N. (1985) Sinusoidal efflux of glutathione in the perfused rat liver. J. Clin. Invest. 75:258–265.
21. Aw, T. Y., Ookhtens, M., Ren, C., and Kaplowitz, N. (1986) Kinetics of GSH efflux from freshly isolated rat hepatocytes. Am. J. Physiol. 250:G236–243.
22. Lu, S., Garcia-Ruiz, C., Kuhlenkamp, J., Ookhtens, M., Salas-Prato, M., and Kaplowitz, N. (1990) Hormonal regulation of hepatic GSH efflux: Studies in cultured rat hepatocytes and perfused liver. J. Biol. Chem. 265:16088–16095.
23. Aw, T. Y., Ookhtens, M., Kuhlenkamp, J. and Kaplowitz, N. (1987) Trans-stimulation and driving forces for GSH transport in sinusoidal membrane vesicles from from rat liver. Biochem. Biophys. Res. Commun. 143:377–382.
24. Ookhtens, M., Fernandez-Checa, J., Lyon, I., and Kaplowitz, N. (1988) Inhibition of GSH efflux in the perfused rat liver and isolated hepatocytes by organic anions and bilirubin: Kinetics, sidedness and molecular forms. J. Clin. Invest. 82:608–616.
25. Aw, T. Y., Ookhtens, M., and Kaplowitz, N. (1984) Inhibition of glutathione efflux from isolated hepatocytes by methionine. J. Biol. Chem. 259:9355–9358.
26. Aw, T. Y., Ookhtens, M., and Kaplowitz, N. (1986) Mechanism of inhibition of glutathione efflux from isolated rat hepatocytes by methionine. Am. J. Physiol. 251:G354–G361.

27. Kaplowitz, N., Fernandez-Checa, J., and Ookhtens, M. (1989) Hepatic GSH transport. In: Glutathione Centennial: Molecular Perspectives, and Clinical Implications (Tanaguchi, N., Higashi, T., Sakamoto, Y., and Meister, A., eds.), pp. 395–406. Academic Press, Inc.

27a. Yi, J. R., Lu, S., Fernandez-Checa, J., and Kaplowitz, N. (1995) Expression cloning of the cDNA for a polypeptide associated with rat hepatic sinusoidal reduced glutathione transport: characteristics and comparison with the canalicular transporter. Proc. Natl. Acad. Sci. USA 92:1495–1499.

28. Fernandez-Checa, J., Ookhtens, M., and Kaplowitz, N. (1987) Effect of chronic ethanol feeding on rat hepatocytic glutathione: compartmentation, efflux, and response to incubation with ethanol. J. Clin. Invest. 80:57–62.

29. Lew, H., Pyke, S., and Quintanilha, A. (1985) Changes in the glutathione status of plasma, liver and muscle following exhaustive exercise in rats. FEBS Lett. 185:262–266.

30. Lauterburg, B. H., Adams, J. D., and Mitchell, J. R. (1984) Hepatic glutathione homeostasis in the rat: Efflux accounts for glutathione turnover. Hepatology 4:586–590.

31. Jaeschke, H. (1992) Enhanced sinusoidal glutathione efflux during endotoxin-induced oxidant stress in vivo. Am. J. Physiol. 263:G60–68.

32. Fernandez-Checa, J., Ookhtens, M., and Kaplowitz, N. (1993) Selective induction by phenobarbital of the electrogenic transport of GSH and organic anions in rat liver canalicular membrane vesicles. J. Biol. Chem. 268:10836–10841.

33. Fernandez-Checa, J., Ren, C., Aw, T. Y., Ookhtens, M., and Kaplowitz, N. (1988) Effect of membrance potential and cellular ATP on GSH efflux from isolated rat hepatocytes. Am. J. Physiol. 18:G403–G408.

34. Sies, H., and Graf, P. (1985) Hepatic thiol and glutathione efflux under the influence of vasopressin, phenylephrine and adrenaline. Biochem. J. 226:545–549.

35. Sato, C., Liu, J. H., Tang, L., Sakai, Y. Yauehi, T., Izumi, N., Liu, J., Takano, T., and Marumo, F. (1992) Possible involvement of protein kinase C and calcium in GSH efflux from Hep G2 cells. Life Sci. 51:2057–2063.

36. Ballatori N., and Truong, A. (1990) Cholestasis, altered junctional permeability, and inverse changes in sinusoidal and biliary glutathione release by vasopressin and epinephrine. Mol. Pharmacol. 38:64–71.

37. Lu, S. C., Ge, J. L., Huang, H. Y., Kuhlenkamp, H., and Kaplowitz, N. (1993) Thiol-disulfide effects on hepatic glutathione transport: Studies in cultured rat hepatocytes and perfused livers. J. Clin. Invest. 92:1188–1197.

38. Lu, S. C., Kuhlenkamp, J., Ge, J. L., Sun, W. M., and Kaplowitz, N. (1994) Specificity and directionality of thiol effects on sinusoidal GSH transport in the rat liver. Mol. Pharmacol. 46:578–585.

11

Oxidative Stress and Mitochondrial Permeability Transition

Donald J. Reed
Environmental Health Sciences Center, Oregon State University, Corvallis, Oregon

I. INTRODUCTION

Studies on the mechanisms of oxygen-related modifications of cellular molecules and processes have led to a greater understanding of oxidative stress–induced events. In mammalian cells, events can occur that lead to the loss of control of endogenous oxidative processes in fulfilling the cellular requirement for molecular oxygen. Oxidative stress may occur to an extent that ranges from a minor to a major contribution to overall effects at the cellular level. Several mechanisms of oxidative stress exist in which the status of cellular integrity and protective systems are challenged by the production of reactive oxygen species. This chapter will discuss oxidative stress (1) as related to the metabolism of molecular oxygen and the need for antioxidant defense systems that are dependent on glutathione and vitamin E. The continual formation of reactive oxygen species is important for normal physiological functions. However, when generated in excess they can be toxic and even more so in the presence of transition metal ions such as iron or copper. For a review see Ref. 2.

Under normal conditions, endogenous oxidative stress does not necessarily place biological tissues and cells at risk since protective systems are present and functioning. However, protective systems can be overwhelmed in various ways leading to a high degree of oxidative stress. It is the excess generation of reactive oxygen species within tissues that can cause damage to cellular constituents. In these instances, oxygen metabolism is involved with the formation of oxygen-containing reactive intermediates, many of which can undergo a cycling process between oxidation and reduction. This process is known as redox cycling and causes oxidative stress in cells. For example, certain chemicals that are known to undergo redox cycling by interaction with molecular oxygen can cause oxidative stress to such a degree that they play a major role in chemically induced toxicity. Such oxidative stress can cause oxidation of cellular constituents such as low molecular weight thiol-containing compounds, including glutathione, protein thiols, as well as other functional groups on macromolecules, and peroxidation of lipids and other susceptible cellular constituents. It is now known that many chemicals can undergo bioactivation to both form biological reactive intermediates that bind to macromolecules and also enhance the formation of oxygen radicals with a concomitant oxidative stress. The evolution of bioactivation processes that form biological reactive intermediates is thought to have necessitated the concomitant evolution of cellular protection systems for cell survival in an oxygen-containing atmosphere. All tissues and cells contain systems for detoxification of biological reactive intermediates and to prevent or limit cellular damage due to oxidative stress events. In many of these cases, disruption of intracellular calcium homeostasis is implicated in the injury, and conditions have been identified in which disrupted Ca^{2+} homeostasis is involved with oxidative stress. Whether oxidative stress results in loss of Ca^{2+} homeostasis or disrupted Ca^{2+} homeostasis causes oxidative stress is not known. However, the two processes appear related in some way.

Toxic processes have reversible and irreversible features that are a consequence of the interplay of bioactivation and oxidative stress with the cellular protective systems that prevent or limit cellular toxicity and possibly repair. Reversible toxicity occurs even with chemicals known as "safe" chemicals. Irreversible toxicity may cause cell death regardless of what antidotal or preventive measures are taken after exposure of cell or tissues to a dose of toxic chemical sufficient to prevent cell survival via free exchange between the intracellular constituents and the surrounding milieu. We review here oxidative stress and the mechanisms by which cells are protected by preventing or limiting cellular damage.

II. OXIDATIVE STRESS

Major contributors to oxygen toxicity are the sources of increased oxidative stress that include increased endogenous superoxide generation, loss of integrity of heme proteins or metalloproteins to enhance the contribution of transition metals to superoxide production, and inadequate antioxidant defenses based on glutathione and vitamins C and E (Table 1). The protective effect of the interaction of vitamin E, ascorbic acid, and glutathione against oxidative damage has been reviewed by Reed (3).

Table 1 Factors Responsible for Initiating Origin of Oxidative Stress

Superoxide and hydrogen peroxide
Redox cycling agents
Transition metals
Excess consumption of antioxidants

Specific sources of reactive oxygen species involved in pathogenic oxidative stress are shown in Table 2. Without mitochondrial antioxidant defense systems such as the glutathione redox cycle, aerobic metabolism, an essential process for life in many species, would likely be impossible due to the endogenous formation of reactive oxygen species (ROS) in association with mitochondrial functions. It is estimated that about 90% of the total O_2 consumed by mammalian species is delivered to mitochondria where a four-electron reduction of molecular oxygen to H_2O by the respiratory chain is coupled to ATP synthesis (4,5). Nearly 4% of mitochondrial O_2 is incompletely reduced, due to leakage of electrons along the respiratory chain, forming ROS such as the superoxide radical (O_2), hydrogen peroxide (H_2O_2), singlet oxygen, and hydroxyl radical (HO•) (4,5).

The superoxide radical undergoes disproportionation to H_2O_2 and O_2 enzymatically via the Mn-containing superoxide dismutase as well as nonenzymatically, although at a rate approximately four orders of magnitude less

Table 2 Specific Sources of Reactive Oxygen Species Involved in Pathogenic Oxidative Stress

Endogenous
Mitochondrial electron transport system
Stimulated phagocytic cells
Nonphagocytic cells being promoted or during metabolism of xenobiotics
Induction of prooxidant enzymes including xanthine oxidase
Quinone semiquinone redox cycling agents
Induction of fatty acid CoA oxidases by peroxisome proliferators
Ischemia-reperfusion
Inhibition of antioxidant enzymes
Exogenous
Pollutants
Pesticides
Drugs
Ionizing radiation

at pH 7.4 (6). Richter (7) calculated that during normal metabolism, one rat liver mitochondrion produces 3×10^7 superoxide radicals per day. It is estimated that superoxide and hydrogen peroxide steady-state concentrations are in the picomolar and nanomolar range, respectively (8). Jones et al. (9) has estimated the hepatocyte steady-state H_2O_2 concentration to be <25 μM. If endogenously formed H_2O_2 is not detoxified to H_2O, formation of the HO• by a metal (iron)–catalyzed Haber Weiss [Eq. (1a)] or Fenton [Eq. (1b)] reaction can occur (10). The HO• species is one of the most reactive and short-lived biological radicals and has the potential to initiate lipid peroxidation of biological membranes (10,11), although not as effectively as other radicals including the ROO• radical (12). Unless termination reactions occur, the process of lipid peroxidation will propagate, resulting in potentially high levels of oxidative stress. Therefore, detoxification of endogenously produced H_2O_2 is critical for redox maintenance of mitochondrial as well as cellular homeostasis.

$$Fe^{+3} + O_2\bullet + H_2O_2 \rightarrow Fe^{2+} + O_2 + OH^- + HO\bullet \quad (1a)$$

$$Fe^{+2} + H_2O_2 \rightarrow Fe^{+3} + HO\bullet + OH^- \quad (1b)$$

Goldstein et al. (13) have reviewed the Fenton reagents and concluded that the oxidizing intermediates formed via the Fentonlike reactions in vivo are not possible to identify. Also, the question of whether an •OH radical is formed rather than a metal in a higher oxidation state that is capable of causing biological damage is nearly impossible to determine.

Sohal (14) has concluded that there is a variation in the sites of superoxide or hydrogen peroxide generation among mitochondria from different tissues and species. This is a result of the rate of mitochondrial superoxide production being dependent on at least three variables: (1) ambient oxygen concentration, (2) levels of autoxidizable respiratory carriers, especially ubiquinone, and (3) the redox state of the autoxidizable carriers (4,8,14–16).

Paraidathathu et al. (17) have shown that mitochondria from hypoxic rat heart tissue produced less reactive oxygen species than normoxic tissue. However, they found that in the presence of calcium the reactive oxygen species of mitochondria from hypoxic rats increased to that of normoxic rats. These workers concluded that hypoxic injury may be a result of the combination of production of reactive oxygen species and increased intracellular calcium levels by mechanisms that may not be mutually exclusive.

Since mitochondria contain the enzymes and cofactors necessary for the glutathione (GSH)/glutathione disulfide (GSSG) redox cycle (18) but do not contain catalase (19), we may assume that a primary function of mitochondrial glutathione (GSH) is in the detoxification of endogenously produced H_2O_2. This redox cycle requires the enzymes GSH peroxidase (selenium-containing) and GSSG reductase along with the cofactors GSH and NADPH (Fig. 1). In addi-

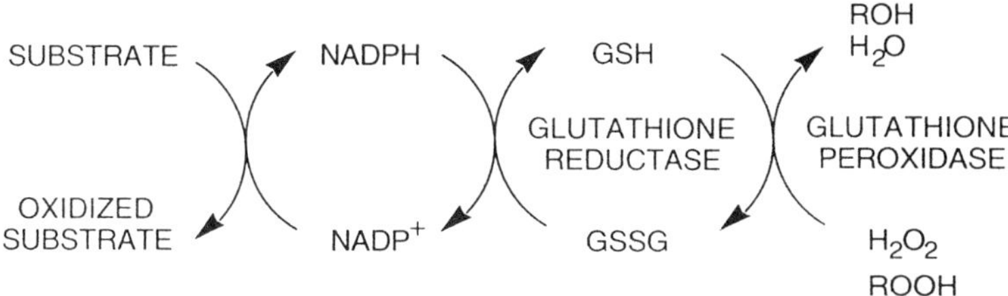

Figure 1 Glutathione and the glutathione redox cycle.

tion to detoxifying H_2O_2, GSH also protects protein sulfhydryls from oxidation (20).

Stimulated phagocytic cells including polymorphonuclear leukocytes (PMNs) and macrophages can generate large quantities of superoxide as a mechanism for destruction of foreign cells. Such an oxidative burst requires that these cells maintain defenses against their own oxygen toxicity. It is known that an NADPH oxidase provides the catalysis for the rapid consumption of molecular oxygen (21,22). The respiratory burst response to appropriate stimuli utilizes the enzyme respiratory burst oxidase to generate superoxide as a precursor for the formation of highly reactive oxidizing agents, including oxidized halogen, oxidizing radicals, and singlet oxygen (for a review, see Ref. 23).

The oxidative stress of superoxide or related oxidants may serve as biological signals when cell function is modified by exposure to oxidizing agents. It has been speculated that enzymes that produce small amounts of superoxide and hydrogen peroxide could do so for signaling purposes in widely distributed biological systems (23). An example is the hydrogen peroxide activation of NFκB, a transcription factor that induces the formation of proteins related to the inflammatory response (24).

Nonphagocytic cells can be promoted by agents such as TPA that can induce oxidative events that may even involve the formation of xanthine oxidase by the conversion of the dehydrogenase to an oxidase (25). Exogenous redox cycling agents include the drug adriamycin and the herbicide paraquat. Such agents have the potential of converting large quantities of oxygen to superoxide by undergoing one-electron redox cycling, the reduced form of the redox agent being able to provide a one-electron reduction of oxygen.

III. PEROXIDATION EVENTS AND ANTIOXIDANTS

Diquat (DQ), a bipyridyl herbicide that is hepatotoxic, is a model compound for studies of redox cycling. Diquat utilizes molecular oxygen to generate large amounts of superoxide anion radical and hydrogen peroxide within cells. Evidence to support the protective role of the glutathione redox cycle is that the toxicity of diquat in vivo and in vitro appears to be mediated by redox cycling and is greatly enhanced by prior treatment with 1,3-bis(2-chloroethyl)-1-nitrosourea (BCNU) (26), which is an inhibitor of glutathione reductase (27). Superoxide anion radical, which can reduce ferric iron to ferrous iron, is produced during diquat toxicity (28). Desferrioxamine, which chelates intracellular iron in the ferric state with an affinity constant of 10^{31}, provides considerable protection against diquat-induced toxicity. Therefore, hydrogen peroxide and transition metals have been suggested as major contributors to diquat toxicity. Even though the hydroxyl radical or a related species seems the most likely ultimate toxic product of the hydrogen peroxide/ferrous iron interaction, scavengers of hydroxyl radical afford only minimal protection. The high degree of reactivity of hydroxyl radicals assures, however, that the site of interaction with cellular components is within close proximity (a few angstroms) of the site of generation of the radical. Much remains to be understood about the mechanism of cytotoxicity by the various redox cycling agents, including the quinones, menadione, and adriamycin.

Interaction of antioxidants provides essential protection against the damaging effects of an oxygen-based metabolism (Table 3) (for a review, see Ref. 29). Complementary antioxidant systems in cytoplasmic and membrane compartments provide efficient cellular defense against oxidative injury (1). α-Tocopherol (αTH) protects cellular membranes against oxidative damage (30–32), and dietary αTH deficiency enhances the susceptibility of biological membranes to oxidative damage in vitro (32–36), and in vivo (36–38). Nonetheless, clinical manifestations of αTH deficiency in humans are poorly defined and are usually secondary to other disease states (30,39). Diplock (40) and Rice-Evans and Diplock (41) have reviewed the status of antioxidant nutrients and disease prevention. This section will focus on the interaction of antioxidants for protection against oxidative damage.

Lipid peroxidation, which can cause extensive damage to subcellular organelles and biomembranes, has been demonstrated to occur in isolated mitochondria, lysosomes, microsomes (42), and nuclei (43). Lipid peroxidation is an important biological consequence of oxidative cellular damage that can be

Table 3 Protection Against Oxygen Toxicity

Superoxide scavengers
Transition metal chelation
Enhancement of antioxidant levels
Enhancement of energy sources for antioxidant maintenance

measured by several noninvasive techniques in humans (44). The destruction of unsaturated fatty acids, which occurs in lipid peroxidation, has been linked with altered membrane structure (45) and enzyme inactivation (46,47). In addition to lipid hydroperoxides and lipid radicals (48), lipid peroxidation generates activated oxygen species such as hydroxyl radicals (49) and superoxide anions (50). The decomposition of peroxidized polyunsaturated fatty acids also generates reactive carbonyl compounds such as malondialdehyde (MDA) and hydroxyalkenals (51).

Biochemical regulation of cellular αTH status is only beginning to be understood. Several studies have addressed the hypothesis that soluble cellular antioxidants, particularly ascorbic acid, maintain membrane αTH levels by regenerating αTH from its oxidation products. Indeed, the reduction of the αTH semiquinone radical by ascorbic acid was reported to proceed in homogeneous solution with a bimolecular rate constant of 1.55×10^6 M^{-1} s^{-1} (52). Synergistic antioxidant protection by αTH and ascorbic acid has been demonstrated in lard (53), homogeneous solution (54), micellar (55,56), and liposome suspensions (57–60). Other reports have provided direct evidence for the αTH-sparing action of ascorbic acid (61,62). Although these data are consonant with a hypothesis of reductive regeneration of αTH, the reaction of ascorbic acid as well as αTH with free radicals complicates interpretation and may account for the additive or synergistic antioxidant effects observed (63). In biological lipid peroxidation, soluble antioxidants also may exert prooxidant effects via reduction of metals (64). The role of soluble antioxidants thus appears complicated. αTH is principally consumed by lipid oxy-radicals* rather than by oxygen radical initiators. Lipid oxy-radical propagation is effectively inhibited by αTH at concentrations above a threshold level. The celebrated Fenton reaction [Eqs. (1b)–(4)] (64) can participate in lipid peroxidation events:

$$Fe^{2} + H_2O_2 \rightarrow Fe^{3+} + \bullet OH + OH^- \quad (1b)$$

$$LH + \bullet OH \rightarrow L\bullet + H_2O \quad (2)$$

$$L\bullet + O2 \rightarrow LOO\bullet \quad (3)$$

$$LOO\bullet + LH \rightarrow LOOH + L\bullet \quad (4)$$

A Fe^{2+} and lipid hydroperoxide–dependent pathway also occurs (64,65):

$$Fe^{2+} + LOOH \rightarrow Fe^{3+} + LO\bullet + OH^- \quad (5)$$

$$LO\bullet + LH \rightarrow LOH + L\bullet \quad (6)$$

*The term "lipid oxy-radicals" refers to both lipid peroxyl radicals (LOO•) and lipid alkoxyl radicals (LO•). The former species are assumed to predominate during uninhibited lipid oxy-radical propagation. However, relative contributions of both species to αTH depletion may depend on αTH concentration (vide infra).

Both ascorbic acid and GSH act as prooxidants when incubated with αTH-free liposomes and Fe^{2+}/H_2O_2. The prooxidant effect of ascorbic acid, but not of GSH, is reversed by incorporation of αTH into the liposomes. αTH levels in excess of the threshold are effectively maintained by ascorbic acid, but GSH antagonizes the antioxidant effects of the αTH/ascorbic acid combination. Effective antioxidant protection by αTH appeared to be due to efficient reaction with lipid oxy-radical in the bilayer rather than to interception of initiating oxygen radicals. At concentrations above a threshold level of approximately 0.2 mol% (mol αTH/mol phospholipid × 100), αTH completely suppressed lipid oxy-radical propagation, which was measured as malondialdehyde production. Both ascorbic acid and glutathione, alone or in combination, can enhance lipid oxy-radical propagation (66). αTH, incorporated into membranes such as in liposomes at concentrations above its threshold protective level, can reverse the prooxidant effects of 0.1–1.0 mM ascorbic acid but not those of glutathione. Ascorbic acid also prevented αTH depletion. The combination of ascorbic acid and subthreshold levels of αTH only temporarily suppresses lipid oxy-radical propagation and does not maintain the αTH level. Glutathione antagonizes the antioxidant action of the αTH/ascorbic acid combination regardless of αTH concentration. These observations indicate theat membrane αTH status can control the balance between pro- and antioxidant effects of ascorbic acid. These studies also provide direct evidence that ascorbic acid interacts directly with components of the phospholipid bilayer (66).

The remarkable antioxidant efficiency of the αTH/ascorbic acid combination may be due in part to efficient radical scavenging by these agents in hydrophobic and hydrophilic environments, respectively. Liebler et al. (66) have shown that ascorbic acid also regenerates αTH from its oxidation products. In addition, αTH completely reversed the prooxidant effects of 0.1 or 1.0 mM ascorbic acid. The inhibition of lipid oxy-radical propagation coincided with maintenance of αTH. The ability of ascorbic acid, a relatively polar and lipid insoluble molecule, to interact directly with components of the lipid bilayer has been questioned (63). However, the data of Scarpa et al. (62) indicate direct reaction between ascorbic acid and liposomal oxy-radicals.

The limited antioxidant capacity of the αTH/ascorbic acid combination at subthreshold αTH concentrations is noteworthy. αTH at 0.1 mol% does not suppress lipid oxy-radical propagation in liposomes unless 0.5–1.0 mM ascorbic acid is present. Even so, αTH status was not maintained—the rate of αTH consumption was merely decreased. The threshold level thus represents a point of no return—propagation reactions cannot be effectively suppressed, nor can the αTH level be maintained. Ascorbic acid, even at 1 mM, cannot maintain αTH levels because the rate of reaction of the αTH semiquinone radical with lipid oxy-radicals exceeds the rate of its reduction to αTH.

Oxidative challenge in liposomes may mimic biological oxidative injury, in which oxygen radicals are generated due to the presence of copper and iron (64,67). In the absence of αTH, the availability of Fe^{2+} determines the extent of SPC liposome peroxidation. Reduction of Fe^{3+} by ascorbic acid or GSH maintains an active Fe^{2+} pool and enhances both the rate and extent of lipid oxy-radical propagation. The concentration-dependent prooxidant effect of GSH is thus easily explained as described by Miller et al. (67). The prooxidant action of ascorbic acid, however, was maximal at the lowest concentration studied (0.1 mM), and concentrations above 1 mM temporarily inhibited lipid peroxidation (66). Inhibition due to scavenging of hydroxyl radicals—presumably the initiating species in this model—by ascorbic acid is unlikely. GSH produces no such inhibition despite being a superior hydrogen atom donor (68). On the other hand, ascorbic acid at higher concentrations may react directly with lipid oxy-radicals at the lipid-water interface. Because GSH displays no antioxidant activity with liposomes, it does not appear to react with lipid oxy-radicals to a significant extent or regenerate αTH from its oxidation products. Both reactions may be kinetically unfavorable for GSH (68). The notable antagonism of the αTH/ascorbic acid antioxidant combination by GSH may involve at least two mechanisms. First, GSH chelation and reduction of iron enhances and sustains the oxidative challenge (69). Second, the glutathione radicals thus produced are themselves rapidly reduced by ascorbic acid (68). The consumption of ascorbic acid would consequently exceed that expected on the basis of its antioxidant functions alone.

Ascorbic acid inhibits lipid peroxidation by regenerating αTH rather than by reacting directly with lipid oxy-radicals. These studies represent one of the first demonstrations of a "sparing" effect of ascorbic acid upon αTH in lipid bilayers under conditions where significant reaction between ascorbic acid and radicals can be unequivocally ruled out. Furthermore, ascorbic acid (at 0.1 mM) offers no antioxidant protection when αTH is present at a subthreshold level. This observation supports a regeneration hypothesis but contradicts the view that ascorbic acid reacts independently with lipid oxy-radicals to spare αTH. Liebler et al. (70) provided evidence that, in vitro or in liposomes, ascorbate does not reduce α-tocopherone directly but reacts instead with the tocopherone cation. Slow rates of these reactions led these workers to suggest that biochemical catalysis would be required to complete a two-electron αTH redox cycle in biological membranes.

A central conclusion is that the balance between pro- and antioxidant effects of ascorbic acid is critically dependent upon the bilayer concentration of αTH relative to the threshold level. Ascorbic acid cannot maintain αTH at subthreshold levels, presumably because the rate of αTH consumption exceeds the rate of its reductive regeneration from a radical intermediate by ascorbic acid.

The effectiveness of the αTH/ascorbic acid combination is thus intimately related to the threshold effect. The liposome studies indicate that the balance between pro- and antioxidant effects of ascorbic acid ultimately depends on the αTH status of the lipid bilayer. The results further imply that maintenance of membrane αTH status may be an essential function of cellular ascorbic acid in vivo. If so, then NADH-dependent reduction of semidehydroascorbate (71) may subject αTH recycling to a limited degree of metabolic control. Attempts to manipulate tissue αTH levels by modifying dietary ascorbic acid intake have met with limited success (63). However, ascorbic acid treatment enhanced lipid peroxidation in αTH-deficient rats (36), and Litov and coworkers concluded that in the absence of αTH, lipid oxy-radical propagation in cellular membranes proceeds unchecked by ascorbic acid. This conclusion is essentially identical to that of the liposome studies; ascorbic acid, at concentrations below 1 mM, did not function as an antioxidant unless the αTH concentration exceeded the threshold level.

Damage to membrane lipids and the associated alterations in bulk properties of membranes are the basis for chemical-induced loss of cell viability. Horrobin (72) has focused on the possibility that loss of membrane essential fatty acids is the basis for cell injury rather than the accumulation of toxic lipid peroxidation products. Possibly in agreement with that hypothesis, lipid peroxidation and further oxidative modification of low-density lipoprotein (LDL) is thought to increase its atherogenicity (73–75). These workers have shown that ascorbic acid oxidation products, including dehydro-L-ascorbic acid, protect human LDL against atherogenic modification and that anti- rather than prooxidant activity was observed for ascorbic acid in the presence of transition metal ions (73).

The failure of GSH to function as an antioxidant in liposomes despite mounting evidence for such a role in biological systems (76,77) is significant. Enzymic participation in the antioxidant action of GSH is thought to be essential, and the results of these studies reinforce that view. Evidence for protein- and GSH-dependent antioxidant protection of cell membranes has been reported (76,78,79), but evidence for the participation of αTH in GSH-dependent protection is somewhat equivocal. Further study of GSH-dependent antioxidant activity in biological membranes may reveal a system whose function complements that of αTH. One possible mechanism by which GSH would stimulate lipid peroxidation is by reducing iron and thus promoting Fe^{2+}-dependent peroxidation. This is not a new suggestion; it is well supported by previously published work (80,81). Moreover, evidence that thiols may directly reduce iron and stimulate peroxidation has come from Aust and colleagues (69,82).

Vitamin E participates in the inhibition of lipid peroxidation of membranes by breaking of the chain propagation (83) by reactions that appear to include scavenging of lipid hydroperoxyl radicals. Murphy et al. (84) have shown that the protection by ascorbate and glutathione against microsomal lipid peroxidation is dependent on vitamin E. Loss of vitamin E is related to both chemiluminescence and thiobarbituric-reactive material formation, which increased when the αTH content of the microsomes had decreased to about 0.38 nmol/mg protein, or about 86% of the initial content (84).

Dietary supplementation with vitamin E resulted in a 10- to 20-fold increase of vitamin E in the rat liver mitochondrial membrane (85). Treatment of submitochondrial particles with an oxidizing system composed of lipoxygenase and arachidonic acid provided evidence that reduction of the tocopheroxyl radical could occur via NADH, succinate, and reduced cytochrome *c*–linked oxidation. Thus, the electron-transport system is proposed to have an important physiological role in recycling vitamin E by reduction of the tocopheroxyl radical to prevent its accumulation and vitamin E consumption.

A limited number of studies have focused on the susceptibility of the cell nucleus to lipid peroxidation. The nuclear membrane regulates the transport of mRNA into the cytoplasm and aids in the process of nuclear division. DNA is also frequently associated with certain regions of the nuclear membrane (86), and it seems likely that nuclear membrane peroxidation may disrupt many of these critical functions. The proximity of the nuclear membrane to DNA could also contribute to the interaction of DNA with reactive compounds generated in lipid peroxidation. Several studies indicate that such lipid peroxidation products can alter the structure and function of DNA (87–90). This fact is of importance since hydroxyl radicals diffuse an average of only 60 Å before reacting with cellular components (91). Assays for 8-hydroxy-2′-deoxyguanosine as a biomarker of oxidative DNA damage include in vivo studies with urine samples (92). Nuclear peroxidation may also increase interactions between more stable peroxidation products and DNA. The cytosolic enzymes aldehyde dehydrogenase (93), glutathione transferase (94), and glutathione peroxidase (95) have all been shown to metabolize various reactive lipid peroxidation products. Such cytosolic enzymes may metabolize peroxidation products generated throughout the cell before they diffuse into the nucleus and interact with DNA.

Endogenous αTH levels in isolated rat liver nuclei have been measured and found to be 0.045 mol% (96). This value corresponds to 970 polyunsaturated fatty acid (PUFA) moieties to one molecule of αTH in the nuclear membrane. These values are higher than values reported for rat liver microsomes (3313) (35) and mitochondria (2100) (97). A threshold level of 0.085 mol% for the prevention of NADPH-induced lipid peroxidation was established for isolated nuclei. That value could be lowered to 0.040 mol% when 1 mM GSH was added to assist in the inhibition of lipid peroxidation. The ability of GSH to enhance αTH-dependent protection against nuclear lipid peroxidation appears to be mainly by a "sparing" effect on the near-threshold level of αTH in the nuclear membrane.

Thiol groups are well known to be important for normal protein functions, and increasing evidence supports the vital importance of these thiols for cell viability during cytotoxic events. Membrane-bound enzymes are damaged during lipid peroxidation, and evidence of vitamin E protection strongly supports a free radical mechanism for protein damage via oxidative stress (98,99). Oxidative stress can cause loss of protein function by damaging amino acid residues other than cysteine, including methionine, tryptophan, and histidine. An important aspect of such damage is that lipid peroxidative events can amplify free radical processes, which propagate chain reactions. Failure to terminate free radical processes with a chain-breaking antioxidant, such as vitamin E, can lead to 4–10 propagation events per initiation, and thus each initiation is amplified (100). Products of lipid peroxidation are toxic and lipid peroxidation events can compromise GSH-dependent detoxication systems. Since the reduction of lipid hydroperoxides by GSH utilizes NADPH for the regeneration of GSH from GSSG, the rate of NADPH production can be limiting during oxidative stress. Therefore, GSSG may be transported from the liver when not reduced due to limited levels of NADPH. Decreased availability of NADPH and GSH can impair other GSH-dependent detoxication pathways, including metabolism of hydrogen peroxide, decreased protection of thiols in protein (101), and decreased reaction with free radicals. Thus, energy-dependent processes involving NADPH, GSH, and thiols in proteins appear to be critically involved in cellular homeostasis during drug-induced toxicity.

The prevention of nitrofurantoin-induced cytotoxicity in isolated hepatocytes by fructose appears to relate to maintenance of the cellular glutathione pool by limiting GSSG efflux from these cells. Silva et al. (102) concluded that fructose induced ATP depletion (20% of control) to such an extent that ATP-dependent GSSG efflux was prevented, thus protecting the cells. This protection is achieved by the retained GSSG being reduced back to GSH, limiting the nitrofurantoin-induced oxidative stress. In addition, a combination of calcium release from intracellular stores and inhibition of calcium efflux appears to result in a marked and sustained increase in cytosolic concentration of calcium associated with surface blebbing and increased activity of certain calcium-dependent enzymes.

IV. THE PROTECTIVE ROLE OF GLUTATHIONE

Mammalian cells have evolved a major protective system to minimize injurious events that result from biotransformation of xenobiotics and normal oxidative products of cellular metabolism. This system is dependent upon the unique tripeptide glutathione. Depending on the cell type, the intracellular concentration of GSH is in the range of 0.5–10 mM (for a review, see Ref. 103). Concentrations in mammalian liver are 4–8 mM with nearly all of the glutathione being present as reduced glutathione, and less than 5% of the total glutathione being present as oxidized glutathione (glutathione disulfide) with minor fractions being mixed disulfides of GSH and other cellular thiols and minor amounts of thioethers from endogenous conjugation reactions (104). Overall, the GSH content of various organs and tissues represents at least 90% of the total non-protein, low molecular weight thiols. GSH can be depleted directly by conjugation with electrophiles and indirectly by the addition of inhibitors of GSH biosynthesis and regeneration (Fig. 2). The liver appears unique in that GSH is synthesized in vivo with the sulfur of cysteine supplied by methionine via the cystathionine pathway (105). Cysteine formed by this pathway as well as the low level of cysteine that is transported into liver cells serve as substrate for γ-glutamylcysteine synthase to form γ-glutamylcysteine, which is then coupled with glycine via GSH synthetase (EC 6.3.2.3) to complete the synthesis of GSH. Most of the other cell types with few exceptions must depend entirely on the uptake of cysteine because of the lack of the cystathionine pathway for cysteine synthesis and in turn GSH synthesis. Inhibition of γ-glutamylcysteine synthetase (EC

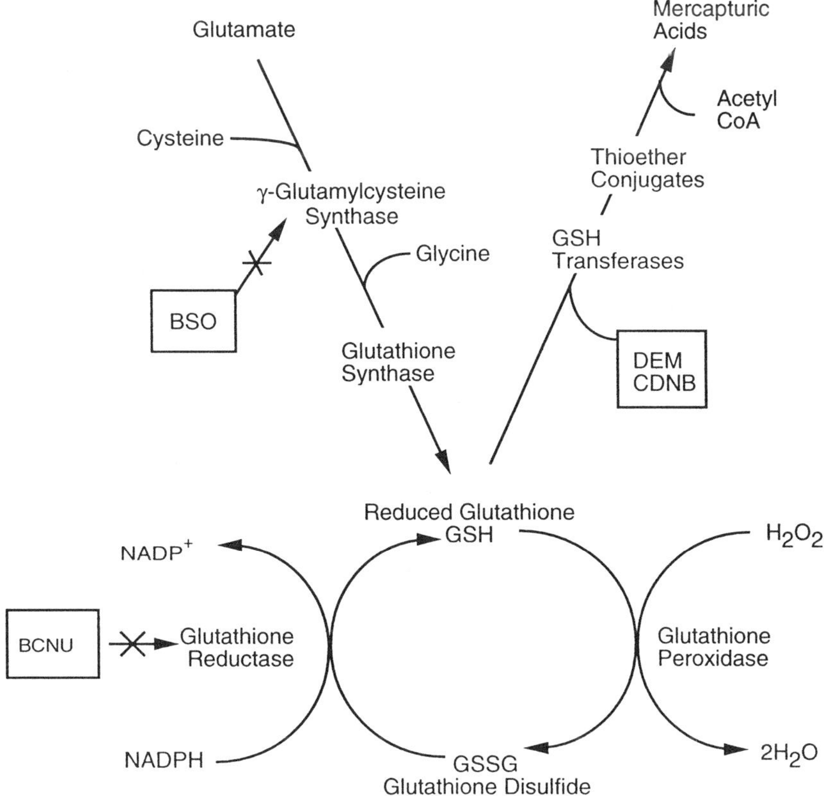

Figure 2 Biosynthesis and depletion of GSH and the glutathione redox cycle.

6.3.2.2), the cytosolic rate-limiting enzyme, by buthionine sulfoximine (BSO) diminishes the rate of GSH synthesis but does not cause depletion of mitochondrial GSH during cytosolic GSH depletion. The two main functions of GSH are (1) as a nucleophilic "scavenger" of numerous compounds and their metabolites, via enzymatic and chemical mechanisms, converting electrophilic centers to thioether bonds, and (2) as a substrate in the GSH peroxidase-mediated metabolism of hydroperoxides to the corresponding alcohols. Alternatively, GSH content in isolated hepatocytes can be augmented by the addition of its sulfur-containing precursors, such as cysteine or methionine, to maximize biosynthesis. Enhanced GSH synthesis has major beneficial effects during severe GSH depletion caused by consumption of GSH during xenobiotic metabolism. Cystine has a sparing effect on the requirement of the essential amino acid methionine in the rat (106). This observation is in agreement with the undirectional process of trans-sulfuration in which methionine sulfur and serine carbon are utilized in cysteine biosynthesis via the cystathionine pathway.

The cystathionine pathway is of major importance to pathways of drug metabolism in the liver that involve GSH or cysteine or both. Depletion of GSH by rapid conjugation can increase synthesis of GSH to rates as high as 2–3 μmol/h/g wet liver tissue (107). The cysteine pool in the liver, which is about 0.2 μmol/g, has an estimated half-life of 2–3 min at such high rates of synthesis of GSH. Although the cystathionine pathway appears to be highly responsive to the need for cysteine biosynthesis in the liver, the organ distribution of the pathway is limited. In mammals, such as rats, the liver is the main site of cysteine biosynthesis, which occurs via the cystathionine pathway. Maintenance of high concentrations of GSH in the liver, in association with high rates of secretion into plasma and extensive extracellular degradation of GSH and GSSG, supports the concept that liver GSH is a physiological reservoir of cysteine (108,109).

In vivo treatment of rats with AT-125, an inhibitor of γ-glutamyl transpeptidase, prevents degradation of GSH in plasma leading to massive urinary excretion of GSH (110). This treatment also lowers the hepatic content of GSH because it inhibits recycling of cysteine to the liver. A physiological decrease in interorgan recycling of cysteine to the liver for synthesis of GSH may also account in part for the decrease of hepatic GSH during starvation and for the marked diurnal variation in concentration of GSH in liver. The efflux of liver GSH and metabolism of the resulting plasma GSH and GSSG appears to help ensure a continuous supply of plasma cysteine. This cysteine pool should in turn minimize the degree of fluctuation of GSH concentrations within the various body organs and cell types that require cysteine, cystine, or both, rather than methionine for synthesis of GSH.

V. OXIDATIVE STRESS AND THE GLUTATHIONE REDOX CYCLE

The intracellular concentration of GSH in isolated hepatocytes has been examined under conditions that result in oxidative stress. Production of malondialdehyde, which is an index of lipid peroxidation and occurs during oxidative stress, can be stimulated by addition of a glutathione depletor, diethyl maleate. This observation suggests that intracellular concentrations of GSH under these conditions are important for membrane and cellular integrity.

A major endogenous protective system against reactive oxygen species is the glutathione redox cycle (for a review, see Ref. 111). GSH depletion to about 20–30% of the normal glutathione level can impair the cell's defense against the toxic actions of both biological reactive intermediates and reactive oxygen species and may lead to cell injury and death. Endogenous oxidative stress, which is a normal physiological process, is a consequence of aerobic metabolism that occurs mostly in the mitochondria of eukaryotic cells. H_2O_2, if not decomposed, can lead to the formation of the very reactive hydroxyl radical and cause the formation of lipid hydroperoxides that can damage membranes, nucleic acid, and proteins and alter their functions.

GSH is maintained in a redox couple with GSSG within the cell and is regenerated by GSH reductase (EC 1.6.4.2), a cytosolic NADPH-dependent enzyme. Inhibition of this enzyme, and hence of GSH regeneration, with 1,3-bis(2-chloroethyl)-1-nitrosourea also depletes intracellular GSH (Fig. 2). It is apparent that a major protective role against the reactive drug intermediates, which are generated by bioreduction and cause oxidative stress by redox cycling, is provided by the ubiquitous glutathione redox cycle (Figs. 1 and 2). This cycle utilizes NADPH- and, indirectly, NADH-reducing equivalents in the mitochondrial matrix as well as the cytoplasm to provide GSH by the glutathione reductase–catalyzed reduction of GSSG. When the glutathione redox cycle is functioning at maximum capacity to eliminate hydrogen peroxide, a regulatory effect is imposed on other NADPH-dependent pathways.

Glutathione peroxidase is a selenium-dependent enzyme that is extremely specific for glutathione and is capable of rapidly detoxifying hydrogen peroxide and certain hydroperoxides. Selenium-dependent glutathione peroxidase activity is the result of the expression of multiple isozymes. Four isozymes have been characterized: (1) the classical cellular glutathione peroxidase, GSHPx-1, (2) the phospholipid hydroperoxide glutathione peroxidase, GSHPX, (3) the plasma glutathione peroxidase, GSHPx-P, and (4) GSHPx-GI (112). Weitzel and Wendel (113) have reported findings that phospholipid hydroperoxide–glutathione peroxidase activity regulates the activity of 5-lipoxygenase by regulating the tone of endogenous hydroperoxides.

Ursini et al. (114) has purified an interfacial glutathione peroxidase. This enzyme has been shown to reduce linoleic acid hydroperoxides, cumene hydroperoxide, *tert*-butyl hydroperoxide, and hydrogen peroxide. However, this enzyme, which does not conjugate CDNB with GSH (114), displays glutathione peroxidase activity toward cumene hydroperoxide, hydrogen peroxide, and lipid hydroperoxides and is distinct from the classical glutathione peroxidase (115). Evidence suggests that the enzyme is interfacial in character and can interact directly with liposomes to reduce phospholipid hydroperoxides (114). The addition of this protein to microsomal incubation mixtures inhibited lipid peroxidation (114). Substrate specificities indicate that this enzyme is distinct from the nuclear glutathione transferase (96).

Lipid hydroperoxides located in biological membranes increase the potential for lipid peroxidation. If this potential is to be diminished, these peroxides must be removed from the membrane environment or be reduced to lipid alcohols. If glutathione peroxidase activity is associated with the phospholipid bilayer of the nuclear membrane, such an association may contribute to the ability of the peroxidase to reduce lipid hydroperoxides. Since lipid hydroperoxides can initiate lipid peroxidation, the reduction of these compounds can contribute to the inhibition of peroxidation. The association of a glutathione dependent peroxidase with membranes may encourage the reduction of lipid hydroperoxides located within lipid bilayers.

Glutathione protection of isolated rat liver nuclei against lipid peroxidation is abolished by exposing isolated nuclei to the glutathione transferase inhibitor S-octylglutathione (96). S-Octylglutathione also inhibited nuclear glutathione transferase activity and glutathione peroxidase activity. A large percentage of the glutathione transferase activity associated with isolated nuclei was solubilized with 0.3% Triton X-100. Studies suggest that this treatment removes nuclear membranes while preserving the integrity of the remaining nucleus.

Extraction of nuclei with 1 M NaCl accompanied with brief sonication failed to solubilize the peroxidase activity. Extraction with 0.3% Triton X-100, however, solubilized the GSH-dependent peroxidase activity. Electron microscopy studies conducted by Dabeva et al. (116) indicate that treatment with even higher concentrations of Triton X-100 removed the outer nuclear membrane but preserved the integrity of the remaining nucleus. Based on this information, it appears that the peroxidase activity is associated with the nuclear membrane. This activity in conjunction with GSH may contribute to the inhibition of lipid peroxidation in nuclear membranes and thereby preserve the integrity of this important membrane system. Increasing evidence suggests that this inhibition of peroxidation may in turn protect the structure and function of DNA.

VI. ORGANELLE GLUTATHIONE

Compartmentation of glutathione has been demonstrated in that separate pools of glutathione exist in the cytoplasm from that in the mitochondria (Fig. 3). A separate pool has been proposed for the nucleus, but that finding is in dispute. Mitochondrial GSH functions as a discrete pool separate from cytosolic GSH. A report by Jocelyn (117) demonstrated that mitochondrial GSH is imperme-

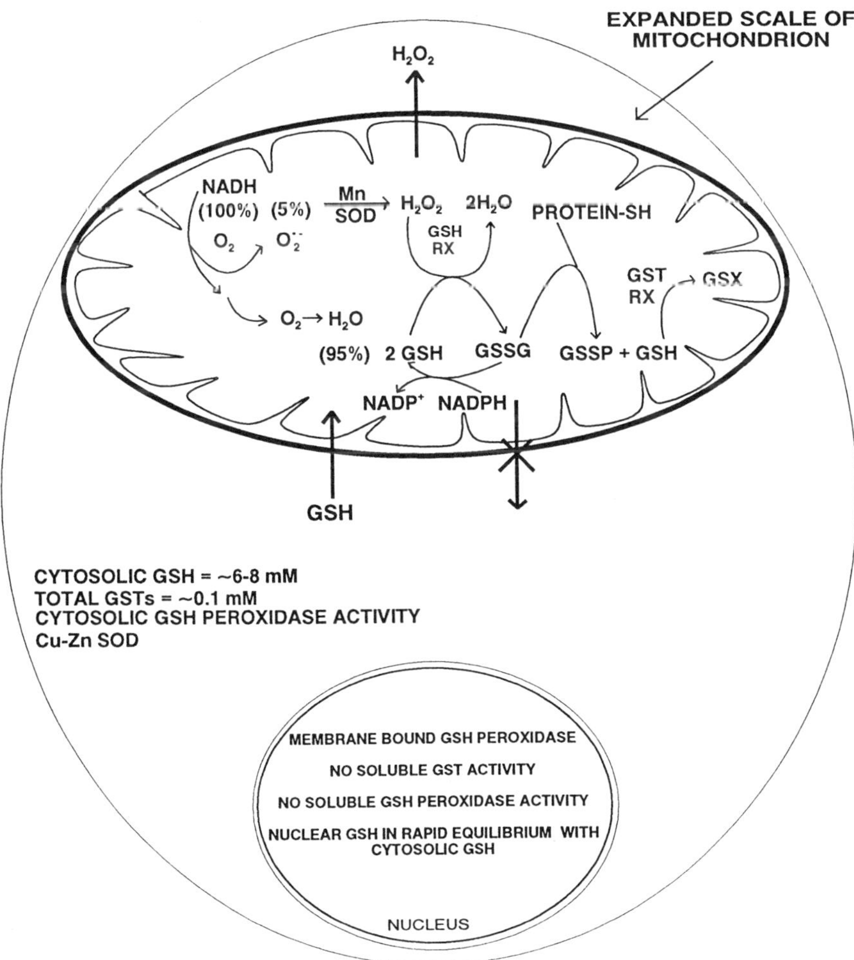

Figure 3 Cellular protective systems and organelle glutathione.

able to the inner membrane following isolation of mitochondria, and Wahlländer et al. (118) reported the concentration of mitochondrial GSH (10 mM) to be higher than cytosolic GSH (7 mM). Studies by Meredith and Reed (119) demonstrated different rates of GSH turnover in the cytosol and mitochondria, confirming the existence of separate intracellular GSH pools. The ratio of GSH:GSSG in mitochondria is approximately 10:1 under normal (untreated) conditions. Unlike cytosolic GSSG, mitochondrial GSSG is not effluxed from the soluble compartment as reported by Olafsdottir et al. (120). This study demonstrated that during oxidative stress induced with *t*-butyl hydroperoxide, GSSG is accumulated in the mitochondrial matrix and eventually reduced back to GSH. However, as the redox state of the mitochondria increased, an increase in protein mixed disulfides was also observed. This study concluded that mitochondria are more sensitive to redox changes in GSH:GSSG than the cytosol and therefore mitochondria may be more susceptible to the damaging effects of oxidative stress. These findings suggest that under certain experimental conditions irreversible cell injury due to oxidative challenge may result from irreversible changes in mitochondrial function.

Studies addressing mitochondria as target organelles of certain types of irreversible cell injury, as related to oxidative stress as well as disrupted Ca^{2+} homeostasis, represents an area of intense investigation (121–125). GSH-dependent protection against lipid peroxidation has been demonstrated in mitochondria, nuclei, microsomes, and cytosol of rat liver. Lipid peroxidation induced in mitochondria is inhibited by respiratory substrates such as succinate, which leads indirectly to reduction of ubiquinone to ubiquinol. The latter is a potent antioxidant (126–128). The essential factor in preventing accumulation of lipid peroxides and lysis of membranes in mitochondria, however, is glutathione peroxidase (18). Although the prevention of free radical attack on membrane lipids may occur by an electron shuttle that utilizes vitamin E and GSH in microsomes, similar activity may not be capable of inhibiting peroxidation in mitochondria (78,129). Instead, mitochondrial GSH transferase(s) may prevent lipid peroxidation in mitochondria by a nonselenium glutathione-dependent peroxidase activity. Three GSH transferases have been isolated from the mitochondrial matrix (130), and nearly 5% of the mitochondrial outer membrane protein consists of microsomal glutathione transferase (131). GSH transferase in the outer mitochondrial membrane could provide the GSH-dependent protection of mitochondria by scavenging lipid radicals by a mechanism that requires vitamin E and is abolished by bromosulfophthalein (131).

Since there is growing experimental evidence that oxidative stress is a factor in the neuropathology of several adult neurodegenerative disorders, many studies are being directed toward understanding of the oxidative stress associated with these diseases (132). For example, Adams et al. (133) have examined brain oxidative stress induced by *t*-butyl hydroperoxide and have concluded from studies with 2- and 8-month-old mice that aging make the brain more susceptible to oxidative damage.

The pathogenesis of cell injury and death during oxidative stress may be caused by calcium ion–induced permeability transition of the mitochondrial inner membrane. A feature of oxidative stress–induced cell injury is morphological and functional changes in mitochondria (134–136). Mitochondrial dysfunction, a critical event in anoxia/reoxygenation injury of liver sinusoidal endothelial cells, appears due to the permeability transition since acidosis protects against lethal oxidative injury (137) as does a combination of cyclosporin A (CsA) plus trifluoperazine (138).

VII. MITOCHONDRIAL PERMEABILITY TRANSITION

Several decades of studies with isolated mitochondria from a variety of tissues indicate that under certain experimental conditions, the mitochondrial inner membrane, which is normally impermeable to solutes, becomes permeable (139–142). Although the mechanism(s) by which this occurs remains controversial, this process is frequently referred to as mitochondrial permeability transition (142,143). The permeability transition is characterized as a nonspecific and as a specific (144) Ca^{2+}-dependent inner membrane pore, which presents itself in mitochondria following treatment with Ca^{2+} and a second agent, termed an inducing agent. Many inducing agents have been identified, and they vary greatly in both structure and function (142) (Table 4). It is thought that these agents, in the presence of Ca^{2+}, act through a common mechanism during the permeability transition. Examples of inducing agents include inorganic phosphate, cytosolic factor, fatty acids, heavy metals, organic sulfhydryl reagents, and oxidants such as *t*-butyl hydroperoxide (141,142). Several identifiable events of

Table 4 Permeability Transition in Mitochondria

Cell types	Nature of change	Inducing agents
Liver, kidney, heart, adrenal cortex	Inner membrane becomes permeable to ions and molecules of mw <1200	Initial calcium loading
		Secondary Agents:
		High calcium
		Phosphate ion
	Cyclosporin A is a potent inhibitor	Sulfhydryl reagents
		Hydroperoxides
		Quinones
		Various drugs
		Heavy metals

the permeability transition have been characterized in the last several years. These include inner membrane permeability to small ions and solutes with molecular weights < 1200 daltons, large amplitude swelling, loss of coupled functions, and sensitivity to cyclosporin A (CsA) (145–148). In addition, loss of matrix proteins via the inner membrane pore has also been reported (149). During permeability transition, Ca^{2+} as well as other ions are rapidly released from the mitochondrial matrix, presumably via the inner membrane pore. Following inner membrane permeability and the release of matrix solutes, a colloidal osmotic pressure arises in the mitochondrial matrix due to the high concentration of proteins, which are slow to equilibrate (142). In order to correct the osmotic imbalance, the entrance of H_2O results in massive swelling of the mitochondria (142).

Mitochondrial swelling under these conditions is termed large amplitude swelling. Although secondary to inner membrane permeability and solute release, large amplitude swelling occurs within a short time (3–10 min) and is easily detected by monitoring light scatter of mitochondrial suspensions at 540 nm. Monitoring of mitochondrial swelling is a simple assay and is often utilized as an indicator of permeability transition. Different experimental conditions, result, however, in different patterns and times of swelling, some with shorter or longer lag periods (148).

Studies by Fournier et al. (150) showed with isolated mitochondria that CsA treatment promoted retention of accumulated Ca^{2+}. Subsequent studies by Crompton et al. (151) were designed to examine the effect of CsA on pore opening since the loss of Ca^{2+} observed by Fournier et al. (150) was possibly due to pore opening. This study revealed that CsA is a potent inhibitor of the permeability transition when the inducing agents inorganic phosphate or *t*-butyl hydroperoxide were added in the presence of Ca^{2+}. Broekemeier et al. (148) tested CsA as an inhibitor of the inner membrane pore in the presence of several different inducing agents and found that very low concentrations of CsA protected against permeability transition. These studies demonstrated the potency of CsA in preventing the permeability transition, as inhibition of inner membrane permeability is observed with concentrations as low as 100 pmol CsA/mg mitochondrial protein.

Studies by Halestrap and Davidson (152) have focused on the mechanism of pore formation and have proposed that the adenine nucleotide translocase is the putative pore structure. They hypothesize that the "m" conformation of this inner membrane protein is modified to the "c" conformation by the binding of negative effectors such as Ca^{2+}, cyclophilin, and inducing agents, resulting in a protein conformation that can no longer function as an adenine nucleotide translocase but as a nonspecific pore. Atractyloside and bongkrekic acid are known modifiers of the translocase and have been helpful in proposing this mechanism. This study also demonstrates that both ATP and ADP are positive effectors of the putative permeability transition pore.

Valle and coworkers (153) have shown that oxidative stress induced by *t*-butyl hydroperoxide is mediated by reactive oxygen species capable of oxidative attack on inner membrane protein thiols. In the presence of Ca^{2+}, the membrane pore may undergo opening as a result of alterations of protein thiols due to the generation of reactive oxygen species from added oxidant and one electron reduction of oxygen by the mitochondrial electron transport system. Petronilli et al. (154) have concluded that the oxidation-reduction state of vicinal cysteinyl residues of the inner membrane has a critical role in pore opening with an increase in the probability of pore opening with increasing degree of thiol oxidation.

Richter et al. (144) propose that pore formation is not required for Ca^{2+} release and suggest that ADP-ribosylation of inner membrane proteins is the responsible mechanism of permeability transition. They hypothesize that uptake of Ca^{2+} and the consequent recycling of Ca^{2+} results in NAD(P)H oxidation followed by the hydrolysis of NAD+ forming a nicotinamide and an ADP-ribose moiety. Subsequent ADP-ribosylation of critical proteins results in inner membrane permeability allowing the specific release of Ca^{2+}.

The present understanding of permeability transition is that the pore structure is an allosteric inner membrane protein, perhaps the adenine nucleotide translocase, with several different regulatory sites, which are affected by Ca^{2+}, Pi, oxidants, sulfhydryl reagents, heavy metals, cyclophilin, ADP, ATP, atractyloside, and bongkrekic acid. The sites that are modified may determine whether permeability transition occurs. It is hypothesized that inhibition of permeability transition with CsA results from the binding of cyclophilin, a matrix protein *cis-trans* isomerase thought to be involved with protein folding (155). Whether the process of permeability transition is physiological remains an intriguing question.

Several methods are available to detect/measure mitochondrial permeability transition in isolated mitochondria. Crompton and Costi (147) have developed a method that exploits two aspects of the permeability transition: (1) the entry of radiolabeled sucrose into the mitochondrial matrix, a normally impermeable solute to the mitochondrial inner membrane and (2) the reversible nature of the pore. By adding a chelator of Ca^{2+}, such as EDTA, or CsA, a potent pore inhibitor, the inner membrane nonspecific pore can be closed. This allows for the entrapment of sucrose into the matrix, which can then be quantified.

Release of added Ca^{2+} or other ions from mitochondria and its inhibition by CsA are also indicative of permeability transition (142,148). As previously discussed, loss of absorbance at 540 nm of mitochondrial suspensions is frequently used as an indicator of permeability transition in isolated mitochondria.

However, since large-amplitude swelling is a secondary event, which does not always accompany permeability transition (156), monitoring of solute movement may be a more sensitive parameter of permeability changes.

Mitochondrial GSH (307 Da) release may be a useful and sensitive endogenous indicator of permeability transition under the conditions we have examined (156–158). Studies from our laboratory with rat liver mitochondria have shown that during permeability transition induced with 70 μM Ca^{2+} and 3 mM Pi, mitochondrial GSH is rapidly and completely released from the matrix and recovered extramitochondrially as GSH (158). This release is completely prevented by addition of 0.5 μM CsA, indicating that release occurs via the Ca^{2+}-dependent, CsA-sensitive inner membrane pore (158). GSH release under these conditions parallels Ca^{2+} release in that both of these molecules are almost completely released during a 5-min incubation. Mitochondrial swelling, a secondary process, is somewhat slower, however, and this process results in a substantial loss of mitochondrial density, which is prevented by the addition of CsA.

An interesting finding is that 3 mM Pi alone induces a permeability transition by a CsA-sensitive mechanism. Mitochondria undergo large-amplitude swelling with Pi alone, although not as extensive as in mitochondria treated with both Ca^{2+} and Pi. However, complete GSH release occurs with Pi treatment alone, although at a slower rate than with the addition of both Ca^{2+} and Pi. This release and swelling appears due to the presence of 6–10 nmol Ca^{2+}/mg protein in the mitochondrial preparations. The effect of Pi is abolished following the addition of EGTA. Although monitoring of swelling indicates that permeability transition is occurring, the finding suggests permeability transition is not as extensive relative to Ca^{2+} and Pi treatment. However, without exogenously added Ca^{2+}, monitoring of Ca^{2+} release is not a sensitive indicator of permeability transition, as endogenous Ca^{2+} levels are very low and therefore a significant change in Ca^{2+} levels is not observed.

Although swelling is detected when Pi is added alone, it is less than maximal swelling detected by light scatter in the presence of 70 μM Ca^{2+} and 3 mM Pi. However, electron microscopy of samples treated with Pi alone or Pi plus Ca^{2+} reveals indistinguishable levels of large-amplitude swelling, with varied times (158). Thus, monitoring of Ca^{2+} release or swelling is not always a sensitive indicator of the degree of permeability transition. Monitoring release of a highly concentrated endogenous solute, such as GSH release by a CsA-sensitive mechanism, may be a sensitive indicator of inner membrane permeability transition.

Mitochondria treated with Ca^{2+} and Pi in the presence of metabolic inhibitors of electron transport or ATP synthesis do not undergo mitochondrial large amplitude swelling as detected by light scatter and electron microscopy (158). However, monitoring of matrix GSH indicates that GSH is released from the matrix into the extramitochondrial environment. Treatment with 0.5 μM CsA prevents this release of GSH, suggesting that release occurred via the Ca^{2+}-dependent inner membrane pore. The rate and extent of GSH release varies with the different conditions, however, release of this endogenous molecule may be a useful marker of mitochondrial permeability transition.

Studies by Richter et al. (144) propose that oxidation of pyridine nucleotides and the subsequent hydrolysis is the mechanism responsible for Ca^{2+} release. These studies treated mitochondria with the oxidant *t*-butyl hydroperoxide as the inducing agent in the presence of Ca^{2+}. Measurement of total pyridine nucleotides in our model, in the presence of Pi rather than an oxidant, revealed that pyridine nucleotides undergo extensive oxidation very early during permeability transition, suggesting the involvement of endogenous oxidative stress (156). The loss of reduced pyridine nucleotides was accounted for by the oxidized product, indicating that hydrolysis of oxidized pyridine nucleotides does not occur to a significant degree and therefore does not appear to be involved with Ca^{2+} release under these conditions. In these experiments, GSSG did not exceed untreated control values, which was approximately 0.5 nmol GSSG/mg mitochondrial protein.

Conditions of permeability transition greatly diminished mitochondrial energy status as indicated by the redox ratios of NAD+/NADH, NADP+/NADPH, and the energy charge. Treatment with 0.5 μM CsA maintains pyridine nucleotide ratios very similar to untreated controls while increasing slightly the energy charge (156). Oxidation of pyridine nucleotides in the presence of Ca^{2+} and Pi suggest that oxidative stress may be involved in this model of permeability transition. Although increased levels of GSSG were not detected, possibly due to a very active glutathione redox cycle or lipid peroxidation in this system, oxidation of pyridine nucleotides is consistent with an oxidative stress. This is the first indication that permeability transition may be intricately connected with oxidative stress. If this type of injury occurs at the cellular level, irreversible cellular injury may occur if mitochondria cannot repair this damage. The physiological consequences of permeability transition and possible relationship to oxygen toxicity are currently under investigation.

ACKNOWLEDGMENTS

Support in part for this work was by United States Public Health Service Grants ES01978 and ES00210.

REFERENCES

1. Sies, H. (1985) Introduction. In: Oxidative Stress (Sies, H., ed.), pp. 1–7. Academic Press, Orlando, FL.

2. Halliwell, B., and Gutteridge, J. M. C. (1990) Role of free radicals and catalytic metal ions in human disease: An overview. In: Methods in Enzymology (Packer, L., and Glazer, A. N., eds.), pp. 1-85. Academic Press, San Diego, CA.
3. Reed, D. J. (1992) Interaction of vitamin E, ascorbic acid, and glutathione in protection against oxidative stress. In: Vitamin E in Health and Disease (L. Packer, ed.), pp. 269-281. Marcel Dekker, Inc., New York.
4. Chance, B., Sies, H., and Boveris, A. (1979) Hydroperoxide metabolism in mammalian organs. Physiol. Rev. 59:527-605.
5. Cadenas, E. (1989) Biochemistry of oxygen toxicity. Annu. Rev. Biochem. 58:79-110.
6. Fridovich, I. (1974) Superoxide dismutases. Adv. Enzymol. Relat. Areas Mol. Biol. 41:35-97.
7. Richter, D. (1988) Do mitochondrial DNA fragments promote cancer and aging? FEBS Lett. 241:1-5.
8. Forman, H. J., and Boveris, A. (1982) Superoxide radical and hydrogen peroxide in mitochondria. In: Free Radicals in Biology, Vol. V (Pryor, W. A., ed.), pp. 65-90. Academic Press, New York.
9. Jones, D. P., Eklow, L., Thor, H., ad Orrenius, S. (1981) Metabolism of hydrogen peroxide in isolated hepatocytes: Relative contributions of catalase and glutathione peroxidase in decomposition of endogenously generated H_2O_2. Arch. Biochem. Biophys. 210:505-516.
10. Pryor, W. A. (1986) Oxy-radicals and related species: Their formation, lifetimes, and reactions. Annu. Rev. Physiol. 48:657-667.
11. Bindoli, A. (1988) Lipid peroxidation in mitochondria. Free Rad. Biol. Med. 5:247-261.
12. Dix, T. A., and Aikens, J. (1993) Mechanisms and biological relevance of lipid peroxidation initiation. Chem. Res. Toxicol. 6:2-18.
13. Goldstein, S., Meyerstein, D., and Czapski, G. (1993) The Fenton reagents. Free Rad. Biol. Med. 15:435-445.
14. Sohal, R. S. (1993) Aging, cytochrome oxidase activity, and hydrogen peroxide release by mitochondria. Free Rad. Biol. Med. 14:583-588.
15. Turrens, J. F., and McCord, J. M. (1990) Mitochondrial generation of reactive oxygen species. In: Free Radicals, Lipoproteins, and Membrane Lipids (Paulet, A. C., Douste-Blazy, L., and Paoletti, R., eds.), pp. 203-212. Plenum Press, New York.
16. Boveris, A., and Cadenas, E. (1982) Production of superoxide in mitochondria. In: Superoxide Dismutase, II (Oberley, L. W., ed.), pp. 15-30. CRC Press, Boca Raton, FL.
17. Paraidathathu, T., de Groot, H., and Kehrer, J. P. (1992) Production of reactive oxygen by mitochondria from normoxid and hypoxic rat heart tissue. Free Rad. Biol. Med. 13:289-297.
18. Flohe, L., and Schlegel, W. (1971). Glutathione peroxidase. IV. Intracellular distribution of the glutathione peroxidase system in the rat liver. Hoppe-Seyler's Z. Physiol. Chem. 352:1401-1410.
19. Neubert, D., Wojtszak, A. B., and Lehninger, A. L. (1962) Purification and enzy-

matic identity of mitochondrial contraction factors I and II. Proc. Natl. Acad. Sci. USA 48:1651–1658.
20. Vignais, P. M., and Vignais, P. V. (1973) Fuscin, an inhibitor of mitochondrial SH-dependent transport-linked functions. Biochem. Biophys. Acta 325:357–374.
21. Babior, B. M. (1984) Oxidant from phagocytes: Agents of defense and destruction. Blood 64:959–966.
22. Babior, B. M. (1987) The respiratory burst oxidase. Trends Biol. Sci. 12:241–242.
23. Chanock, S. J., El Benna, J., Smith, R. M., and Babior, B. M. (1994) The respiratory burst oxidase. J. Biol. Chem. 269:24519–24522.
24. Schreck, R., Rieber, P. and Baeuerle, P. A. (1991) Reactive oxygen intermediates as apparently widely used messengers in the activation of the NF-kappa B transcription factor and HIV-1. EMBO J. 10:2247–2258.
25. Frenkel, K. (1992) Carcinogen-mediated oxidant formation and oxidative DNA damage. Pharmacol. Ther. 53:127–166.
26. Sandy, M. S., Moldéus, P., Ross, D., et al. (1987) Cytotoxicity of the redox cycling compound diquat in isolated hepatocytes: Involvement of hydrogen peroxide and transition metals. Arch. Biochem. Biophys. 259:29.
27. Reed, D. J. (1985) Cellular defense mechanisms against reactive metabolites. In: Bioactivation of Foreign Compounds (Anders, M. W., ed.), pp. 71–108. Academic Press, Orlando, FL.
28. Keberle, H. (1964) The biochemistry of desferrioxamine and its relation to iron metabolism. Ann. NY Acad. Sci. 119:758.
29. Di Mascio, P., Murphy, M. E., and Sies, H. (1991) Antioxidant defense systems: The role of carotenoids, tocopherols, and thiols. Am. J. Clin. Nutr. 53:194S–200S.
30. Machlin, L., ed. (1980) Vitamin E: A Comprehensive Treatise. Marcel Dekker, New York.
31. Burton, G. W., Joyce, A., and Ingold, K. U. (1983) Is vitamin E the only lipid-soluble, chain-breaking antioxidant in human blood plasma and erythrocyte membranes? Arch. Biochem. Biophys. 221:281–290.
32. Tappel, A. L. (1972) Vitamin E and free radical peroxidation of lipids. Ann. NY Acad. Sci. 203:12–28.
33. McCay, P. B., Poyer, J. L., Pfeifer, P. M., May, H. E., and Gilliam, J. M. (1971) A function for α-tocopherol: Stabilization of the microsomal membrane from radical attack during TPNH-dependent oxidations. Lipids 6:297–306.
34. Kornbrust, D. J., and Mavis, R. D. (1980) Relative susceptibility of microsomes from lung, heart, liver, kidney, brain and testes to lipid peroxidation: Correlation with vitamin E content. Lipids 15:315–322.
35. Sevanian, A., Hacker, A. D., and Elsayed, N. (1982) Influence of vitamin E and nitrogen dioxide on lipid peroxidation in rat lung and liver microsomes. Lipids 17:269–277.
36. Litov, R. E., Matthews, L. C., and Tappel, A. L. (1981) Vitamin E protection against in vivo lipid peroxidation initiated in rats by methyl ethyl ketone peroxide as monitored by pentane. Toxicol. Appl. Pharmacol. 59:96–106.
37. Herschberger, L. A., and Tappel, A. L. (1982) Effect of vitamin E on pentane exhaled by rats treated with methyl ethyl ketone peroxide. Lipids 17:686–691.

38. Dillard, C. J., Kunert, K. J., and Tappel, A. L. (1982) Effects of vitamin E, ascorbic acid and mannitol on alloxan-induced lipid peroxidation in rats. Arch. Biochem. Biophys. 216:204–212.
39. Muller, D. P. R., Lloyd, J. K., and Wolff, O. H. (1983) Vitamin E and neurological function: Abetalipoproteinaemia and other disorders of fat absorption. In: Biology of Vitamin E, pp. 106–121. Pitman Books, London.
40. Diplock, A. T. (1991) Antioxidant nutrients and disease prevention: An overview. Am. J. Clin. Nutr. 53:189S–193S.
41. Rice-Evans, C. A., and Diplock, A. T. (1993) Current status of antioxidant therapy. Free Rad. Biol. Med. 15:77–96.
42. Tappel, A. L. (1973) Lipid peroxidation damage to cell components. Fed. Proc. 32:1870–1874.
43. Tirmenstein, M. A., and Reed, D. J. (1988) Characterization of glutathione-dependent inhibition of lipid peroxidation of isolated rat liver nuclei. Arch. Biochem. Biophys. 261:1–11.
44. Pryor, W. A., and Godber, S. S. (1991) Noninvasive measures of oxidative stress status in humans. Free Rad. Biol. Med. 10:177–184.
44. Plaa, G. L., and Witschi, H. (1976) Chemicals, drugs, and lipid peroxidation. Annu. Rev. Pharmacol. Toxicol. 16:125–141.
45. Eichenberger, K., Bohni, P., Winterhalter, K. H., Kawato, S., and Richter, C., Microsomal lipid peroxidation causes an increase in the order of the membrane lipid domain. FEBS Lett. 142:59–62.
46. Baker, S. P., and Hemsworth, B. A., Effect of mitochondrial lipid peroxidation on monoamine oxidase. Biochem. Pharmacol. 27:805–806.
47. Thomas, C. E., and Reed, D. J. (1988) Effect of extracellular Ca^{++} omission on isolated hepatocytes. I. Induction of oxidative stress and cell injury. J. Pharmacol. Exp. Ther. 245:493–500.
48. Porter, N. A. (1984) Chemistry of lipid peroxidation. Methods Enzymol. 105:273–282.
49. O'Brien, P. J., and Hawco, F. J. (1978) Hydroxyl-radical formation during prostaglandin formation catalyzed by prostaglandin cyclo-oxygenase. Biochem. Soc. Trans. 6:1169–1171.
50. Svingen, B. A., O'Neal, F. O., and Aust, S. D., The role of superoxide and singlet oxygen in lipid peroxidation. Photochem. Photobiol. 28:803–809.
51. Brambilla, G., Sciaba, L., Faggin, P., Maura, A., Marinari, U. M., Ferro, M., and Esterbauer, H. (1986) Cytotoxicity, DNA fragmentation and sister-chromatid exchange in Chinese hamster ovary cells exposed to the lipid peroxidation product 4-hydroxynonenal and homologous aldehydes. Mutat. Res. 171:169–176.
52. Packer, J. E., Slater, T. F., and Willson, R. L. (1979) Direct observation of a free radical interaction between vitamin E and vitamin C. Nature (London) 278:737–738.
53. Golumbic, C., and Mattill, H. A. (1941) Antioxidants and the autoxidation of fats. XIII. The antioxygenic action of ascorbic acid in association with tocopherols, hydroquinones and related compounds. J. Am. Chem. Soc. 63:1279–1280.
54. Niki, E., Saito, T., and Kamiya, Y. (1983) The role of vitamin C as an antioxidant. Chem. Lett. 631–632.

55. Tappel, A. L., Brown, W. D., Zalkin, H., and Maier, V. P. (1961) Unsaturated lipid peroxidation catalyzed by hematin compounds and its inhibition by vitamin E. J. Am. Oil Chem. Soc. 38:5–9.
56. Barclay, L. R. C., Bailey, A. M. H., and Kong, D. (1985) The antioxidant activity of α-tocopherol-bovine serum albumin complex in micellar and liposome autoxidations. J. Biol. Chem. 260:15809–15814.
57. Leung, H. W., Vang, M. J., and Mavis, R. D. (1981) The cooperative interaction between vitamin E and vitamin C in suppression of peroxidation of membrane phospholipids. Biochim. Biophys. Acta 664:266–272.
58. Fukuzawa, K., Takase, S., and Tsukatani, H. (1985) The effect of concentration on the antioxidant effectiveness of α-tocopherol in lipid peroxidation induced by superoxide free radicals. Arch. Biochem. Biophys. 240:117–120.
59. Doba, T., Burton, G. W., and Ingold, K. U. (1985) Antioxidant and co-antioxidant activity of vitamin C. The effect of vitamin C, either alone or in the presence of vitamin E or a water-soluble vitamin E analogue, upon the peroxidation of aqueous multi-lamellar phospholipid liposome. Biochim. Biophys. Acta 835:298–303.
60. Niki, E., Kawakami, A., Yamamoto, Y., and Kamiya, Y. (1985) Oxidation of lipids. VIII. Synergistic inhibition of oxidation of phosphatidylcholine liposome in aqueous dispersion by vitamin E and vitamin C. Bull. Chem. Soc. Jpn. 58:1971–1975.
61. Bascetta, E., Gunstone, F. D., and Walton, J. C. (1983) Electron spin resonance study of the role of vitamin E and vitamin C in the inhibition of fatty acid oxidation in a model membrane. Chem. Phys. Lipids 33:207–210.
62. Scarpa, M., Rigo, A., Maiorino, M., Ursini, F., and Gregolin, C. (1984) Formation of α-tocopherol radical and recycling of α-tocopherol by ascorbate during peroxidation of phosphatidylcholine liposomes. An electron paramagnetic resonance study. Biochim. Biophys. Acta 801:215–219.
63. McCay, P. B. (1985) Vitamin E: Interactions with free radicals and ascorbate. Ann. Rev. Nutr. 5:323–340.
64. Aust, S. D., Morehouse, L. A., and Thomas, C. E. (1985) Role of metals in oxygen radical reactions. J. Free Rad. Biol. Med. 1:3–25.
65. Svingen, B. A., Buege, J. A., O'Neal, F. O., and Aust, S. D. (1979) The mechanism of NADPH-dependent lipid peroxidation. The propagation of lipid peroxidation. J. Biol. Chem. 254:5892–5899.
66. Liebler, D. C., Kling, D. S., and Reed, D. J. (1986) Antioxidant protection of phospholipid bilayers by α-tocopherol. J. Biol. Chem. 261:12114–12119.
67. Miller, D. M., Buettner, G. R., and Aust, S. D. (1990) Transition metals as catalysts of "autoxidation" reactions. Free Rad. Biol. Med. 8:95–108.
68. Willson, R. L. (1983) Free radical protection: Why vitamin E, not vitamin C, beta carotene or glutathione? In: Biology of Vitamin E, pp. 19–44. Pitman Books, London.
69. Bucher, J. R., Tien, M., Morehouse, L. A., and Aust, S. D. (1983) Redox cycling and lipid peroxidation: The central role of iron chelates. Fundam. Appl. Toxicol. 3:222–226.
70. Liebler, D. C., Kaysen, K. L., and Kennedy, T. A. (1989) Redox cycles of vi-

tamin E: Hydrolysis and ascorbic acid dependent reduction of 8a-(alkyldioxy) tocopherones. Biochemistry 28:9772–9777.
71. Ito, A., Hayashi, S., and Yoshida, T. (1981) Participation of a cytochrome b5-like hemoprotein of outer mitochondrial membrane (OM cytochrome b) in NADH-semidehydroascorbic acid reductase activity of rat liver. Biochem. Biophys. Res. Commun. 101:591–598.
72. Horrobin, D. F. Is the main problem in free radical damage caused by radiation, oxygen, and other toxins, the loss of membrane essential fatty acids, rather than the accumulation of toxic materials? Med. Hypotheses 35:23–26.
73. Retsky, K. L., Freeman, M. W., and Frei, B. (1993) Ascorbic acid oxidation product(s) protect human low density lipoprotein against atherogenic modification. J. Biol. Chem. 268:1304–1309.
74. Steinberg, D., Parthasarathy, S., Carew, T. E., Khoo, J. C., and Witztum, J. L. (1989) Beyond cholesterol. Modifications of low-density lipoprotein that increase its atherogenicity. N. Engl. J. Med. 320:915–924.
75. Schwartz, C. J., Valente, A. J., Sprague, E. A., Kelley, J. L., and Nerem, R. M. (1991) The pathogenesis of atherosclerosis: An overview. Clin. Cardiol. 14:1–16.
76. Haenen, G. R. M. M., and Bast, A. (1983) Protection against lipid peroxidation by a microsomal glutathione-dependent labile factor. FEBS Lett. 159:24–28.
77. McCay, P. B., Brueggemann, G., Lai, E. K., and Powell, S. R. (1989) Evidence that α-tocopherol functions cyclically to quench free radicals in hepatic microsomes. Requirement for glutathione and a heat-labile factor. Ann. NY Acad. Sci. 570:32–45.
78. Reddy, C. C., Sholz, W. W., and Thomas, C. B., et al. (1982) Vitamin E-dependent reduced glutathione inhibition of rat liver microsomal lipid peroxidation. Life Sci. 31:571.
79. Hill, K. E., and Burk, R. F. (1984) Influence of vitamin E and selenium on glutathione-dependent protection against microsomal lipid peroxidation. Biochem. Pharmacol. 33:1065–1068.
80. Misra, H. P. (1974) Generation of superoxide free radical during the autoxidation of thiols. J. Biol. Chem. 249:2151–255.
81. Rowley, D. A., and Halliwell, B. (1982) Superoxide-dependent formation of hydroxyl radicals in the presence of thiol compounds. FEBS Lett. 138:33–36.
82. Tien, M., Bucher, J. R., and Aust, S. D. (1982) Thiol-dependent lipid peroxidation. Biochem. Biophys. Res. Commun. 107:279–285.
83. Niki, E., Yamamoto, Y., Komuro, E., and Sato, K. (1991) Membrane damage due to lipid oxidation. Am. J. Clin. Nutr. 53:201S–205S.
84. Murphy, M. E., Scholich, H., Wefers, H., and Sies, H. (1989) Alpha-tocopherol in microsomal lipid peroxidation. Ann. NY Acad. Sci. 570:480–486.
85. Maguire, J. J., Wilson, D. S., and Packer, L. (1989) Mitochondrial electron transport-linked tocopheroxyl radical reduction. J. Biol. Chem. 264:21462–21465.
86. Franke, W. W. (1974) Structure, biochemistry, and functions of the nuclear envelope. Int. Rev. Cytol. (Suppl. 4):71–236.
87. Akasaka, S. (1986) Inactivation of transforming activity of plasmid DNA by lipid peroxidation. Biochim. Biophys. Acta 867:201–208.

88. Ueda, K., Kobayashi, S., Morita, J., and Komano, T. (1985) Site-specific DNA damage caused by lipid peroxidation products. Biochim. Biophys. Acta 824:341–348.
89. Brawn, K., and Fridovich, I. (1981) DNA strand scission by enzymically generated oxygen radicals. Arch. Biochem. Biophys. 206:414–419.
90. Mukai, F. H., and Goldstein, B. D. (1976) Mutagenicity of malonaldehyde, a decomposition product of peroxidized polyunsaturated fatty acids. Science 191:868–869.
91. Roots, R., and Okada, S. (1975) Estimation of life times and diffusion distances of radicals involved in x-ray-induced DNA strand breaks of killing of mammalian cells. Radiat. Res. 64:306–320.
92. Shigenaga, M. K., and Ames, B. N. (1991) Assays for 8-hydroxy-2′-deoxyguanosine: A biomarker of in vivo oxidative DNA damage. Free Radical Biol. Med. 10:211–216.
93. Hjelle, J. J., and Petersen, D. R. (1983) Metabolism of malondialdehyde by rat liver aldehyde dehydrogenase. Toxicol. Appl. Pharmacol. 70:57–66.
94. Alin, P., Danielson, U. H., and Mannervik, B. (1985) 4-Hydroxyalk-2-enals are substrates for glutathione transferase. FEBS Lett. 179:267–270.
95. Christophersen, B. O. (1968) Formation of monohydroxy-polyenic fatty acids from lipid peroxides by a glutathione peroxidase. Biochim. Biophys. Acta 164:35–46.
96. Tirmenstein, M. A., and Reed, D. J. (1989) Effects of glutathione on the α-tocopherol-dependent inhibition of nuclear lipid peroxidation. J. Lipid Res. 30:959–965.
97. Gruger, E. H., and Tappel, A. L. (1971) Reactions of biological antioxidants: III. Composition of biological membranes. Lipids 6:147–148.
98. Dean, R. T., and Cheeseman, K. H. (1987) Vitamin E protects against free radical damage in lipid environments. Biochem. Biophys. Res. Commun. 148:1277.
99. Thomas, C. E., and Reed, D. J. (1990) Radical-induced inactivation of kidney Na^+, K^+-ATPase: Sensitivity to membrane lipid peroxidation and the protective effect of vitamin E. Arch. Biochem. Biophys. 281:96–105.
100. McCay, P. B., Lai, E. K., Powell, S. R., et al. (1976) Vitamin E functions as an electron shuttle for glutathione-dependent "free radical reductase" activity in biological membrane. Fed. Proc. Fed. Am. Soc. Exp. Biol. 45:1729.
101. Pascoe, G. A., and Reed, D. J. (1987) Relationship between cellular calcium and vitamin E metabolism during protection against cell injury. Arch. Biochem. Biophys. 253:287.
102. Silva, J. M., McGirr, L., and O'Brien, P. J. (1991) Prevention of nitrofurantoin-induced cytotoxicity in isolated hepatocytes by fructose. Arch. Biochem. Biophys. 289:313–318.
103. Reed, D. J. (1990) Glutathione: Toxicological implications. Annu. Rev. Pharmacol. Toxicol. 30:603–631.
104. Kosower, N. S., and Kosower, E. (1979) Glutathione status of cells. Int. Rev. Cytol. 54:289.
105. Reed, D. J., and Beatty, P. W. (1980) Biosynthesis and regulation of glutathione. Toxicological implications. In: Reviews in Biochemical Toxicology (Hodgson, E., Bend, J. R., and Philpot, R. M., eds.), p. 213. Elsevier Press, New York.

106. Womack, M., Kremmer, K. S., Rose, W. C. (1937) The relation of cysteine and methionine to growth. J. Biol. Chem. 121:403.
107. White, I. N. H. (1976) Role of liver glutathione in acute toxicity of retrorsine to rats. Chem. Biol. Interact. 13:333.
108. Tateishi, N., Higashi, T., Naruse, A., et al. (1977) Rat-liver glutathione. Possible role as a reservoir of cysteine. J. Nutr. 107:51.
109. Higashi, T., Tateishi, N., Naruse, A., et al. (1977) Novel physiological role of liver glutathione as a reservoir of L-cysteine. J. Biochem. 82:117.
110. Reed, D. J., and Ellis, W. W. (1982) Influence of γ-glutamyl transpeptidase inactivation on the status of extracellular glutathione and glutathione conjugates. In: Biological Reactive Intermediates, IIA (Snyder, R., Parke, C. V., Kocsis, J. J., et al., eds.), p. 75. Plenum Press, New York.
111. Reed, D. J. (1986) Regulation of reductive processes by glutathione. Biochem. Pharmacol. 35:7–13.
112. Chu, F. F., Doroshow, J. H., and Esworthy, R. S. (1993) Expression, characterization, and tissue distribution of a new cellular selenium-dependent glutathione peroxidase, GSHPx-GI. J. Biol. Chem. 268:2571–2576.
113. Weitzel, F., and Wendel, A. (1992) Selenoenzymes regulate the activity of leukocyte 5-lipoxygenase via the peroxide tone. J. Biol. Chem. 268:6288–6292.
114. Ursini, F., Maiorino, M., and Gregolin, C. (1985) The selenoenzyme phospholipid hydroperoxide glutathione peroxidase. Biochim. Biophys. Acta 839:62–70.
115. Nakamura, W., Hosoda, S., and Hayashi, K. (1974) Purification and properties of rat liver glutathione peroxidase. Biochim. Biophys. Acta 358:251–261.
116. Dabeva, M. D., Petrov, P. T., Stoykova, A. S., and Hadjiolov, A. A. (1977) Contamination of detergent purified rat liver nuclei by cytoplasmic ribosomes. Exp. Cell Res. 108:467–471.
117. Jocelyn, P. C. (1975) Some properties of mitochondrial glutathione. Biochim. Biophys. Acta 396:427–436.
118. Wahländer, A., Sobell, S., Sies, H., Linke, I., and Müller, M. (1979) Hepatic mitochondrial and cytosolic glutathione content and the subcellular distribution of GSH-S-transferase. FEBS Lett. 97:138–140.
119. Meredith, M. J., and Reed, D. J. (1982) Status of the mitochondrial pool of glutathione in the isolated hepatocyte. J. Biol. Chem. 257:3747–3753.
120. Olafsdottir, K., and Reed, D. J. (1988) Retention of oxidized glutathione by isolated rat liver mitochondria during hydroperoxide treatment. Biochim. Biophys. Acta 964:377–382.
121. Nohl, H., deSilva, D., and Summer, K. H. (1989) 2,3,7,8,-Tetrachlorodibenzo-p-dioxin induces oxygen activation associated with cell respiration. Free Rad. Biol. Med. 6:369–374.
122. Chacon, E., and Acosta, D. (1991) Mitochondrial regulation of superoxide by Ca^{2+}: An alternate mechanism for the cardiotoxicity of doxorubicin. Toxicol. Appl. Pharmacol. 107:117–128.
123. Turrens, J. F., Beconi, M., Barilla, J., Chavez, U. B., and McCord, J. M. (1991) Mitochondrial generation of oxygen radicals during reoxygenation of ischemic tissues. Free Rad. Res. Commun. 12–13:681–689.
124. Cleeter, M. W., Cooper, J. M., and Schapira, A. H. (1992) Irreversible inhibi-

tion of mitochondrial complex I by 1-methyl-4-phenylpyridinium: Evidence for free radical involvement. J. Neurochem. 58:786–789.
125. Schulze-Osthoff, K., Bakker, A. C., Vanhaesebroeck, B., Beyaert, R., Jacob, W. A., and Fiers, W. (1992) Cytotoxic activity of tumor necrosis factor is mediated by early damage of mitochondrial functions. Evidence for the involvement of mitochondrial radical generation. J. Biol. Chem. 267:5317–5323.
126. Takayanagi, R., Takeshige, K., and Minakami, S. (1980) NADH- and NADPH-dependent lipid peroxidation in bovine heart submitochondrial particles. Dependence on the rate of electron flow in the respiratory chain and an antioxidant role of ubiquinol. Biochem. J. 192:853.
127. Bindoli, A., Cavallini, L., and Jocelyn, P. (1982) Mitochondrial lipid peroxidation by cumene hydroperoxide and its prevention by succinate. Biochim. Biophys. Acta 681:496.
128. Mészaros, L., Tihanyi, K., and Horvath, I. (1982) Mitochondrial substrate oxidation-dependent protection against lipid peroxidation. Biochim. Biophys. Acta 731:675.
129. McCay, P. B., Gibson, D. D., Fong, K. L., et al. (1988) Effect of glutathione peroxidase activity on lipid peroxidation in biological membranes. Biochim. Biophys. 261:1.
130. Kraus, P. (1980) Resolution, purification and some properties of three glutathione transferases from rat liver mitochondria. Hoppe-Seyler's Z. Physiol. Chem. 361:9.
131. Morgenstern, R., Lundqvist, G., Andersson, G., et al. (1984) The distribution of microsomal glutathione transferase among different organelles, different organs, and different organisms. Biochem. Pharmacol. 33:3609.
132. Coyle, J. T., and Puttfarcken, P. (1993) Oxidative stress, glutamate, and neurodegenerative disorders. Science 262:689–695.
133. Adams, J. D., Jr., Wang, B., Klaidman, L. K., LeBel, C. P., Odunze, I. N., and Shah, D. (1993) New aspects of brain oxidative stress induced by tert-butylhydroperoxide. Free Rad. Biol. Med. 15:195–202.
134. Takeyama, N., Matsuo, N., and Tanaka, T. (1993) Oxidative damage to mitochondria is mediated by the Ca^{2+}-dependent inner membrane permeability transition. Biochem. J. 294:719–725.
135. Carini, R., Parola, M., Dianzani, M. U., and Albano, E. (1992) Mitochondrial damage and its role in causing hepatocyte injury during stimulation of lipid peroxidation by iron nitriloacetate. Arch. Biochem. Biophys. 297:110–118.
136. Masaki, N., Kyle, M. E., Serroni, A., and Farber, J. L. (1989) Mitochondrial damage as a mechanism of cell injury in the killing of cultured hepatocytes by tert-butyl hydroperoxide. Arch. Biochem. Biophys. 270:672–680.
137. Bronk, S. F., and Gores, G. J. (1991) Acidosis protects against lethal oxidative injury of liver sinusoidal endothelial cells. Hepatology 14:150–157.
138. Fujii, Y., Johnson, M. E., and Gores, G. J. (1994) Mitochondrial dysfunction during anoxia/reoxygenation injury of liver endothelial cells. Hepatology 20:177–185.
139. Malamed, S., and Recknagel, R. O. (1959) The osmotic behavior of the sucrose-inaccessible space of mitochondrial pellets from rat liver. J. Biol. Chem. 234:3027–3030.

140. Lehninger, A. L. (1962) Water uptake and extrusion by mitochondria in relation to oxidative phosphorylation. Phys. Rev. 42:467–517.
141. Fagian, M. M., Pereira-da-Silva, L., Martins, I. S., and Vercesi, A. E. (1990) Membrane protein thiol cross-linking associated with the permeabilization of the inner mitochondrial membrane by Ca^{2+} plus prooxidants. J. Biol. Chem. 265:19955–19960.
142. Gunter, T. E., and Pfeiffer, D. R. (1990) Mechanisms by which mitochondria transport calcium. Am. J. Physiol. 258:C755–C786.
143. Szabö, I., and Zoratti, M. (1992) The mitochondrial megachannel is the permeability transition pore. J. Bioenerg. Biomembr. 24:111–117.
144. Richter, C., Theus, M., and Schlegel, J. (1990) Cyclosporine A inhibits mitochondrial pyridine nucleotide hydrolysis and calcium release. Biochem. Pharmacol. 40:779–782.
145. Al-Nasser, I., and Crompton, M. (1986) The reversible Ca^{2+}-induced permeabilization of rat liver mitochondria. Biochem. J. 239:19–29.
146. Petronilli, V., Cola, C., and Bernardi, P. (1993) Modulation of the mitochondrial cyclosporin A-sensitive permeability transition pore. II. The minimal requirements for pore induction underscore a key role for transmembrane electrical potential, matrix pH, and matrix Ca^{2+}. J. Biol. Chem. 268:1011–1016.
147. Crompton, M., and Costi, A. (1988) Kinetic evidence for a heart mitochondrial pore activated by Ca^{2+}, inorganic phosphate and oxidative stress. A potential mechanism for mitochondrial dysfunction during cellular Ca^{2+} overload. Eur. J. Biochem. 178:489–501.
148. Broekemeier, K. M., Dempsey, M. E., and Pfeiffer, D. R. (1989) Cyclosporin A is a potent inhibitor of the inner membrane permeability transition in liver mitochondria. J. Biol. Chem. 264:7826–7830.
149. Igbavboa, U., Zwizinski, C. W., and Pfeiffer, D. R. (1989) Release of mitochondrial matrix proteins through a Ca^{2+}-requiring, cyclosporin-sensitive pathway. Biochem. Biophys. Res. Commun. 161:619–625.
150. Fournier, N., Ducet, G., and Crevat, A. (1987) Action of cyclosporine on mitochondrial calcium fluxes. J. Bioenerg. Biomembr. 19:297–303.
151. Crompton, M., Ellinger, H., and Costi, A. (1988) Inhibition by cyclosporin A of a Ca^{2+}-dependent pore in heart mitochondria activated by inorganic phosphate and oxidative stress. Biochem. J. 255:357–360.
152. Halestrap, A. P., and Davidson, A. M. (1990) Inhibition of Ca2(+)-induced large-amplitude swelling of liver and heart mitochondria by cyclosporin is probably caused by the inhibitor binding to mitochondrial-matrix peptidyl-prolyl cis-trans isomerase and preventing it interacting with the adenine nucleotide translocase. Biochem. J. 268:153–160.
153. Valle, V. G., Fagian, M. M., Parentoni, L. S. Meinicke, A. R., and Vercesi, A. E. (1993) The participation of reactive oxygen species and protein thiols in the mechanism of mitochondrial inner membrane permeabilization by calcium plus prooxidants. Arch. Biochem. Biophys. 307:1–7.
154. Petronilli, V., Costantini, P., Scorrano, L., Colonna, R., Passamonti, S., and Bernardi, P. (1994) The voltage sensor of the mitochondrial permeability transi-

tion pore is tumed by the oxidation-reduction state of vicinal thiols. J. Biol. Chem. 269:16638–16642.
155. Griffiths, E. J., and Halestrap, A. P. (1991) Further evidence that cyclosporin A protects mitochondria from calcium overload by inhibiting a matrix peptidyl-prolyl cis-trans isomerase. Implications for the immunosuppressive and toxic effects of cyclosporin. Biochem. J. 274:611–614.
156. Savage, M. K., and Reed, D. J. (1994) Oxidation of pyridine nucleotides and depletion of ATP and ADP during calcium- and inorganic phosphate-induced mitochondrial permeability transition. Biochem. Biophys. Res. Commun. 200:1615–1620.
157. Savage, M. K., Jones, D. P., and Reed, D. J. (1991) Calcium- and phosphate-dependent release and loading of glutathione by liver mitochondria. Arch. Biochem. Biophys. 290:51–56. 1991.
158. Savage, M. K., and Reed, D. J. (1994) Release of mitochondrial glutathione and calcium by a cyclosporin A-sensitive mechanism occurs without large amplitude swelling. Arch. Biochem. Biophys. 315:142–152.

12

Phospholipid Hydroperoxide Glutathione Peroxidase: More Than an Antioxidant Enzyme?

Matilde Maiorino, Antonella Roveri, and Carlo Gregolin
Università di Padova, Padova, Italy

Fulvio Ursini
Università di Udine, Udine, Italy

I. SELENIUM AND SELENOENZYMES

In plants that accumulate selenium, the element is absorbed as selenates and selenites. Selenium is found mainly in nonprotein amino acids, such as methylselenocystine, selenocystine, selenohomocystine, and selenocystathionine. These amino acids can be transformed into a volatile form from which selenium may be released (1). Non–selenium-accumulating plants growing in selenium-rich soil also contain selenium, although to a lower extent, but there it is present mainly as selenomethionine in proteins (2).

Selenium has been considered a poison for hundreds of years. Marco Polo described selenium toxicity in beasts of burden eating selenium-accumulating plants of the genus *Astragalus*. Moreover, the definition of a specific toxic syndrome was achieved only in the first half of the twentieth century. Nevertheless, in 1957 it was discovered by Klaus Schwartz that the "poison" selenium was required to prevent a nutritionally induced liver necrosis (3). In 1989, following the identification of selenoenzymes, a Recommended Dietary Allowance (RDA) was established. Thus, the "poison" became a beneficial and necessary trace element.

Dietary intake of selenium in animals is achieved through ingestion of inorganic salts and organic compounds, mainly from plants and animal proteins. From these different sources selenium converges through metabolic pathways that are not completely elucidated into specific proteins. The function of these selenoproteins apparently accounts for all biological functions of the oligo-element. In all known proteins in which selenium is specifically incorporated, it is found as catalytically competent selenocysteine. Selenomethionine can replace methionine in proteins rich in this amino acid, but this appears to be a random substitution devoid of physiological—and possibly toxicological—effects.

In all selenoenzymes the specific catalysis is driven by the selenol moiety. Due to its pK, which is lower than that of thiols, selenocysteine is highly nucleophilic at physiological pH. Selenocysteine has been found in three bacterial enzymes (glycine reductase, formate dehydrogenase, and a hydrogenase) in the glutathione peroxidase family, in tetraiodothyronine 5′-deiodinase, and in the plasma glycosylated selenoprotein P, the enzymatic activity of which has no known function (4).

The discovery, published in 1973 (5), that glutathione peroxidase (GPx), the unusual peroxidase described by Mills in 1957 (6), contains one selenium atom per monomer opened the field of selenium biochemistry. The triad selenium/glutathione peroxidase/antioxidant protection accounts almost completely for the great body of scientific reports on the biology of selenium.

The study of antioxidant activity present in the cytosol of different tissues led to the discovery in 1982 of the second selenoperoxidase: phospholipid hydroperoxide glutathione peroxidase (PHGPx) (7). The purification of GPx activity in plasma, previously thought to be a result of contamination from red blood cells, led to the discovery in 1987 of the plasma glycosylated glutathione peroxidase (pGPx) (8,9). The most recent member of the superfamily of glutathione peroxidases has been identified in the gastrointestinal tract of rodents (GI-GPx) by gene technology (10). With the exception of PHGPx, which is a monomer of 19 kDa, glutathione peroxidases are tetrameric, each monomer weighing 22 kDa.

The occurrence in liver cytosol of a thermolabile factor that in the presence of glutathione (GSH) inhibits lipid peroxidation was described by McCay and coworkers in 1976 (11). Reports on the catalytic effect of GPx on fatty acid hydroperoxides suggested at first that this enzyme could account for the observed antiperoxidant activity. Other groups postulated that the peroxidase activity of some glutathione transferases accounted for this antiperoxidant effect. Eventually, McCay presented clear chromatographic evidence that the "factor" described by him was not connected with any known peroxidase or transferase (12).

A protein accounting for the inhibition of lipid peroxidation was purified from pig heart cytosol in our laboratory in the late 1970s and reported in 1982 (7). The purification of this protein, initially named the peroxidation inhibiting protein (PIP), was carried out by following the inhibition of peroxidation in a simple system during purification. Peroxidation was induced by the Fe^{3+}-triethylenetetramine–iron complex in phosphatidylcholine liposomes containing trace amounts of peroxidized phospholipids. The process was followed by measuring oxygen uptake. Preincubation of the system with GSH and cytosol fractions blocked peroxidation, indicating the presence of the antioxidant protein. The simplicity of this procedure, requiring no more than a few minutes to check the antioxidant activity of a sample, had dramatic impact on the success of the purification of PIP.

In a coupled spectrophotometric assay with NADPH and glutathione reductase, the homogeneous protein showed glutathione peroxidase activity on phospholipid hydroperoxides, as confirmed by TLC, HPLC, and mass spectroscopic analysis of substrates and products. Thus, the peroxidation inhibiting protein was renamed phospholipid hydroperoxide glutathione peroxidase (E.C. 1.11.1.12) (13,14).

II. KINETICS

The kinetics of the selenium-dependent peroxidase reaction were studied in detail with GPx (see Ref. 15 for a review) before the discovery of the other members of the glutathione peroxidase family. The recognition that the same mechanism applies to GPx, PHGPx (16), and pGPx (17) suggests that during evolution a divergence toward specific enzymatic activities took place, while the catalytic properties of the common ancestor were conserved. This concept is in remarkable agreement with structural information from sequence analysis of these proteins (see below).

A. Kinetic Mechanism

All selenium-dependent peroxidases catalyze the redox reaction where a hydroperoxide is reduced and two glutathione molecules are oxidized to the corresponding disulfide:

$$\mathrm{ROOH} + 2\mathrm{GSH} \rightarrow \mathrm{ROH} + \mathrm{H_2O} + \mathrm{GSSG}$$

The currently accepted mechanism for this reaction (Fig. 1) involves the formation of a highly oxidizing species of selenium (selenenic acid) and a selenodisulfide intermediate.

The apparently similar reaction catalyzed by selenium-independent GSH transferases is instead a transferase reaction, which takes place through a nucleophilic displacement leading first to a two-electron oxidation of sulfur of glutathione and eventually to a disulfide:

$$\mathrm{ROOH} + \mathrm{GSH} \rightarrow \mathrm{ROH} + \mathrm{GS\text{-}OH}$$

$$\mathrm{GS\text{-}OH} + \mathrm{GSH} \rightarrow \mathrm{GSSG} + \mathrm{H_2O}$$

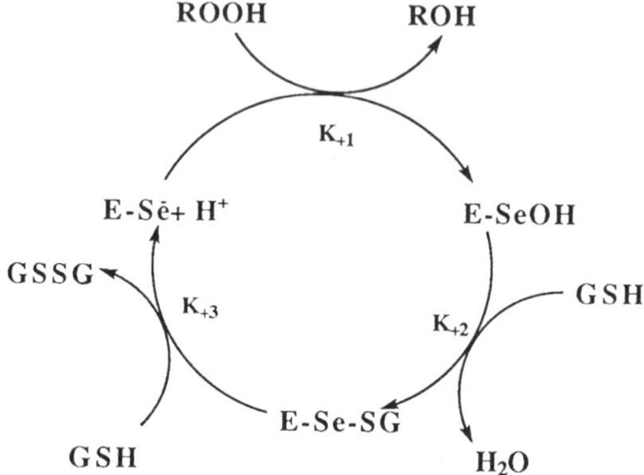

Figure 1 The ping-pong catalytic mechanism proposed for glutathione peroxidases.

It is worth noting that this reaction, although faster when catalyzed by transferases, takes place at a reasonable rate in the absence of the enzyme and accounts for the chemical oxidation of thiols in the presence of hydrogen peroxide.

Several hemoproteins, by interacting with hydrogen peroxide, yield the high oxidation state of iron analogous to compound II of heme-peroxidases (18), the chemical reactivity of which accounts for the oxidation of several compounds (19). Due to the poor specificity for the electron donor, this reaction sequence, which can be formally described as a peroxidatic cycle, is potentially damaging in a biological environment where antioxidants are consumed (20) and lipid peroxidation initiated (21).

From these considerations it can be concluded that although the overall net reaction appears to be the same when the peroxidatic cycle is catalyzed by selenium peroxidases, GSH transferases, or heme peroxidases, a major biological advantage is attained when the reaction is catalyzed by selenoenzymes. When selenium GSH peroxidases are involved in the peroxidatic reaction, no freely diffusable, highly oxidizing intermediate is produced. Furthermore, the oxidizing potential at the active site is specifically oriented to the appropriate donor substrate, while this is not the case when the reaction is catalyzed by the heme peroxidases.

The experimental data obtained using GPx and PHGPx fit a simplified form of the general Dalziel equation:

$$\frac{[\mathrm{Eo}]}{\mathrm{v}} = \frac{\Phi_1}{[\mathrm{ROOH}]} + \frac{\Phi_2}{[\mathrm{GSH}]}$$

A plot of the above equation, as well as the Lineweaver-Burk plot utilizing one substrate, produces parallel lines at different concentrations of the second substrate (13). This describes a ping-pong kinetic mechanism with unlimited maximum velocity, where the apparent V_{max} and K_m for one substrate are dependent on the concentration of the other substrate. In other words, these unusual kinetics appear as a sequence of biomolecular steps with no accumulation of enzyme-substrate complexes (15).

The steady-state kinetic analysis of GPx reaction was characterized by measuring the initial rate of hydrogen peroxide consumption, while for PHGPx the analysis was characterized by a single decay component with complete consumption of the hydroperoxide substrate (13), since neither reverse reaction nor product inhibition takes place. In both cases, experimental data confirmed the following equations:

$$\Phi_1 = \frac{1}{k_{+1}}$$

and

$$\Phi_2 = \frac{1}{k_{+2}} + \frac{1}{k_{+3}}$$

where k_{+1} is the rate constant for the oxidation of the enzyme in the presence of a hydroperoxide and k_{+2} and k_{+3} are the rate constants for the two reductive steps in the presence of GSH. Using this kinetic approach, the rate constant for the interaction of different selenium peroxidases with different substrates has been calculated (14).

B. The Catalytic Cycle, Substrates, and Inhibitors

The above-described kinetic mechanism depicts a catalytic cycle where a reduced form of selenium at the active site, probably the dissociated form of the selenol, is first oxidized upon interaction with the peroxidizing substrate to selenenic acid, which is then reduced by a suitable hydrogen donor. The involvement of a selenolate anion in the catalytic cycle has been suggested by the inhibitory effect of iodoacetate (13). Iodoacetate, indeed, inhibits selenoperoxidases apparently by alkylating the active site. This takes place only in the presence of donor substrates, which reduce the possible higher oxidation states of selenium. Moreover, since the alkylation occurs at neutral pH, where selenols but not thiols are dissociated, the concept that a selenolate is actually involved in the catalysis seems plausible.

The rate constant of the interaction between the reduced form of selenoenzymes and the peroxidizing substrate in the first part of the catalytic cycle has been calculated from the steady-state kinetic analysis and has been used for studying substrate specificity. While GPx reacts faster than PHGPx with hydrogen peroxide and the same rate constant has been calculated for linoleic acid hydroperoxide, phospholipid hydroperoxides, which react with a rate constant of approximately 10^6 M^{-1} s^{-1} with PHGPx, are not substrates for GPx (14). PHGPx also reduces hydroperoxides of cholesterol and cholesterol esters. The specificity of PHGPx for large lipophilic substrates has been confirmed in membranes and lipoproteins where all hydroperoxides are reduced (22).

In the second part of the catalytic cycle, the donor substrate, usually glutathione, produces first a mixed selenodisulfide and then a disulfide, yielding the selenium at the active site in the reduced form of the enzyme. The kinetic analysis of GPx and PHGPx indicates that this "reductive" part of the catalytic cycle is rate limiting, and this is particularly relevant in the case of the monomeric enzyme. This is an important aspect of the forthcoming discussion on the actual specificity of PHGPx for the donor substrate.

The specificity of GPx for glutathione is rather high (15). Minimal modifications of this substrate completely abolish catalytic activity. On the other hand, this specificity is not apparent for PHGPx. This has been shown by the kinetics of inhibition in the presence of iodoacetate and in the presence of mercaptoethanol, indicating that the latter reduces the active site (13). Moreover, for the enzyme recovered from rat testis, the specific activity of the peroxidase reaction acting on hydrogen peroxide was higher when dithiothreitol was used instead of glutathione, and mercaptoethanol and cysteine also showed significant activity (23). These catalytic differences among selenoenzymes have been discussed in terms of modifications of certain key amino acids around the active site (see below).

Some disulfides inhibit selenoperoxidases, apparently slowing down the last reductive step. The evidence that oxidized lipoate behaves as a noncompetitive inhibitor with respect to glutathione fits the hypothesis that the disulfide reverses the reductive part of the catalytic cycle, leading to a lower steady-state concentration of the reduced form of the enzyme. Interestingly, the K_i for oxidized lipoate is much lower for PHGPx than for GPx (0.15 vs. 2 mM), which is possibly related to the higher lipophilicity of the active site of PHGPx. In agreement with this, oxidized glutathione, which inhibits GPx, does not affect significantly PHGPx (24).

III. STRUCTURE

The similarities in terms of catalytic mechanisms and molecular weights for PHGPx and the monomeric form of GPx left open the possibility that the monomeric enzyme could be a variant of the more prevalent tetrameric peroxidase. However, amino acid, cDNA, and genomic sequencing studies showed that

PHGPx is actually a distinct gene product. Comparison of PHGPx and GPx amino acid sequence revealed a rather poor homology between the two selenoproteins (Fig. 2). On the other hand, the presence of some clusters of notable similarity revealed a possible common ancestral origin (25). Thus, the two selenoproteins belong to a superfamily identified by misfit analysis of several analog proteins. Furthermore, the presence of the conserved clusters permitted modeling of the PHGPx structure on the basis of the elucidated three-dimensional structure of intracellular bovine GPx. The modeling indicated that the conserved residues are positioned around the active site, represented by the selenocysteine moiety, such that some peculiar features of PHGPx could be explained.

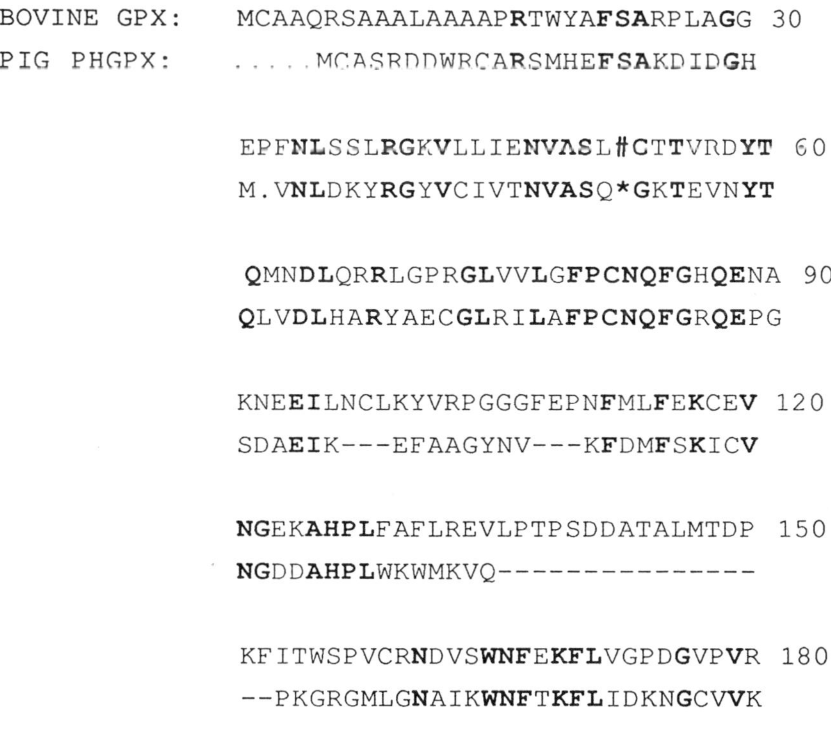

Figure 2 Alignment of pig PHGPx with bovine cytosolic GPx. Position numbering refers to bovine GPx. *Indicates the selenocysteine residue. Amino acids conserved are shown in bold.

A. Amino Acid Sequence

The primary structure of pig heart PHGPx was obtained by overlapping peptide sequences, obtained by cyanogen bromide cleavage and tryptic digestion, along with cDNA sequences obtained from a PCR amplified pig heart and pig parathyroid gland cDNA library (26,27). The screening of two different cDNA libraries was necessary in order to obtain the complete primary sequence. In fact, a first screening of a pig heart library failed to give clones containing the presumed PHGPx 5′ terminal coding region. In addition, the PHGPx N-terminus was found to be resistant to Edman degradation. On the other hand, an appropriate cDNA fragment extending 42 nucleotides upstream from the supposed ATG start codon was obtained from the porcine parathyroid library. This also overlapped with the N-terminal sequence of PHGPx, which was eventually elucidated from a different enzyme preparation, despite the poor sequencing yields (see below).

The primary structure of the protein started with position 3 of the cDNA-deduced amino acid sequence. Based on this, PHGPx is determined to be a 170 amino acid protein, but composed, at least in one of its forms, of 168 residues.

The selenocysteine residue was identified as a carboxymethyl derivative during amino acid analysis. The presence of this unusual amino acid was further confirmed by the presence of the TGA codon within the open reading frame of the cDNA. This is a common feature of selenoproteins from both prokaryotics and eukaryotics, wherein TGA was replaced as a stop codon by selenocysteine.

Interestingly, within the sequence, a tyrosine residue was found to be modified by phosphorylation (see below). Another still undefined derivatization(s) of the PHGPx peptide chain seems to take place based on the following: (1) the molecular mass of 19,257 daltons, calculated on the basis of the amino acid sequence, is lower than the mass measured by laser desorption mass spectroscopy (19,672 daltons), and (2) the yield during N-terminal sequencing is dramatically low in comparison of the amount of protein undergoing Edman degradation. This suggests the presence of an unprotected N-terminus in a copurified minor portion of the protein sample, while this residue is blocked in the bulk of the PHGPx.

B. Genomic Sequence

The sequence of the genomic DNA encoding PHGPx was derived from overlapping fragments isolated from a pig liver genomic DNA library (27). While the GPx gene-coding area contains two exons, the coding area of the PHGPx gene was found to be composed of seven exons, all intron/exon transitions complying with the GT/AG rule.

By screening of the PHGPx genomic DNA sequence for putative regulatory elements, multiple consensus sequences for estrogen and progesterone/glucocorticoid were tentatively identified. Furthermore, the PHGPx gene reveals the presence of cyclic nucleotide-responsive elements, a CCAAT/enhancer-binding protein consensus sequence, AP-3 AP-2 consensus sequences, SP1 sites, and an NF-1 site. The presence of these boxes in the genomic sequence seems to suggest that PHGPx is not likely to be a constitutive protein, while its expression would be part of a more complex system, possibly eliciting a still undefined physiological response.

Nevertheless, the elucidation of PHGPx cDNA and genomic sequences added further support to the hypothesis that a common mechanism enables mammalian cells to recognize UGA as the triplet for selenocysteine insertion in the growing peptide chain. In fact, the screening of the secondary structure of PHGPx mRNA revealed the possibility of a stem loop structure in the 3′UTR, as it was observed to occur in the mRNA of the other mammalian selenoproteins. Eventually, by comparing the 3′UTR of type I iodothyronine 5′-deiodinase, GPx, selenoprotein P (28), and PHGPx, it was possible to identify a consensus that is tentatively proposed to be an identification site(s) for the still uncharacterized factor(s) involved in selenocysteine insertion (27).

C. Homology Considerations

By comparing pig PHGPx and intracellular bovine GPx amino acid sequences, an overall homology close to 30% was found. This level of homology was much lower than that occurring among other mammalian intracellular glutathione peroxidases (27). However, despite this low homology, it was possible to recognize some clusters of pronounced similarity between the sequences of GPx and PHGPx (Fig. 2). According to the x-ray structure for intracellular bovine GPx, the conserved sequences corresponded largely to those parts contributing largely to the immediate environment of GPx active site, thus suggesting a very similar shape to the core of the two enzymes: (1) a pleated sheet adjacent to the selenocysteine (positions 47–53 of the bovine GPx sequence), (2) the area around glutamine 87 in the turn between the β-sheet at positions 74–80 and the helix at positions 91–96, and (3) tryptophan 165 at the beginning of a turn followed by a β-sheet (positions 168–172). Moreover, the residues of selenocysteine, glutamine 87, and tryptophan 165 were found to be conserved in homologous positions in PHGPx (Fig. 2). The presence of these residues brings to mind an almost identical catalytic core in the active site of PHGPx and intracellular GPx and supports the view of the three residues forming a characteristic triad important for an efficient redox catalysis, as proposed for bovine GPx (27,29). The evidence that the same triad exists in homologous positions for GPx and pGPx adds further support to this hypothesis (30). On the other hand, all residues involved in the binding of glutathione to the bovine GPx active site (arginines 57, 103, 184, and 185) were found mutated or deleted in the PHGPx sequence.

The analysis of clusters distant from the catalytic core revealed deletions in the PHGPx sequence, possibly causing a substantial short-cut of the subunit interaction loop, together with the elimination of a small helix also involved in subunit interaction in bovine cellular GPx. As a consequence, both the interfaces responsible for dimerization and tetramerization in bovine cellular GPx are missing in PHGPx, which is, in fact, a monomeric protein.

Interestingly, the highly conserved parts of PHGPx sequence are restricted to only two out of the seven exons of the PHGPx gene (exons 3 and 5), suggesting that the low overall homology with cellular GPx possible originated from shuffling of some unrelated exons into the PHGPx gene at a later stage of evolution. By comparing PHGPx sequences with the corresponding sequences of other members of the GPx superfamily, a dendrogram was constructed based on mutation statistics as a measure of relatedness among the members of the superfamily. Despite the fact that only the conserved third exon of PHGPx was used for comparison, PHGPx appears to have branched off very early from the phylogenetic path for intracellular cytosolic and pGPx species, the latter forming a separate clade with cysteine-containing homologs. Furthermore, PHGPx appeared more closely related to GPx of *Schistosoma mansoni* or to the cysteine-containing analog of *Nicotiana tabaccum* than to any vertebrate GPx. On this basis, PHGPx must be considered a phylogenetically old achievement of the GPx superfamily, from which it diverged very early, possibly before the appearance of mammals.

D. Structure/Catalytic Mechanism Relationships

The modeling of PHGPx with the elucidated three-dimensional structure of intracellular bovine GPx provided an explanation for the structural basis for some peculiar features of PHGPx and GPx catalysis. The identity of the catalytic mechanism of GPx and PHGPx fits the homology of the active site, highlighting the role of the catalytic center, represented by the conserved triad of selenocysteine, glutamine 87, and tryptophan 165. A comparison of their structures accounted for the peculiar substrate specificity of PHGPx toward phospholipid and cholesterol hydroperoxides. The deletion observed in the PHGPx sequence, by cutting the subunit interaction loop, produces a more freely accessible active site and allows interaction with more complex hydroperoxide substrates such as those integrated into lipid bilayers. To further support this observation, it should be mentioned that in both enzymes, computer modeling revealed that the catalytic center resides in a flat depression of the molecular surface composed of hydrophobic residues similarly oriented within the molecule.

The finding that PHGPx is reduced less specifically than GPx by different thiols is also supported by structural data. Residues involved in glutathione binding in GPx are mutated or deleted in the PHGPx sequence. Moreover, the absence of these positively charged residues, holding glutathione in the correct orientation to the selenocysteine moiety, could theoretically account for the lower rate constant by one order of magnitude (k_{+2} + k_{+3} in Fig. 1) for the reductive step of the catalytic cycle of PHGPx. This observation poses the interesting question as to whether glutathione is the actual donor substrate of PHGPx in vivo and envisages the possibility that this selenoenzyme could be involved physiologically in the oxidation of other peculiar and still unidentified thiol moieties.

IV. PHGPx AS AN ANTIOXIDANT ENZYME

The demonstration that the antiperoxidant effect of the "cytosolic factor" described by McCay was actually due to PHGPx present in liver cytosol highlighted the relevance of lipid hydroperoxides in the overall mechanism of lipid peroxidation and provided a major insight into the identification of the physiological mechanism of antioxidant protection. Lipid hydroperoxides play a dual role in lipid peroxidation, being both a major product of lipid peroxidation, as well as a substrate for initiation reactions. These initiations are referred to as "hydroperoxide dependent" or "secondary initiations."

The reduction of lipid hydroperoxides to the corresponding hydroxy derivatives, while completely preventing hydroperoxide-dependent peroxidation, slows down lipid peroxidation as initiated by a hydroperoxide-independent mechanism. Under these conditions, in fact, the hydroperoxide-dependent mechanism usually predominates as soon as some hydroperoxides are produced. The following section will address the mechanism of lipid peroxidation from the standpoint of antioxidant effects arising from the reduction of lipid hydroperoxides as catalyzed by PHGPx.

A. Mechanisms of Initiation of Lipid Peroxidation

The homolysis of a C-H bond of a fatty acid in the presence of oxygen, leading to oxygen addition to the carbon-centered radical, accounts for the initiation of lipid peroxidation. This reaction is facilitated when the C-H bond dissociation energy is lower than that of a "nonactivated" bond (94 kcal/mol), as in the case of the *bis*-allylic C-H bond of a polyunsaturated fatty acid (85 kcal/mol). This is why under physiological conditions polyunsaturated fatty acids are largely subject to the free radical peroxidative process.

Several strong oxidants are capable of initiating primary reactions involving hydrogen abstraction, such as hydroxy radicals ($HO^{\bullet}$), perhydroxy radicals ($HOO^{\bullet}$), thiyl radicals ($RS^{\bullet}$), and possibly iron-dioxygen complexes (31). The relative physiological relevance of these initiation reactions has been the subject of considerable debate in that under various experimental conditions the thermodynamics and the kinetics of each individual initiating reaction or the protective effect of antioxidant systems eliminating these initiating species differ considerably (for reviews see Refs. 32 and 33).

Although oxidants such as peroxy (ROO•) and alkoxy (RO•) radicals are also capable of extracting the hydrogen atom from a divinyl-methane structure, these reactions do not fit the requirement for the definition of primary initiation. In fact, when a peroxy radical, produced by oxygen addition to a fatty acid containing a carbon-centered radical, reacts with a fatty acid, the reaction is more correctly referred to as a propagation reaction. On the other hand, when either peroxy or alkoxy radicals are produced from a preexisting hydroperoxide, the mechanism of the initiation must be regarded as hydroperoxide dependent. The mechanism of lipid hydroperoxide formation by singlet oxygen or lipoxygenases is considered a primary initiation, although lipoxygenases are actually activated by hydroperoxides.

Lipid hydroperoxides, produced by the above primary initiation mechanisms, are the substrate for secondary initiations. Thus, they can be viewed as "storing" oxygen radicals, which, when liberated, would lead to uncontrolled oxidations and to cell damage (34). Although the dissociation energy of the LO-OH bond (44 kcal/mol) is lower than that of LOO-H bond (90 kcal/mol), the homolytic decomposition leading to hydroxy and alkoxy radicals is unlikely to occur at physiological temperature. The O-O bond energy weakens following hydrogen bonding with alcohols and acids as well as with other hydroperoxides (35). However, this mechanism of "molecule-assisted homolysis" has little chance of being relevant in vivo, where other reactions are much faster. The one-electron redox transition is generally recognized as a relevant pathway of decomposition of lipid hydroperoxides leading not only to secondary products but also to initiation of peroxidative chain reactions (34,35).

The oxidative pathway, producing peroxy radicals, could take place in the presence of free radicals that can abstract a hydrogen atom from a hydroperoxide. For such a reaction to occur, thermodynamic requirements are such that the adduct of the oxidizing free radical and hydrogen have a dissociation energy higher than that of the OO-H bond. This dissociation energy is similar to that of the methylene bond in polyunsaturated fatty acids, and, as such, the reaction of an oxidizing free radical with lipid hydroperoxides is unlikely to occur, at least under physiological conditions where the concentration of hydroperoxides is much lower than that of unsaturated fatty acids. These thermodynamic and kinetic considerations minimize the actual role of a free radical–mediated decomposition of hydroperoxides producing peroxy radicals. On the other hand, the formation of peroxy radicals from lipid hydroperoxides by electron transfer to a transition metal ion complex is a well-established mechanism of initiation, where the kinetics as well as the thermodynamics of the reaction depend on the redox potential of the complex and the nature of metal chelation (33).

Reductive decomposition reactions involving transfer of an electron from a reduced transition metal to a hydroperoxide are generally much faster than the oxidative mechanisms described above. In this case, the product is an alkoxy radical, which can abstract a hydrogen atom from a molecule containing a X-H bond, the dissociation energy of which is lower than that of the O-H bond of the hydroperoxide. The dissociation energy of the O-H bond being higher than that of the OO-H bond (104 vs. 90 kcal/mol) accounts, in thermodynamic terms, for the higher reactivity of alkoxy radicals (35). Hydrogen transfer from an adjacent fatty acid to an alkoxy radical is, therefore, favorable for initiating lipid peroxidation from a preexisting hydroperoxide. This reaction, however, is actually in competition with alternative pathways. As an alkoxy radical is usually adjacent to the conjugated double bond, the molecule can undergo a cyclization leading to a epoxyallylic radical, which upon oxygen addition forms an epoxide-containing peroxy radical (36). Other alternative pathways are disproportion of alkoxy radicals or β-scission to form aldehydes and new peroxy radicals.

In summary, reductive decomposition of lipid hydroperoxides produces new peroxidative chain reactions and an array of degradation products, such as aldehydes, all potentially dangerous for biological structures. All the reaction mechanisms for initiation of lipid peroxidation require, directly or indirectly, traces of transition metal complexes, the concentration and the reactivity of which are highly controlled in the cellular environment. However, the majority of iron in a cellular milieu is in the form of hemoproteins, the reactivity of which plays a major role in the initiation of lipid peroxidation. An example of primary initiation catalyzed by a hemoprotein is the formation of a hypervalent oxoferry complex by oxidation of metmyoglobin in the presence of hydrogen peroxide. This complex is capable of oxidizing several hydrogen donors, thus completing a peroxidatic cycle. It also effectively initiates lipid peroxidation by abstracting a hydrogen atom from a polyunsaturated fatty acid (18–21).

More complex mechanisms exist for free radical generation from hemoproteins or hematin and phospholipid hydroperoxides, the most likely mechanism appearing to be homolytic cleavage leading to alkoxy radicals (36). Although this would also imply the formation of hypervalent states of iron (such as in the case of hydrogen peroxide), which have not so far been demonstrated, the above mechanism is supported by product analysis showing the typical pattern of products derived from alkoxy radicals (21,36).

B. Microsomal Lipid Peroxidation

Enzymatically induced microsomal lipid peroxidation was described by Hochstein and Ernster in 1963, and its nature is still in some respects puzzling

(32). The first—and possibly most challenging—consideration is that microsomal lipid peroxidation requires physiological compounds at physiological concentrations. This highlights the role of an extremely efficient antioxidant system, the search for which long frustrated researchers. It was found to be difficult to decrease the rate of microsomal lipid peroxidation using the usual physiological antioxidants. It has been the experience of almost all laboratories working in the field since the 1970s that superoxide dismutase, catalase, glutathione peroxidase, and the so-called hydroxyl radical scavengers failed to inhibit the peroxidative chain reaction initiated by NADPH (or ascorbate) and iron-ADP (or ATP). Vitamin E, when inserted in membranes at a concentration 10 times higher than the physiological concentration, just decreased peroxidation rate without preventing it (37). On the other hand, PHGPx, acting synergistically with vitamin E, completely prevented microsomal lipid peroxidation, thus answering, at least in terms of a plausible mechanism, the original question as to why lipid peroxidation is not as devastating in vivo, although microsomes are in the presence of NADPH, ADP, and possibly traces if iron (37).

The analysis of the conditions determining the ability of an iron complex to initiate microsomal lipid peroxidation showed that the iron complex must be enzymatically reduced in the presence of NADPH, it must react with oxygen, and eventually it must undergo a "slow autoxidation" (31). This was an empirical concept introduced to differentiate the behavior of the Fe^{2+} EDTA–O_2 complex, which immediately releases superoxide without initiating peroxidation, from the Fe^{2+} ADP–O_2 complex, which appears more stable and induces peroxidation. The suggested initiating species, accounting for almost all experimental observations under different conditions, was the perferryl ion $Fe^{2+} - O_2 \leftrightarrow Fe^{3+} - O_2^{-}$. The thermodynamic considerations described by Koppenol and Liebman (38), however, argue against the involvement of this species.

The simplest mechanism of microsomal lipid peroxidation, accounting for experimental evidence as well as for kinetic and thermodynamic constraints, indicates that the reductive decomposition of preformed lipid hydroperoxides by iron complexes—in turn reduced by specific enzymes or ascorbate—is a significant event. However, the observation that PHGPx does not prevent lipid peroxidation in vitamin E–depleted microsomes or in liposomes supplemented with iron/ascorbate appears to rule out the possibility that in microsomal lipid peroxidation only hydroperoxide-dependent initiations take place (although there is general agreement that this reaction is actually relevant once lipid peroxidation is started).

C. Antioxidant Effect of PHGPx and Synergism with Vitamin E

The catalytic specificity of PHGPx toward lipid hydroperoxides accounts a priori for an antioxidant effect in all conditions where initiating free radicals are produced by decomposition of preexisting hydroperoxides. Nevertheless, the application of this in vitro concept to a cellular environment requires consideration of the competing reactions involving lipid hydroperoxides and kinetic parameters of the enzymes involved: the reduction of hydroperoxides by PHGPx, on the one hand, and the cleavage of hydroperoxides by metal complexes (producing radical species that initiate peroxidation), on the other. The rate constant for the interaction of PHGPx with lipid hydroperoxides is high, thus supporting an efficient antioxidant effect. If hydroperoxide-dependent initiations took place—escaping from PHGPx control—chain-breaking antioxidants would minimize the oxidative event by slowing propagation. In this process, the chain-breaking antioxidant follows PHGPx in the priority order of antioxidant mechanisms. This is not always the case. For instance, when lipid peroxidation is initiated by an iron complex and a source of reducing equivalents (ascorbate or NADPH and the microsomal reductase), the initiation is apparently independent of the hydroperoxide and PHGPx plays a minor antioxidant effect. Also, vitamin E, present in native or artificial membranes, slows the lipid peroxidation rate without preventing peroxidation. Only the simultaneous occurrence of vitamin E, PHGPx, and glutathione decreases peroxidation to undetectable levels (37). The rationale of this synergism can be given in the following terms (37):

1. Vitamin E does not quench the initiator (likely the perferryl species), while scavenging lipid peroxy radicals, thus playing a chain-breaking effect.
2. From lipid hydroperoxides, produced by the above reaction, ferrous iron produces alkoxy radicals, which react almost equally quickly with lipids (producing new chain reactions) and vitamin E, which is cooxidized without playing any significant antioxidant role.
3. The reduction of lipid hydroperoxides by PHGPx prevents secondary initiations, thus sparing vitamin E.

In these antioxidant events, the activity of PHGPx follows the chain-breaking reaction of vitamin E. These considerations offer a plausible mechanism for the synergistic effect of vitamin E and selenium in terms of the catalytic capacity of the enzyme.

Taken together, the two peroxidation systems (hydroperoxide dependent and hydroperoxide independent) are likely to account for the large majority of oxidative events taking place in membranes. This view also highlights the significance of a two-step antioxidant system, which does not necessarily follow a strict priority order: the scavenging of peroxy radicals (a one-electron process) and the reduction of hydroperoxides (a two-electron process). The reciprocal interdependence of the aforementioned antioxidant systems seems to optimize the antioxidant protection of the biological environment.

V. BEYOND ANTIOXIDANT PROTECTION

PHGPx has been discovered and purified following peroxidation-inhibiting activity, and its antioxidant capacity was completely accounted for by the peroxidase activity. However, recent studies on the sequence of the structural gene, tissue distribution, hormonal control, binding to specific membranes in some tissues, and modulation of the activity (possibly through posttranscriptional modifications) suggest that the physiological function of PHGPx could surpass antioxidant protection.

A housekeeping nature, or just an induction by oxidative stress, would be expected for an antioxidant enzyme, but not a complex mechanism of regulation, such as that which is becoming progressively apparent. In the following section the most recent knowledge about the physiology of PHGPx will be summarized, bearing in mind that the most relevant physiological effect of this oxidoreductase is still to be identified.

A. Tissue Distribution and Presence of Catalytically Inactive Forms

In different tissues PHGPx activity is partitioned between a soluble and a membrane-bound component (23,39). The rank order of soluble activity is as follows: testis > adrenals > kidney > ovary > thyroid > liver > heart > lung = brain. For membrane-bound enzyme the rank order of the specific activity is as follows: testis >> liver = heart = lung. This tissue distribution is much more reminiscent of endocrine function than of antioxidant protection.

The membrane-bound form has been purified from rat testis mitochondrial membranes, in parallel with the soluble form (23). The two forms proved identical in respect to catalytic and kinetic properties as well in terms of structure as analyzed by peptide fingerprint. The membrane-bound form can be released by the simultaneous presence of ionic strength and thiols (unpublished observations). This indicates a possible binding to specific proteins in the membranes.

When the content of PHGPx in the soluble and membrane compartments of different tissues was analyzed by an immunoenzymatic test, the rank order of the relative concentration in tissues and membranes grossly diverged from the rank order of specific activities (39). Although the enzyme protein content in endocrine tissues such as testis, adrenals, and thyroid still prevailed over that of other tissues, PHGPx appeared substantially expressed also in membranes of tissues such as liver, muscle, and brain, where the activity, if any, resulted at the lower limit of detectability. Taken together, these facts indicate that PHGPx activity can be modulated differently in various tissues.

B. Possible Posttranscriptional Modifications

The above discrepancy between PHGPx content and activity suggests that the catalytic activity could be posttranscriptionally modulated. As a matter of fact,

PHGPx is phosphorylated in vitro by a tyrosine kinase of the *src* family (tyrosine kinases nonreceptor but bound to the membrane) called Fgr (M. Maiorino et al., unpublished). Moreover, using an ELISA test using monoclonal antibodies against phophotyrosine, it has been shown that 10% of membrane-bound and soluble PHGPx purified from rat testis mitochondria is actually phosphorylated (23). Up to now, clear-cut proof that typrosine phosphorylation actually affects the catalytic activity is missing. However, the enzyme purified from rat liver mitochondria (which is not active) fails to react with the antiphosphotyrosine antibody, which recognizes the active enzyme from rat testis (F. Ursini et al., unpublished). This observation strongly suggests the conclusion that tyrosine phosphorylation is actually a key modulator of the activity.

C. Search for a Role

It is apparent from enzymological and structural data that the enzyme catalyzes the reduction of hydroperoxy groups (primarily, but not necessarily, in membranes) using thiols as donor substrates with a specificity for thiols not restricted to glutathione as in the case of the tetrameric peroxidase. Several pieces of evidence indicate that PHGPx is involved in a wide relational system. PHGPx is under endocrine control, not only as suggested by the high specific activity of the membrane-bound enzyme in endocrine tissues, but also as shown by the fact that PHGPx is expressed in rat testes only after puberty, disappears after hypophysectomy, and is partially restored by gonadotropin treatment (40). In addition, the activity is very likely modulated by tyrosine phosphorylation (reminiscent of a signal transduction pathway), and PHGPx-binding structures exist in membranes.

The above characteristics have prompted a search for new functions of PHGPx beyond mere cellular protection afforded through antioxidant effect. This search must obviously deal with physiological effects of either specific hydroperoxides or specific disulfides. As far as the hydroperoxides are concerned, PHGPx would play its physiological function switching off a signal brought about by these compounds, while in the case of disulfides the enzyme, in the presence of a suitable oxidant, would be the effector of the thiol/disulfide transition on the target.

Several examples for either possibility can be found. For the first mechanism to be physiologically relevant, the first condition is that lipid hydroperoxides be generated in a specific manner, and a counteracting system must operate to abate the physiological signal originating from lipid hydroperoxides (34). The best known example of this is the reticulocyte 15-lipoxygenase, which appears to be active on intact phospholipids. This enzyme is expressed during erythrocyte differentiation and is instrumental in the destruction of mitochondria and other intracellular membranes (41). An important consideration concerning this peroxidation-dependent destruction of cellular membranes is that the peroxidation process does not account per se for the complete disappearance of membranes and subsequent reactions must follow that facilitate elimination of hydroperoxide-bearing membranes. The plausible enzymatic step after membrane hydroperoxidation is the hydrolysis of the affected lipids by a specific phospholipase A_2. Once activated, the phospholipase is capable of hydrolyzing unoxidized phospholipids as well (34). The formation of the initial hydroperoxide species could merely be a signal marking membrane loci for destruction. The enzymatic reduction of membrane hydroperoxides would be the counteracting stop signal. It is tempting to speculate that such a mechanism requiring a lipoxygenase, phospholipase, and PHGPx could apply also to other tissues during differentiation. A controlled phospholipid peroxidation could be involved in other regulatory mechanisms. The 5-lipoxygenase of macrophages, for instance, is controlled by a "peroxide tone" in membranes, in turn regulated by PHGPx (42). The control of cellular proliferation may be another major physiological effect of membrane hydroperoxides (43), possibly through the activity of the aforementioned enzymes.

The oxidation of specific protein thiols to disulfides is the second appealing possibility for a new physiological function of PHGPx. Examples of proteins whose activity is modulated in this way are the protein systems involved in calcium fluxes between different cellular compartments. Slow calcium channels (44), IP_3 receptors (45), mitochondrial pore (46), and sarcoplasmic reticulum (47) are notable examples. The recent observation that PHGPx binds to chromatin in rat testis nuclei (F. Ursini et al., unpublished) suggests a possible involvement of the enzyme in physiologically relevant thiol disulfide transitions, such as those taking place during condensation-relaxation of chromatin, or in the specific conformation of zinc finger structures (48–50).

VI. CONCLUSION

Almost 15 years after the discovery that the antioxidant protein factor present in the cell sap is actually a new selenoenzyme, the elucidation of protein and genomic sequence, far from providing the bottom line on the mechanism of antioxidant protection, opens another, possibly more difficult series of questions. If PHGPx would be just an antioxidant enzyme, why is it so carefully regulated, by a mechanism reminiscent of the signal transduction pathway? Why is it mainly expressed in endocrine tissues? Is the main physiological function related to the reduction of hydroperoxides or to the oxidation of thiols? Is the synergism with vitamin E, the tissue distribution of which roughly overlaps that of PHGPx, just related to an antioxidant protection, or are they both sensors-effectors of a signaling system working through either mono or bi-electronic redox transitions?

ACKNOWLEDGMENT

The authors would like to thank Professor Leopold Flohé (and his groups in Aachen and Braunschweig, who did the majority of the work for sequencing protein and DNA) for his useful suggestions and constructive criticism. It is from those discussions and productive input that exciting new ideas and perspective are born.

REFERENCES

1. Frankenberger, W. T., and Karlson, J. (1992) Dissipation of soil selenium by microbial volatilisation. In: Biochemistry of Trace Metals (Adriano, D. C., ed.), pp. 365–381. Lewis Publishers, Boca Raton, FL.
2. Olson, O. E., Novacek, E. J., Whithead, E. I., and Palmer, I. S. (1970) Investigations on selenium in wheat. Phytochemistry 9:1181–1188.
3. Schwarz, K., and Foltz, C. M. (1957) Selenium as an integral part of Factor 3 against dietary liver degeneration. J. Am. Chem. Soc. 79:3292–3293.
4. Stadtman, T. C. (1991) Biosynthesis and function of selenocysteine-containing enzymes. J. Biol. Chem. 266:16257–16260.
5. Flohé, L., Günzler, W. A., and Schock H. H. (1973) Glutathione peroxidase: A selenoenyzme. FEBS Lett., 32:132–134.
6. Mills, G. C. (1957) Hemoglobin catabolism I. Glutathione peroxidase, an erythrocyte enzyme which protects hemoglobin from oxidative breakdown. J. Biol. Chem. 229:189–197.
7. Ursini, F., Maiorino, M., Valente, M., Ferri, L., and Gregolin, C. (1982) Purification from pig liver of a protein which protects liposomes and biomembranes from peroxidative degradation and exhibits glutathione peroxidase activity on phosphatidyl-choline liposomes. Biochim. Biophys. Acta 710:197–211.
8. Mattipati, K. R., and Marnett, L. J. (1987) Characterization of the major hydroperoxide reducing activity of human plasma. Purification and properties of a selenium dependent glutathione peroxidase. J. Biol. Chem. 262:17398–17403.
9. Takahashi, K. T., Avissar, N., Within, J., and Cohen, H. (1987) Purification and characterization of human plasma glutathione peroxidase: A selenoprotein distinct from the known cellular enzyme. Arch. Biochem. Biophys. 256:677–686.
10. Chu, F. F., Doroshow, J. H., and Esworthy, S. (1993) Expression, characterization and tissue distribution of a new cellular selenium dependent glutathione peroxidase. GSHPx-GI J. Biol. Chem. 268:2571–2576.
11. McCay, P., Gibson, D. D., Fong, K. L., and Hornbrook, R. (1976) Effect of glutathione peroxidase activity on lipid peroxidation in biological membranes. Biochim. Biophys. Acta 431:459–468.
12. Gibson, D. D., Hornbrook, R., and McCay, P. (1980) Glutathione dependent inhibition of lipid peroxidation by a soluble, heat labile factor in animal tissues. Biochim. Biophys. Acta 620:572–582.
13. Ursini, F., Maiorino, M., and Gregolin, C. (1985) The selenoenzyme phospholipid hydroperoxide glutathione peroxidase. Biochim. Biophys. Acta 839:62–70.

14. Maiorino, M., Gregolin, C., and Ursini F. (1990) Phospholipid hydroperoxide glutathione peroxidase. Methods Enzymol. 186:448–457.
15. Flohé, L. (1989) The selenoprotein glutathione peroxidase. In: Glutathione: Chemical, Biomedical and Medical Aspects. Part A (Dolphin, D., Poulson, R., and Avramovic O., eds.), pp. 643–731. J. Wiley & Sons, New York.
16. Maiorino, M., Roveri, A., Gregolin, C., and Ursini, F. (1986) Different effects of Triton X-100, deoxycholate and fatty acids on the kinetics of glutathione peroxidase and phospholipid hydroperoxide glutathione peroxidase. Arch. Biochem. Biophys. 251:600–605.
17. Yamamoto, Y., and Takahashi K. (1993) Glutathione peroxidase isolated from plasma reduces phospholipid hydroperoxides. Arch. Biochem. Biophys. 305:541–545.
18. George, P., and Irvine, D. H. (1952) The reaction between metmyoglobin and hydrogen peroxide. Biochem. J. 52:511–517.
19. Giulivi, C., and Cadenas, E. (1994) Ferrylmyoglobin: Formation and chemical reactivity toward electron donating compounds. Methods Enzymol. 233:189–202.
20. Giulivi, C., Romero, F. J., and Cadenas, E. (1992) The interaction of Trolox C, a water soluble vitamin E analog, with ferrylmyoglobin: Reduction of the oxoferryl moiety. Arch. Biochem. Biophys. 299:302–312.
21. Maiorino, M., Ursini, F., and Cadenas, E. (1993) Reactivity of metmyoglobin towards phospholipid hydroperoxides. Free Rad. Biol. Med. 16:661–667.
22. Thomas, J. P., Maiorino, M., Ursini, F., and Girotti, A. W. (1990) Protective action of phospholipid hydroperoxide glutathione peroxidase against membrane-damaging lipid peroxidation: In situ reduction of phospholipid and cholesterol hydroperoxides. J. Biol. Chem. 265:454–461.
23. Roveri, A., Maiorino, M., Nisii, C., and Ursini, F. (1994) Purification and characterization of phospholipid hydroperoxide glutathione peroxidase from rat testis mitochondrial membranes. Biochim. Biophys. Acta 1208:211–221.
24. Maiorino, M. (1994) Effect of a-lipoic acid on Se-dependent glutathione peroxidases. In: Biological Oxidants and Antioxidants (L. Packer and E. Cadenas, eds.), pp. 60–75. Hippocates Verlag, Stuttgart.
25. Flohé, L., Aumann, K. D., Brigelius-Flohé, R., Schomburg, D., Straβburger, W., and Ursini, F. (1993) A structural comparison of the glutathione peroxidases. In: Active Oxygens, Lipid Peroxides and Antioxidants (Yagi, K., ed.), pp. 299–311. CRC Press, Boca Raton, FL.
26. Schuckelt, R., Brigelius-Flohé, R., Maiorino, M., Roveri, A., Reumkens, J., Straβburger, W., Ursini, F., Wolf, B., and Flohé L. (1991) Phospholipid hydroperoxide glutathione peroxidase is a selenoenzyme distinct from the classical glutathione peroxidase as evident from cDNA and amino acid sequencing. Free Rad. Res. Commun. 14:343–361.
27. Brigelius-Flohé, R., Aumann, K. D., Blöcker, H., Gross, G., Kiess, M., Klöppel, K. D., Maiorino, M., Roveri, A., Schuckelt, R., Ursini, F., Wingender, E., and Flohé, L. (1994) Phospholipid hydroperoxide glutathione peroxidase: Genomic DNA, cDNA and deduced amino acid sequence. J. Biol. Chem. 269:7342–7348.
28. Hill, K. E., Lloyd, S. R., and Burk, R. F. (1993) Conserved nucleotide sequences

in the open reading frame and 3′ untranslated region of selenoprotein P mRNA. Proc. Natl. Acad. Sci. USA 90:537–541.
29. Epp, O., Landstein, R., and Wendel, A. (1983) The refined structure of the selenoenzyme glutathione peroxidase at 0.2 nm resolution. Eur. J. Biochem. 133:51–69.
30. Ursini, F., Maiorino, M., Brigelius-Flohé, R., Aumann, K. D., Roveri, A., Schomburg, D., and Flohè, L. (1995) The diversity of glutathione peroxidases. Methods Enzymol. (in press).
31. Ursini, F., Maiorino, M., Hochstein, P., and Ernster, L. (1989) Microsomal lipid peroxidation: Mechanism of initiation. Free Rad. Biol. Med. 6:31–36.
32. Ernster, L. (1993) Lipid peroxidation in biological membranes: Implications and mechanisms. In: Active Oxygens, Lipid Peroxides and Antioxidants (Yagi, K., ed.), pp. 1–38. CRC Press, Boca Raton, FL.
33. Samokyszyn, V. M., Miller, D. M., Reif, D. W., and Aust, S. D. (1990) Iron-catalyzed lipid peroxidation. In: Membrane Lipid Oxidation (Vigo-Pelfrey, C., ed.), pp. 101–127. CRC Press, Boca Raton, FL.
34. Ursini, F., Maiorino, M., and Sevanian, A. (1991) Membrane hydroperoxides. In: Oxidative Stress: Oxidants and Antioxidants (Sies, H., ed.), pp. 319–336. Academic Press, London.
35. Terao, J. (1990) Reactions of lipid hydroperoxides. In: Membrane Lipid Oxidation (Vigo-Pelfrey, C., ed.), pp. 219–238. CRC Press, Boca Raton, FL.
36. Dix, T. A., and Marnett, L. J. (1985) Conversion of linoleic acid hydroperoxide to hydroxy, keto, epoxyhydroxy and trihydroxy fatty acids by hematin. J. Biol. Chem. 260:5351–5357.
37. Maiorino, M., Coassin, M., Roveri, A., and Ursini, F. (1989) Microsomal lipid peroxidation: Effect of vitamin E and its functional interactions with phospholipid hydroperoxide glutathione peroxidase. Lipids 24:721–726.
38. Koppenol, W. H., and Liebman, J. F. (1984) The oxidizing nature of hydroxyl radical. A comparison with the ferryl ion. J. Phys. Chem. 88:99–105.
39. Roveri, A., Maiorino, M., and Ursini, F. (1994) Enzymatic and immunological measurements of soluble and membrane bound PHGPx. Methods Enzymol. 233:202–212.
40. Roveri, A., Casasco, A., Maiorino, M., Dalan, P., Calligaro, A. and Ursini, F. (1992) Phospholipid hydroperoxide glutathione peroxidase of rat testis: Gonadotropin dependency and immunocytochemical identification. J. Biol. Chem. 267:6142–6146.
41. Rapoport, S. M., Schewe, T., Weisner, R., Halangk, W., Ludwig, P., Janicke-Höhne, M., Tannert, C., Hiebsch, C., and Klatt, D. (1979) The lipoxygenase of reticulocytes. Eur. J. Biochem. 96:545–561.
42. Weitzel, F., and Wendel A. (1993) Selenoenzymes regulate the activity of leukocyte 5-lipoxygenase via the peroxide tone. J. Biol. Chem. 268:6288–6292.
43. Cornwell, D. G., and Morisaki, N. (1984) Fatty acid paradoxes in the control of cell proliferation: Prostaglandins, lipid peroxides and cooxidation reactions. In: Free Radicals in Biology, Vol. VI (Pryor, W. A., ed.), pp. 95–148. Academic Press, New York.

44. Roveri, A., Coassin, M., Maiorino, M., Zamburlini, A., van Amsterdam, F. T. M., Ratti, E., and Ursini, F. 1992 Effect of hydrogen peroxide on calcium homeostasis in smooth muscle cells. Arch. Biochem. Biophys. 297:265–270.
45. Hilly, M., Pietri-Rouxel, F., Coquil, J. F., Guy, M., Maugler, J. P. (1993) Thiol reagents increase the affinity of the inositol 1,4,5 triphosphate receptor. J. Biol. Chem. 268:16488–16494.
46. Petronilli, V., Costantini, P., Scorrano, L., Colonna, R., Passamonti, S., Bernardi, P. (1994) The voltage sensor of the mitochondrial permeability transition pore is tuned by the oxidation reduction state of vicinal thiols. J. Biol. Chem. 269:16638–16642.
47. Salama, G., Abramson, J. J., and Pike, G. K. (1992) Sulphydryl reagents trigger Ca^{2+} release from the sarcoplasmic reticulum of skinned rabbit psoas fibres. J. Physiol. London 454:389–420.
48. Walker, J., Chen, T. A., Sterner, R., Berger, M., Winston, F., and Allfrey, V. G. (1990) Affinity chromatography of mammalian and yeast nucleosomes. Two modes of binding of trascriptionally active mammalian nucleosomes to organomercurial-agarose columns, and contrasting behaviour of the active nucleosomes of yeast. J. Biol. Chem. 265:5736–5746.
49. Chen-Cleland, T. A., Boffa, L. C., Carpaneto, E. M., Mariani, M. R., Valentin, E., Mendez, E., and Allfrey, V. G. (1993) Recovery of transcriptionally active chromatin restriction fragments by binding to organomercurial-agarose magnetic beads. A rapid and sensitive method for monitoring changes in higher order chromatin structure during gene activation and repression. J. Biol. Chem. 268:23409–23416.
50. Knoepfel, L., Steinköhler, C., Carrì, M. T., and Rotilio, G. (1994) Role of zinc-coordination and of the glutathione redox couple in the redox susceptibility of human trascription factor Sp1. Biochem. Biophys. Res. Commun. 201:871–877.

13

Biochemical and Clinical Aspects of Extracellular Glutathione and Related Thiols

Masayasu Inoue, Yukiko Minamiyama, Hidenori Yu, Seiichi Yamamasu, Keiko Inoue, and Eisuke F. Sato
Osaka City University Medical School, Osaka, Japan

I. INTRODUCTION

Glutathione (GSH) is one of the most abundant thiols in living organisms and is synthesized exclusively in a cytosolic compartment in its reduced form. Metabolism of GSH and related compounds occurs via interorgan and intraorgan cycles in which liver, kidney, and small intestine play critical roles (Fig. 1). The liver is the major organ that exports GSH in plasma and bile (1,2); excretion of GSH in both compartments occurs via carrier-mediated mechanisms (2,3). The major fraction of plasma GSH (60–70%) is extracted by glomerular filtration (15–20%) and nonfiltrating peritubular mechanism (45–50%). GSH and related tripeptides filtered by the glomerulus rapidly undergo lumenal degradation by γ-glutamyltransferase (GGT) and peptidases that hydrolyze the cysteinylglycine bond (3). GGT and the peptidases are also involved in peritubular handling of GSH and its S-conjugates (4-6). Although Na-dependent uptake has been shown in isolated plasma membrane vesicles from kidney and other tissues (7,8), K_m values for GSH uptake (3–4 mM) are much larger than the GSH levels in plasma and extracellular compartments (5–25 μM), and, hence, its physiological significance remains to be elucidated. Despite a short half-life for plasma GSH (2–3 min) as compared with those of cytosolic compartments in liver and kidney (3 and 0.5 h, respectively), loading doses of GSH failed to increase GSH levels in these tissues of normal and GGT-inactivated animals during the period when most of the administered GSH disappeared from the circula-

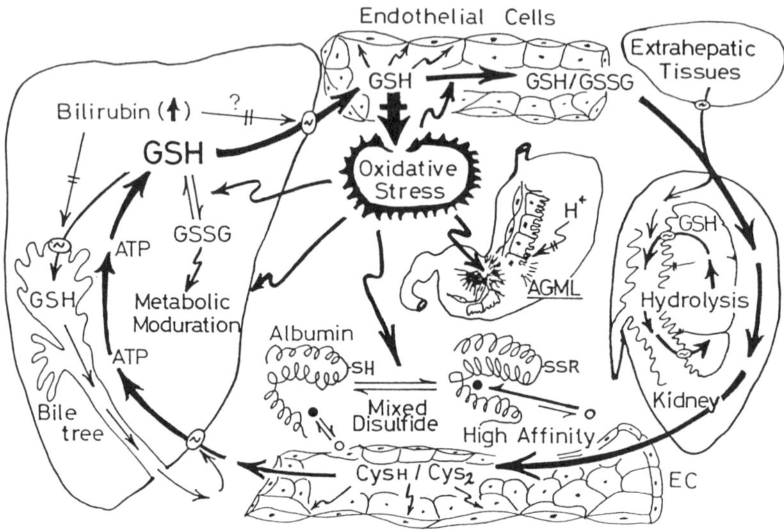

Figure 1 Interorgan and intraorgan cycles of GSH.

tion (5,9,10). Kinetic analysis using radiolabeled GSH revealed that tissue-associated radioactivity was accounted for predominantly by its constituent amino acids. In addition to hepatic GSH transport systems, carrier-mediated excretion of glutathione disulfide (GSSG) and GS-conjugates of xenobiotics has been reported in liver and other tissues (11–16). Since GGT and peptidases that hydrolyze cysteinyl-glycine are highly enriched on the apical membrane surface of kidney, small intestine, and bile-tree cells, lumenal GSH and related tripeptides are also degraded to their constituent amino acids (Fig. 1). This chapter describes the dynamic aspects of the metabolism, pharmacokinetics, and therapeutic potentials of GSH and related thiols.

II. PHYSIOLOGICAL ROLES OF EXTRACELLULAR THIOLS

Kinetic analysis by L-buthionine sulfoximine (BSO), a potent inhibitor of GSH synthase, revealed that GSH turnover in liver and kidney of male Wistar rats (age 8 weeks) occurs 4 and 24 times a day, respectively. GSH synthesis and its extracellular secretion depend on membrane potential and/or ATP, and reabsorption of its constituent amino acids (glutamate, cysteine, and glycine) depends on Na-gradient across the membranes (3,17). Thus, significantly large amounts of energy are consumed to maintain the GSH cycles.

A. Protein Mixed-Disulfide Formation

The presence of such energy-consuming GSH cycles and the occurrence of GSH, GSSG, cysteine, and cystine in the circulation might suggest physiological significance of extracellular thiols. In fact, when incubated with freshly prepared plasma, rapid oxidation of GSH to GSSG occurred by some oxygen-dependent and EDTA-inhibitable mechanism. At the same time, significant amounts of glutathione (and cysteine) formed mixed disulfides with plasma proteins. Analysis using radioactive GSH and cysteine revealed that they bound predominantly to the Cys34 of albumin (18,19). Small amounts of glutathione and cysteine also bound to plasma proteins other than albumin. Analysis using specific antiserum revealed that α_2-macroglobulin was responsible for about 30% of mixed disulfides in nonalbumin fractions. GSH and cysteine also formed mixed disulfides with circulating albumin particularly when animals were challenged with oxidative stress (18,19). In fact, amounts of nonmercaptalbumin increased in patients with various diseases, such as renal failure (20). However, in normal subjects, most albumin circulates in its mercapt form. Thus, when injected in normal animals, nonmercaptalbumin rapidly decreased with concomitant increase in mercaptalbumin. The rate of conversion of nonmercaptalbumin to mercaptalbumin significantly decreased in rats with acute inflammation and in totally viscelectomized animals (18), suggesting that the redox state of albumin and other plasma proteins might be regulated by interorgan metabolism of GSH. Biochemical analysis revealed that the affinity of albumin for hydrophobic organic acids increases significantly when this protein forms mixed disulfide with GSH. The affinity of glutathione-albumin mixed disulfide for hydrophobic anions was higher than that of cysteine albumin mixed disulfide (18,19). Similarly, the affinity of albumin for free fatty acids and heavy metals increased significantly after mixed disulfide formation. Interestingly, nucleophilic nature of Cys34 increased markedly if the binding site(s) on albumin was occupied by such ligands (18,19). Thus, the redox state of Cys34 and the ligand-binding activity of albumin seem to be regulated by way of mixed-disulfide formation (Fig. 2). Under oxidative stress, GSH is easily oxidized to GSSG, which forms nonmercaptalbumin, thereby increasing the binding capacity for free fatty acids generated under oxygen stress.

B. Regulation of Circulatory Status by Thiols

Since significant amounts of extracellular glutathione and cyst(e)ine might occur in the circulation, the interorgan cycle of GSH is also important for protecting extrahepatic tissues, including vascular endothelial cells and blood cells, from oxygen toxicity. In vivo analysis revealed that oxidative stress increased vascular permeability and reduced the volume of effectively circulating plasma by some mechanisms, which was normalized either by GSH (21) or a long-acting

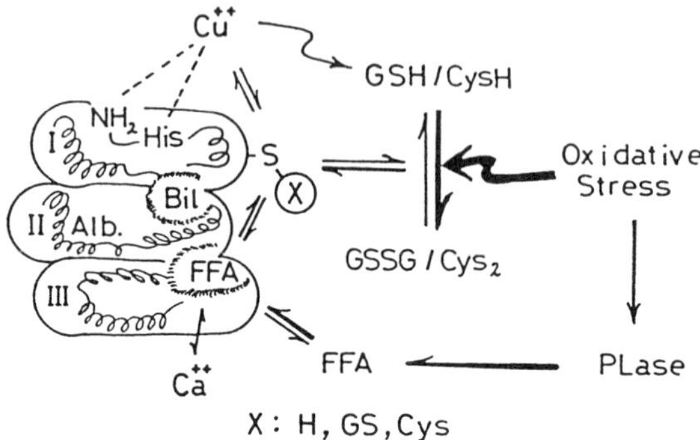

Figure 2 Regulation of albumin by mixed disulfide formation. Oxidative stress stimulates phospholipase activity and mixed disulfide formation, thereby potentiating the binding capacity of albumin for organic anions. S, Cys34; X, GS- or cys-; FFA, free fatty acids; Bil, bilirubin; PLase, phospholipase.

SOD (22–24). It should be noted that vascular endothelial cells have two enzymes, xanthine oxidoreductase and nitric oxide synthase, which generate superoxide and nitric oxide (NO) radicals, respectively. Intracellularly synthesized NO diffuses out of cells and binds to heme-containing proteins, such as hemoglobin and guanylate cyclase, thereby modulating biological functions. NO in and around resistance arteries was first recognized as an endothelium-derived relaxing factor (EDRF), which decreases vascular resistance by increasing cGMP levels in smooth muscles (25–28). Under physiological concentrations of oxygen, the NO radical is fairly unstable and is oxidized to nitrate with a half-life of about 5 s. NO also reacts with superoxide radical, thereby generating highly reactive peroxynitrite (ONOO–). Hence, vascular concentrations of bioactive NO would be decreased if superoxide production were increased in and around endothelial cells. Nakazono et al. (29) reported that targeting SOD to vascular endothelial cells selectively decreased the blood pressure of spontaneously hypertensive rats (SHR) and deoxycorticosterone/NaCl-induced hypertensive rats. Selective dismutation of superoxide radicals around vascular walls by endothelially targeted SOD (30–32) would increase the lifespan of de novo synthesized NO, and thereby maximize its depressor effect. Thus, the ratio of NO to superoxide in and around vascular walls is important in determining the blood pressure and circulatory status of animals. NO also reacts rapidly with various thiols, including GSH and cysteine, thus forming nitrosothiols. These nitrosothiols have potent depressor activity, as does NO. Under physiological pH and oxygen concentrations, the stability of nitrosothiols is significantly greater than that of NO, and, hence, the depressor activity of GS-NO and cys-NO was retained much longer than NO (Fig. 3). Since GSH is degraded extracellularly,

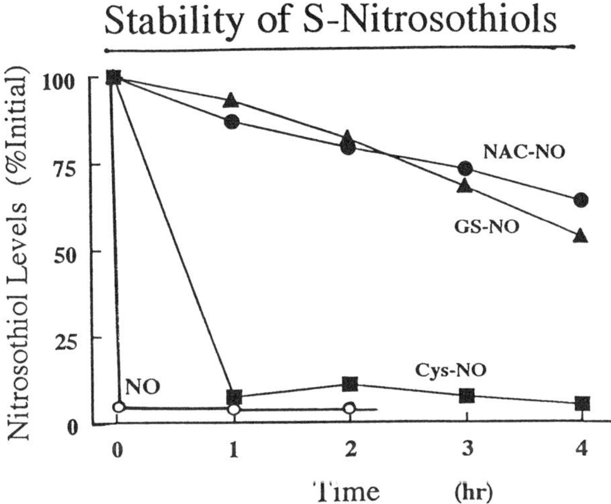

Figure 3 Stability of NO and various nitrosothiols. NO and various forms of nitrosothiols are incubated in 0.1 M phosphate buffer, pH 7.4, at 37°C. At indicated times, nitrosothiols remaining in the medium were determined spectroscopically. When injected i.v. into rats, these nitrosothiols reveal marked depressor activity.

significant amounts of cyst(e)ine also appear in the circulation and enter various cells, including vascular endothelial cells and smooth muscle cells, via active transport systems. Thus, the interorgan cycle of GSH also plays an important role in regulating the circulatory status by way of modulating the metabolism of NO in and around vascular walls (Fig. 4).

C. Thiol Status in AIDS Patients

It is known that GSH and related thiols play critical roles in the regulation of ligand-induced blastogenesis and immunological functions of lymphocytes. HIV infection is associated with altered levels of GSH in cells and extracellular fluids (33). Markedly decreased plasma cystine and cysteine concentrations have also been reported with HIV-infected patients at all stages of the disease and in SIV-infected rhesus macaques (34). Levels of GSSG in pulmonary lavage fluid were higher in AIDS patients with *Pneumocystis carinii* pneumonia than in control subjects (33), suggesting that these patients suffer from oxygen toxicity in their lungs. The intact immune system appears to require a delicate balance between prooxidant and antioxidant conditions, and this balance is obviously disturbed in HIV infection and may contribute to the pathogenesis of AIDS. Elevated glutamate levels are also found in AIDS patients.

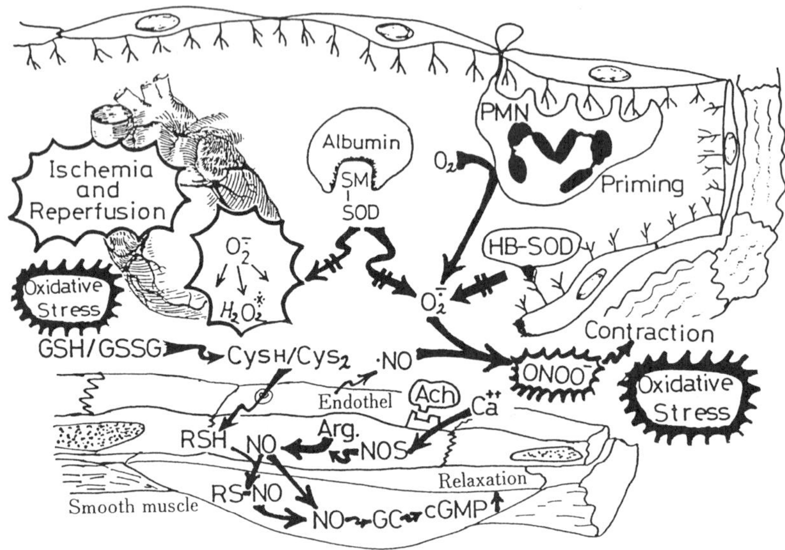

Figure 4 Regulation of NO metabolism and vascular resistance by biothiols. Effect of thiols on the metabolism of NO. NOS, NO synthase; Arg., L-arginine; RSH, GSH or cysteine; RS-NO, nitrosothiols; GC, guanylate cyclase; SM-SOD, long-acting SOD bound to albumin; HB-SOD, SOD having high affinity for heparin sulfates on vascular endothelial cells.

It should be noted that mitogenically stimulated human peripheral blood lymphocytes and T-cell clones have weak transport activity for cystine but strong transport activity for cysteine. Cysteine is represented at the lowest concentration among all protein-forming amino acids in the blood plasma. Proliferative responses of mitogenically stimulated lymphocytes and T-cell clones and the activation of cytotoxic T cells in allogenic mixed lymphocyte cultures are strongly influenced by small variations in extracellular cysteine concentration, even in the presence of relatively high and approximately physiological concentrations of cystine. The extracellular supply of cysteine influences strongly the intracellular level of GSH and the activity of the transcription factor NF-κB that regulates the expression of immunologically relevant genes. Cysteine can be substituted by *N*-acetylcysteine but not by cystine (35).

Cysteine seems to play an important role as a regulatory mediator between macrophages and lymphocytes. Cell culture experiments show that a three- to five-fold elevation of the extracellular glutamate concentration causes a substantial decrease of the intracellular cysteine and glutathione content of murine peritoneal macrophages (36). A combination of low plasma glutamate and high cystine levels was found to be correlated with high CD4+ T-cell numbers even in

HIV-negative healthy human individuals (37). Glutamate and cystine levels also correlated with $CD4^+$ T-cell numbers in chimpanzees. Murine peritoneal macrophages, human peripheral blood monocytes, and murine fibroblastoid cells (L-cells) consume cystine and release cysteine into the extracellular space. The relatively high cystine transport activity of the antigen-presenting macrophages provides these cells with a "cysteine-pumping" function that allows the antigen-binding T cells in their vicinity to shift to the antioxidant state (38). Eck and Droge (36) suggested a possible inhibition of cystine uptake by increased glutamate. Although the elevated levels of plasma glutamate in AIDS patients have been postulated to aggravate the cysteine deficiency by inhibiting the transport activity for cystine, direct evidence that low levels of cellular thiols are due to such competition of the two amino acids in AIDS patients is lacking. It should be noted that glutamate inhibits the feedback inhibition of γ-glutamylcysteine synthase by GSH (39,40), and, hence, high levels of cellular glutamate increases GSH levels in cells and plasma (41). Thus, although elevated levels of glutamate may possibly inhibit the rate of cystine uptake by lymphocytes and other cells, simple competition of the two amino acids may not result in systemic decrease in GSH and related thiols, particularly under steady-state conditions in vivo. When T cells do not receive sufficient amounts of cysteine, cellular GSH levels and rate of DNA synthesis activity decrease and the cells may suffer from various manifestations of oxidative damage.

HIV-1 proviral DNA contains two binding sites for the transcription factor NF-κB. HIV-1–infected individuals have abnormally high levels of tumor necrosis factor alpha (TNFα) and abnormally low plasma cysteine levels (42). Thus, oxidative stress might activate HIV replication through the activation of a factor that binds to a DNA-binding protein, NF-κB, which in turn stimulates HIV gene expression by acting on the promoter region of the viral long terminal repeat. TNFα, an essential cytokine produced by activated macrophages, is also involved in the activation of HIV infection through similar mechanisms (43). TNFα-mediated cytotoxicity of cells may be related to the generation of intracellular hydroxyl radicals. Oxidative stress in HIV-associated disease progression might seem to accelerate the consumption of GSH during HIV infection and progression. Cysteine or *N*-acetylcysteine raised the intracellular GSH level and inhibited HIV-1 replication in persistently infected Molt-4 and U937 cells. However, inhibition of HIV-1 replication appears not to be directly correlated with GSH levels. Cysteine and *N*-acetylcysteine also inhibit NF-κB activity as determined by electrophoretic mobility shift assays and chloramphenicol acetyltransferase gene expression under control of NF-κB–binding sites in uninfected cells. Thus, the cysteine deficiency in HIV-infected individuals may cause an overexpression of NF-κB–dependent genes and enhance HIV replication.

It should be noted that AIDS-like symptoms are not found in HIV-infected chimpanzees and SIV-infected African green monkeys (37). This fact may provide important clues about the pathogenic mechanism of AIDS and the resistance of HIV-infected persons and SIV-infected rhesus macaques with marked decrease in cysteine, cystine, and glutathione levels and elevated plasma glutamate concentrations. Plasma levels of vitamin E, polyunsaturated fatty acids of phospholipids (PUFA-PL), lipoperoxides as well as erythrocyte glutathione peroxidase activity (GSH-Px) were evaluated in 200 migrants from developing countries, some of which were at high risk for serious infectious diseases (44,45). HIV-1 and syphilis infections were also investigated. About 57% of subjects (n = 114) had blood levels of vitamin E, PUFA-PL, and GSH-Px significantly lower than those of normal healthy individuals (n = 30), while lipoperoxides values were unchanged. Eight from this group were found to be HIV-1 positive, and five were TPHA positive. In contrast, the remaining 86 migrants did not show any signs of infection and their blood parameters were normal. These results show that factors such as widespread poverty, inadequate housing, malnutrition, insufficient access to medical care, and psychological stress are strictly correlated to the reduction of blood parameters that are critical for the normal cell function of mammalian cells. The lack of vitamin E and GSH-Px, which are considered major protective molecules against lipoperoxidation damage in vivo, has been involved in several human diseases. Low blood levels of vitamin E, GSH-Px, and particularly PUFA-PL may play a pathogenic role in the onset and development of AIDS and other infections. In this connection, a deficiency of these blood parameters in patients with AIDS (n = 50) and in 32% of HIV-seronegative intravenous drug abusers (n = 100) has been reported. Reduced blood levels of vitamin E, GSH-Px, and particularly polyunsaturated fatty acids may be added to the list of risk factors favoring the onset and the development of AIDS.

The serum from AIDS patients also had four to five times higher catalase activity than that from healthy control subjects (46). Moreover, serum catalase (but not glutathione peroxidase) activity increased progressively with advancing HIV infection (i.e., AIDS > symptomatic infection > asymptomatic infection > controls). Serum catalase activity correlated with serum LDH activity but did not appear to be a consequence of hemolysis, since RBC fragility and serum haptoglobin levels were comparable in HIV-infected and control subjects. Increases in serum catalase activity may reflect and/or compensate for systemic glutathione and other antioxidant deficiencies in HIV-infected individuals. The elevated activity of serum catalase and LDH may result from oxidative tissue injury caused by systemic decrease of GSH and related thiols.

D. Thiol Status and Apoptosis

Apoptosis or programmed cell death is a type of death occurring in various physiological processes. Data suggest that apoptosis may play a critical role in AIDS pathogenesis and that an increase in cellular levels of reactive oxygen species can be associated with activation of previously latent HIV virus. TNF, a cytokine capable of inducing oxygen free radicals and apoptosis, appears also to be involved in HIV activation. The presence of positive correlation between the percentage of apoptotic cells, GSH depletion, and an increase of p24 antigenemia suggested that pretreatment with *N*-acetylcysteine might decrease apoptosis in HIV-infected U937 cells (47). In this context, it has been reported that T-cell colony-forming cell (CFC) is impaired in HIV infection by a thiol-inhibitable mechanism. Levy et al. (48) reported the effect of the thiol-depleting reagents BSO, cyclohexene-1-one, and copper phenanthroline on T-CFC formation and cell cycle progression in HIV-positive and control subjects. All three reagents inhibited T-CFC formation and cell cycle progression (48). Thus, colony formation by cells from HIV-positive subjects was more sensitive to the effects of thiol depletion. Oxidation of membrane thiols, as well as depletion of intracellular GSH, might inhibit T-CFC formation and cell cycle progression for mitogen-stimulated cells in culture. Analysis by a fluorescence-activated cell sorter (FACS) revealed that CD4 and CD8 T cells with high intracellular GSH levels are selectively and significantly depleted early during HIV infection and lost as the infection progresses (49). Maintenance of intracellular GSH levels has been implicated in blocking cytokine-stimulated HIV replication in vitro, in both acute and latent infection models. Subsets of human peripheral blood mononuclear cells differ substantially in GSH levels. FACS analysis also revealed that B cells have the lowest GSH levels, T cells are intermediate, and monocytes and macrophages have the highest levels (49). Furthermore, GSH levels subdivide the CD4 and CD8 T-cell subsets into two classes each: high- and low-GSH cells, which cannot be distinguished by cell size or by currently known surface markers. *N*-Acetylcysteine blocks cytokine-stimulated HIV replication in an acutely infected T-cell line and in acutely infected peripheral blood mononuclear cells from normal individuals. *N*-Acetylcysteine also inhibits stimulated HIV expression in chronically infected monocyte and T-cell lines, which are used as models for latent infection in AIDS (50). Thus, GSH, *N*-acetylcysteine, or other types of GSH precursors may possibly be considered for the treatment of patients with HIV-1 infection to normalize the thiol status and lymphocyte functions, thereby inhibiting apoptosis and opportunistic infection.

III. PHARMACOKINETICS AND THERAPEUTIC POTENTIAL OF EXTRACELLULAR THIOLS

Because of the unique properties of GSH to regulate redox states and to detoxify electrophilic xenobiotics, this tripeptide has been used for the treatment of various diseases including hepatitis and drug poisoning. Despite the fact that the administered GSH is rapidly degraded to its constituent amino acids, GSH cannot be replaced by cysteine predominantly because of the instability and toxicity of a high dose of this amino acid. Under physiological conditions, cysteine but not GSH is rapidly oxidized, which suggests that the reducing activity of cysteine is much greater than that of GSH. The fairly stable nature of GSH is due to masking of the cysteinyl amino group by γ-glutamyl linkage. Analogously, masking of the α-amino group by *N*-acetylation also stabilizes the thiol group of cysteine. Since *N*-acetylcysteine and its S-conjugates behave as organic anions rather than amino acids, they enter cells via the organic anion transport system. When injected intravenously, *N*-acetylcysteine rapidly disappears from the circulation with a half-life of a few minutes and accumulates in the kidney and other tissues via active transport systems for organic acids (10,51,52). Since renal tubule cells are highly enriched with acylase, significant fractions of the administered *N*-acetylcysteine would be hydrolyzed in the kidney and the resulting cysteine returns to the circulation. Because small intestine, liver, and other tissues also have transport systems for organic anions, orally administered *N*-acetylcysteine is taken up by the intestine, transferred to the portal circulation and captured by the liver and other organs. Since the liver also has high acylase activity, *N*-acetylcysteine accumulated in this organ is hydrolyzed to cysteine. In contrast, orally administered GSH is hydrolyzed by GGT and peptidases in the intestine, and the resulting cysteine is taken up by epithelial cells and transferred to the liver. Thus, the pharmacokinetic behavior of GSH is different from that of *N*-acetylcysteine. Such a pharmacokinetic difference should be considered in the clinical use of the two thiols.

A. AIDS

In the healthy lung, the oxidant burden is balanced by the local antioxidant defenses. However, both an increased oxidant burden and/or decreased antioxidant defenses may reverse the physiological oxidant-antioxidant balance in favor of oxidants, leading to lung injury. In fact, several lung disorders are believed to be characterized by an increase in alveolar oxidant burden, potentially depleting alveolar and lung GSH. Low GSH has been linked to abnormalities in the lung surfactant system and the interaction between GSH and antiproteases in the epithelial lining fluid of patients. The increase of intracellular reactive oxygen species is believed to correlate with the activation of NF-κB. There is now sufficient data to strongly implicate free radical injury in the genesis and maintenance of several lung disorders in humans (53). This information is substantial and will help the development of clinical studies examining a variety of inflammatory lung disorders. Thus, it is rational to consider whether it is possible to safely augment thiol status in the epithelial lining fluid of HIV-seropositive individuals. Because GSH and *N*-acetylcysteine reverse TNFα toxicity both in cells and in animals and are well-known drugs that can be administered without known toxicity in humans, these thiols might have a therapeutic potential with AIDS patients. In this context, the protective effect of *N*-acetylcysteine against oxidant lung injury was investigated in a model of acute immunological alveolitis in the rat. Intrapulmonary immune complex deposition in rat lungs caused pulmonary damage associated with a marked decrease in 5-hydroxytryptamine uptake, a marker of endothelial cell function (54). Oral administration of *N*-acetylcysteine effectively prevented the endothelial cell injury of the lung caused by deposition of antigen/antibody complex. *N*-Acetylcysteine reduced the susceptibility of the lung to free radical–induced damage by potentiating the antioxidant defense systems. Although intravenously administered GSH and related thiols circulate over the surface of endothelial cells in the lung, pulmonary uptake of these thiols is low. In contrast, inhalation of aerosol is the efficient way to deliver various drugs including GSH and *N*-acetylcysteine. In fact, GSH can be administered directly to the respiratory epithelial surface by aerosol and is fully functional as an antioxidant both in vitro and in vivo (55). In pulmonary diseases such as idiopathic pulmonary fibrosis or following HIV infection, GSH aerosol therapy not only normalizes deficient GHS levels in the lung, but is capable of favorably influencing cellular events such as oxidant release by pulmonary inflammatory cells. It might be possible to use antioxidants to reverse the imbalance between oxidants and antioxidants at the site of oxidant generation and/or oxidative injury to prevent the progressive tissue damage in lung disorders characterized by high oxidant states.

Pneumocystis carinii pneumonia is a debilitating disease of the lung that can accompany HIV infection. Inhalation of aerosol is the efficient way to deliver GSH and *N*-acetylcysteine. Since GGT activity in the lung is low, the inhaled GSH might be localized within this organ for a fairly long time. To test the possible benefit of modulating the pulmonary thiol status of AIDS patients, GSH was delivered by aerosol to 14 HIV-seropositive individuals, and the glutathione levels in lung epithelial lining fluid were compared before and at 1, 2, and 3 h after aerosol administration (56). Before treatment, total glutathione concentrations in the epithelial lining fluid were approximately 60% of controls. After 3 days of twice daily doses of 600 mg each reduced glutathione, total glutathione levels in the epithelial lining fluid increased and remained in the normal range for at least 3 h after treatment. Strikingly, even though >95% of the glutathione in the aerosol was in its reduced form, the percentage of GSSG in epithelial lining fluid increased from 5% to about 40% 3 h after treatment, probably reflecting the use of GSH as an antioxidant in vivo. No adverse effects were observed. It is feasible and safe to use aerosolized GSH and/or *N*-acetylcysteine to augment the deficient GSH levels of the lower respiratory tract of HIV-seropositive individuals. Based on such findings, evaluation of GSH and related thiols in improving host defense in HIV-seropositive individuals is under current investigation.

B. Ischemia-Reperfusion and Transplantation

In the course of cardiac transplantation, donor hearts undergo a four-step sequence of events—arrest, cold storage, global ischemia during implantation, and reperfusion—during which myocardial damage occurs. The functional recovery of these hearts could be improved by exposure to some preservation solutions throughout this sequence. The formulation of solutions were designed principally based on the following concept: prevention of cell swelling (by high concentrations of myocardium-specific impermeants, such as mannitol), calcium overload (ionic manipulations), and oxidative damage (antioxidants including GSH) and enhancement of anaerobic energy production (such as glutamate). Menasche et al. (57) reported that the maximal rate of ventricular pressure increase and left ventricular compliance of rat hearts, which were subjected to cardioplegic arrest, cold storage for 5 h, global ischemia at 15°C for 1 h, and normothermic reperfusion for an additional hour, were retained by addition of GSH but not *N*-acetylcysteine. They suggested that extracellularly added GSH might play a critical role in protecting the myocardium from reactive oxygen species generated during the storage and/or reperfusion.

Kobayashi et al. (58) reported the beneficial effects of both GSH and γ-glutamylcysteine ester in a model of normothermic hepatic ischemia-reperfusion injury. Although, γ-glutamylcysteine ester increased glutathione levels in the liver, high doses of GSH failed to increase it. Thus, the protective action of these thiols, particularly GSH, might operate extracellularly. Although reactive oxygen species have been assumed to occur intracellularly and, hence, the role of GSH is seen as a cosubstrate for glutathione peroxidase, direct evidence for a pathologically relevant oxidative stress in hepatocytes during reperfusion is lacking. Moreover, even depletion of hepatic GSH by 90% before inducing an intracellular oxidant stress failed to enhance liver injury (59). Kawamoto et al. (60,61) demonstrated a beneficial effect of a long-acting SOD, which circulates bound to albumin, and suggested that the pathologically relevant superoxide radicals during reperfusion might take place in the hepatic vasculature. In this context, Jaeschke (62) reported that intravenously injected GSH is oxidized to GSSG more rapidly in the plasma of postischemic animals than in controls. Administration of GSH and related thiols leads to significantly high cysteine levels in plasma (63), and cysteine and methionine stimulate the hepatic turnover and efflux of glutathione (see Fig. 1). Thus, GSH and related thiols should be considered for clinical use for attenuating ischemia-reperfusion injury and oxygen toxicity caused by transplantation of the various organs.

C. Drug Poisoning

There are numerous data about the therapeutic potential of GSH for patients with oxidative tissue injury caused by drug intoxication and irradiation (64). Since paraquat is a potent redox cycler and generates superoxide radicals, administration of this drug results in severe lung injury. Ando et al. (65) reported that intravenous administration of paraquat markedly increased GSSG levels in the liver and increased the rate of its secretion in the bile, thereby enhancing the interorgan cycle of GSH. Administration of a loading dose of GSH significantly decreased the serum levels of asparate aminotransferase and LDH. Hoffer et al. (66) also tested the effect of *N*-acetylcysteine against oxidative lung damage in paraquat-intoxicated rats. Administration of *N*-acetylcysteine to paraquat-intoxicated animals did not change the glutathione status of the lungs but suppressed the paraquat-induced release of chemoattractants for neutrophils in the bronchoalveolar fluid. The infiltration of inflammatory cells in bronchoalveolar lavage was significantly reduced by *N*-acetylcysteine 24 h after paraquat intoxication. *N*-Acetylcysteine has a protective effect against oxidative lung damage by delaying inflammation and prevents the paraquat-induced reduction of superoxide anion production by stimulated alveolar macrophages. However, this thiol failed to affect the survival rate of paraquat-intoxicated rats. Ogino et al. (67) reported the effect of GSH on a patient who took 500 ml of paraquat. This amount of paraquat is 30–40 times higher than its lethal dose and is sufficient to kill healthy subjects within a few days. The patient was given mixtures of antioxidants, such as GSH, vitamin E, and desferoxamine (to decrease the Fenton reaction caused by Fe^{2+}) and his pO_2 was controlled at around 50–60 mm Hg. Biochemical analysis of the patient blood revealed markedly low levels of LDH and transaminases for at least 3 weeks. Unfortunately, the patient died of pneumonia on day 23. Although the critical effect of GSH administration alone on paraquat-intoxicated patients is not clear at present, a combination of antioxidants and treatment that prevent the generation of reactive oxygen species seems to have beneficial effect. The therapeutic potential of GSH, related thiols, and other antioxidants should be studied more precisely.

D. Stress, Gastric Injury, and Apoptosis

Strong stress has been known to impair not only gastrointestinal function but also immunological response of human subjects. Recent studies reveal that reactive oxygen species also underlie the pathogenesis of stress-induced gastric mucosal injury. In fact, stress-induced gastric mucosal injury could be inhibited by a long-acting SOD, which circulates bound to albumin and accumulates in injured tissues (21–24). Thus, oxidative stress caused by the superoxide radical plays a critical role in the pathogenesis of stress-induced gastric injury. To test whether the GSH cycle also plays an important role in protecting the stomach from oxidative stress, dynamic aspects of GSH status were determined in water-immersion–restraint rats. Interestingly, hepatic levels of total glutathione markedly decreased by stress predominantly through inhibition of de novo synthesis of GSH (21). Thus, the amounts of extracellular GSH derived from the liver would be significantly decreased in stress-loaded animals. The mucosal injury is markedly decreased by a loading dose of GSH. This treatment significantly increases GSH levels in plasma but not in liver and stomach. Thus, extracellular thiols seem to play important roles in protecting gastric mucosa from oxidative stress. Recent studies revealed that stress-loading induced a marked apoptosis of lymphocytes in the thymus. Fragmentation of thymocyte DNA was affected by intravenous administration of either GSH or *N*-acetylcysteine (68). The therapeutic potential of GSH, related thiols, and other antioxidants for clinical handling of stress-induced tissue injury, including apoptosis of lymphocytes, should be studied further.

REFERENCES

1. Meister, A. (1985) Glutathione. Ann. Rev. Biochem. 52:711–760.
2. Inoue, M. (1985) Interorgen metabolism and transport of glutathione and related compounds. In: Renal Biochemistry (Kinne, R., ed.), pp. 225–269, Elsevier.
3. Inoue, M., and Hirota, M. (1989) Dynamic aspects of GSH metabolism during oxidative stress. In: Glutathione Centennial (Sakamoto, Y., Higashi, T., Taniguchi, N., and Meister, A., eds.), pp. 381–394, Academic Press, New York.
4. Anderson, M. E., Bridges, R. J., and Meister, A. (1980) Direct evidence for inter-organ transport of glutathione and non-filtrating renal mechanism for glutathione utilization involves γ-glutamyltranspeptidase. Biochem. Biophys. Res. Commun. 96:848–853.
5. Inoue, M., Shinozuka, S., and Morino, Y. (1986) Mechanism of renal peritubular extraction of plasma glutathione: The catalytic activity of contralumenal γ_i-glutamyltransferase is prerequisite to the apparent peritubular extraction of plasma glutathione. Eur. J. Biochem. 157:605–609.
6. Späte r, H., Poruchynsky, M., Quintana, N., Inoue, M., and Novikoff, A. (1982) Immunocytochemical localization of γ-glutamyl transferase in rat kidney with protein A-horseradish peroxidase. Proc. Natl. Acad. Sci. USA 79:3547–3550.
7. Lash, L. H., and Jones, D. P. (1984) Renal glutathione transport: Characteristics of the sodium-dependent system in the basal-lateral membranes. J. Biol. Chem. 259:14508- 14514.
8. Lash, L. H., Hagen, T. M., and Jones, D. P. (1984) Exogenous glutathione protects intestinal epithelial cells from oxidative injury. Proc. Natl. Acad. Sci. USA 83:4641–4645.
9. Hahn, R., Wendel, A., and Flohe, L. (1978) The fate of extracellular glutathione in the rat. Biochim. Biophys. Acta 539:324–337.
10. Inoue, M., Okajima, K., and Morino, Y. (1982) Metabolic coordination of liver and kidney in mercapturic acid biosynthesis in vivo. Hepatology 2:311–316.
11. Inoue, M., Kinne, R., Tran, T., and Arias, I. M. (1984) Glutathione transport across hepatocyte plasma membrane: Analysis using isolated rat liver sinusoidal membrane vesicles. Eur. J. Biochem. 138:491–495.
12. Inoue, M., Kinne, R., Tran, T., Akerboom, T., Sies, H., and Arias, I. M. (1984)

Biliary transport of glutathione S-conjugate by rat liver canalicular membrane vesicles. J. Biol. Chem. 259:4998–5002.

13. Inoue, M., Kinne, R., Tran, T., and Arias, I. M. (1983) The mechanism of biliary secretion of reduced glutathione: Analysis of transport process in isolated rat liver canalicular membrane vesicles. Eur. J. Biochem. 134:467–471.
14. Akerboom, T., Inoue, M., Sies, H., Kinne, R., and Arias, I. M. (1984) Biliary transport of glutathione disulfide studied with isolated rat liver canalicular membrane vesicles. Eur. J. Biochem. 141: 211–215.
15. Fernandez-Checa, J. C., Takikawa, H., Horie, T., Ookhtens, M., and Kaplowitz, N. (1992) Canalicular transport of reduced glutathione in normal and mutant Eisai hyperbilirubinemic rats. J. Biol. Chem. 267:1667–1673.
16. Jaeschke, H. (1992) Enhanced sinusoidal glutathione efflux during endotoxin-induced oxidant stress in vivo. Am. J. Physiol. 263:G60–68.
17. Inoue, M., and Morino, Y. (1985) Direct evidence for the role of the membrane potential in glutathione transport by renal brush border membranes. J. Biol. Chem. 260:326–331.
18. Inoue, M., Saito, Y., Hirata, E., Morino, Y., and Nagase, S. (1987) Regulation of redox states of plasma proteins by metabolism and transport of glutathione and related compounds. J. Protein Chem. 6:207–225.
19. Inoue, M. (1989) Dynamic aspects of protein-mixed disulfide formation. In: Coenzymes and Cofactors (Dolphin, D., Poulson, R., Abramovic, O., ed.), pp. 613–644. John Wiley & Sons, New York.
20. Sogami, M., Nagaoka, S., Era, S., Honda, M., and Noguchi, K. (1984) Resolution of human mercapt- and nonmercaptalbumin by high-performance liquid chromatography. Int. J. Peptide Protein Res. 24:96–103.
21. Hirota, M., Inoue, M., Sakamoto, Y., Mori, K., Morino, Y., and Akagi, M. (1989) Inhibition of stress-induced gastric injury by glutathione. Gastroenterology 97:853–859.
22. Ogino, T., Inoue, M., Ando, Y., Maeda, M., Awai, M., and Morino, Y. (1988) Chemical modification of superoxide dismutase: Extension of plasma half life of the enzyme through reversible binding to the circulating albumin. Int. J. Protein Peptide Res. 32:153–159.
23. Inoue, M., Ebashi, I., Watanabe, N., and Morino, Y. (1989) Synthesis of a SOD derivative that circulates bound to albumin and accumulates in tissues whose pH is decreased. Biochemistry 28:6619–6624.
24. Hirota, M., Inoue, M., Ando, Y., and Morino, Y. (1990) Inhibition of stress-induced gastric mucosal injury by a long acting superoxide dismutase that circulates bound to albumin. Arch. Biochem. Biophys. 280:269–273.
25. Palmer, R., Ferrige, A., and Moncada, S. (1987) Nitric oxide release accounts for the biological activity of endothelium-derived relaxing factor. Nature 327:524–526.
26. Sakuma, I., Stuehr, D., Gross, S., Nathan, C., and Levi, L. (1988) Identification of arginine as a precursor of endothelium-derived relaxing factor. Proc. Natl. Acad. Sci. USA 85:8664–8667.
27. Ress, D., Palmer, R., and Moncada, S. (1989) Role of endothelium-derived nitric oxide in the regulation of blood pressure. Proc. Natl. Acad. Sci. USA 86:3375–3378.

28. Vanhoutte, P. M. (1989) Endothelium and control of vascular function. Hypertension 13:658–667.
29. Nakazono K., Watanabe, N., Matsuno, K., Sasaki, J., Sato, T., and Inoue, M. (1991) Does superoxide underlie the pathogenesis of hypertension? Proc. Natl. Acad. Sci. USA 88:10045–10048.
30. Inoue, M., Watanabe, N., Sasaki, J., Tanaka, Y., and Amachi, T. (1990) Inhibition of oxygen toxicity by targeting superoxide dismutase to endothelial cell surface. FEBS Lett. 269:89–92.
31. Inoue, M., Watanabe, N., Sasaki, J., Matsuno, K., Tanaka, Y., Hatanaka, H., and Amachi, T. (1991) Expression of a hybrid Cu,Zn-type superoxide dismutase which has high affinity for heparin-like proteoglycans on vascular endothelial cells. J. Biol. Chem. 266:16409–16414.
32. Inoue, M. (1993) Targeting SOD by gene and protein engineering. Methods Enzymol. 233:213–221.
33. Adams, J. D., Jaresko, G. S., Louie, S. G., Klaidman, L. K., Kennedy, D., Sharma, O., and Boylen, C. T. (1993) Pneumocystis carinii pneumonia in HIV infected patients: Effects of the diseases on glutathione and glutathione disulfide. J. Med. 24:337–352.
34. Droge, W., Eck, H. P., and Mihm, S. (1992) HIV-induced cysteine deficiency and T-cell dysfunction—a rationale for treatment with N-acetylcysteine. Immunol. Today. 13:211–214.
35. Droge, W., Eck, H. P., Gmunder, H., and Mihm, S. (1991) Modulation of lymphocyte functions and immune responses by cysteine and cysteine derivatives. Am. J. Med. 91: S140–144.
36. Eck, H. P., and Droge, W. (1989) Influence of the extracellular glutamate concentration on the intracellular cyst(e)ine concentration in macrophages and on the capacity to release cysteine. Biol. Chem. Hoppe. Seyler 370:109–113.
37. Droge, W., Murthy, K. K., Stahl-Hennig, C., Hartung, S., Plesker, R., Rouse, S., Peterhans, E., Kinscherf, R., Fischbach, T., and Eck, H. P. (1993) Plasma amino acid dysregulation after lentiviral infection. AIDS Res. Hum. Retroviruses 9:807–809.
38. Droge, W., Eck, H. P., Gmunder, H., and Mihm, S. (1991) Requirement for prooxidant and antioxidant states in T cell mediated immune responses. Klin. Wochenschr. 69:1118–1122.
39. Jackson, R. C. (1969) Study on the enzymology of glutathione metabolism in human erythrocytes. Biochem. J. 111:309–315.
40. Richman, P., and Meister, A. (1982) Regulation of γ-glutamylcysteine synthetase by nonallosteric feedback inhibition by glutathione. J. Biol. Chem. 250:1422–1426.
41. Maeda, Y., Inaba, M., and Taniguchi, N. (1983) Increase of Na,K-ATPase activity, glutamate, and aspartate uptake in dog erythrocytes associated with hereditary high accumulation of GSH, glutamate, glutamine and aspartate. Blood 61:493–499.
42. Mihm, S., Ennen, J., Pessara, U., Kurth, R., and Droge, W. (1991) Inhibition of HIV-1 replication and NF-kappa B activity by cysteine and cysteine derivatives. AIDS 5:497–503.
43. Baruchel, S., and Wainberg, M. A. (1992) The role of oxidative stress in disease

progression in individuals infected by the human immunodeficiency virus. J. Leukocyte Biol. 52:111–114.
44. Passi, S., Morrone, A., Picardo, M., De-Luca, C., Bartoli, F., Zurlo, A., and Ippolito, F. (1990) Blood levels of vitamin E, polyunsaturated fatty acids of phospholipids, lipoperoxides, glutathione peroxidase activity and serological screening for syphilis and HIV in immigrants from developing countries. G. Ital. Dermatol. Venereol. 125:487–491.
45. Passi, S., De-Luca, C., Picardo, M., Morrone, A., and Ippolito, F. (1990) Blood deficiency values of polyunsaturated fatty acids of phospholipids, vitamin E and glutathione peroxidase as possible risk factors in the onset and development of acquired immunodeficiency syndrome. G. Ital. Dermatol. Venereol. 125:125–130.
46. Leff, J. A., Oppegard, M. A., Curiel, T. J., Brown, K. S., Schooley, R. T., and Repine, J. E. (1992) Progressive increases in serum catalase activity in advancing human immunodeficiency virus infection. Free Rad. Biol. Med. 13:143–149.
47. Malorni, W., Rivabene, R., Santini, M. T., and Donelli, G. (1993) N-Acetylcysteine inhibits apoptosis and decreases viral particles in HIV-chronically infected U937 cells. FEBS Lett. 327:75–78.
48. Levy, E. M., Wu, J., Salibian, M., and Black, P. H. (1992) The effect of changes in thiol subcompartments on T-cell colony formation and cell cycle progression: Relevance to AIDS. Cell Immunol. 140:370–380.
49. Roederer, M., Staal, F. J., Osada, H., and Herzenberg, L. A. (1991) CD4 and CD8 T cells with high intracellular glutathione levels are selectively lost as the HIV infection progresses. Immunology 3:933–937.
50. Roederer, M., Raju, P. A., Staal, F. J., Herzenberg, L. A., and Herzenberg, L. A. (1991) N-Acetylcysteine inhibits latent HIV expression in chronically infected cells. AIDS Res. Human Retroviruses 7:563–567.
51. Inoue, M., Okajima, K., and Morino, Y. (1981) Renal transtubular transport of mercapturic acid in vivo. Biochim. Biophys. Acta 64:122–128.
52. Okajima, K., Inoue, M., Itoh, K., Nagase, S., and Morino, Y. (1985) Role of plasma albumin in renal elimination of a mercapturic acid: Analysis in normal and mutant analbuminemic rats. Eur. J. Biochem. 150:195–199.
53. Morris, P. E., and Bernard, G. R. (1994) Significance of glutathione in lung disease and implications for therapy. Am. J. Med. Sci. 307:119–127.
54. Sala, R., Moriggi, E., Corvasce, G., and Morelli, D. (1993) Protection by N-acetylcysteine against pulmonary endothelial cell damage induced by oxidant injury. Eur. Respir. J. 6:334–336.
55. Buhl, R., and Vogelmeier, C. (1994) Therapy of lung diseases with anti-oxidants. Pneumologie 48:50–56.
56. Holroyd, K. J., Buhl, R., Borok, Z., Roum, J. H., Bokser, A. D., Grimes, G. J., Czerski, D., Cantin, A. M., and Crystal, R. G. (1993) Correction of glutathione deficiency in the lower respiratory tract of HIV seropositive individuals by glutathione aerosol treatment. Thorax 48:985–989.
57. Menasche, P., Pradier, F., Grousset, C., Peynet, J., Mouas, C., Bloch, G., and Piwnica, A. (1993) Improved recovery of heart transplants with a specific kit of preservation solutions. J. Thorac. Cardiovasc. Surg. 105:353–363.

58. Kobayashi, H., Kurokawa, T., and Kitahara, S. (1992) The effect of γ-glutamylcysteine ethyl ester, a prodrug of glutathione, on ischemia-reperfusion-induced liver injury in rats. Transplantation 54:414–419.
59. Jaeschke, H. (1993) The therapeutic potential of glutathione in hepatic ischemia-reperfusion injury. Transplantation 56:256–257.
60. Kawamoto, S., Inoue, M., Tashiro, S., Morino, Y., and Miyauchi, Y. (1990) Inhibition of ischemia and reflow-induced liver injury by an SOD derivative that circulates bound to albumin. Arch. Biochem. Biophys. 277:160–165.
61. Kawamoto, S., Tashiro, S., Miyauchi, Y., and Inoue, M. (1990) Mechanism for enterohepatic injury caused by circulatory disturbance of hepatic vessels in the rat. Proc. Soc. Exp. Biol. Med. 198:629–635.
62. Jaeschke, H. (1990) Glutathione disulfide formation and oxidant stress during acetaminophen-induced hepatotoxicity in mice in vivo: The protective effect of allopurinol. J. Pharm. Exp. Ther. 255:935–941.
63. Jaeschke, H. (1990) Glutathione disulfide as index of oxidant stress in rat liver during hypoxia. Am J. Physiol. 258:G499–505.
64. Dolphin, R., Poulson, R., Avramovic, O., eds. Coenzymes and Cofactors. John Wiley, New York.
65. Ando, Y., Inoue, M., Hirata, H., and Morino, Y. (1988) Dynamic aspects of glutathione metabolism in paraquat-treated rats. Biochem. Life Sci. Adv. 7:259–263.
66. Hoffer, E., Avidor, I., Benjaminov, O., Shenker, L., Tabak, A., Tamir, A., Merzbach, D., and Taitelman, U. (1993) N-Acetylcysteine delays the infiltration of inflammatory cells into the lungs of paraquat-intoxicated rats. Toxicol. Appl. Pharmacol. 120:8–12.
67. Ogino, T., Ando, Y., and Inoue, M. (1992) Molecular mechanism of paraquat toxicity. In: Active Oxygens and Diseases (Inoue, M., ed.), pp. 643–662. Gakkai Shutsupan Center, Tokyo.
68. Watanabe, N., Yu, H., Takahashi, K., Inoue, A., and Inoue, M. (1994) Inhibition of stress-induced apoptosis of thymocytes by glutathione and related thiols in vivo. (submitted).

14

Glutathionyl Specificity of Thioltransferases: Mechanistic and Physiological Implications

John J. Mieyal, Usha Srinivasan, and David W. Starke
Case Western Reserve University School of Medicine, Cleveland, Ohio

Stephen A. Gravina
Howard Hughes Medical Institute, Rockefeller University, New York, New York

Paul A. Mieyal
New York Medical College, Valhalla, New York

I. INTRODUCTION

A. Overview

Sulfhydryl biochemistry plays a remarkably broad and important role in cell biology. The redox status of cysteine sulfhydryl groups dictates the native structure and activity of many enzymes, receptors, protein transcription factors, and transport proteins. The ubiquitous sulfhydryl-containing tripeptide glutathione (GSH) and enzymes that utilize GSH play a central role in cellular defense against chemical oxidants, free radicals, and electrophilic species that perturb the sulfhydryl status. Other sulfhydryl-containing cofactors like coenzyme A and lipoic acid also are vital to many metabolically important enzyme reactions.

Much research has been focused on characterizing the physical and catalytic properties of a class of sulfhydryl enzymes called thiol-disulfide oxidoreductases (TDOR) that likely are involved in cellular sulfhydryl homeostasis through their catalysis of thiol-disulfide interchange reactions. The TDOR enzyme systems include thioltransferase (glutaredoxin), thioredoxin, and protein disulfide isomerase, along with their associated reductase enzymes (glutathione disulfide reductase and thioredoxin reductase) and their common cofactor NADPH. The TDOR systems have been shown to catalyze reduction of disulfide bonds in a wide variety of protein and nonprotein substrates under particular conditions. A preponderance of previous work has contributed to the impression that these enzyme systems have overlapping substrate specificities and possibly redundant physiological functions. However, current information on the substrate specificity and redox potentials of these enzymes, and on the redox states of cellular compartments, indicates characteristic differences among the TDOR systems that may be physiologically important. These differences suggest separate and/or synergistic functions rather than redundancy of action. This chapter emanates from a recent discovery of unusual selectivity of thioltransferase for glutathione-containing disulfide substrates. The evidence for this extraordinary property of thioltransferase that distinguishes it from thioredoxin is reviewed. The implications with respect to the mechanism of catalysis by thioltransferase and the potential physiological inferences are then discussed in separate sections of this chapter. For additional perspectives the reader is referred to a number of other reviews that have considered the TDOR enzymes (1–5).

B. Historical Perspective and Comparative Properties of the TDOR Enzymes

The net reaction catalyzed by thioltransferase is appropriately depicted as a thiol-disulfide exchange reaction involving nucleophilic displacement reactions rather than single electron transfer reactions that would involve radical intermediates. Accordingly, the original name "transhydrogenase" that was applied to the enzyme activity from rat liver (6) was replaced by the name "thioltransferase" because the latter more accurately represents the nature of the reaction catalyzed (7). In another context, Holmgren discovered an enzyme that catalyzed GSH-dependent turnover of ribonucleotide reductase in a mutant of *Escherichia coli* that lacked thioredoxin and named it "glutaredoxin" (8). Subsequent to those earlier studies, "thioltransferase" and "glutaredoxin" enzymes from a variety of organisms and mammalian tissues have been isolated and characterized, and a high degree of homology among the amino acid sequences has been noted for these proteins across species and tissues (see below) (9,10). This has led to the supposition that "thioltransferase" and "glutaredoxin" simply represent alternative names for the same family of enzymes (1). To facilitate the descriptions and discussions in this chapter, the name thioltransferase (TTase) will be used in most cases.

1. *Thioltransferases and Thioredoxins*

Thioltransferases and thioredoxins comprise a class of low molecular weight (ca. 12 kDa) and remarkably heat-stable enzymes found throughout nature. Although primarily localized to the cytosol, some association with membrane fractions has been indicated (11). Thought to be responsible for catalysis of reduction of disulfide bonds in a variety of protein and nonprotein substrates, the TDOR enzymes appear to play key roles in cell economy, especially in the proliferative stage and in response to oxidative stress. Thioltransferases and thioredoxins have been implicated in a wide variety of reactions involving disulfide interchange, ranging from reductive denaturation of insulin to modulation of enzyme and receptor activities and gene expression (1–5). Thioltransferases and thioredoxins are classified generally as TDOR enzymes, because thiol groups on the respective enzymes are believed to participate in the catalysis of reduction of disulfide substrates. Thus, a characteristic structural feature of these enzymes is an active-site pair of cysteine residues that can exist in reduced (dithiol) and oxidized (intramolecular disulfide) forms. In the oxidized state, thioltransferase and thioredoxin enzymes are protected against inactivation by thiol-selective electrophilic reagents like iodoacetate, because in their disulfide forms the active sites have lost their nucleophilic character and do not react with these reagents. The reduced enzymes are inactivated by iodoacetate with characteristic pH profiles and stoichiometries that reflect unusually low pK_a values for one of the two cysteine-SH groups at the respective active sites (thioredoxin-C-32, $pK_a = 6.7$; thioltransferase-C-22, $pK_a = 3.5$). These unusual pK_as relative to free cysteine ($pK_a = 8.3$) have been attributed to stabilization of the thiolate anion either by ion pairing or hydrogen bonding (12–15).

As noted above, a high degree of homology has been observed among the mammalian thioltransferase enzymes that have been sequenced (Table 1). The human RBC enzyme showed at least 80% homology with all of the other mammalian sequences published, i.e., pig liver, calf thymus, and rabbit bone marrow. The hRBC enzyme was much less homologous with the *E. coli* enzyme (22%). We have cloned human brain TTase and expressed it at high levels in *E. coli* (19). The cDNA sequence corresponds to the amino acid sequence of hRBC TTase (Table 1), and the activity of the recombinant enzyme matches that of natural hRBC TTase (19). According to the sequence alignment shown in Table 1, 23 residues are either identical or represent conservative replacements in common for all six enzymes, including yeast thioltransferase and *E. coli* glutaredoxin. Besides the CPY(F)C active site sequence, most of the other consensus involves hydrophobic residues, except for the two aspartate residues (D58 and D84 in the human sequence), that may play a role in activating the Cys22 thiol or stabilizing the thioltransferase-S-S-glutathione intermediate (see Sec. II). In this regard, the residue in the *E. coli* enzyme that corresponds to D84 was identified to interact electrostatically with the glutathionyl moiety in the mixed disulfide according to NMR analysis of the C14S glutaredoxin-SSG protein (20). Remarkably, all other residues identified as defining the region of interaction of the GS-moiety with the glutaredoxin protein are coincident with or adjacent to one of the 23 residues identified as identical in the six aligned sequences.

Table 1 Amino Acid Sequences of Thioltransferases from Various Species and Tissues[a]

			Active Site	
	1	10	20	30
1.	Ac-A--QEFV	NCKIQPGKVV	VFI--KPTCPYC	RRAQE-I LSQ
2.	Ac-A--QAFV	NSKIQPGKVV	VFI--KPTCPYC	RKTQE-LLSQ
3.	Ac-A--QEFV	NSKIQPGKVV	VFI--KPTCPYC	RKTQE-ILSQ
4.	Ac-A--QAFV	NSKIQPGKVV	VFI--KPTCPFC	RKTQE-LLSQ
5.	VSQETVAHV	KDLIGQKEVF	VFAAKTYCPYC	KATL STLFQE
6.		MQTV	I F-GRSGCPYC	VRAKD-LAEK
	40	50	60	70
1.	LP IKQGLLE	FVDI TATDHT	N EIQDYLQQL	TGAR--TVP R
2.	LPFKQGLLE	FVDI TAAGN I	SEIQDYLQQL	TGAR--TVPR
3.	LPFKQGLLE	FVDITATSDM	SEIQDYLQQL	TGAR--TVPR
4.	LPFKEGLLE	FVDITATSDT	NEIQDYLQQL	TGAR--TVPR
5.	LNVPKSKAL	VLELDEMSNG	SEIQDALEEI	SGQK--TVPN
6.	LSNERDDFQ	YQYVDIRAEG	ITKED-LQQK	AGKPVETVPQ
	80		90	100
1.	VFIGKDCIG	GCSDLVSLQ	QS-GELLTRLK	QIGAL-Q
2.	VFIGQECIG	CTDLVNMH	ER-GELLTRLK	QMGAL-Q
3.	VFLGKDCIG	GCSDLIAMQ	EK-GELLARLK	EMGAL RQ
4.	VFIGKECIG	GCTDLESMH	KR-GELLTRLQ	QIGALK
5.	VYINGKHIG	GNSDLETLK	KNGKLAEILK	PV
6.	IFVDQQHIG	GYTDFAAWV	K--- --ENLD	A

[a]Sequences were aligned according to the method of Higgins and Sharp (147).

	Homology (%)
1. HRBC thioltransferase (10)	100
2. Calf thymus glutaredoxin (9)	80
3. Rabbit bone marrow glutaredoxin (16)	83
4. Pig liver thioltransferase (17)	82
5. Yeast thioltransferase (13)	ca. 29
6. *E. coli* glutaredoxin (18)	ca. 22

Despite the many similarities between the thioltransferase and thioredoxin enzymes described above, there are characteristic differences. Thioredoxins have essentially no sequence homology with the thioltransferase proteins, and their respective antibodies do not cross-react, although the global conformations of thioredoxin and thioltransferase are considered to be similar according to x-ray data and modeling studies (2,3). The oxidized forms of thioltransferase and thioredoxin enzymes are reducible by dithiothreitol (DTT), but thioltransferases are distinguished from thioredoxins by their ability to be reduced by GSH. Moreover, oxidized thioredoxins are recycled by coupling to a specific flavoprotein thioredoxin reductase that does not reduce thioltransferase. The thioltransferase enzymes are recycled by coupling to GSH and a different flavoprotein, glutathione disulfide (GSSG) reductase, which does not mediate reduction of thioredoxin. Accordingly, the standard assay for thioltransferase has relied on coupling GSSG formation to NADPH utilization by GSSG reductase. As discussed in the next section, this approach has led to the general concept that thioltransferase does not discriminate among disulfide substrates. In fact, thioltransferase apparently displays a high degree of selectivity for glutathionyl disulfides and catalyzes their reduction more efficiently than does thioredoxin, which does not display a similar catalytic preference (21,22). This finding has many implications regarding potential physiological functions (see Sec. III).

2. *Protein Disulfide Isomerase*

Protein disulfide isomerase (PDI) is a TDOR enzyme believed to be responsible for mediating the correct posttranslational pairing of cysteine residues into intramolecular disulfides by catalyzing repeated oxidations and reductions (23,24). The PDI enzyme contains two domains with vicinal Cys residues (Cys-Gly-His-Cys), analogous to the active sites of thioredoxin (Cys-Gly-Pro Cys) and thioltransferase (Cys-Pro-Tyr-Cys). PDI is distinguished from the thioredoxin and thioltransferase enzymes by its size (MW = 57 kDa) and by its localization (lumen of the endoplasmic reticulum of eukaryotic cells). In addition, PDI is reducible both by GSH with GSSG reductase (like thioltransferase) and by thioredoxin reductase (25), but the latter seems not to be physiologically relevant. Thus, Lundstrom and Holmgren (26) interpreted the high redox potential of PDI to mean that it would be a good oxidant of nascent protein thiols, but by the same token the reduced form of PDI would not be reoxidized by $NADP^+$ via thioredoxin reductase.

3. *Relative Redox Potentials of the TDOR Enzymes*

Table 2 displays the relative redox potentials for the various TDOR enzymes, along with values for active site peptides (27) corresponding to a number of the enzymes. Three things are evident from these data. First, the redox potential of the intact protein usually differs from that of the active site peptide, indicating environmental influences on the active site dithiol moiety; however, Sandberg et al. (29) reported that the redox potentials of the disulfide bond appear to be similar in folded and unfolded *E. coli* glutaredoxin. Second, there is disparity in the values measured for PDI (25,26), warranting reevaluation. Third, despite the variation in the PDI value, it is clear that the redox potential of mammalian PDI is substantially higher than the redox potential for *E. coli* thioredoxin, and it is likely that mammalian thioredoxin will mimic the *E. coli* enzyme. Thus, the relationship among the redox potentials leads to the expectation that thioredoxin usually would act to reduce disulfides, whereas PDI (and DsbA) would promote oxidation, provided that the redox potentials of the cellular

Table 2 Redox Potentials of Thiol-Disulfide Oxidoreductase Enzymes

Enzyme	Redox potential (V)	
	Active site peptide	Intact protein
Bacterial thioredoxin	-0.190[a]	-0.270[b]
Mutant thioredoxin (P34H)		-0.235[b]
Bacterial glutaredoxin	-0.215[a]	(-0.230)[c]
Bacterial PDI (DsbA)		-0.089[d]
Mammalian thioredoxin		Unknown[e]
Mammalian thioltransferase		Unknown[e]
Mammalian PDI	-0.205[a]	-0.110[f]
	-0.205[a]	-0.175[b]

[a]From Ref. 27.
[b]From Ref. 26.
[c]By analogy to T4 glutaredoxin, which was actually measured (28).
[d]From Ref. 32.
[e]Redox potentials for eukaryotic thioredoxins and thioltransferases (glutaredoxins) have not been reported.
[f]From Ref. 25.

compartments where these enzyme work lie somewhere in between those of thioredoxin and PDI. Since the cellular distribution of thioredoxin may include the endoplasmic reticulum (11), it is conceivable that PDI and thioredoxin may act synergistically in catalyzing isomerization of protein disulfides, analogous to the DsbA/DsbB sytem of *E. coli* (30–32). Whether or not thioltransferase also may be involved in protein processing remains to be tested (see Sec. III. D).

Although useful in a broad sense, the thermodynamic relationship among redox potentials is not a sufficient criterion for distinguishing the various potential physiological functions of the TDOR enzymes, as discussed in several contexts in the following sections.

II. CHARACTERIZATION OF THIOLTRANSFERASE DISULFIDE SUBSTRATE SPECIFITY AND MECHANISM

A. Catalytic Properties of Thioltransferase

Early studies (33–35) reported catalysis by thioltransferase of glutathione-dependent reduction of a wide variety of disulfides when measured spectrophotometrically by a coupled assay. This standard assay relies on coupling the formation of the final product GSSG to oxidation of NADPH by the GSSG reductase (GRase) reaction (see Scheme I). As long as excess GSSG reductase is added, the coupled assay effectively measures the overall rate of formation of GSSG in 1:1 correspondence to the observed NADPH oxidation. As indicated by the equations, a nonenzymatic reaction occurs in the absence of thioltransferase, and this rate is subtracted from the overall rate to give the thioltransferase-mediated rate. Previous studies (35) also were interpreted to suggest that thioltransferase could catalyze both the first step and the second step in the two-step sequential formation of GSSG associated with disulfide reduction:

$$\mathrm{RSSR' + GSH \rightleftharpoons RSH + GSSR'}$$

$$\mathrm{GSSR' + GSH \rightleftharpoons GSSG + R'SH}$$

$$\mathrm{GSSG + NADPH \overset{GRase}{\rightleftharpoons} 2GSH + NADP}$$

Scheme I

According to the coupled assay, thioltransferase was shown to catalyze the formation of GSSG from various protein disulfides as well as low molecular weight disulfides in the presence of GSH. Thus, many putative disulfide substrates for thioltransferase have been reported, including cystine, cystamine, CoA-disulfide, hydroxyethyldisulfide, cysteinyl-bovine serum albumin mixed disulfide (BSA-SSCysteine), insulin, oxidized ribonucleotide reductase, and various glutathione-containing mixed disulfides (15,35,36). These data suggested that thioltransferase was capable of catalyzing reduction of both glutathione-containing as well as non–glutathione-containing disulfides, although catalysis of the first step of reduction of non–glutathione-containing disulfides (Scheme I) cannot be observed directly by the standard coupled assay. Thus thioltransferase had been characterized as a relatively nonspecific thiol-disulfide oxidoreductase. This led to the use of hydroxyethyl disulfide (HEDS), cysteine thiosulfate, and cystine as prototype substrates in the coupled assay system (2,14,35,37).

Furthermore, a general mechanism was proposed for catalysis of disulfide reduction by thioltransferase whereby the intramolecular disulfide form of the enzyme is an intermediate in the overall reaction (14). Glutathione would then function simply to reduce the enzyme disulfide and form GSSG to complete the cycle (Fig. 1). This mechanism would be expected to generate a ping-pong kinetic pattern (parallel lines on 1/V vs. 1/S plots) when one or the other of the substrates (i.e., disulfide RSSR′ or GSH) is varied while the other is added at several fixed concentrations.

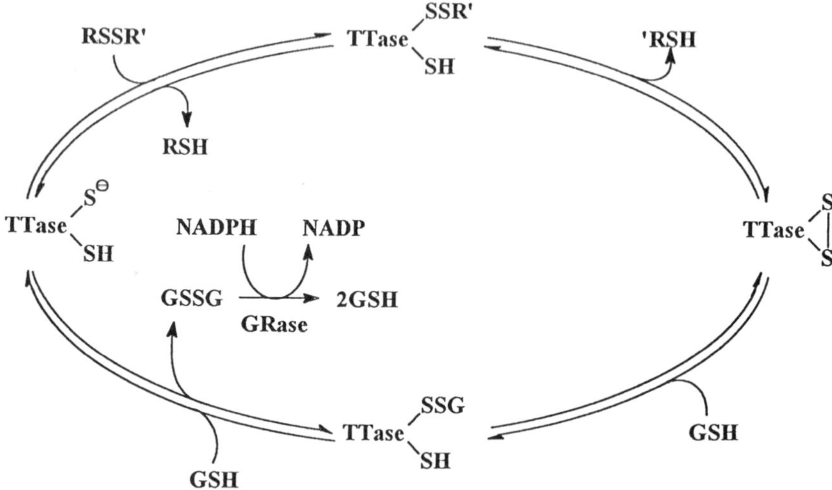

Figure 1 Previous hypothetical scheme for TTase catalysis of GSH-dependent disulfide reduction coupled to GSSG reductase (GRase). This scheme depicts the necessary involvement of the intramolecular disulfide form of TTase in the catalytic cycle. Kinetic studies (15) and mutagenesis studies (38) do not support this mechanism.

Instead of the ping-pong pattern predicted by Figure 1, a sequential kinetic pattern (intersecting lines, Fig. 2) was observed via the coupled assay with GSH and the non–glutathione-containing prototype disulfide substrate HEDS, suggesting that GSH played a more specific role in the reaction scheme than simply reducing the enzyme disulfide (15). Moreover, site-directed mutagenesis studies in which the second cysteine at the active site of pig liver thioltransferase was changed to a serine showed the mutant to have complete catalytic competence for GSH-dependent reduction of disulfide substrates (38). Thus, the intramolecular disulfide form of the enzyme (Fig. 1, right) does not appear to be a required intermediate (see Sec. II.C).

One alternative suggested by the kinetic pattern of Figure 2 was a mechanism whereby thioltransferase would form a ternary complex involving both the disulfide substrate and GSH before commitment to catalysis (15). This possibility led us to develop radiolabel and HPLC assays in order to examine the early stages of the thioltransferase reaction scheme. For the radiolabel assay protein mixed disulfides were prepared with radiolabel on the nonprotein thiol moiety, and the radiolabeled protein-S-SR* was incubated with GSH in the absence or presence of thioltransferase. Reactions were quenched with trichloroacetic acid (TCA), and the supernatants were analyzed for TCA-soluble radioactivity re-

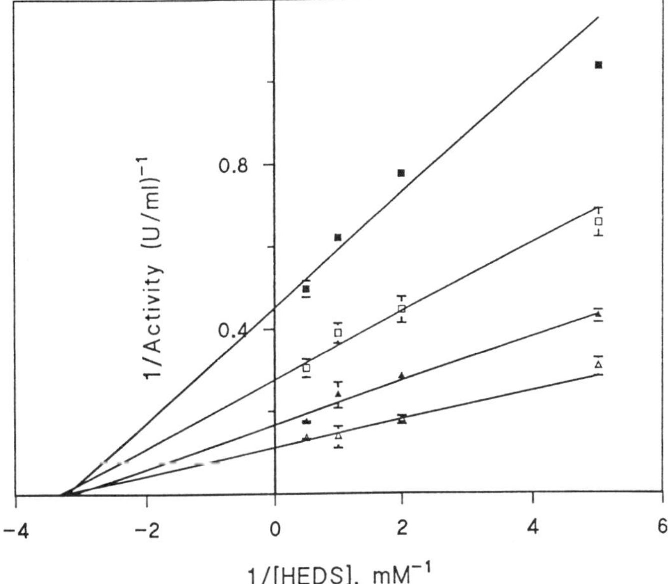

Figure 2 Two substrate kinetics of hRBC thioltransferase: dependence on [HEDS] at various [GSH]. Reactions (1 ml) were allowed to proceed at 25°C in 0.1 M potassium phosphate, pH 7.5, 0.2 mM NADPH, 2 units GRase, and 0.01 units hRBC thioltransferase, at various concentrations of HEDS and several constant GSH concentrations. The formation of GSSG was measured by the spectrophotometric assay. The GSH concentrations were as follows: 0.1 mM (closed rectangles); 0.2 mM (open rectangles); 0.5 mM (closed triangles); 1.0 mM (open triangles). Data are presented on a 1/V vs. 1/[S] plot. Each data point represents the mean of three separate experiments ± SE (15). Analogous patterns of intersecting lines were obtained for the converse experiment (i.e., varied [GSH], several constant HEDS concentrations), and analogous results were obtained when rat liver thioltransferase was substituted for hRBC TTase (15).

leased by reductive dethiolation of the protein-S-SR*. Thus, nonenzymatic and enzyme-mediated rates of dethiolation were measured (21). Initial results with bovine serum albumin-S-S-cysteine [^{14}C] showed no increase in the rate of radiolabel release in the presence of thioltransferase, even though reduction of the albumin-S-S-cysteine disulfide was apparently catalyzed by thioltransferase when measured by GSSG formation according to the standard spectrophotometric coupled assay. These data were not consistent with a seqential mechanism involving a ternary complex, and instead led us to hypothesize that thioltransferase catalyzes only the second step in the two-step reaction that leads to GSSG formation:

$$\text{BSA-S-SCys} + \text{GSH} \rightleftharpoons \text{BSA-SH} + \text{GS-SCys} \quad \text{(radiolabel assay)}$$

$$\text{GS-SCys} + \text{GSH} \xrightleftharpoons{\text{TTase}} \text{Cys-SH} + \text{GSSG}$$

$$\text{GSSG} + \text{NADPH} \xrightleftharpoons{\text{GRase}} \text{NADP}^+ + 2\ \text{GSH} \quad \text{(spectrophotometric assay)}$$

Scheme II

The results with BSA-SSCysteine also suggested a broader interpretation, namely, that thioltransferase catalysis of disulfide reduction is specific for glutathione-containing disulfides. The experimental basis for acceptance of this generalized conclusion is delineated next.

B. Evidence for Glutathionyl-Disufide Selectivity of Thioltransferase

Scheme III depicts possible reactions of GSH with protein mixed disulfides (23). Equations (a) and (b) show the potential radiolabel distributions when the disulfide is labeled on the nonprotein moiety. Equations (c) and (d) show the unlabeled disulfide reacting with radiolabeled GSH. The data presented below show that the GSH thiol reacts preferentially with the nonprotein sulfur [Eqs. (a) and (c)] and that thioltransferase apparently catalyzes the reaction only when R is a glutathionyl moiety.

$$\text{Protein–SSR}^* + \text{GSH} \rightleftharpoons \begin{cases} \text{Protein–SH} + \text{GSSR}^* & \text{(a)} \\ \text{or} & \\ \text{Protein–SSG} + {}^*\text{RSH} & \text{(b)} \end{cases}$$

$$\text{Protein–SSR} + {}^*\text{GSH} \rightleftharpoons \begin{cases} \text{Protein–SH} + {}^*\text{GSSR} & \text{(c)} \\ \text{or} & \\ \text{Protein–SSG}^* + \text{RSH} & \text{(d)} \end{cases}$$

Scheme III

1. *Reduction of Hemoglobin Mixed Disulfides as Prototypes*

Native hemoglobin has one sulfhydryl group on each β-subunit (cys-β93) that can form mixed disulfides with other thiol compounds. Thus, it is a well-defined and convenient model protein substrate for studying catalysis of disulfide reduction, and previous data indicated that TTase catalyzed dethiolation of the hemoglobin-S-S-glutathione mixed disulfide Hb-SSG (37). Comparison of the rates of GSH-dependent reduction of Hb-SSG [^{35}S] and Hb-SSCysteine [^{14}C] in the presence and absence of the enzyme showed that TTase did catalyze dethiolation of the glutathionyl substrate (Fig. 3A), but TTase (up to 0.6 units/ml) did

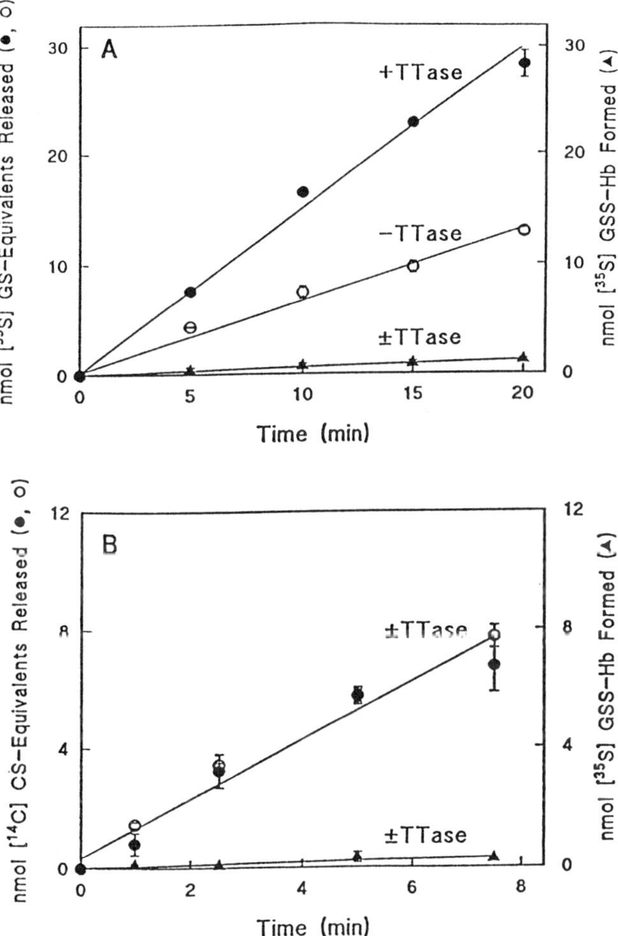

Figure 3 Time dependence of dethiolation of [^{35}S] oxyHb-SSGlutathione and [^{14}C] oxyHb-SSCysteine by GSH ± TTase. (A) Time dependence of dethiolation of [^{35}S] oxyHb-SSG by GSH ± TTase. Reaction mixtures (1 ml) at 30°C contained 0.1 M potassium phosphate, pH 7.5, 0.15 mM [^{35}S] oxyHb-SSG, 0.5 mM GSH, 0.2 mM NADPH, and 2 units/ml yeast GSSG reductase, in the absence or presence of 0.64 units/ml of hRBC TTase. Dethiolation of [^{35}S] oxyHb-SSG and incorporation of [^{35}S] GSH into unlabeled oxyHb-SSG were measured as described by Gravina and Mieyal (21). Open circles correspond to samples without TTase, closed circles represent TTase-containing samples. Triangles represent the incorporation of [^{35}S] GSH into Hb. The minimum detectable radiolabel incorporation would correspond to $<0.1\%$ of the theoretical maximum incorporation. Data have been normalized to the amount of product formed per 1 ml of reaction mixture. Each data point represents the mean of three separate experiments ± SE. Where error bars are not easily apparent, they occur within the symbols. (B) Time dependence of dethiolation of [^{14}C] oxyHb-SSCysteine by GSH ± TTase. Experiments were carried out as described under (A) above, except that [^{14}C] oxyHb-SSCysteine replaced [^{35}S] oxyHb-SSG. (From Ref. 21.)

not catalyze dethiolation of the cysteinyl mixed disulfide (Fig. 3B). When unlabeled Hb-SSG was reacted with radiolabeled GSH, negligible incorporation of [^{35}S] GSH into precipitated Hb (Fig. 3, closed triangles) was observed under the same conditions where substantial release of radioactivity from the labeled mixed disulfides of Hb was measured, confirming that GSH attacks on the nonprotein sulfur moiety of the disulfide substrate as depicted in Scheme III [Eqs. (a) and (c)]. Hb-SSCysteamine, another nonglutathione mixed disulfide, was also tested (Fig. 4). HPLC was used to monitor the reaction, because radiolabeled cysteamine was not available. GSH-dependent dethiolation occurred as observed by appearance of the Hb-SH peak, and no peak corresponding to Hb-SSG appeared. Thus, the only detectable reaction with GSH was conversion of Hb-SSCysteamine to Hb-SH, again showing preferential attack by GSH on the nonprotein S-atom of the disulfide. No catalysis by TTase was observed (see Fig. 4), even in the presence of relatively large amounts of the enzyme (0.6–1.6 units/ml).

2. *Comparative Study of Protein Disulfides*

To examine the selectivity of thioltransferase more thoroughly, three different sets of protein mixed disulfides were used, and both hRBC and rat liver thioltransferase were studied. Thioltransferase catalyzed the dethiolation of the [^{35}S] glutathione mixed disulfides of papain (Papain-S-SG [^{35}S]), bovine serum albumin (BSA-S-SG [^{35}S]), and hemoglobin (Hb-S-SG [^{35}S]). In contrast, thioltransferase did not catalyze GSH-dependent dethiolation of any of the non–GS-containing mixed disulfides, namely, Hb-S-SCysteine [^{14}C], papain-S-SCysteine [^{14}C], BSA-S-SCysteine [^{14}C], and Hb-S-SCysteamine. A maximum value for the potential catalytic activity of TTase for RS-SCysteine substrates may be estimated as $<2.5\%$ of its activity for RS-SG substrates as follows: no catalysis of BSA-SSCysteine by TTase was observed even at an enzyme concentration 40 times greater than that at which catalysis of BSA-SSG reduction was easily observed.

Table 3 summarizes the initial rate data for the various pairs of protein-SSCys and protein-SSG substrates assayed by release of radiolabel and by the spectrophotometric coupled assay where possible. For each pair, an accelerated rate of dethiolation with TTase was detected by the radiolabel assay only for the GS-containing derivatives. As expected, there was good agreement between the radiolabel and spectrophotometric assays with the glutathione-containing substrates. In contrast, with the cysteine-containing substrates, the two assays did not agree. In the absence of TTase, release of the radiolabeled cysteinyl moiety occurred more rapidly than formation of GSSG, consistent with the two-step reaction depicted in Scheme II, where the second step is rate limiting. According to this scheme, the catalysis of GSSG formation observed by the spectrophotometric assay requires a preenzymatic step. This requirement predicts that

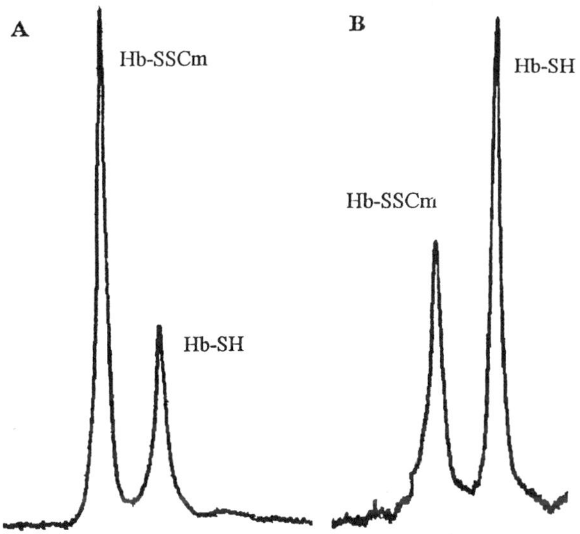

Figure 4 Time course of hemoglobin-SSCysteamine dethiolation by GSH measured by HPLC. The reaction mixture contained 0.1 M potassium phosphate, pH 7.5, 2.0 mM GSH, 0.2 mM NADPH, and 2 units/ml GSSG reductase, in the presence of 1.6 units/ml TTase at 30°C (specific activity 30 units/mg). The reactions were initiated by the addition of Hb-SSCysteamine (Hb-SSCm) to 0.15 mM. At various times, 10 μL of the reaction mixture was diluted 1:6 in deionized water and products were analyzed by anion exchange HPLC. Panels A and B show the product distribution at 30 and 120 min, respectively (21), and mean values ± SE from three determinations are tabulated below. Values of nmol/ml of Hb-SH formed at each time were calculated by multiplying fractional peak area (ΔHb-SH peak area/sum of Hb-SSCm + Δ Hb-SH peak areas) by the concentration of Hb-SSCysteamine in the original reaction mixture.

Time (min)	Hb-SH (nmol ml^{-1})	
	-TTase	+TTase
30	18.5 ± 1.0	19.5 ± 3.0 (hRBC) 17.3 ± 0.8 (rat liver)
120	52.5 ± 2.3	52.5 ± 2.3

non–GS-containing disulfides should display a lag phase in the spectrophotometric assay, but GS-containing mixed disulfides should not. Figure 5 shows a comparison of the initial time courses for GSH-dependent GSSG formation from BSA-SSCysteine and BSA-SSG measured spectrophotometrically. At equal concentrations of mixed disulfide, only BSA-SSCysteine showed a lag phase. These

Table 3 Rates of Dethiolation for Various Protein Mixed Disulfides ± Thioltransferase[a]

	Radiolabel assay		Spectrophotometric assay	
Substrate	Dethiolation rate (nmol/min/ml)	TTase added (units/ml)	GSSG formation rate (nmol/min/ml)	TTase added (units/ml)
Papain-SSG	7.2	none	7.7	none
	20.7	0.008	20.0	0.008
Papain-SSCys	65.3	none	3.0	none
	65.6	0.008–1.1	12.0	0.008
BSA-SSG	3.5	none	2.4	none
	6.2	0.016	5.7	0.016
BSA-SSCys	27.2	none	6.6	none
	27.6	0.01–0.64	13.2	0.008
oxyHb-SSG	0.64	none	N/A[b]	N/A
	1.4	0.64	N/A	N/A
oxyHb-SSCys	1.0	none	N/A	N/A
	0.94	0.64	N/A	N/A

[a]Rate values represent the slopes of regression analyses of the linear phases of product formation vs. time graphs except for papain-SSCys, where the values correspond to the extent of radioactivity released in 1 min. At least three determinations were made at each time point. All regression values were ≥0.98. Reaction mixtures at 30°C contained 0.5 mM GSH, 0.15 mM protein mixed disulfide, 0.2 mM NADPH, 2 units/ml GSSG reductase, and 0.1 M potassium phosphate pH 7.5 (except papain mixed disulfides were run in 0.1 M Tris HCl, pH 7.5). The spectrophotometric assay measures Δ A_{340nm} coupled to GSSG formation, and the radiolabel assay measures TCA-nonprecipitable counts (21).
[b]N/A, not applicable; spectrophotometric assays were not performed with the hemoglobin substrates, because the heme absorbance interferes with accurate determination of A_{340} changes.

results suggest that the GS moiety is incorporated into a mixed disulfide substrate before TTase catalysis occurs. Thus, with the protein-SSCysteine compounds, Cysteine-SSG is formed nonenzymatically and would be the actual substrate for TTase catalysis (Scheme II, step 2) as measured by the spectrophotometric assay. Accordingly, Cysteine-SSG was synthesized and found to be a superior substrate for TTase with a low apparent K_M (35 μM) and high turnover number (2200 min^{-1}) (21).

The data described above document a restrictive substrate specificity for thioltransferase involving a preference for mixed disulfides containing the glutathionyl moiety. We specifically focused on cysteine, GSH, and cysteamine among the various nonprotein thiols that might be considered physiologically relevant compounds of protein-SSR mixed disulfides, because the common features of their structures provided a stringent test of specificity (21). Although cysteamine occurs at much lower concentrations in cells than GSH, nevertheless it has been considered as a potential mediator of intracellular protein thiol-

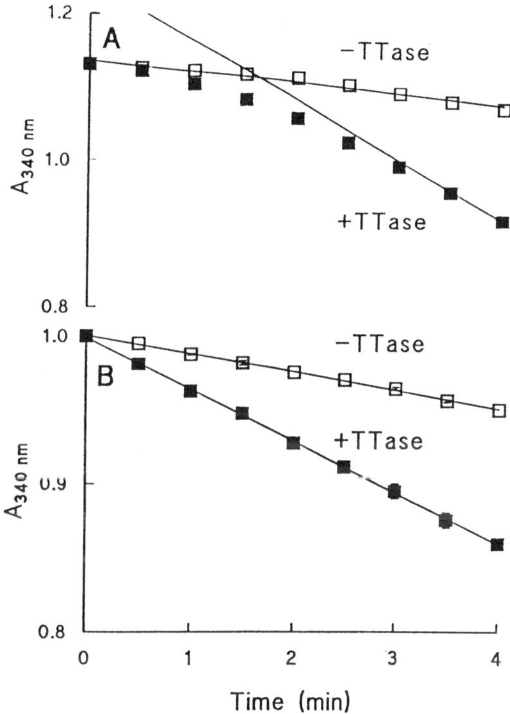

Figure 5 Initial time courses of GSH-dependent reduction of BSA-SSCysteine and BSA-SSGlutathione measured by the spectrophotometric coupled assay. (A) Initial time course for TTase-catalyzed GSSG formation from BSA-SSCysteine and GSH. Reaction mixtures (1 ml) at 30°C contained 0.1 M potassium phosphate, pH 7.5, 0.15 mM BSA-SSCysteine, 0.5 mM GSH, 0.2 mM NADPH, 2 units/ml yeast GSSG reductase, and 0.016 units/ml of hRBC TTase. Reactions were initiated by addition of BSA-SSCysteine with a mixing device and followed spectrophotometrically. The line represents extrapolation of the later linear portion of the time course (not shown completely). Data have been normalized to the amount of product formed per 1 ml of reaction mixture. Each data point represents the mean of three separate experiments ± SE. Where error bars are not easily apparent, they occur within the symbols. (B) Initial time courses for GSSG formation from BSA-SSG and GSH ± TTase. The upper line (open rectangles) corresponds to the time course in the absence of TTase, and the lower line (closed rectangles) corresponds to the time course in the presence of TTase. Experiments were performed as described in (A) above, except that BSA-SSG replaced BSA-SSCysteine. (From Ref. 21.)

disulfide exchange (5). Also, previously reported data from a study of total cytosolic protein from rat liver were interpreted to reflect thioltransferase-catalyzed reduction of cysteamine-containing disulfides (39); however, initial rates of dethiolation were not measured, so unambiguous interpretation is difficult.

Therefore, we examined the possibility that a cysteamine-containing disulfide could be a substrate for thioltransferase in two ways. First, the rates of GSH-dependent dethiolation of a discrete protein-cysteamine mixed disulfide Hb-SSCysteamine were measured in the presence and absence of thioltransferase, and this study revealed no catalysis by TTase (Fig. 4; Table 3). Second, the GSH-dependent reduction of cysteamine monitored by the spectrophotometric assay displayed a distinct lag phase analogous to Figure 5A. Indeed, results analogous to Figure 5 were found for other pairs of substrates, e.g., CoA-SSCoA vs. CoA-SSG, where the non–GS-containing disulfide only displayed a lag phase (data not shown). Such lag phases have been attributed previously to buildup of the GSSG substrate for the reporter enzyme GSSG reductase (35). No lag phases, however, were observed for any of the GS-containing substrates, including Cysteine-SSG, that were tested at concentrations as low as 5 μM using spinach GSSG reductase, which is highly specific for GSSG (21,22). Thus, the more likely explanation for appearance of a lag phase in the GSSG reductase–coupled assay of GSH-dependent reduction of non–GS-containing disulfides is the requirement for accumulation of a GS-containing substrate for the primary enzyme thioltransferase.

Rat liver thioltransferase exhibited the same catalytic characteristics as hRBC thioltransferase, i.e., only the GS-containing disulfides served as substrates (21). Thus, a selectivity for GS-containing disulfide substrates seems to be a general property of mammalian thioltransferases. Moreover, a recent arcticle stated that the preferred substrate in the hydroxyethyldisulfide assay for *E. coli* glutaredoxin was the glutathione-containing mixed disulfide, although no data were provided (40). Thus, there is an indication that the glutathionyl selectivity of thioltransferase (glutaredoxin) may extend to prokaryotes, but further study is necessary to determine whether it is a universal property of this class of enzymes (see Sec. II.D).

3. *Kinetics of Dethiolation of Glutathione-Containing Mixed Disulfides by hRBC Thioltransferase*

A wide range in catalytic efficiency (V_{max}/K_M) was observed for thioltransferase-mediated dethiolation of GS-containing substrates in the presence of 0.5 mM GSH (Table 4). Whereas thioltransferase displayed the highest efficiency for the small molecule Cysteine-SSG, its catalytic efficiency was substantially lower for the various protein-SSG substrates. The progression from Cysteine-SSG to oxyHb-SSG covers more than a 1500-fold range in efficiency, probably reflecting steric influences. For example, the difference between metHb-SSG and oxyHb-SSG likely illustrates the effect of steric hindrance by the substrate protein that interferes with accessibility to the enzyme active site. Each β-subunit of hemoglobin contains a cys-β93 moiety next to his-β92 that is ligated to the iron atom of the heme. Whereas the ferrous atom of oxyHb (diameter ~ 1.91 Å)

Table 4 Kinetics of TTase Catalysis of GSH-Dependent Dethiolation of Glutathionyl Mixed Disulfides[a]

Substrate	$K_{M\,(app)}$[b] (mM)	$V_{max\,(app)}$[b,c] (min^{-1})	V_{max}/K_M ($mM^{-1}\,min^{-1}$)
Spectral assay			
GSSCysteine	0.035	2200	63,000
Radiolabel assay			
BSA-SSG	0.44	1800	4,100
metHb-SSG	0.33	133	400
oxyHb-SSG	0.63	27	40

[a]Reaction mixtures at 30°C contained 0.1 M potassium phosphate, pH 7.5, 0.5 mM GSH, 0.2 mM NADPH, 2 units/ml GSSG reductase, in the presence or absence of hRBC TTase. The spectrophotometric assay measures the loss of NADPH at A_{340nm} coupled to GSSG formation, and the radiolabel assay mesaures TCA soluble counts (21).

[b]$K_{M(app)}$ and V_{max} values were determined by nonlinear fit of velocity vs. [substrate] curves.

[c]Apparent turnover numbers at 0.5 mM GSH were calculated by multiplying the V_{max} (μmol/min/unit TTase) by the specific activity (110 units/mg) and molecular weight (11.3 mg/μmol) of pure hRBC TTase.

fits into the porphyrin plane (~2.02 Å) and pulls his-β92 and cys-β93 toward the interior of the protein, the ferric iron atom of metHb (~2.06 Å) is 0.3 Å out of the porphyrin plane and makes the cys-β93 more solvent exposed (41). Predictably, the efficiency of dethiolation of metHb-SSG was more than 9-fold greater than that of oxyHb-SSG (Table 4). Extending this steric consideration to BSA-SSG suggests that the cysteinyl-S-S-glutathionyl moiety in that case is even more accessible to thioltransferase, because the efficiency of dethiolation was more than 10-fold greater than that of metHb-SSG (Table 4) (see Sec. II.C.4).

4. *Thioredoxin Catalysis of Protein Mixed Disulfide Dethiolation*

In contrast to thioltransferase, *E. coli* thioredoxin did not discriminate between the GS-containing and non–GS-containing disulfide substrates, e.g., the rates of dethiolation of Hb-SSG and Hb-SSCysteine by *E. coli* thioredoxin were similar (Table 5). Thus, the selectivity for the glutathionyl moiety appears to be a property characteristic of thioltransferase. An initial estimate of the relative efficiency of thioltransferase and thioredoxin for dethiolation of protein-SSG substrates was made from data with metHb-SSG as substrate, and this comparison favors thioltransferase. Thus 0.5 μM hRBC thioltransferase and 5.0 μM *E. coli* thioredoxin catalyzed dethiolation of metHb-SSG at rates of 6.9 and 1.5 nmol min^{-1} ml^{-1}, respectively, showing a nearly 50-fold greater turnover num-

Table 5 Rates of Dethiolation for Various Protein Mixed Disulfides ± Thioredoxin[a]

Substrate	Radiolabel assay (nmol/min/ml)	Thioredoxin (μM)	Spectrometric assay (nmol/min/ml)
Papain -SSCys	12.9	5	12.1
BSA-SSCys	8.0	0.5	8.3
metHb-SSG	1.5	5	N/A[b]
metHb-SSCys	1.2	5	N/A

[a]Rate values represent the slopes of regression analyses of the linear phases of product formation vs. time graphs. The values represent the mean of at least two determinations. In all cases the individual rate values deviated by 10% or less relative to the mean. Reactions were performed at 30°C in 0.1 M potassium phosphate pH 7.5 (except papain-SSCys; 0.1 M Tris HCl pH 7.5), 0.2 mM NADPH, 0.15 mM protein mixed disulfides, 0.1 μM *E. coli* thioredoxin reductase ± *E. coli* thioredoxin. No reactions occurred in the absence of thioredoxin. Spectrophotometric assay measures the loss of NADPH at A_{340nm}, and the radiolabel assay measures TCA-soluble counts (21).
[b]N/A, not applicable; spectrophotometric assays were not performed with the hemoglobin substrates, because the heme absorbance interferes with accurate determination of A_{340} changes.

ber for thioltransferase (i.e., 14 min^{-1} vs. 0.3 min^{-1}). Although this comparison suggests that thioltransferase is the more effective catalyst of dethiolation of protein-SSG substrates, a more systematic study in which the relative K_M and V_{max} values are obtained for a number of different substrates is necessary to establish their relative efficiencies. A number of studies have compared the effectiveness of the thioltransferase and thioredoxin systems in other contexts (42–46). For example, Holmgren (42) reported that glutaredoxin was about 10 times more effective than thioredoxin as the dithiol reductant of oxidized ribonucleotide reductase, and Mannervik et al. (43) reported that the thioltransferase system was more effective in reducing small molecules, but the thioredoxin system reduced insulin more effectively. Flamigni et al. (45) reported that oxidized ornithine decarboxylase could be reactivated by both the thioltransferase and thioredoxin systems. Yoshitake et al. (46) interpreted their data to mean that thioltransferase was more effective at reactivating sulfhydryl enzymes that were modified by mixed disulfide formation, whereas thioredoxin more effectively repaired enzymes whose thiol groups were oxidized to the sulfenic or sulfinic acids. In most of these examples, however, the actual nature of the substrates being acted upon by the thioltransferase and thioredoxin enzymes was not characterized directly and substrate and enzyme concentrations were not varied, so they retain ambiguity with regard to relative substrate specificity and relative efficiency.

C. Mechanistic Interpretation of Thioltransferase Catalysis

Figure 6 represents a catalytic scheme for thioltransferase-catalyzed disulfide reduction that is consistent with current data. The central portion of this scheme depicts the simplest mechanism for TTase catalysis when a GS-containing disulfide and GSH are the cosubstrates. According to this scheme TTase is interconverted between the reduced thiolate form and a TTase-SSG intermediate. The occurrence of a TTase-SSG intermediate has broad implications regarding the potential physiological roles of thioltransferase-mediated glutathionyl transfer reactions in homeostasis and regulation (see Sec. III).

This reaction scheme (Fig. 6) predicts that when a GS-containing disulfide and GSH are the substrates for thioltransferase, a ping-pong pattern of two

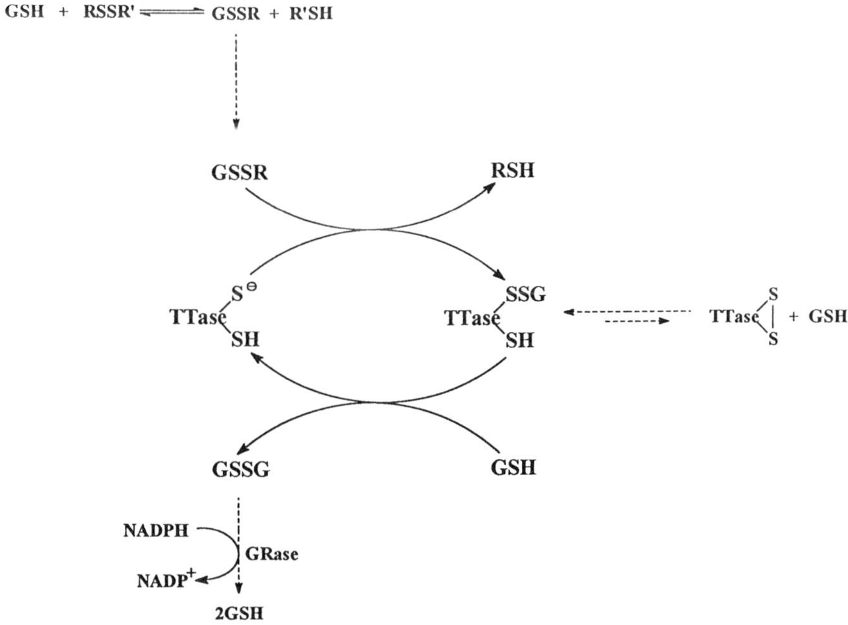

Figure 6 Mechanism of thioltransferase action. The central portion of this scheme depicts the simplest mechanism for TTase catalysis when a GS-containing disulfide and GSH are the cosubstrates. TTase is interconverted between the reduced thiolate form and a TTase-SSG intermediate. The upper portion of the figure shows the preenzymatic formation of GSSR consistent with a sequential kinetic pattern when a non–GS-containing substrate is present (15). At the right is the possible formation of the intramolecular disulfide form of TTase that is readily converted to TTase-SSG by GSH. Coupling with GSSG reductase is shown at the bottom.

substrate kinetics (parallel line pattern) should be observed. Therefore the variation in dethiolation rates were measured as a function of the concentration of the GS-containing disulfide substrate at several fixed concentrations of GSH, and vice versa. Cysteine-SSG and metHb-SSG were tested separately as the disulfide substrate (Fig. 7). As shown, both experiments gave parallel line patterns with hRBC TTase, and analogous results were observed with rat liver TTase (21,22), supporting the concept that a common catalytic mechanism applies to the thioltransferases (Fig. 6). The upper portion of Figure 6 represents the necessary preenzymatic step for a non–GS-containing disulfide in which GSH reacts

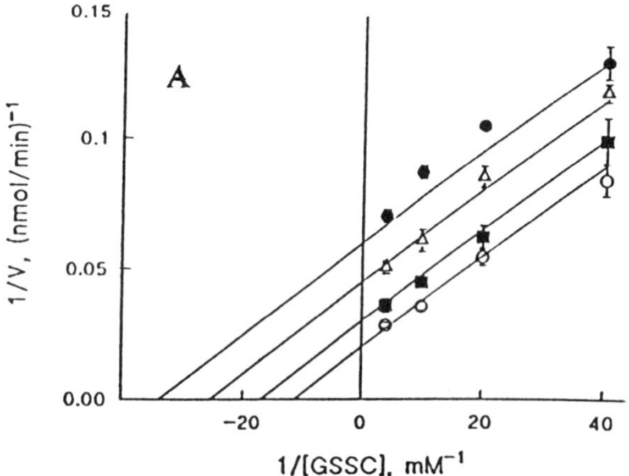

Figure 7 Two-substrate kinetics: dependence of hRBC TTase activity on [GSSCysteine] at various GSH concentrations. Rates of GSSG formation were measured in the standard spectrophotometric assay as a function of GSSCysteine concentration at several fixed GSH concentrations (21). The GSSCysteine concentrations are shown; the GSH concentrations were as follows: 0.5 mM (solid circles); 0.75 mM (open triangles); 1.0 mM (solid rectangles); 1.5 mM (open circles). Data have been normalized to the amount of product formed per 1 ml of reaction mixture. Each data point represents the mean of four separate experiments ± SE. Where error bars are not easily apparent, they occur within the symbols. (B) Two-substrate kinetics: dependence of hRBC TTase activity on [MetHb-SSG] at various GSH concentrations. Rates of GSSG formation were measured in the standard radiolabel assay as a function of Hb-SSG [^{35}S] concentration at several fixed GSH concentrations (22). The Hb-SSG concentrations are shown; the GSH concentrations were as follows: 0.25 mM (solid circles); 0.5 mM (open triangles); 1.0 mM (solid rectangles); 2.0 mM (open circles). Data have been normalized to the amount of product formed per 1 ml of reaction mixture. Each data point represents the mean of three separate experiments ± SE. Where error bars are not easily apparent, they occur within the symbols.

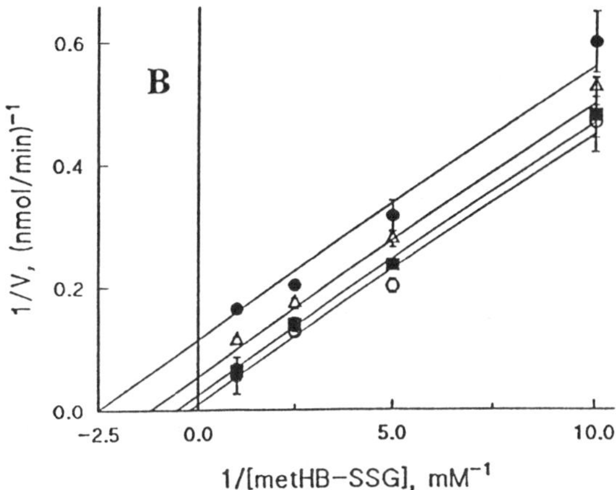

directly to form a GS-containing disulfide as the actual substrate for thioltransferase. This preenzymatic reaction is consistent with the sequential pattern of two substrate kinetics (Fig. 2) observed for non–GS-containing disulfides and GSH, as described above. Although thioltransferase could form an intramolecular disulfide (Fig. 6, right), high intracellular concentrations of GSH would drive the equilibrium toward TTase-SSG and reduced TTase. Completion of the cycle in the reducing direction would be facilitated also by coupling with GSSG reductase, which converts GSSG back to GSH as shown at the bottom of Figure 6.

1. Basis for Thiolate Reactivity of the Active Site Cysteine

The first step in thioltransferase catalysis appears to be nucleophilic attack on the disulfide substrate by the Cys-22 thiolate moiety, releasing the first thiol product. This step is facilitated by the low apparent pK_a (3.5) of this cysteine (13,15,38). Previous studies aimed at identifying the amino acid residues involved in thiolate formation at the active site of pig liver TTase focused on ion pairing between Cys-22 and neighboring positively charged Arg-26 and Lys-27 residues (38). Residues were changed by site-directed mutagenesis to produce two single mutants, R26V and K27E, and the corresponding double mutant. The R26V and K27E mutants had 32% and 67% catalytic activity relative to wild type as measured by the spectrophotometric coupled assay. Titration of pH-dependent inactivation by iodoacetate of the mutant K27E showed little change in the apparent pK_a of Cys-22, whereas the R26V mutant showed no inactivation from pH 2.5–8.5, indicating a dramatic change in pK_a or a change in accessibility of the Cys-22 thiol moiety. Although these data seem to support the salt bridge model for thiolate formation, the authors described solubility problems for the Arg mutant. Moreover, some increase in the rate of inactivation by iodoacetate should have been noted between pH 7.0 and 8.5 (assuming a pK_a for Cys-22 of ~8.5, i.e., a typical value for a cysteine residue that is not ion-paired). Nevertheless, Wells et al. have provided cogent arguments in favor of the ion-pairing hypothesis on the basis of predicted and observed changes in pI with selective mutations of the pig liver thioltransferase (4).

An alternative interpretation of the reactivity of Cys-22, however, is hydrogen bonding of the thiol proton of Cys-22 to a carboxylic residue with a typical pK_a of 3.5 (13,15). Since hydrogen bonds are resistant to ionic strength effects, whereas ion pairs are more sensitive, we studied pH-dependent iodoacetamide inactivation of thioltransferase at ionic strengths from 0.055 to 2 M (15). An analogous study showed papain inactivation by chloroacetamide to be ionic strength sensitive, consistent with an ion pair mechanism of thiolate stabilization (47). In contrast we found no effect of ionic strength on the apparent pK_a of thioltransferase inactivation. Furthermore, sequence analysis shows an aspartic acid residue at position 58 (position 47 for *E. coli*) which is conserved in thioltransferases from *E. coli* to humans. This Asp residue was recently shown by NMR to be near the active site thiol of *E. coli* glutaredoxin (48). Further sequence analysis shows that the homologous Arg-26 and Glu-27 residues of the mammalian thioltransferases are replaced by Glu and Ala or Val and Arg in the yeast and *E. coli* enzymes, respectively, and that the Arg of *E. coli* glutaredoxin is not near the active site thiol (48). Accordingly, further study is needed to resolve the ambiguity regarding the mechanism of thiolate stabilization at the active sites of the thioltransferases.

2. *Evidence for Intramolecular Disulfide and Mixed Disulfide Forms of TTase*

Thioltransferases contain a characteristic dithiol tetrapeptide at their active sites (9,10). Nucleophilic attack by one of the active site cysteines on the disulfide substrate releases the first thiol product, and the enzyme can then form an intramolecular disulfide with the second cysteine and release the second thiol product as depicted in Figure 1 (14). Treatment of reduced TTase with [^{14}C] cystine protected TTase from inactivation by iodoacetate without incorporation of radioactivity, suggesting formation of an intramolecular disulfide on TTase (14). Although a TTase intramolecular disulfide is chemically possible, its participation in the typical catalytic mechanism is doubtful for the reasons already reviewed above. Instead, a TTase-SSG intermediate seems more likely (Fig. 6, central cycle). Consistent with this concept, Eriksson et al. (49) reported that incubation of a partially purified preparation of oxidized TTase with [^{35}S] GSH resulted in incorporation of radiolabel into a TTase derivative that was isolable by ion exchange chromatography, suggesting formation of TTase-SSG [^{35}S]. In this context also, Yang and Wells (50) offered an alternative catalytic mechanism for the mutant TTase lacking the second cysteine residue. The intramolecular disulfide form of TTase in their usual mechanism was replaced by TTase-SSG. Moreover, a recent article on *E. coli* glutaredoxin (40) suggested a ping-pong mechanism analogous to that in Figure 6.

As an initial approach to obtain direct evidence for the TTase-SSG intermediate, we conducted preliminary experiments in which reduced hRBC TTase was treated with Cysteine-SSG [^{35}S] in the absence of GSH; then aliquots of the reaction mixture were precipitated periodically with TCA over a time course of 3 min and radioactivity in the TTase-containing pellets was measured. Rapid incorporation of radiolabel into TTase was observed ($\leq$30 s), followed by release of the label. These data are consistent with rapid formation of TTase-SSG [^{35}S] followed by conversion to the TTase-(S-S) intramolecular disulfide (22). We have also treated reduced TTase with [^{35}S] GSSG in the absence of GSH, followed by addition of excess iodoacetamide, then gel chromatography. Coelution of some of the radiolabel with TTase in this case suggested isolation of a TTase-SSG [^{35}S] derivative that was stabilized by alkylation of the Cys-25 residue (51). *E. coli* thioredoxin treated in the same fashion with [^{35}S] GSSG and quenched with iodoacetamide in less than 5 s did not incorporate any radiolabel, even though a substantial fraction of the thioredoxin was oxidized to the iodoacetamide-resistant intramolecular disulfide form by this brief exposure to GSSG. These experiments reveal a clear difference between the reactivity of the active site dithiol moieties of TTase and thioredoxin with GSSG and support the concept of a TTase-SSG intermediate that is formed rapidly and is sufficiently stable to be observed and trapped. Although these several lines of evidence support the intermediate role of TTase-SSG in the catalytic scheme (Fig. 6), further study is necessary to isolate and characterize this derivative and establish its kinetic competence. For example, evidence for the kinetic competence of a TTase-SSG intermediate would be obtained from measuring the rates of transfer of the GS moiety to an acceptor RSH; i.e., if a TTase-SSG intermediate is involved in reversible catalysis of transfer of the glutathionyl moiety, then the rate of transfer of a GS moiety from TTase-SSG to an acceptor RSH must be at least as fast as the overall rate of the catalyzed reaction.

3. *Alternative Reducing Substrates*

Figure 6 shows the reaction of TTase-SSG with GSH to complete the catalytic cycle. Although GSH is the probable physiological reducing substrate because of its intracellular abundance relative to other thiols, any thiol should be able to replace GSH as the reductive regenerator of reduced TTase according to the principle of microscopic reversibility; i.e., other thiols besides GSH should be capable of acting as the second substrate because this is equivalent to the reverse reaction (Scheme IV, right).

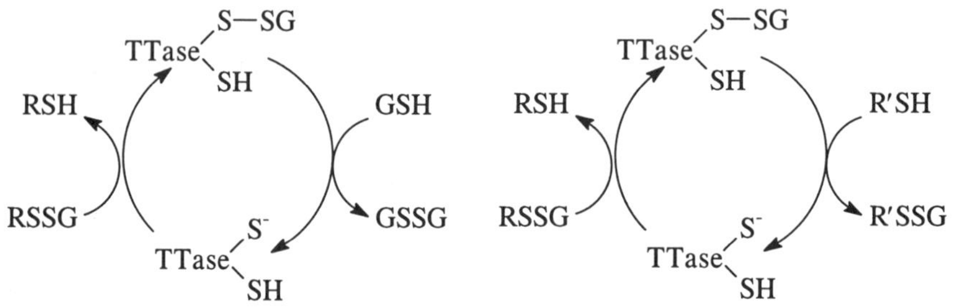

TTase(S–SG)(SH) RSH GSH RSSG GSSG TTase(S⁻)(SH)

TTase(S–SG)(SH) RSH R′SH RSSG R′SSG TTase(S⁻)(SH)

Scheme IV

Consistent with this interpretation, Axelsson and Mannervik (36) reported that a variety of thiol compounds could substitute for GSH in supporting TTase catalysis of release of radiolabel from rat liver cytosolic proteins that had been reacted with radiolabeled GSSG, including CoA, cysteamine, cysteine, dithiothreitol, mercaptoethanol, penicillamine, and thioglycolate. In addition, we demonstrated that TTase does catalyze the reverse of GSH-dependent reduction of hydroxyethyldisulfide, i.e., β-mercaptoethanol–dependent reduction of GSSG; as expected the same position of equilibrium is reached in the forward and reverse directions in the absence and presence of TTase (15). Using a discrete disulfide substrate BSA-SSG [^{35}S], we have confirmed the broad tolerance of TTase for reducing substrates, with notable exceptions. Thus, TTase-mediated dethiolation of BSA-SSG [^{35}S] was observed when GSH was replaced by cysteine, 3-mercaptopropionic acid, β-mercaptoethanol, and trifluoroethanethiol, but two aromatic thiols—ergothionine and *p*-nitrothiolphenol—did not support catalysis, although they reacted directly with the BSA-SSG (22,52).

4. *pH Profiles Indicative of the Rate-Limiting Step in TTase-Catalyzed Disulfide Reduction*

As described above, the apparent pK_a for the active site thiol of the thioltransferases is about 3.5, but the pH dependence of TTase-catalyzed disulfide reduction with GSH as the cosubstrate displays an inflection point near pH 8–8.5 (15,52,53). The distinct separation between these pH profiles suggests that the reaction of the active site cysteine of TTase with the disulfide substrate is not involved in the rate-determining step of the overall reaction. Furthermore, the similarity of the reaction pH profile with the titration of the GSH thiol moiety suggested that the rate-limiting step involves nucleophilic attack by the glutathionyl thiolate species. Since the previously reported pH profiles utilized the non–GS-containing disulfide hydroxyethyldisulfide (HEDS), it was not clear whether the pH dependence of this reaction might reflect the titration of GSH necessary for the preenzymatic formation of the GS-containing substrate for TTase, or whether the reaction of GSH with the TTase-SSG intermediate was rate determining (Fig. 6). To resolve this ambiguity we substituted the GS-containing substrate Cysteine-SSG for HEDS and repeated the pH rate profile. The results with Cysteine-SSG were identical to those with HEDS, suggesting that the regeneration of reduced TTase from the TTase-SSG intermediate is the rate-limiting step of the overall reaction (52). This would explain why most nonprotein substrates tested previously by the spectrophotometric assay displayed similar apparent turnover numbers—within a factor of two (15). Even closer agreement is noticed when GS-containing substrates are compared, which avoids the complication of the lag phase for preenzymatic formation of GS-disulfides from non–GS-containing disulfides. Thus, Cysteine-SSG, CoA-SSG, and BSA-SSG gave turnover numbers of 2200, 1830, and 1800 min^{-1}, respectively (15, 21). The comparison breaks down when the glutathionyl mixed disulfides of Hb are included (metHb-SSG and oxyHb-SSG gave turnover numbers of 133 and 27 min^{-1}) (Table 4), most likely indicating a change in rate-limiting step duc to difficult accessibility for interaction of the TTase-active site with the Hb-SSG disulfide.

To test the interpretation concerning the rate-determining step further, pH profiles for TTase-catalyzed dethiolation of BSA-SSG [^{35}S] were measured by the radiolabel assay with several alternative reducing substrates whose pK_a values range on both sides of the pK_a of GSH (pK_a = 8.7). Thus, trifluoroethanethiol has a pK_a of 7.5, and β-mercaptoethanol has a pK_a of 9.8. If the pH profile of the enzyme-catalyzed reaction simply reflects titration of the thiol moiety of the compound that converts the TTase-SSG intermediate back to the TTase-S^- form, then the pH-rate profiles for TTase catalysis of BSA-SSG dethiolation should correspond to the titration of the respective thiol groups of the reducing substrates. The data in Table 6 document this correspondence. The inflection points (apparent pK_as) for the pH-dependent variation in both the nonenzymatic and the TTase-catalyzed dethiolation rates rates parallel the pK_a values of the respective thiol compounds (52). A difference of about 0.4 pH units was noted between the pK_a values obtained by pH meter titration at room temperature and those from the pH-rate profiles at 30°C, probably reflecting, at least in part, the temperature coefficient for thiolate formation in each case.

D. Basis for the Selectivity of Thioltransferase for Glutathionyl Disulfides

The distinction between the substrate selectivities of thioltransferase and thioredoxin conceivably relates to one or more of several factors: (1) differences in redox potentials, (2) a glutathionyl-recognition site on thioltransferase but not

Table 6 Variation of pH Profile for TTase-Catalyzed BSA-SSG Reduction with pK_a of Thiol Substrate[a]

Thiol substrate	Apparent pK_a of thiol moiety[b]	Apparent pK_a of nonenzymatic reaction[c]	Apparent pK_a of TTase-catalyzed reaction[c]
Glutathione	8.7	8.1 (0.93)	8.3 (0.97)
β-mercaptoethanol	9.8 (0.92)[d]	9.3 (0.93)	9.4 (0.8)
Trifluoroethanethiol	7.5 (0.91)	7.0 (0.92)	7.0 (0.92)

[a]S-Carboxymethyl bovine serum albumin (Sigma) was derivatized at the amino terminus with *N*-hydroxysuccinimide–activated 3-mercaptopropionic acid disulfide, then the disulfide was converted to the mixed disulfide with [^{35}S]-glutathione to give radiolabeled BSA-NH-CO-CH_2-CH_2-S-*SG (52). The pH dependence of dethiolation of the BSA-S*SG was assessed by incubating 0.1 mM BSA-SSG for 10 min. at 30°C with the various thiol compounds in the appropriate buffers [MES for pH 4–7, HEPPSO for pH 7–8, AMPSO for pH 8–10, and CAPS for pH 10–12 (Sigma)], in the absence (nonenzymatic) and presence of partially purified hRBC thioltransferase (0.08–0.17 units). The ionic strength was adjusted to 0.3 M in all cases with NaCl, and the reactions were initiated with the thiol substrate (0.5–4 mM). Reactions were terminated periodically by addition of two volumes of ice cold 20% trichloroacetic acid followed by centrifugation, and an aliquot of the supernatant was measured for released radiolabel.

[b]The apparent pK_a for thiolate formation for GSH is a standard reported value, and data for the pK_a values of β-mercaptoethanol and trifluoroethanethiol were obtained conductometrically with a pH meter by titration with 1 M NaOH, 0.3 M ionic strength, at room temperature.

[c]Initial rates of dethiolation were determined and plotted as a function of pH in order to generate the pH profiles for the nonenzymatic and TTase-catalyzed dethiolation reactions for deriving the corresponding apparent pK_a values.

[d]Apparent pK_a values were calculated by nonlinear least-square fitting of the data to a modified form of the Hill equation (54), and correlation coefficients (shown in parentheses) were calculated using a curve-fitting program (148).

on thioredoxin, (3) steric constraints on the protein substrates that differentially limit access of the respective active sites of thioltransferase and thioredoxin, and (4) the nature and stability of covalent enzyme-substrate intermediates. The latter two possibilites seem more likely, as discussed below.

1. Redox Potential Differences

It is unlikely that the substrate selectivity of thioltransferase could be based on differences in the redox potentials of the protein-Cys-S-SR mixed disulfides that were tested as substrates. The protein-cysteinyl moiety was common to all of them, and the redox potentials for the RS moieties [e.g., glutathione (205 mV); cysteamine (203 mV) (55)] are too similar to account for the observed selectivity.

2. Glutathionyl Recognition Site

The classical interpretation of substrate selectivity for an enzyme involves a substrate-recognition site comprised of a compatible conformational arrangement of amino acid side chains that are conducive to the binding and/or bond-making and bond-breaking events associated with catalysis. In this regard there have been a number of studies that have addressed the question of a glutathione site on the thioltransferase (glutaredoxin) enzymes, but these have shortcomings that limit their straightforward interpretation, including the faulty premise that GSH itself has selective affinity for the enzyme (see Sec. II.C.3).

Two nonmammalian thioltransferases have been examined for "glutathione-binding sites" on the respective proteins. A picture of a proposed interaction between reduced GSH and the active site of T4 glutaredoxin was derived from molecular modeling and site-directed mutagenesis experiments (56). This model appears to have several limitations. First, it was based on the assumption that the intramolecular disulfide form of the enzyme is a true intermediate in the reaction scheme where GSH interacts. Although reduction of this form may be a prerequisite for maintaining a sufficient quantity of the enzyme in an efficient catalytic cycle, it would appear not to be directly involved in the catalytic cycle as described above (Fig. 6). Second, "binding" of GSH was assessed indirectly according to changes in the kinetic parameters for thioltransferase activity measured with the nonglutathionyl substrate hydroxyethyldisulfide. Third, the magnitude of the changes associated with removal of proposed key stabilization interactions were not very large; e.g., with the substitutions for His-12, the K_M changes were within a factor of two and k_{cat} within a factor of 3, so that catalytic efficiency was diminished only by a factor of 2–5. Fourth, since the changes were in the region of the active site thiolate (Cys-14), changes in its environment could affect catalysis without affecting the binding of GSH per se.

In a different study, NMR data for the Cys-12 glutathionyl mixed disulfide derivative of the C14S mutant of *E. coli* glutaredoxin were analyzed for interactions of the glutathionyl moiety with protein residues (20). The GS-moiety was described as localized on the surface of the protein in a cleft bounded by the amino acid residues Y13, T58, V59, Y72, T73, and D74, which appear to participate in H-bonds and electrostatic interactions with the GS-moiety. The active site locale of this mixed disulfide form of the C14S mutant was interpreted to resemble the intramolecular disulfide form (C12-C14) of the protein more than the reduced dithiol form (20), and it could be inferred from the authors' discussion that they interpreted their results as well as our characterization of the glutathionyl-disulfide selectivity of TTase (21) as being indicative of a special affinity of the thioltransferase enzymes for reduced glutathione. Instead, as evidenced by the study of alternative reducing substrates, there appears not to be a selectivity of TTase for GSH even though it is likely that GSH is the physiological reducing substrate. This behavior is analogous to the situation with the glutathione peroxidase enzyme. Like TTase (21,22), glutathione peroxidase displays ping-pong kinetics for its oxidized and reduced substrates (57). Also like TTase, the dependence of the glutathione peroxidase reaction on GSH concentration over the attainable concentration range showed little evidence of saturation, and other thiols can substitute for GSH (57). In addition, no radioactivity was found associated with glutathione peroxidase after incubation with radiolabeled GSH, followed by gel filtration (57). Likewise we found no binding of reduced thioltransferase to an affinity column in which GSH in its reduced form is attached to a spacer arm via the terminal amino group of the tripeptide (58). In contrast, glutathione-S-transferase binds avidly to this type of column (59). Moreover, neither S-methyl glutathione nor S-nitroso glutathione (up to 5 mM) appear to be inhibitors of mammalian thioltransferase (60), although Hoog et al. (61) reported competitive inhibition of *E. coli* glutaredoxin by S-substituted glutathione analogs. There appears to be a remarkable difference between enzymes like glutathione-S-transferase and glyoxylase, which utilize bound GSH as the primary nucleophile for adduction of the second substrate, and enzymes like thioltransferase and glutathione peroxidase, which utilize GSH to regenerate reduced enzyme from a covalent enzyme intermediate formed by primary reaction of an enzyme nucleophile with the oxidized substrate (disulfide or peroxide, respectively) (1). Despite the various indications of low affinity for GSH, it is nevertheless believed that GSH is the physiological thiol substrate for thioltransferase because of its abundance relative to other thiols.

3. *Steric Constraints as the Basis for TTase Selectivity*

Selectivity for protein mixed disulfides containing glutathione vs. those containing cysteine or cysteamine as the nonprotein thiol may relate to the accessibility of the disulfide bond to the thiolate moiety at the active site of TTase. The greater bulk of the GS-tripeptide moiety may expose the S-S bond of the protein mixed disulfide more to the thioltransferase active site than do the cysteinyl or cysteaminyl moieties, which may be accommodated more readily into protein crevices. This explanation, however, would seem not to account for the catalytic discrimination that thioltransferase displays with small disulfides, e.g., Cys-S-SG vs. Cys-S-S-Cys, where the non–GS-containing disulfides display a lag phase in the coupled assay (21). Hence, a further consideration is necessary.

4. *Stability of the TTase-SSR Intermediate as the Determinant of Selectivity*

Figure 8 is a generic representation of the TTase reaction scheme in which GSH is not necessarily involved as the second substrate. Here the central intermediate is the mixed disulfide form of the TTase enzyme. If this intermediate reacts with a thiol compound in solution, the thiol-disulfide exchange reaction is completed. If instead the second thiol near the active site of the TTase reacts with the mixed disulfide intramolecularly, then the intramolecular disulfide form of TTase is formed, which must be reduced with two moles of thiol compound in order to be recruited back into the cycle. TTase cannot catalyze the thiol-disulfide exchange reaction if it is tied up in a nonproductive interchange be-

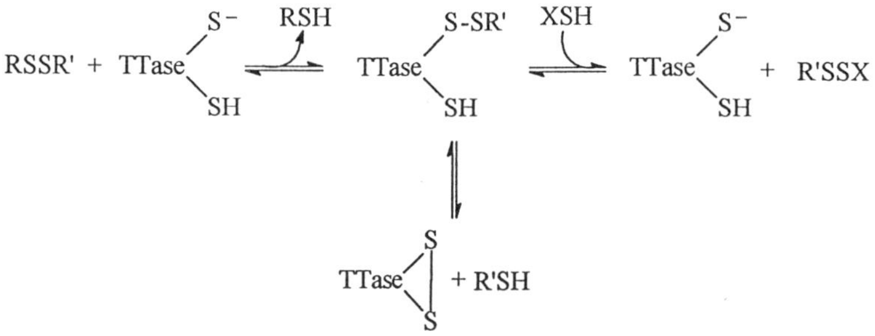

$$\mathrm{RSSR'} + \mathrm{TTase}\begin{matrix}\mathrm{S^-}\\ \mathrm{SH}\end{matrix} \underset{}{\overset{\mathrm{RSH}}{\rightleftharpoons}} \mathrm{TTase}\begin{matrix}\mathrm{S\text{-}SR'}\\ \mathrm{SH}\end{matrix} \overset{\mathrm{XSH}}{\rightleftharpoons} \mathrm{TTase}\begin{matrix}\mathrm{S^-}\\ \mathrm{SH}\end{matrix} + \mathrm{R'SSX}$$

$$\Updownarrow$$

$$\mathrm{TTase}\begin{matrix}\mathrm{S}\\ |\\ \mathrm{S}\end{matrix} + \mathrm{R'SH}$$

Figure 8 Generic representation of the thioltransferase reaction scheme. This scheme depicts a generalized reaction for thioltransferase in which GSH is not necessarily involved. Here the central intermediate is the mixed disulfide with the TTase enzyme. If this intermediate reacts with a thiol compound in solution, the thiol-disulfide exchange reaction is completed. If instead the second thiol near the active site of the TTase reacts with the mixed disulfide intramolecularly, then the intramolecular disulfide form of TTase is formed, which must be reduced with two moles of thiol compound in order to be recruited back into the cycle.

tween the TTase (S-S) intramolecular disulfide and the TTase(SH)-SSR′ or TTase(SH)-SSX forms, because TTase(SH)-S^- must be regenerated. Hence the efficiency of the thiol-disulfide exchange cycle is dependent on the relative stabilities of the various TTase (SH)-SSR′ species; i.e., the nature of the R′S moiety affects the relative rates of intramolecular versus intermolecular reaction of thiols with the TTase mixed disulfide. The selectivity of TTase for GS-containing disulfides would be explained if the conformation of the TTase(SH)-SSG intermediate relative to the TTase(SH)-SSCysteine intermediate, for example, were such that the neighboring thiol was oriented in a less favorable position or otherwise deactivated for intramolecular nucleophilic displacement of the R′S moiety. Consistent with this interpretation, reaction of [^{14}C] cystine with the TTase enzyme was reported to give the intramolecular disulfide form of the enzyme [TTase-(S-S)] without evidence for the mixed disulfide intermediate TTase(SH)-S-SCysteine [^{14}C] in the time frame of the experimental manipulations (14). In contrast, observations of reactions of TTase with radiolabeled glutathionyl disulfides, [^{35}S] GSSG and [^{14}C] Cysteine-SSG, suggest a relatively more stable TTase-SSG intermediate, as described above. Accordingly, one can argue that the selective catalysis of reduction of GS-containing disulfides by TTase is related to the greater stability of the TTase-SSG mixed disulfide compared to TTase-SS-cysteinyl or cysteaminyl mixed disulfides; i.e., the glutathionyl mixed disulfide on TTase is maintained for attack by the acceptor thiol rather than attack by the second thiol on the enzyme.

5. *Conclusion*

In the case of protein mixed disulfide substrates, the key determinants of glutathionyl mixed disulfide selectivity by thioltransferase seem to be accessibility of the disulfide bond to the thiolate active site of the enzyme and the selective attack on the nonprotein side of the S-S bond. If TTase attacked directly on the other side, then catalysis of protein-SSR dethiolation would not be selective. According to this logic, the active site thiolate of thioredoxin, which does not discriminate among protein disulfide substrates, is more penetrating than that of thioltransferase. In the case of nonprotein substrates, TTase in fact may act on either side of the disulfide, but only certain TTase(SH)-SSR intermediates (i.e., TTase-SSG) may be sufficiently stable to lead to overall catalysis rather than futile cycling. As indicated by our studies of mixed disulfide substrates for thioltransferase (21) and the modeling of the T4 (56) and *E. coli* (20) glutaredoxin-binding sites, it would seem reasonable to propose an enzyme site that recognizes the glutathionyl moiety; however, it is more likely that this site would recognize the GS-moiety of GS-SR disulfides rather than reduced GSH, because the enzyme reaction is not selective for GSH as the reducing substrate. Even more likely than "recognition" of the GS-moiety for noncovalent binding is the concept that stabilization of the GS-moiety of the TTase(SH)-SSG covalent intermediate is the basis for the glutathionyl specificity of TTase. Additional studies are necessary to test this hypothesis.

E. Possible Exceptions to the Glutathionyl Selectivity of TTase

It is always difficult to ascertain whether a high degree of selectivity in enzymatic reactions represents a universal property of the enzyme or whether exceptions may exist that reveal a broader versatility of the enzyme. To examine this question regarding the glutathionyl specificity of TTase, possible exceptions are considered under three categories: (1) apparent catalysis by TTase of GSH-dependent dethiolation of non–GS-containing protein mixed disulfides, (2) apparent catalysis by TTase of GSH-dependent reduction of intramolecular disulfides, and (3) apparent catalysis by TTase of GSH-dependent reduction of nondisulfide substrates.

1. *Dethiolation of Non–GS-Containing Protein Mixed Disulfides*

Previously thioltransferase was reported to catalyze the reduction of protein-SSCysteamine mixed disulfides in solutions containing the proteins from rat liver cytosol preparations that were treated with cystamine (39). These results are difficult to interpret, however, because initial rates of dethiolation were not reported. In a more recent article, Terada et al. (62) reported TTase catalysis of reactivation of cystamine-inactivated glutathione-S-transferase (GSTase) and TTase catalysis of release of radioactivity from GSTase that had been treated with radiolabeled cystine. In this case interpretation of the apparent catalysis of GSTase-SSR dethiolation/reactivation by TTase is complicated by the fact that GSH was present in all of the experiments, allowing the possibility of preformation of GS-containing substrates for TTase. The observed TTase catalysis could represent direct action of the TTase enzyme on the GSTase-SSR "substrate," and thus it would be a true exception to the GS specificity. Alternatively, the redox kinetics for the reactions of the mixed disulfides of GSTase with GSH may be such that TTase would accelerate formation of the active GSTase-SH form by still reacting preferentially with GS-containing disulfides and drawing the equilibrium away from the GSTase-SSR form, as indicated by the stepwise reaction in Scheme V:

$$\text{GSTase-S-SR} + \text{GSH} \rightleftharpoons \text{GSTase-SH} + \text{GS-SR}$$

$$\text{GS-SR} + \text{TTase-S}^- \rightleftharpoons \text{TTase-SSG} + \text{RSH}$$

$$\text{GSH} + \text{TTase-SSG} \rightleftharpoons \text{TTase-S}^- + \text{GSSG}$$

$$[\text{GSTase-SH} + \text{GSSG} \rightarrow \text{No covalent modification}]$$

Scheme V

Thus, TTase catalysis might be due to accelerated conversion of glutathionyl mixed disulfide products (GSSR) to GSSG, because GSSG does not react with GSTase-SH to form an inactivated GSTase-SSG covalent complex according to the report of Terada et al. (62). In order to obviate the ambiguity of the interpretation, we repeated the inactivation of GSTase by cystamine, but instead of providing GSH as the reducing substrate for reactivation of GSTase-S-SCysteamine, we substituted cysteamine because oxidized TTase can be reduced by thiols other than GSH (as described above). Hence, if TTase were to act directly on the GSTase-S-SCysteamine mixed disulfide, catalysis should be evident because cysteamine is able to recycle the oxidized enzyme. In fact, although cysteamine did mediate reactivation of GSTase-S-SCysteamine in the absence of TTase, no acceleration of the rate of reactivation was observed in the presence of TTase unless GSH was also added (Fig. 9). Therefore, the observed catalysis of reactivation of GSTase-SSR appears to be consistent with the glutathionyl specificity of TTase, rather than an exception.

2. *Reduction of Intramolecular Disulfides of Proteins and Peptides*

Glutaredoxin was discovered in *E. coli* mutants lacking thioredoxin, according to its ability to catalyze the GSH-dependent reduction of the intramolecular disulfide form of ribonucleotide reductase (RRase) and thereby support deoxyribonucleotide synthesis. Certain mammalian thioltransferases have been shown to support RRase catalysis, while others have been reported as incompetent in that reaction (4) (see Sec. III.A). TTase and thioredoxin also have been compared according to their relative abilities to catalyze reduction of the intramo-

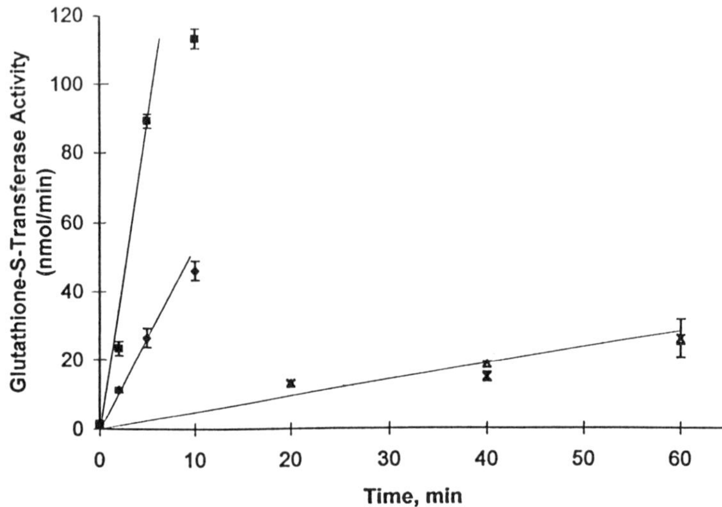

Figure 9 Reactivation of glutathione-S-transferase-SSCysteamine. Glutathione-S-transferase (GSTase π), 0.1 ml of 10 U/ml, was treated with 1 mM DTT for 30 min at room temperature and then purified using a spin column (1-cc syringe containing Sephadex G-25, equilibrated with 0.2 M Tris-HCl, pH 8.0). This sample was then treated with 1mM cystamine for 1.5 h at room temperature. GSTase was >90% inactivated after this treatment (based on comparative yields of GSTase activity from the same spin column). Aliquots of this cystamine-inactivated GSTase (20 μl) were then treated with 2 mM cysteamine in the absence (Δ) or presence (×) of 8 μM thioltransferase (lowest line); or with 2 mM glutathione in the absence (◆) or presence (■) of 8 μM thioltransferase (middle and upper lines, respectively). GSTase activity was measured (and unit activity defined) by adaptation of a standard assay that uses 2 mM glutathione and 1mM 1-chloro-2,4-dinitrobenzene as substrates (63).

lecular disulfide bonds of insulin (43). In both of these reactions it is not known whether the GSH-dependent catalysis by TTase represents direct interaction of the TTase enzyme with the intramolecular disulfide or initial nonenzymatic reaction of GSH with the intramolecular disulfide to form a GS-containing mixed disulfide as the actual substrate for TTase. It may be discovered that the distinction among mammalian species regarding TTase catalysis of the ribonucleotide reductase reaction lies in the relative ability of the oxidized forms of the respective RRases to react with GSH to form the corresponding RRase-SSG substrates for TTase. Further study is necessary to resolve these questions.

3. *Reduction of Dehydroascorbate and Arsenate*

Wells et al. (64) reported that thioltransferase from pig liver and human placenta catalyzes the GSH-dependent reduction of dehydroascorbic acid (oxidized vitamin C) with the concomitant formation of GSSG, and we have observed this activity with hRBC TTase also (51). This reaction is mechanistically intriguing, because it represents the first example of TTase catalysis of a reaction involving a nondisulfide substrate. Subsequent to the original report, the mechanism for dehydroascorbate (DHA) reduction was proposed to involve direct attack of TTase on the DHA molecule (4) (Fig. 10, left). Since nonenzymatic reduction of DHA by GSH occurs in the absence of TTase (51,64), this indicates that GSH can react directly with DHA and suggests an alternative mechanism for TTase catalysis (Fig. 10, right). Accordingly, the first reaction in this scheme involves the formation of a GS adduct with the dehydroascorbate (Fig. 10, right), analogous to the GS-containing intermediate formed in the glyoxylase-catalyzed reduction of glyoxal (1). TTase would then act upon the DHA-SG intermediate in a fashion analogous to the reaction scheme presented in Figure 6. Further studies are necessary to distinguish the alternative mechanisms shown in Figure 10. For example, it is critical to determine whether a DHA-SG intermediate is involved, and whether the mutant of TTase lacking the second thiol group (C25S) is still capable of catalyzing the dehydroascorbate reduction reaction despite its inability to form an intramolecular disulfide.

Bacterial glutaredoxin has been shown to catalyze the GSH-dependent reduction of arsenate (65). This reaction is analogous to TTase catalysis of dehyroascorbate reduction, and in like fashion a mechanism has been proposed that

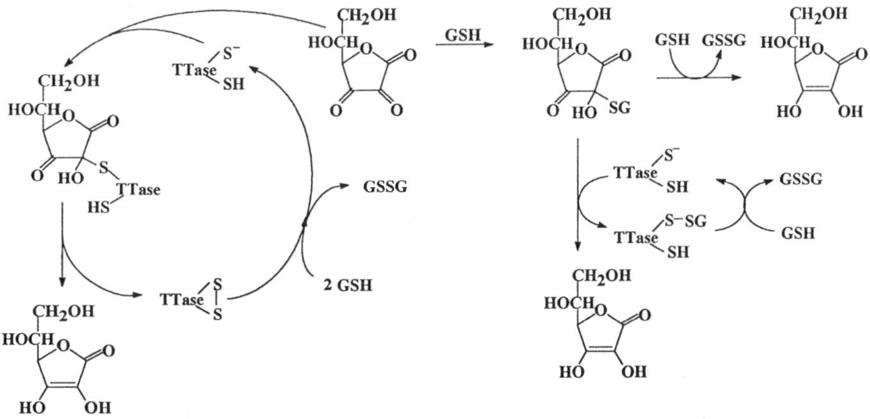

Figure 10 Alternative mechanisms of TTase catalysis of GSH-dependent reduction of dehydroascorbic acid (DHA). At the left is shown the direct attack of TTase on DHA resulting in formation of the intramolecular disulfide form of TTase. At the right is shown the reaction of TTase with a DHA-SG adduct resulting in formation of the TTase-SG intermediate analogous to the mechanism of TTase catalysis of disulfide reduction shown in Figure 6.

involves direct attack of the glutaredoxin enzyme on the arsenate molecule (65) (analogous to Fig. 10, left). Similarly, it is likely that GSH may react directly with arsenate and the glutaredoxin catalytic scheme may be more like that depicted by the right side of Figure 10. Further study is necessary to distinguish these possibilities.

4. *Conclusion*

Although several apparent exceptions to the glutathionyl specificity of thioltransferase catalysis can be gleaned from various reported studies as described above, all of these can be explained by alternative interpretations that involve nonenzymatic preformation of GS-containing substrates for TTase. Moreover, since GSH is the preponderant nonprotein thiol in cells, it is appropriate to focus in the final section of this chapter on the potential physiological implications of the glutathionyl specificity of thioltransferase.

III. POTENTIAL PHYSIOLOGICAL FUNCTIONS

A. Overview

Although much work has been dedicated to elucidating the structure and catalytic properties of thioltransferase, the physiological functions of thioltransferase are not well established. Many proteins have essential thiols, which are critical for function, and it is expected that the thioltransferase system, and the related thioredoxin system, are involved in homeostatic maintenance of intracellular thiols and possibly in regulation of sulfhydryl enzymes and receptors (1–5).

As described above, ribonucleotide reductase oscillates between dithiol and disulfide forms in catalyzing formation of deoxyribonucleotides. Turnover of ribonucleotide reductase by the glutaredoxin system in *E. coli* was calculated to be more efficient than that by the thioredoxin system (3), and thioltransferase has been shown to support ribonucleotide reductase in T4-phage and in calf thymus (66,67). Previous work with a mutant *E. coli* that expresses neither glutaredoxin nor thioredoxin led to the conclusion that the enzymes are essential for sulfate reduction but apparently not for DNA synthesis (68). A follow-up study with the double mutant, however, revealed additional isozymes of glutaredoxin as well as an elevated level of ribonucleotide reductase that would support DNA synthesis and thus could account for the viability of these mutants (69). There has been some difficulty, however, in establishing the role of the TDOR enzymes in conjunction with ribonucleotide reductase action in mammals. Thus, rabbit bone marrow thioltransferase and thioredoxin and pig liver thioltransferase have been reported to be ineffective co-catalysts for the rabbit ribonucleotide reductase reaction (4,16,70). Hence additional study is necessary to establish the physiological roles of the TDOR enzymes in DNA synthesis.

Pyruvate kinase, phosphofructokinase, and GSTase are a few examples of enzymes that are inactivated under oxidizing conditions and reactivated by TTase (37,62,71,72). A more comprehensive listing of sulfhydryl enzymes whose activities are altered in a reversible fashion by thiol modification can be obtained from previous reviews (1,4,5). In most cells GSH is the major intracellular nonprotein thiol and would be expected to form the majority of any mixed disulfides under oxidative stress. Under certain conditions, the formation of a glutathione mixed disulfide could serve as a "tag" for reduction by the TTase system. Under other conditions, relative redox potentials and the GSH/GSSG redox ratio could support TTase catalysis of the reverse reaction, where TTase would function to glutathionylate thiol proteins and thereby modulate their function or protect them from irreversible oxidative damage. In one study mitochondrial Ca^{2+} permeability was linked to GSH/GSSG status independent of the redox status of the pyridine nucleotides, and it was speculated that this might involve catalysis by thioltransferase (73). A difficulty with this interpretation, however, is the lack of reported documentation for TTase in mitochondria (4). Another context where the role of thioltransferase (or other TDOR enzymes) needs to be examined is the aging process in the lens tissue of the eyes, where structural proteins do not turn over and accumulation of protein-mixed disulfides has been implicated in cataract formation (74,75).

Despite considerable attention to posttranslational modification of cysteine residues by thiol-disulfide redox cycling as a potential regulatory mechanism, the redox potentials of only a few proteins have been studied in detail, and most studies of thiol-disulfide oxidoreductase enzymes have emphasized catalysis of disulfide reduction rather than catalysis of reversible thiol-disulfide interchange (5,23,76).

It is conceivable that thioltransferase might also participate in sulfhydryl protein processing. For example, exposed Cys-SH groups on cytosolic domains of membrane proteins that are threaded through the lumen of the endoplasmic reticulum during synthesis might be delivered to the cytosol as protein-S-SG derivatives, because the redox buffer of the endoplasmic reticulum is more oxidizing than the cytosol (77). Hence, cytosolic thioltransferase may catalyze deglutathionylation to reexpose the SH groups.

As described in the previous section, there is evidence that the TTase system with GSH can catalyze the reduction of the nondisulfide compound dehydroascorbate (oxidized vitamin C) (64), and the authors suggested that thioltransferase may be central to ascorbic acid turnover in vivo. Since maximal rates were reported but no enzyme concentrations were given, it is not possible to determine relative turnover numbers for the thioltransferase enzymes that were tested. Thus it is not known if rates of thioltransferase-mediated reduction would correspond to physiological rates of dehydroascorbic acid turnover. In this context also, it is conceivable that other nondisulfide compounds such as nitroso and halo-ethyl xenobiotics, which are known to form adducts with GSH, may also be metabolized by the TTase system (see below). There are multiple considerations of potential physiological functions of the thioltransferase system, and these are discussed with a particular focus on the glutathionyl specificity of thioltransferase in more detail below.

B. Repair and Protection of Protein Thiols

1. *Oxidation of Sulfhydryls*

Aerobic cells require oxygen for energy metabolism, but suffer the potential consequences of hydrogen peroxide, superoxide, and hydroxyl radicals as reactive toxic byproducts. At critical concentrations activated oxygen species may cause sulfhydryl oxidation, lipid peroxidation, enzyme inactivation, DNA damage, and cell lysis. The intracellular milieu is generally a reducing environment due to high concentrations of GSH (1–10 mM), but oxidative stress can shift the thiol-disulfide equilibrium (5,23), so that sulfhydryls may be reversibly oxidized to disulfides or sulfenates. In most cells GSH is the major intracellular nonprotein thiol and would be expected to form the majority of mixed disulfides. Cells have evolved efficient scavenger systems, such as superoxide dismutase and glutathione peroxidase, to remove activated oxygen species nearly completely. Nevertheless, certain cellular processes, pathophysiological states, and xenobiotics may overwhelm these systems and produce oxidative stress. For example, tissue damage associated with myocardial infarction has been linked to generation of activated oxygen during reperfusion of blood after ischemia (78–80). During an inflammatory response, white blood cells kill invading bacteria by digestion with oxidants that may also injure neighboring host cells. The fact that cells can recover from oxidative stress suggests there are systems to repair oxidative damage, and thioltransferase is expected to participate in such repair. Since much of the oxidative damage is attributed to functional modification of sulfhydryl proteins by disulfide bond formation, TTase is thought to act as a cellular repair enzyme by catalyzing reduction of disulfide bonds.

2. *Sulfhydryl Status in Red Blood Cells*

Most cells are susceptible to oxidative injury, but red blood cells (RBCs) are perhaps more at risk because of their high iron content and exposure to high oxygen tension. Their membranes are rich in polyunsaturated fatty acids susceptible to lipid peroxidation, and they experience unusual stress while traversing capillaries. Many RBC proteins have SH groups that are sensitive to oxidative modifications that alter their functional properties. For example, modification of the cys-β93-SH groups at the conformationally responsive dimer interface of hemoglobin alters O_2 and heme binding (81). Similarly, oxidation of SH groups on metabolic control enzymes like phosphofructokinase inactivate them and interfere with RBC energy balance (23). Oxidation of SH groups on membrane proteins like spectrin and Band 3 alters deformability and permeability of the RBC membranes (82,83).

As part of the characterization of hRBC thioltransferase, we investigated its ability to catalyze reactions in vitro that would represent sulfhydryl homeostatic reactions within RBCs. Figure 11 illustrates the relative ability of GSH alone or with thioltransferase, each coupled with GSSG reductase, to effect reduction of the hemoglobin-glutathione disulfide adduct (Hb-SSG), a known product of oxidative stress in RBCs. It is evident that thioltransferase is much more effective at reducing Hb-SSG than is GSH alone (Fig. 11). In order to estimate the contribution of TTase to the total Hb-SSG dethiolase activity of RBCs, we compared the activity of hemolysate (containing a known amount of TTase) to

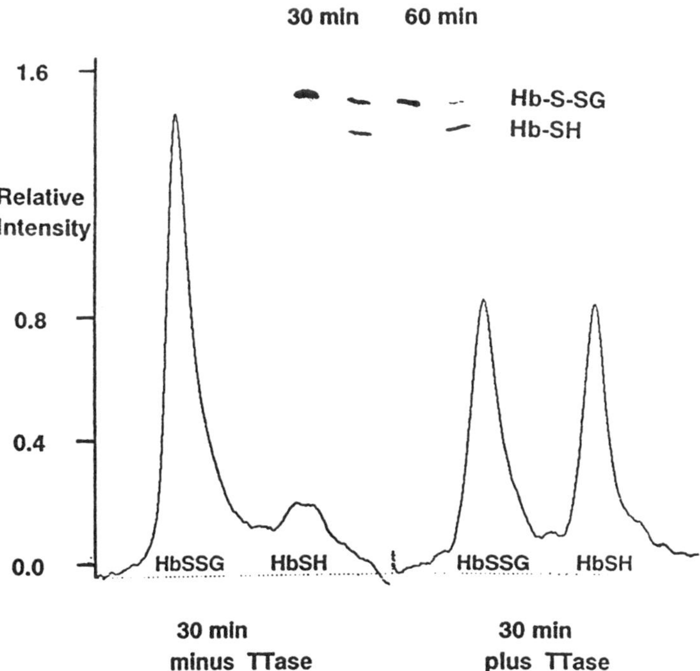

Figure 11 Catalysis of HbSSG reduction by purified hRBC thioltransferase. Time-dependent formation of Hb-SH from Hb-SSG by GSH plus GSSG reductase and NADPH, in the absence (left) and presence (right) of hRBC thioltransferase was measured by isoelectric focusing whereby the separate bands for Hb-SH and Hb-SSG (see gel photo) were quantitated by densitometry as shown by the tracings at 30 min. There is a marked enhancement in the reduction of HbSSG when TTase is present. (From Ref. 37.)

Table 7 Relative Dethiolation of [^{35}S] OxyHb-SSG by Hemolysate and Isolated Thiol-transferase[a]

Conditions	% Total cpm in supernate
Hb-SSG alone	0.0[b]
Hb-SSG + 100 mM DTT	100[b]
Hb-SSG + 2 mM GSH + GSSG reductase (GRase, 2 U/ml)	3.3
Hb-SSG + 0.7 μM pure TTase + 2 mM GSH + GRase	11.7
Hb-SSG + hRBC lysate (0.7 μM TTase)[c] + 2 mM GSH + GRase	12.2
Hb-SSG + 2 mM ascorbate	0.5

[a]Reaction mixtures (25 μl) containing 25 mM potassium phosphate, pH 7.5, 0.2 mM NADPH, and 100 μM oxyHb-SSG with the components indicated were incubated for 10 min at 30°C, then quenched with TCA.
[b]TCA-soluble radioactivity released by 100 mM DTT (100%) corresponded to total precipitable [^{35}S] GS-equivalents in the TCA pellet in the absence of reducing agents.
[c]Lysate was dialyzed to remove GSH and NADPH, TTase activity in lysates was determined as described (37) and concentrated by ultrafiltration to achieve the TTase concentration indicated.

that of pure TTase, along with GSH and GSSG reductase. The data of Table 7 indicate that equivalent amounts of pure TTase and TTase-containing hemolysate have nearly identical dethiolation properties. According to this comparison, the TTase system alone appears to account for the total enzymatic activity of RBCs for reduction of oxyHb-SSG [^{35}S].

Recently Wells et al. (64) suggested that TTase might function physiologically as an ascorbic acid turnover enzyme, i.e., that ascorbate might be the direct reductant of disulfides and that TTase would serve to recycle the oxidized dehydroascorbate. As an initial test of this hypothesis, we measured Hb-SSG [^{35}S] reduction by ascorbate also. Table 7 shows little radioactivity release from Hb-SSG [^{35}S] by 2 mM ascorbate, i.e., little direct reduction of the disulfide by ascorbate occurred. This result does not support an intermediate role for ascorbate in TTase catalysis of disulfide reduction.

The sulfhydryl enzyme phosphofructokinase represents another target of oxidative stress in RBCs and a potential substrate for hRBC TTase. To model this possibility, hRBC PFK was inactivated in a concentration-dependent manner by methylmethanethiosulfonate (MMTS) in the presence of the physiological concentration of GSH in RBCs (2 mM). Then additional GSH was added in the absence or presence of various amounts of hRBC TTase (or DTT) to study reactivation of the PFK enzyme. GSH alone was ineffective, whereas TTase displayed a concentration-dependent catalysis of PFK reactivation (37). Although DTT could also reactivate the PFK, TTase was 1200 times more efficient on a molar basis (e.g., 0.4 μM TTase = 500 μM DTT), consistent with its catalytic function.

Cytochrome b_5 reductase is the key enzyme in the primary pathway of methemoglobin (Hb^{3+}) reduction involving NADH and cytochrome b_5 (84). Since HbO_2 in circulating RBCs is oxidized spontaneously to Hb^{3+}, the NADH-methemoglobin reductase system is important in maintaining oxygen transport capability. Hackett et al. (85) showed that cytochrome b_5 reductase has a catalytically essential thiol moiety (Cys-283). Thus, under oxidative stress cytochrome b_5 reductase may be susceptible to inactivation by disulfide formation that could be reversed by TTase. As an initial test of this hypothesis, we found that treatment of cytochrome b_5 reductase with GSSG completely inactivated the enzyme's ability to reduce $K_3Fe(CN)_6$ (an alternate electron acceptor). When the GSSG-inactivated cytochrome b_5 reductase was treated with 0.5 mM GSH alone, about 30% of the original activity was regained. When hRBC TTase (0.01 units) was also added, about 70% of the activity was restored in the same 10-min time period (22). Although these studies are preliminary, they suggest that cytochrome b_5 reductase may be another physiological substrate for hRBC TTase.

The above observations for three different RBC proteins suggest a consistent pattern of thiol inactivation and regeneration whereby the primary mechanism for reduction of disulfide-modified proteins involves catalysis by TTase (15,21,22,37). This interpretation supplants the traditional belief that protein disulfide reduction in RBCs was accomplished by a nonenzymatic reaction with GSH to form GSSG, and then coupled to GSSG reductase.

Terada et al. (86) also reported the purification of hRBC TTase and its ability to reactivate phosphofructokinase. In addition they reported that TTase levels are similar in RBCs of various species, and they described initial results that implicate TTase in the repair of disulfide-modified thiol groups on RBC membrane proteins. Thus, RBC membrane preparations that were pretreated with diamide and showed a loss of thiol content regained monobromobimane-derivatized thiol groups when treated with TTase and GSH. Although specific membrane proteins could not be identified as the substrates for TTase from this study, it is likely that protein-SSG derivatives were involved according to the TTase specificity characterized in the previous section. Consistent with this concept, earlier work showed that GS-containing mixed disulfides were formed with particular membrane proteins when erythrocytes containing prelabeled [^{14}C] GSH were treated with diamide (87). In another article, diamide and *t*-butyl hydroperoxide were reported to decrease the activity of TTase and increase the concentration of protein mixed disulfides in rat erythrocytes in the absence of glucose, and these changes were reversed by glucose (88). The latter observation is consistent with the influence of NADPH-generating capacity and GSSG reductase activity on the position of equilibrium for reactions catalyzed by thioltransferase (15); i.e., decreased reduction of GSSG would led to increased levels of protein-SSG.

3. Potential Protective Role of Thioltransferase-Catalyzed Disulfide Formation

As indicated earlier, most attention on the TDOR enzymes has been focused on their ability to mediate reduction of disulfides rather than the reverse reaction. During oxidative stress, however, the redox milieu is shifted and the GSSG:GSH ratio is increased. Under such oxidative conditions TTase may catalyze formation of protein-glutathione mixed disulfides rather than reduction. In this way TTase may mediate "glutathione masking" of critical SH proteins under oxidative stress to protect them from irreversible oxidative damage. In extreme cases such as glucose-6-phosphate dehydrogenase deficiency in RBCs where the cells are energy competent but cannot produce adequate NADPH, exposure to oxidants results in high levels of GSSG that cannot be reduced effectively, and these cells have been shown to export GSSG by an active process (89). RBCs treated with H_2O_2 as a model of oxidative stress display alterations in cell shape, decreased membrane deformability, hemolysis, and increased recognition by anti-IgM antibodies (90). These changes have been ascribed to crosslinking of Hb to spectrin by disulfide bond formation in vitro and ex vivo. Although the concentrations of H_2O_2 used to induce these cellular changes were high (0.3–0.8 mM), this study reflects the possible consequences of disulfide bond formation under oxidative stress. This study further suggested that the damaging cellular changes could be diminished by pretreatment with the sulfhydryl blocking agent *N*-ethylmaleimide. In like manner TTase-catalyzed formation of reversible protein-SSG disulfides could play a protective role during oxidative stress. Conversely if certain protein-SSG species represent intermediates in the formation of protein-S-S-protein crosslinks, then TTase might facilitate the deleterious effects of oxidative stress by catalyzing formation of such intermediates. If certain protein-protein interactions shield disulfides from the intracellular milieu, then GSH might not be able to react with those disulfides. Consequently, TTase would also be prevented from reversing the protein disulfide crosslinking even when reducing equivalents were replaced. In this way, the slow buildup of disulfide-bonded proteins may lead to erythrocyte death. Hence, identification of GSH/TTase-resistant disulfide-bonded proteins may lead to a better understanding of the mechanism of erythrocyte lysis. It is evident that further work is required to distinguish among the various ways that thioltransferase may participate in membrane sulfhydryl homeostasis in RBCs.

4. Potential Role of RBCs and Thioltransferase in Systemic Detoxification

The concentration of reactive thiol groups on Hb (cys-β93) is 8 mM in erythrocytes, whereas GSH concentration is 2 mM. Although GSH is typically considered the primary thiol reductant of the cell, the higher concentration of reactive Hb-SH may in fact be a first line of defense against harmful bloodborne substances. Thus, Hb cys-β93 may act as a "buffer" against oxidative stress. This hypothesis is consistent with the concept that erythrocytes, in addition to their primary function as gas transporters, may also play a role as "circulating detoxifying packets" (89). Under conditions of oxidative stress, thioltransferase could act by forming Hb-SSG to dissipate the oxidative potential and protect other sensitive sulfhydryls from oxidative modification. As reducing equivalents were replenished, TTase would catalyze reformation of Hb-SH. In this context, RBCs have been considered to contribute also to the protection of other cells. For example, endothelial cells constitute the surface lining of the entire vascular system from the heart to the smallest capillary, and they control the passage of materials into and out of the blood stream. The endothelial cells are the sites of formation of atherosclerotic plaques, and they are the sites of injury in inflammatory reactions. Previous studies showed that addition of RBCs could mitigate the effects of H_2O_2 on isolated endothelial cells (91,92), consistent with the interpretation that RBCs may serve as an adjunct to the internal protective mechanisms of these cells against the oxidative effects of inflammation and ischemia-reperfusion. The authors suggested a simple scavenging mechanism based on the facile permeability of H_2O_2 into cells. But this seems too restrictive, as if the RBCs acted only as a packet of catalase and GSH peroxidase. Not considered was the possibility that the oxidizing potential of H_2O_2 might be transmitted intracellularly also via reactions with sulfhydryl groups. Such a mechanism was offered as an interpretation of the intracellular changes in the GSH:GSSG ratio observed after treatment of RBCs with the impermeable sulfhydryl oxidant dithiobisnitrobenzoate (DTNB) (93). This speculative mechanism may involve TTase as the mediator of thiol-disulfide interchange that would propagate the extracellular oxidative signal to the intracellular thiols (see Sec. III.C). A later study by Toth et al. (94) indicated that RBCs from smokers had enhanced protective capacity that correlated with increased levels of GSH and catalase, but they acknowledged that the antioxidant capacity of the RBCs could be contributed by some other scavenging mechanism that they had not measured. As discussed above, Hb-SH coupled to the thiotransferase system would be a likely candidate for such a role.

5. *Evidence for Dethiolation Activity of Thioltransferase (and Thioredoxin) in Various Cells*

Besides red blood cells, the thiol-disulfide oxidoreductase enzymes have been implicated in the repair and protection of sulfhydryl proteins in various other contexts. Miller et al. used two glutathionylated protein mixed disulfide substrates (phosphorylase *b*–SSG and carbonic anhydrase III–SSG) to survey various tissues for what they called NADPH-dependent and GSH-dependent dethiolase activities (95). They reported a decreasing amount of total dethiolase activity in rat cells in the order neutrophils > skeletal muscle > liver, heart. The NADPH-dependent activities that they measured probably reflect a combination of the activities of the thioltransferase and thioredoxin systems since endogenous GSH was likely present. The BCNU-independent but GSH-dependent activity that they described probably reflects the combination of nonenzymatic dethiolation by GSH plus turnover of TTase by GSH in the absence of GSSG reductase [i.e., bis-chloroethylnitrosourea (BCNU) is a selective inactivator of GSSG reductase; Sec. III. E]. Thus, several experimental ambiguities make interpretation of these dethiolase studies difficult with regard to attributing the activities to the thioltransferase and/or thioredoxin systems. For example, the tissue preparations were not freed of endogenous GSH before testing the activities in the absence and presence of added GSH, and only nanomolar concentrations of the protein-SSG substrates were used to measure "enzymatic" activities. If the enzyme concentration is in excess of the substrate concentration, this would preclude assessment of actual catalysis involving turnover of the enzyme [e.g., when pure *E. coli* thioredoxin was used as a model, 30 μM of the thioredoxin enzyme was combined with only 14 nM of the carbonic anhydrase-SSG [^{3}H] substrate (95)].

A more recent study was designed to test the relative ability of the TTase and thioredoxin systems to repair proteins that may be damaged in different ways by oxidative stress (46). The glycolytic enzymes phosphofructokinase (PFK) and glyceraldehyde-3-phosphate dehydrogenase (G3PD) were used as representatives of oxidatively sensitive sulfhydryl enzymes that are naturally co-localized in the cytosol with the thiol-disulfide oxidoreductase systems. Treatments of PFK with GSSG, and G3PD with H_2O_2 and tetrathionate, were interpreted to give PFK-SSG, G3PD-S-OH, and G3PD-S-SO_3^- as the corresponding inactivated/reactivatable forms of the enzymes, respectively. Comparison of equimolar amounts of the TTase and thioredoxin enzymes near their estimated physiological concentrations indicated that TTase was much more effective than thioredoxin for reactivating PFK-SSG, as predictable from the substrate selectivity of TTase described above. Thioredoxin, however, was better for reactivating the two types of modified G3PD. Thus, the two enzyme systems may act in concert to repair intracellularly modified sulfhydryl moieties. A previous publication reported inhibition of the reactivation of oxidized G3PD in intact endothelial cells by 13-*cis*-retinoic acid (96), a putative selective inhibitor of thioredoxin reductase (97). This result was interpreted as evidence for the oxidative repair role of the thioredoxin system in vivo (96). In a similar fashion BCNU, azelaic acid, and sodium arsenite were used with erythroleukemia cells (in conjunction with in vitro studies of the isolated enzymes) to implicate the thioltransferase and thioredoxin systems in the reactivation of oxidatively inactivated ornithine decarboxylase in situ (45).

6. *Thioltransferase as a Target for Therapeutic Agents*

As a consequence of a single amino acid substitution on hemoglobin (Glu β-6 Val), sickle cell Hb is prone to polymerize under hypoxic conditions. Thus, agents that increase the oxygen affinity of sickle Hb have been shown to inhibit the polymerization reaction (98). Thiol reagents that modify cys-β93 and thereby alter the oxygen binding of sickle Hb have been shown to decrease fiber formation in vitro and ex vivo, including GSSG, cystamine, iodoacetamide, 2,2′-dithiodipyridine, *N*-ethylmaleimide, and 4-aminophenyl disulfide (99–101). The design of the thiol-directed antisickling agents has focused on adding an exogenous compound that can cross the red cell membrane and react with the cys-β93-SH moiety of Hb. This approach has been problematic because the concentrations of reactive disulfides needed for modification of Hb also elicit toxic side effects (100). Moreover, disulfide modified hemoglobins, if reactive with GSH, would be expected to be substrates for TTase, so that the therapeutic value of the potential disulfide drugs might be countered by the action of TTase. Thus, selective inhibition of TTase that would allow accumulation of Hb-SSG in the sickle RBCs might be considered as an alternative or adjunct therapy of the disease.

In another therapeutic context, Mizoguchi et al. considered the effects of antiinflammatory drugs on the thioltransferase in white blood cells (102). They reported that a wide variety of antiinflammatory agents inhibited TTase preparations isolated from bovine leukocytes, erythrocytes, and liver when tested in vitro using the standard spectrophotometric assay. In an attempt to confirm some of these observations, we (60) observed no inhibition of TTase (bovine liver or hRBC) by indomethacin or mefenamic acid at concentrations of these agents up to 10 times the IC_{50} values reported by Mizoguchi et al. The basis for this discrepancy is not clear; however, the natural absorbance of the putative inhibitors in the region of 340 nm likely interferes with the spectrophotometric assay of TTase, which is dependent on linear Beer's law behavior for accurate measurement of the time-dependent changes in NADPH absorbance.

7. *Conclusion*

Certainly much more study is necessary to delineate the relative contributions of thioltransferase and thioredoxin to sulfhydryl homeostasis in vivo. Since mild to severe oxidative stress may be imposed on cells by various insults, including ischemia-reperfusion, inflammatory responses, and chemotherapy, different oxidative challenges may confront the thioldisulfide oxidoreductase systems, and the activities of the TDOR enzymes may be decreased, or possibly increased by induction, under various conditions. Hence systematic studies need to be conducted that would involve generation of various levels of oxidative stress in situ to alter the redox status of the cells incrementally [extending from some of the studies cited above (45,46,96)]. These studies should include assessment of the endogenous levels of the TDOR enzymes as well as manipulation of the TDOR enzyme levels by selective enhancement of gene expression and by gene knockout experiments.

C. Potential Regulatory Function of Thiol-Disulfide Interchange

In the previous section various examples were discussed from the point of view of repair whereby thioltransferase-catalyzed disulfide reduction reactions led to restoration of the functional properties of sulfhydryl proteins. For the most part, those studies represented extreme cases of oxidant exposure and ideal conditions for reductive reversal by thioltransferase, and they were so designed deliberately to identify possible physiological events. The perspective of this section is to distinguish sulfhydryl-linked regulatory changes from maintenance and repair processes as potential physiological roles for catalysis by thioltransferase (and thioredoxin).

Although the concept of functional modulation by thiol-disulfide interconversion often has been considered as a potential regulatory mechanism akin to phosphorylation/dephosphorylation, sufficient documentation is lacking that would establish such a mechanism for any process in eukaryotic cells (for critical reviews, see Refs. 5,23). Nevertheless there have been many studies in which the activities of functionally important proteins have been altered by thiol oxidation and restored by disulfide reduction, and these results have been interpreted to reflect the possibility for regulation in situ. For example, the ability of reduced thioredoxin to restore glucocorticoid-binding activity to the oxidized glucocorticoid receptor is cited as evidence for redox regulation of the receptor (103); and both thioltransferase and thioredoxin were reported to enhance the binding of L-triiodothyronine to its hepatic nuclear receptors (104). More recently the effects of sulfhydryl reagents were interpreted to implicate thiol-disulfide interchange in IP_3 receptor modulation connected with Ca^{2+} ionization in liver cells, but TDOR enzyme involvement was not considered in that case (105). Thioredoxin has been implicated as part of a putative redox control mechanism that modulates the DNA-binding activity of the transcription factors Fos/Jun and NF-κB (106,107). In the latter study, differential effects of GSSG were reported such that a low level of GSSG appeared to activate the transcription factor NF-κB, while higher concentrations interfered with its binding to DNA. These data were interpreted to reflect the possibility that a certain amount of GSSG may be required for activation and nuclear translocation of NF-κB, while an excess of GSSG may inhibit NF-κB binding to DNA (107). Although the authors suggested that both effects may be counterregulated by thioredoxin, the potential involvement of NF-κB-SSG suggests that thioltransferase might be the preferred enzyme mediator due to its glutathionyl selectivity. The stepwise sensitivity to GSSG may also indicate more than one SH group on NF-κB with different redox potentials or kinetic properties. A similar biphasic response to increasing oxidative conditions was associated with changes in the responsiveness of guanylate cyclase to nitric oxide (108). In both cases the intermediate activated forms were not characterized, and they may be either mixed disulfides with GSH or intramolecular disulfides, or both.

Previous work of Gilbert et al. (23,76,109,110) showed that variations in redox buffer status can alter the activities of key metabolic enzymes like phosphofructokinase, fatty acid synthetase, and HMG-CoA reductase. There are multiple factors to consider regarding which cellular sulfhydryl proteins might be affected by changes in the redox milieu, how rapidly the changes would occur, and how long the effect would persist. These factors include the magnitude of the change in GSH:GSSG ratio, the relative redox potentials of the proteins, and whether the approach to new steady state is enzyme catalyzed. At least two conditions are necessary in order for protein-thiol/disulfide interchange to serve as a regulatory mechanism, namely: (1) the redox potential of the critical protein should be near the range of GSH:GSSG ratios in vivo, and (2) enzyme catalysis should be observed, because typical rates of nonenzymatic reactions of protein -SH groups with GSSG under physiological conditions are quite slow relative to the time range expected for regulatory changes (minutes to hours) (76).

The concept of regulation is supported by observations of both inactivation and activation associated with thiol oxidation of particular proteins. In this context, fructose-1,6-biphosphatase (FBPase) and phosphofructokinase (PFK) represent enzymes that catalyze opposing reactions that are central to the control of glycolysis and gluconeogenesis. While PFK is inactivated by thiol oxidation, FBPase is activated by it, and these changes are reversed by disulfide reducing agents (5). Moreover, recent studies have reported that thioltransferase catalyzes the GSH-dependent reactivation of oxidized PFK (37,46,86) and the GHS-dependent deactivation of oxidized FBPase (111) (see below).

The potential involvement of thioltransferase in cellular regulatory phenomena was first suggested by Mannervik and Axelsson (39). In their report, progress curves were presented for the GSSG-dependent glutathionylation (formation of protein-SSG [^{3}H] and GSH-dependent deglutathionylation (release of GSH [^{3}H]) of cytosolic proteins from rat liver in the absence or presence of purified cytoplasmic thioltransferase from rat liver. There was a clear indication of catalysis of both the forward and reverse reactions, but the progress curves in the absence and presence of TTase did not tend toward the same equilibrium positions as would be expected, and the apparent rate enhancement by TTase for formation of the protein-SSG (~7.5-fold) (Fig. 12A) was more than two times greater than the rate enhancement for the reverse dethiolation reaction (~3-fold) (Fig. 12B). In our study of a prototype reaction for thioltransferase catalysis (hydroxyethyldisulfide + GSH ↔ GSSG + β-mercaptoethanol), we found that the enhancement by thioltransferase (~3-fold) was about

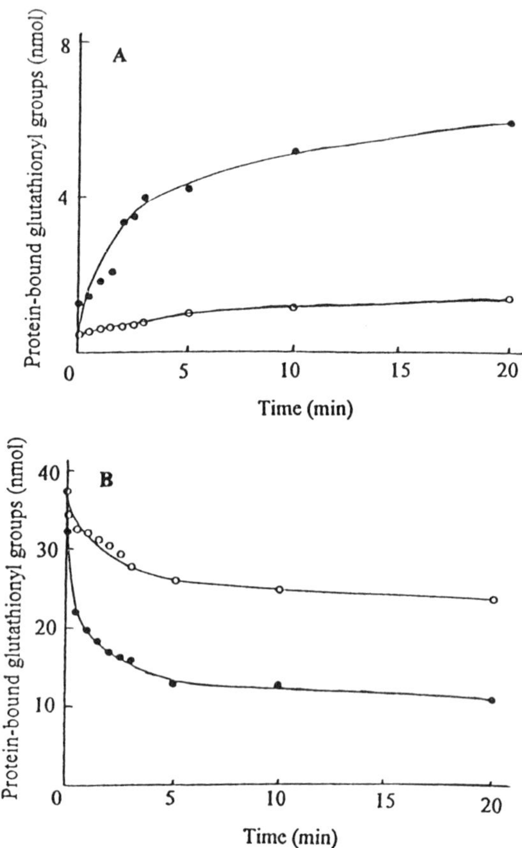

Figure 12 (A) Progress curves for the formation of protein-glutathione mixed disulfides (protein-SSG [^{3}H]) as monitored by incorporation of radioactivity. A post-Sephadex G-75 fraction of DTT-reduced cytosolic protein (11 mg/ml) from rat liver homogenate was allowed to react with 0.3 mM GSSG [^{3}H] in the absence (○) and presence (●) of partially purified rat liver cytoplasmic thioltransferase (0.6 units/ml) in a total volume of 0.4 ml. (B) Progress curves for the decomposition of protein-glutathione mixed disulfides (protein-SSG [^{3}H]) as monitored by loss of protein bound radioactivity. Protein-SSG [^{3}H] (1 mg/ml) was incubated with 0.5 mM GSH in the absence (○) and presence (●) of partially purified thioltransferase (1.2 units/ml) in a total volume of 0.2 ml. (From Ref. 39.)

the same for the forward and reverse reactions, and the position of equilibrium was identical in the absence and presence of enzyme and approached from either direction (Fig. 13).

None of the above-mentioned examples of potential regulation of activities by thiol-disulfide interchange included documented cases of catalysis of both the

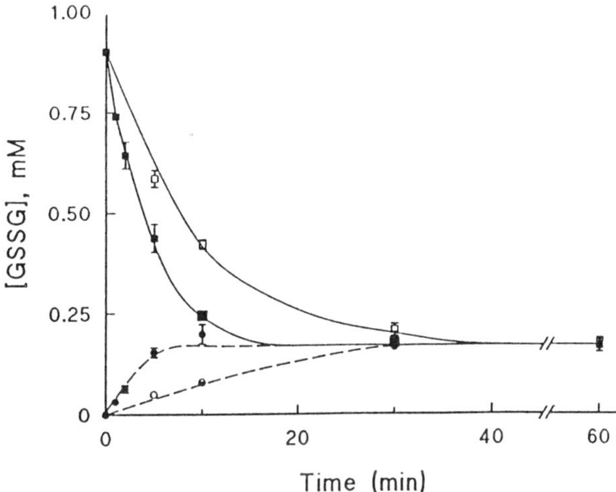

Figure 13 TTase catalysis of reversible thiol-disulfide interchange-approach to equilibrium. For one set of experiments (lower, dotted lines), the initial reaction mixtures contained 0.9 mM hydroxyethyldisulfide and 1.8 mM GSH ± RBC TTase (0.01 units), 30°C, pH 7.5. For the other set (upper lines), 0.9 mM GSSG and 1.8 mM mercaptoethanol were added ± TTase. The solid symbols in each case refer to the "plus-TTase" samples. Note the acceleration by enzyme to the same position of equilibrium. (From Ref. 15.)

forward and reverse directions of activity modification. Although the study of reversible transfer of GS equivalents to and from bulk cytosolic proteins provides a conceptual precedent for the approach (39), discrete proteins were not identified in that case, and there was no connection to modulation of cellular functions. In this regard, phosphofructokinase is an appropriate primary model enzyme to study, because its redox potential is near the range of physiological GSH:GSSG ratios (23), and catalytic reactivation of the oxidized enzyme has been studied, as cited above (37,46,86). The unique selectivity of TTase as a GS-transfer catalyst (described in the previous section) prompted us to begin studies of its ability to catalyze reversible interconversions (protein-SH/protein-SSG) of potentially physiologically relevant proteins, including phosphofructokinase. We and others have observed thioltransferase catalysis of reactivation of oxidized PFK-SSG and found it to be much more effective than thioredoxin (46,112). Characterizing catalysis of the forward reaction, however, has been problematic; i.e., we did not observe a significant enhancement of the rate of inactivation of PFK by GSSG in the presence of TTase (or thioredoxin) (112). This paradoxical result may be consistent with the nonenzymatic characterization of the reaction by Walters and Gilbert, who observed that inactivation of PFK by GSSG is complicated by the biphasic nature of the process whereby two of the SH groups of PFK are converted to mixed disulfides by GSSG, but the kinetics and thermodynamics of the two reactions differ substantially (109). Thus, the initial phase of oxidation of PFK by GSSG is kinetically rapid but thermodynamically disfavored so that a high GSSG:GSH ratio is required, whereas the kinetically slow phase is favored thermodynamically; i.e., it occurs at lower GSSG:GSH ratios. Therefore, further study is necessary to segregate the two steps of the inactivation reaction. For example, if the scheme developed by Walters and Gilbert is accurate (109), it should be possible to trap two different carboxymethyl-S-PFK-SSG modified forms of PFK by capturing one, then the other, of the PFK-SH groups as the iodoacetate derivative; i.e., during the rapid formation of the one GS-mixed disulfide and the rapid reduction of the other. These hybrids would give two different carboxymethyl-S-PFK-SH forms whose monophasic reactions with GSSG could be studied in the absence and presence of TTase.

Since thioltransferase displays selective catalysis of glutathione transfer between protein thiols and disulfides, it is conceivable that TTase may function to glutathionylate proteins to modulate function (Fig. 14). Thus, reversible action by TTase (or in conjunction with thioredoxin) to mediate glutathionylation/deglutathionylation would be analogous to the kinase/phosphatase cycles of phosphorylation/dephosphorylation. The hypothetical scheme in Figure 14 shows TTase as the intermediary in a signal transduction pathway that modulates the activity of intracellular enzymes or receptors. In this scheme an extracellular signal (e.g., disulfide-containing hormone) would bind to its receptor and cause intracellular exposure of a disulfide bond, which would react with GSH to form a glutathionyl mixed disulfide. TTase then would transfer the GS moiety from the receptor-SSG to glutathionylate critical proteins, altering their functions. Turnover of the TTase-SSG intermediate could be coupled to GSSG reductase as described earlier, or thioredoxin with its different redox potential might participate by serving as the reductant. The latter situation could also link the transduction pathway to calcium fluxes, because the activity of thioredoxin reductase appears to be altered by calcium binding (113).

A concept related to the scheme shown in Figure 14 has been presented in a different context (93). In that case, spin-echo NMR studies of intact erythrocytes were interpreted to reflect transduction of both oxidizing and reducing equivalents from impermeant DTNB and GSH, respectively, across the RBC membrane, resulting in changes in the intracellular redox state (93,114). DTNB caused changes in sulfhydryl status within the cell, including diminution of GSH and ergothionine levels. The intracellular changes were analyzed to be larger than expected from the measured stoichiometry of modified exofacial-SH groups. The authors offered a provocative hypothesis whereby the external oxidant effects could be transmitted into the internal environment via thiol/disulfide exchange involving the intramembrane glucose transporter and the inward-facing

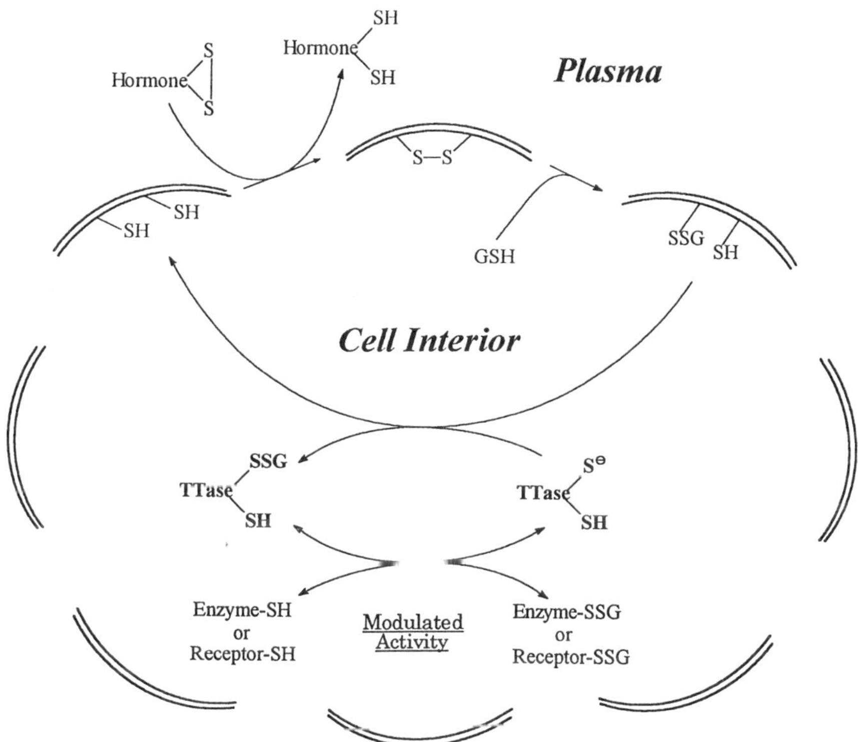

Figure 14 Hypothetical signal transduction mediated by thioltransferase. The scheme depicts a possible mechanism whereby thioltransferase would act as the catalyst in the transduction of a regulatory signal from an extracellular hormone, which utilizes glutathione as the second messenger.

cytoskeletal element spectrin, leading to spectrin-SSG accumulation (93). No data were provided to support the details of their model, but it represents a parallel concept to the hypothetical scheme of Figure 14 which depicts the potential role of TTase as a catalyst of such oxidant-initiated signal transduction involving protein-SSG formation.

A speculative mechanism akin to the concept of Figure 14 has been implied also for the action of follicle-stimulating hormone (FSH), which contains a tetrapeptide sequence identical to the active site sequence of thioredoxin and displays thioredoxin-like catalytic activity (115). Thus, it was suggested that the redox-active dithiol present within a receptor-binding domain of the beta-subunit of FSH may have a physiological role in receptor binding or signal transduction.

The conceptual framework for regulation of physiological events by thiol-disulfide interchange continues to be broadened by more examples of physiologically important molecules whose activity is changed by modification of thiol status in a chemically competent and reversible fashion. Evidence is lacking, however, for kinetically competent modulation of activities by alterations of thiol status that are physiologically relevant. Most of the reports that are used to support the concept of regulation are better viewed as support for homeostatic repair mechanisms. Thorough studies have been carried out by Gilbert and his associates, but the thermodynamic feasibility rather than the kinetic feasibility was emphasized. Clearly enzyme intervention must occur if regulation is to be a significant consideration. Equally critical is the scarcity of documentation of functional changes in thiol-disulfide status for specific proteins in situ in intact tissues. This is a challenging task that represents the frontier for advancement of knowledge in this area.

D. Participation of TDOR Enzymes in Protein Processing

Disulfide reduction may be an important initial step in the degradation of proteins whose native structures are maintained by an intramolecular disulfide lattice. Consistent with this concept, reduction of a specific disulfide bond was shown to be required for ubiquitin conjugation and degradation of lysozyme in vitro (116). The catalytic ability of TTase and thioredoxin to reduce disulfide bonds in various proteins has led to the speculation that one of their physiological functions may be to initiate degradation of certain hormones and proteins (117). As alluded to earlier, it is conceivable that cytosolic thioltransferase might also be involved in completing the processing of transmembrane proteins. It is likely that the cytosolic domains of membrane proteins are threaded through the lumen of the endoplasmic reticulum (ER) during synthesis of the entire protein. Such domains that have exposed Cys-SH groups without other Cys-SH residues in conformational proximity would be delivered to the cytosol as protein-S-SG derivatives, because the redox buffer of the ER is shifted toward oxidation; i.e., the GSH:GSSG ratio of the ER apparently approaches 1:1, in contrast to the cytosol, where the GSH:GSSG ratio exceeds 100:1 (77). Thus, cytosolic thioltransferase may complete the protein processing of the cytosolic domains by catalyzing deglutathionylation to reexpose the SH groups.

Appropriate pairing of cysteine residues to form disulfide bonds is an important stage in posttranslational formation of the native structures of many proteins. Catalysis of this process has been attributed primarily to protein disulfide isomerase (PDI), a TDOR enzyme distinct from thioltransferase and thioredoxin (23,24), as described in Section I.

Renaturation of reduced, denatured ribonuclease (RNase) has been used as a typical model for protein-disulfide processing (118,119). In one study Lyles and Gilbert found that PDI at concentrations about the same as that of the substrate RNase induced disulfide bonds, and the PDI was reoxidized apparently by the RNase disulfides in the absence of other cofactors (118). They argued that PDI concentration in the ER may approach the millimolar range, providing amounts of the "enzyme" in situ that would be stoichiometric with substrate proteins. In a companion study, catalytic amounts of PDI were recycled between redox states in appropriate GSH/GSSG redox buffers (119). From these examples at least two alternative modes of action of PDI are suggested, namely, stoichiometric redox cycling and catalytic thiol-disulfide interchange facilitated by GSH/GSSG. Besides these two possibilities, two other potential mechanisms of enzyme-assisted disulfide pairing warrant consideration on the basis of other studies.

By analogy to the glutathionyl specificity of thioltransferase as depicted by the catalytic scheme of Figure 6, it is conceivable that PDI catalyzes the formation of disulfides by transferring the glutathionyl moiety from GSSG to form a HS-protein-SSG mixed disulfide that is then poised to form a protein-(S-S) intramolecular disulfide with a conformationally near Cys residue by displacement of GSH (see Fig. 15). Consistent with thioltransferase-like behavior, PDI is capable of catalyzing GSH-dependent disulfide reduction coupled to GSSG reductase, and PDI displays sequential two-substrate kinetics for non-glutathionyl disulfide substrates (120). The hypothetical scheme for PDI (Fig. 15) is based also on interpretation of studies of the processing of mutant forms of lysozyme (122,123). Taniyama et al. (122) found that a mutant of human lysozyme (C77A), in which the usual C77-C95 disulfide bond was prevented, was secreted by yeast partly as the lysozyme-C95-S-SG mixed disulfide. They viewed this protein-SSG derivative as representing a typical intermediate in normal disulfide formation in the ER that had been stabilized because the C77-SH was missing. In a follow-up study, Hayano et al. (123) showed that PDI in a redox buffer (2 mM GSH/0.2 mM GSSG) displaced GSH from the lysozyme-SSG "intermediate" more efficiently than did thioredoxin; however, each "enzyme" (PDI and thioredoxin) was present at twice the concentration of the protein-S-SG substrate.

In a recent study using a 28-residue disordered peptide containing only two cysteines as a model substrate for PDI, catalysis of formation of glutathionyl mixed disulfides with each of the cysteines was observed, but PDI did not catalyze the reaction of cystamine with the reduced peptide. PDI also did not catalyze interchange of the GS moiety between the two cysteine residues in the GS-S-peptide-SH hybrid. However, PDI did catalyze intramolecular disulfide formation from both the HS-protein-SSG and the HS-protein-SSCysteamine derivatives (124). These results show some similarity to the GS selectivity of TTase, and they also display some paradoxical lack of catalysis reminiscent of what we observed in our initial studies of TTase involvement in the inactiva-

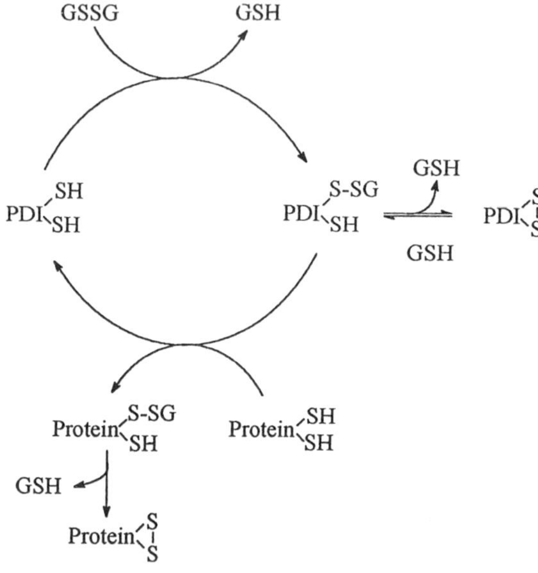

Figure 15 Hypothetical reaction scheme for PDI catalysis. This reaction scheme is analogous to that derived kinetically for TTase (Fig. 6). PDI is predicted to operate in the oxidizing direction, however, showing a specific recognition for GSSG (rather than protein-S-SG) and displaying little specificity for the protein dithiol substrates. The latter supposition is consistent with the report of Morjana and Gilbert (121), which showed that PDI was not very sensitive to a variety of peptides and proteins tested as potential inhibitors in an attempt to probe the substrate recognition site of PDI.

tion of PFK by GSSG (see previous section). Certainly additional studies are necessary to unravel the detailed mechanism of PDI catalysis.

Another consideration (see Sec. I) is the concept that protein-disulfide processing by PDI may be coupled with other TDOR enzymes in a fashion analogous to the protein disulfide isomerase system that is being characterized in *E. coli* (30–32,125). Bardwell et al. (30,31) discovered mutants of *E. coli* that are deficient in protein disulfide bond formation, and thereby identified two proteins DsbA and DsbB that are required for this process. The periplasm in *E. coli* is likened to the endoplasmic reticulum in eukaryotes, and DsbA has a redox potential that favors oxidation of thiols (32) analogous to PDI (25,27) (see Table 2). Thus, DsbA and PDI are expected to oxidize dithiols to disulfides (125). The companion protein DsbB is suggested to serve by reoxidizing DsbA. The mechanism of redox cycling of the DsbB has not been identified, but it is thought to be linked to redox processes in the cytoplasm. Thus a network of enzymes may combine to accomplish protein disulfide processing in *E. coli*. By analogy, the TDOR enzymes of eukaryotes, which display different redox potentials, may act synergistically also. Consistent with this hypothesis, the cellular distribution of thioredoxin apparently includes the ER (11), suggesting coexistence with PDI. The redox potential of thioltransferase and its possible role in protein processing remain to be determined.

E. Potential Role of TDOR Enzymes in Cancer

Under chemotherapy, the intracellular environment of cancer cells and normal cells are confronted with alkylating and quinone agents like BCNU, doxorubicin, etc., which generate electrophiles, radicals, oxidants, and peroxides (126). Glutathione and various enzymes associated with sulfhydryl biochemistry have been viewed as the main line of defense against these insults. Breakdown of this defense in cancer cells and reinforcement in normal cells is the ideal of effective therapy. It is also likely that changes in the levels of these enzymes and their substrates and cofactors may contribute to chemotherapeutic resistance in cancer cells. The key enzymes in this context include glutathione-S-transferase (GSTase), thioltransferase (TTase), glutathione peroxidase (GPXase), glutathione reductase (GRase), thioredoxin (TRx), thioredoxin reductase (TRase), and glucose-6-phosphate dehydrogenase (G6PD). As reviewed below, much attention has been given to changes in GSH levels in cancer and normal cells associated with chemotherapy, and antineoplastic-dependent inhibition and induction of some of the GSH-utilizing enzymes have been studied. The TDOR enzymes thioltransferase and thioredoxin and their associated reductases, however, generally have not been included in this context despite their multifaceted involvement in cellular physiology, especially in proliferating cells under chemical stress. Moreover, cross-inhibition of the various enzymes by typical anticancer agents or concomitant induction associated with resistance have not been studied systematically.

1. Influence of TDOR Enzymes on Anticancer Agents

TDOR enzymes occur throughout nature and play key roles in cell economy. They are required for DNA synthesis in conjunction with ribonucleotide reductase–mediated production of the deoxyribonucleotides, and they are implicated in a wide variety of other disulfide interchange reactions. A role in overall protein synthesis has also been indicated (127). A protein closely related to thioredoxin and associated with plasma membranes was suggested to serve as an essential growth factor (128). The growth-promoting factor adult T-cell–derived factor (ADF) is also closely related to thioredoxin, and it has been detected in many malignant tissues (129). As described in the preceding section, thioredoxin has been implicated as part of the redox control for enhancing the DNA-binding activity of the transcription factors Fos/Jun and NF-κB

(106,107), and it is expected that thioltransferase also may participate in this process because of the likely involvement of protein-SSG mixed disulfides.

TDOR enzymes have also been implicated in the repair of key sulfhydryl enzymes that may be damaged by oxidation or alkylation associated with chemotherapy. Phosphofructokinase is reactivated by TTase (37,46,86). Ornithine decarboxylase, the first committed step in synthesis of polyamine mediators of cell proliferation, is inactivated by oxidation or alkylation, and the inactivated enzyme can be reactivated by TTase and thioredoxin (45). The family of GSTases are characterized as important scavengers of electrophiles, a detoxification process. In contrast, GSTase has been implicated also in catalytic formation of the toxic episulfonium species (130). Certain isozymes of GSTase have been shown to be susceptible to oxidative and radical-mediated inactivation, and thioltransferase was shown to catalyze reactivation of the modified GSTase (62,72). Ribonucleotide reductase, thioredoxin reductase, and GSSG reductase are all inactivated by β-chlorethyl–containing nitrosoureas like BCNU (131,132), and covalently inactivated thioredoxin reductase apparently can be reactivated by thioredoxin (described below) (132). The preceding list of sulfhydryl proteins that play critical roles in proliferating cells and may be targets of chemotherapy is certainly not comprehensive. Other examples of important proteins whose activity can be reversibly altered by sulfhydryl oxidation/reduction include the cytoskeletal proteins tubulin and actin; HMG-CoA reductase, the first committed step in cholesterol synthesis; glucose-6-phosphate dehydrogenase and GSSG reductase, vital for maintaining NADPH and GSH, respectively; and lipoamide dehydrogenase, required for production of acetyl-CoA, which is linked to mitochondrial ATP production (5,23). Definitive studies on reactivation of these other proteins by TDOR enzymes, however, have not been reported.

2. *Effects of Anticancer Agents on TDOR Enzymes*

Many cancer chemotherapeutic agents act by damaging rapidly replicating DNA by alkylation or oxidative processes, which may be protected by GSH and TDOR enzymes (for review of GSH role in cancer chemotherapy, see Ref. 133). Hence, direct damage of the TDOR enzymes may play a key role in the cytotoxic effects of various cancer chemotherapeutic agents and relate to the efficacy or side effects of the drugs. Recently pig liver TTase has been shown under anaerobic conditions to be strongly inhibited by the chemotherapeutic agents cisplatin and carboplatin in vitro (134). Our laboratory is also currently testing the effects of chemotherapeutic agents on thioltransferase. Figure 16 depicts results of a study of the relative sensitivities of GSSG reductase and TTase to inactivation by BCNU. We found that TTase is much less sensitive to BCNU than is GSSG reductase; i.e., 50% inactivation of an equivalent amount of TTase required about 100 times more BCNU than did GSSG reductase (135). Thus, the reductive capacity of the TTase system is more likely to be decreased by

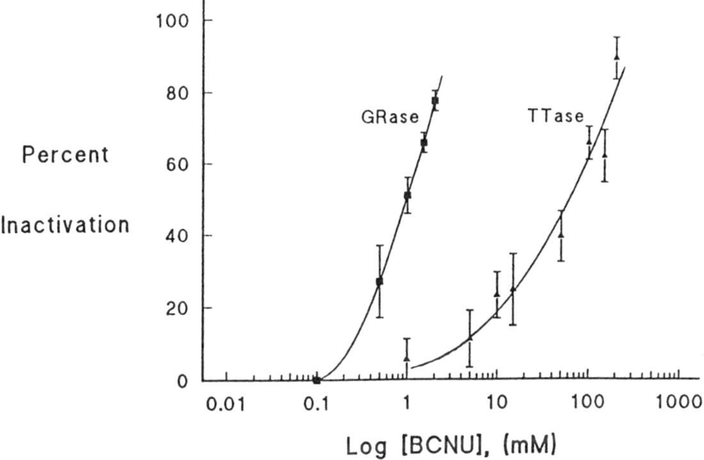

Figure 16 Inactivation of GSSG reductase and thioltransferase by BCNU. The enzymes were pretreated with the appropriate reducing agent (0.2 mM NADPH for GRase and 0.2 mM DTT for TTase) to ensure that they were in reduced forms that would be susceptible to alkylation. Then aliquots of each enzyme solution were diluted to 0.5 μM in separate reaction mixtures containing potassium phosphate buffer, pH 7.5, and various concentrations of BCNU as indicated. The reaction mixtures were incubated at room temperature for 30 min, and then aliquots were diluted into the appropriate assay mixtures and the remaining enzyme activity was determined. Controls representing 100% activity were carried through the same procedure in the absence of BCNU. The small amount of DTT in the reaction mixtures with TTase and BCNU could not account for the low potency of BCNU toward TTase, because the residual DTT concentration was insignificant relative to the BCNU concentrations.

BCNU inactivation of GSSG reductase rather than by direct inactivation of TTase. This selective effect of BCNU on the reductase could divert the action of TTase in the oxidative direction to glutathionylate proteins. All of the enzymes that have been reported to be sensitive to BCNU inhibition contain a pair of cysteine residues at their active sites [e.g., GSSG reductase, thioredoxin reductase, ribonucleotide reductase (131,132)]. This particular feature is common also to TTase and thioredoxin. The relative insensitivity of TTase to BCNU, if true also for thioredoxin, would suggest that the dithiol nature of their active sites per se is not a sufficient determinant for sensitivity of the various reductase enzymes to inactivation by BCNU, and further study is necessary to understand the selective toxicity. Schallreuter et al. reported that chloroethyl-thioether adducts were formed from radiolabeled fotemustine (an analog of BCNU) at the active sites of the inactivated reductase enzymes, and that the adducts could be displaced via thiol-mediated beta-elimination reactions yielding chloride and ethylene (132). The release of radiolabel from the thioredoxin reductase adduct was mediated by reduced thioredoxin, and the adduct was released from GSSG reductase by GSH. Although it was not clear from this report how much of the inactivation of the respective enzymes was actually related to the particular adduct described, these data suggest a broader repair role for the TDOR enzymes, potentially involving beta-elimination reactions besides dethiolation of protein disulfides. Figure 17 depicts β-elimination reactions initiated by reduced thioredoxin and GSH that would lead to reactivation of thioredoxin reductase (TRase) and glutathione reductase (GRase), respectively. A potential role for TTase catalysis is also shown in the GRase reactivation reaction. Much more experimentation is necessary to test the various aspects of these hypothetical schemes.

In another study, doxorubicin was found to inactivate thioredoxin reductase in vitro, suggesting that this effect is contributory to its growth inhibitory action (136). This study did not examine the effects of doxorubicin on TTase, thioredoxin, or GSSG reductase. Since the net anticancer effect may be related not only to disruption of DNA processing, but also to interference with defense and repair systems, one might expect quinones like doxorubicin to cause inactivation of the components of the TDOR enzyme systems and other GSH enzymes as well.

3. *Drug Resistance*

Although amplification of P-glycoprotein and increased drug efflux has been associated with many cases of multiple drug resistance (MDR), alternative mechanisms also contribute importantly, e.g., amplification of target enzymes, increased cellular detoxification processes, increased DNA repair systems, and diminution in drug-activating enzymes. In this regard, previous studies have shown correlations between elevated levels of detoxification enzymes like GSTase in cells resistant to anticancer agents (137). For example, a human breast cancer cell line displayed a correlation between the level of resistance to doxorubicin and increase in activity of an unusual GSTase with high organic peroxidase activity (138). In a model study high levels of human GSTase were expressed in yeast cells, resulting in decreased sensitivity to chlorambucil and doxorubicin (139). In another case, development of antineoplastic resistance to doxorubicin and other quinones was mimicked by the novel introduction of GSTase and GPXase into MCF-7 cells by transient mechanical disruption of the cell membrane, without any changes in the phenotype or genotype of the cells (140). Cole et al. (141) reported isolation of an MDR human small-cell lung cancer line that had shown no increase in P-glycoprotein and a sixfold lower concentration of GSH but elevated levels of GSTase and GSSG reductase. The common feature of these various studies is the indication that it is possible for drug resistance to develop by changes in the levels of the GSH-utilizing enzymes

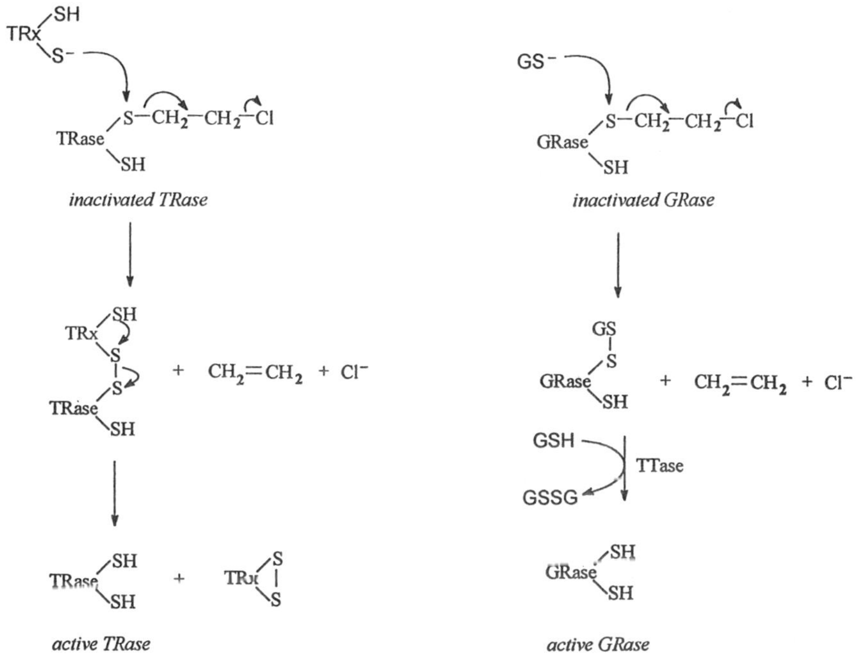

Figure 17 Hypothetical schemes for reactivation of chloroethylated adducts of thioredoxin reductase and glutathione reductase. The scheme at the left depicts a beta-elimination reaction mediated by reduced thioredoxin (TRx) in which the chloroethyl adduct is displaced from thioredoxin reductase (TRase) and an intermediate TRx-S-S-TRase interprotein disulfide is formed with concomitant release of ethylene and chloride. Active TRase is generated from this intermediate as shown, along with oxidized TRx. NADPH and active TRase would be necessary to recycle the oxidized TRx. The scheme at the right depicts a beta-elimination reaction intiated by reduced glutathione in which the chloroethyl adduct is displaced from glutathione reductase (GRase) and an intermediate GRase-S-SG mixed disulfide is formed with concomitant release of ethylene and chloride. Active GRase is generated from this intermediate by a second reaction involving GSH, and this step could be catalyzed by thioltransferase (TTase). NADPH and active GRase would be necessary to recycle the GSSG. These schemes were generated from interpretation of the study of fotemustine inactivation of TRase and GRase by Schallreuter et al. (132). Certainly additional experimentation is necessary to test the various details of these schemes.

without concomitant changes in GSH concentration or enhancement of transport processes for extruding the drugs or drug adducts. Although good quantitative correlation between enzyme level and degree of resistance was found in some cases, often the degree of resistance exceeded the changes in specific enzyme activities. This disparity suggests additional or synergistic adaptive mechanisms.

Whereas GSTase protects cellular targets by forming GSH adducts with alkylating agents and GPXase detoxifies peroxides generated by redox active anticancer compounds, TTase and thioredoxin are expected to serve as repair enzymes that can reverse the damage from oxidizing agents and radical generators. In adapting to a variety of chemical insults, one might expect cells to increase not only the protective factors but also the repair factors. Hence it is likely that increases in repair enzymes like TTase and thioredoxin would contribute to synergistic resistance. Pertinent to this possibility is evidence for the inducibility of these enzymes in other contexts.

4. *Inducibility of the TDOR Enzymes*

TTase and GSSG reductase, like GSTase, are inducible at least in liver by a variety of agents including phenobarbital and the dietary antioxidant agent BHA (2,3-*t*-butyl-hydroxyanisole) (142). Hence there is a precedent for modulation of the level of TTase-linked activities, which relates to the potential for development of resistance to chemotherapeutic agents that put stress on the GSH network. Likewise, there is evidence for inducibility of the thioredoxin system. From a developmental standpoint, the activity of the thioredoxin system may be more dependent on the level of thioredoxin reductase than on that of thioredoxin in neonatal rat level (143). In other studies it has been noted that thioredoxin reductase, thioredoxin, and ribonucleotide reductase are elevated in rapidly dividing cells (144). Also, mouse and human melanoma cells in culture have higher levels of (membrane-associated) thioredoxin reductase activities compared to normal melanocytes (145). It would seem that the ultimate redox support for the TDOR systems, namely, production of NADPH, would also increase under conditions of greater demand. Although not usually measured, in one report of mouse leukemia cells resistant to doxorubicin, G6PD (which produces NADPH) was nearly doubled (146).

5. *Conclusion*

Under conditions of combination chemotherapy (simultaneous or sequential), the nonspecific variety of chemical insults may affect all sulfhydryl groups to some extent. The consequences of SH modification on different proteins then depend on the status of each protein with regard to its cellular concentration, its role in rate-determining or committed steps in metabolic pathways, the state of proliferation of the cell, etc. The net effect on each of these processes also depends on the status of the sulfhydryl defense and repair systems.

The finding of a rather strict substrate specificity for thioltransferase, requiring a glutathionyl disulfide, provides a fundamental distinction between the mechanisms of action of thioltransferase and thioredoxin, because thioredoxin operates completely independently of GSH in reduction of disulfides. This distinction is important in the context of chemotherapy, because diminution of GSH and inactivation of GSSG reductase would disable the thioltransferase system even if the thioltransferase protein itself were not inactivated. These effects, however, would not disable the thioredoxin system. Thus it is important to understand the relative efficiencies of the thioltransferase and thioredoxin systems in catalyzing similar reactions that are important to proliferating cells, as well as to distinguish their catalytic mechanisms and sensitivities to inhibition by anticancer agents.

ACKNOWLEDGMENTS

Grateful recognition is given to our former coworkers who contributed to some of the work reviewed in this article: Susanna Beshai, Jennifer Grant, and Brenda Nieslanik. We also thank Dr. G. David McCoy for critical review of the manuscript prior to submission. The work was supported in part by grants from the American Heart Association (881186 and 91014450) and from the American Cancer Society.

REFERENCES

1. Mannervik, B., Carlberg, L., and Larson, K. (1989) Glutathione: General review of mechanism of action. In: Coenzymes and Cofactors Vol. III, Part A, (Dolphin, D., Paulson, R., and Amramovic, O., eds.), pp. 475–516. John Wiley & Sons, New York.
2. Holmgren, A. (1985) Thioredoxin. Annu. Rev. Biochem. 54:237–271.
3. Holmgren, A. (1989) Thioredoxin and glutaredoxin systems. J. Biol. Chem. 264:13963–13966.
4. Wells, W. W., Yang, Y., Deits, T. L., and Gan, Z. R. (1993) Thioltransferases. Adv. Enzymol. Relat. Areas Mol. Biol. 66:149–201.
5. Ziegler, D. M. (1985) Role of reversible oxidation-reduction of enzyme thiols-disulfides in metabolic regulation. Annu. Rev. Biochem. 54:305–329.
6. Racker, E. (1955) Glutathione-homocystine transhydrogenase. J. Biol. Chem. 217:867–974.
7. Askelof, P., Axelsson, K., Eriksson, S., and Mannervik, B. (1974) Mechanism of action of enzymes catalyzing thiol-disulfide interchange. Thioltransferases rather than transhydrogenases. FEBS Lett. 38:263–267.
8. Holmgren, A. (1976) Hydrogen donor system for Escherichia coli ribonucleoside-diphosphate reductase dependent upon glutathione. Proc. Natl. Acad. Sci. USA 73:2275–2279.
9 Papayannopoulos, I. A., Gan, Z. R., Wells, W. W., and Biemann, K. (1989) A revised sequence of calf thymus glutaredoxin. Biochem. Biophys. Res. Commun. 159:1448–1454.
10. Papov, V. V., Gravina, S. A., Mieyal, J. J., and Biemann, K. (1994) The primary structure and properties of thioltransferase (glutaredoxin) from human red blood cells. Protein Sci. 3:428–434.
11. Rozell, B., Holmgren, A., and Hansson, H. A. (1988) Ultrastructural demonstration of thioredoxin and thioredoxin reductase in rat hepatocytes. Eur. J. Cell Biol. 46:470–477.

12. Kallis, G. B., and Holmgren, A. (1980) Differential reactivity of the functional sulfhydryl groups of cysteine-32 and cysteine-35 present in the reduced form of thioredoxin from *Escherichia coli*. J. Biol. Chem. 255:10261–10265.
13. Gan, Z. R., Sardana, M. K., Jacobs, J. W., and Polokoff, M. A. (1990) Yeast thioltransferase—the active site cysteines display differential reactivity. Arch. Biochem. Biophys. 282:110–115.
14. Gan, Z. R., and Wells, W. W. (1987) Identification and reactivity of the catalytic site of pig liver thioltransferase. J. Biol. Chem. 262:6704–6707.
15. Mieyal, J. J., Starke, D. W., Gravina, S. A., and Hocevar, B. A. (1991) Thioltransferase in human red blood cells: kinetics and equilibrium. Biochemistry 30:8883–8891.
16. Hopper, S., Johnson, R. S., Vath, J. E., and Biemann, K. (1989) Glutaredoxin from rabbit bone marrow. Purification, characterization, and amino acid sequence determined by tandem mass spectrometry. J. Biol. Chem. 264:20438–20447.
17. Gan, Z. R., and Wells, W. W. (1987) The primary structure of pig liver thioltransferase. J. Biol. Chem. 262:6699–6703.
18. Hoog, J. O., Jornvall, H., Holmgren, A., Carlquist, M., and Persson, M. (1983) The primary structure of *Escherichia coli* glutaredoxin. Distant homology with thioredoxins in a superfamily of small proteins with a redox-active site cystine disulfide/cysteine dithiol. Eur. J. Biochem. 136:223–232.
19. Chrestensen, C. A., Eckman, C., Starke, D. W., and Mieyal, J. J. (1995) Cloning and Expression of Human Thioltransferase (TTase) in *E. coli*. ASBMB/ACS Joint Meeting, San Francisco, May, 1995.
20. Bushweller, J. H., Billeter, M., Holmgren, A., and Wuthrich, K. (1994) The nuclear magnetic resonance solution structure of the mixed disulfide between *Escherichia coli* glutaredoxin (C14S) and glutathione. J. Mol. Biol. 235:1585–1597.
21. Gravina, S. A., and Mieyal, J. J. (1993) Thioltransferase is a specific glutathionyl mixed disulfide oxidoreductase. Biochemistry 32:3368–3376.
22. Gravina, S. A. (1993) Characterization and kinetic mechanism of thioltransferase. Ph.D. thesis. Case Western Reserve University, Cleveland, OH.
23. Gilbert, H. F. (1990) Molecular and cellular aspects of thiol-disulfide exchange. Adv. Enzymol. Relat. Areas Mol. Biol. 63:69–172.
24. Bassuk, J. A., and Berg, R. A. (1989) Protein disulphide isomerase, a multifunctional endoplasmic reticulum protein. Matrix 9:244–258.
25. Hawkins, H. C., Blackburn, E. C., and Freedman, R. B. (1991) Comparison of the activities of protein disulphide-isomerase and thioredoxin in catalysing disulphide isomerization in a protein substrate. Biochem. J. 275:349–353.
26. Lundstrom, J., and Holmgren, A. (1993) Determination of the reduction-oxidation potential of the thioredoxin-like domains of protein disulfide-isomerase from the equilibrium with glutathione and thioredoxin. Biochemistry 32:6649–6655.
27. Siedler, F., Rudolph, B. S., Doi, M., Musiol, H. J., and Moroder, L. (1993) Redox potentials of active-site bis(cysteinyl) fragments of thiol-protein oxidoreductases. Biochemistry 32:7488–7495.
28. Nilsson, O., Tapia, O., and van Gunsteren, W. F. (1990) Structure and fluctuations of bacteriophage T4 glutaredoxin modelled by molecular dynamics. Biochem. Biophys. Res. Commun. 171:581–588.

29. Sandberg, V. A., Kren, B., Fuchs, J. A., and Woodward, C. (1991) *Escherichia coli* glutaredoxin: cloning and overexpression, thermodynamic stability of the oxidized and reduced forms, and report of an N-terminal extended species. Biochemistry 30:5475–5484.
30. Bardwell, J. C., McGovern, K., and Beckwith, J. (1991) Identification of a protein required for disulfide bond formation in vivo. Cell 67:581–589.
31. Bardwell, J. C., Lee, J. O., Jander, G., Martin, N., Belin, D., and Beckwith, J. (1993) A pathway for disulfide bond formation in vivo. Proc. Natl. Acad. Sci. USA 90:1038–1042.
32. Wunderlich, M., Otto, A., Seckler, R., and Glockshuber, R. (1993) Bacterial protein disulfide isomerase: efficient catalysis of oxidative protein folding at acidic pH. Biochemistry 32:12251–12256.
33. Chang, S. H., and Wilken, D. R. (1966) Participation of the unsymmetrical disulfide of coenzyme A and glutathione in an enzymatic sulfhydryl-disulfide interchange. I. Partial purification and properties of the bovine kidney enzyme. J. Biol. Chem. 241:4251–4260.
34. Nagai, S., and Black, S. (1968) A thiol-disulfide transhydrogenase from yeast. J. Biol. Chem. 243:1942–1947.
35. Axelsson, K., Eriksson, S., and Mannervik, B. (1978) Purification and characterization of cytoplasmic thioltransferase (glutathione:disulfide oxidoreductase) from rat liver. Biochemistry 17:2978–2984.
36. Axelsson, K., and Mannervik, B. (1980) General specificity of cytoplasmic thioltransferase (thiol:disulfide oxidoreductase) from rat liver for thiol and disulfide substrates. Biochim. Biophys. Acta 613:324–336.
37. Mieyal, J. J., Starke, D. W., Gravina, S. A., Dothey, C., and Chung, J. S. (1991) Thioltransferase in human red blood cells: purification and properties. Biochemistry 30:6088–6097.
38. Yang, Y. F., and Wells, W. W. (1991) Identification and characterization of the functional amino acids at the active center of pig liver thioltransferase by site-directed mutagenesis. J. Biol. Chem. 266:12759–12765.
39. Mannervik, B., and Axelsson, K. (1980) Role of cytoplasmic thioltransferase in cellular regulation by thiol-disulphide interchange. Biochem. J. 190:125–130.
40. Bushweller, J. H., Aslund, F., Wuthrich, K., and Holmgren, A. (1992) Structural and functional characterization of the mutant *Escherichia coli* glutaredoxin (C14S) and its mixed disulfide with glutathione. Biochemistry 31:9288–9293.
41. Perutz, M. F., Fersht, A. R., Simon, S. R., and Roberts, G. C. (1974) Influence of globin structure on the state of the heme. II. Allosteric transitions in methemoglobin. Biochemistry 13:2174–2186.
42. Holmgren, A. (1979) Glutathione-dependent synthesis of deoxyribonucleotides. Characterization of the enzymatic mechanism of *Escherichia coli* glutaredoxin. J. Biol. Chem. 254:3672–3678.
43. Mannervik, B., Axelsson, K., Sundewall, A. C., and Holmgren, A. (1983) Relative contributions of thioltransferase-and thioredoxin-dependent systems in reduction of low-molecular-mass and protein disulphides. Biochem. J. 213:519–523.
44. Park, E. M., and Thomas, J. A. (1988) S-thiolation of creatine kinase and glycogen phosphorylase b initiated by partially reduced oxygen species. Biochim. Biophys. Acta 964:151–160.

45. Flamigni, F., Marmiroli, S., Caldarera, C. M., and Guarnieri, C. (1989) Involvement of thiol transferase- and thioredoxin-dependent systems in the protection of 'essential' thiol groups of ornithine decarboxylase. Biochem. J. 259:111–115.
46. Yoshitake, S., Nanri, H., Fernando, M. R., and Minikami, S. (1994) Possible differences in the regenerative roles played by thioltransferase and thioredoxin for oxidatively damaged proteins. J. Biochem. 116:42–46.
47. Chaiken, I. M., and Smith, E. L. (1969) Reaction of chloroacetamide with the sulfhydryl group of papain. J. Biol. Chem. 244:5087–5094.
48. Xia, T. H., Bushweller, J. H., Sodano, P., Billeter, M., Bjornberg, O., Holmgren, A., and Wuthrich, K. (1992) NMR structure of oxidized *Escherichia coli* glutaredoxin: Comparison with reduced E. coli glutaredoxin and functionally related proteins. Protein Sci. 1:310–321.
49. Eriksson, S., Askelof, P., Axelsson, K., and Mannervik, B. (1974) Evidence for a glutathione-dependent interconversion of two forms of a thioltransferase from rat liver catalyzing thiol-disulfide interchange. Acta Chem. Scand. B. 28:931–936.
50. Yang, Y. F., and Wells, W. W. (1991) Catalytic mechanism of thioltransferase. J. Biol. Chem. 266:12766–12771.
51. Mieyal, P. A., and Mieyal, J. J. (1993) Unpublished observations.
52. Srinivasan, U., Mieyal, P. A., and Mieyal, J. J. (1995) pH profiles indicative of the rate limiting step in thioltransferase catalyzed disulfide reduction. Biochemistry (submitted).
53. Yang, Y. F., and Wells, W. W. (1990) High-level expression of pig liver thioltransferase (glutaredoxin) in *Escherichia coli.* J. Biol. Chem. 265:589–593.
54. Markley, J. L. (1974), Observation of histidine residues in proteins by means of nuclear magnetic resonance spectroscopy. Acc. Chem. Res. 8:70–80.
55. Keire, D. A., Strauss, E., Guo, W., Noszal, B., and Rabenstein, D. L. (1992) Kinetics and equilibria of thiol/disulfide interchange reactions of selected biological thiols and related molecules with oxidized glutathione. J. Org. Chem. 57:123–127.
56. Nikkola, M., Gleason, F. K., Saarinen, M., Joelson, T., Bjornberg, O., and Eklund, H. (1991) A putative glutathione-binding site in T4 glutaredoxin investigated by site-directed mutagenesis. J. Biol. Chem. 266:16105–16112.
57. Flohe, L., and Gunzler, W. A. (1974), Glutathione peroxidase. In: Glutathione, Proc. 16th Conf. German Soc. Biol. Chem. (Flohe, L., Benohr, H., Sies, H., Waller, H. D., and Wendel, A., eds.). Academic Press, New York, pp. 132–144.
58. Suhr, L. A., and Mieyal, J. J. (1992) Unpublished observations.
59. Mozer, T. J., Tiemeier, D. C., and Jaworski, E. G. (1983) Purification and characterization of corn glutathione S-transferase. Biochemistry 22:1068–1072.
60. Srinivasan, U. (1995) Studies on the properties and selective inhibition of thioltransferase. Ph.D. thesis, Case Western Reserve University, Cleveland, OH. In preparation.
61. Hoog, J. O., Holmgren, A., D'Silva, C., Douglas, K. T., and Seddon, A. P. (1982) Glutathione derivatives as inhibitors of glutaredoxin and ribonucleotide reductase from *Escherichia coli*. FEBS Lett. 138:59–61.
62. Terada, T., Maeda, H., Okamoto, K., Nishinaka, T., Mizoguchi, T., and Nishihara, T. (1993) Modulation of glutathione S-transferase activity by a thiol/disulfide exchange reaction and involvement of thioltransferase. Arch. Biochem. Biophys. 300:495–500.

63. Habig, W. H., Pabst, M. J., and Jakoby, W. B. (1974) Glutathione S-transferases. The first enzymatic step in mercapturic acid formation. J. Biol. Chem. 249:7130–7139.
64. Wells. W. W., Xu, D. P., Yang, Y. F., and Rocque, P. A. (1990) Mammalian thioltransferase (glutaredoxin) and protein disulfide isomerase have dehydroascorbate reductase activity. J. Biol. Chem. 265:15361–15364.
65. Gladysheva, T. B., Oden, K. L., and Rosen, B. P. (1994) Properties of the arsenate reductase of plasmid R773. Biochemistry 33:7288–7293.
66. Holmgren, A. (1978) Glutathione-dependent enzyme reactions of the phage T4 ribonucleotide reductase system. J. Biol. Chem. 253:7424–7430.
67. Luthman, M., Eriksson, S., Holmgren, A., and Thelander, L. (1979) Glutathione-dependent hydrogen donor system for calf thymus ribonucleoside-diphosphate reductase. Proc. Natl. Acad. Sci. USA 76:2158–2162.
68. Russel, M., Model, P., and Holmgren, A. (1990) Thioredoxin or glutaredoxin in Escherichia coli is essential for the sulfate reduction but not for deoxyribonucleotide synthesis. J. Bacteriol. 172:1923–1929.
69. Aslund, F., Ehn, B., Miranda-Viuete, Pueyo, C., and Holmgren, A. (1994) Two additional glutaredoxins exist in *Escherichia coli*: Glutaredoxin 3 is a hydrogen donor for ribonucleotide reductase in a thioredoxin/glutaredoxin 1 double mutant. Proc. Natl. Acad. Sci. USA 91:9813–9817.
70. Höpper, S., and Iurlano, D. (1983) Properties of a thioredoxin purified from rabbit bone marrow which fails to serve as a hydrogen donor for the homologous ribonucleotide reductase. J. Biol. Chem. 258:13453–13457.
71. Axelsson, K., and Mannervik, B. (1983) An essential role of cytosolic thioltransferase in protection of pyruvate kinase from rabbit liver against oxidative inactivation. FEBS Lett. 152:114–118.
72. Shen, H. X., Tamai, K., Satoh, K., Hatayama, I., Tsuchida, S., and Sato, K. (1991) Modulation of class Pi glutathione transferase activity by sulfhydryl group modification. Arch. Biochem. Biophys. 286:178–182.
73. Beatrice, M. C., Stiers, D. L., and Pfeiffer, D. R. (1984) The role of glutathione in the retention of Ca^{2+} by liver mitochondria. J. Biol. Chem. 259:1279–1287.
74. Cui, X. L., and Lou, M. F. (1993), The effect and recovery of long-term H_2O_2 exposure on lens morphology and biochemistry. Exp. Eye Res. 57:157–167.
75. Dickerson, J. E., and Lou, M. F. (1993), A new mixed disulfide species in human cataractous and aged lenses. Biochim. Biophys. Acta 1157:141–146.
76. Gilbert, H. F. (1984) Redox control of enzyme activities by thiol/disulfide exchange. Methods Enzymol. 107:330–351.
77. Hwang, C., Sinskey, A. J., and Lodish, H. F. (1992) Oxidized redox state of glutathione in the endoplasmic reticulum. Science 257:1496–1502.
78. Jolly, S. R., Kane, W. J., Bailie, M. B., Abrams, G. D., and Lucchesi, B. R. (1984) Canine myocardial reperfusion injury. Its reduction by the combined administration of superoxide dismutase and catalase. Circ. Res. 54:277–285.
79. Halliwell, B., and Gutteridge, J. M. (1984) Oxygen toxicity, oxygen radicals, transition metals and disease. Biochem. J. 219:1–14.
80. Lesnefsky, E. J., Dauber, I. M., and Horwitz, L. D. (1991) Myocardial sulfhydryl pool alterations occur during reperfusion after brief and prolonged myocardial ischemia in vivo. Circ. Res. 68:605–613.

81. Craescu, C. T., Poyart, C., Schaeffer, C., Garel, M. C., Kister, J., and Beuzard, Y. (1986) Covalent binding of glutathione to hemoglobin. II. Functional consequences and structural changes reflected in NMR spectra. J. Biol. Chem. 261:14710–14716.
82. Hebbel, R. P. (1986) Erythrocyte antioxidants and membrane vulnerability. J. Lab. Clin. Med. 107:401–404.
83. Wagner, G. M., Lubin, B. M., and Chiu, D.T.-Y. (1988) Oxidative damage to red blood cells. In: Cellular Antioxidant Defense Mechanisms, Vol. I (Chow, C. K., ed.), pp. 185–195. CRC Press, Boca Raton, FL.
84. Passon, P. G., and Hultquist, D. E. (1972) Soluble cytochrome b_5 reductase from human erythrocytes. Biochim. Biophys. Acta 275:62–73.
85. Hackett, C. S., Novoa, W. B., Ozols, J., and Strittmatter, P. (1986) Identification of the essential cysteine residue of NADH-cytochrome b_5 reductase. J. Biol. Chem. 261:9854–9857.
86. Terada, T., Oshida, T., Nishimura, M., Maeda, H., Hara, T., Hosomi, S., Mizoguchi, T., and Nishihara, T. (1992) Study on human erythrocyte thioltransferase: comparative characterization with bovine enzyme and its physiological role under oxidative stress. J. Biochem. Tokyo 111:688–692.
87. Haest, C. W., Kamp, D., and Deuticke, B. (1981) Topology of membrane sulfhydryl groups in the human erythrocyte. Demonstration of a non-reactive population in intrinsic proteins. Biochim. Biophys. Acta 643:319–326.
88. Terada, T., Nishimura, M., Oshida, H., Oshida, T., and Mizoguchi, T. (1993) Effect of glucose on thioltransferase activity and protein mixed disulfides concentration in GSH-depleting reagents treated rat erythrocytes. Biochem. Mol. Biol. Int. 29:1009–1014.
89. Beutler, E. (1983) Active transport of glutathione disulfide from erythrocytes. In: Functions of Glutathione: Biochemical, Physiological, Toxicological, and Clinical Aspects, pp. 65–74. Raven Press, New York.
90. Snyder, L. M., Fortier, N. L., Leb, L., McKenney, J., Trainor, J., Sheerin, H., and Mohandas, N. (1988) The role of membrane protein sulfhydryl groups in hydrogen peroxide-mediated membrane damage in human erythrocytes. Biochim. Biophys. Acta 937:229–240.
91. Toth, K. M., Clifford, D. P., Berger, E. M., White, C. W., and Repine, J. E. (1984) Intact human erythrocytes prevent hydrogen peroxide-mediated damage to isolated perfused rat lungs and cultured bovine pulmonary artery endothelial cells. J. Clin. Invest. 74:292–295.
92. Tsan, M. F., and White, J. E. (1988) Red blood cells protect endothelial cells against H_2O_2-mediated but not hyperoxia-induced damage. Proc. Soc. Exp. Biol. Med. 188:323–327.
93. Reglinski, J., Hoey, S., Smith, W. E., and Sturrock, R. D. (1988) Cellular response to oxidative stress at sulfhydryl group receptor sites on the erythrocyte membrane. J. Biol. Chem. 263:12360–12366.
94. Toth, K. M., Berger, E. M., Beehler, C. J., and Repine, J. E. (1986) Erythrocytes from cigarette smokers contain more glutathione and catalase and protect endothelial cells from hydrogen peroxide better than do erythrocytes from nonsmokers. Am. Rev. Respir. Dis. 134:281–284.

95. Miller, R. M., Park, E. M., and Thomas, J. A. (1991) Reduction (dethiolation) of protein mixed-disulfides; distribution and specificity of dethiolating enzymes and N,N′-bis(2-chlorethyl)-N-nitrosourea inhibition of an NADPH-dependent cardiac dethiolase. Arch. Biochem. Biophys. 287:112–120.
96. Fernando, M. R., Nanri, H., Yoshitake, S., Nagata, K. K., and Minakami, S. (1992) Thioredoxin regenerates proteins inactivated by oxidative stress in endothelial cells. Eur. J. Biochem. 209:917–922.
97. Schallreuter, K. U., and Wood, J. M. (1989) The stereospecific suicide inhibition of human melanoma thioredoxin reductase by 13-cis-retinoic acid. Biochem. Biophys. Res. Commun. 160:573–579.
98. Manning, J. M. (1991) Covalent inhibitors of the gelation of sickle cell hemoglobin and their effects on function. Adv. Enzymol. Relat. Areas Mol. Biol. 64:55–91.
99. Hassan, W., Beuzard, Y., and Rosa, J. (1976) Inhibition of erythrocyte sickling by cystamine, a thiol reagent. Proc. Natl. Acad. Sci. USA 73:3288–3292.
100. Garel, M. C., Domenget, C., Galacteros, F., Martin, C. J., and Beuzard, Y. (1984) Inhibition of erythrocyte sickling by thiol reagents. Mol. Pharmacol. 26:559 565.
101. Garel, M. C., Domenget, C., Caburi, M. J., Prehu, C., Galacteros, F., and Beuzard, Y. (1986) Covalent binding of glutathione to hemoglobin. I. Inhibition of hemoglobin S polymerization. J. Biol. Chem. 261:14704–14709.
102. Mizoguchi, T., Nishnaka, T., Uchida, G., Mizuta, J., Uchida, H., Terada, T., and Toya, H. (1993) Inhibition of bovine leukocyte thioltransferase by anti-inflammatory drugs and antihistaminic drugs. Biol. Pharm Bull. 16:840–842.
103. Grippo, J. F., Holmgren, A., and Pratt, W. B. (1985) Proof that the endogenous, heat-stable glucocorticoid receptor-activating factor is thioredoxin. J. Biol. Chem. 260:93–97.
104. Takagi, S., Bhat, G. B., Hummel, B. C., and Walfish, P. G. (1989) Thioredoxin and glutaredoxin enhance the binding of L-triiodothyronine to its hepatic nuclear receptors. Biochem. Cell Biol. 67:477–480.
105. Bird, G. S., Burgess, G. M., and Putney, J. J. (1993) Sulfhydryl reagents and cAMP-dependent kinase increase the sensitivity of the inositol 1,4,5-trisphosphate receptor in hepatocytes. J. Biol. Chem. 268:17917–17923.
106. Xanthoudakis, S., Miao, G., Wang, F., Pan, Y. C., and Curran, T. (1992) Redox activation of Fos-Jun DNA binding activity is mediated by a DNA repair enzyme. EMBO J. 11:3323–3235.
107. Galter, D., Mihm, S., and Droge, W. (1994) Distinct effects of glutathione disulfide on the nuclear transcription factors kB and the activator protein-1. Eur. J. Biochem. 221:639–648.
108. Wu, X. B., Brune, B., von Appen, F., and Ullrich, V. (1992) Reversible activation of soluble guanylate cyclase by oxidizing agents. Arch. Biochem. Biophys. 294:75–82.
109. Walters, D. W., and Gilbert, H. F. (1986) Thiol/disulfide exchange between rabbit muscle phosphofructokinase and glutathione. Kinetics and thermodynamics of enzyme oxidation. J. Biol. Chem. 261:15372–15377.
110. Cappel, R. E., and Gilbert, H. F. (1993) Oxidative inactivation of 3-hydroxy-3-

methylglutaryl-coenzyme A reductase and subunit cross-linking involve different dithiol/disulfide centers. J. Biol. Chem. 268:342–348.

111. Terada, T., Hara, T., Yazawa, H., and Mizoguchi, T. (1994) Effect of thioltransferase on the cystamine-activated fructose-1,6-biphosphatase by its redox regulation. Biochem. Mol. Biol. Int. 32:239–244.
112. Nieslanik, B., Starke, D. W., Grant, J., and Mieyal, J. J. (1994) Unpublished observations.
113. Schallreuter, K. U., Pittelkow, M. R., and Wood, J. M. (1989) EF-hands calcium binding regulates the thioredoxin reductase/thioredoxin electron transfer in human keratinocytes. Biochem. Biophys. Res. Commun. 162:1311–1316.
114. Ciriolo, M. R., Paci, M., Sette, M., De, M. A., Bozzi, A., and Rotilio, G. (1993) Transduction of reducing power across the plasma membrane by reduced glutathione. A 1H-NMR spin-echo study of intact human erythrocytes. Eur. J. Biochem. 215:711–718.
115. Grasso, P., Santa, C. T., Boniface, J. J., and Reichert, L. J. (1991) A synthetic peptide corresponding to hFSH-beta-(81-95) has thioredoxin-like activity. Mol. Cell. Endocrinol. 78:163–170.
116. Dunten, R. L., Cohen, R. E., Gregori, L., and Chau, V. (1991) Specific disulfide cleavage is required for ubiquitin conjugation and degradation of lysozyme. J. Biol. Chem. 266:3260–3267.
117. Axelsson, K., and Mannervik, B. (1980) A possible role of cytoplasmic thioltransferase in the intracellular degradation of disulfide-containing proteins. Acta Chem. Scand. B. 34:139–140.
118. Lyles, M. M., and Gilbert, H. F. (1991) Catalysis of the oxidative folding of ribonuclease A by protein disulfide isomerase: pre-steady-state kinetics and the utilization of the oxidizing equivalents of the isomerase. Biochemistry 30:619–625.
119. Lyles, M. M., and Gilbert, H. F. (1991) Catalysis of the oxidative folding of ribonuclease A by protein disulfide isomerase—dependence of the rate on the composition of the redox buffer. Biochemistry 30:613–619.
120. Gilbert, H. F. (1989) Catalysis of thiol/disulfide exchange: single-turnover reduction of protein disulfide-isomerase by glutathione and catalysis of peptide disulfide reduction. Biochemistry 28:7298–7305.
121. Morjana, N. A., and Gilbert, H. F. (1991) Effect of protein and peptide inhibitors on the activity of protein disulfide isomerase. Biochemistry 30:4985–4990.
122. Taniyama, Y., Seko, C., and Kikuchi, M. (1990) Secretion in yeast of mutant human lysozymes with and without glutathione bound to cysteine 95. J. Biol. Chem. 265:16767–16771.
123. Hayano, T., Inaka, K., Otsu, M., Taniyama, Y., Miki, K., Matsushima, M., and Kikuchi, M. (1993) PDI and glutathione-mediated reduction of the glutathionylated variant of human lysozyme. FEBS Lett. 328:203–208.
124. Darby, N. J., Freedman, R. B., and Creighton, T. E. (1994) Dissecting the mechanism of protein disulfide isomerase: catalysis of disulfide bond formation in a model peptide. Biochemistry 33:7937–7947.
125. Zapun, A., Bardwell, J. C., and Creighton, T. E. (1993) The reactive and destabilizing disulfide bond of DsbA, a protein required for protein disulfide bond formation in vivo. Biochemistry 32:5083–5092.

126. Sinha, B. K. (1989) Free radicals in anticancer drug pharmacology. Chem. Biol. Interact. 69:293–317.
127. Hunt, T., Herbert, P., Campbell, E. A., Delidakis, C., and Jackson, R. J. (1983) The use of affinity chromatography on 2′5′ ADP-sepharose reveals a requirement for NADPH, thioredoxin and thioredoxin reductase for the maintenance of high protein synthesis activity in rabbit reticulocyte lysates. Eur. J. Biochem. 131: 303–311.
128. Martin, H., and Dean, M. (1991) Identification of a thioredoxin-related protein associated with plasma membranes. Biochem. Biophys. Res. Commun. 175: 123–128.
129. Nakamura, H., Masutani, H., Tagaya, Y., Yamauchi, A., Inamoto, T., Nanbu, Y., Fujii, S., Ozawa, K., and Yodoi, J. (1992), Expression and growth-promoting effect of adult T-cell leukemia-derived factor. Cancer 69:2091–2097.
130. Humphreys, W. G., Kim, D. H., Cmarik, J. L., Shimada, T., and Guengerich, F. P. (1990) Comparison of the DNA-alkylating properties and mutagenic responses of a series of S-(2-haloethyl)-substituted cysteine and glutathione derivatives. Biochemistry 29:10342–10350.
131. Ahmad, T., and Frischer, H. (1985) Active site-specific inhibition by 1,3-bis(2-chloroethyl)-1-nitrosourea of two genetically homologous flavoenzymes: Glutathione reductase and lipoamide dehydrogenase. J. Lab. Clin. Med. 105:464–471.
132. Schallreuter, K. U., Gleason, F. K., and Wood, J. M. (1990) The mechanism of action of the nitrosourea anti-tumor drugs on thioredoxin reductase, glutathione reductase and ribonucleotide reductase. Biochim. Biophys. Acta 1054:14–20.
133. Russo, A., Carmichael, J., Friedman, N., DeGraff, W., Tochner, Z., Glatstein, E., and Mitchell, J.B. (1986) The roles of intracellular glutathione in antineoplastic chemotherapy. Int. J. Radiat. Oncol. Biol. Phys. 12:1347–1354.
134. Wells, W. W., Rocque, P. A., Xu, D. P., Yang, Y., and Deits, T. L. (1991) Interactions of platinum complexes with thioltransferase(glutaredoxin), in vitro. Biochem. Biophys. Res. Commun. 180:735–741.
135. Beshai, S., Mieyal, P. A., and Mieyal, J. J. (1992) Unpublished observations.
136. Mau, B. L., and Powis, G. (1990) Inhibition of thioredoxin reductase (E.C. 1.6. 4.5.) by antitumor quinones. Free Radic. Res. Commun. 8:365–372.
137. Tew, K. D., and Clapper, M. L. (1988), Glutathione S-transferases and anticancer drug resistance. Bristol-Myers Cancer Symp. 9 (Mech. Drug Resist. Neoplast. Cells):141–159.
138. Batist, G., Tulpule, A., Sinha, B. K., Katki, A. G., Myers, C. E., and Cowan, K. H. (1986) Overexpression of a novel anionic glutathione transferase in multidrug-resistant human breast cancer cells. J. Biol. Chem. 261:15544–15549.
139. Black, S. M., Beggs, J. D., Hayes, J. D., Bartoszek, A., Muramatsu, M., Sakai, M., and Wolf, C. R. (1990) Expression of human glutathione S-transferases in Saccharomyces cerevisiae confers resistance to the anticancer drugs adriamycin and chlorambucil. Biochem. J. 268:309–315.
140. Doroshow, J. H. (1986) Role of hydrogen peroxide and hydroxyl radical formation in the killing of Ehrlich tumor cells by anticancer quinones. Proc. Natl. Acad. Sci. USA 83:4514–4518.

141. Cole, S. P., Downes, H. F., Mirski, S. E., and Clements, D. J. (1990) Alterations in glutathione and glutathione-related enzymes in a multidrug-resistant small cell lung cancer cell line. Mol. Pharmacol. 37:192–197.
142. Di Simplicio, P., Jensson, H., and Mannervik, B. (1989) Effects of inducers of drug metabolism on basic hepatic forms of mouse glutathione transferase. Biochem. J. 263:679–685.
143. Demarquoy, J., Fairand, A., Vaillant, R., and Gautier, C. (1991) Development and hormonal control of thioredoxin and the thioredoxin-reductase system in the rat liver during the perinatal period. Experientia 47:497–500.
144. Hansson, H. A., Holmgren, A., Rozell, B., and Stemme, S. (1986). In: Thioredoxin and Glutaredoxin Systems (A. Holmgren, C.-I. Branden, H. Jornvall, and B.-M. Sjoberg, eds.), pp. 134–141. Raven Press, New York.
145. Schallreuter, K. U., and Wood, J. M. (1987) Azelaic acid as a competitive inhibitor of thioredoxin reductase in human melanoma cells. Cancer Lett. 36: 297– 305.
146. Nair, S., Singh, S. V., Samy, T. S., and Krishan, A. (1990) Anthracycline resistance in murine leukemic P388 cells. Role of drug efflux and glutathione related enzymes. Biochem. Pharmacol. 39:723–728.
147. Higgins, D. G., and Sharp, P. M. (1988) CLUSTAL: A package for performing multiple sequence alignment on a microcomputer. Gene 73:237–244.
148. Miller, A. R. (1981) Pascal Programs for Scientists and Engineers, pp. 159–163. SYBEX Publications.

15

Lipoic Acid–Requiring Proteins: Recent Advances

Mulchand S. Patel and Nataraj N. Vettakkorumakankav
State University of New York at Buffalo, Buffalo, New York

I. INTRODUCTION

Since its isolation and characterization in 1957 by Reed et al. (1), lipoic acid (1,2-dithiolane-3-valeric acid) has been shown to be part of several enzyme systems playing vital roles in intermediary metabolism (for a review see Ref. 2). The enzymatic pathways leading to the synthesis of lipoic acid in cells are still unknown; however, several intermediate steps have been identified by isotope-labeling experiments in bacteria and preparations of rat liver (3,4). These experiments have demonstrated that octanoic acid serves as the immediate precursor for the 8-carbon fatty acid chain and cysteine appears to be the source of sulfur (3,5). Two genes unique to the lipoic acid biosynthetic pathway have been identified: Lip A and Lip B (6). The Lip A gene has been cloned and encodes a polypeptide of mass 35 kDa (7) and is suggested to play a role in the thiolation of octanoic acid as octanoic acid was unable to restore growth of Lip A mutants (6). More recently, a lipoate biosynthetic locus has been identified in *Saccharomyces cerevisiae* by complementation of mutants (8). From sequence comparisons, it appears as if this gene is responsible for insertion of sulfur atoms in lipoic acid and that this enzyme might be mitochondrial (8). Lipoic acid can be transported by an energy-dependent process that is capable of concentrating up to 100 times the normal amount of lipoic acid in cells (9). The quantitation of lipoic acid in bacterial or mammalian cells is a difficult process due to the relatively low concentrations of lipoic acid in vivo (0.5–25 ng/g dry weight) (10).

Various quantitation procedures including chemical, gas chromatography, bioasssay, and competitive binding immunoassay are available for lipoic acid measurements (10–12). The principle of the bioassay procedure relies on bacterial strains dependent on lipoic acid for growth.

II. COVALENT ATTACHMENT OF R-LIPOIC ACID

The biological activity of lipoic acid requires its covalent attachment to the N^6-amino group of lysine residues of several enzymes. The attachment of lipoic acid to the proteins requires the participation of lipoate protein–activating enzymes (lipoate protein ligase) whose activity has been demonstrated in both prokaryotes and eukaryotes (13).

The attachment of lipoate to enzymes proceeds in two steps in *Escherichia coli*:

1. Enzyme + Lipoate + ATP = Enzyme-lipoyl-adenylate + PPi
2. Enzyme-lipoyl-adenylate + Apoprotein = Holoprotein + AMP + Enzyme

The above reactions are catalyzed by a single enzyme in *E. coli*. However, more recently, two distinct lipoate-protein ligase activities (LPL-A and LPL-B) have been detected in *E. coli* based on the ability to lipoylate purified, overexpressed, lipoyl apo-domains of dihydrolipoamide acetyltransferase (14). In mammalian tissues, two distinct activities are required for the lipoylation of proteins: lipoyl-activating enzyme to form a lipoyl-AMP intermediate and lipoyl transferase activity (15). These activities are located in the mitochondria of mammalian cells (16). Compared to the attachment of lipoate moiety to proteins, very little is known about the removal of lipoate moieties from proteins. However, lipoamidase activity has been detected in extracts of *Streptococcus faecalis, E. coli,* and baker's yeast (13,17).

III. BIOLOGICAL ROLES OF LIPOIC ACID

There are five lipoate-containing proteins known in eukaryotes where the lipoate moiety is covalently attached to a specific lysine residue(s) (Table 1). These proteins are part of four multienzyme complexes localized in the mitochondria of eukaryotes, three of which are collectively referred to as the α-keto acid dehydrogenase complexes. The fourth is the glycine cleavage system. Most of these complexes are present in prokaryotes as well. These complexes play an important role in the metabolism of carbohydrates, of the branched-chain amino acids, and of glycine (Fig. 1). In the archaebacteria, however, lipoic acid has been detected without the apparent presence of these multienzyme complexes (18).

Table 1 Lipoic Acid–Dependent Biological Processes in Eukaryotes

Complex	Protein	Biological role
PDC	E2-PDC	Oxidation of pyruvate
	protein X	E2-E3 interaction
α-KGDC	E2-α-KGDC	Oxidation of α-keto-glutarate
BCKADC	E2-BCKADC	Oxidation of branched-chain α-keto acids
GCS	H-Protein	Glycine metabolism

PDC, pyruvate dehydrogenase complex; α-KGDC, α-ketoglutarate dehydrogenase complex; BCKADC, branched-chain α-keto acid dehydrogenase complex; GCS, glycine cleavage system; E2, dihydrolipoamide acyltransferase; H-Protein, hydrogen carrier protein.

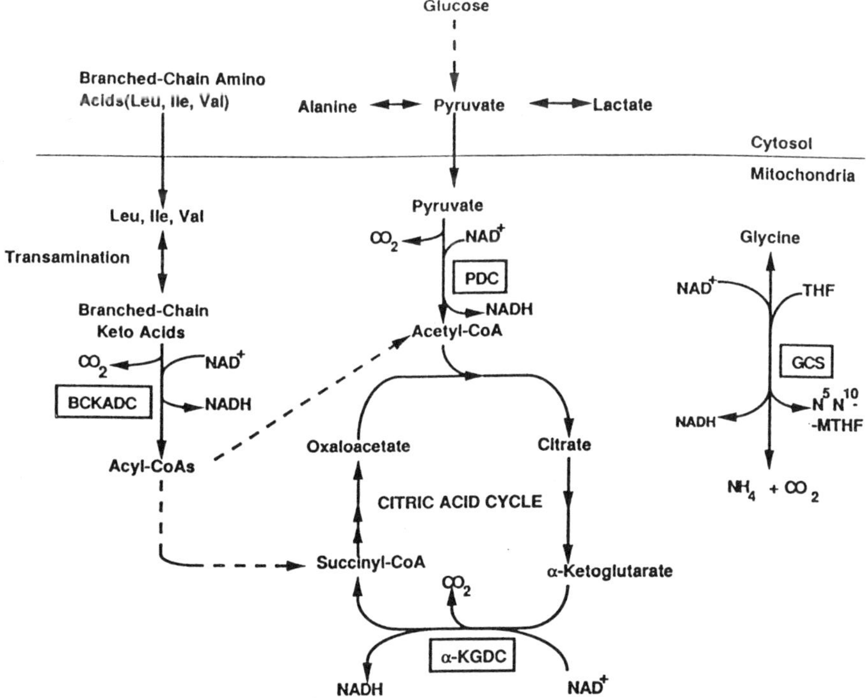

Figure 1 Involvement of lipoic acid–requiring multienzyme complexes in intermediary metabolism. PDC, pyruvate dehydrogenase complex; α-KGDC, α-ketoglutarate dehydrogenase complex; BCKADC, branched-chain α-keto acid dehydrogenase complex; GCS, glycine cleavage system.

In this paper we will review the recent advances in the structure-function relationships of α-keto acid dehydrogenase complexes.

IV. α-KETO ACID DEHYDROGENASE COMPLEXES

The α-keto acid dehydrogenase complexes catalyze the oxidative decarboxylation of pyruvate, α-ketoglutarate, and branched-chain α-keto acids by the coordinated action of atleast three component enzymes shown schematically in Figure 2 (19–21). These multienzyme complexes are pyruvate dehydrogenase complex (PDC), α-ketoglutarate dehydrogenase complex (α-KGDC), and branched-chain α-keto acid dehydrogenase complex (BCKADC). Each of these complexes are composed of multiple copies of α-keto acid dehydrogenase (E1), dihydrolipoamide acyltransferase (E2), and dihydrolipoamide dehydrogenase (E3). The enzyme components E1 and E2 are unique for each complex, while the E3 component is shared by these complexes with the exception of *Pseudomonas*

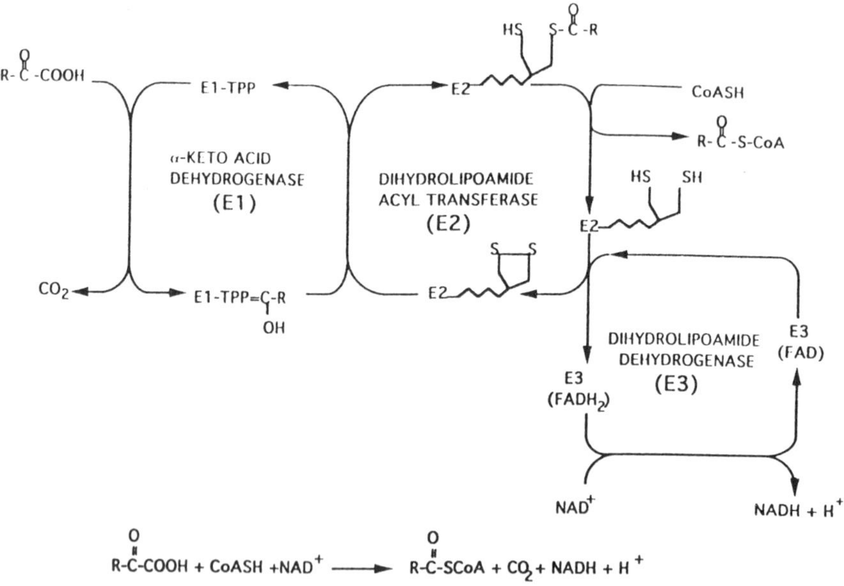

$$R\text{-}\overset{O}{\overset{\|}{C}}\text{-COOH} + CoASH + NAD^+ \longrightarrow R\text{-}\overset{O}{\overset{\|}{C}}\text{-SCoA} + CO_2 + NADH + H^+$$

Figure 2 Schematic representation of the reactions catalyzed by the α-keto acid dehydrogenase complexes. R = $-CH_3$ in pyruvate dehydrogenase complex, CH_2-CH_2-COOH in α-ketoglutarate dehydrogenase complex, and $(CH_3)_2$-CH–, $(CH_3)_2$-CH-CH_2–, or CH_3-$CH_2(CH_3)$-CH– in the case of the branched-chain α-keto acid dehydrogenase complex.

putida, which has three different E3 proteins (22). These complexes differ in their subunit composition and structure as one moves up the evolutionary ladder, which is responsible for the unique regulatory properties of the complexes isolated from different sources, especially the PDC (19–21,23). One of the unique features of eukaryotic PDC is the presence of an additional protein component called protein X (20,21). Protein X is responsible for the interaction between E2 and E3 but also possesses a covalently attached lipoate moiety (20,21).

The three catalytic components catalyze a series of reactions illustrated in Figure 3. These reactions are sequential and highly coordinated. E1 component of these complexes catalyzes both the TPP-dependent decarboxylation of its cognate substrate and the reductive acylation reaction, which results in the transfer of the acyl group in thioester linkage to the covalently attached lipoate moiety of E2 component. E1 component is unable to use free lipoamide and short lipoylated peptide for this reaction, while free lipoyl domains from PDCs serve as excellent substrates, signifying the importance of properly folded lipoyl domains of E2 for reductive acylation by E1 (26). The E2 component also catalyzes the transfer of the acyl group from its acyl-lipoyl moieties to coenzyme A (CoA), resulting in the formation of acyl CoA. In the PDC, the acetyl groups on the E2 component can be transferred to the lipoyl moiety of protein

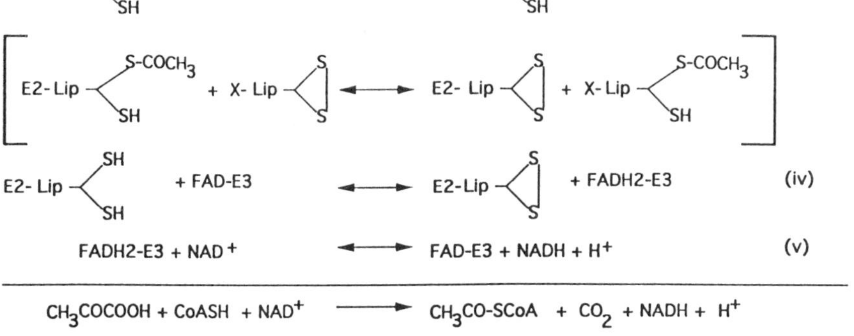

$$CH_3COCOOH + TPP\text{-}E1 \longrightarrow CH_3C(OH)\text{=}TPP\text{-}E1 + CO_2 \quad \text{(i)}$$

$$CH_3C(OH)\text{=}TPP\text{-}E1 + E2\text{-}Lip{<}^{S}_{S}| \longleftrightarrow TPP\text{-}E1 + E2\text{-}Lip{<}^{S\text{-}COCH_3}_{SH} \quad \text{(ii)}$$

$$E2\text{-}Lip{<}^{S\text{-}COCH_3}_{SH} + CoASH \longleftrightarrow E2\text{-}Lip{<}^{SH}_{SH} + CH_3CO\text{-}SCoA \quad \text{(iii)}$$

$$\left[E2\text{-}Lip{<}^{S\text{-}COCH_3}_{SH} + X\text{-}Lip{<}^{S}_{S}| \longleftrightarrow E2\text{-}Lip{<}^{S}_{S}| + X\text{-}Lip{<}^{S\text{-}COCH_3}_{SH}\right]$$

$$E2\text{-}Lip{<}^{SH}_{SH} + FAD\text{-}E3 \longleftrightarrow E2\text{-}Lip{<}^{S}_{S}| + FADH2\text{-}E3 \quad \text{(iv)}$$

$$FADH2\text{-}E3 + NAD^+ \longleftrightarrow FAD\text{-}E3 + NADH + H^+ \quad \text{(v)}$$

$$CH_3COCOOH + CoASH + NAD^+ \longrightarrow CH_3CO\text{-}SCoA + CO_2 + NADH + H^+$$

Figure 3 Partial reactions catalyzed by the pyruvate dehydrogenase complex in eukaryotes. TPP, thiamine pyrophosphate; Lip, lipoic acid moiety covalently attached to the dihydrolipoyl acetyltransferase (E2) and protein X (X) components of this complex; CoASH, coenzyme A.

X. However, protein X cannot transfer these acyl groups to CoA as this protein lacks catalytic activity. The acyl groups, therefore, are shuttled back and forth between the lipoyl moieties of protein X and E2 (Fig. 3). It should be stated that only the R-enantiomers of lipoic acid and dihydrolipoamide serve as substrates for these reactions, while the S-enantiomer acts as an inhibitor (27).

Another role for the E2 component of these complexes is in the assembly of the multimeric complex. E2 components form the icosahedral or octahedral core around which the E1 and E3 components are arranged (19,28,29). Recent x-ray analysis of the *Azotobacter vinelandii* PDC-E2 (30) has provided further insight into the assembly of the *A. vinelandii* PDC-E2. The catalytic domain of the E2 component assembles into oligomers of 24 subunits with a 432 octahedral symmetry. The inner core, composed of the 24 E2 subunits, assemble as trimers, i.e., 8 trimers form the core. The shape of this multisubunit structure is that of a truncated cube with the trimers forming the corners of the cube. It appears as if the formation of the tightly interacting trimers is important for the formation of the cubic core, as the interactions between components of these trimers are most extensive. The determination of this structure and the availability of low-resolution structures of E2 and E3 components and the subsequent generation of a predicted structural model (31) have allowed a better understanding of the complexity of the reactions catalyzed by these complexes.

The structural arrangement of various domains of E2 component has been the subject of extensive study in various laboratories (21,26,32). There are three structural domains in E2 component and, from amino to carboxy termini, these domains are the lipoyl domain, the subunit binding domain, and the catalytic domain responsible for the transfer of the acyl group to CoA (Fig. 4). The sequences linking these domains are rich in charged amino acids, alanine, and proline and are regions with extremely high conformational mobility (21,26,32). The lipoyl domains (approximately 80 amino acids) form a tightly folded structure with the lipoyl moieties attached to a specific lysine residue in the only sharp turn in the antiparallel β-sheet structure (33). The structure of the E3 and E1 binding domain of the *Bacillus stearothermophilus* PDC has been solved using nuclear magnetic resonance spectroscopy, and the structure is extremely ordered for a small polypeptide (34). The structure of native E3/E1 binding domain was determined to be similar to that of a synthetic peptide representing the same region (34). The subunit binding domain is composed of two short, parallel α-helices connected by two overlapping β-turns (34).

There is no obvious correlation between the number of repeating lipoyl domains in the E2 component of PDC and the source or structure of the complex. The E2 component of PDC from *E. coli* and *A. vinelandii* have three repeating lipoyl domains, while the E2 component of PDC from humans has two repeating units and that from yeast has one. Genetic engineering studies that inactivated two of the three lipoyl domains of E2 from *E. coli* PDC had no

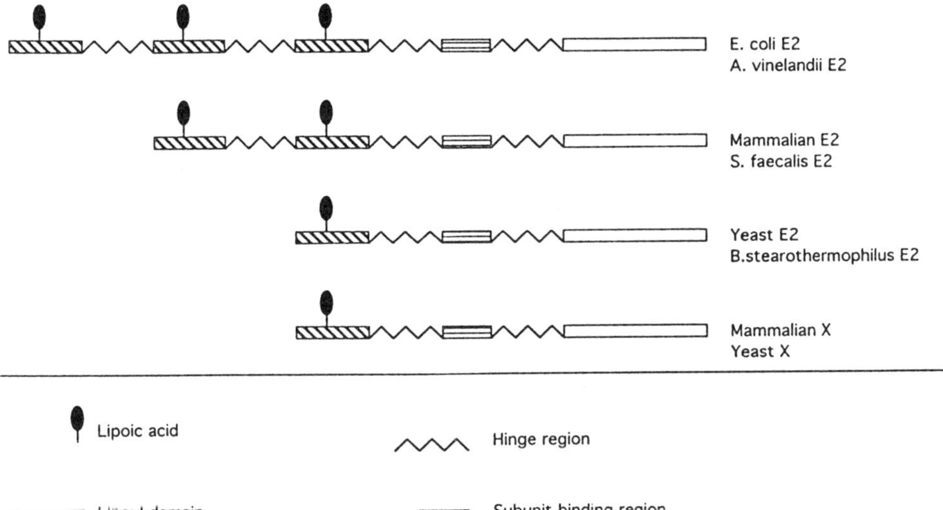

Figure 4 Domain structure of dihydrolipoyl acyltransferase (E2) component of prokaryotes and eukaryotes and the protein X (X) component of eukaryotic pyruvate dehydrogenase complex.

adverse effect on catalysis by the PDC, indicating that one lipoyl domain is sufficient for the normal functioning of this complex (26). Increasing the number of lipoyl domains in *E. coli* PDC from three to nine had no adverse effects on the assembly of E1 and E3 components on the core complex, while catalytic efficiency was markedly reduced in the complexes containing four to nine lipoyl domains (35), suggesting that excess lipoyl domains are in fact detrimental to the overall complex function. This could be due to a lowering in the extent of lipoylation of those domains that do participate in catalysis (35). Overexpressed recombinant lipoyl domains of *E. coli* and *B. stearothermophilus* PDC-E2 and mature bovine H-protein of the glycine cleavage system (GCS) under different conditions have been shown to produce three types of proteins: correctly lipoylated protein, unlipoylated protein, and protein modified by octanoylation (15,36,37). Even when the culture medium was supplemented with 30 μM of lipoate, only 10% of the overexpressed bovine H-protein was lipoylated, while another 10% was presumably modified with an octanoyl group and the remaining 80% were in the unlipoylated form (15). These results indicate that the cellular lipoate content is the limiting factor for the correct lipoylation of overexpressed lipoyl-containing proteins.

The domains of *S. cerevisiae* PDC-E2 have been analyzed by deletion mutagenesis (25), and the results are summarized in Figure 5. Deletion of the lipoyl

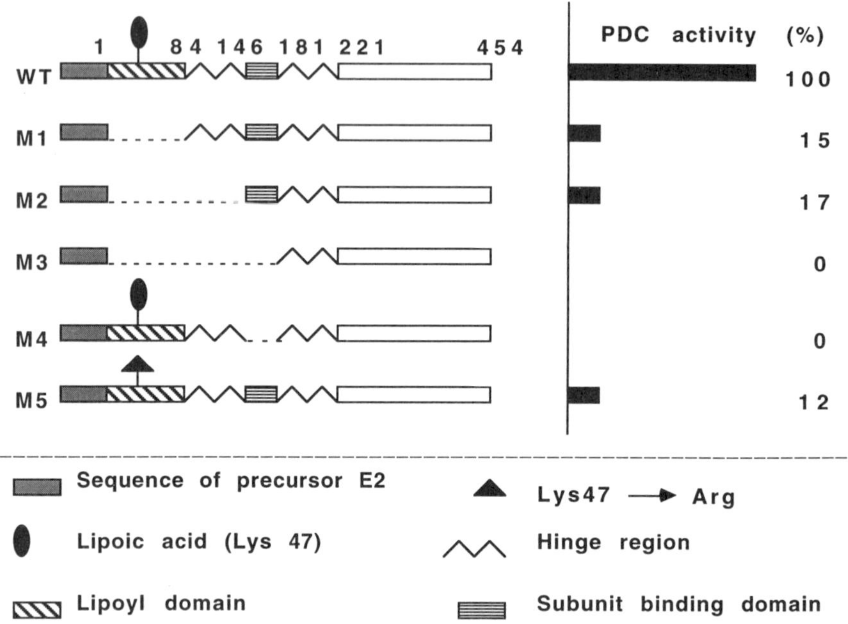

Figure 5 Mutational analysis of E2 (dihydrolipoyl acetyltransferase) component of pyruvate dehydrogenase complex from *S. cerevisiae* (25).

domain leads to approximately 85% reduction of PDC activity with 15% activity being retained perhaps due to the lipoyl domain of PDC protein X participating in the acetyl transfer reaction. Upon site-directed mutagenesis of the lysine residue that is normally lipoylated, the activity of the PDC is reduced by approximately 88%, illustrating the role for this specific lysine residue. Upon deletion of the subunit binding domain, the PDC is completely inactivated once again, illustrating the structural role played by this lipoyl-containing protein in the assembly of PDC (25).

Protein X, as mentioned before, is part of the eukaryotic PDC (21). Protein X is lipoylated at a single site in the lipoyl domain and can participate in the reductive acetylation reaction in the presence of PDC-E1 and pyruvate (Fig. 4) (21,24,38). Site-directed and deletion mutagenesis of yeast protein X revealed that this lipoylation of protein X is not essential for the activity of the PDC, while it can couple with the catalytic domain of E2 devoid of lipoyl domains and retains about 15% of overall complex activity (24). These results are summarized in Figure 6. The presence of protein X is, however, necessary for the assembly of the functional PDC.

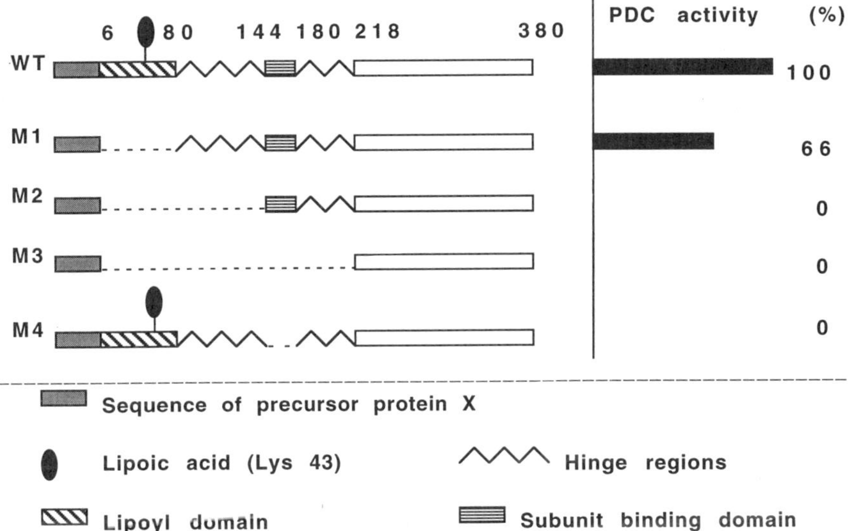

Figure 6 Mutational analysis of the domain structure of protein X of pyruvate dehydrogenase complex from *S. cerevisiae* (24).

E3, a flavoprotein, oxidizes the reduced lipoyl moieties of E2 and transfers the reducing equivalents to NAD^+. This is achieved by a two step reduction as shown in Figure 3. E3 belongs to the family of pyridine nucleotide-dependent disulfide oxidoreductases and has been extensively characterized from several sources (39).

V. GLYCINE CLEAVAGE SYSTEM

The glycine cleavage system (GCS) catalyzes the reversible oxidation of glycine to CO_2, NH_3, 5,10-methylene tetrahydrofolate, and NADH (Fig. 7) and is composed of pyridoxal phosphate–requiring protein (P-protein), hydrogen carrier protein (H-protein), tetrahydrofolate-dependent protein (T-protein), and dihydrolipoamide dehydrogenase (L-protein) (40,41). Glycine is decarboxylated by the P-protein and the methylamine moiety is transferred to the lipoyl group of the H-protein. The release of NH_3 and the formation of 5,10-methylene tetrahydrofolate is catalyzed by the T-protein. The reoxidation of the lipoyl moiety of the H-protein is catalyzed by the L-protein, which is referred to as the E3 component of the α-keto acid dehydrogenase complexes. The primary structures of several H-proteins have been deduced from the nucleotide sequences of corresponding cDNAs and are available for comparison. All of these

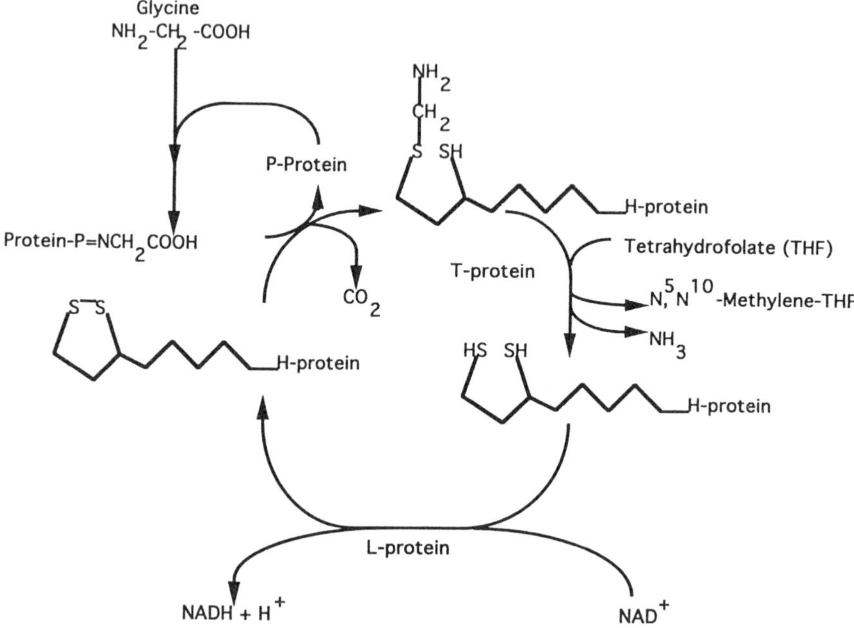

Figure 7 Schematic representation of the reactions catalyzed by the glycine cleavage system. P-protein, T-protein, and H-protein, to pyridoxal phosphate–, tetrahydrofolate-, and lipoic acid–dependent proteins, respectively; L-protein, dihydrolipoamide dehydrogenase (E3) component.

sequences show considerable homology around the lipoyl lysine residues, and this homology can be extended to the corresponding regions of the E2 component and the X protein (Fig. 8) (42). The three-dimensional structure of H-protein of GCS from pea leaf has been predicted by homology modeling using the available three-dimensional structure of lipoyl domains of *B. stearothermophilus* E2 (43). This structure, like the lipoyl domain of E2, is an all β-sheet structure with a similar β-turn on which the posttranslationally modified lysine residue is situated (43).

VI. AMINO ACID SEQUENCE HOMOLOGY IN THE LIPOYLATION SITE

To determine the possible role for lipoylation of several conserved residues in the lipoate attachment site of bovine H-protein, Gly-43, Gly-70, Glu-56, and Glu-63 were replaced by different residues by site-directed mutagenesis. Although Gly-43 is conserved among H-proteins from various sources, it was found not to be essential for proper lipoylation. However, Glu-56 and Gly-70

```
                                        *
Hu/E2P 28  KEGDKINEGDLIAEVETD K ATVGFESLEECYM 59
Ra/E2P 28  KEGEKISEGDLIAEVETD K ATVGFESLEECYM 59
Ye/E2P 29  KEGDQLSPGEVIAEIETD K AQNDFEFQEDGYL 60
Hu/E2B 26  KEGDTVSQFDSICEVQSD K ASVTITSRYDGVI 57
Ra/E2K 25  AVGDAVAEDEVVCEIETD K TSVQVPSPANGII 56
Ec/E2P 22  KVGDKVEAEQSLITVEGD K ASNEVPSPQAGIV 53
Av/E2P 21  KTGDLIEVEQGLVVLESA K ASNEVPSPKAGVV 52
Ec/E2K 25  KPGDAVVRDEVLVEIETD K VVLEVPASADGIL 56
Ye/X   25  KVGEPFSAGDVILEVETD K SQIDVEALDDGKL 56

Hu/H   41  EVGTKLNKQDEFGALESV K AASELYSPLSGEV 72
Bo/H   41  EVGTKLNKQEEFGALESV K AASELYSPLSGEV 72
Pi/H   45  EPGVSVTKGKGFGAVESV K ATSDVNSPISGEV 76
Ch/H   41  EIGTKLNKDDEFGALESV K AASELYSPLTGEV 72
```

Figure 8 Amino acid sequence homology in the region of the lipoylated lysine residues of E2 components of PDC (E2P), BCKADC (E2B), and α-KGDC (E2K); the protein X component of eukaryotic PDC (X); and H protein of GCS (H) from humans (Hu), rat (Ra), yeast (Ye), bovine (Bo), pig (Pi), chicken (Ch), *Escherichia coli* (Ec), and *Azotobacter vinelandii* (Av). The asterisk indicates the lysine residue on the polypeptide that is lipoylated. Numbers at the beginning and end of the sequences refer to the numbers of the amino acids (39,42,50,51).

are essential for proper lipoylation (42). There are no reports of similar studies on the lipoyl domains of E2. Details of lipoylation of specific proteins are presented elsewhere in this volume.

VII. AUTOIMMUNE AND GENETIC DISEASES ASSOCIATED WITH LIPOYL-CONTAINING PROTEINS

One of the major characteristics of primary biliary cirrhosis is the production of antimitochondrial antibodies exhibited by about 90–95% of patients with this disorder. Van de Water et al. (44) demonstrated that the epitope of the major autoantigen in primary biliary cirrhosis corresponds to the lipoyl domain of E2 component of the PDC. It was also demonstrated that the lipoylated form of this autoepitope had higher affinity for the antibodies when compared to the unlipoylated form (44). Detailed analysis of the antigenic capacity of synthetic peptides corresponding to various regions of the E2 component, as evidenced by the ability of these peptides to cross-react with sera from primary biliary cirrhosis patients, has been performed (45). The results indicated that the lipoylated peptide from the lipoyl domain of E2 was recognized by the sera from these patients.

Monospecific antibodies to trifluoroacetylated proteins, obtained from a polyclonal serum of rabbits exposed to the anaesthetic agent halothane, were purified and shown to cross-react with distinct trifluoroacetylated proteins of 52 and 64 kDa proteins from humans not previously exposed to halothane (45). In patients with halothane hepatitis, there was very low or no cross-reactivity of this antibody with the 52 and 64 kDa proteins (46). These results suggested that the trifluoroacetylated proteins, to which the antibody was raised, resemble the epitopes on the 52 and 64 kDa proteins. Further characterization of the 64 kDa protein revealed that this protein was the E2 component of the PDC (46) where the specific lysine residue was trifluoroacetylated. This relationship was established by the isolation of this protein and the sequence analysis of tryptic fragments of this protein and sequence comparisons to the nucleotide-derived amino acid sequence of E2 component of PDC (47). These results indicate the role of lipoylated peptides in the molecular mimicry by trifluoroacetylated proteins and presumably the development of halothane hepatitis. The unique structural feature of the lysine residue is illustrated by the fact that antibodies can be raised to this epitope.

One patient with chronic lactic acidemia has been shown to have deficiency in the E2 component of the PDC as evidenced by enzymatic assays and detection of E2 using anti-E2 antibodies (48). Deficiency of protein X has been reported in two patients with lactic acidemia and one patient with metabolic encephalomyelopathy due to PDC deficiency using the Western blotting technique (48,49). These results are preliminary and, due to the unavailability of nucleotide sequence for the human gene coding for protein X, it has not been possible to characterize the mutation leading to the deficiency of protein X.

VIII. SUMMARY

Significant progress has been made in our understanding of the structure and function relationships of the lipoyl-requiring proteins over the past decade. However, there has been very little progress in our understanding of the enzymes involved in the covalent modification of proteins by lipoylation and delipoylation. More studies are required also to address the issue of specific amino acid interactions of protein X with E2 and E3 components of eukaryotic PDC. The unique structure of the lipoyl domain of these lipoic acid–requiring proteins is further illustrated by the development of antibodies (1) in primary biliary cirrhosis, where the reason for the development of antibodies to E2 of PDC has not been established, and (2) against these proteins in response to halothane exposure (in which case the antigens are trifluoroacetylated proteins). These results indicate a unique role for these proteins in intermediary metabolism and in the assembly and function of multienzyme complexes.

REFERENCES

1. Reed. L. J., DeBusk, B. G., Gunsalus, I. C., and Hornberger, Jr., C. S. (1957) Crystalline α-lipoic acid: A catalytic agent associated with pyruvate dehydrogenase. Science 114:93–94.
2. Patel, M. S., and Smith, R. L. (1994) Biochemistry of lipoic acid containing proteins: Past and present. In: New Strategies in Prevention and Therapy (Schmidt, K., Diplock, A. T., and Ulrich, H., eds.), pp. 65–77. Hippokrates Verlag, Stuttgart, Germany.
3. Dupre, S., Spoto, G., Matarese, R. M., Orlando, M., and Cavalini, D. (1980) Biosynthesis of lipoic acid in the rat: Incorporation of ^{35}S- and ^{14}C-labelled precursors. Arch. Biochem. Biophys. 202:361–365.
4. Eisenberg, M. (1988) Biosynthesis of biotin and lipoic acid. In: *Escherichia coli* and *Salmonella typhimurium*: Cellular and Molecular Biology (Neidhardt, F. C., Ingraham, L., Low, K. B., Magasanik, B., Schaechter, M., and Umbarger, H. E., eds.), pp. 544–550. American Society of Microbiology, Washington, DC.
5. White, R. H. (1980) Stable isotope studies on the biosynthesis of lipoic acid in *Escherichia coli*. Biochemistry 19:15–19.
6. Herbert, A. A., and Guest, J. R. (1968) Biochemical and genetic studies with lysine + methionine mutants of *Escherichia coli*: Lipoic acid and α-ketoglutarate dehydrogenase-less mutants. J. Gen. Microbiol. 53:363–381.
7. Spratt, B. G., Boyd, A., and Stoker, N. (1980) Defective and plaque forming lambda transducing bacteriophage carrying penicillin-binding protein-cell shape genes: Genetic and physical mapping and identification of gene products from the lip-dacA-rodA-pbpA-leuS region of the *Escherichia coli* chromosome. J. Bacteriol. 142:569–581.
8. Sulo, P., and Martin, N. C. (1993) Isolation and characterization of LIP5: Lipoate biosynthetic locus of *Saccharomyces cerevisiae*. J. Biol. Chem. 268:17634–17639.
9. Leach, F. R. (1970) Lipoic acid transport. Methods Enzymol. 18:276–281.
10. White, R. H. (1980) A gas chromatographic method for the analysis of lipoic acid in biological samples. Anal. Biochem. 110:89–92.
11. Herbert, A. A., and Guest, J. R. (1970) Turbidometric and polarographic assays for lipoic acid using mutants of *Escherichia coli*. Methods Enzymol. 18a:269–272.
12. MacLean, A. I., and Bachas, L. G. (1991) Homogenous enzyme assay for lipoic acid based on the pyruvate dehydrogenase complex: A model for an assay using a conjugant with one ligand per subunit. Anal. Biochem. 195:303–307.
13. Reed, L. J. (1957) The chemistry and function of lipoic acids. Adv. Enzymol. 18:319–347.
14. Brookfield, D. E., Green, J., Ali, S. T., Machado, R. S., and Guest, J. R. (1991) Evidence for two protein-lipoylation activities in *Escherichia coli*. FEBS Lett. 295:13–16.
15. Fujiwara, K., Okamura-Ikeda, K., and Motokawa, Y. (1992) Expression of mature bovine H-protein of the glycine cleavage system in *Escherichia coli* and in vitro lipoylation of the apoform. J. Biol. Chem. 267:20011–20016.
16. Griffin, T. A., Wynn, R. M., and Chuang, D. T. (1990) Expression and assembly of mature apotransacylase (E2b) of bovine branched-chain α-keto acid dehydrogenase complex in *Escherichia coli*. J. Biol. Chem. 265:12104–12110.

17. Suzuki, K., and Reed, L. J. (1963) Lipoamidase. J. Biol. Chem. 238:4021–4025.
18. Danson, M. J. (1988) Archaebacteria: The comparative enzymology of their central metabolic pathways. Adv. Microbiol. Physiol. 29:165–231.
19. Reed, L. J. (1974) Multienzyme complex. Acc. Chem. Res. 7:40–46.
20. Yeaman, S. J. (1989) The 2-oxo acid dehydrogenase multienzyme complexes: Recent advances. Biochem. J. 257:625–632.
21. Patel, M. S., and Roche, T. E. (1990) Molecular biology and biochemistry of pyruvate dehydrogenase complexes. FASEB J. 4:3224–3233.
22. Palmer, J. A., Madhusudhan, K. T., Hatter, K., and Sokatch, J. R. (1991) Cloning, sequence and transcriptional analysis of the structural gene for LPD-3, the third lipoamide dehydrogenase of *Pseudomonas putida*. Eur. J. Biochem. 202:231–240.
23. Perham, R. N., Packman, L. C., and Radford, S. E. (1987) 2-Oxo acid dehydrogenase multienzyme complexes: In the beginning and halfway there. Biochem. Soc. Symp. 54:67–81.
24. Lawson, J. E., Behal, R. H., and Reed, L. J. (1991) Disruption and mutagenesis of the *Saccharomyces cerevisiae* PDX1 gene encoding the protein X component of the pyruvate dehydrogenase complex. Biochemistry 30:2834–2839.
25. Lawson, J. E., Niu, X.-D., and Reed, L. J. (1991) Functional analysis of the domains of dihydrolipoamide acetyltransferase from *Saccharomyces cerevisiae*. Biochemistry 30:11249–11254.
26. Perham, R. N. (1991) Domains, motifs and linkers in 2-oxo acid dehydrogenase multienzyme complexes: A paradigm in the design of a multifunctional protein. Biochemistry 30:8501–8512.
27. Yang, Y., and Frey, P. A. (1989) 2-Ketoacid dehydrogenase complexes of *Escherichia coli*: Stereospecificities of the three components for (R)-lipoate. Arch. Biochem. Biophys. 268:465–474.
28. Yeaman, S. J. (1986) The mammalian 2-oxoacid dehydrogenases: A complex family. Trends Biochem. Sci. 11:293–296.
29. Reed, L. J., and Hackert, M. L. (1990) Structure-function relationships in dihydrolipoamide acetyltransferases. J. Biol. Chem. 265:8971–8974.
30. Mattevi, A., Obmolova, G., Schulze, E., Kalk, K. H., Westphal, A. H., DeKok, A., and Hol, W. G. J. (1992) Atomic structure of the cubic core of the pyruvate dehydrogenase multienzyme complex. Science 255:1544–1550.
31. Fuller, C. C., Reed, L. J., Oliver, R. M., and Hackert, M. L. (1979) Crystallization of dihydrolipoyl-transacetylase-dihydrolipoyl dehydrogenase subcomplex and its implications regarding the subunit structure of the pyruvate dehydrogenase complex from *Escherichia coli*. Biochem. Biophys. Res. Commun. 90:431–437.
32. Mattevi, A., DeKok, A., and Perham, R. N. (1992) The pyruvate dehydrogenase multienzyme complex. Curr. Opin. Struct. Biol. 2:877–887.
33. Dardel, F., Laue, E. D., and Perham, R. N. (1991) Sequence-specific 1H-NMR assignments and secondary structure of the lipoyl domain of *Bacillus stearothermophilus* pyruvate dehydrogenase multienzyme complex. Eur. J. Biochem. 201:203–209.
34. Kalia, Y. N., Brocklehurst, S. M., Hipps, D. S., Appella, E., Sakaguchi, K., and Perham, R. N. (1993) The high resolution structure of the peripheral subunit-binding domain of dihydrolipoamide acetyltransferase from the pyruvate dehydrogenase multienzyme complex of *Bacillus stearothermophilus*. J. Mol. Biol. 230:323–341.

35. Machado, R. S., Clark, D. P., and Guest, J. R. (1992) Construction and properties of pyruvate dehydrogenase complexes with up to nine lipoyl domains per lipoate acetyltransferase chain. FEMS Microbiol. Lett. 100:243-248.
36. Ali, S. T., Moir, A. J. G., Ashton, P. R., Engel, P. C., and Guest, J. R. (1990) Octanoylation of the lipoyl domains of the pyruvate dehydrogenase complex in a lipoyl deficient strain of *Escherichia coli*. Mol. Microbiol. 4:943–950.
37. Ali, S. T., and Guest, J. R. (1990) Isolation and characterization of lipoylated and unlipoylated domains of the E2p subunit of the pyruvate dehydrogenase complex of *Escherichia coli*. Biochem. J. 271:139–145.
38. Behal, R. H., Browning, K. S., Hall, T. B., and Reed, L. J. (1989) Cloning and nucleotide sequence of the gene for protein X from *Saccharomyces cerevisiae*. Proc. Natl. Acad. Sci. USA 86:8732–8736.
39. Williams, Jr., C. H. (1992) Lipoamide dehydrogenase, glutathione reductase, thioredoxin reductase, and mercuric ion reductase-a family of flavoenzyme transhydrogenases. In: Chemistry and Biochemistry of Flavoenzymes, Vol. III (Muller, F., ed.), pp. 121–211. CRC Press, Boca Raton, FL.
40. Fujiwara, K., Okamura, K., and Motokawa, Y. (1979) Hydrogen carrier protein from chicken liver: Purification, characterization and role of its prosthetic group, lipoic acid, in the glycine cleavage system. Arch. Biochem. Biophys. 197:454–462.
41. Kikuchi, G., and Hiraga, K. (1982) The mitochondrial glycine cleavage system: Unique features of glycine decarboxylation. Mol. Cell Biochem. 45:137–149.
42. Fujiwara, K., Okamura-Ikeda, K., and Motokawa, Y. (1991) Lipoylation of H-protein of the glycine cleavage system. FEBS Lett. 293:115–118.
43. Brocklehurst, S. M., and Perham, R. N. (1993) Prediction of the three-dimensional structures of the biotinylated domain from yeast pyruvate carboxylase and of the lipoylated H-protein from pea leaf glycine cleavage system: A new automated method for the prediction of protein tertiary structure. Protein Sci. 2:626–639.
44. Van de Water, J., Gershwin, M. E., Leung, P., Ansari, A., and Coppel, R. L. (1988) The autoepitope of the 74-kD mitochondrial autoantigen of primary biliary cirrhosis corresponds to the functional site of dihydrolipoamide acetyltransferase. J. Exp. Med. 167:1791–1799.
45. Tuaillon, N., Andre, C., Briand, J-P., Penner, E., and Muller, S. (1992) A lipoyl synthetic octadecapeptide of dihydrolipoamide acetyltransferase specifically recognized by anti-M2 autoantibodies in primary biliary cirrhosis. J. Immunol. 148:445–450.
46. Gut, J., Christen, U., Huwyler, J., Burgin, M., and Kenna, J. G. (1992) Molecular mimicry of trifluoroacetylated human liver protein adducts by constitutive proteins and immunochemical evidence for its impairment in halothane hepatitis. Eur. J. Biochem. 210:569–576.
47. Christen, U., Jeno, P., and Gut, J. (1993) Halothane metabolism: The dihydrolipoamide acetyltransferase subunit of the pyruvate dehydrogenase complex molecularly mimics trifluoracetyl-protein adducts. Biochemistry 32:1492–1499.
48. Robinson, B. H., Mackay, N., Petrova-Benedict, R., Ozalp, I., Coskun, T., and Stacpoole, P. (1990) Defects in the E2 lipoyl transacetylase and the X-lipoyl containing component of the pyruvate dehydrogenase complex in patients with lactic acidemia. J. Clin. Invest. 85:1821–1824.

49. Marsac, C., Stansbie, D., Bonne, G., Cousin, J., Jenenson, P., Benelli, C., Jeroux, J. P., and Lindsay, G. (1993) Defect in the lipoyl-bearing protein-X subunit of the pyruvate dehydrogenase complex in 2 patients with encephalomyelopathy. J. Pediatr. 123:915–920.
50. Matuda, S., Nakano, K., Ohta, S., Shimura, M., Yamanaka, T., Nakagawa, S., Titani, K., and Miyata, T. (1992) Molecular cloning of dihydrolipoamide acetyltransferase of the rat pyruvate dehydrogenase complex: Sequence comparison and evolutionary relationship to other dihydrolipoamide acetyltransferases. Biochim. Biophys. Acta 1131:114–118.
51. Nakano, K., Matuda, S., Yamanaka, T., Tsubouchi, S., Nakagawa, S., Titani, K., Ohta, S., and Miyata, T. (1991) Purification and molecular cloning of succinyltransferase of the rat α-ketoglutarate dehyrogenase complex. J. Biol. Chem. 266:19013–19017.

16

Role of Lipoic Acid in the Glycine Cleavage System and Lipoylation of H-Protein

Yutaro Motokawa, Kazuko Fujiwara, and Kazuko Okamura-Ikeda
The Institute for Enzyme Research, The University of Tokushima, Tokushima, Japan

Lipoic acid is a prosthetic group of H-protein of the glycine cleavage system and dihydrolipoamide acyltransferases (E2) of the pyruvate, α-ketoglutarate, and branched-chain α-keto acid dehydrogenase complexes. The lipoyl moiety is attached in amide linkage to the ε-amino group of specific lysine residues of the proteins. This chapter summarizes the role of lipoic acid in the glycine cleavage reaction and properties of the components of the glycine cleavage system. In addition, it deals with the enzyme responsible for the attachment of lipoic acid to H-protein.

I. THE GLYCINE CLEAVAGE SYSTEM

A. Properties of the Reaction Catalyzed by the Glycine Cleavage System and the Role of Lipoic Acid

The glycine cleavage system is composed of three enzymes and a carrier protein. The enzymes are P-protein or glycine dehydrogenase (decarboxylating) (EC 1.4.4.2), T-protein or aminomethyltransferase (EC 2.1.2.10), and L-protein or dihydrolipoamide dehydrogenase (EC 1.8.1.4). The carrier protein is called H-protein, the prosthetic group of which is lipoic acid. The system catalyzes the reversible oxidation of glycine (the glycine cleavage reaction) shown in the following equation:

Glycine + tetrahydrofolate + NAD^+ = CO_2 + NH_3 + 5,10-methylene-tetrahydrofolate + NADH + H^+

In addition to tetrahydrofolate (H_4folate) and NAD^+, pyridoxal phosphate, protein-bound lipoic acid, and FAD are required. The reaction is partitioned into three partial reactions (Fig. 1) (1). The first partial reaction is catalyzed by P-protein, and H-protein serves as a cosubstrate. The partial reaction proceeds via a sequential random mechanism where the carboxyl carbon of glycine is converted to carbon dioxide and the remnant of the glycine molecule is transferred to one of the sulfhydryl groups formed by the reductive cleavage of disulfide in the lipoyl prosthetic group of H-protein (2). The other sulfhydryl group of the lipoyl moiety remains free. The modification of the free sulfhydryl group with *N*-ethylmaleimide abolishes the function of the H-protein loaded with the aminomethyl group as a substrate for the second partial reaction (3). The free sulfhydryl group seems to be essential for the further degradation of glycine. The decarboxylation is not accompanied by the removal of a C-2 hydrogen atom of glycine, and instead both C-2 hydrogens are transferred with the alpha carbon atom to H-protein (4). K_m values for glycine and H-protein are 5.8 mM and 3.4 μM, respectively (2). H-protein with the aminomethyl moiety derived from glycine can be isolated when the reaction mixture is subjected to gel filtration on Sephadex G-100 (1). The reaction is reversible, and we can obtain glycine from CO_2 and H-protein loaded with the aminomethyl group. The free sulfhydryl group of the lipoyl moiety of H-protein again seems to be essential for the reaction back to glycine, since the modification of the sulfhydryl group makes the H-protein–aminomethyl complex inactive as a substrate for P-protein (3).

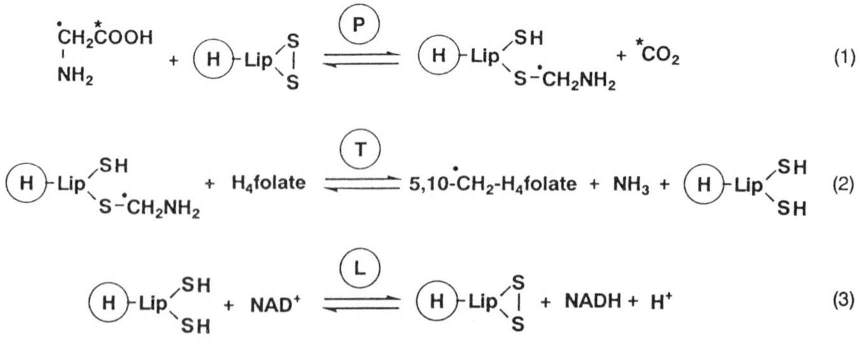

$$\dot{C}H_2\overset{*}{C}OOH(NH_2) + (H)\text{-Lip}\langle S{-}S \rangle \underset{(P)}{\rightleftharpoons} (H)\text{-Lip}\langle SH,\ S{-}\dot{C}H_2NH_2 \rangle + {}^{*}CO_2 \quad (1)$$

$$(H)\text{-Lip}\langle SH,\ S{-}\dot{C}H_2NH_2 \rangle + H_4\text{folate} \underset{(T)}{\rightleftharpoons} 5,10\text{-}\dot{C}H_2\text{-}H_4\text{folate} + NH_3 + (H)\text{-Lip}\langle SH,\ SH \rangle \quad (2)$$

$$(H)\text{-Lip}\langle SH,\ SH \rangle + NAD^+ \underset{(L)}{\rightleftharpoons} (H)\text{-Lip}\langle S{-}S \rangle + NADH + H^+ \quad (3)$$

Figure 1 Reaction scheme for the oxidative degradation of glycine by the glycine cleavage system. P, H, T, and L in the circles represent respective proteins, and Lip represents lipoic acid.

When glycine, $^{14}CO_2$, and H-protein are present in the reaction mixture, P-protein catalyzes the exchange of the carboxyl group of glycine with $^{14}CO_2$ (1). This reaction is a convenient assay method for P-protein and H-protein. For the determination of P-protein activity, an excess of H-protein is added to the reaction mixture. For the assay of H-protein activity, an excess of P-protein is added. The reactive species of CO_2 in the exchange reaction and the reverse reaction is CO_2,but not HCO_3^-. The initial rate of these reactions is higher with CO_2 as a substrate than with HCO_3^- and becomes identical when equilibrium is reached (2).

The second partial reaction is catalyzed by T-protein. The methylene carbon is transferred to H_4folate to give 5,10-methylenetetrahydrofolate (5,10-CH_2-H_4folate), and α-amino group is released as ammonia (3). The formation of these products is stoichiometrical. The lipoyl group of H-protein is left in the reduced form. K_m values for H-protein with the aminomethyl group and H_4folate are 2.2 and 50 μM, respectively. In the absence of H_4folate, formaldehyde and ammonia are produced but the reaction rate is less than 0.05% of that measured in the presence of H_4folate. The formation of the products is stoichiometrical. The results indicate that H_4folate serves not only as an acceptor for the one-carbon unit from glycine but also as a stimulator of the turnover rate presumably by removing formaldehyde from the active site. Kinetic studies reveal that the reverse reaction with ammonia, 5,10-CH_2-H_4folate, and the reduced form of H-protein as substrates proceeds via an ordered ter bi mechanism (5) defined by Cleland (6). H-protein is the first substrate that binds T-protein followed by 5,10-CH_2-H_4folate and ammonia. The order of product release is H_4folate and the methylamine-loaded H-protein. The kinetic studies are carried out under the condition where the aminomethyl moiety attached to H-protein is converted to glycine in the presence of CO_2 and the excess amount of P-protein. To maintain the concentration of reduced H-protein constant, a reducing system for the lipoyl moiety of H-protein consisting of dihydrolipoamide dehydrogenase and NADH is coupled. This coupled assay is suitable for the assay of T-protein.

The third partial reaction, the reoxidation of the lipoyl moiety of H-protein, is catalyzed by L-protein. This reaction is analogous to the reoxidation of the dihydrolipoyl group on acyltransferases of the α-keto acid dehydrogenase complexes catalyzed by dihydrolipoamide dehydrogenase (E3).

Thus, the lipoic acid prosthetic group of H-protein interacts with the active sites of three different enzymes and transfers the intermediate and reducing equivalents in a manner similar to that found for α-keto acid dehydrogenase complexes (7). The lipoyl moiety in the dihydrolipoamide acyltransferases is a part of a flexible lipoyl domain and couples with active sites of the component enzymes of the α-keto acid dehydrogenase complexes. The molecular mass of H-protein is similar to that of the lipoyl domain plus hinge region.

B. Properties of the Individual Components of the Glycine Cleavage System

The glycine cleavage system is ubiquitously distributed among prokaryotes and eukaryotes. In animals, the system is mainly found in liver and situated in the mitochondrial inner membrane (8).

P-protein of vertebrates, pea, and *Escherichia coli* is a homodimer, and each subunit (about 100 kDa) contains a molecule of pyridoxal phosphate. P-protein of some anaerobic bacteria such as *Clostridium acidiurici* and *Eubacterium acidaminophilum* exists as an $\alpha_2\beta_2$ tetramer containing polypeptides of 54–65 kDa (9,10), but the nature of these bacterial P-proteins has not been well characterized. cDNAs for P-protein of human (11), chicken (11), and pea (*Pisum sativum*) (12) and the gene for *E. coli* P-protein (13) have been cloned and characterized. The amino acid sequence of the mature human P-protein (981 amino acids) is 84% and 54% identical to the chicken (970 amino acids) and pea (971 amino acids) counterparts, respectively, and 52% identical to *E. coli* P-protein (956 amino acids) (Fig. 2). Pyridoxal phosphate is attached to a specific lysine residue (e.g., Lys-704 of chicken P-protein) (14). Amino acid sequence around the lysine residue is well conserved among all the P-protein so far examined. Some 50 of the 61 amino acids (e.g., Gly-682 to Pro-742 of human P-protein) are identical, including a run of 25 amino acids, which includes the site of attachment of pyridoxal phosphate (Fig. 2). The pyridoxal phosphate-binding site has the His-Lys(PLP)-X consensus sequence found in pyridoxal phosphate-dependent amino acid decarboxylases and serine hydroxymethyltransferase (15). A region rich in glycine exists seven residues from the pyridoxal phosphate-binding site. The region is found in amino acid decarboxylases and thought to form a hydrophobic pocket in which the cofactor or substrate binds (16). Alternatively, these glycine residues may provide a flexible loop which facilitates the approach of the lipoyl moiety of H-protein to the active site of P-protein.

T-protein is a monomer with a molecular mass of about 40 kDa. This enzyme is a basic protein (pI = 9.8) and forms a 1:1 complex with H-protein (17). The formation of this complex may be due in part to an acidic nature of H-protein (pI = 4.0) (1) and a basic nature of T-protein. cDNAs for T-protein of human (18), bovine (19), chicken (20), and pea (21) and the gene for *E. coli* T-protein (13,22) have been cloned. The amino acid sequence of the mature human T-protein (375 amino acids) is 90, 68, and 53% identical to the bovine (375 amino acids), chicken (376 amino acids), and pea (378 amino acids) counterparts, respectively, and 28% identical to *E. coli* T-protein (363 amino acids) (Fig. 3). There is a highly conserved region (e.g., Gly-229 to Asp-248 of human T-protein), which contains two conserved arginine residues among all the T-protein examined. The region may constitute a part of the active site.

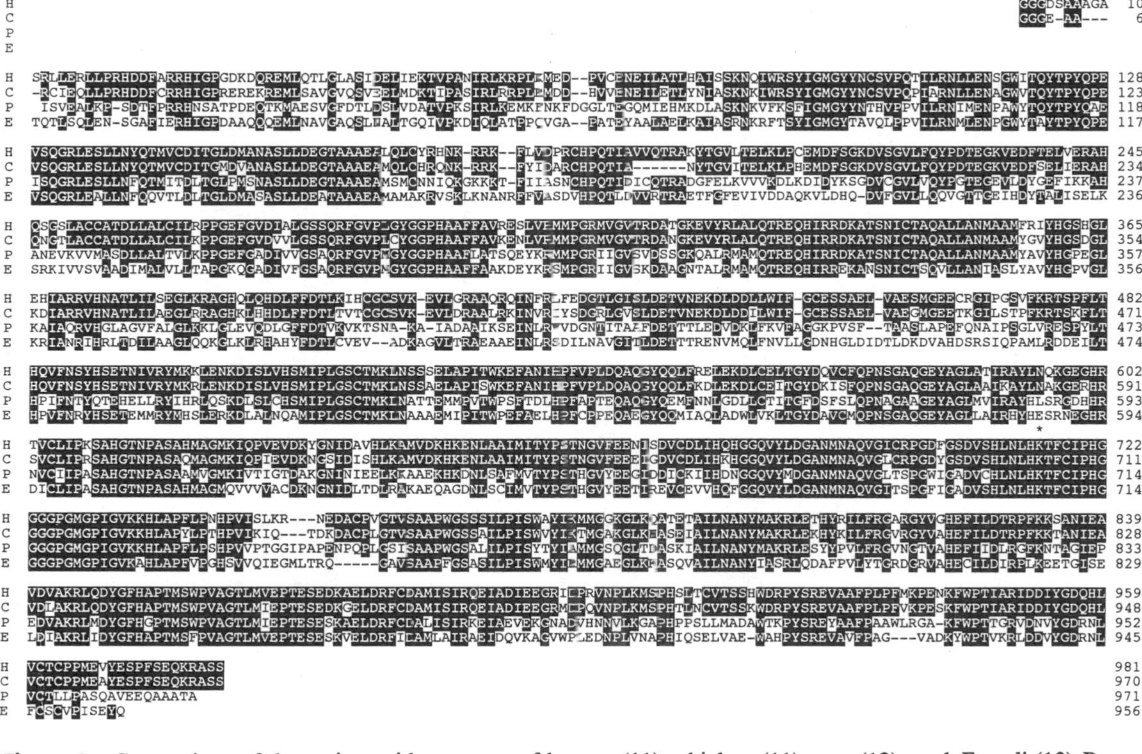

Figure 2 Comparison of the amino acid sequence of human (11), chicken (11), pea (12), and *E. coli* (13) P-protein. The lysine residue involved in pyridoxal phosphate attachment is marked by an asterisk. Amino acid residues identical with human P-protein are indicated by black boxes. H, human; C, chicken; P, pea; E, *E. coli*.

Figure 3 Comparison of the amino acid sequence of human (18), bovine (19), chicken (20), pea (21), and *E. coli* (13,22) T-protein. Amino acid residues identical with human T-protein are indicated by black boxes.

L-protein is a dihydrolipoamide dehydrogenase, a common flavoprotein component of the three α-keto acid dehydrogenase complexes. Dihydrolipoamide dehydrogenase specific to the glycine cleavage system has not been reported. Genes for T-protein, H-protein, and P-protein of *E. coli* constitute an operon (13). The gene for dihydrolipoamide dehydrogenase of *E. coli* is not included in this operon but situated downstream of the *ace* operon that encodes pyruvate dehydrogenase and dihydrolipoamide acetyltransferase (23). The observation suggests that dihydrolipoamide dehydrogenase catalyzes the identical reaction in both the α-keto acid dehydrogenase complexes and the glycine cleavage system and is the same gene product.

H-protein is a monomeric, acidic, and heat-stable protein of about 14 kDa (1). Vertebrate H-protein is composed of 125 amino acids, and lipoic acid is covalently linked to Lys-59 (24). The presence of lipoic acid in the animal H-protein was first determined by paper chromatography of a material obtained from H-protein hydrolyzed with *p*-toluene sulfonic acid (1) and later confirmed by mass spectrometry (25). The binding site of lipoic acid was verified by

Edman degradation and amino acid analysis of a peptide from H-protein whose lipoyl moiety was reduced and modified with [^{14}C]*N*-ethylmaleimide (24). Amino acid sequence around the site of lipoate attachment is conserved among animal, plant, and *E. coli* H-protein. The α-helical structure of the region is predicted (24) by the method of Chou and Fasman (26). cDNAs for H-protein of human (27,28), bovine (29), chicken (30), and plants [*P. sativum* (31,32) and *Arabidopsis thaliana* (33)] and the gene for *E. coli* H-protein (13,34) have been cloned. The amino acid sequence of the mature human H-protein is 97, 86, 45, and 47% identical to the bovine, chicken, pea (131 amino acids), and *A. thaliana* (131 amino acids) counterparts, respectively, and 42% identical to *E. coli* H-protein (128 amino acids) (Fig.4).

```
Human
Bovine
Chicken
P. sativum                                                          SNVL    4
A. thaliana                                                         STVL    4
E. coli                                                             SNVP    4

Human        SVRKFTEKHEWVTTENGIG-TVGISNFAQEALGDVVYCSLPEVGTKLNKQD     50
Bovine       SVRKFTEKHEWVTTENGVG-TVGISNFAQEALGDVVYCSLPEVGTKLNKQE     50
Chicken      SARKFTDKHEWISVENGIG-TVGISNFAQEALGDVVYCSLPEIGTKLNKDD     50
P. sativum   DGLKYAPSHEWVKHEGSVA-TIGITDHAQDHLGEVVFVELPEPGVSVTKGK     54
A. thaliana  EGLKYANSHEWVKHEGSVA-TIGITAHAQDHLGEVVFVELPEDNTSVSKEK     54
E. coli      AELKYSKEHEWLRKEADGTYTVGITEHAQELLGDMVFVDLPEVGATVSAGD     55

                     *
Human        EFGALESVKAASELYSPLSGEVTEINEALAENPGLVNKSCYEDGWLIKMT    100
Bovine       EFGALESVKAASELYSPLSGEVTEINKALAENPGLVNKSCYEDGWLIKMT    100
Chicken      EFGALESVKAASELYSPLTGEVTDINAALADNPGLVNKSCYQDGWLIKMT    100
P. sativum   GFGAVESVKATSDVNSPISGEVIEVNTGLTGKPGLINSSPYEDGWMIKIK    104
A. thaliana  SFGAVESVKATSEILSPISGEIIEVNKKLTESPGLINSSPYEDGWMIKVK    104
E. coli      DCAVAESVKAASDIYAPVSGEIVAVNDALSDSPELVNSEPYAGGWIFKIK    105

Human        LSNPSELDELMSEEAYEKYIKSIEE                             125
Bovine       FSNPSELDELMSEEAYEKYIKSIEE                             125
Chicken      VEKPAELDELMSEDAYEKYIKSIED                             125
P. sativum   PTSPDELESLLGAKEYTKFCEEEDAAH                           131
A. thaliana  PSSPAELESLMGPKEYTKFCEEEDAAH                           131
E. coli      ASDESELESLLDATAYEALLEDE                               128
```

Figure 4 Comparison of the amino acid sequence of human (27,28), bovine (29), chicken (30), pea (31,32), *A. thaliana* (33), and *E. coli* (13,34) H-protein. The lysine residue involved in lipoic acid attachment is marked by an asterisk. Amino acid residues identical with human H-protein are indicated by black boxes.

II. LIPOYLATION OF H-PROTEIN

A. Intramitochondrial Lipoylation of H-Protein

The roles of lipoic acid in the glycine cleavage system and the α-keto acid dehydrogenase complexes have been well studied. However, we know little about the enzymatic attachment of lipoic acid to the ε-amino group of a specific lysine residue of the proteins. In eukaryotes lipoate-bearing proteins are found only in mitochondria. They are synthesized on cytoplasmic ribosomes as a precursor form, transported into mitochondria, and processed to the mature form. It seems likely that the enzyme responsible for attachment of lipoic acid (lipoyltransferase) is located within mitochondria. To obtain direct evidence in support of this assumption, we examined the intramitochondrial lipoylation of H-protein employing the in vitro transcription-translation system (29).

The *Eco*RI fragment of the cDNA for bovine H-protein encoding the full length of the precursor form was subcloned into plasmid pGEM-4Z. The recombinant plasmid DNA was linearized with *Bam*HI. Capped transcripts were produced in the presence of $m^7G(5')ppp(5')G$ from the SP6 promoter, and the purified product was translated in vitro with rabbit reticulocyte lysate in the presence of L-[^{35}S]methionine. A portion of the mixture was reserved, and the product was immunoprecipitated with antibody against chicken H-protein and analyzed by sodium dodecyl sulfate–polyacrylamide gel electrophoresis (SDS-PAGE). The immunoprecipitated product had a molecular mass of approximately 19 kDa (Fig. 5, lane 1). The size is in good agreement with that predicted for the precursor form of H-protein (18,790). The remaining translation mixture was incubated with freshly isolated bovine liver mitochondria. The mixture was further incubated with trypsin and then with soybean trypsin inhibitor. After disruption of mitochondria with Triton X-100, H-protein was immunoprecipitated and analyzed by SDS-PAGE. The product was comigrated with the authentic H-protein purified from bovine liver and resistant to trypsin digestion (Fig. 5, lanes 2 and 3). Clearly, the protein product synthesized in vitro can be imported into mitochondria and processed to its mature size.

The mature form of H-protein synthesized in vitro and processed in mitochondria has a lipoyl group attached to a specific lysine residue (29). The covalent attachment of lipoic acid to the radiolabeled H-protein was confirmed by lysylendopeptidase digestion of the product and resolution of the resultant radioactive peptides by high-performance liquid chromatography (HPLC). Lysylendopeptidase specifically cleaves polypeptides on the carboxyl terminal side of lysine residues but is unable to cleave the peptide bond when the lysine residue is modified (14). The purified H-protein from bovine liver was digested with lysylendopeptidase, and peptides were separated by HPLC. A peptide containing lipoic acid (K5) was identified (Fig. 6A). Peptide K5 consists of 29 amino acids (49–QEEFGALESV**K**AASELYSPLSGEVTEINK-77) including Lys-59

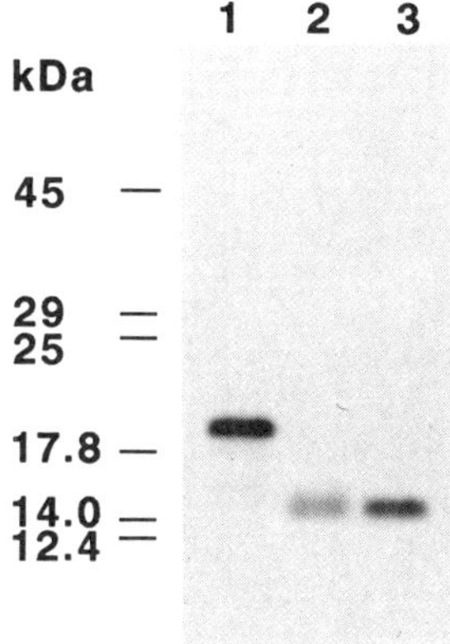

Figure 5 Mitochondrial import and processing of the in vitro-synthesized bovine H-protein precursor. An in vitro translation product (lane 1) was incubated with isolated mitochondria (lane 2). A reaction mixture with mitochondria was further incubated with trypsin (lane 3). All samples were subjected to immunoprecipitation and SDS-PAGE. (From Ref. 29.)

(**K**), to which the lipoic acid prosthetic group is attached. The in vitro–translated precursor form of H-protein labeled with L-[^{3}H]leucine gave no radioactive peak corresponding to the position of peptide K5; instead, two shorter peptides, A and B (Fig. 6B), were indicated when digested with lysylendopeptidase. Edman degradation of these peptides indicated that the two peptides were derived from peptide K5, since significant radioactivity was detected in the cycle where leucine residue was expected to emerge. This indicates that Lys-59 in the precursor form of H-protein is not modified with the lipoyl group. On the other hand, a radioactive peak comigrating with peptide K5 was detected when H-protein incorporated into mitochondria was analyzed as above (Fig. 6C). The results clearly indicate that H-protein is lipoylated after the incorporation into mitochondria and the lipoylation is not a prerequisite for the transport.

Comparison of amino acid sequences around the lipoate-attachment site of H-protein and dihydrolipoamide acyltransferases of the α-keto acid dehydrogenase complexes from various sources indicates the presence of several conserved amino acid residues (Fig. 7). The observation suggests that some features of the primary structure of the region are required for a common function—a carrier of an intermediate to different active sites. Another possibility is that the conserved residues are crucial for the enzymatic attachment of lipoic acid cofactor. To test the latter possibility, the conserved residues of the bovine H-protein were modified individually by site-directed mutagenesis, and each modified H-protein was translated in vitro and incubated with bovine liver mitochondria

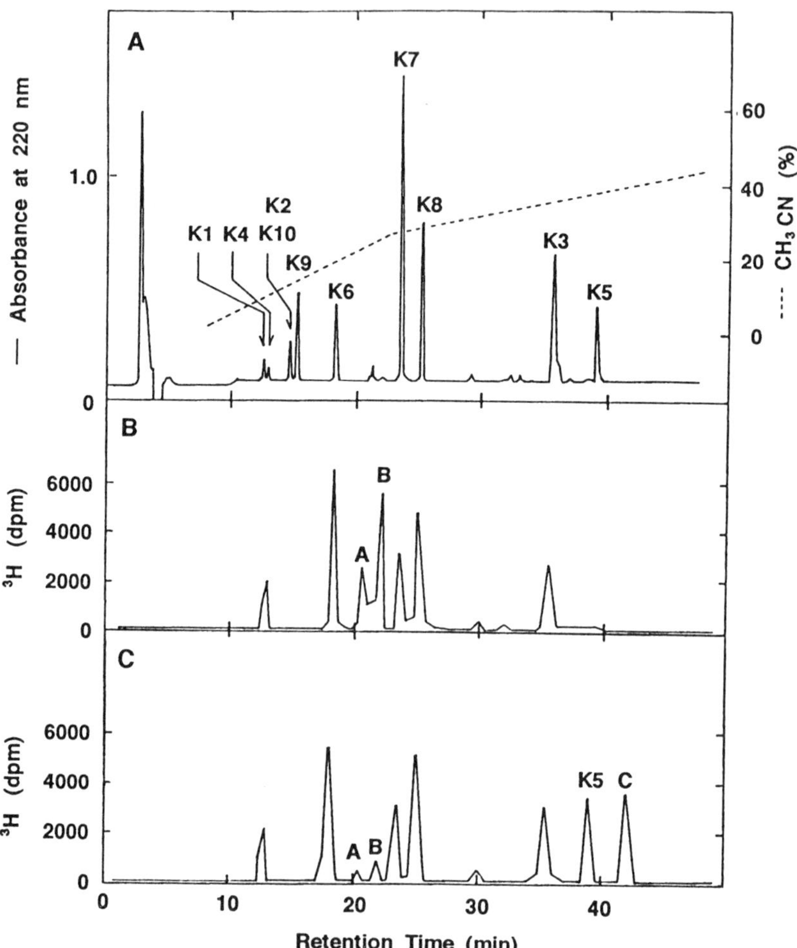

Figure 6 HPLC profile of peptides from bovine H-protein with lysylendopeptidase digestion. (A) Purified H-protein from bovine liver. Peptides were numbered according to their relative location in the sequence of H-protein from the NH_2 terminus. (B) In vitro translated precursor labeled with L-[^{3}H]leucine. (C) Radiolabeled mature H-protein imported into mitochondria. (From Ref. 29.)

(45). All the modified H-proteins were transported into mitochondria and processed to the mature form. Replacement of Glu-56 by glutamine, alanine, or aspartic acid resulted in great decrease in lipoylation. Replacement of Glu-63 by aspartic acid had no effect, and replacement by alanine reduced the lipoylation slightly. The fact that the residues corresponding to Glu-63 are replaced by

```
                                                 *
H-protein  Human                 41 EVGTKLNKQDEFGALESVKAASELYSPLSGEV 72
           Bovine                41 EVGTKLNKQEEFGALESVKAASELYSPLSGEV 72
           Chicken               41 EIGTKLNKDDEFGALESVKAASELYSPLTGEV 72
           Pea                   45 EPGVSVTKGKGFGAVESVKATSDVNSPISGEV 76
           E. coli               46 EVGATVSAGDDCAVAESVKAASDIYAPVSGEI 77

E2p        Human liver           28 KEGDKINEGDLIAEVETDKATVGFESLEECYM 59
           Rat heart             28 KEGEKISEGDLIAEVETDKATVGFESLEECYM 59
           S. cerevisiae         29 KEGDQLSPGEVIAEIETDKAQMDFEFQEDGYL 60
           E. coli               22 KVGDKVEAEQSLITVEGDKASMEVPSPQAGIV 53
           A. vinelandii         21 KTGDLIEVEQGLVVLESAKASMEVPSPKAGVV 52
E2k        Rat heart             25 AVGDAVAEDEVVCEIETDKTSVQVPSPANGII 56
           E. coli               25 KPGDAVVRDEVLVEIETDKVVLEVPASADGIL 56
           A. vinelandii         24 KPGEPVKRDELIVDIETDKVVMEVLAEADGVI 55
E2b        Human and
             bovine liver        26 KEGDTVSQFDSICEVQSDKASVTITSRYDGVI 57
           P. putida             25 KVGDIIAEDQVVADVMTDKATVEIPSPVSGKV 56
                                                 ^
```

Figure 7 Comparison of amino acid sequence surrounding the attachment site of lipoic acid. The sequences of human (27,28), bovine (29), chicken (30), pea (31,32), and *E. coli* (13,34) H-proteins are compared to the sequences of dihydrolipoamide acetyltransferase (E2p) of human liver (35), rat heart (36), *Saccharomyces cerevisiae* (37), *E. coli* (38), and *Azotobacter vinelandii* (39), to the sequences of dihydrolipoamide succinyltransferase (E2k) of rat heart (40), *E. coli* (41), and *A. vinelandii* (42), and to the sequences of human (43) and bovine (43) liver and *Pseudomonas putida* (44) acyltransferases of the branched-chain α-keto acid dehydrogenase complex (E2b). The asterisk indicates the residue involved in lipoic acid attachment. The conserved amino acid residues which were chosen for mutagenesis are indicated by black boxes. The numbers on the right and left refer to the positions of the amino acids of these proteins.

aspartic acid or other amino acids in pea and *E. coli* H-protein and several acyltransferases indicates that Glu-63 is not responsible for lipoylation. Gly-43 seems to be not essential for lipoylation, since the replacement of Gly-43 by asparagine or serine did not reduce the lipoylation rate significantly. On the other hand, replacement of Gly-70 by asparagine or serine reduced the lipoylation rate. These results demonstrate that the well-conserved amino acid residues, Glu-56 and Gly-70, are important, if not crucial, for the efficient lipoylation of H-protein. The region where the lipoyllsyine residue (Lys-59) is situated is predicted to form an α-helical structure and Glu-56 and Lys-59 are expected to reside at the same side of the helix (24). Possibly Glu-56 interacts with lipoyltransferase and the anionic charge may play a role in the interaction. Gly-70 is predicted to be situated in a β-turn structure (24). Presence of glycine at position 70 may be required to facilitate the lipoylation of the specific lysyl residue by providing a flexible loop. Interestingly, a glycine residue corresponding to Gly-70 is also found in biotin enzymes (46).

B. Properties of Lipoyltransferase

Studies on lipoyltransferase have been hindered by the lack of an appropriate acceptor protein. Preparation of substrate amounts of apoprotein is essential for elucidation of the mechanism of lipoic acid attachment. For this purpose, bovine apoH-protein was expressed at high levels (25) through the use of a phage T7 overexpression system (47). H-protein was chosen because it is a monomeric, heat-stable, and small protein that is easy to handle. Nucleotide sequence encoding the mitochondrial targeting presequence was deleted from cDNA for the full-length bovine H-protein by site-directed mutagenesis. The 0.6-kilobase pair fragment encoding the mature form of bovine H-protein was subcloned into the *Nde*I/*Bam*HI site of the expression vector pET-3a. *E. coli* BL21 (DE3)pLysS transformed with the expression vector was grown in a medium supplemented with 25 μM isopropyl-β-D-thiogalactopyranoside at 30°C for 24 h. The cultured cells contained mature bovine H-protein as a soluble form at a level of about 10% of the total bacterial protein. When the cells were cultured in medium containing 30 μM lipoate, about 10% of the recombinant protein expressed was the lipoylated active form, 10% was an inactive octanoylated form, and the remaining 80% was the unlipoylated apoform. Little of the H-protein was lipoylated in *E. coli* cultured without added lipoate. ApoH-protein has been purified to homogeneity from cells cultured without lipoate.

Reed and coworkers showed that lipoyl-AMP is required to activate apopyruvate dehydrogenase complex in *Streptococcus faecalis* and *E. coli* (48). Chemically synthesized lipoyl-AMP is used as a lipoyl donor for the assay of lipoyltransferase. The assay mixture for lipoyltransferase contains apoH-protein, lipoyl-AMP, enzyme preparation, bovine serum albumin (BSA), and phosphate buffer. After incubation at 37°C, the reaction is stopped by heating. Lipoylated H-protein is then assayed by the glycine-$^{14}CO_2$ exchange reaction. The sensitivity of the method is such that we can detect lipoylation of apoH-protein in amounts down to 2 pmol (49).

Lipoyltransferase that catalyzes the transfer of lipoyl group from lipoyl-AMP to apoH-protein has been purified from bovine liver mitochondria (49). The postmitochondrial supernatant has no lipoyltransferase activity. Mitochondrial extract was chromatographed on a hydroxylapatite column. Two active peaks emerged from the column. Lipoyltransferase in the peak eluting at higher phosphate concentration has been purified to homogeneity and named lipoyltransferase II. Lipoyltransferase contained in the peak eluting at lower phosphate concentration has been purified and named lipoyltransferase I, but the final product still contained a minor component. Gel permeation chromatography showed that the molecular mass of native form of both lipoyltransferases is 40 kDa. SDS-PAGE of lipoyltransferase II gave a single protein band corresponding to a molecular mass of 40 kDa. The preparation of lipoyltransferase I consisted of a major band at 40 kDa and a minor band at 61 kDa on SDS-PAGE. These results indicate that the lipoyltransferases are monomeric enzymes.

The lipoylation reaction is absolutely dependent on lipoyl-AMP, apoH-protein, and lipoyltransferase. Lipoate plus ATP cannot replace lipoyl-AMP, indicating that lipoyltransferase has no lipoate-activating activity that produces lipoyl-AMP from lipoate and ATP. Lipoate-activating enzyme has been partially purified from bovine liver (50). Lipoyl-AMP inhibits the reaction at concentrations greater than 50 μM. The inhibition is overcome by the addition of 0.2 mg/ml BSA, and the reaction proceeds hyperbolically up to at least 200 μM lipoyl-AMP. Ovalbumin and cytochrome *c* cannot substitute BSA. BSA binds nonspecifically to some hydrophobic molecules including fatty acids and polyprenyl chains (51,52). Likewise, BSA may interact with lipoyl-AMP to prevent the binding of lipoyl-AMP to the site other than the active site of the enzyme. Both lipoyltransferase activities require phosphate ion as reported with the partially purified *E. coli* enzymes that lipoylate lipoyl apodomains of bacterial pyruvate dehydrogenase complex (53). The activity of lipoyltransferase I and II is maximal at pH 7.9 in potassium phosphate buffer.

Kinetic constants were obtained by varying the concentration of one substrate with the constant concentration of the other substrate (49). The reaction exhibited typical Michaelis-Menten kinetics with respect to both substrates. $K_m^{app.}$ values obtained with lipoyltransferase I and II for lipoyl-AMP at 2 μM apoH-protein were 13 and 16 μM and for apoH-protein at 0.1 mM lipoyl-AMP were 0.29 and 0.17 μM, respectively. $V_{max}^{app.}$ values of lipoyltransferase I and II were 135 and 144 nmol/min/mg of protein, respectively. Thus, properties of lipoyltransferase I and II are not distinguishable kinetically and physically. Further studies are required to elucidate the difference of these lipoyltransferases.

The early work on the lipoylation of the apopyruvate dehydrogenase complex of *E. coli* and *S. faecalis* showed that excess octanoyl-AMP inhibits the lipoylation (48). The lipoylation of H-protein catalyzed by lipoyltransferase I or II was inhibited by analogs of lipoyl-AMP, including hexanoyl-, octanoyl-, and decanoyl-AMP, to a similar extent (49). When 40 μM lipoyl-AMP was employed as substrate, lipoylation was inhibited to about 30–40% by the addition of 40 μM of these acyl-AMPs. The result suggests that both lipoyltransferases have affinities for these acyl-AMPs similar to that for lipoyl-AMP. Moreover, both lipoyltransferases are able to catalyze the transfer of the acyl groups from these acyl-AMPs to apoH-protein as determined by nondenaturing PAGE. When modified at Lys-59, H-protein moves faster than apoH-protein because the apo form has an additional positive charge on the unmodified Lys-59 (Fig. 8) (49). 2-Propylpentanoic acid, an antiepileptic drug, inhibits the glycine cleavage reaction (54). 2-Propylpentanoyl-AMP inhibited the lipoylation only slightly (about 20% inhibition), and the propylpentanoyl group was transferred to apoH-protein, although poorly (Fig. 8). Octanoylation of a specific lysine residue that should be lipoylated has been observed when the mature bovine H-protein (25) and lipoyl domains of bacterial dihydrolipoamide acetyltransferase component (55,56) were overexpressed in *E. coli*. The octanoylation was confirmed by mass spectroscopic analysis. The mature form of bovine H-protein translated in vitro and processed in mitochondria gave a peptide not obtained from the authentic H-protein (peptide C in Fig. 5C) when digested with lysylendopeptidase and analyzed by HPLC (29). The retention time of peptide C is similar to that of octanoylated peptide from bovine H-protein overexpressed in *E. coli* on HPLC (25). These results indicate that peptide C may be an octanoyl peptide and that octanoylation is not restricted in *E. coli*. However, no acylated H-protein other than lipoylated H-protein has been found in any organism. The presence of a large amount of H-protein or lipoyl domain to be lipoylated in mitochondria or *E. coli* is presumably the principal cause of the mismodification with an octanoyl group, although the precise mechanism involved is not known.

Lipoyltransferases I and II are active when the nascent apoH-protein translated in vitro in a rabbit reticulocyte lysate was employed as an acceptor protein (49). cDNA encoding the precursor or the mature form of H-protein were transcribed and translated in vitro, and the translated products were subjected

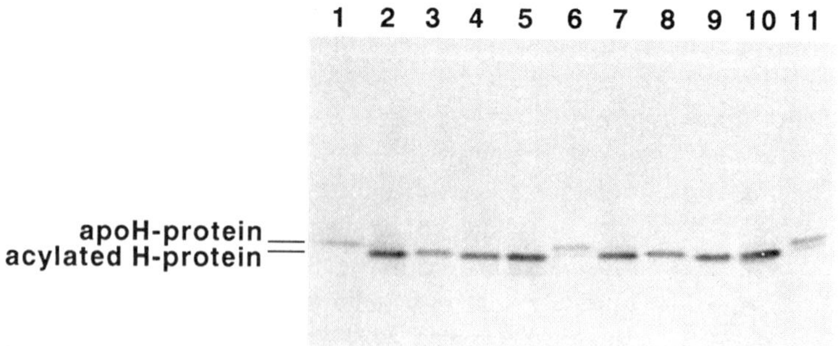

Figure 8 Lipoylation and acylation of apoH-protein. Reactions were carried out without enzyme (lane 1), with lipoyltransferase I (lanes 2–6), and with lipoyltransferase II (lanes 7–11). Products were subjected to nondenaturing PAGE, electrotransferred onto Immobilon-P, and detected with antibody against chicken H-protein. Lanes 1, 2, and 7, with lipoyl-AMP; lanes 3 and 8, with hexanoyl-AMP; lanes 4 and 9, with octanoyl-AMP; lanes 5 and 10, with decanoyl-AMP; lanes 6 and 11, with 2-propylpentanoyl-AMP. (From Ref. 49.)

to the lipoylation experiment. The precursor and mature forms were both modified with the lipoyl group, indicating that the loosely folded apoH-protein is able to accept the lipoyl moiety.

We have not tested the possibility that lipoyltransferases purified from bovine liver can lipoylate apoacyltransferase of the pyruvate, α-ketoglutarate, or branched-chain α-keto acid dehydrogenase complex. The presence of two enzymes responsible for the attachment of lipoic acid to apolipoyl domains of the pyruvate dehydrogenase complex has been reported in *E. coli* (53). A striking difference between bovine lipoyltransferases and *E. coli* enzymes is that the latter are able to use lipoate in the presence of MgATP in place of lipoyl-AMP, while bovine enzymes are not. It is interesting to speculate whether these lipoate-transferring enzymes from bovine and *E. coli* can lipoylate various protein substrates or are specific to the limited protein substrate.

REFERENCES

1. Fujiwara, K., Okamura, K., and Motokawa, Y. (1979) Hydrogen carrier protein from chicken liver: Purification, characterization, and role of its prosthetic group, lipoic acid, in the glycine cleavage reaction. Arch. Biochem. Biophys. 197:454–462.
2. Fujiwara, K., and Motokawa, Y. (1983) Mechanism of the glycine cleavage reaction: Steady state kinetic studies of the P-protein-catalyzed reaction. J. Biol. Chem. 258:8156–8162.
3. Fujiwara, K., Okamura-Ikeda, K., and Motokawa, Y. (1984) Mechanism of the glycine cleavage reaction: Further characterization of the intermediate attached to H-protein and of the reaction catalyzed by T-protein. J. Biol. Chem. 259:10664–10668.
4. Fujiwara, K., Okamura-Ikeda, K., Ohmura, Y., and Motokawa, Y. (1986) Mechanism of the glycine cleavage reaction: Retention of C-2 hydrogens of glycine on the intermediate attached to H-protein and evidence for the inability of serine hydroxymethyltransferase to catalyze the glycine decarboxylation. Arch. Biochem. Biophys. 251:121–127.
5. Okamura-Ikeda, K., Fujiwara, K., and Motokawa, Y. (1987) Mechanism of the glycine cleavage reaction: Properties of the reverse reaction catalyzed by T-protein. J. Biol. Chem. 262:6746–6749.
6. Cleland, W. W. (1970) Steady state kinetics. In: The Enzymes, 3rd ed., Vol. 2 (Boyer, P. D., ed.), pp. 1–65. Academic Press, New York.
7. Reed, L. J., and Hackert, M. L. (1990) Structure-function relationships in dihydrolipoamide acyltransferases. J. Biol. Chem. 265:8971–8974.
8. Motokawa, Y., and Kikuchi, G. (1971) Glycine metabolism in rat liver mitochondria: V. Intramitochondrial localization of the reversible glycine cleavage system and serine hydroxymethyltransferase. Arch. Biochem. Biophys. 146:461–466.
9. Gariboldi, R. T., and Drake, H. L. (1984) Glycine synthase of the purinolytic bacterium *Clostridium acidiurici*: Purification of the glycine-CO_2 exchange system. J. Biol. Chem. 259:6085–6089.

10. Freudenberg, W., and Andreesen, J. R. (1989) Purification and partial characterization of the glycine decarboxylase multienzyme complex from *Eubacterium acidaminophilum*. J. Bacteriol. 171:2209–2215.
11. Kume, A., Koyata, H., Sakakibara, T., Ishiguro, Y., Kure, S., and Hiraga, K. (1991) The glycine cleavage system: Molecular cloning of the chicken and human glycine decarboxylase cDNAs and some characteristics involved in the deduced protein structures. J. Biol. Chem. 266:3323–3329.
12. Turner, S. R., Ireland, R., and Rawsthorne, S. (1992) Cloning and characterization of the P subunit of glycine decarboxylase from pea (*Pisum sativum*). J. Biol. Chem. 267:5355–5360.
13. Okamura-Ikeda, K., Ohmura, Y., Fujiwara, K., and Motokawa, Y. (1993) Cloning and nucleotide sequence of the gcv operon encoding the *Escherichia coli* glycine-cleavage system. Eur. J. Biochem. 216:539–548.
14. Fujiwara, K., Okamura-Ikeda, K., and Motokawa, Y. (1987) Amino acid sequence of the phosphopyridoxyl peptide from P-protein of the chicken liver glycine cleavage system. Biochem. Biophys. Res. Commun. 149:621–627.
15. Bossa, F., Barra, D., Martini, F., Schirch, L. V., and Fasella, P. (1976) Serine transhydroxymethylase from rabbit liver: Sequence of a nonapeptide at the pyridoxal-5′-phosphate-binding site. Eur. J. Biochem. 70:397–401.
16. Moore, R. C., and Boyle, S. M. (1990) Nucleotide sequence and analysis of the speA gene encoding biosynthetic arginine decarboxylase in *Escherichia coli*. J. Bacteriol. 172:4631–4640.
17. Okamura-Ikeda, K., Fujiwara, K., and Motokawa, Y. (1982) Purification and characterization of chicken liver T-protein, a component of the glycine cleavage system. J. Biol. Chem. 257:135–139.
18. Hayasaka, K., Nanao, K., Takada, G., Okamura-Ikeda, K., and Motokawa, Y. (1993) Isolation and sequence determination of cDNA encoding human T-protein of the glycine cleavage system. Biochem. Biophys. Res. Commun. 192:766–771.
19. Okamura-Ikeda, K., Fujiwara, K., Yamamoto, M., Hiraga, K., and Motokawa, Y. (1991) Isolation and sequence determination of cDNA encoding T-protein of the glycine cleavage system. J. Biol. Chem. 266:4917–4921.
20. Okamura-Ikeda, K., Fujiwara, K., and Motokawa, Y. (1992) Molecular cloning of a cDNA encoding chicken T-protein of the glycine cleavage system and expression of the functional protein in *Escherichia coli*: Effect of mRNA secondary structure in the translational initiation region on expression. J. Biol. Chem. 267:18284–18290.
21. Bourguignon, J., Vauclare, P., Merand, V., Forest, E., Neuburger, M., and Douce, R. (1993) Glycine decarboxylase complex from higher plants: Molecular cloning, tissue distribution and mass spectrometry analyses of the T protein. Eur. J. Biochem. 217:377–386.
22. Stauffer, L. T., Ghrist, A., and Stauffer, G. V. (1993) The *Escherichia coli* gcvT gene encoding the T-protein of the glycine cleavage enzyme system. DNA Sequence 3:339–346.
23. Stephens, P. E., Lewis, H. M., Darlison, M. G., and Guest, J. R. (1983) Nucleotide sequence of the lipoamide dehydrogenase of *Escherichia coli* K12. Eur. J. Biochem. 135:519–527.

24. Fujiwara, K., Okamura-Ikeda, K., and Motokawa, Y. (1986) Chicken liver H-protein, a component of the glycine cleavage system: Amino acid sequence and identification of the N^{ε}-lipoyllysine residue. J. Biol. Chem. 261:8836–8841.
25. Fujiwara, K., Okamura-Ikeda, K., and Motokawa, Y. (1992) Expression of mature bovine H-protein of the glycine cleavage system in *Escherichia coli* and in vitro lipoylation of the apoform. J. Biol. Chem. 267:20011-20016.
26. Chou, P. Y., and Fasman, G. D. (1978) Prediction of the secondary structure of proteins from their amino acid sequence. Adv. Enzymol. Relat. Areas Mol. Biol. 47:45–148.
27. Fujiwara, K., Okamura-Ikeda, K., Hayasaka, K., and Motokawa, Y. (1991) The primary structure of human H-protein of the glycine cleavage system deduced by cDNA cloning. Biochem. Biophys. Res. Commun. 176:711–716.
28. Koyata, H., and Hiraga, K. (1991) The glycine cleavage system: Structure of a cDNA encoding human H-protein, and partial characterization of its gene in patients with hyperglycinemias. Am. J. Hum. Genet. 48:351–361.
29. Fujiwara, K., Okamura-Ikeda, K., and Motokawa, Y. (1990) cDNA sequence, in vitro synthesis, and intramitochondrial lipoylation of H-protein of the glycine cleavage system. J. Biol. Chem. 265:17463–17467.
30. Yamamoto, M., Koyata, H., Matsui, C., and Hiraga, K. (1991) The glycine cleavage system: Occurrence of two types of chicken H-protein mRNAs presumably formed by the alternative use of the polyadenylation consensus sequences in a single exon. J. Biol. Chem. 266:3317–3322.
31. Kim, Y., and Oliver, D. J. (1990) Molecular cloning, transcriptional characterization, and sequencing of cDNA encoding the H-protein of the mitochondrial glycine decarboxylase complex in peas. J. Biol. Chem. 265:848–853.
32. Macherel, D., Lebrun, M., Gagnon, J., Neuburger, M., and Douce, R. (1990) cDNA cloning, primary structure and gene expression for H-protein, a component of the glycine-cleavage system (glycine decarboxylase) of pea (*Pisum sativum*) leaf mitochondria. Biochem. J. 268:783–789.
33. Srinivasan, R., and Oliver, D. J. (1992) H-Protein of the glycine decarboxylase multienzyme complex. Complementary DNA encoding the protein from *Arabidopsis thaliana*. Plant Physiol. 98:1518–1519.
34. Stauffer, L. T., Steiert, P. S., Steiert, J. G., and Stauffer, G. V. (1991) An *Escherichia coli* protein with homology to the H-protein of the glycine cleavage enzyme complex from pea and chicken liver. DNA Sequence 2:13–17.
35. Thekkumkara, T. J., Ho, L., Wexler, I. D., Pons, G., Liu, T.-C., and Patel, M. S. (1988) Nucleotide sequence of a cDNA for the dihydrolipoamide acetyltransferase component of human pyruvate dehydrogenase complex. FEBS Lett. 240:45–48.
36. Matuda, S., Nakano, K., Ohta, S., Shimura, M., Yamanaka, T., Nakagawa, S., Titani, K., and Miyata, T. (1992) Molecular cloning of dihydrolipoamide acetyltransferase of the rat pyruvate dehydrogenase complex: Sequence comparison and evolutionary relationship to other dihydrolipoamide acyltransferases. Biochim. Biophys. Acta 1131:114–118.
37. Niu, X.-D., Browning, K. S., Behal, R. H., and Reed. L. J. (1988) Cloning and nucleotide sequence of the gene for dihydrolipoamide acetyltransferase from *Saccharomyces cerevisiae*. Proc. Natl. Acad. Sci. USA 85:7546–7550.

38. Stephens, P. E., Darlison, M. G., Lewis, H. M., and Guest, J. R. (1983) The pyruvate dehydrogenase complex of *Escherichia coli* K12: Nucleotide sequence encoding the dihydrolipoamide acetyltransferase component. Eur. J. Biochem. 133:481–489.
39. Hanemaaijer, R., Janssen, A., de Kok, A., and Veeger, C. (1988) The dihydrolipoyltransacetylase component of the pyruvate dehydrogenase complex from *Azotobacter vinelandii*: Molecular cloning and sequence analysis. Eur. J. Biochem. 174:593–599.
40. Nakano, K., Matuda, S., Yamanaka, T., Tsubouchi, H., Nakagawa, S., Titani, K., Ohta, S., and Miyata, T. (1991) Purification and molecular cloning of succinyltransferase of the rat α-ketoglutarate dehydrogenase complex: Absence of a sequence motif of the putative E3 and/or E1 binding site. J. Biol. Chem. 266:19013–19017.
41. Spencer, M. E., Darlison, M. G., Stephens, P. E., Duckenfield, I. K., and Guest, J. R. (1984) Nucleotide sequence of the sucB gene encoding the dihydrolipoamide succinyltransferase of *Escherichia coli* K12 and homology with the corresponding acetyltransferase. Eur. J. Biochem. 141:361–374.
42. Westphal, A. H., and de Kok, A. (1990) The 2-oxoglutarate dehydrogenase complex from *Azotobacter vinelandii*: 2. Molecular cloning and sequence analysis of the gene encoding the succinyltransferase component. Eur. J. Biochem. 187:235–239.
43. Lau, K. S., Griffin, T. A., Hu, C.-W. C., and Chuang, D. T. (1988) Conservation of primary structure in the lipoyl-bearing and dihydrolipoyl dehydrogenase binding domains of mammalian branched-chain α-keto acid dehydrogenase complex: Molecular cloning of human and bovine transacylase (E2) cDNAs. Biochemistry 27:1972–1981.
44. Burns, G., Brown, T., Hatter, K., and Sokatch, J. R. (1988) Comparison of the amino acid sequences of the transacylase components of branched chain oxoacid dehydrogenase of *Pseudomonas putida*, and the pyruvate and 2-oxoglutarate dehydrogenases of *Escherichia coli*. Eur. J. Biochem. 176:165–169.
45. Fujiwara, K., Okamura-Ikeda, K., and Motokawa, Y. (1991) Lipoylation of H-protein of the glycine cleavage system: The effect of site-directed mutagenesis of amino acid residues around the lipoyllysine residue on the lipoate attachment. FEBS Lett. 293:115–118.
46. Browner, M. F., Taroni, F., Sztul, E., and Rosenberg. L. E. (1989) Sequence analysis, biogenesis, and mitochondrial import of the α-subunit of rat liver propionyl-CoA carboxylase. J. Biol. Chem. 264:12680–12685.
47. Studier, F. W., Rosenberg, A. H., Dunn, J. J., and Dubendorff, J. W. (1990) Use of T7 RNA polymerase to direct expression of cloned genes. Methods Enzymol. 185:60–89.
48. Reed, L. J., Leach, F. R., and Koike, M. (1958) Studies on a lipoic acid-activating system. J. Biol. Chem. 232:123–142.
49. Fujiwara, K., Okamura-Ikeda, K., and Motokawa, Y. (1994) Purification and characterization of lipoyl-AMP:N^{ε}-lysine lipoyltransferase from bovine liver mitochondria. J. Biol. Chem. 269:16605–16609.
50. Tsunoda, J. N., and Yasunobu, K. T. (1967) Mammalian lipoic acid activating enzyme. Arch. Biochem. Biophys. 118:395–401.

51. Richieri, G. V., Anel, A., and Kleinfeld, A. M. (1993) Interactions of long-chain fatty acids and albumin: Determination of free fatty acid levels using the fluorescent probe ADIFAB. Biochemistry 32:7574–7580.
52. Ohnuma, S., Koyama, T., and Ogura, K. (1991) Purification of solanesyl-diphosphate synthase from *Micrococcus luteus*: A new class of prenyltransferase. J. Biol. Chem. 266:23706–23713.
53. Brookfield, D. E., Green, J., Ali, S. T., Machado, R. S., and Guest, J. R. (1991) Evidence for two protein-lipoylation activities in *Escherichia coli*. FEBS Lett. 295:13–16.
54. Kochi, H., Hayasaka, K., Hiraga, K., and Kikuchi, G. (1979) Reduction of the level of the glycine cleavage system in the rat liver resulting from administration of dipropylacetic acid: An experimental approach to hyperglycinemia. Arch. Biochem. Biophys. 198:589–597.
55. Ali, S. T., Moir, A. J. G., Ashton, P. R., Engel, P. C., and Guest, J. R. (1990) Octanoylation of the lipoyl domains of the pyruvate dehydrogenase complex in a lipoyl-deficient strain of *Escherichia coli*. Mol. Microbiol. 4:943–950.
56. Dardel, F., Packman, L. C. and Perham, R. N. (1990) Expression in *Escherichia coli* of a sub-gene encoding the lipoyl domain of the pyruvate dehydrogenase complex of *Bacillus stearothermophilus*. FEBS Lett. 264:206–210.

17

α-Lipoic Acid and N-Acetylcysteine: New Concepts for Old Drugs

Aalt Bast and Guido R. M. M. Haenen
Leiden/Amsterdam Center for Drug Research, Vrije Universiteit, Amsterdam, The Netherlands

I. INTRODUCTION

α-Lipoic acid liver is used as a protective agent against damage resulting from hepatotoxic compounds, peripheral polyneuropathies, and as adjuvant during treatment of polyneuritis and other consequences of vitamin B deficiency.

N-acetylcysteine is prescribed as a mucolytic agent and is claimed to reduce bronchial mucus viscosity by cleavage of disulfide bridges between glycopeptides. Somewhat later *N*-acetylcysteine was also used as rescue agent after paracetamol intoxication.

The faces of both drugs changed over the last years since their antioxidant activity was raised. Here the antioxidant profile of both compounds is described.

The clinical use of both compounds is still in line with the original information leaflets, namely, α-lipoic acid is applied in diabetic-induced polyneuropathy and *N*-acetylcysteine in lung disease, especially in chronic obstructive pulmonary disease (COPD). It should be realized, however, that the distinct antioxidant properties of both biothiols offer broader clinical perspectives.

II. ENDOGENOUS PROTECTION AGAINST LIPID PEROXIDATION

The peroxidation of membrane lipids is one of the major mechanisms of injury in cells subjected to oxidative stress. As a result of lipid peroxidation, all membrane functions can be inactivated. The barrier function of the membrane is lost, since pores in the membrane are formed (1). Also the activity of membrane-bound enzymes is reduced by lipid peroxidation (2,3), and signal transduction over the membrane is impaired (4). Beside this, cytotoxic products, like hydroxyalkenals, are formed during lipid peroxidation (3).

Aerobic cells are equipped with an elaborate defense system against oxidative stress. In the membrane the antioxidant vitamin E is present, while the cytosol contains the antioxidants glutathione (GSH) and vitamin C. An increasing amount of evidence indicates that the combination of two antioxidants provides a more efficient line of defense against lipid peroxidation than the sum of the two antioxidants working apart. An example for such a synergy is found in the antioxidant activity of GSH.

The hydrophilic antioxidant GSH is not an efficient scavenger of free radicals, especially when the radicals are formed in the lipid membrane (5). With respect to the GSH-dependent defense against lipid peroxidation, much attention has been given to the GSH peroxidases. The peroxidases catalyze the reduction of hydroperoxides (ROOH) into their corresponding, less reactive alcohols (ROH) at the expense of GSH, which is oxidized to oxidized glutathione (GSSG).

$$\mathrm{ROOH + 2\ GSH \xrightarrow{GSH-px} ROH + H_2O + GSSG}$$

Under normal conditions, GSH is regenerated from GSSG by the GSH reductase, which uses NADPH as cofactor.

Much emphasis has been placed on the idea that the detoxification of phospholipid hydroperoxides by the GSH peroxidases, either alone or in combination with phospholipase A_2, is the major—if not only—GSH-dependent protection mechanism against lipid peroxidation (6,7). However, there are several indications that none of the GSH peroxidases are involved in the GSH-dependent protection against microsomal lipid peroxidation (3,8).

More direct evidence exists for the participation of vitamin E in the GSH-dependent protection against lipid peroxidation. In control liver microsomes from guinea pigs, GSH induces a concentration-dependent protection against lipid peroxidation (Fig. 1A). This effect is specific for GSH, since other thiols like *N*-acetyl-L-cysteine (L-NAC) do not protect (Fig. 1A). The GSH-dependent protection was absent in control microsomes that were heated (100°C) for 60 s, or in trypsin-treated microsomes (not shown).

Extraction of the microsomes with acetone reduced the vitamin E content from 0.20 ± 0.01 nmol/mg protein of 0.025 0.001 nmol/mg protein. In the microsomes extracted with acetone, the GSH-dependent protection was absent (Fig. 1B). The extraction with acetone had no effect on the amount of polyunsaturated fatty acid of the microsomes (not shown). Reintroduction of vita-

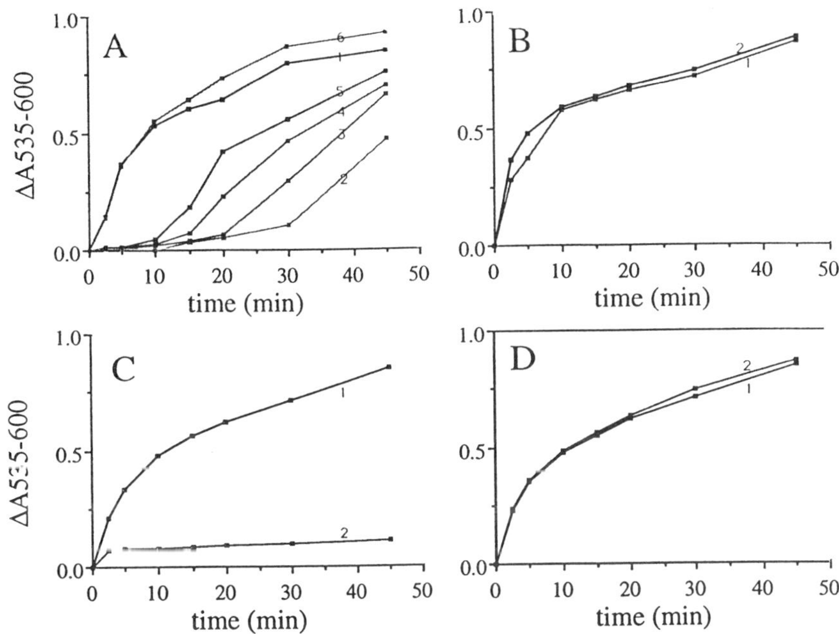

Figure 1 Time course of lipid peroxidation in guinea pig liver microsomes. The experiments were performed as reported previously (2,3). (A) Control microsomes, (B) microsomes extracted with acetone, and (C) microsomes extracted with acetone where vitamin E was subsequently reintroduced to the control level were used. (D) The microsomes of C pretreated with trypsin were used. In all cases lipid peroxidation was induced by 0.2 mM ascorbate and 10 μM Fe^{2+}. Further additions were: none (curves 1), 1 mM GSH (curves 2), 0.5 mM GSH (curve 3), 0.2 mM GSH (curve 4), 0.1 mM GSH (curve 5), and 1 mM L-NAC (curve 6). Lipid peroxidation was measured with the thiobarbituric acid test as described previously (2) and was expressed as the absorbance at 535 nm vs. 600 nm ($\Delta A_{535-600}$).

min E in the membrane to the control level (0.197 ± 0.003 nmol/mg protein) restored the GSH-dependent protection (Fig. 1C). In the extracted microsomes where the vitamin E level was restored to the control level, heating (not shown) or trypsin treatment (Fig. 1D) destroyed the GSH-dependent protection again.

As shown by these results, liver microsomes from guinea pigs are protected against lipid peroxidation by GSH. The effect of GSH was more pronounced in guinea pig liver microsomes than in rat or mouse liver microsomes (9). Species differences in the potency of GSH to protect against lipid peroxidation have been reported previously (10); however, no explanation for this difference has been given. We found that this difference is not due to a different vitamin

E content of the microsomes, nor to a different activity of the microsomal GSH transferase in the different species (9). The specificity for GSH indicates that a phospholipid hydroperoxide GSH peroxidase is not involved, since thiols other than GSH are accepted by this enzyme as cofactor, while these thiols do not give protection.

The most likely explanation for the GSH-dependent protection is the regeneration of vitamin E (Vit E) from the vitamin E radicals (Vit E$^{\bullet}$) by GSH, a reaction catalyzed by a heat-labile factor that functions as a free radical reductase (2).

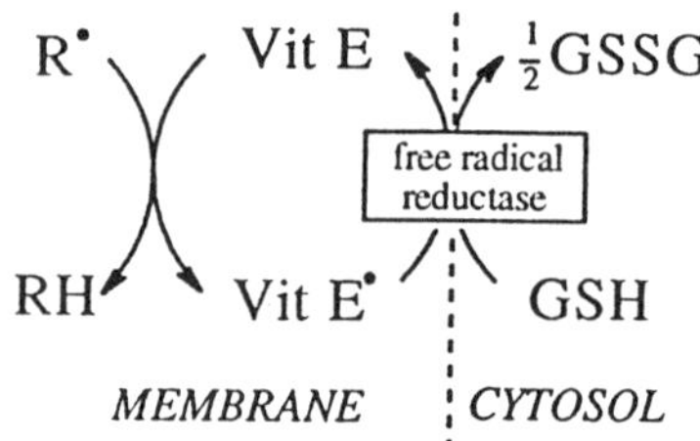

This explains the synergistic effect of vitamin E and GSH in the protection against lipid peroxidation.

III. ANTIOXIDANT ACTIVITY OF L-NAC AND DIHYDROLIPOATE

In the prevention of radical-induced damage, the thiol-containing drugs like L-NAC and dihydrolipoate are used. To understand the effectiveness of these drugs, the interplay with the endogenous defense mechanisms against lipid peroxidation has to be investigated. As shown in Figure 1A, L-NAC did not protect against lipid peroxidation in liver microsomes. This indicates that the protection by L-NAC via the free radical reductase can only be achieved via the synthesis of GSH, since L-NAC is a precursor of GSH. L-NAC stimulated Fe^{2+}/ascorbate-induced lipid peroxidation, whereas it had no effect on Fe^{2+}-induced lipid peroxidation. The pro-oxidant effect of L-NAC is similar to the pro-oxidant effect of sodium 2-mercaptoethanesulfonate (11) and probably the result of the indirect reduction of iron via the regeneration of ascorbate.

Dihydrolipoate itself does not induce lipid peroxidation, but the combination of dihydrolipoate with either Fe^{3+} or Fe^{2+} induces lipid peroxidation (Fig. 2). The effect of dihydrolipoate is concentration-dependent (Fig. 3). Increasing the concentration of dihydrolipoate increases the final extent of lipid peroxidation, and with concentrations of dihydrolipoate above 0.5 mM, a lag time is induced. Addition of 0.5 mM dihydrolipoate to an incubation system containing Fe^{2+} and 0.2 mM vitamin C induces a short lag time (Fig. 3).

The effects of dihydrolipoate are comparable to the dual role ascorbate has on lipid peroxidation. In a low concentration, ascorbate acts as a pro-oxidant,

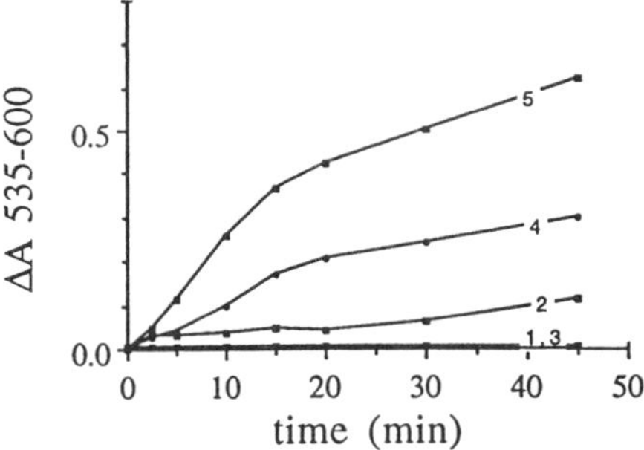

Figure 2 Time course of lipid peroxidation in control rat liver microsomes. Lipid peroxidation was induced by 10 μM Fe^{3+} (1), 10 μM Fe^{2+} (2), 0.5 mM dihydrolipoate (3), 10 μM Fe^{3+} and 0.5 mM dihydrolipoate (4), or 10 μM Fe^{2+} and 0.5 mM dihydrolipoate (5).

whereas higher concentration of ascorbate protect against lipid peroxidation (12). It has been stated the the ratio of Fe^{2+}/Fe^{3+} is the determining factor for the initiation of lipid peroxidation reactions (13,14). High concentrations of ascorbate might inhibit lipid peroxidation due to a reductive maintenance of iron exclusively in the Fe^{2+} form, thus preventing the Fe^{2+} to Fe^{3+} conversion that is needed for inducing lipid peroxidation (14).

In microsomal lipid peroxidation induced by Fe^{2+}/ascorbate, dihydrolipoate displayed antioxidant activity in a concentration of 0.5 mM since a lag phase is observed. Dihydrolipoate in a concentration of 0.5 mM in an incubation system with only Fe^{2+} without ascorbate only stimulated lipid peroxidation. Apparently, the pro- and antioxidant effects of dihydrolipoate and vitamin C

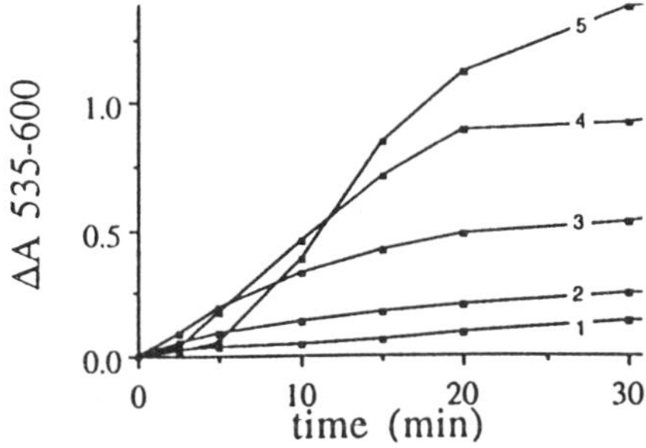

Figure 3 Modulation of lipid peroxidation in control rat liver microsomes by dihydrolipoate. The concentrations of dihydrolipoate for curves 1–5, respectively, were 0, 0.25, 0.5, 1.0, and 2.0 mM. All reactions were started with the addition of 10 μM Fe^{2+}.

proceed via a similiar mechanism, and the effects of both compounds are additive. This effect of dihydrolipoate is comparable to the previously reported effect of L-cysteine (15).

The presented results show that dihydrolipoate has no prominent effect on microsomal lipid peroxidation. This again illustrates that the free radical reductase is rather selective for GSH. Also GSSG does not protect against Fe^{2+}/ascorbate-induced lipid peroxidation. However, the combination of GSSG and dihydrolipoate provides a more pronounced protection against lipid peroxidation than dihydrolipoate alone (Fig. 4). Heating the microsomes in boiling water for 60 s abolished the protective effect of either GSH alone, or the combination of GSSG and dihydrolipoate (not shown). This suggests that the protective effect of the combination of GSSG and dihydrolipoate proceeds via GSH and the free radical reductase. In order to verify this possibility, the formation of GSH from GSSG and dihydrolipoate was determined. As shown in Table 1, there was indeed an extensive but incomplete reduction of GSSG to GSH by dihydrolipoate. This implies that dihydrolipoate (DHLA) can indirectly protect against lipid peroxidation via the free radical reductase by the reduction of GSSG to GSH. Lipoate (LA) formed in this reaction can be reduced to dihydrolipoate by lipoamide dehydrogenase using NADH as cofactor. In this enzymatic reaction, only the R isomer of lipoate is reduced. L-NAC potentiates the GSH-dependent protection by acting as a GSH-precursor.

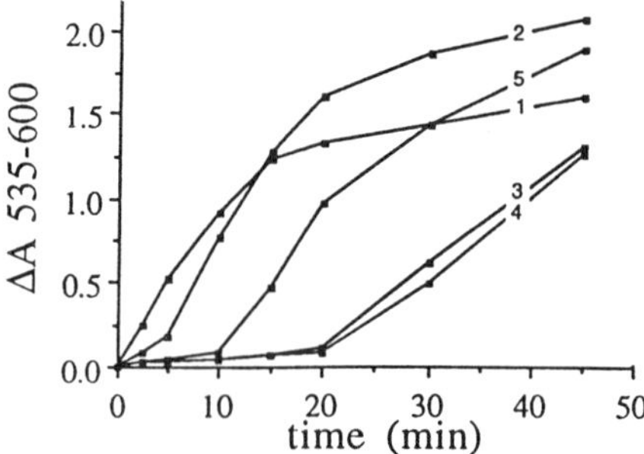

Figure 4 Time course of lipid peroxidation in control rat liver microsomes. Hepatic microsomes were incubated with 0.2 mM ascorbate and (1) no addition, (2) 0.5 mM dihydrolipoate, (3) 1 mM GSH, (4) 0.5 mM dihydrolipoate and 1 mM GSH, (5) 0.5 mM dihydrolipoate and 0.5 mM GSSG. Addition of 0.5 mM GSSG alone or 0.5 mM lipoate alone had no effect, and the result was identical to time course (1). Addition of 0.5 mM lipoate and 1 mM GSH did not differ from the effect obtained with 1 mM GSH alone (3). The latter time courses are not indicated for the sake of clarity. All reactions were started by addition of 10 μM Fe^{2+}.

Table 1 Reduction of GSSG to GSH by Dihydrolipoate

Condition	GSH (mM)	GSSG (mM)
0.5 mM GSSG	0.00 ± 0.01	0.51 ± 0.02
0.5 mM dihydrolipoate + 0.5 mM GSSG	0.42 ± 0.02	0.27 ± 0.02

Incubations were performed for 45 min at 37°C in Tris-HCl/KCl buffer. Data represent the mean ± SEM of three experiments.

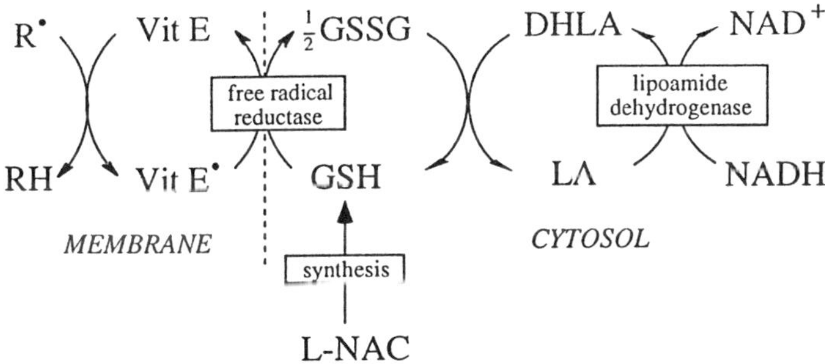

IV. REACTIVE OXYGEN SPECIES AND LUNG INJURY

Histological examination of asthmatic lungs shows shedding and loss of the epithelial layer (16). Clinically, asthma is characterized by pulmonary inflammation, reversible airway obstruction, and a hyperreactivity of the lungs to several stimuli such as histamine or cold air. During the inflammation process, macrophages and mast cells become activated and release chemotactic products. Neutrophils and eosinophils are then attracted to the inflamed pulmonary tissue (17). Activated macrophages, neutrophils, and eosinophils produce $O_2^{-\bullet}$ by the membrane-bound enzyme NADPH oxidase. The $O_2^{-\bullet}$ are further converted to other reactive oxygen species (ROS), like H_2O_2, •OH and HOCl, and may damage proteins and polyunsaturated fatty acids.

Interestingly, incubation of lung tissue with ROS (18) will also lead to a destruction of the epithelial layer (Fig. 5) comparable to what is found in asthmatic lung tissue.

Epithelial denudation will lead to a loss of barrier function and will result in a greater penetration of allergens and exposure of underlying sensory nerve endings. The long-term effect of epithelial injury may include a proliferative

(a)

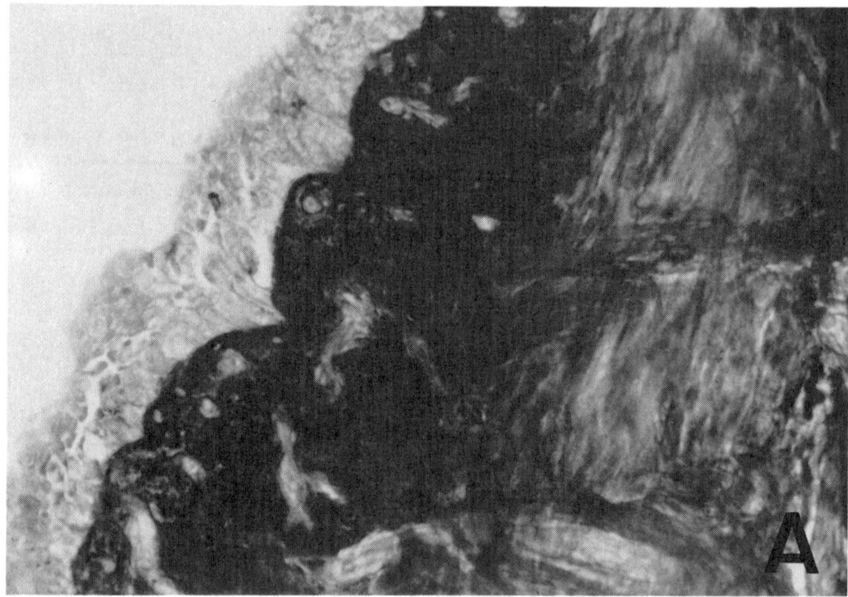

Figure 5 Trachea preparations were obtained from male Dunkin-Hartley guinea pigs (350–450 g). The trachea was dissected free from surrounding tissue and cut in longitudinal direction opposite the smooth muscle. After a 30-min incubation period with various concentrations of HOCl (a: control; b: 10^{-4} M HOCl; c: 10^{-3} M HOCl) and a 30-min washing procedure with six intermediate changes of buffer solution, the tracheal strips were fixed in a 25% formalin, 5% concentrated acetic acid in saturated picric acid solution. The tissue was subsequently embedded in paraffin, sectioned at 7 μm, and stained with cezocarmine G, aniline blue, and orange G, as described (19). Histological stainings are presented with an original magnification of 200×. Epithelial destruction is observed after incubation with HOCl.

response of columnar epithelial cells in the mucosa which may result from release of autocrine growth factors, e.g., epidermal growth factor receptor stimulants, transforming growth factor α (TGFα), epidermal growth factor itself, or the peptide endothelin-1 (ET_1). The mitogens will give proliferation of the underlying fibroblasts and airway smooth muscle cells. The radius of the resulting thickened airway will reduce to a greater extent as response to a certain degree of smooth muscle shortening than the normal airways (20).

Besides inducing epithelial destruction, ROS have a variety of other actions in lung tissue. ROS increase mucus secretion (21), stimulate the synthesis of platelet activating factor (22), and lead to degranulation of mast cells to release the bronchoconstricting agents histamine or serotonin (23,24).

(b)

(c)

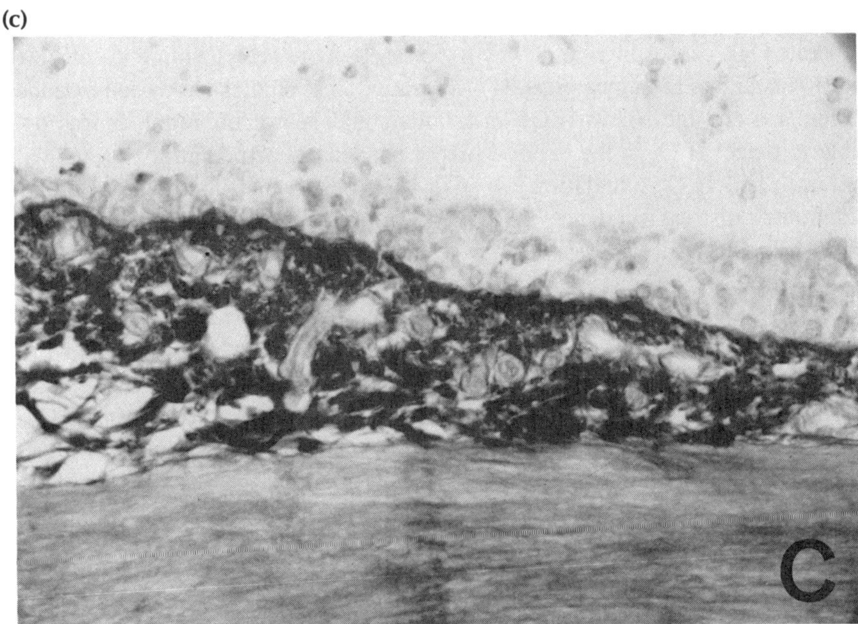

Airway hyperreactivity in patients with pulmonary diseases may be the result of an increased responsiveness to contractile stimuli like muscarinic agonists and/or a decreased potency of stimuli to relax smooth muscle. Comparison of the effects of HOCl on muscarinic and β-adrenoceptor responses (Fig. 6) indicates that the β-adrenoceptor response is more susceptible to HOCl treatment (total destruction at 10^{-2} M HOCl) than the muscarinic receptor response (still 20% contraction at 10^{-2} M HOCl). Apparently, HOCl changes the balance (functional antagonism) between muscarinic and β-adrenergic receptor response of guinea pig tracheal strips in favor of the muscarinic response. Similar data have been obtained for other ROS (25,26).

A hypofunction of the β-adrenergic receptor response has been observed in asthmatic lungs and parallels the increase in airway hyperreactivity (27). β-Adrenergic receptor responses not only relax the airway smooth muscle, but also attenuate the release of chemical spasmogens from mast cells, improve mucociliary transport, inhibit release of acetylcholine from cholinergic nerves, and reduce airway edema (27). The β-adrenergic receptor response of guinea pig airways is also impaired after antigen challenge (28,29), which is a widely used model to investigate allergic asthma. An involvement of ROS was suggested (29) because antioxidants could prevent the ovalbumine-induced airway hyperresponsiveness and the concomitantly observed impairment of β-adrenoceptor function. The data further implied that xanthine oxidase could be regarded as a major source for ROS in this model in guinea pigs (29).

It has been found that inhalation of xanthine and xanthine oxidase (which will generate $O_2^{-\bullet}$) causes in vivo airway hyperreactivity to acetylcholine in cats (30), and to histamine in guinea pigs (31). Recently, these studies have been extended to humans and the role of ROS in asthma could be corroborated. It was possible to detect H_2O_2 in the expired breath condensate of pediatric patients with asthma (32). H_2O_2 exhalation has also been observed in patients with adult respiratory distress syndrome (ARDS) (33).

V. THIOLS AND LUNG FUNCTION

The lung protects itself against ROS through an array of antioxidants in the epithelial lining fluid (ELF). Among the ELF antioxidants are catalase, vitamin E, vitamin C, ceruloplasmin, transferrin, lactoferrin, and GSH (34). The GSH levels in ELF are at least 100 times higher than normal plasma GSH levels. Various lung pathologies are accompanied by a decreased ELF GSH concentration. Patients with idiopathic pulmonary fibrosis (IPF), a chronic inflammatory lung disease, show a marked ELF GSH deficiency (35). Also in patients suffering from cystic fibrosis (36) or ARDS (37), ELF GSH levels are lower than normal.

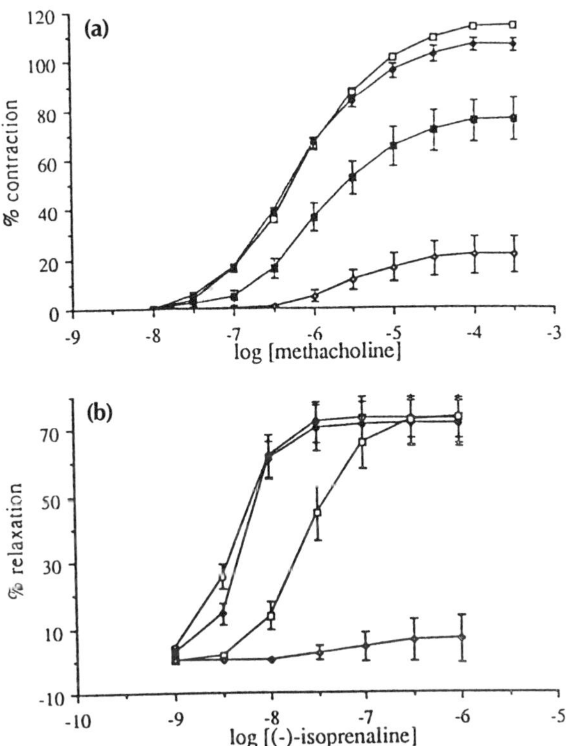

Figure 6 The effect of different concentrations of HOCl on the (a) metacholine (contraction) and (b) (–)isoprenaline (relaxation) concentration response curve (CRC) in guinea pig tracheal strips. Changes in length of the smooth muscle were isotonically recorded with a contraweight of 0.5 g. After incubation with HOCl, equipotent metacholine concentrations were used for the precontraction that preceded the (–)isoprenaline CRCs. After control (□), HOCl 10^{-4} M (◆), 10^{-3} (■), and 10^{-2} (◇), the strips in the case of (–)isoprenaline CRCs were precontracted with 3×10^{-7} M, 3×10^{-7} M, 10^{-6} M, and 10^{-4} M metacholine, respectively. The maximal effect of (–)isoprenaline (78 ± 8%) was expressed as a percentage of the precontraction. The maximal effect of metacholine was related to the curve obtained before incubation with HOCl (or control).

Recently, it was reported that also symptom-free human immunodeficiency virus (HIV)-seropositive persons experience low ELF GSH levels (38). A low GSH concentration in the ELF may result in an imbalanced oxidant/antioxidant ratio and may lead to further lung damage. Moreover, decreased GSH-levels may also have direct functional consequences. In vitro studies showed that GSH (in the concentration range normally found in ELF) suppressed fibroblast proliferation (39). In a group of smokers with chronic bronchitis or chronic obstructive pulmonary disease a correlation between reduced 1-s forced expiratory volume (FEV_1) and glutathione levels in the bronchoalveolar lavage was found (40). Lung GSH depletion in mice gave type II cell lamellar body swelling and disintegration which is associated with abnormalities in the lung surfactant system (41). Because the presence of inflammatory cells on the epithelial surface might play a vital role in host defense (e.g., in cystic fibrosis where high numbers of neutrophils on the airway epithelial surface are found), it is not feasible to decrease the number of inflammatory cells. A better therapeutic strategy seems to be protection of the epithelial layer by augmenting the antioxidant levels, e.g., GSH (42).

L-NAC is an old drug that has long been marketed as a mucolytic agent. The mucolytic action of L-NAC is by direct cleavage of disulfide bridges between glycopeptides with a resulting decrease in the viscosity. Currently, its therapeutic effect is primarily ascribed to its ability to act as a GSH-precursor. Indeed, L-NAC might act by raising intracellular concentrations of L-cysteine and hence of GSH. ELF GSH-levels increased in patients with IPF when they were treated with *N*-acetyl-L-cysteine (3 dd 600 mg p.o. for 5 days) (43). Intravenous L-NAC treatment during 72 h improved systemic oxygenation and reduced the need for ventilatory support in patients with mild to moderate acute lung injury (44). Patients with ARDS undergo severe oxidative stress indicated by a decrease in protein thiols (45). Lung lavages of patients with ARDS are characterized by increased numbers of neutrophils and interleukin-1α (IL-1). Intratracheal administration of IL-1 in rats gave severe lung damage that could effectively be treated by post-insult administration of NAC (46). Long-term treatment with L-NAC decreases the number of exacerbations in COPD patients (47). This therapeutic effect of L-NAC has been associated with decreased intrabronchial bacterial numbers observed during L-NAC treatment (48).

VI. *N*-ACETYLCYSTEINE AND α-LIPOIC ACID: DIRECT OR INDIRECT ACTING ANTIOXIDANTS?

Many data suggest the replenishment of cellular glutathione as the principal mechanism of action of L-NAC (vide supra). In a recent study the effects of L-NAC and *N*-acetyl-D-cysteine (D-NAC) were compared for their protective effect in a hyperoxic-induced lung damage model in rats (49). D-NAC will not act as GSH-precursor. Nevertheless, both NAC isomers, given intravenously, prevented the increase of lung edema to the same extent. Intragastric administration even showed that only the D-form of NAC was active. The latter was explained by a lack of deacetylation of D-NAC, enabling the compound to pass intact through the intestinal mucosa. The study seems to indicate that in this model, NAC has a direct protective action and does not act via glutathione (49). Indeed, NAC has some direct scavenging activity since it reacts with HOCl (50,51), with •OH and slowly with H_2O_2 (50). In humans a direct action of NAC on lung tissue is less likely since orally administered L-NAC does not penetrate into bronchoalveolar lavage fluid (52).

The direct scavenging effect of both NAC and α-lipoic acid deserves some further attention. Neutrophils contain the enzyme myeloperoxidase, that catalyses the reaction between H_2O_2 and Cl^-, in which the powerful oxidant HOCl is formed. Neutrophils derive their bactericidal action among other things from HOCl. Surrounding macromolecules are, however, also vulnerable. A target for HOCl is α-$_1$-antiprotease, an important inhibitor of serine proteases such as elastase. HOCl oxidizes a critical methionine residue of α_1-antiprotease to a sulfoxide with consecutive loss of activity of the elastase-inhibitory capacity. Moreover, activated neutrophils excrete elastase. An imbalance between elastase and anti-elastase may occur during oxidative stress and may lead to elastase-dependent hydrolysis of elastin and eventually to emphysema. NAC, dihydrolipoic acid and GSH are equally active as HOCl scavengers and equipotent in protecting α_1-antiprotease against activation (51). Interestingly, not only dihydrolipoic acid, but also the oxidized form lipoic acid are good HOCl scavengers which contrasts the findings with GSH and GSSG, where only the reduced form (GSH) can act as HOCl-scavenger. In analogy with GSH, the action of the dithiol dihydrolipoic acid can of course be expected. The HOCl scavenging activity of the oxidized form (lipoic acid) has been explained by the existence of ring strain in the five-membered ring, which contains an intramolecular S-S bridge (51,53).

Unbalanced thiol homeostasis in the cell may further unsettle the delicate intracellular interplay between thiols, ATP, and Ca^{2+} (54). A sustained increase in intracellular Ca^{2+} may set off a series of cytotoxic events. These include activation of Ca^{2+}-phospholipases. This may lead to release of arachidonic acid and subsequent formation of vaso- and bronchoconstricting agents like thromboxane A_2. This process seems to be important especially in ROS-induced lung injury (55).

A promising new field of research is the redox control of the transcription factor activating protein-1 (AP-1), which is a heterodimer of Fos and Jun proteins, the products of the protooncogens of c-fos and c-jun (56). Induction of AP-1 activity occurs at low GSH-levels (56). L-NAC inhibits the induction of AP-1 binding (56). Interestingly, steroids (widely used drugs in lung diseases) counteract the chronic inflammatory effects of cytokines in a similar way (i.e., a direct interaction between the transcription factor AP-1 and the glucocorticoid receptor or a counteracting effect of the glucocorticoid receptor on the AP-1 induced gene transcription at the level of DNA binding may occur) (57).

REFERENCES

1. Crompton, M., Costi, A., and Hayat, L. (1987) Evidence for the presence of a reversible, Ca^{2+}-dependent pore activated by oxidative stress in heart mitochondria. Biochem. J. 245:915–918.
2. Haenen, G. R. M. M., and Bast, A. (1983) Protection against lipid peroxidation by a microsomal glutathione-dependent heat labile factor. FEBS Lett. 159:24–28.
3. Haenen, G. R. M. M., Tai Tin Tsoi, J. N. L., Vermeulen, N.P. E., Timmerman, H., and Bast, A. (1987) 4-Hydroxy-2,3-*trans*-nonenal stimulates microsomal lipid peroxidation by reducing the glutathione dependent protection. Arch. Biochem. Biophys. 259:449–456.
4. Haenen, G. R. M. M., Van Dansik, P., Vermeulen, N. P. E., Timmerman, H., and Bast, A. (1988) The effect of hydrogen peroxide on β-adrenoceptor function in the heart. Free Rad. Res. Comms. 4:243–249.
5. Barclay, L. R. C. (1988) The cooperative antioxidant role of glutathione with a lipid-soluble and water soluble antioxidant during peroxidation of liposomes initiated in the aqueous phase and in the lipid phase. J. Biol. Chem. 263:16138–16142.
6. Ursini, F., Maiorino, M., Valente, M., Ferri, L., and Gregolin, C. (1982) Purification from pig liver of a protein which protects liposomes and biomembranes from peroxidative degradation and exhibits glutathione peroxidase activity on phosphatidylcholine hydroperoxides. Biochim. Biophys. Acta 710:197–211.
7. Tan, K. H., Meyer, D. J., Belin, J., and Ketterer, B. (1984) Inhibition of microsomal lipid peroxidation by glutathione and glutathione transferases B and AA. Biochem. J. 220:243–252.
8. McCay, P. B. (1985) Vitamin E, interactions with free radicals and ascorbate. Ann. Rev. Nutr. 5:323–340.
9. Bastiaans, H. M. M., Haenen, G. R. M. M., and A. Bast (1990) Interplay between ascorbic acid, α-tocopherol, glutathione and lipoic acid in the protection against microsomal lipid peroxidation. Eur. J. Pharmacol. 183:2436–2437.
10. Murphy, M. E., and Kehrer, J. P. (1988) Lipid peroxidation factors in liver and muscle of rat, mouse, and chicken. Arch. Biochem. Biophys. 268:585–593.
11. Bast, A., Haenen, G. R. M. M., and Savenije-Chapel, E. M. (1987) Inhibition of rat hepatic microsomal lipid peroxidation by mesna via glutathione. Arzneim. Forsch. 37:1043–1045.
12. Bast, A, Haenen, G. R. M. M., and Doelman, C. J. A. (1990) Oxidants and antioxidants: State of the art. Am. J. Med. 91:3C-S3–3C-13S.
13. Braugler, J. M., Duncan, L. A., and Chase, R. L. (1986) The involvement of iron in lipid peroxidation. J. Biol. Chem. 261:10282–10289.
14. Minotti, G., and Aust, S. D. (1987) The requirement for iron (III) in the initiation of lipid peroxidation by iron (II) and hydrogen peroxide. J. Biol. Chem. 262:1098–1104.
15. Haenen, G. R. M. M., Vermeulen, N. P. E., Timmerman, H., and Bast, A. (1989) Effect of several thiols on lipid peroxidation in rat liver microsomes. Chem.-Biol. Interact. 71:202–212.
16. Houston, J. C., DeNavasquez, S., and Trounce, J. R. (1953) A clinical and pathological study of fatal cases of status asthmaticus. Thorax 8:207–213.

17. Doelman, C. J. A., and Bast, A. (1990) Oxygen radicals in lung pathology. Free Rad. Biol. Med. 9:381–400.
18. Doelman, C. J. A., Oosterom, W. C., and Bast, A. (1990) Regulation of sympathetic and parasympathetic receptor responses in the rat trachea by epithelium: Influence of mechanical and chemical removal of epithelium. J. Pharm. Pharmacol. 42:831–836.
19. Gurr, E. (1962) Staining Animal Tissue. London Hill Ltd., London.
20. Stewart, A. G., Tomlinson, P. R., and Wilson, J. (1993) Airway wall remodelling in asthma: A novel target for the development of anti-asthma drugs. Trends Pharmacol. Sci. 14:275–279.
21. Adler, K. B., Holden-Stauffer, W. J., and Repine, J. E. (1990) Oxygen metabolites stimulate release of high-molecular weight glycoconjugates by cell and organ cultures of rodent respiratory epithelium via an arachidonic acid-dependent mechanism. J. Clin. Invest. 85:75–85.
22. Lewis, M. S., Whatley, R. E., Cain, P., McIntyre, T. M., Prescott, S. M., and Zimmerman, G. A. (1988) Hydrogen peroxide stimulates the synthesis of platelet activating factor by endothelium and induces endothelial cell-dependent neutrophil adhesion. J. Clin. Invest. 82:2045–2055.
23. Mannioni, P. F., Giannella, E., Palmerani, B., Pistelli, A., Gambassi, F., Bain-Sacchi, T., Bianchi, S., and Mansini, E. (1988) Free radicals as endogenous histamine releasers. Agents Actions 23:129–142.
24. Szarek, J. L., and Schmidt, N. L. (1990) Hydrogen peroxide-induced potentiation of contractile responses in isolated rat airways. Am. J. Physiol. 258:L232–L237.
25. Van der Vliet, A., and Bast, A. (1992) Effect of oxidative stress on receptors and signal transmission. Chem.-Biol. Interact. 85:95–116.
26. Kramer, K., Doelman, C. J. A., Timmerman, H., and Bast, A. (1987) A disbalance between β-adrenergic and muscarinic reponses caused by hydrogen peroxide in rat airway in vitro. Biochem. Biophys. Res. Commun. 145:357–362.
27. Lulich, K. M., Goldie, R. G., and Paterson, J. W. (1988) Beta-adrenoceptor function in asthmatic bronchial smooth muscle. Gen. Pharmacol. 19:307–311.
28. Daffonchio, L., Abbracchio, M. P., Di Luca, M., Hernandez, A., Amadeo, L., Cattabeni, F., and Omini, C. (1990) β-Adrenoceptor desensitization induced by antigen challenge in guinea-pig trachea. Eur. J. Pharmacol. 178:21–27.
29. Ikuta, N., Sugiyama, S., Takagi, K., Satake, T., and Ozawa, T. (1992) Implications of oxygen radicals on airway hyperresponsiveness after ovalbumin challenge. Am. Rev. Respir. Dis. 145:561–565.
30. Katsumata, D., Ichinose, M., Miura, M., Kimura, K., Inoue, H., and Takishima, T. (1988) Reactive oxygen exposure produces airway hyperresponsiveness. Am. Rev. Respir. Dis. 137:A285.
31. Nakano, E., and Misawa, M. (1990) In vivo effect of superoxide anion on guinea-pig airways. Eur. J. Pharmacol. 183:1178.
32. Dohlman, A. W., Black, H. R., and Royall, J. A. (1993) Expired breath hydrogen peroxide is a marker of acute airway inflammation in pediatric patients with asthma. Am. Rev. Respir. Dis. 148:955–960.
33. Kietzmann, D., Kahl, R., Müller, M., Burchardi, H., and Kettler, D. (1993)

Hydrogen peroxide in expired breath condensate of patients with acute respiratory failure and with ARDS. Intensive Care Med. 19:78–81.

34. Cantin, A. M., and Bégin, R. (1991) Glutathione and inflammatory disorders of the lung. Lung 169:123–138.
35. Cantin, A. M., Hubbard, R. C., and Crystal, R. G. (1989) Glutathione deficiency in the epithelial lining fluid of the lower respiratory tract in idiopathic pulmonary fibrosis. Am. Rev. Respir. Dis. 139:370–372.
36. Roum, J. H., Buhl, R., McElvaney, N. G., Borok, Z., Hubbard, R. C., Chernick, M., Cantin, A. M., and Crystal, R. G. (1990) Cystic fibrosis is characterized by a marked reduction in glutathione levels in pulmonary epithelial lining fluid. Am. Rev. Respir. Dis. 141: A87.
37. Pacht, E. H., Timerman, A. F., Lykens, M. G., and Merole, A. J. (1991) Deficiency of alveolar fluid glutathione in patients with sepsis and the adult respiratory distress syndrome. Chest 100:1397–1403.
38. Buhl, R., Jaffe, H. A., Holroyd, K. J., Wells, F. B., Mastrangeli, A., Saltini, C., Cantin, A. M., and Crystal, R. G. (1989) Systemic glutathione deficiency in symptom-free HIV-seropositive individuals. Lancet 2:1294–1298.
39. Cantin, A. M., Larivée, P., and Bégin, R. (1990) Extracellular glutathione suppresses human lung fibroblast proliferation. Am. J. Respir. Cell. Mol. Biol. 3:79–85.
40. Linden, M., Rasmussen, J. B., Piitulainen, E., Tunek, A., Larson, M., Tegner, H., Venge, P., Laitinen, L. A., and Brattsand, R. (1993) Airway inflammation in smokers with nonobstructive and obstructive chronic bronchitis. Am. Rev. Respir. Dis. 148:1226–1232.
41. Mårtensson, J., Jain, A., Frayer, W., and Meister, A. (1989) Glutathione metabolism in the lung: Inhibition of its synthesis leads to lamellar body and mitochondrial defects. Proc. Natl. Acad. Sci. USA 86:5296–5300.
42. Crystal, R. G. (1991) Oxidants and respiratory tract epithelial injury: Pathogenesis and strategies for therapeutic intervention. Am. J. Med. 91 (Suppl. 3C):39–44.
43. Meyer, A., and Magnussen, H. (1991) Der Effekt von oralem N-Acetylcystein auf die Glutathion-konzentration in der bronchoalveolären Lavage von Patienten mit fibrosierenden Lungerkrankungen. Med. Klin. 86, 279–283.
44. Suter, P. M., Domenighetti, G., Schaller, M.-D., Lavarrière, M.-C., Ritz, R., and Perret, C. (1994) N-Acetylcysteine enhances recovery from acute lung injury in man. A randomized, double-blind, placebo-controlled clinical study. Chest 105:190–194.
45. Quinlan, G. J., Evans, T. W., and Gutteridge, J. M. C. (1994) Oxidative damage to plasma proteins in adult respiratory distress syndrome. Free Rad. Res. 20:289–298.
46. Leff, J. A., Wilke, C. P., Hybertson, B. M., Shanley, P. F., Beechler, C. J., and Repine, J. E. (1993) Post insult treatment with N-acetyl-L-cysteine decreases IL-1-induced neutrophil influx and lung leak in rats. Am. J. Physiol. 265:L501–L506.
47. Boman, G., Bäcker, U., Larsson, S., Melander, B., and Wåhlander, L. (1983) Oral acetylcysteine reduces exacerbation rate on chronic bronchitis: Report of a trial organized by the Swedish Society for Pulmonary Diseases. Eur. J. Respir. Dis. 64:405–415.

48. Riise, G. C., Larsson, S., Larsson, P., Jeansson, S., and Andersson, B. A. (1994) The intrabronchial microbial flora in chronic bronchitis patients: a target for N-acetylcysteine therapy? Eur. Respir. J. 7:94–101.
49. Särnstrand, B. (1992) Is N-acetylcysteine a free radical scavenger in vivo? The effect of N-acetylcysteine in oxygen-induced lung injury. Eur. Respir. Rev. 2:11–15.
50. Auroma, O. I., Halliwell, B., Hoey, B. M., and Butler, J. (1989). The antioxidant action of N-acetylcysteine: its reaction with hydrogen peroxide, hydroxyl radical, superoxide, and hypochlorous acid. Free Rad. Biol. Med. 6:593–597.
51. Haenen, G. R. M. M., and Bast, A. (1991) Scavenging of hypochlorous acid by lipoic acid. Biochem. Pharmacol. 42:2244–2246.
52. Cotgreave, I. A., Eklund, A., Larsson, K., and Moldéus, P. W. (1987) No penetration of orally administered N-acetylcysteine into bronchoalveolar lavage fluid. Eur. J. Respir. Dis. 70:73–77.
53. Biewenga, G. P., De Jong, J., and Bast, A. (1994) Lipoic acid favors thiosulfinate formation after hypochlorous acid scavenging: A study with lipoic acid derivatives. Arch. Biochem. Biophys. 312:114–120.
54. Bast, A. (1993) Oxidative stress and calcium homeostasis. In: DNA and Free Radicals (Halliwell, B., and Aruoma, O. I., eds.), pp. 95–108. Ellis Horwood, New York.
55. Olafsdottir, K., Ryrfeldt, Å., Atzori, L., Berggren, M., and Moldéus, P. W. (1991) Hydroperoxide-induced broncho- and vasoconstriction in the isolated rat lung. Exp. Lung Res. 17:615–627.
56. Bergelson, S., Pinkus, R., and Daniel, V. (1994) Intracellular glutathione levels regulate fos/jun induction and activation of glutathione S-transferase gene expression. Cancer Res. 54:36–40.
57. Barnes, P. J., and Adcock, I. (1993) Anti-inflammatory actions of steroids: Molecular mechanisms. Trends Pharmacol. Sci. 14:436–441.

18

Inhibition of Xanthine Oxidase by Lipoic Acid

Silvia Alvarez and Alberto Boveris
University of Buenos Aires, Buenos Aires, Argentina

I. INTRODUCTION

Lipoic acid (α-lipoic acid; 1,2-dithiolane-3-pentanoic acid) is widely distributed in animals and plants as a cofactor of different multienzyme complexes involved in the oxidative decarboxilation of pyruvic and α-ketoglutaric acids. Addition of exogenous lipoic acid is essential for the growth of certain microorganisms, but so far no effects of deficiency in humans or animals are known (1). Additionally, lipoic acid protects vitamin E reductase activity in microsomes; a combination of lipoic acid and tocopherol results in an enhancement of vitamin E efficacy (2).

Lipoic acid has been proposed for therapeutic use in liver cirrhosis, diabetes mellitus, atherosclerosis, and polyneuritis (3,4). Because the free radical–mediated peroxidation of phospholipids may be a common factor in the development of the above-mentioned pathologies, studies of the antioxidant effect of lipoic acid have been carried out in various systems (5–8). Moreover, the reduced form, α-dihydrolipoic acid, can act as a potent sulfhydryl reductant and metal chelator (8).

McCord et al. have described how xanthine dehydrogenase rapidly converts into xanthine oxidase in nonperfused ileum (9,10). Moreover, it has been postulated that xanthine may accumulate during ischemia as a result of ATP degradation and that upon reoxygenation of the tissue, xanthine oxidase will markedly increase superoxide anion and hydrogen peroxide production. This process in turn will initiate a chain of free radical reactions leading to cell damage (9,10). The xanthine oxidase inhibitor allopurinol prevents damage during ischemia reperfusion stress, as should be expected if xanthine oxidase were involved in this process (10,11).

Results from this laboratory on the inhibition of intestine chemiluminescence overshoot upon reperfusion suggested an inhibitory effect of lipoic acid on the conversion of xanthine dehydrogenase to oxidase (12,13). The present work focuses on the in vivo and in vitro inhibition of lipoic acid on xanthine oxidase activity by studying the oxidation of xanthine to uric acid and describes lipoic acid as a competitive inhibitor with respect to xanthine of the enzyme.

II. IN VIVO INHIBITORY EFFECT

Lipoic acid (50, 100, 150 mg/kg body weight) was administered to Wistar rats, 180–200 g, and fed ad libitum 1 h before sacrifice. Xanthine oxidase activity from intestine homogenate was determined as a function of time. The doses injected prevented by 40–52% ($p < 0.01$) the xanthine dehydrogenase-to-xanthine oxidase conversion after 20 min of incubation at 10°C of the intestine homogenate (Fig. 1). Lipoic acid was also shown to be effective in inhibiting reperfusion chemiluminescence overshoot in rat intestine when administered in a dose of 100 mg/kg 1 h before the experiment. The time course of reperfusion chemiluminescence as well as the effect of lipoic acid are shown in Figure 2.

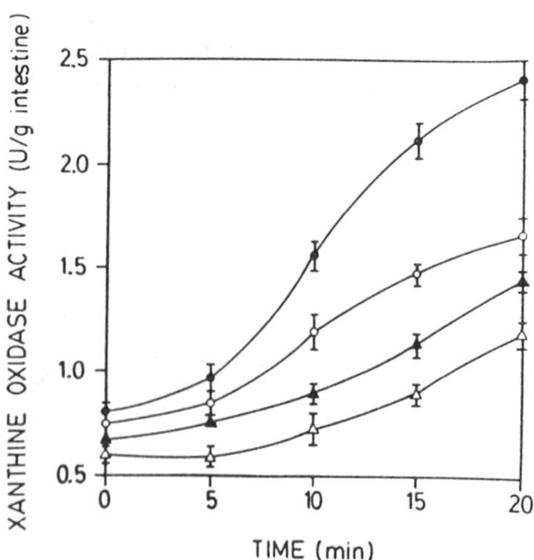

Figure 1 In vivo effect of lipoic acid on the xanthine oxidase activity of rat intestine. (●), No additive. Lipoic acid doses: (○), 50 mg/kg weight; (▲), 100 mg/kg weight; (Δ), 150 mg/kg weight.

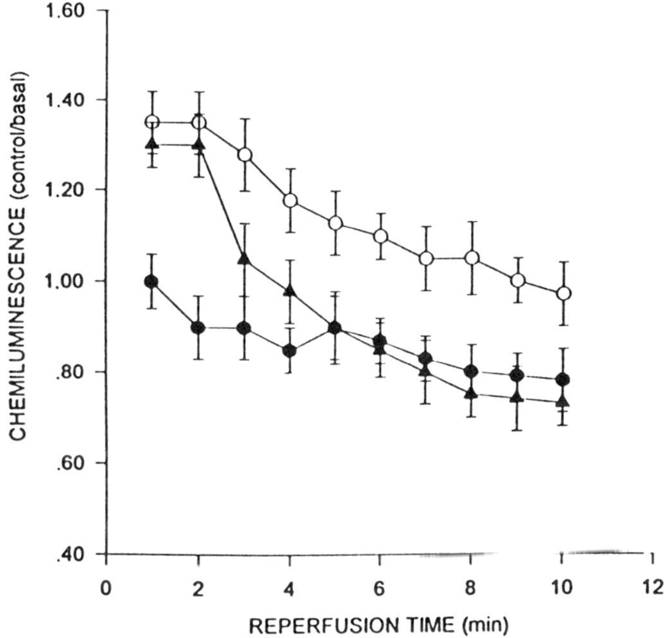

Figure 2 In vivo chemiluminescence of rat intestine: (●) spontaneous emission, (○), after reperfusion following 3-min ischemia, and (Δ) after reperfusion following 5-min ischemia pretreated with lipoic acid. (From Ref. 12.)

III. IN VITRO INHIBITORY EFFECT

Buttermilk xanthine oxidase activity was inhibited by lipoic acid and by its reduced form, dihydrolipoic acid. A dose-dependent inhibition of activity was observed; the reduced metabolite was slightly more potent than the oxidized form. Dihydrolipoic and lipoic acid, both at 3 mM, produced inhibitions of xanthine oxidase activity of 50% and 42%, respectively ($p < 0.01$) (Fig. 3).

The conversion of xanthine dehydrogenase to oxidase and the xanthine oxidase activity were inhibited when 1.5 and 3.0 mM were added to the supernatant of rat liver homogenate (Fig. 4). Xanthine dehydrogenase-to-oxidase conversion was completely inhibited. Moreover, after 20 min of incubation at 10°C, decreases in xanthine oxidase activity of 8% ($p < 0.02$) and 21% ($p < 0.01$) were obtained with 1.5 and 3.0 mM lipoic acid. Dihydrolipoic acid at 1.5 and 3.0 mM in the same experimental conditions inhibited xanthine oxidase activity by 17 and 29% ($p < 0.01$), respectively (Fig. 4). Partial inhibition of xanthine oxidase activity was observed when 3 mM dihydrolipoic acid was added to intestine homogenate supernatant (Fig. 5).

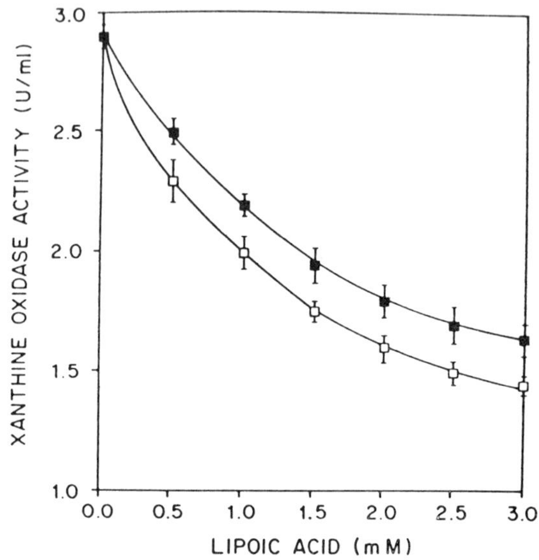

Figure 3 In vitro effect of (■) lipoic acid and (□) dihydrolipoic acid on buttermilk xanthine oxidase activity.

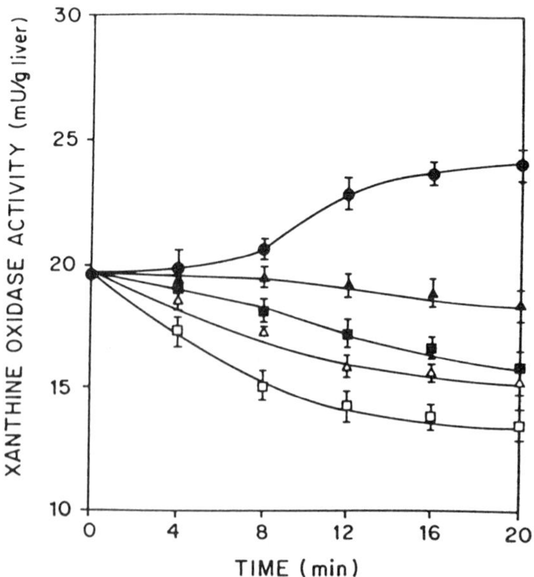

Figure 4 In vitro effect of (▲) 1 mM and (Δ) 3 mM lipoic acid and (■) 1 mM and (□) 3 mM dihydrolipoic acid on the xanthine oxidase activity of rat liver homogenate. (●), No additive.

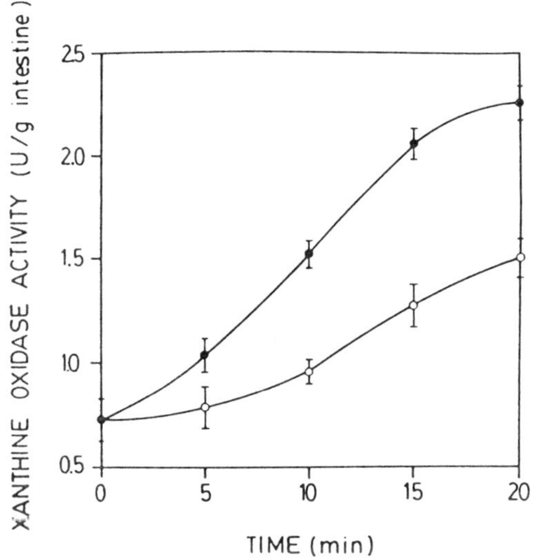

Figure 5 In vitro effect of (○) 3 mM dihydrolipoic acid on the xanthine oxidase activity of rat intestine homogenate. (●), No additive.

IV. LIPOIC ACID AS XANTHINE OXIDASE COMPETITIVE INHIBITOR

Lineweaver-Burk and Dixon plots of xanthine oxidase activity with variable amounts of xanthine and lipoic acid indicated that lipoic acid inhibited xanthine oxidase activity in competition with its substrate (Fig. 6). The K_m value calculated from Figure 5 is 6.45 ± 0.70 μM, and the K_i value of lipoic acid calculated from Fig. 6 is 1.66 ± 0.20 mM.

V. DISCUSSION

Lipoic acid may be potentially useful in the treatment of pathophysiological conditions, such as a variety of liver and intestine diseases or dysfunctions in which lipid peroxidation appears to be involved (14). However, the mechanisms of this protection are not well understood. Experimental evidence suggests that the observed protection may be potentially due to the antioxidant action of lipoic acid as inhibitor of xanthine oxidase (15). The observed in vivo effect of lipoic acid in inhibiting xanthine oxidase in intestine homogenates requires a lipoic acid concentration in the 1–3 mM range (Figs. 1 and 4), which seems consistent with the injected doses that should yield a whole body concentration

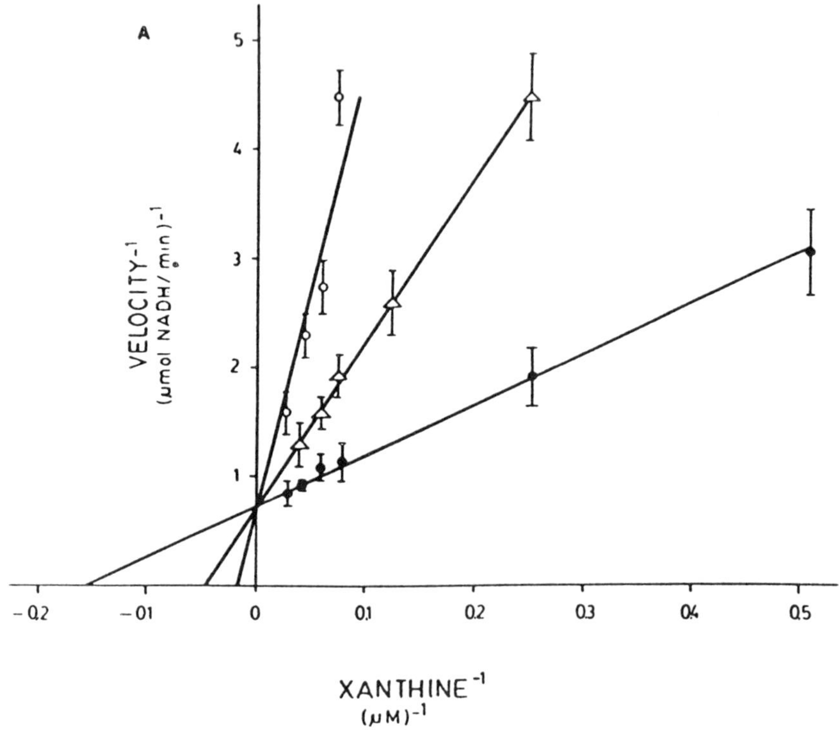

Figure 6 (A) Double reciprocal plot (Lineweaver-Burk) using different concentrations of xanthine in the presence of (Δ) 1 mM and (○) 3 mM lipoic acid. (●), No additive. (B) Simple reciprocal plot (Dixon) using different concentrations of lipoic acid and (○) 17 μM, (●) 21 μM, and (Δ) 25 μM xanthine.

in the 0.5–1.0 mM range, slightly increased in the intestine due to the intraperitoneal route of administration. In order to determine the type of inhibition produced by lipoic acid on xanthine oxidase activity, variable amounts of lipoic acid and xanthine were utilized to determine xanthine oxidase reaction rates. Lineweaver-Burk and Dixon plots gave evidence of competitive inhibition. Lipoic acid (K_i = 1.7 mM) can be understood as a xanthine oxidase inhibitor of low specificity, as is guanidinium (K_i = 8 mM) (15), in contrast to the highly specific allopurinol. This indicates that the active center has a relatively wide crevice that accommodates unrelated molecules (xanthine, allopurinol, lipoic acid, guanidinium).

Recent evidence seems to indicate that the xanthine oxidase catalytic center has negatively and positively charged sites that bind guanidinium cations and

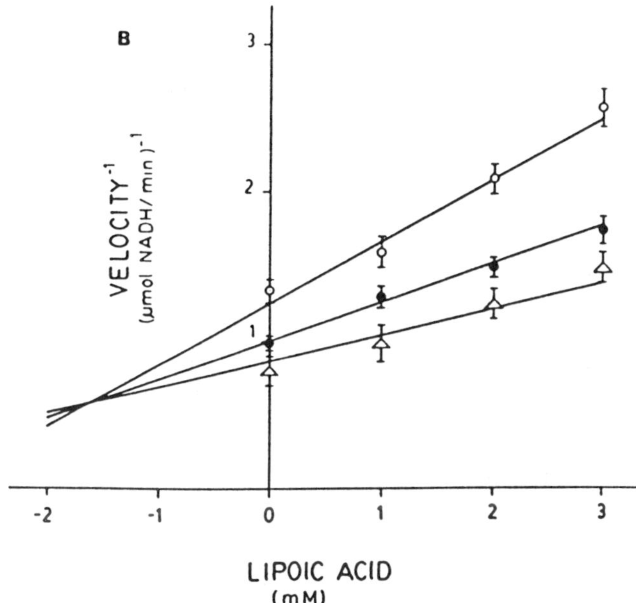

chloride or bromide anions (15). The positively charged site may constitute the binding site for the carboxylic group of lipoic acid.

Hausladen and Fridovich have aroused interest in xanthine oxidase inhibitors (competitive with respect to xanthine) with the aim of improving the superoxide dismutase assay; competitive inhibitors at a fixed substrate concentration should increase the univalent oxygen reduction by xanthine oxidase (15).

VI. CONCLUDING REMARKS

The basic information provided here is important for understanding the pharmacokinetics and in vivo antioxidant action of lipoic acid. Studies of the reduction of molecular oxygen by xanthine oxidase to superoxide anion (univalent pathway) or to hydrogen peroxide (divalent pathway) remain to be done, with the objective of fully characterizing the inhibitory mechanism of lipoic acid. The possibility exists that lipoic acid may be reduced at the active center to dihydrolipoic acid, which may act in a low turnover cycle involving a one-electron reduction step to thiyl radical and an oxygen-reoxidation step yielding superoxide anion.

ACKNOWLEDGMENTS

The authors wish to thank Dr. Enrique Cadenas and Dr. Rafael Radi for helpful discussion. This work was supported by grants from the University of Buenos Aires and CONICET.

REFERENCES

1. Mattulat, A. (1992) Determination of the lipoic acid content of animal tissue. In: International Thioctic Acid Workshop (Schmidt, K., and Ulrich, H., eds.), pp. 69–73. Universimed Verlag GmbH, Frankfurt.
2. Bast, A., and Haenen, G. R. (1990) Regulation of lipid peroxidation by glutathione and lipoic acid: involvement of liver microsomal vitamin E free radical reductase. In: Antioxidants in Therapy and Preventive Medicine (Emerit, I., Packer, L., and Auclair, C., eds.), pp. 111–116. Plenum Press, New York.
3. Altenkirch, H., Stoltenburg, Didinger, G., Wagner, H. M., Herrman, J., and Walter, G. (1990) Effect of lipoic acid in hexacarbon-induced neuropathy. Neurotoxicol. Teratol. 12:619–622.
4. Wagh, S. S., Natraj, C. V., and Menon, K. K. (1987) Mode of action of lipoic acid in diabetes. J. Biosci. 11:59–74.
5. Servinova, E., Reznick, S. K., and Packer, L. (1992) Thioctic acid against ischemia-reperfusion injury in the isolated perfused Lagendorff heart. Free Rad. Res. Commun. 17:49–58.
6. Kieglstein, J., Peruche, B., Prehn, J, Nuglisch, J., and Karkoutly, C. (1992) In: International Thioctic Acid Workshop (Schmidt, K. and Ulrich, H., eds), pp. 89–102. Universimed Verlag GmbH, Frankfurt.
7. Bast, A., and Haenen, G. R. (1988) Interplay between lipoic acid and glutathione in the protection against microsomal lipid peroxidation. Biochim. Biophys. Acta 963:558–561.
8. Scheer, B., and Zimmer, G. (1993) Dihyrolipoic acid prevents hipoxic/reoxygenation and peroxidative damage in rat heart mitochondria. Arch. Biochem. Biophys. 302:385–390.
9. McCord, J. M. (1985) Oxygen derived free radicals in postischemic tissue injury. N. Engl. J. Med. 312:159–163.
10. Roy, R. S., and McCord, J. M. (1983) Superoxide and ischemia: conversion of xanthine dehydrogenase to xanthine oxidase. In: Oxyradicals and Their Scavenger Systems (Greenwald, R., and Cohen, G., eds.), pp. 143–145. Elsevier Science, New York.
11. Gonzalez Flecha, B., Reides, C., Cutrin, J. C., Llesuy, S., and Boveris, A. (1993) Oxidative stress produced by suprahepatic occlusion and reperfusion. Hepatology 18:881–889.
12. Roldan, E. J. A., Pinus, C. R., Turrens, J. F., and Boveris, A. (1989) Chemiluminescence of ischaemic and reperfused intestine in vivo. Gut 30:184–187.
13. Roldan, E. J. A., and Boveris, A. (1993) Short-term ischemic intestine reperfusion oxidative-stress. Antioxidative effect of drugs. In: Pathophysiology of Reperfusion Injury (Dipak, K. D., ed.), pp. 491–508. CRC Press, Boca Raton, FL.

14. Boveris, A., Roldan, E. J. A., and Alvarez, S. (1994) Effect of lipoic acid on short-term ischemia-reperfusion in rat intestine. In: Biological Oxidants and Antioxidants: New Developments in Research and Health Effects (Packer, L., and Cadenas, E., eds.), pp. 81–89. Hippokrates Verlag, Stuttgart.
15. Hausladen, A., and Fridovich, I. (1993) Competitive inhibition of xanthine oxidase by guanidinium: dependence upon monovalent anions and effects on production of superoxide. Arch. Biochem. Biophys. 304:479–482.

19

Effects of Thiols on Cigarette Smoke–Induced Biomolecular Damage

Jason P. Eiserich, Charles A. O'Neill, Carroll E. Cross, and Albert van der Vliet
University of California, Davis, California

Barry Halliwell
University of London, King's College, London, England

Abraham Z. Reznick
The Bruce Rappaport Faculty of Medicine, Technion, Haifa, Israel

I. INTRODUCTION

Cigarette smoking is associated with a variety of pulmonary and cardiovascular disorders including emphysema, atherosclerosis, and cancer (1–3). However, the exact chemical/biochemical mechanisms underlying cigarette smoke (CS)–induced effects on biological systems are incompletely understood. CS can be simplistically divided into two phases: a gas phase and a tar phase. Important chemical constituents of the gas phase include highly noxious aldehydes including acetaldehyde, formaldehyde, acrolein, and crotonaldehyde, as well as methanol, isoprene, and nitric oxide (4). It is these and other CS constituents inhaled into the lung, including abundant alkyl, alkoxyl, and peroxyl radical species (5), that can be expected to directly react with constituents of the respiratory tract lining fluids (RTLFs) that cover the respiratory tract epithelial cells (RTECs) (6–8). It has been proposed that free radicals in CS, or generated by exposure to it (e.g., by CS-dependent activation of phagocytes), may be major contributory factors to CS-related diseases (9,10).

Since RTLFs are not easily obtainable in an undiluted and easily characterized form, we have used human plasma as a surrogate model to study further the interactions between CS and complex biological systems, which include antioxidants, proteins, and lipids (10–12). Using this model, we have examined the effect of gas-phase CS on plasma antioxidants, focusing on the relationships of antioxidant depletion with CS-induced modifications of plasma proteins and lipids. Although this model has important limitations, perhaps most notably in that RTLFs contain mucus, surfactant, and much higher levels of reduced glutathione (GSH), we have rationalized that CS reactions with plasma would be fairly representative of the types of biomolecular interactions CS might have with RTLFs.

The major findings of our work have shown that gas-phase CS depletes plasma aqueous-phase antioxidants, especially ascorbic acid (AA). We have observed that CS is capable of modifying plasma proteins, as measured by protein-SH loss and protein carbonyl formation. We also observed that, in general, lipid-soluble antioxidants were depleted to a lesser extent than aqueous-phase antioxidants and that these processes were accompanied by the formation of small amounts (<1 μM) of lipid hydroperoxides (10). Importantly, we have shown that AA could protect against CS-induced lipid peroxidation and that reduced GSH and dihydrolipoic acid (DHLA) appear to protect plasma proteins from CS-induced modification. The aim of the present chapter is to review the known and speculative interactions of CS with biological macromolecules and to discuss the role and possible mechanisms whereby selected thiol compounds may ameliorate subsequent CS-induced biomolecular modifications.

II. EFFECTS OF GAS-PHASE CS ON PLASMA CONSTITUENTS

A. Effects on Plasma Antioxidants

To explore the effects and relative reactivity of various antioxidants, we exposed plasma [15–20 ml, a volume that approximates the estimated RTLF volume in humans (13)] to CS at a rate of one puff every 20 min according to the method described by Frei et al. (10). When human plasma was exposed to sequential "puffs" of gas-phase CS for a period of 3 h, AA was completely consumed at between 1 and 2 h (Fig. 1). Other aqueous-phase antioxidants in plasma, such as protein-SH and uric acid, disappeared at a slower rate. However, because of the relatively larger concentrations of these compounds, they are still considered significant antioxidant species reacting with CS. As also shown in Figure 1, ubiquinol-10 was rapidly utilized, but α-tocopherol, the major lipid-phase antioxidant, was decreased by only 20–25% at the end of CS exposure (3 h). Thus, it is readily apparent that CS has different effects on various plasma antioxidant components; AA and ubiquinol-10 are rapidly depleted, whereas uric acid and α-tocopherol appear relatively less sensitive or more selectively reactive with certain CS components and/or secondary reaction products derived from CS. Consumption of ubiquinol-10 and α-tocopherol appears to occur only after AA was nearly completely utilized.

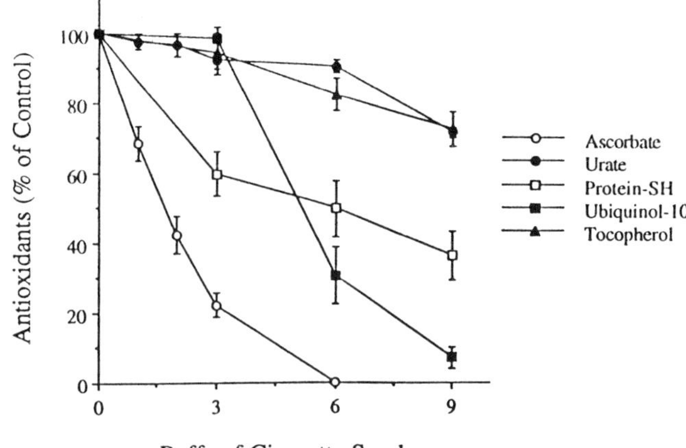

Figure 1 Depletion of antioxidants in plasma exposed to gas-phase cigarette smoke (CS). Human plasma (20 ml) in a 500-ml filter flask was exposed to puffs of CS as previously described (10). Levels of endogenous plasma antioxidants are shown as percentages of their initial concentrations, which were: ascorbic acid, 61 ± 20 μM; uric acid, 304 ± 27 μM; protein -SH, 524 ± 37 μM; ubiquinol-10, 0.9 ± 0.1 μM; α-tocopherol, 29 ± 9 μM. Values are mean ± SD obtained from three separate experiments.

B. Effect of CS on Lipid Peroxidation and Lipoprotein Electrophoretic Mobility

CS-induced lipid hydroperoxide (LOOH) formation was measured by HPLC using chemiluminescence detection, which is capable of detecting nM quantities of LOOH (14). After 2–3 h of CS exposure (6–9 "puffs"), low (approx. 200 nM) levels of LOOH could be detected. Supplementation of plasma with additional amounts of AA prior to CS exposure delayed the initiation of CS-induced formation of LOOH, whereas depletion of AA (by treatment of plasma with ascorbate oxidase prior to CS exposure) resulted in an immediate initiation of LOOH formation (12). However, it should be emphasized that relatively low levels of LOOH could be detected and that they formed before the complete depletion of α-tocopherol. Unlike ubiquinol-10, α-tocopherol is not able to prevent LOOH formation, as it acts as a chain-breaking antioxidant. The fact that α-tocopherol is still present suggests that gas-phase CS leads only to small amounts of lipid peroxidation initiation products. In addition, we observed no lipid hydroperoxide formation when plasma was exposed to whole (unfiltered) CS. However, since our only basis for determining the extent of lipid peroxidation was by LOOH formation, we cannot rule out the possibility that LOOHs are, in fact, formed but are rapidly decomposed by transition metals present in the CS tar.

Plasma low-density lipoproteins (LDL) exhibited slight yet consistent increases in the anodic electrophoretic mobility when exposed to CS, as measured by gel electrophoresis (10). Isolated LDL (1 mg protein/ml in PBS) exposed to gas-phase CS for 3 h exhibited a twofold increase in electrophoretic mobility, and this change was not inhibited by the presence of EDTA (1 mM) and desferrioxamine (1 mM), suggesting a mechanism independent of metal-catalyzed oxidation (15).

C. Effect on Protein Modification as Measured by Protein-SH Losses and Formation of Protein Carbonyls

A standard measure of protein oxidation/modification is the formation of protein-bound carbonyl groups, which are determined by reaction with 2,4-dinitrophenylhydrazine (DNPH) and subsequent spectrophotometric detection (16). Basal levels of carbonyl groups present in human plasma proteins range between 0.5 and 0.9 nmol/mg protein. Exposure of plasma to "puffs" of CS (3 h) caused a depletion of protein-SH groups (Fig. 1) and a linear increase in protein carbonyls to levels of 5–7 nM of carbonyl/mg protein. Addition of AA did not alter the rate at which protein carbonyls were formed over this exposure period, suggesting a mechanism of damage different from that responsible for lipid peroxidation. We also showed that CS-induced protein modification probably did not include iron-catalyzed oxidations, as the presence of desferrioxamine had little influence on accumulation of protein carbonyls induced by gas-phase CS (11).

D. Effect of Gas-Phase CS on the Activity of Plasma Enzymes

Effects of gas-phase CS on human plasma creatine kinase (CK) and pure forms of CK were examined. Plasma CK activity was decreased by 40% after nine "puffs" of CS. Purified rabbit muscle CK activity was lowered by 82% after treating solutions of rabbit CK with similar quantities of CS. Since CK is a thiol-dependent enzyme (17), it is likely that some modification of protein-SH groups within the active site of CK was responsible for the inactivation of this enzyme by CS. Other plasma enzymes such as lactate dehydrogenase (LDH), aspartate aminotransferase (AAT), and γ-glutamyl transferase (GGT) were not readily affected by exposure of plasma to this level of gas-phase CS under our exposure conditions. McCall et al. (18) have shown that exposure of plasma to gas-phase CS causes a dramatic loss of activity in lecithin-cholesterol acyltransferase (LCAT), which is an important component of reverse cholesterol transport and, like CK, contains protein-SH group(s), which are important for its functional activity.

III. EFFECTS OF THIOLS ON CS-INDUCED MODIFICATIONS OF PLASMA ANTIOXIDANTS, PROTEINS, AND LIPIDS

A. Effects of GSH and DHLA on CS-Induced Ascorbic Acid and α-Tocopherol Depletions

As shown above, plasma AA is depleted relatively quickly upon exposure to CS. Addition of reduced GSH (1 mM) did not change the rate of loss of endogenous AA, but DHLA (1 mM) appeared to cause a small decrease in the initial rate of AA depletion. It should be noted that exposure of plasma to air also caused a depletion of AA to about 60% of its initial value after 3 h and that DHLA also appears to decrease this air-induced AA loss. The results suggest that DHLA, but not GSH, may be capable of protecting or regenerating AA in both air and in CS-exposed plasma.

Most of the α-tocopherol in plasma is contained within the lipoprotein fractions, especially the LDL. Exposure of plasma to CS for 3 h (nine "puffs") caused only a relatively small (20–25%) decrease in levels of α-tocopherol. Addition of GSH (1 mM) to the plasma did not alter the rate of α-tocopherol oxidation, but DHLA (1 mM) appeared to inhibit the loss of α-tocopherol to a small extent. This observation could conceivably be due to the recycling of α-tocopherol by AA, which in turn may be regenerated by DHLA. However, DHLA is also capable of attenuating the loss of α-tocopherol in the absence of AA. The effectiveness of DHLA, however, appears quite minimal as the concentrations required (1 mM) to achieve such beneficial effects are extremely high relative to endogenous plasma antioxidant levels. Of note, DHLA was ineffective at inhibiting ubiquinol-10 depletion by CS both in the presence and absence of AA.

B. Effect of GSH and DHLA in Decreasing Lipid Peroxidation

After ascorbic acid depletion, plasma that had been exposed to the gas phase of CS exhibited increased levels of lipid hydroperoxides, indicating that at least some lipid peroxidation was induced by CS. Addition of DHLA (1 mM) inhibited the formation of detectable lipid hydroperoxides (Fig. 2). Supplementation of plasma with GSH (1 mM) also appeared to afford some protection of plasma lipids against CS-induced oxidation, although less so than DHLA.

C. Effects of GSH and DHLA on Protein Modification and Enzyme Activity

As illustrated in Figure 3, the addition of DHLA and GSH (1 mM concentrations) inhibited CS-induced protein carbonyl formation. The results show that

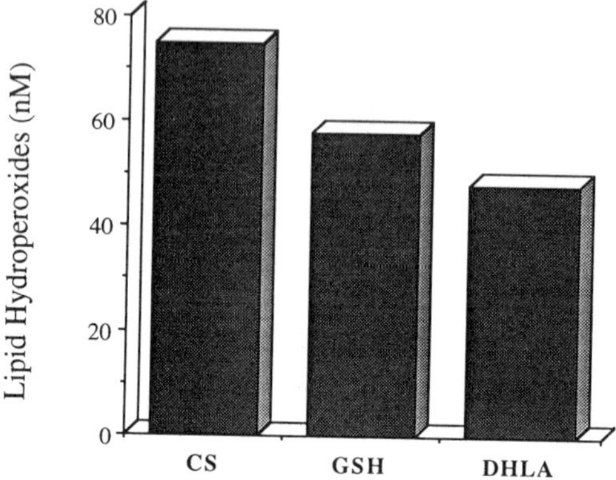

Figure 2 Inhibitory effects of GSH and DHLA (1 mM) on the formation of lipid hydroperoxides in plasma exposed to six "puffs" of gas-phase CS. Levels of lipid hydroperoxides are the mean of two separate experiments.

both thiols can protect proteins from accumulating protein carbonyl groups when exposed to gas-phase CS. Plasma creatine kinase (CK) has an essential-SH group at its active site (17), oxidation of which causes a loss of activity when plasma is incubated at 37°C under air, and DHLA prevents this loss of activity (19). Exposure of plasma to CS accelerated the loss of enzyme activity, and addition of DHLA, GSH, or dithiothreitol (DTT) to plasma exposed to CS surprisingly

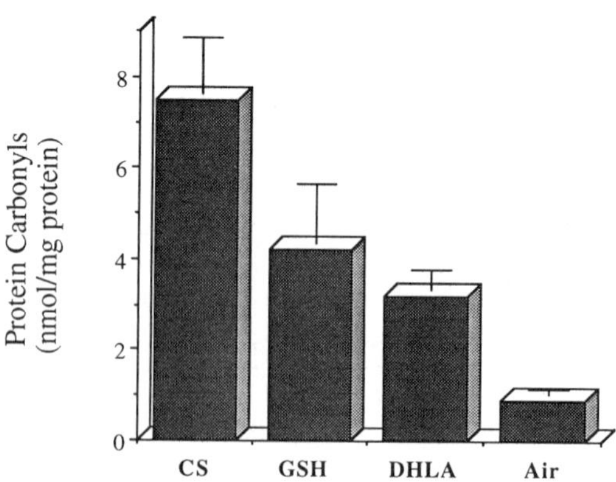

Figure 3 Formation of protein carbonyls in plasma exposed to nine puffs of gas-phase CS and its inhibition by GSH and DHLA (1 mM). Control samples were exposed to puffs of room air instead of CS. Values are mean ± SD obtained from three separate experiments.

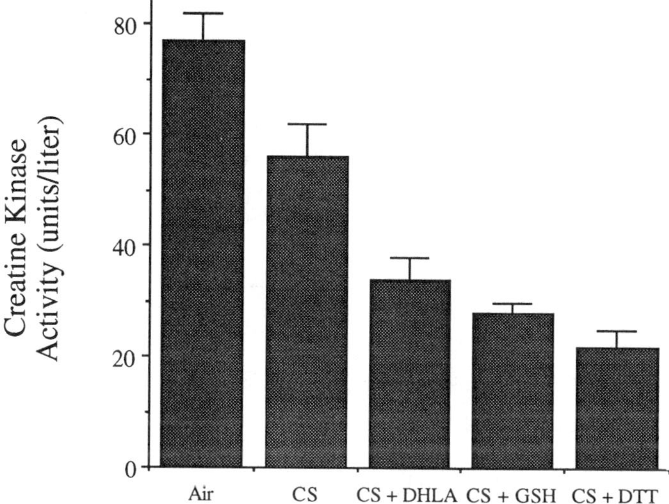

Figure 4 Effects of DHLA, GSH, and DTT (1 mM) on human plasma CK activity after exposure to 9 puffs of CS. Values are expressed as mean ± SD obtained from three separate experiments.

exacerbated, rather than inhibited, CS-induced loss of CK activity (Fig. 4). The relatively large abundance of various radicals in CS is thought to induce the formation of reactive thiyl radicals of protein-SH, DHLA, GSH, and DTT, which, when combined, lead to the formation of mixed disulfides of the protein and nonprotein moieties. However, this process of S-thiolation is reversible, unlike the irreversible formation of CS-induced α,β-unsaturated aldehyde adducts with protein-SH groups, and may actually serve as a protective mechanism.

IV. MECHANISMS OF CS-INDUCED BIOMOLECULAR DAMAGE AND THIOL PROTECTION

A. Modification of Plasma Constituents by Aldehydes Found in CS

Aldehydes have been shown to be important bioactive constituents of gas-phase CS. They are reactive toward nucleophilic sites in proteins, including but not limited to the amino acids cysteine and lysine, which possess reactive-SH and -NH_2 groups, respectively. We evaluated the effect of CS-derived aldehydes on plasma proteins to determine the possibility that this may be an important mechanism of CS-induced protein damage.

Preliminary studies showed that a mixture of three aldehydes added to plasma (acetaldehyde, acrolein, and propanal) in quantities known to be present in CS caused a significant increase in protein carbonyl formation (11). In subsequent studies (20), incubation of the α,β-unsaturated aldehydes acrolein and crotonaldehyde with human plasma significantly increased protein carbonyl formation during the first hour of exposure. Addition of aldehydes to plasma did not cause a depletion of antioxidants such as ascorbic acid and α-tocopherol and did not induce plasma lipid peroxidation. This strongly suggests that reactions of α,β-unsaturated aldehydes with proteins could conceivably account for much (if not all) of the carbonyl formation induced by gas-phase CS. We propose that the majority of CS-induced protein carbonyl formation is due to the reactions of α,β-unsaturated aldehydes with protein-SH and -NH_2 groups yielding stable Michael adducts.

B. Effects of DHLA and GSH on Formation of Protein Carbonyls by α, β-Unsaturated Aldehydes

Addition of 1 mM GSH to plasma prior to the addition of α,β-unsaturated aldehydes decreased the formation of protein carbonyls by almost 50% (11). Recently we showed that a mixture of saturated aldehydes and α,β-unsaturated aldehydes, in amounts reported to be present in the mainstream smoke of one reference cigarette, caused an increase in protein carbonyl formation. A concentration-dependent increase in carbonyl formation was observed with varying levels of crotonaldehyde and acrolein. Acrolein at 80 μM induced carbonyl formation to a level of 5.1 nmol carbonyl/mg protein. DHLA (1 mM) or GSH (1 mM) had similar effects in attenuating the formation of protein carbonyls (11).

Exposure of plasma to CS causes a marked decrease of plasma protein-SH groups. α,β-Unsaturated aldehydes, such as acrolein, are known to readily react with protein -SH groups by Michael addition (21,22), resulting in a protein-bound carbonyl group. The α,β-unsaturated aldehyde acrolein, when added to plasma, decreased the levels of protein-SH groups in a concentration-dependent manner and produced a corresponding increase in protein carbonyl groups. Addition of 80 μM acrolein to plasma leads to a net loss of plasma protein-SH of 329 μM. This amount of acrolein results in a net protein carbonyl formation of 4.5 nmol/mg protein, corresponding to a concentration of 315 μM. The values for protein-SH loss and protein carbonyl formation indicate a nearly 1:1 stoichiometry, as observed by a close correlation between protein-SH depletion and carbonyl formation on a molar basis. The effects of acrolein, at concentrations ranging from 10 μM to 10 mM, on the loss of protein-SH and -NH_2 groups and subsequent protein carbonyl and Schiff base (imine) formation are shown

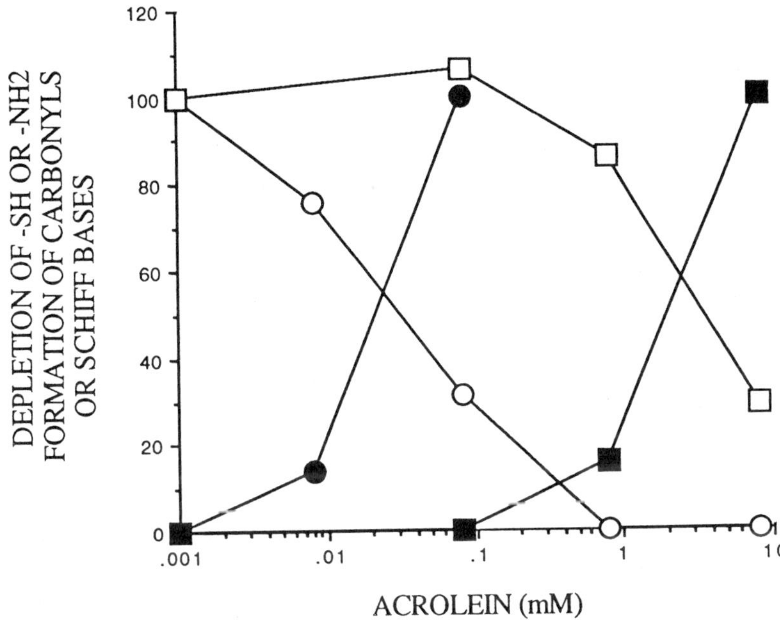

Figure 5 Depletion of protein-SH (O) and -NH_2 (□) groups and subsequent formation of protein carbonyls (●) and Schiff bases (■) in plasma incubated with various concentrations of acrolein. Measurements were made after 1 h of incubation and represent the mean of two separate experiments. For -SH and -NH_2 levels, values are expressed as a percentage of that measured in unexposed plasma. Protein carbonyls and Schiff base levels are expressed as a percentage of that found at the highest concentrations of acrolein used (10 mM).

in Figure 5. Acrolein appears to react selectively with protein-SH groups at low concentrations, and only upon complete depletion of protein-SH groups by high levels of acrolein do -NH_2 groups react with acrolein yielding Schiff bases. Therefore, at concentrations of acrolein relevant to those found in CS, protein-SH groups appear to be the primary targets of reactivity and subsequent protein modification. However, in chronic cigarette smoking, depletions of -SH groups in respiratory tract proteins potentially occur, and if so, protein -NH_2 groups are likely secondary targets. α,β-Unsaturated aldehydes can also react with amine groups of side chains of amino acids through Michael addition yielding protein carbonyl groups, analogous to protein-SH reactivity. As mentioned earlier, it should be emphasized that Michael addition reactions, unlike reactions leading to Schiff base formation, are not reversible and may represent important, "nonrepairable" modifications of proteins by CS.

C. Nitration and Oxidation of *N*-Acetyltyrosine by Gas-Phase CS: A Potential Free Radical Mechanism of Protein Modification

An important free radical in CS is nitric oxide (nitrogen monoxide, •NO), which is present in mainstream gas-phase CS at up to 1000 ppm and probably represents one of the greatest exogenous sources of •NO to which humans are exposed. To evaluate the possibility that nitrogen oxides in CS play a role in its damaging effects on proteins, we exposed *N*-acetyltyrosine, a representative model for tyrosine residues in a peptide sequence, to puffs of gas-phase CS. Acid hydrolysis of *N*-acetyltyrosine following CS exposure liberated free tyrosine and its modification products. We observed that both 3-nitrotyrosine and dityrosine were formed, as determined by HPLC with in-line UV and fluorescence detection (23). The formation of both 3-nitrotyrosine and dityrosine after six "puffs" of CS (5-min incubations between puffs) and its inhibition by GSH, ascorbic acid, and uric acid is shown in Figure 6.

GSH, ascorbic acid, and uric acid at 100 μM concentrations, which approximate those found in human (13) RTLFs, inhibited the formation of 3-nitrotyrosine by 35, 70, and 85%, respectively. Similar trends were observed for the inhibition of dityrosine formation by these antioxidants. It is of interest to note that GSH and ascorbic acid were completely depleted after three and six "puffs," respectively. However, uric acid was only depleted by approximately 20% after six "puffs" of CS. Addition of dithiothreitol (DTT) (1 mM) to the GSH sample upon completion of the CS exposure resulted in only a 3% recovery of GSH, whereas nearly 100% recovery of GSH was observed when the authentic oxidized form (GSSG) was treated in a similar fashion. This indicates that GSH was probably not oxidized to GSSG or nitroso-glutathione (GSNO), as these species are readily converted to GSH by reductants such as DTT. The fate of GSH was more likely a result of its reaction with aldehydes in CS and may explain its relatively low inhibitory activity compared to ascorbic acid and uric acid.

Figure 7 summarizes the potential mechanisms by which components of gas-phase CS may modify proteins in the respiratory tract, either in the RTLFs or on the surface of respiratory tract cells. Reactions of CS-derived saturated and α,β-unsaturated aldehydes with -SH and -NH_2 groups of proteins lead to the formation of Schiff bases and Michael addition products, contributing significantly to the overall modification of proteins by CS. Gas-phase CS is also potentially capable of modifying proteins through free radical mechanisms involving thiyl and tyrosyl radicals, subsequently leading to the formation of mixed disulfides with low molecular thiols, intramolecular disulfide linkages, dityrosine moieties, and even nitration products including nitrotyrosine. It may be that reactions of aldehydes with proteins represent the major mechanism involved in protein modification by gas-phase CS.

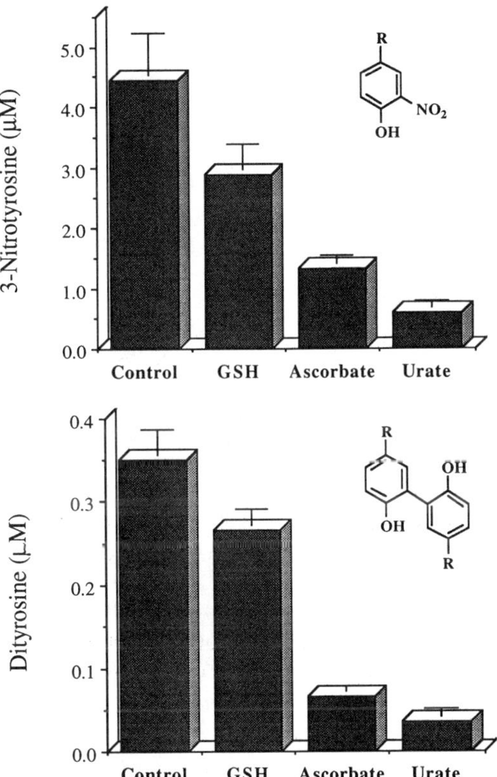

Figure 6 Formation of 3-nitrotyrosine (top) and dityrosine (bottom) by gas-phase CS and its inhibition by GSH, ascorbate, and urate (all at 100 μM). Solutions of *N*-acetyltyrosine (1 mM) in 100 mM KH_2PO_4 were exposed to 6 puffs of CS, hydrolyzed in 6 M HCl at 105°C for 3 h and the levels of the products determined by HPLC. Values are expressed as mean ± SD of three separate experiments.

V. POTENTIAL PHYSIOLOGICAL RELEVANCE

There is increasing evidence that thiols, including GSH, are important respiratory tract lining fluid (RTLF) antioxidants (6–8). Decreased levels of GSH have been found in the RTLFs of such diverse lung disorders as cystic fibrosis (24), HIV infection (25), and idiopathic pulmonary fibrosis (26). The RTLFs provide an "antioxidant" screen, protecting the underlying RT epithelial cells against both inhaled environmental oxidants and endogenous oxidants produced by phagocytes contained within the RTLFs (6–8). Not surprisingly, it has been suggested that therapeutic administrations of thiol compounds (e.g., by aerosol) could

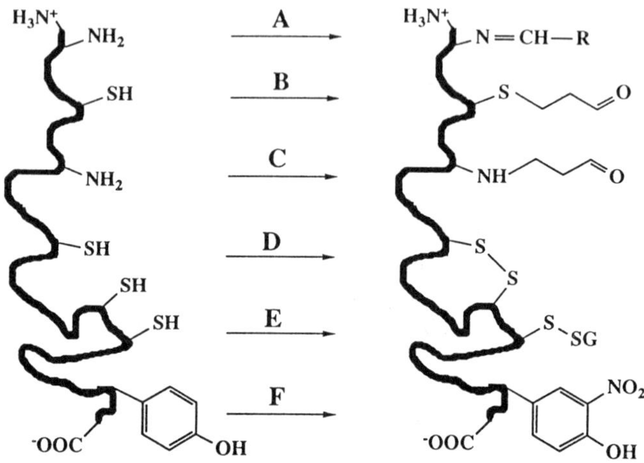

Figure 7 Schematic representation of the possible interactions/reactions of gas-phase cigarette smoke components with proteins leading to a variety of structural modifications. Reactions are as follows: (A) Schiff base formation resulting from the interaction of saturated aldehydes with -NH_2 groups; (B) and (C) Michael addition products from the reaction of -SH and -NH_2 groups with α,β-unsaturated aldehydes (acrolein); (D) oxidant-induced formation of disulfides; (E) formation of mixed disulfides with nonprotein thiols (glutathione); and (F) nitration of tyrosine by nitrogen oxides present in CS.

augment the RTLF "antioxidant" screen, perhaps ameliorating, to some degree, the disease processes (27).

However, some degree of prudence is indicated before this therapeutic approach is generally accepted. Thiols can be toxic in vitro, e.g., they can reduce transition metal ions, thereby participating in Fenton chemistry with the production of reactive oxygen species such as •OH and H_2O_2 (28,29). This might be expected to be a particular problem in clinical diseases where proteases could be expected to be present in RTLFs, thus liberating biologically complexed iron [e.g., cystic fibrosis (30)] or in circumstances where catalytic transition metals could be present on inhaled particulates [e.g., pneumoniosis (31), air pollutants (32), or cigarette smoke (33)]. Thiols can react with other reactive oxygen (or nitrogen) species to form reactive and potentially toxic thiyl and/or oxysulfur radicals (34).

Another possible concern is that the optimal function of a number of RTEC plasma membrane proteins, including receptors, signal transduction pathways, and ion-conducting channels, may require a balance between reduced and oxidized thiols to maintain membrane function. Excessive amounts of reducing agents (e.g., thiols administered at mM levels) may influence the integrity of disulfide bonds within plasma membrane proteins, thereby altering the binding, kinetic, and/or permeation properties of these proteins. In fact, a delicate balance between reduced sulfhydryls and oxidized disulfides is required for the optimal function of many enzymes, including the NADH-ascorbate free radical reductase (35). We have shown that incubation of plasma with DHLA and GSH (1 mM) results in a drastic increase of protein-SH groups (Fig. 8) and is concomitant with a relatively rapid loss of DHLA and GSH. These thiol compounds, when added to plasma, probably reduce disulfide linkages of proteins liberating free -SH groups. It has been proposed that plasma GSH serves to mobilize compounds bound by disulfide linkage to plasma proteins to form GSSG and other low molecular weight derivatives of glutathione such as mixed disulfides with cysteine and other thiols (36).

VI. SELECTED UNSETTLED ISSUES

CS consists of several thousands of both identified and unidentified compounds, many of which can be expected to react with biomolecular species in the respiratory tract. It is thus no surprise that a myriad of biopathological responses to CS have been described.The aforementioned studies have addressed selected

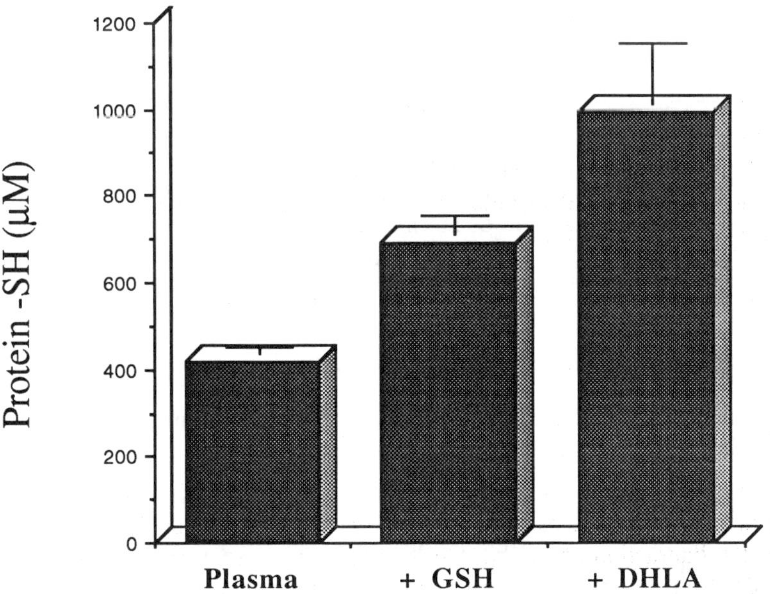

Figure 8 Effects on the concentrations of plasma protein-SH groups by the addition of GSH and DHLA (1 mM). Levels of protein-SH were determined after 20-min incubation of plasma either alone or in the presence of the thiols. Values are expressed as mean ± SD of three separate experiments.

studies of mainstream gas-phase CS responses in human plasma, focusing upon antioxidant depletion and its relationship to protein modification and lipid hydroperoxide formation. However, the smoker is exposed to a matrix that may be quite different from that which passes the Cambridge filter used in our studies. CS particulates, or tar phase, which are deposited in the respiratory tract, may contribute directly to damaging biomolecules and may act synergistically with the gas phase of CS producing an increased oxidant "burden" on the respiratory tract.

Our studies have exposed biological extracellular fluids to CS for relatively long periods of time (up to 20 min), which may not directly reflect the true exposure experienced by cigarette smokers. However, the inhaled components of CS, including a variety of aldehydes and free radicals such as •NO and peroxyl radicals, are probably absorbed into the RTLFs and continue to react with the biomolecules in these fluids and the underlying epithelial cells for an unkown period of time. CS that is inhaled into the more distal regions of the lung may reach poorly ventilated areas and, hence, increase its residence time and potential reactivity. The time-dependent formation of "different" reactants in CS is poorly understood, and aging of CS, as in the case of passive smoking (secondhand smoke), is thought to have different reactivity and, hence, different toxicity compared to mainstream smoke. Can •NO be converted to a more reactive species in either the gas phase or as it is deposited in the RTLFs, and does this change the reactivity and damage imposed by CS? If so, what compounds are formed in aging smoke, and what are these "new" components reacting with in the lung? Of the thousands of possible species present in CS, which are primarily responsible for lipid and protein modifications? Are the radical species, abundant aldehydes, a combination of the two, or secondary products responsible for the deleterious effects of CS? These and many other questions must be answered in order to begin to unravel the mechanisms by which CS causes its deleterious effects on biological systems.

VII. SUMMARY AND CONCLUSIONS

The studies described in this chapter were undertaken to further our understanding of the factors and mechanisms by which components of gas-phase CS cause modifications and/or oxidations of plasma components. The experimental findings indicate that perhaps a major mechanism of CS-induced damage to proteins is by the interaction of α,β-unsaturated aldehydes (e.g., acrolein and crotonaldehyde) with proteins resulting in the formation of irreversible Michael addition products with protein-SH and -NH_2 groups. The formation of nitrated and oxidized forms of tyrosine by CS strongly suggests that free radical reactions of nitrogen oxides and other reactive oxygen species may also play key roles in CS-induced protein modification. These types of modifications may result in altered activity of critical enzymes, membrane receptors, and transport proteins (such as membrane ion channels) and would be expected to interfere with cell signaling pathways, possibly including those involving tyrosine phosphorylation. The biothiols DHLA and GSH, which react readily with CS aldehydes and reactive oxygen and nitrogen species, afford protection from CS-induced damage to proteins and to a lesser extent lipids, thus implicating these biothiols as possible agents that may neutralize certain adverse effects of CS. However, using thiols as therapeutic agents may upset the delicate redox balance of the "finely tuned" biological system and could theoretically exacerbate an already existing oxidative stress situation.

ACKNOWLEDGMENTS

The authors would like to thank Balz Frei for contributing some of the work and data included in this review and Lester Packer and Hans Tritschler for their support and encouragement.

REFERENCES

1. Carp, H., and Janoff, A. (1978) Possible mechanisms for emphysema in smokers: in vitro suppression of serum elastase-inhibitory capacity by fresh cigarette smoke and its prevention by antioxidants. Am. Rev. Respir. Dis. 118:617–621.
2. Evans, M. D., and Pryor, W. A. (1994) Cigarette smoking, emphysema, and damage to α1-proteinase inhibitor. Am. J. Physiol. 266:L593–L611.
3. Haapanen, A., Koskenvuo, M., Kaprio, J., Kesaniemi, Y. A., and Keikkila, K. (1989) Carotid arteriosclerosis in identical twins discordant for cigarette smoking. Circulation 80:10–16.
4. Cueto, R., and Pryor, W. A. (1994) Cigarette smoke chemistry: Conversion of nitric oxide to nitrogen dioxide and reactions of nitrogen oxides with other smoke components as studied by Fourier transform infrared spectroscopy. Vib. Spectrosc. 7:97–111.
5. Pryor, W. A., and Stone, K. (1993) Oxidants in cigarette smoke. Radicals, hydrogen peroxide, peroxynitrate, and peroxynitrite. Ann. NY Acad. Sci. 686:12–28.
6. Slade, R., Stead, A. G., Graham, J. A., and Hatch, G. E. (1985) Comparison of lung antioxidant levels in humans and laboratory animals. Am. Rev. Respir. Dis. 131:742–746.
7. Slade, R., Crissman, K., Norwood, J., and Hatch, G. (1993) Comparison of antioxidant substances in bronchoalveolar lavage cells and fluid from humans, guinea pigs, and rats. Exp. Lung Res. 19:469–484.
8. Cross, C. E., van der Vliet, A., O'Neill, C. A., Louie, S., and Halliwell, B. (1994) Oxidants, antioxidants, and respiratory tract lining fluids. Environ. Health Perspect. 102 (Suppl. 10):185–191.
9. Church, D. F., and Pryor, W. A. (1985) Free radical chemistry of cigarette smoke and its toxicological implications. Environ. Health Perspect. 64:111–126.

10. Frei, B., Forte, T. M., Ames, B. N., and Cross, C. E. (1991) Gas phase oxidants of cigarette smoke induce lipid peroxidation and changes in lipoprotein properties in human blood plasma. Protective effects of ascorbic acid. Biochem. J. 277:133–138.
11. Reznick, A. Z., Cross, C. E., Hu, M.-L., Suzuki, Y. J., Khawaja, S., Safadi, A., Mochnik, P. A., Packer, L., and Halliwell, B. (1992) Modification of plasma proteins by cigarette smoke as measured by protein carbonyl formation. Biochem. J. 286:607–611.
12. Cross, C. E., O'Neill, C. A., Reznick, A. Z., Hu, M.-L., Marcocci, L., Packer, L., and Frei, B. (1993) Cigarette smoke oxidation of human plasma constituents. Ann. NY Acad. Sci. 686:72–90.
13. Hatch, G. E. (1992) Comparative biochemistry of airway lining fluid. In: Treatise on Pulmonary Toxicology. Comparative Biology of the Normal Lung (Parent, R. A., ed.), pp. 617–632. CRC Press, Boca Raton, FL.
14. Frei, B., Yamamoto, Y., Niclas, D., and Ames, B. N. (1988) Evaluation of an isoluminol chemiluminescence assay for the detection of hydroperoxides in human blood plasma. Anal. Biochem. 175:120–130.
15. Davis, P. A., O'Neill, C., van der Vliet, A., Packer, L., and Cross, C. E. (1994) Investigations of the possible roles of cigarette smoke-induced oxidative stress in processes related to atherosclerosis. In: Biological Oxidants and Antioxidants (Packer, L., and Cadenas, E., eds.), pp. 213–228. Hippokrates Verlag, Stuttgart.
16. Levine, R. L., Garland, D., Oliver, C. N., Amici, A., Climent, I., Lenz, A.-G., Ahn, B.-W., Shaltiel, S., and Stadtman, E. R. (1990) Determination of carbonyl content in oxidatively modified proteins. Meth. Enzymol. 186:464–485.
17. Watts, D. C. (1973) Creatine kinase (adenosine 5'-triphosphate-creatine-phosphotransferase). In: The Enzymes (Boyer, P. D., ed.), pp. 383–455. Academic Press.
18. McCall, M. R., van den Berg, J. J. M., Kuypers, F. A., Tribble, D. L., Krauss, R. M., Knoff, L. J., and Forte, T. M. (1994) Modification of LCAT activity and HDL structure. New links between cigarette smoke and coronary heart disease risk. Arterioscler. Thromb. 14:248–253.
19. Scott, B. C., Aruoma, O. I., Evans, P. J., O'Neill, C., van der Vliet, A., Cross, C. E., Tritschler, H., and Halliwell, B. (1994) Lipoic and dihydrolipoic acids as antioxidants. A critical evaluation. Free Rad. Res. 20:119–133.
20. O'Neill, C. A., Halliwell, B., van der Vliet, A., Davis, P. A., Packer, L., Tritschler, H., Strohman, W. A., Rieland, T., Cross, C. E., and Reznick, A. Z. (1994) Aldehyde-induced protein modifications in human plasma: protection by glutathione and dihydrolipoic acid. J. Lab. Clin. Med. 124:359–370.
21. Esterbauer, H., Schauer, R. J., and Zollner, H. (1991) Chemistry and biochemistry of 4-hydroxynonenal, malonaldeldehyde and related aldehydes. Free Rad. Biol. Med. 11:81–129.
22. Witz, G. (1989) Biological interactions of α,β-unsaturated aldehydes. Free Rad. Biol. Med. 7:333–349.
23. Eiserich, J. P., Vossen, V., O'Neill, C. A., Halliwell, B., Cross, C. E., and van der Vliet, A. (1994) Molecular mechanisms of damage by excess nitrogen oxides: Nitration of tyrosine by gas-phase cigarette smoke. FEBS Lett. 353:53–56.

24. Roum, J. H., Buhl, R., McElvaney, N. G., Borok, Z., and Crystal, R. G. (1993) Systemic deficiency of glutathione in cystic fibrosis. J. Appl. Physiol. 75: 2419–2424.
25. Buhl, R., Holroyd, K. J., Cantin, A. M., Jaffe, H. A., Wells, F. B., Saltini, C., et al. (1989) Systemic glutathione deficiency in symptom-free HIV-seropositive individuals. Lancet 2:1294–1298.
26. Cantin, A. M., Hubbard, R. C., and Crystal, R. G. (1989) Glutathione deficiency in the epithelial lining fluid of the lower respiratory tract in idiopathic pulmonary fibrosis. Am. Rev. Respir. Dis. 139:380–372.
27. Crystal, R. G. (1991) Oxidants and respiratory tract epithelial injury: Pathogenesis and strategies for therapeutic intervention. Am. J. Med. 91 (Suppl. 3C):395–445.
28. Huston, P., Espenson, J. H., and Bakac, A. (1992) Reactions of thiyl radicals with transition-metal complexes. J. Am. Chem. Soc. 114:9510–9516.
29. Held, K. D., and Biaglow, J. E. (1994) Mechanisms for the oxygen radical-mediated toxicity of various thiol-containing compounds in cultured mammalian cells. Radiat. Res. 139:15–23.
30. Mohammed, J. R., Mohammed, B. S., Pawluk, L. J., Bucci, D. M., Baker, N. R., Davis, W. B. (1988) Purification and cytotoxic potential of myeloperoxidase in cystic fibrosis sputum. J. Lab. Clin. Med. 112:711 720.
31. Janssen, Y. M. W., Van Houten, B., Borm, P. J. A., and Mossman, B. T. (1993) Biology of disease: Cell and tissue responses to oxidative damage. Lab. Invest. 69:261–274.
32. Pritchard, R. J., and Ghio, A. J. (1994) Humic acid-like substances are present in air pollution particles. Am. J. Respir. Crit. Med. 149:A840.
33. Ghio, A. J., Stonehuerner, J., and Quigley, D. R. (1994) Humic-like substances in cigarette smoke condensate and lung tissue of smokers. Am. J. Physiol. 266:L382–388.
34. Munday, R. (1989) Toxicity of thiols and disulphides: Involvement of free-radical species. Free Rad. Biol. Med. 7:659–673.
35. Villalba, J. M., Canalejo, A., Buron, M. I., Cordoba, F., and Navas, P. (1993) Thiol groups are involved in NADH-ascorbate free radical reductase activity of rat liver plasma membrane. Biochem. Biophys. Res. Comm. 192:707–713.
36. Anderson, M. E., and Meister, A. (1980) Dynamic state of glutathione in blood plasma. J. Biol. Chem. 255:9530–9533.

20

Regulation of Gene Expression by Lipoic Acid

Yuichiro J. Suzuki and Lester Packer
University of California, Berkeley, California

Masashi Mizuno
Kobe University, Kobe, Japan

Heinz Ulrich
ASTA Medica AG, Frankfurt, Germany

I. INTRODUCTION

Lipoic acid is an essential cofactor of four multienzyme complexes localized in the mitochondria. Three of these complexes are collectively referred to as the α-keto acid dehydrogenase complexes, which catalyze the oxidative decarboxylation of pyruvate, α-ketoglutarate, and the branched-chain α-keto acids (1). The fourth is the glycine cleavage system, which catalyzes the reversible oxidation of glycine, forming CO_2, NH_4^+, and a methylene group, which is accepted by tetrahydrofolate (2). In these reactions, lipoic acid is covalently linked to specific lysine residues of the proteins and functions as the lipoamide in the oxidized form to transfer the carbon fragment while being reduced in the process. The standard oxidation-reduction potential (E_0') of the dihydrolipoic acid/lipoic acid pair is quite low as determined (at pH 7) to be −0.32 V by polarography and −0.29 V from the equilibrium with NAD in the presence of dihydrolipoic acid dehydrogenase (3).

In addition to its well-known role in metabolism, other functions of lipoic acid have been proposed and may contribute to cell homeostasis. Lipoic acid and its reduced form, dihydrolipoic acid, were recently identified as antioxidants, which eliminate biological oxidants and attenuate cellular damage (see Chap. 22). The antioxidant properties of lipoic acid and dihydrolipoic acid include scavenging of oxygen radicals (4) and recycling of endogenous antioxidants (5). Furthermore, the redox regulation of functional proteins involved in cell signaling and gene transcription may be a novel physiological function of lipoic acid. Exogenously administered lipoic acid has also been found to modulate gene expression, indicating a potential use of lipoic acid in prevention and therapy for various diseases.

II. REDOX REGULATION OF TRANSCRIPTION FACTORS BY DIHYDROLIPOIC ACID: PHYSIOLOGICAL IMPLICATIONS

The regulation of protein activity through redox modulation of thiols is a well-recognized phenomenon (6). Activities of many mammalian DNA-binding proteins are also regulated through redox mechanisms, and redox regulation of DNA-protein interactions plays an important role in mechanisms of actions of transcription factors for controlling gene transcription. NF-κB DNA-binding activity, which regulates gene expression of those involved in inflammatory processes as well as human immunodeficiency virus (7), is stimulated by reducing agents such as dithiothreitol, 2-mercaptoethanol, and thioredoxin, and it is abolished by the exposure to sulfhydryl-modifying agents, diamide, and *N*-ethylmaleimide (8–11). Site-directed mutagenesis studies have identified important sulfhydryl groups for DNA-binding activity such as Cysteine-62 in the p50 subunit of NF-κB (10,12). Similarly, the DNA-binding activity of Fos-Jun heterodimeric complex (AP-1) has been shown to be regulated by a redox mechanism (13). Fos and Jun can be converted to an inactive state by chemical oxidation or modification of a conserved cysteine residue located in the DNA-binding domains of Fos (Cysteine-154) and Jun (Cysteine-272) flanked by basic amino acids. The DNA-binding activity of Fos and Jun can be enhanced by mutation of the cysteine to serine or by treatment with high concentrations of reducing agents including reduced thioredoxin. One ubiquitous 37 kDa protein, redox factor-1 (Ref-1), has been proven to participate in physiological regulation of AP-1 (14,15). Steroid receptors such as glucocorticoid receptor (16) and progesterone receptor (17) also require reduced sulfhydryl groups for DNA-binding activity. Grippo et al. (18) identified the glucocorticoid receptor–activating factor, which enhances the glucocorticoid binding to receptor, as thioredoxin. Thioredoxin is a small ubiquitous protein, which functions to transfer electrons in various biochemical processes (19).

Dihydrolipoic acid and dihydrolipoamide can reduce thioredoxin (20,21). Gleason and Holmgren (20) pointed out that, since lipoic acid is a good reductant of thioredoxin in vitro, a lipoic acid–coupled system might function under normal conditions in *E. coli*. Spector et al. (22) proposed that thioredoxin may be a physiological electron acceptor from mitochondrial dihydrolipoamide in mammalian systems. Thioredoxin has been suggested to play an important role in cell regulation by enhancing the activities of transcription factors including

NF-κB (11), AP-1 (13), and glucocorticoid receptor (18). Thus, mitochondrial dihydrolipoamide may participate in gene regulation by donating electrons to thioredoxin, which in turn donates electrons to transcription factors.

In vitro electrophoretic mobility shift assays demonstrate that dihydrolipoic acid can directly enhance the DNA-binding activities of transcription factors. When nuclear extracts (from Jurkat cells) were isolated and the DNA-binding assays were performed without the use of DTT (23), NF-κB did not bind to radiolabeled oligonucleotide, which contains binding sites for this transcription factor (Fig. 1). This is consistent with the previous finding that in vitro DNA-binding activity of NF-κB requires reducing factors (9). The addition of DTT or dihydrolipoic acid in the binding reaction mixtures for the electrophoretic mobility shift assay restored the DNA-binding activity (Fig. 1). Similarly, the inhibition of NF-κB DNA-binding activity by exposure to a thiol-oxidizing agent, diamide (8,9), was blocked by dihydrolipoic acid (Fig. 2). Dihydrolipoic acid also blocked the diamide-induced inhibition of DNA binding activity of AP-1 (Fig. 3), another transcription factor whose DNA binding has been shown to be regulated through a redox mechanism (13). Redox regulation of transcrip-

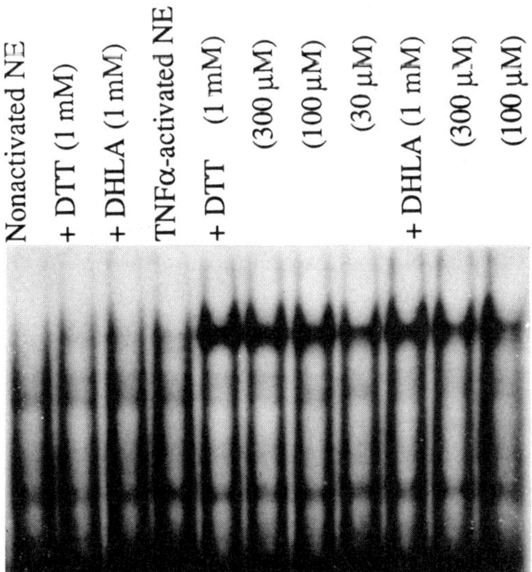

Figure 1 Restoration of the DNA-binding activity of activated NF-κB by DTT and dihydrolipoic acid. Jurkat cells (1×10^6 cells/ml) were activated with TNFα (25 ng/ml). Nuclear extracts (NE) were prepared in the absence of DTT in buffer C (23). Extracts were then incubated with DTT or dihydrolipoic acid (DHLA) in the binding reaction mixture for EMSA containing ^{32}P-labeled oligonucleotide probe for NF-κB.

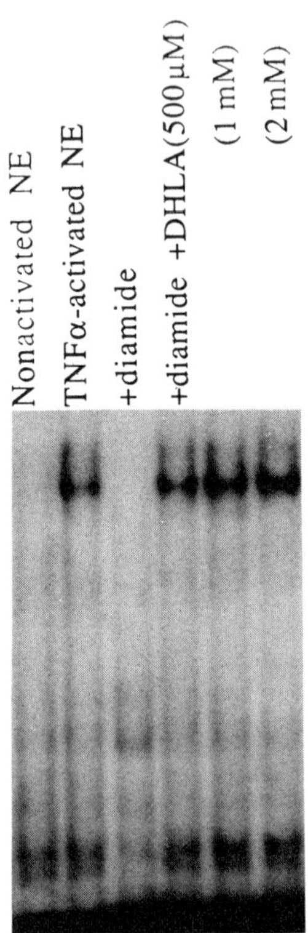

Figure 2 Prevention of diamide-induced inhibition of NF-κB DNA-binding activity by dihydrolipoic acid. Jurkat cells (1×10^6 cells/ml) were activated with TNFα (25 ng/ml). Nuclear extracts were prepared and then incubated with diamide (100 μM) in the absence or presence of DHLA in the binding reaction mixture for EMSA containing DTT (0.5 mM) and ^{32}P-labeled oligonucleotide probe for NF-κB.

tion factors appears to be a common phenomenon; the DNA-binding activity of vitamin D receptor was not exhibited in the absence of reducing factors, and this was restored by the addition of dihydrolipoic acid (Fig. 4).

III. MODULATION OF GENE EXPRESSION BY LIPOIC ACID: PHARMACOLOGICAL SIGNIFICANCE

Modulation of gene expression by targeting components of cell signaling has significant clinical implications in cancer, AIDS, and inflammation. Lipoic acid

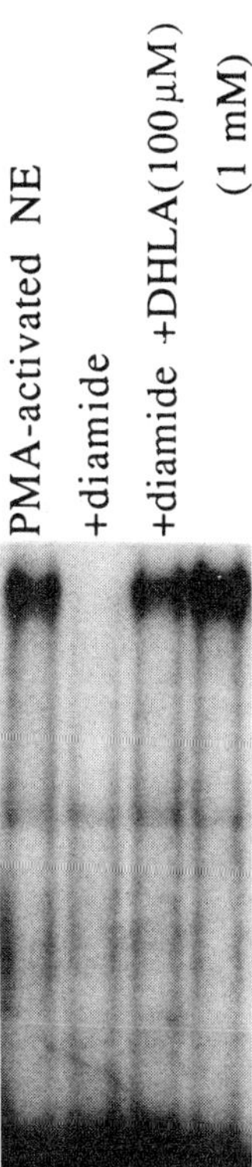

Figure 3 Prevention of diamide-induced inhibition of AP-1 DNA-binding activity by dihydrolipoic acid. Jurkat cells (1×10^6 cells/ml) were activated with PMA (50 ng/ml). Nuclear extracts were prepared and then incubated with diamide (100 μM) in the absence or presence of DHLA in the binding reaction mixture for EMSA containing DTT (0.5 mM) and ^{32}P-labeled oligonucleotide probe for AP-1.

can alter the course of the expression of various genes and thus resultant biological processes by (1) exerting antioxidant actions against the regulation of gene expression mediated by oxidative stress and (2) directly inhibiting the DNA-binding activity of transcription factors.

A. Antioxidant Actions Against Gene Expression by Oxidative Stress

As lipoic acid has been shown to exert potent reactive oxygen-scavenging actions (4), it is conceivable that the administration of lipoic acid would block the activation of transcription factors that are controlled by oxidative stress. Accord-

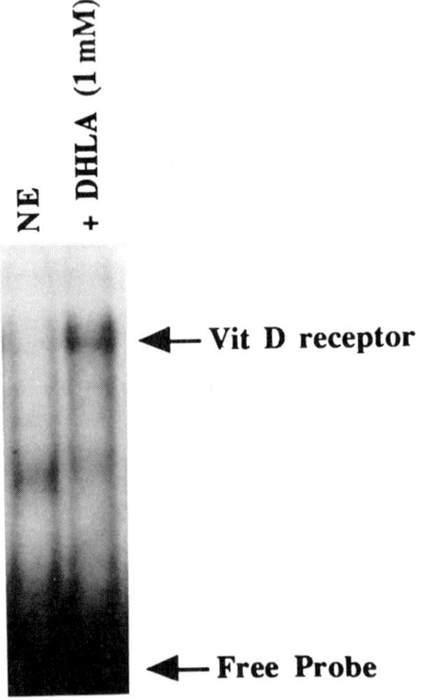

Figure 4 Restoration of the DNA-binding activity of vitamin D receptor by dihydrolipoic acid. Nuclear extract was prepared from Jurkat cells (1×10^6 cells/ml) in the absence of DTT in buffer C (23). Extracts were then incubated with DHLA in the binding reaction mixture for EMSA containing ^{32}P-labeled oligonucleotide probe for DRE (32).

ingly, we found that the administration of lipoic acid attenuates the activation of NF-κB (24) and the induction of c-*fos* protooncogene expression (25).

Activation of NF-κB transcription factor by various stimuli including tumor necrosis factor-α (TNFα) and phorbol 12-myristate 13-acetate (PMA) has been shown to be inhibited by antioxidants (26), suggesting that reactive oxygen species are used as second messengers in cell-signaling processes (27). Preincubation of Jurkat (human T) cells with α-lipoic acid inhibited the activation of NF-κB induced by TNFα or PMA as shown in Figure 5 (24). It has been hypothesized that the action of α-lipoic acid involves the intracellular formation of dihydrolipoic acid, which in turn scavenges reactive oxygen species, thus alleviating signal transduction (24). The intracellular reduction of α-lipoic acid to dihydrolipoic acid in Jurkat cells has been demonstrated using HPLC (28). Furthermore, the hypothesis is supported by the observations that the direct

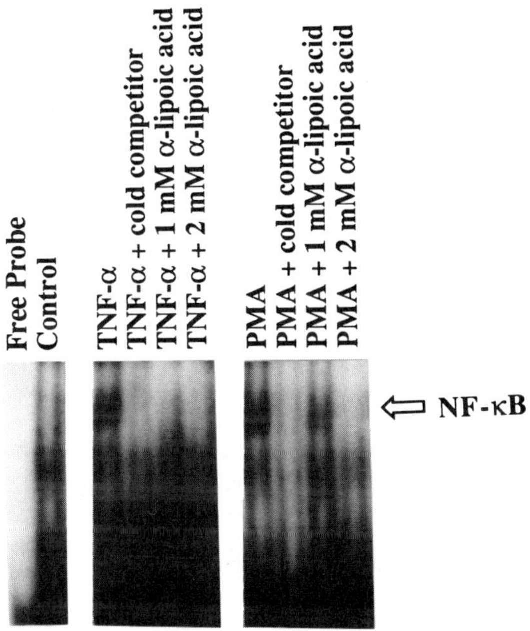

Figure 5 Effects of α-lipoic acid on NF-κB activation induced by TNFα or PMA. Jurkat cells (1×10^6 cells/ml) were incubated for 2 h with various concentrations of α-lipoic acid followed by incubation with TNFα (25 ng/ml) or PMA (50 ng/ml) for 4 h. Lane 1, free probes; lane 2, untreated control; lane 3, TNFα; lane 4, TNFα plus unlabeled competitor (1000-fold excess); lane 5, TNFα plus α-lipoic acid (1 mM); lane 6, TNFα plus α-lipoic acid (2 mM); lane 7, PMA; lane 8, PMA plus unlabeled competitor (1000-fold excess); lane 9, PMA plus α-lipoic acid (1 mM); lane 10, PMA plus α-lipoic acid (1 mM). (From Ref. 24.)

addition of dihydrolipoic acid to the culture medium of Jurkat cells inhibits NF-κB activation (29).

Cell signaling for the expression of protooncogene c-*fos*, which is a precursor for AP-1 transcription factor, has also been shown to be regulated by reactive oxygen species (30). A 5-day incubation of Jurkat cells with dihydrolipoic acid was found to attenuate PMA-induced expression of c-*fos* mRNA, as shown in Figure 6 (25). This suggests that the antioxidant action of dihydrolipoic acid modulates c-*fos* expression. α-Lipoic acid (oxidized form), however, in this experimental condition did not attenuate, but rather potentiated the c-*fos* expression induced by PMA. The long-term incubation thus appears to exhibit different

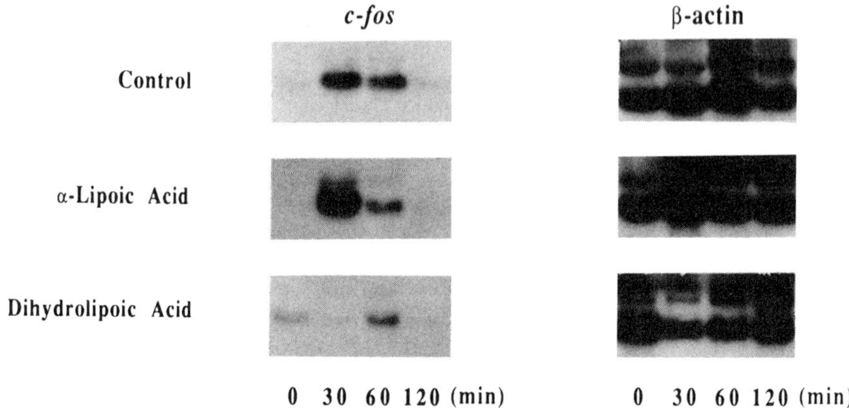

Figure 6 The effects of α-lipoic acid and dihydrolipoic acid on gene expression of c-*fos* mRNA. Jurkat cells (1×10^4) were pretreated with α-lipoic acid (0.2 mM) or dihydrolipoic acid (0.2 mM) for 5 days. Pretreated cells were incubated with PMA (0.5 μM), and total RNA was isolated after 30, 60, and 120-min treatment. RNA was then electrophoresed, blotted, and hybridized with ^{32}P-labeled oligonucleotide probe for c-*fos* or β-actin. (From Ref. 25.)

phenomena as the intracellular reduction of lipoic acid may be limited by the duration of incubation.

B. Inhibition of DNA–Transcription Factor Interactions

In addition to a likely role of α-lipoic acid, i.e., to exert reactive oxygen-scavenging actions in inhibiting NF-κB activation, we found that α-lipoic acid can inhibit the NF-κB DNA-binding activity in vitro. As shown in Figure 7, the addition of α-lipoic acid in the binding reaction mixtures for the electrophoretic mobility shift assays resulted in a concentration-dependent inhibition of DNA-binding activity of NF-κB in nuclear extracts isolated from Jurkat cells, in contrast to the potentiating effect of dihydrolipoic acid on DNA–transcription factor interactions, as discussed above. We further found that α-lipoic acid inhibits whereas dihydrolipoic acid potentiates NF-κB activation when it is induced by okadaic acid (Fig. 8), an NF-κB activator that is insensitive to antioxidant inhibition (31). Dihydrolipoic acid thus inhibits NF-κB activation by antioxidant-sensitive activators such as TNFα by eliminating reactive oxygen species—a component of cell signaling. α-Lipoic acid does the same either di-

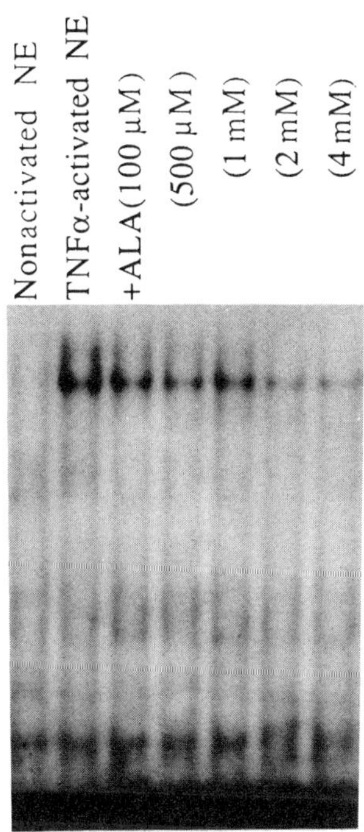

Figure 7 Inhibition of NF-κB binding by α-lipoic acid. Jurkat cells were activated with TNFα. Nuclear extracts were prepared (24) and then incubated with α-lipoic acid (ALA) in the binding reaction mixture for EMSA containing DTT (0.5 mM) and ^{32}P-labeled oligonucleotide probe for NF-κB.

rectly or by forming dihydrolipoic acid, and it also directly inhibits the DNA-binding activity of activated form of NF-κB.

IV. CONCLUDING REMARKS

This chapter has reviewed recent advances in understanding the role of lipoic acid in modulating gene expression. Although experimental evidence supporting the possible involvement of lipoic acid in physiological regulation of gene transcription is still in its infancy, its effectiveness in pharmacological modulation of gene expression is quite possible. Natural compounds that can influence the gene expression processes are attractive in managing various diseases by supporting the actions of more aggressive agents without risk of toxicity. As it is now clear that lipoic acid can modulate gene transcription by various inducers, it is of interest to identify genes that are directly induced by lipoic acid.

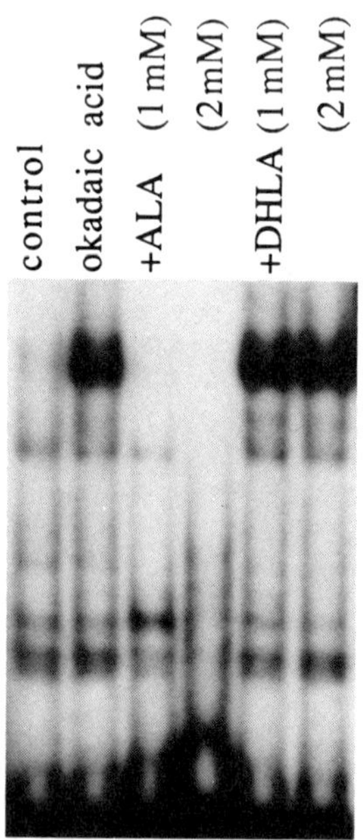

Figure 8 Effects of α-lipoic acid and dihydrolipoic acid on okadaic acid–induced NF-κB activation. Jurkat cells (1×10^6 cells/ml) were pretreated for 30 min with α-lipoic acid or dihydrolipoic acid and then incubated with okadaic acid (50 ng/ml) for 4 h. Nuclear extracts were prepared and then incubated in the binding reaction mixture for EMSA containing DTT (0.5 mM) and ^{32}P-labeled oligonucleotide probe for NF-κB. The control lane contained the nuclear extract from cells without okadaic acid stimulation.

ACKNOWLEDGMENTS

Supported by the National Institutes of Health (CA47597), this work was done during the tenure of a research fellowship from the American Heart Association, California affiliate, to YJS.

REFERENCES

1. Reed, L. J. (1974) Multienzyme complex. Acc. Chem. Res. 7:40–46.
2. Kikuchi, G., and Hiraga, K. (1982) The mitochondrial glycine cleavage system: Unique features of the glycine decarboxylation. Mol. Cell Biochem. 45:137–149.
3. Jocelyn, P. C. (1967) The standard redox potential of cysteine-cystine from the thiol-disulphide exchange reaction with glutathione and lipoic acid. Eur. J. Biochem. 2:327–331.
4. Suzuki, Y. J., Tsuchiya, M., and Packer, L. (1991) Thioctic acid and dihydrolipoic

acid are novel antioxidants which interact with reactive oxygen species. Free Rad. Res. Comms. 15:255–263.
5. Kagan, V. E., Shvedova, A., Serbinova, E., Khan, S., Swanson, C., Powell, R., and Packer, L. (1992) Dihydrolipoic acid—a universal antioxidant both in the membrane and in the aqueous phase. Biochem. Pharmacol. 44:1637–1649.
6. Ziegler, D. M. (1985) Role of reversible oxidation-reduction of enzyme thiols-disulfides in metabolic regulation. Annu. Rev. Biochem. 54:305–329.
7. Baeuerle, P. (1991) The inducible transcription activator NF-κB: Regulation by distinct protein subunits. Biochem. Biophys. Acta 1072:63–80.
8. Toledano, M. B., and Leonard, W. J. (1991) Modulation of transcription factor NF-κB binding activity by oxidation-reduction in vitro. Proc. Natl. Acad. Sci. USA 88:4328–4332.
9. Molitor, J. A., Ballard, D. W., and Greene, W. C. (1991) κB-Specific DNA binding proteins are differentially inhibited by enhancer mutations and biological oxidation. New Biol. 3:987–996.
10. Matthews, J. R., Wakasugi, N., Virelizier, J.-L., Yodoi, J., and Hay, R. T. (1992) Thioredoxin regulates the DNA binding activity of NF-κB by reduction of a disulphide bond involving cysteine 62. Nucleic Acid Res. 20:3821–3830.
11. Hayashi, T., Ueno, Y., and Okamoto, T. (1993) Oxidoreductive regulation of nuclear factor κB. Involvement of a cellular reducing catalyst thioredoxin. J. Biol. Chem. 268:11380–11388.
12. Toledano, M. B., Ghosh, D., Trinh, F., and Leonard, W. J. (1993) N-Terminal DNA-binding domains contribute to differential DNA-binding specificities of NF-κB p50 and p65. Mol. Cell Biol. 13:852–860.
13. Abate, C., Patel, L., Rauscher III, F. J., and Curran, T. (1990) Redox regulation of Fos and Jun DNA-binding activity in vitro. Science 249:1157–1161.
14. Xanthoudakis, S., and Curran, T. (1992) Identification and characterization of Ref-1, a nuclear protein that facilitates AP-1 DNA-binding activity. EMBO J. 11: 653–665.
15. Xanthoudakis, S., Miao, G., Wang, F., Pan, Y.-C. E., and Curran, T. (1992) Redox activation of Fos-Jun DNA binding activity is mediated by a DNA repair enzyme. EMBO J. 11:3323–3335.
16. Silva, C. M., and Cidlowski, J. A. (1989) Direct evidence for intra- and intermolecular disulfide bond formation in the human glucocorticoid receptor. Inhibition of DNA binding and identification of a new receptor-associated protein. J. Biol. Chem. 264:6638–6647.
17. Peleg, S., Schrader, W. T., and O'Malley, B. W. (1989) Differential sensitivity of chicken progesterone receptor forms to sulfhydryl reactive reagents. Biochemistry 28:7373–7379.
18. Grippo, J. F., Holmgren, A., and Pratt, W. B. (1985) Proof that the endogenous, heat-stable glucocorticoid receptor-activating factor is thioredoxin. J. Biol. Chem. 260:93–97.
19. Holmgren, A. (1989) Thioredoxin and glutaredoxin systems. J. Biol. Chem. 264:13963–13966.
20. Gleason, F. K., and Holmgren, A. (1988) Thioredoxin and related protein in procaryotes. FEMS Microbiol. Rev. 54:271–298.

21. Holmgren, A. (1979) Thioredoxin catalyzes the reduction of insulin disulfides by dithiothreitol and dihydrolipoamide. J. Biol. Chem. 254:9627–9632.
22. Spector, A., Huang, R.-R. C., Yan, G.-Z., and Wang, R.-R. (1988) Thioredoxin fragment 31-36 is reduced by dihydrolipoamide and reduces oxidized protein. Biochem. Biophys. Res. Commun. 150:156–162.
23. Suzuki, Y. J., and Packer, L. A. (1995) Redox regulation of DNA-protein interactions by biothiols. Meth. Enzymol. 252 (in press).
24. Suzuki, Y. J., Aggarwal, B. B., and Packer, L. (1992) α-Lipoic acid is a potent inhibitor of NF-κB activation in human T cells. Biochem. Biophys. Res. Commun. 189:1709–1715.
25. Mizuno, M., and Packer, L. (1994) Effects of α-lipoic acid and dihydrolipoic acid on expression of proto-oncogene c-fos. Biochem. Biophys. Res. Commun. 200:1136–1142.
26. Staal, F. J. T., Roederer, M., Herzenberg, L. A., and Herzenberg, L. A. (1990) Intracellular thiols regulate activation of nuclear factor κB and transcription of human immunodeficiency virus. Proc. Natl. Acad. Sci. USA 87:9943–9947.
27. Schreck, R., Rieber, P., and Baeuerle, P. A. (1991) Reactive oxygen intermediates as apparently widely used messengers in the activation of the NF-κB transcription factor and HIV-1. EMBO J. 10:2247–2258.
28. Handelman, G. J., Han, D., Tritschler, H., and Packer, L. (1994) α-Lipoic acid reduction by mammalian cells to the dithiol form, and release into the culture medium. Biochem. Pharmacol. 47:1725–1730.
29. Packer, L., and Suzuki, Y. J. (1994) Structural consequences of NF-κB inhibition by natural antioxidants: α-lipoic acid and vitamin E. In: Oxidative Stress, Cell Activation and Viral Infection (Pasquier, C., Olivier, R. Y., Auclair, C., and Packer, L., eds.), pp. 113–130. Birkhäuser Verlag, Basel.
30. Crawford, D., Zbinden, I., Amstad, P., and Cerutti, P. (1988) Oxidant stress induces the proto-oncogenes c-fos and c-myc in mouse epidermal cells. Oncogene 3:27–32.
31. Suzuki, Y. J., Mizuno, M., and Packer, L. Signal transduction for NF-κB activation. Proposed location of antioxidant-inhibitor step. J. Immunol. 153:5008–5015.
32. Sone, T., Kerner, S., and Pike, W. (1991) Vitamin D receptor interaction with specific DNA. Association as a 1,25-dihydroxyvitamin D_3-modulated heterodimer. J. Biol. Chem. 266:23296–23305.

21

Therapeutic Effects of α-Lipoic Acid on Diabetic Neuropathy

Dan Ziegler and F. Arnold Gries
Diabetes Research Institute, Heinrich Heine University, Düsseldorf, Germany

I. INTRODUCTION

Near-normoglycemia is generally accepted as the primary goal of causal treatment of diabetic neuropathy (5,9). However, sometimes relatively long periods of near-normal glycemic control (up to several years) are needed to beneficially influence nerve dysfunction in diabetic patients (15). Hence, additional long-term drug treatment is often necessary to maintain a patient's quality of life.

Potential forms of treatment that have emerged from the putative concepts on the pathogenesis of diabetic neuropathy include the reduction of increased flux through the polyol pathway using aldose reductase inhibitors (18,29), substitution of *myo*-inositol (10), inhibition of the accumulation of advanced glycation end products on nerve and vessel proteins by aminoguanidine (16), correction of depletion of neurotrophic factors by nerve growth factor substitution (11), elimination of endoneurial hypoperfusion resulting in hypoxia (8) by vasodilators (22), and correction of alterations in essential fatty acid metabolism by γ-linolenic acid (12). However, these specific therapeutic approaches are currently not available in clinical routine. Although symptomatic treatment using drugs, such as tricyclic antidepressants, is effective, it may be of limited value due to frequent adverse reactions (19).

Recents studies provide evidence that increased free radical formation (oxidative stress) may play an important role in the pathogenesis of diabetic neuropathy (17). Administration of physiological antioxidants such as glutathione and butylated hydroxytoluene resulted in partial or complete prevention of experimental diabetic neuropathy (2,4). The antioxidant α-lipoic acid (24) has been shown to increase the intracellular content of glutathione (3), to inhibit glycosylation and structural modification of albumin in vitro (25), to induce nerve fiber sprouting (7), and to prevent nerve-conduction slowing in diabetic rats (30) (M. Nagamatsu et al., Diabetes Care, 1995, in press). For these reasons, α-lipoic acid may be suitable for the treatment of diabetic neuropathy.

Previous clinical trials that studied the therapeutic value of α-lipoic acid in diabetic patients with neuropathy have had controversial results (6,13,23). In the present study we examined the effects of high-dose, initially parenteral treatment with α-lipoic acid as compared with vitamin B_1 on clinical and neurophysiological parameters of peripheral and autonomic nerve function in patients with symptomatic peripheral neuropathy.

II. STUDY DESIGN AND PATIENTS

The study was a single-blind, 1:1 randomized, parallel-group comparison. An initial 3-week in-patient parenteral phase of α-lipoic acid (Thioctacid™, ASTA Medica, Frankfurt am Main, Germany) infused i.v. in 0.9% saline using 2×300 mg/day and Vitamin B_1 (Benerva™, Hoffmann-LaRoche, Grenzach-Wyhlen, Germany) injected i.m. using 2×200 mg/day was followed by a 12-week out-patient oral phase using identical doses.

After approval of the Heinrich Heine University Ethical Committee and informed consent had been obtained, 23 patients who met the following inclusion criteria entered the study:

1. Insulin-treated diabetic patients or those receiving diet alone or additional oral hypoglycemic agents, classified according to the proposals of the National Diabetes Data Group (21)
2. Age between 18 and 70 years
3. Symptomatic symmetrical distal diabetic neuropathy in presence of abnormal nerve conduction velocity.

Patients with one of the following exclusion criteria were excluded: causes of neuropathy other than diabetes (e.g., chronic alcohol abuse), treatment with drugs potentially influencing peripheral or autonomic nerve function, peripheral vascular disease with intermittent claudication, proliferative retinopathy, serum creatinine > 1.5 mg/dl, $HbA_1 > 13\%$, ketotic derangement and/or blood glucose levels > 250 mg/dl during the week prior to the study, and pregnancy or unsafe anticonception.

The clinical data for the patients studied are shown in Table 1. There were no significant differences between patients assigned to α-lipoic acid and vitamin B_1 treatment with regard to the parameters listed.

Table 1 Clinical Data for Patients[a] at Randomization

	α-Lipoic acid	Vitamin B_1
Number	11	12
Male/Female	6/5	5/7
Age (years)	60.0 (27–70)	56.5 (39–70)
Duration of diabetes (years)	16.0 (7–37)	15.0 (1–20)
IDDM/NIDDM	3/8	3/9
$HbA_{1a\text{-}c}$ (%)	11.5 (8.2–13.0)	11.3 (8.3–13.0)
Mean blood glucose (mg/dl)	179 (130–297)	197 (77–267)

[a]Median (range) or number of patients.

III. METHODS

A. Peripheral Nerve Function

Nerve-conduction measurements were performed with a Schwarzer-Picker EMG 2000 electromyograph (Munich, Germany) using surface electrodes. During each investigation local skin temperature was controlled using a temperature probe (Digimed H11S, ttw, Waldkirch) and maintained at 33–34°C (28). All measurements were carried out in the right upper and lower limbs. Motor nerve conduction velocity (NCV) was determined in the median and peroneal nerves. Sensory NCV was measured orthodromically in the median nerve and antidromically in the sural nerve. Sensory potentials were averaged from at least 60 signals (28).

The three major symptoms of peripheral neuropathy (pain, paresthesias, and numbness) in the upper and lower extremities were assessed using 10-cm horizontal analog scales. In the six scales used in total, 0 cm signified that the particular symptom was not present, while 10 cm was considered as the maximal subjective symptom perception.

B. Autonomic Nerve Function

Cardiac autonomic function was evaluated using the Neurocard-Analyzer II (Argustron, Mettmann, Germany). The coefficient of R-R interval variation (CV) was calculated from 150 consecutive beats recorded after a 10-min rest in a supine position (20).

Pupillary function was assessed using a computer-based TV-Pupillometer System, Series 1000 (G+W Applied Science Laboratories, Waltham, MA) (28). During each session six pupil reflexes were registered on the monitor, four of which without noise were evaluated. The computer calculated and plotted the pupillary dilation velocity (PDV = ratio of the half-maximal redilation and the time interval between maximal miosis and half-maximal redilation) and the pupillary reflex latency (PRL). Pupillary function tests were measured on each eye, and the results were expressed as the mean of both values.

The age-related normal ranges and day-to-day reproducibility of the nerve function tests have been published (27,28). Neurophysiological parameters and subjective symptoms were assessed at baseline (week 0) as well as after 3 and 15 weeks.

C. Metabolic Parameters

Glycated hemoglobin (HbA_{1a-c}) was determined using HPLC (Diamat, Bio-Rad, Munich, Germany). The mean ± SD in healthy subjects is 6.4 ± 0.6%. Capillary blood glucose was measured using the hexokinase method (ACP 5040 autoanalyzer, Eppendorf, Hamburg, Germany). Mean blood glucose was calculated as the mean of five values (fasting, postprandial, before lunch, before supper, and at bedtime).

D. Statistical Analysis

Data showing normal distribution were expressed by the arithmetical mean ± SEM. Descriptive variables were given as median (range). The nerve function indices were anayzed by adjusting the follow-up values to the baseline values (week 0). The resulting differences from baseline were analyzed between the groups using the *t*-test. The medians and absolute frequencies were tested using the Wilcoxon-Mann-Whitney U-test and chi-squared test, respectively. Significance levels were fixed uniformly at $\alpha = 0.05$.

IV. RESULTS

The metabolic parameters are shown in Table 2. No significant differences with regard to HbA_1 and mean blood glucose between the treatment groups was noted

Table 2 Glycated Hemoglobin (HbA_{1a-c}) and Mean Blood Glucose (MBG) Levels

	Week 0	Week 3	Week 15
HbA_{1a-c} (%)			
α-Lipoic acid	11.3 ± 0.5	10.0 ± 0.4	9.4 ± 0.6
Vitamin B_1	11.4 ± 0.4	10.3 ± 0.4	9.3 ± 0.3
MBG (mg/dl)			
α-Lipoic acid	187 ± 13	169 ± 19	156 ± 20
Vitamin B_1	182 ± 15	163 ± 10	185 ± 11

Mean ± SEM.

at baseline and during the study. There was a significant decrease in HbA_1 in both groups after 3 and 15 weeks as compared with baseline ($p < 0.05$), whereas mean blood glucose remained unchanged.

At baseline (week 0) no significant differences between the groups were observed regarding the intensity of pain, paresthesias, and numbness as assessed using the horizontal analog scales. The degree of changes in intensity of pain and paresthesias on the analog scales for the lower limbs after 3 and 15 weeks as compared with baseline are shown in Figure 1. After 3 weeks of parenteral treatment with α-lipoic acid, a significant reduction of both pain and paresthesias was noted as compared with vitamin B_1 ($p < 0.05$). These effects were preserved following the 12-week phase of oral treatment ($p < 0.05$). In the upper limbs there was a significant decrease in paresthesias ($p < 0.05$), but not pain after 3 and 15 weeks in patients treated with α-lipoic acid compared with those given vitamin B_1 (Fig. 2). There were no significant differences between the groups during the study regarding numbness in the upper and lower limbs. The changes in the sum scores for pain, paresthesias, and numbness in the lower and upper limbs are illustrated in Figure 3. During both the parenteral and oral phases, a significant reduction in neuropathic symptom scores was seen in patients treated with α-lipoic acid as compared with those who received vitamin B_1 ($p < 0.05$). Significant differences regarding motor and sensory NCV as well as autonomic nerve function tests were observed neither between the groups

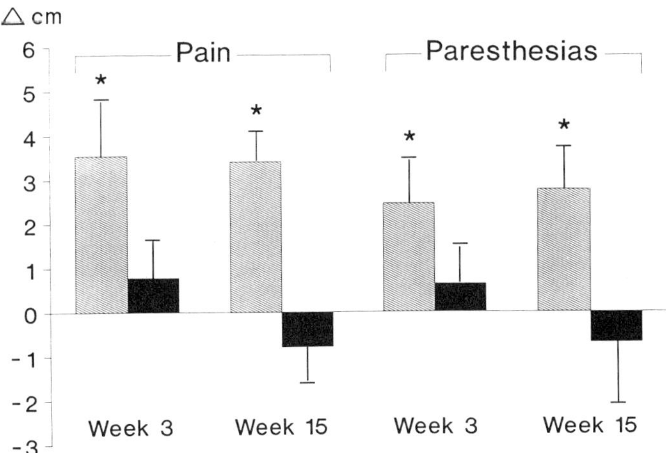

Figure 1 Changes in intensity of pain and paresthesias in the lower extremities after 3 and 15 weeks vs. baseline. Data shown are the absolute changes on the horizontal analog scales in cm (positive values correspond with improvement). □, α-Lipoic acid ($n = 11$); ■, vitamin B_1 ($n = 12$). *$p < 0.05$ vs. vitamin B_1.

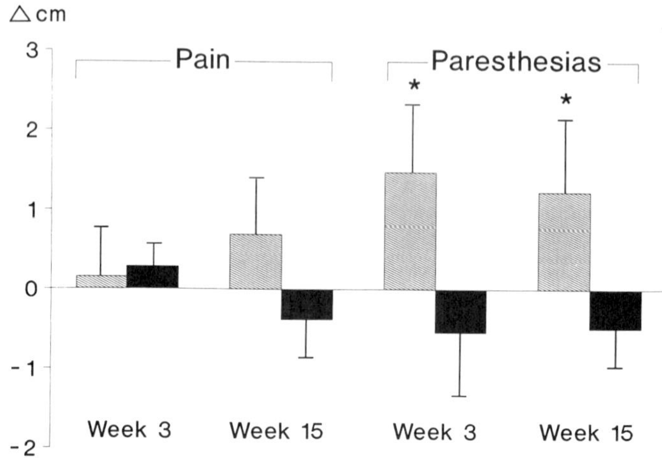

Figure 2 Changes in intensity of pain and paresthesias in the upper extremities after 3 and 15 weeks vs. baseline. Data shown are the absolute changes on the horizontal analog scales in cm (positive values correspond with improvement). □, α-Lipoic acid ($n = 11$); ■, vitamin B_1 ($n = 12$). *$p < 0.05$ vs. vitamin B_1.

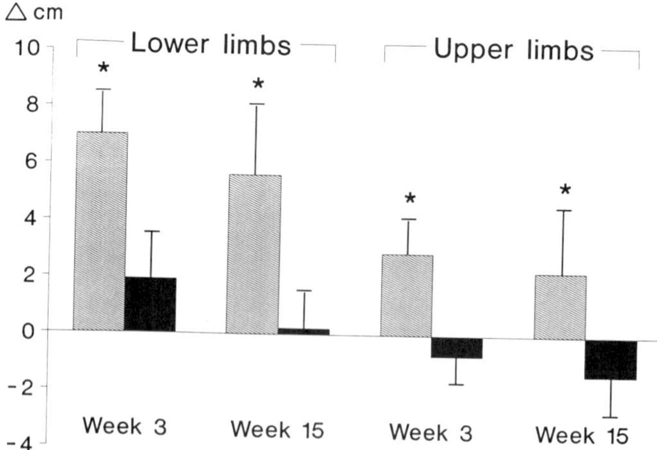

Figure 3 Changes in sum scores for pain, paresthesias, and numbness in the lower and upper extremities after 3 and 15 weeks vs. baseline. Data shown are the absolute changes on the horizontal analog scales in cm (positive values correspond with improvement). □, α-Lipoic acid ($n = 11$); ■, vitamin B_1 ($n = 12$). *$p < 0.05$ vs. vitamin B_1.

Table 3 Nerve Conduction Velocity (NCV), Coefficient of R-R Interval Variation (CV) at Rest and Pupillary Dilation Velocity (PDV)

	Week 0	Week 3	Week 15
Motor NCV (m/s)			
Median MNCV			
α-Lipoic acid	46.4 ± 0.5	46.4 ± 1.0	45.9 ± 0.9
Vitamin B_1	44.5 ± 1.4	45.3 ± 1.7	43.9 ± 1.4
Peroneal MNCV			
α-Lipoic acid	38.9 ± 1.2	38.3 ± 1.4	36.8 ± 1.2
Vitamin B_1	39.7 ± 2.0	40.3 ± 1.8	36.6 ± 1.8
Sensory NCV (m/s)			
Median SNCV			
α-Lipoic acid	46.0 ± 2.5	46.9 ± 2.7	47.4 ± 2.1
Vitamin B_1	41.2 ± 2.7	42.9 ± 3.1	41.3 ± 3.1
Sural SNCV			
α-Lipoic acid	35.2 ± 2.4	34.8 ± 2.4	33.1 + 2.2
Vitamin B_1	37.9 ± 1.7	37.5 ± 1.3	37.0 ± 1.7
Autonomic nerve function			
VK (%)			
α-Lipoic acid	4.10 ± 1.45	4.29 ± 1.52	4.94 ± 1.60
Vitamin B_1	2.36 ± 0.28	2.85 ± 0.35	3.05 ± 0.52
PDV (mm/s)			
α-Lipoic acid	0.61 ± 0.06	0.66 ± 0.07	0.73 ± 0.09
Vitamin B_1	0.83 ± 0.08	0.90 ± 0.08	0.91 ± 0.08

Mean ± SEM.

nor in comparisons to baseline (Table 3). During the treatment period there were no adverse reactions related to the drugs.

V. DISCUSSION

In this study we demonstrated a significant reduction in painful and paresthetic symptoms associated with diabetic neuropathy during a 3-week high-dose parenteral treatment period using α-lipoic acid as compared with vitamin B_1 administration. This effect was preserved after a further 12-week oral phase, whereas the quantitative indices of peripheral and autonomic nerve function remained unchanged within the 15-week period.

Previous studies did not provide convincing data regarding the efficacy of α-lipoic acid in diabetic peripheral neuropathy. In a double-blind controlled study over 12 weeks, Jörg et al. (13) treated 35 Type 1 diabetic patients with clinical or electrophysiological signs of neuropathy with α-lipoic acid (600 mg/day p.o.) or a vitamin B combination (600 mg/day p.o.), respectively. No significant therapeutic effects were demonstrated for any of the parameters measured (sensory NCV, nerve action potentials to single and double stimuli, relative refractory period, and nerve ischemia resistance). However, neuropathic symptoms were not assessed in their study. Delcker et al. (6) used a design similar to the one used in the present study. They administered 400 mg α-lipoic acid (n = 14) or 130 mg of a vitamin B combination (n = 11) over 1 month i.v., followed by an oral treatment phase for 6 months. In that study, likewise, no changes in NCV were noted, and the neuropathic signs also remained unaffected. The evaluation of the analog scales showed a reduction in pain intensity after 7 months as compared with baseline, but no differences between the groups were observed. In a 3-week double-blind study Sachse and Willms (23) used a relatively low oral dose of 300 mg/day in 10 diabetic patients. They could not detect any differences between drug and placebo treatment for subjective symptoms, NCV, and vibration perception threshold.

In line with these studies, we did not demonstrate any effect of α-lipoic acid on motor and sensory NCV within 15 weeks of treatment. Likewise, the indices of autonomic nerve function were unchanged. However, since persistent neurophysiological changes and autonomic dysfunction are to be regarded as the consequence of a chronic process, it cannot be readily expected that a relatively short-term treatment with α-lipoic acid over weeks or months would result in substantial improvement of advanced nerve function deficits. In fact, studies addressing the question as to whether diabetic neuropathy may be improved or its progression delayed following the initiation of long-term (near-) normoglycemia have shown that in some cases several years are needed to achieve a beneficial effect on neurophysiological parameters (15,26). Thus, in Type 1 diabetic patients who were normoglycemic following pancreas transplantation, a significant effect on peripheral neuropathy was observed only after 42 months (15). Long-term improvement of cardiac autonomic dysfunction may require even longer periods of (near-)normoglycemia (15,26).

In the present study both groups showed a significant decrease in HbA_1 after 3 and 15 weeks as compared with the baseline values. This effect is due to the optimized glycemic control during the initial 3-week in-patient phase. As, however, there were at no time any differences between the groups regarding HbA_1 and mean blood glucose, the effects observed cannot be explained by an influence of glycemic control.

Due to a potential risk of inducing anaphylactic shock following i.v. administration of vitamin B_1, the substance was initially given i.m. Hence, in view of the different application modes during the initial phase and the single-blind study design, a possible bias of the results cannot be ruled out. The persisting neuropathic symptoms at the end of the oral phase argue against any significant influence of the different means of administration in the initial phase.

The exact mechanisms by which α-lipoic acid acts regarding its effect on pain and paresthesias has yet to be established, but at least two potential hypotheses are worthy of discussion. First, α-lipoic acid may induce a dose-dependent sprouting of neurites in cultured neuroblastoma cells (7). Changes in membrane fluidity that are mediated by the sulfhydryl groups of the substance are thought to be responsible for this effect. In experimentally induced acrylamide neuropathy there is a marked reduction of the sprouting phenomenon, and distal neuropathy occurs due to the depletion of substances containing sulfhydryl groups, such as glutathione, in the axon (14). Spontaneous sprouting and integrity of the membrane at the nerve terminal can be maintained by administration of α-lipoic acid in vivo and in vitro. Moreover, in animal experiments the substance promotes regeneration after partial denervation (14). Such a protective effect of α-lipoic acid has also been demonstrated in hexacarbon-induced neuropathy (1). The second approach is based on the property of α-lipoic acid to act as a radical scavenger. Recent studies showed increased oxygen free radical activity in the sciatic nerve in diabetic rats (17), and experimental diabetic neuropathy can be prevented by administration of antioxidants (2,4). It is conceivable that the initial diabetes-related changes in the nerve are mediated by oxidative stress, which, on a long-term basis, could result in progressive neuronal damage and would therefore be of pathogenetic relevance. In light of this, a potential rationale is given for treating diabetic neuropathy using antioxidants such as α-lipoic acid. The value of this therapy must be verified by long-term studies, particularly addressing the question of a possible effect on objective neurophysiological parameters.

ACKNOWLEDGMENTS

We thank Ms. M. Behler and Ms. C. Gottschalk for excellent technical assistance and Prof. Dr. H. Reinauer and his team for the clinical chemistry measurements. This study was supported by ASTA Medica, Frankfurt am Main, Germany.

REFERENCES

1. Altenkirch, H., Stoltenburg-Didinger, G., Wagner, H. M., Herrmann, J., and Walter, G. (1990) Effects of lipoic acid in hexacarbon-induced neuropathy. Neurotoxicol. Teratol. 12:619–622.
2. Bravenboer, B., Kappelle, A. C., Hamers, F. P. T., van Buren, T., Erkelens, D. W., and Gispen, W. H. (1992) Potential use of glutathione for the prevention and treatment of diabetic neuropathy in the streptozotocin-induced diabetic rat. Diabetologia 35:813–817.
3. Busse, E., Zimmer, G., Schopohl, B., Kornhuber, B. (1992) Influence of α-lipoic acid on intracellular glutathione in vitro and in vivo. Drug Res. 42:829–831.

4. Cameron, N. E., Cotter, M. A., Maxfield, E. K. (1993) Anti-oxidant treatment prevents the development of peripheral nerve dysfunction in streptozotocin-diabetic rats. Diabetologia 36:299–304.
5. Committee on Health Care Issues, American Neurological Association (1986) Does improved control of glycemia prevent or ameliorate diabetic polyneuropathy? Ann. Neurol. 19:288–290.
6. Delcker, A., Fischer, P.-A., Ulrich, H. (1989) Randomisierte Studie Thioctsäure gegenüber Vitamin-B-Kombinationspräparat bei Patienten mit diabetischer Polyneuropathie unter besonderer Berücksichtigung des peripheren Nervensystems. In: Neue Biochemische, Pharmakologische und Klinische Erkenntnisse zur Thioctsäure (Borbe, H. O., and Ulrich, H., eds.), pp. 335–344. pmi Verlag, Frankfurt.
7. Dimpfel, W., Spüler, M., Pierau, F.-K., and Ulrich, H. (1990) Thioctic acid induces dose-dependent sprouting of neurites in cultured rat neuroblastoma cells. Dev. Pharmacol. Ther. 14:193–199.
8. Dyck, P. J. (1989) Hypoxic neuropathy: Does hypoxia play a role in diabetic neuropathy? Neurology 39:111–118.
9. Greene, D. A., Sima, A. A. F., Albers, J. W., and Pfeifer, M. A. (1990) Diabetic neuropathy. In: Diabetes Mellitus (Rifkin, H., and Porte, D. eds.), pp. 710–755. Elsevier, New York.
10. Gregersen, G. (1987) Myo-inositol supplementation. In: Diabetic Neuropathy (Dyck, P. J., Thomas, P. K., Asbury, A. K., Winegrad, A. I., and Porte, D., eds.), pp. 188–189. Saunders, Philadelphia.
11. Hellweg, R., and Hartung, H. D. (1990) Endogenous levels of nerve growth factor are altered in experimental diabetes mellitus: A possible role for NGF in the pathogenesis of diabetic neuropathy. J. Neurosci. Res. 26:258–267.
12. Jamal, G. A. (1990) Pathogenesis of diabetic neuropathy: The role of the n-6 essential fatty acids and their eicosanoid derivatives. A new hypothesis. Diabetic Med. 7:574–579.
13. Jörg, J., Metz, F., Scharafinski, H. (1988) Zur medikamentösen Behandlung der diabetischen Polyneuropathie mit der Alpha-Liponsäure oder Vitamin B-Präparaten. Nervenarzt 59:36–44.
14. Kemplay, S., Martin, P., and Wilson, S. (1988) The effects of thioctic acid on motor nerve terminals in acrylamide-poisoned rats. Neuropathol. Appl. Neurobiol. 14:275–288.
15. Kennedy, W. R., Navarro, X., Goetz, F. C., Sutherland, D. E. R., and Najarian, J. S. (1990) Effects of pancreatic transplantation on diabetic neuropathy. N. Engl. J. Med. 322:1031–1037.
16. Kihara, M., Schmelzer, J. D., Poduslo, J. F., Curran, G. L., Nickander, K. K., and Low, P. A. (1991) Aminoguanidine effects on nerve blood flow, vascular permeability, electrophysiology, and oxygen free radicals. Proc. Natl. Acad. Sci. USA 88:6107–6111.
17. Low, P. A., and Nickander, K. K. (1991) Oxygen free radical effects in sciatic nerve in experimental diabetes. Diabetes 40:873–877.
18. Masson, E. A., and Boulton, A. J. M. (1990) Aldose reductase inhibitors in the treatment of diabetic neuropathy. A review of the rationale and clinical evidence. Drugs 39:190–202.

19. Max, M. B., Lynch, S. A., Muir, J., Shoaf, S. E., Smoller, B., and Dubner, R. (1992) Effects of desipramine, amitriptyline, and fluoxetine on pain in diabetic neuropathy. N Engl J Med 326, 1250-1256.
20. Morguet, A., and Springer, H. J. (1981) Microcomputer-based measurement of beat-to-beat intervals and analysis of heart rate variability. Med. Progr. Technol. 8:77–82.
21. National Diabetes Data Group (1979) Classification and diagnosis of diabetes mellitus and other categories of glucose intolerance. Diabetes 28:1039–1057.
22. Robertson, S., Cameron, N. E., and Cotter, M. A. (1992) The effect of the calcium antagonist nifedipine on peripheral nerve function in streptozotocin-diabetic rats. Diabetologia 35:1113–1117.
23. Sachse, G., and Willms, B. (1980) Efficacy of thioctic acid in the therapy of peripheral diabetic neuropathy. In: Aspects of Autonomic Neuropathy in Diabetes (Gries, F. A., Freund, H. J., Rabe, F., and Berger, H., eds.), pp. 105–108. Horm. Metab. Res., Suppl. Series 9, Thieme, Stuttgart, New York.
24. Suzuki, Y. J., Tsuchiya, M., and Packer, L. (1991) Thioctic acid and dihydrolipoic acid are novel antioxidants which interact with reactive oxygen species. Free Rad. Res. Comms. 15:255–263.
25. Suzuki, Y. J., Tsuchiya, M., and Packer, L. (1992) Lipoate prevents glucose-induced protein modifications. Free Rad. Res. Comms. 17:211–217.
26. Ziegler, D., Dannehl, K., Wiefels, K., and Gries, F. A. (1992) Differential effects of near-normoglycaemia for 4 years on somatic nerve dysfunction and heart rate variation in Type 1 diabetic patients. Diabetic Med. 9:622–629.
27. Ziegler, D., and Gries, F. A. (1990) Reproducibility of electrophysiologic tests. Diabete Metab. 16:373–376.
28. Ziegler, D., Mayer, P., Mühlen, H., and Gries, F. A. (1991) The natural history of somatosensory and autonomic nerve dysfunction in relation to glycaemic control during the first 5 years after diagnosis of Type 1 (insulin-dependent) diabetes mellitus. Diabetologia 34:822–829.
29. Ziegler, D., Mayer, P., Rathmann, W., and Gries, F. A. (1991) One-year treatment with the aldose reductase inhibitor, ponalrestat, in diabetic neuropathy. Diabetes Res. Clin. Pract. 14:63–74.
30. M. Nagamatsu et al. Diabetes Care. (In press.)

22

Antioxidant Properties and Clinical Implications of α-Lipoic Acid

Lester Packer and Eric H. Witt
University of California, Berkeley, California

Hans Jürgen Tritschler, Klaus Wessel, and Heinz Ulrich
ASTA Medica AG, Frankfurt, Germany

I. INTRODUCTION

α-Lipoic acid (Fig. 1) was first isolated by Reed and coworkers as an acetate-replacing factor (1,2). As lipoamide, it is a cofactor in the multienzyme complexes that catalyze the oxidative decarboxylation of α-keto acids such as pyruvate, α-ketoglutarate, and branched-chain α-keto acids (3). Although it was tentatively classified as a vitamin after its isolation (1,4), it was later found to be synthesized by animals and humans (5). The complete enzyme pathway that is responsible for its de novo synthesis has not yet been elucidated. Octanoate appears to be the immediate precursor for the 8-carbon fatty acid chain, and cysteine is the source of sulfur (6).

It is becoming increasingly clear that, in addition to its metabolic role, α-lipoic acid and its reduced form, dihydrolipoic acid (DHLA) (Fig. 1), are also powerful antioxidants. This paper will review the antioxidant properties of these compounds and the preventive and therapeutic implications and applications that arise from these antioxidant properties.

Many criteria must be considered when evaluating the antioxidant potential of a compound. Some of these concern chemical and biochemical aspects, such as specificity of free radical quenching, metal-chelating activity, interaction with other antioxidants, and effects on gene expression. Other criteria that are important when considering preventive or therapeutic applications include absorption and bioavailability, concentration in tissues and cells, and extracellular fluid location.

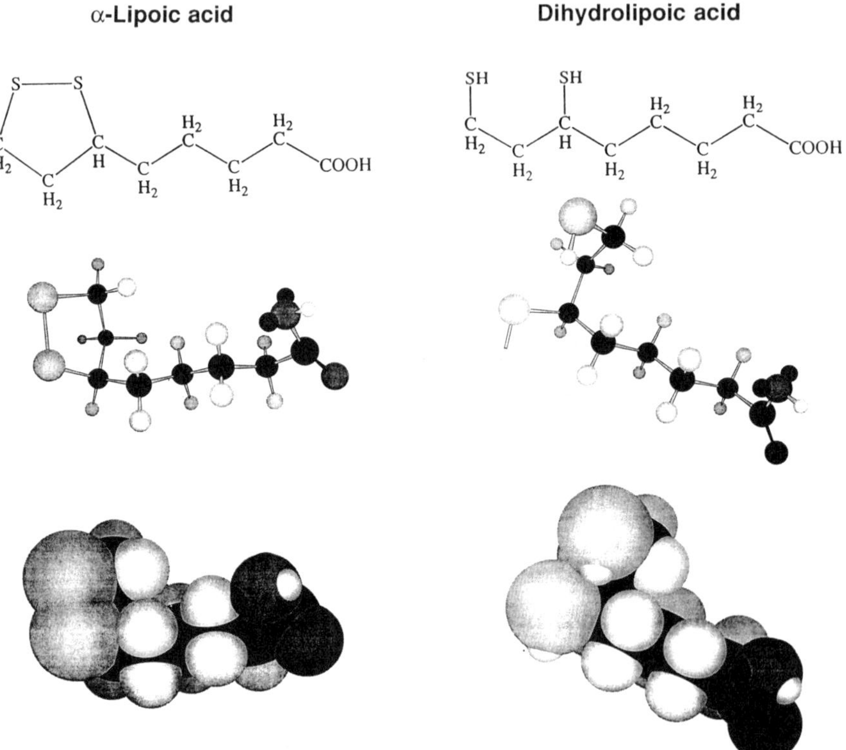

Figure 1 The structures of α-lipoic acid and dihydrolipoic acid, shown as chemical structures (top), ball-and-stick models (middle), and space-filling models (bottom). The naturally occurring R-enantiomers are shown. Synthetic α-lipoic acid is a racemic mixture of the R- and the S-enantiomer.

It is not necessary to meet all of these criteria in order to be a potent antioxidant. For example, vitamin E acts only in the membrane or lipid domains, its dominant action is to quench lipid peroxyl radicals, it has little or no activity against radicals in the aqueous phase, yet it is considered one of the central antioxidants of the body. Epidemiological studies are confirming its role in the prevention of numerous oxidant-related diseases, such as heart disease (7,8).

The α-lipoic acid/dihydrolipoic acid redox pair fulfills a number of the above criteria and has been called a "universal antioxidant" (9). α-Lipoic acid is readily absorbed from the diet. It is probably rapidly converted to DHLA in many tissues, as recent advances in assay technique have made evident (10,11). One or both of the components of the redox pair effectively quench a number of free radicals in both lipid and aqueous domains. Both DHLA and α-lipoic acid have metal-chelating activity. DHLA acts synergistically with other antioxidants, indicating that it is capable of regenerating other antioxidants from their radical or inactive forms. Finally, there is evidence that DHLA and α-lipoic acid may have effects on regulatory proteins and on genes involved in normal growth and metabolism.

Because of these antioxidant attributes, a number of experimental and clinical studies have been carried out that show α-lipoic acid to be useful or potentially useful as a therapeutic agent in such conditions as diabetes, ischemia-reperfusion injury, heavy metal poisoning, radiation damage, neurodegeneration, and HIV infection.

II. ANTIOXIDANT ACTIONS

An antioxidant function for α-lipoic acid was suggested as early as 1959 by Rosenberg and Culik (4), who observed that administration of α-lipoic acid prevented scurvy symptoms in vitamin C–deficient guinea pigs as well as preventing symptoms of vitamin E deficiency in rats fed a diet lacking α-tocopherol. It is only recently, however, that the specific effects of α-lipoic acid and DHLA in free radical quenching, metal chelation, antioxidant recycling, and gene expression have been investigated.

A. ANTIOXIDANT ACTIONS OF LIPOIC ACID

There is general agreement about the antioxidant properties of α-lipoic acid (Table 1). It scavenges hydroxyl radicals (12) with a rate constant of 4.7×10^{10} $M^{-1}s^{-1}$ (13); this is an essentially diffusion-limited reaction rate. It reacts efficiently with hypochlorous acid; 50 μM almost completely abolished the inactivation of α_1-antiproteinase by 50 μM HOCl (13,14). α-Lipoic acid has also been reported to scavenge singlet oxygen in at least four different systems (15–19). Although it reacts with hydrogen peroxide under nonphysiological conditions (30% H_2O_2 in acetone) (20), it does not appear to scavenge hydrogen peroxide under more physiological conditions (13). Two separate studies have also failed to show that α-lipoic acid can scavenge superoxide radical (12,13).

The situation regarding the ability of α-lipoic acid to scavenge peroxyl radicals is not as clear-cut. When lipid- or water-soluble peroxyl radical generators were used to generate peroxyl radicals, no reaction with α-lipoic acid was detected (9), whereas peroxyl radical ($CCl_3O_2\cdot$) generated by radiolysis of CCl_4 and propan-2-ol reacted with α-lipoic acid with a rate constant of 1.8×10^8 $M^{-1}s^{-1}$ (13). Although the explanation for the disagreement may lie in the different methodologies of the two groups, further work is necessary to clarify this question.

Table 1 Antioxidant Effects of α-Lipoic Acid

Oxidant	Scavenged by α-lipoate?	Test system	Ref.
Superoxide radical	No	O_2^- generated by xanthine–xanthine oxidase and detected by ESR using spin traps.	12
	No	O_2^- generated by hypoxanthine–xanthine oxidase and detected by O_2^--dependent cytochrome *c* reduction.	13
Hydrogen peroxide	No	H_2O_2 added directly and detected by peroxidase-based assay system	13
Hydroxyl radical	Yes	$OH^{\bullet}$ generated by H_2O_2 + $FeSO_4$ and detected by ESR using spin traps or by chemiluminescence using luminol	12
	Yes	$OH^{\bullet}$ generated by H_2O_2 + $FeCl_3$ + ascorbate and detected by deoxyribose degradation; rate constant for reaction = 4.71×10^{10} $M^{-1}s^{-1}$	13
Hypochlorous acid	Yes	HOCl added directly and detected by effect on α_1-antiproteinase activity	14
	Yes	Same	13
Peroxyl radical	No	Peroxyl radicals generated in aqueous phase by thermal decomposition of 2,2′-azobis(2-amidinopropane)-dihydrochloride and detected by fluorescence quenching of phycoerythrin; peroxyl radicals generated in lipids (liposomes or microsomes) by thermal decomposition of 2,2′-azobis (2,4 dimethylvaleronitrile)	9
	Yes	$CCl_3O_2\cdot$ generated by linear accelerator and detected spectrophotometrically; rate constant of reaction 1.8×10^8 $M^{-1}s^{-1}$	13
Singlet oxygen	Yes	$^1O_2$ generated by rubrene autoperoxidation in air-saturated benzene when stimulated at 546 nm and detected by	15

		following disappearance of rubrene spectrophotometrically; rate constant of the reaction = 1×10^8 $M^{-1}s^{-1}$; DHLA not tested	
	Yes	$^1O_2$ generated by photosensitized oxidation of methylene in chloroform or methanol; reaction detected by analyzing oxidation products of the methyl ester of LA; DHLA not tested.	16
	Yes	$^1O_2$ generated by thermolysis of endoperoxide and detected by infrared chemiluminescence; rate constant of the reaction = 1.38×10^8 $M^{-1}s^{-1}$	17
	Yes	$^1O_2$ generated by thermolysis of endoperoxide and detected by single-strand DNA breaks	18,19
Transition metals	Chelator	α-Lipoic acid formed stable complexes with Cu^{2+}, Mn^{2+}, and Zn^{2+} in aqueous solution; bisnorlipoate and tetranorlipoate formed more stable complexes.	23
	Chelator	Same as system to test $^1O_2$; when EDTA was added, the protective effect of LA decreased, suggesting that it was at least partially due to metal chelation	19
	Possible chelator	LA decreased Cd^{2+} toxicity in hepatocytes, but much less effectively than DHLA; the authors speculate that LA is taken up and converted to DHLA, which is the compound exerting the protective effect	21
	No chelation	In rat liver microsomes $+FeSO_4$ + ascorbate, LA had no effect on accumulation of TBARS	22
	Chelator	LA decreased site-specific iron-induced degradation of deoxyribose,suggesting that it chelated the iron	13
	Chelator	Decreased Cu^{2+} -induced oxidation of ascorbate, increased partitioning of Cu^{2+} into octanol, inhibited Cu^{2+} -induced lipid peroxidation.	24

α-Lipoic acid may also exert an antioxidant effect in biological systems through transition metal chelation. It reduces Cd^{2+}-induced toxicity in isolated hepatocytes (21) and may chelate iron (13,19), although in a rat liver microsomal system in which lipid peroxidation was induced by $FeSo_4$ (22), α-lipoic acid did not inhibit peroxidation. α-Lipoate forms stable complexes with Mn^{2+}, Cu^{2+}, and Zn^{2+}, with the complex being almost entirely with the carboxylate group (23). The effects of α-lipoic acid on copper have recently been extended to oxidation situations. α-Lipoic acid was effective in preventing Cu^{2+}-catalyzed ascorbic acid oxidation, increased the partitioning of Cu^{2+} into *n*-octanol from aqueous solution, and inhibited Cu^{2+}-catalyzed liposomal peroxidation (24). These observations indicate that lipoic acid is a copper chelator.

B. Antioxidant Actions of Dihydrolipoic Acid

With a redox potential of –0.32 V (25) for the DHLA/α-lipoic acid redox pair, dihydrolipoic acid is a potent reductant [for comparison, the redox potential of the GHS/GSSG pair is –0.24 V (26)]. DHLA will reduce GSSG to GSH, but GSH is incapable of reducing α-lipoic acid to DHLA (27).

Like α-lipoic acid, DHLA is a potent antioxidant, although there is uncertainty as to its effects (Table 2). DHLA scavenges hypochlorous acid (14) and peroxyl radicals (9,13,28). DHLA has been shown to have both antioxidant (12) and prooxidant (13) effects in systems in which hydroxyl radical was generated; however, the prooxidant effects, if real, are probably due to DHLA's effects on iron (see below). It does not appear to react with hydrogen peroxide (13,29) or with singlet oxygen (17–19).

However, there is disagreement among studies as to whether DHLA is also capable of scavenging superoxide radical and whether it acts as an antioxidant or a prooxidant in its interactions with iron ion. In a study in which O_2^{-} was generated by xanthine–xanthine oxidase and detected by ESR using DMPO spin trap (12), DHLA was found to react with superoxide radical with a rate constant of 3.3×10^5 $M^{-1}s^{-1}$ (Fig. 2). These results were confirmed by this group in a later study (28) using the same system for generating superoxide radical and competition with epinephrine oxidation to assess DHLA's scavenging ability. In this study a rate constant of 7.3×10^5 $M^{-1}s^{-1}$ was found, in good agreement with the previous work. In contrast to these studies, another group, using a similar superoxide-generating system (hypoxanthine–xanthine oxidase) and detecting superoxide by O_2^{-}-dependent nitro-blue tetrazolium reduction, found no scavenging effect for DHLA (13). It is difficult to explain the discrepancy in the results of the two groups, and further work is clearly indicated.

Perhaps the most crucial question surrounding DHLA is whether it acts as a prooxidant under certain circumstances, either as a reductant for transition metals (especially iron) or by regenerating ascorbate, which is known to reduce

Table 2 Antioxidant Effects of Dihydrolipoic Acid

Oxidation	Scavenged by DHLA?	Test system	Ref.
Superoxide radical	Yes	O_2^- generated by xanthine–xanthine oxidase and detected by ESR using spin traps; sulfhydryl content of DHLA decreased during reaction and $[H_2O_2]$ increased; rate constant of reaction = $3.3 \times 10^5\ M^{-1}\ s^{-1}$	12
	Yes	$O_2^{\cdot -}$-generated by xanthine–xanthine oxidase and detected by epinephrine oxidation: rate constant of reaction = $7.3 \times 10^5\ M^{-1}\ s^{-1}$	28
	No	O_2^- generated by hypoxanthine–xanthine oxidase and detected by O_2^- dependent nitro-blue tetrazolium reduction	13
Hydrogen peroxide	No	H_2O_2 added directly and detected by iron-thiocyanate	29
	No	H_2O_2 added directly; reaction monitored by measuring thiol content of DHLA	13
Hydroxyl radical	Yes	$OH^{\bullet}$ generated by H_2O_2 + $FeSO_4$ and detected by ESR using spin traps	
	No (promoted oxidation)	$OH^{\bullet}$ generated by H_2O_2 + $FeCl_3$ + ascorbate and detected by deoxyribose degradation; prooxidant effect postulated due to reduction of ferrous ion and/or regeneration of ascorbate by DHLA	13
Hypochlorous acid	Yes	HOCl added directly and detected by effect on α_1-antiproteinase activity	14
	Yes	Same	13
Peroxyl radical	Yes	Peroxyl radicals generated in aqueous phase by thermal decomposition of 2,2′-azobis(2-amidinopropane)-dihydrochloride and detected by fluorescence quenching of phycoerythrin; peroxyl radicals generated in lipids (liposomes or microsomes) by thermal decomposition of 2,2′-azobis (2,4 dimethylvaleronitrile); stoichiometry: 1.5 mol peroxyl radicals quenched per mol DHLA	9

(*continued*)

Table 2 Continued

Oxidation	Scavenged by DHLA?	Test system	Ref.
	Yes	Peroxyl radicals generated in liposomes by AMVN and detected by fluorescence decay of parinaric acid	28
	Yes	$CCl_3O_2\cdot$ generated by linear accelerator and detected spectrophotometrically; rate constant of reaction $2.7 \times 10^7\ M^{-1}\ s^{-1}$	13
Singlet oxygen	No	$^1O_2$ generated by thermolysis of endoperoxide and detectd by infrared chemiluminescence	17
	Yes(?)	$^1O_2$ generated by thermolysis of endoperoxide and detectd by single-strand DNA breaks; the protective effect of DHLA may not be due to direct quenching of $^1O_2$	18,19
Transition metals	Chelation	DHLA decreased Cd^{2+} toxicity in isolated hepatocytes; also decreased TBARS in Cd^{2+} exposed hepatocytes, indicating that, in the absence of added iron, DHLA's activity is overall antioxidant against peroxidation	21
	Chelation	DHLA bound iron from ferritin in the ferrous and ferric states	32,33
	Prooxidant	In $FeCl_3$-EDTA + H_2O_2 system, accelerated deoxyribose degration: accelerated reaction also when no EDTA in system, suggesting that its iron-binding is not as effective as its iron-reduction; however, no prooxidant effect when DHLA was used alone in an Fe (III)-bleomycin-DNA system	13

Prooxidant	Rat liver microsomes + $FeSO_4$ ± ascorbate; DHLA accelerated peroxidation as measured by TBARS	22
No prooxidant effect	No electron transfer from DHLA to Fe^{3+} as measured by Fe^{2+}-phenanthroline complex; also, DHLA did not potentiate formation of OH. in Fe SO_4-H_2O_2 system, as measured by ESR (DMPO spin trap)	12
Some prooxidant effect	In lipid peroxidation in microsomes, induced by AMVN, in the absence of iron (chelated by deferoxamine), DHLA decreased peroxidation (as measured by TBARS) about 50%; in the presence of iron, DHLA did not decrease peroxidation (although there was no prooxidant effect; i.e., the antioxidant and prooxidant effects of DHLA balanced in this system); in this system Tetranor-DHLA greatly increased (5×) peroxidation in the presence of iron	28
Prooxidant	Micromolar DHLA, in the presence of Cu^{2+} ions, caused single-strand nicks in pSP64 plasmid DNA; other ions were tested in the same system and had no effect: Co^{2+}, Cr^{3+}, Fe^{3+}, Fe^{2+}, Ni^{2+}, Mn^{2+}, and Zn^{2+}; other thiols also found to cleave DNA in this system in the presence of Cu^{2+}	34

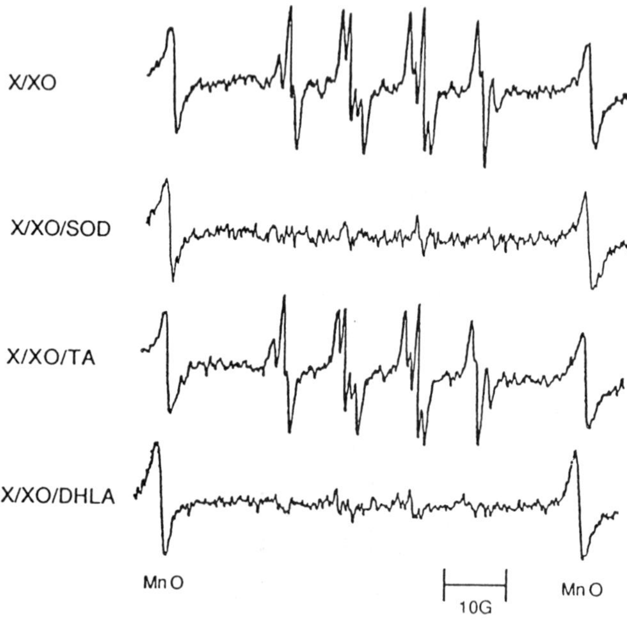

Figure 2 ESR study of the effects of α-lipoic acid and dihydrolipoic acid on superoxide anion radicals generated by xanthine plus xanthine oxidase. DMPO spin adducts generated by 5 mM xanthine (X) plus 100 μg xanthine oxidase (XO) in the presence of 40 μM DETAPAC in 150 mM KH_2PO_4-KOH (pH 7.4) in total volume of 1 ml. [DHPO] = 45 mM; [SOD] = 200U; [α-lipoic acid (TA)] = 5 mM; [dihydrolipoic acid (DHLA)] = 5 mM. ESR signals were recorded 3 min after the initiation of xanthine oxidase reaction.

iron. DHLA binds both Fe^{2+} and Fe^{3+} and may reduce bound Fe^{3+} and Fe^{2+} (30,31). DHLA also appears to be capable of removing iron from the storage protein ferritin in both the ferrous and ferric states (32,33), though it does not appear capable of removing iron from heme (13). DHLA was found to act as prooxidant in the peroxidation of rat liver microsomes (22) and in the production of hydroxyl radicals (13). In addition, DHLA, as well as a number of other thiols and dithiols, has also been shown to induce single-strand breaks in plasmid DNA in the presence of Cu^{2+}, but not in the presence of Fe^{2+} or Fe^{3+} (34). This is in contrast to the antioxidant Cu^{2+}-chelating effect of α-lipoic acid (24).

On the other hand, in another system in which iron was used to generate hydroxyl radical, the effect of DHLA was clearly antioxidant (12), and in this study no electron transfer from DHLA to Fe^{3+}, as measured by formation of Fe^{2+}-phenanthroline complex, was observed. In Cd^{2+} toxicity of isolated hepatocytes, addition of DHLA decreased TBARS (21), so at least in this system, the antioxidant effect of DHLA against lipid peroxidation was greater than any prooxidant effect.

Hence, the question of whether DHLA acts a prooxidant in biological systems is yet to be resolved. The interaction of DHLA with other antioxidants may negate a prooxidant effect in a physiological system. α-Lipoic acid itself may counteract any prooxidant effect of DHLA (13).

C. Influences on Other Antioxidants

DHLA appears to be able to regenerate other antioxidants, such as ascorbate and (indirectly) vitamin E, from their radical forms. Vitamin E is the major chain-breaking antioxidant, which protects membranes from lipid peroxidation (35). Vitamin E exists in biological membranes in a low molar ratio to unsaturated phospholipids, usually less than 0.1 nmol/mg of membrane protein, or, in other words, one molecule/1000–2000 membrane phospholipid molecules, which are the main target of oxidation in membranes. Lipid peroxyl radicals can be generated in membranes at the rate of 1–5 nmol/mg of membrane protein/min, yet destructive oxidation of membrane lipids does not normally occur, nor is vitamin E rapidly depleted. Furthermore, deficiency states for vitamin E are remarkably difficult to induce in adult animals. These apparent paradoxes can be explained by "vitamin E recycling," in which the antioxidant ability of vitamin E is continuously restored by other antioxidants. Those antioxidants that recycle vitamin E are vitamin C, ubiquinols, and thiols (Fig. 3) (36,37).

Evidence for vitamin E recycling by DHLA has come from a number of studies. DHLA protects against microsomal lipid peroxidation, but only in the presence of vitamin E (38); α-lipoic acid was not effective in either system. DHLA could have exerted its effect in this system by directly reducing tocopheroxyl radical or by reducing other antioxidants (such as ascorbate), which then regenerated vitamin E. There may be a weak direct interaction between DHLA and the tocopheroxyl radical; we have found that the presence of DHLA reduces the tocopheroxyl radical ESR signal in liposomes exposed to UV light (unpublished data). The use of UV light, which directly produces tocopheroxyl radicals, eliminates the possibility of iron chelation or other antioxidant effects of DHLA, and the use of liposomes eliminates the possibility of DHLA acting through other antioxidants. Therefore, in this system it seems likely that DHLA, which partitions mainly in the aqueous phase, is directly reducing tocopheroxyl radicals at the membrane/water interface. However, the effect is weak, and the major recycling of vitamin E by DHLA in biological systems probably occurs through other antioxidants. Current evidence points to the notion that DHLA can recycle vitamin E via glutathione (22,39), vitamin C (9,40,41), ubiquinol (41a,41b), NADPH, or NADH (9).

In fact, protection of other antioxidants by α-lipoic acid was suggested as early as 1959 by Rosenberg and Culik (4), who stated, with startling prescience,

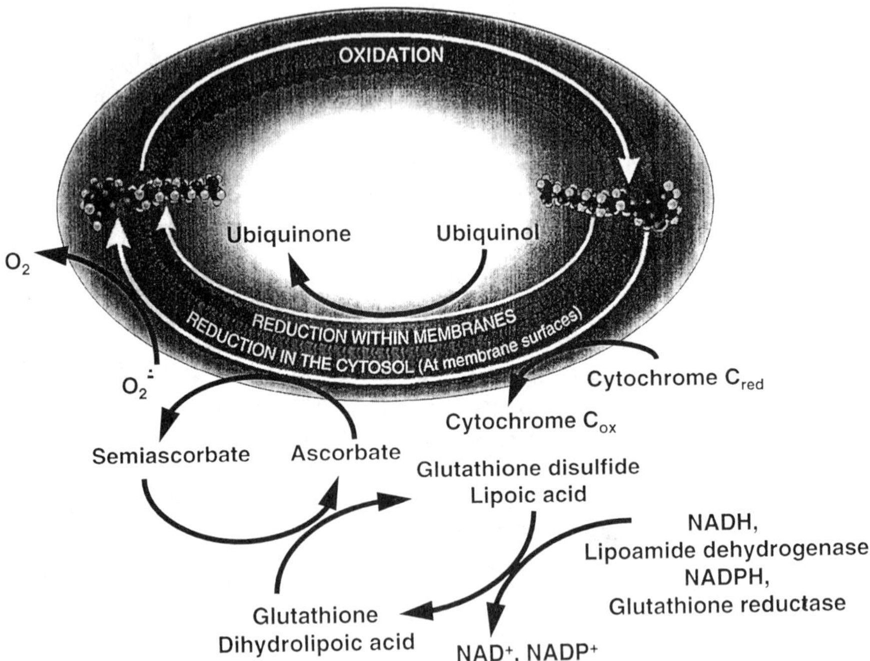

Figure 3 Vitamin E recycling. The tocopheroxyl radical, formed during oxidation of α-tocopherol (shown in the membrane), may be reduced back to α-tocopherol by a number of compounds, including ubiquinol, cytochrome *c*, and ascorbate. Ascorbate can be regenerated through reaction with thiols such as glutathione or lipoic acid. These can be returned to their reduced forms by various mechanisms, drawing on the reducing power of NADH or NADPH.

that "α-lipoic acid, and even more so its dihydro derivative into which it is converted rapidly after entering cellular metabolism, might act as an antioxidant for ascorbic acid and tocopherols." In the studies of Rosenberg and Culik, α-lipoic acid was found to prevent symptoms of both vitamin E and vitamin C deficiency. Recently, we have found similar protective effects of α-lipoic acid administration in tocopherol-deficient hairless mice (42) (Fig. 4). Such results are consistent with a recycling of tocopherol and/or ascorbate by α-lipoic acid, but could also be explained by the ability of α-lipoic acid to spare ascorbate and vitamin E through its separate but overlapping radical-scavenging effects.

α-Lipoic acid also causes an increase of 30–70% in intracellular glutathione in murine neuroblastoma and melanoma cell lines, as well as tissues of supplemented mice (43). These results have been confirmed in human Jurkat cell lines, in which intracellular concentrations of GSH increase approximately 50% 5 h

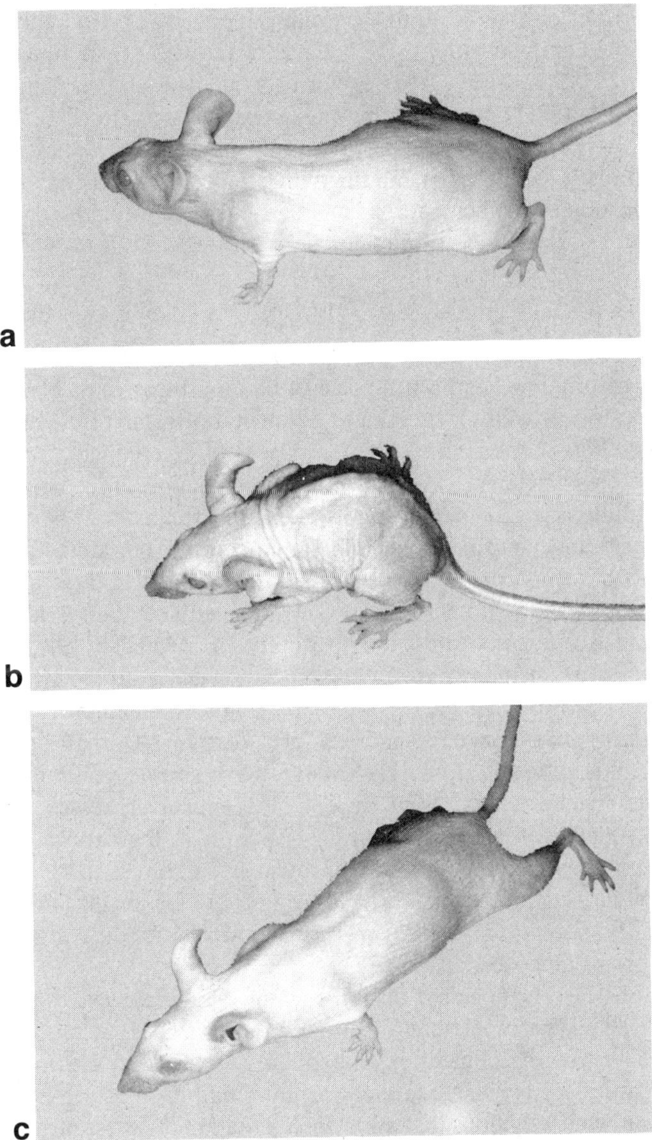

Figure 4 Adult 12-week-old hairless mice after 6 weeks of (a) normal control diet, (b) vitamin E–deficient diet, and (c) vitamin E–deficient, α-lipoate–supplemented diet. The animal on the vitamin E–deficient diet shows symptoms of vitamin E deficiency with muscular dystrophy and weight loss.

after addition of α-lipoic acid to the culture medium (D. Han and G. Handelman, personal communication). Such elevations in GSH cannot be explained by reduction of GSSG, since GSSG is normally present at less than 10% of the concentration of GSH (44). These intriguing observations are yet to be explained.

Thus, it appears that α-lipoic acid and DHLA act as antioxidants not only directly, through radical quenching and metal chelation, but indirectly as well, through recycling of other antioxidants and possibly through induction of increased intracellular levels of glutathione.

D. Redox Effects on Proteins and Influence on Protein Folding

Thiolation of proteins has been reported to be a protective mechanism against oxidative stress, as well as affecting the function of some thiol-containing proteins (44a). Glutathione is the most abundant thiol in mammalian cells (44) and may be the primary agent involved in redox regulation of protein thiols under normal conditions. α-Lipoic acid and DHLA do not appear to be present in the unbound state under normal physiological conditions, but after dietary supplementation both forms appear in various tissues in the unbound from (42). DHLA also has a lower redox potential than GSH, as well as having a lower molecular weight. These factors lead to the possibility that administration of exogenous α-lipoic acid may influence intracellular function not only through antioxidant actions but also through affecting the redox status of thiol-containing proteins, such as thioredoxin, enzymes, and transport proteins. Dihydrolipoate and dihydrolipoamide can reduce thioredoxin (47,48), a small ubiquitous protein that functions to transfer electrons in various biochemical processes (49). Spector et al. (50) proposed that thioredoxin may be a physiological electron acceptor from lipoamide. Glucose transport has also been shown to be enhanced by DHLA in a number of systems (76–78), and it is thought that this stimulation may be due to reduction of sulfhydryl groups involved in the regulation of insulin-stimulated glucose transport (81).

DHLA has also been found to influence physiologically important proteins through mechanisms other than disulfide reduction. DHLA reduces metmyoglobin and ferrylmyoglobin to oxymyoglobin (45). DHLA also elicits glandular kallikrein-induced prolactin proteolysis by acting upon prolactin to refold the molecule into conformations that are glandular kallikrein substrates (46). Hence, DHLA may exert an influence over intracellular metabolism by means other than antioxidant effects, but the physiological significance of this remains to be elucidated.

E. Effects on Gene Expression

There has recently been a great deal of interest in the effects of oxidants and antioxidants on signal transduction and gene expression in cell growth and differentiation. In this regard α-lipoic acid and DHLA have been investigated in terms of their effect on the transcription factors NF-κB (51), which regulates the expression of genes that regulate normal response, such as inflammatory response, as well as the genes of human immunodeficiency virus type 1 (HIV-1) (52). The effects of α-lipoic acid and DHLA on the expression of c-*fos* have also been studied.

NF-κB is regulated through redox mechanisms (53,54), and sulfhydryl groups such as Cys62 in the p50 subunit are involved this regulation (55). There are two steps in the process of NF-κB activation both of which may be influenced by a thiol antioxidant such as α-lipoic acid. Early steps involve the activation of NF-κB and dissociation from an inhibitory subunit, IκB. These are apparently at least partly under redox control, with oxidation stimulating activation and dissociation. The binding of activated NF-κB to DNA involves cysteine residues whose redox status is also important, with reduced cysteine apparently enhancing binding. Hence, the effects of thiol-containing antioxidants can be complex. For example, overexpression of thioredoxin inhibits TPA-induced activation of NF-κB (55a), whereas in an in vitro system thioredoxin enhanced binding of NF-κB to DNA (55,55b). Similarly, incubation of human Jurkat T cells in medium containing 4 mM α-lipoic acid completely inhibits NF-κB activation induced by tumor necrosis factor α or phorbol myristate 13-acetate (56). Recently, the DNA-binding activity of NF-κB was found to be enhanced by DHLA and inhibited by α-lipoic acid, and inhibition of NF-κB DNA binding induced by a nonreducing environment or by exposure to a thiol-oxidizing agent, diamide, was reversed by DHLA (57), so the interplay of these compounds and their effects on NF-κB in a physiological cellular environment is difficult to predict.

The effects of both α-lipoic acid and DHLA on the expression of the growth-regulating gene, c-*fos*, have also been investigated (58). Jurkat T cells preincubated with either α-lipoic acid or DHLA were exposed to TPA, an activator of c-*fos* expression. Cells incubated in DHLA exhibited less c-*fos* mRNA expression compared to controls, whereas those preincubated with α-lipoic acid exhibited greater expression compared to controls. Active oxygen species are produced by cultured cells stimulated by TPA (59–61), and superoxide production is enhanced in TPA-stimulated leukocytes (62). The suppression of c-*fos* expression by DHLA but not by α-lipoic acid may be due to the fact that DHLA scavenges superoxide, whose production may be involved in enhancing c-*fos* expression, whereas α-lipoic acid does not.

Apoptosis is another process in which intracellular oxidation is thought to play a role. In rat thymocytes exposed to methylprednisolone or etoposide, inducers of apoptosis, preincubation with DHLA or lipoamide inhibited apoptosis. Lipoic acid had no effect (S. Orrenius, personal communication).

While it appears that DHLA and/or α-lipoic acid may influence gene expression at one or more levels, the exact mechanisms and significance have yet to be elucidated. This area holds great promise.

III. EXPERIMENTAL AND CLINICAL THERAPEUTIC STUDIES

α-Lipoic acid administration has been shown to be effective in preventing pathology in various experimental models in which reactive oxygen species have been implicated. Before considering these studies, however, it is necessary to discuss whether α-lipoic acid is absorbed as a dietary supplement, to what degree it is taken up by tissues, whether it is reduced to DHLA, and whether it is metabolized to shorter-chain homologs.

A. The Fate of Exogenously Supplied α-Lipoic Acid

α-Lipoic acid is absorbed from the diet, transported to the tissues, and taken up by cells, where a large proportion is rapidly converted to DHLA. This has been shown by a number of studies. The work of Rosenberg and Culik (4) showed that α-lipoic acid, supplemented in the diet of rats and guinea pigs, protected against the symptoms of vitamin E or vitamin C deficiency. In later experiments (63), animals were fed a diet supplemented with α-lipoic acid for 12 days, then liver, skin, and brain homogenates were tested for susceptibility to lipid peroxidation; α-lipoate decreased peroxidation by 50–79%.

Experiments using radiolabeled α-lipoic acid indicate that the compound is absorbed. When ^{14}C-labeled lipoic acid was administered to rats, either as an i.p. injection or orally, via stomach tube, 80% of the administered radioactivity was either excreted or found in the tissues (64). α-Lipoic acid administered to a variety of cell and tissue systems appears in the medium as DHLA (65). In a more recent study (10), lipoic acid was added to the culture medium for human fibroblasts or Jurkat T cells at concentrations from 1 to 4 mM. The concentrations of both α-lipoic acid and DHLA in the cells and in the culture medium were determined by HPLC with electrochemical detection at various times up to 2 h. The intracellular concentration of DHLA in the Jurkat cells reached 1.5 mM within 10 min. It was also found that the cells released DHLA into the medium.

The results indicate that normal mammalian cells are capable of taking up α-lipoic acid, reducing it to DHLA, and releasing DHLA. Hence, the effects of both α-lipoic acid and DHLA may be present both intracellularly and extracellularly when α-lipoic acid alone is administered extracellularly. This has important implications in, for example, the use of α-lipoic acid supplementation to prevent LDL oxidation.

Finally, a recent study that repeated and extended the experiments of Rosenberg and Culik (4) involving vitamin E–deficient rats confirm that α-lipoic acid is absorbed and is converted to DHLA in tissues (42). Hairless mice were used as a model for studying the effects of E deficiency with or without α-lipoic acid supplementation, because they display obvious symptoms of deficiency within 5 weeks (Fig. 4). Mice were fed either a normal diet, a vitamin–E deficient diet, or an E-deficient diet supplemented with 1.65 g α-lipoic acid/kg diet. α-Lipoic acid supplementation completely prevented symptoms of vitamin E deficiency. Both α-lipoic acid and DHLA (unbound) were measured after 5 weeks on the various diets in liver, kidney, heart, and skin. Only the mice fed the diet supplemented with α-lipoic acid displayed any unbound α-lipoic acid or DHLA in these tissues. Total α-lipoic acid was highest in heart (3.42 ± 2.20 nmol/g wet weight) and lowest in liver (0.60 ± 0.33 nmol/g). In all tissues DHLA was detected as well as α-lipoic acid and represented 21–45% of the total. Metabolites of α-lipoic acid were not measured.

The absorption and reduction of α-lipoic acid in humans is not well studied; endogenous plasma levels of α-lipoic acid were 1–25 ng/ml and of DHLA 33–145 ng/ml in six healthy volunteers (66), but plasma levels of α-lipoic acid and DHLA in individuals supplemented with α-lipoic acid remain to be thoroughly investigated.

The extent to which exogenously supplied α-lipoic acid is metabolized to shorter-chain homologs, which may have different effects than α-lipoic acid itself, is also not known. In studies of administration of radiolabeled α-lipoic acid to rats (67), much of the excreted α-lipoic acid was in altered forms, including the β-oxidation products bisnorlipoate, tetranorlipoate, and β-hydroxy-bisnorlipioc acid. It is not known what proportion of administered α-lipoic acid is converted to these metabolites; in one study of α-lipoic acid administered to cultured T lymphocytes, no formation of shorter-chain length β-oxidation products of lipoic acid was observed over 2 h (10). Hence, the question of how much α-lipoic acid is converted to its shorter-chain homologs is still unanswered. These metabolites may also play a significant role in the observed effects of treatment with α-lipoic acid.

B. Diabetes

Many of the complications induced by diabetes, including polyneuropathy and cataract formation, appear to be mediated by oxygen free radical generation (68); diabetic patients have elevated serum levels of thiobarbituric acid–reactive substances compared to nondiabetics (69). α-Lipoic acid has potential preventive or ameliorative effects in both Type I and Type II diabetes (70–73), hence the relationship between the role of oxidative stress in diabetes-induced complications and the antioxidant properties of α-lipoic acid and dihydrolipoic acid is of particular interest. α-Lipoic acid has also been shown to have promise for preventing glycation reactions. Glycation of proteins may play a role in pathologies accompanying the disease.

1. *Type I Diabetes*

Type I diabetes results from β-cell destruction by immunological or inflammatory attack. One animal model for this type of diabetes is the nonobese diabetic mouse. When diabetes development was accelerated in nonobese diabetic mice by cyclophosphamide administration, 60% of the mice developed diabetes within 1–3 weeks (74). If α-lipoic acid was given (10 mg/kg, i.p.) for 10 days before and 10 days after cyclophosphamide administration, only 30% of the mice developed diabetes. This effect could be due to suppression of nitric oxide release by macrophages, which has been shown to occur in the presence of DHLA (75), as well as to scavenging of other reactive oxygen species released by inflammatory cells, which attack the pancreatic β-cells. The large suppression of diabetic induction by α-lipoic acid should prompt further research in this area.

2. *Type II Diabetes*

Most Type II diabetics are hyperinsulinemic, hence no insulin therapy is warranted. A number of studies have therefore examined other means of increasing glucose uptake. As skeletal muscle tissue is the major sink in the body for glucose following a meal, agents that enhance glucose uptake by skeletal muscles are potentially useful in the long-term treatment of Type II diabetes.

α-Lipoic acid enhances glucose utilization in isolated rat diaphragm (76), heart (77), and cultured myotubes (78). In animal models, using the obese Zucker rat as an animal model of insulin resistance in obesity, it was demonstrated that α-lipoic acid treatment increased the uptake of glucose (as 2-deoxyglucose) in the absence or presence of insulin in epitrochlearis muscles by over 50% (79).

Human studies (80) have also demonstrated an effect of α-lipoic acid in Type II diabetes. α-Lipoic acid administration (1000 mg, i.v.) enhanced insulin-stimulated whole body glucose disposal by about 50%. It has been suggested that the exogenously supplied α-lipoic acid may bring endogenous levels of α-lipoic acid, known to be low in diabetic animals (70) back to normal. Alternatively, α-lipoic acid may react with cellular sulfhydryl groups, believed to be involved in the regulation of insulin-stimulated glucose transport (81). The effect may also be due to LA's antioxidant function. At present it is not possible to distinguish among these possibilities.

3. *Glycation Reactions*

Glycation of proteins may be an underlying factor in a number of the pathologies of diabetes (82), and free radicals may be involved in the process (68). Although several mechanisms have been postulated for the pathogenesis of chronic diabetic complications, protein glycation and oxidation by glucose (glycoxidation) are plausible working hypotheses (68,83,84).

α-Lipoic acid decreased the inhibition of enzyme activity in lysozyme and the extent of glycosylation in serum albumin that was incubated in glucose (85). In another study, (86) it was found that noncovalent hydrophobic binding of α-lipoic acid to albumin was involved in its protective effects (Fig. 5) Since α-lipoate has a hydrophobic carbon chain, it is likely that α-lipoate binds to albumin by hydrophobic interactions similar to fatty acids. Glycated human serum

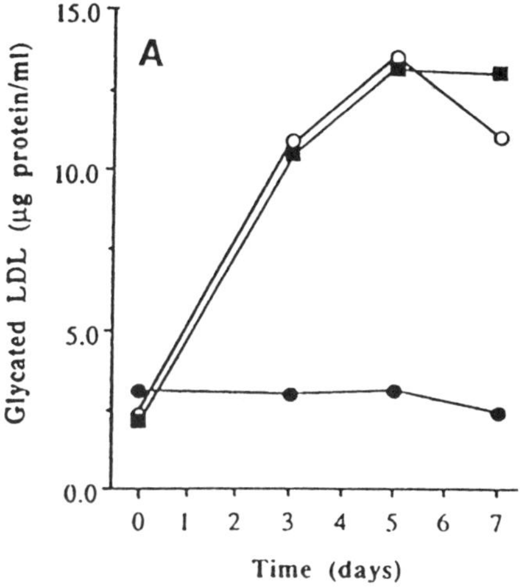

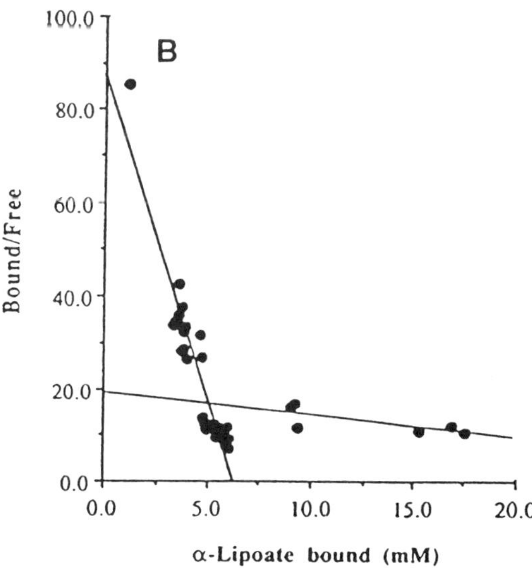

Figure 5 Effect of α-lipoate on LDL glycation. LDL (1 mg protein/ml) was incubated with 200 mM glucose in the presence (open circles) or absence (filled squares) of α-lipoate (20 mM) in Hepes-saline (pH 7.4) at 37°C. At the indicated times, glycated LDL was separated by affinity chromatography and protein concentrations measured. Values are averages of duplicate observations. LDL was also incubated without glucose as a negative control (filled circles). (B) Scatchard plot of binding of α-lipoate to BSA. BSA (1 mM) was incubated with various concentrations of α-lipoate in Hepes-saline (pH 7.4) at 37°C for 18 h. The sample was centrifuged in a centricon 30, and the α-lipoate concentration in the filtrate was measured by absorption spectroscopy ($\varepsilon = 150\ M^{-1}\ cm^{-1}$).

albumin binds fewer fatty acids than does nonglycated albumin (87), suggesting that glycation sites are adjacent to fatty acid–binding sites. In an early step of albumin glycation, α-lipoic acid may protect BSA from glycation by masking the glycation site through hydrophobic binding. Although covalent interaction with serum albumin has been reported to inhibit the protein glycation, noncovalent interaction of Diclofenac (Voltaren) with human serum albumin was also reported to inhibit glucose attachment to human serum albumin (88). α-Lipoic acid may act in a similar manner.

Diabetic patients have a higher frequency of atherosclerosis than nondiabetic individuals (89), which may be related to glycation. α-Lipoic acid was found not to protect LDL from glycation in vitro (86) (Fig. 5). These findings also emphasize the significance of the hydrophobicity of α-lipoate in protection against glycation. Schepkin et al. (90) also reported using NMR techniques that α-lipoate binds to albumin but not to LDL.

However, although α-lipoic acid was not found to protect lipoic acid in the short term, the long-term effects of glycation involve further reaction to Amadori products and eventually the formation of advanced glycosolated end products (AGE). These steps may involve oxidative mechanisms. It is worth noting in this regard that α-lipoic acid has been found to be protective of human LDL exposed to oxidative stress (40,91) (Fig. 6). Hence, over the long term, α-lipoic acid may help protect diabetics from the atherosclerosis that is a common component of the disease.

4. *Polyneuropathy*

Endoneural blood flow and oxygen tension are reduced in experimental diabetic neuropathy. This process of endoneural hypoxia is associated with an increase of oxidative stress and an impairment of the nerve conduction velocity (92).

Antioxidants such as glutathione have been shown to prevent neuropathy in animal models (93), and α-lipoic acid induces sprouting of neurites in culture (94). Clinical trials have been carried out on the effects of α-lipoic acid supplementation on diabetic neuropathy, and its use for this condition is approved in Germany.

In one placebo-controlled double-blind study (95), 21 days of intravenous administration of α-lipoic acid (200 mg daily) caused alleviation of some clinical symptoms, but objective measures, such as vibration sense and nerve conduction velocity, showed no change with i.v. lipoic acid administration. In a longer, single-blind study, α-lipoic was compared to vitamin B_1 as treatment for diabetic neuropathy (96). α-Lipoic acid (600 mg/day) or vitamin B_1 (400 mg/day) was administered intravenously and intramuscularly, respectively, to diabetics for 3 weeks, followed by 12 weeks of oral administration of the same dose. Pain and parathesia were reduced in the α-lipoic acid patients when compared with the vitamin B_1 patients, but again in this study no improvement in

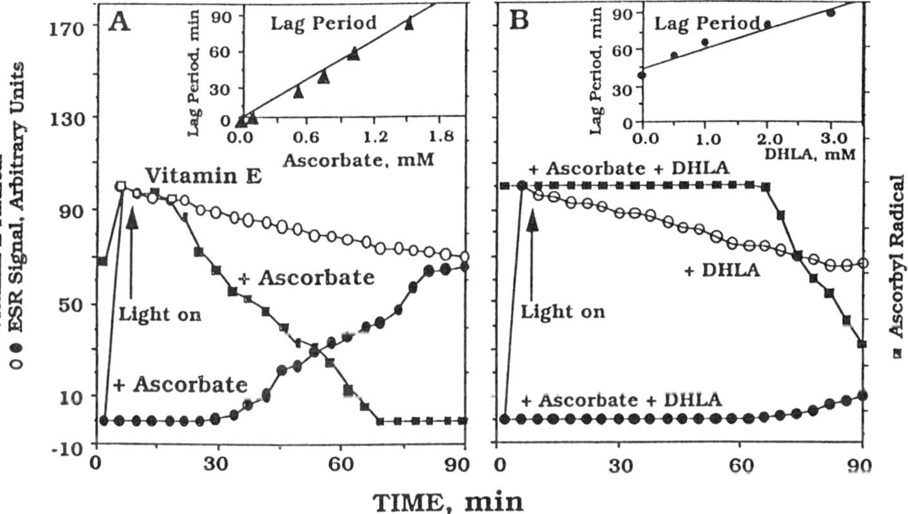

Figure 6 Time course of UV-induced vitamin E chromanoxyl and ascorbyl ESR signals in LDL suspensions. (A) Effect of 500 μM ascorbate. (Inset) Dependence of the lag period (during which the vitamin E chromanoxyl radical ESR signal was not observed) on the concentration of ascorbate. (B) Effect of 2M dihydrolipoic acid plus 500 μM ascorbate. (Inset) dependence of the lag period (during which vitamin E chromanoxyl radical ESR signal was not observed) on the concentration of dihydrolipoic acid (the ascorbate concentration was 500 μM). Incubation medium (100 μL) contained LDL samples with endogenous vitamin E (6.2 nmol/mg protein, 14 mg protein/ml), ascorbate (500 μM in phosphate buffer), pH 7.4, 25°C. All values are given as a percentage of the maximum magnitude obtained. Vitamin E chromanoxyl radicals were generated by irradiation with UV light (290–400 nm).

motor or sensory nerve conduction velocity could be demonstrated due to lipoic acid administration. In a nonblinded study in which several parameters were studied (97), diabetic patients (both Type I and Type II) were given oral supplements of either α-lipoic acid (600 mg/day for 2 weeks, then 300 mg/day for 10 weeks), α-tocopherol (1575 IU daily), or selenium (100 μg daily), and compared to controls (no supplements, no placebo) after 12 weeks. Blood malondialdehyde decreased from 14.4 to 10.9 μmol/l in the α-lipoic acid group. Similar reductions were seen in the α-tocopherol and selenium groups. No change was seen in the control group. Albuminuria decreased 50, 46, and 26% in the α-lipoic acid–, α-tocopherol–, and selenium-supplemented groups, respectively. There was subjective improvement in neurological symptoms in all supplemented groups but not in the control group. In addition, while retinopathy worsened in five of nine in the control group, there was only one case of worsening of retinopathy in any of the antioxidant-supplemented groups. This study showed that mitigating oxidation, whether by α-lipoic acid supplementation or supplementation with other antioxidants, improved biochemical and functional parameters in diabetics; however, the study suffered from lack of placebo control and lack of double blind.

In general, it is not surprising that the short-term studies cited here have shown no objective improvement in neurological functioning, since such improvement would be expected to occur over months or years, not weeks. However, the improvement of subjective neurological symptoms, observed by blinded evaluators, indicates that objective effects might appear given longer trials.

5. *Cataracts*

Cataract formation is a common diabetic complication. α-Lipoic acid supplementation decreased cataract formation, caused by BSO-induced inhibition of GSH synthesis, 60% in newborn rats, while sparing lens ascorbate, tocopherol, and GSH (98). Significant protection of lens ascorbate, tocopherol, and glutathione was also observed in the lipoate-supplemented rats compared to nonsupplemented ones. In vitro diabetic cataractogenesis in rat lens cell cultures exposed to high concentrations of glucose was prevented by α-lipoic acid (99) (Fig. 7).

Our group (100) has hypothesized that α-lipoic acid prevents oxidative stress in diabetic conditions by sparing vitamin C. Since vitamin C and glucose share the same carrier in non–insulin-dependent tissues, the elevated blood glucose in diabetes competitively inhibits the cell entry of vitamin C, resulting in localized intracellular vitamin C deficiency (Fig. 8). Exogenously supplemented α-lipoic acid utilizes other transport systems to enter the cells, is converted to dihydrolipoate, and recycles vitamin C. This would explain its protective effect in diabetic cataractogenesis, as well as other complications of diabetes.

C. Ischemia-Reperfusion Injury

Ischemia-reperfusion injury occurs when a burst of free radicals is produced during reoxygenation of tissue that has become hypoxic. It is important in cardiac tissue (especially with the introduction of clot-dissolving drugs for the treatment of heart attack) and in brain. Agents that prevent ischemia-reperfusion injury may therefore prove important in the treatment of cardiac infarct, during open-heart surgery, and in the treatment of stroke and other conditions that cause interruption of blood flow to the brain.

In vitro, DHLA prevents ischemia-reperfusion–induced changes in fluidity and polarity of rat heart mitochondria (101). Several studies extend these experiments to more closely match in vivo situations. Hearts from rats fed α-lipoic acid, then exposed to ischemia-reperfusion, exhibited higher levels of vitamin E in heart tissue, improved postischemic left ventricular functional

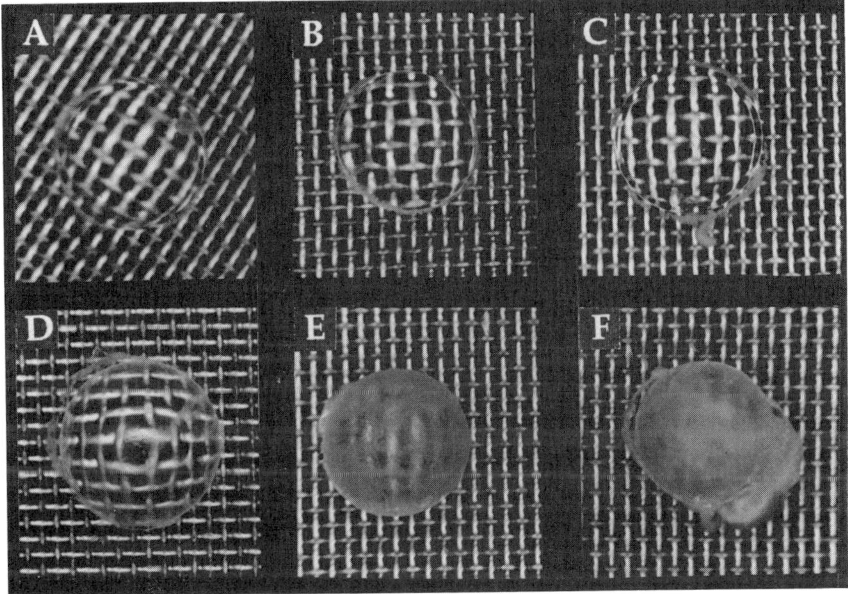

Figure 7 Effect of α-lipoate on glucose-induced lens opacity. Lenses as seen through a dissection microscope after 8 days of incubation. Rat lenses with intact capsules were incubated in Modified Eagle Medium 199, with (A) normal glucose (5.56 mM) and 1 mM R-α-lipoic acid, (B) normal glucose (5.56 mM) and 1 mM S-α-lipoic acid, (C) normal glucose (5.56 mM) and 1 mM racemic α-lipoic acid, (D) elevated glucose (55.6 mM) and 1 mM R-α-lipoic acid, (E) elevated glucose (55.6 mM) and 1 mM S-α-lipoic acid, (F) elevated glucose (55.6 mM) and 1 mM racemic α-lipoic acid.

recovery, and decreased lipid peroxidation and lactate dehydrogenase leakage (a marker of membrane damage) compared to control hearts (102). The protective effects of dihydrolipoic acid against rat myocardial ischemia-reperfusion injury are dependent on vitamin E, suggesting that α-lipoic acid functions in this system by regenerating tocopherol from the tocopheroxyl radical (103). Assadnazari et al. (104), using a working heart system in an NMR magnet, found that DHLA added to the reperfusion buffer accelerated the recovery of aortic flow during reperfusion. DHLA also appeared to increase ATP synthesis in the heart.

The brain is another area where ischemia-reperfusion injury can have serious consequences, e.g., in head trauma, subarachnoid hemorrhage, stroke, or cardiac arrest. Prehn et al., in a study with middle cerebral artery occlusion in mice, demonstrated that in this model of focal ischemia, treatment with DHLA, but not α-lipoic acid, reduced the size of the infarct (105,106). The authors also noted a hypoglycemic effect at the doses used (100 mg/kg body weight).

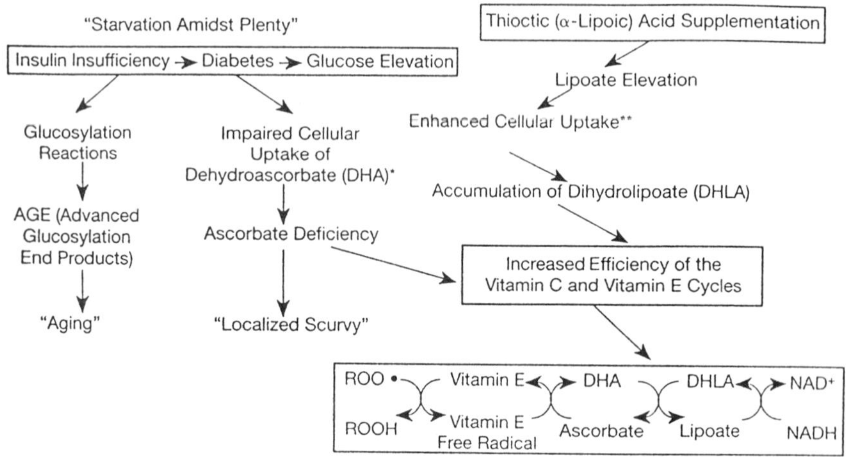

Figure 8 Schematic diagram of the pathobiochemical situation in diabetes mellitus and the possible correction by oral administration of α-lipoic acid. Since glucose competitively inhibits the absorption of ascorbic acid into the cells, α-lipoic acid increases the efficiency of the vitamin C cycle and thereby also activates the vitamin E cycle. *Since glucose and dehydroascorbate use the same sugar transporter system, high glucose competitively inhibits vitamin C uptake. **Lipoate uses fatty acid transporter system.

In other ischemia-reperfusion systems, Boveris et al. (107) reported that α-lipoic acid administration prevented rat intestinal short-term ischemia-reperfusion induced overshoot of chemiluminescence, which is indicative of an increased steady-state level of singlet oxygen and an increased rate of lipid peroxidation. Boveris and coworkers also observed that α-lipoic acid can inhibit xanthine oxidase activity; xanthine oxidase is postulated to play an important role in the mechanism of ischemia-reperfusion injury.

D. Liver Diseases

α-Lipoic acid is often used for therapy in conditions that involve liver pathology, especially mushroom poisoning and alcoholic liver degeneration. There is, however, little evidence that α-lipoic acid is useful in either of these conditions.

Although case reports describe complete recovery from mushroom (Amanita) poisoning in patients treated with α-lipoic acid (108,109), 10–50% of victims recover without α-lipoic acid treatment. In animal studies, α-lipoic acid has proved completely ineffective as a treatment for Amanita poisoning, and some groups recommend its elimination from a treatment regimen for this condition (110).

Several studies indicated that α-lipoic acid administration might be beneficial in liver disease, especially alcoholic liver disease (111–113), but these studies suffered from lack of control groups, lack of statistical analysis, or the use of other treatments in addition to α-lipoic acid. In a controlled, double-blind, long-term study, α-lipoic acid had no effect on the course of alcohol-related liver disease (114). Hence, its use for the treatment of alcoholic liver disease cannot be recommended.

E. NF-κB Activation, HIV, and AIDS

Oxidative stress may play a role in several aspects of HIV infection, including activation of virus replication (115), immunosuppression (116,117), and tumor initiation and promotion (118). HIV-infected patients have been reported to be deficient in various antioxidants (119–121).

A pilot study was recently conducted on the effects of α-lipoic acid supplementation (150 mg three times daily for 14 days) in 12 HIV-positive individuals (122). Plasma ascorbate and glutathione increased, while markers of plasma lipid peroxidation decreased. Also, the number of T-helper cells increased in six of nine patients, and the T-helper:T-suppressor cell ratio improved in 6 of 10 patients. Also, in cultured cells, α-lipoic acid and dihydrolipoic acid prevented HIV replication (123) and the activation of NF-κB transcription factor (56), which are regulated by oxidative stress. These studies suggest that further research is warranted on the effect of α-lipoic acid in HIV infection.

F. Neurodegenerative Diseases

Oxidative stress may be greater in neurological tissues because of their constant high rate of oxygen consumption and high mitochondrial density. Mitochondria inevitably produce free radicals as "byproducts" of normal oxidative metabolism (124,125), which damage mitochondrial DNA (126). Defective proteins coded for by the damaged DNA can lead to synthesis of mitochondria in which components of the electron transport chain preceding the damaged protein become reduced, leading to greater free radical production (127–129) and more mitochondrial damage, in a vicious cycle. Such a vicious cycle may be responsible, in part, for neurodegenerative diseases. Support for this view comes from the observation of high degrees of oxidative damage as well as damaged mitochondria in tissues from patients with neurodegenerative disease (129–134). Therapy or prevention would therefore logically involve antioxidant treatment. α-Lipoic acid is an especially good candidate because of the variety of its antioxidant functions.

In mice, α-lipoic acid (100 mg/kg body weight for 15 days) improved performance in an open-field memory test—in fact, the α-lipoic acid–treated animals performed better than young animals 24 h after the first test (though the difference was not significant) (135). Treatment with α-lipoic acid did not improve memory in young animals. The authors concluded that α-lipoic acid's free radical–scavenging ability may improve *N*-methyl-D-aspartate receptor density, leading to improved memory. Recently, Greenamyre et al. (136) observed that the intraperitoneal administration of α-lipoic acid or DHLA reduced by half the rat striatum lesions induced by excitotoxins that affect NMDA receptors, which may lead to calcium influx and generation of nitric oxide and other free radicals (137).

Another way in which mitochondria may be important in neurodegeneration is through alterations in their effects on calcium homeostasis. In this regard, the recent report of Christof Richter (personal communication) that α-lipoic acid inhibits mitochondirial calcium transport may be relevant to its beneficial effects in neurodegenerative disorders noted by Greenamyre (136).

Ischemia-reperfusion injury, induced by head trauma, subarachnoid hemorrhage, stroke, or cardiac arrest, is another potential source of free radical damage to neuronal structures. The protective effect of α-lipoic acid in this condition has already been discussed (see Sec. III.C). In addition, α-lipoic acid treatment of rats exposed to inhalation of *n*-hexane (constant exposure to 700 ppm) delayed the onset of severe neuropathy by 50% (138).

G. Radiation Injury

Irradiation is known to produce a cascade of free radicals, and antioxidant compounds have long been used to treat irradiation injury. α-Lipoic acid, but not DHLA, protected against radiation injury to hematopoietic tissues in mice and increased the LD_{50} from 8.67 to 10.93 Gy (139). α-Lipoic acid administration to murine neuroblastoma cells (43) increased cell survival after irradiation from 2% to about 10%, and the effect correlated with an increase of the intracellular GSH/GSSG ratio induced by α-lipoate. These results extended to mice irradiated with 8 Gy; α-lipoic acid administration (16 mg/kg) increased survival rates from 35% in untreated animals to 90% in α-lipoic acid–treated animals.

A recent study examined the effects of 28 days of antioxidant treatment on a variety of blood and urinary parameters in children living in areas affected by the Chernobyl nuclear accident who are continuously exposed to low-level radiation (140). Treatment with α-lipoic acid alone lowered blood peroxidation values to the same level seen in non–radiation-exposed children; α-lipoic acid + vitamin E treatment further lowered blood peroxidation to below-normal values; vitamin E alone was without effect. Urinary excretion of radioactive metabolites was also lowered by α-lipoic acid but not by vitamin E, presumably due to chelation by α-lipoic acid. Liver and kidney functions were also normalized by α-lipoic acid treatment.

H. Cigarette Smoke Effects on Plasma Proteins

It is known that CS contains a number of free radical species (141,142). Since many of the diseases caused by smoking involve, at least in part, free radical–mediated processes, it has been proposed that the free radicals in CS contribute to smoking-related diseases. The effects of CS on lung-lining fluids have been investigated using plasma as a model system. Freshly obtained human plasma is exposed to "puffs" of gas-phase or whole CS, and antioxidants and markers of oxidative damage are measured. In this system, ascorbic acid is consumed first, followed by protein thiol groups (143,144). Lipid hydroperoxides as well as protein carbonyls (a marker of protein oxidation) appear (143,145). Exogenously supplied DHLA is protective against CS-induced oxidation of antioxidants, proteins, and lipids (144). DHLA also partially protected polymorphonuclear leukocytes from the damaging effects of CS (146). This effect may be due to scavenging of oxidants in the aqueous or lipid phases or to regeneration of ascorbic acid that has been converted to ascorbyl radical as it is oxidized by CS components. In this regard, DHLA could play a role in minimizing the pathological consequences of smoking.

I. Heavy Metal Poisoning

The possible chelating effects of α-lipoic acid, together with its antioxidant effects, make it a good candidate for the treatment of heavy metal poisoning. It may be especially effective against arsenite, cadmium, and mercury. Grunert (147) demonstrated that α-lipoic acid administration completely protected mice and dogs from arsenite poisoning if the ratio of α-lipoic acid to arsenite was at least 8:1. This was true even if the α-lipoic acid was administered after severe symptoms of poisoning were already seen. Similar complete protection was seen for Hg^{2+} poisoning in mice, although at low doses (2:1 molar ratio) α-lipoic acid appeared to potentiate the lethal effect of Hg^{2+}. α-Lipoic acid was ineffective against lead and gold poisoning in mice in this study. Exposure of isolated hepatocytes to α-lipoic acid or DHLA resulted in the amelioration of cadmium^{2+}-induced membrane damage, lipid peroxidation, and the depletion of cellular glutathione (21). These findings were extended to a rat model in which α-lipoic acid at a dose of 30 mg completely prevented cadmium-induced lipid peroxidation in brain, heart, and testes (148). In this study it also completely abolished cadmium-induced decreases in the activities of Ca^{2+}, Na^{+}-, and Mg^{+}-ATPases in these organs. α-Lipoic acid administration to rats also increased biliary excretion of injected Hg^{2+} (the major route of excretion) 12- to 37-fold (149) but dramatically decreased biliary excretion of Cd^{2+}, methyl mercury, Zn^{2+}, and Cu^{2+}. It is not clear whether this was due to chelation of these metals in the circulation by α-lipoic acid or other effects of α-lipoic acid. α-Lipoic acid administration has also been found to greatly increase the rate of elimination of radiomercury in rabbits (150). Hence, α-lipoic acid may be of clinical value in the treatment of mercury and cadmium poisoning. Its effect on other heavy metals is not yet completely clear.

J. Side Effects

Neither animal nor human studies to date have shown serious side effects with administration of α-lipoic acid (151). The LD_{50} is approximately 400–500 mg/kg following intravenous administration in rats and 400–500 mg/kg after oral dosing in dogs. In long-term oral supplementation at doses sufficient to reduce body weight gain, no functional or laboratory adverse effects were seen in animals (151). There is no evidence of carcinogenic or teratogenic effects, but it is recommended that pregnant women avoid taking supplemental α-lipoic acid until more data are available (151). In humans, side effects include allergic skin reactions and possible hypoglycemia in diabetic patients as a consequence of improved glucose utilization with high doses of α-lipoic acid (151).

IV. CONCLUSIONS

α-Lipoic acid and its reduced form, DHLA, quench a variety of reactive oxygen species, inhibit reactive oxygen generators, and spare other antioxidants. Experimental as well as clinical studies point to the usefulness of α-lipoic acid as a therapeutic agent for such diverse disease conditions as myocardial and cerebral ischemia-reperfusion injury, heavy metal poisoning, radiation damage, diabetes, neurodegenerative diseases, and HIV infection. High doses of α-lipoic acid are approved in Germany for treatment of diabetic polyneuropathy. Furthermore, the interesting antioxidant properties of α-lipoic acid and its interaction with other important antioxidants like vitamin E, ascorbate, and glutathione will provide a fertile field for continued research.

ACKNOWLEDGMENTS

Research for this work was supported by the National Institutes of Health (CA47597), Council for Tobacco Research (33446R2), and the State of California Tobacco Related Disease Research Program through the University of California (#4RT-0065). We thank Dr. Y. J. Suzuki for preliminary discussions at the outset of this project.

REFERENCES

1. Reed, L. J., DeBusk, B. G., Gunsalus, I. C., and Hornberger, Jr., C. S. (1951) Crystalline α-lipoic acid: A catalytic agent associated with pyruvate dehydrogenase. Science 114:93–94.

2. Reed, L. J. (1957) The chemistry and function of lipoic acids. Adv. Enzymol. 18:319–347.
3. Reed, L. J. (1974) Multienzyme complex. Acc. Chem. Res. 7:40–46.
4. Rosenberg, H. R., and Culik, R. (1959) Effect of α-lipoic acid on vitamin C and vitamin E deficiencies. Arch. Biochem. Biophys. 80:86–93.
5. Carreau, J. P. (1979) Biosynthesis of lipoic acid via unsaturated fatty acids. Meth. Enzymol. 62:152–158.
6. Dupre, S., Spoto, G., Matarese, R. M., Orlando, M., and Cavallini, D. (1980) Biosynthesis of lipoic acid in the rat: incorporation of ^{35}S- and ^{14}C-labeled precursors. Arch. Biochem. Biophys. 202:361–365.
7. Rimm, E. B., Stampfer, M. J., Ascherio, A., Giovannucci, E., Colditz, G. A., and Willett, W. C. (1993) Vitamin E consumption and the risk of coronary heart disease in men. N. Engl. J. Med. 328:1450–1456.
8. Stampfer, M. J., Hennekens, C. H., Manson, J. E., Colditz, G. A., Rosner, B., and Willett, W. C. (1993) Vitamin E consumption and the risk of coronary disease in women. N. Engl. J. Med. 328:1444–1449.
9. Kagan, V. E., Shvedova, A., Serbinova, E., Khan, S., Swanson, C., Powell, R., and Packer, L. (1992) Dihydrolipoic acid—a universal antioxidant both in the membrane and in the aqueous phase. Biochem. Pharmacol. 44:1637–1649.
10. Handelman, G. J., Han, D., Tritschler, H., and Packer, L. (1994) α-Lipoic acid reduction by mammalian cells to the dithiol form and release into the culture medium. Biochem. Pharmacol. 47:1725–1730.
11. Podda, M., Han, D., Koh, B., Fuchs, J., and Packer, L. (1994) Conversion of lipoic acid to dihydrolipoic acid in human keratinocytes. Clin. Res. 42:41A.
12. Suzuki, Y. J., Tsuchiya, M., and Packer, L. (1991) Thioctic acid and dihydrolipoic acid are novel antioxidants which interact with reactive oxygen species. Free Rad. Res. Comms. 15:255–263.
13. Scott, B. C., Aruoma, O. I., Evans, P. J., O'Neill, C., van der Vliet, A., Cross, C. E., Tritschler, H., and Halliwell, B. (1994) Lipoic and dihydrolipoic acids as antioxidants. A critical evaluation. Free Rad. Res. 20:119–133.
14. Haenen, G. R., and Bast, A. (1991) Scavenging of hypochlorous acid by lipoic acid. Biochem. Pharmacol. 42:2244–2246.
15. Stevens, B., Perez, S. R., and Small, R. D. (1974) The photoperoxidation of unsaturated organic molecules-IX. Lipoic acid inhibition of rubrene autoperoxidation. Photochem. Photobiol. 19:315–316.
16. Stary, F. E., Jindal, S. J., and Murray, R. W. (1975) Oxidation of α-lipoic acid. J. Org. Chem. 40:58–62.
17. Kaiser, S., Di Mascio, P., and Sies, H. (1989) Lipoat und Singulettsauerstoff. In: Thioctsäure (Borbe, H. O., and Ulrich, H., eds.), pp. 69–76. pmi Verlag GmbH, Frankfurt.
18. Devasagayam, T. P., Di Mascio, P., Kaiser, S., and Sies, H. (1991) Singlet oxygen induced single-strand breaks in plasmid pBR322 DNA: the enhancing effect of thiols. Biochim. Biophys. Acta 1088:409–412.
19. Devasagayam, T. P. A., Subramanian, M., Pradhan, D. S., and Sies, H. (1993) Prevention of singlet oxygen-induced DNA damage by lipoate. Chem.-Biol. Interact. 86:79–92.

20. Saito, I., and Fukui, S. (1967) Studies on the oxidation products of lipoic acid. J. Vitaminol. 13:115–121.
21. Müller, L., and Menzel, H. (1990) Studies on the efficacy of lipoate and dihydrolipoate in the alteration of cadmium^{2+} toxicity in isolated hepatocytes. Biochim. Biophys. Acta 1052:386–391.
22. Bast, A., and Haenen, G. R. (1988) Interplay between lipoic acid and glutathione in the protection against microsomal lipid peroxidation. Biochim. Biophys. Acta 963:558–561.
23. Sigel, H. Prijs, B., McCormick, D. B., and Shih, J. C. H. (1978) Stability of binary and ternary complexes of α-lipoate and lipoate derivatives with Mn^{2+}, Cu^{2+}, and Zn^{2+} in solution. Arch. Biochem. Biophys. 187:208–214.
24. Ou, P., Tritschler, H. J., and Wolff, S. P. (1994) Thioctic (Lipoic) acid: A therapeutic metal-chelating antioxidant? Biochem. Pharmacol. (in press).
25. Searls, R. L., and Sanadi, D. R. (1960) α-Ketoglutaric dehydrogenase. 8. Isolation and some properties of a flavoprotein component. J. Biol. Chem. 235: 2485–2491.
26. Scott, I., Duncan, W., and Ekstrand, V. (1963) Purification and properties of glutathione reductase of human erythrocytes. J. Biol. Chem. 238:3928–3939.
27. Jocelyn, P. C. (1967) The standard redox potential of cysteine-cystine from the thiol-disulphide exchange reaction with glutathione and lipoic acid. Eur. J. Biochem. 2:327–331.
28. Suzuki, Y. J., Tsuchiya, M., and Packer, L. (1993) Antioxidant activities of dihydrolipoic acid and its structural homologues. Free Rad. Res. Commun. 18:115–122.
29. Haenen, G. R., de Rooij, B. M., Vermeulen, N. P. E., and Bast, A. (1990) Mechanism of the reaction of ebselen with endogenous thiols: Dihydrolipoate is a better cofactor than glutathione in the peroxidase activity of ebselen. Mol. Pharmacol. 37:412–422.
30. Bonomi, F., Werth, M. T., and Kurtz, Jr., D. M. (1985) Assembly of $[Fe_nS_n(SR)_4]^{2-}$ (n = 2,4) in aqueous media from irons, salts, thiols, and sulfur, sulfide, or thiosulfate plus rhodanese. Inorg. Chem. 24:4331–4335.
31. Bonomi, F., Pagani, S., Cariati, F., Pozzi, A., Crisponi, G., Cristiani, F., Nurchi, V., Russo, U., and Zanoni, R. (1992) Synthesis and characterization of iron derivatives of dihydrolipoic acid and dihydrolipoamide. Inorgan. Chim. Acta 195:109–115.
32. Bonomi, F., and Pagani, S. (1986) Removal of ferritin-bound iron by DL-dihydrolipoate and DL-dihydrolipoamide. Eur. J. Biochem. 155:295–300.
33. Bonomi, F., Cerioli, A., and Pagani, S. (1989) Molecular aspects of the removal of ferritin-bound iron by DL-dihydrolipoate. Biochim. Biophys. Acta 994: 180–186.
34. Reed, C. J., and Douglas, K. T. (1989) Single-strand cleavage of DNA by Cu(II) and thiols: A powerful chemical DNA-cleaving system. Biochim. Biophys. Res. Comm. 162:1111–1117.
35. Burton, G. W., and Ingold, K. U. (1981) Autoxidation of biological molecules. 1. The antioxidant activity of vitamin E and related chain-breaking phenolic antioxidant in vitro. J. Am. Chem. Soc. 103:6472–6477.

36. Sies, H. (1993) Strategies of antioxidant defense. Eur. J. Biochem. 215:213–219.
37. Packer, L. (1992) New horizons in vitamin E research—the vitamin E cycle, biochemistry and clinical applications. In: Lipid-Soluble Antioxidants: Biochemistry and Clinical Applications (Ong, A. S. H., and Packer, L., eds.), pp. 1–16. Birkhauser Verlag, Boston.
38. Scholich, H., Murphy, M. E., and Sies, H. (1989) Antioxidant activity of dihydrolipoate against microsomal lipid peroxidation and its dependence on α-tocopherol. Biochim. Biophys. Acta 1001:256–261.
39. Bast, A., and Haenen, G. R. (1990) Regulation of lipid peroxidation of glutathione and lipoic acid: involvement of liver microsomal vitamin E free radical reductase. In: Antioxidant in Therapy and Preventive Medicine (Emerit, I., Packer, L., and Auclair, C., eds.), pp. 111–116. Plenum Press, New York.
40. Kagan, V. E., Serbinova, E. A., Forte, T., Scita, G., and Packer, L. (1992) Recycling of vitamin E in human low density lipoproteins. J. Lipid Res. 33: 385–397.
41. Constantinescu, A., Han, D., and Packer, L. (1993) Vitamin E recycling in human erythrocyte membranes. J. Biol. Chem. 268:10906–10913.
41a. Gotz M. E., Dirr, A., Burger, R., Janetzky, B., Weinmuller, M., Chan, W. W., Chen, S. C., Reichmann, H., Rausch, W. D., and Riederer, P. (1994) Effect of lipoic acid on redox state of coenzyme Q in mice treated with 1-methyl-4-phenyl-1,2,3,6-tetrahydropyridine and diethyldithiocarbamate. Eur. J. Pharmacol. 266: 291–300.
41b. Kagan, V., Serbinova, E., and Packer, L. (1990) Antioxidant effects of ubiquinones in microsomes and mitochondria are mediated by tocopherol recycling. Biochem. Biophys. Res. Comm. 169:851–857.
42. Podda, M., Tritschler, H. J., Ulrich, H., and Packer, L. (1994) α-Lipoic acid supplementation prevents symptoms of vitamin E deficiency. Biochem. Biophys. Res. Commun. 204:98–104.
43. Busse, E., Zimmer, G., Schopohl, B., and Kornhuber, B. (1992) Influence of alpha-lipoic acid on intracellular glutathione in vitro and in vivo. Arzneimittel-Forsch. 42:829–831.
44. Halliwell, B., and Gutteridge, J. M. C. (1989) Free Radicals in Biology and Medicine. Clarendon Press, Oxford.
44a. Thomas, J. A., Chai, Y. C., and Jung, C. H. (1994) Protein S-thiolation and dethiolation. Meth. Enzymol. 233:385–395.
45. Romero, F. J., Ordonez, I., Arduini, A., and Cadenas, E. (1992) The reactivity of thiols and disulfides with different redox states of myoglobin. Redox and addition reactions and formation of thiyl radical intermediates. J. Biol. Chem. 267:1680–1688.
46. Hatala, M. A., DiPippo, V. A., and Powers, C. A. (1991) Biological thiols elicit prolactin proteolysis by glandular kallikrein and permit regulation by biochemical pathways linked to redox control. Biochemistry 30:7666–7672.
47. Gleason, F. K., and Holmgren, A. (1988) Thioredoxin and related protein in procaryotes. FEMS Microbiol. Rev. 54:271–297.
48. Holmgren, A. (1979) Thioredoxin catalyzes the reduction of insulin disulfides by dithiothreitol and dihydrolipoamide. J. Biol. Chem. 254:9627–9632.

49. Holmgren, A. (1989) Thioredoxin and glutaredoxin systems. J. Biol. Chem. 264: 13963–13966.
50. Spector, A., Huang, R.-R. C., Yan, G.-Z., and Wang, R.-R. (1988) Thioredoxin fragment 31-36 is reduced by dihydrolipoamide and reduces oxidized protein. Biochem. Biophys. Res. Commun. 150:156–162.
51. Hayashi, T., Ueno, Y., and Okamoto, T. (1993) Oxidoreductive regulation of nuclear factor κB. Involvement of a cellular reducing catalyst thioredoxin. J. Biol. Chem. 268:11380–11388.
52. Baeuerle, P. A. (1991) The inducible transcription activator NF-κB: Regulation by distinct protein subunits. Biochim. Biophys. Acta 1072:63–80.
53. Toledano, M. B., and Leonard, W. J. (1991) Modulation of transcription factor NF-κB binding activity by oxidation-reduction in vitro. Proc. Natl. Acad. Sci. USA 88:4328–4332.
54. Molitor, J. A., Ballard, D. W., and Greene, W. C. (1991) κB-Specific DNA binding proteins are differentially inhibited by enhancer mutations and biological oxidation. New Biol. 3:987–996.
55. Matthews, J. R., Wakasugi, N., Virelizier, J.-L., Yodoi, J., and Hay, R. T. (1992) Thioredoxin regulates the DNA binding activity of NF-κB by reduction of a disulphide bond involving cysteine 62. Nucleic Acid Res. 20:3821–3830.
55a. Schenk, H., Klein, M., Erdbrugger, W., and Droge, W. (1994) Distinct effects of thioredoxin and antioxidant on the activation of transcription factors NF-κB and AP-1. Proc. Natl. Acad. Sci. USA 91:1672–1676.
55b. Hayashi, T., Ueno, Y., and Okamoto, T. (1993) Oxidoreductive regulation of nuclear factor kappa B. Involvement of a cellular reducing catalyst thioredoxin. J. Biol. Chem. 268:11380–11308.
56. Suzuki, Y. J., Aggarwal, B. B., and Packer, L. (1992) α-Lipoic acid is a potent inhibitor of NF-κB activation in human T cells. Biochem. Biophys. Res. Commun. 189:1709–1715.
57. Suzuki, Y. J., Ulrich, H., and Packer, L. (1994) Redox regulation of NF-κB DNA binding activity by dihydrolipoate and α-lipoate. Arch. Biochem. Biophys. (submitted).
58. Mizuno, M., and Packer, L. (1994) Effects of α-lipoic acid and dihydrolipoic acid on expression of the proto-oncogene c-fos. Biochem. Biophys. Res. Commun. 220: 1136–1142.
59. Fischer, S. M., and Adams, L. M. (1985) Suppression of tumor promoter-induced chemiluminescence in mouse epidermal cells by several inhibitors of arachidonic acid metabolism. Cancer Res. 45:3130–3136.
60. Bonser, R. W., Dawson, J., Thompson, N. T., Hodson, H. F., and Garland, L. G. (1986) Inhibition of phorbol ester stimulated superoxide production by 1-oleoyl-2-acetyl-sn-glycerol (OAG), fact or artefact? FEBS Lett. 209:134–138.
61. Shibanuma, M., Kuroki, T., and Nose, K. (1987) Effects of the protein kinase C inhibitor H-7 and calmodulin antagonist W-7 on superoxide production in growing and resting human histiocytic leukemia cells (U937). Biochem. Biophys. Res. Commun. 144:1317–1323.
62. Semba, M., Inui, N., and Yamada, M. A. (1992) Inhibitory effects of anti-

promoters and radical scavengers on the promotion process in two-stage cell transformation. Tohoku J. Exp. Med. 168:133–136.
63. Packer, L. (1992) New horizons in antioxidant research: action of the thioctic acid/dihydrolipoic acid couple in biological systems. In: Thioctsäure 2. International Thioctic Acid Workshop (Schmidt, K., and Ulrich, H., eds.), pp. 35–44. Universimed Verlag GmbH, Frankfurt.
64. Harrison, E. H., and McCormick, D. B. (1974) The metabolism of dl-[1,6-^{14}C] lipoic acid in the rat. Arch. Biochem. Biophys. 160:514–522.
65. Peinado, J., Sies, H., and Akerboom, T. P. M. (1989) Hepatic lipoate uptake. Arch. Biochem. Biophys. 273:389–395.
66. Teichert, J., and Preiss, R. (1992) HPLC-methods for determination of lipoic acid and its reduced form in human plasma. Int. J. Clin. Pharmacol. Ther. Toxicol. 30:511–512.
67. Spence, J. T., and McCormick, D. B. (1976) Lipoic acid metabolism in the rat. Arch. Biochem. Biophys. 174:13–19.
68. Wolff, S. P., Jiang, Z.-Y., and Hunt, J. V. (1991) Protein glycation and oxidative stress in diabetes mellitus and ageing. Free Rad. Biol. Med. 10:339–352.
69. Kuklinksi, B., and Pietschmann, A. (1991) Oxidativer Streβ und Altern. Z. Geriatrie 4:224–226.
70. Natraj, C. V., Gandhi, V. M., and Menon, K. K. G. (1984) Lipoic acid and diabetes: Effect of dihydrolipoic acid administration in diabetic rats and rabbits. J. Biosci. 6:37–46.
71. Gandhi, V. M., Wagh, S. S., Natraj, C. V., and Menon, K. K. G. (1985) Lipoic acid and diabetes II: Mode of action of lipoic acid. J. Biosci. 9:117–127.
72. Wagh, S. S., Gandhi, V. M., Natraj, C. V. and Menon, K. K. G. (1986) Lipoic acid and diabetes—Part III: Metabolic role of acetyl dihydrolipoic acid. J. Biosci. 10:171–179.
73. Wagh, S. S., Natraj, C. V., and Menon, K. K. G. (1987) Mode of action of lipoic acid in diabetes. J. Biosci. 11:59–74.
74. Faust, A., Burkart, V., Ulrich, H., Weischer, C. H., and Kolb, H. (1994) Effect of lipoic acid on cyclophosphamide-induced diabetes and insulitis in non-obese diabetic mice. Int. J. Immunopharmacol. 16:61–66.
75. Burkart, V., Koike, T., Brenner, H. H., Imai, Y., and Kolb, H. (1993) Dihydrolipoic acid protects pancreatic islet cells from inflammatory attack. Agents Actions 38:60–65.
76. Haugaard, N., and Haugaard, E. S. (1970) Stimulation of glucose utilization by thioctic acid in rat diaphragm incubated in vitro. Biochim. Biophys. Acta 222:583–586.
77. Singh, H. P. P., and Bowman, R. H. (1970) Effect of D,L-alpha lipoic acid on the citrate concentration and phosphofructokinase activity of perfused hearts from normal and diabetic rats. Biochem. Biophys. Res. Commun. 41:555–561.
78. Bashan, N., Burdett, E., Guma, A., and Klip, A. (1993) Effect of thioctic acid on glucose transport. In: The Role of Antioxidants in Diabetes Mellitus (Gries, F. A., and Wessel, K., eds.), pp. 221–229. PMI Verlag, Frankfurt.
79. Henriksen, E. J., Jacob, S., Tritschler, H., Wellel, K., Augustin, H. J., and Dietze, G. J. (1994) Chronic thioctic acid treatment increases insulin-stimulated

glucose transport activity in skeletal muscle of obese Zucker rats. Diabetes (Suppl. 1):122A.
80. Jacob, S., Henriksen, E. J., Tritschler, H. J., Clancy, D. E., Simon, I., Schiemann, A. L., Ulrich, H., Jung, I., Dietze, G. J., and Augustin, H. J. (1994) Thioctic acid enhances glucose-disposal in patients with Type 2 diabetes. Hormon. Metab. (submitted).
81. Henriksen, E. J., and Holloszy, J. O. (1990) Effects of phenylarsine oxide on stimulation of glucose transport in rat skeletal muscle. Am. J. Physiol. 258:C648–C653.
82. Renold, A. E., Mintz, D. H., Muller, W. A., and Cahill, Jr., G. F. (1978) Diabetes mellitus. In: The Metabolic Basis of Inherited Diseases (Stanbury, J. B., Wyngaarden, J. B., and Fredrickson, D. S., eds.), pp. 80–109. McGraw-Hill, New York.
83. Baynes, J. W. (1991) Role of oxidative stress in development of complications in diabetes. Diabetes 40:405–412.
84. Brownlee, M., Cerami, A., and Vlassara, H. (1988) Advanced glycosylation end products in tissue and the diabetic basis of diabetic complications. N. Engl. J. Med. 318:1315–1321.
85. Suzuki, Y. J., Tsuchiya, M., and Packer, L. (1992) Lipoate prevents glucose-induced protein modifications. Free Rad. Res. Commun. 17:211–217.
86. Kawabata, T., and Packer, L. (1994) Alpha-lipoate can protect against glycation of serum albumin, but not low density lipoprotein. Biochem. Biophys. Res. Commun. 203:99–104.
87. Keita, Y., Kratzer, W., Worner, W., and Rietbrock, N. (1993) Effect of free fatty acids on the binding kinetics at the benzodiazepine binding site of glycated serum albumin. Int. J. Clin. Pharmacol. Ther. Toxicol. 31:337–342.
88. van Boekel, M. A., van den Bergh, P. J., and Hoenders, H. J. (1992) Glycation of human serum albumin: Inhibition by Diclofenac. Biochim. Biophys. Acta 1120: 201–204.
89. Lyons, T. (1992) Lipoprotien glycation and its metabolic consequences. Diabetes 41 (Suppl. 2):67–73.
90. Schepkin, V., Kawabata, T., and Packer, L. (1994) NMR study of lipoic acid binding to bovine serum albumin. Biochem. Mol. Biol. Int. 33:879–886.
91. Kagan, V., Freisleben, H.-J., Tsuchiya, M., Forte, T., and Packer, L. (1991) Generation of probucol radicals and their reduction by ascorbate and dihydrolipoic acid in human low density lipoproteins. Free Rad. Res. Commun. 15:265–276.
92. Low, P. A., and Nickander, K. K. (1991) Oxygen free radical effects in sciatic nerve in experimental diabetes. Diabetes 40:873–877.
93. Bravenboer, B., Kappelle, A. C., Hamers, F. P. T., van Buren, T., Erkelens, D. W., and Gispen, W. H. (1992) Potential use of glutathione for the prevention and treatment of diabetic neuropathy in the streptozotocin-induced diabetic rat. Diabetologia 35:813–817.
94. Dimpfel, W., Spuler, M., Pierau, F.-K., and Ulrich, H. (1990) Thioctic acid induces dose-dependent sprouting of neurites in cultured rat neuroblastoma cells. Dev. Pharmacol. Ther. 14:193–199.

95. Sachse, G., and Willms, B. (1980) Efficacy of thioctic acid in the therapy of peripheral diabetic neuropathy. In: Aspects of Autonomic Neuropathy in Diabetes: International Symposium of the German Diabetes Association, Dusseldorf, May, 1978. (Gries, F. A., Freund, H. J., Rabe, F., and Berger, H., eds.), Stuttgart, New York: G. Thieme pp. 105–108.
96. Ziegler, D., Mayer, P., Muhlen, H., and Gries, F. A. (1993) Effeckte einer Therapie mit α-Liponsaure gegenüber Vitamin B_1 bei der diabetischen Neuropathie. Diab. Stoffw. 2:443–448.
97. Kähler, W., Kuklinski, B., Ruhlmann, C., and Plotz, C. (1993) Diabetes mellitus—a free radical-associated disease. Effects of adjuvant supplementation of antioxidants. In: The Role of Anti-oxidants in Diabetes Mellitus: Oxygen Radicals and Anti-oxidants in Diabetes (Gries, F. A., and Wessel, K., eds.), pp. 33–53. pmi Verl.-Gruppe, Frankfurt am Main.
98. Maitra, I., Serbinova, E., Tritschler, H., and Packer, L. (1995) α-Lipoic acid prevents buthionine sulfoximine-induced cataract formation in newborn rats. Free Rad. Biol. Med. 18:823–829.
99. Kilic, F., Packer, L., and Trevethick, J. R. (1994) Modeling cortical cataractogenesis 17: In vitro effect of α-lipoic acid on glucose-induced lens membrane damage, a model of diabetic cataractogenesis. Exp. Eye Res. (submitted).
100. Packer, L. (1993) The role of anti-oxidative treatment of diabetes mellitus. Diabetologia 36:1212–1213.
101. Scheer, B., and Zimmer, G. (1993) Dihydrolipoic acid prevents hypoxic/reoxygenation and peroxidative damage in rat heart mitochondria. Arch. Biochem. Biophys. 302:385–390.
102. Serbinova, E., Khwaja, S., Reznick, A. Z., and Packer, L. (1992) Thioctic acid protects against ischemia-reperfusion injury in the isolated perfused Langendorff heart. Free Rad. Res. Commun. 17:49–58.
103. Haramaki, N., Packer, L., Assadnazari, H., and Zimmer, G. (1993) Cardiac recovery during post-ischemic reperfusion is improved by combination of vitamin E with dihydrolipoic acid. Biochem. Biophys. Res. Commun. 196:1101–1107.
104. Assadnazari, H., Zimmer, G., Freisleben, H. J., Werk, W., and Leibfritz, D. (1993) Cardioprotective efficiency of dihydrolipoic acid in working rat hearts during hypoxia and reoxygenation. ^{31}P nuclear magnetic resonance investigations. Arzneimittel-Forsch. 43:425–432.
105. Prehn, J. H., Peruche, B., Karkoutly, C., Roβberg, C., Mennel, H. D., and Krieglstein, J. (1990) Dihydrolipoic acid protects neurons against ischemic/hypoxic damage. In: Pharmacology of Cerebral Ischemia (Krieglstein, J., and Oberpichler, H., eds.), pp. 357–362. Wissenschaftliche Verlagsgesellscheaft GmbH, Stuttgart.
106. Prehn, J. H., Karkoutly, C., Nuglisch, J., Peruche, B., and Krieglstein, J. (1992) Dihydrolipoate reduces neuronal injury after cerebral ischemia. J. Cereb. Blood Flow Metab. 12:78–87.
107. Boveris, A., Roldan, E. J. A., and Alvarez, S. (1994) Effect of lipoic acid on short-term ischemia-reperfusion in rat intestine. In: Biological Oxidants and Antioxidants. (Packer, L. and Cadenas, E., eds.), pp. 77–85. Hippokrates Verlag, Stuttgart.

108. Teutsch, C., and Brennan, R. W. (1978) Amanita mushroom poisoning with recovery from coma: A case report. Ann. Neurol. 3:177–179.
109. Plotzker, R., Jensen, D. M., and Payne, J. A. (1982) Case report: Amanita virosa acute hepatic necrosis: Treatment with thioctic acid. Am. J. Med. Sci. 283:79–82.
110. Vesconi, S., Langer, M., Iapichino, G., Costantino, D., Busi, C., and Fuime, L. (1985) Therapy of cytotoxic mushroom intoxication. Crit. Care Med. 13:402–406.
111. Bani, G. (1970) Sperimentazione clinica di un complesso coenzymatico in epatopazienti. Cl. Terap. 52:267–280.
112. Colombi, A., Thölen, H., and Huber, F. (1969) Einfluss von Coenzym A, Nicotinamid-adenin-dinucleotid (NAD), α-Liponsäure und Cocarboxylase auf die akute Hepatitis (Doppelblindversuch). Int. J. Clin. Pharmacol. 2:133–138.
113. Möller, E., and Schmitt, R. (1976) Ein Beitrag zur Behandlung chronischer Lebererkrankungen. Med. Klin. 71:1831–1835.
114. Marshall, A. W., Graul, R. S., Morgan, M. Y., and Sherlock, S. (1982) Treatment of alcohol-related liver disease with thioctic acid: A six month randomised double-blind trial. Gut 23:1088–1093.
115. Legrand-Poels, S., Vaira, D., Pincemail, J., van de Vorst, A., and Piette, J. (1990) Activation of human immunodeficiency virus type 1 by oxidative stress. AIDS Res. Hum. Retrovir. 6:1389–1397.
116. Freed, B. M., Rapoport, R., and Lempert, N. (1987) Inhibition of early events in the human T-lymphocyte response to mitogens and alloantigens by hydrogen peroxide. Arch. Surg. 122:99–104.
117. Aune, T. M., and Pierce, C. W. (1981) Conversion of soluble immune response suppressor to macrophage-derived suppressor factor by peroxide. Proc. Natl. Acad. Sci. USA 78:5099–5103.
118. Cerutti, P. A. (1985) Prooxidant status and tumor promotion. Science 227:375–381.
119. Buhl, R., Jaffe, H. A., Holroyd, K. J., Wells, F. B., Mastrangeli, A., Saltini, C., Cantin, A. M., and Crystal, R. G. (1990) Glutathione deficiency and HIV. Lancet 335:546.
120. Dworkin, B. M., Wormser, G. P., Axelrod, F., Pierre, N., Schwarz, E., Schwartz, E., and Seaton, T. (1990) Dietary intake in patients with acquired immunodeficiency syndrome (AIDS), patients with AIDS-related complex, and serologically positive human immunodeficiency virus patients: correlations with nutritional status. Jpn. J. Parent. Enteral Nutr. 14:605–609.
121. Bohl, R., Jaffe, H. A., Holroyd, K. J., Wells, F. B., Mastranseli, A., Saltini, C., Cantin, A. M., and Crystal, R. G. (1989) Systemic glutathione deficiency in symptom-free HIV-seropositive individuals. Lancet 2:1294–1298.
122. Fuchs, J., Schofer, H., Milbradt, R., Freisleben, H. J., Buhl, R., Siems, W., and Grune, T. (1993) Studies on lipoate effects on blood redox state in human immunodeficiency virus infected patients. Arzneimittel-Forsch. 43:1359–1362.
123. Baur, A., Harrer, T., Peukert, M., Jahn, G., Kalden, J. R., and Fleckenstein, B. (1991) Alpha-lipoic acid is an effective inhibitor of human immuno-deficiency virus (HIV-1) replication. Klin. Wochenschr. 69:722–724.
124. Boveris, A., and Chance, B. (1973) The mitochondrial generation of hydrogen

peroxide: General properties and the effect of hyperbaric oxygen. Biochem. J. 134:707–716.
125. Nohl, H., and Jordan, W. (1986) The mitochondrial site of superoxide formation. Biochem. Biophys. Res. Commun. 138:533–539.
126. Richter, C., Park, J. W., and Ames, B. N. (1988) Normal oxidative damage to mitochondrial and nuclear DNA is extensive. Proc. Natl. Acad. Sci. USA 85: 6465–6467.
127. Takeshige, K., and Minakami, S. (1979) NADH- and NADPH-dependent formation of superoxide amines by bovine heart submitochondrial particles. Biochem. J. 180:129–135.
128. Zaccarato, F., Cavallini, L., Deana, R., and Alexandre, R. (1988) Pathways of hydrogen-peroxide generation in guinea-pig cerebral cortex mitochondria. Biochem. Biophys. Res. Commun. 154:727–734.
129. Saggu, H., Cooksey, J., Dexter, D., Wells, F. R., Lees, A., Jenner, P., and Marsden, C. D. (1989) A selective increase in particulate superoxide dismutase activity in parkinsonian substantia nigra. J. Neurochem. 53:692–697.
130. Dexter, D. T., Carayon, A., Javoy-Agid, F., Agid, Y., Wells, F. R., Daniel, S. E., Lees, A. J., Jenner, P., and Marsden, C. D. (1991) Alterations in the levels of iron, ferritin and other trace metals in Parkinson's disease and other neurodegenerative diseases affecting the basal ganglia. Brain 114:1953–1975.
131. Jenner, P., Dexter, D. T., Sian, J., Schapira, A. H. V., and Marsden, C. D. (1992) Oxidative stress as a cause of nigral cell death in Parkinson's disease and incidental Levy body disease. Ann. Neurol. 32:S87–S97.
132. Shoffner, J. M., Watts, R. L., Juncos, J. L., Torroni, A., and Wallace, D. C. (1991) Mitochondrial oxidative phosphorylation defects in Parkinson's disease. Ann. Neurol. 30:332–339.
133. Volicer, L., and Crino, P. B. (1990) Involvement of free radicals in dementia of the Alzheimer type: A hypothesis. Neurobiol. Aging 11:567–571.
134. Wallace, D. C., Singh, G., Lott, M. T., Hodge, J. A., Schurr, T. G., Lezza, A. M. S., and Elsas, L. J. (1988) Mitochondrial DNA mutation associated with Leber's hereditary optic neuropathy. Science 242:1427–1430.
135. Stoll, S., Hartmann, H., Cohen, S. A., and Müller, W. E. (1993) The potent free radical scavenger α-lipoic acid improves memory in aged mice: Putative relationship to NMDA receptor deficits. Pharmacol. Biochem. Behav. 46:799–805.
136. Greenamyre, J. T., Garcia-Osuna, M., and Greene, J. G. (1994) The endogenous cofactors, thioctic acid and dihydrolipoic acid, are neuroprotective against NMDA and malonic acid lesions of striatum. Neurosci. Lett. 171:17–20.
137. Choi, D. W. (1988) Glutamate neurotoxicity and diseases of the nervous system. Neuron 1:623–634.
138. Altenkirch, H., Stoltenburg-Didinger, H., Wagner, M., Hermann, J., and Walter, G. (1990) Effects of lipoic acid in hexacarbon-induced neuropathy. Neurotoxicol. Teratol. 12:619–622.
139. Ramakrishnan, N., Wolfe, W. W., and Catravas, G. N. (1992) Radioprotection of hematopoietic tissues in mice by lipoic acid. Radiation Res. 130:360–365.
140. Korkina, L. G., Afanas'ef, I. B., and Diplock, A. T. (1993) Antioxidant therapy

in children affected by irradiation from the Chernobyl nuclear accident. Biochem. Soc. Trans. 21:314S.

141. Pryor, W. A. (1987) Cigarette smoke and the involvement of free radical reactions in chemical carcinogenesis. Br. J. Cancer (Suppl. 85):19–23.
142. Church, D. F., and Pryor, W. A. (1985) Free-radical chemistry of cigarette smoke and its toxicological implications. Environ. Health Perspect. 64:111–126.
143. Frei, B., Forte, T. M., Ames, B. N., and Cross, C. E. (1991) Gas phase oxidants of cigarette smoke induce lipid peroxidation and changes in lipoprotein properties in human blood plasma: Protective effects of ascorbic acid. Biochem. J. 277: 133–138.
144. Cross, C. E. ,O'Neill, C. A., Reznick, A. Z., Hu, M., Packer, L., and Frei, B. (1993) Cigarette smoke oxidation of human plasma constituents. Ann. NY Acad. Sci. 686:72–90.
145. Reznick, A. Z., Cross, C. E., Hu, M.-L., Suzuki, Y. J., Kwaja, S., Safadi, A., Motchnik, P. A., Packer, L., and Halliwell, B. (1992) Modifications of plasma proteins by cigarette smoke as measured by protein carbonyl formation. Biochem. J. 286:607–611.
146. Tsuchiya, M., Thompson, D. F., Suzuki, Y. J., Cross, C. E., and Packer, L. (1992) Superoxide formed from cigarette smoke impairs polymorphonuclear leukocyte active oxygen generation activity. Arch. Biochem. Biophys. 299:30–37.
147. Grunert, R. R. (1960) The effect of DL-α-lipoic acid on heavy-metal intoxication in mice and dogs. Arch. Biochem. Biophys. 86:190–194.
148. Sumathi, R., Devi, V. K., and Varalakshmi, P. (1994) DL-α-lipoic acid protection against cadmium-induced tissue lipid peroxidation. Med. Sci. Res. 22:23–25.
149. Gregus, Z., Stein, A. F., Varga, F., and Klaassen, C. D. (1992) Effect of lipoic acid on biliary excretion of glutathione and metals. Toxicol. Appl. Pharmacol. 114:86–96.
150. Cardinale, A., and De Simone, G. F. (1968) The use of sulphydryl containing drugs and chelating agents to reduce renal irradiation during ^{203}Hg chlormerodrin scintigraphy. J. Nucl. Biol. Med. 12:121–127.
151. Asta-Medica, α-Lipoic acid prescribing information. Asta Medica AG, Frankfurt, Germany.

Index

About the Editors

LESTER PACKER is Professor of Molecular and Cell Biology and Director, Membrane Bioenergetics Group, University of California, Berkeley. Dr. Packer is the author of over 500 published articles and coeditor of *Oxidative Stress in Dermatology, Vitamin E in Health and Disease*, with Jürgen Fuchs, and *Retinoids: Progress in Research and Clinical Applications*, with Maria A. Livrea (all titles, Marcel Dekker, Inc.). He is President of the Oxygen Club of California, President-Elect of the International Society of Free Radical Research, and a member of the Oxygen Society, the Society of Investigative Dermatology, and the American Society for Biochemistry and Molecular Biology, among other organizations. Dr. Packer received the B.S. (1951) and M.S. (1952) degrees in biology and chemistry from Brooklyn College, New York, and the Ph.D. degree (1956) in microbiology and biochemistry from Yale University, New Haven, Connecticut.

ENRIQUE CADENAS is a Professor in and Chairman of the Department of Molecular Pharmacology and Toxicology in the School of Pharmacy and a Professor of Biochemistry in the School of Medicine, University of Southern California, Los Angeles. The author of over 140 professional publications, Dr. Cadenas serves on the editorial board of numerous journals. He is Treasurer of both the International Society for Free Radical Research and the Oxygen Society, Vice-President of the Oxygen Club of California, and a member of the American Chemical Society, the American Society for Photobiology, the

Biochemical Society (United Kingdom), and the European Society for Photobiology, among others. Dr. Cadenas received the M.D. degree (1973) and the Ph.D. degree (1977) in biochemistry from the University of Buenos Aires School of Medicine, Argentina.

Emerging Targets in Antibacterial and Antifungal Chemotherapy

Emerging Targets in Antibacterial and Antifungal Chemotherapy

Edited by
Joyce A. Sutcliffe and
Nafsika H. Georgopapadakou

Chapman and Hall
New York London

First published in 1992 by
Chapman and Hall
an imprint of
Routledge, Chapman & Hall, Inc.
29 West 35 Street
New York, NY 10001-2291

Published in Great Britain by
Chapman and Hall
2-6 Boundary Row
London

Printed in the United States of America

Library of Congress Cataloging in Publication Data

Emerging targets in antibacterial and antifungal chemotherapy /
[edited by] Joyce A. Sutcliffe, Nafsika H. Georgopapadakou.
p. cm.
Includes bibliographical references and index.
ISBN 0-412-02711-9
1. Pharmaceutical microbiology. 2. Antibiotics. 3. Microbial sensitivity tests. I. Sutcliffe, Joyce A. II. Georgopapadakou, Nafsika H., 1950–
QR46.5.E44 1991
615'.329—dc20 91-15763
CIP

British Library cataloguing in publication data also available.

To our parents: Tasso and Elli Georgopapadakou
Ralph and Hazel Stanfield

Contributors

KEITH BARRETT-BEE (16,21), *Bioscience 1 Department, ICI Pharmaceuticals, Mereside, Alderley Park, Macclesfield SK10 4TG, United Kingdom*

RICHARD A. CALDERONE (20), *Department of Microbiology, Georgetown School of Medicine, Washington, D.C. 20007, U.S.A.*

WARREN M. CASEY (15), *Department of Microbiology, North Carolina State University, Raleigh, NC 27695, U.S.A.*

JOANNA CLANCY (11), *Central Research Division, Pfizer Inc., Groton, CT 06340, U.S.A.*

JACK COLEMAN (10), *Department of Biochemistry and Molecular Biology, Louisiana State University Medical Center, 1901 Perdido St., New Orleans, LA 70112, U.S.A.*

EVA CZYZEWSKA (17), *Department of Chemistry, Simon Fraser University, Burnaby, B.C. V5A 1S6, Canada*

ASIS DAS (4), *Department of Microbiology, University of Connecticut Health Center, Farmington, CT 06030, U.S.A.*

STEPHEN P. DENYER (14), *Department of Pharmacy, Brighton Polytechnic, Moulescomb, Brighton, Sussex BN 4GJ, United Kingdom*

JOSEPH DEVITO (4), *Department of Microbiology, University of Connecticut Health Center, Farmington, CT 06030, U.S.A.*

GEORGE L. DRUSANO (2), *Division of Infectious Disease, University of Maryland School of Medicine, 10 S. Pine St., Baltimore, MD 21201, U.S.A.*

W. STEPHEN FARACI (8), *Central Research Division, Pfizer Inc., Groton, CT 06340, U.S.A.*

D. P. FIGGITT (14), *Department of Pharmaceutical Sciences, University of Nottingham, Notthingham NG7 2RD, United Kingdom*

MAKIKO FUKAYAMA (20), *Department of Microbiology, Georgetown University School of Medicine, Washington, D.C. 20007, U.S.A.*

NAFSIKA H. GEORGOPAPADAKOU (9,18), *Department of Chemotherapy (58/2), Roche Research Center, Nutley, NJ 07110-1199, U.S.A.*

SARAH K. HIGHLANDER (12), *Department of Microbiology and Immunology, Bay-*

lor College of Medicine, Houston, TX 77030, U.S.A.

KUNIMOTO HOTTA (1), *National Institute of Health of Japan, Tokyo, Japan*

AKIRA IMADA (1), *Biology Research Laboratories, Takeda Chemical Industries, Osaka, Japan*

SUZANNE JACKOWSKI (6), *Department of Biochemistry, St. Jude Children's Research Hospital, 332 N. Lauderdale, Memphis, TN 38105, U.S.A.*

DAVID E. JACKSON (14), *Department of Pharmaceutical Sciences, University of Nottingham, Nottingham NG7 2RD, United Kingdom*

JOSEPH A. KOVACS (22), *Critical Care Medicine Department, Clinical Center, National Institutes of Health, Bethesda, MD 20892, U.S.A.*

GREGG Y. LIPSCHIK (22), *Critical Care Medicine Department, Clinical Center, National Institutes of Health, Bethesda, MD 20892, U.S.A.*

R. TODD LORENZ (15), *Department of Microbiology, North Carolina State University, Raleigh, NC 27695, U.S.A.*

JOE LUTKENHAUS (5), *Department of Microbiology, Molecular Genetics & Immunology, University of Kansas Medical Center, 39th and Rainbow Blvd., Kansas City, KS 66103, U.S.A.*

CHARLES S. MCHENRY (3), *Department of Biochemistry, Biophysics and Genetics, University of Colorado Medical School, 4200 E. Ninth Ave., Denver, CO 80262, U.S.A.*

RAJEEV MISRA (7), *Department of Microbiology, Arizona State University, Tempe, AZ 85287-2701, U.S.A.*

FRANCIS C. NEUHAUS (9), *Department of Biochemistry, Molecular & Cell Biology, Northwestern University, 2153 Sheridan Road, Evanston, IL 60208, U.S.A.*

A. CAMERON OEHLSCHLAGER (17), *Department of Chemistry, Simon Fraser University, Burnaby, B.C. V5A 1S6, Canada*

LEO W. PARKS (15), *Department of Microbiology, North Carolina State University, Raleigh, NC 27695, U.S.A.*

CATHERINE P. REESE (11), *Central Research Division, Pfizer Inc., Groton, CT 06340, U.S.A.*

NEIL S. RYDER (16), *Sandoz Forschungsinstitut, Gesellschaft M.B.H., Brunner Strasse 59, A1235 Wien, Austria*

JOHN F. RYLEY (21), *ICI Pharmaceuticals, Mereside, Alderley Park, Macclesfield, Cheshire SK10 4TG, United Kingdom*

THOMAS SILHAVY (7), *Department of Molecular Biology, Princeton University, Princeton, NJ 08544-1014, U.S.A.*

JASON SPARKOWSKI (4), *Department of Microbiology, University of Connecticut Health Center, Farmington, CT 06030, U.S.A.*

JAN S. TKACZ (19), *Department of Fermentation Microbiology, Merck Sharp & Dohme Research Laboratories, P.O. Box 2000, Rahway, NJ 07065-0900, U.S.A.*

THOMAS J. WALSH (13), *Section of Infectious Diseases, National Cancer Institute, Bldg. 10 Rm 13N-240, Bethesda, MD 20892, U.S.A.*

FREDERICK WARREN (4), *Creative BioMolecules, Hopkinton, MA 01748, U.S.A.*

GEORGE M. WEINSTOCK (12), *Department of Biochemistry and Molecular Biology, University of Texas Medical School, Houston, TX 77225, U.S.A.*

Contents

Preface

The prevalence of bacterial and fungal infections is reflected in the large market share captured by antimicrobial agents (20% of the total pharmaceutical sales, worldwide). Recent advances in medicine, such as transplantation and the use of prosthetic devices, have resulted in an increased need for effective antiinfective therapy. The treatment of chronic diseases (rheumatoid arthritis, diabetes, asthma, cancer) and AIDS patients is often undermined by recalcitrant secondary bacterial or fungal infections. Although many antibiotics are available to treat bacterial infections, they are often less efficacious because of resistance development. For fungal infections, on the other hand, a wide array of treatment choices is not currently available. New antimicrobial agents, with different structures and mechanisms of action, are clearly desirable, as they are likely to be immune to existing resistance mechanisms, and thus add to the armamentarium of the infectious disease physician.

The impetus for this collection of reviews was two seminars that highlighted potentially new antibacterial and antifungal targets. The seminars were organized for and presented at the 1989 American Society for Microbiology meeting. Since bacterial and fungal diseases represent some of the same challenges, we decided to combine approaches for drug discovery in these two areas. The coverage of novel targets is far from exhaustive; rather, only a selected number of worthwhile approaches has been encompassed. By focusing basic biological concepts on emerging chemotherapeutic targets, we hope that this volume will fill a gap between general texts on antimicrobial drugs and highly specialized reference books.

For many authors, writing a review in their area of expertise is not difficult to contemplate, but our contributors were presented with a more difficult task—to survey their areas from a chemotherapeutic perspective and to present their ideas in a readable format for a multidisciplinary audience. In cases where biochemical pathways are well defined, such as fatty acid or lipid A biosynthesis (chapters 6 and 10), specific enzymes that are unique to bacteria (acetoacetyl-ACP synthase) or rate-limiting (UDP-GlcNac acyltransferase) were highlighted. In other cases, where the biochemistry is less advanced, suggestions for specific targets were made on the basis of genetic analysis. In still other cases, where specific targets are yet to be defined, empirical approaches were discussed, such as the availability of strains that allow screening methodology to be established.

The modern era of antimicrobial therapy began in the 1930s with the launch of Prontosil, the first sulfonamide to be synthesized and developed as an antibacterial. Together with the successful development of penicillin as a systemic antibacterial agent in the 1940s, it triggered an intensive search for new antimicrobials, especially from fermentation sources. Early drug discoveries were largely the result of serendipity; they produced substances (chloramphenicol, gentamicin, polymyxin, tetracycline, amphotericin) with mechanisms of action that were initially unknown. Our expanding knowledge and insight into how microbes establish niches, survive, and grow have refined our approaches. Further, advances in recombinant DNA technology have strengthened the ability to enrich for drugs with specific modes of action. Yet despite our improved abilities in discovering drugs, we are humbled by an older, more pervasive force—natural selection. Selective pressure from antibiotic use has resulted in diminished potency and spectrum limitations in clinically useful antibiotics. Microbial resistance is the dark side, anthropocentrically speaking, of Darwinian evolution. A historical perspective on the more recent approaches of antibiotic discovery is given by Imada and Hotta (chapter 1).

Of antibacterials, β-lactams continue to be the most successful class and account for approximately two-thirds of the worldwide antibiotic market. An update on recent developments and new insights for this drug class is provided by Neuhaus and Georgopapadakou (chapter 9). Ergosterol synthesis inhibitors are the most widely used antifungals (chapter 16) and drugs with this mechanism of action are discussed by Barrett-Bee and Ryder.

Since the bacterial cell wall is a structure unique to procaryotes, the biosynthetic enzymes for this giant macromolecule represent promising antibacterial targets. Faraci (chapter 8) details the early steps in peptidoglycan synthesis, providing background and rationale useful to the selection of a target enzyme.

The remarkable medical and commercial success of quinolones has rekindled interest in nucleic acid synthesis and cell division. The identification of multiple, unexploited targets in RNA and DNA synthesis is captured in reviews by Das *et al.* (chapter 4) and McHenry (chapter 3), respectively. The roles of the many players involved in bacterial cell division, with special consideration for the essential role of FtsZ protein, are described by Lutkenhaus (chapter 5).

Some antibiotics are ineffective because they never reach targets within the cell; drugs may be excluded from entry into the cell by cellular membranes or effluxed by cellular pumps. Reese and Clancy (chapter 11) describe several screening strategies designed to expand the spectrum and/or enhance the potency of well-characterized antibiotics by disrupting the gram-negative cell envelope.

The barrier function of the outer membrane is also expected to be compromised by a lipid A inhibitor; potential targets in the biosynthetic pathway of this essential macromolecule are discussed by Coleman (chapter 10).

The protein secretion pathway, and appropriate methodology and screening tactics to enrich for novel inhibitors, are reviewed by Misra and Silhavy (chapter 7). An inhibitor of protein secretion might also be expected to effect the localization of certain virulence factors; this approach, as well as other approaches and an evaluation of numerous virulence targets, is discussed in the review by Highlander and Weinstock (chapter 12).

Because both fungi and mammals are eukaryotes, the selective eradication of pathogenic fungi poses a difficult challenge. Among the most attractive targets are components of the fungal cell wall, a structure totally lacking in mammalian cells. Two key enzymes involved in cell wall biosynthesis, chitin and glucan synthases, are reviewed by Georgopapadakou (chapter 18) and Tkacz (chapter 19), respectively.

Ergosterol, the major sterol in fungi, is not only a structural membrane component but also an essential player in fungal growth and development. The physiological outcome of inhibiting ergosterol synthesis is discussed by Parks *et al.* (chapter 15), while potential targets in the ergosterol biosynthetic pathway and rationally designed sterol inhibitors are covered by Oehlschlager and Czyzewska (chapter 17).

Less exploited antifungal targets include type II topoisomerase, an enzyme essential for DNA replication (chapter 14) and virulence factors such as acidic proteinase and adhesins elaborated by *Candida albicans* (chapter 20). A problem that is generic to antifungal discovery approaches is the applicability of existing *in vitro* and *in vivo* screens for evaluation of new antifungals; this area is critically reviewed by Ryley and Barrett-Bee (chapter 21).

The dramatic increase in the incidence of *Pneumocystis carinii* pneumonia, secondary to the spread of AIDS, and reclassification of this pathogen as a fungus make a review of this organism and its chemotherapeutic targets especially timely (chapter 22).

We have been fortunate to have enlisted enthusiastic colleagues who, we believe, have successfully met the challenge of writing reviews in their respective areas of expertise from a chemotherapeutic perspective. We sincerely thank all the authors for their patience and fortitude and our editor, Greg Payne, for his gentle encouragement, quiet efficiency, and uplifting humor.

A successful response to a challenge provokes a fresh challenge in its turn.

—Arnold Toynbee

Antibacterial Chemotherapy

1

Historical Perspectives of Approaches to Antibiotics Discovery

Akira Imada and *Kunimoto Hotta*

I. Introduction

The first 20 years in the half century of antibiotic search and discovery yielded many successful antibiotics. The chances of discovering new useful antibiotics, however, declined, sharply thereafter, precipitating strategic changes. Mechanism-based or target-directed screening, utilization of rare microorganisms and unusual culture conditions, and introduction of gene technology to modify antibiotic productivity have been among the major changes; these topics will be reviewed in this chapter.

A. *History of Antibiotics until 1970*

1. *Antibacterial Agents*

Penicillin was discovered in 1929 by Fleming (19) and its therapeutic efficacy was demonstrated by Chain *et al.* in 1940 (10). The clinical effectiveness of penicillin triggered a systematic search for antibacterial agents. Early work by Waksman demonstrated that actinomycetes are a treasure box of antibiotics; consequently, most of the clinically important antimicrobial agents were discovered as actinomycete products by the late 1950s: streptomycin (82), chloramphenicol (15), chlortetracycline (14), neomycin (100), oxytetracycline (18), erythromycin (58), leucomycin (27), oleandomycin (87), cycloserine (26), kanamycin (98) and rifamycin (85). Cephalosporin C (64) was also discovered in this period as a fungal product. The antibiotics discovered in the next decade number considerably fewer; *e.g.*, lincomycin (53), gentamicin (101), and fosfomycin (29). These antibiotics are also growth products of actinomycetes. In parallel with the screening efforts, chemical modifications of penicillins, cephalosporins, and aminoglycosides began in the 1960s, resulting in the synthesis of many useful derivatives.

2. *Antifungal Agents*

The history of antifungal agents is nearly as old as that of antibacterial agents. The first antifungal antibiotic, giseofulvin, was isolated from *Penicillium griseoflavum* and chemically characterized in 1939 (77). However, in sharp contrast to antibacterial agents, antifungal agents with favorable chemotherapeutic index and wide spectrum have been elusive. Several compounds were introduced clinically despite their low chemotherapeutic index: actinomycete products such as nystatin (28), amphotericin B (22), and azalomycin F (3); fungal products such as variotin (93) and siccanin (39); and a bacterial product, pyrrolnitrin (6). Of these, only amphotericin B has been used widely for systemic infections. Its use, however, is limited to life-threatening fungal infections due to severe adverse reactions.

B. *Need for a Breakthrough*

The rapid decline in the discovery of useful antibiotics at the end of the 1960s inevitably stimulated the need for strategic changes. In 1971, Conover (11) polled pharmaceutical companies worldwide, soliciting changes that might improve the chances of discovering novel antibiotics. The results of his survey are shown in Tables 1.1 and 1.2.

Table 1.1. Ranking of approaches to increased antibacterial microbial metabolite discovery 1970–1979

Approach	Rank[a]	Percent judging approach fruitful[b]
Application of new techniques for collecting, storing, and processing soil samples; for isolating and growing potential antibiotic-producing microorganisms; and for detecting new antibiotics	1	67
Examination of genera of microorganisms that have received relatively little attention thus far in the search for antibiotics	2	58
Examination of marine microorganisms	3	53
Examination of terrestrial microorganisms that grow under unusual or extreme environmental conditions	3	42
Examination of microorganisms that grow in the presence of pathogens	4	33

(Note): The introduction of numerous semisynthetic penicillins and cephalosporins since the 1971 survey confirmed that the design of structural analogs has proven profitable (Table 1.2). Another favored approach, the isolation of new microbial metabolites, however, has not proven nearly as successful. Only a few β-lactam compounds of microbial origin and their synthetic derivatives have enriched the antibiotic marketplace.

[a] Derived from individual rankings by use of a weighted scoring system.

[b] For the discovery of useful new antibacterial antibiotics in the next decade.

Table 1.2. Ranking of discovery potentital of approaches to antibacterial drug discovery

Approach	Rank[a]	Percent judging approach fruitful
Preparation of structural analogs of useful antibiotics by chemical or other means	1	83
Isolation of new microbial metabolites	1	92
Preparation of structural analogs of toxic or poorly efficacious antibiotics by chemical or other means	2	42
Directed synthesis of organic compounds based upon a biochemical rationale	3	[b]
Empirical screening of organic compounds unrelated to existing antibiotics	4	[b]

[a] Derived from individual rankings by use of a weighted scoring system.

[b] 65% of respondents judged that the search for new microbial metabolites will represent a better approach to the discovery of antibacterial drugs having novel structure, mode of action, and range of efficacy.

Antibiotic researchers of the late 1960s thought that improving the diversity of culture collections and application of new screening technology were most likely to enhance antibiotic discovery efforts (Table 1.1). In this chapter, we review the progress made in both of these areas.

C. Screening Technology

The screening of microbial metabolites includes four processes: (1) the isolation and (2) cultivation of microorganisms and (3) the detection and (4) identification of the active components. We shall call the techniques for discovering new, useful metabolites "screening technology." In this section we discuss detection techniques.

In the search for pharmacologically active substances, synthetic chemicals and microbial fermentation broths are traditionally the two major sources of test substances. An obvious contrast between these two sources is that broths are complex, containing compounds of unknown structure, while synthetic compounds, with known structure, are more simple and homogeneous. Additionally, active principles in microbial broths are usually present in small amounts while the majority of synthetic compounds can be readily supplied in larger amounts. The advantage to screening microbial samples for desired activity is that structurally novel, pharmacologically active compounds can be discovered. However, because microbial broths are more complex, the screens need to possess three features: simplicity, sensitivity, and specificity.

Simplicity: Screening for antimicrobial substances in culture broths usually requires the growth and cultivation of thousands of microbial strains. Therefore, assays that are simple to run and are compatible with whole fermentation broths are preferred.

Sensitivity: Extensive searches for antibiotics have been conducted worldwide for over 40 years. To discover lead compounds that have escaped previous screens, the methodology needs to be sensitive enough to detect antibiotics with weak activity and those produced at low concentrations.

Specificity: For the sake of efficiency, methodology that selectively enriches for the antibiotics with a specific mode of action is desirable at an early stage of screening.

Technological aspects of the detection methods should satisfy these three S's. An example that typifies these concepts is a screen for β-lactam antibiotics that uses supersensitive mutants. This screen compares the antibacterial susceptibility of the supersensitive mutants to their resistant parental strains by using simple agar plate diffusion assays. The use of isogenic pairs affords the sensitivity and specificity that neither strain alone would impart.

In addition to the three S's, a fourth S, *sensibility,* is also important—as important as the most sophisticated software (38). The researcher's awareness of what might be successful is also important.

In contrast to random screening, screening technology that has emerged since the 1970s is more target-directed (see below). Use of robotic equipment and computers to collect and sort data and miniaturization of assays are characteristic components of modern screening technology. Readers are referred to Nolan and Cross (66) for a review of other basic concepts for screening.

II. Screening Technology for Antibacterial Agents

Conventional screening programs for antibacterial agents have been geared to enrich for antibiotics with broad spectra and potency against pathogenic bacteria. Wild-type strains of gram-positive and gram-negative bacteria (*Staphylococcus, Escherichia, Salmonella, Pseudomonas,* and *Mycobacterium*) as the test organisms were sensitive enough to discover the classic antibiotics described in section I. Because of the emergence of resistant bacteria and an increase in the number of known antibiotics, however, innovative and sophisticated screens were needed by the middle of 1960s. As information concerning the mode of action, the basis of resistance mechanisms, and selective toxicity of antibiotics became available, target-directed or mechanism-based screens were established.

A. β-lactam Antibiotics

Screens for β-lactam antibiotics represent mechanism-based screens with high sensitivity, high specificity, and simplicity (2). They usually employ one or more of the following: (1) selective activity against β-lactam hypersensitive mutants; (2) sensitivity of test substances to hydrolysis by β-lactamase; (3) inhibition or (4) induction of β-lactamase; (5) induction of morphological changes in the test organism; (6) inhibition of DD-carboxypeptidase activity; and (7) inhibition of cell wall synthesis. By using one or more of these screening assays, various novel β-lactam antibiotics have been discovered (Table 1.3).

Prior to 1970 and establishment of these screens, penicillins and cephalosporins were reported as fungal metabolites (*Penicillium* and *Cephalosporium*) exclusively. In 1971, however, the first β-lactam antibiotics (cephamycins or 7-methoxycephalosporins) of nonfungal origin were reported as the metabolites of a *Streptomyces* species (60). Later, it was shown that a wide variety of *Streptomyces* species could produce cephamycins (89); methodology that used bacterial strains that were hypersusceptible to β-lactams (≤ 1 μg/ml) or were more sensitive to morphological change helped to detect the active components. Additionally, the degradation of antimicrobial activity by β-lactamases helped to characterize the principles.

A screening strategy that used bacterial strains hypersusceptible to β-lactam antibiotics discovered the nocardicins, the first monocyclic β-lactam antibiotics

Table 1.3. Differential assay systems targeting β-lactam antibiotics

Antibiotic	Producer[b]	Hyper-sensit. mutant	Sen.	β-lactamase		Ind.	Morpho-logical change	Cell wall Inh.
				Res.	Inh.			
Penicillins	*Penicillium*							
Cephalosporins	*Cephalosporium*							
Cephamycins	*Streptomyces*	+	+				+	
7α-Formylamino-cephems								
Cephabacins	*Lysobacter*	+	+	+	+		+	
SQ28516	*Flavobacterium*					+		
Norcardicins	*Actinosynnema*	+	+					
Formadicins	*Flexibacter*	+	+	+	+		+	
Clavams	*Streptomyces*				+			
Carbapenems								
Thienamycin	*Streptomyces*							+
Olivanic acid	*Streptomyces*				+			
PS-5	*Streptomyces*		+	+				
SQ-27860	*Serratia*					+		
Monobactams								
Sulfazecin	*Pseudomonas*	+	+				+	
SQ26445[c]	*Gluconobacter*					+		
SQ26180	*Chromobacterium*					+		
EM5400	*Agrobacterium*					+		
Lactivicin[d]	*Empedobacter*	+	+				+	+

[a] +: Used as screen.

[b] Only first-reported organisms are listed. In cases of bacterial products, some other bacterial genera were also reported as producers in the first reports. The nocardicin producer was formerly classified as *Nocardia*.

[c] SQ26445=sulfazecin.

[d] Lactivicin is not a β-lactam, but it has similar structure and activity to those of β-lactams.

(1), and thienamycin, the first carbapenem (42). β-lactam-supersensitive mutants can be derived from *Pseudomonas aeroginosa* (48, 72), *Escherichia coli* (1), *Comamonas terrigena* (71), and *Staphylococcus aureus* (43). It should also be noted that a wide variety of fungal genera other than *Penicillium* and *Cephalosporium* can produce penicillins and cephalosporins (47). Using a similar approach, Imada *et al.* discovered the first bacterially derived β-lactam antibiotics, sulfazecin and isosulfazecin, in 1981 (37). Independently, Sykes *et al.* (91) discovered the *N*-sulfonated monocyclic β-lactam antibiotics and proposed that this class be called monobactams. The latter authors subsequently reported that they had used the β-lactamase induction assay in *Bacillus licheniformis* to discover the monobactams; this system can detect β-lactam antibiotics in concentrations as low as 1 ng/ml (92). Clavulanic acid is the first clavam discovered by a subtle method used to detect β-lactamase inhibitors (9).

Other β-lactam-directed screens include inhibition of *Actinomadura* DD-carboxypeptidase activity (20, 83) and autolysin-triggering activity (78).

The discovery of lactivicin by β-lactam-directed screening methodology is notable (67). This bacterial product does not have the classical β-lactam nucleus, but it can be categorized with β-lactam antibiotics because it binds to penicillin-binding proteins of various bacteria and is susceptible to degradation by β-lactamases.

B. Cell Wall Inhibitors other than β-lactam Antibiotics

Efficacy and low toxicity of β-lactam antibiotics boosted efforts toward establishing screening methodology for cell wall inhibitors. Omura *et al.* (73, 75) established a screening system based on differential activity of microbial broths against *Bacillus subtilis* and *Mycoplasma,* the organism that lacks cell wall. Broths active against only *B. subtilis* that specifically inhibited the incorporation of [^{3}H]diaminopimelic acid (DAP) without inhibiting [^{14}C]leucine incorporation were selected for further evaluation. This differential assay system resulted in the discovery of new cell wall inhibitors, including the glycopeptides, azureomycins, and izupeptins (88). Known compounds like 3-aminoglucose, D-cycloserine, ristocetin, and penicillin G were also uncovered with this screening strategy.

To screen for glycopeptide antibiotics that inhibit the lipid cycle involved in peptidoglycan synthesis, a differential assay that was highly specific, sensitive, and simple was successfully designed by Rake *et al.* (79). In their screening program, actinomycete strains were isolated by using media containing vancomycin and cultivated under conditions suitable for glycopeptide production. For the differential assay, a vancomycin-resistant *Staphylococcus aureus* strain was used as the test organism in combination with diacetyl-L-lysyl-D-alanyl-D-alanine, a tripeptide analog of the glycopeptide receptor. Subsequently, new glycopeptide antibiotics such as aridicins and kibdelins were discovered.

Some other novel screens based on the mode of action of vancomycin were reported recently. O'Sullivan *et al.* (76) established an assay system using two agar plates seeded with *B. subtilis;* one plate is supplemented with a cell wall preparation of *S. aureus* that contains a variety of potential binding determinants to vancomycin. Agents with smaller inhibitory zones on the cell wall-containing plate could be distinguished, resulting in the discovery of a novel antibiotic, lysobactin, from *Lysobacter*. Mahoney *et al.* (52), on the other hand, established a peptide-binding chromogenic assay. This assay employed a pentapeptide (L-alanyl-D-isoglutaminyl-L-lysyl-D-alanyl-D-alanine)–bovine serum albumin conjugate immobilized on the wall of microtiter wells. The binding of vancomycin–alkaline phosphatase to the peptide could be monitored spectrophotometrically; the presence of glycopeptides inhibited the binding, thereby reducing the amount of phosphatase reporter activity. This assay system is simple, sensitive, and specific for glycopeptides.

The use of hypersensitive mutants to select for cell wall synthesis inhibitors other than β-lactams was also reported (44, 45). *Staphylococcus aureus* mutants hypersensitive to D-cycloserine or fosfomycin resulted in the discovery of TA-243 or α-D-glucose–1-phosphate, respectively, as inhibitors of cell wall synthesis.

Other screens to detect inhibitors of cell wall biosynthetic enzymes include an assay for alanine racemase and/or D-alanine:D-alanine ligase (99) and an assay that enriches for activities synergistic with D-cycloserine (50).

C. *Inhibitors of Protein and Nucleic Acid Syntheses*

Inhibitors of bacterial protein synthesis include such structurally diverse, clinically useful antibiotics as macrolides, aminoglycosides, tetracyclines, and chloramphenicol. A general assay for all protein synthesis inhibitors is the incorporation of radiolabeled amino acid (*e.g.*, leucine) into protein (TCA-insoluble fraction). In this assay, however, inhibitors of nucleic acid synthesis will also be detected because of their polar effect on protein synthesis. Thus, secondary screens are needed to distinguish specific inhibitors. An interesting assay for protein biosynthetic inhibitors was reported by Naveh *et al.* (63). They used a dark variant of a luminous bacterium, *Photobacterium leiognathi;* luminescence in this mutant strain could be induced with DNA-intercalating agents such as acridine dyes. By using a rapid bioluminescence system, *de novo* protein-dependent induction of luminescence could be monitored and protein synthesis inhibitors detected at concentrations of 0.1–0.3 μg/ml.

Several screens designed to enrich for specific classes of protein synthesis inhibitors were also reported. Yao and Mahoney (104) established a gentamicin enzyme-linked immunosorbent assay (ELISA) that detected aminoglycoside antibiotics selectively, with no cross-reactivity to nonaminoglycosides. In this assay, microtiter wells were coated with gentamicin antibody and fermentation broths that competed for a gentamicin–alkaline phosphatase conjugate were sought. This technique is applicable to other antibiotic classes if appropriate antibody and antibiotic-enzyme conjugates are available. Another simple approach specific for aminoglycoside antibiotics is the use of actinomycete strains with different aminoglycoside resistance patterns as test organisms; more details will be described in section IV (70). Additionally, a mutant of *Klebsiella pneumoniae* hypersensitive to aminoglycoside antibiotics has been used successfully to find new aminoglycosides (68).

Inhibitors of nucleic acid synthesis can be categorized into four types based on their mode of action: (1) compounds that directly interact with DNA, (2) inhibitors of topoisomerases, (3) inhibitors of DNA synthesis, and (4) inhibitors of RNA synthesis.

Agents that interact primarily with DNA often have antitumor activity; mechanism-based screens for inhibitors of bacterial nucleic acid synthesis have been used to discover antitumor antibiotics. For example, DNA repair-deficient-bacterial strains (*rec*−) can detect microbial metabolites with potential antitumor activity because they are more sensitive to agents interacting with DNA than isogenic *rec*+ strains (55). Matsuda *et al.* (57) used a *recA* mutant of *Escherichia coli* in a medium supplemented with or without paraquat to enrich for agents interacting with DNA. Paraquat increases the intracellular level of superoxide dismutase, providing the *recA* mutant with increased resistance to antitumor agents like anthracyclines; these latter agents produce oxygen radicals and cause DNA strand scission. This approach revealed an unexpected aspect of D-cycloserine; the drug produces oxygen radicals and cleaves DNA strands (56). Steinberg *et al.* (90) detected DNA synthesis inhibitors in a *dim* mutant of *Photobacterium leiognathi* by using a bioluminescence assay.

Currently, one of the most clinically useful classes of antibacterials is the fluoroquinolones, synthetic compounds that inhibit bacterial DNA gyrase (topoisomerase II). Although no microbial products specifically inhibiting the A subunit of gyrase have been reported, novobiocin and coumermycin are natural products that inhibit the B subunit (21, 33). Additionally, topoisomerase inhibitors of eucaryotic topoisomerases (camptothecin, etoposide) were found in natural product screens. The great clinical success of the fluoroquinolones will undoubtedly stimulate assay systems targeting DNA topoisomerases.

Omura *et al.* (74) established a differential assay using *Enterococcus faecium* to select novel inhibitors of folate metabolism. *Enterococcus faecium* is deficient in some enzymes essential for folate biosynthesis and requires a culture medium supplemented with folate, pteroate, and leucovorin for growth. This requirement was exploited to search for inhibitors of the folate biosynthetic pathway by detecting compounds active against the organism grown in a medium containing a limited amount of pteroate, but inactive when *E. faecium* is grown in medium containing thymidine. This approach resulted in the discovery of diazaquinomycins A and B and AM–8042.

DNA and RNA synthesis inhibitors are elaborated elsewhere (see chapters 3 and 4).

D. Other Agents

Screening systems that target other essential cellular metabolic functions like cell division, membrane transport and integrity, fatty acid synthesis, lipopolysaccharide biosynthesis, and systems that identify agents that reduce the virulence of bacteria will be described in later chapters.

III. Screening Technology for Antifungal Agents

The development of efficacious antifungal agents has not been as successful as that of antibacterial agents. The main reason is that fungi and mammals are both eukaryotes and their metabolic pathways are not significantly different; consequently, it is difficult to find lethal, selective targets. However, the need for antifungal agents is on the rise due to the increase of fungal infections in immunocompromised hosts (see chapters 13 and 22).

The most attractive antifungal targets are essential components of the cell wall or membrane and will be briefly described in this section. For more detail, the reader is referred to other chapters in the book.

A. Agents Affecting Fungal Membranes

Amphotericin B is the only naturally occurring antifungal antibiotic that is used worldwide in systemic infections. Its site of action is the fungal membrane (13). Ergosterol, the major sterol in yeast, has been identified as the binding site for amphotericin B and other polyene macrolide antibiotics (95). These antibiotics also bind to cholesterol, the major sterol in mammalian membranes. The differential binding to ergosterol and cholesterol provides a basis for a mechanism-based screen to discover new agents with better selective toxicity (65). Some differential assays to enrich for polyene macrolides can be found in the literature: antagonism of antifungal activity by ergosterol (106) or interference of biological activity by fatty acids (36) or Tween 80 (30). Recently, Etienne *et al.* (17) isolated mutants of *Saccharomyces cerevisiae* resistant to both polyene macrolides and cycloheximide; the resistant strains were used to eliminate polyene antibiotics in search of new nonpolyene antifungal agents.

Recently, synthetic imidazole and triazole compounds such as ketoconazole, intraconazole, and fluconazole have been developed to treat systemic fungal infections. As will be reviewed later (chapter 16), these compounds inhibit the P450-mediated demethylation of lanosterol to ergosterol, thereby decreasing the major sterol of the fungal membrane. The P450-mediated demethylation step in ergosterol synthesis represents a viable target for microbial antifungal products, though no natural inhibitors of this step have thus far been reported.

Ryder and Dupont (81) showed that squalene epoxidase, another enzyme involved in ergosterol synthesis, can also be a viable target for antifungal agents;

the fungal enzyme is 1,000 times more sensitive to allylamines than the mammalian enzyme.

B. Cell Wall Inhibitors

A major difference between plants (including fungi) and animals is the presence of a cell wall in the former. The enzymes that constitute the cell wall biosynthetic pathway, therefore, are targets for antifungal agents that should be selectively toxic.

The fungal cell wall contains glucans and chitin as its major structural components. Several inhibitors of chitin synthase and 1,3-β-glucan synthase have been described; these compounds have agricultural rather than medical applications. Brillinger (8) proposed the use of an *in vitro* chitin synthase system to screen for insecticides. This approach could also be used to look for antifungal agents.

Isono (41) used chitin synthase from *Pyricularia oryzae* and 1,3-β-glucan synthase and mannan synthase from *Saccharomyces cerevisiae* as enzyme sources for *in vitro* screens; UPD-N-acetylglucosamine, UDP-glucose, and UDP-mannose were radiolabeled substrates, respectively. Cell wall synthesis was monitored by thin-layer chromatography of polymer fractions. Neopolyoxins A, B, and C (49) were discovered as new chitin synthase inhibitors and lipopeptins (96) and neopeptins A, B, and C (97) as glucan and mannan synthase inhibitors, respectively. Neopolyoxins were active not only against *P. oryzae* but also against clinically important strains of *Candida* and *Aspergillus;* however, the chemical instability of neopolyoxins in weakly acidic solutions hindered their development. Masson *et al.* (54) also described a screening method for chitin synthase inhibitors that uses permeabilized cells of *Saccharomyces cerevisiae*. Other examples of chitin and glucan synthase assays will be reviewed in chapters 18 and 19.

Selitrennikoff (84) reported an elegant screening method for cell wall inhibitors. He isolated a temperature-sensitive, osmotically fragile mutant of *Neurospora crassa* that grows and divides freely as protoplasts at 37°C. Upon a temperature shift to 22°C, the majority of cells regenerate cell wall and form mycelium. Agents which prevent this regeneration, but not protoplast growth or division, were sought. In operation, paper discs containing test samples were placed on an agar plate seeded with protoplasts. Incubation at 22°C permitted cell wall regeneration and growth. Plates were examined for indications of zones where cell wall regeneration, but not growth, had been inhibited. Compounds that inhibited only regeneration, resulting in hazy inhibition zones containing clumps of large protoplasts, were easily identified. The method is simple, easy to use, and readily adaptable for large-scale screens; however, some antibiotics other than fungal cell wall inhibitors also produce hazy zones of inhibitions. Although the method is not specific for fungal cell wall inhibitors, an understanding of the nature of the false-positive compounds provides important clues that could be used to weed out these samples early.

C. Agents Inducing Morphological Changes

Dähn *et al.* (12) reported the discovery of a polyoxin analog active against *Candida* in their morphology screen that used *Botrytis cinerea* and *Mucor hiemalis*. Beppu (7) developed a screening method based on the observation of morphological changes in fungi and yeasts (see also ref. 24). The latter investigations grew *Candida albicans* or *Mucor racemosus* on nutrient agar plates and added paper discs impregnated with microbial broths to the surface of the seeded agar. Cell morphology, including dimorphism, was microscopically observed around the discs. Microbial broths inducing morphological changes were detected with a 20% frequency, while those inducing dimorphism were not detected. The active principles were identified mainly as families of polyene macrolides, polyethers (ionophores), antimycin (respiratory chain inhibitors), streptothricins (protein synthesis inhibitors), polyoxins (chitin synthase inhibitors), and validamycin (antifungal antibiotic for agricultural use). Thus, although the morphological method might be assessed as having rather low specificity, structurally novel agents like leptomycin A and B and phosphazinomycin were discovered after screening only 2,000 soil isolates (24, 25).

D. Gap between In Vitro *and* In Vivo *Activities*

One of the issues regarding the search for clinically active antifungal agents is the poor correlation between *in vitro* antifungal activity and *in vivo* therapeutic efficacy. This gap has been found not only with naturally occurring agents but also with synthetic azole derivatives (see chapter 2). The reason for this is unknown. The establishment of an *in vitro* screen that can predict *in vivo* efficacy is most urgently needed.

IV. Selection and Manipulation of Antibiotic Producers

A. Target Organisms as Screening Source

A wide variety and large number of actinomycetes, bacteria, and fungi have been exploited in the search for antibacterial and antifungal agents. Waksman introduced streptomycetes into his systematic screening program in the early 1940s and these organisms turned out to be excellent sources of antibiotics. Subsequently, they have become the major resources for antibiotic screening worldwide. Until 1974, *Streptomyces* produced about 95% of the 2,000 antibiotics of actinomycete origin, including many medically important ones. As the number of antibiotics increased, however, it became increasingly difficult to discover new, useful ones. Consequently, microbiologists turned their attention to "rare" actinomycetes. As shown in Figure 1.1, "rare" actinomycetes (see section I and Table 1.1) as antibiotic sources increased in importance and produced about 30%

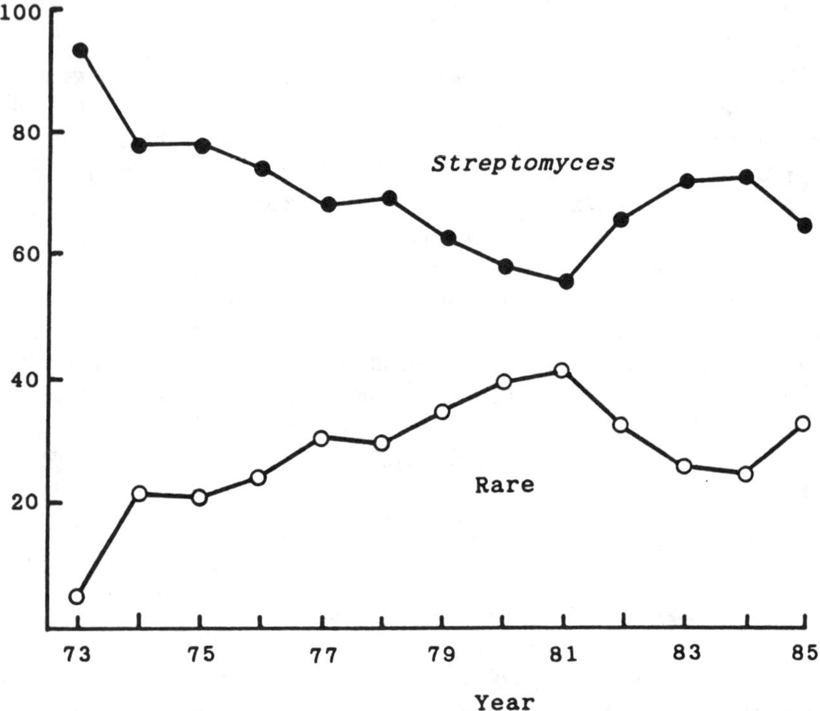

Figure 1.1 Ratio between *Streptomyces* and rare actinomycetes that yielded new antibiotics (data mainly based on the 1973–1985 issues of the *Journal of Antibiotics*).

of the 1,500 antibiotics of actinomycete origin reported in the 10 years following 1974.

There is no doubt that rare actinomycetes are important sources for antibiotic screening. These organisms have provided novel antibacterial and antifungal antibiotics: *e.g.*, a novel carbapenem AB–110-D by *Kitasatosporia* (61) and an antifungal antibiotic, benanomicin, by *Actinomadura* (94). The word "rare" for actinomycetes was first used by Lechevalier and Lechevalier in 1967 (51) to describe strains of the actinomycete genera other than *Streptomyces* that were less frequently encountered by conventional isolation. Rare actinomycetes now include about 40 genera such as *Nocardia, Micromonospora, Actinoplanes,* and so on (102). It should be noted, however, that the major antibiotic source is still *Streptomyces*. *Streptomyces* strains may be regarded as rare or unusual if their incidence of isolation is very low; *S. kanamyceticus* (kanamycin producer) and *S. cattleya* (thienamycin producer) are such examples.

It has been shown that antibiotic production is not specific to species, but to strain (70). Thus, the selection of microorganisms based only on taxonomic characterization is no longer predictive. To increase the probability of new antibiotic discovery, it is necessary and helpful to establish characterization methods based on genetic and/or biochemical properties. Additionally, enrichment for unexplored microorganisms that have resisted most isolation efforts will be increasingly important. Otherwise, discovery of new antibiotics may only result from screening a large number of microorganisms. Thus, from the point of view of antibiotic production we should redefine rare antinomycetes as metabolically, biochemically, or genetically rare.

In this context, the following examples should be noted. In searching for new aminoglycoside antibiotic (AG) producers, Hotta *et al.* (34) devised a method to isolate actinomycetes in media containing AGs, followed by the characterization of AG-resistance patterns of the isolates. This approach was based on the fact that AG producers generally have individual, multiple AG-resistance patterns depending on the type of AGs they produce. Subsequently, this approach yielded actinomycetes that produced a wide variety of aminoglycosides, including new ones at a rate of about 10%. The glycopeptide-directed screening program established by Rake *et al.* (79) is another example (see section II B). Their approach yielded a diverse group of actinomycetes (vancomycin-resistant isolates) which produced glycopeptides at a rate of 2.2% (44 out of 1935 cultures). It is notable that new glycopeptides, kibdelins and aridicins, are produced by a new genus, *Kibdelosporangium,* and were discovered multiple times, at a rate of one out of every 320 cultures. A total target-directed screening system for carbapenem antibiotics was also reported by Nakamura *et al.* (62). The system involved (1) isolation of soil organisms in a medium containing both amoxicillin and clavulanate, (2) cultivation in a selected medium suitable for carbapenem production, and (3) a carbapenem-specific agar diffusion assay. As a result of this selection methodology, 20% of the organisms produced carbapanem antibiotics and a novel carbapenem antibiotic, AB–110-D, was discovered.

In order to keep discovering new antibiotics, criteria for discriminating novel or unexplored microorganisms from the more routine strains need to be established. Because antibiotic production is strain-specific, the criteria should be based on genetic and/or biochemical properties rather than taxonomic properties. New approaches of isolation technology may result in a breakthrough.

B. Cultivation Conditions

Since production of novel antibiotics may require different physiological conditions and/or genetic backgrounds, manipulating cultivation conditions is one of the key points to the discovery of novel antibiotics. This includes the isolation and cultivation of wild-type soil isolates as well as laboratory-derived mutants or recombinants.

Actinomycetes are usually cultivated in various liquid media with shaking at 180–250 rpm at 27–28°C for antibiotic production. Cultivation media usually consist of 1–3% carbon and 0.5–2% nitrogen sources in combination with 5–15 mM inorganic phosphate. Cations such as Ca^{+2}, Mg^{+2}, and Na^{+} are supplemented at 10–15 mM and trace elements like Co^{+2}, Cu^{+2}, Fe^{+2}, and Zn^{+2} can be added. Since regulatory mechanisms for antibiotic production are so varied, no single medium can be satisfactory for the production of all types of antibiotics. The regulatory mechanisms include inhibition by carbon and nitrogen catabolites, phosphate, and the antibiotic product. The induction by endogenous bioregulators such as A-factor has also been reported (46). Therefore, media with varied composition are useful in screens for diverse antibiotics. Some antibiotics are produced under unusual conditions which might not favor the production of many other antibiotics. Thus, pyrrolnitrin (6), nocardicin (1) and phosphonic acid antibiotics (73) were discovered in media containing 100–200 mM inorganic phosphate; the production of most antibiotics is negatively regulated by such high concentrations of inorganic phosphate. Aplasmomycin (69) was found when actinomycetes were examined for differential antibiotic productivity in nutritionally rich and poor media. The antibiotic was produced only when the rich medium was diluted 16-fold. Cryomycin (105) was found by incubating a facultatively psychrophilic subspecies of *S. griseus* at a temperature ranging from 0 to 18°C while incubation of *S. lavendulae* under an unusual intensive shaking led to the discovery of chlorocarcin, mimosamycin, and saframycin (4, 5).

C. Genetic Manipulation

It has been shown that the substrate specificity of the biosynthetic enzymes in an antibiotic pathway is relatively flexible, allowing the accumulation of structurally related antibiotic metabolites in the fermentation broths (70). This permissive substrate specificity lends itself to the addition of precursors and sometimes results in the formation of new antibiotics, either directly or by mutasynthesis (mutational biosynthesis). Interspecific crosses, cell fusion, and gene manipulations have also provided new antibiotics. The genetic manipulations used for generating new antibiotics are exemplified below.

1. Mutasynthesis

A blocked mutant of *S. fradiae* that requires 2-deoxystrepatamine (DOS) to biosynthesize neomycin was found to incorporate exogenous DOS analogues into antibiotic molecules without further modification (86). The new antibiotics that were produced are called hybrimycins. Blocked mutants that grow without special medium supplementation but require supplementation for antibiotic biosynthesis are called "idiotrophs." Antibiotic production using idiotrophs is called "mutasynthesis" by Rinehart (80) and "mutational biosynthesis" by Nagaoka and Demain (59). New mutasynthetic antibiotics include derivatives of aminoglycosides, novobiocin, and platenomycin, a macrolide (70).

2. *Cell Fusion*

Techniques for the fusion of actinomycete protoplast cells have been developed and used to obtain mutants, heterokaryons, or recombinant organisms that produce new antibiotics. Indolizomycin was the first antibiotic obtained by using this approach (23, 103). Two strains of aminoglycoside producers, *S. griseus* (streptomycin (SM) producer) and *S. tenjimariensis* (istamycin (IS) producer), were chosen for fusion studies, and the differences in their native resistance patterns to SM and kanamycin (KM) were used as selective markers. After fusion, clones that had acquired resistance to both SM and KM were selected; one of them was found to produce a novel antibiotic named indolizomycin. Interestingly, indolizomycin is not a hybrid antibiotic but is novel, with no structural similarity to either SM or IS. The clone was taxonomically identified as *S. griseus*. Biochemical and genetic characterization (35, 40) suggested that the indolizomycin production was not due to genetic recombination but to the activation of *S. griseus* genes that were cryptic.

3. *Gene Cloning and Heterologous Expression*

Cloning of genes involved in antibiotic biosynthesis from actinomycetes is now possible and regulatory elements for gene expression, coding sequences, and genetic arrangements are recognizable. Generally, the biosynthetic genes are clustered in association with the self-resistance genes (32).

The generation of a mederrhodin A–producing *Streptomyces* strain was the first example of a new antibiotic produced by gene manipulation (31). This system was based on the structural similarity between actinorhodin and medermycin that are produced by *S. coelicolor* A3(2) and *Streptomyces* sp. AM7161, respectively. Gene segments from the former strain were cloned into a vector plasmid, plJ922, and introduced into the medermycin producer; a recombinant strain that produced mederrhodin was obtained. Recently, a novel hybrid macrolide antibiotic, isovaleryl-spiramycin, was produced in *S. ambofaciens* by introduction of a cloned carbomycin (a macrolide) biosynthetic gene, *carE*, from *S. thermotolerans* (16).

V. Conclusion and Future Prospects

It is well recognized that discovery of novel useful antibiotics is becoming more and more difficult. Additionally, it has been argued that no more new antibacterials are needed. We have witnessed, however, a recent burst of methicillin-resistant *S. aureus* (MRSA) and the rapid emergence of quinolone-resistant *P. aeruginosa* and other bacteria in the clinic. These facts suggest that the battle against bacteria is never-ending. Screening efforts should continue to find new antibacterial leads to enable us to cope with the emergence of resistant bacteria and with newly introduced problem pathogens. The increase in immunocompromised patients has changed the pattern of infectious diseases encountered clinically; consequently, immunostimulants as adjunct antiinfective therapeutic agents are needed. There is also a pressing need for better antifungal agents and for antiviral agents in this latter context.

We have emphasized the necessity of simplicity, sensitivity, and specificity in screening technology and the importance of the use of unexplored (rare) microorganisms as screening sources. Genetic manipulation was also suggested as a way to enrich for new antibiotics. Based on these fundamentals, we believe that sophisticated modern screening systems can be constructed.

The knowledge of bacterial and fungal pathogenicity is expected to provide new insight for new targets. When a gene product is identified as a pathogenic factor, we can focus our efforts on the regulation of expression as well as the function of the gene. More sophisticated methods in molecular biology, cell biology, genetic engineering, and "receptology" (understanding of a receptor at a molecular level) can be recruited to set up novel screening strategies.

References

1. Aoki, H., H. Sakai, M. Kohsaka, T. Konomi, J. Hosoda, Y. Kubouchi, E. Iguchi, and H. Imanaka. 1976. Nocardicin A, a new monocyclic β-lactam antibiotic. I. Discovery, isolation and characterization. J. Antibiotics **29:** 492–500.

2. Aoki, H., K. Kunugita, J. Hosoda, and H. Imanaka. 1977. Screening of new and novel β-lactam antibiotics. Japanese J. Antibiotics (Suppl.) **30:** 207–217.

3. Arai, M. 1960. Azalomycins B and F, two new antibiotics. II. Properties of azalomycins B and F. J. Antibiotics **13A:** 51–56.

4. Arai, T., K. Yazawa, Y. Mikami, A. Kubo, and K. Takahashi, 1976. Isolation and characterization of satellite antibiotics, mimosamycin and chlorocarcins from *Streptomyces lavendulae,* streptothricin source. J. Antibiotics **29:** 398–407.

5. Arai, T., K. Takahashi, and A. Kubo, 1977. New antibiotics, saframycins A, B, C, D and E. J. Antibiotics **30:** 1015–1018.

6. Arima, K., H. Imanaka, M. Kousaka, A. Fukuda, and G. Tamura. 1965. Studies on pyrrolnitrin, a new antibiotic. I. Isolation and properties of pyrrolnitrin. J. Antibiotics **18A:** 201–204.

7. Beppu, T. 1982. Screening of antifungal agents using morphological abnormality as an index, p. 129–137. *In* K. Arima, H. Umezawa, S. Fukui, and D. Mizuno (ed.), Medicine and microbial production, Part I. Gakkai Shuppan Center, Tokyo.

8. Brillinger, G.U. 1979. Chitin synthase from fungi: A test model for substances with insecticidal properties. Arch. Microbiol. **121:** 71–74.

9. Brown, A.G., D. Butterworth, M. Cole, G. Hanscomb, J.D. Hood, C. Reading, and G.N. Rolinson. 1976. Naturally-occurring β-lactamase inhibitors with antibacterial activity. J. Antibiotics **29:** 668–669.

10. Chain, E., H.W. Florey, A.D. Gardner, N.G. Heatley, M.A. Jennings, J. Orr-Ewing, and A.G. Sanders. 1940. Penicillin as a chemotherapeutic agent. Lancet **II:** 226–228.

11. Conover, L.H. 1971. Drug discovery from microbiological sources, p. 33–80. *In* R.F. Gould (ed.), Drug Discovery (Advances in Chemistry, series 108). American Chemical Society, Washington, D.C.

12. Dähn, U., H. Hagenmaier, H. Höhne, W.A. König, G. Wolf, and H. Zähner. 1976. Metabolic products of microorganisms. 154. Nikkomycin, a new inhibitor of fungal chitin synthesis. Arch. Microbiol. **107:** 143–160.

13. de Kruyff, B., and R.A. Demel. 1974. Polyene antibiotic-sterol interactions in membranes of *Acholeplasma laidlawii* cells and lecithin liposomes. III. Molecular structure of polyene antibiotic-cholesterol complexes. Biochem. Biophys. Acta. **339:** 57–70.

14. Duggar, B.M. 1948. Aureomycin: a product of the continuing search for new antibiotics. Ann. N.Y. Acad. Sci. **51:** 177–181.

15. Ehrlich, J., G.R. Bartz, R.M. Smith, and D.A. Joslyn. 1947. Chloromycetin, a new antibiotic from a soil actinomycete. Science **106:** 417.

16. Epp, J.K., M.L.B. Huber, J.R. Turner, T. Goodson, and B.E. Schoner. 1989. Production of a hybrid macrolide antibiotic in *Streptomyces ambofaciens* and *Streptomyces lividans* by introduction of a cloned carbomycin biosynthetic gene from *Streptomyces thermotolerans*. Gene **85:** 293–301.

17. Etienne, G., E. Armau, and G. Tiraby. 1990. A screening method for antifungal substances using *Saccharomyces cerevisiae* strains resistant to polyene macrolides. J. Antibiotics **48:** 199–206.

18. Finlay, A.C., G.L. Hobby, S.Y. Pan, P.P. Regna, J.B. Routien, D.B. Seeley, G.M. Shull, B.A. Sobin, I.A. Solomons, J.W. Vinson, and J.H. Kane. 1950. Terramycin, a new antibiotic. Science **111:** 85.

19. Fleming, A. 1929. On the antibacterial action of culture of a *Penicillium,* with special reference to their use in the isolation of *B. influenzae*. Brit. J. Exp. Pathol. **10:** 226–236.

20. Frere, J.M., D. Klein, and J.M. Ghuysen, 1980. Enzymatic method for rapid and sensitive determination of β-lactam antibiotics. Antimicrob. Agents and Chemother. **18:** 506–510.

21. Gellert, M., M.H. O'Dea, T. Itoh, and J.I. Tomizawa. 1976. Novobiocin and coumermycin inhibit DNA supercoiling catalyzed by DNA gyrase. Proc. Natl. Acad. Sci. U.S.A. **73:** 4474–4478.

22. Gold, W., H.A. Stout, J.F. Pagano, and R. Donovick. 1956. Amphotericin A and B, antifungal antibiotics produced by a *Streptomycetes*. I. *In vitro* studies, p. 579–586. *In* H. Welch and F. Marti-Ivanez (ed.), Antibiotics Annual. Medical Encyclopedia, Inc., New York.

23. Gomi, S., D. Ikeda, H. Nakamura, H. Naganawa, F. Yamashita, K. Hotta, S. Kondo, Y. Okami, H. Umezawa, and Y. Iitaka. 1984. Isolation and structure of a new

antibiotic, indolizomycin, produced by a strain SK2–52 obtained by interspecies fusion treatment. J. Antibiotics **37:** 1491–1494.

24. Gunji, S., K. Arima, and T. Beppu. 1983. Screening of antifungal antibiotics according to activities inducing morphological abnormalities. Agr. Biol. Chem. **47:** 2061–2069.

25. Hamamoto, T., S. Gunji, H. Tsuji, and T. Beppu. 1983. Leptomycins A and B, new antifungal antibiotics. I. Taxonomy of the producing strains and their fermentation, purification and characterization. J. Antibiotics **36:** 639–645.

26. Harned, R.L., P.H. Hidy, and E.K. LaBaw. 1955. Cycloserine. I. A preliminary report. Antibiotics & Chemotherapy **5:** 204.

27. Hata, T., Y. Sano, N. Ohki, Y. Yokoyama, A. Matsumae, and S. Ito. 1953. Leucomycin, a new antibiotic. J. Antibiotics **6A:** 87–89.

28. Hazen, E.L., and R. Brown. 1950. Two antifungal agents produced by a soil actinomycete. Science **112:** 423.

29. Hendlin, D., E.O. Stapley, M. Jackson, H. Wallick, A.K. Miller, F.J. Wolf, T.W. Miller, L. Chaiet, F.M. Kahan, E.L. Folz, H.B. Woodruff, J.M. Mata, S. Hernandez, and S. Mochales. 1969. Phosphonomycin, a new antibiotic produced by a strain of *Streptomyces*. Science **166:** 122–123.

30. Hickey, R.J. 1953. The antagonism between the antifungal antibiotic, ascosin, and some long chain, unsaturated fatty acids. Arch. Biochem. Biophys. **46:** 331–336.

31. Hopwood, D.A., F. Malpartida, H.M. Kieser, H. Ikeda, J. Duncan, I. Fujii, B.A.M. Rudd, H.G. Floss, and S. Omura. 1985. Production of "hybrid" antibiotics by genetic engineering. Nature **314:** 642–644.

32. Hopwood, D.A. 1988. Towards an understanding of gene switching in *Streptomyces,* the basis of sporulation and antibiotic production. Proc. R. Soc. Lond. **B235:** 121–138.

33. Hooper, D.C., J.S. Wolfson, G.L. Michugh, M.B. Winters, and M.N. Swartz. 1982. Effect of novobiocin, coumermycin A1, chlorobiocin, and their analogs on *Escherichia coli* DNA gyrase and bacterial growth. Antimicrob. Agents Chemother. **22:** 662–671.

34. Hotta, K., A. Takahashi, Y. Okami, and H. Umezawa. 1983. Relationship between antibiotic resistance and antibiotic productivity in actinomycetes which produce aminoglycoside antibiotics. J. Antibiotics **36:** 1789–1791.

35. Hotta, K., F. Yamashita, Y. Okami, and H. Umezawa. 1985. New antibiotic-producing streptomycetes, selected by antibiotic resistance as a marker. II. Features of a new antibiotic-producing clone obtained after fusion treatment. J. Antibiotics **38:** 64–69.

36. Iannitelli, R.C., and M. Ikawa. 1980. Effect of fatty acids on action of polyene antibiotics. Antimicrob. Agents Chemother. **17:** 861–864.

37. Imada, A., K. Kitano, K. Kintaka, M. Muroi, and M. Asai. 1981. Sulfazecin and isosulfazecin, novel β-lactam antibiotics of bacterial origin. Nature **289:** 590–591.

38. Imada, A., and H. Okazaki. 1987. Takeda's efforts to discover β-lactam antibiotics from bacteria: Ten years of experience, p. 3–12. *In* R.E. Cape, M.I. Goldberg, T. Hata, and K. Maeda (ed.), Antibiotic Research and Biotechnology. Japan Antibiotics Research Association, Tokyo, Japan.

39. Ishibashi, K. 1962. Studies on antibiotics from *Helminthosporium* sp. fungi. VII. J. Antibiotics **15:** 161–167.

40. Ishikawa, J., Y. Koyama, S. Mizuno, and K. Hotta. 1988. Mechanisms of increased kanamycin-resistance generated by protoplast regeneration of *Streptomyces griseus*. II. Mutational gene alteration and gene amplification. J. Antibiotics **41:** 104–112.

41. Isono, K. 1985. Inhibitors of fungal cell walls, p. 195–216. *In* Antibiotics: development into new fields (ed., The Japanese Agricultural Scientific Society). Asakura Shoten, Tokyo.

42. Kahan, J.S., F.M. Kahan, R. Goegelman, S.A. Currie, M. Jackson, E.O. Stapley, T.W. Miller, A.K. Miller, D. Hendlin, S. Mochales, S. Hernandez, H.B. Woodruff, and J. Birnbaum. 1979. Thienamycin, a new β-lactam antibiotic. I. Discovery, taxonomy, isolation and physical properties. J. Antibiotics **32:** 1–12.

43. Kamogashira, T. 1988. Some characteristics of a hypersensitive mutant to β-lactam antibiotics derived from a strain of *Staphylococcus aureus*. Agric. Biol. Chem. **52:** 1841–1843.

44. Kamogashira, T., M. Sugawara, and M. Takano. 1988. Isolation of a fosfomycin-hypersensitive mutant and production of a D-glucose-1-phosphate from *Bacillus* sp. BA–3796 screened by its use. J. Ferment. Technol. **66:** 649–655.

45. Kamogashira, T., and S. Takegata. 1988. A screening method for cell wall inhibitors using a D-cycloserine hypersensitive mutant. J. Antibiotics **41:** 803–806.

46. Khokhlov, A.S. 1988. Results and perspectives of actinomycete autoregulators studies, p. 338–345. *In* Y. Okami, T. Beppu, and H. Ogawara (ed.), Biology of Actinomycetes '88. Japan Scientific Societies Press, Tokyo, Japan.

47. Kitano, K., K. Kintaka, S. Suzuki, K. Kitamoto, K. Nara, and Y. Nakao. 1975. Screening of microorganisms capable of producing β-lactam antibiotics. J. Ferment. Technol. **53:** 327–338.

48. Kitano, K., K. Nara, and Y. Nakao. 1977. Screening for β-lactam antibiotics using a mutant of *Pseudomonas aeruginosa*. Japanese Antibiotics (Suppl.) **30:** 239–245.

49. Kobinata, K., M. Uramoto, N. Nishii, H. Kusakabe, G. Nakamura, and K. Isono. 1980. Neopolyoxins A, B and C, new chitin synthetase inhibitors. Agr. Biol. Chem. **44:** 1709–1711.

50. Kuroda, Y., M. Okuhara, T. Goto, E. Iguchi, M. Kohsaka, H. Aoki, and H. Imanaka. 1980. FR–900130, a novel amino acid antibiotic. I. Discovery, taxonomy, isolation and properties, J. Antibiotics **33:** 125–131.

51. Lechevalier, H.A., and M.P. Lechevalier. 1967. Biology of actinomycetes. Ann. Rev. Microbiol. **21:** 71–100.

52. Mahoney, D.F., D.K. Baisden, and R.C. Yao. 1989. A peptide binding chromogenic assay for detecting glycopeptide antibiotics. J. Ind. Microbiol. **4:** 43–47.

53. Mason, D.J., A. Dietz, and C. DeBoer. 1963. Lincomycin, a new antibiotic. I. Discovery and biological properties, p. 554–559. *In* J.C. Sylvester (ed.), Antimicrobial Agents and Chemotherapy 1962. American Society for Microbiology, Ann Arbor, Michigan.

54. Masson, J.M., V. Guillermet, and F. LeGoffic. 1987. Chitin synthase as a target for the design of antifungal agents. Eur. J. Med. Chem. **22:** 377–381.

55. Masui, I., S. Murakawa, and T. Takahashi. 1980. Attempt of microbial mutagen screening with Rec$^-$ assay: isolation of chartreusin. Agric. Biol. Chem. **44:** 919–920.

56. Matsuda, Y., M. Kitahara, K. Maeda, and H. Umezawa. 1982. A method of screening for antibiotics producing oxygen radicals. J. Antibiotics **35:** 928–930.

57. Matsuda, Y., M. Kitahara, K. Maeda, and H. Umezawa. 1982. Some evidence for interaction of D-cycloserine with DNA. J. Antibiotics **35:** 893–899.

58. McGuire, J.H., R.L. Bunch, R.C. Anderson, H.E. Boaz, E.H. Flynn, H.M. Powell, and J.W. Smith. 1952. Itomycin (erythromycin—Lilly). Antibiotics & Chemotherapy **2:** 281–283.

59. Nagaoka, K., and A.L. Demain. 1975. Mutational biosynthesis of a new antibiotic, streptomutin A, by an idiotroph of *Streptomyces griseus*. J. Antibiotics **28:** 627–635.

60. Nagarajan, R., L.D. Boeck, M. Gorman, R.L. Hamill, C.E. Higgens, M.M. Hoehn, W.M. Stark, and J.G. Whitney. 1971. β-lactam antibiotics from *Streptomyces*. J. Amer. Chem. Soc. **93:** 2308–2310.

61. Nakamura, Y., K. Ishii, E. Ono, M. Ishihara, T. Kohda, Y. Yokogawa, and H. Shibai. 1988. A novel naturally occurring carbapenem antibiotic, AB–110-D, produced by *Kitasatosporia papulosa* novo sp. J. Antibiotics **41:** 707–711.

62. Nakamura, Y., E. Ono, T. Kohda, and H. Shibai. 1989. Highly targeted screening system for carbapenem antibiotics. J. Antibiotics **42:** 73–83.

63. Naveh, A., I. Potasman, H. Bassan, and S. Ulitzur. 1984. A new rapid and sensitive bioluminescence assay for antibiotics that inhibit protein synthesis. J. Appl. Bacteriol. **56:** 457–463.

64. Newton, G.G.F., and E.P. Abraham. 1955. Cephalosporin C, a new antibiotic containing sulphur and D-aminoadipic acid. Nature **175:** 548.

65. Nisbet, L.J., and N. Porter. 1989. The impact of pharmacology and molecular biology on the exploitation of microbial products, p. 309–342. *In* S. Baumberg, I. Hunter, and M.J. Rhodes (ed.), Microbial Products: New Approaches. Cambridge University Press, Cambridge, U.K.

66. Nolan, R.D., and T. Cross. 1988. Isolation and screening of actinomycetes, p. 1–32. *In* M. Goodfellow, S.T. Williams, and M. Mordarski (ed.), Biotechnology in Actinomycetes. Academic Press, London, England.

67. Nozaki, Y., N. Katayama, H. Ono, S. Tsubotani, S. Harada, H. Okazaki, and Y. Nakao. 1987. Binding of a non-β-lactam antibiotic to penicillin-binding proteins. Nature **325:** 179–180.

68. Numata, K., H. Yamamoto, M. Hatori, T. Miyaki, and H. Kawaguchi. 1986. Isolation of an aminoglycoside hypersensitive mutant and its application in screening. J. Antibiotics **39:** 994–1000.

69. Okami, T., T. Okazaki, T. Kitahara, and H. Umezawa. 1976. Studies on marine microorganisms. V. A new antibiotic, aplasmomycin, produced by a streptomycete isolated from shallow sea mud. J. Antibiotics **29:** 1019–1025.

70. Okami, Y., and K. Hotta. 1988. Search and discovery of new antibiotics, p. 33–67. *In* M. Goodfellow, S.T. Williams, and M. Mordarski (ed.), Biotechnology in Actinomycetes. Academic Press, London, England.

71. Okamura, H., A. Koki, M. Sakamoto, K. Kubo, Y. Mutoh, Y. Fukagawa, K. Kouno, Y. Shimauchi, T. Ishikura, and J. Lein. 1979. Microorganisms producing new β-lactam antibiotics. J. Ferment. Technol. **57:** 265–272.

72. Okumura, M., Y. Kuroda, T. Goto, M. Okamoto, H. Tarano, M. Kohsaka, H. Aoki, and H. Imanaka. 1980. Studies of new phosphonic acid antibiotics. I. FR–900098, isolation and characterization. J. Antibiotics **33:** 13–17.

73. Omura, S., H. Tanaka, R. Oiwa, T. Nagai, Y. Koyama, and Y. Takahashi. 1979. Studies on bacterial cell wall inhibitors. VI. Screening method for the specific inhibitors of peptidoglycan synthesis. J. Antibiotics **32:** 978–984.

74. Omura, S., M. Murata, K. Kimura, S. Matsukura, T. Nishihara, and H. Tanaka. 1985. Screening for new antifolates of microbial origin and a new antifolate, AM 8402. J. Antibiotics **38:** 1016–1024.

75. Omura, S. 1986. Philosophy of new drug discovery. Microbiol. Rev. **50:** 259–279.

76. O'Sullivan, J., J.E. McCullough, A.A. Tymiak, D.R. Kirsch, W.H. Trejo, and P.A. Principe. 1988. Lysobactin, a novel antibacterial agent produced by *Lysobacter* sp. I. Taxonomy, isolation and partial characterization. J. Antibiotics **41:** 1740–1744.

77. Oxford, A.E., H. Raistrick, and P. Simonart. 1939. XXIX. Studies in the biochemistry of microorganisms. LX. Griseofluvin, $C_{17}H_{17}O_6Cl$, a metabolic product of *Penicillium griseofulvum* Dierckx. Biochem. J. **33:** 40–248.

78. Patel, M.V. 1985. An agar plate method for the screening of antibiotics triggering autolytic enzymes. J. Antibiotics **38:** 527–529.

79. Rake, J.B., R. Gerber, R.J. Mehta, D.J. Newman, Y.K. Oh, C. Phelen, M.C. Shearer, R.D. Sitrin, and L.J. Nisbet. 1986. Glycopeptide antibiotics: A mechanism-based screen employing bacterial cell wall receptor mimetic. J. Antibiotics **39:** 58–67.

80. Rinehart, K.L. Jr. 1977. Mutasynthesis of new antibiotics. Pure and Applied Chemistry **49:** 1361–1384.

81. Ryder, N.S., and M.C. Dupont. 1985. Inhibition of squalene epoxidase by allylamine antimycotic compounds. A comparative study of the fungal and mammalian enzymes. Biochem. J. **230:** 765–770.

82. Schats, A., and S.A. Waksman. 1944. Effect of streptomycin and other antibiotic substances upon *Mycobacterium tuberculosis* and related organisms. Proc. Soc. Exp. Biol. Med. **57:** 244–248.

83. Schindler, P.W., W. Konig, S. Chatterjee, and B.N. Ganguli. 1986. Improved screening for β-lactam antibiotics: A sensitive, high-throughput assay using DD-carboxypeptidase and a novel chromophore-labeled substrate. J. Antibiotics **39:** 53–57.

84. Selitrennikoff, C.P. 1983. Use of a temperature-sensitive, protoplast forming *Neurospora crassa* strain for the detection of antifungal antibiotics. Antimicrob. Agents Chemother. **23:** 757–765.

85. Sensi, P., P. Margalith, and M.T. Timbal. 1959. Rifamycin, a new antibiotic. Preliminary report. Il Farmaco, Ed. Sci. **14:** 146.

86. Shier, W.T., P. C Schaefer, D. Gottlieb, and K.L. Rinehart, Jr. 1974. Use of mutants in the study of aminocyclitol antibiotic biosynthesis and preparation of the hybrimycin C complex. Biochemistry **13:** 5073–5078.

87. Sobin, B.A., A.R. English, and W.O. Celmer. 1955. PA105, a new antibiotic, p. 827–830. *In* H. Welch and F. Marti-Ibannez (ed.), Antibiotics Ann.—1954/1955. Medical Encyclopedia, Inc., New York, NY.

88. Spiri-Nakagawa, P., Y. Fukushi, K. Maebashi, N. Imamura, Y. Takahashi, Y. Tanaka, H. Tanaka, and S. Omura. 1986. Izupeptins A and B, new glycopeptide antibiotics produced by an actinomycete. J. Antibiotics **39:** 1719–1723.

89. Stapley, E.O., M. Jackson, S. Hernandez, S.B. Zimmerman, S.A. Currie, S. Mochales, J.M. Mata, H.B. Woodruff, and D. Hendlin. 1972. Cephamycins, a new family of β-lactam antibiotics. I. Production by actinomycetes, including *Streptomyces lactamdurans* sp. n. Antimicrob. Agents Chemother. **2:** 122–131.

90. Steinberg, D.A., G.A. Patterson, R.J. White, and W.M. Maiese. 1985. The stimulation of bioluminescence in *Photobacterium leignathi* as a potential for antitumor agents. J. Antibiotics **38:** 1401–1407.

91. Sykes, R.B., C.M. Cimarusti, D.P. Bonner, K. Bush, D.M. Floyd, N.H. Georgopapadakou, W.H. Koster, W.C. Liu, W.L. Parker, P.A. Principle, M.L. Rathnum, W.A. Slusarchyk, W.H. Trejo, and J.S. Wells. 1981. Monocyclic β-lactam antibiotics produced by bacteria. Nature **291:** 489–491.

92. Sykes, R.B., and J.S. Wells. 1985. Screening for β-lactam antibiotics in nature. J. Antibiotics **38:** 119–121.

93. Takeuchi, S., H. Yonehara, and H. Umezawa. 1959. Studies on variotin, a new antifungal antibiotic. I. Preparations and properties of variotin. J. Antibiotics **12A:** 195–200.

94. Takeuchi, T., T. Hara, H. Naganawa, H. Okada, M. Hamada, H. Umezawa, S. Gomi, M. Sezaki, and S. Kondo. 1988. New antifungal antibiotics, benanomicins A and B from an actinomycete. J. Antibiotics **41:** 807–811.

95. Thomas, A.H. 1986. Suggested mechanisms for the antimycotic activity of the polyene antibiotics and the N-substituted imidazoles. J. Antimicrob. Chemother. **17:** 269–279.

96. Tsuda, K.T. Kihara, M. Nishii, G. Nakamura, K. Isono, and S. Suzuki. 1980. A new antibiotic, lipopeptin A. J. Antibiotics **33:** 247–248.

97. Ubukata, M., M. Uramoto, J. Uzawa, and K. Isono. 1986. Structure and biological activity of neopeptins A, B and C, inhibitors of fungal cell wall glucan synthesis, Agr. Biol. Chem. **50:** 351–356.

98. Umezawa, H., M. Ueda, K. Maeda, K. Yagishita, S. Kondo, Y. Okami, R. Utahara, Y. Osato, K. Nitta, and T. Takeuchi. 1957. Production and isolation of a new antibiotic, kanamycin. J. Antibiotics **10A:** 181–188.

99. Vicario, P.P., B.G. Green, and H. Katzen. 1987. A single assay for simultaneously testing effectors of alanine racemase and/or D-alanine: D-alanine ligase. J. Antibiotics **40:** 209–216.

100. Waksman, S.A., and H.A. Lechevaier. 1949. Neomycin, a new antibiotic active against streptomycin-resistant bacteria, including tuberculosis organisms. Science **109:** 305–307.

101. Weinstein, M.J., G.M. Luedemann, E.M. Oden, G.H. Wagman, J.P. Rosselet, J.A. Marquez, C.T. Coniglio, W. Charney, H.L. Herzog, and J. Black. 1963. Gentamicin, a new antibiotic complex from *Micromonospora*. J. Med. Chem. **6:** 463–464.

102. Williams, S.T., M. Goodfellow, and G. Alderson. 1989. Suprageneric classification of *Actinomycetes,* p. 2333–2339. *In* S.T. Williams (ed.), Bergey's Manual of Systematic Bacteriology, Vol. 4. Williams and Wilkins, Baltimore, Maryland.

103. Yamashita, F., K. Hotta, S. Kurasawa, Y. Okami, and H. Umezawa. 1985. New antibiotic-producing streptomycetes, selected by antibiotic resistance as a marker. I. New antibiotic production generated by protoplast fusion treatment between *Streptomyces griseus* and *S. tenjimariensis*. J. Antibiotics **38:** 58–63.

104. Yao, R.C., and D.F. Mahoney. 1984. Enzyme-linked immunosorbent assay for the detection of fermentation metabolites: aminoglycoside antibiotics. J. Antibiotics **37:** 1462–1468.

105. Yoshida, N., Y. Tani, and K. Ogata. 1972. Cryomycin, a new peptide antibiotic produced only at low temperature. J. Antibiotics **25:** 653–659.

106. Zygmunt, W.A., and P.A. Tavormina. 1966. Steroid interference with antifungal activity of polyene antibiotics. Appl. Microbiol. **14:** 865–869.

2

Bacterial Pathogens for the 1990s: A Case for New Drug Development

George L. Drusano

I. Introduction

The decade of the Eighties was marked by an unprecedented acceleration in the development of new antimicrobial compounds. The early part of the decade was dominated by the introduction of multiple new beta-lactams, both extended-spectrum penicillins (*e.g.*, mezlocillin, piperacillin, and azlocillin) and third-generation cephalosporins (cefotaxime, moxalactam, cefoperazone, ceftazidime, ceftizoxime, ceftriaxone, *etc.*). The mid to latter part of the decade saw the introduction of carbapenem antibiotics (imipenem/cilastatin), enzyme inhibitors combined with older beta-lactams, such as ticarcillin/clavulanate and ampicillin/sulbactam, and last, but certainly not least, fluoroquinolone antimicrobials.

These compounds have extremely broad spectra of activity and are active at concentrations previously seen only with gram-positive organisms and older beta-lactam antibiotics (*e.g.*, *Streptococcus pneumoniae* and aqueous penicillin G). Even problem pathogens, such as *Pseudomonas aeruginosa* and *Bacteroides fragilis,* are contained within the spectrum of a number of these drugs. Given this, one may reasonably ask if there is a need for further drug development in the area of serious bacterial infections. Antimicrobial drug development is costly, diverts resources from the development of other classes of compounds and, overall, drives up the cost of drug acquisition. If there is no clinical need, the pharmaceutical industry would be wise to use its resources elsewhere.

It is the thesis of this chapter, however, that patient needs in the Nineties will require the development of new antibacterial compounds; both classical agents, such as beta-lactams, and nontraditional compounds with different sites of action are necessary to provide an acceptable standard of care.

As opposed to other areas of infectious disease therapy, where new pathogens dominate the need for continued drug development (see chapters 13 and 22), the needs for new antibacterial chemotherapy revolve, for the most part, around familiar pathogens.

In addition, there will be an increasing need for these compounds in terms of the numbers of patients to be treated. This increase in infected patients with resistant pathogens is caused by a number of different factors. Many of the pathogens to be discussed below are often seen in the setting of nosocomial infections.

Improvements in multiple areas of medicine have resulted in the creation of new populations of patients who are immunosuppressed and who require relatively long periods of hospitalization, putting them at high risk for nosocomial infections. Examples can be seen in the aggressive regimens common for the therapy of neoplastic disease, the increasing numbers and kinds of organ transplant patients, often with profound immunosuppression, and the large population of patients infected with the human immunodeficiency virus.

Patients with different kinds of catheters and prosthetic devices also are a group in which unusual and drug-resistant pathogens are common.

Finally, travel is both common and rapid. Highly resistant organisms which have arisen in a restricted locale now may be rapidly disseminated worldwide. Indeed, single-resistance plasmids have been successfully tracked in their transnational sojourn (25).

The recent advances seen with continuing development of carbapenem-type antimicrobials (meropenem), cephalosporins with poor substrate affinity for Sykes class I enzymes (cefepime and cefpirome), quinolone antimicrobials (ciprofloxacin, ofloxacin, enoxacin, temafloxacin, and fleroxacin), and 14–15-membered ring macrolides (azithromycin and clarithromycin) represent the recognition within the pharmaceutical industry that, while having come a long way from penicillin, the fight continues.

II. Emerging Resistant Pathogens

A. Gram-Positive Cocci

1. Enterococci

Among gram-positive cocci, there have been major changes in susceptibility patterns. This has been seen most dramatically among the enterococci and the staphylococci. Three classes of agents have traditionally been employed for the therapy of serious enterococcal infections (*i.e.*, where bactericidal activity is required, such as endocarditis): penicillins, especially aqueous penicillin G and ampicillin, vancomycin, and aminoglycosides. Over the course of the last 4–5 years, major resistance has been documented among these organisms to each of the aforementioned classes of drugs. Murray (22), among others, has demonstrated that high-level penicillin and ampicillin resistance has occurred. The mechanism is the straightforward acquisition of a beta-lactamase-bearing plasmid, allowing hydrolysis of these relatively unstable compounds. This caused less concern than it might, because clinicians knew that vancomycin was always available for therapy in the event a beta-lactamase-positive strain was isolated.

However, strains of enterococci resistant to vancomycin have recently been described (15). Common to almost all of these strains is the elaboration of a 35–39 kDa cell wall protein (1). It has been speculated that this protein, possibly a DD-carboxypeptidase, binds the glycopeptide before it can bind to its intended target site, UDP-muramyl-pentapeptide (39). These strains often possess a large (~1 Md [Megadaltons]) plasmid. This might cause concern that widespread vancomycin resistance would result, if the plasmid could easily be passed among different species. To date, however, all attempts at filter mating have been unsuccessful in passing the plasmid. Nonetheless, the loss of both major classes of cell-wall-active antibiotics effective against enterococci serves to point to a major need for drug development in the next decade.

Enterococci are relatively unique in uniformly displaying tolerance to cell-wall-active antibiotics (9). Further, there are circumstances, such as endocarditis, when bacterial killing is required. To obtain bactericidal activity, clinicians have traditionally employed a combination of a cell-wall-active antimicrobial and an aminoglycoside. Moellering and Weinberg (21) have demonstrated that one obtains markedly improved uptake of the aminoglycoside when the organism has been preexposed or is simultaneously exposed to a cell-wall-active drug. This presupposes, however, that one has an active aminoglycoside antibiotic available.

The last several years have revealed a marked increase in aminoglycoside resistance in this genus (28). This has been particularly marked for gentamicin, the drug most often employed clinically and the one that displays perhaps the most marked synergy. Aminoglycoside resistance should be differentiated as essentially all strains are "low-grade" aminoglycoside resistant, usually with a minimal inhibitory concentration (MIC) to gentamicin of less than 50–100 μg/ml. However, clinically important resistance is seen when these organisms have very high MICs (>500–>2,000 μg/ml, depending upon the investigation), where one does not see synergy, even in the presence of full therapeutic concentrations of both cell-wall-active drug and the aminoglycoside. This "high-level" resistance has been described in pockets around the country, occasionally accounting for 30–40% of the isolates from some hospital laboratories (41).

Increasing drug resistance among enterococci appears to be a major problem. It is likely that strains will be described with simultaneous resistance to all major therapeutic agents currently in clinical use, highlighting the need for continued drug development. Neither quinolones nor newer macrolides hold much promise for this problem; in the former, little therapeutic success has been evident and, the latter, most strains are already macrolide-resistant.

2. Staphylococci

Staphylococci are common clinical pathogens, accounting for a large percentage of infections from both community and nosocomial sources. *Staphylococcus aureus* has been the traditional organism to which antimicrobial chemotherapy has been directed. With increasing numbers of patients living for long periods of time with medical devices in place, and with increasing numbers of immunocompromised patients, coagulase-negative staphylococci, particularly *S. epidermidis* and *S. hemolyticus,* are now recognized as pathogens in their own right (27, 35).

Over the last 10 years, resistance among strains of *S. aureus* to nafcillin and the isoxazolyl penicillins has become distressingly common in American hospitals (23). This resistance has been seen mostly among nosocomial isolates, but community cases arising among abusers of intravenous drugs have been described (7, 20). Indeed, only vancomycin, among the current clinically available antimicrobial agents, provides reliable activity against multiply resistant isolates.

With the introduction of the fluoroquinolones, there was hope that this oral class of agents would provide reliable activity against methicillin-resistant staphylococci. While fluoroquinolones are active in the test tube at relatively low concentrations (ca. 1 μg/ml as an MIC_{90} for ciprofloxacin), rapid emergence of resistance has been seen clinically (29). Although newer quinolones are under development with greater activity against gram-positive pathogens, we must have a conservative view of these agents, since resistance among staphylococci appears to be at the level of altered DNA gyrase (36) and any new agents must have more than an eight-fold increase in activity to be truly promising.

Several reports have described resistance to vancomycin in *S. hemolyticus*. In a well-documented case, a vancomycin-resistant strain arose after repeated exposures to this drug. A foreign body probably served as a reservoir for this pathogen (35).

It is clear that gram-positive cocci, which have been thought of as "easy-to-treat" pathogens because of the success achieved with classical beta-lactams, are emerging as a major challenge for the next decade in both community-acquired and nosocomial infection settings. New antimicrobial agents are required that are not susceptible to prevalent resistance mechanisms.

B. Gram-negative Bacilli

1. Enterobacteriaceae, Pseudomonas, Bacteroides

The identification of TEM-type beta-lactamases as the cause of resistance of some gram-negative organisms to early broad-spectrum penicillins (*e.g.*, ampicillin) led to the search for compounds with greater intrinsic stability relative to this group of enzymes (see chapter 9). Cephalosporin molecules with a bulky aminothiazolyl-oxyimino side chain sterically hinder the entry of the molecule into the active site of TEM-type enzymes, thus preventing hydrolysis of the beta-lactam bond and preserving microbiological activity.

This side chain was a common feature of all the early "third-generation" cephalosporins. Testing of these compounds against *Enterobacteriaceae* with

Sykes class I chromosomal beta-lactamases initially displayed promising results. This was probably due to the use of nonphysiological concentrations of drug in the beta-lactamase assays. The large concentrations allowed easy measurement, but because of the Michaelis-Menten kinetics of drug hydrolysis, the amount hydrolyzed was such a small fraction of the total drug that these drugs appeared stable to these enzymes. When more appropriate concentrations were tested (1 μM, the approximate steady-state concentration in the periplasm), it became clear that significant hydrolysis of these drugs occurred by means of Sykes class I enzymes (38).

Normark and colleagues have demonstrated that these chromosomal enzymes, which are normally inducible, can become constitutive, high-level producers (stably derepressed for beta-lactamase production) of beta-lactamase (16). This finding has documented clinical implications. Sanders and Moellering have demonstrated cases of failure due to selection of stably derepressed mutants by exposure to weakly inducing beta-lactams (33). Sanders and Sanders reviewed this issue and found that when patients infected with a Sykes class I–bearing organism were examined, resistant mutants could be isolated from >20% of the patients (34). In addition, 50% of the colonized patients had a clinical failure or relapse of infection.

Organisms producing beta-lactamase constitutively cause nosocomial infections and are well recognized by clinicians. Examples include *Pseudomonas aeruginosa, Enterobacter cloacae, Serratia marcescens, Citrobacter* species, and indole-positive *Proteus,* among others.

Once stably derepressed for beta-lactamase production, these organisms can be cross-resistant to a wide variety of beta-lactam antibiotics. Data are presented in Figure 2.1 in which more than 200 sequential isolates carrying a Sykes class I enzyme from patients in medical intensive-care units had susceptibilities determined to a wide variety of beta-lactam antibiotics (data presented at the Third Decennial International Conference on Nosocomial Infections, Atlanta, Georgia, August 1990). Sensitivity or resistance to one beta-lactam (in this case, ceftazidime) is highly correlated with sensitivity or resistance to the other beta-lactams. This implies that when organisms become stably derepressed for beta-lactamase production, MICs increase to virtually all the beta-lactams. (Penems and carbapenems may be exceptions.) Obviously, then, these pathogens will remain troublesome clinical problems, and new drug development is necessary to provide additions to the physician's therapeutic armamentarium.

Among the chromosomal beta-lactamase-bearing organisms, *Pseudomonas aeruginosa* represents a special and most troubling case. All of what has been written above regarding stable derepression certainly applies to *Pseudomonas*. In addition, penem and carbapenem antibiotics, which tend to retain activity against other species and even against the stably derepressed mutants, often are ineffective when the pathogen is *Pseudomonas*. Resistance in this pathogen is mediated by a different mechanism. Quinn *et al.*, Buschler *et al.*, and Lynch *et al.* demon-

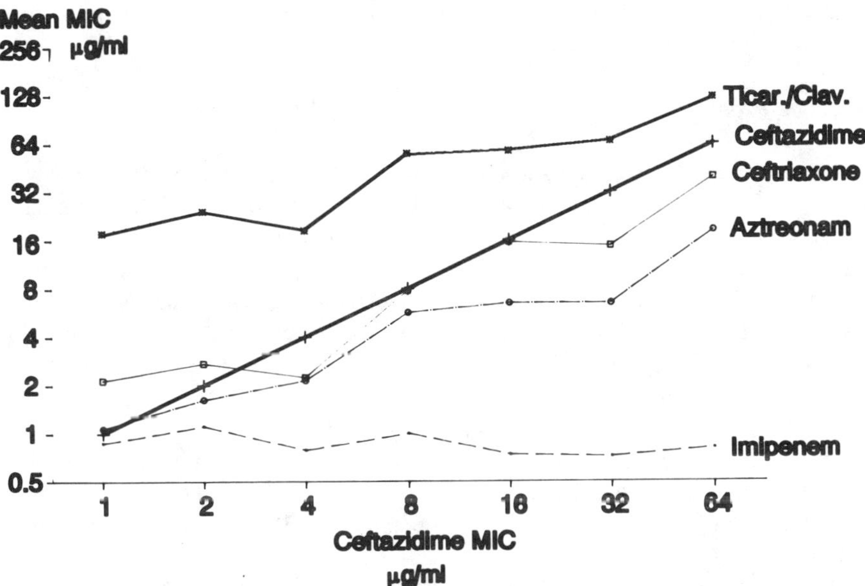

Figure 2.1. Plot of the mean minimal inhibitory concentrations (MICs) to ceftazidime of 241 isolates of *Enterobacteriaceae* bearing a Sykes class I chromosomal beta-lactamase against the mean MICs of various other members of the beta-lactam class of antibiotics. Good correlation is noted for all drugs except the carbapenem class.

strated that resistance was associated with the loss of an outer membrane protein (31, 4, 17). Lynch *et al.* showed that the loss of this porin protein was associated with a decrease in the transport of radiolabelled imipenem into the resistant organisms (17). Nikaido and colleagues demonstrated that this porin was $ompD_2$ and that it possessed specific carbapenem binding sites (37). *Pseudomonas* remains a persistent clinical problem, even for our newly introduced antimicrobial agents.

Among the fluoroquinolones currently available, or about to be available, ciprofloxacin has the greatest intrinsic gram-negative activity, especially for *Pseudomonas aeruginosa*. Yet clinical failures are well noted as resistant organisms rapidly emerge during therapy. The mechanism for resistance is clear, resulting from an altered A subunit of DNA gyrase. Organisms producing this altered subunit have approximately an eight-fold increase in MIC to all the new fluoroquinolones. For ciprofloxacin, where the MIC_{90} for *P. aeruginosa* is 0.5–1.0 μg/ml, organisms resistant by this mechanism will have an increase in MIC values to 4–8 μg/ml. As maximal oral doses of ciprofloxacin produce transient peaks of only 3.0–4.0 μg/ml, these isolates will be clinically resistant to the drug, since the MIC is, by and large, above pharmacologically achievable levels of drug (11).

Pseudomonas aeruginosa remains a problem pathogen. Extended-spectrum penicillins, third-generation cephalosporins, penem and carbapenem antibiotics, and the new fluoroquinolones are not always effective clinically.

C. The Problem of Beta-lactamases

Several new developments in the last few years also point to development of resistance to our newer agents (particularly beta-lactams) among organisms such as *E. coli* and *K. pneumoniae,* often associated with community-acquired infections (Table 2.1).

Enzyme inhibitors, such as potassium clavulanate and penicillanic acid sulfones (*e.g.*, sulbactam, tazobactam), were developed primarily as inhibitors of plasmid-mediated enzymes. These beta-lactamase inhibitors have been very successful in expanding the spectrum of older beta-lactams (aminopenicillins such as ampicillin, amoxicillin, and ticarcillin), with proven safety profiles. Organisms have been described recently, however, that are resistant to combination therapy. These organisms are resistant because they contain plasmids of high copy number and produce very large amounts of beta-lactamase (32), saturating the number of

Table 2.1. Gram-negative bacilli bearing specific resistance mechanisms: resistance to classes of beta-lactam antibiotics

		Type of resistance mechanism		
	Sykes class I chromosomal beta-lactamase	High-copy-number plasmids	Extended-TEM beta-lactamase	Zinc metalo-enzyme[a] beta-lactamase
Extended-spectrum penicillins[b]	Stably derepressed mutants resistant	Resistant	Resistant	Most resistant
Cepahlosporins and monobactams	Stably derepressed mutants resistant	Sensitive	Resistant[c]	Most resistant[d]
Carbapenems	Sensitive	Sensitive	Sensitive	Resistant
Beta-lactam plus enzyme inhibitor	Resistant	Resistant	Sensitive	Most resistant[e]
Poor physiologic substrate cephalosporins[f]	Sensitive	Sensitive	Resistant	Most resistant

[a] All *Xanthomonas maltophilia* and very rare cases of *Bacteroides fragilis*.

[b] Examples are ticarcillin, piperacillin, mezlocillin, and azlocillin.

[c] Resistance is often specific (*i.e.*, hydrolyzing ceftazidime, but not ceftriaxone or vice versa).

[d] Ceftazidime may represent the most active of the cephalosporins against *X. maltophilia*.

[e] Ticarcillin/clavulanante may have the most activity among extant combinations of beta-lactam/beta-lactamase inhibitors.

[f] Examples are cefepime and cefpirome.

molecules of inhibitor. Since the enzyme inhibitors are stably bound to beta-lactamase, unbound (free) enzyme can hydrolyze the aminopenicillin.

Even more disturbing are reports of "extended TEM-type" beta-lactamases. As indicated above, the third-generation cephalosporins obtain their stability to TEM-type enzymes through steric hindrance of entry of the molecule into the active site of the beta-lactamase. Point mutations can allow these folded proteins to open, however, allowing entry of drug molecules (18). The result is that drugs such as cefotaxime, ceftazidime, ceftriaxone, aztreonam, ceftizoxime, *etc.*, can be hydrolyzed. Over 17 of these enzymes (the number is steadily rising) have been described. An interesting, but confusing, aspect of the extended TEM-type beta-lactamases is that the enzymes have different substrate specificities; some are "cefotaximases" while others are "ceftazidimases," "aztreonamases," *etc.* Consequently, the clinical microbiology laboratory is faced with the problem of having to test a large number of different drugs to determine the specific susceptibilities of an isolate. Further, hospitals may need to carry more drugs on their formularies than desired.

Also problematic is that these enzymes are often on plasmids with a broad host range, transmittable virtually to any gram-negative isolate. Their presence is suspect when the isolate is an *E. coli* resistant to aztreonam or ceftazidime. However, if the organism contains a Sykes class I enzyme, the presence of the extended TEM-type beta-lactamase may be overlooked, because resistance would be thought to result from a stably derepressed beta-lactamase. The substrate specificity of the extended TEM enzyme and its sensitivity to drug combinations with clavulanate and penicillanic acid sulfones can provide the clues to its presence, however.

Thus for gram-negative organisms there is a need for new drug development. Virtually all the advances attained over the last two decades have been at least partially attenuated by the acquisition (or at least amplification) of newly recognized resistance mechanisms.

Stable derepression of chromosomal beta-lactamases often causes problems for many of our beta-lactams with organisms seen in the intensive care unit. Production of very large amounts of beta-lactamase by plasmids of high copy number can defeat our enzyme inhibitors and can show up even in community isolates. Extended TEM enzymes can defeat many of our newer drugs and, being on plasmids, can be found in virtually any organism in the community as well as in the hospital. *Pseudomonas* continues to represent a major challenge, becoming resistant to virtually all our agents, including carbapenems and fluoroquinolones.

III. Pathogens of Increasing Clinical Significance

A. Leuconostoc, Bacteroides, Mycobacteria, *and* Xanthomonas *Species*

There are other pathogens emerging that are worrisome as their frequency of isolation is increasing. Among the gram-positives, *Leuconostoc* species have been identified recently as causing serious infection in hospitalized patients (13). This organism is vancomycin-resistant, again pointing to the need for new drug development for these organisms.

Among the gram-negative isolates, more reports are surfacing with *Xanthomonas maltophilia* as a cause of serious nosocomial infection (8). This organism has been endowed with multiple resistance mechanisms, including a resilient outer membrane with low permeability for many drugs (40) as well as a completely unique beta-lactamase, a planar zinc metalloenzyme (2). This metalloenzyme can hydrolyze any beta-lactam, including the penems and carbapenems, and is not well inhibited by beta-lactamase inhibitors (designed for folded-domain-type beta-lactamases) (24).

Bacteroides fragilis, a common pathogen in multiple clinical settings, but especially in serious infections of the abdomen, has traditionally been rather easy to treat. Drugs such as chloramphenicol, metronidazole, cefoxitin, cefotetan, ampicillin/sulbactam, ticarcillin/clavulanate, and imipenem/cilastatin have been the mainstays of therapy. This pathogen is fighting back, acquiring multiple mechanisms of resistance (3) and frequencies of resistance to cefoxitin and cefotetan of 16–20% (6). A few strains have acquired zinc metallo-beta-lactamases, causing them to be resistant to imipenem/cilastatin as well as the enzyme-inhibitor drugs. There have even been reports of metronidazole resistance (3, 14). Clearly, new discovery efforts would be welcome for this pathogen.

The epidemic of the human immunodeficiency virus has brought us the resurgence of an old pathogen, *Mycobacterium* (National Nosocomial Infection Survey). The number of cases of *mycobacterial* infection has increased dramatically, with 23,000 cases being reported in 1989, resulting in a provisional death total of 1,750 (*Proceedings,* NIH Workshop on Future Directions in Tuberculosis Research, Bethesda, Maryland, December 12–14, 1990). While the resistance patterns of the *M. tuberculosis* cases remain fairly typical, a large number of cases of mycobacterial disease are caused by the multiply drug-resistant avium-intracellulare complex. Attempts to design rational multidrug combinations for therapy of infections with this group of organisms have had disappointing clinical results (5). Once again, this should be an area of focus for discovery efforts.

IV. Conclusions and Future Directions

The pharmaceutical industry has already begun to respond to some of the needs highlighted above. New classes of drugs, such as lipopeptides (10), have been developed which are active against methicillin-resistant staphylococci and vancomycin-resistant enterococci. In addition, when high enough concentrations can be reached, this class of agents is bactericidal against the enterococci as a single agent. New fluoroquinolones, highly active against gram-positive species, are currently under development, while carbapenem antibiotics with useful activity against methicillin-resistant staphylococci are on the drawing board. Novel glycopeptides with improved activity against vancomycin-resistant enterococci are being evaluated.

For gram-negatives, new cephalosporins are being developed (cefepime and cefpirome) that have poor affinity for Sykes class I enzymes, making them very useful for stably derepressed, multiply antibiotic-resistant mutant organisms (30). Unfortunately, the "extended-TEM" beta-lactamases hydrolyze these drugs, and so these represent only a partial solution. Other agents, however, particularly the beta-lactamase inhibitors (clavulanate, sulbactam, tazobactam, BRL 42715, *etc.*), inhibit these enzymes, at least for the present. The problem of *Pseudomonas* is being partially addressed by drugs, such as meropenem, that have improved activity, even against $ompD_2$-deleted mutants (19), and drugs like catechol-cephalosporins that are actively transported via iron-scavenging pathways. The problem of organisms bearing zinc metalloenzyme beta-lactamases may be solved only with the development of novel antibiotic classes.

We are also witnessing the emergence of new agents that are adjunctive to antimicrobials, designed to improve survivorship in severely ill patients with septic shock. Monoclonal antibodies directed against lipopolysaccharide, such as HA–1A and E–5 (42), may have a place in our therapeutic armamentarium in the future. Drugs that are IL–1 receptor antagonists may also prove useful (26). Perhaps even more exciting is the recognition of the central role of tumor necrosis factor (TNF) as a mediator in seriously ill patients. The results of phase I studies employing monoclonal antibodies to TNF will be fascinating, indeed, as scientists argue whether TNF is "good" or "bad" for septic patients.

While tremendous advances have been made by our medicinal chemists to produce large numbers of new agents, more work needs to be done. Continuing pressure from the use of these agents—sometimes inappropriate use (12)—has allowed selection of organisms, both gram-positive and gram-negative, with resistance mechanisms that provide them an advantage in an environment such as the hospital. In contrast to other pathogens, such as fungi and viruses, where new pathogens are arising that require new and improved therapies, the challenge to be met in the next decade and for the foreseeable future with bacterial pathogens is the design of novel compounds, perhaps with new molecular targets, to inhibit familiar organisms that are resistant to currently available agents.

References

1. al-Obeid, S., E. Collatz, and L. Gutmann. 1990. Mechanism of resistance to vancomycin in *Enterococcus faecium* D366 and *Enterococcus faecalis* A256. Antimicrob. Agents Chemother. **34:**252–56.
2. Bicknell, R., E.L. Emanuel, J. Gagnon, and S.G. Waley. 1985. The production and molecular properties of the zinc beta-lactamase of *Pseudomonas maltophilia* IID 1275. Biochem. J. **229:**791–97.
3. Brogan, O., P.A. Garnett, and R. Brown. 1989. *Bacteroides fragilis* resistant to metronidazole, clindamycin and cefoxitin. J. Antimicrob. Chemother. **23:**660–62.

4. Buscher, K.H., W. Cullmann, W. Dick, and W. Opferkuch. 1987. Imipenem resistance in *Pseudomonas aeruginosa* resulting from diminished expression of an outer membrane protein. Antimicrob. Agents Chemother. **31:**703–08.

5. Chiu, J., J. Nussbaum, S. Bozzette, J.G. Tilles, L.S. Young, J. Leedom, P.N. Heseltine, and J.A. McCutchan. 1990. Treatment of disseminated *Mycobacterium avium* complex infection in AIDS with amikacin, ethambutol, rifampin, and ciprofloxacin. Ann. Intern. Med. **113:**358–61.

6. Cornick, N.A., G.J. Cuchural Jr., D.R. Snydman, N.V. Jacobus, P. Iannini, G. Hill, T. Cleary, J.P. O'Keefe, C. Pierson, and S.M. Finegold. 1987. The antimicrobial susceptibility patterns of the *Bacteroides fragilis* group in the United States, 1987. J. Antimicrob. Chemother. **25:**1011–19.

7. Crane, L.R., D.P. Levine, M.J. Zervos, and G. Cummings. 1986. Bacteremia in narcotic addicts at the Detroit Medical Center. I. Microbiology, epidemiology, risk factors, and empiric therapy. Rev. Infect. Dis. **8**: 364–73.

8. Elting, L.S., and G.P. Bodey. 1990. Septicemia due to *Xanthomonas* species and non-*aeruginosa Pseudomonas* species: increasing incidence of catheter-related infections. Medicine (Baltimore) **69**: 296–306.

9. Fabbri, A., G. Manno, A. Tacchella, M.L. Belli, and C. Palmero. 1986. Susceptibility of enterococci. I. Inhibitory and bactericidal activity of several chemoantibiotics against *Streptococcus faecalis* and *Streptococcus faecium*. Chemioterapia **5:**302–08.

10. Fasching, C.E., L.R. Peterson, J.A. Moody, L.M. Sinn, and D.N. Gerding. 1990. Treatment evaluation of experimental staphylococcal infections: comparison of beta-lactam, lipopeptide, and glycopeptide antimicrobial therapy. J. Lab. Clin. Med. **116:**697–706.

11. Forrest, A., M. Weir, K.I. Plaisance, G.L. Drusano, J. Leslie, and H.C. Standiford. 1988. Relationships between renal function and disposition of oral ciprofloxacin. Antimicrob. Agents Chemother. **32:**1537–40.

12. Frieden, T.R., and R.J. Mangi. 1990. Inappropriate use of oral ciprofloxacin. J. Am. Med. Assoc. **264:** 1438–40.

13. Handwerger, S., H. Horowitz, K. Coburn, A. Kolokathis, and G.P. Wormser. 1990. Infection due to *Leuconostoc* species: six cases and review. Rev. Infect. Dis. **12:**602–10.

14. Hickey, M.M., U.M. Davies, J. Dave, M. Vogler, and R.A. Wall. 1990. Metronidazole resistant *Bacteroides fragilis* infection of a prosthetic hip joint. J. Infect. **20:** 129–33.

15. Johnson, A.P., A.H. Uttley, N. Woodford, and R.C. George. 1990. Resistance to vancomycin and teicoplanin: an emerging clinical problem. Clin. Microbiol. Rev. **3:**280–91.

16. Lindberg, F., and S. Normack. 1986. Contribution of chromosomal beta-lactamases to beta-lactam resistance in enterobacteria. Rev. Infect. Dis. **8**(Suppl. 3):S292–304.

17. Lynch, M.J., G.L. Drusano, and H.L. Mobley. 1987. Emergence of resistance to imipenem in *Pseudomonas aeruginosa*. Antimicrob. Agents Chemother. **31:**1892–96.

18. Mabilat, C., S. Goussard, W. Sougakoff, R.C. Spencer, and P. Courvalin. 1990. Direct sequencing of the amplified structural gene and promoter for the extended-broad-spectrum beta-lactamese TEM–9 (RHH–1) of *Klebsiella pneumoniae*. Plasmid **23:**27–34.

19. Margaret, B.S., G.L. Drusano, and H.C. Standiford. 1989. Emergence of resistance to carbapenem antibiotics in *Pseudomonas aeruginosa*. J. Antimicrob. Chemother. **24**(Suppl A):161–67.

20. McGowan, J.E. Jr. 1988. Gram-positive bacteria: spread and antimicrobial resistance in university and community hospitals in the USA. J. Antimicrob. Chemother. **21**(Suppl C): 49–55.

21. Moellering, R.C. Jr., and A.N. Weinberg. 1971. Studies on antibiotic syngerism against enterococci. II. Effect of various antibiotics on the uptake of 14 C-labeled streptomycin by enterococci. J. Clin. Invest. **50:**2580–84.

22. Murray, B.E. 1990. The life and times of the *Enterococcus*. Clin Microbiol. Rev. **3:**46–65.

23. Mylotte, J.M., C. McDermott, and J.A. Spooner. 1987. Prospective study of 114 consecutive episodes of *Staphylococcus auerus* bacteremia. Rev. Infect. Dis. **9:**891–907.

24. Neu, H.C., G. Saha, and N.X. Chin. 1989. Resistance of *Xanthomonas maltophilia* to antibiotics and the effect of beta-lactamase inhibitors. Diagn. Microbiol. Infect. Dis. **12**:283–85.

25. O'Brien, T.F., M.P. Pla, K.H. Mayer, H. Kishi, E. Gilleece, M. Syvanen, and J.D. Hopkins. 1985. Intercontinental spread of a new antibiotic resistance gene on an epidemic plasmid. Science **230:**87–88.

26. Ohlsson, K., P. Bjork, M. Bergenfeldt, R. Hageman, and R.C. Thompson. 1990. Interleukin–1 receptor antagonist reduces mortality from endotoxin shock. Nature **348:**550–52.

27. Oren, I., and D. Merzbach. 1990. Clinical and epidemiological significance of species identification of coagulase-negative staphylococci in a microbiological laboratory. Isr. J. Med. Sci. **26:**125–28.

28. Patterson, J.E., and M.J. Zervos. 1990. High-level gentamicin resistance in *Enterococcus:* microbiology, genetic basis, and epidemiology. Rev. Infect. Dis. **12:**644–52.

29. Peterson, L.R., J.N. Quick, B. Jensen, S. Homann, S. Johnson, J. Tenquist, C. Shanholtzer, R.A. Petzel, L. Sinn, and D.N. Gerding. 1990. Emergence of ciprofloxacin resistance in nosocomial methicillin-resistant *Staphylococcus aureus* isolates. Resistance during ciprofloxacin plus rifampin therapy for methicillin-resistant *S. auerus* colonization. Arch. Intern. Med. **150:** 2151–55.

30. Phelps, D.J., D.D. Carlton, C.A. Farrell, and R.E. Kessler. 1986. Affinity of cephalosporins for beta-lactamases as a factor in antibacterial efficacy. Antimicrob. Agents Chemother. **29:**845–48.

31. Quinn, J.P., E.J. Dudek, C.A. DiVincenzo, D.A. Lucks, and S.A. Lerner. 1986. Emergence of resistance to imipenem during therapy for *Pseudomonas aeruginosa* infections. J. Infect. Dis. **154:**289–94.

32. Sanders, C.C., J.P. Iaconis, G.P. Bodey, and G. Samonis. 1988. Resistance to ticarcillin-potassium clavulanate among clinical isolates of the family *Enterobacteriaceae:* Role of PSE–1 beta-lactamase and high levels of TEM–1 and SHV–1 and problems with false susceptibility in disk diffusion tests. Antimicrob. Agents Chemother. **32:**1365–1369.

33. Sanders, C.C., R.C. Moellering Jr., R.R. Martin, R.L. Perkins, D.G. Strike, T.D. Gootz, and W.E. Sanders Jr. 1982. Resistance to cefamandole: a collaborative study of emerging clinical problems. J. Infect. Dis. **145:**118–25.

34. Sanders, W.E. Jr., and C.C. Sanders. 1988. Inducible beta-lactamases: clinical and epidemiologic implications for use of newer cephalosporins. Rev. Infect. Dis. **10:**830–38.

35. Schwalbe, R.S., J.T. Stapleton, and P.H. Gilligan. 1987. Emergence of vancomycin resistance in coagulase-negative staphylococci. N. Engl. J. Med. **316:**927–31.

36. Sreedharan, S., M. Oram, B. Jensen, L.R. Peterson, and L.M. Fisher. 1990. DNA gyrase *gyrA* mutations in ciprofloxacin-resistant strains of *Staphylococcus aureus:* close similarity with quinolone resistance mutations in *Escherichia coli*. J. Bacteriol. **172:**7260–62.

37. Trias, J., and H. Nikaido. 1990. Outer membrane protein D2 catalyzes facilitated diffusion of carbapenems and penems through the outer membrane of *Pseudomonas aeruginosa*. Antimicrob. Agents Chemother. **34:**52–57.

38. Vu, H., and H. Nikaido. 1985. Role of beta-lactam hydrolysis in the mechanism of resistance of a beta-lactamase-constitutive *Enterobacter cloacae* strain to expanded-spectrum beta-lactams. Antimicrob. Agents Chemother. **27:**393–98.

39. Watanakunakorn, C. 1984. Mode of action and in-vitro activity of vancomycin. J. Antimicrob. Chemother. **14** (Suppl D): 7–18.

40. Yamazaki, E., J. Ishii, K. Sato, and T. Nakae. 1989. The barrier function of the outer membrane of *Pseudomonas maltophilia* in the diffusion of saccharides and beta-lactam antibiotics. FEMS Microbiol. Lett. **51:**85–89.

41. Zervos, M.J., M.S. Terpenning, D.R. Schaberg, P.M. Therasse, S.V. Medendorp, and C.A. Kauffman. 1987. High-level aminoglycoside-resistant enterococci. Colonization of nursing home and acute care hospital patients. Arch. Intern. Med. **147:**1591–94.

42. Ziegler, E.J., J.A. McCutchan, J. Fierer, M.P. Glauser, J.C. Sadoff, H. Douglas, and A.I. Braude. 1982. Treatment of gram-negative bacteremia and shock with human antiserum to a mutant *Escherichia coli*. N. Engl. J. Med. **307:**1225–30.

3

DNA Replication

Charles S. McHenry

I. Introduction

The DNA replication apparatus of bacteria presents numerous attractive, largely unexplored targets for new antibacterial chemotherapeutic agents. Replication in bacteria is a complex process; in *E. coli,* over 30 proteins are required. It has been established biochemically and genetically that many replication genes encode products that are essential for replication (Table 3.1) Temperature-sensitive mutations in each of these genes are conditionally lethal. Thus, blocking any of these essential proteins with chemotherapeutic agents should block cell proliferation. Because the replication system of *E. coli* is well characterized, I will focus on it throughout this review. Similar enzymes and genes have also been identified in widely divergent bacteria including gram-positive species.[1]

II. General Features of DNA Replication in *E. coli*

DNA replication in bacteria is a highly regulated process that is coordinated with other processes required for cell growth and division (59, 87). The principal decision to proceed with a round of replication occurs at the level of initiation. Once initiation occurs, replication proceeds bidirectionally from a unique origin until the forks meet on the opposing side of the circular bacterial chromosome or encounter replication termination sequences (Figure 3.1).

Initiation of a round of replication requires binding of an initiator protein to specific sequences found within the unique origin. In concert with other proteins, the structure of the origin is opened to permit the assembly of a primosome, an apparatus that migrates with the fork, splitting the two strands of the helix apart

[1]For example, the catalytic subunit of DNA polymerase III from *B. subtilis* and *E. coli* can be aligned and shows 27% identity (37). The *B. subtilis dnaH* and *dnaE* genes are analogous to the *E. coli* initiation gene *dnaA* (103) and primase *dnaG* (134).

Table 3.1. *Essential DNA replication proteins and sources of overproducing strains*

Gene	Protein-encoded map position	Reference to overproducer
dnaA	83	47
dnaB	92	6
dnaC	99	57
dnaT	99	78
dnaG	67	141
ssb	92	72
dnaE	4	76
dnaN	83	51
dnaX	11	131
dnaQ	5	117
gyrA	48	94, 36
gyrB	83	94, 36
lig	52	107
polA	87	92

while interacting with a priming enzyme. This latter enzyme generates primers for discontinuous replication on the lagging strand.

DNA polymerases cannot initiate synthesis *de novo;* they can only elongate preexisting chains in the 5′→3′ direction. Their inability to initiate synthesis explains the necessity of separate priming enzymes to generate the initiating

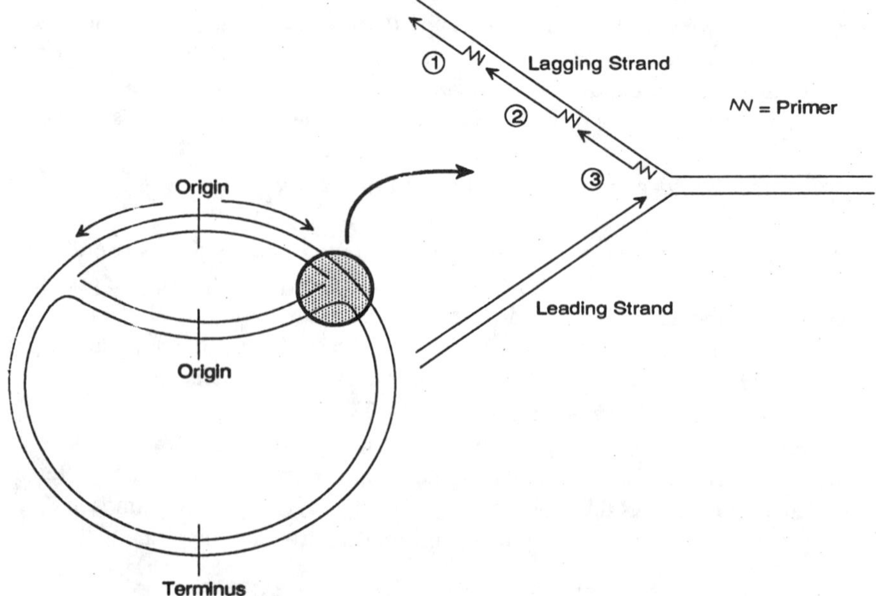

Figure 3.1. General features of *E. coli* replication. Arrows indicate the direction of fork movement and polymerase movement in the chromosome and replication fork illustrations, respectively.

oligonucleotides for further extension by the DNA polymerases. Synthesis of regularly spaced primers on the lagging strand also leads to a solution of the problem caused by DNA being antiparallel and DNA polymerases being unidirectional. The replicative DNA polymerase, synthesizing DNA continuously in the 5′→3′ direction on the leading strand template, tracks behind the helicase and jumps between successively synthesized primers on the lagging strand extending them 5′→3′ to synthesize DNA discontinuously. These discontinuous fragments are subsequently processed and joined to form high-molecular-weight DNA (Fig. 3.1).

III. Enzymology of Replication Fork Movement

A. DNA Polymerase III Holoenzyme

The DNA polymerase III holoenzyme is the major replicative enzyme of *E. coli*. It contains a DNA polymerase core plus auxiliary subunits that confer upon it the special properties that distinguish replicative complexes from repair polymerases (80). These properties include a rapid elongation rate and a corresponding high rate of 3′→5′ proofreading, high processivity, ability to utilize a long single-stranded template coated with single-stranded DNA binding protein (SSB[2]), ability to jump over obstacles created by annealed oligonucleotides, resistance to physiological levels of salt, ability to interact with other proteins of the replicative apparatus, and ability to coordinate replication through an asymmetric dimeric structure.

The DNA polymerase III holoenzyme can be biochemically resolved into a series of successively simpler polymerase forms (Table 3.2). Correlation of less complex forms with their biochemical properties has contributed to our understanding of the role of the polymerase auxiliary proteins in the replicative reaction.

Table 3.2. DNA Polymerase III Forms

Form	Subunit composition	Processivity[a]	Reference
Holoenzyme	$\alpha, \varepsilon, \theta, \tau, \gamma, \delta, \delta', \chi, \psi, \beta$	>5,000, 150,000	23, 95
Reconstituted holoenzyme-ε	$\alpha, \gamma, \delta, \delta', \chi, \psi, \beta$	1–3,000	42
Pol III*	$\alpha, \varepsilon, \theta, \tau, \gamma, \delta, \delta', \chi, \psi$	200	24
Pol III'	$\alpha, \varepsilon, \theta, \tau$	60	24
Pol III (core)	$\alpha, \varepsilon, \theta$	10	23

[a] Number of nucleotides inserted per binding-catalysis-dissociation step.

[2]The abbreviations used are: SSB, single-stranded DNA binding protein; DNAC, 3-decynoyl-N-acetylcysteamine.

1. DNA Polymerase III Core

The DNA polymerase III holoenzyme contains a DNA polymerase III core that is composed of three subunits: α, ε, and θ, of 130,000, 27,500, and 10,000 daltons, respectively (83). α contains the polymerase activity while the $3' \rightarrow 5'$ exonucleolytic editing activity resides within ε (117). The structural genes for α and ε are *dnaE* and *dnaQ* (*mutD*), respectively. Conditionally lethal mutations provide genetic evidence for a central role of these two proteins in the replication process (28, 46, 21). Neither a function nor a gene for θ has been established.

The core DNA polymerase III can efficiently replicate nuclease-activated DNA that contains short gaps (60, 71). While it is even more active than complex forms of DNA polymerase III in this system, it is inert in more natural replication systems described below. In part, this is due to the limited processivity of DNA polymerase III. It only inserts 10–15 nucleotides per binding-catalysis-dissociation step, several orders of magnitude less than the DNA polymerase III holoenzyme (23, Table 3.2). It is inhibited by physiological salt, spermidine, and SSB (14, 17, 31, 23).

Specific inhibitors of *E. coli* DNA polymerase III core have not yet been found. Brown and colleagues, however, have developed an excellent series of inhibitors that block DNA polymerase III of *Bacillus subtilis* by inducing formation of a ternary polymerase-inhibitor-template-primer complex (12). These inhibitors, 6-(arylamino)- and 6-(arylazo)-pyrimidines, base pair with pyrimidines in the template and lock the polymerase in a dead-end complex, effectively blocking further polymerization (16, 75).

2. DNA Polymerase III′ and DNA Polymerase III*

DNA polymerase III′ is a complex of the core DNA polymerase III and the τ subunit (81). DNA polymerase III′ can be distinguished from DNA polymerase III core by its increased processivity (*ca.* 60, Table 3.2) and stimulation by physiological concentrations of spermidine (81, 24). DNA polymerase III dimerizes upon binding τ, forming DNA polymerase III′ (81). Since the higher forms of DNA polymerase III also contain τ, they have also been presumed to form dimers, but this has not been rigorously demonstrated.

DNA polymerase III* is a complex of DNA polymerase III′ and the $\gamma\delta$ complex (γ, δ, δ', ψ, χ), with a processivity of *ca* 300 (24). It is stimulated by SSB, a protein that inhibits both DNA polymerase III and III′.

3. DNA Polymerase III Holoenzyme

The DNA polymerase III holoenzyme is composed of DNA polymerase III* plus the β subunit. It is the only polymerase III form that can function in reconstituted natural replication systems that include *E. coli* SSB and other replication proteins. The DNA polymerase III holoenzyme is the most processive form (23). It is able to replicate all long single-stranded templates tested without dissociating and remains stably bound to the template for periods up to 40 minutes (23, 49). This suggests that, if the dynamic affinity of the active polymerase for template was the same as that observed for the arrested complex, the holoenzyme is sufficiently processive to replicate the entire leading strand of the *E. coli* chromosome without ever dissociating. Recently the ability of the DNA polymerase III holoenzyme to synthesize 150,000-base-pair DNA *in vitro* has been demonstrated (95).

For holoenzyme to form a highly processive complex, it must first form an initiation complex with primed DNA in a reaction that is ATP-dependent (140, 49) (Figure 3.2). ATP is hydrolyzed to ADP and orthophosphate. 2′(3′)Trinitrophenyl ATP blocks initiation complex function, ATP binding by the holoenzyme, and the primed DNA-dependent ATPase activity of the DNA polymerase III holoenzyme (102). Initiation complex formation can be conveniently monitored by creation of a complex that is resistant to inhibition by anti-β IgG (49). β, an essential component of the elongation complex, presumably is buried in the initiation complex, sterically precluding antibody attachment. Initiation complexes can be isolated by gel filtration. Addition of the four required dNTPs to these complexes results in their rapid conversion to complete duplex molecules without polymerase dissociation (49).

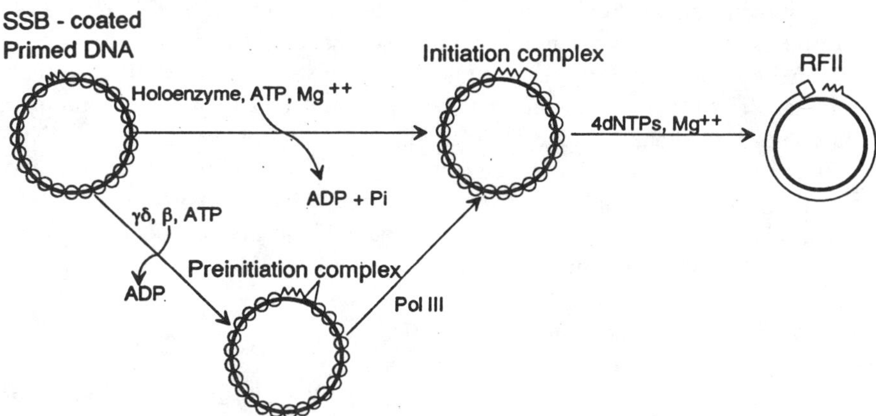

Figure 3.2. Intermediates in the DNA polymerase III holoenzyme-catalyzed replicative reaction. Initiation complexes can either be formed directly in a concerted reaction starting with holoenzyme and ATP or through formation of a preinitiation complex with the $\gamma\delta$ complex and β in the presence of ATP. The core DNA polymerase III can interact with the preinitiation complex to form a highly processive initiation complex. Alternatively, the τ and δ or δ' subunits can substitute for the $\gamma\delta$ complex. (See text.)

4. *Function of β and $\gamma\delta$ Complex*

In experiments involving isolated holoenzyme subassemblies, the $\gamma\delta$ complex functions to transfer β onto a template primer in an ATP-dependent reaction (136, 100). Alternatively, τ can interact with δ or δ' to transfer β to DNA in an analogous ATP-dependent reaction (101). This is the same ATP-dependent step that accompanies initiation complex formation with the intact DNA polymerase III holoenzyme (Fig. 3.2). The core DNA polymerase III can interact with the preinitiation complex formed by this reaction to assemble a highly processive replicative complex. Presumably, the ATP-dependent step in initiation complex formation between DNA polymerase III holoenzyme and primed templates also involves transfer of β to the template and altered interactions with DNA polymerase III, but these reactions may occur in a concerted fashion (80). It is not known if complexes formed in the multistage reaction differ from those formed with intact holoenzyme.

If added in vast excess, β alone can convert core DNA polymerase III to a highly processive enzyme (17, 63). Presumably, this occurs by a mass action effect, bypassing the need for the $\gamma\delta$ complex-dependent transfer reaction. This $\gamma\delta$ complex-independent reaction is inhibited by SSB and physiological ionic strength.

5. *Role of the τ Subunit*

Both the τ and γ subunits are products of the *dna*X gene (58, 97). They share 47,500 daltons of identical amino-terminal sequence. After translation of all but the final two amino acids, γ is generated by a site-specific frameshifting mechanism into a reading frame that contains a stop codon (25). τ contains an extra 23,600-dalton domain that interacts strongly with DNA. τ, when added to DNA polymerase III, increases its processivity; γ, by itself, does not (24).[3] Presumably the strong τ-DNA interaction is responsible for this increase in processivity. Both τ and γ have ATP binding sites and may function in analogous roles in different halves of the DNA polymerase III holoenzyme (38, 131).

6. *The Asymmetric Dimer Hypothesis*

Early studies designed to determine the role of ATP in formation of an initiation complex between holoenzyme and primed single-stranded DNA suggested an asymmetry in holoenzyme function. The ATP analog, ATPγS, will substitute for

[3]This increase in processivity is due solely to τ-template interactions. Since the δ subunits were missing from these experiments, neither τ nor γ could transfer β to the complex in an ATP-dependent reaction, resulting in much higher processivities.

ATP in supporting initiation complex formation, but only partially. In spite of its greater efficacy at low concentrations, ATPγS will only support one-half as much initiation complex formation as ATP (50, 82). ATPγS also triggers dissociation of initiation complexes formed in the presence of ATP. In a nucleotide-free state, one-half of the complexes are dissociated by the addition of ATPγS.

When the ATPγS observation was first made, we proposed that ATPγS was unveiling a functional asymmetry (50, 82). We proposed that the two halves of holoenzyme were functionally different and that this functional asymmetry was used by the cell to solve the asymmetric functional problem of polymerases at the replication fork (Figure 3.3). One polymerase, the leading strand polymerase, once bound to a primer at the origin, need not dissociate until the entire *E. coli* chromosome has ben replicated. The lagging strand polymerase must bind to a primer, synthesize an Okazaki fragment, release, and bind to the next primer each second (each cycle in Figure 3.3). It would appear to be advantageous for the cell to contain an asymmetric polymerase with a very high processivity on the leading strand and a decreased processivity on the lagging strand. This would

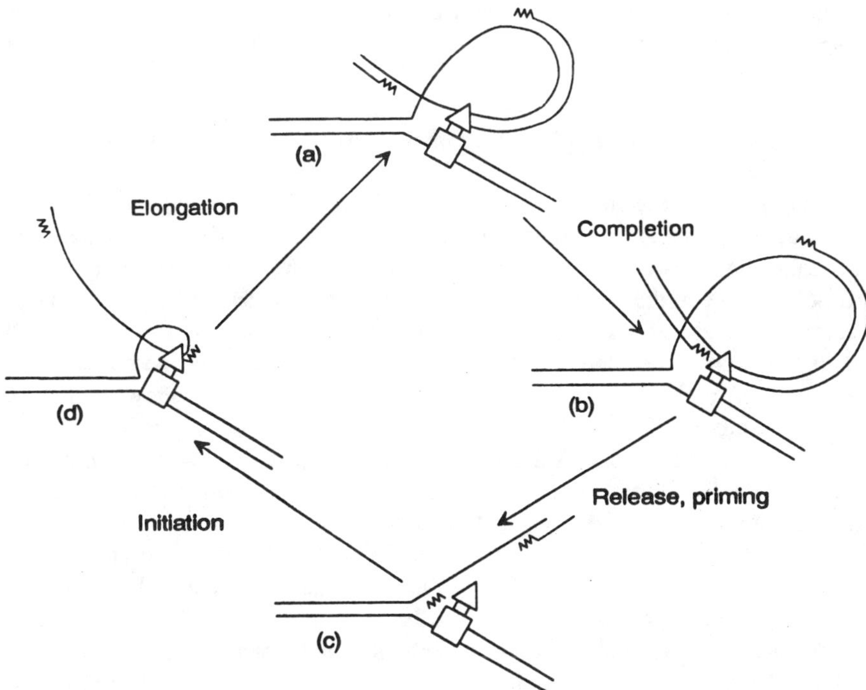

Figure 3.3 Coordinated action of the two halves of an asymmetric dimeric polymerase during coupled leading and lagging strand synthesis. The labels elongation, completion, release, and initiation refer to the actions of the lagging strand polymerase during Okazaki fragment synthesis.

permit rapid cycling each second during Okazaki fragment synthesis, while guaranteeing that the replicative complex will remain associated with the replication fork.

The presence of τ and γ within the holoenzyme may provide a molecular explanation for functional asymmetry (38). Both proteins reside within the same holoenzyme assemblies and compete with one another in reconstitution experiments, suggesting that they may occupy equivalent sites in the two halves of holoenzyme (77). Perhaps the τ subunit is located in the leading strand half of holoenzyme. Its tight template binding domain would be expected to confer greater processivity on the leading strand polymerase. By similar logic, the γ subunit would be expected to be located in the lagging strand half, permitting the rapid cycling between Okazaki fragment primers due, in part, to its weaker interactions with DNA. The ATPγS result is consistent with either γ or τ using ATPγS productivity to support initiation complex formation. This hypothesis has not yet been tested however. Lagging strand cycling may also be facilitated by communication between polymerase halves. Cooperative interactions between holoenzyme halves could be used to coordinate leading with lagging strand replication. It has been demonstrated that the two ATP binding sites that are responsible for initiation complex formation within holoenzyme interact with strong positive cooperativity (84).

7. *Structure of the DNA Polymerase III Holoenzyme*

Central to our understanding of the mechanism used for coordinated replication of leading and lagging strands and the mechanism of rapid cycling between Okazaki fragments is the determination of the structure of the DNA polymerase III holoenzyme at the level of subunit arrangements and subunit-DNA contacts. Our present understanding of these structures is primitive and is primarily limited to information gained through functional interactions detected and suppressor mutations. A preliminary model of the holoenzyme structure is shown in Figure 3.4. Upon mixing, the α and ε subunits form a complex (76a). Genetic confirmation of this interaction comes from mutations in *dnaQ* (structural gene for ε) that suppress *dnaE* (α) mutations (79). It is generally assumed that this type of suppressor mutation arises by modification of the structure of a subunit that interacts directly with the suppressed gene product. The DNA polymerase III core (α ε θ complex) is isolable, so θ must minimally interact with either α or θ, but we don't know which (83). τ can be isolated in a complex with the core DNA polymerase III, but the contacted subunit is unknown (81). Mutations have been isolated in *dnaX* (structural gene for τ and γ) that suppress *dnaN* (β) (22). Consistent with a τ-β and γ-β interaction is the finding that γ-δ or τ-(δ or δ') can transfer β to primed DNA to form a preinitiation complex (101). Suppressor data indicate an interaction between β and α (62). Reinforcing this conclusion, we

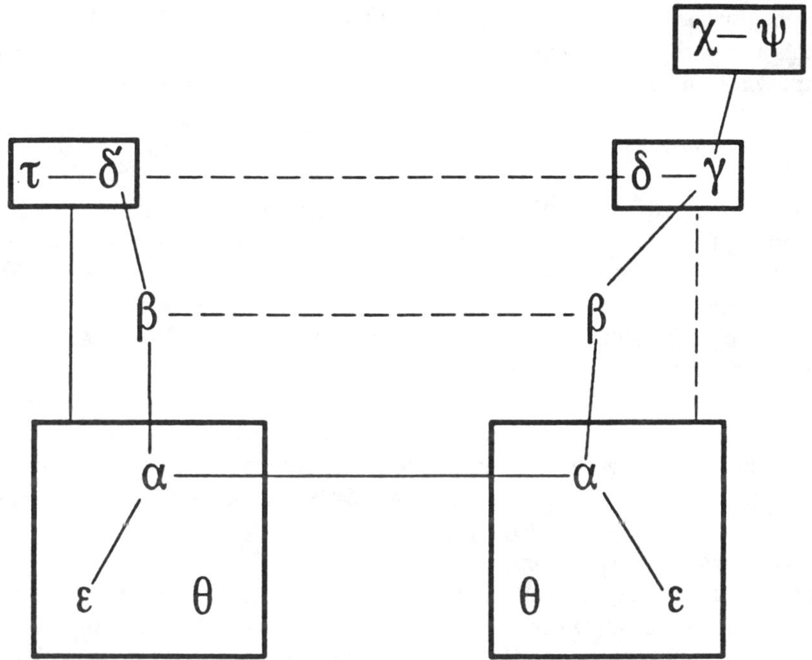

Figure 3.4. Minimal structure arrangement of subunits of the DNA polymerase III holoenzyme. Solid lines indicate known interactions. Lines between subunits indicate subunit-subunit interactions. Lines to boxes indicate subunit-complex interactions. Dotted lines indicate less certain interactions. See text for details.

also know that β can interact with and increase the processivity of core DNA polymerase III (63). Genetic evidence for α-α interaction and DNA polymerase III holoenzyme being a dimer derives from the discovery of *dna*E interallelic complementation by the Moses laboratory (13). From the studies of O'Donnell, it is known that δ can interact with γ and that δ or δ' can interact with τ (101). δ and δ' form an isolable complex. χ and ψ have been isolated in a complex with γ (101).

My laboratory has been developing fluorescence energy transfer as a method to measure subunit-subunit distances and subunit-template-primer distances. We know that β forms a dimer when it is free by itself in solution and that under physiological conditions, Mg^{++} induces dissociation (32, 30). Thus the aggregation state of β in active holoenzyme is uncertain. Fluorescent reporter groups located on cys_{333} of β first revealed this dissociation. We know that cys_{333}s are on the distal side of the β dimer and that a conformational change accompanying dissociation is transmitted to these distal sites (32). Fluorescence energy transfer from a fluorophore located on the third nucleotide from the 3'-end of a primer to an acceptor located on cys_{333} placed this distance at *ca*. 65 Å (33). Continuation of this approach coupled with subunit-subunit cross-linking and subunit-template-primer cross-linking should permit the holoenzyme structure to be established.

B. dnaG *Primase*

DNA synthesis in all cellular systems examined to date is primed by RNA-containing fragments called primers. These primers are synthesized by a specialized polymerase that can use single-stranded DNA templates to generate short oligonucleotides that can be efficiently transferred to the appropriate DNA polymerase. In *E. coli,* the *dnaG* protein is the priming enzyme (9). *dnaG* primase is a 60,000-dalton protein that is a monomer in solution (139, 112), but may function as a multimer as supported by its reaction kinetics (126). Normally, the primase requires additional primosomal proteins to function, but some single-stranded bacteriophages permit primase to recognize them directly. For example, *dnaG* primase synthesizes a 29-nucleotide RNA primer at the origin of bacteriophage G4 single-stranded DNA (10). SSB is required (9, 146). The primase recognition site forms a complex tertiary structure (123).

Although primase must initiate a chain with a ribonucleotide, dNTPs can be inserted in place of rNTPs, but it cannot efficiently elongate a deoxynucleotide-terminated chain (113, 85). Thus, *in vivo,* the primers synthesized on G4 are probably considerably shorter than the 29-mer. Although primase exhibits a strict requirement for use of a ribonucleotide for the first nucleotide incorporated, the initiating nucleotide can contain a variety of modifications in the triphosphate moiety. ATP, adenosine tetraphosphate, ADP, ATPγS, and ADPNP can all serve as initiating nucleotides (113).

2′-deoxy–2′-azido CTP and the 2′-deoxy–2′-azido derivatives of ATP, GTP, and UTP are effective inhibitors of *E. coli dnaG* primase (111). Inhibition results from incorporation of the analog resulting in shortened primers that cannot be extended. Primase lacks a proofreading exonuclease, so there is no efficient mechanism to remove the incorporated inhibitor. It is not known if *dnaG* primase remains tightly bound to the primer, blocking elongation, or if the 3′→ 5′ proofreading exonuclease of holoenzyme removes the primer.

C. Primosome

For most replicons, the *dnaG* primase cannot directly recognize priming sites, but must interact with the primosome for instructions to generate primer. The primosome is a combined helicase and mobile promoter that migrates along the lagging strand in the direction of fork movement. The primosome minimally contains the *dnaB* protein that can be loaded onto the template in a site-specific reaction by several different mechanisms. One mechanism, provided by *dnaC* protein, n, n′, n″, and i proteins is discussed below in this section. Alternative mechanisms include the *dnaA* protein-mediated process at the *E. coli* origin also discussed in this chapter and the λ, O, and P protein-mediated mechanism at the bacteriophage λ origin (89).

Incubation of *dnaB* protein, *dnaC* protein, *dnaT* protein (i), n, n′ (factor Y), and n″ with ϕX174 DNA, SSB, and ATP leads to formation of a complex that can be isolated by gel filtration. The complex supports the replicative reaction with only the addition of the *dnaG* primase and the DNA polymerase III holoenzyme and their nucleoside triphosphate substrates (135). This complex, termed a primosome, minimally contains the *dnaB* protein and protein n′. The primosome can move 5′→ 3′ along the lagging strand template in the direction of fork movement and transiently interact with *dnaG* protein to trigger primer formation (86, 4). Given the similarity of these reactions to those catalyzed by *dnaB* protein alone, *dnaB* protein is probably responsible for both of these activities.

1. dnaB *Protein*

The *dnaB* protein is a hexamer with a monomeric subunit size of 52,000 daltons (110, 99). The protein can hydrolyze any natural ribonucleoside triphosphate, but prefers purines (86, 109). The Km for ATP is 0.6 mM. Both single-stranded and double-stranded DNA serve as an effector of the ATPase reaction, although the properties of the two reactions differ (3).

Under nonphysiological conditions in the absence of SSB, the *dnaB* protein can function to facilitate priming by *dnaG* primase in the absence of any other replication proteins. The *dnaB* protein accomplishes this by binding nonspecifically to single-stranded DNA and transiently interacting with *dnaG* primase to trigger it to initiate primer synthesis (2). Coating of the template strand with SSB precludes interactions of *dnaB* protein without the aid of other proteins (2). This data supported the earlier model, with *dna*B protein functioning as a mobile promoter within the primosome, presenting a protein signal for the initiation of RNA synthesis, analogous to the DNA promoter signal that specifies the initiation of transcription for classical RNA polymerases (86).

dnaB protein also functions as a helicase and is probably the major helicase responsible for replication fork movement (66). It requires a 3′-OH extension of the displaced strand of more than 40 bases (66) before it can initiate the displacement reaction. Alone, the *dnaB* protein can unwind 1,000 bases of helix in under 30 seconds (66). It can also function efficiently with the DNA polymerase III holoenzyme to catalyze strand displacement synthesis at rates approximating the rate of fork movement *in vivo* (*ca.* 730 nucleotides/second at 30°C; 95). The reaction proceeds at the same rate with or without the other primosomal components, although primosome formation is more efficient than *dnaB* loading without the assistance of ancillary proteins. SSB inhibits helicase action if added before *dnaB* protein, but stimulates the reaction if added afterwards (66). SSB probably blocks *dnaB* protein binding to the fork, but stimulates strand displacement afterwards. This suggests that *dnaB* protein is processive, since SSB would inhibit *dnaB* protein after it dissociates the first time. Mok and Marians have obtained DNA products as large as 150,000 base pairs in an apparently processive reaction driven by the *dnaB* helicase (95).

2. *n′ Protein (factor Y)*

Protein n′ (factor Y) is a ϕX174-dependent ATPase (137, 122). In its native state, it is a single polypeptide chain of 76,000 daltons (121). It recognizes a 55-nucleotide fragment that can form a stable hairpin structure of 44 nucleotides (122). The n′ protein can recognize this sequence on SSB-coated DNA and displaces 30–180 molecules of SSB in an ATP-dependent reaction (120). Similar n′ effector sites are found near the origin of colE1 and pBR322 and can function as origins (147, 148). Correlation of n′ ATPase effector sites and functional origins implies that it is this recognition that leads to assembly of the primosome (124).

Recent studies have revealed that n′ functions as a helicase that moves 3′–5′ along the template, counter to the direction of the *dnaB* helicase (68, 65). It has been proposed that this protein functions to modulate the activities of the primosome on a loop formed on the lagging strand in dimeric-coupled leading and lagging strand models for replication (68). Supporting this notion, Marians and colleagues have recently demonstrated that the direction of primosome movement can be in either the *dnaB* helicase or n′ direction, depending on the concentration and type of nucleotides in the reaction mixture (69). This multienzyme bidirectional helicase requires all of the primosomal proteins and a primosome assembly site recognized by n′.

3. dnaC *Protein*

The *dnaC* protein, a monomer of 28,000 daltons (56, 98), interacts with the *dnaB* protein in the presence of ATP, forming a $dnaB_6$-$dnaC_6$ complex (138, 55, 133). Both *dnaC* and *dnaB* proteins bind nucleotides (8). Complexing of *dnaC* with *dnaB* protein inhibits the DNA-stimulated ATPase activity of *dnaB* protein (138, 55). A complex of *dnaB* and *dnaC* in the presence of ATPγS complex forms, but is functionally inactive (133). It has been proposed that ATP is hydrolyzed when the $dnaC_6$-$dnaB_6$ complex enters the primosome, concomitant with *dnaC* protein release (133). Consistent with this mechanism, in the λ replication system, *dnaC* protein is replaced by the λP proteins. The λP-*dnaB* complex is inactive at the origin until acted upon by the heat-shock proteins, *dnaJ* and *dnaK,* actively separating *dnaB* protein from the complex (1).

D. DNA Gyrase

As helicases split the two strands of the DNA helix apart in preparation for priming and replication, positive superhelical turns are induced ahead of the fork. If not relaxed, this superhelicity would stop fork progression. This problem is solved in *E. coli* by DNA gyrase, a special type II topoisomerase that can not only relax positive superhelical tension but use the energy of ATP hydrolysis to induce negative superhelicity (29, 93, 41, 20). This enzyme maintains the *E. coli* chromosome under negative superhelical tension, predisposing it to unwinding and replication. DNA gyrase is the target of numerous antibacterial agents, including nalidixic acid and novobiocin (93, 130, 129). As it has been the subject of several reviews, it will not be addressed further here (108a, 133a, 133b, 142a).

IV. Initiation of Replication

Replication initiates bidirectionally from a unique site (*oriC*) at 83 minutes on the *E. coli* chromosome (143). The origin was cloned by virtue of its ability to confer autonomous replication on a selectable marker lacking a functional origin (143). Further studies defined a 245-base-pair segment required for origin function. The availability of a well-defined plasmid-containing origin sequence enabled an *oriC*-specific *in vitro* replication system to be established. The initial crude system was inhibited by rifampicin and nalidixic acid and was defective when prepared from *dnaB* or *dnaA* mutant strains (27). The *dnaA* gene product, not required for earlier reconstituted systems, was known through genetic studies to be required for *oriC* function.

By using *oriC* plasmid replication as an assay, a purified system was established that required the following proteins for efficient synthesis: the *dnaA* protein, *dnaB* protein, *dnaC* protein, *dnaG* protein, SSB, DNA gyrase, DNA polymerase III holoenzyme, and low levels of the *E. coli* histone-like protein, HU.

The origin serves to initially bind the *dnaA* protein-ATP complex to a cluster of 9 base pairs (11, 15, 90). Four *dnaA* protein binding sites occur in the right-most region of the origin. *In vivo* footprinting studies indicate that three participate in DNA binding; perhaps the structure formed precludes the fourth site from interacting (115) (W. Messer, personal communication). All four *dnaA* protein binding sites are required for function *in vivo* (11). Upon *dnaA* protein binding, higher concentrations of ATP and high temperature (38°C) are required to induce a conformational change resulting in melting a series of AT-rich 13-mer sequences found in the left-most portion of the origin (11). Thus, origin function depends on a complex architecture and conformational changes. Part of the origin function derives from specialized structures in the DNA. The 13-mer sequences on the left side of the origin will melt in the absence of added protein under specific conditions *in vitro* (61). If the left-most 13-mer is mutated, the two right-most sequences will not unwind either. The left-most sequence can be replaced by other sequences of reduced helical stability that do not conform to the 13-mer consensus sequence. Kowalski and Eddy have termed the left-most element a DUE (DNA unwinding element) (61). The two other 13-mer elements cannot be replaced, presumably due to specific protein–nucleic acid interactions. Consistent with this notion, the *dnaA* protein is thought to interact with the 13-mer sequences after an initial interaction with the 9-mers (145).

The histone-like protein HU is needed for optimal origin function, but is not absolutely required (19). Its level determines the requirement for other replication proteins. (See below.) The *dnaB* and *dnaC* proteins in the presence of SSB form the primosome, just as they do in the n′ (factor Y)-dependent replication of φX174. In the plasmid *oriC* systems, there is no requirement for the other primosomal proteins, n, n′, and n″ or the *dnaT* protein. The *dnaG* protein and holoenzyme perform the priming and elongation roles that they do in other reconstituted reactions (132, 35). In *E. coli,* the composition of the primosome may change at a site distal to the origin. This would permit entry of n′ in concert with the n, n″, and *dnaT* protein, presumably when the fork encounters an n′ (factor Y) primosome assembly site (108). Such sequences have been detected in the *E. coli* chromosome downstream from the origin. Such a switching mechanism would be analogous to a switch from DNA polymerase I to DNA polymerase III holoenzyme and primosome-dependent replication as seen 400 nucleotides downstream from the pBR322 origin (91, 114, 125). The lethal phenotype of *dnaT* (structural gene for protein i) mutants suggests a role for these proteins in normal *E. coli* replication (64, 78).

The initial crude *oriC* system was sensitive to rifampicin, implying a role for RNA polymerase (27). RNA polymerase is required for efficient replication under a variety of conditions, including the absence of HU or the presence of high levels of HU, inadequately supercoiled template, temperatures less than 27°C, or the presence of topoisomerase I (7). With the requirement for two RNA polymerases, the classical enzyme and primase, the question of which enzyme supplied primers caused confusion. In the λ phage system, RNA polymerase serves to overcome the negative effects of physiological levels of HU, thus allowing transcriptional activation of the λ origin (88). The product of the RNA polymerase reaction does not prime, but serves to perturb the structure so that the other replication proteins can act. The same situation occurs in *E. coli*. The promoters for RNA polymerase near the origin can either point into or away from the origin and can even be terminated with the nonextendable nucleotide, cordecepin,[4] and still function to activate (7). The possibility that specific RNA polymerase-replication protein interactions are required was eliminated by successful substitution of the promoters with those of foreign polymerases.

The *oriC* system just described is not absolutely specific for *oriC*-containing

[4]This result must be treated with caution since the 3′→5′ exonuclease of the DNA polymerase III holoenzyme can excise cordecepin, activating these primers (34).

plasmids. A variety of other plasmids can also be replicated. Using specificity for *oriC*-containing plasmids as an assay, two additional proteins were purified, topoisomerase I and RNAse H (52, 104). Presumably, they play a role in destabilizing RNA primers synthesized at sites other than *oriC*. These two proteins apparently play this role *in vivo* also. Genetic inactivation of either *rnh* or *topA*, the structural genes for RNAse H nd topoisomerase I, suppresses defects in *dnaA*, permitting initiation at alternative origins (73, 104).

The *dnaA* protein binds ATP tightly and hydrolyzes it slowly to ADP, which is released very slowly (119, 118). The ADP form of the enzyme is inactive. Acidic phospholipids that contain unsaturated fatty acids are active in effecting ADP release (118, 144). A fluid membrane is required for *dnaA* protein activation. Anesthetics like fluphenazine disrupt membrane structure and block phospholipid-dependent *dnaA* protein activation (118). The inhibitor 3-decynoyl-N-acetylcysteamine (DNAC) blocks unsaturated fatty acid synthesis by interfering with β-hydroxydecanoyl thioester dehydrase (39, 53). This compound exerts its antibacterial activity by blocking initiation of new rounds of replication (26). Phospholipids prepared from DNAC treated cells will not efficiently support *dnaA* protein activation unless the cells were supplemented with an unsaturated fatty acid during growth. This suggests that membrane content and structure could play a positive role in regulating the initiation of DNA replication. That the effect is exerted through the *dnaA* protein may be merely correlative. To determine whether blockage of *dnaA* protein activation is the key cause of DNAC lethality, viable DNAC-resistant mutants that mapped to the *dnaA* gene would need to be isolated.

A negative role for phospholipid regulation of *dnaA* protein is also possible. Phospholipids can bind to and inhibit the nucleotide-free form of *dnaA* protein (118, 144). Consistent with this role, origins associated with the cell membrane for the first 10 minutes following a round of replication have been found (106). This is the period in which new rounds of replication are suppressed. Thus, membrane association may be responsible for blocking new rounds of replication. There is no strong evidence that this association takes place through the *dnaA* protein though.

V. Termination of Replication

Replication in *E. coli* proceeds bidirectionally away from *oriC* until the forks encounter termination sequences located between 28 and 36 minutes on the *E. coli* genetic map (18, 43). The terminus consists of two sequences that are specific for forks moving in only one direction. They consist of a 23-nucleotide sequence that is bound by the protein product of the *tus* gene (45). The 23-nucleotide termination sequence and *tus* protein are sufficient to block termination *in vitro* in an orientation-specific fashion (44, 74, 67). The termination sequences and *tus* are not essential for *E. coli* viability (40) and, therefore, do not offer a promising target for antimicrobial development.

VI. Screens for Inhibitors of the Replication Apparatus

A. General Perspective

The DNA replicative machinery of bacteria constitutes an attractive, largely unexplored family of targets for development of new antibiotics. The total mass of the proteins participating in initiation and fork elongation listed in this review exceeds 1.7 million daltons.[5] These present not only a number of defined catalytic sites but numerous other domains and clefts that may be able to accommodate inhibitory compounds. The size of the "replisome" approaches that of a ribosome. Recalling the large number of useful antibiotics that do not resemble substrates that attack this latter target, one should be encouraged to search for a variety of novel compounds that inhibit the replicative apparatus. Thus, one might expect to find useful antibiotics that do not resemble either DNA or nucleotides, an area to which past replication inhibitor development work has been largely confined.

Genetics tells us that no replacement exists for at least 14 members of the replication complex. Mutations of any of these genes block cellular DNA replication, leading to cell death. Many of the additional components for which no gene has been identified may also be essential.

The complexity of DNA replicative reactions in terms of their multiple labile components and the mandatory integrity of their primers, templates, and products has dictated the use of purified reconstituted systems for nearly all useful mechanistic studies. These systems avoid interfering cellular proteases and nucleases. They also lack a variety of repair reactions that could obscure a positive result in a drug screen since they catalyze macromolecular nucleotide incorporation, the basis of most rapid assays suitable for screens. Soluble systems also permit detection of hydrophilic inhibitors that might be excluded in whole-cell screens. Useful hydrophilic compounds, once their importance is recognized, might be appropriately modified to overcome membrane permeability barriers.

The availability of strains that overproduce replication proteins (Table 3.1) and production of replication enzymes and templates by a number of commercial sources[6] should facilitate development of purified reconstitution systems.

---

[5]Only one of each subunit is accounted for regardless of the stoichiometry in the complex. Proteins include *dnaB*, *dnaC*, n, n′, n″, i, *dnaG*, DNA polymerase III holoenzyme, DNA gyrase, SSB, *dnaA*, HU, RNAseH, topoisomerase I, DNA polymerase I, DNA ligase, and RNA polymerase.

[6]Boehringer Mannheim Biochemicals (PO Box 50414, Indianapolis, IN 46250) supplies *E. coli* RNAse H, RNA polymerase, DNA polymerase I, and DNA ligase. USB (26111 Miles Rd., Cleveland, OH 44122) sells *E. coli* DNA polymerase I, RNAse H, DNA ligase, and SSB. Stratagene (11099 North Torrey Pines Road, La Jolla, CA 92037) offers DNA polymerase III holoenzyme as part of a kit. The DNA polymerase III holoenzyme and numerous other replication proteins including initiation and priming components are available from Enzyco, Inc. (PO Box 18247–180, Denver, CO, 80218).

The most convenient way to monitor DNA replication is by incorporation of acid-soluble deoxynucleoside triphosphates into acid-insoluble DNA. Although seemingly a crude assay, with adequate attention to the combination and levels of proteins and the use of appropriate templates, the reactions become very specific, enabling the screening for inhibitors of different replication proteins simultaneously. A premix of replication proteins could be prepared and aliquoted in microtiter plates; after the addition of an agent to be screened, the reactions could be initiated with a limiting component such as the radioactive triphosphate. After incubation, acid is added to precipitate the product and the reaction mix is filtered over a glass fiber filter, collecting the precipitate and allowing radioactivity to be quantitated. These reactions have been miniaturized to the 25-μl scale already and should be very amenable to robotics. With the use of microtiter-based filtration systems and filter mats and methodology that requires little or no scintillation fluid, it should be possible to economically perform a thousand or more assays per day.

B. *Assays for Inhibitors of Fork Movement*

1. *DNA Polymerase III Holoenzyme*

The DNA polymerase III holoenzyme can be directly assayed by providing an oligonucleotide-primed single-stranded circle in the presence of SSB (34). This assay would screen for the holoenzyme's ability to form a functional initiation complex in an ATP-dependent reaction and to rapidly extend the primer in several seconds to generate a full-length circular product. This rapid, highly processive reaction minimally is dependent on the α, γ, δ, τ and ε subunits of the holoenzyme (128).

2. *Holoenzyme-primase-coupled System*

Use of a single-stranded G4 template and the addition of the *dnaG* primase permits the simultaneous screening for inhibition of primase and the DNA polymerase III holoenzyme (146, 48). *dnaG* primase recognizes a specific site on SSB-coated G4 and generates a primer that is transferred to the holoenzyme. We have evidence that the *dnaG* protein may become part of the elongation complex if both primase and holoenzyme are present during the priming reactions (31).

3. *Holoenzyme-primase-*dnaB *protein System*

An assay that would screen for specific *dnaG-dnaB* protein interactions is provided by the general priming system (2). In this assay, SSB must be deleted since SSB blocks *dnaB* protein interactions with DNA unless more specific "loading systems" such as those mediated by n′ (factor Y), *dnaA* protein, or the λ O and P proteins are present. In the general priming system, any single-stranded

DNA lacking a G4 type origin can be used. *dnaB* protein interacts with DNA and forms a complex that can trigger priming by *dnaG* primase and enable elongation by the DNA polymerase III holoenzyme.

4. Complete Primosome-primase-holoenzyme Systems

An assay that would screen for the entire n′ (factor Y)-dependent primosome on single-stranded DNA is provided by ϕX174 DNA templates (85, 5). Specificity for all of the primosome components is enforced by inclusion of SSB. This system requires n′ to recognize a primosome assembly site. In an ATP-dependent reaction in concert with proteins n, n″, *dnaT*, and *dnaC*, *dnaB* is transferred to the template where it can migrate, transiently interacting with *dnaG* protein and triggering a priming event. As with the other systems, elongation is mediated by the DNA polymerase III holoenzyme.

A similar assay that includes all of the above features plus an assay for the helicase activity of *dnaB* protein is provided by a model template developed by Marians and colleagues (95). A synthetic oligonucleotide containing the primosome assembly site is annealed to a single-stranded circle. The primosome assembly site is contained in the 5′-noncomplementary tail of the oligonucleotide, since the primosome can only assemble efficiently on single-stranged DNA. The primer is elongated by the DNA polymerase III holoenzyme until it replicates the full circle. Normally, the holoenzyme pauses at this site, but if the *dnaB* helicase is loaded by the n′-dependent primosome, it will progress ahead of the holoenzyme displacing the product strand and setting up a rolling circle whose tails approach 150,000 bases in length (95). If *dnaG* primase is included, coupled synthesis on the lagging strand takes place; only a twofold reduction in incorporation would result, however, if primase was inhibited. Thus, the preceding assays are more sensitive in monitoring primase inhibition.

5. Preliminary Determination of Bacterial Selectivity In Vitro

The availability of SV40 *in vitro* mammalian replication systems that presumably use the same elongation apparatus as mammalian chromosomes provides a relatively inexpensive preliminary means to determine whether inhibitors are selective for bacterial targets. The mammalian systems require the α and δ polymerases, requisite replication factors, and the SV40 origin-specific T antigen (54, 142, 70, 127). This system could be set up and candidate inhibitors tested to determine whether they also inhibit the mammalian replication apparatus. This would provide a means to determine which inhibitors to pursue in further tests.

C. Screens for Inhibitors of Initiation

1. dnaA *Protein-dependent Assays*

Plasmids containing *oriC* can be used to drive a variety of increasingly complex *dnaA* protein-dependent reactions. The simplest and most practical would be one that contained low, carefully titrated HU levels so that RNA polymerase and the specificity factors topoisomerase I and RNAse H could be deleted. This assay would include *dnaA* protein, SSB, *dnaB* protein, *dnaC* protein, *dnaG* primase, DNA gyrase, and DNA polymerase III holoenzyme (132, 105).

No system is available that assays both the *dnaA* protein and n′ (factor Y)-dependent pathways for primosome assembly although it is possible that a transition between these two complexes occurs shortly after chromosomal initiation.

D. Crude Systems for Screening DNA Replication Inhibitors

Several crude systems for testing the effects of DNA replication inhibitors exist. These are inferior to the purified reconstituted systems for reasons already discussed but present the opportunity to test inhibitors in a more biological context. These systems suitably adapted to bacteria other than *E. coli* may present a means to test inhibitors initially detected in the purified *E. coli* system for their general efficacy on other bacteria.

1. Permeabilized Cell Assay

The permeabilized cell system employs cells permeabilized to small hydrophilic molecules by treatment with toluene (96). Since nucleotide precursors of replication become permeable to these cells, incorporation of radioactivity can be measured as in the soluble systems. The template is the *E. coli* chromosome. Only elongation functions are monitored since this system does not permit initiation of replication. These cells should also be permeable to most small-molecule inhibitors of replication.

2. Cellophane Disk Assay

The cellophane disk system also monitors elongation activities on preexisting *E. coli* replication forks. Cells are lysed gently in a concentrated nondisrupted state on cellophane disks that float on a solution containing radioactive triphosphate precursors (116). This system has the advantage of being able to add macromolecules, including antibodies, that can be used to validate the requirement for a specific protein in a reaction.

Crude unfractionated extracts do not work well in supporting replicative reactions unless supplemented with high concentrations of a portion of the required replication proteins. Presumably, this is due to release of inhibitory substances including proteases and nucleases and to fragmentation of the chromosome during lysis.

E. Determination of the Target of Inhibition

Due to the multiple components of the replication systems described, additional experiments will need to be performed to identify the target of inhibition. Most useful approaches for addressing this issue require the purified *in vitro* systems since the crude systems do not offer experimental control.

In the simplest case, an inhibitor would bind to a replication protein free in solution by itself and prevent it from entering the replication complex. The target of inhibition could be identified by determining the protein that is most effective in overcoming the inhibition when added in excess. Of course, all the components in a system should be tested individually and attention should be paid to the possibility that a component that is not the direct target could provide partial protection by pulling equilibria, drawing the target into a complex resistant to inhibition. If the inhibitor is competing with one of the small-molecule participants in the replicative reaction, inhibition could be overcome by increasing substrate concentration.

In cases where the inhibitor acts by binding to a component of a complex that includes the template primer and forms a dead-end complex, analysis becomes more difficult because inhibition cannot be overcome with additional protein. The first step in addressing this problem is to determine which stage of the replicative reactions is blocked. Each of the intermediates of a reaction could be reconstituted, isolated by gel filtration, and tested for its ability to proceed upon addition of the missing components in the absence of inhibitor. For example, in the n′ (factor Y)-directed reaction, it could be determined whether an inhibitor blocks primosome formation, priming, initiation complex formation, or the elongation reaction. A reciprocal approach would be to reconstitute the various intermediates without inhibitor and to determine whether the overall reaction is still sensitive to inhibitor if it is added.

A complementary approach would be to examine which proteins are bound to the template primer in fully inhibited reactions. This could be accomplished by isolating template primer free of unbound protein (gel filtration through Biogel A5m), denaturation of the complex, resolution of bound proteins by SDS polyacrylamide gel electrophoresis, and identification of those proteins present by direct staining or immunoblotting (49). Once adequate structural information is available for an inhibitor, it might be possible to radiolabel it and identify its target by chemical or photo-crosslinking approaches.

A genetic approach to target identification would be provided by isolating inhibitor-resistant mutants and determining the map location of the resistance locus. A correlation between *in vivo* and *in vitro* resistance could be made with the candidate protein isolated from the resistant mutant.

VII. Summary

The *E. coli* replication system, including nearly all of the initiation, elongation, and termination reactions, became very well defined over the last 15 years. All the required proteins are available in pure form, many from overproducers, and their principal contribution to the replicative reaction are understood. From genetics, we know with certainty that at least 14 replication proteins are required for cell survival. Biochemical and genetic studies tell us that all or most bacteria share homologous replication machinery. With our advanced knowledge of the biochemistry of DNA replication, the time is ripe to make concerted efforts to find novel compounds that selectively interfere with bacterial DNA replication, yielding antibacterial agents.

References

1. Alfano, C., and R. McMacken. 1989. Ordered assembly of nucleoprotein structures at the bacteriophage λ replication origin during the initiation of DNA replication. J. Biol. Chem. **264:** 10699–10708.
2. Arai, K.I., and A. Kornberg. 1979. A general priming system employing only *dnaB* protein and primase for DNA replication. Proc. Natl. Acad. Sci. USA 76: 4308–4312.
3. Arai, K.I., and A. Kornberg. 1981. Mechanism of *dnaB* protein action: ATP hydrolysis by *dnaB* protein dependent on single-stranded or double-stranded DNA. J. Biol. Chem. **256:** 5253–5259.
4. Arai. K.I., and A. Kornberg. 1981. Unique primed start of phage ϕX174 DNA replication and mobility of the primosome in a direction opposite chain synthesis. Proc. Natl. Acad. Sci. USA **78:** 69–73.
5. Arai, K.I., R. McMacken, S.I. Yasuda, and A. Kornberg. 1981. Purification and properties of *Escherichia coli* protein: a prepriming protein in phage ϕX174 DNA replication. J. Biol. Chem. **256:** 5281–5286.
6. Arai, K.I., S.I. Yasuda, and A. Kornberg. 1981. Mechanism of *dnaB* protein action: crystallization and properties of *dnaB* protein, an essential replication protein in *Escherichia coli*. J. Biol. Chem. **256:** 5247–5252.
7. Baker, T.A., and A. Kornberg. 1988. Transcriptional activation of initiation of replication from the *E. coli* chromosomal origin: an RNA-DNA-hybrid near *oriC*. Cell **55:** 113–123.
8. Biswas, S.B., and E.E. Biswas. 1987. Regulation and *dnaB* function in DNA replication in *Escherichia coli* by *dnaC* and λP gene products. J. Biol. Chem. **262:** 7831–7838.

9. Bouché, J.P., K. Zechel, and A. Kornberg. 1975. *DnaG* gene product: a rifampicin resistant RNA polymerase initiates the conversion of a single-stranded *coli* phage DNA to its duplex replicative form. J. Biol. Chem. **250:** 5995–6001.

10. Bouché, J.P., L. Rowen, and A. Kornberg. 1978. The RNA primer synthesized by primase to initiate phage G4 DNA replication. J. Biol. Chem. **253:** 765–769.

11. Bramhill, D., and A. Kornberg. 1988. Duplex opening by *dnaA* protein at novel sequences in initiation of replication at the origin of the *Escherichia coli* chromosome. Cell **52:** 743–756.

12. Brown, N.C. 1971. Inhibition of bacterial DNA replication by 6-(*p*-hydroxyphenyl-azo)-uracil: differential effect on repair and semi-conservative synthesis in *Bacillus subtilis*. J. Mol. Biol. **59:** 1–16.

13. Bryan, S., M. Hagensee, and R.E. Moses. 1988. DNA polymerase III is required for mutagenesis, p. 305–313. *In* R. Moses and W. Summers (ed.), DNA replication and mutagenesis. American Society for Microbiology, Washington, D.C.

14. Burgers, P.M.J., and A. Kornberg. 1982. ATP activation of DNA polymerase III holoenzyme from *Escherichia coli:* initiation complex stoichiometry and reactivity. J. Biol. Chem. **257:** 11474–11478.

15. Cleary, J., D. Smith, N. Harding, and J. Zyskind. 1982. Primary structure of the chromosomal origins (*oriC*) of *Enterobacter aerogenes* and *Klebsiella pneumoniae:* comparisons and evolutionary relationships. J. Bacteriol. **150:** 1467–1471.

16. Coulter, C.L., and N.R. Cozzarelli. 1975. Crystal structure of an inhibitor and a model for inhibition of replicative DNA synthesis. J. Mol. Biol. **91:** 329–344.

17. Crute, J.J., R.J. LaDuca, K.O. Johanson, C.S. McHenry, and R.A. Bambara. 1983. Excess β subunit can bypass the ATP requirement for highly processive synthesis by the II *Escherichia coli* DNA polymerase III holoenzyme. J. Biol. Chem. **258:** 11344–11349.

18. deMassey, B.J., S. Bejar, J. Louarn, J.-M. Louarn, and J.-P. Bouché. 1984. Two terminator loci separated by five minutes on the *Escherichia coli* chromosome cause unidirectional inhibition of replication. Proc. Natl. Acad. Sci. USA **84:** 1759–1763.

19. Dixon, N.E., and A. Kornberg. 1984. Protein HU in the enzymatic replication of the chromosomal origin of *Escherichia coli*. Proc. Natl. Acad. Sci. USA **81:** 424–428.

20. Drlica, K. 1984. Biology of bacterial deoxyribonucleic acid topoisomerases. Microbiol. Rev. **48:** 273–289.

21. Echols, H., E. Lu, and P.M.J. Burgers. 1983. Mutator strains of *Escherichia coli, mutD* and *dnaQ* with defective exonucleolytic editing by DNA polymerase III holoenzyme. Proc. Natl. Acad. Sci. USA **80:** 2189–2192.

22. Engstrom, J., A. Wong, and R. Maurer. 1986. Interaction of DNA polymerase III γ and β subunits *in vivo* in *Salmonella typhimurium*. Genetics **113:** 499–516.

23. Fay, P.J., K.O. Johanson, C.S. McHenry, and R.A. Bambara. 1981. Size classes of products synthesized processively by DNA polymerase III and DNA polymerase III holoenzyme of *Escherichia coli*. J. Biol. Chem. **256:** 976–983.

24. Fay, P.J., K.O. Johanson, C.S. McHenry, and R. Bambara. 1982. Size classes of products synthesized processively by two subassemblies of *Escherichia coli* DNA polymerase III holoenzyme. J. Biol. Chem. **257:** 5692–5699.

25. Flower, A.M., and C.S. McHenry. 1990. The γ subunit of DNA polymerase III holoenzyme of *Escherichia coli* is produced by ribosomal frameshifting. Proc. Natl. Acad. Sci. USA **87:** 3713–3717.

26. Fralick, J.A., and K.G. Lark. 1973. Evidence for the involvement of unsaturated fatty acids in initiating chromosome replication in *Escherichia coli*. J. Mol. Biol. **80:** 459–475.

27. Fuller, R.S., J.M. Kaguni, and A. Kornberg. 1981. Enzymatic replication of the origin of the *Escherichia coli* chromosome. Proc. Natl. Acad. Sci. USA **78:** 7370–7374.

28. Gefter, M.L., Y. Hirota, T. Kornberg, J.A. Wechsler, and C. Barnoux. 1971. Analysis of DNA polymerases II and III in mutants of *Escherichia coli* thermosensitive for DNA synthesis. Proc. Natl. Acad. Sci. USA **68:** 3150–3153.

29. Gellert, M., K. Mizuuchi, M.H. Odea, and H.A. Nash. 1976. DNA gyrase, an enzyme that introduces superhelical turns into DNA. Proc. Natl. Acad. Sci. USA. **73:** 3872–3876.

30. Griep, M.A., and C.S. McHenry. 1988. The dimer of the β subunit of *Escherichia coli* DNA polymerase III holoenzyme is dissociated into monomers upon binding magnesium-II. Biochemistry **27:** 5210–5215.

31. Griep, M.A., and C.S. McHenry. 1989. Glutamate overcomes the salt inhibition of DNA polymerase III holoenzyme. J. Biol. Chem. **264:** 11294–11301.

32. Griep, M., and C. McHenry. 1990. Dissociation of the DNA polymerase III holoenzyme β_2 subunits is accompanied by a conformational change at distal cysteines$_{333}$. J. Biol. Chem. **265:** 20356–20363.

33. Griep, M., and C.S. McHenry. 1991. Fluorescence energy transfer between the primer and the β subunit of DNA polymerase III holoenzyme. Submitted for publication.

34. Griep, M., J. Reems, M. Franden, and C. McHenry. 1990. Reduction of the potent DNA polymerase III holoenzyme $3' \rightarrow 5'$ exonuclease activity by template-primer analogs. Biochemistry **29:** 9006–9014.

35. Grossman, A.D., D.B. Straus, W.A. Walter, and C.A. Gross. 1987. σ^{32} synthesis can regulate the synthesis of heat shock proteins in *Escherichia coli*. Genes Dev. **1:** 179–184.

36. Hallett, P., A.J. Grimshaw, D.B. Wigley, and A. Maxwell. 1990. Cloning of the DNA gyrase genes under *tac* promoter control—overproduction of the gyrase protein-A and B-protein. Gene **93:** 139–142.

37. Hammond, R., M. Barnes, S. Mack, J. Mitchener, and N. Brown. 1991. *Bacillus subtilis* DNA polymerase III: complete sequence, overexpression, and characterization of the *polC* gene. Gene **98:** 29–36.

38. Hawker, J.R., and C.S. McHenry. 1987. Monoclonal antibodies specific for the τ subunit of the DNA polymerase III holoenzyme of *Escherichia coli*. Used to demonstrate that τ is the product of the *dnaZX* gene and that both it and the γ, the *dnaZ* gene

product, are integral components of the same enzyme assembly. J. Biol. Chem. **262:** 12711–12727.

39. Helmkamp, G.M., Jr., R.R. Rando, D.J.H. Brock, and K. Bloch. 1968. β-hydroxydecanoyl thioester dehydrase. J. Biol. Chem. **243:** 3329–3231.

40. Henson, J.M., B. Kopp, and P.L. Kuempel. 1984. Deletion of 60 kilobase pairs of DNA from the *ter*C region of the chromosome of *Escherichia coli*. Mol. Gen. Genet. **193:** 263–268.

41. Higgins, N.P., C.L. Peebles, A. Sugino, and N.R. Cozzarelli. 1978. Purification of subunits of *Escherichia coli* DNA gyrase and reconstitution of enzymatic activity. Proc. Natl. Acad. Sci. USA **75:** 1773–1777.

42. Hill, T.M., A.J. Pelletier, M.L. Tecklenburg, and P.L. Kuempel. 1988. Identification of the DNA sequence from the *E. coli* terminus region that halts replication forks. Cell **55:** 459–466.

43. Hill, T.M., J.M. Henson, and P.L. Kuempel. 1987. The terminus region of the *Escherichia coli* chromosome contains two separate loci that exhibit polar inhibition of replication. Proc. Natl. Acad. Sci. USA **84:** 1754–1758.

44. Hill, T.M., and K.J. Marians. 1990. *Escherichia coli tus* protein acts to arrest the progression of DNA replication forks *in vitro*. Proc. Natl. Acad. Sci. USA **87:** 2481–2485.

45. Hill, T.M., M.L. Tecklenburg, A.J. Pelletier, and P.L. Kuempel. 1989. *tus,* the trans-acting gene required for termination of DNA replication in *Escherichia coli,* encodes a DNA-binding protein. Proc. Natl. Acad. Sci. USA **86:** 1593–1597.

46. Horiuchi, T., H. Maki, and M. Sekiguchi. 1978. A new conditional lethal mutator (*dnaQ49*) in *Escherichia coli* K12. Mol. Gen. Genet. **163:** 277–283.

47. Hwang, D.S., and J.M. Kaguni. 1988. Purification and characterization of the *dnaA–46* gene product. J. Biol. Chem. **263:** 10625–10632.

48. Johanson, K.O., and C.S. McHenry. 1980. Purification and characterization of the β subunit of the DNA polymerase III holoenzyme of *Escherichia coli*. J. Biol. Chem. **255:** 10984–10990.

49. Johanson, K.O., and C.S. McHenry. 1982. The β subunit of the DNA polymerase III holoenzyme becomes inaccessible to antibody after formation of an initiation complex with primed DNA. J. Biol. Chem. **257:** 12310–12315.

50. Johanson, K.O., and C.S. McHenry. 1984. Adenosine 5′-*O*-(3-thiotriphosphate) can support the formation of an initiation complex between the DNA polymerase III holoenzyme and primed DNA. J. Biol. Chem. **259:** 4589–4595.

51. Johanson, K.O., T.E. Haynes, and C.S. McHenry. 1986. Chemical characterization and purification of the β subunit of the DNA polymerase III holoenzyme from an overproducing strain. J. Biol. Chem. **261:** 11460–11465.

52. Kaguni, J.M., and A. Kornberg. 1984. Topoisomerase I confers specificity in enzymatic replication of the *Escherichia coli* chromosomal origin. J. Biol. Chem. **259:** 8578–8583.

53. Kass, L.R. 1968. The antibacterial activity of 3-decynoyl-N-acetylcysteamine. J. Biol. Chem. **243:** 3223–3228.

54. Kelly, T.J. 1988. SV40 DNA replication. J. Biol. Chem. **263:** 17889–17892.

55. Kobori, J.A., and A. Kornberg. 1982. The *Escherichia coli dnaC* gene product: properties of the *dnaB-dnaC* protein complex. J. Biol. Chem. **257:** 13770–13775.

56. Kobori, J.A., and A. Kornberg. 1982. The *Escherichia coli dnaC* gene product: purification, physical properties and role in replication. J. Biol. Chem. **257:** 13763–13769.

57. Kobori, J.A., and A. Kornberg. 1982. The *Escherichia coli dnaC* gene product: overproduction of the *dnaC* proteins of *Escherichia coli* and *Salmonella typhimurium* by cloning into a high copy number plasmid. J. Biol. Chem. **257:** 13757–13762.

58. Kodaira, M., S.B. Biswas, and A. Kornberg. 1983. The *dnaX* gene encodes the DNA polymerase III holoenzyme τ subunit, precursor of the γ subunit, the *dnaZ* gene product. Mol. Gen. Genet. **192:** 80–86.

59. Kornberg, A. 1980. *In* A. Kornberg (ed.), DNA replication, p. 1–724. W.H. Freeman and Company, San Francisco.

60. Kornberg, T., and M.L. Gefter. 1972. Deoxyribonucleic acid synthesis in cell-free extracts. J. Biol. Chem. **247:** 5369–5375.

61. Kowalski, D., and M.J. Eddy. 1989. The DNA unwinding element: a novel cis-acting component that facilitates opening of the *Escherichia coli* replication origin. EMBO J. **8:** 4335–4344.

62. Kuwabara, N., and H. Uchida. 1981. Functional cooperation of the *dnaE* and *dnaN* gene products in *Escherichia coli*. Proc. Natl. Acad. Sci. USA **78:** 5764–5767.

63. LaDuca, R.J., J.J. Crute, C.S. McHenry, and R.A. Bambara. 1986. The β subunit of the *Escherichia coli* DNA polymerase III holoenzyme interacts functionally with the catalytic core in the absence of other subunits. J. Biol. Chem. **261:** 7550–7557.

64. Lark, C.A., J. Riazi, and K.G. Lark. 1978. *dnaT*-dominant conditional lethal mutation affecting DNA replication in *Escherichia coli*. J. Bacteriol. **136:** 1008–1017.

65. Lasken, R.S., and A. Kornberg. 1988. The primosomal protein n′ of *Escherichia coli* is a DNA helicase. J. Biol. Chem. **263:** 5512–5518.

66. Lebowitz, J.H., and R. McMacken. 1986. The *Escherichia coli dnaB* replication protein is a DNA helicase. J. Biol. Chem. **261:** 4738–4748.

67. Lee, E.H., A. Kornberg, H. Masumi, T. Kobayashi, and T. Horiuchi. 1989. *Escherichia coli* replication termination protein impedes the action of helicases. Proc. Natl. Acad. Sci. USA **86:** 9104–9108.

68. Lee, M.S., and K.J. Marians. *Escherichia coli* replication (factor Y): a component of the primosome can act as a DNA helicase. Proc. Natl. Acad. Sci. USA **84:** 8345–8349.

69. Lee, M.S., and K.J. Marians. 1989. The *Escherichia coli* primosome can translocate actively in either direction along a DNA strand. J. Biol. Chem. **264:** 14531–14542.

70. Lee, S.H., T. Eki, and J. Hurwitz. 1989. Synthesis of DNA containing the SV40 origin of replication by the combined action of DNA polymerases α and δ. Proc. Natl. Acad. Sci. USA **86:** 7361–7365.

71. Livingston, D.M., D.C. Hinkle, and C.C. Richardson. 1975. Deoxyribonucleic acid polymerase III of *Escherichia coli*. J. Biol. Chem. **250:** 461–469.

72. Lohman, T.M., J.M. Green, and R.S. Beyer. 1986. Large-scale overproduction and rapid purification of the *Escherichia coli ssb* gene product expression of the *ssb* gene under λP_L control. Biochemistry **25:** 21–25.

73. Louarn, J., J.P. Bouché, J. Patte, and J.M. Louarn. 1984. Genetic inactivation of topoisomerase I suppresses a defect in initiation of chromosome replication in *Escherichia coli*. Mol. Gen. Genet. **195:** 170–174.

74. MacAllister, T., G.S. Khatri, and D. Bastia. 1990. Sequence-specific and polarized replication termination: *in vitro* complementation of extracts of *tus*-negative *Escherichia coli* by purified *ter* protein and analysis of termination intermediates. Proc. Natl. Acad. Sci. USA **87:** 2828–2832.

75. Mackenzie, J.M., N.M. Neville, G.E. Wright, and N.C. Brown. 1973. Hydroxyphenylazopyrimidines: characterization of the active forms and their inhibitory action on a DNA polymerase from *Bacillus subtilis*. Proc. Natl. Acad. Sci. USA **70:** 512–516.

76. Maki, H., and A. Kornberg. 1985. The polymerase subunit of DNA polymerase III of *Escherichia coli:* purification of the α subunit devoid of nuclease activities. J. Biol. Chem. **260:** 12987–12992.

76a. Maki, H., and A. Kornberg. 1987. Proofreading by DNA polymerase III of *Escherichia coli* depends on cooperative interaction of the polymerase and exonuclease subunits. Proc. Natl. Acad. Sci. USA **84:** 4389–4392.

77. Maki, H., S. Maki, and A. Kornberg. 1988. DNA polymerase III holoenzyme of *Escherichia coli:* the holoenzyme is an asymmetric dimer with twin active sites. J. Biol. Chem. **263:** 6570–6578.

78. Masai, H., M.W. Bond, and K.I. Arai. 1986. Cloning of the *Escherichia coli* gene for primosomal protein: the relationship to *dnaT,* essential for chromosomal DNA replication. Proc. Natl. Acad. Sci. USA **83:** 1256–1260.

79. Maurer, R., B.C. Osmond, and D. Botstein. 1984. Genetic analysis of DNA replication in bacteria *dnaB* mutations that suppress *dnaC* mutations and *dnaQ* mutations that suppress *dnaE* mutations in *Salmonella typhimurium*. Genetics **108:** 25–38.

80. McHenry, C. 1988. DNA polymerase III holoenzyme of *Escherichia coli*. Annu. Rev. Biochem. **57:** 519–550.

81. McHenry, C.S. 1982. Purification and characterization of DNA polymerase III': identification of τ as a subunit of the DNA polymerase III holoenzyme. J. Biol. Chem. **257:** 2657–2663.

82. McHenry, C.S., and K.O. Johanson. 1984. DNA polymerase III holoenzyme of *Escherichia coli:* an asymmetric dimeric replicative complex containing distinguishable leading and lagging strand polymerases, p. 315–319. *In* U. Hübscher and S. Spadari (ed.), Proteins involved in DNA replication. Plenum Publishing Corp., New York.

83. McHenry, C.S., and W. Crow. 1979. DNA polymerase III of *Escherichia coli* purification and identification of subunits. J. Biol. Chem. **254:** 1748–1753.

84. McHenry, C.S., R. Obserfelder, K. Johanson, H. Tomasiewicz, and A. Franden. 1987. Structure and mechanism of the DNA polymerase III holoenzyme, p. 47–62. *In* R. McMacken and T.J. Kelly (ed.), DNA replication and recombination. Alan R. Liss, Inc., New York.

85. McMacken, R., and A. Kornberg. 1978. A mulltienzyme system for priming the replication of ϕX174 viral DNA. J. Biol. Chem. **253:** 3313–3319.

86. McMacken, R., K. Ueda, and A. Kornberg. 1977. Migration of *Escherichia coli dnaB* protein on the template DNA strand as a mechanism in initiating DNA replication. Proc. Natl. Acad. Sci. USA **74:** 4190–4194.

87. McMacken, R., L. Silver, and C. Georgopoulos. 1987. DNA replication, p. 564–612. *In* J.L. Ingraham, K.B. Low, B. Magasanik, M. Schaechter, and H.E. Umbarger (ed.), *Escherichia coli* and *Salmonella typhimurium,* vol. 1. American Society for Microbiology, Washington, D.C.

88. Mensa-Wilmot, K., K. Carroll, and R. McMacken. 1989. Transcriptional activation of bacteriophage λ DNA replication *in vitro:* regulatory role of histone-like protein HU of *Escherichia coli*. EMBO J. **8:** 2393–2402.

89. Mensa-Wilmot, K., R. Seaby, C. Alfano, M.S. Wold, B. Gomes, and R. McMacken. 1989. Reconstitution of a nine-protein system that initiates bacteriophage λ DNA replication. J. Biol. Chem. **264:** 2853–2861.

90. Messer, W. 1987. Initiation of DNA replication in *Escherichia coli*. J. Bacteriol. **169:** 3395–3399.

91. Minden, J.S., and K.J. Marians. 1985. Replication of pBR322 DNA *in vitro* with purified proteins: requirement for topoisomerase I in the maintenance of template specificity. J. Biol. Chem. **260:** 9316–9325.

92. Minkley, E.G., Jr., A.T. Leney, J.B. Bodner, M.M. Panicker, and W.E. Brown. 1984. *Escherichia coli* DNA polymerase I: construction of a *polA* plasmid for amplification and an improved purification scheme. J. Biol. Chem. **259:** 10386–10392.

93. Mizuuchi, K., M.H. Odea, and M. Gellert. 1978. DNA gyrase subunit structure and ATPase activity of the purified enzyme. Proc. Natl. Acad. Sci. USA **75:** 5960–5963.

94. Mizuuchi, K., M. Mizuuchi, M.H. Odea, and M. Gellert. 1984. Cloning and simplified purification of *Escherichia coli* DNA gyrase A and B proteins. J. Biol. Chem. **259:** 9199–9201.

95. Mok, M., and K.J. Marians. 1987. The *Escherichia coli* preprimosome and *dnaB* helicase can form replication forks that move at the same rate. J. Biol. Chem. **262:** 16644–16654.

96. Moses, R.E., and C.C. Richardson. 1970. Replication and repair of DNA in cells of *Escherichia coli* treated with toluene. Proc. Natl. Acad. Sci. USA **67:** 674–681.

97. Mullin, D.A., C.L. Woldringh, J.M. Henson, and J.R. Walker. 1983. Cloning of the *Escherichia coli dnaZX* region and identification of its products. Mol. Gen Genet. **192:** 73–79.

98. Nakayama, N., M.W. Bond, A. Miyajima, J. Kobori, and K.I. Arai. 1987. Structure of *Escherichia coli dnaC:* identification of a cysteine residue possibly involved in association with *dnaB* protein. J. Biol. Chem. **262:** 10475–10480.

99. Nakayama, N., N. Arai, M.W. Bond, Y. Kaziro, and K. Arai. 1984. Nucleotide sequence of *dnaB* and the primary structure of the *dnaB* protein from *Escherichia coli*. J. Biol. Chem. **259:** 97–101.

100. O'Donnell, M.E. 1987. Accessory proteins bind a primed template and mediate rapid cycling of DNA polymerase III holoenzyme from *Escherichia coli*. J. Biol. Chem. **262:** 16558–16565.

101. O'Donnell, M., and P.S. Studwell. 1990. Total reconstitution of DNA polymerase III holoenzyme reveals dual accessory protein clamps. J. Biol. Chem. **265:** 1179–1187.

102. Oberfelder, R., and C.S. McHenry. 1987. Characterization of 2′ 3′ trinitrophenyl-ATP as an inhibitor of ATP-dependent initiation complex formation between the DNA polymerase III holoenzyme and primed DNA. J. Biol. Chem. **262:** 4190–4194.

103. Ogasawara, N., S. Moriya, K. von Meyenburg, F.G. Hansen, and H. Yoshikawa. 1985. Conservation of genes and their organization in the chromosomal replication origin region of *Bacillus subtilis* and *Escherichia coli*. EMBO J. **4:** 3345–3350.

104. Ogawa, T., G.G. Pickett, T. Kogoma, and A. Kornberg. 1984. RNAse H confers specificity in the *dnaA*-dependent initiation of replication at the unique origin of the *Escherichia coli* chromosome *in vivo* and *in vitro*. Proc. Natl. Acad. Sci. USA **81:** 1040–1044.

105. Ogawa, T., T.A. Baker, A. Van Der Ende, and A. Kornberg. 1985. Initiation of enzymatic replication at the origin of the *Escherichia coli* chromosome: contributions of RNA polymerase and primase. Proc. Natl. Acad. Sci. USA **82:** 3562–3566.

106. Ogden, G.B., M.J. Pratt, and M. Schaechter. 1988. The replicative origin of the *E. coli* chromosome binds to cell membranes only when hemimethylated. Cell **54:** 127–135.

107. Panasenko, S.M., J.R. Cameron, R.W. Davis, and I.R. Lehman. 1977. Five hundred-fold overproduction of DNA ligase after induction of a hybrid λ lysogen constructed *in vitro*. Science **196:** 188–189.

108. Parada, C., and K. Marians. 1991. J. Biol. Chem., in press.

108a. Radl, S. 1990. Structure-activity relationships in DNA gyrase inhibitors. Pharmac. Ther. **48:** 1–17.

109. Reha Krantz, L.J., and J. Hurwitz. 1978. The *dnaB* gene product of *Escherichia coli:* single-stranded DNA dependent ribonucleotide triphosphatase activity. J. Biol. Chem. **253:** 4051–4057.

110. Reha Krantz, L.J., and J. Hurwitz. The *dnaB* gene product of *Escherichia coli:* purification homogeneity and physical properties. J. Biol. Chem. **253:** 4043–4050.

111. Reichard, P., L. Rowen, R. Eliasson, J. Hobbs, and F. Eckstein. 1978. Inhibition of primase: the *dnaG* protein of *Escherichia coli* by 2′-deoxy- 2′-azido CTP. J. Biol. Chem. **253:** 7011–7016.

112. Rowen, L., and A. Kornberg. 1978. Primase, the *dnaG* protein of *Escherichia coli:* an enzyme which starts DNA chains. J. Biol. Chem. **253:** 758–764.

113. Rowen, L., and A. Kornberg. 1978. A ribonucleotide deoxyribonucleotide primer synthesized by primase. J. Biol. Chem. **253:** 770–774.

114. Sakakibara, Y., and J. Tomizawa. 1974. Replication of ColE1 plasmid DNA in cell extracts: selective synthesis of early replicative intermediates. Proc. Nat. Acad. Sci. USA **71:** 1403–1407.

115. Samitt, C.E., F.G. Hansen, J.F. Miller, and M. Schaechter. 1989. *In vivo* studies of *dnaA* binding to the origin of replication of *Escherichia coli*. EMBO J. **8:** 989–994.

116. Schaller, H., B. Otto, V. Nusslein, J. Hof, R. Herrmann, and F. Bonhoeffer. 1972. Deoxyribonucleic acid replication *in vitro*. J. Mol. Biol. **63:** 183–200.

117. Scheuermann, R.H., and H. Echols. 1984. A separate editing exonuclease for DNA replication: the ε subunit of *Escherichia coli* DNA polymerase III holoenzyme. Proc. Natl. Acad. Sci. USA **81:** 7747–7751.

118. Sekimizu, K., and A. Kornberg. 1988. Cardiolipin activation of *dnaA* protein: the initiation protein of replication in *Escherichia coli*. J. Biol. Chem. **263:** 7131–7135.

119. Sekimizu, K., D. Bramhill, and A. Kornberg. 1987. ATP activates *dnaA* protein in initiating replication of plasmids bearing the origin of the *Escherichia coli* chromosome. Cell **50:** 259–266.

120. Shlomai, J., and A. Kornberg. 1980. A prepriming DNA replication enzyme of *Escherichia coli:* actions of protein n′: a sequence-specific DNA-dependent ATPase. J. Biol. Chem. **255:** 6794–6798.

121. Shlomai, J., and A. Kornberg. 1980. A prepriming DNA replication enzyme of *Escherichia coli:* purification of protein n′: a sequence-specific DNA-dependent ATPase. J. Biol. Chem. **255:** 6789–6793.

122. Shlomai, J., and A. Kornberg. 1980. An *Escherichia coli* replication protein that recognizes a unique sequence within a hairpin region in phage ϕX174 DNA. Proc. Natl. Acad. Sci. USA **77:** 799–803.

123. Sims, J., and E.W. Benz, Jr. 1980. Initiation of DNA replication by the *Escherichia coli dnaG* protein: evidence that tertiary structure is involved. Proc. Natl. Acad. Sci. USA **77:** 900–904.

124. Soeller, W.C., and K.J. Marians. 1982. Deletion mutants defining the *Escherichia coli* replication (factor Y) effector site sequences in pBR322 DNA. Proc. Natl. Acad. Sci. USA **79:** 7253–7257.

125. Staudenbauer, W. 1977. Replication of the ampicillin resistance plasmid RSF1030 in extracts of *Escherichia coli:* separation of the replication cycle into early and late stages. Mol. Gen. Genet. **156:** 27–34.

126. Stayton, M.M., and A. Kornberg. 1983. Complexes of *Escherichia coli* primase with the replication origin of G4 phage DNA. J. Biol. Chem. **258:** 13205–13212.

127. Stillman, B. 1989. Initiation of eukaryotic DNA replication *in vitro*. Annu. Rev. Cell Biol. **5:** 197–245.

128. Studwell, P.S., and M. O'Donnell. 1990. Processive replication is contingent on the exonuclease subunit of DNA polymerase III holoenzyme. J. Biol. Chem. **265:** 1171–1178.

129. Sugino, A., C.L. Peebles, K.N. Kreuzer, and N.R. Cozzarelli. 1977. Mechanism of action of nalidixic acid: purification of *Escherichia coli nalA* gene product and its relationship to DNA gyrase and a novel nicking-closing enzyme. Proc. Natl. Acad. Sci. USA **74:** 4767–4771.

130. Sugino, A., N.P. Higgins, P.O. Brown, C.L. Peebles, and N.R. Cozzarelli. 1978. Energy coupling in DNA gyrase and the mechanism of action of novobiocin. Proc. Natl. Acad. Sci. USA **75:** 4838–4842.

131. Tsuchihashi, Z., and A. Kornberg. 1989. ATP interactions of the τ and γ subunits of DNA polymerase III holoenzyme of *Escherichia coli*. J. Biol. Chem. **264:** 17790–17795.

132. Van Der Ende, A., T.A. Baker, T. Ogawa, and A. Kornberg. 1985. Initiation of enzymatic replication at the origin of the *Escherichia coli* chromosome primase as the sole priming enzyme. Proc. Natl. Acad. Sci. USA **82:** 3954–3958.

133. Wahle, E., R.S. Lasken, and A. Kornberg. 1989. The *dnaB-dnaC* replication protein complex of *Escherichia coli:* formation and properties. J. Biol. Chem. **264:** 2463–2468.

133a. Wang, J.C. 1985. DNA topoisomerases. Annu. Rev. Biochem. **54:** 665–697.

133b. Wang, J.C. 1987. Recent studies of DNA topoisomerases. Biochem. Biophys. Acta **909:** 129.

134. Wang, L.-F., C.W. Price, and R.H. Doi. 1985. *Bacillus subtilis dnaE* encodes a protein homologous to DNA primase. J. Biol. Chem. **260:** 3368–3372.

135. Weiner, J.H., R. McMacken, and A. Kornberg. 1976. Isolation of an intermediate which precedes *dnaG* RNA polymerase participation in enzymatic replication of bacteriophage ϕX174 DNA. Proc. Natl. Acad. Sci. USA **73:** 752–756.

136. Wickner, S. 1976. Mechanism of DNA elongation catalyzed by *Escherichia coli* DNA polymerase III, *dnaZ* protein and DNA elongation factor I and factor III. Proc. Natl. Acad. Sci. USA **73:** 3511–3515.

137. Wickner, S., and J. Hurwitz. 1975. Association of phage ϕX174 DNA dependent ATPase activity with an *Escherichia coli* protein replication (factor Y) required for *in vitro* synthesis of phage ϕX174 DNA. Proc. Natl. Acad. Sci. USA **72:** 3342–3346.

138. Wickner, S., and J. Hurwitz. 1975. Interaction of *Escherichia coli dnaB* and *dnaCD* gene products *in vitro*. Proc. Natl. Acad. Sci. USA **72:** 921–925.

139. Wickner, S., M. Wright, and J. Hurwitz. 1973. Studies on *in vitro* DNA synthesis purification of the *dnaG* gene product from *Escherichia coli*. Proc. Natl. Acad. Sci. USA **70:** 1613–1618.

140. Wickner, W., and A. Kornberg. 1973. DNA polymerase III* requires ATP to start synthesis on a primed DNA. Proc. Natl. Acad. Sci. USA **70:** 3679–3683.

141. Wold, M.S., and R. McMacken. 1982. Regulation of expression of the *Escherichia coli dnaG* gene and amplification of the *dnaG* primase. Proc. Natl. Acad. Sci. USA **79:** 4907–4911.

142. Wold, M.S., D.H. Weinberg, D.M. Virshup, J.J. Li, and T.J. Kelly. 1989. Identification of cellular proteins required for SV40 DNA replication. J. Biol. Chem. **264:** 2801–2809.

142a. Wolsson, J.S., and D.C. Hooper. 1989. Fluoroquinolone antimicrobial agents. Clin. Microbiol. Rev. **2:** 378–424.

143. Yasuda, S., and Y. Hirota. 1977. Cloning and mapping of the replication origin of *Escherichia coli*. Proc. Natl. Acad. Sci. USA **74:** 5458–5462.

144. Yung, B.Y.M., and A. Kornberg. 1988. Membrane attachment activates *dnaA* protein, the initiation protein of chromosome replication in *Escherichia coli*. Proc. Natl. Acad. Sci. USA **85:** 7202–7205.

145. Yung, B.Y.M., and A. Kornberg. 1989. The *dnaA* initiator protein binds separate domains in the replication origin of *Escherichia coli*. J. Biol. Chem. **264:** 6146–6150.

146. Zechel, K., J.P. Bouché, and A. Kornberg. 1975. Replication of phage G4: a novel and simple system for the initiation of DNA synthesis. J. Biol. Chem. **250:** 4684–4689.

147. Zipursky, S.L., and K.J. Marians. 1980. Identification of two *Escherichia coli* (factor Y) effector sites near the origins of replication of the plasmids ColE1 and pBR322. Proc. Natl. Acad. Sci. USA **77:** 6521–6525.

148. Zipursky, S.L., and K.J. Marians. 1981. *Escherichia coli* (factor Y) sites of plasmid pBR322 can function as origins of DNA replication. Proc. Natl. Acad. Sci. USA **78:** 6111–6115.

4

RNA Synthesis in Bacteria: Mechanism and Regulation of Discrete Biochemical Events at Initiation and Termination

Asis Das, Joseph DeVito, Jason Sparkowski, and *Frederick Warren*

I. Introduction

Transcription, the synthesis of RNA from DNA, is the first step in gene expression. It is a multistep biochemical process. RNA polymerase, the enzyme that catalyzes transcription, locates and binds a specific promoter sequence on the template DNA, unwinds a small region surrounding a specific start site, initiates the synthesis of an RNA chain, elongates the chain to a definite stop site, and finally releases the completed transcript and dissociates itself from the template DNA. RNA polymerase is composed of a battery of distinct polypeptide subunits whose function requires accurate interaction not only with each other but also with the template DNA, four ribonucleotide precursor molecules, the emerging RNA chain, and numerous other ancillary subunits that influence one or more individual steps of transcription. The overall chemical mechanism of the basic transcription process is identical in procaryotic and eucaryotic organisms, with certain features of the individual components of the transcription apparatus conserved from bacteria to man. However, the two types of organisms differ more in the individual protein components and the ancillary proteins than they show identity. This diversity makes the transcription apparatus of pathogenic bacteria an ideal target for developing novel antibacterial drugs.

The primary goal of this chapter is to encourage attempts to develop new antibacterial agents directed toward the transcription apparatus. The available antibiotics known to inhibit transcription selectively are limited to rifampicin and streptolydigin. In addition to their use as antibacterial agents, these drugs have been enormously helpful as tools in basic research on transcription. The likelihood of finding additional antibiotics is very good because RNA synthesis involves several distinct biochemical steps and the transcription apparatus is built from several essential subunit polypeptides. The review that follows summarizes our current state of knowledge of RNA synthesis in *Escherichia coli,* the model bacterium with which the majority of genetic and biochemical studies have been carried out. A number of previous articles and monographs have described many of these aspects in greater depth and detail (1, 27, 54, 55, 63, 79, 86, 90, 113, 116, 129, 131, 137, 139, 142–146, 149, 151, 156, 180, 184, 190–192, 194). We hope this summary makes evident how the identification of new inhibitors of RNA synthesis could help test the current models and allow for new experimental approaches toward a further molecular understanding. To this end, we present a few simple strategies that may help identify new therapeutic agents.

II. Chemical and Biological Views of RNA Synthesis

RNA (ribonucleic acid) is a polymer of ribonucleotides. All natural RNA molecules are made up of four different ribonucleoside monophosphates (ribonucleotides). A ribonucleoside contains a purine (adenine and guanine) or a pyrimidine (cytosine and uracil) moiety joined to ribose sugar. As in DNA, individual nucleotides in the RNA polymer are joined by a phosphodiester bridge utilizing the 5′ and 3′ hydroxyl groups of the ribose moieties and arranged sequentially in a 5′ to 3′ direction. Enzymatic synthesis of RNA is template-directed. The enzyme RNA polymerase utilizes ribonucleoside 5′-triphosphates (rNTP) as the precursors to synthesize RNA molecules by sequential joining of the ribonucleotides. The linear assembly of the nucleotides is strictly dependent upon the template DNA sequence: one of the two strands of the DNA double helix serves as the template for the synthesis of a complementary RNA chain according to the Watson-Crick base-pairing rule, *i.e.*, adenine pairing with uracil and guanine with cytosine. In polymerization, the nucleotides are joined sequentially starting at the 5′-triphosphate end and ending at the 3′-hydroxyl end. Each phosphodiester bond formation results from the concomitant hydrolysis and release of a pyrophosphate moiety.

There are three major classes of natural RNAs produced in every cell type: (1) thousands of different messenger RNA molecules that encode distinct cellular proteins, (2) ribosomal RNAs that are incorporated into the two subunits of ribosomes involved in protein synthesis, and (3) a set of transfer RNA molecules that are required for the sequential recognition of codons in mRNA and insertion of respective amino acids into the polypeptide chain. A small number of additional RNA molecules are also produced in the cell for more specialized functions (184): these include various RNAs that constitute small nuclear ribonucleoprotein particles (snRNPs) involved in RNA splicing, and also a small RNA in the signal recognition particle required for the subcellular localization and extracellular transport of proteins.

The genome of all organisms is expressed as small transcription units. There are signals in the DNA template known as the promoter and the terminator that dictate where RNA synthesis starts and where it ends. Thus, transcription consists of three major steps: specific initiation at a promoter, elongation, and termination at a specific site. To initiate transcription, RNA polymerase must be able to locate and recognize the promoter signal on the DNA double helix (binding), melt or denature the double helix at the start region (formation of open complex) so that the catalytic center of the enzyme [*i.e.*, nucleotide binding site(s)] is accurately juxtaposed to the template strand of the transcription start site, and finally produce a dinucleotide by choosing and joining the first two nucleotides of the mRNA to be synthesized. In commencing elongation, the polymerase has to remain tightly bound to the DNA in the ternary transcription complex (i.e., DNA–RNA polymerase–RNA complex), melt the double helix ahead of it, and begin a processive polymerization. During this transition from the initiation mode to the elongation mode, the polymerase must release the promoter DNA segment (promoter clearance) to allow a second round of initiation. While transcription of some genes initiates every second, initiation at others occurs only once in the cell cycle. Thus, strong promoters (signals for highly transcribed genes) must favor all three events of initiation (*i.e.*, binding, formation of open complex, and promoter clearance) with highest efficiency or rate and, as might be predicted, one or more of these steps must be rate-limiting with weak promoters. During elongation of an RNA chain, the polymerase must not dissociate from the ternary complex nonspecifically; rather, it must recognize a termination signal, cease further polymerization, release the RNA chain, and dissociate from the DNA to begin a new round of RNA synthesis.

Transcription of all genes in bacteria is carried out by a single DNA-dependent RNA polymerase enzyme (20, 27, 146). In contrast, distinct forms of RNA polymerases in eucaryotic cells are involved in the synthesis of mRNA, rRNA, and tRNA (146). Template-directed RNA synthesis is also catalyzed by replicases encoded by certain viruses that utilize an RNA genome to synthesize complementary RNA and by the primase enzyme that produces RNA primers for DNA replication (184). In addition, RNA molecules can also be synthesized independent of a template. The enzyme polynucleotide phosphorylase, which normally carries out the degradation of RNA molecules in the cell, can synthesize RNA (homopolymers and mixed polymers) from ribonucleoside diphosphate precursors (184). In fact, before the discovery of RNA polymerase, polynucleotide phosphorylase was thought to be involved in gene transcription.

Not all genes in a cell are expressed at the same level and not all genes are expressed at the same time in the cell's life. Certain cells produce a characteristic set of proteins that are not expressed in another cell type of the same organism. Cells regulate RNA synthesis as one of the means to achieve this differential pattern of gene expression. In fact, it is the regulation of transcription that plays the most predominant role in the regulation of gene expression during development and differentiation. Transcription is regulated at all three basic steps and the control is mediated by both genetic determinants (*i.e.*, cis-acting sites on DNA) and cellular factors (*i.e.*, trans-acting proteins and small molecules). These modulators dictate whether to start transcription or not, and whether to continue transcription or terminate it. There are activators that turn specific genes "on" and repressors that turn specific genes "off." Likewise, certain proteins promote termination of RNA synthesis preventing the expression of downstream genes, while others prevent termination so that the downstream genes are transcribed. As we discuss below, RNA polymerase serves as the basic RNA synthetic machinery whose activity is modulated by an interplay of these regulatory signals and factors. It is this interplay that ultimately manifests most of the temporal and spatial pattern of gene expression in all organisms.

III. The Anatomy of *E. coli* RNA Polymerase

The multisubunit structure of RNA polymerase was elucidated by the purification of the *E. coli* enzyme (20, 27). One of the most exciting earlier findings on RNA polymerase was that it consisted of two functionally separable entities: a catalytic "core" and a σ subunit (21). The core enzyme is composed of two α subunits (Mr 37 kd, 329 aa [amino acid]), a β subunit (Mr 151 Kd, 1,342 aa), and a β' subunit (Mr 155 kd, 1,470 aa), encoded by the *rpoA*, *rpoB*, and *rpoC* genes, respectively (20). All three genes are essential for the growth of *E. coli*. The *rpoA* gene maps at 73 min on the *E. coli* chromosome. Curiously, it is part of a polycistronic operon that encodes ribosomal proteins S13, S11, S4, and L17. The *rpoB* and *rpoC* genes map at 90 min; again, these genes are part of a polycistronic operon that encodes ribosomal proteins L10, L7, and L12 (Figure 4.1). An intergenic transcriptional terminator, located between the *rplL* and *rpoB* cistrons, is thought to be a modulator of the expression of *rpoB* and *rpoC*. The core enzyme contains an additional subunit named omega (Mr. 10 kdd, 90 aa). This subunit is encoded by the *rpoZ* gene, which seems to be dispensable for *E. coli* growth (61).

The core enzyme, by itself, is unable to carry out faithful transcription of natural genes (17). Its association with a σ subunit (17) produces the RNA polymerase holoenzyme complex that not only initiates RNA synthesis at a natural promoter, but also carries out all of the subsequent steps of transcription. As this initial discovery suggested, the σ subunit directs the RNA polymerase to a specific promoter signal, and as described later, it plays a pivotal role in transcription initiation.

The pioneering work of Losick and Pero and of Chamberlin and co-workers has shown that the gram-positive bacterium *Bacillus* and its phages encode multiple σ factors that govern transcription initiation from different gene sets (79). This led to the unifying hypothesis of "the σ-cascade" in temporal control of gene expression, particularly in endospore formation in *Bacillus* (123). Initially, it appeared that *E. coli* might contain a single σ protein, σ^{70} (Mr 70 Kd, 613 aa). However, subsequent studies in a number of laboratories have changed this view. *E. coli* encodes a number of additional σ factors. These include (1) σ^{32}, the product of the *htpR* (*rpoH*) gene, required for the expression of heat-shock genes

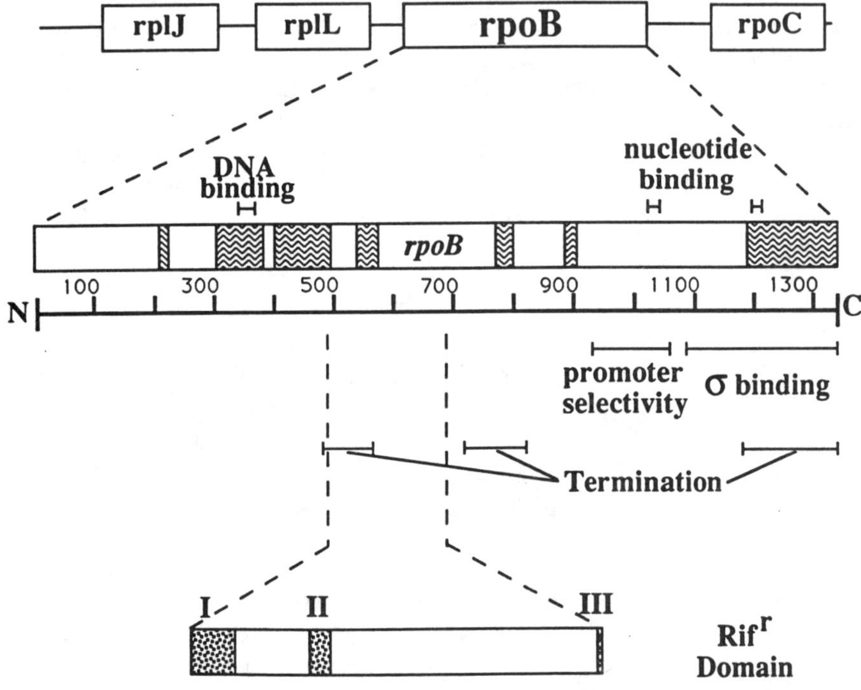

Figure 4.1. Functional domains of the RNA polymerase β subunit. Barred regions indicate functional domains, as inferred from genetic and/or biochemical studies detailed in the text. Scale depicts amino acid length (in increments of 100 residues) from the N, or amino, terminus to the C, or carboxy, terminus.

(73), (2) σ^{54}, the product of *ntrA* (*rpoN*), required for the expression of nitrogen-fixation genes (82, 154), and (3) σ^{E}, involved in transcription of a second class of heat-shock genes (for instance, *htrA*) (44). A distinct σ factor might also be required for expression of genes involved in chemotaxis and motility (M. Chamberlin, personal communication), and it would not be surprising to see this list growing.

σ^{70}, the major σ factor essential for viability of *E. coli,* is encoded by the *rpoD* gene (20). It maps at 67 min on the *E. coli* chromosome and is a part of a polycistronic operon that encodes the ribosomal protein S21 and the DnaG primase involved in DNA replication. These genes are regulated by multiple promoters, an intergenic terminator, and at the level of mRNA stability.

The β subunit of the core is thought to carry out all the basic aspects of RNA synthesis (Figure 4.1). It binds ribonucleotides (68) as well as rifampicin and streptolydigin (25, 122), both of which inhibit the catalytic polymerization process. In each cycle of polymerization, the core enzyme has to bind and couple a nucleotide to the 3′-hydroxyl-end of the nascent RNA, translocating forward by one nucleotide on the template so that the next cycle of binding and coupling can be performed. As we shall describe later, processive polymerization occurs within an 18-base-pair (bp) DNA bubble that has a 12-bp DNA-RNA hybrid (58). Thus, the active site of the core is likely to contain a DNA-RNA helicase domain that keeps unwinding one DNA-RNA bp at a time, about 12 bp upstream from the polymerization site as the polymerization domain keeps adding a DNA-RNA bp. If the β subunit is responsible for these activities, it should be in direct contact with DNA and RNA as well as with the ribonucleotide precursors. Indeed, the β subunit can be cross-linked with photo-reactive nucleoside triphosphate analogs (68). It can also be cross-linked with the DNA template containing bromodeoxyuridine (164), as well as with the nascent RNA containing photoreactive nucleotide(s) incorporated solely at the 5′-end, at random internal sites, or exclusively at the 3′-end (15, 40, 76). As might be expected, mutations in β affect all aspects of transcription (Figure 4.1): the binding of the σ subunit (64), promoter recognition (65), kinetics of transcript elongation and termination (48, 94, 114), and interaction with termination and antitermination factors (34, 74, 95, 169).

Much less is known about the function of the other subunits of the core. The β' subunit is thought to bind DNA. Certain mutations in *rpoC* encoding β' are known to affect termination and interaction with NusA protein (92, 96). RNA polymerase is a metalloenzyme containing two Zn^{2+} atoms. An individual Zn^{2+} atom is bound by each of β and β' subunits (170, 189). No specific function has yet been ascribed to the α subunit, though certain mutations in the *rpoA* gene, which encodes α, are known to alter gene expression in specific ways (172). The fact that α is an essential protein in the cell clearly suggests that it plays some critical role in the assembly and function of RNA polymerase. In fact, the current view of the sequential pathway of RNA polymerase assembly suggests an initiator role for the α subunit (196).

The β and β' subunits of *E. coli* share amino acid sequence homology with the largest subunits of RNA polymerases from yeast, fruit fly, and mouse (4, 13, 174). Thus, it seems likely that the structure of the catalytic center of RNA polymerase might be conserved across species. Previous chemical cross-linking studies have outlined a subunit topology of RNA polymerase holoenzyme (81). More recently, a three-dimensional structure of the enzyme has been predicted based on electron-microscopic analysis of two-dimensional crystals on positively charged lipid layers (31, 32). RNA polymerase is seen as an irregularly shaped molecule of about 100 × 100 × 160 Å in size. The three-dimensional structure derived by electron imaging reveals a cleft in RNA polymerase similar to the active-site cleft of DNA polymerase I that was determined by X-ray diffraction (136). This cleft of RNA polymerase is about 25 Å in diameter and about 55 Å in length and is appropriate for binding about 16 bp of B-form DNA double helix. Since RNA polymerase is expected to interact with an 18-bp DNA bubble during processive elongation (58, 192), it is attractive to imagine that the cleft represents the active site of the enzyme. However, during promoter recognition and initiation, RNA polymerase interacts with a larger (50–60 bp) DNA segment which is at least (170 Å in length (162, 164). It is possible that DNA bending at the promoter allows the proper fitting of the DNA to the active site cleft required for the initiation phase of the polymerization reaction (109).

Upon initiation of transcription, a battery of additional proteins interacts with the core RNA polymerase to modulate elongation and termination (9, 54, 55, 88, 192). These factors, considered as the elongation subunits of RNA polymerase, include NusA, NusB, NusE (the S10 ribosomal protein), and NusG (36, 37, 42, 62, 70, 88). Collectively, these subunits interact with elongating RNA polymerase leading to the formation of a transcriptosome. In addition to these common subunits, there may be operon-specific or gene-specific elongation subunits in the cell. These include the antitermination proteins, namely, phage λ N and Q gene products, factors that engage RNA polymerase and modify it to a termination-resistant form (9, 69). It is attractive to imagine that proteins with analogous function are involved in antitermination control of some cellular genes. However, to date, no such cellular factor has been reported in *E. coli*. Factors such as Rho and Tau (18, 148), known to cause the termination and release of nascent RNA chains at specific genetic sites, may also attach to the transcriptosome early and function later when RNA polymerase has moved to the respective sites of action for these factors. We imagine that the transcription apparatus in the cell may not be a unitary complex, but rather a collection of different sets of complexes that are engaged in transcription of various sets of genes.

IV. Mechanism of Initiation

The initiation of RNA synthesis has the potential to be the target for many antimicrobial agents as it entails a number of discrete biochemical steps (129, 192): promoter-σ interaction, regional melting of DNA duplex, and multiple interaction of the active center of RNA polymerase (both β and β' subunits) with DNA template, ribonucleotides, and the nascent RNA. In fact, initiation is inhibited by the two most widely used antimicrobial agents known to affect RNA synthesis, rifampicin and streptolydigin (126, 127). In this section, we will describe the nature of the promoter sequence, the interaction of promoter with the holoenzyme, and the structure and function of the σ subunit in promoter-recognition and initiation. We will then discuss how both positive and negative modulators of transcription are thought to influence the discrete steps of initiation.

A. Determinants of a Bacterial Promoter

The nature of the promoter sequence recognized by RNA polymerase was revealed from DNA sequence analysis (141). DNA sequences of a large number of promoters have allowed the derivation of a consensus sequence for bacterial promoters (78, 151). Centered around the -10 position with respect to the start site (Figure 4.2) is a conserved hexamer sequence classically known as the Pribnow box (141). In addition to the conserved Pribnow box (5′-TATAAT consensus), a second conserved element is found centered around the −35 position (5′-TIGACA consensus). Promoters recognized by σ^{70} all contain these two elements with a spacer of about 17 bp (78, 86, 151). As might be predicted, the promoters recognized by σ^{32} and σ^{38} contain a different set of consensus sequences (86). In addition, the position of the two sequence elements with respect to the start sites of transcription vary greatly from one class of promoters to the other.

The functional significance of the promoter elements has been elucidated both genetically and biochemically. A large number of up- and down-promoter mutations isolated by Susskind and co-workers (195) have defined the nucleotide residues critical for promoter-recognition. Similarly, biochemical studies performed by McClure and colleagues and others have provided clues on the function of these conserved elements in promoter-recognition and function (129, 132). Generally speaking, there is a good correlation between the degree of identity with the consensus sequence and the functional strength of a promoter both *in vivo* and *in vitro* (86). Mutations enhancing the strength of a promoter result from the substitution of a nonconsensus bp by the consensus bp. Conversely, promoter-down mutations are due to the conversion of a consensus bp to a nonconsensus one.

B. Interaction of RNA Polymerase with Promoter

The initiation process can be operationally divided into three steps as outlined in Figure 4.2 (27, 86, 129, 180): (1) binding of the promoter by RNA polymerase forming a "closed" promoter–RNA polymerase complex (R.Pc), (2) isomerization of the closed complex to "open" complex (R.Po) in which the DNA duplex is melted around the start site, and (3) formation of a small RNA transcript and clearance of the promoter. Preceding this scheme of events is the location of the promoter sequence by RNA polymerase.

The core enzyme is capable of binding DNA nonspecifically (salt-sensitive binding) *in vitro*. The addition of σ factor reduces this nonspecific binding by the core enzyme. This favors the idea that the holoenzyme is assembled prior to the search for the promoter. One model for promoter-searching proposes a tracking mechanism (181, 192) in which the RNA polymerase binds DNA nonspecifically and then slides along DNA until a promoter is located, where it forms a metastable "closed" complex that isomerizes to the stable (heparin-resistant) "open-promoter" complex. Thus, at first approximation, the strength of a promoter is determined by the combined rates at which the two specific complexes are formed. The isomerization is a slow step and it is generally rate-limiting in the overall initiation process (128). Whereas mutations in the −35 region can affect both of these rates, mutations in the −10 region seem to affect primarily the rate of isomerization.

The frequency of initiation at a promoter is indirectly influenced by the rate of promoter-clearance. The σ factor, which binds to core RNA polymerase revers-

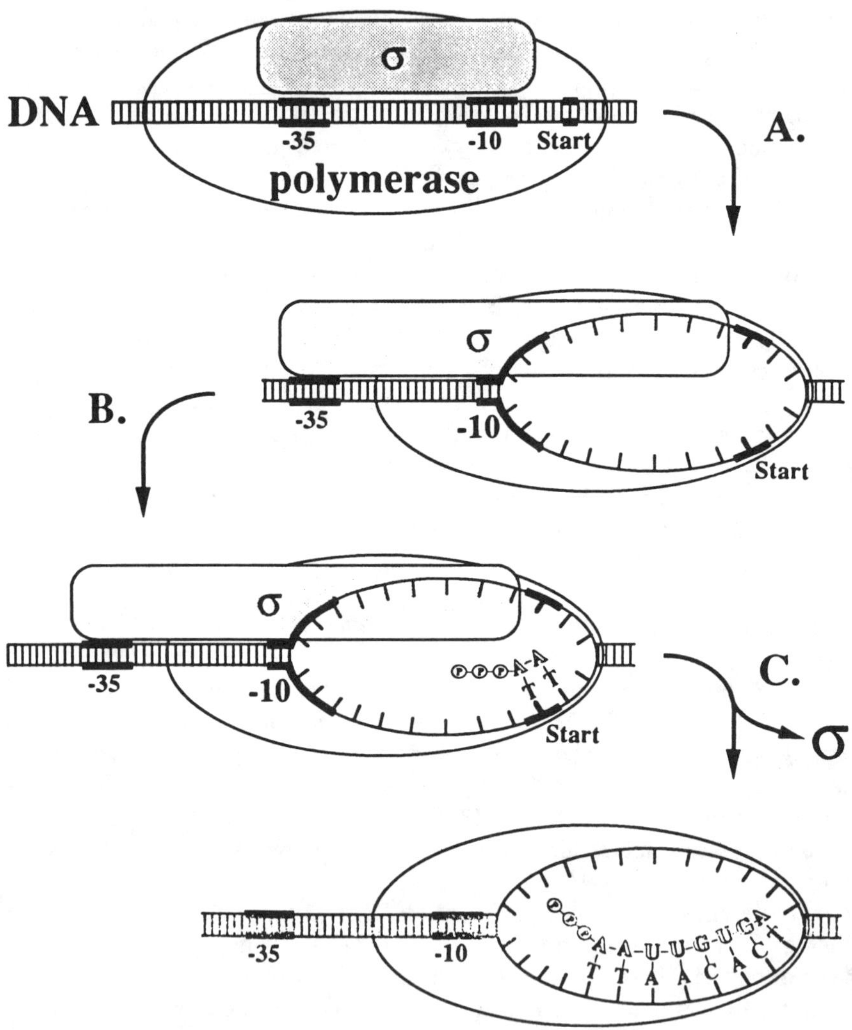

Figure 4.2. The transcription initiation process. A. Isomerization from closed complex (R.Pc) to open complex (R.Po). B. Transition from closed complex to an initiation competent holoenzyme. C. σ release and promoter clearance.

ibly to form the initiation-competent holoenzyme, dissociates from the transcription complex after synthesis of a short transcript (about eight nucleotides). Release of σ factor is thought to be an indication of promoter-clearance, since as discussed below σ is believed to recognize the promoter-sequence directly. Because the concentration of σ is limiting in the cell, the released σ factor recycles in new rounds of initiation by reforming a holoenzyme complex and occupying the promoter cleared from the previous round of initiation. Thus, a faster intrinsic rate of promoter-clearance would increase the overall strength of a promoter.

Curiously, *in vitro* transcription from many promoters has been shown to undergo "abortive" initiation, a process in which RNA polymerase carries out many cycles of synthesizing and releasing short (dinucleotide and larger) transcripts before clearing the promoter and engaging a stable elongation mode (24, 97, 104, 126). This reiterated synthesis of various-length oligonucleotides per full-length RNA molecule occurs in the presence of heparin, demonstrating that the RNA polymerase does not dissociate from the template. In the presence of rifampicin, the reiterative synthesis of abortive products is restricted to mostly dinucleotides, and to some extent trinucleotides, depending on the promoter (24, 104, 126). Thus, rifampicin primarily inhibits the formation of the second phosphodiester bond, preventing initiation. Presumably, many of the initial promoter-contacts are not broken and the melted topology of the start region is maintained during cycles of synthesis and release of the abortive products. Thus, at the early phase of elongation of RNA chains, RNA polymerase translocation is in an equilibrium. Later, this changes to a state that favors a processive polymerization reaction; when a sufficiently large (8–10-mer) oligoribonucleotide has been synthesized, promoter-contacts are broken and σ factor is released from the complex. At this point, RNA synthesis is completely resistant to rifampicin.

It is not known whether abortive initiation takes place *in vivo*. Given that it does, abortive initiation would decrease the promoter-strength. Studies on the pL promoter of phage λ suggest that the sequences in the initial transcribed region might influence promoter-clearance and conversion of RNA polymerase to the stable elongation mode (106). The pL promoter is one of the strongest promoters *in vivo*. Yet, its −10 hexamer differs significantly from the consensus and RNA polymerase binds pL inefficiently (15–30-fold reduced) compared to other promoters of similar strength. Variants of pL that are more homologous to the consensus bind polymerase efficiently; however, the *in vivo* strength of these variants is significantly reduced compared to the natural pL promoter. Knaus and Bujard discovered that a high promoter-strength of these variants could be restored if they replaced the downstream sequence (+4 to +20) with that derived from a typical consensuslike promoter (PN25 from phage T5 of *E. coli*) (106). This suggests that although the pL promoter binds RNA polymerase less efficiently, its strength is ultimately determined by a more favorable promoter-clearance and transition into a stable elongation mode. Likewise, promoters with a consensus −10 sequence bind polymerase well. However, to be strong promoters, these might have evolved to contain certain downstream sequence elements that favor transition to the productive elongation mode. Clearly, the precise mechanisms of these events and their universality remain to be elucidated.

C. Discrete Domains of the σ Factor for Promoter Recognition and Core Binding

Since all σ factors have at least one common function, i.e., to bind the core RNA polymerase, they should share at least one common structural motif. As sequences of several σ factors became available, it was clear that they have several discernable sequence similarities. According to Helmann and Chamberlin (79), the σ factors can be aligned at four separate regions, called regions 1 through 4 (Figure 4.3). Region 2 can be further subdivided into four separate homology regions, 2.1 through 2.4. It is this region that shows significant homology to the yeast TFIID factor (87), the TATA-box binding protein that is thought to be a functional analog of the procaryotic σ factor. Region 2 also shares homology to the human RAP30 protein, a factor isolated by affinity chromatography with RNA polymerase II. RAP30, together with another subunit called

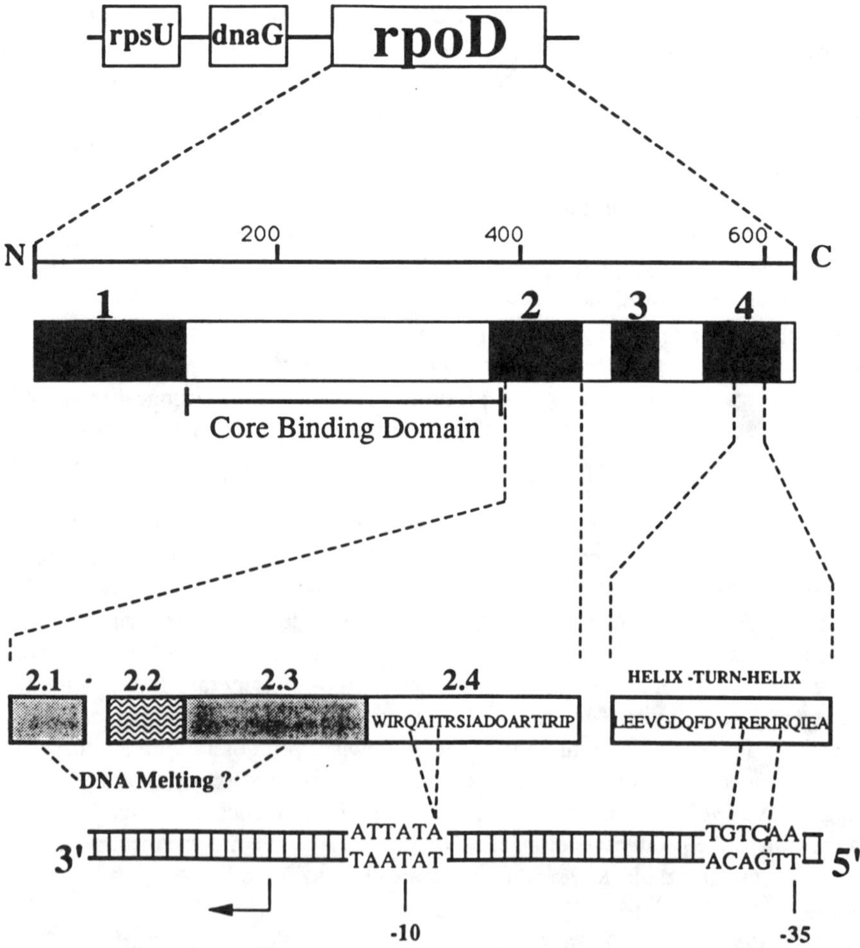

Figure 4.3. Functional domains of *E. coli* σ 70. Enlarged regions 2.4 and 4 show the respective amino sequences in these regions. Dashed lines below indicate contact between specific amino acids and the A.T. bp at position −12, and the two G residues in the −35 region, as detailed in the text. Scale depicts amino acid length in increments of 200 residues.

RAP70, constitutes a DNA helicase activity (167). Initially, region 2.2 was thought to be involved in core binding. However, recent biochemical studies by Lesley and Burgess demonstrate that the core binding requires a 30-aa segment that precedes region 2.2 and partially overlaps with region 2.1 (119).

Conceptually, the σ subunit may directly recognize one or both of the promoter elements (175) or it may activate a promoter-recognition site on the core polymerase by a conformational change (190). The σ subunit may also act in a second capacity: by binding the promoter, it may actively melt the DNA at the start site to form the open complex. There is now excellent evidence suggesting that discrete domains of σ are involved in the direct recognition and binding of the promoter-elements (59, 102, 154, 161, 182, 198). Since a number of replication initiators from procaryotes and eucaryotes are known to induce strand denaturation upon binding the specific recognition sequence at the origin (16, 188), it is likely that the transcription initiator proteins also perform a similar duplex-opening function (79, 154).

A number of laboratories have now isolated suppressor mutations in σ factor which can compensate for mutations at -35 and -10 hexamers (59, 102, 161, 182, 198). Mutations in three different σ factors have been reported to date: σ70 of *E. coli,* its *Bacillus* homolog, called σA, and a minor σ factor involved in sporulation of *Bacillus,* called σH. The suppressor mutations implicate two discrete regions of σ factor in the direct recognition of each of the two conserved elements of the promoter (see Figure 4.3). Suppressors of -35 hexamer mutations map in region 4, the carboxy-terminal conserved segment that shows a striking similarity to the helix-turn-helix motif of known DNA-binding proteins (137). Suppressors of the -10 hexamer mutants have been located in the central segment called region 2.4. Though this region does not show a particularly striking similarity to the known DNA binding motifs, an α-helical structure of the region can be predicted. Three highly conserved Ile residues spaced three amino acids apart from each other would lie on one side of the predicted helix, perhaps forming an interior hydrophobic pocket. Facing outward on the helix would be those residues which are not always conserved, including two that are changed in the suppressor mutants. It is these two residues that are specifically hypothesized to make direct contact with the Pribnow box hexamer at the -12 position (Figure 4.3).

The isomerization of the "closed" complex to the "open" transcription complex involves the melting of 12 bp of DNA surrounding the transcription start site, from -9 to $+3$ (105, 158, 162). There is also a change in the conformation of RNA polymerase as evident from altered protease sensitivity and resistance to dissociation from the DNA template by treatment with polyanions such as heparin (147, 155). Helmann and Chamberlin (79) have proposed that the conserved regions 2.3 and 2.1 of σ might be involved in the unwinding of DNA, based on the facts that (1) these regions are rich in aromatic amino acids whose side chains may be involved in stacking between nucleic acid bases and that (2) this feature is conserved in certain eucaryotic proteins (for instance, snRNP proteins) that bind RNA and single-stranded DNA. In addition, the aromatic residues in these subregions are flanked by basic amino acids that may play a role in charge neutralization during the melting process (79).

Recently, Sasse-Dwight and Gralla have proposed a mechanism for DNA melting by the σ54 protein (154). σ54 has some features that are common to σ70 and its homologs, but other features reminiscent of the eucaryotic transcription factors such as Fos and Jun. These authors suggest the role of an intramolecular "leucine-zipper" domain in positioning an acidic, "strand-opening" domain near the DNA melting region. Specifically, they hypothesize that the acidic domain could strip shielding cations from DNA and thereby force the two highly negatively charged strands to melt out due to repulsion. This model is provocative. Nevertheless, the solution to the precise mechanism of isomerization will require future study and a combination of many experimental approaches. The isolation of novel antibiotics inhibiting any aspect of the isomerization process would be highly rewarding.

V. Regulation of Initiation

Both positive and negative control of gene expression have been shown to involve transcriptional regulators. To date, many transcriptional regulators have been discovered from bacteria as well as yeast and higher eucaryotes. A regulator may be responsible for controlling only a specific gene or an operon. Other regulators act globally, allowing control of many groups of genes required for a certain developmental pathway. Multiple sets of genes can be controlled temporally, one set expressed prior to the other in a program of development or differentiation. In addition, genes can be spatially controlled, meaning that a given gene or a set of genes is expressed in one cell type but not another. Given these multiple kinds of gene control, it is remarkable that both negative and positive regulators of transcription act by a set of simple and common mechanisms (1, 86, 129, 143, 145, 146, 156, 180). Repressors bind specific DNA sequences near the promoter to inhibit initiation. In contrast, activators increase the frequency of initiation by binding to specific DNA sequences.

Repressors inhibit either the initial binding of the promoter by RNA polymerase or the subsequent isomerization process. The mechanism seems to depend on the location(s) of the repressor binding site: upstream or downstream of the promoter region, in between the −35 and −10 hexamers, or overlapping with either or both promoter elements. In some instances, repression is mediated by operators that lie at a great distance from the promoter sequence; interaction between the two sites bound by the repressor causes the intervening DNA to form a loop (1, 156).

Likewise, activators stimulate the initial binding by polymerase or the subsequent isomerization process. Generally, the activator binding sites are located upstream of the −35 hexamer; however, there is at least one example where the activator binds at a far upstream site (134, 140, 153). Though the effect of activator binding on DNA structure is not clearly understood, some regulators (for instance, the CAP protein required for the activation of *lac* and other sugar utilization operons) appear to bend DNA. A common mechanism for activators is thought to involve a direct functional contact between the activator and the RNA polymerase (143). Clearly such a contact would increase the rate of association of RNA polymerase with the promoter. However, in addition, this binding could cause a conformational change required for the isomerization step (140, 153, 154, 171).

Most activators and repressors bind DNA by a common mechanism (137). The binding sites on DNA consist of two symmetric half-sites at which two monomers bind. The regulatory proteins have a specific, functionally separable, DNA binding domain that is constituted by a "helix-turn-helix" motif (137). They also have a separate dimerization domain which allows the monomers to associate in solution. In addition, some regulators (the *lac* repressor, for instance) form tetramers so that two dimers bound at distantly located operator sites can interact, causing DNA looping (108, 125).

Those regulators that are modulated by small effector molecules (for instance, cAMP for the CAP protein) have an additional domain for effector binding. Sometimes, the effector binding activates the regulator to its DNA binding conformation (the cAMP-CAP complex). Other regulators (the *lac* repressor, for example) become incapable of recognizing DNA upon binding of the effector molecule.

Generally, gene expression is modulated by a combination of independent control mechanisms, providing the system a greater flexibility. The same gene or a set of genes is frequently controlled by both positive and negative regulators. Control might be exerted not only at the level of transcription initiation, but also by termination and posttranscriptional mechanisms (80).

A. *Negative Regulation*

The simplest mechanism for negative regulation of initiation by a repressor is one in which it binds the promoter region to exclude binding of RNA polymerase. Extensive genetic , biochemical, and X-ray crystallographic studies of the Cro and CI repressors of λ and related phages have provided a molecular understanding of the underlying principles of the protein–nucleic acid interaction (135, 137). The DNA binding domain of these repressors consists of two helices, one of which recognizes and binds the operator sequence directly. The diversity of these recognition helices in different repressors enables sequence-specific recognition of each respective operator sequence. Certain residues in the recognition helix and the operator site are most critical for recognition and binding, while others influence the overall binding strength. The dimer formation mediated by a discrete domain in these repressors is required for stable repressor-operator interaction. The binding affinity is also influenced in a cooperative manner by the presence of a second operator that may be located nearby or at a distance. Cooperativity in binding at nearby sites requires that the repressors be bound to the same face of the DNA double helix, suggesting that the repressor dimers touch each other.

The structure of the promoter-operator region of the λ rightward operon at which CI and Cro bind is shown in Figure 4.4. There are three sets of operators, called oR1, oR2, and oR3, that are bound by CI and Cro with opposite affinities: CI binds oR1 with highest affinity and Cro binds oR3 most efficiently (144). CI acts as a repressor for lytic growth of phage λ, a process that requires transcription from the pR promoter (80). Thus, the high affinity of CI for oR1 is consistent with the physiological role of CI. In contrast, Cro acts to repress lysogeny, a process that requires the synthesis of CI. We describe below how CI autogenously controls its own synthesis by acting as a specific transcriptional activator for the pRM promoter. The high affinity of Cro for oR3 is again consistent with its role as the repressor for transcription from the pRM promoter.

B. Mechanism of Repression

How does binding of a repressor affect initiation? Does repressor binding prevent the interaction of polymerase with promoter, the isomerization step, or

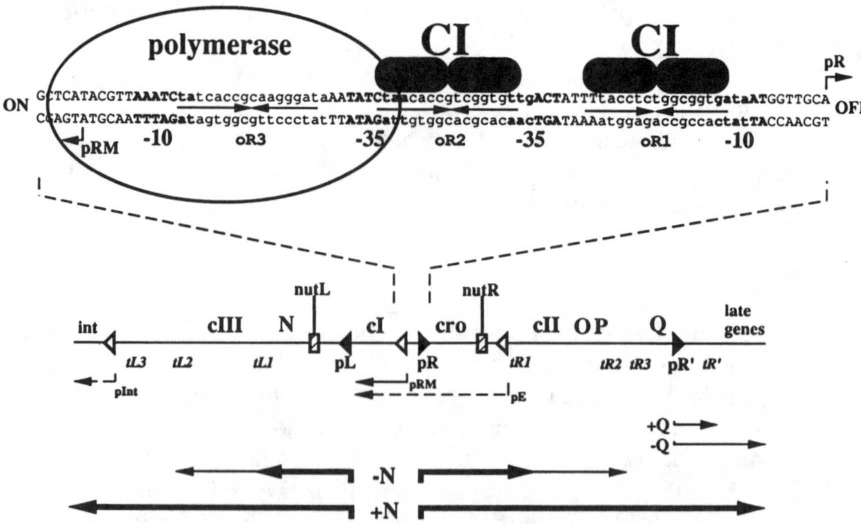

Figure 4.4. Control circuits in phage λ. Enlarged region (top) shows the rightward operator-promoter, indicating how the binding of CI repressor at oR1 and oR2 activates pRM and represses pR. Solid triangles denote promoters which do not require activators for function, while open triangles indicate promoters whose function is dependent upon activating factors. Dashed arrows represent cII-dependent transcripts.

the clearance from the promoter and transition into a stable elongation mode? This is an important question because the operator sites are not always placed at the same position with respect to the promoter. Many *E. coli* genes that are induced upon DNA damage and are repressed by the LexA repressor contain the LexA binding site at various locations with respect to the promoter. In some promoters, the *lexA* operator is overlapping with either the −35 hexamer or the −10 hexamer. At other promoters, the *lexA* operator maps upstream of the −35 hexamer, in between the −35 and −10 hexamers, or downstream of the −10 hexamer. As shown in Figure 4.4, the oR2 operator overlaps with the −35 hexamer of pR, and the oR1 operator is partly overlapping the −10 region. Indeed, the binding of CI repressor to oR1 and oR2 directly reduces the binding of RNA polymerase at the pR promoter (77). In contrast, although the Arc repressor of *Salmonella* phage P22 binds as its operator located between the −10 and −35 region of P_{ant}, it does not prohibit RNA polymerase binding to any great extent; rather, it significantly reduces the isomerization process (179). It is possible that a repressor bound upstream of the −35 region may interfere with the required binding of an activator or an RNA polymerase-activator interaction. A repressor bound downstream of −10 in the early transcribed region may increase abortive initiation and prevent RNA polymerase from entering a stable elongation mode. Though repressors are generally known not to affect RNA polymerase activity during elongation, it is conceivable that an operator-bound repressor can act as a road block to transcript elongation from an upstream promoter. Recent *in vitro* work by Steege and co-workers has provided evidence for this possibility (138).

C. Control from a Distance

The activity of a promoter can be controlled by operators that lie at a distance from the promoter (1, 2, 5, 30, 43, 72, 91, 156). Three operons of *E. coli (gal, ara,* and *deo*) show this unusual arrangement of the operator sites with respect to the promoter (1, 5, 156). In each case, repressor molecules bound at two distant sites interact with each other forming a DNA loop between the sites (5, 43, 72, 125). If the two sites are located nearby, within about 100 bp, it is important that the binding sites be spaced in multiples of 10 bp so that proteins bound on the same side of the DNA double helix can interact with each other (43). When there is a greater distance, there is little spatial constraint because of the flexibility of the intervening DNA.

With regard to the mechanism of repression by DNA looping, it is interesting that the *lac* operon also contains distant operator sites besides the one that is positioned near the transcription start site (108). The *lac* repressor has long been known to form a tetramer. Evidently, the dimer-dimer interaction strengthens occupancy of each operator by a single repressor dimer. However, it is not clear whether the DNA looping itself contributes to the repressor's action. The DNA looping may distort the structure of the promoter to prevent RNA polymerase binding. Alternatively, RNA polymerase may be able to bind but be unable to isomerize or enter into the stable elongation mode.

D. Positive Regulation

Activators that have been characterized in greater detail are the CAP protein of *E. coli,* required for transcription of a number of sugar utilization operons including *lac, mal,* and *ara* (19), and the CII and CI proteins of phage λ required for the establishment and maintenance of the lysogenic state, respectively (191). All three proteins bind DNA near the −35 region at a specific recognition site and increase the rate or extent of open complex formation by RNA polymerase.

1. DNA Binding.

Each activator protein binds DNA as a dimer, utilizing a helix-turn-helix motif in the protein and two discrete, almost identical sequences on the DNA double helix. The sites of monomer binding are oriented in an inverted manner for both CAP and CI. In contrast, the CII binding site consists of a direct repeat of the monomer recognition sites (191). Whereas the CI and CII proteins bind DNA without the aid of a small effector, the CAP protein requires cAMP to bind DNA. cAMP is known to cause an allosteric change in the protein that activates the DNA binding domain; cAMP-independent mutants of CAP have been isolated that are thought to be folded spontaneously in the DNA binding conformation (60).

2. The Affected Step of Initiation.

The initial hypothesis on the mechanism of activators proposed that they increase the binding affinity of RNA polymerase for promoters. Indeed, the CAP protein functions by increasing the rate of formation of the closed promoter complex, but it does not seem to influence the rate of isomerization to the open complex. However, the CI protein greatly influences the rate of isomerization without affecting the initial binding step. The CII protein has features common to each of the two distinct paradigms of transcriptional activators: it influences each of the two individual steps of open complex formation.

3. RNA Polymerase-activator Interaction.

An attractive hypothesis for transcription activation proposes that activation is mediated by a direct contact between RNA polymerase and the activator protein. Initially, this idea stemmed from the isolation of a "positive control" (PC) mutation in λcI encoding the CI protein (83). As the name implies, these mutations cause a defect in the activator function without affecting the DNA binding capacity of the activator.

One of the PC mutations altered a position in the CI protein speculated to be in close proximity to RNA polymerase. *In vitro* analysis of the mutant protein had shown that it behaved like the wild-type protein in DNA binding. The CI protein binds at three sites surrounding the divergent promoters pRM and pR on the λ chromosome (Figure 4.4). The binding of CI at the oR1 site represses the pR promoter (for the maintenance of the lysogenic state) and binding at oR2 causes activation of the pRM promoter. Hawley and McClure found that the positive control mutant protein repressed the pR promoter normally; however, compared to over 10-fold stimulation by the wild-type protein, the mutant protein increased the isomerization at pRM promoter only about twofold (77, 86).

4. *Converting a Repressor to an Activator.*

That the site of CI affected by positive control mutations is indeed involved in transcription activation was demonstrated elegantly by the "helix-swap" experiments done by Bushman and Ptashne (22). They altered the Cro repressor protein, converting it to an activator for the pRM promoter. The Cro protein, like CI, utilizes a helix-turn-helix motif to bind the same set of oR sequences, repressing both the pRM and pR promoters. The function of Cro differs from that of CI in two respects. First, whereas CI binds oR1 more tightly than oR2 and oR3, Cro binds oR3 more tightly than oR2 and oR1. This discrimination for the binding sites enables Cro to act as the lysogenic repressor and CI as the lytic repressor in the two opposite modes of phage λ development. Second, the binding of CI repressor at oR2 causes activation of the pRM promoter, whereas the binding of Cro at either oR3 or oR2 leads to the repression of the pRM promoter (See Figure 4.4). Distinct amino acid residues in the recognition helices of the two proteins and distinct nucleotide residues in the oR1 and oR3 sites allow CI and Cro to discriminate between these sites (84). However, molecular modeling suggested that the helix-turn-helix motifs of the two proteins lie in a nearly identical position on DNA (84). This analogous alignment of the two proteins was the key to the success of Bushman and Ptashne in converting the Cro protein into an activator by simply placing the activator region of CI at the equivalent location within Cro.

In order to test the function of the hybrid Cro protein that contained substitutions of four residues, the oR3 site was inactivated so that there was no repression of pRM. Due to an unexpected increase in the binding affinity of the mutant Cro for oR1, the oR2 site was also mutated (but not the −35 and −10 hexamers of pRM) such that the mutant Cro bound oR2 and oR1 with equal affinity. The mutant Cro protein could stimulate pRM transcription by about fivefold. In a subsequent study, Bushman *et al.* (23) have shown that among the four acidic residues that form the activating patch on helix 2 of the CI protein, the most critical determinant of CI-RNA polymerase interaction is a single glutamic acid residue. The role of an acidic patch in activation appears to be a widely utilized mechanism in gene control: two yeast activators (GAL4 and GCN4) also contain an acidic activator domain functionally separable from the respective DNA binding domains (143).

E. Role of DNA Structure

Another hypothesis on the mechanism of activators is that they alter DNA conformation to stimulate transcription. The binding of the CAP protein with its recognition site induces a bend in the DNA. However, this protein-induced bending has not been correlated with activation. Certain promoter regions—for instance, *hisT* of *Salmonella,* and *rrnB* of *E. coli*—are thought to have unusual structures in the DNA as they show aberrant mobility on polyacrylamide gels. Evidence that such a structure may contribute to transcription initiation came from Bossi and Smith, who showed that a 3-bp deletion at −71 to −73 position of the *hisT* promoter reduces the promoter function about 2.5-fold both *in vivo* and *in vitro* and allowed the promoter DNA fragment to migrate normally on polyacrylamide gels (14). It is as yet unclear exactly how these unusual structures in the DNA affect transcription initiation.

F. Control by DNA Methylation

Methylation of DNA by *dam* methylase is known to regulate a number of genes in *E. coli*. Methylation directly influences the rate of transcription initiation of the transposase gene of Tn10. The transposase promoter contains a *dam* methylation site in the −10 hexamer. As measured by fusion to the *lacZ* gene, the transposase promoter is about tenfold more active in a *dam*− host compared to the wild type (150). Similarly, the methylated promoter shows a tenfold-reduced activity *in vitro* compared to the unmethylated DNA (150). An interesting aspect of *dam* methylation control is that methylation plays a negative role. Since hemimethylated DNA is formed only transiently after the passage of the replication fork, a methylation-controlled gene would be transcribed momentarily after each round of replication and will remain silent throughout the remainder of the cell cycle. Indeed, the DnaA protein required for the initiation of DNA replication is regulated by *dam* methylation, and it would not be surprising if a number of cell cycle proteins follow this mode of gene regulation.

G. Control by DNA Supercoiling

The superhelicity of DNA may have a general role in transcription initiation, elongation, and termination due to topological constraints during transcription. There are a number of promoters that have been reported to be affected dramatically by the extent of supercoiling of the temperate DNA. Among these are the three genes, *gyrA*, *gyrB*, and *topA*, which are responsible for maintaining the superhelical density of the bacterial chromosome. Menzel and Gellert have shown that the expression of DNA gyrase encoded by the *gyrA* and *gyrB* genes is increased about tenfold upon treatment of *E. coli* with nalidixic acid, an inhibitor of DNA gyrase, or during growth of a ts *gyr* mutant at the restrictive temperature (130). This homeostatic control ensures that if the superhelical density is lowered, the synthesis of gyrase increases in order to maintain the proper superhelicity. As might be expected according to a homeostatic control, the activity of the *topA* promoter is decreased about two- to threefold in the temperature-sensitive *gyr* mutant strain (176).

VI. Transduction of Environmental Signals

The activation of genes negatively regulated by a repressor requires that, upon receiving an appropriate environmental signal, the repressor does not occupy the operator. In case of CI and LexA, this signal is induced by DNA-damaging agents. DNA damage activates the RecA protein to facilitate cleavage of the repressor molecules, thereby preventing them from binding DNA. Initially, the RecA protein was thought to be a specific protease, but recent studies from the Mount laboratory suggest that LexA undergoes a self-cleavage reaction that is promoted by RecA. The repressors of the *lac* and *gal* operons each bind a specific sugar (inducer) which prevents the repressor from binding DNA. There is a second class of repressors—for instance, the *trp* repressor—which binds DNA only when it is bound by tryptophan. When the cell is limiting for tryptophan, the ligand-free repressor molecule no longer binds the operator, clearing the way for RNA polymerase to engage the promoter.

A. *How is an Activator Activated?*

As mentioned above, the CAP protein requires cAMP to bind DNA. Thus, the level of cAMP in the cell, which is negatively controlled by glucose, appropriately determines whether genes for other sugar utilization are expressed or not. The expression of the *ara* operon of *E. coli* is induced by arabinose (inducer) and repressed by fucose (anti-inducer). The activation of the *araBAD* promoter requires cAMP-CAP and, in addition, the AraC protein that is transcribed from an adjacent promoter oriented divergently. The AraC protein acts as both a negative and a positive regulator. It does not require an effector molecule to bind DNA, but must bind the inducer to cause activation. There are two operators at which the AraC protein binds to bring about its dual effect (75). Initially, it was thought that the binding of inducer and the anti-inducer causes a differential affinity of AraC for these operators. However, this turned out not to be true. It appears that when arabinose is absent, the DNA bound by AraC and cAMP-CAP does not bind or retain RNA polymerase. When arabinose is present, all three proteins appear to form a complex with DNA.

B. Protein Phosphorylation Induced by Environmental Signals

A general mechanism for converting an inactive regulatory protein to its active form involves protein phosphorylation (90). These systems involve a transmitter and a receiver component to transduce an appropriate environmental signal in controlling gene expression (107). The transmitter, for example, the NtrB protein that controls genes for nitrogen fixation and assimilation, is a protein kinase whose activity is modulated by starvation for nitrogen or ammonia (101, 133). The receiver is NtrC, which upon phosphorylation by NtrB kinase becomes an activator for a number of nitrogen-regulated genes, for example, the *glnA* gene that encodes glutamine synthetase (101, 133). The phosphorylation of NtrC is not required for binding to its target site near the *glnA* promoter. Rather, the phosphorylated NtrC must interact with RNA polymerase to facilitate open complex formation (140). The transcription of *glnA* requires σN, the product of *rpoN* or *ntrA* gene (82, 134, 140, 154). The holoenzyme binds the promoter but is unable to form an open complex without phosphorylated NtrC (134, 140). This suggests that the promoter melting function of σN is induced upon its interaction with NtrC (154). It should be noted that the NtrC binding site can function like a typical eucaryotic enhancer element. Recent work has shown that NtrC bound at distant sites can contact σN-RNA polymerase complex at the *glnA* promoter by looping of the intervening DNA (171). By using catenated DNA circles, it has been shown quite elegantly that the function of the enhancer site is solely to tether the activator in the vicinity of the target RNA polymerase (185).

At least one regulatory protein of *E. coli* that undergoes signal-induced protein phosphorylation acts as a transcriptional antiterminator (6). It seems likely that signal-induced phosphorylation would be a common mechanism of modulating regulatory proteins, including essential cellular growth factors.

VII. Transcript Elongation

Once RNA synthesis begins and the growing RNA chain reaches a length of approximately 10 bases, a new biochemical event occurs—the release of σ factor from the holoenzyme and conversion of polymerase into a stable mode (192). Recent studies of Bowser and Hanna indicate that the σ factor loses contact with the 3′-end of nascent RNA after synthesis of a tetranucleotide (15). Precisely how and when the σ factor dissociates from the transcription complex may differ from promoter to promoter. Upon promoter-clearance, RNA polymerase becomes processive and can synthesize the entire RNA transcript by the sequential addition of nucleotides (at 30–50 nucleotides/second) without dissociating from the DNA or releasing the elongating RNA. This transition to the elongation mode can be correlated with a change in the conformation of polymerase, altering its interaction with the DNA template. The elongation complex is highly resistant to salt (119). It can be purified from nucleotides and stored for weeks at room temperature without loss of activity (7, 120). Whereas at initiation, RNA polymerase protects about 70 bp of DNA from digestion by nuclease, only about 30 bp of DNA are protected during processive elongation (192). In addition, there is also a change in the topology of the transcription bubble (192): (1) the DNA:RNA hybrid grows to approximately 12 bp in length, equivalent to approximately one complete turn of an A-form DNA double helix (76, 110), and (2) the DNA bubble increases from 12 bp at initiation to approximately 18 bp during stable elongation (58). Also, the loss of σ factor now allows the association of other polymerase-associated proteins that play important roles in the regulation of transcript elongation and accurate termination. (See below.)

As RNA polymerase moves down the template DNA adding nucleotides to the growing 3′-end of the nascent transcript, the 18-bp DNA bubble and the 12-bp RNA:DNA helix move along with RNA polymerase. Thus, there is a continuous denaturation and renaturation of the leading and lagging edges of the DNA bubble. Likewise, there is a continuous formation and breaking of RNA:DNA bp, respectively, at the leading and lagging edges of the RNA:DNA helix. How RNA polymerase solves these topological problems while maintaining extreme fidelity of nucleotide incorporation is not clear. It seems likely that RNA polymerase has an activity that denatures the lagging edge of the DNA;RNA helix, with a site on RNA polymerase directly contacting the melted strands so that the 18-bp bubble is maintained.

A. *Proteins Associated with Elongating RNA Polymerase*

The studies of cellular factors involved in the antitermination of transcription mediated by the N protein of phage λ have identified several *E. coli* proteins that interact with RNA polymerase during the elongation phase of the transcription cycle. One of these proteins, NusA, binds directly to the core RNA polymerase after σ dissociates from the holoenzyme (70). This association most likely involves a direct interaction with β as mutations in *rpoB* can suppress *nusA* mutations in an allele-specific manner (95, 169). NusA reduces the rate of elongation by core polymerase *in vitro*. One specific function of NusA in modulating elongation is to enhance RNA polymerase pausing at discrete sites (103, 112, 115). Core RNA polymerase pauses at specific sites on DNA (29, 100, 112, 115); some of these pause sites are G/C-rich and others have interrupted palindromic sequences that can form hairpin structures in the RNA. We discuss below the importance of RNA hairpins as signals for pausing and termination. NusA enhances pausing at some but not all sites. In addition to reducing the rate of elongation, NusA promotes termination and release of RNA chains at certain termination sites (46). Paradoxically, NusA is also a required cofactor for antitermination mediated by λ N and Q proteins (69, 186). Thus, NusA is the essential elongation subunit of RNA polymerase that couples both termination and antitermination factors.

The roles of other Nus factors (NusB, NusE, and NusG) capable of forming a complex with elongating RNA polymerase (88), have not yet been elucidated. We should note that the rate of polymerization *in vivo* is considerably greater than the rate at which the core RNA polymerase elongates RNA chains in the presence of NusA *in vitro*. This means that at least one function of the other proteins that associate with polymerase is to enhance the rate of polymerization. There may be a specific cellular factor that neutralizes the action of NusA, thereby enhancing the rate of RNA chain elongation. An inhibitor of this hypothesized factor would certainly affect termination at selective termination sites and may block or significantly reduce cell growth.

VIII. Transcription Termination

RNA polymerase must stop transcription at predetermined intergenic sites to maintain proper expression of genes as discrete transcription units. Indeed, discrete signals for terminating RNA synthesis are located at the ends of genes and operons. However, sometimes terminators are present within a transcription unit, upstream of genes, or even within genes. Many studies of transcription termination have been devoted to these sites since they act as regulatory signals in the control of transcription (11, 54, 113, 139, 151, 192, 194). It is through these studies that the mechanism for suppressing transcription termination and activating downstream genes, antitermination, has been discovered.

A. Sequencing Determinants of Bacterial Terminators

The primary signal for termination by RNA polymerase resides immediately upstream of the site where RNA synthesis stops, though there may be some contribution of the downstream sequence at certain sites (54, 118, 139, 151, 192, 194). The termination signals have been defined primarily by down mutations, and also, by cloning as a separable genetic entity to reduce expression of a downstream reporter gene. Terminators may be classified on the basis of whether they are recognized by the core RNA polymerase in a purified transcription reaction *in vitro*.

Factor-independent terminators have two features, each contributing to termination. They have a G/C-rich region that constitutes an inverted palindrome followed by a run of six to eight A residues on the template strand. Generally speaking, mutations in the palindrome that decrease base-pairing capacity reduce termination efficiency (29, 113, 151, 194). Conversely, mutations optimizing base-pairing capacity enhance termination. The interrupted palindrome allows the formation of a hairpin structure which acts as the signal for termination

(Figure 4.5). Evidence supporting this hypothesis stems from studies where ribonucleotide analogs that specifically affect RNA hairpin formation affect termination efficiency (45, 124, 194). Though no significant consensus sequence is discernible from a collection of many terminators, there seems to be a bias for a G-rich sequence immediately preceding the run of U's in the RNA. This bias may facilitate the formation of the hairpin in an unknown way, or it may provide a specific structural feature of the terminator hairpin that is recognized by RNA polymerase as the pause or termination signal.

Though "factor-independent" terminators are recognized by RNA polymerase in the absence of any ancillary factors *in vitro,* the efficiency of termination at certain sites is greater *in vivo* than *in vitro* and factors known to influence termination (Rho and NusA) can increase the efficiency of some of these termina-

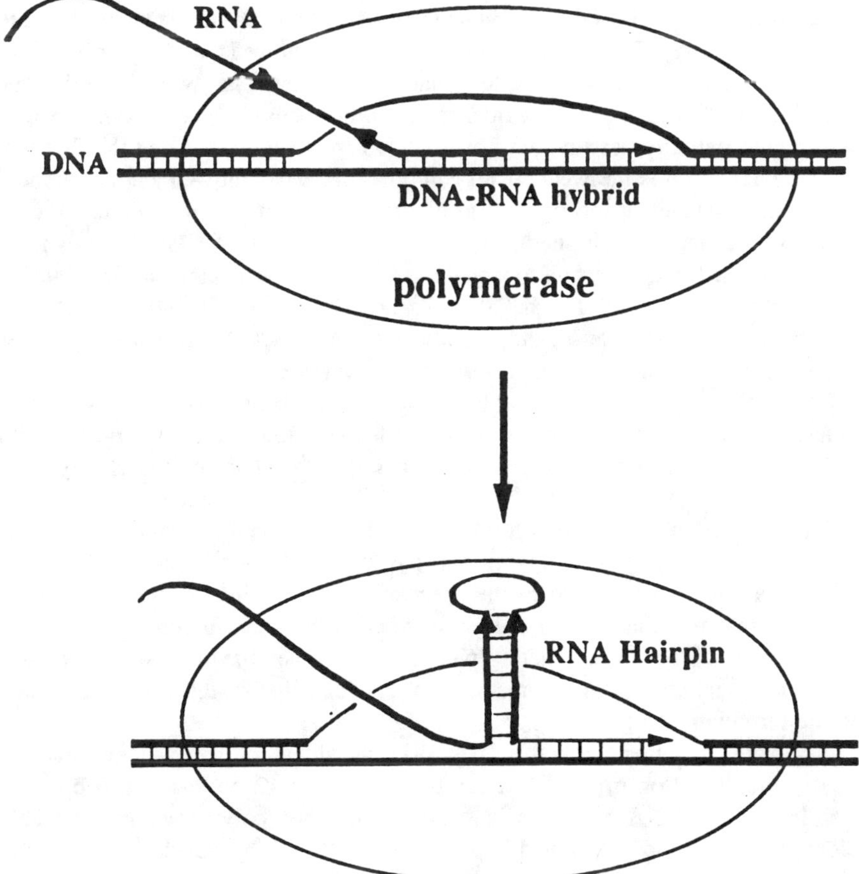

Figure 4.5. A model for factor-independent termination. Transcription of the terminator region (dark arrows) allows formation of an RNA hairpin, leading to termination.

tors. There is a second class of terminators that are unable to function without Rho, *in vitro* or *in vivo* (11, 54, 139, 151). These terminators differ markedly from the simple, factor-independent terminators in two respects: (1) there is no stretch of A residues and a hairpin does not always precede the site of termination, and (2) a long upstream region that is high in C content but low in G's and relatively free of secondary structure is a determinant of the termination signal.

B. Mechanism of Factor-independent Termination

A terminator signal should suffice for three distinct biochemical events: (1) pausing of RNA polymerase, (2) release of RNA chain from the transcription bubble, and (3) dissociation of RNA polymerase from DNA. Evidence that the run of A's in the template strand may contribute to the release of nascent RNA came from an analysis of mutants in which the length of the stretch was reduced to four A's (45, 192). Transcription of the mutant site showed that RNA polymerase paused at the site, but did not terminate as efficiently as with the wild-type site. This finding, along with evidence that other pause sites can be mapped to interrupted palindromes that are not followed by a run of A's and the fact that dA:rU interaction is the weakest bp combination, have allowed a simple model on factor-independent termination to emerge. According to this model (Figure 4.5), the nascent RNA immediately upstream of the run of U's denatures from the DNA and forms a hairpin to signal a reversible transcription pausing. Before the hairpin melts to reform the normal transcription bubble and release RNA polymerase from the pause, the remainder of the DNA:RNA hybrid denatures spontaneously and falls off the transcription complex.

This simple model predicts that the efficiency of a simple terminator would be determined by the relative stability of the hairpin and that all hairpins should induce pausing. Though these predictions generally seem to be true, clearly there are exceptions. The context of the hairpin, the loop sequence, and the stem sequence itself all seem to contribute to the efficiency of termination at one terminator or another. Future work will tell us how these different determinants of the terminator contribute to the mechanism of termination. One important question that remains to be resolved is whether the hairpin actively holds the polymerase captive by interacting with a cleft in the enzyme and if there is a conformational change in polymerase that converts it from the elongation mode to the termination mode.

According to Yager and von Hippel (192), the signal for pause and termination lies in the reduction of the RNA:DNA hybrid from 12 bp to about 6 bp. The stability of the DNA bubble, which maintains RNA polymerase in the stable elongation mode, is determined in part by the DNA:RNA hybrid. The reduction of the hybrid distorts the active site to inhibit polymerization and also makes the DNA bubble thermodynamically unstable. Thus, hairpin formation ultimately leads to the collapse of the bubble, releasing both the nascent RNA and the RNA polymerase from the DNA. In the premise of this thermodynamic model, one could imagine that the context and the precise sequence of the palindrome might indeed contribute to the rate of hairpin formation and stability, thus influencing termination efficiency. Alternatively, these sequences could influence a different aspect of the process that contributes to termination independent of the hairpin formation.

C. Mechanism of Rho-dependent Termination

The Rho protein, a hexamer of a 48-Kd polypeptide encoded by the essential *rho* gene of *E. coli,* maps at min 84 on the *E. coli* chromosome (33, 54, 148, 192). Rho is an RNA-dependent ATPase and a DNA:RNA helicase. Both ATP hydrolysis and RNA binding are essential for the termination activity of Rho. Rho protein was discovered by Roberts based on its activity at the tR1 and tL1 terminators of phage λ (148). Since then, many Rho-dependent sites have been uncovered not only at the ends of bacterial operons but also within operons that regulate transcription of downstream genes (38). Rho-dependent termination occurs at several discrete sites in the tR1 terminator region (115). It is clear that a region of at least 80 nucleotides (nt) upstream of the first termination site is required for Rho function (26, 28). As first suggested from observations on transcriptional polarity in bacterial operons, translation of this region inhibits Rho function, presumably by interfering with Rho-RNA interaction (D. Court and M. Zuber, personal communication).

The upstream region in the RNA seems to be the site where Rho binds in order to act as a transcript release factor (11, 54, 139, 192). Rho may not be able to engage the transcription complex or access the transcription bubble unless it binds the upstream region. The upstream region of tR1 contains two specific sites, called rutA and rutB, that seem to be required for Rho binding (28). Rho has two RNA binding sites that are occupied in succession and binding of the second site triggers ATP hydrolysis, without involving a phosphorylated intermediate (11, 54). Rho bound to ADP, the product of ATP hydrolysis, has a different conformation compared to its ATP-bound form. A combination of this conformational change and the energy of ATP hydrolysis may allow the protein to migrate toward RNA polymerase, which is presumably stalled at the site of termination due to the intrinsic nature of the termination site. There, Rho can extract the RNA from the transcription bubble through its DNA:RNA helicase activity (17). If RNA polymerase escapes the first termination site before Rho can catch up, termination can still be facilitated at one of subsequent favorable sites.

It should be noted that Rho may have to displace RNA polymerase prior to extracting RNA or use RNA polymerase as an anchor to physically extract RNA from the RNA:DNA duplex. Both models suggest an interaction of Rho with RNA polymerase. Indeed, certain mutations in *rpoB* can compensate for *rho* mutations in an allele-specific manner (34, 94, 169), and *in vitro* studies have shown that the mutant polymerase becomes susceptible to the action of one mutant Rho protein but not another (34). Rho may not be able to bind RNA polymerase without the aid of the NusA protein (157). Recently, specific alleles of *rpoB* have been described that compensate for both *nusA* and *rho* mutations in an allele-specific manner (95, 169).

IX. Regulation of Termination: Attenuation

Conceptually, termination can be regulated positively and negatively by a number of distinct mechanisms: (1) influencing RNA hairpin formation, (2) regulating the level of termination factors and their interaction with the transcription complex, and (3) modifying RNA polymerase so that its elongation behavior is altered or its capacity to interact with termination signals and factors is altered. The studies of "attenuation" (premature termination within a transcription unit) in several operons involved in biosynthesis of amino acids and nucleotides have shown that the formation of the terminator hairpin can be modulated positively or negatively depending on the level of a specific amino acid or nucleotide in the cell (113, 194). In the case of attenuation in the *trp* operon, it was discovered that the formation of the terminator hairpin can be controlled by a complementary upstream RNA sequence. The ability of this competing RNA sequence to interact with the terminator hairpin sequence and disrupt the terminator structure is modulated by translation of the upstream region which, in turn, is modulated by the level of tryptophan. When the level of tryptophan is low, the ribosome stalls at two tandem trp condons, facilitating interaction of the competitor RNA sequence with the terminator sequence. This favors readthrough and thus expression of the *trp* operon. When tryptophan is abundant, the ribosome continues translation through these codons and prevents the interaction of the competitor RNA with the terminator. The formation of terminator hairpin causes termination and prevents *trp* expression.

The level of Rho in the cell is controlled by autogenous regulation (8). The mechanism of action of Psu protein of phage P4, which seems to suppress Rho-dependent termination in general (55, 111, 173), may suggest that the cell has an anti-Rho function for modulating Rho activity. Such a function has not been discovered. Recent studies on the regulation of the *bgl* operon have led to the discovery of an antitermination protein (BglG) that acts by binding to a specific RNA sequence overlapping the terminator hairpin (89). Presumably, the binding of BglG to its target prevents the formation of the terminator hairpin, though an interaction of BglG with RNA polymerase has not been excluded. The RNA binding activity of BglG is modulated by phosphorylation involving a second protein of the operon, BglF, a signal-transducing protein kinase that transports and phosphorylates β-glucosides as well (6, 89). In the absence of β-glucosides, low levels of BglG in the cell remain in the inactive phosphorylated form. When a β-glucoside is present, BglF dephosphorylates BglG, converting the protein to an active antiterminator.

Direct disruption of the terminator hairpin upon binding by a specific RNA binding protein might be a general mechanism of antitermination since attenuation of the *trp* operon of *Bacillus* is also controlled by a trans-acting protein (*mtrA* gene product) that shows sequence similarity with a known RNA binding protein of phage T4 (66).

X. Operon-specific Antitermination

Antitermination plays a major role in the temporal control of gene expression of phage λ (55). Lambda is a temperate phage that can either lysogenize the host upon infection or undergo lytic development to produce progeny phages. The N gene product, an operon-specific antitermination protein that suppresses both Rho-dependent and Rho-independent terminators (67), is a positive regulator of both lysogenic and lytic mode of λ development (Figure 4.1). The lysogenic mode requires a high-level expression of CII, a protein that activates transcription of both the CI repressor gene as well as the integrase gene, and CIII, a protein that stabilizes CII by protecting it from proteolysis by the cellular Hfl protease (53, 55). Lytic growth requires the expression of Q protein that, like N, is an operon-specific antiterminator essential for transcription of the late genes (55, 149). N directly controls the expression of CIII from the pL promoter, and CII and Q from the pR promoter, by suppressing terminators that act as negative regulators for expression of each of these three genes (Figure 4.4).

A. *Antitermination by λ N Protein*

The N protein suppresses both Rho-dependent and Rho-independent terminators on the λ genome (67). The action of N on terminators is dependent upon an upstream, cis-acting site, *nut,* that is located between the promoter and the first terminator within both pL and pR operons. The *nut* site, defined by cis-dominant mutations (152), is a distinct genetic entity that can function when placed upstream of virtually any terminator, factor-dependent or independent, of phage and bacterial origin (39, 117, 183). The cloned *nut* site allows N protein, supplied in trans, to overcome multiple terminators placed thousands of bp away from the *nut* site (183). The *nut* site contains three genetic elements: boxA, boxB, and boxC (55). Of these, both boxA and boxC sequences are conserved in the *nut* sites of other lambdoid phages that utilize specific N gene products to allow antitermination in the respective genomes (49). The boxB sequence, an interrupted palindrome that can form a hairpin in the mRNA, is unique in each of these phages, and serves as the genome specificity element for N action (49, 117). Recent *in vitro* work has shown that the N protein binds the boxB RNA hairpin directly in a sequence-specific manner (S. Chattopadhyay, J. Garcia, and A. Das, in preparation). An arginine motif, conserved in the amino termini of each of the lambdoid phage N proteins, is responsible for this sequence-specific recognition of RNA hairpins (117).

Antitermination by N is also dependent on a set of trans-acting host factors, NusA–G, defined primarily by recessive mutations in the host that affect antitermination (35–37, 42, 50–53, 62, 88). As mentioned above, NusA binds RNA polymerase directly (70). Greenblatt, Chamberlin, and their co-workers have shown that NusA also binds N and Rho directly, indicating its pivotal role in coupling termination and antitermination factors to RNA polymerase (71, 157). Immunochemical studies have shown that N, NusA, NusB, NusE (S10 ribosomal protein), and NusG form a complex with RNA polymerase once it has transcribed past the *nut* site (9, 88) (S. Barik, B. Ghosh, and A. Das, unpublished results; J. Greenblatt, personal communication). According to the current working hypothesis (187), the *nut* site sequence in the nascent mRNA acts as the anchor to bind N and bring the protein in contact with the downstream RNA polymerase (Figure 4.6). The following observations lend support to this hypothesis: (1) N directly binds boxB RNA in the absence of any host factor, and the boxB mutants defective in N binding *in vitro* are also defective in antitermination *in vivo* (S. Chattopadhyay, J. Mena-Garcia, and A. Das, in preparation); (2) the boxB hairpin in the nascent mRNA is essential for N binding to the transcription complex and the modification of RNA polymerase to a termination-resistant form (187); (3) an excess of boxB hairpin added *in vitro* inhibits N-dependent antitermination by sequestering N protein (J. DeVito and A. Das, unpublished results); (4) truncated N gene products lacking an essential carboxy-terminal domain act as trans-dominant mutants, binding boxB and excluding the wild-type protein from capturing RNA polymerase (116); (5) the boxB hairpin can act to deliver N protein to RNA polymerase when the latter has moved hundreds of bp away from the *nut*

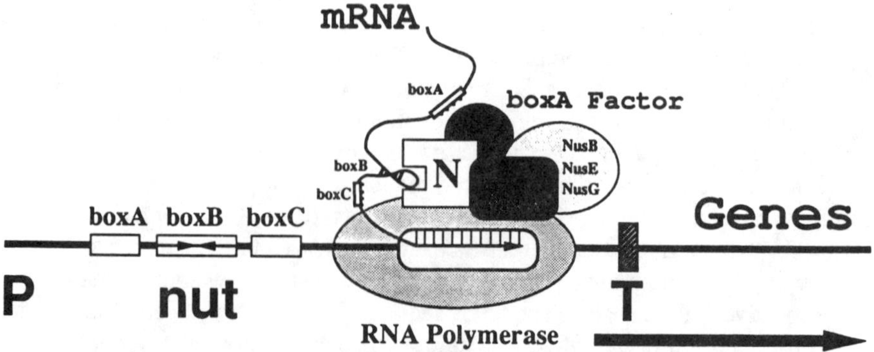

Figure 4.6. The antitermination complex formed by the λ N protein. Bold arrow indicates the direction of transcription from the promoter (**P**) to and through the terminator (**T**). The respective protein and nucleic acid components are described in the text.

site (187); and finally, (6) once the boxB hairpin has been synthesized, the *nut* site DNA is no longer necessary for N to modify RNA polymerase (187).

N-dependent antitermination has been reconstituted *in vitro* with only one host factor, NusA, that binds both N as well as RNA polymerase (186). In this minimal system, a variety of phage and bacterial terminators placed downstream of the *nut* site are efficiently suppressed by N. However, the inefficient suppression of the same terminators placed farther distal to the *nut* site suggests that the core antitermination complex that forms in presence of N and NusA is metastable (W. Whalen and A. Das, in preparation). The efficiency of suppression of distal terminators can be increased either by increasing the concentration of N protein alone, or by the presence of additional host factors—NusB, NusE, NusG, and at least one other unidentified host factor (W. Whalen, B. Ghosh, and A. Das, unpublished results; J. Greenblatt, personal communication). These observations lead to the hypothesis that the additional Nus factors act in an ancillary capacity to stabilize the core complex, endowing processivity in antitermination over the long transcription units on the phage genome. Some of these ancillary Nus factors have additional functions in the direct recognition of boxA and boxC and also in bestowing differential specificity in suppressing different kinds of terminators (56, 116, 117, 187).

The boxA element is conserved not only within the *nut* sites of lambdoid phages but also in the leader region of the *rrn* operons that are also regulated by an antitermination mechanism. (See below.) Whereas mutations in boxB abolish antitermination by N, boxA mutations (including deletions) only reduce the extent of antitermination (117, 197). In fact, boxA is not required for antitermination in the minimal system (J. DeVito, W. Whalen, and A. Das, unpublished results), indicating that it plays an ancillary role in antitermination by λ N protein. Recent genetic studies by Lazinski have led to the discovery that boxA provides a second, genome-independent mechanism for polymerase capture by N antiterminator. The key experiment in support of this conclusion is that a consensus boxA sequence can suppress mutations in either boxB or in N that completely abolish the sequence-specific interaction of boxB with N (116). Since boxA is not required for boxB-N interaction, and since boxA does not increase the affinity of boxB-N interaction (S. Chattopadhyay, J. Garcia, and A. Das, in preparation), we have hypothesized that boxA is directly recognized by a Nus factor which in turn binds N. Though NusA was initially thought to bind boxA based on indirect genetic evidence, direct physical studies have failed to detect such an interaction (41, 177). Friedman and colleagues have recently reported that high-level transcription of a *nut* site containing the consensus boxA sequence can inhibit antitermination by N and that an increased expression of NusB in the cell overcomes this inhibition (56). These results are consistent with the hypothesis that boxA binds a host factor. It remains to be seen whether NusB directly binds boxA or interacts with an unidentified factor that recognizes boxA.

The boxA binding factor may be an essential component of the cell that allows antitermination in cellular operons utilizing analogs of λ N protein. The dual mechanism of capture involving the recognition of conserved boxA by a common factor and the recognition of unique boxB elements by genome-specific antiterminators provides a mechanism for regulating the extent and specificity of antitermination. In λ, for instance, we can imagine that the boxA factor could play a role in the decision between lytic and lysogenic growth. The cell undergoing lysogeny will have a low level of the Hfl protease so that CII is produced at a maximum rate. The same cell may also have a low level of the boxA factor, preventing the formation of a stable antitermination complex so that the lytic activator protein Q is not expressed.

How does N actually suppress termination? As a subunit of RNA polymerase, N must somehow change the response of polymerase to termination signals and factors, locking it in a stable elongation mode. One simple possibility is that N enhances the intrinsic rate of chain elongation. In that event, RNA polymerase could transcribe through the terminator so rapidly that the terminator hairpin is not formed before polymerase escapes the terminator region. A second possibility, not necessarily mutually exclusive from the first, is that N occupies or modifies a cleft in polymerase that might directly interact with termination factors and/or RNA hairpins (9, 186). According to this scenario, N would sterically exclude the interaction of polymerase with terminator hairpin and factors that promote termination. A third possibility is that N actively maintains an extended DNA bubble by stabilizing the RNA:DNA duplex, by inhibiting the RNA:DNA duplex melting activity of RNA polymerase at the lagging edge of the transcription bubble, or by directly destabilizing the RNA hairpin (186).

B. Antitermination by λ Q Protein

Like the N protein, the Q protein is an operon-specific antiterminator that utilizes a promoter-proximal engagement site to capture RNA polymerase and modify it to a termination-resistant form (55, 150). The site required for Q's action, *qut,* overlaps with the later promoter of λ, pR′. Unlike the *nut* site, the *qut* site is inseparable from the promoter (193). Elegant *in vitro* work from the Roberts laboratory has shown that RNA polymerase undergoes a significant pausing soon after initiation from the pR′ promoter at site +16 (69). Q acts upon this paused polymerase specifically and accelerates the polymerase not only through this pause site but also others that are encountered downstream, enabling polymerase to read through terminators (69). Though there is no direct evidence, we imagine that Q becomes a subunit of RNA polymerase at the +16 site. Antitermination by Q is extremely processive and, unlike the case of N, the single host factor NusA seems to be sufficient for processive antitermination (10, 69, 193) (S. Barik and A. Das, unpublished results). Both mutational studies of *qut* as well as nuclease protection analysis suggest that Q recognizes distinct features of the *qut* site DNA (193) (W. Yarnell and J. Roberts, personal communication).

Pausing at +16 might be essential for three reasons: (1) to juxtapose polymerase with the Q-recognition site which is likely to include DNA sequences between the −10 and −35 hexamers, (2) to endow polymerase with the conformation that Q recognizes and binds, and (3) to hold polymerase captive until Q captures it. All of these features are quite different from those required by N to capture RNA polymerase. In this regard, it is interesting that the life cycle of λ does not demand that Q action be regulated. Rather, once a decision to lytic development has been made, the Q protein should function at an optimal rate to allow rapid coordinate expression of the head and tail proteins and the host-cell lysis functions.

C. Antitermination in the rRNA Operons

The ribosomal RNAs are encoded by seven *rrn* operons whose expression is subject to an antitermination mechanism (131). The leader region in these operons contains sequence elements that show similarity to the *nut* site of λ and its relatives (121). Of the three elements, both boxA and boxC sequences are conserved, as is the case in the lambdoid phages. Remarkably, the boxA element alone has been found to be sufficient for suppression of Rho-dependent terminators cloned downstream of the leader region (3, 12, 85). However, strong Rho-independent terminators are not suppressed by the *rrn* leader (3), indicating that there is terminator specificity in boxA-dependent antitermination. The boxB sequences of the seven *rrn* operons which lie upstream of the boxA sequence are not well conserved (12), and it is not clear if and how the boxB sequence plays a role in antitermination within the *rrn* operons. Though NusB is thought to participate in antitermination in *rrn* operons (160) (H. Schriner and C. Squires, personal communication), the details of the antitermination mechanism remain to be elucidated. In particular, as in antitermination in λ, the factor that recognizes boxA remains to be identified. Moreover, it remains to be determined whether *rrn* antitermination requires a cellular analog of λ N protein that might recognize the *rrn* boxB sequences directly.

XI. Antimicrobial Agents Affecting RNA Synthesis

Very few effective antimicrobial agents are currently available whose action is specifically targeted toward components of the transcription apparatus. It is interesting that the two that have had some success, rifampicin and streptolydigin, both interact with and bind to the β subunit of RNA polymerase. Even though they both exert their effect by interacting with β, their modes of action are different.

Rifampicin is a semisynthetic derivative of rifamycin B that is produced by *Streptomyces mediterranei*. This antibiotic has been especially useful in the treatment of tuberculosis. Rifampicin is one of the most potent inhibitors of bacterial RNA polymerase, with 50% inhibition of the *E. coli* enzyme at levels as low as 2×10^{-8} M (57). Its action is mediated by a stoichiometric binding to RNA polymerase. Rifampicin does not block the binding of RNA polymerase to the template (178), suggesting that different sites in polymerase are involved in binding of rifampicin and DNA. Mechanistic studies (24, 97, 126, 159, 165) have shown that RNA polymerase can synthesize a dinucleotide, and on occasion a trinucleotide, in the presence of rifampicin. Thus rifampicin does not prevent isomerization, nucleotide binding, and formation of the first phosphodiester bond. Rather, it greatly inhibits the subsequent addition of the third nucleotide and almost abolishes the addition of the fourth nucleotide. Presumably, rifampicin inhibits the initial translocation step(s) by simple steric hindrance. The differential inhibition on the second and subsequent phosphodiester bond formation might be attributed to the precise sequence of the start site, which may not be a single site for some promoters. Curiously, a mutation in a conserved region of *rpoB* (K1051R) shows a dominant-negative phenotype and alters RNA polymerase so that it is able to catalyze the formation of the first phosphodiester bond, but is defective in carrying out the subsequent phosphodiester bond formation (99). This is virtually the same phenotype seen when wild-type polymerase is treated with rifampicin.

Rifampicin does not inhibit the stable elongation process (24, 104); RNA polymerase becomes resistant to inhibition by rifampicin once it has produced a tetranucleotide or larger transcript. It is conceivable that the nascent transcript occupies the site bound by rifampicin or that the conformational change associated with σ dissociation and transition to the stable elongation mode might abolish the binding of rifampicin to RNA polymerase.

Bacteria rapidly develop resistance to rifampicin, usually by mutation in the *rpoB* gene. The frequency of spontaneous Rif^R mutants varies, but has been reported to be approximately 1×10^{-10} for *Mycobacterium tuberculosi,* and as high as 1×10^{-8} for *E. coli* (47). The rifampicin binding domain of *rpoB* has been defined by an exhaustive collection of Rif^R mutations (93). The mutations are clustered in three discrete locations. Some of the mutations are recessive, conditional lethal. These and others that do not affect the viability of the cell are often associated with an altered termination property (48, 94). Some increase the intrinsic termination efficiency while others decrease it. Certain Rif^R alleles alter the interaction of β with Rho and NusA proteins, compensating for *rho* and *nusA* mutations in an allele-specific manner (34, 74, 95, 169). One such Rif^R allele, ts8, caused by a F522 substitution, has a conditional lethal phenotype that is apparently due to the defect of the β subunit in interaction with a hitherto-unidentified transcription factor required for growth of *E. coli* at high temperature (168, 169).

Although streptolydigin also inhibits the activity of RNA polymerase by interacting with β, there are several major differences between its mechanism of action and that of rifampicin. The most important difference is that streptolydigin inhibits the RNA chain polymerization reaction, slowing the rate of phosphodiester bond formation at any step in the RNA synthetic cycle, including abortive initiation (127, 128). Compared to rifampicin, streptolydigin binds weakly to RNA polymerase and can dissociate from polymerase upon simple dilution of the reaction mixture (25). Genetic mapping of streptolydigin-resistant mutations in *rpoB* and recombination of these alleles with Rif^R alleles demonstrate that the binding site of streptolydigin differs from that of rifampicin (166).

XII. Screening for New Inhibitors of RNA Synthesis

Clearly, there are a number of proteins and even additional sites on the β subunit that might be the target of novel antimicrobial drugs inhibiting RNA synthesis. For instance, it may be possible to find specific inhibitors directed against each of the three distinct domains of σ that are likely to be involved in binding the core enzyme, recognition of the promoter, and melting the start site. It is likely that there are essential growth regulators of bacteria which are required for transcription activation of specific developmental genes or even housekeeping genes such as the tRNA and rRNA genes. These are obviously targets for a second set of drugs. Finally, since RNA chain elongation and termination are essential processes of overall RNA synthesis, elongation and termination factors are the targets for yet another class of drugs.

A. A Simple Screening Method

The modern genetic techniques available to today's microbiologist should allow the engineering of plasmids or phage vectors suitable for the rapid testing of compounds that affect some step of RNA synthesis. The simplest approach amenable to the screening of a large number of test substances is to monitor expression of a reporter gene whose activity is measured easily. Two reporters suitable for this purpose come to mind: (1) the *lacZ* gene of *E. coli,* whose product can hydrolyze a chromogenic β-galactoside substrate (ONPG) (163), and (2) the RS gene cluster of phage λ, whose products cause a rapid lysis of the bacterial culture (D. Lazinski and A. Das, in preparation).

The reporter genes, *lacZ* and λ*RS,* can be fused to the *lac* promoter-operator region in a plasmid vector suitable for the bacterial strain, along with a copy of the *lac* repressor gene (Figure 4.7A). Bacterial transformed with the plasmid can be grown in culture and then small aliquots of the culture can be treated with various test substances in microtiter dishes. After a brief exposure to these potential inhibitors, IPTG can be added and incubation continued for about 2 hours. IPTG will inactivate the *lac* repressor, and those substances that do not affect transcription or translation will allow the expression of *lacZ* and the *RS* genes. The expression of *RS* gene cluster will cause cell lysis, detectable by a significant drop in optical density. After recording the optical density, a mixture of chloroform and SDS can be added to each well, followed by the addition of

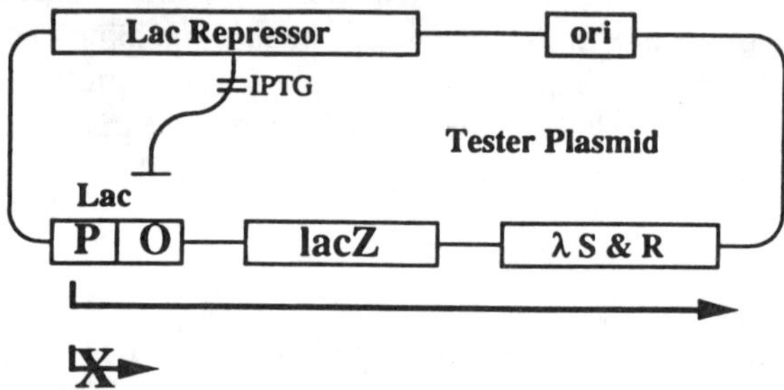

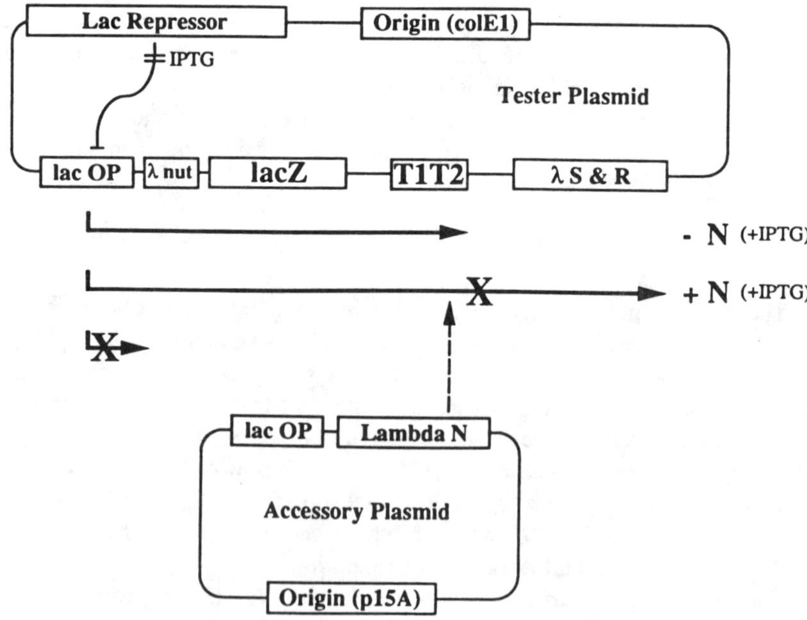

Figure 4.7. Plasmids for screening of novel antibiotics. A. Single-tester plasmid for screening general inhibitors of RNA synthesis. B. Two-plasmid system for testing inhibitors of transcriptional initiation, termination, and antitermination.

ONPG. Any culture expressing *lacZ* will hydrolyze ONPG, producing a yellow color. An inhibitor of transcription or translation is expected to block cell division, prevent cell lysis and remain colorless upon addition of ONPG. The combined use of the two reporters is ideal for avoiding system-specific inhibition by test substances.

An identical strategy may be applied to screen compounds which inhibit termination or antitermination of transcription by engineering terminator and antiterminator sequences upstream of the reporter genes. In our study of antitermination, the strong terminator cluster T1–T2 of *E. coli rrnB* operon was placed between the *lac* promoter and the *RS* genes. These terminators are strong enough to keep RS expression down, and cells harboring the plasmid grow normally even after complete induction of the *lac* promoter with IPTG. When the *nut* site of phage λ is inserted between the promoter and the terminators in the tester plasmid and the cells are provided with λ N protein encoded by a second compatible plasmid (Figure 4.7B), sufficient expression of RS genes ensues, resulting in cell lysis.

The two systems can be combined to contain the *lacZ* gene upstream of the terminators followed by the *RS* genes so that inhibitors of RNA synthesis and the inhibitors of termination or antitermination processes may be screened in a single test (Figure 4.7B). Inhibitors of transcription will prevent cell lysis and remain colorless when ONPG is added. Inhibitors of termination, tested in the absence of N, will allow cell lysis and produce a yellow color due to ONPG hydrolysis. Conversely, inhibitors of antitermination, tested with the N protein expressed, will prevent cell lysis, but produce a yellow color. Since some components of the transcription apparatus—for instance, NusA—are involved in both termination and antitermination, certain drugs could inhibit both processes. However, as the outcome of a lack of termination is the same as active antitermination, a specific inhibitor of antitermination is the one that blocks cell lysis both in the presence and in the absence of N.

After the initial screening process, the active compounds need to be specifically tested for direct inhibition of RNA synthesis. The simplest approach would be to monitor the RNA synthetic capacity of a cell-free lysate in the presence and absence of the test drugs, employing a plasmid DNA as the template. The *in vitro* assay, performed under conditions optimum for transcription and most refractory to RNA degradation, is quite convenient and simple, as it would utilize a set of extracts prepared by proven methods (36). The same extract may be used with a different set of standard reagents to test if the drug inhibits translation rather than transcription.

Following the initial characterization of compounds that specifically affect the process of RNA synthesis, a battery of pathogenic bacterial species could be tested for sensitivity or resistance to these drugs. For drugs that show promise to be effective as specific antimicrobial agents, more studies could be done to determine which component of the RNA synthetic machinery is affected. Genetic studies can be performed to determine the frequency of revertants to the compound and also to map the target gene. Once the target proteins are identified, a more detailed biochemical characterization of the drug-protein interaction could be carried out. Finally, a rational drug design strategy could be implemented to attempt to produce more effective derivatives of the compound.

Acknowledgments

This work is dedicated to the memory of F.W.'s mother, Julia M. Warren, who passed away during its preparation.

Some of the experiments described here were supported by grants from the National Institutes of Health, the National Science Foundation, and the American Cancer Society, and by an Established Investigator Award from the American Heart Association to A.D.

References

1. Adhya, S. 1989. Multipartite genetic control elements: communication by DNA loop. Ann. Rev. Genet. **23:**227–250.
2. Adhya, S., and W. Miller. 1979. Modulation of the two promoters of the galactose operon of *Escherichia coli*. Nature **279:**4923–494.
3. Albrechtson, B., C.L. Squires, S. Li, and C. Squires. 1990. Antitermination of characterized transcriptional terminators by the *Escherichia coli rrnG* leader region. J. Mol. Biol. **213:**123–134.
4. Allison, L.A., M. Moyle, M. Shales, and C.J. Ingles. 1985. Extensive homology among the largest subunits of eukaryotic and prokaryotic RNA polymerases. Cell **42:**599–610.
5. Amouyal, M., L. Mortensen, H. Bue, and K. Hammer. 1989. Single and double loop formation when *deoR* repressor binds to its natural operator sites. Cell **58:** 545–551.
6. Amster-Choder, O., and A. Wright. 1990. Regulation of activity of a transcriptional anti-terminator in *E. coli* by phosphorylation *in vito*. Science **249:**540–542.
7. Arndt, K.M., and M.J. Chamberlin. 1990. RNA chain elongation by *E. coli* RNA polymerase: factors affecting the stability of elongating ternary complexes. J. Mol. Biol. **213:**79–108.
8. Barik, S., P. Bhattacharya, and A. Das. 1985. Autogenous regulation of transcription termination factor Rho. J. Mol. Biol. **182:**495–508.
9. Barik, S., B. Ghosh, W. Whalen, D. Lazinski, and A. Das. 1987. An antitermination protein engages the elongating transcription apparatus at a promoter-proximal recognition site. Cell **50:**885–899.
10. Barik, S., and A. Das. 1990. An analysis of the role of host factors in transcription antitermination *in vitro* by the Q protein of coliphage lambda. Mol. Gen. Genet. **222:**152–156.
11. Bear, D.G., and D.S. Peabody. 1988. The *E. coli* Rho protein: an ATPase that terminates transcription. TIBS **13:**343–347.
12. Berg, K.L., C. Squires, and C.L. Squires. 1989. Ribosomal RNA operon antitermination: function of leader and spacer region *boxB-boxA* sequences and their conservation in diverse microorganisms. J. Mol Biol. **209:**345–358.

13. Biggs, J., L.L. Searles, and A.L. Greenleaf. 1985. Structure of the eukaryotic transcription apparatus: features of the gene for the largest subunit of *Drosophilia* RNA polymerase II. Cell **42:**611–621.

14. Bossi, L., and D.M. Smith. 1984. Conformation change in the DNA associated with an unusual promoter mutation in a tRNA gene of *Salmonella*. Cell **39:**643–652.

15. Bowser, C.A., and M.M. Hanna. 1991. The sigma subunit of *Escherichia coli* RNA polymerase loses contacts with the 3′ end of the nascent RNA after synthesis of a tetranucleotide on Poly [d(A-T)]. J. Mol. Biol., in press.

16. Bramhill, D., and A. Kornberg. 1988. Duplex opening by *DnaA* protein at novel sequences in initiation of replication at the origin of *E. coli* chromosome. Cell **52:**743–755.

17. Brennan, C., A. Dombroski, and T. Platt. 1987. Transcription termination factor Rho is an RNA-DNA helicase. Cell **48:**945–952.

18. Briat, J.F., and M.J. Chamberlin. 1984. Identification and characterization of a new transcriptional termination factor from *E. coli*. Proc. Natl. Acad. Sci. USA **81:**7373–7377.

19. Buc, H. 1986. Mechanism of activation of transcription by the complex formed between cyclic AMP and its receptor in *Escherichia coli*. Biochem. Soc. Trans. **14:**196–199.

20. Burgess, R.R., B. Erickson, D.L.M. Gentry, D. Hager, S. Lesley, M. Strickland, and N. Thompson. 1987. Bacterial RNA polymerase subunits and genes, p. 3–16. *In* W.S. Reznikoff, R.R. Burgess, J.E. Dahlberg, C.A. Gross, M.T. Record, and M.P. Wickens (ed.), RNA polymerase and the regulation of transcription. Elsevier, New York.

21. Burgess, R.R., A.A. Travers, J.J. Dunn, and E.K.F. Bautz. 1969. Factor stimulating transcription by RNA polymerase. Nature **211:**43–46.

22. Bushman, F.D., and M. Ptashne. 1988. Turning λ Cro into a transcriptional activator. Cell **54:**191–197.

23. Bushman, F.D., C. Shang, and M. Ptashne. 1989. A single glutamic acid residue plays a key role in the transcriptional activation function of lambda repressor. Cell **58:**1163–1171.

24. Carpousis, A.J., and J.D. Gralla. 1980. Cycling of ribonucleic acid polymerase to produce oligonucleotides during initiation *in vitro* at the *lacUV5* promoter. Biochemistry **19:**3245–3253.

25. Cassani, G., R.R. Burgess, H.M. Goodman, and L. Gold. 1971. Inhibition of RNA polymerase by streptolydigin. Nature **230:**197–200.

26. Ceruzzi, M., and J.P. Richardson. 1985. Interaction of Rho protein with λ*cro* and T7 D111 early gene transcripts, p. 161–170. *In* R. Calender and L. Gold (ed.), Sequence specificity in transcription and translation. Alan R. Liss, Inc., New York.

27. Chamberlin, M.J., 1976. RNA polymerase—an overview, p. 17–67. *In* R. Losick and M. Chamberlin (ed.), RNA polymerase. Cold Spring Harbor Laboratory, Cold Spring Harbor, N.Y.

28. Chen, C.A., G.R. Galluppi, and J.P. Richardson. 1986. Transcription termination at λ tR1 is mediated by interaction of Rho with specific single-stranded domains near the 3′ end of cro RNA. Cell **46:**1023–1028.

29. Court, D., C. Brady, M. Rosenberg, D.L. Wulff, M. Behr, M. Mahoney, and S. Izumi. 1980. Control of transcription termination: a *rho*-dependent termination site in bacteriophage lambda. J. Mol. Biol. **138:**231–254.

30. Dandanell, G., and K. Hammer. 1987. Long-range cooperativity between gene regulatory sequences in a prokaryote. Nature **325:**823–826.

31. Darst, S.A., H.O. Ribi, D.W. Pierce, and R.D. Kornberg. 1988. Two dimensional crystals of *Escherichia coli* RNA polymerase holoenzyme on positively charged lipid layers. J. Mol. Biol. **203:**269–273.

32. Darst, S.A., E.W. Kubalek, and R.D. Kornberg. 1989. Three-dimensional structure of *Escherichia coli* RNA polymerase holoenzyme determined by electron crystallography. Nature **340:**730–732.

33. Das, A., D. Court, and S. Adhya. 1976. Isolation and characterization of conditional lethal mutants of *Escherichia coli* defective in transcription termination factor Rho. Proc. Natl. Acad. Sci. USA **73:**1959–1963.

34. Das, A., C. Merril, and S. Adhya. 1978. Interaction of RNA polymerase and Rho in transcription termination: coupled ATPase. Proc. Natl. Acad. Sci. USA **75:**4828–4832.

35. Das, A., M. Gottesman, J. Wardwell, P. Trisler, and S. Gottesman. 1983. A mutation in the *Escherichia coli rho* gene that inhibits the N activity of phage lambda. Proc. Natl. Acad. Sci. USA **80:**5530–5534.

36. Das, A., and K. Wolska. 1984. Transcription antitermination *in vitro* by lambda N gene product: requirement for a phage *nut* site and the products of host *nusA*, *nusB* and *nusE* genes. Cell **38:**165–173.

37. Das, A., B. Ghosh, S. Barik, and K. Wolska. 1985. Evidence that ribosomal protein S10 itself is a cellular component necessary for transcription antitermination by phage λ N protein. Proc. Natl. Acad. Sci. USA **82:**4070–4074.

38. de Crombrugghe, B., S. Adhya, M. Gottesman, and I. Pastan. 1973. Effect of Rho on transcription of bacterial operons. Nature (New Biol) **241:**260–264.

39. de Crombrugghe, B., M. Mudryj, R. DiLauro, and M. Gottesman. 1979. Specificity of the bacteriophage lambda N gene product (p*N*): *nut* sequences are necessary and sufficient for the antitermination by p*N*. Cell **18:**1145–1151.

40. Dissinger, S., and M.M. Hanna. 1990. Active site labeling of *Escherichia coli* transcription elongation complexes with 5-[(4-azidophenacyl)thio] uridine 5′-triphosphate. J. Biol. Chem. **265:**7662–7668.

41. Dissinger, S., and M.M. Hanna. 1991. RNA-protein interactions in a *nusA*-containing *E. coli* transcription complex paused at an RNA hairpin. J. Molec. Biol. *218:* 1–15.

42. Downing, W.L., S.L. Sullivan, M.E. Gottesman, and P.P. Dennis. 1990. Sequence and transcriptional pattern of the essential *Escherichia coli secE-nusG* operon. J. Bacteriol. **172:**1621–1627.

43. Dunn, T.M., S. Haber, S. Ogden, and R.F. Schleif. 1984. An operator at −280 base pairs that is required for repression of *araBAD* operon promoter: addition of DNA helical turns between the operator and promoter cyclically hinders repression. Proc. Natl. Acad. Sci. USA **81:**5017–5020.

44. Erirkson, J.W., and C.A. Gross. 1989. Identification of the σ^E subunit of *Escherichia coli* RNA polymerase: a second alternate σ factor involved in high-temperature gene expression. Genes Dev. **3:**1462–1471.

45. Farnham, P.J., and T. Platt. 1980. A model for transcription termination suggested by studies on the *trp* attenuator *in vitro* using base analogs. Cell **20:**739–748.

46. Farnham, P.J., J. Greenblatt, and T. Platt. 1982. Effects of nusA protein on transcription termination in the tryptophan operon of *Escherichia coli*. Cell **29:**945–951.

47. Farr, B., and G.L. Mandell. 1982. Rifampin. Med. Clin. N. Am. **66(1):**157–168.

48. Fisher, R.F., and C. Yanofsky. 1983. Mutations of the beta subunit of RNA polymerase alter both transcription pausing and transcription termination in the trp operon region. J. Biol. Chem. **258:**8146–8150.

49. Franklin, N.C. 1985. Conservation of genome form but not sequence in the transcription antitermination determinants of bacteriophages λ, 21 and P22. J. Mol. Biol. **181:**5–84.

50. Friedman, D.I., and L.S. Baron. 1974. Genetic characterization of a bacterial locus involved in the activity of the N function of phage λ. Virology **58:**141–148.

51. Friedman, D.I., M. Baumann and L.S. Baron. 1976. Cooperative effects of bacterial mutations affecting λ N gene expression. 1. Isolation and characterization of a *nusB* mutant. Virology **73:**119–127.

52. Friedman, D.I., A.T. Schauer, M.R. Baumann, L.S. Baron, and S.L. Adhya. 1981. Evidence that ribosomal protein S10 participates in the control of transcription termination. Proc. Natl. Acad. Sci. USA **78:**1115–1118.

53. Friedman, D.I., E.R. Olson, C. Georgopoulos, K. Tilly, I. Herskowitz, and F. Banuett. 1984. Interactions of bacteriophage and host macromolecules in the growth of bacteriophage λ. Microbiol. Rev. **48:**299–325.

54. Friedman, D.I., M.J. Imperiale, and S.L. Adhya. 1987. RNA 3′ end formation in the control of gene expression. Annu. Rev. Genet. **21:**453–488.

55. Friedman, D.I. 1988. Regulation of phage gene expression by termination and antitermination of transcription, p. 263–319. *In* R. Calender (ed.), The bacteriophages vol. 2. Plenum Publishing Corp., New York.

56. Friedman, D.I., E.R. Olsen, L.L. Johnson, D. Alessi, and M.G. Craven. 1990. Transcription-dependent competition for a host factor: the function and optimal sequence of the phage lambda *boxA* transcription antitermination signal. Genes and Develop. **4:**2210–2222.

57. Gale, E.F., E. Cundliffe, P.E. Reynolds, M.H. Richmond, and M.J. Waring. 1981. Inhibitors of nucleic acid synthesis, p. 370–401. *In* Gale, et al. (ed.) The molecular basis of antibiotic action. John Wiley and Sons Ltd., New York.

58. Gamper, H.B., and J.E. Hearst. 1982. A topological model for transcription based on unwinding angle analysis of *E. coli* RNA polymerase binding, initiation and ternary complexes. Cell **29:**81–90.

59. Gardella, T., H. Moyle, and M.M. Susskind. 1989. A mutant *Escherichia coli* σ70 subunit of RNA polymerase with altered promoter specificity. J. Mol. Biol. **206:**579–590.

60. Garges, S., and S. Adhya. 1985. Sites of allosteric shift in the structure of the cyclic AMP receptor protein. Cell **41:**745–751.

61. Gentry, D.R., and R.R. Burgess. 1989. *rpoZ,* encoding the omega subunit of *Escherichia coli* RNA polymerase, is in the same operon as *spoT*. J. Bacteriol. **171:**1271–1277.

62. Ghosh, B., and A. Das. 1984. NusB: A protein factor necessary for transcription antitermination *in vitro* by phage lambda *N* gene product. Proc. Natl. Acad. Sci. USA **81:**6305–6309.

63. Gilbert, W. 1976. Starting and stopping sequences for the RNA polymerase, p. 193–205. *In* R. Losick and M. Chamberlin (ed.), RNA polymerase. Cold Spring Harbor Laboratory, Cold Spring Harbor, N.Y.

64. Glass, R.E., A. Honda, and A. Ishihama. 1986. Genetic studies on the β subunit of *Escherichia coli* RNA polymerase. IX. The role of the carboxy-terminus in enzyme assembly. Mol. Gen. Genet. **203:**487–491.

65. Glass, R.E., S.T. Jones, V. Nene, T. Nomura, N. Fujita, and A. Ishihama. 1986. Genetic studies on the β subunit of *Escherichia coli* RNA polymerase. VIII. Localization of a region involved in promoter selectivity. Mol. Gen. Genet. **203:** 487–491.

66. Gollnick, P., S. Ishino, M.I. Kuroda, D.J. Henner, and C. Yanofsky. 1990. The *mtr* locus is a two-gene operon required for transcription attenuation in the *trp* operon of *Bacillus subtilis*. Proc. Natl. Acad. Sci. **87:**8726–8730.

67. Gottesman, M., S. Adhya, and A. Das. 1980. Transcription antitermination by bacteriophage lambda N gene product. J. Mol. Biol. **140:**57–65.

68. Grachev, M.A., E.A. Lukhtanov, A.A. Mustaev, E.F. Zaychikov, M.N. Abdukayumov, I.V. Rabinov, V.I. Richter, Y.S. Skolov, and P.G. Christyakov. 1989. Studies on the functional topography of *Escherichia coli* RNA polymerase. A method for localization of the sites of affinity labelling. Eur. J. Biochem. **180:**577–585.

69. Grayhack, E.J., X. Yang, L.F.J. Lau, and J.W. Roberts. 1985. Phage lambda gene Q antiterminator recognizes RNA polymerase near the promoter and accelerates it through a pause site. Cell **42:**259–269.

70. Greenblatt, J., and J. Li. 1981. Interaction of the sigma factor and the *nusA* gene protein of *E. coli* with RNA polymerase in the initiation-termination cycle of transcription. Cell **24:**421–428.

71. Greenblatt, J., and J. Li. 1981. The *nusA* gene protein of *Escherichia coli;* its identification and a demonstration that it interacts with the gene *N* transcription antitermination protein of bacteriophage lambda. J. Mol. Biol. **147:**11–23.

72. Griffith, J., A. Hochschild, and M. Ptashne. 1986. DNA loops induced by cooperative binding of λ repressor. Nature **322:**750–752.

73. Grossman, A.D., J.W. Erikson, and C.A. Gross. 1984. The *htpR* gene product of *E. coli* is a sigma factor for heat-shock promoters. Cell **32:**151–159.

74. Guarente, L.P., and J. Beckwith. 1978. Mutant RNA polymerase of *E. coli* terminates transcription in strains making defective *rho* factor. Proc. Natl. Acad. Sci. USA **75:**294–297.

75. Hamilton, E.P., and N. Lee. 1988. Three binding sites for AraC protein are required for autoregulation of *araC* in *Escherichia coli*. Proc. Natl. Acad. Sci. USA **85:**1749–1753.

76. Hanna, M.M., and C.F. Mears. 1983. Topography of transcription: The path of the leading end of nascent transcript through the *Escherichia coli* transcription complex. Proc. Natl. Acad. Sci. USA **80:**4238–4242.

77. Hawley, D.K., A.D. Johnson, and W.R. McClure. 1985. Functional and physical characterization of transcription initiation complexes in the bacteriophage λ O_R region. J. Biol. Chem. **260:**8618–8626.

78. Hawley, D.K., and W.R. McClure. 1983. Compilation and analysis of *Escherichia coli* promoter DNA sequences. Nucleic Acids Res. **11:**2237–2255.

79. Helmann, J.D., and M.J. Chamberlin. 1988. Structure and function of bacterial sigma factors. Ann. Rev. Biochem. **57:**839–872.

80. Hendrix, R.W., J.W. Roberts, R.W. Stahl, and R.A. Weisberg. 1983. *Lambda II*. Cold Spring Harbor Laboratory, Cold Spring Harbor, N.Y.

81. Hillel, Z., and C.-W. Wu. 1977. Subunit topography of RNA polymerase from *Escherichia coli*. A cross-linking study with bifunctioanl reagents. Biochemistry **16:**3334–3342.

82. Hirschman, J., P.-K. Wong, K. Keener, and K. Kustu. 1985. Products of nitrogen regulating genes *ntrA* and *ntrC* of enteric bacteria activate *glnA* transcription *in vitro:* evidence that the *ntrA* product is a σ factor. Proc. Natl. Acad. Sci. USA **82:**7525–7529.

83. Hochschild, A., N. Irwin, and M. Ptashne. 1983. Repressor structure and the mechanism of positive control. Cell **32:**319–325.

84. Hochschild, A., J. Douhan III, and M. Ptashne. 1986. How λ repressor and λ cro distinguish between O_R1 and O_R3. Cell **47:**807–816.

85. Holben, W.E., and E.A. Morgan. 1984. Antitermination of transcription from an *Escherichia coli* ribosomal RNA promoter. Proc. Natl. Acad. Sci. USA **81:**6789–6793.

86. Hoopes, B.C., and W.M. McClure. 1987. Strategies in regulation of transcription initiation, p. 1231–1240. *In* F.C. Neidhardt, J.L. Ingraham, K.B. Low, B. Magasanik, M. Schaechter, and H.E. Umberger (ed.), *Escherichia coli* and *Salmonella typhimurium:* cellular and molecular biology. American Society for Microbiology, Washington, D.C.

87. Horikoshi, M., C.K. Wang, H. Fujii, J.A. Cromlish, P.A. Weil, and R.G. Roeder. 1989. Cloning and structure of a yeast gene encoding a general transcription initiation factor TFIID that binds to the TATA box. Nature (London) **341:**299–303.

88. Horwitz, R.J., J. Li, and J. Greenblatt. 1987. An elongation control particle containing the *N* gene transcriptional antitermination protein of bacteriophage lambda. Cell **51:**631–641.

89. Houman, F., M.R. Diaz-Torres, and A. Wright. 1990. Transcriptional antitermination in the *bgl* operon of *E. coli* is modulated by a specific RNA binding protein. Cell **62:**1153–1163.

90. Igo, M.M., J.M. Slauch, and T.J. Silhavy. 1990. Signal transduction in bacteria: kinases that control gene expression. New Biol. **2:**5–9.

91. Irani, M., L. Orosz, and S. Adhya. 1983. A control element within a structural gene: the *gal* operon of *Escherichia coli*. Cell **32:**783–788.

92. Ito, K., E. Kohji, and Y. Nakamura. 1991. Genetic interaction between β' subunit of RNA polymerase and arginine-rich domain of *Escherichia coli* NusA protein. J. of Bacteriol., in press.

93. Jin, D.J., and C.A. Gross. 1988. Mapping and sequencing of mutations in the *Escherichia coli rpoB* gene that lead to rifampicin resistance. J. Mol. Biol. **202:**245–263.

94. Jin, D.J., W.A. Walter, and C.A. Gross. 1988. Characterization of the termination phenotypes of rifampicin resistant mutants. J. Mol. Biol. **202:**245–263.

95. Jin, D.J., M. Cashel, D.L. Friedman, Y. Nakamura, W.A. Walter, and C.A. Gross. 1988. Effects of rifampicin resistant *rpoB* mutations on antitermination and interaction with *nusA* in *Escherichia coli*. J. Mol. Biol. **204:**247–261.

96. Jin, D.J., and C.A. Gross. 1989. Three *rpoBC* mutations that suppress the termination defects of *rho* mutants also affect the functions of *nusA* mutants. Mol. Gen. Genet. **216:**269–275.

97. Johnston, D.E., and W.R. McClure. 1976. Abortive initiation of *in vitro* RNA synthesis on bacteriophage λ DNA, p. 101–126. *In* R. Losick and M. Chamberlin (ed.), RNA polymerase. Cold Spring Harbor Laboratory, Cold Spring Harbor, N.Y.

98. Kajitani, M., and A. Ishihama. 1983. Determination of the promoter strength in the mixed transcription system: promoters of lactose, tryptophan and ribosomal protein L10 operons from *Escherichia coli,* Nucleic Acids Res. **11:**671–686.

99. Kashlev, M., J. Lee, K. Zalenskaya, V. Nikiforov, and A. Goldfarb. 1990. Blocking of the initiation to elongation transition by a transdominant RNA polymerase mutation. Science **248:**1006–1009.

100. Kassavetis, G.A., and M.J. Chamberlin. 1981. Pausing and termination of transcription within the early region of bacteriophage T7 DNA *in vitro*. J. Biol. Chem. **256:**2777–2786.

101. Keener, J., and S. Kustu. 1988. Protein kinase and phosphoprotein phosphatase activities of nitrogen regulatory proteins NTRB and NTRC of enteric bacteria: roles of the conserved amino-terminal domain of NTRC. Proc. Natl. Acad. Sci. USA **85:**4976–4980.

102. Kenney, T.J., K. York, P. Youngman, and C.P. Moran. 1989. Genetic evidence that RNA polymerase associated with σ^A factor uses a sporulation-specific promoter in *Bacillus subtilis*. Proc. Natl. Acad. Sci. USA **86:**9109–9113.

103. Kingston, R.E., and M.J. Chamberlin. 1981. Pausing and attenuation of *in vitro* transcription in the *rrnB* operon of *E. coli*. Cell **27:**523–531.

104. Kinsella, L., C.-Y. Hsu, W. Schulz, and D. Dennis. 1982. RNA polymerase: Correlation between transcript length, abortive product synthesis, and formation of a stable ternary complex. Biochemistry **21:**2719–2723.

105. Kirkegaard, K., H. Buc, A. Spassky, and J.C. Wang. 1983. Mapping of single-stranged regions in duplex DNA at the sequence level: single-strand specific cytosine methylation in RNA polymerase-promoter complexes. Proc. Natl. Acad. Sci. USA **80:**2544–2548.

106. Knaus, R., and H. Bujard. 1988. P_L of coliphage lambda: an alternative solution for an efficient promoter. EMBO **7:**2919–2923.

107. Kofoid, E.C., and J.S. Parkinson. 1988. Transmitter and receiver modules in bacterial signaling proteins. Proc. Natl. Acad. Sci. USA **85:**4981–4985.

108. Kramer, H., M. Niemoller, M. Amouyal, B. Revet, B. von Wilcken-Bergmann, and B. Muller-Hill. 1987. *lac* repressor forms loops with linear DNA carrying two suitably spaced *lac* operators. EMBO **6:**1481–1491.

109. Kuhnke, G., H. Fritz, and R. Ehring. 1987. Unusual properties of promoter-up mutations in the *Escherichia coli* galactose operon and evidence suggesting RNA polymerase-induced DNA binding. EMBO **6:**507–513.

110. Kumar, S.A., and J.S. Krakow. 1975. Studies on the product binding site of the *Azotobacter vinelandii* ribonucleic acid polymerase. J. Biol. Chem. **250:**2878–2884.

111. Lagos, R., R.-Z. Jiang, S. Kim, and R. Goldstein. 1986. An extrachromosomal gene product which antagonizes Rho-dependent termination of a bacterial operon. Proc. Natl. Acad. Sci. USA **83:**9561–9565.

112. Landick, R., and C. Yanofsky. 1987. Isolation and structural analysis of the *Escherichia coli trp* leader paused transcription complex. J. Mol. Biol. **196:**363–377.

113. Landick, R., and C. Yanofsky. 1987. Transcription attenuation, p. 1276–1301. *In* F.C. Neidhardt, J.L. Ingraham, K.B. Low, B. Magasanik, M. Schaechter, and H.E. Umbarger (ed.), *Escherichia coli* and *Salmonella typhimurium:* cellular and molecular biology. Amer. Soc. Microbiol., Washington, D.C.

114. Landick, R., J. Stewart, and D.N. Lee. 1990. Amino acid changes in conserved regions of the β-subunit of *Escherichia coli* RNA polymerase alter transcription pausing and termination. Genes and Develop. **4:**1623–1636.

115. Lau, L.F., J.W. Roberts, and R. Wu. 1983. RNA polymerase pausing and transcript release at the Lambda *tR1* terminator *in vitro*. J. Biol. Chem. **258:**9391–9397.

116. Lazinski, D. 1989. The molecular determinants involved in the sequence-specific recognition of RNA signals by bacteriophage antiterminators. Ph.D. Dissertation, University of Connecticut, Storrs.

117. Lazinski, D., E. Grzadzielska, and A. Das. 1989. Sequence-specific recognition of RNA hairpins by bacteriophage antiterminators requires a conserved arginine-rich motif. Cell **59:**207–218.

118. Lee, D.N., L. Phung, J. Stewart, and Landick, R. 1990. Transcription pausing by *E. coli* RNA polymerase is modulated by downstream DNA sequences. J. Biol. Chem. **265:**15145–15153.

119. Lesley, S.A., and R.R. Burgess. 1989. Characterization of the *Escherichia coli* transcription factor α^{70}: localization of a region involved in the interaction with core polymerase. Biochemistry **28:** 7728–7734.

120. Levin, J.R., B. Krummel, and M.J. Chamberlin. 1987. Isolation and properties of transcribing ternary complexes of *Escherichia coli* RNA polymerase positioned at a single template base. J. Mol. Biol. **196:**173–174.

121. Li, S.C., C.L. Squires, and C. Squires. 1984. Antitermination of *E. coli* rRNA transcription is caused by a control region segment containing lambda *nut*-like sequences. Cell **38:**851–860.

122. Lill, U.I., and G.R. Hartmann. 1973. On the binding of rifampicin to the DNA-directed RNA polymerase from *Escherichia coli*. Eur. J. Biochem. **38:**336–345.

123. Losick, R., and J. Pero. 1981. Cascades of sigma factors. Cell **25:**582–584.

124. Lynn, S.P., C.E. Bauer, K. Chapman, and J.F. Gardner. 1985. Identification and characterization of mutants affecting transcription termination at the threonine operon attenuator. J. Mol. Biol. **183:**529–541.

125. Mandal, N., W. Su, R. Harber, S. Adhya, and H. Echols. 1990. DNA looping in cellular repression of transcription of the galactose operon. Genes and Develop. **4:**410–418.

126. McClure, W.R., and C.L. Cech. 1978. On the mechanism of rifampicin-inhibition of RNA synthesis. J. Biol. Chem. **253:**8949–8956.

127. McClure, W.R. 1980. On the mechanism of streptolydigin inhibition of *Escherichia coli* RNA polymerase. J. Biol. Chem. **255:**1610–1616.

128. McClure, W.R. 1980. Rate-limiting steps in RNA chain initiation. Proc. Natl. Acad. Sci. USA **77:**5634–5638.

129. McClure, W.R. 1985. Mechanism and control of transcription initiation in prokaryotes. Ann. Rev. Biochem. **54:**171–204.

130. Menzel, R., and M. Gellert. 1983. Regulation of the genes for *E. coli* DNA gyrase: homeostatic of DNA supercoiling. Cell **34:**105–113.

131. Morgan, E. 1986. Antitermination mechanisms in rRNA operons of *Escherichia coli*. J. Bact. **168:**1–5.

132. Mulligan, M.E., D.K. Hawley, R. Entriken, and W.R. McClure. 1984. *Escherichia coli* promoter sequences predict *in vitro* RNA polymerase selectivity. Nucleic Acids Res. **9:**789–800.

133. Ninfa, A.J., and B. Magasanik. 1986. Covalent modification of the *glnG* product, NRI_{I}, by the *glnL* product, NR_{II}, regulates the transcription of the *glnALG* operon in *Escherichia coli*. Proc. Natl. Acad. Sci. USA **83:**5909–5913.

134. Ninfa, A.J., L.J. Reitzer, and B. Magasanik. 1987. Initiation of transcription at the bacterial *glnAp2* promoter by purified *E. coli* components is facilitated by enhancers. Cell **50:**1039–1046.

135. Ohlendorf, D.H., W.F. Anderson, R.G. Fisher, Y. Takeda, and B.W. Matthews. 1982. The molecular basis of DNA-protein recognition inferred from the structure of cro repressor. Nature **298:**718–723.

136. Ollis, D.L., P. Brick, R. Hamlin, N.G. Xuong, and T.A. Steitz. 1985. Structure of the large fragment of *Escherichia coli* DNA polymerase I complexes with dTMP. Nature **313:**762–766.

137. Pabo, C., and R. Sauer. 1984. Protein-DNA recognition Ann. Rev. Biochem. **53:**293–321.

138. Pavco, P.A., and D.A. Steege, 1990. Elongation by *Escherichia coli* RNA polymerase is blocked *in vitro* by a site specific DNA binding protein. J. Biochem. **265:**9960–9969.

139. Platt, T. 1986. Transcription termination and the regulation of gene expression. Annu. Rev. Biochem. **55:**339–372.

140. Popham, D.L., D. Szeto, J. Keener, and S. Kustu. 1989. Function of a bacterial activator protein that binds to transcriptional enhancers. Science **243:**629–635.

141. Pribnow, D. 1975. Nucleotide sequence of an RNA polymerase binding site at an early T7 promoter. Proc. Natl. Acad. Sci. USA **72:**784–788.

142. Ptashne, M. 1988. How eukaryotic transcriptional enhancers work. Nature **355:**683–689.

143. Ptashne, M. 1989. How gene activators work. Sci. Amer. **260:**41–47.

144. Ptashne, M., A. Jeffrey, A.D. Johnson, R. Maurer, B.J. Meyer, C.O. Pabo, T.M. Roberts, and R.T.Sauer. 1980. How the λ repressor and cro work. Cell **19:**1–11.

145. Raibaud, O., and M. Schwartz. 1984. Positive control of transcription in bacteria. Annu. Rev. Genet. **18:**173–206.

146. Reznikoff, W.S., R.R. Burgess, J.E. Dahlberg, C.A. Gross, M.T. Record, and M.P. Wickens. 1987. RNA polymerase and the regulation of transcription. Elsevier, New York.

147. Rhodes, G., and M.J. Chamberlin. 1974. Ribonucleic acid chain elongation by *Escherichia coli* ribonucleic acid polymerase I. Isolation of ternary complexes and the kinetics of elongation. J. Biol. Chem. **249:**6675–6683.

148. Roberts, J.W. 1969. Termination factor for RNA synthesis. Nature **224:**1168–1174.

149. Roberts, J.W. 1988. Phage lambda and the regulation of transcription termination. Cell **52:**5–6.

150. Roberts, D, B.C. Hoopes, W.R. McClure, and N. Kleckner. 1985. IS10 transposition is regulated by DNA adenine methylation. Cell **43:**117–130.

151. Rosenberg, M., and D. Court. 1979. Regulatory sequences involved in the promotion and termination of transcription. Ann. Rev. Genet. **13:**319–353.

152. Salstrom, J.S., and W. Szybalski. 1978. Coliphage λ*nutL*−*:* a unique class of mutants defective in the site of gene *N* product utilization for antitermination of leftward transcription. J. Mol. Biol. **124:**195–221.

153. Sasse-Dwight, S., and J.D. Gralla. 1988. Probing the *Escherichia coli glnALG* upstream activation mechanism *in vivo*. Proc. Natl. Acad. Sci. USA **85:**8934–8938.

154. Sasse-Dwight, S., and J.D. Gralla. 1990. Role of eukaryotic-type functional domains found in the prokaryotic enhancer receptor factor σ^{54}. Cell **62:**945–954.

155. Schafer, R., R. Kramer, W. Zillig, and H. Cudny. 1973. On the initiation of transcription by DNA-dependent RNA polymerase from *Escherichia coli*. Eur. J. Biochem. **33:**207–214.

156. Schleif, R. 1988. DNA looping. Science **240:**127–128.

157. Schmidt, M.C., and M.J. Chamberlin. 1984. Binding of *rho* factor to *Escherichia coli* RNA polymerase mediated by nusA protein. J. Biol. Chem. **259:**15000–15002.

158. Schmitz, A., and D.J. Galas. 1979. The interaction of RNA polymerase and *lac* repressor with the *lac* control region. Nucleic Acids Res. **6:**111–137.

159. Schulz, W., and W. Zillig. 1981. Rifampicin inhibition of RNA synthesis by destabilization of DNA-RNA polymerase-oligonucleotide complexes. Nucleic Acids Res. **24:**6889–6906.

160. Sharrock, R.A., R.L. Gourse, and M. Nomura. 1985. Defective antitermination of rRNA transcription and derepression of rRNA and tRNA synthesis in the *nusB5* mutant of *Escherichia coli*. Proc. Natl. Acad. Sci. USA **82:**5275–5279.

161. Seigele, D.A., J.C. Hu, W.A. Walter, and C.A. Gross. 1989. Altered promoter recognition by mutant forms of the σ^{70} subunit of *Escherichia coli* RNA polymerase. J. Mol. Biol. **206:**591–604.

162. Siebenlist, U., R.B. Simpson, and W. Gilbert. 1980. *E. coli* RNA polymerase interacts homologously with two different promoters. Cell **20:**269–281.

163. Silhavy, T.J., and J.R. Beckwith. 1985. Uses of *lac* fusions for the study of biological problems. Microbiol. Rev. **49:**398–418.

164. Simpson, R.B. 1979. The molecular topography of RNA polymerase-promoter interaction. Cell **18:**277–285.

165. Sippel, A.E., and G.R. Hartmann. 1970. Rifampicin resistance of RNA polymerase in the binary complex with DNA. Eur. J. Biochem. **16:**152–157.

166. Sokolova, E., M. Ovadis, Z. Gorlenko, and R. Khesin. 1970. Localization of a streptolydigin resistant mutation in the *E. coli* chromosome and effects of streptolydigin in T2 phage development in *stl-r* and *stl-s* strains of *E. coli*. Biochem. Biophys. Res. Commun. **41:**870–876.

167. Sopta, M., Z.F. Burton, and J. Greenblatt. 1989. Structure and associated DNA-helicase activity of a general transcription initiation factor that binds to RNA polymerase. Nature (London) **341:**410–414.

168. Sparkowski, J., and A. Das. 1990. The nucleotide sequence of *greA*, a suppressor gene that restores growth of an *Escherichia coli* RNA polymerase mutant at high temperature. Nucleic Acids Res. **18:**6443.

169. Sparkowski, J., and A. Das. 1991. Simultaneous gain and loss of functions caused by a single amino acid substitution in the beta subunit of *Escherichia coli* RNA polymerase: allelic suppression and conditional lethality. Genetics, accepted for publication.

170. Speckhard, D.C., F.Y.H. Wu, and C.-W. Wu. 1977. Role of the intrinsic metal in RNA polymerase from *Escherichia coli*. *In vivo* substitution of tightly bound zinc with cobalt. Biochemistry **16:**5228–5234.

171. Su, W., S. Porter, S. Kustu, and H. Echols. 1990. DNA-looping and enhancer activity: association between DNA-bound NtrC activator and RNA polymerase at the bacterial glnA promoter. Proc. Natl. Acad. Sci. USA **87:**5504–5508.

172. Sunshine, M., and B. Sauer. 1975. A bacterial mutation blocking P2 phage late gene expression. Proc. Natl. Acad. Sci. USA **72:**2770–2774.

173. Sunshine, M., E. Six, K. Barrett, and C. Calender. 1976. Relief of P2 bacteriophage amber mutant polarity by the satellite bacteriophage P4. J. Mol. Biol. **106:**673–682.

174. Sweetster, D., M. Nonet, and R.A. Young. 1987. Prokaryotic and eukaryotic RNA polymerases have homologous core subunits. Proc. Natl. Acad. Sci. USA **84:**1192–1996.

175. Travers, A.A., R. Buckland, M. Goman, S.S.G. Le Grice, and J.G. Scaife. 1978. A mutation affecting the sigma subunit of RNA polymerase changes transcriptional specificity. Nature **273:**354–358.

176. Tse-Dinh, Y.-C. 1985. Regulation of the *Escherichia coli* DNA topoisomerase I gene by DNA supercoiling. Nucleic Acids Res. **13:**4751–4764.

177. Tsugawa, A., T. Kurihara, M. Zuber, D. Court, and Y. Nakamura. 1985. *E. coli* NusA protein binds *in vitro* to an RNA sequence immediately upstream of the *boxA* signal of bacteriophage lambda. EMBO **4:**2337–2342.

178. Umezawa, H., S. Mizuno, H. Yamazaki, and K. Nitta. 1968. Inhibition of DNA-dependent RNA synthesis by rifamycins. J. Antibiot. (Tokyo), Ser. A. **21:**234–236.

179. Vershon, A.K., S.-M. Liao, W.R. McClure, and R.T. Sauer. 1987. Interaction of the bacteriophage P22 arc repressor with operator DNA.J. Mol. Biol. **195:**323–331.

180. von Hippel, P.H., D.G. Bear, W.D. Morgan, and J.A. McSwiggen. 1984. Protein-nucleic acid interactions in transcription: a molecular analysis. Ann. Rev. Biochem. **53:**389–446.

181. von Hippel, P.H., A. Revzin, C.A. Gross, and A.C. Wang. 1974. Non-specific DNA binding of genome regulating proteins as a biological control mechanism: 1. The *lac* operon: Equilibrium aspects. Proc. Natl. Acad. Sci. USA **71:**4808–4812.

182. Waldburger, C., T. Gardella, R. Wong, and M. Susskind. 1990. Changes in conserved region 2 of *Escherichia coli* σ70 affecting promoter recognition. J. Mol. Biol. **215:**267–276.

183. Warren, F. and A. Das. 1984. Formation of termination-resistant transcription complex at phage lambda *nut* locus: effects of altered translation and a ribosomal mutation. Proc. Natl. Acad. Sci. USA **81:**3612–3616.

184. Watson, J.D., N. Hopkins, J.W. Roberts, J.A. Steitz, and A.M. Weiner. 1987. Molecular biology of the gene. Fourth edition. The Benjamin/Cummings Publishing Co., Menlo Park, California.

185. Wedel, A., D.S. Weiss, D. Popham, P. Droge, and S. Kustu. 1990. A bacterial enhancer functions to tether a transcriptional activator near a promoter. Science **248:**486–490.

186. Whalen, W., B. Ghosh, and A. Das. 1988. NusA protein is necessary and sufficient *in vitro* for phage λ*N* gene product to suppress a ρ-independent terminator placed downstream of *nutL*. Proc. Natl. Acad. Sci. USA **85:**2494–2498.

187. Whalen, W., and A. Das. 1990. Action of an RNA site at a distance: role of the *nut* genetic signal in transcription antitermination by phage λ*N* gene product. New Biol. **2:**975–991.

188. Wold, M.S., J.J. Li, and T. Kelly. 1987. Initiation of simian virus 40 DNA replication *in vitro:* large-tumor-antigen- and origin-dependent unwinding of the template. Proc. Natl. Acad. Sci. USA **84:**3643–3647.

189. Wu, C.-W., F.Y.H. Wu, and D.C. Speckhard. 1977. Subunit location of the intrinsic divalent metal ions in RNA polymerase from *Escherichia coli*. Biochemistry **16:**5449–5954.

190. Wu, F., L. Yarbrough, and C.-W. Wu. 1976. Conformational transition of *Escherichia coli* RNA polymerase induced by the interaction of sigma subunit with core enzyme. Biochemistry **15:**3254–3258.

191. Wulff, D.L., and M. Rosenberg. 1983. Establishment of repressor synthesis, p. 53–73. *In* R.W. Hendrix, J.W. Roberts, R.W. Stahl, and R.A. Weisberg (ed.), *Lambda II*. Cold Spring Harbor Laboratory, Cold Spring Harbor, N.Y.

192. Yager, T.D. and P.H. von Hippel. 1987. Transcript elongation and termination in *Escherichia coli,* p. 1241–1275, *In* F.C. Neidhardt, J.L. Ingraham, K.B. Low, B. Magasanik, M. Schaechter and H.E. Umbarger (ed.), *Escherichia coli* and *Salmonella typhimurium:* cellular and molecular biology. Amer. Soc. Microbiol., Washington, D.C.

193. Yang, X., C.M. Hart, E.J. Grayhack, and J.W. Roberts. 1987. Transcription antitermination by phage lambda gene Q protein requires a DNA segment spanning the RNA start site. Genes and Dev. **1:**217–226.

194. Yanofsky, C. 1981. Attenuation in the control of expression of bacterial operons. Nature (London) **289:**751–758.

195. Youderian, P., S. Bouvier, and M. Susskind. 1982. Sequence determinants of promoter activity. Cell **30:** 843–853.

196. Yura, T., and A. Ishihama. 1979. Genetics of bacterial RNA polymerases. Ann. Rev. Genet. **13:**59–97.

197. Zuber, M., T.A. Patterson, and D.L. Court. 1987. Analysis of *nutR:* a site required for transcription antitermination in phage lambda. Proc. Natl. Acad. Sci. USA **84:**4515–4518.

198. Zuber, P., J.M. Healy, H.L. Carter III, S. Cutting, C.P. Moran, Jr., and R. Losick. 1989. Mutation changing the specificity of an RNA polymerase sigma factor. J. Mol. Biol. **206:**605–614.

5

Bacterial Cell Division

Joe Lutkenhaus

The process of cell division is necessary for the growth of all cells. The mechanism by which eubacteria divide has so far shown little homology to the mechanisms employed by eukaryotic cells. Although this lack of similarity may be a result of our lack of knowledge about these processes it may also be that different mechanisms are involved. If so, it would make cell division a selective target for antimicrobial agents.

I. Cell Division in Bacteria

The following remarks about bacterial cell division come mostly from studies with *E. coli* since this organism has been more intensely studied; however, the principles should be generally applicable even though the details will surely vary.

A. Physiology of Cell Division

The bacterial cell cycle, consisting of chromosome replication followed by cell division, is normally well coordinated such that very few DNA-less cells are produced (76). How these two cell-cycle events are spatially and temporally coordinated has not been fully resolved. Several possibilities have been suggested for positive coupling between these two events and negative coupling systems are known. (See below.) The positive coupling models suggest that some event of the chromosome replication cycle, such as initiation or termination, acts as a signal for the cell division cycle (81). In a novel approach to this question Berlander and Nordstrom (16) manipulated chromosome replication by substituting the chromosomal origin with a plasmid origin in which initiation can be manipulated by temperature. Increasing the frequency of initiation did not dramatically affect cell size although DNA content per cell was increased. Provided nucleoid segregation was occurring normally, this study would argue that the replication cycle and the division cycle are independent of each other and run in parallel; however, both events may be initiated with a common signal that is coupled to the growth rate.

Donachie and Begg (49) have noted that cells divide at the same cell length independent of growth rate or cell shape and have postulated that the attainment of this minimal cell length is necessary to segregate newly replicated nucleoids. These studies suggest that nucleoid segregation acts as a signal for cell division or is at least a requirement for cell division. Studies done on DNA segregation mutants suggest that nucleoid segregation does not regulate the frequency of division but may be involved in localization of the division site (70). Mulder and Woldringh (104) have suggested that the nucleoid can negatively influence cell division, although a mechanism for this is unknown.

Studies examining residual cell division after DNA synthesis is blocked provide clear evidence that only cells that have completed chromosome replication can divide. Inhibitors of protein synthesis give the same result. Thus division, once initiated, can be completed in the absence of any additional DNA or protein synthesis.

B. Morphological Aspects

1. Periseptal and Polar Annuli

Cell division in bacteria occurs by the ingrowth of the cell wall at defined sites in a process that results in a constriction or formation of a septum (27, 28). Through the examination of serial sections of plasmolyzed cells by electron microscopy a structure related to the division event has been observed. Plasmolysis of bacterial cells leads to a shrinkage of the cytoplasm, revealing sites where the cytoplasmic membrane appears attached to the cell wall. It is these attachments (zones of adhesion) that may play a role in the division process. In cells of all lengths, sites of attachment exist at midcell which form two continuous, circumferential rings, referred to as periseptal annuli (56, 96). These periseptal annuli are thought to flank the future division site and segregate it from the rest of the cell. New periseptal annuli appear to arise from the periseptal annuli at midcell and are laterally displaced to one-fourth and three-fourths cell lengths to be used in the next division cycle (33). The periseptal annuli persist as polar annuli after division is completed (reviewed in 41).

The morphological appearance of these zones of adhesion suggests they play a role in the cell division process. Membrane vesicles isolated from sucrose gradients with a buoyant density between cytoplasmic and outer membrane have been identified which may have arisen from zones of adhesion (31). Such vesicles appear as hybrid vesicles based upon analysis of their protein content; however, it has not yet been possible to biochemically identify any unique components. Perhaps morphological analysis of cell division mutants will reveal deficiencies in the formation of such sites and suggest which genes are involved.

2. *Septal Attachment Site*

Examination of septating cells by high-resolution electron microscopy revealed that the leading edge of the septum had a characteristic appearance in which the cytoplasmic membrane appears attached to the cell wall (97). The incorporation of new wall material during septum formation, monitored through diaminopimelic acid (DAP) labelling, occurs predominantly at the leading edge of the septum and presumably coincides with the septal attachment site (142). Although such synthetic activity can be detected as increased labelling, no change in the composition of the peptidoglycan can be detected during septation (42). Completion of septation results in a birth scar which can be seen as an indentation of the cell wall following plasmolysis (97). This structure is only seen at one pole so it must not persist. So far no specific components of the septal attachment site have been identified, but this structure likely contains one or more of the penicillin binding proteins, particularly PBP3.

3. *Penicillin-induced Lysis*

Growing cells are more sensitive to penicillin lysis than nongrowing cells and the lysis occurs at division sites (27, 48, 124). This lysis is preceded by bulge formation which can be inhibited by blocking cell division through induction of the SOS response or through the use of mutations in the cell division gene *ftsZ*. The interpretation of this result is that the initiation of cell division is followed by a step that has increased sensitivity to β-lactams.

C. *Genetic Aspects*

1. *Identification of Cell Division Genes*

Genes involved in cell division (summarized in Table 5.1) have been identified through the isolation of conditional lethal mutations that result in filament formation at the nonpermissive condition. These genes have been designated *fts* for filamenting temperature sensitive. In general two approaches have been used. The first, used by Ricard and Hirota (118), simply involved screening temperature-sensitive conditional lethals for filament formation when grown at the nonpermissive temperature. A modification of this approach by Donachie and co-workers using a temperature-sensitive amber suppressor (92) resulted in the isolation of conditional lethals that were ambers. A second approach used by a number of workers (12, 55, 135, 138) involves mutagenesis and enrichment for filament-forming mutants by filtration after time allotted for expression of the mutant phenotype. This latter approach requires filament formation to be reversible or at least nonlethal during the period at the nonpermissive temperature. These

Table 5.1. Genes that affect cell division in E. coli

Gene	Map location	Comment
ftsl	2	Codes for penicillin-binding protein 3; specifically required for septum formation; also designated *pbpB, sep*
ftsZ(sulB)	2	Functions in initiation of cell division-rate limiting; target for cell division inhibitors, SulA, MinCD, MinC-DicB, SfiC; overproduction leads to a minicell phenotype; cytoplasmic protein
ftsA	2	Component of septum, interacts with PBP3, located in cytoplasmic membrane
ftsQ	2	Required for septation
ftsW	2	Required for septation, probable integral membrane protein that may interact with PBP3
ftsH	69	Can be suppressed by overproduction of PBP3
ftsYEXS	76	*ftsY* gene is homologous to 54K component of eukaryotic signal recognition particle; ftsE is homologous to a family of transport proteins
envA	2	Essential gene; affects cell separation and permability
minB	26	Consists of the *minCDE* genes; affects placement of the septum; inactivation of this locus results in a minicell phenotype; *minCD* encodes an inhibitor of cell division that acts only at the cell poles in the presence of *minE*
sulA(sfiA)	22	Inducible inhibitor of cell division; part of the SOS response, prevents formation of anucleate cells following interuption of DNA replication
dicB	35	Coinhibitor of cell division with *minC*
dicF	35	Small RNA that can inhibit division
sfiC	28	Component of genetic element *e14*; inhibits cell division following DNA damage in strains carrying the element
lon	10	Encodes a protease (La) that degrades SulA
era	55	Essential ras-like protein; loss of *era* results in filamentation
dnaK	0	Heat-shock gene; required for DNA segregation and cell division
pbpA	15	Codes for PBP2; also designated *mrdA*; required for maintenance of cell shape
rodA	15	Codes for an integral membrane protein that is homologous to FtsW; required for cell shape
mreB	71	Affects sensitivity to mecillinam; round cell morphology; negatively affects ftsl expression; slight homology to FtsA

approaches have led to the identification of a number of genes thought to be involved in cell division.

Early it was noted that interfering with DNA metabolism led to filament formation. Inhibition of DNA replication or segregation leads to filament formation through a number of different coupling mechanisms (discussed below). Thus, it was realized that mutations that lead to filament formation may affect cell division indirectly. For example, the *ftsB* mutation is allelic to *nrdB* and filamentation is due to limiting precursors for DNA synthesis (84, 128). Mutations that affect protein secretion can also lead to filament formation (114) probably as a direct result of the improper localization of many important peptidoglycan biosynthetic enzymes and other cell division proteins. (See below.) It is worth noting that *azi* resistance mutations that can lead to filamentous growth lie within the *secA* gene, an essential component of the protein secretion machinery (Oliver, personal communication). Mutations that affect phospholipid biosynthesis can also lead to filamention by disrupting the normal function of the membrane.

2. fts *Genes*

The genetics of bacterial cell division was initiated with the studies of Hirota and co-workers (118). They screened a heavily mutagenized collection of colonies for morphological alterations at the nonpermissive temperature and identified a number of genes involved in cell division. Although no other extensive studies have been conducted, other workers have identified additional genes that fall into this category. A noteworthy feature is that many of the *fts* genes are located in a large cluster at 2 min on the *E. coli* genetic map (Figure 5.1).

a) *ftsI (pbpB, sep)*. The *ftsI* gene codes for one of the four high-molecular-weight penicillin binding proteins, PBP3, that has both transpeptidase and trans-glycosylase activities *in vitro* (73). The amino acid sequence suggests that the protein is anchored to the cytoplasmic membrane through a hydrophobic sequence near the amino terminus and that the enzymatically active domains are present in the periplasmic space (107). This possibility is consistent with recent fusion studies which find only one export signal located near the amino-terminal end (24). Recent studies revealed that PBP3 undergoes posttranslational modification involving cleavage at its carboxy-terminal end (65, 106). The function of this cleavage is not known. Genetic studies and studies done with antibiotics specific for PBP3, such as furazlocillin (22), indicate that this protein is specifically required for septation (125). Since *ftsI* mutants or cells treated with furazlocillin appear to undergo normal cell elongation, the transpeptidase activity of PBP3 does not appear essential for this aspect of cell growth. Although some redundancy exists with respect to the functions of the penicillin binding proteins, none appears to be able to replace the role of PBP3 in septation.

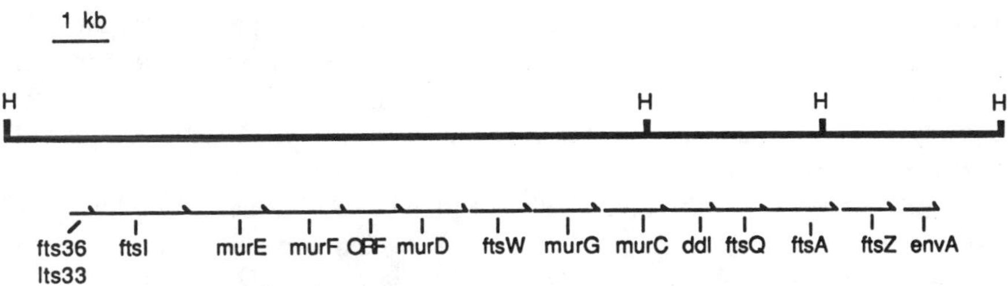

Figure 5.1. The 2-min gene cluster. This region contains five *fts* genes interspersed with genes required for assembling the UDP-N-acetylmuramyl pentapeptide precursor (*murEFDGC and ddl*).

b) *ftsZ*. The *ftsZ* gene was discovered when it was determined that not all previously identified *ftsA* mutations were in the same gene (94). The *ftsZ* gene codes for a stable 40,000-dalton protein that is found mostly in the cytoplasm although a small portion is found in the membrane fraction (80, 144). This gene is of interest for several reasons. Increased expression of this gene leads to increased septation activity which is seen as a minicell phenotype (141). This phenotype has not been reported for any of the other *fts* genes, although this needs to be critically examined. In addition, genetic and biochemical studies have indicated that the FtsZ protein is the target of several division inhibitors (91). These inhibitors are all encoded by genes on the bacterial chromosome that appear to be expressed under various stress conditions or play a role in septal site selection. Recently, we have confirmed that *ftsZ* is an essential cell division gene. This was expected because a thermosensitive lethal mutation had been isolated in the gene. Inhibiting the expression of *ftsZ* results in a rapid inhibition of cell division with eventual cell lysis (21).

c) *ftsA*. The *ftsA* gene was the first identified cell division gene and shows up most frequently in mutant isolation procedures that use size selection (47). The filaments are quite characteristic with indentations along the filament at presumed division sites. The *ftsA* gene codes for a 46,400-dalton protein that localizes to the cytoplasmic membrane although it has no signal sequence and shows no exceptional hydrophobicity (32, 116). Synthesis of this protein for only a short period just prior to cell division is sufficient for division to occur (52). Several studies have indicated that this protein may play a direct role in septation. The study of a temperature-sensitive allele that encodes a protein that is irreversibly denatured suggests that FtsA may be a component of the septum (131). Division sites bypassed during growth at the nonpermissive temperature cannot be immediately reused upon transfer to permissive temperature possibly due to the incorporation of the mutant FtsA protein at these septal sites. An allele of *ftsA* also prevents β-lactam antibiotics from binding to PBP3 at the nonpermissive temperature, suggesting that PBP3 and FtsA may interact directly (132).

d) *ftsQ*. The *ftsQ* gene maps immediately upstream of *ftsA* within the large 2-min cluster (12). This gene product is very poorly expressed, presumably due to the lack of a recognizable ribosome binding site, and has only recently been identified (126). Inspection of the amino acid sequence of the gene product reveals an amino-terminal sequence similar in length and hydrophobicity to that present in PBP3 (120, 145). This suggests that this protein is also anchored to the cytoplasmic membrane through this hydrophobic region with most of the protein extending into the periplasmic space.

e) *ftsW*. The *ftsW* gene is a recently identified gene that lies within the large 2-min gene cluster (74). The gene product shares extensive homology with the products of the *rodA* gene and the *spoVE* gene of *B. subtilis* (72). All of these proteins are thought to be integral membrane proteins. Since the *rodA* gene product appears to be required for function of PBP2 it is tempting to speculate that FtsW might function in cooperation with PBP3. At present, however, there is no direct evidence to support this possibility.

f) *fts YEX* operon and *ftsS*. Analysis of *ftsE* mutations led to the discovery of an operon consisting of three genes mapping near the *rpoH* gene (61). Conditional lethal mutations have been identified in two of these genes although some doubt has been raised about *ftsE* as a cell division gene since filamentation of an *ftsE* mutant is growth rate dependent (129). The *ftsY* gene is homologous to the SRP54 protein, a component of eukaryotic secretion machinery, and its function in *E. coli* remains to be elucidated (17, 121). The FtsE protein has homology to a family of nucleotide binding proteins, all involved in transport. A possibility is that the proteins of this gene complex are involved in the transport of some septal components. The *ftsS* gene has not been studied in detail but is closely linked to the *ftsYEX* operon.

3. *Gene Organization and Expression*

Several of the well-characterized *fts* genes lie within a large cluster of genes at 2 min on the *E. coli* genetic map (Figure 5.1). These genes all have the same transcriptional orientation and are tightly clustered with some of them overlapping (47, 72, 120, 145). Despite this tight juxtapositioning, many of the genes can be expressed from individual promoters as well as additional upstream promoters. For example, the *ftsZ* gene is expressed from at least six different promoters that are scattered throughout a 2-kb region within the coding sequences of upstream genes (95, 120, 127, 145). Also, an additional promoter contributes to *ftsZ* expression as cells enter stationary phase (2). The only transcriptional terminator identified within this 18-kb region is located at the end of the *envA* gene (5).

One possibility for the regulation of periodic events is through periodic gene expression. However, transcription of penicillin binding protein genes and *ftsZ* is continuous (119, 143). This would suggest that the division cycle is regulated through modulation of the activity of proteins rather than through changes in gene expression.

Experiments in which the gene dosage or gene expression was altered indicate that the level of some of the *fts* gene products must be maintained within a fairly narrow range for normal morphology. An increase in FtsZ of severalfold results in a minicell phenotype whereas a slight decrease results in filamentation (141) (Dai and Lutkenhaus, submitted). How the *ftsZ* gene is regulated is not completely known, although its regulation has been examined under several different conditions. The *ftsZ* gene is not autoregulated and does not respond to mutational or antibiotic inhibition of cell division (45, 119) (Dai and Lutkenhaus, submitted).

The expression of *ftsZ*, as a fraction of total protein synthesis, decreases with increasing growth rate; the consequence is that the amount of FtsZ per cell or septum is relatively constant (2). During the cell cycle the *ftsZ* gene is expressed at a constant rate that doubles at a unique point in the cell cycle (119) coincident with initiation of DNA replication. The expression of *ftsZ* appears linked to DNA replication through the DnaA protein (102). Two *dnaA* binding sites are present within the *ftsA* gene and there is a third in the *ftsQ* gene. DnaA may exert its influence through action at these sites. Whether *ftsZ* gene expression is responding directly to DnaA remains to be determined.

Increased expression of other *fts* genes does not appear to affect cell morphology as readily. Increased gene dosage of *ftsL, ftsQ,* or *ftsA* has little affect; however, if *ftsA* expression is further increased in an expression vector cell division is inhibited (139) (Lutkenhaus, unpublished).

4. *Fts Complex?*

It has been suggested several times that some of the *fts* gene products form a complex (44). Interaction among *fts* gene products has been inferred from genetic studies and therefore is tentative until direct evidence is obtained. One observation is that introduction of the *ftsM1* mutation or a DNA fragment containing the *ftsI* gene on a multicopy plasmid can affect survival of *ftsZ, ftsA,* or *ftsQ* mutants (44, 82). Another is that introduction of mutant alleles of *ftsZ* or *ftsI* in the presence of wild-type alleles can lead to filamentation (140). Such a dominant negative phenotype may be due inactive complex formation or to these products titrating out some other essential component. These studies suggest that the balance of various gene products is important but give little information as to possible interactions.

A genetic approach to looking for interactions among gene products, which has been exploited in several systems, involves looking for extragenic suppressors. So far there has been little success in applying this to cell division genes, although there have been some attempts. A mutation in *rodA* has been reported to suppress a temperature-sensitive mutation in the *ftsI* gene (13). However, this suppression is not due to protein-protein interaction, but is due to a modification of the peptidylglycan due to an increase in the amount of PBP5 (11). A specific mutation in *ftsA* has been shown to affect the ability of PBP3 to bind penicillin, suggesting a possible interaction (132). This possibility needs to be further investigated.

It has been reported that the synthesis of the flagellar components requires several functional *fts* genes (110). In various *fts* mutants transcription of the flagellar genes is switched off at the nonpermissive temperature. This is an intriguing observation that remains to be explained.

5. *Other Genes Affecting Cell Division*

Many other genes affect cell division, but do not fall into the *fts* class because they are not essential or because temperature-sensitive mutations have not been isolated within them.

a) *min* genes. One of the oldest-known mutations that affects cell division is the *minB1* mutation (1, 38). This mutation results in the production of minicells and nucleated cells with a heterogeneous cell length distribution due to the presence of elongated cells. The minicells lack chromosomes but appear to be produced by the normal cell division machinery operating frequently at the cell poles (57). Analysis of the *minB* locus has shown that it consists of three genes, *minC, minD,* and *minE,* that act in concert to prevent the division potential from being used at the cell poles (39). The products of *minC* and *minD* act together to inhibit cell division and have the capacity, when in excess of MinE, to inhibit division at all potential sites. When *minE* is expressed at the normal physiological level the inhibition is limited to the cell poles, but if *minE* is overexpressed, relative to *minCD,* the MinCD inhibitor is completely suppressed.

b) *envA*. The *envA* gene is located within the 2-min gene cluster just downstream of *ftsZ*. A mutation in this gene (which is not conditional) leads to sensitivity to antibiotics and the growth of cells in chains (111). At least one suppressor of *envA* has been identified which increases the protein content of the outer membrane and suppresses the permeability defect (64). One attractive possibility, that *envA* might code for N-acetylmuramyl-L-amidase, has been ruled out. The *envA* gene is known to be essential and inhibiting its expression leads to short chains of rounded cells which lyse, while an increased gene dosage appears harmful to cells (5, 51).

c) heat-shock genes. The heat-shock genes are known to influence cell division since mutations in several heat-shock genes can give a filamentous phenotype following thermal stress (46). Improper regulation of the heat-shock response leads to filamentation; failure to turn on the heat-shock genes, as in *rpoH* mutants, results in filamentous growth at high temperatures (133), and failure to turn off the heat-shock response as in *dnaK* mutants leads to filamentous growth (47). Deletion of the *dnaK* gene results in cells that are filamentous and can only grow in a very narrow temperature range; suppressor mutations are rapidly acquired (25). Filamentation due to the loss of *dnaK* can be suppressed by increasing the level of FtsZ or by increasing the gene dosage of a nonessential gene, *dksA* (25, 83). The loss of *dnaK* also affects DNA metabolism, most notably segregation (26), but at least one *dnaK* mutation primarily affects initiation (122). The products of the heat-shock genes are known to be involved in protein folding and unfolding and proteolysis; the absence of these latter functions from the cell might certainly be expected to affect the assembly of any complex required for cell division. The *lon* gene has also been classified as a heat-shock gene and plays a role in resumption of cell division following recovery from the SOS response. (See below.)

d) *ftsM* gene. The *ftsM* gene was originally identified as a cell division gene tightly linked to *leu*. Subsequent analysis has shown that the *ftsM1* mutation is actually a missense suppressor, located in the minor tRNA, *serU,* that would compete for leucine tRNAs to insert serine at leucine (UUG) codons (86). The presence of this missense suppressor leads to a thermosensitive filamentation phenotype as well as some phenotypes reminiscent of *lon*—such as sensitivity to UV and shifts in growth rate. One possibility is that serine for leucine insertion in one or a few key proteins is responsible for the observed effects. It is interesting that increasing the gene dosage of *ftsZ* can suppress these phenotypes, but the possible explanations remain numerous.

6. Are there Additional Cell Division Genes?

Genetic analysis has so far identified a number of genes that are intimately involved in cell division; however, it is certainly possible that many more genes remain to be discovered. Even the most complete screening by Hirota and coworkers (118) failed to turn up such genes as *ftsI, ftsA,* and others. It is likely that screening heavily mutagenized strains introduces multiple mutations some of which may prevent cell growth at the nonpermissive temperature and a filamentous phenotype from being revealed. When mutants are selected based upon their increased length, mutations are usually found in genes in the 2 min region with most of these mutations occurring in the *ftsA* gene (46). Thus, more extensive searches would have to be conducted or novel selections utilized.

Recently, Ishino *et al.* (74) used mutagenized P1 lysates to probe the 2 min region further. This approach resulted in the identification of the *ftsW* gene, which had not been detected in earlier studies utilizing filament selection that had turned up the *ftsQ* gene (12). In addition, this approach resulted in the isolation of two mutations, one resulting in lysis and a second resulting in filamentation, that may lie within a short open reading frame just upstream of *ftsI*. In this case localized mutagenesis of a region known to have cell division genes allowed additional cell division genes to be discovered. In contrast localized mutagenesis of the *ftsYEX* region failed to turn up conditional lethal mutations in *ftsY* even though the gene is essential (62). This is consistent with temperature-sensitive mutations being rare in some genes.

The *era* gene has been characterized from *E. coli* and is similar to the eukaryotic RAS family of genes; the gene product has been shown to bind GTP *in vitro* (100). The *era* gene is essential for growth and diluting the gene product from the cell results in a reduced growth rate and filamentation, suggesting that the Era protein may function in cell division.

Schmid *et al.* (123) extensively mutagenized *S. typhimurium* to isolate temperature-sensitive mutations to identify essential genes. They found only 115 essential targets of which 17% showed some degree of filamentation at the nonpermissive temperature. It is not known what fraction showed abnormal nucleoid segregation as this would indicate the fraction affected in DNA metabolism. This study points out two things: (1) the small number of essential genes found in this type of study (the 18-kb segment in Figure 5.1 alone has 12 essential genes) and (2) the relative high frequency of filamenting mutants.

II. DNA Replication and Cell Division

Inhibition of DNA replication in any one of a number of different ways results in inhibition of cell division. The inhibition is due to the induction of specific inhibitors of cell division and by a less-well-understood mechanism. Inhibition of DNA segregation does not result in complete inhibition of cell division but can lead to abnormal location of the septum and the production of DNA-less cells. It has also been proposed that termination of chromosomal replication can give a positive signal to initiate septation (81), although more recent studies appear to rule this out (16).

A. The SOS-response-mediated Inhibition of Cell Division

Many kinds of DNA damage lead to induction of the SOS response in *E. coli* (35, 88). This response consists of the increased expression of at least 15 genes associated with DNA repair. These genes are induced as a result of the activation of RecA by DNA damage which in turn mediates cleavage of the LexA protein, repressor of the SOS response genes. Associated with SOS induction is inhibition of cell division, which is lethal in *lon* mutants. The inhibition of cell division is due to the induction of one of the SOS genes, the *sulA* (*sfiA*) gene, which alone is sufficient to inhibit cell division (68, 69). The SulA inhibitor is very unstable and rapidly disappears once its synthesis ceases; however, in *lon* mutants *SulA* is stabilized, resulting in filamentous death (103). The *sulA* gene can be disrupted without a noticeable effect on the physiology of the cell in the absence of DNA damage. The expression of *sulA* is negligible under normal growth conditions but increases dramatically following induction of the SOS-stress response. These properties of *SulA,* inducibility and lability, make it an efficient transient inhibitor following DNA damage. An additional inducible inhibitor is present in strains of *E. coli* carrying the genetic element *e14*. In these strains the *sfiC* gene is induced by DNA damage and inhibits cell division (36).

B. Non-SOS-mediated Inhibition of Cell Division

It is clear that SulA (and SfiC) is not the only means of inhibiting cell division following DNA damage although it is the most effective (29). This can be seen in mutants where the SOS response is rendered noninducible (*recA* or *lexA* mutations) or where the *sulA* (and *sfiC*) gene is disrupted (76). In such mutants cell division is reduced, but not blocked, following SOS-inducing treatments. Little is known about this mechanism of inhibition. It is possible that some aspect of chromosome replication, perhaps segregation, is needed for a positive signal for cell division (81), 104) or that another unidentified inhibitor is activated. From the studies of Jaffe *et al*. (76), it is clear that this second coupling mechanism is not as efficient as the SOS mechanism since DNA-less cells are produced after inhibition of DNA synthesis in mutants unable to synthesize SulA. For this uncoupled cell division to occur, an intact cAMP-CRP (cAMP Receptor Protein) system must be present (76). In its absence little division occurs.

C. DNA Segregation and Cell Division

Several DNA partition mutants have been reported; however, further examination reveals that they are affected in DNA metabolism, which affects segregation. In temperature-sensitive DNA gyrase mutants, DNA segregation does not take place; however, net DNA synthesis can continue for some time at the nonpermissive temperature and the DNA can be visualized as a large body, usually near the center of the cell (115). Under such conditions cell division still occurs but septa are placed at sites that are displaced from the DNA mass (70, 105). As a result, DNA-less cells are produced of various sizes. What is interesting is that the frequency of septation appears normal for the DNA and mass increase, so division is not immediately inhibited, but the localization of the division site is uncoupled (70, 71).

Recently, Hiraga and colleagues (67) utilized a novel screen to look for mutants that produce DNA-less cells. Besides *min* mutants, they isolated mutations at the *tolC* locus and *mukB*. Such mutants produce a significant amount of normal-sized DNA-less cells. In such mutants it appears that cell division occurs in the absence of proper nucleoid segregation. The *tolC* locus codes for an outer membrane protein and mutations at this locus are known to be quite pleiotropic, altering sensitivity to colicins and detergents; however, these phenotypes can be separated from the effect on segregation. Recently, it has been reported that *tolC* mutations can affect DNA supercoiling, probably indirectly, and it may be this property of *tolC* mutations that affects nucleoid segregation and production of DNA-less cells (53). The *mukB* gene represents a newly identified gene. It encodes a large protein consisting of 1,534 amino acids with a structure reminiscent of myosin (108). Perhaps the MukB protein can provide force for chromosome segregation.

The *min* mutant has been described primarily as a mutant that has abnormal localization of the septal site. Recent analysis of the segregation of DNA in *minB* mutants reveals that segregation may also be affected. One study found abnormal nucleoid segregation in 10–20% of the cells (77) and another study found that nucleoid segregation was retarded in all cells (105). This raises the possibility that minicell formation is the result of inefficient DNA segregation. However, *minB* can not be absolutely required as *minB* deletion mutants grow quite well.

These various observations on DNA segregation and septation suggest that interference in DNA segregation results in an uncoupling of segregation and division. In the absence of proper DNA segregation, division still takes place but the localization of the septal site is perturbed. The septal site is no longer located between the replicated nucleoids but is placed in an anucleate region which may be somewhat random or be possibly preferentially at the poles (70, 104, 105).

This also points out that cells can survive with a fairly high frequency of improper septal localization.

III. Cell Shape and Cell Division

In an organism with a rigid cell wall, cell shape and cell division are intimately connected. Disruption of the shape of the cell will ultimately delay or inhibit division. In bacteria the peptidoglycan is the shape-determining component of the cell wall and the shape appears to result from the synthetic activities of the penicillin binding proteins (PBPs). It has been suggested that there may be two competing activities; one for cell elongation and one for cell division. β-lactams lead to cell death by their inhibition of one or more of the various PBPs; the mechanism can be quite complex depending upon the affinities for the various PBPs.

A. The High-molecular-weight Penicillin Binding Proteins

The PBPs are present in the cytoplasmic membrane and covalently bind β-lactams (125). In *E. coli* there are at least seven distinct PBPs. The four higher-molecular-weight proteins are essential for cell growth and are thought to be the lethal targets of the β-lactams whereas the lower-molecular-weight PBPs are not essential. The role of the higher-molecular-weight PBPs in the growth of the cell was elucidated through the use of genetics and antibiotics (125). The results can be summarized as follows: PBP1a and PBP1b are necessary for overall growth and inhibition of their activity leads to lysis; PBP2 is necessary for maintenance of rod shape and inhibition of its activity leads to coccal-shaped cells; PBP3 is required for septation and inhibition of its activity leads to filamentation.

Although the postulated roles for the high-molecular-weight PBPs are supported by several lines of evidence, their role in cell growth may not be as simple as summarized above. For example, although inhibition of PBP1a and PBP1b leads to cell lysis, the lytic event appears to be coupled to cell division; lysis is initiated in the cell cycle coincident with the initiation of septation (43). This may indicate a crucial role for these PBPs in initiation of septation. It has also been observed that inhibition of these PBPs along with inhibition of PBP2 leads to immediate lysis. One possibility is that PBP2 can fill in for the PBP1s during cell elongation but not during septation.

The high-molecular-weight PBPs show regions of homology that extend over their transglycosylase (amino terminal) and transpeptidase (carboxy terminal) domains (3). Each of these PBPs contains a hydrophobic segment near their amino terminal end that acts as a signal sequence. This sequence may not be cleaved and could therefore act not only to transport the protein to the periplasm but also to anchor it to the cytoplasmic membrane. Also of note is the length of sequence that precedes the hydrophobic segment; for most secreted proteins this portion of the signal sequence is quite short and contains a few positive charges. PBP1b contains a highly charged segment that is 63 amino acids in length, while PBP2 and PBP3 contain segments of 20 and 23 residues, respectively. PBP1a, an exception, contains a sequence that is only five residues in length. It is tempting to speculate that these segments, located on the cytoplasmic side of the membrane, may interact with cytoplasmic components that may regulate their activity. The amino-terminal-charged portion of the FtsQ signal sequence is 24 residues in length (145). Thus, the FtsQ protein resembles PBP2 and PBP3 in its localization and may play some role in peptidoglycan metabolism.

There is some evidence that the PBPs are not randomly distributed in the cytoplasmic membrane. Fractionation of membrane vesicles by several techniques shows that the higher-molecular-weight PBPs are present in vesicles distinct from the lower-molecular-weight PBPs (87).

1. PBP3

The role of PBP3 in cell division is very well established. It does not appear to be involved in initiation of division but is required during the subsequent steps (113). PBP3 biochemistry is complex because of a number of posttranslational modifications that it undergoes. Although it contains the consensus sequence for modification and processing of lipoproteins, only a small fraction appears to be modified along this route. Essentially all of the protein is proteolytically processed at the carboxyl-terminal end and a temperature-sensitive mutant has been identified in which this processing is defective (65). However, it is not clear whether the temperature sensitivity of this mutant is due to the defect in processing. The function of the posttranslational modifications remains to be elucidated. The *ftsH* mutant can be suppressed by increasing the gene dosage of *ftsI* (54). Perhaps the *ftsH* mutant is deficient in PBP3 localization or stabilization—defects which can be suppressed by an increased level of PBP3.

2. PBP2

The target of the antibiotic mecillinam is PBP2 and treatment of cells with this antibiotic results in round cells which eventually lyse (125). Resistance to mecillinam is quite complex and can be obtained by mutation at several different loci besides *pbpA* (the gene for PBP2). These loci include *rodA, lov,* the *mre* locus, and genes involved in the cAMP-CRP complex (23, 112, 136). PBP2 in conjunction with RodA is required for cell elongation; PBP2 has transpeptidase and transglycosylase activities and RodA appears to be required for the latter activity (75). Temperature-sensitive mutations in the *pbpA* gene result in spherical cell morphology but are not always lethal (112). Complete inactivation of the *pbpA* gene results in inhibition of cell division as well as cell elongation, suggesting that PBP2 may have a role in both processes (112). Interestingly, the *pbpA* gene is not essential when combined with a *cya* or *lov* mutation, suggesting that these mutations modify the peptidoglycan machinery such that the requirement for PBP2 is bypassed. Another possibility is that PBP2 is only essential at fast growth rates and the decrease in growth rate that accompanies *cya* or *lov* mutations makes PBP2 dispensable.

The cAMP-CRP complex has a pleiotropic affect on cell division. In the absence of cAMP cells grow slower but cellular morphology is not drastically affected (37). In the *fic–1* mutant, however, cell division is inhibited by cAMP and perhaps even in wild-type cells when cAMP levels are sufficiently elevated (134). In the absence of cAMP, filamentation of the *ftsZ84* (Ts) mutant is reduced and the production of anucleate cells following inhibition of DNA synthesis does not occur (76). These results, along with dispensability of PBP2 in a *cya* mutant, suggest that a gene or genes regulated by cAMP modifies cell wall physiology in some way.

Mutations at the *mre* locus can either increase or decrease sensitivity to mecillinam (136). The *mre* locus is complex and consists of a number of genes, one of which, *mreB*, has homology to *ftsA* over the latter halves of the two genes (46). Deletion of the *mre* locus is not lethal but results in round cell morphology which is associated with higher levels of PBP3 (137). It is not known if the higher levels of PBP3 directly affect cellular morphology, however.

B. Cell Elongation vs. Cell Division

Schwarz *et al.* (124) suggested that *E. coli* may contain two distinct peptidoglycan synthetic systems: one for elongation and one for septation. A number of observations suggest that these two systems may be competing and alternating during the cell cycle, although it has not been possible to detect biochemical differences in the peptidoglycan (42). Inhibition of cell elongation by mecillinam in some division mutants can stimulate division activity (30). Elevation of PBP5 induces a round cell morphology (101) and can suppress the temperature sensitivity of several *ftsI* mutations (although cell division is still dependent on the residual PBP3 activity), indicating that this increase may favor the septation pathway (11). This indirect suppression of *ftsI* (Ts) is explained as follows: since PBP5 is a D-alanine carboxypeptidase I, elevating its level will increase the proportion of tetrapeptides to pentapeptides and ultimately of tripeptides to pentapeptides (due to carboxypeptidase II activity) in the peptidoglycan. This alteration of the peptidoglycan favors cell division because the tripeptide is thought to be a favored acceptor for transpeptidation by PBP3, whereas the pentapeptide is a favored acceptor for cell elongation by PBP2. If this hypothesis is correct it means that variations in activity of carboxypeptidases, especially carboxypeptidase II, may be instrumental in switching from cell elongation to cell division. It has been reported that carboxypeptidase II activity varies during the cell cycle (9). The gene for this enzyme has not been isolated.

C. Studies on Cell Shape, Division, and Segregation

The interplay between shape and division has been investigated by combining a cell shape mutation with various division mutations. Such double mutants give characteristic shapes at the nonpermissive temperature (10). Cell shape mutations (*pbpA* [Ts] or *rodA* [Ts]) cause cells to grow and divide as spheres. Adding an *ftsQ* (*Ts*), *ftsA* (*Ts*), or *ftsI* (*Ts*) mutation results in swollen rods with indentations along their length as though division was initiating but unable to be completed. In contrast, if an *ftsZ* (*Ts*) mutation is present, the cells grow into elipsoids as though division cannot be initiated. The same result is achieved if the SOS response is induced (thought to block *ftsZ* function) in a cell shape mutant. These studies suggest that *ftsZ* functions at an earlier step than the other cell division genes examined.

Investigation into nucleoid segregation during a rod-to-sphere transition suggests that maintenance of cell shape is necessary to ensure orderly nucleoid segregation (50). During the transition between cell shapes, segregation and division are delayed for several generations until the spherical cells reach a diameter that is about the same length at which rod-shaped cells initiate division. It appears that during the transition, nucleoid segregation and division are delayed due to the shape alterations. The mechanisms involved are unknown.

IV. Inhibitor Genes of Cell Division

The expression of any one of several genes can lead to inhibition of cell division. In addition, an inhibitor of cell division is involved in the process of selecting septal sites. These inhibitors appear to act by inhibiting the activity of FtsZ (Figure 5.2).

A. sulA (sfiA)

The *sulA* gene was identified by mutations that could suppress the sensitivity of *lon* mutants to SOS-inducing conditions (59, 60, 63, 78). Such mutations likely inactivate the *sulA* gene since disruptions of the gene also suppress *lon*. These observations led to the hypothesis that *sulA* encoded an inhibitor of cell division that was activated by DNA damage. This has been confirmed by showing that expression of *sulA* is part of the SOS response (68) and that expression of *sulA* under the control of the *lac* promoter is sufficient to inhibit cell division (69). One important feature of inhibition by SulA is that it is completely reversible (99). This was shown by adding an inhibitor of protein synthesis to filaments formed as a result of *sulA* induction and finding that these filaments, after a short lag, divide into normal-sized cells. Blocking protein synthesis leads to the rapid disappearance of SulA through proteolysis. The observation that the filaments divide indicates that SulA does not irreversibly inactivate any component of the

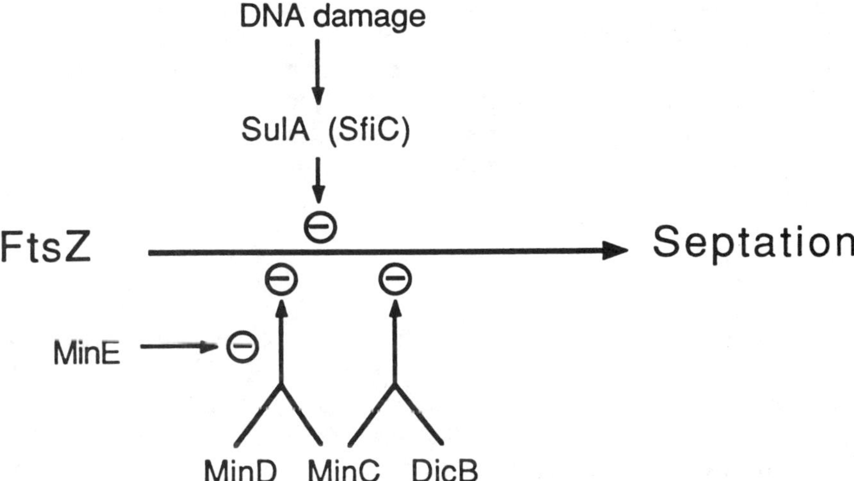

Figure 5.2. Inhibition of cell division. FtsZ is involved in the earliest step of cell division and can be inhibited by several inhibitors. The inhibitors include *sulA*, which is induced by DNA damage; *minCD*, which, in the presence of *minE*, inhibits cell division at the cell poles; and *dicB*, which can form an inhibitor with *minC*.

septation machinery. Possibly SulA can directly block septation or SulA causes some enzymatic modification that can be readily reversed.

B. sfiC

The *sfiC* locus was discovered when it was noted that some strains filamented in the presence of DNA damage even if *sulA* was inactivated (36). This inhibition was characterized by its independence of *lon* and its irreversibility. The comparison of strains containing or lacking the inhibitor led to the discovery that the locus was contained on an excisable genetic element designated *e14* (98). This is a large genetic element (14 kb) that excises inefficiently from the chromosome following DNA damage. It appears to behave to some extent like a cryptic prophage, which has led to the hypothesis that the *sfiC* locus may resemble the λ*kil* gene.

C. minCD

Elucidation of the *minB* locus revealed that it consists of three genes, two of which, *minCD*, function together to inhibit cell division (39). The inhibitory nature of the *minCD* genes is revealed when they are overexpressed, resulting in extensive filamentation. Under physiological conditions, the function of the *minB* locus is to ensure that division takes place at midcell by inhibiting polar sites.

This role of *minB* is supported by deletion of the entire *minB* locus, which results in the classic minicell phenotype, random placement of the septum at potential sites. The specificity of inhibition at the cell poles by *minB* is dependent upon *minE,* the third member of the *minB* locus, since cell division is inhibited at all sites in its absence. How this site selectivity is achieved is not clear although MinE can antagonize MinCD since an excess of MinE results in the minicell phenotype.

D. dicB *and* dicF

The *dic* locus was discovered as a result of looking for temperature-sensitive mutations that mapped near the chromosomal terminus of DNA replication (14). What initially looked like an *fts* gene turned out to be a repressor gene (*dicA*) of an operon that contains two loci, *dicB* and *dicF,* that inhibit cell division when expressed. A clue as to the potential mechanism of one of these inhibitors was the minicell phenotype of suppressors of *dicB* expression (85). Thus, inactivation of the *minB* locus alleviated inhibition of division mediated by the *dicB* gene, indicating that inhibition of cell division by *dicB* was in some way dependent on the *minB* locus. Subsequent analysis revealed that *dicB* can substitute for *minD* to form an active inhibitor with *minC;* the *minC-dicB* complex is *minE* resistant (40). The regulatory region of the *dicB* operon is complex and is similar to P22 phage with the *dicA* gene being similar to the C2 immunity repressor (15). It is not known what physiological stress might induce this locus to inhibit cell division.

The second inhibitor present in the *dicB* operon is *dicF,* which consists of a small, untranslated RNA. This RNA is processed from a larger transcript and may interact with some other RNA (possibly as an antisense RNA) to inhibit division; the mechanism by which it inhibits cell division is unknown, however.

E. Inhibition Mechanism

A possible mechanism by which *sulA* might inhibit cell division was suggested by the isolation of suppressors of *lon's* sensitivity to DNA damage. Two classes of *lon* suppressors were isolated. One, *sulA,* mapped in a nonessential gene which turned out to be an inhibitor of cell division. The second, *sulB* (*sfiB*), appeared to map in an essential gene since some *sulB* mutants had a filamentous phenotype at high temperatures (63). This suggested a model in which the *SulA* inhibitor might interact with an essential cell component and that *sulB* mutations altered that component such that it was resistant to SulA. Subsequent results lend support to this model. Two *sulB* mutations were located within the essential cell division gene, *ftsZ,* and two *sfiB* mutations also appeared to map in this gene (79, 90). Recently, a number of these *sulB* mutations were sequenced and their *ftsZ* location was confirmed (19). These mutations appear scattered within the *ftsZ* gene and have been redesignated *ftsZ* (Rsa—resistant to *SulA*).

Biochemical evidence for an interaction between FtsZ and SulA has come from experiments conducted to examine the half-life of SulA. It has been observed that the half-life of SulA is longer in the presence of FtsZ than in the presence of FtsZ114 (Rsa) or FtsZ9 (Rsa) (22, 84). Thus, an interaction between SulA and FtsZ might shield SulA from proteolytic degradation by Lon; this interaction does not occur between SulA and FtsZ114 (Rsa). Another possible interpretation, that the mutant form of FtsZ promotes degradation of SulA, does not appear to be correct (Bi, and Lutkenhaus, unpublished).

Recently, novel *ftsZ* (Rsa) mutations were isolated by mutagenesis of *ftsZ* cloned on a low copy vector and selected for by suppression of *lon* (19). In this procedure a second copy of *ftsZ* is present on the chromosome, thereby allowing survival of more drastic changes in *ftsZ*. At least one of the mutations isolated by this approach, *ftsZ3*, cannot support cell growth and requires the presence of a wild-type copy of *ftsZ*. Cells containing both of these alleles grow best when the wild-type allele outnumbers the mutant allele. Nonetheless this allele of *ftsZ* can still confer the Rsa phenotype (in the presence of a wild-type copy) and the cells appear normal in the presence of SulA as long as the mutant allele is not in excess. If the two alleles were independent and SulA inhibited the wild-type FtsZ, that would leave cells with the mutant FtsZ3 (Rsa) protein for cell growth. Therefore, to explain this normal morphology in the presence of SulA it has been proposed that the active form of FtsZ is a multimer and the two allele products form a mixed multimer in which the presence of the wild-type FtsZ allows cell growth and the presence of FtsZ3 (Rsa) confers resistance to SulA.

In addition to mutational alterations of *ftsZ* that can suppress *lon* sensitivity to DNA damage, overexpression of the wild-type *ftsZ* can also suppress this aspect of the *lon* phenotype (93). An increase in FtsZ of four- to sevenfold can completely suppress most *lon* mutants for sensitivity to an SOS inducer like nitrofurantoin although this increase in FtsZ is sufficient only to partially suppress a null allele of *lon*. This compensation of FtsZ for SulA can also be seen in another way. A multicopy plasmid containing *sulA* cannot be introduced into *lon* mutants; however, if the level of FtsZ is raised severalfold the plasmid can be introduced (93).

Recently another gene has been discovered that can also influence the FtsZ-SulA interaction. A mutation, designated *cfcA1*, was isolated as a suppressor of a *dnaB* (Ts) mutation, and found to suppress *lon* and partially suppress filamentation due to a temperature-sensitive *ftsZ* allele (109). It was postulated that the *cfcA* gene may interact with *ftsZ*.

Those *ftsZ* (Rsa) mutations that have been tested can also suppress the inhibition by *sfiC*. One class of *ftsZ* (Rsa) mutations (*sfiB*) was isolated in conditions that expressed *sfiC*. It may be that all *ftsZ* (Rsa) mutations are resistant to *sfiC*.

Overexpression of *ftsZ* can suppress SulA and can also suppress inhibition mediated by either the *minCD* inhibitor or the *minC-dicB* inhibitor (40). An increase of FtsZ of four- to fivefold can suppress the inhibitory effect of *minCD* expressed from the *lac* promoter (20). The *ftsZ* (Rsa) mutations can also suppress inhibition by *minCD*. Although most of the *ftsZ* (Rsa) mutations are not completely resistant to *minCD* they can suppress *minCD* lethality at a level where the wild-type FtsZ cannot. Two of the *ftsZ* (Rsa) mutations appear to be very resistant to *minCD* since there is no detectable morphological effect of *minCD* induction. In addition, these alleles of *ftsZ* produce a minicell phenotype even though the level of FtsZ is not significantly increased.

Since two different inhibitors, *sulA* and *minCD*, both appear to inhibit cell division through a common target, this raised the possibility that these inhibitors might act through each other. This is clearly not the case. Either inhibitor can act in the absence of the other, but both inhibitors act directly on FtsZ (20, 40) despite a lack of sequence similarity. If FtsZ is only functional as a multimer, then the mode of action of these inhibitors may be to prevent multimerization.

V. Division Potential

Rod-shaped cells such as *E. coli* appear to grow in cell length and initiate division at a time coincident with completion of chromosome replication. Cells appear to initiate division at the same cell length independent of the growth rate (49, 50). From these physiological studies it is obvious cells divide once per cell doubling, but it is not clear what dictates the frequency of division or what the limiting component is. It is certainly possible that it is the formation of the nascent site that is rate-limiting under normal conditions rather than the completion of chromosome replication. However, the realization, through the study of minicell and DNA segregation mutants, that the cell poles can under some circumstances act as division sites means that one can study limiting factors other than site formation.

The observation that an increase in the level of FtsZ can induce minicell formation suggested that FtsZ may be rate-limiting for division (140). Since minicell formation in the presence of increased FtsZ was not accompanied by an increase in cell length, it indicated that more than one division event was occurring per cell cycle—one at the medial site and one at the cell pole (Figure 5.3).

In the *minB* mutant it appears that the formation of division sites is also not limiting; longer cells are present in the population that presumably have bypassed internal sites and the poles are available for septation. An explanation for the presence of these elongated cells is that division potential is limited and that using it at the cell poles occurs at the expense of the internal sites (130). Thus, the increase in the average cell length of the minicell mutant is a reflection of the frequency with which minicells are produced. If this model is correct, increasing the division potential in the minicell mutant could compensate for the availability of the polar sites and reduce the average cell length. This is what is observed when the level of FtsZ is increased severalfold; the average cell length is reduced

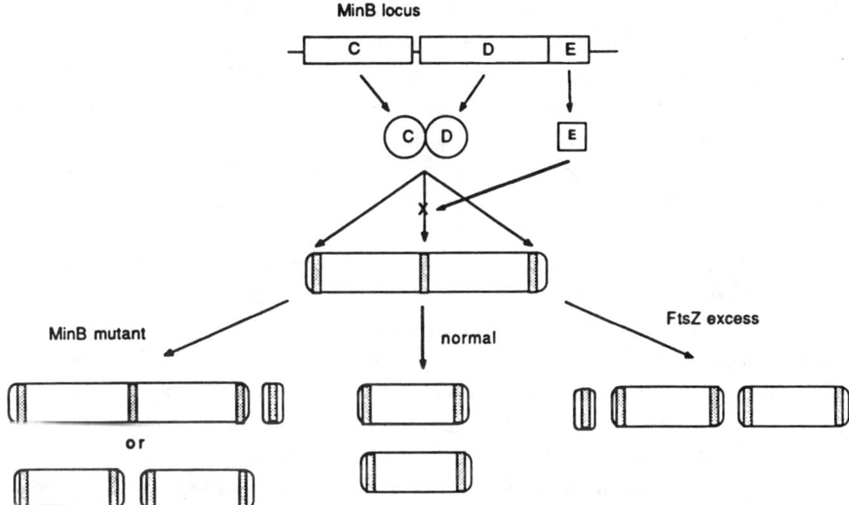

Figure 5.3. Frequency of division and site selection. In the normal situation the MinCD inhibitor, in the presence of MinE, blocks polar sites and division occurs at the medial site. In the absences of *min* function division can take place at any site, but division potential is limited so only one division occurs. Excess FtsZ can overcome *min* inhibition at the poles and increase the division potential such that more than one division can occur.

to normal and, at higher levels of FtsZ, to even less than normal (18). This is consistent with FtsZ being rate-limiting under these conditions. It is possible that increasing the level of FtsZ may result in the increase of other gene products that would otherwise be rate-limiting.

VI. Evolution of Cell Division Genes

A. PBPs

It would appear that PBPs are present in all bacteria that have rigid cell walls. PBPs have been demonstrated either by directly screening for them or by inference from the sensitivity of organisms to β-lactam antibiotics. Although PBPs in other organisms have not been studied in detail, the prevalence of these proteins in most eubacteria and their absence in eukaryotes make them selective targets for antimicrobial agents.

B. Other Cell Division Genes

Cell division mutants isolated in *S. typhimurium* and *B. subtilis* have not been characterized to any great extent genetically. However, evidence for an FtsZ protein in many diverse bacterial species has been found by Corton *et al.* (34)

using antisera against *E. coli* FtsZ. Subsequently, Beall *et al.* (8) cloned the *ftsZ* homolog from *B. subtilis* and found that one of the better-characterized cell division mutations of *B. subtilis, ts1,* mapped in this gene (66). The *ftsZ* gene product from *B. subtilis* is almost 50% identical to its *E. coli* counterpart and is among one of the more conserved genes between these two species (8). Although the *ftsZ* gene is quite widely distributed among the eubacteria, antisera against the *E. coli* FtsZ did not detect an antigen in six different species of *Mycoplasma,* which lack rigid cell walls, or in several eukaryotic species (Beall and Lutkenhaus, unpublished observations). Although such negative results can hardly be considered conclusive, it is possible that the presence of *ftsZ* coincides with the presence of a rigid cell wall containing peptidoglycan.

In *B. subtilis* a homolog of the *E. coli ftsA* gene was found just upstream of *ftsZ* as it is in *E. coli* (8). These gene products have 35% amino acid identity. A temperature-sensitive sporulation mutation, *spo279* (Ts), is located in this gene and affects the expression of at least some *spoII* genes (Leighton, personal communication). Examination of the effects of this *spo279* (Ts) mutation shows that vegetative division is also affected; cells are filamentous, but viability is not dramatically affected (8). It appears likely that the *ftsA* gene in *B. subtilis* may be involved in vegetative division and also in some aspect of sporulation. Recent analysis of the *ftsZ* gene in *B. subtilis* revealed that it is also required for sporulation (7). Reducing the expression of *ftsZ* blocks sporulation by blocking formation of the asymmetric septum, suggesting that FtsZ is required for initiation of this septation event as well.

The large cluster of genes found at 2 min in *E. coli* is not preserved intact in *B. subtilis*. Although the *ftsA* and *ftsZ* genes retain their tandem arrangement in the two organisms, the immediate flanking genes have no homology (6, 8). Using Southern analysis investigators have observed that the arrangement of the 2-min cluster of *E. coli* is preserved in other enteric bacteria and in some other gram negatives as well (Corton and Lutkenhaus, unpublished).

C. Inhibitors of Cell Division

The *sulA* gene has been characterized from several enteric species, indicating that this nonessential gene confers a selective advantage under certain conditions (58). The *sulA* genes from these different species are highly homologous except at the 5′-end of the coding sequence. An SOS-like response is present in *B. subtilis* which includes inhibition of cell division (89). In contrast to *E. coli,* however, where the inhibitor gene is part of the SOS regulon, the inhibition of cell division in *B. subtilis* is observed following DNA damage but is not part of the SOS regulon. It remains to be determined whether this organism has an inducible inhibitor of cell division homologous to *sulA*.

Minicells have been observed in organisms other than *E. coli,* suggesting that the capacity for polar division events may be a relatively common feature of rod-shaped bacteria (117). This implies that the *min* genes may also be conserved, although this remains to be determined.

VII. Summary and Future Prospects

A. Summary

The process of cell division is quite complex, involving temporal and spatial control such that the division site is located at the midpoint of the cell between segregated chromosomes. Studies to date show that chromosomal segregation influences the localization of the division site but does significantly affect the frequency of septation; DNA-less cells are readily produced in the absence of normal chromosome partitioning. In contrast, inhibition of DNA replication results in inhibition of division with little production of DNA-less cells. Thus chromosomal segregation is coupled to site localization and chromosome replication is coupled to site utilization. The *minB* locus, encoding a division inhibitor, MinCD, and a protein that regulates the inhibitor complex, MinE, affects site localization, but not utilization. Localization is effected by inhibiting polar sites but not the site at midcell.

The process of cell division involves the inward growth of the cell wall at the division site following a priming or triggering event. The triggering event is not known but it may involve FtsZ, which functions in the earliest known step committed to cell division. The level of FtsZ regulates the frequency of septation. Inhibition of FtsZ activity by mutation, by induction of the SOS response, or by limiting its expression prevents the appearance of penicillin-sensitive sites or any attempts at constriction. After the FtsZ step the other *fts* gene products participate along with PBP3, which plays an essential and unique role in completion of the septum.

B. Future Prospects

The biosynthesis of the cell wall has been one of the major targets for antibiotics. Synthetic efforts have primarily targeted the penicillin binding proteins. As the three-dimensional structure of these proteins, especially the active site, is elucidated it will further enhance drug design. The latter approach has been hampered because of the low abundance of the high molecular weight PBPs in the cell and their insolubility. However, a soluble form of PBP3 has recently been overproduced, purified, and refolded to bind β-lactams (4). Cloning of the structural gene and overproduction of soluble proteins should be applicable to other PBPs providing sufficient materials to guide rational drug design.

The biochemical activities of the *fts* gene products other than PBP3 remain to be discovered. Nonetheless, they should prove to be good candidates for antimicrobials as they appear to be widespread among the eubacteria and perhaps not present in eukaryotic cells. Discovery of drugs that inhibit any of these proteins could be aided by strains that contain the gene product in excess as these strains would be expected to have increased resistance to such inhibitors. Strains containing these genes on multicopy plasmids are available, but care would have to be exercised as overexpression of some *fts* genes is deleterious. For example, small increases in FtsZ result in minicell formation, but larger increases (>7-fold) result in filamentation and cell death. Overexpression of *ftsA* also results in filamentation and cell death. Perhaps inhibitors of these genes could protect from overproduction lethality. Utilization of *fts* (Ts) mutants might also be of benefit as these mutants might show increased sensitivity to inhibitors at the permissive temperature where their function may already be impaired. Utilization of mutants such as *lon* and *ftsM1* which show sensitivity to division inhibition may also be of some value since they would be expected to show increased sensitivity to inhibitors.

At present it looks as if the *ftsZ* gene is a target of several cell division genes that are endogenous to the cell. Several of these, *sulA*, *sfiC*, and *dicB*, are not normally expressed and are expressed only in response to cellular stress or by mutational activation. Of these type of inhibitors only *sulA* appears to be widespread among eubacteria; the other two are not even present in all *E. coli* strains. Another inhibitor, *minCD*, appears to play a role in the normal physiology, being involved in division site selection. This inhibitor may also be conserved among bacteria, but this remains to be determined. Resistance to these inhibitors can be elicited by either an increased level of FtsZ or by *ftsZ* mutations that overcome the inhibition. Such strains may be useful in screens designed to enrich for inhibitors of FtsZ or MinE since they would be expected to be more resistant. Inhibitors of MinE should remove the topological constraint upon MinCD, allowing this inhibitor to block division at all sites.

Acknowledgments

Research in the author's laboratory is supported by Public Health Service grant GM29764 from the National Institutes of Health and by a grant from the Wesley Foundation.

The author wishes to thank the members of his laboratory for numerous discussions.

References

1. Adler, H.I., W. Fisher, A. Cohen, and A. Hardigree. 1967. Miniature *E. coli* cells deficient in DNA. Proc. Natl. Acad. Sci. USA **57**:321–326.
2. Aldea, M., T. Garrido, J. Pla, and M. Vicente, 1990. Division genes in *Escherichia coli* are expressed coordinately to cell septum requirements by gearbox promoters. EMBO J. **8**:3923–3931.

3. Asoh, S., H. Matsuzawa, F. Ishinmo, J.L. Strominger, M. Matsuhashi, and T. Ohta. 1986. Nucleotide sequence of the *pbpA* gene and characteristics of the deduced amino acid sequence of penicillin-binding protein 2 of *Escherichia coli* K12. Eur. J. Biochem. **160:**231–238.
4. Bartholome–De Belder, J., M. Nguyen-Disteche, N. Houba-Herin, J.M. Ghuysen, I.N. Maruyama, H. Hara, Y. Hirota, and M. Inouye. 1988. Overexpression, solubilization and refolding of a genetically engineered derivative of the penicillin-binding protein 3 of *Escherichia coli* K12. Mol. Microbiol. **2:**519–525.
5. Beall, B., and J. Lutkenhaus. 1987. Sequence analysis, transcriptional organization, and insertional mutagenesis of the *envA* gene of *Escherichia coli*. J. Bacteriol. **169:**5408–5415.
6. Beall, B., and J. Lutkenhaus. 1989. Nucleotide sequence and insertional inactivation of a *Bacillus subtilis* gene that affects cell division, sporulation, and temperature sensitivity. J. Bacteriol. **171:**6821–6834.
7. Beall, B., and J. Lutkenhaus. 1991. FtsZ in *Bacillus subtilis* is required for vegetative septation and asymmetric septation during sporulation. Genes and Develop. **5:**447–455.
8. Beall, B., M. Lowe, and J. Lutkenhaus. 1988. Cloning and characterization of *Bacillus subtilis* homologs of *Escherichia coli* cell division genes *ftsZ* and *ftsA*. J. Bacteriol. **170:**4855–4864.
9. Beck, B.D., and J.T. Park. 1976. Activity of three murein hydrolases during the cell division cycle of *Escherichia coli* K–12 as measured in toluene-treated cells. J. Bacteriol. **126:**1250–1260.
10. Begg, K.J., and W.D. Donachie. 1985. Cell shape and division in *Escherichia coli:* experiments with shape and division mutants. J. Bacteriol. **163:**615–622.
11. Begg, K.J., A. Takasuga, D.H. Edwards, S.J. Dewar, B.G. Spratt, H. Aduchi, T. Ohta, H. Matsuzawa, and W.D. Donachie. 1990. The balance between different peptidoglycan precursors determines whether *Escherichia coli* cells will elongate or divide. J. Bacteriol. **172:**6697–6703.
12. Begg, K.J., G.F. Hatfull, and W.D. Donachie, 1980. Identification of new genes in a cell envelope–cell division gene cluster in *Escherichia coli:* cell division gene *ftsQ*. J. Bacteriol. **144:**435–437.
13. Begg, K.J., B.G. Spratt, and W.D. Donachie, 1986. Interaction between membrane proteins PBP3 and RodA is required for normal cell shape and division in *Escherichia coli*. J. Bacteriol. **167:**1004–1008.
14. Bejar, S., and J.P. Bouche. 1985. A new dispensable genetic locus of the terminus region involved in control of cell division in *Escherichia coli*. Mol. Gen. Genet. **201:**146–150.
15. Bejar, S., F. Bouche, and J.P. Bouche. 1988. Cell division inhibition gene *dicB* is regulated by a locus similar to lambdoid bacteriophage immunity loci. Mol. Gen. Genet. **212:**11–19.
16. Berlander, R., and Kurt Nordstrom. Chromosome replication does not trigger cell division in *E. coli*. Cell **60:**365–374.

17. Bernstein, H.D., M.A. Poritz, K. Strub, P.J. Hoben, S. Brenner, and P. Walter. 1989. Model for signal sequence recognition from amino-acid sequence of 54K subunit of signal recognition particle. Nature **340:**482–486.

18. Bi, E., and J. Lutkenhaus. 1990. FtsZ regulates the frequency of cell division in *Escherichia coli*. J. Bacteriol. **172:**2765–2768.

19. Bi, E., and J. Lutkenhaus, 1990. Analysis of *ftsZ* mutations that confer resistance to the cell division inhibitor, *sulA*. J. Bacteriol. **172:**5602–5609.

20. Bi, E., and J. Lutkenhaus. 1990. Interaction between the *min* locus and *ftsZ*. J. Bacteriol. **172:**5610–5616.

21. Bi, E., K. Dai, S. Subbarao, B. Beall, and J. Lutkenhaus. 1991. FtsZ and cell division. Research in Microbiol. (in press).

22. Botta, G.A., and J.T. Park. 1981. Evidence for the involvement of penicillin-binding protein 3 in murein synthesis during septation but not during cell elongation. J. Bacteriol. **145:**333–340.

23. Bouloc, P., A. Jaffe, and R. D'Ari. 1989. The *Escherichia coli lov* gene product connects peptidoglycan synthesis, ribosomes and growth rate. EMBO J. **8:**317–323.

24. Bowler, L.D., and B.G. Spratt. 1989. Membrane topology of penicillin binding protein 3 of *Escherichia coli*. Mol. Microbiol. **3:**1277–1286.

25. Bukau, B., and G.C. Walker. 1989. Cellular defects caused by deletion of the *Escherichia coli dnaK* gene indicate roles for heat shock protein in normal metabolism. J. Bacteriol. **171:**2337–2346.

26. Bukau, B., and G.C. Walker. 1989. *ΔdnaK52* mutants of *Escherichia coli* have defects in chromosome segregation and plasmid maintenance at normal growth temperatures. J. Bacteriol. **171:**6030–6038.

27. Burdett, I.D.J., and R.G.E. Murray. 1974. Septum formation in *Escherichia coli:* characterization of septal structure and the effects of antibiotics on cell division. J. Bacteriol. **119:**303–324.

28. Burdett, I.D.J., and R.G.E. Murray. 1974. Electron microscope study of septum formation in *Escherichia coli* strains B and B/r during synchronous growth. J. Bacteriol. **119:**1039–1056.

29. Burton, P., and I.B. Holland. 1983. Two pathways of divison inhibition in UV-irradiated *E. coli*. Mol. Gen. Genet. **190:**128–132.

30. Canepari, P., G. Botta, and G. Satta. 1984. Inhibition of lateral wall elongation by mecillinam stimulates cell division in certain cell division conditional mutants of *Escherichia coli*. J. Bacteriol. **157:**130–133.

31. Chakraborti, A.S., K. Ishidate, W.R. Cook, J. Zrike, and L.I. Rothfield. 1986. Accumulation of a murein-membrane attachment site fraction when cell division is blocked in *lkyD* and *cha* mutants of *Salmonella typhimurium* and *Escherichia coli*. J. Bacteriol. **168:**1422–1429.

32. Chon, Y., and R. Gayda. 1988. Studies with FtsA-LacZ protein fusions reveal FtsA-located inner-outer membrane junctions. Biochem. Biophys. Res. Commun. **152:**1023–1030.

33. Cook, W.R., T.J. MacAlister, and L.I. Rothfield. 1986. Compartmentalization of the periplasmic space at division sites in gram-negative bacteria. J. Bacteriol. **168:**1430–1438.

34. Corton, J.C., J.E. Ward, Jr., and J. Lutkenhaus. 1987. Analysis of cell division gene *ftsZ (sulB)* from gram-negative and gram-positive bacteria. J. Bacteriol. **169:**1–7.

35. D'Ari, R. 1985. The SOS system. Biochimie **67:**343–347.

36. D'Ari, R., and O. Huisman. 1983. Novel mechanism of cell division inhibition associated with the SOS response in *Escherichia coli*. J. Bacteriol. **156:**243–250.

37. D'Ari, R., A. Jaffe, P. Bouloc, and A. Robin. 1988. Cyclic AMP and cell division in *Escherichia coli*. 1988. J. Bacteriol. **170:**65–70.

38. Davie, E., K. Syndor, and L.I. Rothfield. 1984. Genetic basis of minicell formation in *Escherichia coli* K–12. J. Bacteriol. **158:**1202–1203.

39. de Boer, P.A.J., R.E. Crossely, L.I. Rothfield. 1989. A division inhibitor and a topological specificity factor coded for by the minicell locus determine proper placement of the division septum in *E. coli*. Cell **56:**641–649.

40. de Boer, P.A.J., R.E. Crossley, and L.I. Rothfield. 1990. Central role for the *Escherichia coli minC* gene product in two different cell division-inhibition systems. Proc. Natl. Acad. Sci. USA **87:**1129–1133.

41. de Boer, P.A.J., W.R. Cook, and L.I. Rothfield. 1990. Bacterial cell division. Ann. Rev. Genet. **24:**249–274.

42. deJonge, B.L.M., F.B. Wientjes, I. Jurida, F. Driehuis, J.T.M. Wouters, and N. Nanninga. 1989. Peptidoglycan synthesis during the cell cycle of *Escherichia coli:* composition and mode of insertion. J. Bacteriol. **171:**5783–5794.

43. del Portillo, F.G., M.A. de Pedro, D. Joseleau-Petit, and R. D'Ari. 1989. Lytic response of *Escherichia coli* cells to inhibitors of penicillin-binding proteins 1a and 1b as a timed event related to cell division. J. Bacteriol. **171:**4217–4221.

44. Descoteaux, A., and G.R. Drapeau. 1987. Regulation of cell division in *Escherichia coli* K–12: probable interactions among proteins FtsQ, FtsA and FtsZ.J. Bacteriol. **169:**1938–1942.

45. Dewar, S.J., V. kagen-Zur, K.J. Begg, and W.D. Donachie. 1989. Transcriptional regulation of cell division gene in *Escherichia coli*. Mol. Microbiol. **3:**1371–1377.

46. Doi, M., M. Wachi, F. Ishino, S. Tomika, M. Ito, Y. Sakagami, A. Suzuki, and M. Matsuhashi. 1988. Determinations of the DNA sequence of the *mreB* gene and of the gene products of the *mre* region that function in formation of the rod shape of *Escherichia coli* cells. J. Bacteriol. **170:**4619–4624.

47. Donachie, W.D., and A.C. Robinson. 1987. Cell division: parameter values and the process, p. 1578–1593. *In* F.C. Neidhardt, J.L. Ingraham, K.B. Low, B. Magasanik, M. Schaechter, and H.E. Umbarger (ed.), *Escherichia coli* and *Salmonella typhimurium:* cellular and molecular biology. American Society for Microbiology, Washington, D.C.

48. Donachie, W.D., and K.J. Begg, 1970. Growth of the bacterial cell. Nature (London) **227:**1220–1224.

49. Donachie, W.D., and K.J. Begg. 1989. Cell length, nucleoid separation, and cell division of rod-shaped and spherical cells of *Escherichia coli*. J. Bacteriol. **171:**4633–4639.

50. Donachie, W.D., and K.J. Begg. 1989. Chromosome partition in *Excherichia coli* requires postreplication protein synthesis. J. Bacteriol. **171:**5405–5409.

51. Donachie, W.D., K.J. Begg, and N.F. Sullivan. 1984. Morphogenes of *Escherichia coli*, p. 27–62. *In* R. Losick and L. Shapiro (ed.), *Microbial development*. Cold Spring Harbor Laboratory, Cold Spring Harbor, N.Y.

52. Donachie, W.D., K.J. Begg, J.F. Lutkenhaus, G.P.C. Salmond, E. Martinez-Salas, and M. Vicente. 1979. Role of the *ftsA* gene product in control of *Escherichia coli* cell division. J. Bacteriol. **140:**388–394.

53. Dorman, C.J., A.S. Lynch, N.N. Bhriain, and C.F. Higgins. 1989. DNA supercoiling in *Escherichia coli: topA* mutations can be suppressed by DNA amplifications involving the tolC locus. Mol. Microbiol. **3:**531–540.

54. Ferreira, L.C.S., W. Keck, A. Betzner, and U. Schwarz. *In vivo* cell division gene product interaction in *Escherichia coli* K–12. J. Bacteriol. **169:**5776–5781.

55. Fletcher, G., C.A. Irwin, J.M. Henson, C. Fillingim, M.M. Malone, and J.R. Walker. 1978. Identification of the *Escherichia coli* cell division gene sep and organization of the cell division–cell envelope genes in the *sep-mur-ftsA-envA* cluster as determined with specialized transducing lambda bacteriophages. J. Bacteriol. **133:**91–100.

56. Foley, M., J.M. Brass, J. Birmingham, W.R. Cook, P.B. Garland, C.F. Higgins, and L.I. Rothfield. 1989. Compartmentalization of the periplasm at cell division sites in *Escherichia coli* as shown by fluorescence photobleaching experiments. Mol. Microbiol. **3:**1329–1336.

57. Frazer, A.C., and R. Curtiss III. 1975. Production, purification and utility of bacterial minicells. Cur. Top. Microbiol. Immunol. **69:**1–84.

58. Freundl, R., G. Braun, N. Honore, and S.T. Cole. 1987. Evolution of the enterobacterial *sulA* gene: a component of the SOS system encoding an inhibitor of cell division. Gene **52:**31–40.

59. Gayda, R.C., L.T. Yamamoto, and A. Markovitz. 1976. Second-site mutations in *capR (lon)* strains of *Escherichia coli* K–12 that prevent radiation sensitivity and allow bacteriophage lambda to lysogenize. J. Bacteriol. **127:**1208–1216.

60. George, J., M. Castellazzi, and G. Buttin. 1975. Prophage induction and cell division in *E. coli*. III. Mutations *sfiA* and *sfiB* restore division in *tif* and *lon* strains and permit the expression of mutator properties of *tif*. Mol. Gen. Genet. **140:**309–322.

61. Gill, D.R., G.F. Hatfull, and G.P.C. Salmond. 1986. A new cell division operon in *Escherichia coli*. Mol. Gen. Genet. **205:**134–145.

62. Gill, D.R., and G.P.C. Salmond. 1990. The identification of the *Escherichia coli ftsY* gene product: an unusual protein. Mol. Microbiol. **4:**575–583.

63. Gottesman, S., E. Halpern, and P. Trisler. 1981. Role of *sulA* and *sulB* in filamentation by *lon* mutants of *Escherichia coli* K–12. J. Bacteriol. **148:**265–273.

64. Grundstrom, T.S., S. Normark, and K. Magnusson. 1980. Overproduction of outer membrane protein suppresses envA-mediated hyperpermeability. J. Bacteriol. **144:**884–890.

65. Hara, H, Y. Nishimura, J.-I. Kato, H. Suzuki, H. Nagasawa, A. Suzuki, and Y. Hirota. 1989. Genetic analyses of processing involving C-terminal cleavage in penicillin-binding protein 3 of *Escherichia coli*. 1989. J. Bacteriol. **171:**5882–5889.

66. Harry, E.J., and R.G. Wake. 1989. Cloning and expression of a *Bacillus subtilis* division initiation gene for which a homolog has not been identified in another organism. J. Bacteriol. **171:**6835–6839.

67. Hiraga, S., H. Niki, T. Ogura, C. Ichinose, H. Mori, B. Ezaki, and A. Jaffe. 1989. Chromosome partitioning in *Escherichia coli:* novel mutants producing anucleate cells. J. Bacteriol. **171:**1496–1505.

68. Huisman, O., and R. D'Ari. 1981. An inducible DNA replication–cell division coupling mechanism in *E. coli*. Nature (London) **290:**797–799.

69. Huisman, O., R. D'Ari, and S. Gottesman. 1984. Cell division control in *Escherichia coli:* specific induction of the SOS function SfiA protein is sufficient to block septation. Proc. Natl. Acad. Sci. USA **81:**4490–4494.

70. Hussain, K., K.J. Begg, G.P.C. Salmond, and W.D. Donachie, 1987. ParD: a new gene code for a protein required for chromosome partitioning and septum localization. Mol. Microbiol. **1:**73–81.

71. Hussain, K., E.J. Elliott, and G.P.C. Salmond. 1987. The ParD− mutant of *Escherichia coli* also carries a $gyrA_{am}$ mutation: the complete sequence of *gyrA*. Mol. Microbiol. **1:**259–273.

72. Ikeda, M., T. Sato, M. Wachi, H.K. Jung, F. Ishino, Y. Kobasyashi, and M. Matsuhashi. 1989. Structural similarity among *Escherichia coli* FtsW, RodA proteins and *Bacillus subtilis* SpoVE protein, which function in cell division, cell elongation, and spore formation, respectively. J. Bacteriol. **171:**6375–6378.

73. Ishino, F., and M. Matsuhashi. 1981. Peptidoglycan synthetic activities of highly purified penicillin-binding protein 3 in *Escherichia coli:* a septum-forming reaction sequence. Biochem. Biophys Res. Commun. **101:**905–911.

74. Ishino, F., H.K. Jung, M. Ikeda, M. Doi, M. Wachi, and M. Matsuhashi. 1989. New mutations *fts–36, lts–33* and *ftsW* clustered in the *mra* region of the *Escherichia coli* chromosome induce thermosensitive cell growth and division. J. Bacteriol. **171:**5523–5530.

75. Ishino, F., W. Park, S. Tomioka, S. Tamaki, I. Takase, K. Kunugita, H. Matsuzawa, S. Asoh, T. Ohta, B.G. Spratt, and M. Matsuhashi. 1986. Peptidoglycan synthetic activities in membranes of *Escherichia coli* caused by overproduction of penicillin-binding protein 2 and RodA protein. J. Biol. Chem. **261:**7024–7031.

76. Jaffe, A., R. D'Ari, and V. Norris, 1986. SOS-independent coupling between DNA replication and cell division in *Escherichia coli*. J. Bacteriol. **165:**66–71.

77. Jaffe, A., R. D'Ari, and S. Hiraga. 1988. Minicell-forming mutants of *Escherichia coli:* production of minicells and anucleate rods. J. Bacteriol. **170:**3094–3101.

78. Johnson, B.F. 1977. Fine structure mapping and properties of mutations suppressing the *lon* mutation in *Escherichia coli* K–12 and B strains. Genet. Res. **30:**273–286.

79. Jones, C.A., and I.B. Holland. 1984. Inactivation of essential genes, *ftsA, ftsZ,* suppresses mutations at *sfiB*, a locus mediating division inhibition during the SOS response in *E. coli*. EMBO J. **3:**1181–1186.

80. Jones, C.A., and I.B. Holland. 1985. Role of the SfiB (FtsZ) protein in division inhibition during the SOS response in *E. coli:* FtsZ stabilizes the inhibitor SfiA in maxicells. Proc. Natl. Acad. Sci. USA **82:**6045–6049.

81. Jones, N.C., and W.D. Donachie, 1973. Chromosome replication, transcription and cell division in *Escherichia coli*. Nature (London) **243:**100–103.

82. Jung, H.K., F. Ishino, and M. Matsuhashi. Inhibition of growth of *ftsQ, ftsA,* and *ftsZ* mutant cells of *Escherichia coli* by amplification of a chromosomal region encompassing closely aligned cell division and cell growth genes. J. Bacteriol. **171:**6379–6382.

83. Kang, P.J., and E.A. Craig. 1990. Identification and characterization of a new *Escherichia coli* gene that is a dosage-dependent suppressor of a *dnaK* deletion mutation. J. Bacteriol. **172:**2055–2064.

84. Kren, B. and J.A. Fuchs. 1987. Characterization of the *ftsB* gene as an allele of the *nrdB* gene in *Escherichia coli*. J. Bacteriol. **169:**14–18.

85. Labie, C., F. Bouche, and J.-B. Bouche. 1989. Isolation and mapping of *Escherichia coli* mutations conferring resistance to division inhibition protein DicB.J. Bacteriol. **171:**4315–4319.

86. Leclerc, G., C. Sirard, and G.R. Drapeau. 1989. The *Escherichia coli* cell division mutation *ftsM1* is in *serU*. J. Bacteriol. **171:**2090–2095.

87. Leidenix, M.J., G.H. Jacoby, T.A. Henderson, and K.D. Young. 1989. Separation of *Escherichia coli* penicillin-binding proteins into different membrane vesicles by agarose electrophoresis and sizing chromatography. J. Bacteriol. **171:**5680–5686.

88. Little, J.W., and D.W. Mount. 1982. The SOS regulatory system of *Escherichia coli*. Cell **29,** 11–22.

89. Love, P.E., and R.E. Yasbin. 1984. Genetic characterization of the inducible SOS-like system of *Bacillus subtilis*. J. Bacteriol. **160:**910–920.

90. Lutkenhaus, J.F. 1983. Coupling of DNA replication and cell division: *sulB* is an allele of *ftsZ*. J. Bacteriol. **154:**1339–1346.

91. Lutkehaus, J. 1990. Regulation of cell division in *E. coli*. Trends in Genetics **6:**22–25.

92. Lutkenhaus, J.F., and W.D. Donachie. 1979. Identification of the *ftsA* gene product. J. Bacteriol. **154:**1088–1094.

93. Lutkenhaus, J.F., B. Sandjanwala, and M. Lowe. 1986. Overproduction of FtsZ suppresses sensitivity of *lon* mutants to division inhibition. J. Bacteriol. **166:**756–762.

94. Lutkenhaus, J.F., H. Wolf-Watz, and W.D. Donachie. 1980. Organization of genes in the *fts-A-envA* region of the *Escherichia coli* genetic map and identification of a new *fts* locus (*ftsZ*). J. Bacteriol. **142:**615–620.

95. Lutkenhaus, J.F., and H.C. Wu. 1980. Determination of transcriptional units and gene products from the *ftsA* region of *Escherichia coli*. J. Bacteriol. **143:**1281–1288.

96. MacAlister, T.J., Macdonald, B., and L.I. Rothfield. 1983. The periseptal annulus: an organelle associated with cell division in gram-negative bacteria. Proc. Natl. Acad. Sci. USA. **80:**1372–1376.

97. MacAlister, T.J., W.R. Cook, R. Weigand, and L.I. Rothfield. 1987. Membrane-murein attachment at the leading edge of the division septum: a second membrane-murein structure associated with morphogenesis of the gram-negative bacterial division septum. J. Bacteriol. **169:**3945–3951.

98. Maguin, E., H. Brody, C.W. Hill, and R. D'Ari. 1986. SOS-associated division inhibition gene *sfiC* is part of excisable element *e14* in *Escherichia coli*. J. Bacteriol. **168:**464–466.

99. Maguin, E., J. Lutkenhaus, and R. D'Ari. 1986. Reversibility of SOS-associated division inhibition in *Escherichia coli*. J. Bacteriol. **166:**733–738.

100. March, P.E., C.G. Lerner, J. Ahnn, X. Cui, and M. Inouye. 1988. The *Escherichia coli* Ras-like protein (Era) has GTPase activity and is essential for growth. Oncogene **2:**539–544.

101. Markiewicz, Z., J.K. Broome-Smith, U. Schwarz, and B.G. Spratt. 1982. Spherical *E. coli* due to elevated levels of D-alanine carboxypeptidase. Nature (London) **297:**702–704.

102. Masters, M., T. Paterson, A.G. Popplewell, T. Owen-Hughes, J.H. Pringle, and K.J. Begg. 1989. The effect of DnaA protein levels and the rate of initiation at *oriC* on transcription originating in the *ftsQ* and *ftsA* genes: *in vivo* experiments. Mol. Gen. Genet. **216:**475–483.

103. Mizusawa, S., and S. Gottesman. 1983. Protein degradation in *Escherchia coli:* the *lon* gene controls the stability of SulA protein. Proc. Natl. Acad. Sci. USA **80:**358–362.

104. Mulder, E., and C.L. Woldringh. 1989. Actively replicating nucleoids influence positioning of division sites in *Escherichia coli* filaments forming cells lacking DNA.J. Bacteriol. **171:**4303–4314.

105. Mulder, E., M. El'Bouhali, E. Pas, and C.L. Woldringh. 1990. The *Escherichia coli minB* mutation resembles *gyrB* in defective nucleoid segregation and decreased negative supercoiling of plasmids. Mol. Gen. Genet. **221:**87–93.

106. Nagasawa, H., Y. Sakagami, A. Suzuki, H. Suzuki, H. Hara, and Y. Hirota. 1989. Determination of the cleavage site involved in C-terminal processing of penicillin-binding protein 3 of *Escherichia coli*. J. Bacteriol. **171:**5890–5893.

107. Nakamura, M., I.N. Maruyama, M. Soma, J. Kato, H. Suzuki, and Y. Hirota. 1983. On the process of cellular division in *Escherichia coli:* nucleotide sequence of the gene for penicillin-binding protein 3. Mol. Gen. Genet. **191:**1–9.

108. Niki, H., A. Jaffe, R. Imamura, T. Ogura, and S. Hiraga. 1991. The new gene *mukB* codes for a 177 kDa protein with coiled-coil domains involved in chromosome partitioning in *E. coli*. EMBO J. **10:**183–193.

109. Nishimura, A. 1989. A new gene controlling the frequency of cell division per round of DNA replication in *Escherichia coli*. Mol. Gen. Genet. **215:**286–293.

110. Nishimura, A., and Y. Hirota. 1989. A cell division regulatory mechanism controls the flagellar regulon in *Escherichia coli*. Mol. Gen. Genet. **216:**340–346.

111. Normark, S., H.G. Boman, and E. Matsson. 1969. Mutant of *Escherichia coli* with anomalous cell division and ability to decrease episomally and chromosomally mediated resistance to ampicillin and several other antibiotics. J. Bacteriol. **97:**1334–1342.

112. Ogura, T., P. Bouloc, H. Niki, R. D'Ari, S. Hiraga, and A. Jaffe. 1989. Penicillin-binding protein 2 is essential in wild type *Escherichia coli* but not in *lov* or *cya* mutants. J. Bacteriol. **171:**3025–3030.

113. Olijhoek, A.J.M., S. Klencke, E. Pas, N. Nanninga, and U. Schwarz. Volume growth, murein synthesis, and murein cross-linking during the division cycle of *Escherichia coli* PA3092. 1982. J. Bacteriol. **152:**1248–1254.

114. Oliver, D., and J. Beckwith. 1981. *E. coli* mutant pleiotropically defective in the export of secreted proteins. Cell **25:**765–772.

115. Orr, E., N.F. Fairweather, I.B. Holland, and R.H. Pritchard. 1979. Isolation and characterization of a strain carrying a conditional lethal mutation in the *cou* gene of *Escherichia coli* K–12. Mol. Gen. Genet. **177:**103–112.

116. Pla, J., A. Dopazo, and M. Vicente. 1990. The native form of FtsA, a septal protein of *Escherichia coli,* is located in the cytoplasmic membrane. J. Bacteriol. **172:**5097–5102.

117. Reeve, J.N., N.H. Mendelson, S.I. Coyne, L.L. Hallock, and R.M. Cole. 1973. Minicells of *B. subtilis*. J. Bacteriol. **114:**860–873.

118. Ricard, M., and Y. Hirota. 1973. Process of cellular division in *Escherichia coli:* physiological study on thermosensitive mutants defective in cell division. J. Bacteriol. **116:**314–322.

119. Robin, A., D. Joseleau-Petit, and R. D'Ari. 1990. Transcription of the *ftsZ* gene and cell division in *Escherichia coli*. J. Bacteriol. **172:**1392–1399.

120. Robinson, A.C., D.J. Kenan, G.F. Hatfull, N.F. Sullivan, R. Spiegelberg, and W.D. Donachie. 1984. DNA sequence and transcriptional organization of essential cell division genes *ftsQ* and *ftsA* of *Escherichia coli:* evidence for overlapping transcriptional organization of the *ddl ftsQ* region. J. Bacteriol. **160:**546:555.

121. Romisch, K., J. Webb, J. Herz, S. Prehn, R. Frank, M. Vingron, and B. Dobberstein. 1989. Nature (London) **340:**478–482.

122. Sakakibara, Y. 1988. The *dnaK* gene of *Escherichia coli* functions in initiation of chromosome replication. J. Bacteriol. **170:**972–979.

123. Schmid, M.B., N. Kapur, D.R. Isaacson, P. Lindroos, and C. Sharpe. 1989. Genetic analysis of temperature-sensitive lethal mutants of *Salmonella typhimurium*. Genet. **123:**625–633.

124. Schwarz, U., A. Asmus, and H. Frank. 1969. Autolytic enzymes and cell division of *Escherichia coli*. J. Mol. Biol. **41:**419–429.

125. Spratt, B.G. 1975. Distinct penicillin binding proteins involved in the division, elongation and shape of *Escherichia coli*. Proc. Natl. Acad. Sci. USA **72:**2999–3003.

126. Storts, D.R., O.M. Aparicio, J.M. Schoemaker, and A. Markovitz. 1989. Overproduction and identification of the *ftsQ* gene product, an essential cell division protein in *Escherchia coli* K–12. J. Bacteriol. **171:**4290–4297.

127. Sullivan, N.F., and W.D. Donachie. 1984. Overlapping functional units in a cell division gene cluster in *Escherichia coli*. J. Bacteriol. **158:**1198–1201.

128. Taschner, P.E.M., J.G.J. Verest, and C.L. Woldringh. 1987. Genetic and morphological characterization of *ftsB* and *nrdB* mutants of *Escherichia coli*. J. Bacteriol. **169:**19–25.

129. Taschner, P.E.M., P. Huls, E. Pas, and C.L. Woldringh. 1988. Division behavior and shape changes in isogenic *ftsZ, ftsQ, ftsA, pbpB,* and *ftsE* cell division mutants of *Escherichia coli* during temperature shift experiments. J. Bacteriol. **170:**1533–1540.

130. Teather, R.M., J.F. Collins, and W.D. Donachie. 1974. Quantal behavior of a diffusible factor which initiates septum formation at potential division sites in *Escherichia coli*. J. Bacteriol. **118:**407–413.

131. Tormo, A., and M. Vicente. 1984. The *ftsA* gene product participates in formation of the *Escherichia coli* septum structure. J. Bacteriol. **157:**779–784.

132. Tormo, A., J.A. Ayala, M.A. de Pedro, M. Aldea, and M. Vicente. 1986. Interaction of FtsA and PBP3 proteins in the *Escherichia coli* septum. J. Bacteriol. **166:**985–992.

133. Tsuchido, T., R.A. VanBogelen, and F.C. Neidhardt. 1986. Heat shock response in *Escherichia coli* influences cell division. Proc. Natl. Acad. Sci. USA **83:**6959–6963.

134. Utsumi, R., M. Noda, M. Kawamukai, and T. Komano. 1989. Control mechanism of the *Escherichia coli* K–12 cell cycle is triggered by the cyclic AMP–cyclic AMP receptor protein complex. J. Bacteriol. **171:**2909–2912.

135. van de Putte, P., J.E. van Dillewijn, and A. Rorsch. 1964. The selection of mutants of *E. coli* with impaired cell division at elevated temperature. Mutat. Res. **1:**121–130.

136. Wachi, M., M. Doi, S. Tamaki, W. Park, S. Nakajima-lijima, and M. Matsuhashi. 1987. Mutant isolation and molecular cloning of the *mre* genes, which determine cell shape, sensitivity to mecillinam, and amount of penicillin-binding proteins in *Escherichia coli*. J. Bacteriol. **169:**4935–4940.

137. Wachi, M., and M. Matsuhashi. 1989. Negative control of cell division by *mreB*, a gene that functions in determining the rod shape of *Escherichia coli* cells. J. Bacteriol. **171:**3123–3127.

138. Walker, J.R., A. Kovarik, J.S. Allan, and R.A. Gustafson. 1975. Regulation of bacterial cell division: temperature sensitive mutants of *Escherichia coli* that are defective in septum formation. J. Bacteriol. **123:**693–703.

139. Wang, H., and R.C. Gayda. 1990. High-level expression of the FtsA protein inhibits cell separation in *Escherichia coli* K–12. J. Bacteriol. **172:**4736–4740.

140. Ward, J.E., and J.F. Lutkenhaus. 1984. A *lacZ-ftsZ* gene fusion is an analog of the cell division inhibitor *sulA*. J. Bacteriol. **157:**815–820.

141. Ward, J.E., and J.F. Lutkenhaus. 1985. Overproduction of FtsZ induces minicells in *E. coli*. Cell **42:**941–949.

142. Wientjes, F.B. and N. Nanninga. 1989. Rate and topography of peptidoglycan synthesis during cell division in *Escherichia coli:* concept of a leading edge. J. Bacteriol. **171:**3412–3419.

143. Wientjes, F.B., T.J.M. Olijhoek, U. Schwarz, and N. Nanninga. 1983. Labelling pattern of major penicillin-binding proteins of *Escherichia coli* during the division cycle. J. Bacteriol. **153:**1287–1293.

144. Yi, Q.-M., and J. Lutkenhaus. 1985. The nucleotide sequence of the essential cell division gene *ftsZ*. Gene **36:**241–247.

145. Yi, Q.-M., S. Rockenbach, J.E. Ward, and J. Lutkenhaus. 1985. Structure and expression of the cell division genes *ftsQ, ftsA,* and *ftsZ*. J. Mol. Biol. **184:**399–412.

6

Fatty Acid Biosynthesis

Suzanne Jackowski

Fatty acid biosynthesis and membrane lipid formation represent major metabolic energy commitments for growing bacterial cells. Some prokaryotic organisms are capable of incorporating exogenous fatty acids into their membranes; however, extracellular fatty acids are not incorporated into lipopolysaccharide acyl moieties in gram-negative bacteria typified by *Escherichia coli*. Thus exogenous fatty acid supplementation cannot compensate for the absence of *de novo* fatty acid production. Since fatty acid biosynthesis is absolutely required for formation of cell membranes and cell walls, a unique inhibitor of this branch of bacterial metabolism would be an effective antibiotic. Antibiotics that target fatty acid biosynthesis are not currently in use. The general perception is that the mammalian and bacterial fatty acid synthase systems are so similar that site-directed inhibitors would block fatty acid formation in both classes of organisms. Recent advances in understanding the *E. coli* fatty acid synthase system indicate that there are discrete differences between mammalian and bacterial synthases that can be exploited to develop antibiotics that specifically inhibit bacterial growth but do not impair mammalian fatty acid production.

Fatty acid synthase systems are divided into two major categories based on their structural organization. (For review see ref. 1.) In type I fatty acid synthase systems, the individual reactions in the pathway are carried out by multifunctional polypeptides. In yeast, the fatty acid synthase complex consists of two unique multifunctional polypeptides, each encoded by a different gene. In mammals, all enzyme activities are located on a single polypeptide chain encoded by a single gene. There are no pathway intermediates released from type I fatty acid synthase complexes, but rather the growing acyl chain remains bound to the multifunctional polypeptide complex until chain elongation is completed. (For review see ref. 2.) Palmitic acid is the primary product formed by the mammalian enzyme and is released from the 4′-phosphopantetheine prosthetic group of the multifunctional polypeptide by an associated thioesterase activity. Unsaturated fatty acids are subsequently synthesized by an aerobic mechanism involving a membrane-bound electron transport system. (For review see ref. 3.)

Type II fatty acid synthase systems are found in bacteria and plants, and each reaction is carried out by a unique protein that can be separated and analyzed individually. The distinct proteins of fatty acid biosynthesis are encoded by different genes whose loci are dispersed throughout the chromosome. The pathway intermediates are ferried from one specific enzyme to another as thioester derivatives of acyl carrier protein (ACP). ACP is a small acidic protein that has a 4′-phosphopantetheine prosthetic group attached via a phosphodiester linkage to Ser-36. (For review see ref. 4.) Pathway intermediates are attached to the terminal sulfhydryl group of ACP as thioesters. The precursors for fatty acid biosynthesis are derived from the acetyl-CoA pool. Malonyl-ACP is required for all of the condensation reactions and is formed by the carboxylation of acetyl-CoA by acetyl-CoA carboxylase followed by transacylation to ACP catalyzed by malonyl transacylase. Acetyl-CoA is the primer for fatty acid biosynthesis. (See below.) The initial condensation reaction is catalyzed by acetoacetyl-ACP synthase and subsequent rounds of elongation are catalyzed by two other condensing enzymes. There are four reactions required to elongate the growing fatty acid by two carbons. First, the β-ketoacyl-ACP synthases condense malonyl-ACP with acyl-ACP to form a β-ketoacyl-ACP. This ketoester is reduced by an NADPH-dependent β-ketoacyl-ACP reductase and a water molecule is then removed by β-hydroxyacyl-ACP dehydrase. The last step is catalyzed by enoyl-ACP reductase to form a saturated acyl-ACP. This sequence of chain-elongation reactions is identical in both the type I and type II systems, except that in the bacterial system, ACP-thioester intermediates dissociate from each individual enzyme. Specific inhibitors of the bacterial pathway can be designed to exploit this small difference in the mechanism of fatty acid synthesis.

Bacteria synthesize unsaturated fatty acids by an anaerobic mechanism as opposed to an aerobic mechanism in mammals. The double bond is introduced into the growing acyl chain by a unique β-hydroxydecanoyl-ACP dehydrase as first described by Bloch and co-workers (5,6); this enzyme is responsible for the key reaction in which the biosynthesis of unsaturated fatty acids diverges from saturated fatty acids. This dehydrase catalyzes the isomerization of *trans*-2- to *cis*-3-decenoyl-ACP and is thus distinct from the other β-hydroxyacyl-ACP dehydrase(s) that participates in the elongation cycle of fatty acid biosynthesis. *cis*-3-Decenoyl-ACP is then elongated to form either *cis*-9-palmitoleic acid or *cis*-11-vaccenic acid by the common reactions of fatty acid synthesis. (See above.) The uniqueness of β-hydroxydecanoyl-ACP dehydrase points to this enzyme as a potential target for antibacterial agents.

I. Specific Inhibitors of β-hydroxydecanoyl Thioester Dehydrase

The dehydrase enzyme that catalyzes both the dehydration and isomerization reactions is specifically and irreversibly inhibited by the acetylenic substrate analog 3-decynoyl-N-acetylcysteamine (Figure 6.1; 3-decynoyl-NAC). (For re-

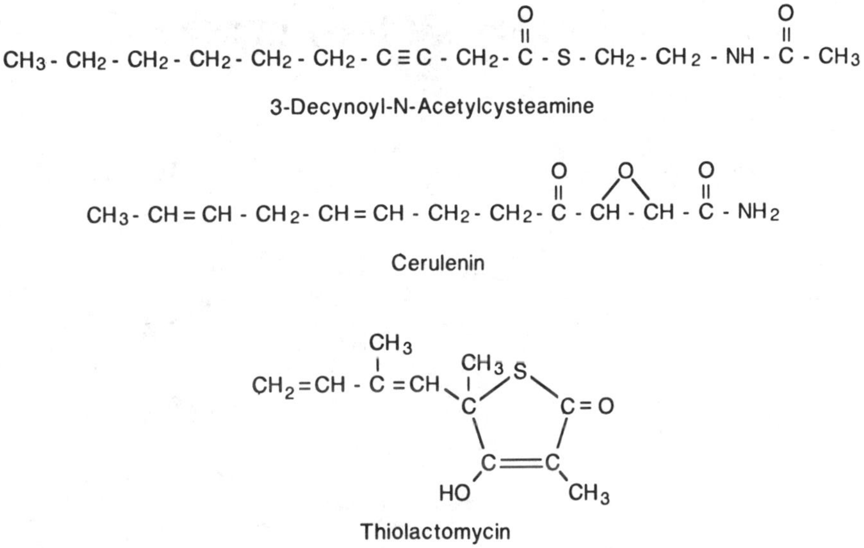

$$CH_3-CH_2-CH_2-CH_2-CH_2-CH_2-C\equiv C-CH_2-\overset{O}{\overset{\|}{C}}-S-CH_2-CH_2-NH-\overset{O}{\overset{\|}{C}}-CH_3$$

3-Decynoyl-N-Acetylcysteamine

$$CH_3-CH=CH-CH_2-CH=CH-CH_2-CH_2-\overset{O}{\overset{\|}{C}}-\overbrace{CH-CH}^{O}-\overset{O}{\overset{\|}{C}}-NH_2$$

Cerulenin

CH3
CH2=CH - C =CH
CH3 S
C
C=O
C=C
HO
CH3

Thiolactomycin

Figure 6.1. Structure of inhibitors of bacterial fatty acid biosynthesis.

view see ref. 6.) The allenic inhibitor, 2,3-decadienoyl-NAC, inhibits dehydrase activity even more effectively (7). 3-Decynoyl-NAC concentrations of 10 to 50 μM completely inhibit bacterial growth (8) but growth inhibition is relieved by addition of unsaturated fatty acids to the medium (8,9). Saturated fatty acid synthesis continues at its normal pace in the presence of 3-decynoyl-NAC and supplies necessary precursors for lipopolysaccharide production. The exogenous unsaturated fatty acids that are capable of relieving the growth inhibition are normally found in the gut and body fluids. Although the dehydrase is required for unsaturated fatty acid synthesis and is unique to the type II systems, one drawback to the use of dehydrase inhibitors as antimicrobial drugs is that they would be ineffective in physiological environments where unsaturated fatty acids are available to rescue the microbes.

II. β-ketoacyl-ACP Synthases as Key Targets for Drug Design

The β-ketoacyl-ACP synthases appear to be the appropriate targets for antibiotic design against dissociated (type II) fatty acid synthase systems typified by *E. coli*. These enzymes catalyze the condensation of malonyl-ACP with the growing acyl chain to form the corresponding β-ketoacyl-ACP (Figure 6.2). The condensation reaction proceeds via a covalent acyl-enzyme intermediate. First, the acyl-ACP binds to the synthase and the acyl moiety is transferred to the active site sulfhydryl of the condensing enzyme. Next malonyl-ACP binds to the synthase which catalyzes the condensation of the acyl-enzyme intermediate with

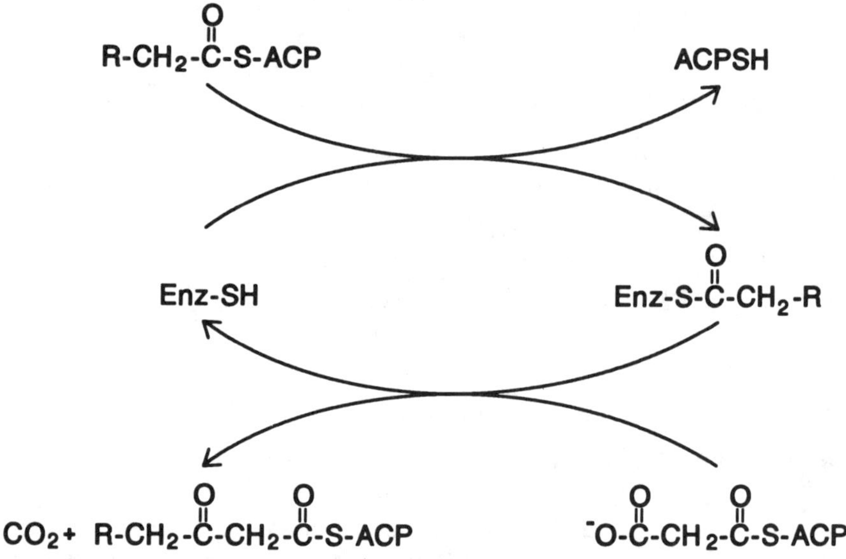

Figure 6.2. Condensation reaction catalyzed by β-ketoacyl-ACP synthases. First, the acyl moiety is transferred from acyl-ACP to the active site sulfhydryl of the β-ketoacyl-ACP synthase. The acyl-enzyme intermediate then condenses with malonyl-ACP to yield a β-ketoacyl-ACP that is two carbons longer than the original acyl chain.

malonyl-ACP. The resulting β-ketoacyl-ACP is then released from the enzyme. β-Ketoacyl-ACP synthases also promote two other reactions, albeit at a much lower rate. In the absence of malonyl-ACP, the synthases catalyze the exchange of acyl moieties between ACP and CoA via the formation of the acyl-enzyme intermediate. In the absence of acyl-ACP, the condensing enzymes will decarboxylate malonyl-ACP to acetyl-ACP.

The β-ketoacyl-ACP synthases are emerging as the key components of the regulatory mechanisms that govern fatty acid biosynthesis in type II fatty acid synthase systems. (For reviews see refs. 10, 11.) There are three condensing enzymes described in *E. coli*. β-Ketoacyl-ACP synthases I and II have distinctly different substrate specificities, antigenic determinants, peptide maps, and amino acid compositions, and their structural genes are physically separated on the *E. coli* chromosome (12). β-Ketoacyl-ACP synthase I is required for the elongation of short-chain unsaturated acyl-ACP (*i.e.*, *cis*-3-decenoyl-ACP), and mutants (*fabB*) lacking synthase I activity are unable to synthesize either palmitoleic or *cis*-vaccenic acids (13). Growth of *fabB* mutants ceases when the unsaturated

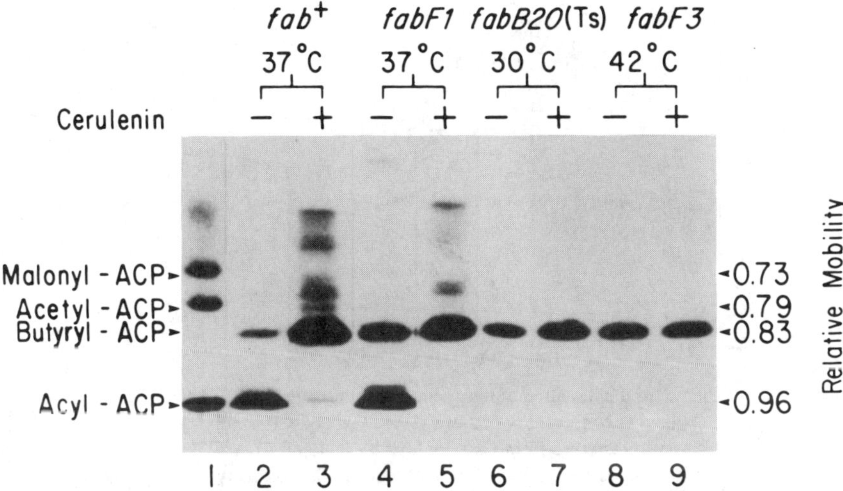

Figure 6.3. Genetic and biochemical evidence for the existence of acetoacetyl-ACP synthase. Extracts were prepared from *E. coli* K-12 strains SJ16 (*fab*$^+$), wild type; SJ86 (*fabF1*), defective in synthase II activity; and CY331 (*fabB20*(Ts) *fabF3*), defective in both synthase I and II. Fatty acid synthase assays were performed at the indicated temperature in either the presence or absence of cerulenin (100 μg/ml). The incorporation of radiolabeled malonyl-CoA into acyl-ACP products was then analyzed by conformationally sensitive gel electrophoresis to resolve the chain lengths of acyl-ACP formed *in vitro* and the bands visualized by fluorography. Reproduced from reference 21.

fatty acid content of the membrane phospholipids falls below 15–20% (14), but these mutant strains are able to grow when cultured in the presence of a variety of unsaturated fatty acids (15). β-Ketoacyl-ACP synthase II is also involved in the regulation of unsaturated fatty acid biosynthesis and is responsible for the temperature-dependent regulation of membrane fatty acid composition. (For review see ref. 16.) Mutants (*fabF*) lacking synthase II activity do not have a growth phenotype, but are unable to synthesize *cis*-vaccenate or regulate their fatty acid composition in response to temperature (17–20). Recently, evidence for the existence of a third condensing enzyme, acetoacetyl-ACP synthase (synthase III), was reported (21). This synthase selectively catalyzes the formation of acetoacetyl-ACP and is genetically and biochemically distinct from synthases I and II (Figure 6.3; 21). In contrast to the other two synthases, synthase III utilizes acetyl-CoA as the acyl primer rather than acyl-ACP (22). Thus fatty acid synthesis draws directly on the acetyl-CoA pool without the need for the transacylation of acetyl-CoA to acetyl-ACP (22). The role of this third condensing enzyme remains to be firmly established, but its position at the beginning of the biosynthetic pathway (Figure 6.4) suggests that it may govern the total rate of fatty acid production.

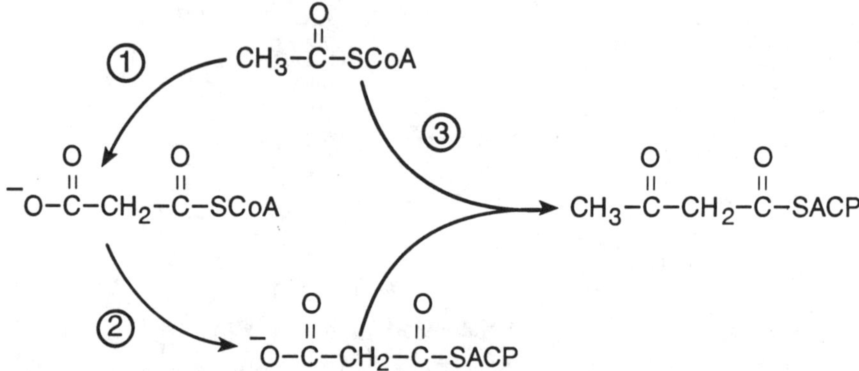

Figure 6.4. Scheme for the initiation of fatty acid biosynthesis in *E. coli*. Acetyl-CoA derived from intermediary metabolism is converted to malonyl-CoA by acetyl-CoA carboxylase (reaction 1). The malonyl moiety is then transferred from CoA to ACP by malonyl-CoA:ACP transacylase (reaction 2). Fatty acid biosynthesis is then initiated by the condensation of acetyl-CoA with malonyl-CoA catalyzed by acetoacetyl-ACP synthase (reaction 3). Reproduced from reference 22.

III. Cerulenin Inhibition of Fatty Acid Biosynthesis

The condensing enzymes are targets for fungal products that effectively inhibit bacterial cell growth. Cerulenin, (2R)(3S)-2,3-epoxy-4-oxo-7,10-dodecadienoly-amide (Figure 6.1), is a fungal product that is a noncompetitive, irreversible inhibitor of β-ketoacyl-ACP synthase I and II activities (23–25) and extremely effective in blocking the growth of a broad spectrum of bacteria. (For reviews see refs. 26 and 27.) Cerulenin blocks β-ketoacyl-ACP synthase activity by covalent modification of the synthase active site (24, 28) and inhibition correlates with the binding of 1 mole of cerulenin per mole of enzyme (24). β-Ketoacyl-ACP synthases I and II contain a fatty acyl and a malonyl-ACP binding site (29–31). Incubation of β-ketoacyl-ACP synthases with acyl-ACP protects the enzymes from cerulenin inhibition. These data strongly support the concept that cerulenin binds to the fatty acyl site of the condensing enzyme. Although cerulenin has proven to be a versatile biochemical tool (for review see 26 and 27), it is not a suitable antibiotic because it is also a potent inhibitor of the condensing enzyme reaction catalyzed by the multifunctional mammalian (type I) fatty acid synthase (23). This observation is not surprising since the type I multifunctional synthases have a fatty acyl binding site analogous to the site on prokaryotic β-ketoacyl-ACP synthases I and II.

Acetoacetyl-ACP synthase (synthase III) is not inhibited by cerulenin (21), indicating that this condensing enzyme lacks the fatty acyl binding site. Treatment of growing *E. coli* cultures with cerulenin results in the accumulation of short-chain acyl-ACPs having eight carbons or less. This suggests that the later condensation steps in the pathway are effectively blocked but that the initial condensation reactions are able to proceed in the presence of the antibiotic (22). Consistent with these observations, β-ketoacyl-ACP synthase III does not catalyze the condensation of long-chain acyl moieties with malonyl-ACP *in vitro* (22). Thus β-ketoacyl-ACP synthase III retains the malonyl-ACP site present in synthases I and II, but does not possess the fatty acyl binding site characteristic of the β-ketoacyl-ACP synthases in both the type I and II fatty acid synthases.

IV. Thiolactomycin, an Inhibitor of β-ketoacyl-ACP Synthases

Thiolactomycin, (4S)(2E,5E)-2,4,6-trimethyl-3-hydroxy-2,5,7-octatriene-4-thiolide (Figure 6.1), is a structurally unique antibiotic that inhibits type II but not type I fatty acid synthases (32–38). The antibiotic is not toxic to mice and affords significant protection against urinary tract and intraperitoneal bacterial infections (35). An analysis of the individual enzymes of type II fatty acid synthase shows that the β-ketoacyl-ACP synthase activity and the acetyl-CoA:ACP transacylase activity are the only activities inhibited by thiolactomycin (38). The observations that malonyl-ACP protects the synthases from thiolactomycin inhibition and that they are competitively inhibited with respect to malonyl-ACP are consistent with thiolactomycin interacting with the malonyl-ACP site on the condensing enzymes rather than the acyl-ACP site. All three condensing enzymes are inhibited by thiolactomycin both *in vivo* and *in vitro* (22).

Some investigators did not observe thiolactomycin inhibition of acetyl transacylase from *E. coli* (39), and in our laboratory thiolactomycin inhibition of acetoacetyl-ACP synthase is not apparent under certain incubation conditions. It is important to understand that thiolactomycin is a reactive chemical species, and care must be taken not to inactivate the antibiotic prior to evaluating its biological activity. Any condition that promotes opening the thiolactone ring will inactivate thiolactomycin. Ring opening is favored by rather mild conditions such as pH > 8.5 and the presence of thiol reducing reagents such as dithiothreitol. Reducing reagents are commonly added to β-ketoacyl-ACP synthase and acetyl transferase assays and if the concentration of reducing agents is too high or if thiolactomycin is preincubated with these sulfhydryl reagents, inhibition of either enzyme activity is not observed. Furthermore, since thiolactomycin inhibition is competitive, proper kinetic analysis is essential. With competitive inhibition the assay conditions (amount of protein, malonyl-ACP concentration, *etc.*) can be adjusted to make the inhibitor appear to be either more or less effective than it actually is. Under appropriate assay conditions and with structurally intact antibiotic, the concentration of thiolactomycin required to inhibit acetoacetyl-ACP synthase (synthase III) *in vitro* is the same as the dose of antibiotic required to inhibit fatty acid formation *in vivo* (22).

In order to investigate the mechanism of thiolactomycin action in more detail, a thiolactomycin-resistant mutant of *E. coli* (strain CDM5) was isolated and characterized (22). The β-ketoacyl-ACP synthase III activity in extracts from strain CDM5 was refractory to thiolactomycin inhibition. In addition, acetyl-CoA:ACP transacylase activity in strain CDM5 was resistant to inactivation by thiolactomycin, suggesting that the acetoacetyl-ACP synthase also catalyzes this transacylation reaction. Verification of this point awaits the purification and biochemical characterization of synthase III, but the idea is supported by the observation that acyl-CoA:ACP transacylase activity is catalyzed by synthases I and II (13,29). The specific activity of acetoacetyl-ACP synthase in extracts from this mutant was tenfold lower than in extracts from its thiolactomycin-sensitive parent, resulting in a marked defect in the ability of strain CDM5 to incorporate acetyl-CoA into fatty acids *in vitro*. However, additional experiments demonstrated that the incorporation of acetyl-CoA into fatty acids is not impaired *in vivo* (unpublished observations). Thus the mutation in strain CDM5 results in the production of an acetoacetyl-ACP synthase that is resistant to thiolactomycin, but is much less stable to *in vitro* manipulations than the wild-type protein. These data point to acetoacetyl-ACP synthase as a target for thiolactomycin inhibition of bacterial fatty acid biosynthesis. The genetic and biochemical characterization of additional mutants will be required to determine whether alterations in β-ketoacyl-ACP synthase III activity are invariably associated with the thiolactomycin-resistant phenotype or whether there are other mechanisms for acquiring thiolactomycin resistance. The genetic characterization of strain CDM5 and the investigation of the genetic alterations in other thiolactomycin-resistant isolates will require the production of more thiolactomycin than is currently available. A preliminary result suggests that two independent mutations in strain CDM5 are required to confer the thiolactomycin-resistant phenotype (unpublished observations).

V. Design of More Effective Antibiotics

Bacterial condensing enzymes possess both an acyl and malonyl-ACP binding site, whereas the corresponding activity in the type I fatty acid synthase system has only the acyl chain site (Table 6.1). Cerulenin binds to the acyl site and thus inhibits both type I and II fatty acid synthase systems. On the other hand,

Table 6.1. Properties of the β-ketoacyl-ACP synthases in type I and type II fatty acid synthase systems

Property	Type I synthase	Type II synthases I and II	Type II synthase III
Acyl chain binding site	Present	Present	Absent
ACP binding site	Absent	Present	Present
Cerulenin	Sensitive	Sensitive	Resistant
Thiolactomycin	Resistant	Sensitive	Sensitive

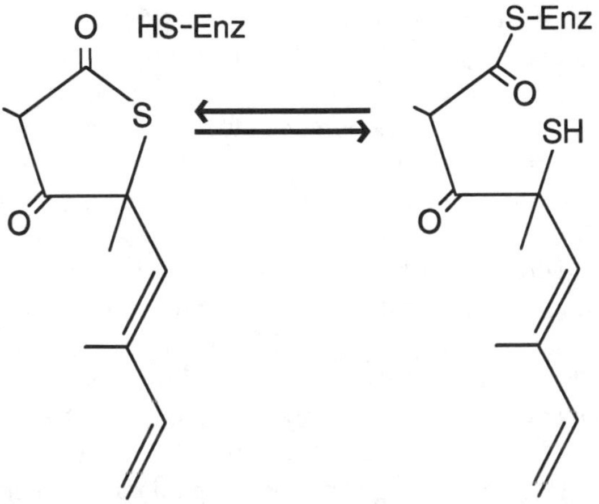

Figure 6.5. Proposed mechanism for the reversible inhibition of β-ketoacyl-ACP synthase reaction by thiolactomycin. Thiolactomycin binding at the malonyl-ACP site on the condensing enzyme places the antibiotic thioester in a favorable position to react with the enzyme active site sulfhydryl. Reduction of the thiolactomycin-enzyme complex with sulfhydryl reagents releases the antibiotic from the enzyme.

thiolactomycin binds at the malonyl-ACP site and selectively blocks type II condensation reactions. The malonyl-ACP site is persistently occupied in the type I fatty acid synthase systems, possibly rendering these multienzyme complexes resistant to thiolactomycin and to future antibiotics that are targeted to the malonyl-ACP site. There are two lines of investigation that should be pursued in the search for more effective type II–specific condensing enzyme inhibitors. First, antibiotic structures should be modified to mimic more closely the structure of malonyl-pantetheine. If the thiolactomycin structure is drawn in the keto rather than the enol form (Figure 6.5), the resemblance to malonyl-pantetheine becomes more obvious. Alterations in this structure that more closely resemble malonate should result in more effective inhibitors due to a tighter association with the enzyme active site. Second, an irreversible inhibitor would be useful. Functional groups need to be introduced into the thiolactomycin structure that would react with amino acid residues proximal to the binding site, thus preventing dissociation of the antibiotic from the enzyme.

VI. Concluding Remarks

Fatty acid and hence membrane phospholipid synthesis constitute an underexploited, but logical, target for the development of new antibiotics that show little or no cross-resistance with other classes of antibiotics. To be effective, these agents must be selective inhibitors of type II fatty acid synthase systems. Acetoacetyl-ACP synthase (β-ketoacyl-ACP synthase III) is unique to the type II systems, making this enzyme a selective target to use in developing inhibitory compounds. It lacks an acyl chain binding site and a screen for potential inhibitors of this enzyme is unlikely to find compounds that also block mammalian fatty acid biosynthesis. The *in vitro* filter disc assay for synthase III (21,22) is quantitative and readily adaptable to hundreds of samples per day. Acetoacetyl-ACP synthase is readily available since its activity can be specifically assayed in cerulenin-treated cell extracts that have been partially purified by ammonium sulfate fractionation and can be stored frozen for months without significant loss of activity (22). Using this acetoacetyl-ACP synthase assay as an activity screen will reveal compounds (like thiolactomycin) that may also inhibit the other β-ketoacyl-ACP synthases. This is an important goal because an antibiotic that blocks the action of multiple condensing enzymes that are required for cell growth would reduce the frequency of appearance of antibiotic-resistant bacteria.

References

1. Vance, D.E. 1976. The rate-limiting reaction catalyzed by a multi-enzyme complex—the fatty acid synthetases. J. Theor. Biol. **59:**409–413.
2. Goodridge, A.G. 1985. Fatty acid synthesis in eucaryotes, p. 143–179 . *In* D.E. Vance, and J.E. Vance (ed.), Biochemistry of Lipids and Membranes. Benjamin/Cummings Publishing Co., Menlo Park, California.
3. Cook, H.W. 1985. Fatty acid desaturation and chain elongation in eucaryotes, p. 181–211. *In* D.E. Vance, and J.E. Vance (ed.), Biochemistry of Lipids and Membranes. Benjamin/Cummings Publishing Co., Menlo Park, California.
4. Prescott, D.J., and P.R. Vagelos. 1972. Acyl carrier protein. Advances Enzymol. **36:**269–311.
5. Bloch, K. 1971. β-Hydroxydecanoyl thioester dehydrase, p. 441–464. *In* P. Boyer (ed.), The Enzymes, 3rd Edition. Academic Press, New York.
6. Helmkamp, G.M., R.R. Rando, D.J.H. Brock, and K. Bloch. 1968. β-Hydroxydecanoyl thioester dehydrase. Specificity of substrates and acetylenic inhibitors. J. Biol. Chem. **243:**3229–3231.
7. Endo, K., G.M. Helmkamp, and K. Bloch. 1970. Mode of inhibition of β-hydroxydecanoyl thioester dehydrase by 3-decynoyl-N-acetylcysteamine. J. Biol. Chem. **245:**4293–4296.
8. Kass, L.R. 1968. The antibacterial activity of 3-decynoyl-N-acetyl-cysteamine. Inhibition *in vivo* of β-hydroxydecanoyl dehydrase. J. Biol. Chem. **243:**3223–3228.
9. Clark, D.P., D. DeMendoza, M.L. Polacco, and J.E. Cronan, Jr. 1983. β-Hydroxydecanoyl thioester dehydrase does not catalyze a rate-limiting step in *Escherichia coli* unsaturated fatty acid synthesis. Biochemistry **22:**5898–5902.
10. Rock, C.O., and J.E. Cronan, Jr. 1985. Lipid metabolism in procaryotes, p. 73–115. *In* D.E. Vance and J.E. Vance (ed.), Biochemistry of Lipids and Membranes. Benjamin/Cummings Publishing Co., Menlo Park, California.

11. Cronan, J.E. Jr., and C.O. Rock. 1987. Biosynthesis of membrane lipids, p. 474–497. *In* F.C. Neidhardt (ed.), *Escherichia coli* and *Salmonella typhimurium:* Cellular and Molecular Biology, Vol. 1. American Society of Microbiology, Washington, D.C.
12. Garwin, J.L., A.L. Klages, and J.E. Cronan, Jr. 1980. Structural, enzymatic, and genetic studies of β-ketoacyl-acyl carrier protein synthases I and II of *Escherichia coli*. J. Biol. Chem. **255:**11949–11956.
13. D'Agnolo, G., I.E. Rosenfeld, and P.R. Vagelos. 1987. Multiple forms of β-ketoacyl-acyl carrier protein synthetase in *Escherichia coli*. J. Biol. Chem. **250:**5289–5294.
14. Jackson, M.B., and J.E. Cronan, Jr. 1978. An estimate of the minimum amount of fluid lipid required for the growth of *Escherichia coli*. Biochim. Biophys. Acta **512:**472–479.
15. Rosenfeld, I.S., G. D'Agnolo, and P.R. Vagelos. 1975. Synthesis of unsaturated fatty acids and the lesion of fabB mutants. J. Biol. Chem. **248:**2452–2460.
16. de Mendoza, D., and J.E. Cronan, Jr. 1983. Thermal regulation of membrane lipid fluidity in bacteria. Trends Biochem. Sci. **8:**49–52.
17. Gelman, E.P., and J.E. Cronan, Jr. 1972. Mutant of *Escherichia coli* deficient in the synthesis of *cis*-vaccenic acid. J. Bacteriol. **112:**381–387.
18. Garwin, J.L., A.L. Klages, and J.E. Cronan, Jr. 1980. β-Ketoacyl-acyl carrier protein synthase II of *Escherichia coli*. J. Biol. Chem. **255:**3263–3265.
19. de Mendoza, D., A.K. Ulrich, and J.E. Cronan, Jr. 1983. Thermal regulation of membrane fluidity in *Escherichia coli*. J. Biol. Chem. **258:**2098–2101.
20. Ulrich, A.K., D. de Mendoza, J.L. Garwin, and J.E. Cronan, Jr. 1982. Genetic and biochemical analyses of *Escherichia coli* mutants altered in the temperature-dependent regulation of membrane lipid composition. J. Bacteriol. **154:**221–230.
21. Jackowski, S., and C.O. Rock. 1987. Acetoacetyl-acyl carrier protein synthase, a potential regulator of fatty acid biosynthesis in bacteria. J. Biol. Chem. **262:**7927–7931.
22. Jackowski, S., C.M. Murphy, J.E. Cronan, Jr., and C.O. Rock. 1989. Acetoacetyl-acyl carrier protein synthase. J. Biol. Chem. **264:**7624–7629.
23. Vance, D.E., I. Goldberg, O. Mitsuhashi, K. Bloch, S. Omura, and S. Nomura. 1972. Inhibition of fatty synthetases by the antibiotic cerulenin. Biochem. Biophys. Res. Commun. **48:**649–656.
24. D'Agnolo, G., I.S. Rosenfeld, J. Awaya, S. Omura, and P.R. Vagelos. 1973. Inhibition of fatty acid synthesis by the antibiotic cerulenin. Specific inactivation of β-ketoacyl-acyl carrier protein synthetase. Biochim. Biophys. Acta **326:**155–166.
25. Kawaguchi, A., H. Tomoda, S. Nozoe, S. Omura, and S. Okuda. 1982. Mechanism of action of cerulenin on fatty acid synthetase. Effect of cerulenin on iodoacetamide-induced malonyl-CoA decarboxylase activity. J. Biochem. (Tokyo) **92:**7–12.
26. Omura, S. 1976. The antibiotic cerulenin, a novel tool for biochemistry as an inhibitor of fatty acid synthesis. Bacteriol. Rev. **40:**681–697.
27. Omura, S. 1981. Cerulenin. Methods Enzymol. **72:**520–532.

28. Kauppinen, S., M. Siggaard-Anderson, and P. von Wettstein-Knowles. 1988. β-Ketoacyl-ACP synthase I of *Escherichia coli:* Nucleotide sequence of the *fabB* gene and identification of the cerulenin binding residue. Carlsberg Res. Commun. **53:**357–370.

29. Greenspan, M.D., A.W. Alberts, and P.R. Vagelos. 1969. Acyl carrier protein. XIII. β-Ketoacyl acyl carrier protein synthetase from *Escherichia coli*. J. Biol. Chem. **244:**6477–6485.

30. Prescott, D.J., and P.R. Vagelos. 1970. Acyl carrier protein. XIV. Further studies on β-ketoacyl acyl carrier protein synthetase from *Escherichia coli*. J. Biol. Chem. **245:**5484–5490.

31. Alberts, A.W., R.M. Bell, and P.R. Vagelos. 1972. Acyl carrier protein. XV. Studies of β-ketoacyl-acyl carrier protein synthetase. J. Biol. Chem. **247:**3190–3198.

32. Oishi, H., T. Noto, H. Sasaki, K. Suzuki, T. Hayashi, H. Okazaki, K. Ando, and M. Sawada. 1982. Thiolactomycin, a new antibiotic. I. Taxonomy of the producing organism, fermentation and biological properties. J. Antibiotics **35:**391–395.

33. Sasaki, H., H. Oishi, T. Hayashi, I. Matsuura, K. Ando, and M. Sawada. 1982. Thiolactomycin, a new antibiotic. II. Structure elucidation. J. Antibiotics **35:**396–400.

34. Noto, T., S. Miyakawa, H. Oishi, H. Endo, and H. Okazaki. 1982. Thiolactomycin, a new antibiotic. III. *In vitro* antibacterial activity. J. Antibiotics **35:**401–410.

35. Miyakawa, S., K. Suzuki, T. Noto, Y. Harada, and H. Okazaki. 1982. Thiolactomycin, a new antibiotic. IV. Biological properties and chemotherapeutic activity in mice. J. Antibiotics **35:**411–419.

36. Hayashi, T., O. Yamamoto, H. Sasaki, A. Kawaguchi, and H. Okazaki. 1983. Mechanism of action of the antibiotic thiolactomycin inhibition of fatty acid synthesis of *Escherichia coli*. Biochem. Biophys. Res. Commun. **115:**1108–1113.

37. Hayashi, T., O. Yamamoto, H. Sasaki, and H. Okazaki. 1984. Inhibition of fatty acid synthesis by the antibiotic thiolactomycin. J. Antibiotics **37:**1456–1461.

38. Nishida, I., A. Kawaguchi, and M. Yamada. 1986. Effect of thiolactomycin on the individual enzymes of the fatty acid synthase system in *Escherichia coli*. J. Biochem (Tokyo) **99:**1447–1454.

39. Lowe, P.N., and S. Rhodes. 1988. Purification and characterization of [acyl-carrier-protein] acetyltransferase from *Escherichia coli*. Biochem. J. **250:**789–796.

7

Protein secretion in bacteria: a chemotherapeutic target?

Rajeev Misra and *Thomas J. Silhavy*

I. Introduction

The *Escherichia coli* cell is composed of four distinct compartments: the cytoplasm, an inner membrane, an outer membrane, and the periplasm, an aqueous compartment sandwiched between the two membranes. All proteins are synthesized in the cytoplasm and a subset of these are exported to noncytoplasmic locations. This process of protein export from the cytoplasm is an essential cellular function performed by a set of proteins discovered primarily through genetic studies. All proteins destined for the outer membrane or periplasmic space are first synthesized as precursors bearing an amino-terminal signal sequence of about 20 amino acids. Cleavage of the signal sequence by signal peptidase at the outer face of the inner membrane precedes proper localization of these exported protein.

Over the last several years, genetic studies of protein translocation across the inner membrane of *E. coli* have led the way in understanding this essential cellular process. (For recent reviews, see ref. 6, 45.) The importance of this work lies in the observation that the export machinery in both bacteria and higher organisms share certain similarities. The translocation machinery of other bacteria, even gram-positives, appears to contain homologous components (51). Moreover, *E. coli* can export various eukaryotic secretory proteins and *vice versa* (52, 55). Therefore, knowledge gained from studying mechanisms of protein export in *E. coli* will be extremely useful in understanding this process in other organisms where genetic manipulations are more difficult.

Despite the apparent similarities of the export process in bacteria and higher organisms, they are not the same. For example, as few as seven cellular proteins may comprise the entire bacterial export machinery, yet the corresponding machinery in mammalian cells is clearly more complex. The bacterial export process and machinery appear to be more similar to those of lower eukaryotes, such as yeast, than those of mammalian cells (45). Thus, it should be possible to develop antibiotics that target components of the one system without affecting the other by exploiting the differences between the protein export processes of the various organisms.

II. Applications of *lacZ* Fusions for Studying Protein Export

lacZ fusions are one of the most widely used tools in studying various biological processes, particularly in bacteria and yeast. *lacZ* gene fusion technology allows the construction of a hybrid gene in which the 5′ portion of the target gene is fused to a large 3′ portion of *lacZ* (specifies β-galactosidase), while retaining partial or full functional activity of both the gene products (49). This technology has been successfully used in developing genetic selections to identify genes whose products are involved in the bacterial export process. In this section we will describe how novel phenotypes of *lacZ* fusion strains have been used to define the mechanisms and components of the bacterial export machinery.

A. Selections Based on Overproduction Lethality

When the amino-terminus of a LacZ fusion is derived from an exported protein, the hybrid molecule can be targeted to the export machinery (4, 17, 18). A useful feature of exported LacZ fusions is that their high level synthesis causes cell death. This lethality is due to an attempt by the cell to engage the hybrid protein in the normal export pathway. The *E. coli* export machinery cannot translocate LacZ sequences effectively. This attempt causes jamming of the export machinery and interference with the export of other proteins, some of which are essential for growth. The extent of the export defect or lethal jamming is dependent on the amount of hybrid protein produced. This lethal jamming feature was successfully exploited by using LacZ fused to maltose-inducible proteins, such as LamB and MalE, which are normally exported to the outer membrane and periplasm, respectively. Thus, strains carrying a MalE-LacZ or LamB-LacZ fusion will not grow on maltose due to the overproduction of fusion proteins. Consequently, such strains are maltose-sensitive (Mal^s). Since lethality is caused by jamming of the export machinery, mutations that prevent entry of the hybrid into the export pathway, or facilitate export of the hybrid, should relieve the lethality.

1. Signal Sequence Mutations

Starting with strains carrying either a MalE-LacZ or LamB-LacZ fusion (thus the strain is Mal^s), maltose-resistant (Mal^r) revertants were isolated (3, 17, 18). A majority of the mutations altered the signal sequence of the hybrid protein and thus blocked its export (5, 16). When these mutations are recombined into the wild-type *malE* or *lamB* gene, they are cis-dominant and block export of the precursor form of molecules (signal sequence still attached), causing their accumutation in the cytoplasm. Analysis of these signal sequence mutations has provided insights regarding the function of the signal sequence and paved the way to an extensive and systematic characterization of this key export signal. (For a review, see ref. 23.)

2. *Suppressors of the Mutant Signal Sequence*

An important application of the signal sequence mutations obtained from the LacZ fusions has been the isolation of extragenic suppressors. It was speculated that mutations in the export machinery could be isolated by searching for extragenic suppressors of a defective signal sequence. Signal sequence mutations in *lamB* prevent LamB from being exported to the outer membrane; therefore, cells bearing such mutations are unable to grow on maltodextrin (an oligomer of four or more glucose units) as the sole carbon source. Thus, signal sequence mutations in *lamB* confer a maltodextrin$^-$ (Dex$^-$) phenotype. By using mutants where the export of LamB was impeded due to a signal sequence mutation, it was possible to identify second-site mutations that suppress this Dex$^-$ phenotype (15, 50). Similarly, strong signal sequence mutations in *malE* confer a Mal$^-$ phenotype (because the *malE* gene product, a periplasmic protein, is required for the maltose transport). Starting with a defective *malE* gene, extragenic suppressor mutations were isolated that conferred a Mal$^+$ phenotype (2, 44). Further analysis of these unlinked suppressor mutations (termed *prl,* for *p*rotein *l*ocalization) revealed the genes, *prlA*, *prlD*, and *prlG*, whose products are thought to participate in protein export.

These *prl* genes were also identified by using selections based on *lacZ* gene fusion technology and were termed *sec* as discussed below. Thus *prlA/secY*, *prlD/secA*, and *prlG/secE* are allelic. This convergence provides convincing evidence for the involvement of these gene products in protein export. It seems likely, given the suppressor phenotype, that the three Prl proteins interact directly with the signal sequence during the export process.

B. *Selection Based on the Lac Phenotype*

Membrane-associated LacZ hybrid proteins have a greatly reduced β-galactosidase activity relative to cytoplasmic fusions (25, 37). This is most likely due to improper assembly of the β-galactosidase tetramer. Therefore, strains carrying such fusions are phenotypically Lac$^-$. It was reasoned that in order for such strains to become Lac$^+$, only a small portion of the hybrid protein need be internalized. Thus, it should be possible to identify mutations that simply decrease rather than abate the export process by selecting "Lac-ups" using the appropriate fusion strains. Two kinds of mutations satisfy this criterion: signal sequence mutations and mutations that affect the functioning of a component of the export machinery.

Starting with a *malE-lacZ* fusion strain that is Lac$^-$ due to localization of the hybrid protein in the membrane, Lac$^+$ revertants were isolated. Oliver and Beckwith (37) imposed an additional criterion on this selection: since the export process is essential for cell survival, they looked for conditional lethal mutations—that is, mutations that confer only a slight export defect at a permissive temperature (which is sufficient to give a Lac-up phenotype), but are lethal at a nonpermissive temperature due to the complete loss of function. By applying these criteria, mutations were isolated that identified *secA*. The *secB* gene was identified in a similar fashion (28). However, this gene product is essential only under certain conditions. (See below.)

In an analogous selection scheme, but with the use of PhoA-LacZ and LamB-LacZ fusions, another essential gene, *secD,* was discovered (21). Subsequent work established that this locus contains two genes termed *secD* and *secF* (22).

C. Selection Based on the Derepression of SecA

The Lac$^+$ selection had an apparent bias that resulted in repeated isolation of mutations in *secA, secB,* or *secD* and *secF*. In order to circumvent this bias, Beckwith and co-workers devised an additional strategy to isolate export-defective mutants. They noted that synthesis of SecA was derepressed if the export machinery was compromised. Thus, expression of a *secA-lacZ* fusion was significantly induced (Lac-up) in the presence of *secA, secY,* or *secD* mutations, or when the export machinery was jammed by a hybrid protein (38, 42, 43). From these observations, it was argued that mutations that compromise the cellular export machinery derepress SecA synthesis. Indeed, by using a *secA-lacZ* fusion in a *secA*$^+$ wild-type background, mutations were identified that resulted in an increased *secA-lacZ* expression (22, 42, 46). These mutations mapped to *secA, secD, secE, secF,* and *secY*. No *secB* mutations were isolated in this screen. This was expected because although *secB* mutations cause an export-defective phenotype, they do not derepress expression of *secA-lacZ* fusion (43).

In summary, six genes were identified by several genetic selections whose products are involved in the bacterial export process. However, none of these schemes yielded mutations in a signal peptidase (leader peptidase; Lep) gene whose product is essential for protein translocation (Table 7.1). An explanation

Table 7.1. Protein components of the E. coli *export machinery*

Proteins	Molecular weight	Cellular location	Function
SecA	102,000	Peripheral membrane, cytoplasm	ATPase, precursor targeting
SecB	16,000	Cytoplasm	Antifolding chaperone
SecD	67,000	Inner membrane	Unknown
SecE	13,500	Inner membrane	Translocator
SecF	35,000	Inner membrane	Unknown
SecY	49,000	Inner membrane	Translocator
Lep	37,000	Inner membrane	Signal sequence processing

for this, as suggested by Schatz and Beckwith (45), is that the above schemes only identify genes whose products function in protein translocation from the cytoplasm. Signal peptidase is not required for translocation and, therefore, mutations that interfere with signal peptidase function would not satisfy these selections.

III. Genes whose Products Are Involved in the Export Process

A. secA (prlD)

secA maps at 2.5 min on the *E. coli* chromosome. The *secA* gene was first identified through mutations conferring a Lac$^+$ phenotype in a strain carrying an export-competent, therefore Lac$^-$, *malE-lacZ* fusion (37). The original *secA* mutation is a temperature-sensitive allele. Strains carrying this *secA*^{ts} allele grow at 30°C but are unable to grow at 42°C. At the semipermissive and nonpermissive temperatures, precursors of many exported proteins accumulate in the cytoplasm. Null mutations in *secA* are lethal, suggesting that SecA is essential for cell growth at all temperatures. *prlD* alleles of *secA* were also isolated as suppressors of the signal sequence mutations (2, 4, 50).

SecA is a 102,000-molecular-weight peripheral membrane protein (47). *In vitro* protein translocation assays have established that SecA hydrolyzes ATP (11, 13, 34). This activity may be involved in the energy-coupling step of protein translocation. SecA most likely interacts with the membrane components of the export machinery, SecE and SecY (26, 35). Direct interaction of SecA with SecB, which is complexed with a precursor protein, has also been established (26). It is apparent from these observations that SecA plays an important role in protein translocation and probably functions to pilot precursor proteins to the membrane export sites.

B. secB

secB maps at 81 min on the *E. coli* chromosome. The *secB* gene was identified utilizing the same selection scheme that resulted in the identification of *secA* (28). However, unlike *secA, secB* is not an essential gene. Strains carrying null alleles of *secB* are viable in minimal but not in rich media; therefore, *secB* is conditionally essential (29). In null mutants, export of only a subset of envelope (outer membrane and periplasm) proteins is affected (29).

The *secB* gene codes for a small, approximately 16,000-molecular-weight, cytoplasmic protein (30, 32). The role of SecB in protein translocation has been established by biochemical studies. SecB has been shown to bind to newly synthesized exported proteins and maintain them in an export-competent conformation (12, 31, 54). Because of this apparent antifolding activity, SecB fits into the category of a molecular chaperone. There is some disagreement with regard to SecB binding sites. Substantial evidence indicates that SecB binds to the mature sequences of proteins (12, 20, 41). However, another report suggests that SecB may also bind to the signal sequence of precursor proteins (53).

C. secD *and* secF

SecD and *secF* map at 9.5 min on the *E. coli* chromosome. The original *secD* alleles were recessive, cold-sensitive, and generally export defective (21). Subsequent analysis of these alleles revealed that several were not present in *secD* but in an adjacent open reading frame representing a new gene, *secF* (22). SecD and SecF are 67,000- and 39,000-molecular-weight polypeptides, respectively. DNA sequence and *phoA* fusion analysis suggest that both SecD and SecF are integral inner-membrane proteins with large periplasmic domains (22) (Johnson and Beckwith, in preparation).

D. secE (prlG)

SecE maps at 90 min on the *E. coli* chromosome (14). The *secE* gene was identified by utilizing two different genetic selections. In one case, it was identified among the recessive, cold-sensitive mutations that derepress the expression of a *secA-lacZ* fusion (42). In a different approach, *secE (prlG)* mutations were isolated as dominant suppressors of a mutant *lamB* signal sequence (50). DNA sequence analysis of several *secE* alleles revealed that the mutations which derepress *secA-lacZ* expression lie in the untranslated region of the mRNA, whereas the *lamB* signal sequence suppressor alleles alter the coding region (Schatz, Bieker, Silhavy, and Beckwith, in preparation).

DNA sequence analysis predicts that SecE is a 13,500-molecular-weight protein. These data, together with *phoA* fusion analysis, suggest that SecE is an integral inner-membrane protein with three membrane-spanning segments (46). Genetic and biochemical studies show that SecE interacts with SecY, and functions in protein translocation (8, 10).

E. secY (prlA)

secY maps at the end of the *spc* ribosomal protein operon, at 72 min of the *E. coli* chromosome (48). Alleles of *secY* (termed *prlA* alleles) were first isolated as suppressors of signal sequence mutations and are the most potent suppressors known (2, 15). Subsequently, other alleles of *secY* were obtained that are conditionally lethal (27) and derepress *secA-lacZ* expression (46). These *secY* alleles cause a general export defect.

SecY is an 49,000-molecular-weight integral inner membrane protein with ten membrane-spanning segments (1) that interacts with SecE and, as noted above, the SecY/E complex functions in translocation to physically move exported proteins across the membrane bilayer (8, 10). In wild-type cells, SecY function is rate-limiting in protein export (7).

IV. A Model for Bacterial Protein Export

A summary of the bacterial export pathway (Figure 7.1), derived from numerous genetic and biochemical studies, is as follows. Precursors of many exported proteins bind with SecB either during or immediately after their synthesis. SecB is believed to prevent the folding of these precursors, keeping them in an export-competent conformation. A subset of precursor proteins do not appear to require SecB. This could either reflect divergent folding properties of the various proteins or the presence of more than one molecular chaperone. Indeed, export of pre-β-lactamase appears to be facilitated by another chaperone, GroEL (9). The signal sequence itself plays an important role with regard to the folding of the precursor. *In vitro* studies have shown that the signal sequence retards protein folding (41). This could facilitate the efficient binding of SecB to the mature portion of the

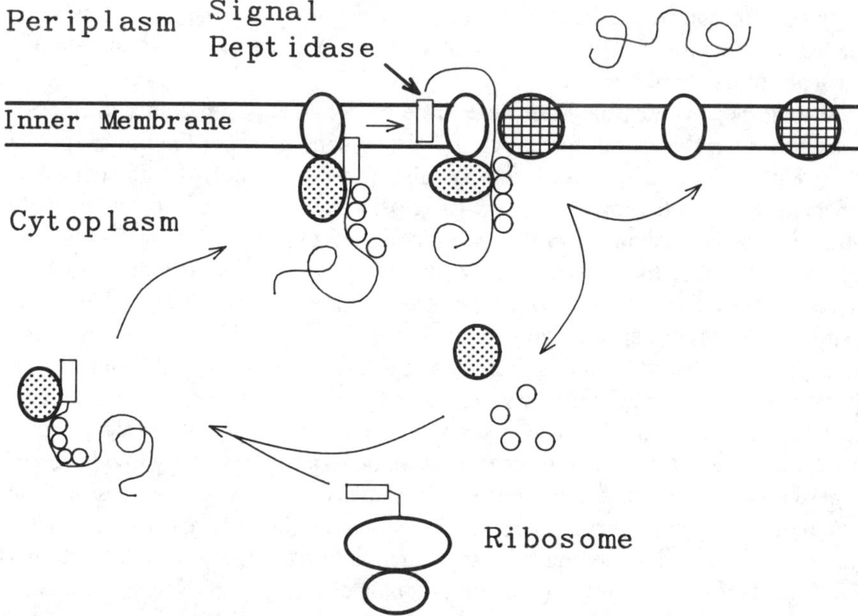

Figure 7.1. The protein translocation pathway in *E. coli*. See text for details. Symbols: small circles, SecB; stippled oval, SecA; open oval, SecE; crosshatched circle, SecY; rectangle, signal sequence; wavy line, mature sequence of an exported protein; thick arrow, cleavage by signal peptidase. SecD and SecF are involved in the translocation event, but the steps at which they participate are unknown and thus not shown. This illustration is not intended to infer any specific stoichiometry between the exported protein and various components of the translocation machinery.

protein. The precursor-SecB complex interacts with SecA. SecA binds with this complex owing to its affinity for both SecB and the signal sequence.

At the membrane, the SecA-SecB-precursor complex interacts with SecY/E; this interaction seems to require acidic phospholipids (35). As noted above, SecY/E is the translocator that facilitates passage of the protein through the membrane bilayer. Translocation is stimulated by a membrane potential (56), and as translocation proceeds, release of SecB from the complex may be promoted by the ATPase activity of SecA (26). Cleavage of the signal sequence by signal peptidase may occur while the precursor is bound to the SecY/E complex. The roles of SecD and SecF are not yet apparent but they may function in later steps in the translocation reaction. These proteins may function to clear the translocator and prepare it for the export of another protein molecule.

V. Scheme to Screen for Compounds that Affect Protein Export

Genetic analysis of the bacterial export process revealed six genes whose products are required for the translocation of exported proteins and are essential for growth. We believe that these genetic strategies can also be exploited in screening for compounds that kill bacteria by targeting the essential proteins of the export machinery. The discovery of such compounds could provide a novel class of antibacterial agents.

Among the several genetic schemes used to isolate export mutants, the selections involving the search for "Lac-ups" are perhaps the most readily adaptable to enrich for export-inhibitory compounds. The practicality of this scheme is reflected by the following features: β-galactosidase is one of the most intensively studied bacterial enzymes; therefore various aspects of its genetics and biochemistry are well understood. Accordingly, many types of indicator plates, such as lactose-MacConkey and lactose-tetrazolium agar, are readily available. The sensitivity of this assay can be significantly increased on a solid medium by using chromogenic substrate analogs such as X-gal (5-bromo-4-chloro-3-indolyl-β-D-galactoside). Moreover, there are various ways of regulating the activity of β-galactosidase, particularly by using an inhibitor called phenylethyl-thio-galactoside. Since the options used to control the amount or to detect very low levels of β-galactosidase are numerous, schemes that involve the use of LacZ fusions can be readily employed. Furthermore, a scheme that involves the use of agar plates can allow for rapid screening of many potentially useful compounds. Alternatively, screens involving the use of microtiter plates can be employed as automated assay systems for β-galactosidase have been developed (36).

As mentioned earlier, there are two genetic screens for Lac-ups that have been used for identifying export defects; one senses internalization of an exported LacZ hybrid protein and the other monitors derepression of *secA-lacZ* fusion. Since these screens have been successfully used to isolate export-defective mutants, it should be possible to exploit the same strategies for screening compounds that interfere with the export process. In practice, the screening method would be identical to that described by Quillardet and Hofnung (40). These authors employed a *sfiA-lacZ* fusion strain to screen for compounds that damage DNA. We suggest a similar strategy, except that the indicator strain would carry either a *malE-lacZ* or a *lamB-lacZ* or a *secA-lacZ* fusion. By this simple spot-test it should be possible to identify compounds that inhibit growth of *E. coli* cells by blocking protein export. It's worth noting that compounds that inhibit growth of *E. coli* by other mechanisms, such as inhibitors of RNA or protein synthesis, will not score as positives in this screen; the Lac-up phenotype requires that the LacZ hybrid protein be made.

Several recent results strengthen our belief that the desired compounds can be identified:

1. Mutations that allow *E. coli* cells to grow in the presence of small quantities of sodium azide (*azi*) map at 2.5 min on the chromosome (33). Recent studies have shown that these mutations are alleles of *secA* (19, 39). As mentioned earlier, SecA possesses an ATPase activity. The enzymatic activity of SecA can be inhibited, both *in vivo* and *in vitro*, by millimolar quantities of sodium azide, thus resulting in the accumulation of precursors of exported proteins. The sodium-azide-resistant mutants possess an altered SecA protein whose enzymatic activity is not inhibited by small quantities of sodium azide (39).
2. The addition of phenethyl alcohol to growing bacterial cells has been shown to cause the accumulation of precursors of exported proteins (24). The mechanism by which phenethyl alcohol interferes with the export process is not understood. Interestingly, however, mutations that confer phenethyl alcohol resistance (*pea*) map close to *secA*, and it has been suggested that *azi* and *pea* are allelic (57).

Thus, small quantities of these compounds could result in the Lac-up phenotype in a strain carrying either *secA-lacZ* or an exported fusion, such as *malE-lacZ* or *lamB-lacZ*. These inhibitors could serve as positive controls in the screening strategy to enrich for novel compounds that interfere with the bacterial export process.

VI. Concluding Remarks

The rapid spread of antibiotic resistance among bacterial populations requires the continuing search for novel antibacterial agents. In *E. coli*, genetic analysis employing *lacZ* fusion strains has identified six genes encoding essential components of the protein export machinery. Each of these six proteins is a potential target site for novel antibacterial agents. These genetic studies predict that compounds that inhibit any one of these proteins can be identified by screening strategies that use the appropriate *lacZ* fusion strains. Since screening methods with *lacZ* fusion strains are well established, requisite screens to find inhibitors of the protein translocation machinery can be readily implemented. Moreover, compounds such as azide or phenethyl alcohol can be detected by these screens; thus, we think it likely that these screens will be successful.

References

1. Akiyama, Y., and K. Ito. 1987. Topology analysis of the SecY, an integral membrane protein involved in protein export in *Escherichia coli*. EMBO J. **6:**3465–3470.
2. Bankaitis, V.A., and P.J. Bassford, Jr. 1985. Proper interaction between at least two components is required for efficient export of proteins to the *Escherichia coli* cell envelope. J. Bacteriol. **161:**169–178.
3. Bassford, P.J., and J. Beckwith. 1979. *Escherichia coli* mutants accumulating the precursor of a secreted protein in the cytoplasm. Nature (London) **277:**538–541.
4. Bassford, P.J. Jr., T.J. Silhavy, and J.R. Beckwith. 1979. Use of gene fusions to study secretion of maltose-binding protein into *Escherichia coli* periplasm. J. Bacteriol. **139:**19–31.
5. Bedouelle, H., P.J. Bassford, Jr., A.V. Flower, I. Zabin, and J. Beckwith. 1980. The nature of mutational alterations in the signal sequence of maltose-binding protein of *Escherichia coli*. Nature (London) **285:**78–81.
6. Bieker, K.L., G.P. Phillips, and T.J. Silhavy. 1990. The *sec* and *prl* genes of *Escherichia coli*. J. Bioenerg. Biomem. **22:**291–310.
7. Bieker, K.L., and T.J. Silhavy. 1989. PrlA is important for the translocation of exported proteins across the cytoplasmic membrane of *Escherichia coli*. Proc. Natl. Acad. Sci. USA **78:**968–972.
8. Bieker, K.L., and T.J. Silhavy. 1990. PrlA (SecY) and PrlG (SecE) interact directly and function sequentially during protein translocation in *E. coli*. Cell **61:**833–842.
9. Bochkareva, E.S., N.M. Lissin, and A.S. Girshovich. 1988. Transient association of newly synthesized unfolded proteins with the heat shock GroEL protein. Nature (London) **336:**254–157.
10. Brundage, L., J.P. Handrick, E. Schiebel, A.J.M. Driessen, and W. Wickner. 1990. The purified *E. coli* integral membrane protein SecY/E is sufficient for reconstitution of SecA-dependent precursor protein translocation. Cell **62:**649–657.
11. Cabelli, R.J., L. Chen, P.C. Tai, and D.B. Oliver. 1988. SecA protein is required for secretory protein translocation into *E. coli* membrane vesicles. Cell **55:**683–692.
12. Collier, D.N., V.A. Bankaitis, J.B. Weiss, and P.J. Bassford, Jr. 1988. The antifolding activity of *secB* promotes the export of the *E. coli* maltose-binding protein. Cell **53:**273–283.
13. Cunningham, K., R., Lill, E. Crooke, M. Rice, and K. Moore. 1989. SecA protein, a peripheral membrane of the *Escherichia coli* plasma membrane, is essential for the functional binding and translocation of proOmpA. EMBO J. **8:**955–959.

14. Downing, W.L., S.L. Sullivan, M.E. Gottesman, and P.P. Dennis. 1990. Sequence and transcriptional pattern of the essential *Escherichia coli secE-nusG* operon. J. Bacteriol. **172:**1621–1627.

15. Emr, S.D., S. Hanley-Way, and T.J. Silhavy. 1981. Suppressor mutations that restore export of a protein with a defective signal sequence. Cell **23:**79–88.

16. Emr, S.D., J. Hedgpeth, J.-M. Clement, T.J. Silhavy, and M. Hofnung. 1980. Sequence analysis of mutations that prevent export of lambda receptor, an *Escherichia coli* outer membrane protein. Nature (London) **285:**82–85.

17. Emr, S.D., M. Schwartz, and T.J. Silhavy. 1978. Mutations altering the cellular localization of the phage lambda receptor of *Escherichia coli*. Proc. Natl. Acad. Sci. USA **75:**5802–5806.

18. Emr, S.D., and T.J. Silhavy, 1980. Mutations affecting localization of *Escherichia coli* outer membrane protein, the bacteriophage lambda receptor. J. Mol. Biol. **141:**63–90.

19. Fortin, Y., P. Phoenix, and G.R. Drapeau. 1990. Mutations conferring resistance to azide in *Escherichia coli* occur primarily in the *secA* gene. J. Bacteriol. **171:**6607–6610.

20. Gannon, P.M., P. Li, and C.A. Kumamoto. 1989. The mature portion of *Escherichia coli* maltose-binding protein (MBP) determines the dependence of MBP on SecB for export. J. Bacteriol. **171:**813–818.

21. Gardel, C., S.A. Benson, J. Hunt, S. Michaelis, and J. Beckwith. 1987. *secD*, a new gene involved in protein export in *Escherichia coli*. J. Bacteriol. **169:**1286–1290.

22. Gardel, C., K. Johnson, A. Jacq, and J. Beckwith. 1990. The *secD* locus of *E. coli* codes for two membrane proteins required for protein export. EMBO J. **9:**3209–3216.

23. Gennity, J., J. Goldstein, and M. Inouye. 1990. Signal peptide mutants of *Escherichia coli*. J. Bioenrg. Biomem. **22:**233–269.

24. Halegoua, S., and M. Inouye. 1979. Translocation and assembly of the outer membrane proteins of *Escherichia coli:* selective accumulation of precursors and novel assembly intermediates caused by phenethyl alcohol. J. Mol. Biol. **130:**39–61.

25. Hall, M.N., M. Schwartz, and T.J. Silhavy. 1982. Sequence information within the *lamB* gene is required for proper routing of the bacteriophage lambda receptor protein to the outer membrane of *Escherichia coli* K-12. J. Mol. Biol. **156:**93–112.

26. Hartl, F.-U., S. Lecker, E. Schiebel, J.P. Hendrick, and W. Wickner. 1990. The binding cascade of SecB to SecA to SecY/E mediates preprotein targeting to the *E. coli* plasma membrane. Cell **63:**269–279.

27. Ito, K., M. Wittekind, M. Nomura, K. Shiba, and T. Yura. 1983. A temperature-sensitive mutant of *E. coli* exhibiting slow processing of exported proteins. Cell **32:**789–797.

28. Kumamoto, C.A., and J. Beckwith. 1983. Mutations in a new gene, *secB*, cause defective protein localization in *Escherichia coli*. J. Bacteriol. **154:**254–260.

29. Kumamoto, C.A., and J. Beckwith. 1985. Evidence for specificity at an early step in protein export in *Escherichia coli*. J. Bacteriol. **163:**267–274.

30. Kumamoto, C.A., L. Chen, J. Fandl, and P.C. Tai. 1989. Purification of the *Escherichia coli secB* gene product and demonstration of its activity in an in vitro protein translocation system. J. Biol. Chem. **264:**2242–2249.

31. Kumamoto, C.A., and P.M. Gannon. 1989. Effects of *Escherichia coli secB* mutations on pre-maltose binding protein conformation and export kinetics. J. Biol. Chem. **263:**11554–11558.

32. Kumamoto, C.A., and A.K. Nault. 1989. Characterization of the *Escherichia coli* protein-export gene *secB*. Gene **75:**167–175.

33. Lederberg, J. 1950. The selection of genetic recombinations with bacterial growth inhibitors. J. Bacteriol. **59:**211–215.

34. Lill, R., K. Cunningham, L.A. Brundage, K. Ito, and D.B. Oliver. 1989. SecA protein hydrolyzes ATP and is an essential component of the protein translocation ATPase of *Escherichia coli*. EMBO J. **8:**961–966.

35. Lill, R., W. Dowhan, and W. Wickner. 1990. The ATPase activity of SecA is regulated by acidic phospholipids, SecY, and the leader and mature domains of precursor proteins. Cell **60:**271–280.

36. Menzel, R. 1989. A microtitre plate-based system for semiautomated growth and assay for bacterial cell's β-galactosidase activity. Anal. Biochem. **181:**40–50.

37. Oliver, D.B., and J. Beckwith. 1981. *E. coli* mutants pleitropically defective in the export of secreted proteins. Cell **25:**2765–2772.

38. Oliver, D.B., and J. Beckwith. 1982. Regulation of a membrane component required for protein secretion in *Escherichia coli*. Cell **30:**311–319.

39. Oliver, D.B., R.J. Cabelli, K.M. Dolan, and G.P. Jarosik. 1990. Azide-resistant mutants of *E. coli* alter the SecA protein, an azide-sensitive component of the protein export machinery. Proc. Natl. Acad. Sci. USA **87:**8227–8231.

40. Quillardet, P., and M. Hofnung. 1985. The SOS chromotest, a colorimetric bacterial assay for genotoxins: procedures. Mutation Res. **147:**65–78.

41. Randall, L.L., T.B. Topping, and S.J.S. Hardy. 1990. No specific recognition of leader peptide by SecB, a chaperone involved in protein export. Science **248:**860–863.

42. Riggs, P.D.A., I. Derman, and J. Beckwith. 1988. A mutation affecting the regulation of a *secA-lacZ* fusion defines a new *sec* gene. Genetics **118:**571–579.

43. Rollo, E.E., and D.B. Oliver. 1988. Regulation of the *Escherichia coli secA* gene by protein secretion defects: analysis of *secA, secB, secD,* and *secY* mutants. J. Bacteriol. **170:**3281–3280.

44. Ryan, J.P., and P.J. Bassford, Jr. 1985. Post-translational export of maltose-binding protein in *Escherichia coli* strains harboring *malE* signal sequence mutations and either prl^{+} or *prl* suppressor alleles. J. Biol. Chem. **260:**14832–14837.

45. Schatz, P.J., and J. Beckwith. 1990. Genetic analysis of protein export in *Escherichia coli*. Ann. Rev. Genet. **24:**215–248.

46. Schatz, P.J., P.D. Riggs, A. Jacq, M.J. Fath, and J. Beckwith. 1989. The *secE* gene encodes an integral membrane protein required for protein export in *E. coli*. Genes Dev. **3:**1035–1044.

47. Schmidt, M.G., E.E. Rollo, J. Grodberg, and D.B. Oliver. 1988. Nucleotide sequence of the *secA* gene and *secA*(ts) mutations preventing protein export in *Escherichia coli*. J. Bacteriol. **170:**3404–3414.

48. Schultz, J., T.J. Silhavy, M.L. Berman, N. Fiil, and S.D. Emr. 1982. A previously unidentified gene in the *spc* operon of *Escherichia coli* K-12 specifies a component of the protein export machinery. Cell **31:**227–235.

49. Silhavy, T.J., and J.R. Beckwith. 1985. Uses of *lac* fusions for the study of biological problems. Microbiol. Rev. **49:**398–418.

50. Stader, J., L.J. Gansheroff, and T.J. Silhavy. 1989. New suppressors of signal-sequence mutations, *prlG,* are linked to the *secE* gene of *Escherichia coli*. Genes Dev. **3:**1045–1052.

51. Suh, J.-W., S.A. Boylan, S.M. Thomas, K.M. Dolan, D.B. Oliver, and C.W. Price. 1990. Isolation of a *secY* homolog from *Bacillus subtilis:* Evidence for a common protein export pathway in Eubacteria. Mol Microbiol. **4:**305–314.

52. Talmadge, K, S. Stahl, and W. Gilbert. 1980. Eukaryotic signal sequence transports insulin antigen in *Escherichia coli*. Proc. Natl. Acad. Sci. USA **77:**3369–3373.

53. Watanabe, M., and G. Blobel. 1989. SecB functions as a cytosolic signal recognition factor for protein export in *E. coli*. Cell **58:**695–705.

54. Weiss, J.B., P.H. Ray, and P.J. Bassford, Jr. 1988. Purified SecB protein of *Escherichia coli* retards folding and promotes membrane translocation of the maltose-binding protein in vitro. Proc. Natl. Acad. Sci. USA **85:**8978–8982.

55. Wiedman, M., A. Huth, and T. Rapoport. 1984. *Xenopus* oocytes can secrete bacterial β-lactamase. Nature (London) **309:**637–639.

56. Yamada, H., H. Tokuda, and S. Mizushima. 1989. Proton motive force-dependent and -independent protein translocation revealed by an efficient *in vitro* assay system of *Escherichia coli*. J. Biol. Chem. **264:**1723–1728.

57. Yura, T, and C. Wada. 1968. Phenethyl alcohol resistance in *Escherichia coli* I. Resistance of strain C600 and its relation to azide resistance. Genetics **59:**177–190.

8

Cytosolic Enzymes in Peptidoglycan Biosynthesis as Potential Antibacterial Targets

W. Stephen Faraci

I. Introduction

The bacterial cell wall provides structural integrity necessary to prevent osmotic-induced cell lysis. The peptidoglycan layer is the primary structural component which gives the cell wall its rigid structure. Peptidoglycan is composed of glycan chains with alternating N-acetylglucosamine and N-acetylmuramic acid residues that are cross-linked through a peptide bridge (89). As shown in Figure 8.1, crosslinking occurs via nucleophilic attack of the meso-diaminopimelate (meso-A_2pm) amino group of one glycan strand on the penultimate D-ala-D-ala group on another glycan strand. Gram-positive bacteria (*e.g.*, *S. aureus*) differ in strand crosslinking in that they utilize a pentaglycine bridge instead of meso-A_2pm. The β-lactam antibiotics irreversibly acylate proteins present in the cell membrane which catalyze transpeptidation and D,D-carboxypeptidation of the peptidoglycan. (For review, see 79.) Inability to crosslink the different glycan chains leads to cell wall degradation by bacterial autolysins, thus leading to cell death.

The initial stages of peptidoglycan biosynthesis occur in the cytoplasm whereby the pentapeptide chain (where pentapeptide = L-Ala-D-Glu-meso-A_2pm-D-Ala-D-Ala) is synthesized on an N-acetylmuramyl uridine diphosphate (UDP-NAcMur) scaffold. N-Acetylmuramyl pentapeptide is subsequently transferred from UDP to an undecaprenyl phosphate carrier at the cell membrane. The use of a lipid carrier allows transportation of the hydrophilic sugar peptide into a hydrophobic membrane environment where it can be attached to the growing peptidoglycan chain (88). The enzymes involved in the biosynthesis of the pentapeptide bridge that forms the covalent crosslink between nascent glycopeptide chains in peptidoglycan are discussed below. Enzymes of the first group (alanine branch) lead to the synthesis of the dipeptide, D-ala-D-ala, from the amino acid precursor L-alanine. Enzymes from the second group (UDP-tripeptide branch)

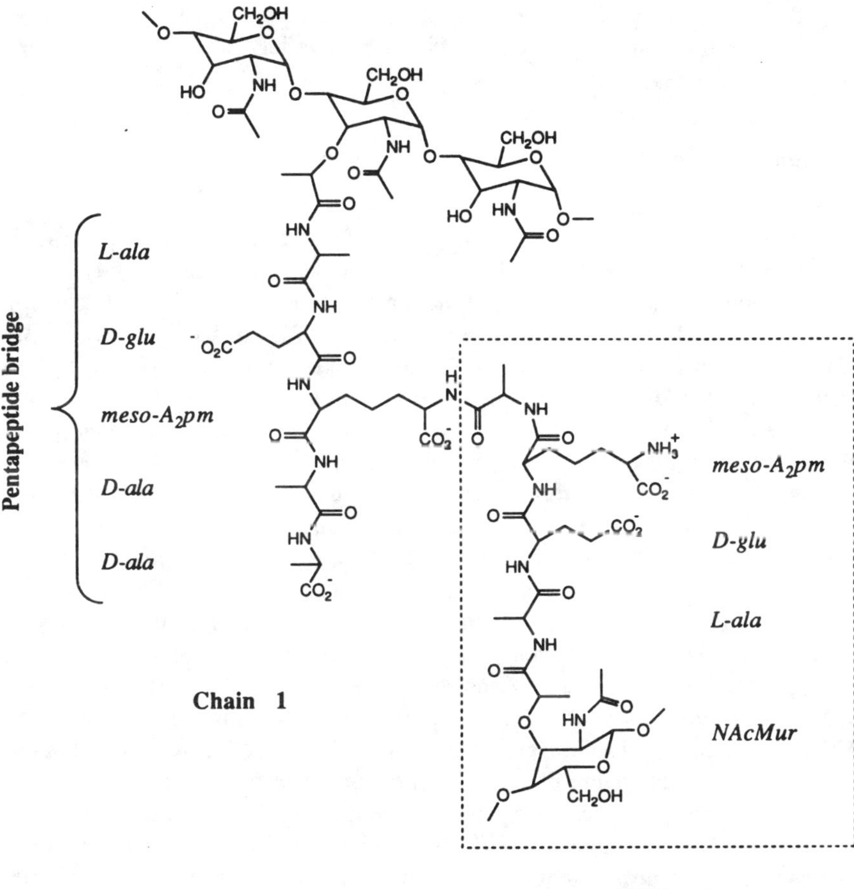

Figure 8.1. Peptidoglycan structure of *E. coli*. Gram-positive bacteria use L-lysine in place of meso-A_2pm and crosslink adjacent glycan chains via a pentaglycine bridge (86).

synthesize the proximal portion of the pentapeptide chain attached to the sugar backbone.

II. Alanine Branch

The alanine branch in the biosynthesis of UDP-MurNac-pentapeptide includes the enzymes which interact with D,L-alanine or D-alanyl-D-alanine. Since D-ala is unique to procaryotes, these enzymes have been the target for antibacterial drug design (54). Recent advances in the understanding of the mechanism of action of these enzymes and their inhibition by a variety of compounds has led to renewed interest in this area.

A. Alanine Racemase

Alanine racemase, the best characterized of the cytosolic enzymes (84, 2, 72), is a pyridoxal phosphate (PLP)-containing enzyme which catalyzes the racemization of alanine. The *in vivo* function of the enzyme is to provide bacteria with a source of D-alanine from the available pool of L-alanine. The mechanism of racemization involves initial transaldimination of substrate to yield the PLP-linked alanine adduct. Abstraction of the α-proton by an enzyme base and subsequent translocation to the opposite site leads to racemization. Shen *et al.* (69) have shown that the amino acid racemase from *Pseudomonas straita* proceeds by a one-base mechanism from internal return studies. However, Adams (2) has proposed a two-base mechanism for alanine racemases isolated from *Pseudomonas putida* and *Bacillus subtilis* due to extensive asymmetry observed in tritium exchange studies and Michaelis-Menten parameters. Faraci and Walsh (24) also observed extensive asymmetry in the racemization of alanine by the enzyme isolated from *Salmonella typhimurium* and have proposed a partial profile for catalysis. For the *Salmonella* enzyme, formation of the L-alanyl-PLP complex is rate-limiting with proton abstraction (re-protonation) not kinetically significant. Although the enzyme reaction profile is asymmetric, no definitive conclusion can be reached as to the nature of catalysis (i.e., one-base or two-base).

The racemase gene has been cloned, overproduced, and purified from *S. typhimurium* (23, 27, 90) and *Bacillus stearothermophilus* (31, 50, 77). Genetic studies in *S. typhimurium* show that two enzymes are present (91). One enzyme, *dadB,* is located on the same operon as D-alanine dehydrogenase and is believed to serve a catabolic function. This gene is inducible and required for cell growth on L-alanine (84). The second enzyme, *alr,* is constitutively expressed and provides D-alanine for the growing peptidoglycan layer (27). Cells are viable if only one gene (either *dadB* or *alr*) is inactivated; however, deletion of both genes is lethal in the absence of exogenous D-alanine. The presence of two genes is also evident in *E. coli* K-12 (92). In *E. coli, dadX* encodes the inducible alanine racemase whereas *alr* codes for the constitutive enzyme. From these studies, it is evident that alanine racemase is an essential enzyme for bacterial growth since deletion of the relevant alleles leads to a loss in cell viability.

Over the past 30 years, a large number of racemase inhibitors have been investigated and the mechanism of inhibition and antibacterial effectiveness of each has been assessed (Table 8.1). Inhibition of alanine racemase by cycloserine (**1**) was first investigated by Roze and Strominger (66) with the *Staphylococcus aureus* enzyme and Lambert and Neuhaus (38) with the *E. coli* enzyme. Competi-

Table 8.1. Inhibitors of the cytoplasmic steps of peptidoglycan synthesis

Compound	Target enzyme	Type of Inhibition	Antibacterial effectiveness	Reference
Cycloserine (1)	Alanine racemase D-ala-D-ala ligase D-amino acid transaminase	Irreversible Irreversible	Broad-spectrum antibacterial; toxic	38, 66, 86, 19, 51, 52, 42, 74
Fluoroalanine (2)	Alanine racemase D-amino acid transaminase	Irreversible Irreversible	broad-spectrum antibacterial; toxic	8, 65, 84, 73, 76
Chloroalanine (3)	Alanine racemase D-amino acid transaminase	Irreversible Irreversible		73,76
Trifluoroalanine (5)	Alanine racemase	Irreversible	No antibacterial activity cell penetration problems	25, 87
1-Aminoethyl phosphonic acid (Ala-P) (6)	Alanine racemase	Slow-binding inhibitor	Broad-spectrum antibacterial as dipeptide	17, 38, 6, 7, 14,
	D-ala-D-ala ligase	Slow-binding inhibitor	L-ala-L-ala-P	20,
	L-ala adding enzyme	Weak competitive inhibitor		4
1-Aminoethyl boronic acid (Ala-B) (7)	Alanine racemase	Slow-binding inhibitor	unknown	18
	D-ala-D-ala ligase	Slow-binding inhibitor		18
Halovinylglycine (9)	Alanine racemase	Irreversible	None; due to permeability problems	78
Aminoethylphosphinic acid (11)	D-ala-D-ala ligase	Slow-binding inhibition	Very weak; due to permeability problems	20, 58
Vinylglycine (14)	D-amino acid transaminase	Irreversible	Unknown	76
Gabaculine (15)	D-amino acid transaminase	Irreversible	Limited to *B. sphaericus* and *S. pidermidis*; toxic	74
3-Chloro-2,6-diamino-pimelic acid (16)	A_2pm epimerase	Competitive inhibitor	Inhibits growth of *E. coli* mutant strains	

tive inhibition is observed with Ki values of roughly 0.5 mM for both enzymes. Wang and Walsh (86) have shown that D-cycloserine is a time-dependent inactiva-

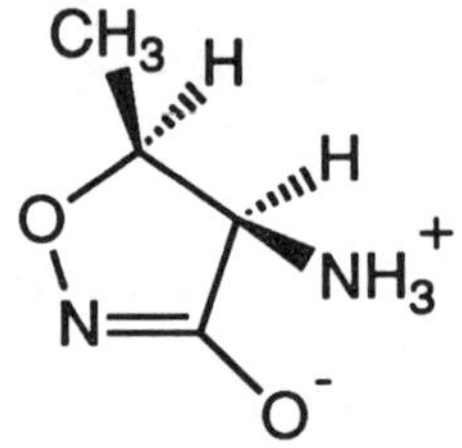

tor of the *E. coli* B alanine racemase. This is in agreement with other studies which show that cycloserine is a time-dependent inactivator of pyridoxal-phosphate-requiring enzymes (74).

Time-dependent inhibition is also observed with enzyme isolated from *S. typhimurium* and *Pseudomonas striata* by β-fluoroalanine (**2**), β-chloroalanine (**3**), and O-acetyl-D-serine (**4**) (65, 8, 84). Processing of the β-substituted alanine analog proceeds by initial transaldimination to form the PLP complex, abstraction of the α-proton, elimination of the β-leaving group, and then release of aminoacrylate as product (Scheme I). Inactivation occurs when, once in every 800

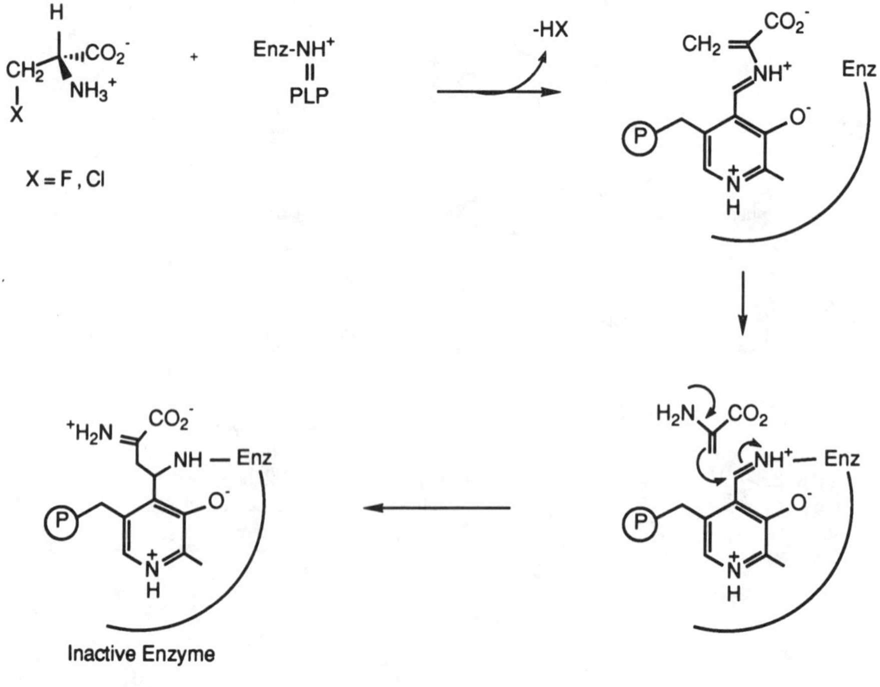

X = F, Cl

Inactive Enzyme

turnover events, nucleophilic attack of released aminoacrylate on the PLP-enzyme leads to an enzyme-PLP-aminoacrolyl complex. This type of inhibition was observed by Metzler and his colleagues in the inactivation of L-aspartate transaminase or glutamate decarboxylase by β-substituted amino acids (40, 80).

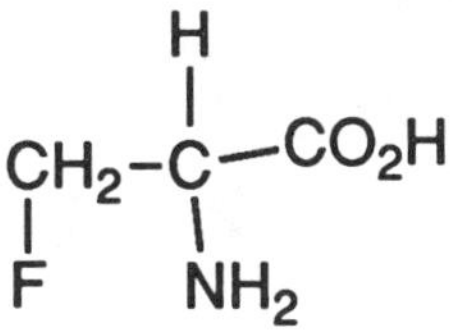

$$\mathrm{F{-}CH_2{-}CH(NH_2){-}CO_2H}$$

2

$$\mathrm{Cl{-}CH_2{-}CH(NH_2){-}CO_2H}$$

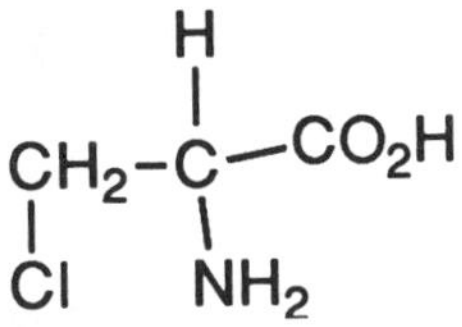

3

$$\mathrm{AcO{-}CH_2{-}CH(NH_2){-}CO_2H}$$

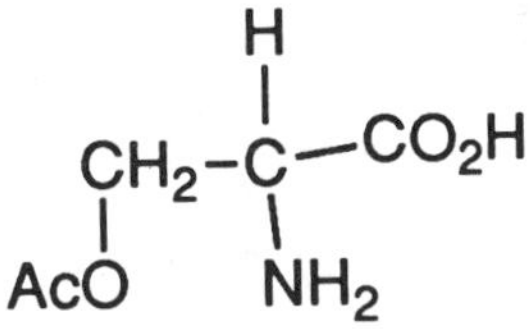

4

Another alanine analog that inhibits alanine racemase is trifluoroalanine (**5**). Wang and Walsh (87) first investigated the inactivation of alanine racemase from *E. coli* and found that trifluoroalanine is an efficient inactivator with an enzyme turnover/inactivation ratio of roughly 10/1. However, a detailed investigation of the inactivation mechanism was difficult due to limited enzyme supplies. Recently, Faraci and Walsh (25) have determined the mechanism of trifluoroalanine inactivation with enzyme isolated from the gram-positive thermophilic bacterium

$$\mathrm{F_3C{-}CH(NH_2){-}CO_2H}$$

5

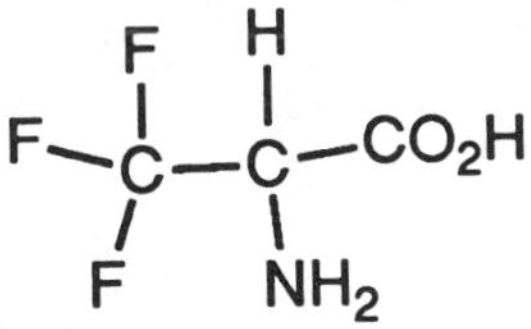

B. stearothermophilus (Scheme II). Inactivation occurs by initial transaldimination to give the trifluoroalanyl-PLP complex. Abstraction of the α-hydrogen is followed by elimination of fluoride to give the electrophilic difluoroeneamine, which undergoes nucleophilic attack by the active site lysine-38. Thus, the enzyme is covalently linked to the alanyl moiety, rendering it useless for further turnover. Precedent for this mechanism is the γ-cystathionase inactivation by trifluoroalanine (70,71); however, the stability of the β-fluoro adduct in the inhibition of racemase appears unique.

A third type of racemase inhibitor is 1-aminoethyl phosphonic acid (Ala-P; **6**). First observed by Dulaney (17) and Lambert and Neuhaus (38), Ala-P was found to competitively inhibit alanine racemase from gram-negative bacteria (*E. coli* and *P. aeruginosa*) and irreversibly inhibit enzyme isolated from gram-positive bacteria (*Staphylococcus aureus* and *Enterococcus faecalis*). The inhibition of alanine racemase from the gram-positive bacteria *E. faecalis* and *B. stearothermophilus* by Ala-P has been investigated in detail by Walsh and co-workers using purified enzyme (6,7). Inhibition is time-dependent and Ala-P can be classified

Scheme II

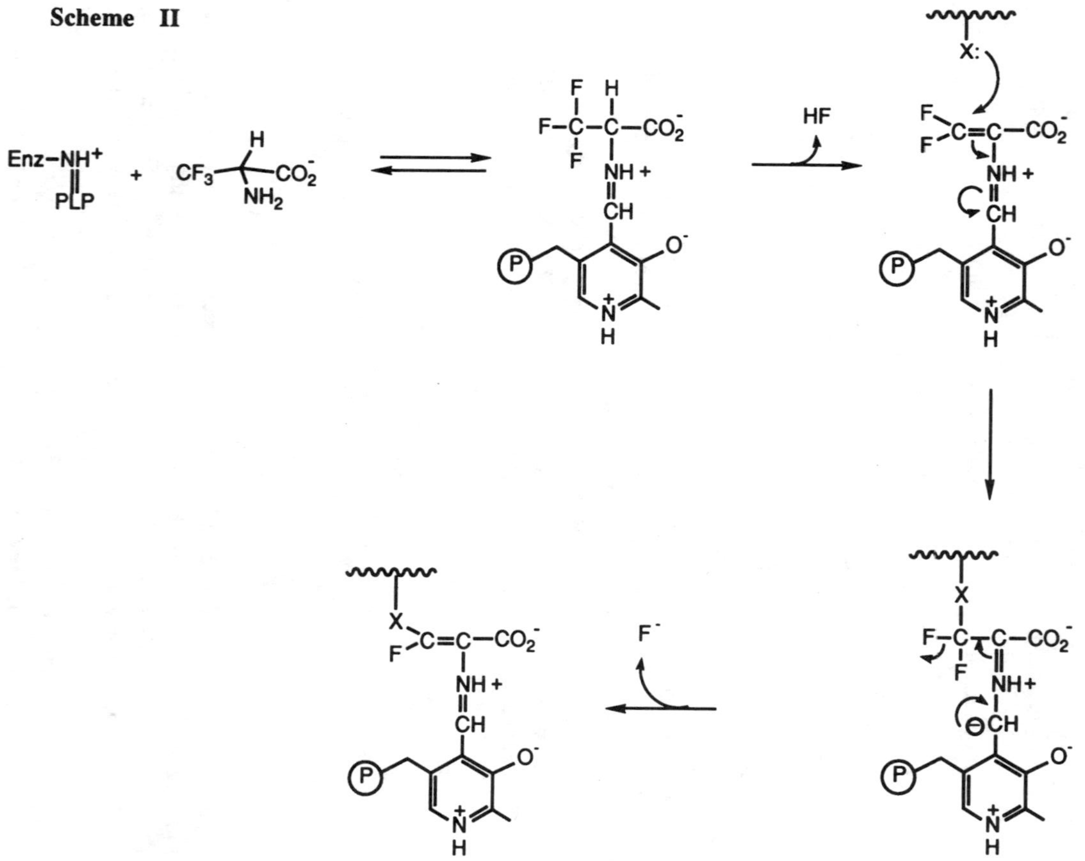

$Enz{-}NH^+{=}PLP + CF_3CH(NH_2)CO_2^-$

HF

F^-

6

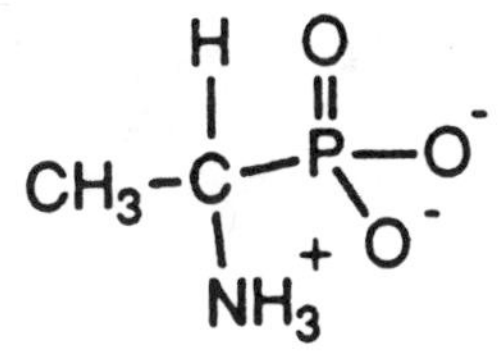

$$CH_3-\underset{NH_3^+}{\overset{H}{C}}-\overset{O}{\overset{\|}{P}}(-O^-)-O^-$$

as a slow-binding inhibitor (47). Kinetic analysis shows inhibition to be a two-step process with Ala-P initially binding to enzyme in a weak, reversible complex (Ki = 1 mM) followed by slow conversion to an enzyme·Ala-P form which dissociates very slowly ($t_{1/2}$ = 25 days). Inhibition is dependent on presence of the dianion, as the phosphinic acid analog is inactive. The structure of the inactive enzyme complex was uncertain as addition of borohydride (which reduces imines to secondary amines) was inconclusive. This question has been resolved due to recent advances in solid-state NMR (29). Using solid-state, magic angle spinning ^{15}N NMR, Copie *et al.* (14) showed that Ala-P forms a protonated Schiff base with pyridoxal-phosphate at the enzyme active site. Apparently, this complex is inaccessible to sodium borohydride for reduction.

It is likely that formation of the slowly dissociating enzyme complex occurs by protein isomerization after initial formation of the enzyme-inhibitor complex. This isomerization restricts access of the active site by outside molecules such as borohydride and probably hinders release of the inhibitor. Badet and Walsh (6) have shown that inhibition occurs equally well with D or L-Ala-P and that α-hydrogen cleavage does not occur. However, slow-binding inhibition occurs *only* in gram-positive racemases; Ala-P inhibits gram-negative racemases in a reversible, competitive fashion. This exemplifies the subtle structural differences that exist in homologous enzymes isolated from gram-positive or gram-negative strains.

Other phosphonic acid inhibitors include aminocyclopropane phosphonate, (1-aminoisopropyl) phosphonate, β-chloro-α-aminoethyl phosphonic acid, and 1-amino-2-propenyl phosphonic acid (22, 6, 82). Aminocyclopropane phosphonate and (1-aminoisopropyl) phosphonate were found to be slow-binding inhibitors of racemase from *Staphylococcus aureus* and *Enterococcus faecalis* while β-chloro-α-aminoethyl phosphonic acid and 1-amino-2-propenyl phosphonic acid were found to be *competitive inhibitors* of alanine racemase from both gram-negative and gram-positive bacteria.

Thus, substitution of the carboxylate of alanine with phosphonate leads to a compound that is a potent inhibitor rather than a substrate. Extension of this principle led to the design of another racemase inhibitor, 1-aminoethyl boronic acid (7) (18). However, 1-aminoethyl boronic acid is not as potent as Ala-P and it is doubtful that the mechanism of inactivation is the same. Inactivation of racemase by 1-aminoethyl boronic acid leads to a change in the absorption spectrum of the enzyme-PLP cofactor, characteristic of pyridoxamine-5-phosphate (PMP) derivatives. Since boron esters form tetrahedral anions, it is possible that nucleophilic attack of the boron hydroxyl group on the aldimine carbon leads to a 5-membered ring which is the inactive enzyme species (**8**).

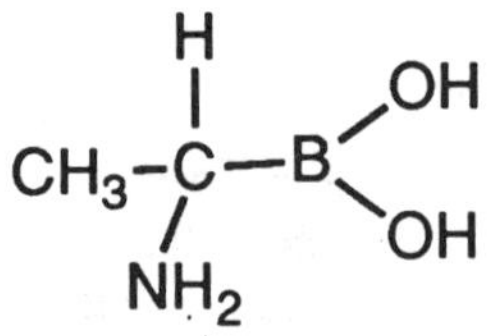

7

8

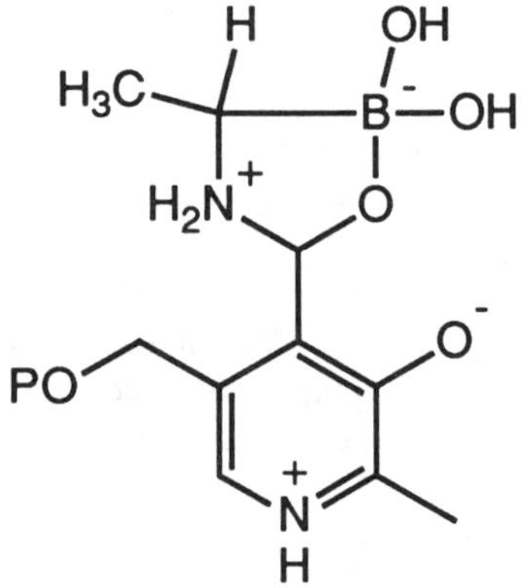

Recent studies by workers at Merck have identified another class of racemase inhibitors, the 3-halovinylglycines (**9**) (59,78). Inactivation of alanine racemase from *E. coli B* by D-chlorovinylglycine is irreversible, with a second-order rate constant for inactivation of $122 M^{-1} sec^{-1}$. This compares favorably to the 3-

9 **(where X = Cl, F)**

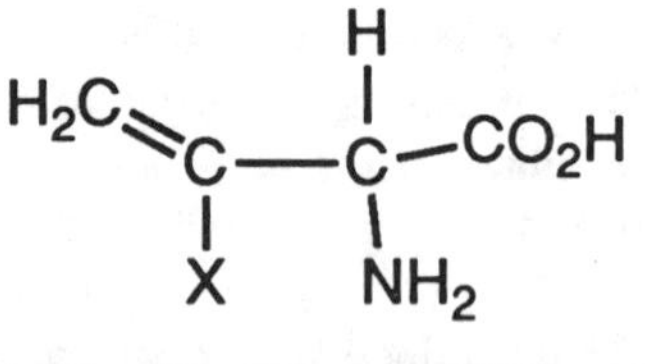

fluoro-D-alanine inactivation rate (8). Inactivation probably occurs via the same mechanism seen for other suicide substrates in which abstraction of the α-proton of D-chlorovinylglycine is followed by elimination of chloride. This generates an electrophilic allene which is able to crosslink the enzyme (Scheme III).

Scheme III

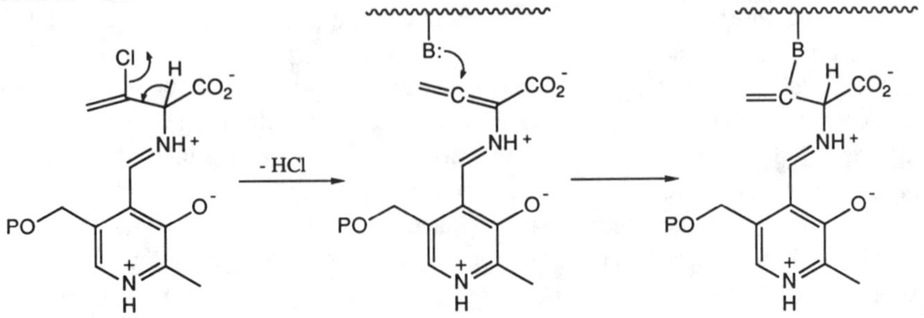

- HCl

B. D-alanine-D-alanine Ligase

The second enzyme in the D-alanine branch of peptidoglycan biosynthesis is the D-ala-D-ala ligase. The enzyme was initially characterized in 1962 by Neuhaus (51,52); D-alanine acts as both donor (N-terminal) and acceptor (C-terminal) in dipeptide formation. There are two kinetically distinct sites which have different K_m values associated with them. Studies have shown that the binding site for the N-terminal D-alanine is highly specific for this amino acid while the C-terminal site is less stringent and accepts a variety of D-amino acids (51,52). Both ATP and magnesium are required for ligase activity. Dipeptide formation probably occurs via phosphorylation of the carboxyl group on the donor D-alanine, thus generating a D-alanyl phosphoryl ester, followed by nucleophilic attack of the amino group on the acceptor D-alanine (Scheme IV). The enzyme is very specific for D-alanine, as L-alanine and other L-amino acids are not processed. Recently,

Scheme IV

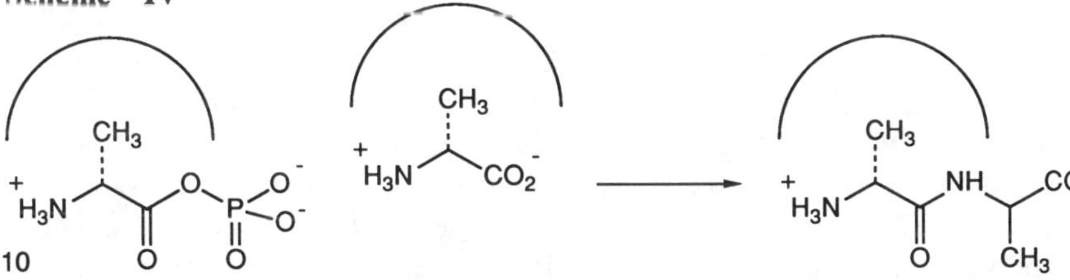

10

Walsh and co-workers (85, 19) have shown that both D-fluoroalanine and trifluoroalanine, which are alanine racemase inhibitors, are also D-ala-D-ala ligase substrates; K_m and k_{cat} values are much poorer as compared to D-alanine. This result explains the paradox in the inhibitory properties of these compounds: at high concentrations, fluoroalanine and trifluoroalanine are incorporated in the cell wall.

The gene encoding the D-Ala-D-Ala-Ligase enzyme, *ddl,* has been cloned and sequenced from *E. coli* (63) and *S. typhimurium* (15). Protein sequence homology is 39%, suggestive of a common ancestral gene. The *S. typhimurium* enzyme has been overproduced (36), providing large quantities of enzyme for X-ray crystallography and other enzymological studies. Daub *et al.* (15) have shown that insertion mutations in *ddl* are not lethal; thus, it is likely that there is a second ligase gene present in *Salmonella*.

Due to the importance of this enzyme in peptidoglycan biosynthesis, the design of inhibitors of D-ala-D-ala ligase are being actively pursued. It is known that cycloserine (**1**), a racemase inhibitor, is also an inhibitor of the ligase (55). Inhibition is reversible and cycloserine competes with D-ala at both the donor (K_i = 22 μM) and acceptor (K_i = 140 μM) sites (53).

Parsons *et al.* (58) have demonstrated that the phosphinate analog, 1-aminoethyl-[2-carboxy-2R-heptyl-1-ethyl]phosphonic acid (**11b**), is an inhibitor of the

D-ala-D-ala ligase from *E. coli*. Duncan and Walsh (20) have shown that inhibition of the *S. typhimurium* enzyme is ATP-dependent and follows slow-binding kinetics with fast k_{on} rates and relatively slow ($t_{1/2}$ of 8 hours) k_{off} rates. Further work (58) demonstrates that a series of phosphinic acid dipeptide analogs are potent inhibitors of the D-ala-D-ala ligase from *Enterococcus faecalis*. Structure-activity relationships of the phosphinic acid analogs shows that the optimal substituent on the amino terminal end is a methyl group while the R_1 substituent contributes to increased activity with bulk (thiomethyl = heptyl > methyl > hydrogen). Sterochemistry at both chiral centers is also very important, as one might expect.

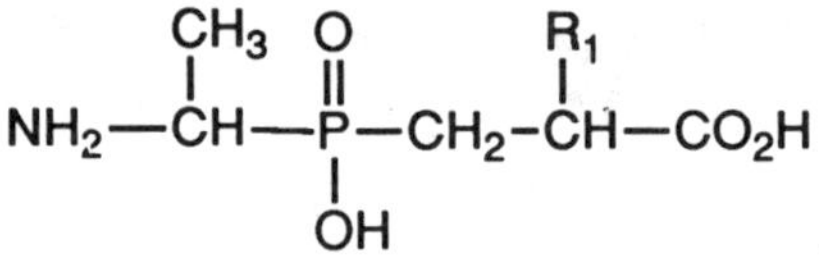

$$NH_2-CH(CH_3)-P(=O)(OH)-CH_2-CH(R_1)-CO_2H$$

R_1	Compound	IC_{50}
thiomethyl	**11a**	4 μM
n-heptyl	**11b**	4 μM
methyl	**11c**	35 μM
hydrogen	**11d**	535 μM

In order to elucidate the structure of the enzyme-inhibitor adduct, McDermott *et al.* (43) used solid-state, magic angle spinning NMR. By measuring ^{31}P chemical-shift anisotropies and ^{31}P-^{31}P dipole couplings, these workers found that inhibition of ligase from *S. typhimurium* by the methyl phosphinic acid (**11c**) occurs by phosphorylation of the phosphinate moiety, thus yielding a phosphoryl phosphinate as the enzyme-inhibitor complex. Precedent for this type of mechanism is the inhibition of glutamine synthetase by methionine sulfoximine and phosphinothricin (13). Even though a phosphoryl phosphinate has been demonstrated in the ligase-inhibitor complex, there is some question as to whether phosphorylation is a requisite of inhibition since noncleavable ATP analogs can substitute for ATP (18).

Other phosphonic acid inhibitors include D-Ala-P (20) and (3-amino-2-oxoalkyl) phosphonic acid (12). Enzyme inhibition by D-Ala-P follows slow-binding kinetics, in which $K_i = 0.5$ mM, $k_{on} = 27\ M^{-1}sec^{-1}$, and the half life for return of activity is 2 minutes. The L-isomer of Ala-P was ineffective. Inhibition of ligase by D-Ala-P was less effective than the alkylamino phosphinate, but the mechanism of inhibition is believed to be the same, as ATP is required for inhibition in both cases. Analogs of the proposed D-alanyl phosphate intermediate (**10** in Scheme IV) were synthesized and found to be modest inhibitors of the ligase. 3-Amino-2-oxobutylphosphonic acid (**12**) and the corresponding aza analog (**13**) were found to inhibit *E. coli* ligase with K_i values of 500 μM and 50 μM, respectively (12). The ability of these compounds to inhibit D-ala-D-ala ligase is consistent with the proposed acyl phosphate displacement mechanism.

Inhibition of ligase by 1-aminoethyl boronic acid (**7**) (Ala-B) was also investigated (18). Inhibition is time-dependent and ATP is required; nonhydrolyzable ATP analogs will not suffice, unlike inhibition with the phosphinate and phos-

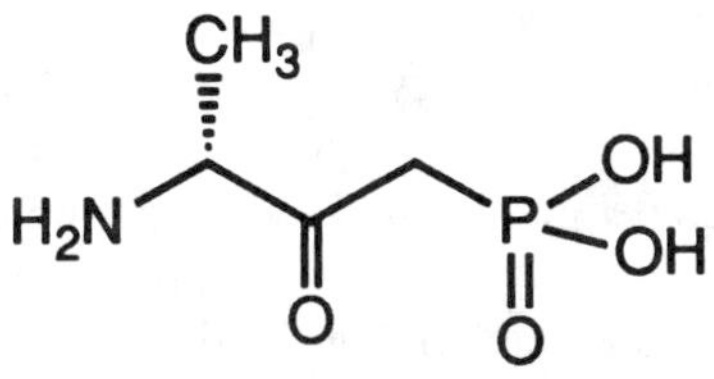

12 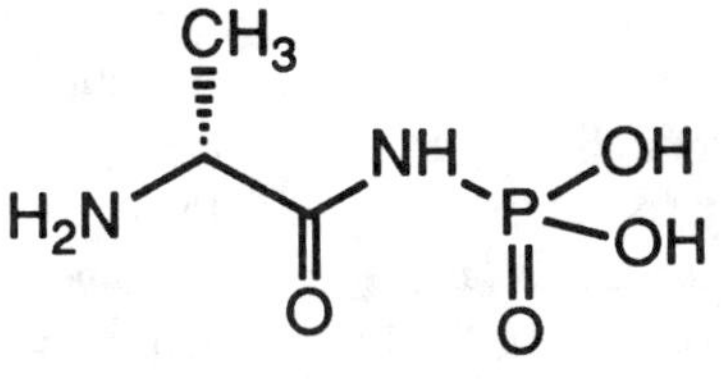 13

phonic acid analogs. Studies show that Ala-B can interact with the free enzyme and ATP to give a enzyme-inhibitor complex with a half-life for return of activity of roughly 2 hours *or* interact with enzyme, ATP, and D-alanine to give an Ala-B/D-alanine/ATP complex that has a half-life for return of enzyme activity of roughly 4.5 days. Thus, Ala-B can interact with the enzyme in a variety of ways, depending on what substrates (*e.g.*, D-alanine, ATP) are present.

It is apparent that inhibitors of the racemase also seem to be either inhibitors or substrates for the ligase. Cycloserine, Ala-P, and Ala-B are all racemase and ligase inhibitors while D-fluoroalanine and trifluoroalanine are substrates (albeit poor) for ligase and inhibitors for racemase. Thus, analogs of D-alanine, a substrate for both enzymes, lead to considerable overlap in the enzyme processing. This could provide synergistic behavior in the case of compounds that inhibit both enzymes but may also abrogate the inhibitory effect of a compound (racemase inhibitor is a ligase substrate). Therefore, substrate mimics must be carefully evaluated as two or more enzymes may be affected.

C. D-alanyl-D-alanine Adding Enzyme

The D-alanyl-D-alanine adding enzyme is an ATP-dependent enzyme which synthesizes UDP-muramyl-L-ala-D-glu-meso-A_2pm-D-ala-D-ala from UDP-muramyl-L-ala-D-glu-meso-A_2pm and D-ala-D-ala. The enzyme from *E. coli* has been partially purified (45) and limited kinetic information has been obtained. Duncan *et al.* (19) have recently purified this enzyme to homogeneity and find that it is a monomeric protein with a molecular weight of 49 kDa. Steady-state kinetic analysis indicates that the K_m for UDP-MurNAc-L-ala-D-glu-meso-A_2pm is 70 μM and the K_m for the second substrate, D-ala-D-ala, is 220 μM; the turnover number (k_{cat}) is 784 min^{-1}. The gene (*murF*) has been cloned and is currently being sequenced. Upon completion of the sequence, the gene can be placed in an overproducing vector to provide sufficient quantities of enzyme for X-ray analysis and enzyme screens to identify novel inhibitors.

Enzymological studies of the D-alanyl-D-alanine adding enzyme show that the enzyme is capable of discriminating between different dipeptides on the basis of the amino acid at the carboxyl end of the dipeptide (18). Thus, the dipeptide substrate may contain other amino acids on the amino terminus but appears to have an absolute requirement for D-alanine at the carboxyl terminus. This is in contrast to the D-ala-D-ala ligase, which has a high degree of specificity for the amino terminal amino acid, but can vary in the carboxyl amino acid (56). Although both enzymes have stringent requirements for D-alanine in at least one position of the dipeptide, the enzyme can be fooled into using D-alanine mimics as substrates. D-fluoroalanine and trifluoroalanine, both racemase inhibitors, are substrates for the ligase (giving the fluorinated dipeptide) and the adding enzyme (giving the fluorinated pentapeptide). Most likely, this precursor could be assembled into the peptidoglycan to provide structural integrity. These observations are consistent with self-reversal of antibacterial action with elevated levels of fluoroalanine (35).

III. UDP-MurNAc Tripeptide Branch

Although less studied than the alanine branch, the biosynthesis of UDP-Mur NAc-L-ala-D-glu-meso-A_2pm (UDP-MurNAc tripeptide) provides a variety of targets for the development of novel antibacterials. These enzymes are involved in the formation of the early parts of the pentapeptide and inhibition of any one of these enzymes should result in poorly crosslinked peptidoglycan, thus inhibiting bacterial growth.

A. The UDP-MurNAc Amino Acid Ligases

Elongation of the pentapeptide chain in peptidoglycan biosynthesis occurs by stepwise addition of the respective amino acid to the UDP-Mur-NAc backbone. Each reaction is catalyzed by a distinct ligase. All three enzymes utilize ATP for peptide-bond formation in a process that is similar to that observed in the D-alanyl-D-alanine adding enzyme. Although all three enzymes are potential targets for antibacterials, very little basic work has been done on enzyme or inhibitor design. Except for 1-aminoethylphosphonic acid, a weak inhibitor of the UDP-MurNAc-alanine ligase (4), there are no known inhibitors of the amino acid adding enzymes involved in peptidoglycan biosynthesis. Thus, these targets represent novel opportunities for drug discovery.

1. UDP-MurNAc-L-alanine Ligase

The first enzyme committed to biosynthesis of the pentapeptide chain is the UDP-MurNAc-L-alanine ligase. Early work by Lugtenberg and van Dam (41), investigating temperature-sensitive mutants of *E. coli,* and Hishinuma *et al.* (30), in which inhibition of this enzyme by glycine was detected, details most of the work done on this enzyme. Atherton *et al.* (4) have shown that 1-aminoethylphosphonic acid weakly inhibits UDP-MurNAc-L-alanine ligase as buildup of UDP-MurNAc occurred at high intracellular concentrations of inhibitor. This, however, appears to be a minor inhibitor for the ligase enzyme; the principal target is alanine racemase.

2. *UDP-MurNAc-L-alanine-D-glutamate Ligase*

This enzyme, also called D-glutamate adding enzyme, is responsible for the addition of D-glutamate to UDP-MurNAc-L-alanine. The enzyme from *E. coli* has been partially purified and has a MW of 54 kDa (45), with K_m values of 280 μM for D-glutamate and 5.5 μM for UDP-MurNAc-L-alanine. Enzyme studies show that the specificity of the UDP-MurNAc-L-ala site does not appear to be strict, as removal of the UMP moiety leads to a compound with reasonable specificity (Km = 200 μM). However, the D-glutamate site is very specific; only 2-amino-4-phosphonobutyric acid showed any inhibition (43% at 10 mM).

3. *UDP-MurNAc-L-alanine-D-glutamate-diaminopimelate Ligase*

The final step in the biosynthesis of the UDP-tripeptide is addition of meso-2,6-diaminopimelate to UDP-MurNAc-L-alanine-D-glutamate by UDP-MurNAc-L-alanine-D-glutamate-diaminopimelate ligase. Abo-Ghalia *et al.* (1) have partially purified the enzyme and prepared substrate analogs to determine the effect on activity. The specificity for UDP-MurNAc-L-alanine-D-glutamate is very strict, similar to other D-amino acid adding enzymes (19, 32, 33, 64, 94). For UDP-MurNAc-L-alanine-D-glutamate-diaminopimelate ligase, no UDP-MurNAc derivative was found to be a substrate, and limited inhibition of enzyme activity by UDP-MurNAc-L-ala, D-lactyl-L-ala-D-glu, propionyl-L-ala-D-glu, and Mur NAc-L-ala-D-glu shows that extensive recognition of the entire substrate molecule must exist.

B. *D-amino Acid Transaminase*

In the biosynthesis of the pentapeptide, D-glutamate is required. Since glutamate naturally occurs as the L-isomer, D-glutamate must be synthesized. D-amino acid transaminase catalyzes the transamination of a wide spectrum of α-amino acids and α-keto acids; however, it is specific for the D-isomer of the amino acid (34). Thus, D-amino acid transaminase may be a good target for antibacterial drug design. The enzyme, purified from *B. subtilis* or *B. sphaericus*, is a dimer of MW 60 kDa, containing two pyridoxal-phosphate molecules per dimer (95). Synthesis of D-glutamate is accomplished by transamination of α-ketoglutarate by D-alanine via a mechanism identical to other PLP-requiring transamination enzymes (83). Manning and colleagues (44, 16) have shown that the enzyme from thermophilic bacteria can undergo mutation of all three cysteine residues without any change in enzyme activity. Thus, none of the cysteine residues are essential for catalysis.

The D-alanine analogues β-fluoro-D-alanine, β-chloro-D-alanine, β-cyano-D-alanine, and O-carbamyl-D-serine are all inhibitors of the transaminase (as well as alanine racemase). Inhibition occurs by elimination of the β-leaving group after transaldimination and subsequent attack of an enzyme nucleophile generating a crosslinked enzyme complex (Scheme V) (73,76). D-cycloserine (**1**) is also a potent, active site-directed inhibitor of D-amino acid transaminase (as well as alanine racemase and D-ala-D-ala ligase). Studies have demonstrated that inhibition is irreversible at pH 8.0 with a Ki of 25 nM (42, 74). Complete activity is restored with dialysis at pH 7.0, however, suggesting that the inactive complex consists of a weak covalent bond or perhaps no covalency at all.

A more specific inhibitor of D-amino acid transaminase is D-vinylglycine (**14** in Scheme VI). Soper *et al.* (76) have shown that D-vinylglycine is an irreversible inactivator of D-amino acid transaminase from *B. subtilis* and *B. sphaericus* (Scheme VI). The number of catalytic events that occur before inactivation is about 450 for the *B. sphaericus* enzyme and 800 for the *B. subtilis* enzyme. The production of 2-keto-3-butenoate during turnover (path 1) leaves the enzyme in the pyridoxamine-P form, which is unable to undergo further reaction with another molecule of vinylglycine (or any other amino acid) in the absence of keto acid. Thus, the enzyme is protected from irreversible enzyme alkylation (path 2) when α-ketoglutarate is not present in the assay.

D,L-Gabaculine (**15**) (5-amino-1,3-cyclohexyldienyl carboxylic acid), a known inhibitor of γ-aminobutyrate transaminase (60,61), also inhibits D-amino acid transaminase. Using the *B.sphaericus* enzyme, Soper and Manning (74) have shown that inactivation is time-dependent with a K_i of 100 μM. Inhibition is blocked by saturating concentrations of either D-alanine or α-ketoglutarate.

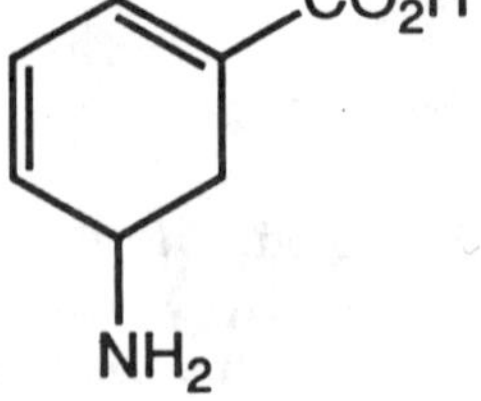

CO_2H

NH_2

15

C. Glutamate Racemase

Although D-amino acid transaminase provides most bacteria with D-glutamate, glutamate racemase is present in some bacteria, such as *Lactobacillus fermenti* (5) and *Pediococcus pentosaceus* (48,49). The enzyme from *Pediococcus pentosaceus* has recently been cloned, expressed in *E. coli,* and purified to homogeneity. Initial studies indicate that the enzyme has a MW of 40 kDa, glutamic acid

Scheme V

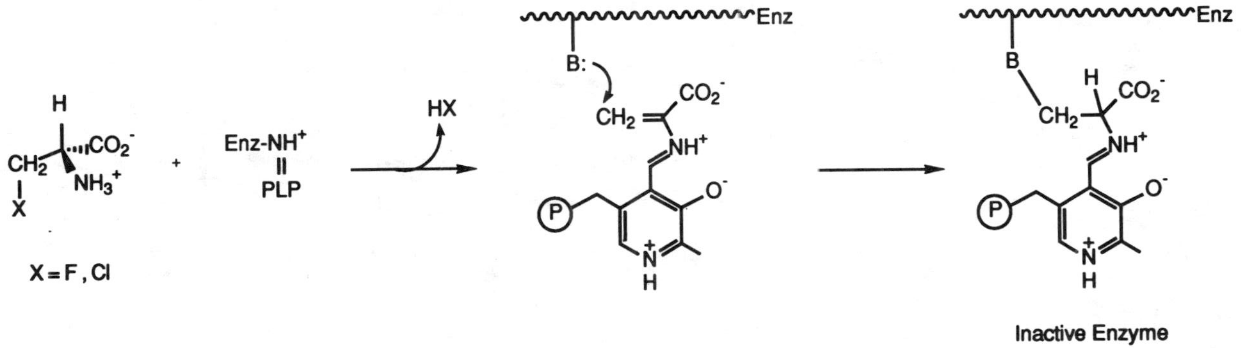

$X = F, Cl$

$Enz\text{-}NH^+ = PLP$

HX

Inactive Enzyme

Scheme VI

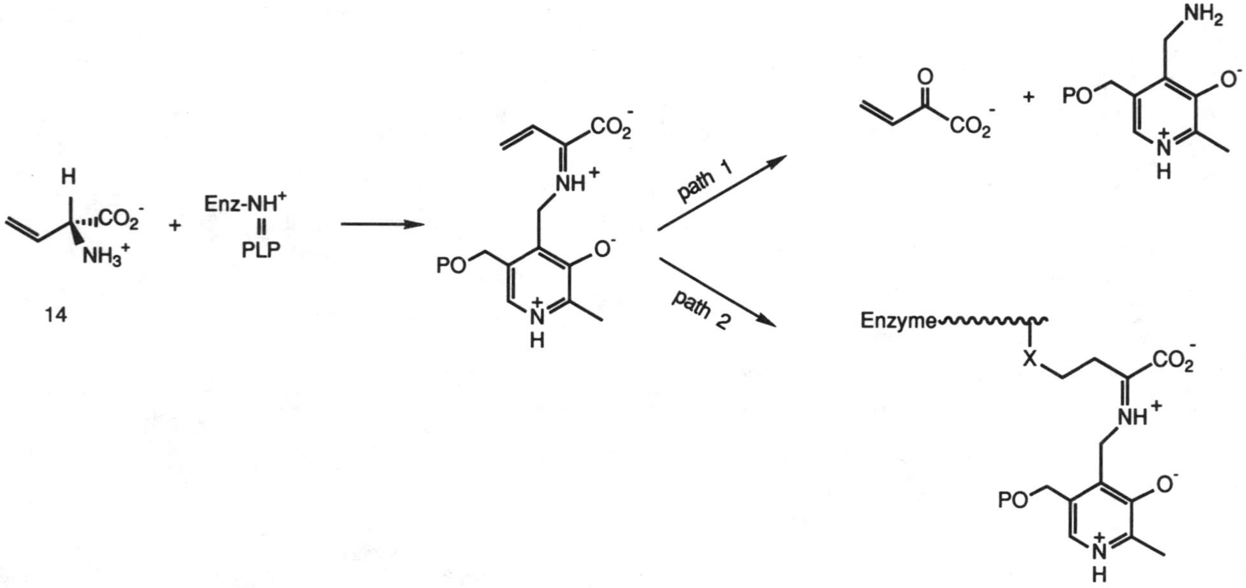

14

Enz-NH$^+$ = PLP

path 1

path 2

Enzyme

is the sole substrate (Km = 14 mM for the D-isomer and 10 mM for the L-isomer), and no cofactors are required. Glutamate racemase is very sensitive to sulfhydryl reagents (e.g., p-chloromercuribenzoate, Ellman's reagent) and reducing thiol is necessary for stability. Thus, this enzyme is similar to diaminopimelic acid epimerase (*vide infra*) and proline racemase (12,26). The enzyme appears exclusively in lactic acid bacteria, such as *Pediococcus, Lactobacillus,* and *Leuconostoc* species (49), and thus does not provide an antibacterial target in clinically important organisms.

D. Diaminopimelic Acid Epimerase

The enzyme diaminopimelic acid epimerase (A_2pm epimerase) catalyzes the epimerization of 2,6-diaminopimelic acid. Diaminopimelic acid is found only in gram-negative bacteria and is involved in crosslinking the peptidoglycan (3,67). In the biosynthesis of meso-A_2pm, L,L-A_2pm is produced and then converted to meso-A_2pm by the epimerase. Wiseman and Nichols (93) have purified the enzyme to homogeneity from *E. coli;* it has a molecular weight of 34 kDa with no requirement for pyridoxal-phosphate or metal ions as cofactors. Tritium exchange experiments suggest that the enzyme utilizes a two-base mechanism for proton translocation. In this way, it resembles proline racemase (11,26). Isotope-effect experiments predict that only one of the proton sites is a thiol and iodoacetamide tritration supports this assertion.

The gene for A_2pm epimerase (*dapF*) from *E. coli* has been identified and *dapF* mutants isolated (62). Strains containing a mutational insertion in *dapF* accumulate large amounts of L,L-A_2pm, confirming that L,L-A_2pm is a physiological substrate for this enzyme; surprisingly, these strains are viable and do not require meso-A_2pm for growth. Although L,L-A_2pm accumulates, meso-A_2pm can be detected, correlating with the A_2pm^+ phenotypes. How is meso-A_2pm synthesized in the absence of a functional *dapF* gene? One possibility is the presence of a second epimerase gene whose activity is undetectable under the assay conditions. However, the large amount of L,L-A_2pm synthesized suggests that the activity of this enzyme would be very poor. A second possibility is that there is another enzyme that, under stressful conditions, can make meso-A_2pm. Meso-A_2pm dehydrogenase, which converts tetrahydrodipicolinate to meso-A_2pm, has been found in some bacteria but not *E. coli* (46). A third possibility is the presence of a nonspecific amino acid racemase that operates with low efficiency, thus making enough meso-A_2pm to satisfy the requirement for peptidoglycan synthesis.

Another study by Baumann et al. (10) has shown that 3-chloro-2,6-diaminopimelate (3-chloro-A_2pm) (**16**) inhibited growth of *E. coli*. These cells can be rescued by exogenously supplying meso-A_2pm. These authors isolated a A_2pm-requiring auxotroph that needed either L,L-A_2pm or meso-A_2pm for growth. Even in the presence of L,L-A_2pm, 3-chloro-A_2pm inhibited growth. Thus, inhibition of the epimerase appears to be lethal, contradicting the studies by Richaud et al. (62). However, these two results can be reconciled if one assumes that multiple epimerases are inhibited by 3-chloro-A_2pm.

3-chloro-A_2pm inhibits enzyme isolated from *E. coli* with a K_i of 200 nM (10). Inhibition by 3-chloro-A_2pm is unusual in that it is actually a very poor substrate and turns over 100 times slower than the normal substrate. Baumann *et al.* (9) propose that enzymic processing of 3-chloro-A_2pm occurs by elimination of HCl followed by cyclization and release of product (Scheme VII); the planar analogue (**17**) probably binds tightly to the enzyme. Thus, 3-chloro-A_2pm is a mechanism-based inhibitor; by itself the compound is innocuous, but on enzymic processing (elimination of HCl) it is transformed into a potent competitive inhibitor. It is interesting to note that although the enzyme catalyzes the epimerization of L,L-A_2pm 100 times faster than the cyclization of 3-chloro-A_2pm to piperidine

Scheme VII

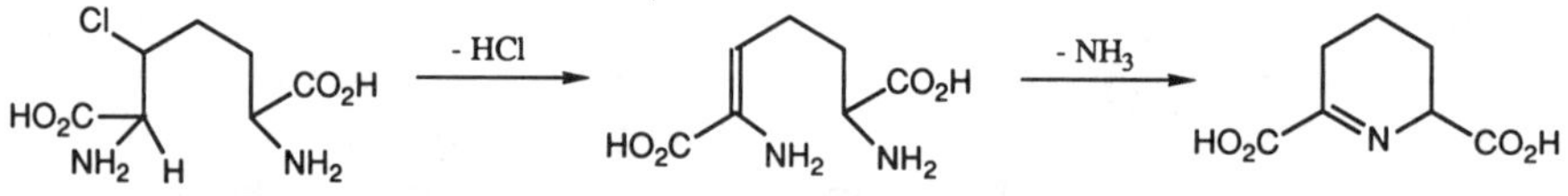

16 —(- HCl)→ 17 —(- NH_3)→

dicarboxylic acid, it is actually more *efficient* in catalyzing the cyclization than the epimerization. This is apparent on analysis of V_{max}Km parameters for both compounds. (Chloro-A_2pm is roughly 20 times better than L,L-A_2pm.)

Other diaminopimelate analogs, such as meso-lanthionine (**18**) and N-hydroxydiaminopimelate (**19**), have been shown to inhibit the *E. coli* enzyme with K_i values of 180 μM and 5.6 μM, respectively (37). Both compounds are competitive inhibitors, with no time-dependent inactivation observed. A potent irreversible inhibitor of the epimerase has recently been reported by Gerhart *et al.* (28). Aziridino-diaminopimelic acid [1-(4-amino-carboxybutyl)aziridine-2-carboxylic acid; **20**] rapidly inactivates A_2pm epimerase with Ki $\geqslant$ 25 μM and $1/k_{inact} \leqslant 10$ seconds.

IV. Conclusions

The majority of scientific effort in the design and synthesis of new antibacterials over the past 30 years has focused on modifications of preexisting antibiotic classes, such as the β-lactam, macrolide, and tetracycline families. Although these classes of antibiotics are clinically effective, resistance is a major concern. One way to circumvent resistance is to discover and develop novel antibacterials, targeting essential enzymes of bacterial pathways for which no known inhibitors exist. The peptidoglycan biosynthetic pathway offers unexploited, essential enzyme targets that have no mammalian counterpart.

Of the enzymes involved in the D-alanine biosynthetic branch, alanine race-

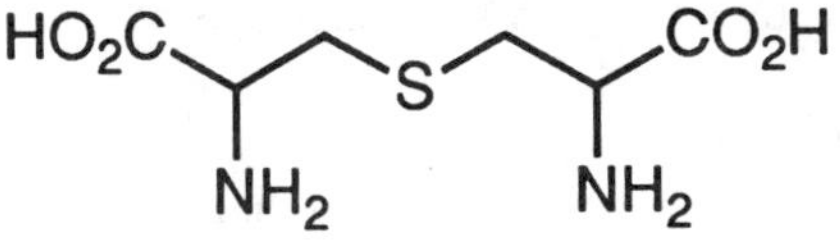

18

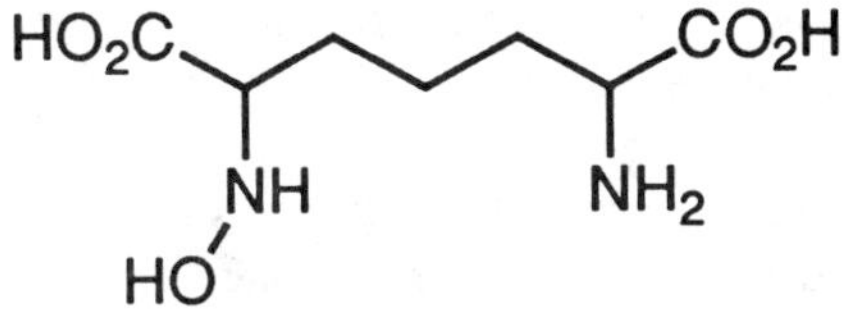

19

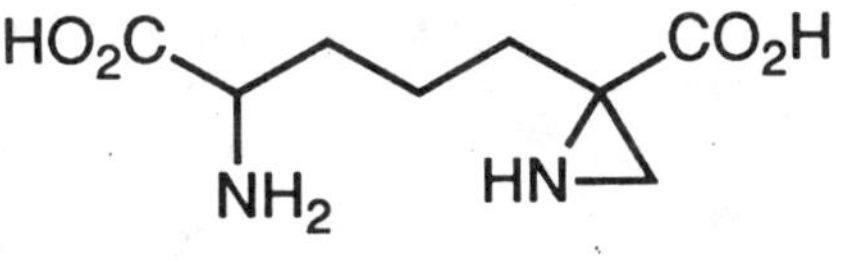

20

mase has been investigated in the most detail. Although alanine racemase activity is essential and a great deal of work has been spent investigating the enzyme mechanism and inhibition, racemase inhibitors are not viable antibiotics. This is partly due to exclusion of inhibitors from the bacterial cytoplasm where the enzyme is located. Ala-P and trifluoroalanine are not transported across the cell membrane while fluoroalanine and cycloserine are transported via the alanine/glycine channel (53). However, both of these latter compounds are toxic; fluoroalanine is metabolized to the highly toxic fluoroacetate in mammalian cells while cycloserine is neurotoxic (53). Ala-P was found to be very safe, but uptake was not observed to any substantial degree. To overcome this obstacle, the Roche group investigated ways to transport the inhibitor across the cell wall to reach its target. The dipeptide L-alanyl-ala-P enters bacteria via a dipeptide transport system and is hydrolyzed by a cytoplasmic aminopeptidase, yielding Ala-P. This approach was very successful in a variety of gram-positive and gram-negative bacterial strains, but some clinically important strains did not transport the dipeptide into the cell. Thus, alanyl-ala-P was not further advanced.

D-alanyl-D-alanyl ligase and the D-ala-D-ala adding enzyme have just recently been cloned and large amounts of these enzymes are available for the first time. The ligase inhibitors synthesized by the Merck group (structures **11a–d**) have been shown to have excellent enzyme inhibitory potency; however, the inability of these compounds to penetrate the cell wall results in limited antibacterial activity (58). The thiomethylphosphinic acid (**11a**) was found to be marginally active against a wide variety of gram-negative and gram-positive bacterial strains, although some synergism with the racemase inhibitor fluoroalanine was noted. Recently, Courvalin *et al.* have shown that the vancomycin-resistant *Enterococcus faecium* strain BM4147 produces a protein (designated VanA) which has sequence similarity to D-ala-D-ala ligase and is inducible by vancomycin (21, 39). VanA is a D,D-dipeptide ligase that is structurally and functionally related to the D-ala-D-ala ligase but has modified substrate specificity (9). Thus, resistance to ligase inhibitors may become widespread as VanA or other ligases with alternate substrate specificity are obtained.

Although the D-ala-D-ala adding enzyme provides an attractive target, obtaining sufficient quantities of enzyme and identifying inhibitors has delayed advancements. This rationale also holds for the amino acid adding enzymes involved in the MurNAc tripeptide branch.

D-amino acid transaminase has been investigated in considerable detail by Manning and co-workers (73–76). The enzyme has been isolated from *Bacillus sphaericus* and *Bacillus subtilis* and the gene has been cloned; deletion mutants have not been described. Inhibitors of the transaminase have been shown to have antibacterial activity; however, none of these compounds inhibits transaminase solely. Although gabaculine specifically inhibits the cell wall enzyme D-amino acid transaminase, it also inhibits a variety of PLP-dependent enzymes, resulting in significant toxicity (68,57). Thus, it is not useful as an antibacterial.

Diaminopimelate epimerase provides bacteria with meso-diaminopimelate, a direct precursor of lysine and an essential component of the cell wall in gram-negative bacteria. The enzyme from *E. coli* has been isolated and purified, and inhibitors have been synthesized. One inhibitor, 3-chloro-2,6-diaminopimelic acid, has also been shown to inhibit growth of *E. coli* (10). However, Richaud *et al.* (62) have shown that cells carrying a deletion of the gene that codes for A_2pm epimerase accumulate L,L-diaminopimelate but do not require meso-diaminopimelate to grow. Thus, the bacteria must have an alternative path to supply meso-A_2pm. The most likely explanation is that two genes are present that code for A_2pm epimerase and both are inhibited by 3-chloro-A_2pm. An additional caution in selecting A_2pm epimerase enzyme as a target is that it is required for peptidoglycan biosynthesis *only* in gram-negative bacteria. Gram-positive bacteria utilize L-lysine for peptidoglycan crosslinking. Thus, a program which develops epimerase inhibitors will not provide a broad-spectrum antibacterial, but rather a gram-negative specific agent.

V. Future Directions

In the design of novel antibacterials, there are a few considerations one must take into account. One concern is the importance of the pathway in bacterial growth; another, obtaining sufficient quantities of the target enzyme so that biochemical and biophysical experiments can be performed; third, that enzymes which perform the same function (*e.g.*, two alanine racemases) must *all* be inhibited. The cytosolic enzymes involved in the biosynthesis of the pentapeptide bridge in the peptidoglycan are all attractive targets; however, only alanine racemase, D-ala-D-ala ligase, D-amino acid transaminase, and diaminopimelate epimerase have been shown to be essential by known inhibitors.

Alanine racemase is the most extensively studied of the cytosolic enzymes and both genetic and biochemical studies have shown that it is essential for bacterial cell wall biosynthesis. Despite decades of work by industrial and academic laboratories, no racemase inhibitor has found its way to the market. In order to overcome the inherent difficulties of these compounds (all are substrate analogs), new compounds must be identified which inhibit the enzyme but are not structural analogs of alanine. Leads can be obtained by screening natural product extracts with purified enzyme. The screening strategy must take into account that two enzymes (*dadB* and *alr* from *Salmonella*) are present and potential leads *must inhibit both enzymes*. A prototypical *in vitro* screening assay has been described by Vicario *et al*. (81) in detecting both racemase and D-ala-D-ala ligase inhibitors. Compounds which inhibit both enzymes can be further characterized for antibacterial activity by using standard minimal inhibitory concentration (MIC) assays. Since bacterial cell wall penetration may be a problem, these secondary screens may be inconclusive. It is possible that novel compounds which inhibit alanine racemase *in vitro* but do not penetrate the cell wall can be modified to retain inhibitory properties and to have increased cellular permeability. Another approach may be the identification of lead compounds which penetrate the cell wall but are weakly inhibitory. This may provide a unique starting point for a semisynthetic approach. One can also envision this strategy for D-ala D-ala ligase and D-amino acid transaminase, where large quantities of enzyme are available for screening. Although irreversible, mechanism-based inhibitors may not be discovered through a screening program, potency may not be an issue for a novel class of compounds. Instead, one may select for good penetration into bacterial cytoplasm that may, in the case of a classical competitive antagonist, slow down peptidoglycan biosynthesis long enough for the bacterial autolysins to degrade the cell wall. This is one way that inhibitors of the cytosolic enzymes involved in pentapeptide bond formation may be a useful addition to existing antibacterials.

Another important aspect in the design of novel antibacterials is that coresistance to known antibiotic classes is not expected. Thus, a structurally unique class of compounds that inhibit an enzyme that has not been previously targeted will complement the numerous antibacterials currently available and may prove useful in the treatment of highly resistant bacterial infections.

References

1. Abo-Ghalia, M., C. Michaud, D. Blanot, and J. van Heijenoort. 1985. Specificity of the uridine-diphosphate-N-acetylmuramyl-L-alanyl-D-glutamate: meso-2,6-diaminopimelate synthetase from *Escherichia coli*. Eur. J. Biochem. **153:**81–87.
2. Adams, E. 1976. Catalytic aspects of enzymatic racemization. Adv. Enzymol. Relat. Areas Mol. Biol. **44:**69–138.
3. Antia, M., D.S. Hoare, and E. Work. 1957. The stereoisomers of α, ε-diaminopimelic acid. Biochem. J. **65:**448–459.
4. Atherton, F.R., M.J. Hall, C.H. Hassall, R.W. Lambert, W.J. Lloyd, and P.S. Ringrose. 1979. Phosphonopeptides as antibacterial agents: mechanism of action of alaphosphin. Antimicrob. Ag. Chemother. **15:**696–705.

5. Ayengar, P., and E. Roberts. 1952. Utilization of D-glutamic acid by *Lactobacillus arabinosus:* glutamic racemase. J. Biol. Chem. **197:**453–459.

6. Badet, B., and C.T. Walsh. 1985. Purification of an alanine racemase from *Streptococcus faecalis* and analysis of its inactivation by (1-aminoethyl)phosphonic acid enantiomers. Biochemistry **24:**1333–1341.

7. Badet, B., K. Inagaki, K. Soda, and C.T. Walsh. 1986. Time-dependent inhibition of *Bacillus stearothermophilus* alanine racemase by (1-aminoethyl)phosphonate isomers by isomerization to noncovalent slowly dissociating enzyme-(1-aminoethyl) phosphonate complexes. Biochemistry **25:**3275–3282.

8. Badet, B., D. Roise, and C.T. Walsh. 1984. Inactivation of the *dadB Salmonella typhimurium* alanine racemase by D and L isomers of β-substituted alanines: kinetics, stoichiometry, active site peptide sequencing, and reaction mechanism. Biochemistry **23:**5188–5194.

9. Baumann, R.J., E.H. Bohme, J.S. Wiseman, M. Vaal, and J.S. Nichols. 1988. Inhibition of *Escherichia coli* growth and diaminopimelic acid epimerase by 3-chlorodiaminopimelic acid. Antimicrob. Ag. Chemother. **32:**1119–1123.

10. Bugg, T.D.H., S. Dutka-Malen, M. Arthur, P. Courvalin, and C.T. Walsh. 1991. Identification of vancomycin resistance protein VanA as a D-alanine:D-alanine ligase of altered substrate specificity. Biochemistry **30:**2017–2021.

11. Cardinale, G.J., and R.H. Abeles. 1968. Purification and mechanism of action of proline racemase. Biochemistry **7:**3970–3978.

12. Chakravarty, P.K., W.J. Greenlee, W.H. Parsons, A.A. Patchett, P. Combs, A. Roth, R.D. Busch, and T.N. Mellin. 1989. (3-Amino-2-oxoalkyl)phosphonic acids and their analogues as novel inhibitors of A-alanine:D-alanine ligase. J. Med. Chem. **32:**1886–1890.

13. Colanduoni, J.A., and J.J. Villafranca. 1986. Inhibition of *Escherichia coli* glutamine synthetase by phosphinotherin. Bioorg. Chem. **14:**163–169.

14. Copie, V., W.S. Faraci, C.T. Walsh, and R.G. Griffin. 1988. Inhibition of alanine racemase by alanine phosphonate: detection of an imine linkage to pridoxal 5′-phosphate in the enzyme-inhibitor complex by solid state ^{15}N nuclear magnetic resonance. Biochemistry **27:**4966–4969.

15. Daub, E., L.E. Zawadzke, D. Botstein, and C.T. Walsh. 1988. Isolation, cloning, and sequencing of the *Salmonella typhimurium ddlA* gene with purification and characterization of its product, D-alanine: D-alanine ligase (ADP forming). Biochemistry **27:**3701–3708.

16. del Pozo, A.M., M. Merola, H. Ueno, J.M. Manning, K. Tanizawa, K. Nishimura., S. Asano, H. Tanaka, K. Soda, D. Ringe, and G.A. Petsko. 1989. Activity and spectroscopic properties of bacterial D-amino acid transaminase after multiple site-directed mutagenesis of a single tryptophan residue. Biochemistry **28:**510–516.

17. Dulaney, E.L. 1970. 1-Aminoethylphosphonic acid, an inhibitor of bacterial cell wall synthesis. J. Antibiotics **23:**567–568.

18. Duncan, K., W.S. Faraci, D.S. Matteson, and C.T. Walsh. 1989. (1-Aminoethyl) boronic acid: a novel inhibitor for *Bacillus stearothermophilus* alanine racemase and

Salmonella typhimurium D-alanine: D-alanine ligase (ADP forming). Biochemistry **28:**3541–3549.

19. Duncan, K., J. van Heijenoort, and C.T. Walsh. 1990. Purification and characterization of the D-alanyl-D-alanine-adding enzyme from *Escherichia coli*. Biochemistry **29:**2379–2386.

20. Duncan, K., and C.T. Walsh. 1988. ATP-dependent inactivation and slow binding inhibition of *Salmonella typhimurium* D-alanine: D-alanine ligase (ADP) by (aminoalkyl)phosphinate and aminophosphonate analogues of D-alanine. Biochemistry **27:**3709–3714.

21. Dutka-Malan, S., C. Molinas, M. Arthur, and P. Courvalin. 1990. The VanA glycopeptide resistance protein is related to D-alanyl-D-alanine ligase cell wall biosynthesis enzymes. Mol. Gen. Genet. **224:**364–372.

22. Erion, M.D., and C.T. Walsh. 1987. 1-Aminocyclopropanephosphonate: time-dependent inactivation of 1-aminocyclopropanecarboxylate deaminase and *Bacillus stearothermophilus:* alanine racemase by slow dissociation behavior. Biochemistry **26:**3417–3425.

23. Esaki, N., and C.T. Walsh. 1986. Biosynthetic alanine racemase of *Salmonella typhimurium:* purification and characterization of the enzyme encoded by the *alr* gene. Biochemistry **25:**3261–2367.

24. Faraci, W.S., and C.T. Walsh. 1988. Racemization of alanine by the alanine racemases from *Salmonella typhimurium* and *Bacillus stearothermophilus:* energetic reaction profiles. Biochemistry **27:**3267–3276.

25. Faraci, W.S., and C.T. Walsh. 1989. Mechanism of inactivation of alanine racemase by β,β,β-trifluoroalanine. Biochemistry **28:**431–437.

26. Fisher, L.M., J.G. Belasco, T.W. Bruice, W.J. Albery, and J.R. Knowles. 1986. Proline racemase: oversaturation and interconversion of enzyme forms. p. 205–216. *In:* P.A. Frey (ed.), Mechanisms of Enzymic Reactions: Stereochemistry. Elsevier Scientific, Amsterdam.

27. Galakatos, N.G., E. Daub, D. Botstein, and C.T. Walsh. 1986. Biosynthetic *alr* alanine racemase from *Salmonella typhimurium:* DNA and protein sequence determination. Biochemistry **25:**3255–3260.

28. Gerhart, F., W. Higgins, C. Tardif, and J. Ducep. 1990. 2-(4-Amino-4-carboxybutyl) aziridine-2-carboxylic acid. A potent irreversible inhibitor of diaminopimelic acid epimerase. Spontanious formation from α-(halomethyl) diaminopimelic acids. J. Med. Chem. **33:**2157–2162.

29. Griffin, R.G., W.P. Aue, R.A. Haberkorn, G.S. Harbison, J.H. Herzfeld, E.M. Menger, M.G. Munowitz, E.T. Olejniczak, D.P. Raleigh, J.E. Roberts, D.J. Ruben, A. Schmidt, S.O. Smith, and S. Vega. 1988. Magic-angle sample spinning. p. 203–266. *In* B. Maraviglia (ed.), Physics of NMR Spectroscopy in Biology and Medicine. Soc. Etal. de Fisicia, Rome.

30. Hishinuma, F., K. Izaki, and H. Takahashi. 1971. Inhibition of L-alanine adding enzyme by glycine. Agric. Biol. Chem. **35:**2050–2058.

31. Inagaki, K., K. Tanizawa, B. Badet, C.T. Walsh, H. Tanaka, and K. Soda. 1986. Thermostable alanine racemase from *Bacillus stearothermophilus:* molecular cloning of the gene, enzyme purification, and characterization. Biochemistry **25:**3268–3274.

32. Ito, E., and J.L. Strominger. 1964. Enzymatic synthesis of the peptide in bacterial uridine nucleotides, J. Biol. Chem. **239:**210–214.

33. Ito, E., and J.L. Strominger. 1973. Enzymatic synthesis of the peptide in bacterial uridine nucleotides. J. Biol. Chem. **248:**3131–3136.

34. Jones, W.M., T.S. Soper, H. Ueno, and J.M. Manning. 1985. D-Glutamate D-amino acid transaminase from bacteria. Methods in Enzymol. **113:**108–113.

35. Kahan, F.M., and H. Kropp. 1975. MK641/MK642 A fixed-ratio combination (1): rationale; The sequential blockade of bacterial cell wall biosynthesis. Abstracts, 15th Interscience Conference on Antimicrobial Agents and Chemotherapy, Washington, D.C., Sept. 1975, No. 103.

36. Knox, J.R., H. Liu, C.T. Walsh, and L.E. Zawadzke. 1989. D-alanine-D-alanine ligase (ADP) from *Salmonella typhimurium*. Overproduction, purification, crystallization and preliminary X-ray analysis. J. Mol. Biol. **205:**461–463.

37. Lam, L.K.P., L.D. Arnold, T.H. Kalantar, J.G. Kelland, P.M. Lane-Bell, M.M. Palcic, M.A. Pickard, and J.C. Vederas. 1988. Analogs of diaminopimelic acid as inhibitors of meso-diaminopimelate dehydrogenase and L,L-diaminopimelate epimerase. J. Bio. Chem. **263:**11814–11819.

38. Lambert, M.P., and F.C. Neuhaus. 1972. Mechanism of D-cycloserine action: alanine racemase from *Escherichia coli* W. J. Bacteriol. **110:**978–987.

39. LeClerq, R., E. Derlot, J. Duval, and P. Courvalin. 1988. Plasmid-mediated resistance to vancomycin and teicoplanin in *Enterococcus faecium*. New Engl. J. Med. **319:**157–161.

40. Likos, J.J., H. Ueno, R.W. Feldhaus, and D.E. Metzler. 1982. A novel reaction of the coenzyme of glutamate decarboxylase with serine O-sulfate. Biochemistry **21:**4377–4386.

41. Lugtenberg, E.J.J., and A.S. van Dam. 1972. Temperature-sensitive mutants of *Escherichia coli* K-12 with low activities of the L-alanine adding enzyme and the D-alanyl-D-alanine adding enzyme. J. Bacteriol. **110:**35–40.

42. Martinez-Carrion, M., and W.T. Jenkins. 1965. D-alanine-D-glutamate transaminase. Inhibitors and the mechanism of transamination of D-amino acids. J. Biol. Chem. **240:**3547–3552.

43. McDermott, A.E., F. Creuzet, R.G. Griffin, L.E. Zawadzke, Q. Ye, and C.T. Walsh. 1990. Rotational resonance determination of the structure of an enzyme-inhibitor complex: phosphorylation of an (aminoalkyl)phosphinate inhibitor of D-alanyl-D-alanine ligase by ATP. Biochemistry. **29:**5767–5775.

44. Merola, M., A.M. del Pozo, H. Ueno, P., Recsei, A. DiDonato, J.M. Manning, K. Tanizawa, Y. Masu, S. Asano, H. Tanaka, K. Soda, D. Ringe, and G.A. Petsko. 1989. Site-directed mutagenesis of the cysteinyl residues and the active site residue of bacterial D-amino acid transaminase. Biochemistry **28:**505–509.

45. Michaud, C., D. Blanot, B. Flouret, and J. van Heijenoort. 1987. Partial purification and specificity studies of the D-glutamate-adding and D-alanyl-D-alanine-adding enzymes from *Escherichia coli* K12. Eur. J. Biochem. **166:**631–637.
46. Misono, H., H. Togawa, T. Yamamoto, and K. Soda. 1979. meso-α,ε-Diaminopimelate D-dehydrogenase: distribution and the reaction product. J. Bacteriol. **137:**22–27.
47. Morrison, J.F., and C.T. Walsh. 1988. The behavior and significance of slow-binding enzyme inhibitors. Adv. Enzymol. Relat. Areas Mol. Biol. **61:**201–301.
48. Nakajima, N., K. Tanizawa, H. Tanaka, and K. Soda. 1986. Cloning and expression in *Escherichia coli* of the glutamate racemase gene from *Pediococcus pentosaceus*. Agric. Biol. Chem. **50:**2823–2830.
49. Nakajima, N., K. Tanizawa, H. Tanaka, and K. Soda. 1988. Distribution of glutamate racemase in lactic acid bacteria and further characterization of the enzyme from *Pediococcus pentosaceus*. Agric. Biol. Chem. **52:**3099–3104.
50. Neidhart, D.J., M.D. Distefano, K. Tanizawa, K. Soda, C.T. Walsh, and G.A. Petsko. 1987. X-ray crystallographic studies of the alanine-specific racemase from *Bacillus stearothermophilus*. J. Biol. Chem. **262:**15323–15326.
51. Neuhaus, F.C. 1962. The enzymatic synthesis of D-alanyl-D-alanine. I. Purification and properties of D-alanyl-D-alanine synthetase. J. Biol. Chem. **237:**778–786.
52. Neuhaus, F.C. 1962. The enzymatic synthesis of D-alanyl-D-alanine. II. Kinetic studies on D-alanyl-D-alanine synthetase. J. Biol. Chem. **237:**3128–3135.
53. Neuhaus, F.C., and W.P. Hammes. 1987. Inhibition of cell wall biosynthesis by analogues of alanine. p. 45–99. *In* D.J. Tipper (ed.), Antibiotic Inhibitors of Bacterial Cell Wall Biosynthesis. Pergamon Press, New York.
54. Neuhaus, F.C., and W.P. Hammes. 1981. Inhibition of cell wall biosynthesis by analogues of alanine. Pharmacol. Ther. **14:**265–319.
55. Neuhaus, F.C., and J.L. Lynch. 1964. The enzymatic synthesis of D-alanyl-D-alanine. III. On the inhibition of D-alanyl-D-alanine synthetase by the antibiotic D-cycloserine. Biochemistry **3:**471–480.
56. Neuhaus, F.C., and W.G. Struve. 1965. Enzymatic synthesis of analogues of the cell wall precursor. I. Kinetics and specificity of uridine diphospho-N-acetylmuramyl-L-alanyl-D-glutamyl-L-lysine:D-alanyl-D-alanine ligase (adenosine diphosphate) from *Streptococcus faecalis* R. Biochemistry **4:**120–131.
57. Palfreyman, M.G., P.J. Schechter, W.R. Buckett, G.P. Tell, and J. Koch-Weser. 1981. The pharmacology of GABA-transaminase inhibitors. Biochem. Pharm. **30:**817–824.
58. Parsons, W.H., A.A. Patchett, H.G. Bull, W.R. Schoen, D. Taub, J. Davidson, P.L. Combs, J.P. Springer, H. Gadebusch, B. Weissberger, M.E. Valiant, T.N. Mellin, and R.D. Busch. 1988. Phosphinic acid inhibitors of D-analyl-D-alanine ligase. J. Med. Chem. **31:**1772–1778.
59. Patchett, A.A., D. Taub, B. Weissberger, M.E. Valiant, H. Gadebusch, N.A. Thornberry, and H.G. Bull. 1988. Antibacterial activities of fluorovinyl- and chlorovinylglycine and several derived dipeptides. Antimicrob. Agents Chem. **32:**319–323.

60. Rando, R.R. 1977. Mechanism of the irreversible inhibition of γ-aminobutyric acid α-ketoglutaric acid transaminase by the neurotoxin gabaculine. Biochemistry. **16:**4604–4610.

61. Rando, R.R., and R.W. Bangerter. 1976. The irreversible inhibition of mouse brain γ-aminobutyric acid (GABA)-α-ketoglutaric acid transaminase by gabaculine. J. Am. Chem. Soc. **98:**6762–6764.

62. Richaud, C., W. Higgins, D. Mengin-Lecreulx, and P. Stragier. 1987. Molecular cloning, characterization, and chromosomal localization of *dapF*, the *Escherichia coli* gene for diaminopimelate epimerase. J. Bacteriol. **169:**1454–1459.

63. Robinson, A.E., D. Kenan, J. Sweeney, and W. Donachie. 1986. Further evidence for overlapping transcriptional units in an *Escherichia coli* cell envelope–cell division gene cluster: DNA sequence and transcriptional organization of the *ddl ftsQ* region. J. Bacteriol. **167:**809–817.

64. Rogers, H.J., H.R. Perkins, and J.B. Ward. 1980. Microbial cell walls and membranes, p. 239–297. Chapman and Hall, London.

65. Roise, D., K. Soda, T. Yagi, and C.T. Walsh. 1984. Inactivation of the *Pseudomonas striata* broad specificity amino acid racemase by D and L isomers of β-substituted alanines: kinetics, stoichiometry, active site peptide, and mechanistic studies. Biochemistry **23:**5195–5201.

66. Roze, U., and J. Strominger. 1966. Alanine racemase from *Staphylococcus aureus:* conformation of its substrates and its inhibitor, D-cycloserine. Mol. Pharmacol. **2:**92–98.

67. Salton, M.R.J. 1964. In The bacterial cell wall, p. 293. Elsevier Scientific, Amsterdam.

68. Schechter, P.J., Y. Tranier, and J. Grove. 1979. Gabaculine and isogabaculine: *in vivo* biochemistry and pharmacology in mice. Life Sciences **24:**1173–1182.

69. Shen, S., H.G. Floss, H. Kumagai, H. Yamada, N. Esaki, K. Soda, S.A. Wasserman, and C. Walsh. 1983. Mechanism of pyridoxal-phosphate-dependent enzymatic amino-acid racemization. J. Chem. Soc. Chem. Commun. 82–83.

70. Silverman, R.B., and R.H. Abeles. 1976. Inactivation of pyridoxal phosphate dependent enzymes by mono and polyhaloalanines. Biochemistry **15:**4718–4723.

71. Silverman, R.B., and R.H. Abeles. 1977. Mechanism of inactivation of γ-cystathionase by β,β,β-trifluoroalanine. Biochemistry **16:**5515–5520.

72. Soda, K., H. Tanaka, and K. Tanizawa. 1986. Mechanism and inhibition of alanine racemase, p. 223–251. *In* D. Dolphin, R. Poulson, and O. Avramovic (eds.), Vitamin B_6 Pyridoxal-Phosphate; Part B. J. Wiley, New York.

73. Soper, T.S., and J.M. Manning. 1978. β-Elimination of β-halo substrates by D-amino acid trasaminase associated with inactivation of the enzyme. Trapping of a key intermediate in the reaction. Biochemistry **17:**3377–3384.

74. Soper, T.S., and J.M. Manning. 1981. Different modes of action of bacterial D-amino acid transaminases. J. Biol. Chem. **256:**4263–4268.

75. Soper, T.S., and J.M. Manning. 1982. Inactivation of pyridoxal-phosphate enzymes by gabaculine. J. Biol. Chem. **257:**13930–13936.

76. Soper, T.S., J.M. Manning, P.A. Marcotte, and C.T. Walsh. 1977. Inactivation of bacterial D-amino acid transaminases by the olefinic amino acid D-vinylglycine. J. Biol. Chem. **252:**1571–1575.

77. Tanizawa, K., A. Ohshima, A. Scheidegger, K. Inagaki, H. Tanaka, and K. Soda. 1988. Thermostable alanine racemase from *Bacillus stearothermophilus:* DNA and protein sequence determination and secondary structure prediction. Biochemistry **27:**1311–1316.

78. Thornberry, N.A., H.G. Bull, D. Taub, W.J. Greenlee, A.A. Patchett, and E.H. Cordes. 1987. 3-Halovinylglycines. Efficient irreversible inhibitors of *E. coli* alanine racemase. J. Am. Chem. Soc. **109:**7543–7544.

79. Tipper, D.J. 1987. Mode of action of β-lactam antibiotics, p. 133–170. *In* Antibiotic Inhibitors of Bacterial Cell Wall Biosynthesis. D.J. Tipper (ed.). Pergamon Press, New York.

80. Ueno, H., J.J. Likos, and D.E. Metzler. 1982. Chemistry of the inactivation of cytosolic aspartate aminotransferase by serine O-sulfate. Biochemistry **21:**4387–4394.

81. Vicario, P.P., B.G. Green, and H.M. Katzen. 1987. A single assay for simultaneously testing effectors of alanine racemase and/or D-alanine: D-alanine ligase. J. Antibiot. **40:**209–216.

82. Vo-Quang, Y., D. Carniato, L. Vo-Quang, A. Lacoste, E. Neuzil, and F. LeGoffic. 1986. (β-Chloro-α-aminoethyl)phosphonic acids as inhibitors of alanine racemase and D-alanine: D-alanine ligase. J. Med. Chem. **29:**148–151.

83. Walsh, C. 1979. Enzymatic reactions requiring pyridoxal phosphate. p. 781–797. *In:* C. Walsh (ed.), Enzymatic Reaction Mechanisms. W.H. Freeman, San Francisco.

84. Walsh, C.T. 1989. Enzymes in the D-alanine branch of bacterial cell wall peptidoglycan assembly. J. Biol. Chem. **264:**2393–2396.

85. Walsh, C.T., E. Daub, L.E. Zawadzke, and K. Duncan. 1988. Enzymes of D-alanine metabolism: cloning, expression, purification, and characterization of *Salmonella typhimurium* D-alanyl-D-alanine ligase, p. 531–540. *In* P. Actor, L. Daneo-Moore, M.L. Higgins, M.R.J. Salton, and G.D. Shockman (eds.), Antibiotic Inhibition of Bacterial Cell Surface Assembly and Function. American Society for Microbiology, Washington, D.C.

86. Wang, E., and C.T. Walsh. 1978. Suicide substrates for the alanine racemase of *Escherichia coli* B. Biochemistry **17:**1313–1321.

87. Wang, E., and C.T. Walsh. 1981. Characteristics of β,β-difluoroalanine and β,β,β-trifluoroalanine as suicide substrates for *Escherichia coli* B alanine racemase. Biochemistry **20:**7539–7546.

88. Ward J.B. 1981. Teichoic and teichuronic acids: biosynthesis assembly and location. Microbiol. Rev. **45:**211–243.

89. Ward, J.B. 1987. Biosynthesis of peptidoglycan: points of attack by wall inhibitors, p. 1–43. *In:* D.J. Tipper (ed.), Antibiotic Inhibitors of Bacterial Cell Wall Biosynthesis. Pergamon Press, New York.

90. Wasserman, S.A., E. Daub, P. Grasafi, D. Botstein, and C.T. Walsh. 1984. Catabolic alanine racemase from *Salmonella typhimurium:* DNA sequence, enzyme purification, and characterization. Biochemistry **23:**5182–5187.

91. Wasserman, S.A., C.T. Walsh, and D. Botstein, 1983. Two alanine racemase genes in *Salmonella typhimurium* that differ in structure and function. J. Bacteriol. **153:**1439–1450.

92. Wild, J., J. Hennig, M. Lobocka, W. Walczak, and T. Klopotowski. 1985. Identification of the *dadX* gene coding for the predominant isozyme of alanine racemase in *Escherichia coli* K12. Mol. Gen. Genet. **198:**315–322.

93. Wiseman, J.S., and J.S. Nichols. 1984. Purification and properties of diaminopimelic acid epimerase from *Escherichia coli*. J. Biol. Chem. **259:**8907–8914.

94. Wyke, A.W., and H.R. Perkins. 1975. The specificity of enzymes adding amino acids in the synthesis of the peptidoglycan precursors of *Corynebacterium poinsettiae* and *Corynebacterium insidiosum*. J. Gen. Microbiol. **88:**159–168.

95. Yonaha, K., K. Misono, T. Yamamoto, and K. Soda. 1975. D-Amino acid aminotransferase of *Bacillus sphaericus*. Enzymologic and spectrometric properties. J. Biol. Chem. **250:**579–587.

9

Strategies in β-lactam Design

Francis C. Neuhaus and *Nafsika Georgopapadakou*

I. Introduction

It has been 50 years since Florey, Chain and their associates first described the treatment of staphylococcal and streptococcal infections with penicillin (49). This epoch-making discovery ushered in a new era of fruitful research culminating in the identification of the bacterial cell wall as the target of this β-lactam (222). Penicillin was found to act by inhibiting the cross-linking enzymes involved in the biosynthesis of peptidoglycan (PG), the major cell-wall polymer (298, 323). For many scientists, the design of β-lactams targeted specifically to these essential enzymes is the ultimate research goal in this area.

In 1965, Tipper and Strominger proposed that penicillin is a structural analog of the donor peptide in the reaction catalyzed by $N^{\varepsilon}-(\text{D}Ala)$[1]-acceptor transpeptidase (298) (Figure 9.1). The transpeptidase forms an acyl-enzyme complex with the donor peptide of nascent peptidoglycan, an intermediate in the assembly of cross-linked polymer. Reaction of this enzyme intermediate with an acceptor peptide leads to the formation of the essential cross-link. Penicillin acylates the transpeptidase, forming a stable penicilloyl-enzyme complex, and thereby inhibits further cross-linking. The structural-analog hypothesis predicted that penicillin-sensitive targets should be detectable as penicillin-binding proteins (PBPs). Experimental verification transformed this prediction from an object of interest to a tool of research; it provided the practical means for identifying and studying the targets of β-lactams.

Although peptidoglycan transpeptidases are considered to be the primary targets of β-lactam action, there is growing realization that bacteriolysis and cell death involve triggering of additional processes. How these processes are initiated

[1]Abbreviations: Unless stated, all abbreviations of amino acid residues denote the L configuration.

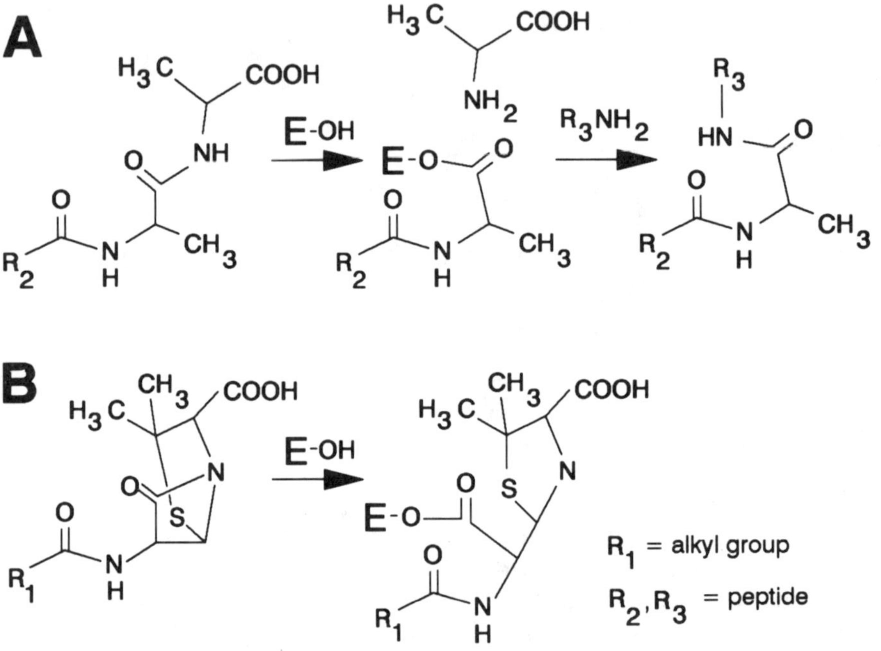

Figure 9.1. Structural analogy between acyl-DAla-DAla, involved in the transpeptidation reaction (A), and benzylpenicillin, involved in enzyme inactivation (B).

in response to inhibition of cell-wall assembly is not well understood. This has important practical implications; in organisms tolerant to penicillin, the inhibition of wall synthesis leads only to growth arrest.

The identification of peptidoglycan transpeptidase as a primary target of penicillin action has provided a major impetus for establishing the three-dimensional structures of these enzymes. From crystallographic, genetic, and kinetic studies, it appears that the vast majority of penicillin-sensitive enzymes are members of a single superfamily that utilizes active-site serine as a nucleophile. Members include the peptidoglycan transpeptidases, DD-carboxypeptidases, and β-lactamases. Structural elements and critical catalytic residues are conserved, while subtle changes in amino-acid sequences in the catalytic site determine whether the enzymes function as β-lactamases, transpeptidases, or DD-carboxypeptidases. The geometry of the active site in each of these enzymes can thus provide clues for designing molecules with optimal fit and maximal acylation efficiency.

Although scientists have produced an impressive array of β-lactams in the past 50 years, there is still an urgent need for new antiinfectives (see chapter 2). Acquisition of resistance to existing β-lactams threatens their effective clinical use and fuels the demand for novel inhibitors of peptidoglycan transpeptidase. Characterizing the receptor targets, the determinants of β-lactam action, and the mechanisms of β-lactam resistance will lead to the design of the next generation of β-lactam antibiotics. Thus, one of the research goals in this area is to utilize microbiological, biochemical, and genetic approaches to build structure/function relationships.

During the past decade, outstanding reviews have been published on *peptidoglycan biosynthesis* (189, 190, 221, 249, 254, 275, 300, 313), the *mechanism of penicillin action* (261, 297, 315, 317); *β-lactams* (42, 82, 195); *chemical reactivity of β-lactams* (218, 219); *PBPs* (100, 108, 112a, 283, 284, 286); *DD-carboxypeptidase/transpeptidase* (93, 109, 110, 144); *structure-activity relationships* (52, 95, 135, 136, 202, 236, 255, 265, 291, 318); *the autolysin-trigger model* (301, 302); and *β-lactamases* (46, 62, 92, 112a, 126, 163). In addition, these basic topics have been covered in the proceedings of recent meetings (1, 19, 43, 124, 261, 308). The objective of this chapter is to review the parameters affecting β-lactam activity and suggest strategies for β-lactam design.

II. Electronic, Steric, and Conformational Determinants of Activity

A. *Minimal β-lactam Structure*

The bicyclic ring systems of penicillin (**1**) and cephalosporin (**2**) consist of a 2-azetidinone ring fused with thiazolidine or dihydrothiazine, respectively. The fused ring systems force the β-lactam nitrogen into a nonplanar conformation. The nonplanarity of nitrogen is expressed by its displacement (h) perpendicular from the plane of atoms 3, 5, and 7 in penicillin (4, 6, and 8 in cephalosporin), which varies from 0.40 Å in benzylpenicillin to 0.24 Å in cephaloridine (78, 96). This nonplanarity has been proposed to hinder amide resonance and thereby contribute to the high chemical reactivity of the β-lactam, a feature thought to be essential for antibiotic action (148).

With the discovery of nocardicins (**3**) in 1976 (11) and other monobactams (**4**) in 1981 (139, 292), the minimal motif for β-lactam action shifted from a bicyclic ring system to the simple 2-azetidinone (β-lactam) ring and an ionizable acidic function within 3.5 Å of the amide nitrogen (**5**). The latter requirement is met by a proximal carboxyl or a sulfonate moiety whose N-S-O bond lengths place the oxygen atoms in a position coincident with the carboxyl of penicillin or cephalosporin. In addition, most β-lactams require a side-chain(s) bearing a hydrogen-bonding group. To achieve acylation of the penicillin-sensitive enzyme(s), a complex mix of steric, conformational, and electronic properties of the β-lactam ligand interacting with the target binding site must come into play. Each of these properties will be examined in the following sections.

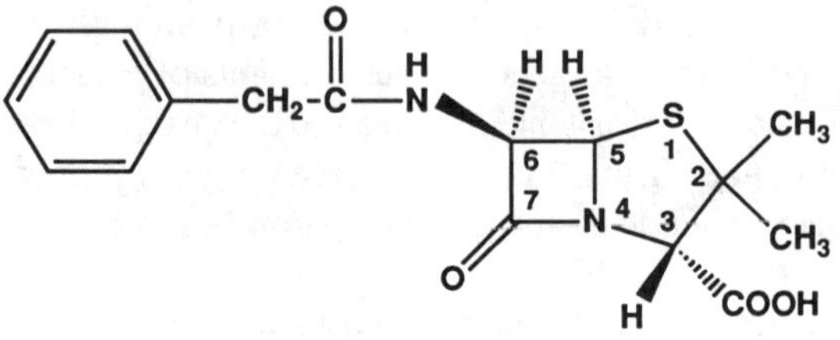

(1) Benzylpenicillin

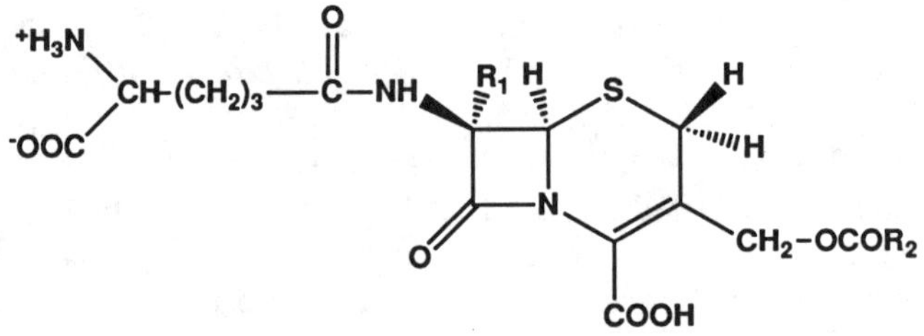

(2) **Cephalosporin C**	$R_1 = H$	$R_2 = -CH_3$
Cephamycin C	$R_1 = OCH_3$	$R_2 = -NH_2$

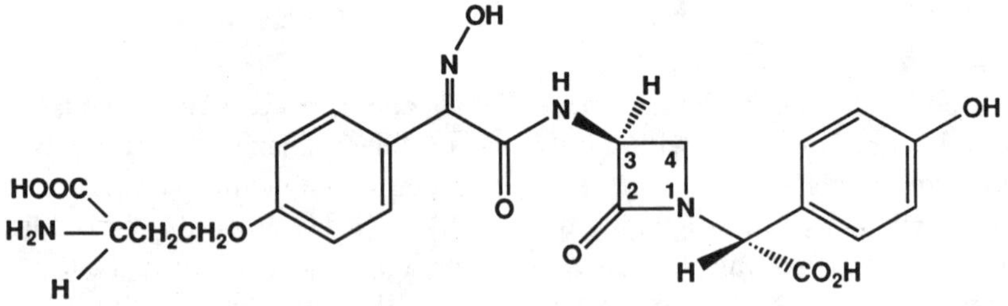

(3) Nocardicin A (monobactam)

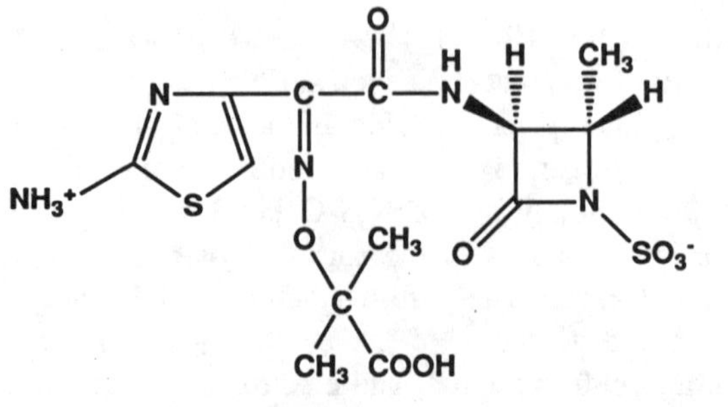

(4) Aztreonam (monobactam)

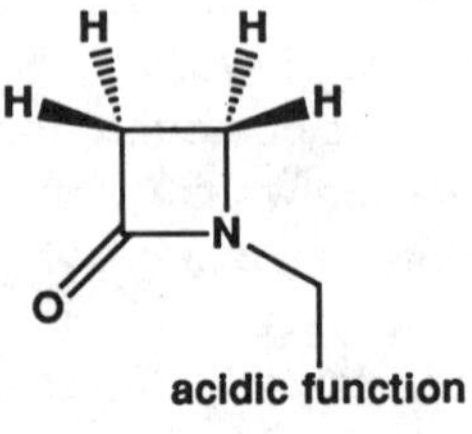

(5) Minimal β-lactam structure

B. Reactivity of the β-lactam Nucleus

The contribution of intrinsic chemical reactivity to the overall antibacterial potency of β-lactams has been a subject of some controversy during the past 15 years. It was proposed that β-lactams have exceptional reactivity, deriving from inhibition of amide resonance and relief of strain in the 2-azetidinone ring. While this proposal is intuitively appealing, it has not been possible to quantify the role of these factors in determining the reactivity of the β-lactam when bound to the enzyme target. Interaction of the β-lactam with the target generates a "binding-site reactivity" (Boyd, personal communication) that may be very different from the intrinsic chemical reactivity of the β-lactam, as measured by acid or base hydrolysis. It is the former that determines biological activity.

In 1970, Sweet and Dahl suggested that benzylpenicillin **(1)** and Δ^3-cephalosporin **(2)** have a shorter carbonyl bond, a longer amide (C-N) bond, and a more pyramidal ring nitrogen when compared with the biologically inactive Δ^2-cephalosporin (289). The lactam nitrogen is nearly planar in the Δ^2-cephalosporin. It was suggested that the pyramidal character of the nitrogen, resulting from strain caused by the bicyclic ring fusion or by electron delocalization through enamine resonance, is the feature that confers the necessary chemical reactivity to the β-lactam.

The hypothesis that biological activity is determined in part by intrinsic chemical reactivity in a class of β-lactams has had experimental support. For example, within the class of thienyl 7-substituted Δ^3-cephems with a given subset of electron-withdrawing groups, biological activity has been correlated with chemical reactivity, a determinant of the transition-state energy (TSE) (33, 35). TSE was used as an approximate quantum-mechanical measure of the electrophilicity of the β-lactam carbonyl (32). Values for TSE are only useful for qualitative comparisons within specific subgroups of β-lactams, *e.g.*, 3-substituted cephems. They also have chemotherapeutic implications; cefaclor **(6;** R=Cl) has a stronger electron-withdrawing group than cephalexin **(6;** R=CH_3) and increased antibacte-

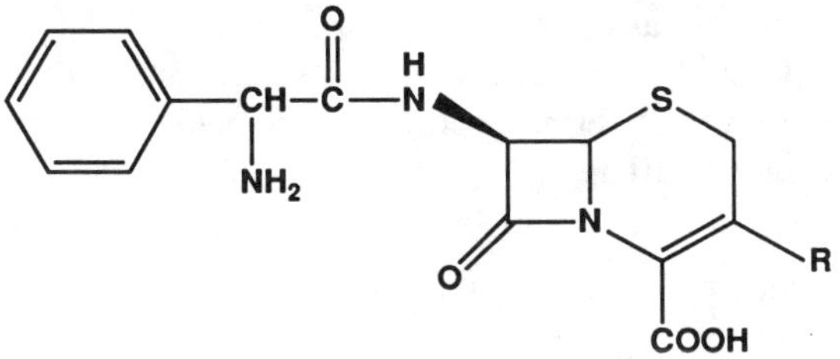

(6) **Cephalexin, R = CH_3**

Cefaclor, R = Cl

rial activity. Nevertheless, correlations between antibacterial activity and chemical reactivity among different classes have been unsuccessful (24, 34, 96, 219).

The importance of the pyramidal character of the ring nitrogen was questioned when three unsubstituted, highly strained β-lactams were examined (228). While each of these β-lactams has a nitrogen that is more pyramidal than that in penicillin, only **(7)** and **(8)** show antibacterial activity, even though the chemical reactivity of the 1-carba-1-penem **(9)** is equivalent to penicillin and the *h* value is 0.54 Å.

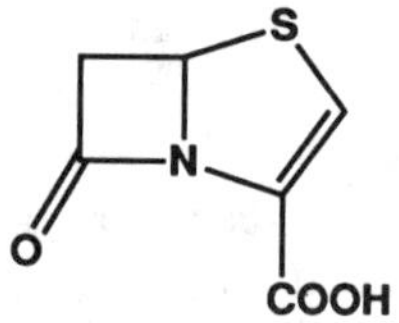

(7) Penem

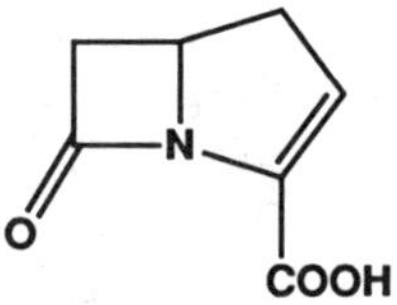

(8) Δ^2-carbapenem

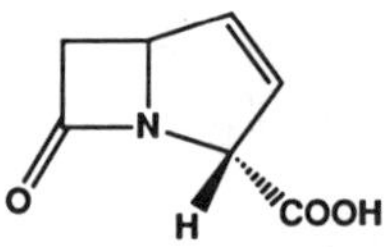

(9) Δ^1-carbapenem

Thus the pyramidal character of nitrogen is not a sufficient condition for biological activity. It is also not a necessary one, but rather a consequence of the electronic structure of the bicyclic β-lactams (34, 57, 218, 228). In monobactams, the nitrogen is only slightly pyramidal (h=0.13 Å), and the reactivity of the β-lactam results in part from the strongly electron-withdrawing sulfonate group (154, 241). Further, in penicillins and cephalosporins, kinetic and ground-state effects do not indicate a significant degree of inhibition of amide resonance (99, 218, 219, 237), nor is there evidence of significant strain energy of the β-lactam being released upon hydrolysis; strained β-lactam systems are not necessarily better antibiotics (218, 237). Nevertheless, a minimal chemical reactivity is required for biological activity (7, 329).

As will be discussed in section III, intrinsic chemical reactivity is only one of several parameters, in addition to "binding fit" and binding-site reactivity, which govern the acylation efficiency of the target enzymes (96, 115). It should be noted that reactivity, as determined by OH^- catalysis, may not be a good model for comparison in structure/activity correlations. Since the DD-peptidases/transpeptidases are atypical serine peptidases, the nucleophile may be ROH instead of RO^- (see section III.B.2).

C. *Conformational Features of the β-lactam*

In addition to binding-site reactivity, one of the primary determinants of biological activity is the binding fit to the target enzyme (113, 115, 172a). A great deal has been deduced about β-lactam conformations from structure/activity studies using minimum inhibitory concentrations (MICs) as a criterion of activity. While these studies are complicated by such factors as permeability and β-lactamase activity (see section IV), they can provide insights into the biologically active conformations of β-lactams and their determinants which are required for interaction with the target(s).

In contrast to the Δ^3-cephem nucleus, the thiazolidine ring of the penam nucleus can exist in two possible conformations. These ring-flip conformations place the carboxyl group in either a pseudoaxial or pseudoequatorial orientation (31, 57) (Figure 9.2). Since this pseudorotation also places the 2β-methyl in a pseudoaxial or equatorial orientation, respectively, the pseudoaxial orientation is also referred to as the "penam-diaxial" and the pseudoequatorial orientation as the "penam-diequatorial" (276).

Cohen (57) compared the molecular geometries of nine β-lactams with fixed conformations and found that all of the biologically active compounds had a pseudoequatorial form, whereas the inactive β-lactams had a pseudoaxial form. In the former group, the distance between the oxygen atom of the β-lactam and the carbon atom of the carboxyl group is 3.6 Å (range, 3.0–3.9 Å) whereas in the latter, 4.2 Å (range, 4.1–4.3 Å). It was concluded that the geometrical features designated by these distances play a key role in the binding of β-lactams to target enzymes (55, 57).

In contrast to the penam nucleus, the penem and carbapenem nuclei are rigid and devoid of any ability to pseudorotate. Thus, these molecular conformations must match the "fit" of the target enzyme to have antibiotic activity.

Based on the proposal that the active conformation of penicillin is pseudoequatorial, Cohen predicted that Δ^2-cephem–4β-carboxylic acid should have biological activity. Although this configuration is opposite to that observed in penicillin, the 4β-COOH of this Δ^2-cephem is in the geometric space of the carboxyl of active β-lactams. This cephem was found to be inactive; thus three-dimensional recognition of this conformation is not the only feature necessary for acylation of the target (58). The importance of the pseudoequatorial orientation of the 3-carboxyl group of benzylpenicillin has been questioned by Wolfe and co-workers

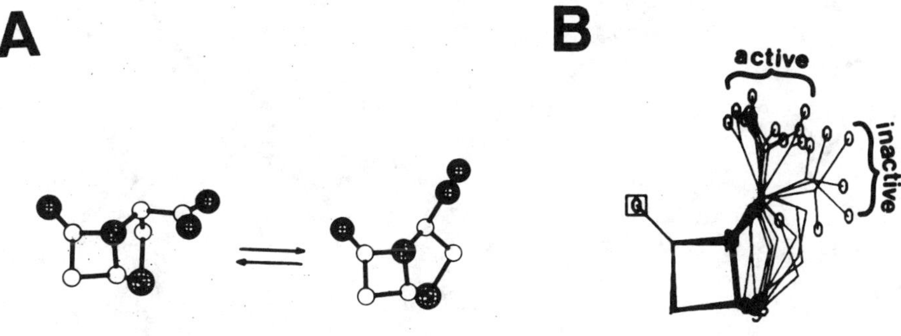

Figure 9.2. Pseudorotation of the penam nucleus (**A**), and geometrical separation between active and inactive structures (**B**). Reprinted with permission (57).

(325, 327) in studies using the active-site peptide, AcVGSVTK-NHMe, of the *Streptomyces* R61 DD-carboxypeptidase as the receptor for penicillin. In their model, the docking of penicillin G to the peptide conformer of this site, which maximizes the proximity of the serine and lysine side chains, positions the serine nucleophile on the convex face of the β-lactam. In this complex, the amide N-H of the side chain is hydrogen-bonded to the *N*-terminal carbonyl oxygen of the peptide and the carboxyl group in the pseudoaxial orientation is closely associated with the protonated ε-amino group of the lysine. On the basis of this model it was suggested that the pseudoaxial conformation of penicillin G is bound preferentially to the active site peptide.

In addition to pseudorotation of the penam nucleus, the second feature of β-lactam conformation involves the acyl side-chain. In the case of benzylpenicillin, the conformers have been classified as (a) compact, (b) extended, and (c) fully extended (242) (Figure 9.3). It was proposed that the active conformation of benzylpenicillin is the compact conformer. Pullman (238) predicted a restricted range of conformational freedom about the C_6-N bond and free rotation at the $PhCH_2$-C bond. Until recently, because of the freedom of rotation around these bonds, it had not been possible to determine the preferred conformation of the acylamido side chain. Synthesis of 6-spiro-epoxypenicillins by Bycroft *et al.* (47) provided the means to restrict the movement of this side-chain and establish the preferred conformation. The 10R isomer (**10**) is biologically active and binds to

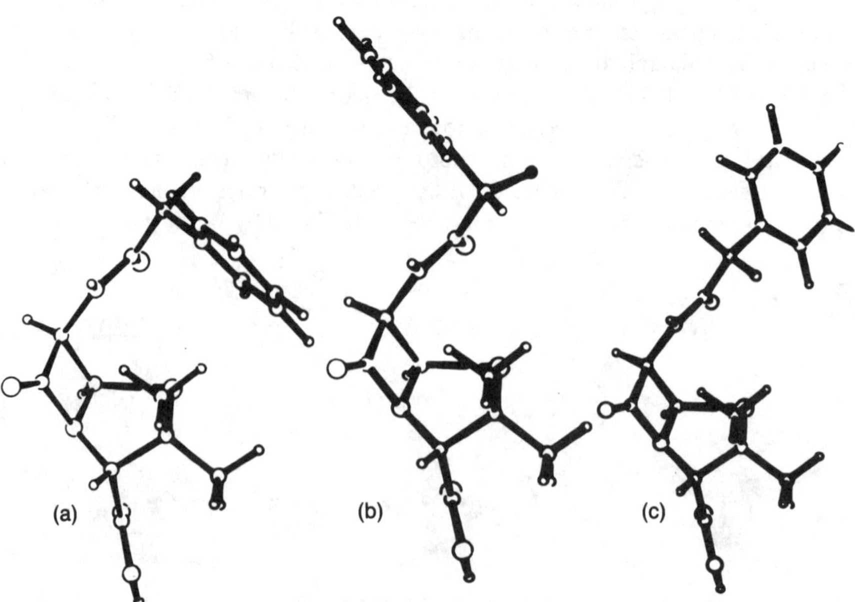

Figure 9.3. Compact (**a**), extended (**b**), and fully extended (**c**) conformations of penicillin. Reprinted with permission (242).

the PBPs of both gram-positive and gram-negative bacteria. The 10S isomer **(11)** is inactive, though it retains some β-lactamase activity.

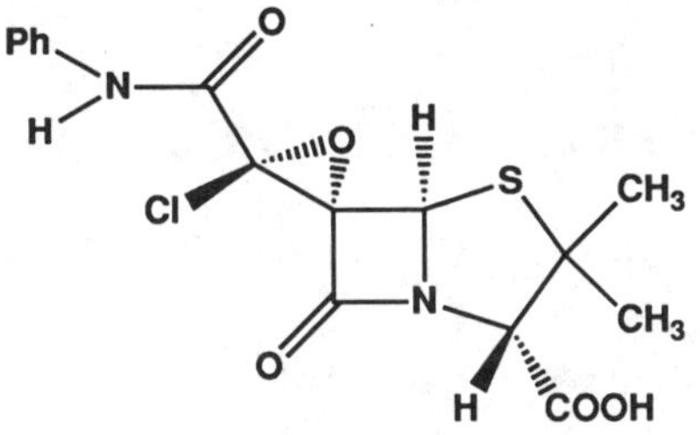

(10) 6-Spiro(10R)epoxypenicillin

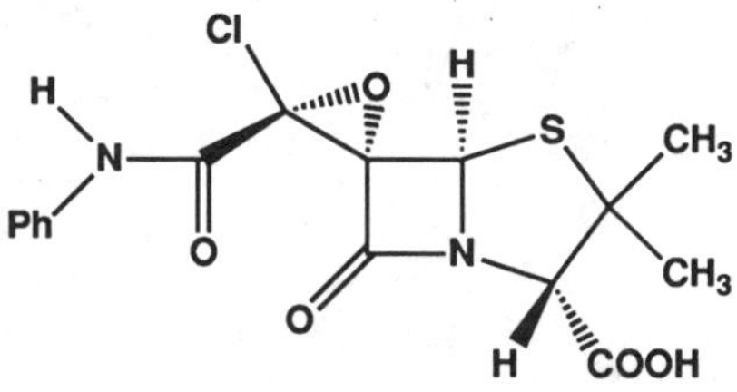

(11) 6-Spiro(10S)epoxypenicillin

Although epoxypenicillin restricts movement around C_6–C_{10}, rotation still occurs at C_{10}–C_{11} and N_{12}–C_{13}. There are 32–33 conformations calculated for each of structures **10** and **11** vs. 58 for benzylpenicillin. Nevertheless, based on this analysis, Shute *et al.* (276) proposed that these derivatives can be considered to be "locked" side-chain analogs of benzylpenicillin. An accessory subsite for the acylamino side chain has thus been defined (Figure 9.4). The basis for this subsite will be further developed in conjunction with the 6(7)β-acylamino substituents.

In addition to the conformational features of the 6(7)β-acyl-amino side-chains, Wolfe *et al.* (325) have drawn attention to the amide N-H function. Alkylation of this nitrogen, *e.g.*, *N*-methylpenicillin, yields an inactive β-lactam. The hydrogen bonding of this function to the DD-peptidase (160) and β-lactamase (193) represents an important interaction with the target binding site (see IIIB.3).

The fused bicyclic ring structure of the β-lactam, together with acyl side-chain, defines two unique faces of the molecule (Figure 9.5), the convex side (α) and the concave side (β). With the carboxyl group in the pseudoaxial orientation, C-2 is flipped up and 2β-methyl is oriented over the β-face **(12).** This conformation has been termed "closed" because of the shielding by the 2β-methyl and can be contrasted with the open conformation **(13)** (56, 155). It was concluded from these studies with the closed form that either the enzyme nucleophile is hindered in its attack on the β-lactam carbonyl or that steric congestion of the tetrahedral intermediate exists.

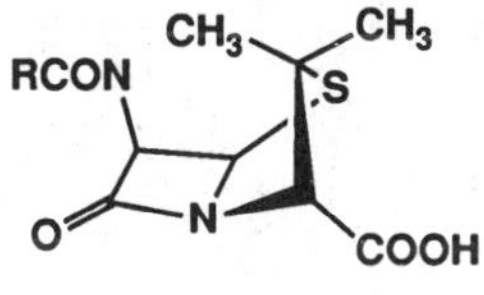

(12) Closed form (pen)

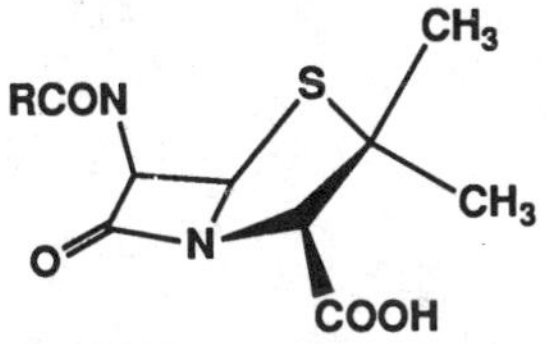

(13) Open form (pen)

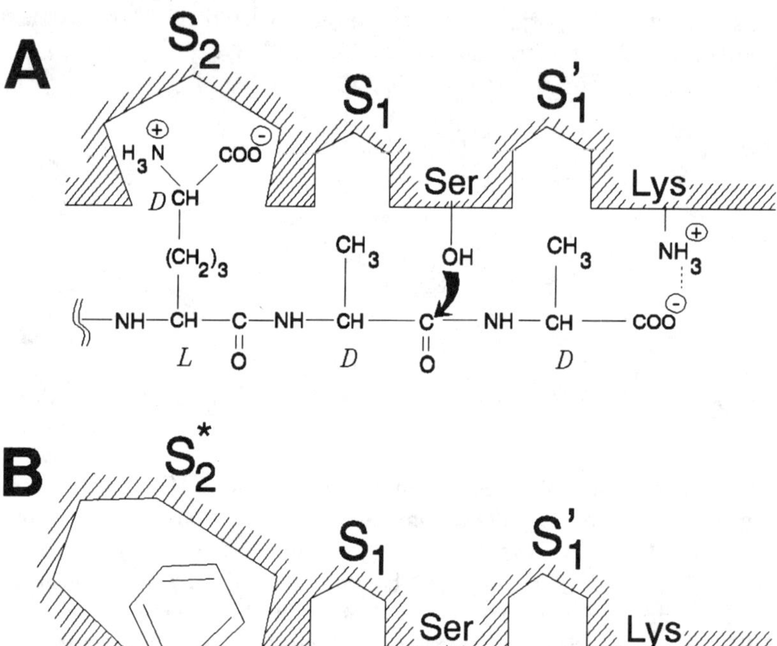

Figure 9.4. Diagrammatic representation of the binding of the acyl-DAla-DAla substrate (**A**) and penicillin (**B**) to the enzyme target. Binding subsites are also shown.

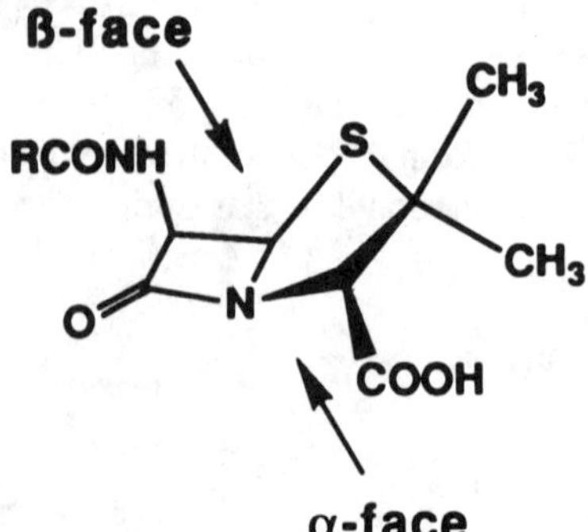

Figure 9.5. Designation of α and β-faces of penicillin.

Keith and co-workers (53, 54, 106, 155) tested this conformational hypothesis by comparing 6β-phenylacetylamino-(2,3)-α-methylenepenicillanic acid and its β-methylene isomer. The α-isomer, which has a sterically unencumbered β-face similar to the open (active) conformation of benzylpenicillin, was biologically more active than the β-isomer (54, 106). The carboxyl carbon–carbonyl oxygen distance is 3.46 Å in the α-isomer and 4.09 Å in the β-isomer (53). In a β-faced attack on the β-lactam of the β-methylenepenam, the nucleophile is hindered in its approach to the carbonyl carbon, while in an α-faced attack with an unhindered approach, the oxygen atom of the tetrahedral intermediate is forced into space occupied by the endo cyclopropyl hydrogen atom. Thus the approach vector of the nucleophile must be open and the possibility of steric congestion of the tetrahedral intermediate must be minimized.

D. Side-chain Substituents

1. 6(7)β-acylamino

The classical β-lactams, penicillins and cephalosporins, require a 6(7)β-acylamino side-chain for potent antibacterial activity. From numerous studies (52, 135), structure/activity relationships (SARs) have emerged that allow one to modify the spectrum of activity, stability to β-lactamases, and potency. The structural-analog hypothesis provides no clues for explaining the profound influence that the 6(7)β-acyl substituents have on the potency of this group of β-lactams. Since these SARs are based on antibacterial activity, it is not generally possible to define a given receptor or target site (see sections III and IV). Nevertheless, they provide evidence for a side-chain-binding subsite, S_2^*, distinct from the S_1/S_1' subsite, that binds the β-lactam nucleus. The presence of the former subsite is strongly supported by the specificity profile of individual PBPs (see section III). It should be noted that the 6(7)-side-chains do not change the chemical reactivity of the β-lactam (140), but are involved specifically in the interaction with the receptor (Figure 9.4) (276).

It might be predicted that subsite S_2^* should be responsible for binding the donor peptide in the transpeptidation reaction. To test this prediction, the antibacterial activity of D-iso-glutamyl-L-lysyl–6β-aminopenicillanic acid was examined (20). This β-lactam was inactive, suggesting that S_2^* does not bind the D-isoglu-Lys moiety. The antibacterial activity of sulfazecin, a D-isoglutamyl-D-alanyl-monobactam, seems to support a minimal role of this subsite in peptide-donor binding (139). It would appear that the donor peptide either does not utilize the S_2^* subsite or only overlaps partly with it. Therefore, it would be appropriate to distinguish between the subsite (S_2^*) that binds the 6(7)β-side chain of β-lactams and the subsite (S_2^*) that binds the peptide donor substrate. This subsite is not defined in the X-ray structure of the acyl-enzyme complex of cephalosporin C and the R61-DD-peptidase (see section III.B).

2. 6(7)α-methoxy

The discovery of cephamycin C **(2)** in 1971 (199) introduced a new class of substituents on the 7α-position of the Δ^3-cephem with increased gram-negative activity. In addition, this substituent confers greatly enhanced β-lactamase stability on the Δ^3-cephem nucleus (119).

Based on the structural-analog hypothesis, Tipper and Strominger (298) predicted that 6α-methyl substitution of the β-lactam ring would enhance the antibacterial properties when compared with penicillin. However, 6α-methylpenicillin had no antibacterial activity (25). In analyzing this observation, Cohen (57a) concluded from modeling studies that it is not possible for the 6(7)α-methyl groups of penams and Δ^3-cephems to occupy the same spatial position as the methyl group of the acyl-DAla-DAla. Thus, it was unexpected to find good antibacterial activity with 7α-methoxy-substituted cephalosporins. In contrast to the 6α-methyl penicillin, the methyl group of 6(7)-methoxy β-lactams can easily occupy the geometric space of the methyl group of the acyl-DAla-DAla (57a). When this substituent was added to ticarcillin, the 6α-methoxy derivative, temocillin **(14),** was found to have good gram-negative activity and high resistance to β-lactamase (170, 278). Cefoxitin **(15),** cefmetazole **(16),** and cefotetan **(17)** are clinically important 7α-methoxy cephalosporins resulting from this research. It should be noted, however, that the effect of this substituent on PBP binding of different cephalosporins is unpredictable, at least in *E. coli* (71, 170). Nevertheless, the success with temocillin stimulated further research on 6(7)α-substituents (180). For example, the 6α-formamido-, 6α-(hydroxymethyl)-, and 6α-isocyano- show many features of the 6α-methoxy group. In contrast to 6(7)β-acylamino substituents, the 6(7)α-groups are all limited in size.

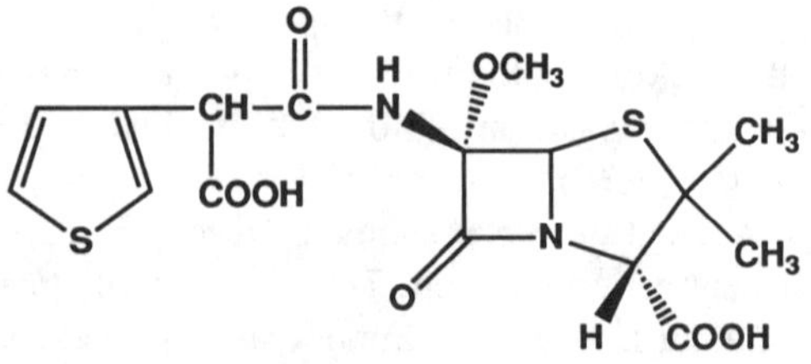

(14) Temocillin

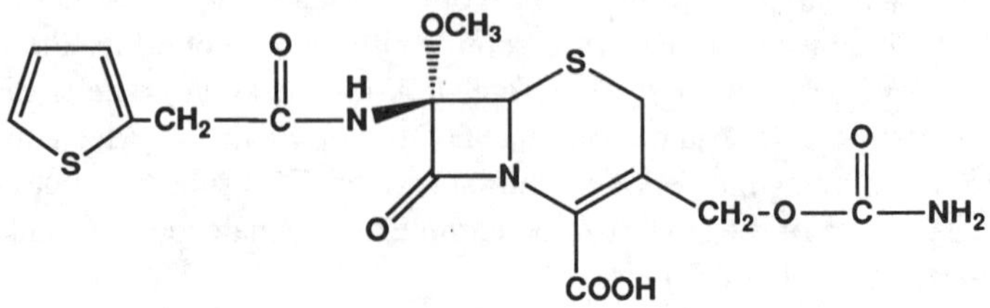

(15) Cefoxitin

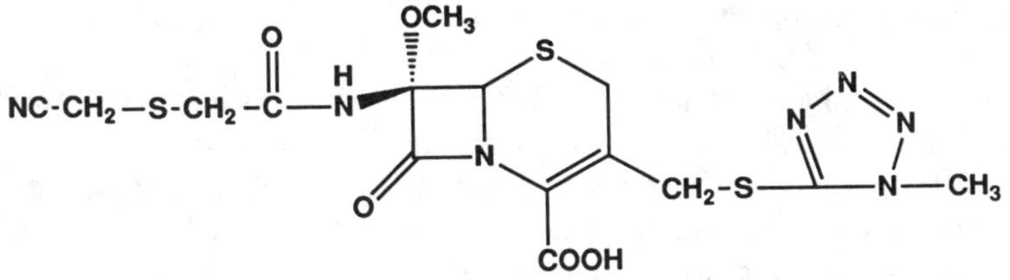

(16) Cefmetazole

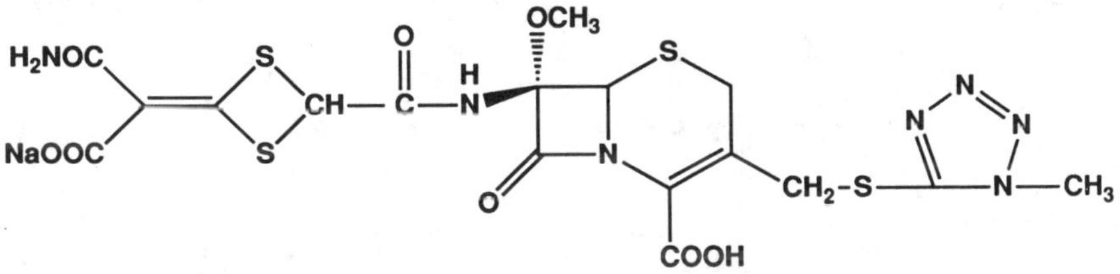

(17) Cofotetan

In early studies, measurements of the pseudo-first-order rate constants of base hydrolysis showed that 6α-methyl-substituted penicillins were less reactive than the parent compounds (141), leading to the conclusion that the substituent effect is steric rather than electronic. In contrast, 7α-methoxy substitution in cephalosporins had no effect on the chemical reactivity. Thus, it is proposed that the 6(7)α-methyl substituent hinders nucleophilic attack on the α-face of the β-lactam, whereas 6(7)α-methoxy has little effect.

3. 6α-alkyl (Δ^2-penems)

The discovery of thienamycin in 1976 provided a new template for the design of β-lactams (153). Synthesis of the amidine derivative gave the more stable and clinically useful imipenem **(18)**. The absence of a large 6β-substituent, the Δ^2 double bond, and the replacement of the sulfur atom by carbon distinguish this class from classical penicillins. The last feature is not essential for activity, as penems (e.g., SCH34343 **(19)**) are similarly active. The 6-hydroxyethylpenems are broad-spectrum antibiotics with good activity against anaerobes and high stability to β-lactamases. The hydroxyethyl substituent is analogous to the 6(7)α-substituents in configuration (83, 244). In 1980 Woodward and colleagues provided the first synthetic carbapenems and recognized that the 6α-substituent was not essential for antibacterial activity (329).

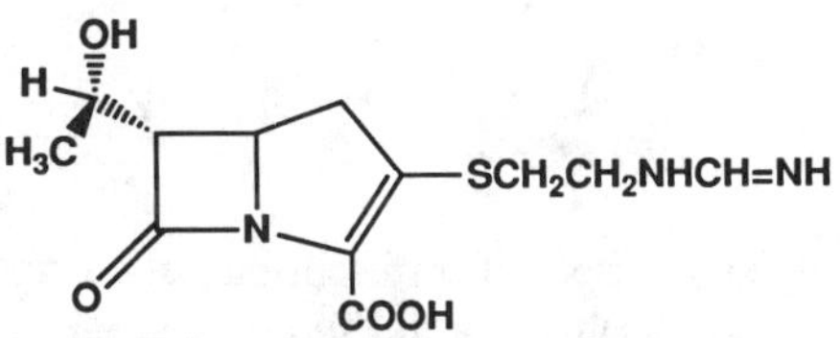

(18) Imipenem (Δ^2-carbapenem)

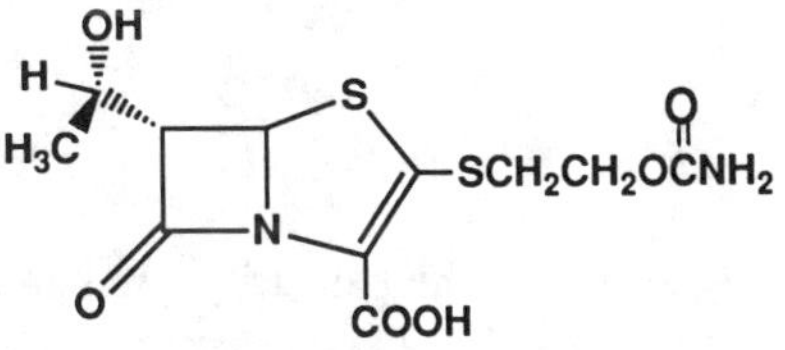

(19) SCH34343 (Δ^2-penem)

4. 3′-substituents (Δ^3-cephems)

3′-Substituents have a significant effect on the antimicrobial activity of cephalosporins (135). Generally, these substituents are either electron-withdrawing (see section II.B) or good leaving groups, such as acetoxy (ccfotaxime, **20**), pyridinium (cephaloridine, **21**), and thio-heterocycle (cefoperazone, **22**). Good 3′ leaving groups are important for β-lactamase stability, as they are eliminated from the acyl-enzyme intermediate and thus stabilize it to deacylation (89, 90). It unlikely that the 3′-substituents interact with a subsite on the target enzyme.

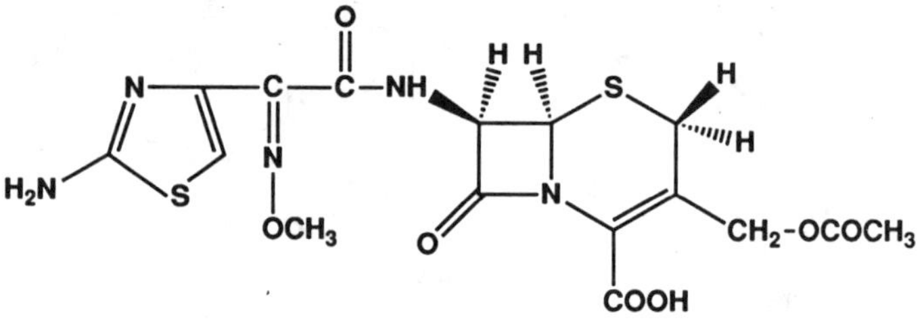

SD20

(20) Cefotaxime

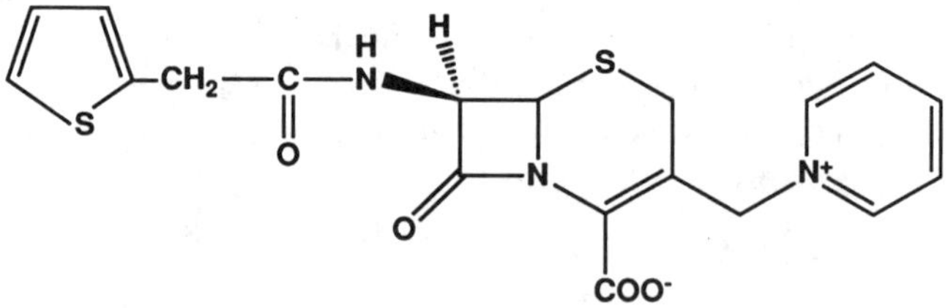

(21) Cephaloridine

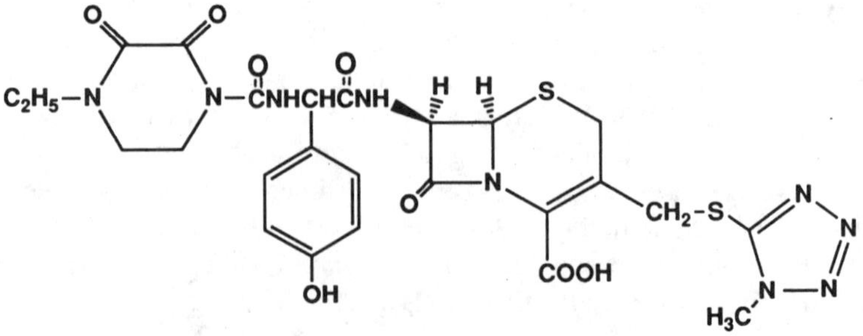

(22) Cefoperazone

E. Other Bicyclic Nuclei—Is a β-lactam Ring Necessary?

Research in the past 15 years has produced a variety of different bicyclic ring systems (**18, 23, 24**) that greatly extend the possibilities for designing new types of transpeptidase/carboxypeptidase inhibitors. In addition, carbacephems, 2-oxa-

and 2-thio isocephems have recently been reported (61a, 61b, 181a). Bicyclic ring analogs of penams and cephems represent a fascinating approach for modifying the properties, including the acylation efficiency, of the β-lactam and hence its biological activity (82).

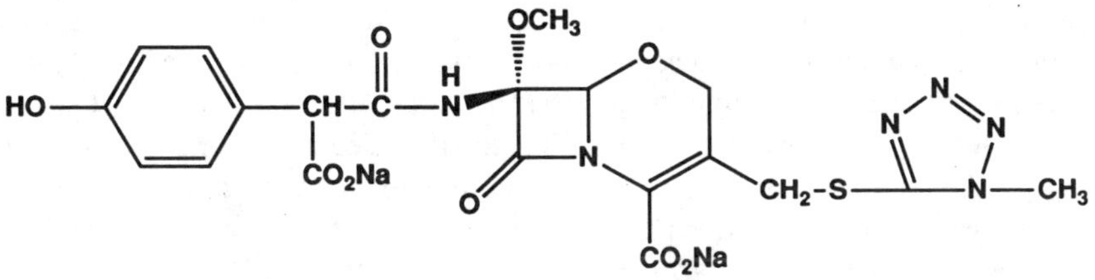

(23) Moxalactam (Δ^3-oxacephem)

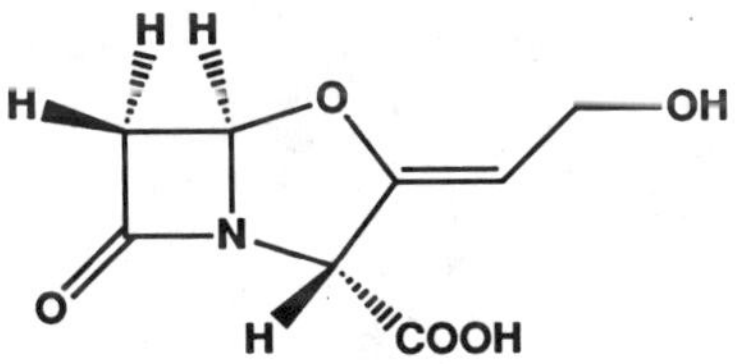

(24) Clavulanic acid (oxapenam)

The discovery of lactivicin **(25)** (130, 215), bicyclic pyrazolidinones **(26)** (151, 294), and γ-lactams **(27)** (7) strongly suggests that the β-lactam ring is not an essential feature (176).

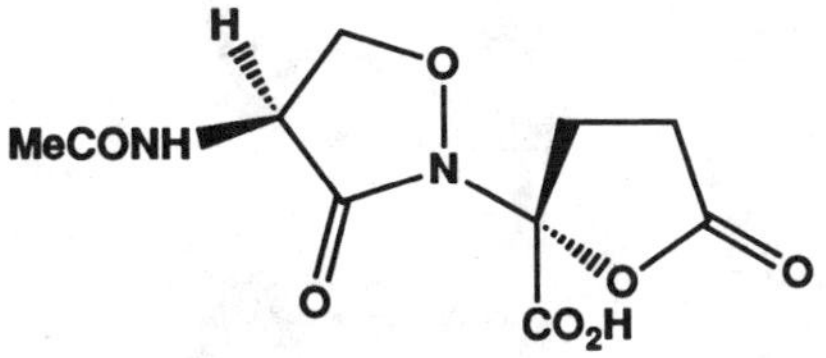

(25) Lactivicin

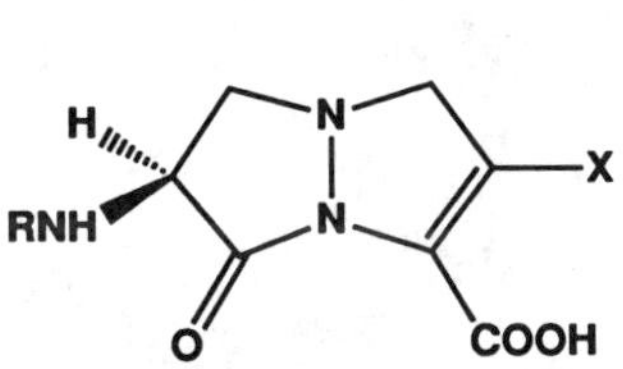

(26) Pyrazolidone

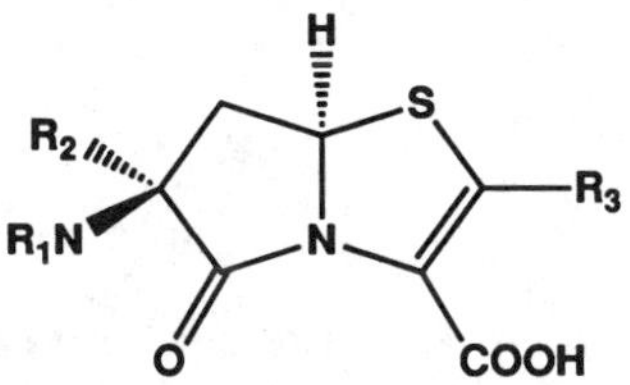

(27) gamma-lactam

The Lilly group (7) has recently described results that predict γ-lactam structures which may possess antibacterial activity. These predictions are based on a comparison of the three-dimensional structure of the γ-lactam derived from molecular orbital calculations with that of cephaloridine **(21).** For example, the 7S-isomer of the bicyclic γ-lactam **(28)** is the best mimic of this cephalosporin (Figure 9.6); it has an MIC of 8 μg/ml against *Streptococcus pyogenes* and *Streptococcus pneumoniae,* whereas the 7R-isomer has an MIC of 64 μg/ml. While the antimicrobial activity is modest, the approach represents a first step in the long-term goal of designing new antibacterials based on altering the β-lactam moiety. The introduction of a nitrogen atom into position 1 of the bicyclic ring of the γ-lactam, to afford the pyrazolidinone **(26),** greatly improved the antibacterial activity and broadened the activity spectrum (294).

(28) 7 (S) isomer

In summary, the five features that represent the basic requirements for β-lactam action are: (1) a β-lactam ring with a carbonyl having a minimum chemical reactivity, (2) a scissile C-N bond, (3) an acidic function with a suitable geometry, (4) an optimal binding "fit" of the β-lactam to the receptor, and (5) an unhindered

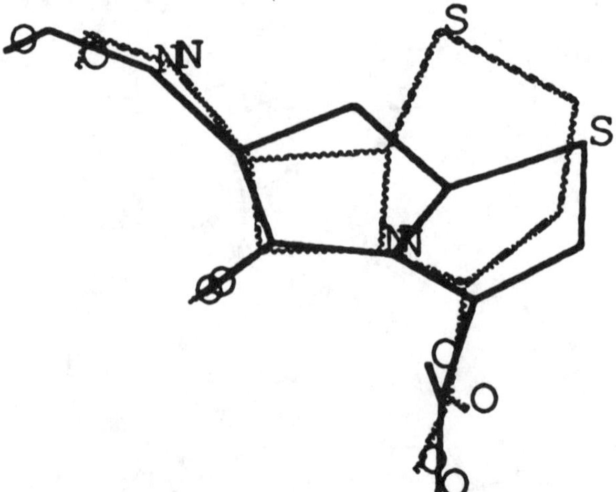

Figure 9.6. View of the 7-(S) γ-lactam isomer (28) fit to cephaloridine structure (**21**). Reprinted with permission (7).

α-face and open conformation of the β-lactam. With regard to the first feature, the requirement for a β-lactam is not absolutely essential. These requirements are completed by the 6(7)-substituents, which facilitate the docking of the β-lactam molecule to its target.

As will be discussed in section III, key features of the inhibitory process are acylation of the target and inability of the leaving group to dissociate. The acylation efficiency is determined by the interaction of the β-lactam and the active-site amino-acid residues of the target enzyme, with the ligand and receptor playing essential and complementing roles (113).

III. Nature of the Receptor Target

A. *Identification of the Lethal Target(s)*

PBPs are cytoplasmic membrane proteins, of molecular weights 35–120 kDa, that covalently bind penicillin and catalyze the terminal stages of peptidoglycan synthesis. They are detected in whole cells or bacterial membranes by the binding of radiolabeled benzylpenicillin. After binding, proteins are fractionated on sodium dodecylsulfate–polyacrylamide gels and fluorographed (282). Binding of a given β-lactam is measured as decreased binding of labeled benzylpenicillin (69). PBPs are ubiquitous in bacteria, with the exception of mycoplasmas which lack the cell wall, and are conventionally numbered in the order of decreasing molecular weight. A survey has revealed a wide variation in both the number and the amounts of PBPs among bacteria (104).

Three complementary approaches have established which of the PBPs are primary targets: (1) the effects of β-lactams on cell growth and morphology have been correlated with their binding affinities for individual PBPs; (2) mutants lacking certain PBPs have been examined for changes in growth and morphology; and (3) mutants resistant to β-lactams have been examined for PBP alterations. The collective evidence strongly suggests that each organism has a multiplicity of penicillin-sensitive targets, each with a characteristic affinity profile for β-lactam(s) (297). Inhibition of one or more of these targets can lead to cell growth arrest and lysis.

Although PG transpeptidases are the primary targets of β-lactam action, there is a growing realization that inhibition of these enzymes is only part of a series of cellular events leading to growth inhibition or cell death (302). For this reason, penicillin-sensitive enzymes (*i.e.*, PBPs) may be viewed as receptors, which initiate a complex set of metabolic events upon binding the β-lactam. Defective cell-wall assembly in penicillin-treated cells triggers autolysin action, secretion of water-soluble PG, and release of lipoteichoic acids and lipids. How defective PG assembly transmits its signal to different cellular systems is not known.

Because the field of PBPs has been extensively reviewed, the discussion that follows will focus primarily on the PBPs from *Escherichia coli* and *Staphylococ-*

cus aureus. The two organisms were chosen because of their intrinsic interest as important pathogens and their ability to illustrate specific points.

1. Escherichia coli

In *E. coli*, there are at least eight PBPs, of molecular weights 91 (PBPs 1a and 1b), 66 (PBP 2), 60 (PBP 3), 49 (PBP 4), 40 (PBPs 5 and 6), and 30 (PBPs 7 and 8) kDa (269, 281, 282, 287). PBPs 1a, 1b, 2, and 3 are bifunctional enzymes, possessing both transglycosylase and transpeptidase activities (142, 201). PBPs 1a and 1b are involved in cell elongation, PBP 2 in initiation of cylindrical wall growth, and PBP 3 in septation. Deletion of the gene for either PBP 1a or 1b does not result in loss of cell viability whereas deletion of both does, suggesting that they can compensate for each other (288). PBP 1b is attached to the plasma membrane by its amino terminus with 90% of the protein in the periplasmic space (84). The predominantly periplasmic location of PBPs 1b, 2 (2), and 3 (28) is consistent with the fact that β-lactams do not require transport through the cytoplasmic membrane to reach their targets. Recently, den Blaauwen *et al.* (74) isolated monoclonal antibodies against epitopes of the transglycosylase and transpeptidase domains of PBP 1b. Only antibodies against the latter prevented the binding of penicillin. The spatial distribution of the epitopes in intact membranes indicates a periplasmic location of the two domains (75, 76). The ability to dissect PBP domains and to selectively inhibit and topologically locate each represents an exciting development in our understanding of peptidoglycan assembly and β-lactam action (44, 285). Further, the ability to modify PBPs, *e.g.*, PBP 5, by site-directed mutagenesis, represents an interesting approach to probing their function (205).

One of the features of β-lactam inhibition in *E. coli* is the distinct affinity profile of each PBP for various β-lactams and the morphological response(s) associated with it (69, 70, 100, 234, 281). Acylation of PBPs 1a and 1b by cephaloridine **(21)** causes lysis, of PBP 2 by mecillinam **(29)** causes the formation of large, osmotically stable, round cells, and of PBP 3 by furazlocillin **(30)**, piperacillin **(31)**, or cefotaxime **(20)** induces filamentation (27, 69, 122, 138, 167, 281, 283). Curiously, with some oxyimino cephalosporins, binding to PBP 3 is obscured by rapid deacylation, and a modification of the binding assay is used (170).

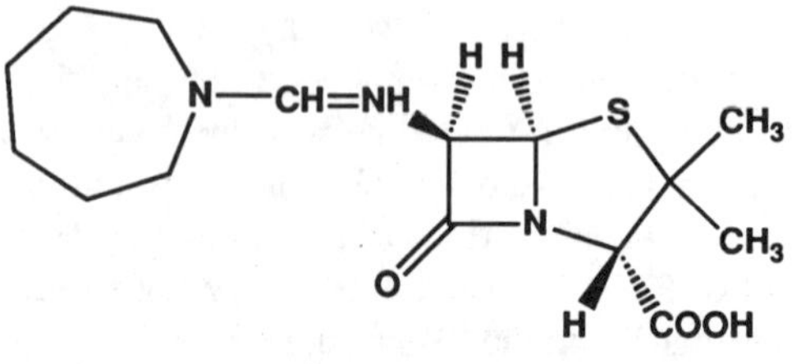

N—CH=NH
H H
S
CH_3
N
CH_3
O
H
COOH

(29) Mecillinam

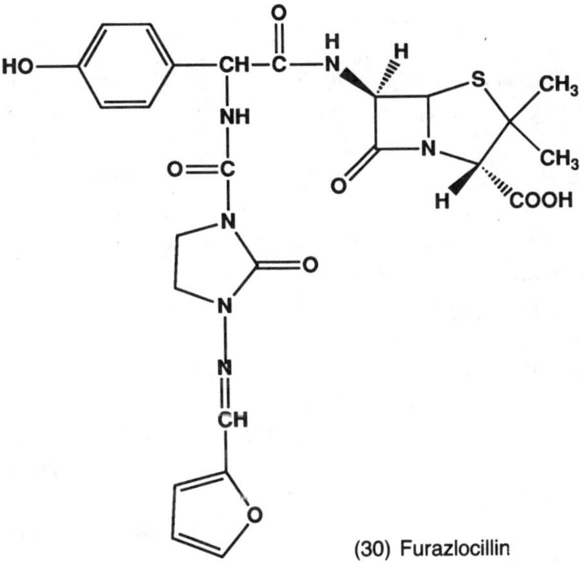

(30) Furazlocillin

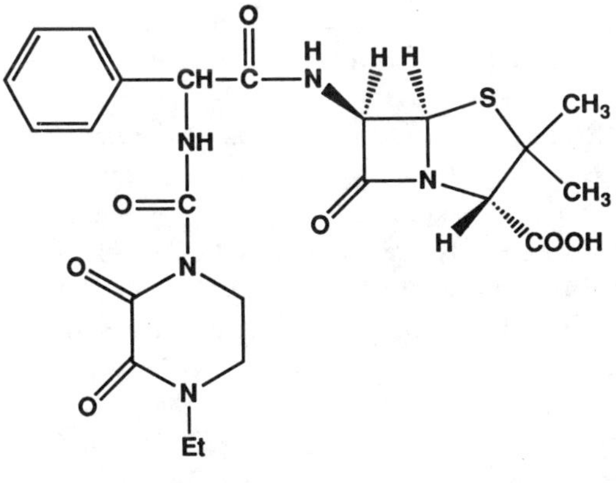

(31) Piperacillin

PBPs 4, 5, and 6 possess endopeptidase and carboxypeptidase activities. PBP 4 has been designated as an endopeptidase and DD-carboxypeptidase; it may also catalyze the synthesis of the novel DAP-DAP cross-bridge (118, 231). Deletion of PBP 4, 5, or 6 does not result in loss of viability under laboratory conditions (39). However, mutants defective in both PBPs 4 and 5 have increased peptidoglycan cross-linking (77). PBP 5 is a periplasmic protein attached to the inner membrane by a short stretch of amino acids. In addition, there are two low-molecular-weight PBPs, 7 (32 kDa) and 8 (29 kDa) (269, 282). PBP 7 binds penems, *e.g.*, imipenem **(18)**, at relatively low concentrations under nongrowing conditions, and has been suggested to be a β-lactam target, as these penems cause lysis of nongrowing cells (306).

2. Staphylococcus Aureus

In *S. aureus,* four PBPs of molecular weights 87, 80, 75, and 41 kDa have been detected (105, 166, 252). Of the four, PBP 1 appears to be the primary transpeptidase and is essential for cell growth (18, 250). PBP 2 may be a transpeptidase (250) that links newly synthesized nascent PG strands to the existing wall when cultures are incubated under nongrowing conditions. By analogy to other organisms, it has been suggested that PBP 3 is a septation-associated transpeptidase (103). PBP 4 is a DD-carboxy peptidase/secondary transpeptidase, as deficient mutants show over 50% reduction in cross-linking (331). In addition, PBP 4 possesses significant penicillinase activity (166). Essential PBPs, and therefore targets of β-lactam action, are PBPs 1, 2, and 3 (18, 67, 68, 102).

In contrast to the bifunctional penicillin-sensitive enzymes of gram-negative organisms, the transglycosylase activities are separate from the penicillin-sensitive transpeptidases in *S. aureus* (223). Thus, the organization of the PG synthetic machinery may be different in gram-positive organisms.

Imipenem **(18)** is a relatively specific inhibitor of PBP 1 (18) and, to a lesser extent, PBP 4 (250). Cefotaxime **(20)** is a specific inhibitor of PBP 2, whereas mecillinam **(29)** is a specific inhibitor of PBP 3 (18, 103, 105, 250). Based on bacteriolysis experiments with imipenem and either cefotaxime or mecillinam, it was suggested that inhibition of PBP 1 and either PBP 2 or PBP 3 is necessary for lysis (18).

In *S. aureus* "Berlin," a correlation does not exist between the simultaneous inhibition of PBPs 1 and 2 (or 3) by penicillin and overall reduction in cross-linking (18, 171). Labischinski *et al.* (171, 248) proposed that penicillin-induced lytic death in *Staphylococci* is mainly the result of a morphogenetic effect of the β-lactam mediated by organelles called murosomes (117). With synchronously growing *S. aureus* SG511, the onset of the lytic event was found to be independent of the stage of the division cycle at which penicillin was added; the cells were always able to perform the next cell division. In the presence of penicillin, wall material is deposited in the wrong cleavage plane, leaving the cells vulnerable to murosome-induced extrusion of cell contents. Initiation of cell separation in the presence of aberrant peptidoglycan deposition and murosome-induced holes triggers the autolytic system resulting in cell lysis (116).

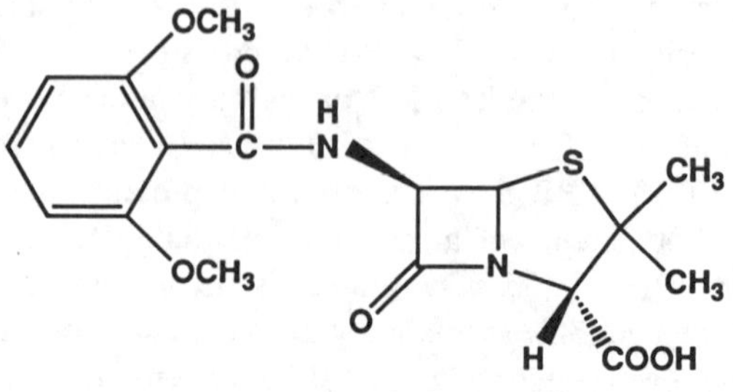

OCH_3 O H C N S CH_3 OCH_3 N CH_3 O H COOH

(32) Methicillin

Methicillin (**32**)-resistant *S. aureus* strains have been isolated from a variety of clinical cases. In each case, the production of an additional PBP, 2′ or 2a (*mecA* locus), with a low affinity for β-lactams was responsible for resistance (132, 182, 251, 303, 307a). PBP 2′ may catalyze a penicillin-insensitive transpeptidation, and has been hypothesized to have evolved from the fusion of a regulatory region for β-lactamase and a PBP-structural gene of unknown origin (22, 280). Induction of PBP 2′ by β-lactams represents a unique resistance mechanism (182, 257, 307). In addition to *mec*, a chromosomal gene, *FemA*, has been implicated in methicillin resistance though not in the production of PBP 2a (22a). Since methicillin resistance is enhanced by growth in the presence of NaCl, by growth at pH 7.0 instead of pH 5.2, and by growth at low temperatures, other factors, e.g. cell wall features, may also play a role in this mechanism (50, 178b).

3. Other Bacteria

Of the major gram-negative pathogens, *Enterobacter* (69), *Klebsiella pneumoniae* (69), and *Salmonella typhimurium* (272) have PBPs very similar to those of *E. coli*. *Proteus* (217), *Serratia marcescens* (104), and *Pseudomonas aeruginosa* (69, 213) PBPs are somewhat different, though still correlatable, to *E. coli* PBPs. *Haemophilus influenzae* has eight PBPs (179), 5 (59 kDa) being associated with resistance (187, 270), while *Neisseria gonorrhoeae* has three PBPs (17, 80), 1 (87 kDa) and 2 (59 kDa) being lethal targets (80, 81). The anaerobe *Bacteroides fragilis* has four PBPs, 1 (100 kDa) and 2 (86 kDa) being probably transpeptidases (107); both are associated with resistance (107, 230).

Of the major gram-positive pathogens, *Staphylococcus epidermidis* has a similar PBP pattern and methicillin-resistance mode to those of *S. aureus* (104, 307). *S. pneumoniae* has five PBPs (321); PBPs 1 (105 kDa) and 2 (95 kDa) are lethal targets for β-lactams (123, 128, 337). *Gaffkya homari* has nine PBPs, of molecular weights 50–200 kDa (330).

Three species of *Enterococci* have a moderately high intrinsic resistance to benzylpenicillin (4–10 μg/ml). *Enterococcus faecalis* has five PBPs (105); PBPs 1 (105 kDa) and 3 (79 kDa) are associated with the resistance found in this organism (105, 320, 322). The related organism *Enterococcus faecium* has a similar PBP pattern and intrinsic resistance (320, 322). *Enterococcus hirae* strains possess seven membrane PBPs ranging in size from 140 to 43 kDa (62b). Two of these, PBPs 2 and 3, are involved in cellular division. In addition, muramidase-2 from walls of this organism binds benzylpenicillin at a site separate from that for muramidase activity (79a). The overproduction of the slow-acylating PBP-5 is considered responsible for the high resistance (92a).

It has been suggested that there may be at least as many target PBPs as there are bacterial pathogens (95). This is of course an exaggeration, as PBP-affinity profiles for individual β-lactams are similar in related organisms (*e.g.*, enterobacteria, *Staphylococci*). Nevertheless, there is no single target PBP common to all bacteria, and thus a universal β-lactam antibiotic may be an elusive goal (108).

B. Stereochemical Constraints of the Receptor Target

1. Geometry of the Binding Site

The DD-carboxypeptidase/transpeptidase (DD-peptidase) from *Streptomyces* sp. R61 and β-lactamases from several sources are the only penicillin-sensitive enzymes whose three-dimensional structures have been established (51, 79, 134, 134a, 158, 160, 193). This structural information is complemented by the amino-acid sequences of 27 penicillin-sensitive enzymes (94). The preliminary crystallographic studies of PBP 1b(δ) from *E. coli* promise additional structural information for β-lactam-sensitive proteins in PG synthesis (143). In addition, derivatives of the *E. coli* PBP 5 with truncated carboxyl terminals have been crystallized, one of which, PBP 5S, yielded enzymatically active crystals suitable for crystallographic analysis (91). Because of the tertiary structural homology of these enzymes to β-lactamases, a consideration of the latter is of interest in defining the geometry of the β-lactam-binding sites (150, 156, 193, 262) and the mechanisms of enzyme action (86, 93, 225).

The β-lactam acylation site of the R61 DD-peptidase has been located by x-ray crystallography (160). Three structural elements determine the site: (1) an α-helix, containing the Ser-62 nucleophile and the conserved Lys-65; (2) an inner β-strand, containing the conserved triad of His-298, Thr-299, and Gly-300; and (3) loop 1, containing Tyr-159 and Phe-164. These structural elements determine a major part of the acylation site for β-lactams and acyl-DAla-DAla. To what extent the binding site for β-lactams is congruent with the binding site for the donor peptide (Figure 9.1) is not known. Molecular simulations combining x-ray data with computational methodologies of the DD-peptidase with waters of solvation in the catalytic site have provided some off the mean interatomic distances within this catalytic site (34a).

X-ray crystallographic studies of the R61 DD-peptidase complexed with β-lactam as the acyl-enzyme provides a unique insight into the enzyme intermediate (**159**, 160). The electron density for the β-lactam, e.g. cephalosporin C (**2**), in the active site indicates that the β-lactam bond is cleaved by the oxygen of Ser-62 and covalently linked at C-8. A rotation of 50° about the C_6-C_7 bond repositions the dihydrothiazine ring; the carboxyl group is consequently placed in the vicinity of Lys-65 and His-298 and within hydrogen-bonding distance (2.9 Å) from the Thr-299 oxygen atom.

The ability to use the stereochemical constraints of the DD-peptidase to model the Lys-DAla-DAla peptide strongly suggests that the peptide dihedral angle is rotated out of planarity to 135° (160). The direction of rotation is toward the conformation existing in β-lactam inhibitors. Thus distortion of the peptide bond of X-DAla-DAla would appear to be an essential first step prior to the formation of the acyl-enzyme intermediate.

2. *Mechanism of acylation*

The design of new antibacterial agents targeted to peptidoglycan transpeptidases requires not only structural information on these enzymes, but also knowledge of the mechanism of binding, acylation, and deacylation:

$$E + I \underset{}{\overset{K}{\rightleftharpoons}} EI \xrightarrow{k_2} EI^* \xrightarrow{k_3} E + P$$

In the above three-step model (93), the first step involves the reversible binding of enzyme and β-lactam (I). The dissociation constant, K, is a measure of binding specificity and reflects the stereochemical constraints of the binding site and stereochemical features of the β-lactam. The first-order rate constants, k_2 and k_3, characterize the irreversible acylation and deacylation steps, respectively. This model can be applied to all penicillin-sensitive enzymes, *i.e.*, transpeptidases, DD-carboxypeptidases, and β-lactamases. The values of k_2 reflect the characteristic binding-site reactivities of β-lactams. In many cases, the individual values of k_2 and K have not been measured. Thus, the second-order rate constant, k_2/K, is used for comparing the acylation efficiency of different β-lactams. To be an effective inactivator, acylation (k_2/K) should be high and deacylation (k_3) low; to be an effective substrate, both k_2/K and k_3 should be high. Mechanistically, deacylation of the penicilloyl-enzyme proceeds either via direct hydrolysis to penicilloic acid, or via a secondary C_5-C_6 cleavage to give phenylglycyl-enzyme which is then hydrolyzed to phenylglycine (93, 315). The former pathway is associated with β-lactamases and most DD-carboxypeptidases, while the latter pathway is associated with peptidoglycan transpeptidases and some DD-carboxypeptidases.

The second-order rate constants (k_2/K) show wide variation. For example, with the R61 DD-peptidase, the values for benzyl penicillin and 6-aminopenicillanic acid (6-APA) are 14,000 and 0.25 $M^{-1}s^{-1}$, respectively, reflecting a 56,000 times faster acylation for benzylpenicillin over 6-APA (93, 96). In contrast, the dissociation constant (K) for a variety of β-lactams ranges only from 0.1 to 13 mM (114). Thus the binding-site reactivity reflects k_2 rather than K.

The ratio of the second-order rate constants for the opening of the β-lactam, $M^{-1}s^{-1}{}_{(enzyme)}/M^{-1}s^{-1}{}_{(OH^-)}$, represents the accelerating effect of the DD-peptidase relative to the intrinsic chemical reactivity of the β-lactam (96). For the R61 DD-peptidase this ratio varies from 1.7 for 6-APA to 38,000 for benzylpenicillin. These results clearly indicate the importance of the interaction of the 6β-acyl substituent of benzylpenicillin with the enzyme in achieving the necessary binding-site reactivity for effective acylation of the enzyme by the β-lactam (96).

There is substantial evidence to indicate that binding of the β-lactam and the donor peptide is a two-step process involving subsites S_1/S_1' and S_2 (Figure 9.4) (114). Subsite S_2 for β-lactams is designated S_2^*, and, on the basis of structure/

activity studies with the R61 enzyme, it is different from subsite S_2 (section II.D) (114, 276). It is assumed that subsite S_1/S_1' is the same for both the β-lactam and the donor peptide. In the first step of binding, the carboxyl and carbonyl of the β-lactam nucleus bind to subsite S_1/S_1', a process that is neither very efficient nor very selective. In the binding of peptide donor, it is the DAla-DAla moiety that binds to subsites S_1/S_1'. In the second step, the lysyl (or DAP) residue of the peptide donor binds to S_2. In the case of β-lactams, the 6(7)β-acylamino side-chains interact with components of the binding subsite S_2^* and correctly orient (dock) the electrophilic carbonyl for nucleophilic attack. The binding of the side-chain serves as the handle through which the β-lactam ring in subsite S_1/S_1' acquires sufficient binding-site reactivity (114). This is accomplished by distorting the β-lactam, straining the structure, polarizing the carbonyl, or by a combination of all three (113a). Tipper (297) focused attention on the secondary interactions as an important feature of the acylation process and concluded that, until the chemistry and determinants of this part of the process are well understood, the design of β-lactam antibiotics will "remain guided by empirical analyses of structure-activity relationships for whole organisms and for specific PBPs."

The mechanism of acylation in *Streptomyces* R61 DD-carboxypeptidase, and presumably penicillin-sensitive enzymes in general, is different from that found in eukaryotic serine proteinases, *e.g.*, chymotrypsin. Since a histidine residue is not involved, other mechanisms of nucleophilic attack have been considered (310a). From theoretical calculations, Wolfe et al (327) concluded that a four-centered transition structure between the OH of the serine residue and the C(7)-N of the β-lactam **(33)** represents the first step in the attack:

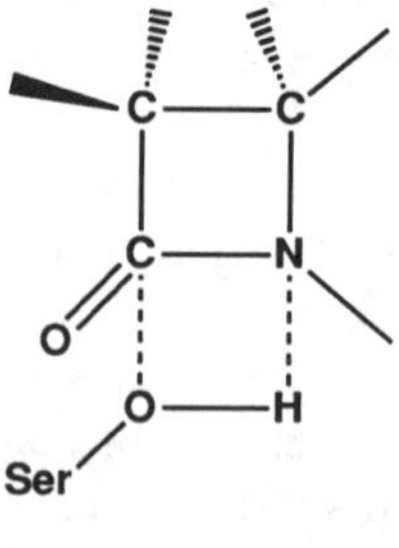

(33) Putative serine * β-lactam complex

An acidic residue (Glu–166) in the R-TEM β-lactamase, acting as a general base, may facilitate the attack of the serine nucleophile (44a, 86, 162). In addition, Glu–166 may also facilitate the deacylation reaction of the acyl-enzyme intermediate, as recent site-directed mutagenesis studies with RTEM and *Bacillus cereus* β-lactamases have indicated (3, 115a).

An important feature for determining the necessary electrophilicity for nucleophilic attack is the oxyanion-binding site (168, 169). In the case of the R61 DD-peptidase, the peptide amide groups of Ser-62 and Thr-301 are thought to provide a part of this site for polarizing the carbonyl (160). A functional group with an apparent pKa of approximately 9.5, the conserved Lys-65, has been proposed to participate in the binding of the C3(4)-carboxylate of β-lactams (310a).

The mechanism of proton transfer in the hydrolysis of the endocyclic amide bond of a β-lactam, as well as the peptide bond of the acyclic peptide donor in the transpeptidation reaction, is essential for understanding the pathway of acyl-enzyme formation, and eventually designing transition-state analogs as potential inhibitors.

A caveat is the penems **(18, 19)**—do they require the steering? Do they have this elusive intrinsic chemical reactivity?

3. *The Effect of Structure on Enzyme Action*

Because of the homology and mechanistic similarity of β-lactamases to DD-carboxypeptidases/transpeptidases, these proteins are grouped into a single superfamily (111, 112, 150). They are all two-domain proteins; one domain has an "all-α" structure and the other has a core of five-stranded β-sheets protected by α-helices (111). Each of these penicillin-interactive proteins possesses seven conserved regions (boxes). Amino-acid replacements in five of these boxes reduce or abolish activity in β-lactamases as well as PBP 3 and PBP 5 of *E. coli* (150). The five boxes occupy critical positions in the three-dimensional structure of the binding site of β-lactamase from *Streptomyces albus*.

As a result of this tertiary structural homology, the three-dimensional coordinates and mechanistic analyses of the β-lactamases can be used for determining some of the geometric constraints that govern acylation of these enzymes as well as other penicillin-interactive enzymes (164). In the case of β-lactamase from *Bacillus licheniformis* 749/C, the C3(4) carboxyl and β-lactam carbonyl are required in the first stage of binding (193). In the three-dimensional structure (Figure 9.7), the amides of Ser-70 (analogous to box II) and Ala-37 attract the carbonyl of the β-lactam while Thr-235 and Lys-234 hydrogen bond to the carboxylate group. With the carboxyl and carbonyl of the β-lactam secured, the β-lactam ring is aligned for an α-face attack by Ser-70. It is hypothesized that the amides of Ser-70 and Ala-37 provide the "oxyanion hole" for stabilizing oxygen during the transition state.

The testing of this model has begun with the aid of site-directed mutagenesis. For example, Lys-234 has been replaced with either Glu-234 or Ala-234. Kinetic analyses of these mutant proteins indicate that Lys-234 is involved in both ground-state and transition-state binding (86). A working hypothesis has been proposed for the action of class A β-lactamases (Figure 9.8). One of its unique features is the role of Glu-166, acting as a general base to facilitate nucleophilic attack of water on the acyl-enzyme intermediate (3). It was shown by Adachi et al. (3) that Glu-166 is not involved in the acylation of the enzyme and, thus, their

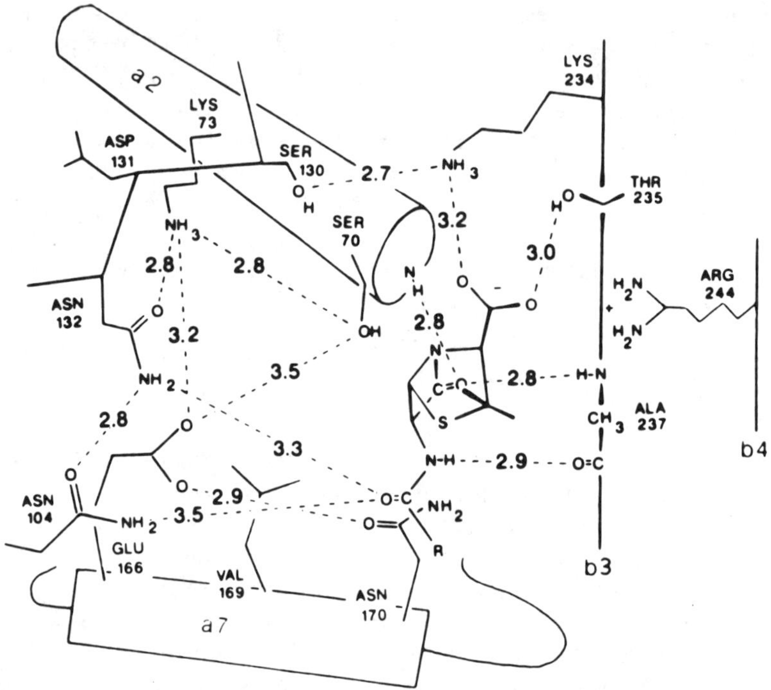

Figure 9.7. Schematic diagram of distances (Å) in penicillin-binding site of the β-lactamase from *B. licheniformis* 749/C. Reprinted with permission (193).

scheme does not involve this residue in facilitating the attack of the serine nucleophile.

The mechanism of acylation/deacylation appears to be essentially identical for the β-lactamase and the R61 DD-peptidase. The difference is in the ability of the DD-peptidase to hinder the approach of a water molecule to the acyl-serine residue. This is thought to be accomplished by Phe-164 which lies very near the α-face of the β-lactam. In contrast, the β-lactamase has a glutamate residue at this location which may facilitate the entry of a water molecule. The mechanism of deacylation is not well understood. Additional changes appear to make the β-lactamase-binding site more hydrophilic than that in the DD-peptidase. As a result, the deacylation constant (k_3) approaches 1,000 s^{-1}. In the DD-peptidase, the k_3 is 10^{-3} s^{-1}. Thus, while there are major differences among these four groups of enzymes, the crystallographic studies have set the groundwork and demonstrated common features of the binding site and the mechanism of acylation. Polarization of the C=O bond, attack at the carbonyl carbon, and protonation of the ring nitrogen are all common features.

Recent studies have revealed several new features of the catalytic mechanism of β-lactamases. Pratt and co-workers (198, 225) recognized that this enzyme

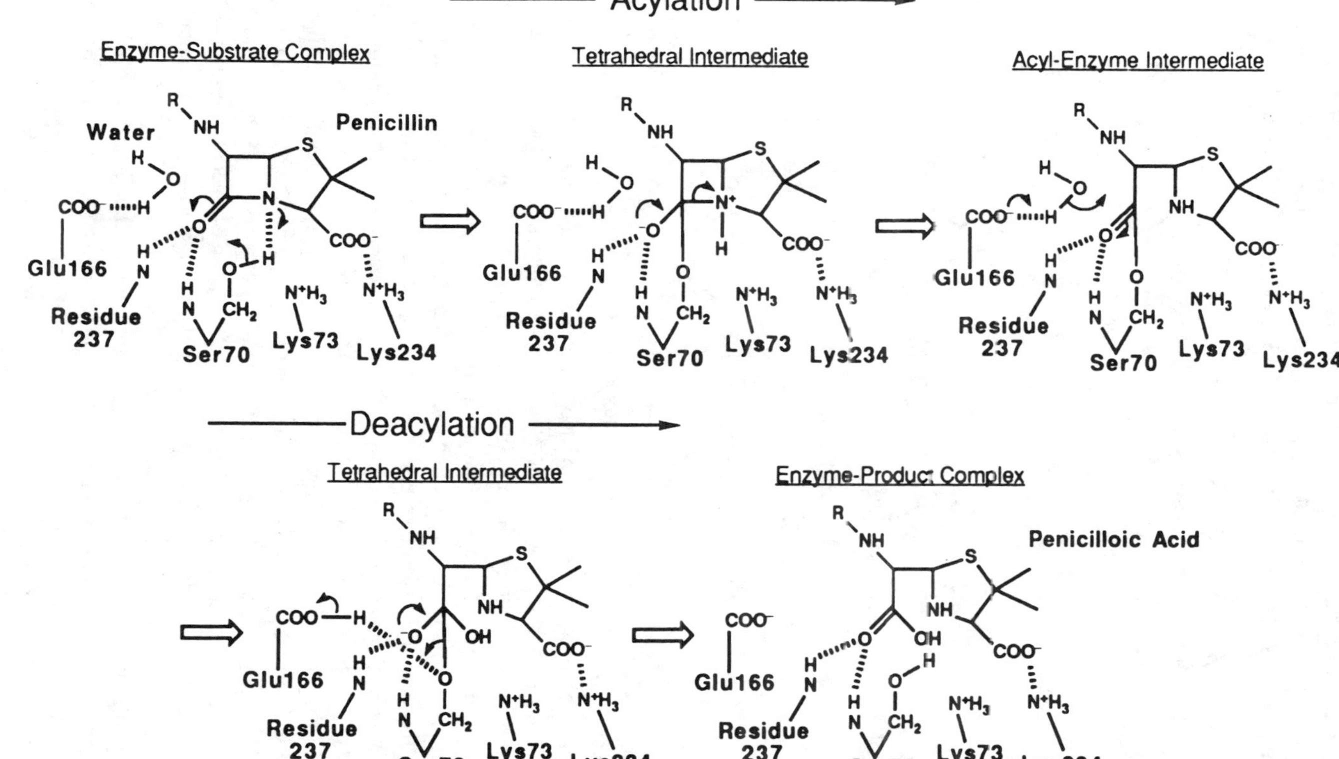

Figure 9.8. A hypothetical mechanism for β-lactamase catalysis. Reprinted with permission (3).

utilizes the depsipeptide (**34**) as a substrate. In the presence of D-phenylalanine, the enzyme catalyzes the aminolysis of (**34**) with the formation of (**35**). Analysis of the steady-state kinetics of the effect of (**35**) and (**36**) indicated that both a

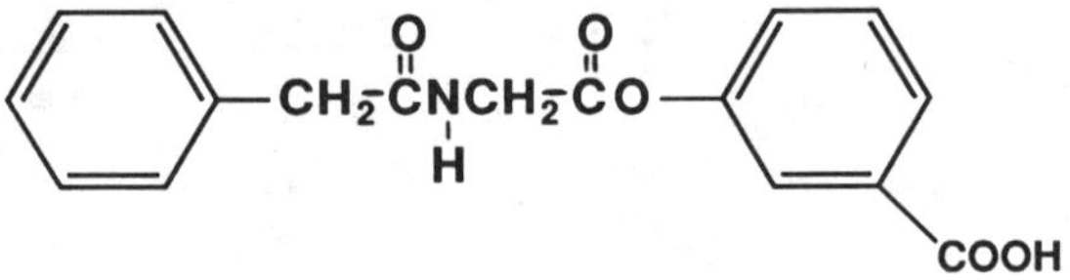

(34) Depsipeptide substrate

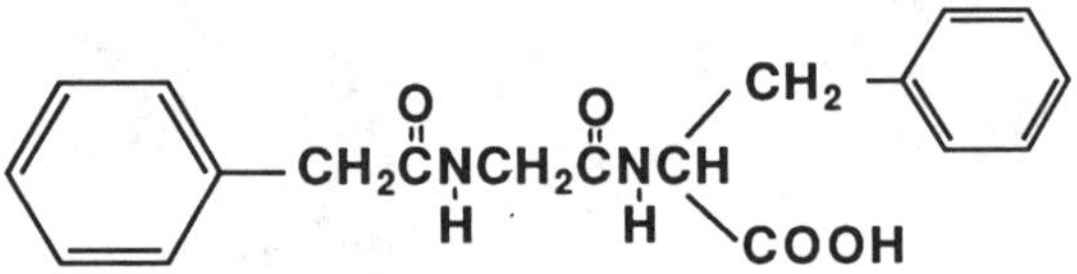

(35) Peptide product

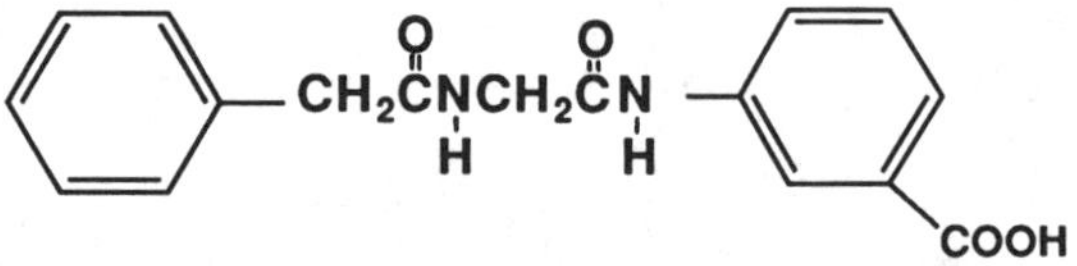

(36) Peptide inhibitor

competitive binding mode and a noncompetitive binding mode exist for each peptide. The important conclusion from these studies is that the β-lactamase probably has two distinct binding sites that can be simultaneously occupied. Because of the evolutionary relationship between DD-peptidases and β-lactamases, it is reasonable to suggest that the two sites represent *donor and acceptor* sites in this penicillin-sensitive enzyme. This proposal may be reconciled with some of the R61 DD-peptidase kinetics that do not appear to be compatible with a single two-stage acyl-enzyme-intermediate model. Nguyen-Distèche *et al*. (203) suggested that the amino acceptor, Gly-Gly or Gly-Ala, modifies the binding of the carbonyl donor Ac_2-Lys-DAla-DAla in the reaction catalyzed by the *Streptomyces* K-15 DD-peptidase. In the presence of these acceptors, both the amide and the ester carbonyl donors are processed without detectable accumulation of the acyl-enzyme. In addition, it was suggested that hydrolysis and transpeptidation do not proceed through identical pathways.

The spatial disposition of the carboxylate function at the C-terminal position and the amide function at the *N*-terminal position with respect to the scissile amide bond has been deduced from comparisons of conformational and crystallographic structural analyses. The molecular electrostatic potential maps reflect the spatial distribution of electron density around the oxygens. The major energy well is associated with the carbonyl of the scissile bond, congruent with the oxyanion hole of the DD-peptidase and β-lactamase (172).

C. Evolution of the Substrate-analog Hypothesis

According to the substrate-analog hypothesis (298, 317), penicillin mimics the acyl-DAla-DAla moiety of the MurNAc-Ala-DGlu-Lys-DAla-DAla donor peptide in the transpeptidation reaction (297). Acylation of the enzyme nucleophile results in the formation of an inactive penicilloyl-enzyme. Thus, penicillin has been proposed to act as (1) an active-site-directed inhibitor (271), (2) a transition-state inhibitor (173), or (3) a mechanism-based inhibitor (suicide substrate).

Evidence so far supports the third proposal; *i.e.*, inactivation of the enzyme targets by β-lactams involves a mechanism-based acylation (93, 110, 317). Because of the endocyclic nature of the scissile amide bond of the β-lactam, the leaving group can not dissociate and remains bound to the site after the amide bond is cleaved. In the transpeptidases and DD-carboxypeptidases, the acyl-enzyme reactivity is poor and hence deacylation is slow. Because the thiazolidine (or dihydrothiazine) ring remains covalently bound, it is proposed that it sterically blocks incoming nucleophiles, *e.g.*, HOH, H_2NOH, or the peptide acceptor of nascent PG (Figure 9.8). However, steric hindrance may not be necessary, as in monobactams the considerably smaller leaving group also remains covalently bound. In the case of penicilloyl enzyme, a secondary C_5-C_6 cleavage results in the release of the thiazolidine group and phenylacetylglycine (110, 316).

In 1975, Rando proposed the induced-strain model of penicillin action (240). It was suggested that binding of the penicillin to the receptor enzyme would increase reactivity by distorting penicillin from its ground-state structure. In the case of penicillins and cephalosporins, the transpeptidase binds the β-lactam in a conformation which requires that the β-lactam approach planarity (see section II.A). This enzyme-induced strain energy could translate into binding-site reactivity. The proposal does not account for the activity of the nearly planar monobactams.

In 1971, Lee suggested that the conformation of the substrate analog, *e.g.*, Gly-DAla-DAla, at its transition state is similar to that of penicillin (173). This is possible if the peptide bond of this analog is twisted 45° out of plane. Using molecular orbital calculations, Boyd (30) concluded that the spatial match between the tetrahedral adduct and penicillin is generally better than that of the peptide and penicillin (31). Thus, a transpeptidase could easily recognize a β-lactam antibiotic as its substrate. The transition-state analog hypothesis has received further support from crystallographic and modeling studies (160). However, it would be expected that a transition-state inhibitor would have a high affinity for the receptor (168). In the penicillin-sensitive enzymes, the dissociation constants (K) for peptide and β-lactam are similar and essentially unrelated to inhibitory activity. For example, with the R61 DD-peptidase, the dissociation constants for the substrate and benzylpenicillin are 14 and 13 mM, respectively (93). Transition-state analog inhibitors of carboxypeptidase A, on the other hand, have dissociation constants in the picomolar range (129). It is clear that β-lactams are not simple affinity-labeling agents or transition-state inhibitors. They bind to the target enzymes and are processed by the normal catalytic mechanism generating an acyl-enzyme intermediate that, because of the endocyclic C5-C6 (C6-C7) bond, cannot dissociate. Thus they have the hallmarks of mechanism-based inhibitors (16a).

D. Relevance of Biochemical Data to β-lactam Design

Boyd and Ott (36, 37) compared data from biochemical and microbiological studies and concluded that the ability to inhibit cell growth by β-lactams in a variety of pathogens is not correlated with their ability to inhibit the DD-peptidase from *Streptomyces* R61. The result is not surprising, since each organism has its own characteristic penicillin-sensitive enzymes, and each enzyme its own specificity profile. Nonetheless, because of the tertiary homology of the transpeptidases, DD-carboxypeptidases, and β-lactamases, we can use structural and mechanistic features common to all β-lactam-sensitive enzymes for the design of new β-lactams or novel antibacterial agents.

In summary, it is now clear that β-lactams are mechanism-based inhibitors, *i.e.*, suicide substrates. The key to the success of these compounds is that the potential leaving group remains covalently linked to the active-site serine, preventing its regeneration. The minimal structure for an inhibitor must incorporate this key element without violating the various specificity determinants, conformational restraints, and reactivity considerations discussed earlier.

The features derived from mechanism common to the vast majority of penicillin-sensitive enzymes are:

- scissile bond
- leaving group unable to dissociate
- carboxyl (or other anion) adjacent to β-lactam
- β-lactam carbonyl in oxyanion hole
- proton acceptor
- free approach vector of nucleophile

Features derived from individual specific targets are:

- acylation rate
- deacylation rate

Acylation and deacylation rates are determined by the binding-site reactivity of the β-lactam and the reactivity of the acyl enzyme, respectively.

The above points emphasize the fact that while one cannot use the geometric constraints of a model enzyme as a template to design the ultimate broad-spectrum β-lactam, there are common structural and mechanistic features that can serve as a conceptual framework for inhibitor design. Balancing the many physical, stereochemical, and biochemical features that lead to growth inhibition in different bacteria is a challenging task. Growth inhibition involves effects on several processes, which collectively translate into traditional structure/activity relationships, *i.e.*, structure/MIC. With the advent of computerized molecular modeling (226), effects on each process can be assessed separately. So far, target multiplicity (section II.A) and accessibility (section IV) have hampered computer-assisted molecular modeling. Nevertheless, the mechanistic approach we have taken in this review may eventually play a significant role in new drug design. Insights on β-lactams designed against specific targets are provided elsewhere (58, 113, 157, 181, 235).

IV. Accessibility of the Receptor Target

Having discussed the determinants of β-lactam activity against the receptor targets, we will now consider the accessibility of the β-lactams to the targets in intact cells. Two factors that influence accessibility are outer-membrane (OM) permeability and β-lactamases.

A. Outer-membrane Permeability

The targets of β-lactam antibiotics reside on the outer face of the cytoplasmic membrane. Thus, these targets are readily accessible in gram-positive bacteria (the cell wall does not constitute a permeability barrier) but not in gram-negative bacteria, where the β-lactam must first pass through the OM. The OM is a protein-rich, *asymmetric* lipid bilayer containing phospholipids in the inner leaflet and lipopolysaccharide (LPS) in the outer leaflet (177, 212). It functions as a molecular sieve, with water-filled channels (pores) formed by 35- to 45-kDa proteins (porins), through which nonspecific transport occurs. In enterobacteria, LPS constitutes the entire outer leaflet of the lipid bilayer, preventing lipophilic compounds from diffusing in. Hydrophilic compounds, such as β-lactam antibiotics, enter the OM mostly through porins.

1. Porins

Porins are major OM proteins which, in enterobacteria at least, form stable trimers (21). *E. coli* K-12 produces two porin species, OmpC (38 kDa) and OmpF (37 kDa), which form channels of 11 Å and 12 Å diameters, respectively, with an exclusion limit of 600–800 kDa. The two porins are reciprocally regulated at the transcriptional level (125). Under physiological conditions, *i.e.*, high osmolarity (300 mosM) and temperature (37°C), production of OmpC (the narrower channel) is favored over production of OmpF. Other *E. coli* strains may produce up to three porins, as do the related *S. typhimurium* (200), *K. pneumoniae,* and *E. cloacae* (266). *Proteus, Morganella, Providencia,* and possibly *Serratia,* on the other hand, may produce a single porin (191).

P. aeruginosa may produce three porins, of molecular weights 70, 46, and 43 kDa (332). The apparent exclusion limit is 250 Da, suggesting that the intrinsic resistance of *P. aeruginosa* to most β-lactams is largely due to the small pores of its OM. With the exception of imipenem (**18**), β-lactams have molecular weights greater than 350 and the porin pathway is probably not significant for this class of compounds. Protein F, a 34-kDa protein suggested to be porin (127), is homologous to OmpA of *E. coli* (328) and may have a structural role in the OM (119a), though very recent functional studies support a porin role (207).

H. influenzae produces a 40-kDa porin with an exclusion limit similar to that of *E. coli* (310). The apparent high susceptibility of this organism to antibiotics might be due to a loose LPS structure allowing nonporin pathways to operate. *N. gonorrhoeae* produces an anion-selective porin (protein I) as its major OM protein (336). *B. fragilis* pores have an exclusion limit similar to that of *P. aeruginosa* (64, 335); its porins have not yet been identified, although its outer-membrane protein profile has been recently reported (224).

The permeability of the porin channels is a function of the size, charge, and hydrophobic nature of the β-lactam molecule. In *E. coli* and related enterobacteria, hydrophilic, cationic molecules are preferred (333). In other gram-negative bacteria, permeability differs from that of *E. coli,* ranging from 100-fold lower in *P. aeruginosa* (333) to tenfold higher in *H. influenzae* (62a).

2. *Other Uptake Pathways*

In addition to the nonspecific porins, the OM possesses specific channels, two of which, the siderophore uptake system and the 45-kDa D2 protein of *P. aeruginosa,* have been implicated in β-lactam permeation. Catechol cephalosporins (**37**) mimic the siderophores involved in iron transport and pass through the OM by using siderophore receptors; they have enhanced activity against enterobacteria and *Pseudomonas* (66, 192, 210, 314). Imipenem (**18**) passes through the outer membrane of *P. aeruginosa* by using a specific sugar channel (D2 protein) (304, 305, 332). Antipseudomonal β-lactams, such as cefoperazone (**22**) and ceftazidime (**38**), may also enter through a nonporin pathway, by removing the magnesium that bridges adjacent LPS molecules.

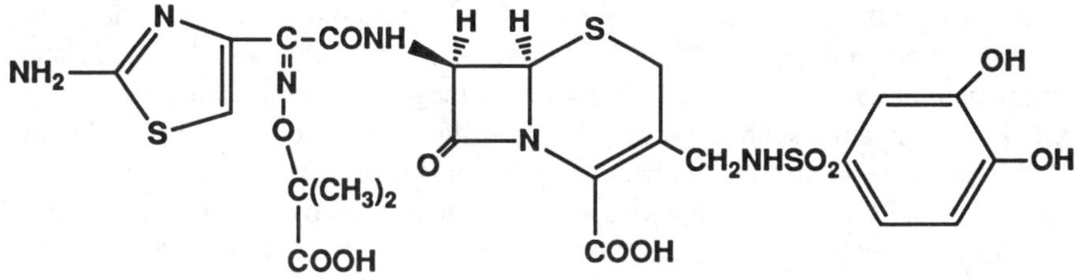

(37) Catechol cephalosporin (ref. 66).

B. The Challenge of β-lactamases

β-Lactamases are largely responsible for bacterial resistance to β-lactam antibiotics, and the development of new β-lactams is closely linked to the emergence of enzymes with altered specificities and expression levels (section V.A) (145, 229, 319). In gram-positive bacteria, the majority of β-lactamases are inducible and exocellular; in gram-negative bacteria, they can be inducible or constitutive and are predominantly periplasmic (108).

β-Lactamase activity can be measured by several assays (reviewed in refs. 46, 46a, 108, and 126), the spectrophotometric (146, 263), microiodometric (214, 264, 268), and acidimetric (259) assays being the most useful ones. Of the three, the first is a continuous assay and highly sensitive, requiring only micromolar amounts of the substrate. In intact cells or crude extracts, β-lactamase activity is commonly assayed by using the chromogenic cephalosporin nitrocefin (216). More recent methods include a colorimetric assay that measures penicilloic acid production (60), and a nitrocefin-competition assay that measures inhibition of nitrocefin hydrolysis (220).

For constructing substrate profiles, β-lactamase activities are normalized to a reference β-lactam. Unfortunately, in early studies, activities were measured as hydrolysis rates under conditions of high substrate concentration (0.1 to 1 mM). This obscured the contribution of substrate affinity, i.e., K_m. A more balanced approach, which takes into account both hydrolysis rate *and* affinity, uses the ratio V_{max}/K_m (termed "physiological efficiency" by Pollock (233)), or v/K_m if values are normalized to a reference β-lactam. For example, cephaloridine **(21)** is more rapidly hydrolyzed by chromosomal cephalosporinases than cefotaxime **(20)**, but has a lower affinity for the enzyme (K_m, >300 μM vs. ~0.3 μM for cefotaxime). Since PBP binding occurs at submicromolar β-lactam concentrations, β-lactamase activity assumes physiological importance at these concentrations. (See section IV.D.)

β-Lactamases have been traditionally classified on the basis of molecular weight, substrate profile, inhibitor profile, isoelectric point, or a combination of the four parameters (17a, 45a, 253a, 293a). As their amino-acid sequence and mechanism of catalysis have become known, they have formed the basis for the ABCD classification scheme, which divides β-lactamases into four broad classes (9, 150). This classification is becoming increasingly attractive, as distinctions between penicillinases and cephalosporinases are blurred by the emergence from the former of extended-spectrum β-lactamases. The ABCD classification will be followed in this review (Table 9.1).

Table 9.1. Major β-lactamases in clinical bacteria

Trivial name	Old class[a]	Original host	Substrate profile	Inhibitor profile
		Class A—serine enzymes, plasmid mediated[b] (~30 kDa)		
PC1		*S. aureus*	pen/ceph[c]	clox/clav/sulb[b]
TEM[e]	IIIa	*E. coli*	pen/ceph	clox/clav/sulb/taz
SHV	IV	*K. pneumoniae*	pen/ceph	clav/sulb/taz
HMS-1		*E. coli*	pen	
ROB-1		*H. influenzae*	pen	clav/sulb
OXA[f]	V	*E. coli*	pen/ceph	clav/sulb/taz
CARB[g]		*P. Aeruginosa*	pen	
K-1	IV	*K. oxytoca*	ceph	
		Class B—metalloenzymes, chromosomal (~22 kDa)		
		B. fragilis	pen/ceph/cpen	
		P. maltophilia	pen/ceph/cpen	
		Class C—serine enzymes, chromosomal (~40 kDa)		
AmpC	I	*E. coli*	ceph	taz
P99	I	*E. cloacae*	ceph	taz
S−A[h]	I	*P. aeruginosa*	ceph	taz
		Class D—serine enzymes, plasmid-mediated[i] (~12 kDa)		
PSE-3	V	*P. aeruginosa*	pen	clav/sulb/taz

[a] After references 253a and 293a.

[b] Except for K1, which is chromosomal.

[c] pen, penicillins; ceph, cephalosporins; cpen, carbapenems

[d] clox, cloxacillin; clav, clavulanic acid; sulb, sulbactam; taz, tazobactam.

[e] Several closely related enzymes. TEM-3 to TEM-11 are mutant forms of TEM-1 or TEM-2 isolated from enterobacteria; they are often referred to by the name of the β-lactam they destroy (*e.g.*, TEM-3 is CTX-1 (cefotaximase) (178a, 226a); TEM-5 is CAZ-1 (ceftazidimase), *etc.*).

[f] Several closely related enzymes (OXA-1 to OXA-7).

[g] Several enzymes. Also called PSE.

[h] Sabath et al. (259).

[i] After Huovinen *et al.* (137).

1. Class A Enzymes

Clinically important enzymes of this class are plasmid-mediated, with the exception of the chromosomal *Klebsiella* enzymes such as the K1 β-lactamase of *Klebsiella pneumoniae* (12, 13, 131, 149). The plasmid-mediated β-lactamases include the staphylococcal enzymes, the TEM, SHV, and HMS enzymes of enterobacteria, the ROB-1 enzyme of *H. influenzae,* and the OXA and PSE (CARB) enzymes of enterobacteria and *Pseudomonas*. They are generally penicillinases, though extended-spectrum β-lactamases have emerged from the *K. oxytoca,* TEM, and SHV enzymes in the past 5 years (229). The trivial names of the enzymes have originated from patients (TEM), inhibitors (SHV), substrates (CARB), organisms (PSE), or researchers themselves (HMS).

S. aureus β-lactamases are divided serologically into four types, A, B, C, and D (253, 256), and are inducible in the majority of wild-type strains. Most studies have used a penicillinase-constitutive strain, PC1, which overproduces the type A enzyme. Since the advent of isoxazolyl penicillins and methicillin (**32**) in the Sixties, staphylococcal β-lactamases have ceased to be of major clinical concern.

The TEM enzyme is the most widely distributed β-lactamase in clinical isolates of gram-negative pathogens, accounting for over two-thirds of β-lactamase producers. It was originally isolated from *E. coli* (72), but has since spread to other enterobacteria and gram-negative bacteria. Other β-lactamases of this class are the SHV (227, 232) and HMS-1 (184) enzymes, originally found in *K. pneumoniae;* the OXA enzymes, found in enterobacteria and *Pseudomonas* (196); the ROB-1 enzyme, found in *H. influenzae* (152, 185, 258); and some of the PSE enzymes of *P. aeruginosa* and enterobacteria (133). The carbenicillin-hydrolyzing PSE-1, PSE-4, CARB-3, and CARB-4 enzymes also belong to this class (26), while the 12-kDa PSE-3 β-lactamase has been placed in a separate class (class D; 137, 150). Some of the newer OXA-type enzymes hydrolyze β-lactamase-stable cephalosporins such as cefotaxime (186).

2. Class B Enzymes

These are chromosomally mediated, zinc metalloenzymes, possessing a thiol group which acts as a ligand to bind zinc. Clinically important enzymes of this class have been found in *Bacteroides* (64, 137a, 334) and *P. maltophilia* (23, 260). Their activity spectrum includes carbapenems and cephamycins, generally regarded as β-lactamase stable (194); they probably function without an acyl-enzyme intermediate. The enzyme from *B. fragilis* has been cloned and sequenced, and is homologous to the *Bacillus cereus* zinc-requiring β-lactamase (10, 243).

3. Class C Enzymes

Chromosomally mediated β-lactamases are either constitutive or inducible, and range from barely detectable amounts to as much as 10% of the cell protein (108). They are typically cephalosporinases and are found in gram-negative bacteria.

Constitutive enzymes include *E. coli ampC* β-lactamase (147) and the enzymes produced by *Shigella* and *Salmonella*. Several strains are known to produce exceptionally high levels of class C enzymes, *e.g.*, *E. cloacae* P99 (120). *Bacteroides* β-lactamases are also mostly constitutive and are usually inhibited by clavulanic acid and sulbactam (85, 277, 296).

Inducible β-lactamases are found in a wide range of bacteria, notably in *Enterobacter, Citrobacter, Acinetobacter,* indole-positive *Proteus, Pseudomonas, Serratia,* and *Yersinia*. Induction (more than 100-fold) can occur with β-lactamase substrates (*e.g.*, ampicillin) or inhibitors (cefoxitin **(15,)** clavulanic acid **(24)**). With substrates, it can result in temporary resistance; with some inhibitors, in antagonism. Selection of mutants overproducing these enzymes constitutively (derepressed mutants (175) is a more serious complication, as it can result in broad, long-term β-lactam resistance (255a; see also section IV.B).

4. Class D Enzymes

These β-lactamases are plasmid-mediated, and include the carbenicillin-hydrolyzing 12-kDa PSE-3 enzyme (137, 149, 183). The oxacillin-hydrolyzing OXA enzymes have been recently suggested to be class A β-lactamases (196).

C. Coping with β-lactamases

1. β-lactamase-resistant β-lactams

A significant number of antibacterial β-lactams are resistant to enzymatic hydrolysis by class A and C β-lactamases. Compounds of this class are 7α-methoxy cephalosporins (cefoxitin **(15)**, cefotetan **(17)**, cefmetazole **(16)**, moxalactam **(19)**), and 6α-methoxy penicillins (temocillin **(14)**); oxyimino cephalosporins (cefotaxime **(20)**, ceftriaxone **(39)**, ceftazidime **(38)**); penems (Sch34343 **(19)**) and carbapenems (imipenem **(18)**, meropenem **(40)**); and monobactams (aztreonam **(4)**, carumonam **(41)**). In the case of monobactams, the 4-substituent provides additional β-lactamase resistance (290).

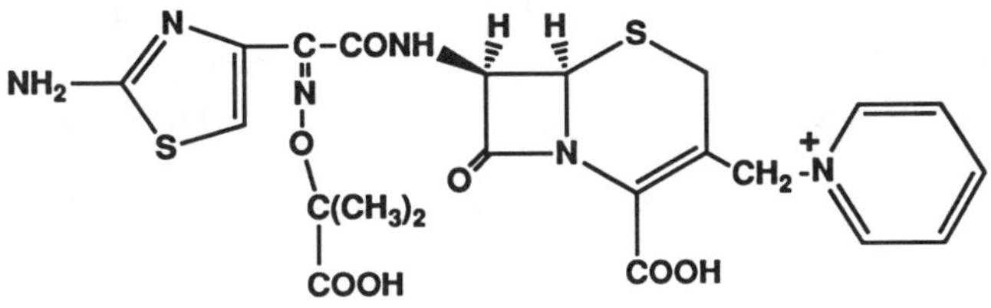

(38) Ceftazidime

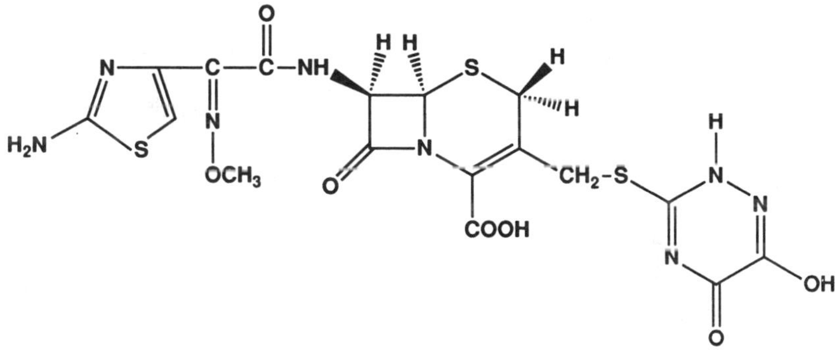

(39) Ceftriaxone

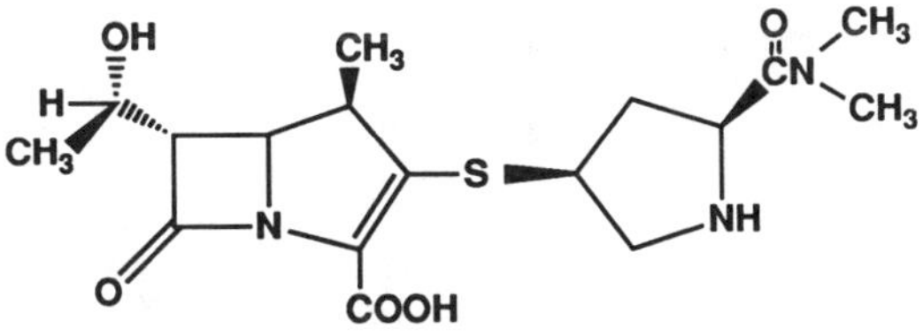

(40) Meropenem (Δ^2-carbapenem)

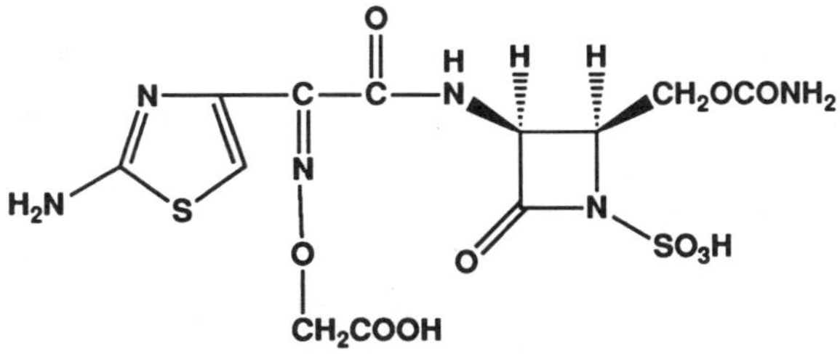

(41) Carumonam (monobactam)

2. *β-Lactamase inhibitors*

These β-lactams lack substantial antibacterial activity but potentiate activity of other β-lactam antibiotics. They include clavulanic acid **(24),** the first mechanism-based inhibitor of β-lactamases to be reported (41, 246); the penam sulfones, sulbactam **(42)** (88) and tazobactam **(43)** (16); 6β-halopenicillanic acids **(44)** (59); 6-(substituted methylene) penams **(45)** (14, 15) and -penems **(46, 47)** (38, 61, 197).

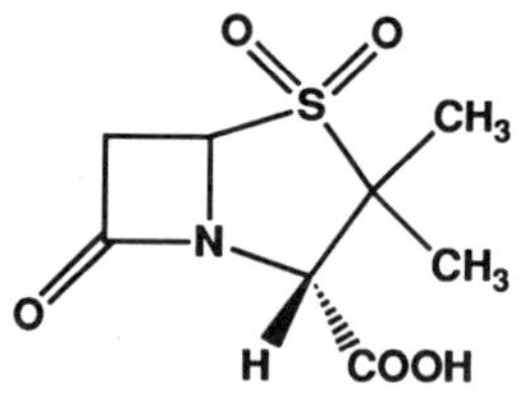

(42) Sulbactam

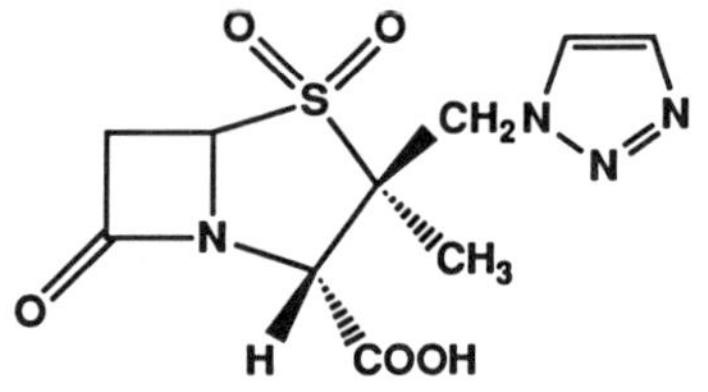

(43) Tazobactam

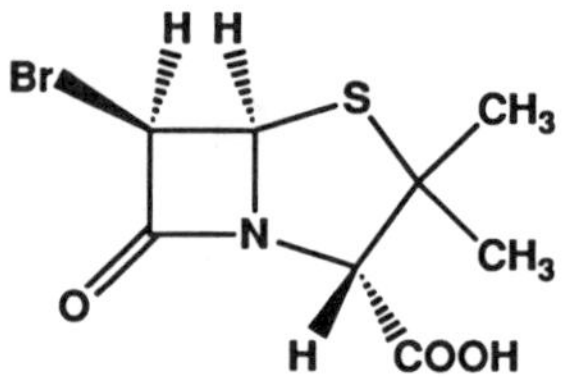

(44) 6beta-Bromopenicillanic acid

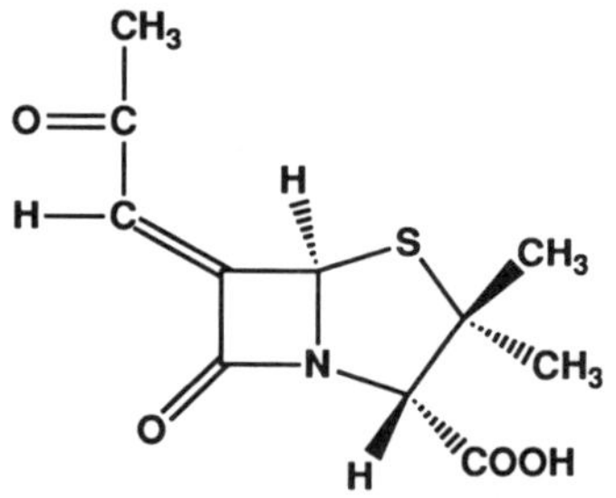

(45) 6-acetylmethylenepenicillanic acid

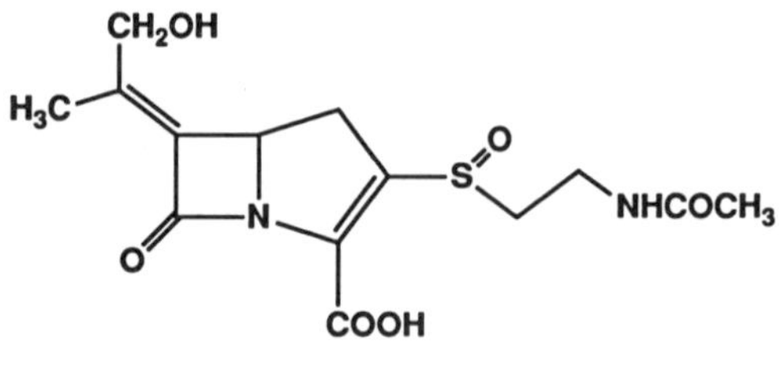

(46) Asparenomycin B

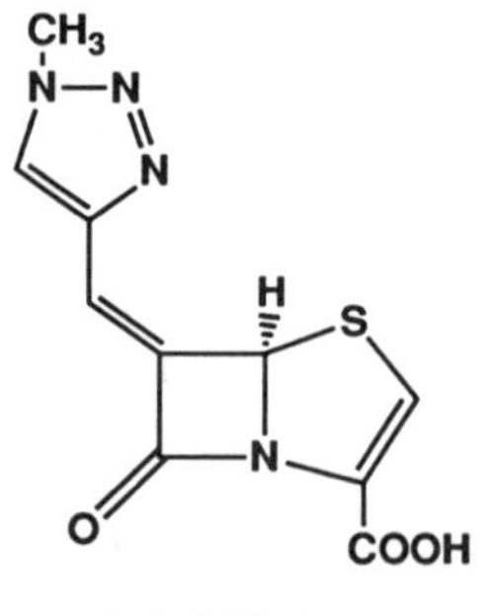

(47) BRL42715

Of these compounds, the first two are currently in clinical use as fixed combinations with established penicillins: clavulanic acid with amoxicillin (Augmentin) and ticarcillin (Timentin); sulbactam with ampicillin, combined (Unasyn) or covalently attached (sultamicillin) (87). Both clavulanic acid and penam sulfones are more inhibitory to class A than class C β-lactamases (65). Since the latter enzymes are increasingly becoming a threat to third-generation cephalosporins, BRL42157, an excellent inhibitor of both classes of enzymes, holds much promise.

There are no known inhibitors of class B β-lactamases other than metal chelators (48).

D. The Interplay between β-lactamases and Permeability

A key aspect of the susceptibility of gram-negative bacteria to β-lactams is the interplay of OM permeability, affinity/turnover for β-lactamases in the periplasmic space, and affinity for target PBPs (208, 293, 312). Although β-lactamases constitute the bulk of β-lactam resistance, it is the combined presence of these enzymes and reduced OM permeability that poses clinical significance and concern. Lowered target affinity is seldom a resistance mechanism in gram-negative bacteria; it has been reported for *H. influenzae* (187) and *N. gonorrhoeae* (80), organisms with relatively permeable OM. The action of β-lactamases invariably involves hydrolysis of β-lactam, *i.e.*, turnover. Covalent binding ("trapping") of the β-lactam by the enzyme probably does not occur, at least not in *E. cloacae* (311) where it was originally reported (295).

The hydrolysis of β-lactams by β-lactamases in the periplasmic space has been exploited in an elegant assay for determining outer-membrane permeability (267, 338). Analysis rests on the assumptions that β-lactam influx across OM occurs by simple diffusion and that β-lactam hydrolysis follows Michaelis-Menten kinetics. Thus, the assay is only applicable when the β-lactam is a substrate for the β-lactamase of the organism being tested. An important consideration is the method of preparation of the cell extracts, as sonication inactivates some β-lactamases, inflating permeability values (188).

An alternative assay involves the measurement of the β-lactam penetration rate in reconstituted vesicles containing porins (209). As the β-lactam enters, the vesicles swell—hence the name "liposome swelling" assay. The swelling can be followed spectrophotometrically by the decrease in the turbidity of the suspension. A major limitation of this assay is that nonphysiological concentrations of β-lactams are used (10 mM *vs.* 0.1 mM for the β-lactamase-based assay).

By using these two assays, extensive structure-permeability relationships for β-lactams have been established in *E. coli* (206, 211). The assays have also been used to assess the OM permeability of β-lactams in other bacteria, with varying degrees of success.

V. Conclusions and Future Directions

A. β-lactam Development and Bacterial Resistance

As stated earlier, the interplay of β-lactam development and β-lactam resistance has been a major driving force in β-lactam research. The early clinical success of benzylpenicillin against *S. aureus* in the Forties was followed by the emergence of penicillinase-associated resistance, leading to the development of penicillinase-resistant penicillins (*e.g.*, methicillin **(32)** and oxacillin). The stability of these latter compounds to penicillinases has been attributed to the steric bulk in the compact conformation (Figure 9.3).

The addition of the amino group in the α-carbon of benzylpenicillin gave rise to ampicillin **(48)**, extending the activity spectrum to gram-negative bacteria (136). Substitution on the α-amino group of ampicillin with 4-ethyl-2,3-dioxo-1-piperazinylcarboxylic acid yielded the acyl derivative, piperacillin **(21)**, with significant antipseudomonal activity. The success of aminopenicillins in the 1960s led to the appearance of plasmid-mediated enzymes in enterobacteria and *Pseudomonas* and the spread of these enzymes to other pathogens (notably, *N. gonorrhoeae* and *H. influenzae*) in the following decade. This in turn provided the impetus for a massive synthetic effort, resulting in the first-generation cephalosporins, such as cephalexin (**6**, R=CH_3).

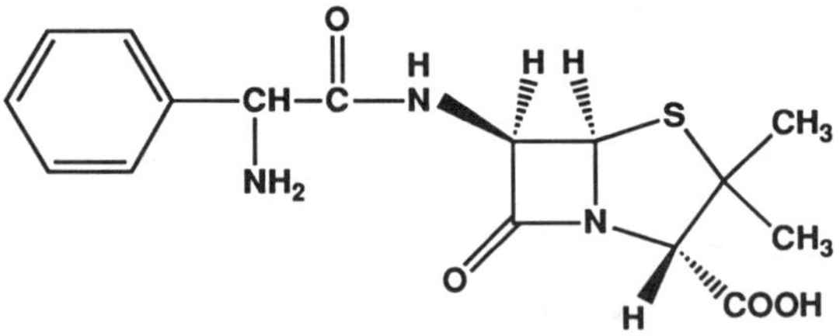

(48) Ampicillin

The 6β-acylamino side-chain, characteristic of most penicillins, was replaced in the 1970s by an acyl group with an N,N-dialkyl formamidine, giving rise to mecillinam **(29)** (178). The compound is very active against many gram-negative bacteria, though relatively inactive against gram-positive bacteria. It illustrates a nonacyl type of modification on the 6-position of penicillin that broadens the antibacterial profile. Due to its limited use, resistance to this compound has not occurred.

The development, success, and widespread use of β-lactam-resistant cephamycins and oxyimino cephalosporins in recent years have led to the emergence of methicillin-resistant *Staphylococci,* and gram-negative organisms that overproduce chromosomal and plasmid-mediated, extended spectrum, β-lactamases. (See chapter 2.)

Even the carbapenems have not escaped unscathed. Their use has revealed an obscure specific channel in the outer membrane of *P. aeruginosa,* whose absence is associated with resistance (45, 239). It has also highlighted a little-known class of β-lactamases in *P. maltophilia* and *B. fragilis* which do not operate via an acyl intermediate, and are thus capable of hydrolyzing this important class of compounds (section IV.B.2).

B. New Developments in β-lactams

The increased understanding of the individual factors involved in β-lactam action has revealed potential targets for β-lactam design: β-lactamases, OM channels, and peptidoglycan transpeptidases. Unfortunately, these components often pose conflicting requirements. For example, the side-chain of ceftazidime **(38)** confers increased β-lactamase stability over cefotaxime **(20)** and ceftriaxone **(39)**, but at a significant loss of binding to *S. aureus* PBPs; hence it is a narrow-spectrum, gram-negative antibacterial agent. Methicillin **(32)** has increased β-lactamase stability over ampicillin **(48)**, but is significantly less permeable; hence it is a narrow-spectrum, antistaphylococcal drug. In a few cases, however, stability to β-lactamase or enhanced OM permeation has not resulted in compromise. Cefoxitin **(15)**, for example, has increased β-lactamase stability *and* PBP binding over cephalothin. Imipenem **(18)** has increased β-lactamase stability *and* PBP binding *and* OM permeability over most cephalosporins. Some catechol-cephalosporins **(37)** have preserved PBP binding while increasing OM penetration. So far, these relationships have emerged *post hoc;* they have not led β-lactam design.

Recent developments in the β-lactam field include interaction with other cell-wall inhibitors, such as phosphonomycin and vancomycin. The former compound acts with β-lactam antibiotics synergistically in *S. aureus* (it suppresses PBP 2′ in MRSA) (309) but antagonistically in *P. aeruginosa* (it suppresses PBP 3 and induces β-lactamase) (247). In the case of vancomycin, resistance in *Enterococci* is associated with the induction of a 39-kDa protein, suggested to be DD-carboxy-peptidase (8, 165, 204, 273). Resistance is suppressed by β-lactam antibiotics that inhibit the DD-carboxy-peptidase (121, 274). More recently, the 39-kDa protein has been purified and identified as a D Ala:D Ala ligase of altered substrate specificity (44b, 83a).

Other developments are the synthesis of 3′-substituted quinolone cephalosporins (4, 5, 6, 73, 73a, 101). In the case of β-lactamase-sensitive cephalosporins with a chemically stable linkage to quinolones, the quinolone component is delivered preferentially to the site of infection. The result can be improved pharmacokinetics and reduced toxicity to the host.

New developments in β-lactam resistance include the alarming proliferation of extended-spectrum, TEM-type β-lactamases (229), the emergence of methicillin-resistant, quinolone-resistant *Staphylococci* (245, 279), the bizarre peptidoglycan formed as a consequence of β-lactam-resistant PBPs in *Pneumococcus* (97, 98), and the growing realization that the bactericidal effect is related to the rate of growth (63); slow-growing bacteria (*e.g.*, in prosthetic devices) are hard to kill (40).

C. Future Prospects

The challenge of β-lactam resistance offers several opportunities for β-lactam design. The universality of PBP 2′ in methicillin-resistant *Staphylococci* (50), the broad spectrum of β-lactam resistance, and the increase in incidence of resistance make this PBP a particularly attractive target. Understanding its expression and elucidating its active site and mechanism of action will be valuable for inhibitor design. Thus, the recent finding of synergy between phosphonomycin and β-lactams is particularly encouraging for new agents against methicillin-resistant staphylococci.

β-Lactamase inhibitors, aimed at class C β-lactamases, are another promising area of research. The recent discovery of methylenepenems, which strongly inhibit both class A and C enzymes, attest to the feasibility of such inhibitors. Although there are as yet no inhibitors of the class B enzymes, the incidence of these chromosomal β-lactamases has been mercifully low. This, of course, is likely to change with the increased use of carbapenems. Incidence in *Bacteroides* and *Pseudomonas* might increase, or they might become mobile and spread beyond these organisms. Thus, there is reason to warrant a search for inhibitors of class B enzymes. The search will undoubtedly be aided by the extensive knowledge on the zinc proteases. But understanding the mechanism of action of the zinc β-lactamases themselves would be most important.

Exploring different substituents at position 3 of cephalosporins, aimed at achieving greater β-lactamase stability to deacylation (section II.D.4), is another valid approach. Increasing the reactivity of the β-lactam (section II.B) may be limited, as this feature is probably responsible for the hypersensitivity reactions, the most common adverse effects with β-lactams.

Fleming's discovery of penicillin in 1928 initiated one of the remarkable sagas in medical history. In 1942, on a Saturday afternoon, Anne Miller received one of the first injections of this miracle drug. By Sunday, her temperature had fallen from 41°C to normal and her streptococcal infection had subsided (161). Yet, penicillin has not become the panacea for bacterial infections. Nature has endowed bacteria with effective mechanisms to circumvent its lethal action: β-lactamases, permeability barriers, resistant targets. Thus research in microbial targeting, receptor interactions, and resistance mechanisms is essential to our continuing quest for new β-lactam antibiotics. The principles outlined in this review present a flexible framework for the design of such compounds.

Acknowledgments

The authors are indebted to Donald Boyd, Stephen Faraci, Michael Heaton, Judith Kelly, James Knox, Rex Pratt, David Pruess, John Roberts, Gerald Shockman, Rabindra Sinha, Alexander Tomasz, Brian Wilkinson, and Saul Wolfe for discussions and manuscripts prior to publication. The research in the laboratory of FN has been supported by grant AI–04615 from the National Institute of Allergy and Infectious Diseases.

References

1. Actor, P., L. Daneo-Moore, M.L. Higgins, M.R.J. Salton, and G.D. Shockman. 1988. Antibiotic inhibition of bacterial cell surface assembly and function, p. 1–657. American Society for Microbiology, Washington, D.C.

2. Adachi, H., T. Ohta, and H. Matsuzawa. 1987. A water-soluble form of penicillin-binding protein 2 of *Escherichia coli* constructed by site-directed mutagenesis. FEBS Lett. **226:**150–154.

3. Adachi, H., T. Ohta, and H. Matsuzawa. 1991. Site-directed mutants, at position 166, of RTEM-1 β-lactamase that form a stable acyl-enzyme intermediate with penicillin. J. Biol. Chem. **266:**3186–3191.

4. Albrecht, H.A., G. Beskid, K.-K. Chan, J.G. Christenson, R. Cleeland, K.H. Deitcher, N.H. Georgopapadakou, D.D. Keith, D.L. Pruess, J. Sepinwall, A.C. Specian Jr., R.L. Then, M. Weigele, K.F. West, and R. Yang. 1990. Cephalosporin 3′-quinolone esters with a dual mode of action. J. Med. Chem. **33:**77–86.

5. Albrecht, H.A., G. Beskid, K.-K. Chan, J.G. Christenson, R. Cleeland, K.H. Deitcher, D.D. Keith, D.L. Pruess, J. Sepinwall, A. Specian Jr., R.L. Then, M. Weigele, and K.F. West. 1988. Dual-action cephalosporins: an idea whose time has come. 28th Intersci. Conf. Antimicrob. Agents Chemother., abstr. 441.

6. Albrecht, H.A., G. Beskid, N.H. Georgopapadakou, D.D. Keith, F.M. Konzelman, D.L. Pruess, P. Rossman, and C.C. Wei. 1990. Dual-action cephalosporins: cephalosporin 3′-quinolone carbamates. 30th Intersci. Conf. Antimicrob. Agents Chemother., abstr. 402.

7. Allen, N.E., D.B. Boyd, J.B. Campbell, J.B. Deeter, T.K. Elzey, B.J. Foster, L.D. Hatfield, J.N. Hobbs, Jr., W.J. Hornback, D.C. Hunden, N.D. Jones, M.D. Kinnick, J.M. Morin, Jr., J.E. Munroe, J.K. Swartzendruber, and D.G. Vogt. 1989. Molecular modeling of γ-lactam analogues of β-lactam antibacterial agents: synthesis and biological evaluation of selected penem and carbapenem analogues. Tetrahedron **45:**1905–1928.

8. Al-Obeid, S., E. Collatz, and L. Gutmann. 1990. Mechanism of resistance to vancomycin in *Enterococcus faecium* D366 and *Enterococcus faecalis* A256. Antimicrob. Agents Chemother. **34:**252–256.

9. Ambler, R.P. 1980. The structure of β-lactamases. Phil. Trans. R. Soc. (Biol) **289:**321–331.

10. Ambler, R.P., M. Daniel, J. Fleming, J.-M. Hermoso, C. Pang, and S.G. Waley. 1985. The amino acid sequence of the zinc-requiring β-lactamase II from the bacterium *Bacillus cereus* 569. FEBS Lett. **189:**207–211.

11. Aoki, H., H. Sakai, M. Kohsaka, T. Konomi, J. Hosoda, Y. Kubochi, E. Iguchi and H. Imanaka. 1976. Nocardicin A, a new monocyclic β-lactam antibiotic. I. Discovery, isolation and characterization. J. Antibiot. **29**:492–500.

12. Arakawa, Y., M. Ohta, N. Kido, Y. Fujii, T. Komatsu, and N. Kato. 1986. Close evolutionary relationship between the chromosomally encoded β-lactamase gene of *Klebsiella pneumoniae* and the TEM β-lactamase gene mediated by R plasmids. FEBS Lett. **207**:69–74.

13. Arakawa, Y., M. Ohta, N. Kido, M. Mori, H. Ito, T. Komatsu, Y. Fujii, and N. Kato. 1989. Chromosomal β-lactamase of *Klebsiella oxytoca,* a new class A enzyme that hydrolyzes broad-spectrum β-lactam antibiotics. Antimicrob. Agents Chemother. **33**:63–70.

14. Arisawa, M., and R.L. Then. 1982. 6-Acetylmethylenepenicillanic acid (Ro 15–903), a potent β-lactamase inhibitor. I. Inhibition of chromosomally and R-factor-mediated β-lactamases. J. Antibiot. **35**:1578–1583.

15. Arisawa, M., and R.L. Then. 1983. Inactivation of TEM-1 β-lactamase by 6-acetylmethylenepenicillanic acid. Biochem. J. **209**:609–615.

16. Aronoff, S.C., M.R. Jacobs, S. Johenning, and S. Yamabe. 1984. Comparative activities of the β-lactamase inhibitors YTR 830, sodium clavulanate, and sulbactam combined with amoxicillin or ampicillin. Antimicrob. Agents Chemother. **26**:580–582.

16a. Ator, M.A., and P.R. Ortiz de Montellano. 1990. Mechanism-based (suicide) enzyme inactivation, p. 213–282. *In* D.S. Sigman and P.D. Boyer (ed.), The enzymes, vol. 19. Mechanism of catalysis. 3rd edition. Academic Press, San Diego.

17. Barbour, A.G. 1981. Properties of the penicillin-binding proteins in *Neisseria gonorrhoeae*. Antimicrob. Agents Chemother. **19**:316–322.

17a. Bauernfeind, A. 1986. Classification of β-lactamases. Rev. Infect. Dis. **8** (Suppl. 5):470–481.

18. Beise, F., H. Labischinski, and P. Giesbrecht. 1988. Role of the penicillin-binding proteins of *Staphylococcus aureus* in the induction of bacteriolysis by β-lactam antibiotics, p. 360–366. *In* P. Actor, L. Daneo-Moore, M.L. Higgins, M.R.J. Salton, and G.D. Shockman (ed.), Antibiotic inhibition of bacterial cell surface assembly and function. American Society for Microbiology, Washington, D.C.

19. Bentley, P.H., and R. Southgate. 1989. Recent advances in the chemistry of β-lactam antibiotics, p. 1–380. Proceedings of the 4th International Symposium. Royal Society of Chemistry, Special Publication No. 70. London.

20. Bentley, P.H., and A.V. Stachulski. 1983. Synthesis and biological activity of some fused β-lactam peptidoglycan analogues. J. Chem. Soc. Perkin Trans. **I**:1187–1192.

21. Benz, R. Porin from bacterial and mitochondrial outer membranes. CRC Crit. Rev. Biochem. **19**:145–190.

22. Berger-Bächi, B., and C. Ryffel. 1990. Control PBP 2′ synthesis in *Staphylococci*. *In* H. Kleinkauf and H. von Döhren (ed.), 50 Years of Penicillin Application, in press, Berlin.

22a. Berger-Bachi, B., L. Barberis-Maino, A. Straessle, and F.H. Kayser. 1989. FemA, a host-mediated factor essential for methicillin resistance in *Staphylococcus aureus:* molecular cloning and characterization. Mol Gen. Genet. **219:**263–269.

23. Bicknell, R., E.L. Emanuel, J. Gagnon, and S.G. Waley. 1985. The production and molecular properties of the zing β-lactamase of *Pseudomonas maltophilia* IID 1275. Biochem. J. **229:**791–797.

24. Blaszczak, L.C., R.F. Brown, G.K. Cook, W.J. Hornback, R.C. Hoying, J.M. Indelicato, C.L. Jordan, A.S. Katner, M.D. Kinnick, J.H. McDonald, III, J.M. Morin, Jr., J.E. Munroe, and C.E. Pasini. 1990. Comparative reactivity of 1-carba-1-dethiacephalosporins with cephalosporins. J. Med. Chem. **33:**1656–1662.

25. Böhme, E.H.W., H.E. Applegate, B. Toeplitz, J.E. Dolfini, and J.Z. Gougoutas. 1971. 6-Methyl penicillins and 7-methyl cephalosporins. J. Am. Chem. Soc. **93:**4324–4326.

26. Boissinot, M., and R.C. Levesque. 1990. Nucleotide sequence of the PSE-4 carbenicillinase gene and correlations with the *Staphylococcus aureus* PC1 β-lactamase crystal structure. J. Biol. Chem. **265:**1225–1230.

27. Botta, G.A., and J.T. Park. 1981. Evidence for involvement of penicillin-binding protein 3 in murein synthesis during septation but not during cell elongation. J. Bacteriol. **145:**333–340.

28. Bowler, L.D., and B.G. Spratt. 1989. Membrane topology of penicillin-binding protein 3 of *Escherichia coli*. Molecular Microbiol. **3:**1277–1286.

29. Deleted.

30. Boyd, D.B. 1979. Conformational analogy between β-lactam antibiotics and tetrahedral transition states of a dipeptide. J. Med. Chem. **22:**533–537.

31. Boyd, D.B. 1982. Theoretical and physicochemical studies on β-lactam antibiotics, p. 437–545. *In* R.B. Morin and M. Gorman (ed.), Chemistry and biology of β-lactam antibiotics: penicillins and cephalosporins, vol. 1. Academic Press, New York.

32. Boyd, D.B. 1983. Quantum mechanics in drug design: methods and applications. Drug Inform. J. **17:**121–131.

33. Boyd, D.B. 1984. Electronic structures of cephalosporins and penicillins. 15. Inductive effect of the 3-position side chain in cephalosporins. J. Med. Chem. **27:**63–66.

34. Boyd, D.B. 1987. Computer-assisted molecular design studies of β-lactam antibiotics, p. 339–356. *In* H. Umezawa (ed.), Frontiers of antibiotic research. Proceedings of the 4th Takeda Science Foundation Symposium on Bioscience, Academic Press, Orlando, FL.

34a. Boyd, D.B., J.D. Snoddy, and H.-S Lin. 1991. Molecular simulations of DD-peptidase, a model β-lactam-binding protein: synergy between X-ray crystallography and computational chemistry. J. Computational Chemistry **12:**635–644.

35. Boyd, D.B., D.K. Herron, W.H.W. Lunn, and W.A. Spitzer. 1980. Parabolic relationships between antibacterial activity of cephalosporins and β-lactam reactivity predicted from molecular orbital calculations. J. Am. Chem. Soc. **102:**1812–1814.

36. Boyd, D.B., and J.L. Ott. 1986a. Lack of relevance of kinetic parameters for exocellular DD-peptidases to cephalosporin MICs. Antimicrob. Agents Chemother. **29:**774–780.

37. Boyd, D.B. and J.L. Ott. 1986b. Examination of model enzyme and penetration systems in relation to antibacterial activity. J. Antibiot. **39:**281–285.

38. Broom, N.J.P., K. Coleman, P.A. Hunter, and N.F. Osborne. 1990. 6-(Substituted methylene)penems, potent broad spectrum inhibitors of bacterial β-lactamase. II. Racemic furyl and thienyl derivatives. J. Antibiot. **43:**76–82.

39. Broome-Smith, J.K. 1985. Construction of a mutant of *Escherichia coli* that has deletions of both the penicillin binding protein 5 and 6 genes. J. Gen. Microbiol. **131:**2115–2118.

40. Brown, M.R., W.D.G. Allison, and P. Gilbert. 1988. Resistance of bacterial biofilms to antibiotics: a growth-rate related effect? J. Antimicrob. Chemother. **22:**777–780.

41. Brown, A.G., D. Butterworth, M. Cole, G. Hanscomb, J.D. Hood, C. Reading, and G.N. Rolinson. 1976. Naturally occurring β-lactamase inhibitors with antibacterial activity. J. Antibiot. **29:**668–669.

42. Brown, A.G., M.J. Pearson, and R. Southgate. 1990. Other β-lactam agents, p. 655–702. *In* C. Hansch, P.G. Sammes, and J.B. Taylor (ed.), Comprehensive medicinal chemistry; the rational design, mechanistic study and therapeutic application of chemical compounds, vol. **2.** Enzymes and other molecular targets. Pergamon Press, Oxford.

43. Brown, A.G., and S.M. Roberts. 1984. Recent advances in the chemistry of β-lactam antibiotics, p. 1–391. Royal Society of Chemistry, Special Publication No. 52. London.

44. Buchanan, C.E. 1981. Topographical distribution of penicillin-binding proteins in *Escherichia coli* membrane. J. Bacteriol. **145:**1293–1298.

44a. Buckwell, S.C., M.I. Page, S.G. Waley, and J.L. Longridge. 1988. Hydrolysis of 7-substituted cephalosporins catalyzed by β-lactamases I and II from *Bacillus cereus* and by hydroxide ion. J. Chem. Soc. Perkin Trans. **II:**1815–1821.

44b. Bugg, T.D.H., S. Dutka-Malen, M. Arthur, P. Courvalin, and C.T. Walsh. 1991. Identification of vancomycin resistance protein VanA as a D-alanine:D-alanine ligase of altered substrate specificity. Biochemistry 30:2017–2021.

45. Büscher, K.-H., W. Cullmann, W. Dick, and W. Opferkuch. 1987. Imipenem resistance in *Pseudomonas aeruginosa* resulting from diminished expression of an outer membrane protein. Antimicrob. Agents Chemother. **31:**703–708.

45a. Bush, K. 1989. Classification of β-lactamases: Groups 1, 2a, 2b, and 2b′. Antimicrob. Agents Chemother. **33:**264–270.

46. Bush, K., and R.B. Sykes. 1984. Interaction of β-lactam antibiotics with β-lactamases as a cause for resistance, p. 1–31. *In* L.E. Bryan (ed.), Antimicrobial drug resistance. Academic Press, New York.

46a. Bush, K., and R.B. Sykes. 1986. Methodology for the study of β-lactamases. Antimicrob. Agents Chemother. **30:**6–10.

47. Bycroft, B.W., R.E. Shute, and M.J. Begley. 1988. Novel β-lactamase inhibitory and antibacterial 6-spiro-epoxypenicillins. J. Chem. Soc., Chem. Commun. **5**:274–276.

48. Cartwright, S.J., and S.G. Waley. 1983. Beta-lactamase inhibitors. Med. Res. Rev. **33**:341–382.

49. Chain, E., H.W. Florey, A.D. Gardner, N.G. Heatley, M.A. Jennings, J. Orr-Ewing, and A.G. Sanders. 1940. Penicillin as a chemotherapeutic agent. Lancet **2**:226–228.

50. Chambers, H.F. 1988. Methicillin-resistant *Staphylococci*. Clin. Microbiol. Rev. **1**:173–186.

51. Charlier, P., O. Dideberg, J.-M Frère, P.C. Moews, and J.R. Knox. 1983. Crystallographic data for the β-lactamase from *Enterobacter cloacae* P99. J. Mol. Biol. **171**:237–238.

52. Christensen, B.G. 1981. Structure-activity relationships in β-lactam antibiotics, p. 101–122. *In* M.R.J. Salton and G.D. Shockman (ed.), β-Lactam antibiotics: mode of action, new developments, and future prospects. Academic Press, New York.

53. Christenson, J., N. Georgopapadakou, D. Keith, K.-C. Luk, V. Madison, R. Mook, D. Pruess, J. Roberts, P. Rossman, C.-C. Wei, M. Weigele, and K. West. 1988a. α-Cyclopropylpenams: the synthesis and activity of a new class of penam antibacterials, p. 33–48. *In* P.H. Bentley and R. Southgate (ed.), Recent advances in the chemistry of β-lactam antibiotics. Proceedings of the 4th international symposium. Royal Society of Chemistry, Special Publication No 70. London.

54. Christenson, J.G., D.L. Pruess, M.K. Talbot, and D.D. Keith. 1988b. Antibacterial properties of (2,3)-α- and (2,3)-β-methylene analogs of penicillin G. Antimicrob. Agents Chemother. **32**:1005–1011.

55. Chung, S.K., and D.F. Chodosh. 1989. Computer graphics/molecular mechanics studies of β-lactam antibiotics. Geometry comparison with x-ray crystal structures. Bull. Korean Chem. Soc., **10**:185–190.

56. Clayden, N.J., C.M. Dobson, L.-Y. Lian, and J.M. Twyman. 1986. A solid-state ^{13}C nuclear magnetic resonance study of the conformational states of penicillins. J. Chem. Soc. Perkin Trans. **II**:1933–1940.

57. Cohen, N.C. 1983. β-Lactam antibiotics: geometrical requirements for antibacterial activities. J. Med. Chem. **26**:259–264.

57a. Cohen, N.C. 1985. Drug design in three dimensions. Advances in Drug Research. **14**:41–145.

58. Cohen, N.C., I. Ernest, H. Fritz, H. Fuhrer, G. Rihs, R. Scartazzini, and P. Wirz. 1987. 183. Are the known Δ^2-cephems inactive as antibiotics because of an unfavourable steric orientation of their 4α-carboxylic group? Synthesis and biology of two Δ^2-cephem-4β-carboxylic acids. Helv. Chim. Acta **70**:1967–1979.

59. Cohen, S.A., and R.F. Pratt. 1980. Inactivation of *Bacillus cereus* β-lactamase I by 6-β-bromopenicillanic acid: mechanism. Biochemistry **19**:3996–4003.

60. Cohenford, M.A., J. Abraham, and A.A. Medeiros. 1988. A colorimetric procedure for measuring β-lactamase activity. Anal. Biochem. **168**:252–258.

61. Coleman, K., D.R.J. Griffin, J.W.J. Page, and P.A. Upshon. 1989. In-vitro evaluation of BRL 42715, a novel β-lactamase inhibitor. Antimicrob. Agents Chemother. **33:**1580–1587.

61a. Cook, G.K., J.H. McDonald, III, W. Alborn, Jr., D.B. Boyd, J.A. Eudaly, J.M. Indelicato, R. Johnson, J.S. Kasher, C.E. Pasini, D.A. Preston, and E.C.Y. Wu. 1989. 3-Quaternary ammonium 1-carba-1-dethiacephems. J. Med. Chem. **32:**2442–2450.

61b. Costerouss, G., S. Gouindambr, J.G. Teutsch. March 1990. U.S. patent 4,908,359.

62. Coulson, A. 1985. β-Lactamase: molecular studies. Biotechnol. Gen. Eng. Rev. **3:**219–253.

62a. Coulton, J.W., P. Mason, and D. Dorrance. 1983. The permeability barrier of *Haemophilus influenzae* type b against β-lactam antibiotics. J. Antimicrob. Chemother. **12:**435–449.

62b. Coyette, J., J.-M. Ghuysen, and R. Fontana. 1980. The penicillin-binding proteins in *Streptococcus faecalis* ATCC 9790. Eur. J. Biochem. **110:**445–456.

63. Cozens, R.M., E. Tuomanen, W. Tosch, O. Zak, J. Suter, and A. Tomasz. 1986. Evaluation of the bactericidal activity of β-lactam antibiotics on slowly growing bacteria cultured in the chemostat. Antimicrob. Agents Chemother. **29:**797–802.

64. Cuchural, G.J., S. Hurlbut, M.H. Malamy, and F.P. Tally. 1988. Permeability to β-lactams in *Bacteroides fragilis*. J. Antimicrob. Chemother. **22:**785–790.

65. Cullmann, W. 1990. Interaction of β-lactamase inhibitors with various β-lactamases. Chemotherapy **36:**200–208.

66. Curtis, N.A.C., R.L. Eisenstadt, S.J. East, R.J. Cornford, L.A. Walker, and A.J. White. 1988. Iron-regulated outer membrane proteins of *Escherichia coli* K-12 and mechanism of action of catechol-substituted cephalosporins. Antimicrob. Agents Chemother. **32:**1879–1886.

67. Curtis, N.A.C., and M.V. Hayes. 1981. A mutant of *Staphylococcus aureus* H deficient in penicillin-binding protein 1 is viable. FEMS Microbiol. Lett. **10:**227–229.

68. Curtis, N.A.C., M.V. Hayes, A.W. Wyke, and J.B. Ward. 1980. A mutant of *Staphylococcus aureus* H lacking penicillin-binding protein 4 and transpeptidase activity in vitro. FEMS Microbiol. Lett. **9:**263–266.

69. Curtis, N.A.C., D. Orr, G.W. Ross, and M.G. Boulton. 1979a. Competition of β-lactam antibiotics for the penicillin-binding proteins of *Pseudomonas aeruginosa, Enterobacter cloacae, Klebsiella aerogenes, Proteus rettgeri,* and *Escherichia coli:* comparison with antibacterial activity and effects upon bacterial morphology. Antimicrob. Agents Chemother. **16:**325–328.

70. Curtis, N.A.C., D. Orr, G.W. Ross, and M.G. Boulton. 1979b. Affinities of penicillins and cephalosporins for the penicillin-binding proteins of *Escherichia coli* K-12 and their antibacterial activity. Antimicrob. Agents Chemother. **16:**533–539.

71. Curtis, N.A.C., G.W. Ross, and M.G. Boulton. 1979c. Effect of 7α-methoxy substitution of cephalosporins upon their affinity for the penicillin-binding proteins

of *E. coli* K12. Comparison with antibacterial activity and inhibition of membrane bound model transpeptidase activity. J. Antimicrob. Chemother. **5**:391–398.

72. Datta, N., and P. Kontomichalou. 1965. Penicillinase synthesis controlled by infectious R-factors in *Enterobacteriaceae*. Nature **208**:239–241.

73. Demuth, T.P., R.E. White, R.A. Tietjen, F.H. Ebetino, W.G. Kraft, J.A. Andersen, C.C. McOsker, M.A. Walling, B.W. Davis, and F.J. Rourke. 1990. C-10 quinolyl-cephem carbamates: synthesis and evaluation of a new class of antibacterial agents. 200th ACS National Meeting, Div. Med. Chem., abstr. 154.

73a. Demuth, T.P., R.E. White, R.A. Tietjen, R.J. Storrin, J.R. Skuster, J.A. Andersen, C.C. McOsker, R. Freedman, and F.J. Rourke. 1991. Synthesis and antibacterial activity of new C-10 quinolyl-cephem esters. J. Antibiot. **44**:200–209.

74. den Blaauwen, T., M. Aarsman, and N. Nanninga. 1990. Interaction of monoclonal antibodics with the enzymatic domains of penicillin-binding protein 1b of *Escherichia coli*. J. Bacteriol. **172**:63–70.

75. den Blaauwen, T., and N. Nanninga. 1990. Topology of penicillin-binding protein 1b of *Escherichia coli* and topography of four antigenic determinants studied by immunocolabeling electron microscopy. J. Bacteriol. **172**:71–79.

76. den Blaauwen, T., F.B. Wientjes, A.H.J. Kolk, B.G. Spratt, and N. Nanninga. 1989. Preparation and characterization of monoclonal antibodies against native membrane-bound penicillin-binding protein 1B of *Escherichia coli*. J. Bacteriol. **171**:1394–1401.

77. De Pedro, M.A., U. Schwarz, U. Nishimura, and Y. Hirota. 1980. On the biological role of penicillin-binding proteins 4 and 5. FEMS Microbiol. Lett. **9**:219–221.

78. Dexter, D.D., and J.M. van der Veen. 1978. Conformations of penicillin G: crystal structure of procaine penicillin G monohydrate and a refinement of the structure of potassium penicillin G.J. Chem. Soc., Perkin Trans. **I**:185–190.

79. Dideberg, O., P. Charlier, J.-P Wéry, P. Dehottay, J. Dusart, T. Erpicum, J.-M. Frère, and J.-M Ghuysen. 1987. The crystal structure of the β-lactamase of *Streptomyces albus G* at 0.3 nm resolution. Biochem. J. **245**:911–913.

79a. Dolinger, D.L., L. Daneo-Moore, and G.D. Shockman. 1989. The second peptidoglycan hydrolase of *Streptococcus faecium* ATCC 9790 covalently binds penicillin. J. Bacteriol. **171**:4355–4361.

80. Dougherty, T.J., A.E. Koller, and A. Tomasz. 1980. Penicillin-binding proteins of penicillin-susceptible and intrinsically resistant *Neisseria gonorrhoeae*. Antimicrob. Agents Chemother. **18**:730–737.

81. Dougherty, T.J., A.E. Koller, and A. Tomasz. 1981. Competition of β-lactam antibiotics for the penicillin-binding proteins of *Neisseria gonorrhoeae*. Antimicrob. Agents Chemother. **20**:109–114.

82. Dürckheimer, W., F. Adam, G. Fischer, and R. Kirrstetter. 1987. Synthesis and biological properties of newer cephem antibiotics, p. 161–192. *In* H. Umezawa (ed.), Frontiers of antibiotic research. Proceedings of the 4th Takeda Science Foundation Symposium on Bioscience. Academic Press, Orlando, FL.

83. Dürckheimer, W., J. Blumbach, R. Lattrell, and K.H. Scheunemann. 1985. Recent developments in the field of β-lactam antibiotics. Angew. Chem. Int. Ed. Eng. **24:**180–202.

83a. Dutka-Malen, S., C. Molinas, M. Arthur, and P. Courvalin. 1990. The VanA glycopeptide resistance protein is related to D-alanyl-D-alanine ligase cell wall biosynthesis enzymes. Mol. Gen. Genet. **224:**364–372.

84. Edelman, A., L. Bowler, J.K. Broome-Smith, and B.G. Spratt. 1987. Use of a β-lactamase fusion vector to investigate the organisation of penicillin-binding protein 1B in the cytoplasmic membrane of *Escherichia coli*. Mol. Microbiol. **1:**101–106.

85. Eley, A., and D. Greenwood. 1986. Characterization of β-lactamases in clinical isolates of *Bacteroides*. J. Antimicrob. Chemother. **18:**325–333.

86. Ellerby, L.M., W.A. Escobar, A.L. Fink, C. Mitchinson, and J.A. Wells. 1990. The role of lysine–234 in β-lactamase catalysis probed by site-directed mutagenesis. Biochemistry **29:**5797–5806.

87. English, A.R., D. Girard, and S.L. Haskell. 1984. Pharmacokinetics of sultamicillin in mice, rats, and dogs. Antimicrob. Agents Chemother. **25:**599–602.

88. English, A.R., J.A. Retsema, A.E. Girard, J.E. Lynch, and W.E. Barth. 1978. CP-45,899, a beta-lactamase inhibitor that extends the antibacterial spectrum of beta-lactams: initial bacteriological characterization. Antimicrob. Agents Chemother. **14:**414–419.

89. Faraci, W.S., and R.F. Pratt. 1986. Interactions of cephalosporins with the *Streptomyces* R61 DD-transpeptidase/carboxypeptidase. Influence of the 3′ substituents. Biochem. J. **238:**309–312.

90. Faraci, W.S., and R.F. Pratt. 1987. Nucleophilic re-activation of the PC1 β-lactamase of *Staphylococcus aureus* and of the DD-peptidase of *Streptomyces* R61 after their inactivation by cephalosporins and cephamycins. Biochem. J. **246:**651–658.

91. Ferreira, L.C.S., U. Schwarz, W. Keck, P. Charlier, O. Dideberg, and J.-M. Ghuysen. 1988. Properties and crystallization of a genetically engineered, water-soluble derivative of penicillin-binding protein 5 of *Escherichia coli* K12. Eur. J. Biochem. **171:**11–16.

92. Fink, A. 1985. The molecular basis of β-lactamase catalysis and inhibition. Pharm. Res. **2:**55–61.

92a. Fontana, R., R. Cerini, P. Longoni, A. Grossato, and P. Canepari. 1983. Identification of a streptococcal penicillin-binding protein that reacts very slowly with penicillin. J. Bacteriol. **155:**1343–1350.

93. Frère, J.-M., and B. Joris. 1985. Penicillin-sensitive enzymes in peptidoglycan biosynthesis. CRC Critical Rev. Microbiol. **11:**299–396.

94. Frère, J.-M., B. Joris, O. Dideberg, P. Charlier, and J.-M. Ghuysen. 1988. Penicillin-recognizing enzymes. Biochemical Society Transactions. **16:**934–938.

95. Frère, J.-M., B. Joris, L. Varetto, and M. Crine. 1988. Structure-activity relationships in the β-lactam family: an impossible dream. Biochem. Pharmacol. **37:**125–132.

96. Frère, J.-M., J.A. Kelly, D. Klein, J.-M. Ghuysen, P. Claes, and H. Vanderhaeghe. 1982. Δ^2- and Δ^3-cephalosporins, penicillinate and 6-unsubstituted penems; intrinsic reactivity and interaction with β-lactamases and D-alanyl-D-alanine-cleaving serine peptidases. Biochem. J. **203:**223–234.

97. Garcia-Bustos, J.F., B.T. Chait, and A. Tomasz. 1988. Altered peptidoglycan structure in a pneumococcal transformant resistant to penicillin. J. Bacteriol. **170:**2143–2147.

98. Garcia-Bustos, J.F., and A. Tomasz. 1990. A biological price of antibiotic resistance: major changes in the peptidoglycan structure of penicillin-resistant pneumococci. Proc. Natl. Acad. Sci. USA **87:**5415–5419.

99. Gensmantel, N.P., D. McLellan, J.J. Morris, M.I. Page, P. Proctor, and G.S. Randahawa. 1980. Mechanisms in the reactions of some β-lactam antibiotics and their derivatives, p. 227–239. *In* G.I. Gregory (ed.), Recent advances in the chemistry of β-lactam antibiotics. Royal Society of Chemistry, Special publication no. 38. London.

100. Georgopapadakou, N.H. 1988. Penicillin-binding proteins. p. 409–431 *In* P.K. Peterson and J. Verhoef (eds.), Antimicrobial agents annual 3. Elsevier Science Publishers BV. Amsterdam.

101. Georgopapadakou, N.H., A. Bertasso, K.K. Chan, J.S. Chapman, R. Cleeland, L.M. Cummings, B.A. Dix, and D.D. Keith. 1989. Mode of action of the dual-action cephalosporin Ro 23-9424. Antimicrob. Agents Chemother. **33:**1067–1071.

102. Georgopapadakou, N.H., L.M. Cummings, E.R. LaSala, J. Unowsky, and D.L. Pruess. 1988. Overproduction of penicillin-binding protein 4 in *Staphylococcus aureus* is associated with methicillin resistance, p. 597–602. *In* P.L. Actor, L. Daneo-Moore, M.L. Higgins, M.R.J. Salton, and G.D. Shockman (ed.), Antibiotic inhibition of bacterial cell surface assembly and function. American Society for Microbiology, Washington, DC.

103. Georgopapadakou, N.H., B.A. Dix, and Y.R. Mauriz. 1986. Possible physiological functions of penicillin-binding proteins in *Staphylococcus aureus*. Antimicrob. Agents Chemother. **29:**333–336.

104. Georgopapadakou, N.H., and F.Y. Liu. 1980a. Penicillin-binding proteins in bacteria. Antimicrob. Agents Chemother. **18:**148–157.

105. Georgopapadakou, N.H., and F.Y. Liu. 1980b. Binding of β-lactam antibiotics to penicillin-binding proteins of *Staphylococcus aureus* and *Streptococcus faecalis:* relation to antibacterial activity. Antimicrob. Agents Chemother. **18:**834–836.

106. Georgopapadakou, N.H., D.A. Russo, A. Liebman, W. Burger, P. Rossman, and D. Keith. 1987. Interaction of (2,3)-methylenepenams with penicillin-binding proteins. Antimicrob. Agents Chemother. **31:**1069–1074.

107. Georgopapadakou, N.H., S.A. Smith, and R.B. Sykes. 1983. Penicillin-binding proteins in *Bacteroides fragilis*. J. Antibiot. **36:**907–910.

108. Georgopapadakou, N.H., and R.B. Sykes. 1983. Bacterial enzymes interacting with β-lactam antibiotics, p. 1–77. *In* A.L. Demain and N.A. Solomon (ed.), Antibiotics

containing the beta-lactam structure II. Handbook of experimental pharmacology, vol. 67/II, Springer-Verlag, Berlin/Heidelberg.

109. Ghuysen, J.-M. 1977. The bacterial DD-carboxypeptidase-transpeptidase enzyme system: a new insight into the mode of action of penicillin, p. 1–162. University of Tokyo Press, Tokyo.

110. Ghuysen, J.-M. 1984. Exploration of active sites of DD-peptidases, p. 115–123. *In* W. Paton, J. Mitchell, and P. Turner (ed.), Proceedings IUPHAR 9th International Congress of Pharmacology, Vol. 1. London.

111. Ghuysen, J.-M. 1988a. Bacterial active-site serine penicillin-interactive proteins and domains: mechanism, structure, and evolution. Rev. Infect. Dis. **10:**726–732.

112. Ghuysen, J.-M. 1988b. Evolution of DD-peptidases and β-lactamases, p. 268–284. *In* P.L. Actor, L. Daneo-Moore, M.L. Higgins, M.R.J. Salton, and G.D. Shockman (ed.), Antibiotic inhibition of bacterial cell surface assembly and function. American Society for Microbiology, Washington, DC.

112a. Ghuysen, J.-M. 1991. Serine β-lactamases and penicillin-binding proteins. Annu. Rev. Microbiol. **45:**37–67.

113. Ghuysen, J.-M., J.-M. Frère, B. Joris, J. Dusart, C. Duez, M. Leyh-Bouille, M. Nguyen-Distèche, J. Coyette, O. Dideberg, P. Charlier, G. Dive, and J. Lamotte-Brasseur. 1989. Inhibition of enzymes involved in bacterial cell wall synthesis, p. 523–572. *In* M. Sandler and H.J. Smith (ed.), Design of enzyme inhibitors as drugs. Oxford University Press, New York.

113a. Ghuysen, J.-M., J.-M. Frère, M. Leyh-Bouille, J. Coyette, J. Dusart, and M. Nguyen-Distèche. 1979. Use of model enzymes in the determination of the mode of action of penicillins and Δ^3-cephalosporins. Ann. Rev. Biochem. **48:**73–101.

114. Ghuysen, J.-M., J.-M Frère, M. Leyh-Bouille, H.R. Perkins, and M. Nieto. 1980. The active centres in penicillin-sensitive enzymes. Phil. Trans. R. Soc. Lond. B289:285–301.

115. Ghuysen, J.-M., J.-M. Frère, M. Leyh-Bouille, M. Nguyen-Distèche, J. Coyette, J. Dusart, B. Joris, C. Duez, O. Dideberg, P. Charlier, G. Dive, and J. Lamotte-Brasseur. 1984. Bacterial wall peptidoglycan, DD-peptidases, and beta-lactam antibiotics. Scand. J. Infect. Dis., Suppl. **42:**17–37 (1984).

115a. Gibson, R.M., H. Christensen, and S.G. Waley. 1990. Site-directed mutagenesis of β-lactamase I. Biochem. J. **272:**613–619.

116. Giesbrecht, P. 1991. On the mechanism of β-lactam action: the majority of *Staphylococci* are killed via murosome-induced wall perforations both under "lytic" and under "non-lytic" doses of penicillin. *In* H. Kleinkauf and H. von Döhren (ed.), 50 Years of Penicillin Application, in press.

117. Giesbrecht, P., H. Labischinski, and J. Wecke. 1985. A special morphogenetic wall defect and the subsequent activity of "murosomes" as the very reason for penicillin-induced bacteriolysis in *Staphylococci*. Arch Microbiol. **141:**315–324.

118. Glauner, B., and U. Schwarz. 1983. The analysis of murein composition with high-pressure-liquid chromatography, p. 29–34. *In* R. Hakenbeck, J.V. Höltje, and H.

Labischinski (ed.), The target of penicillin: the murein sacculus of bacterial cell walls, architecture and growth. Walter de Gruyter and Co., Berlin.

119. Gordon, E.M., and R.B. Sykes. 1982. Cephamycin antibiotics, p. 199–370. *In* R.B. Morin and M. Gorman (ed.), Chemistry and biology of β-lactam antibiotics, vol. 1. Penicillins and cephalosporins. Academic Press, New York.

119a. Gotoh, N., H. Wakebe, E. Yoshihara, T. Nakae, and T. Nishino. 1989. Role of protein F in maintaining structural integrity of the *Pseudomonas aeruginosa* outer membrane. J. Bacteriol. **171**:983–990.

120. Graham, M.N., and T.J. Mantle. 1989. Purification of a class C A-type β-lactamase from a derepressed strain of *Enterobacter cloacae*. Biochem. J. **260**:705–710.

121. Gutmann, L., S. Al-Obeid, D. Billot-Klein, J.F. Acar, E. Collatz, and J. van Heijenoort. 1990. Vancomycin (Van) resistance and synergy between penicillin (pen) involve an inducible carboxypeptidase (CPDase) in Van-resistant *Enterococci* (VRE). 30th Intersci. Conf. Antimicrob. Agents Chemother., abstr. 929.

122. Gutmann, L., S. Vincent, D. Billot-Klein, J.F. Acar, E. Mrèna, and R. Williamson. 1986. Involvement of penicillin-binding protein 2 with other penicillin-binding proteins in lysis of *Escherichia coli* by some β-lactam antibiotics alone and in synergistic lytic effect of amdinocillin (mecillinam). Antimicrob. Agents Chemother. **30**:906–912.

123. Hakenbeck, R., M. Tarpay, and A. Tomasz. 1980. Multiple changes of penicillin-binding proteins in penicillin-resistant clinical isolates of *Streptococcus pneumoniae*. Antimicrob. Agents Chemother. **17**:364–371.

124. Hakenbeck, R., J.-V. Holtje, and H. Labischinski. 1983. The target of penicillin: the murein sacculus of bacterial cell walls; architecture and growth, p. 1–663. Walter de Gruyter and Co., Berlin.

125. Hall, M.N., and T.J. Silhavy. 1979. The *ompB* locus and the regulation of the major outer membrane porin proteins of *Escherichia coli* K-12. J. Mol. Biol. **146**:23–43.

126. Hamilton-Miller, J.M.T., and J.T. Smith (ed). 1979. Beta-lactamases, p. 1–500. Academic Press, New York.

127. Hancock, R.E.W., G.M. Decad, and H. Nikaido. 1979. Identification of the protein producing transmembrane diffusion pores in the outer membrane of *Pseudomonas aeruginosa* PA01. Biochim. Biophys. Acta **554**:323–331.

128. Handwerger, A., and A. Tomasz. 1986. Alterations in kinetic properties of penicillin-binding proteins of penicillin-resistant *Streptococcus pneumoniae*. Antimicrob. Agents Chemother. **30**:57–63.

129. Hanson, J.E., A.P. Kaplan, and P.A. Bartlett. 1989. Phosphonate analogues of carboxypeptidase A substrates are potent transition-state analogue inhibitors. Biochemistry **28**:6294–6305.

130. Harada, S., S. Tsubotani, T. Hida, H. Ono, and H. Okazaki. 1986. Structure of lactivicin, an antibiotic having a new nucleus and similar biological activities to β-lactam antibiotics. Tetrahedron Lett. **27**:6229–6232.

131. Hart, C.A., and A. Percival. 1982. Resistance to cephalosporins among gentamicin-resistant *Klebsiellae*. J. Antimicrob. Chemother. **9**:275–286.

132. Hartman, B.J., and A. Tomasz. 1984. Low-affinity penicillin-binding protein associated with β-lactam resistance in *Staphylococcus aureus*. J. Bacteriol. **158:**513–516.

133. Hedges, R.W., and M. Matthew. 1979. Acquisition by *Escherichia coli* of plasmid-borne beta-lactamases normally confined to *Pseudomonas* spp. Plasmid **2:**269–278.

134. Herzberg, O., and J. Moult. 1987. Bacterial resistance to β-lactam antibiotics: crystal structure of β-lactamase from *Staphylococcus aureus* PC1 at 2.5 Å resolution. Science **236:**694–701.

134a. Herzberg, O. 1991. Refined crystal structure of β-lactamase from *Staphylococcus aureus PC1* at 2.0 A resolution. J. Mol. Biol. **217:**701–719.

135. Hoover, J.R.E. 1983. β-Lactam antibiotics; structure-activity relationships, p. 119–245. *In* A.L. Demain and N.A. Solomon (ed.), Antibiotics containing the beta-lactam structure. II. Handbook of experimental pharmacology, vol. 67/II, Springer-Verlag, Berlin/Heidelberg.

136. Hoover, J.R.E., and G.L. Dunn. 1979. The β-lactam antibiotics, p. 83–172. *In* M.E. Wolff (ed.), Burger's medicinal chemistry. Part II. 4th ed. Wiley and Sons, Inc., New York.

137. Huovinen, P., S. Huovinen, and G.A. Jacoby. 1988. Sequence of PSE-2 β-lactamase. Antimicrob. Agents Chemother. **32:**134–136.

137a. Hurlbut, S., G.J. Cuchural, and F.P. Tally. 1990. Imipenem resistance in *Bacteroides distasonis* mediated by a novel β-lactamase. Antimicrob. Agents Chemother. **34:**117–120.

138. Iida, K., S. Hirata, S. Nakamuta, and M. Koike. 1978. Inhibition of cell division in *Escherichia coli* by a new synthetic penicillin, piperacillin. Antimicrob. Agents Chemother. **14:**257–266.

139. Imada, A., K. Kitano, K. Kintaka, M. Muroi, and M. Asai. 1981. Sulfazecin and isosulfazecin, novel β-lactam antibiotics of bacterial origin. Nature **289:**590–591.

140. Indelicato, J.M., T.T. Norvilas, R.R. Pfeiffer, W.J. Wheeler, and W.L. Wilham. 1974. Substituent effects upon the base hydrolysis of penicillins and cephalosporins. Competitive intramolecular nucleophilic amino attack in cephalosporins. J. Med. Chem. **17:**523–527.

141. Indelicato, J.M., and W.L. Wilham. 1974. Effect of 6-α substitution in penicillins and 7-α substitution in cephalosporins upon β-lactam reactivity. J. Med. Chem. **17:**528–529.

142. Ishino, F., W. Park, S. Tomioka, S. Tamaki, I. Takase, K. Kunugita, H. Matsuzawa, S. Asoh, T. Ohta, B.G. Spratt, and M. Matsuhashi. 1986. Peptidoglycan synthetic activities in membranes of *Escherichia coli* caused by overproduction of penicillin-binding protein 2 and rodA protein. J. Biol. Chem. **261:**7024–7031.

143. Ishino, F., M. Wachi, K.-H. Ueda, Y. Ito, R.A. Nicholas, J.L. Strominger, T. Senda, K. Ishikawa, Y. Mitsui, and M. Matsuhashi. 1988. Crystallization and preliminary crystallographic studies of the high-molecular-weight penicillin-binding protein 1B-δ of *Escherichia coli,* p. 285–291. *In* P. Actor, L. Daneo Moore, M.L. Higgins, M.R.J. Salton, and G.D. Shockman (ed.), Antibiotic inhibition of bacterial

cell surface assembly and function. American Society for Microbiology, Washington, D.C.

144. Izaki, K., M. Matsuhashi, and J.L. Strominger. 1968. Biosynthesis of the peptidoglycan of bacterial cell walls. XIII. Peptidoglycan transpeptidase and D-alanine carboxypeptidase: penicillin-sensitive enzymatic reaction in strains of *Escherichia coli*. J. Biol. Chem. **243:**3180–3192.

145. Jacoby, G.A., and L. Sutton. 1985. β-Lactamases and β-lactam resistance in *Escherichia coli*. Antimicrob. Agents Chemother. **28:**703–705.

146. Jansson, J.A.T. 1965. A direct spectrophotometric assay for penicillin β-lactamase (penicillinase). Biochim. Biophys. Acta **99:**171–172.

147. Jaurin, B., and T. Grundström. 1981. AmpC cephalosporinase of *Escherichia coli* K12 has a different evolutionary origin from that of beta-lactamases of the penicillinase type. Proc. Natl. Acad. Sci. USA **78:**4897–4901.

148. Johnson, J.R., R.B. Woodward, and R. Robinson. 1949. The constitution of the penicillins, p. 440–454. *In* H.T. Clarke, J.R. Johnson, and R. Robinson (ed.), The chemistry of penicillin. Princeton University Press, Princeton, NJ.

149. Joris, B., F. De Meester, M. Galleni, and J.-M. Frère, and J. van Beeumen. 1987. The K1 β-lactamase of *Klebsiella pneumoniae*. Biochem. J. **243:**561–567.

150. Joris, B., J.-M. Ghuysen, G. Dive, A. Renard, O. Dideberg, P. Charlier, J.-M. Frère, J.A. Kelly, J.C. Boyington, P.C. Moews, and J.R. Knox. 1988. The active-site-serine penicillin-recognizing enzymes as members of the *Streptomyces* R61 DD-peptidase family. Biochem. J. **250:**313–324.

151. Jungheim, L.N., S.K. Sigmund, and J.W. Fisher. 1987. Bicyclic pyrazolidinones, a new class of antibacterial agent based on the β-lactam model. Tetrahedron Lett. **28:** 285–288.

152. Juteau, J.-M., and R.C. Levesque. 1990. Sequence analysis and evolutionary perspectives of ROB-1 β-lactamase. Antimicrob. Agents Chemother. **34:**1354–1359.

153. Kahan, J.S., F.M. Kahan, R. Goegelman, S.A. Currie, M. Jackson, E.O. Stapley, T.W. Miller, A.K. Miller, D. Hendlin, S. Mochales, S. Hernadez, H.B. Woodruff, and J. Birnbaum. 1979. Thienamycin, a new β-lactam antibiotic. 1. Discovery, taxonomy, isolation and physical properties. J. Antibiot. **32:**1–12.

154. Kamiya, K., M. Takamoto, Y. Wada, and M. Asai. 1981. Structure of sulfazecin-methanol (1/1). Acta Crystallogr., Sect. B **37:**1626–1628.

155. Keith, D.D., J. Tengi, P. Rossman, L. Todaro, and M. Weigele. 1983. A comparison of the antibacterial and β-lactamase inhibiting properties of penam and (2,3)-β-methylenepenam derivatives: the discovery of a new β-lactamase inhibitor. Conformational requirements for penicillin antibacterial activity. Tetrahedron **39:**2445–2458.

156. Kelly, J.A., O. Dideberg, P. Charlier, J.P. Wery, M. Libert, P.C. Moews, J.R. Knox, C. Duez, C. Fraipont, B. Joris, J. Dusart, J.-M Frère, and J.-M. Ghuysen. 1986. On the origin of bacterial resistance to penicillin: Comparison of a β-lactamase and a penicillin target. Science **231:**1429–1431.

157. Kelly, J.A., J.R. Knox, P.C. Moews, J.-M. Frère, and J.-M. Ghuysen. 1988. Using X-ray diffraction results and computer graphics to design β-lactams. J. Japanese Assoc. Infectious Diseases **62:**182–191.

158. Kelly, J.A., J.R. Knox, P.C. Moews, G.J. Hite, J.B. Bartolone, H. Zhao, B. Joris, J.-M. Frère, and J.-M. Ghuysen. 1985. 2.8 Å structure of penicillin-sensitive D-alanyl carboxypeptidase-transpeptidase from *Streptomyces* R61 and complexes with β-lactams. J. Biol. Chem. **260:**6449–6458.

159. Kelly, J.A., J.R. Knox, P.C. Moews, J. Moring, and H.C. Zhao. 1988b. Molecular graphics: studying β-lactam inhibition in three dimensions, p. 261–267. *In* P. Actor, L. Daneo-Moore, M.L. Higgins, M.R.J. Salton, and G.D. Shockman (ed.), Antibiotic inhibition of bacterial cell surface assembly and function. American Society for Microbiology, Washington, D.C.

160. Kelly, J.A., J.R. Knox, H. Zhao, J.-M. Frère, and J.-M. Ghuysen. 1989. Crystallographic mapping of β-lactams bound to a D-alanyl-D-alanine peptidase target enzyme. J. Mol. Biol. **209:**281–295.

161. Kiester, E. Jr. 1990. Penicillin and a revolution in world health. Smithsonian **21:**172–187.

162. Knap, A.K., and R.F. Pratt. 1991. Inactivation of the RTEM-1 cysteine β-lactamase by iodoacetate. The nature of active-site functional groups and comparisons with the native enzyme. Biochem. J. **273:**85–91.

163. Knowles, J.R. 1985. Penicillin resistance: the chemistry of β-lactamase inhibition. Acct. Chem. Res. **18:**97–104.

164. Knox, J.R., and J.A. Kelly. 1989. Crystallographic comparison of penicillin-recognizing enzymes, p. 46–55 *In* S.M. Roberts (ed.), Molecular recognition: chemical and biochemical problems. Special publication No. 78. Proceedings of an international symposium. Royal Society of Chemistry, London.

165. Knox, J.R., and R.F. Pratt. 1990. Different modes of vancomycin and D-alanyl-D-alanine peptidase binding to cell wall peptide and a possible role for the vancomycin resistance protein. Antimicrob. Agents Chemother. **34:**1342–1347.

166. Kozarich, J.W., and J.L. Strominger. 1978. A membrane enzyme from *Staphylococcus aureus* which catalyzes transpeptidase, carboxypeptidase, and penicillinase activities. J. Biol. Chem. **253:**1272–1278.

167. Kraus, W., and J.-V. Höltje. 1987. Two distinct transpeptidation reactions during murein synthesis in *Escherichia coli*. J. Bacteriol. **169:**3099–3103.

168. Kraut, J. 1977. Serine proteases: structure and mechanism of catalysis. Ann. Rev. Biochem. **46:**331–358.

169. Kraut, J. 1988. How do enzymes work? Science **242:**533–540.

170. Labia, R., P. Baron, J.M. Masson, G. Hill, and M. Cole. 1984. Affinity of temocillin for *Escherichia coli* K-12 penicillin-binding proteins. Antimicrob. Agents Chemother. **26:**335–338.

171. Labischinski, H., H. Maidhof, M. Franz, D. Krüger, T. Sidow, and P. Giesbrecht. 1988. Biochemical and biophysical investigations into the cause of penicillin-induced lytic death of *Staphylococci:* checking predictions of the murosome model, p.

242–257. *In* P. Actor, L. Daneo-Moore, M.L. Higgins, M.R.J. Salton, and G.D. Shockman (ed.), Antibiotic inhibition of bacterial cell surface assembly and function. American Society for Microbiology, Washington, D.C.

172. Lamotte-Brasseur, J., G. Dive, and J.-M. Ghuysen. 1984. On the structural analogy between D-alanyl-D-alanine terminated peptides and β-lactam antibiotics. Eur. J. Med. Chem.-Chim. Ther. **19:**319–330.

172a. Lamotte-Brasseur, J., G. Dive, and J.-M. Ghuysen. 1991. Conformational analysis of β and γ-lactam antibiotics. Eur. J. Med. Chem. **26:**43–50.

173. Lee, B. 1971. Conformation of penicillin as a transition-state analog of the substrate of peptidoglycan transpeptidase. J. Mol. Biol. **61:**463–469.

174. Deleted.

175. Lindberg, F., L. Westman, and S. Normark. 1985. Regulatory components in *Citrobacter freundii ampC* β-lactamase induction. Proc. Natl. Acad. Sci. USA **82:**4620–4624.

176. Lowe, G., and S. Swain. 1984. Do β-lactam antibiotics require a β-lactam ring? p. 209–221. *In* A.G. Brown and S.M. Roberts (ed.), Recent advances in the chemistry of β-lactam antibiotics. Proceedings of the 3rd international symposium. Special publication No 52. Royal Society of Chemistry, London.

177. Lugtenberg, B., and L. van Alphen. 1983. Molecular architecture and functioning of the outer membrane of *Escherichia coli* and other gram-negative bacteria. Biochim. Biophys. Acta **737:**51–115.

178. Lund, F., and L. Tybring. 1972. 6-β-Amidinopenicillanic acids—a new group of antibiotics. Nature New Biology **236:**135–137.

178a. Mabilat, C., and P. Courvalin. 1990. Development of "oligotyping" for the characterization and molecular epidemiology of TEM-derived β-lactamases in members of the family *Enterobacteriaceae*. Antimicrob. Agents Chemother. **34:**2210–2216.

178b. Madiraju, M.V.V.S., D.P. Brunner, B.J. Wilkinson. 1987. Effects of temperature, NaCl, and methicillin on penicillin-binding proteins, growth, peptidoglycan synthesis, and autolysis in methicillin-resistant *Staphylococcus aureus*. Antimicrob. Agents Chemother. **31:**1727–1733.

179. Makover, S.D., R. Wright, and E. Telep. Penicillin-binding proteins in *Haemophilus influenzae*. Antimicrob. Agents Chemother. **19:**584–588.

180. Mansford, K.R.L. 1987. Properties of recent penam antibiotics, p. 149–160. *In* H. Umezawa (ed.), Frontier of antibiotic research symposium in bioscience. Academic Press. Orlando, FL.

181. Marchand-Brynaert, J., Z. Bounkhala-Khrouz, J.S. Carretero, J. Davies, D. Ferroud, B.J. van Keulen, B. Serckx-Poncin, and L. Ghosez. 1989. Synthesis of potential inhibitors of bacterial DD-peptidases, p. 157–170. *In* P.H. Bentley and R. Southgate (ed.), Recent advances in the chemistry of β-lactam antibiotics. Proceedings of the 4th international symposium. Special Publication No 70. Royal Society of Chemistry, London.

181a. Mastalerz, H., M. Menard, V. Vinet, J. Desiderio, J. Fung-Tomc, R. Kessler, and Y. Tsai. 1988. An examination of O-2-isocephems as orally absorbable antibiotics. J. Med. Chem. **31:**1190–1196.

182. Matsuhashi, M., M.D. Song, F. Ishino, M. Wachi, M. Doi, M. Inoue, K. Ubukata, N. Yamashita, and M. Konno. 1986. Molecular cloning of the gene of a penicillin-binding protein supposed to cause high resistance to β-lactam antibiotics in *Staphylococcus aureus*. J. Bacteriol. **167:**975–980.

183. Matthew, M. 1978. Properties of the β-lactamase specified by the *Pseudomonas* plasmid R151. FEMS Microbiol. Lett. **4:**241–244.

184. Matthew, M. 1979. Plasmid-mediated beta-lactamases of gram-negative bacteria: properties and distribution. J. Antimicrob. Chemother. **5:**349–358.

185. Medeiros, A.A., R. Levesque, and G.A. Jacoby. 1986. An animal source for the ROB-1 β-lactamase of *Hemophilus influenzae* type b. Antimicrob. Agents Chemother. **29:**212–215.

186. Medeiros, A.A., M. Cohenford, and G.A. Jacoby. 1985. Five novel plasmid-determined β-lactamases. Antimicrob. Agents Chemother. **27:**715–719.

187. Mendelman, P.M., D.O. Chaffin, T.L. Stull, C.E. Rubens, K.D. Mack, and A.L. Smith. 1984. Characterization of non-β-lactamase-mediated ampicillin resistance in *Haemophilus influenzae*. Antimicrob. Agents Chemother. **26:**235–244.

188. Mett, H., B. Schacher, and L. Wegmann. 1988. Ultrasonic disintegration of bacteria may lead to irreversible inactivation of β-lactamase. J. Antimicrob. Chemother. **22:**293–298.

189. Mirelman, D.E. 1979. Biosynthesis and assembly of cell wall peptidoglycan, p. 115–166. *In* M. Inouye (ed.), Bacterial outer membrane: biogenesis and functions. John Wiley and Sons, New York.

190. Mirelman, D. 1981. Assembly of wall peptidoglycan polymers, p. 67–86. *In* M.R.J. Salton and G.D. Shockman (ed.), β-Lactam antibiotics: mode of action, new developments, and future prospects, Academic Press, New York.

191. Mitsuyama, J., R. Hiruma, A. Yamaguchi, and T. Sawai. 1987. Identification of porins in outer membrane of *Proteus, Morganella,* and *Providencia* spp. and their role in outer membrane permeation of β-lactams. Antimicrob. Agents Chemother. **31:**379–384.

192. Mochizuki, H., H. Yamada, Y. Oikawa, K. Murakami, J. Ishiguro, H. Kosuzume, N. Aizawa, and E. Mochida. 1988. Bactericidal activity of M14659 enhanced in low-iron environments. Antimicrob. Agents Chemother. **32:**1648–1654.

193. Moews, P.C., J.R. Knox, O. Dideberg, P. Charlier, and J.-M. Frère. 1990. β-Lactamase of *Bacillus licheniformis* 749/C. Proteins **7:**156–171.

194. Monks, J., and S.G. Waley. 1988. Imipenem as substrate and inhibitor of β-lactamases. Biochem. J. **253:**323–328.

195. Morin, R.B., and M. Gorman. 1982. Chemistry and biology of β-lactam antibiotics, vol. 1, 2, and 3. Academic Press, New York.

196. Mossakowska, D., N.A. Ali, and J.W. Dale. 1989. Oxacillin-hydrolysing β-lactamases. A comparative analysis at nucleotide and amino acid sequence levels. Eur. J. Biochem. **180:**309–318.

197. Murakami, K., M. Doi, and T. Yoshida. 1982. Asparenomycins A, B and C, new carbapenem antibiotics. V. Inhibition of β-lactamases. J. Antibiot. **35:**39–45.

198. Murphy, B.P., and R.F. Pratt. 1991. *N*-(Phenylacetyl)glycyl-D-aziridine-2-carboxylate, an acyclic amide substrate of β-lactamases: importance of the shape of the substrate in β-lactamase evolution. Biochemistry. **30:**3640–3649.

199. Nagarajan, R., L.D. Boeck, M. Gorman, R.C. Hamill, C.E. Higgins, M.M. Hoehn, W.M. Stark, and J.G. Whitney. 1971. β-Lactam antibiotics from *Streptomyces*. J. Am. Chem. Soc. **93:**2308–2310.

200. Nakae, T., and J. Ishii. 1978. Transmembrane permeability channels in vesicles reconstituted from single species of porins from *Salmonella typhimurium*. J. Bacteriol. **133:**1412–1418.

201. Nakagawa, J., S. Tamaki, S. Tomioka, and M. Matsuhashi. 1984. Functional biosynthesis of cell wall peptidoglycan by polymorphic bifunctional polypeptides. Penicillin-binding protein 1Bs of *Escherichia coli* with activities of transglycosylase and transpeptidase. J. Biol. Chem. **259:**13937–13946.

202. Newall, C.E., and P.D. Hallam. 1990. β-Lactam antibiotics: penicillins and cephalosporins, p. 609–653. *In* C. Hansch, P.G. Sammes, and J.B. Taylor (ed.), Comprehensive medicinal chemistry: the rational design, mechanistic study and therapeutic application of chemical compounds, vol. 2. Enzymes and other molecular targets. Pergamon Press, Oxford.

203. Nguyen-Distèche, M.M. Leyh-bouille, S. Pirlot, J.-M. Frère, and J.M. Ghuysen. 1986. *Streptomyces* K15 DD-peptidase-catalysed reactions with ester and amide carbonyl donors. Biochem. J. **235:**167–176.

204. Nicas, T.I., C.Y.E. Wu, J.N. Hobbs Jr., D.A. Preston, and N.E. Allen. 1989. Characterization of vancomycin resistance in *Enterococcus faecium* and *Enterococcus faecalis*. Antimicrob. Agents Chemother. **33:**1121–1124.

205. Nicholas, R.A., and J.L. Strominger. 1988. Site-directed mutants of a soluble form of penicillin-binding protein 5 from *Escherichia coli* and their catalytic properties. J. Biol. Chem. **263:**2034–2040.

206. Nikaido, H. 1985. Role of permeability barriers in resistance to β-lactam antibiotics. Pharmacol. Ther. **27:**197–231.

207. Nikaido, H., K. Nikaido, and S. Harayama. 1991. Identification and characterization of porins in *Pseudomonas aeruginosa*. J. Biol. Chem. **266:**770–779.

208. Nikaido, H., and S. Normark. 1987. Sensitivity of *Escherichia coli* to various β-lactams is determined by the interplay of outer membrane permeability and degradation by periplasmic β-lactamases: a quantitative treatment. Mol. Microbiol. **1:**29–36.

209. Nikaido, H., and E.Y. Rosenberg. 1983. Porin channels in *Escherichia coli:* studies with liposomes reconstituted from purified proteins. J. Bacteriol. **153:**241–252.

210. Nikaido, H., and E.Y. Rosenberg. 1990. Cir and Fiu proteins in the outer membrane of *Escherichia coli* catalyze transport of monomeric catechols: study with β-lactam antibiotics containing catechol and analogous groups. J. Bacteriol. **172:**1361–1367.

211. Nikaido, H., E.Y. Rosenberg, and J. Foulds. 1983. Porin channels in *Escherichia coli:* studies with β-lactams in intact cells. J. Bacteriol. **153:**232–240.

212. Nikaido, H., and M. Vaara. 1985. Molecular basis of bacterial outer membrane permeability. Microbiol. Rev. **49:**1–32.

213. Noguchi, H., M. Matsuhashi, and S. Mitsuhashi. 1979. Comparative studies of penicillin-binding proteins in *Pseudomonas aeruginosa* and *Escherichia coli*. Eur. J. Biochem. **100:**41–49.

214. Novick, R.P. 1982. Micro-iodometric assay of penicillinase. Biochem. J. **83:**236–240.

215. Nozaki, Y., N. Katayama, H. Ono, S. Tsubotani, S. Harada, H. Okazaki, and Y. Nakao. 1987. Binding of a non-β-lactam to penicillin-binding proteins. Nature **325:**179–180.

216. O'Callaghan, C.H., A. Morris, S.M. Kirby, and A.H. Shingler. 1972. Novel method for detection of β-lactamases by using a chromogenic cephalosporin substrate. Antimicrob. Agents Chemother. **1:**283–288.

217. Ohya, S., M. Yamazaki, S. Sugawara, and M. Mitsuhashi. 1979. Penicillin-binding proteins in *Proteus* species. J. Bacteriol. **137:**474–479.

218. Page, M.I. 1984. The mechanisms of reactions of β-lactam antibiotics. Acc. Chem. Res. **17:**144–151.

219. Page, M.I. 1987. The mechanisms of reactions of β-lactam antibiotics. Adv. Phys. Org. Chem. **23:**165–270.

220. Papanicolaou, G.A., and A.A. Medeiros. 1990. Discrimination of extended-spectrum β-lactamases by a novel nitrocefin competition assay. Antimicrob. Agents Chemother. **34:**2184–2192.

221. Park, J.T. 1987. Murein synthesis, p. 663–671. *In* F.C. Neidhardt, J.L. Ingraham, K.B. Low, B. Magasanik, M. Schaechter, and H.E. Umbarger (ed.), *Escherichia coli* and *Salmonella typhimurium;* cellular and molecular biology, vol. 1. American Society for Microbiology, Washington.

222. Park, J.T., and J.L. Strominger. 1957. Mode of action of penicillin. Biochemical basis for the mechanism of action of penicillin and for its selective toxicity. Science **125:**99–101.

223. Park, W., and M. Matsuhashi. 1984. *Staphylococcus aureus* and *Micrococcus luteus* peptidoglycan transglycosylases that are not penicillin-binding proteins. J. Bacteriol. **157:**538–544.

224. Patrick, S., and D.A. Lutton. 1990. Outer membrane proteins of *Bacteroides fragilis* grown in vivo. FEMS Microbiol. Lett. **71:**1–4.

225. Pazhanisamy, S., and R.F. Pratt. 1989. β-Lactamase-catalyzed aminolysis of depsipeptides: peptide inhibition and a new kinetic mechanism. Biochemistry **28:**6875–6882.

226. Perun, T.J., and C.L. Propst. 1989. Introduction to computer-aided drug design, p. 1–16. *In* T.J. Perun and C.L. Propst (ed.), Computer-aided drug design: methods and applications. Marcel Dekker, New York.

226a. Petit, A., G. Gerbaud, D. Sirot, P. Courvalin, and J. Sirot. 1990. Molecular epidemiology of TEM-3 (CTX–1) β-lactamase. Antimicrob. Agents Chemother. **34:**219–224.

227. Petrocheilou, V., R.B. Sykes, and M.H. Richmond. 1977. Novel R-plasmid-mediated beta-lactamase from *Klebsiella aerogenes*. Antimicrob. Agents Chemother. **12:**126–128.

228. Pfaendler, H.R., J. Gosteli, R.B. Woodward, and G. Rihs. 1981. Structure, reactivity, and biological activity of strained bicyclic β-lactams. J. Am. Chem. Soc. **103:**4526–4531.

229. Philippon, A., R. Labia, and G. Jacoby. 1989. Extended-spectrum β-lactamases. Antimicrob. Agents Chemother. **33:**1131–1136.

230. Piddock, L.J.V., and R. Wise. 1986. Cefoxitin resistance in *Bacteroides* species: evidence indicating two mechanisms causing decreasing susceptibility. J. Antimicrob. Chemother. **19:**161–170.

231. Pisabarro, A.G., F.J. Canada, D. Vazquez, P. Arriaga, and A. Rodriguez-Tebar. 1986. Structural modification of *Escherichia coli* peptidoglycan induced by bicyclomycin. J. Antibiot. **39:**914–921.

232. Pitton, J.S. 1972. Mechanisms of bacterial resistance to antibiotics, p. 15–93. *In* R.H. Adirna (ed.), Review of physiology, vol. 65. Springer-Verlag, Berlin.

233. Pollock, M.R. 1965. Purification and properties of penicillinases from two strains of *Bacillus licheniformis:* a chemical, physicochemical and physiological comparison. Biochem. J. **94:**666–675.

234. Prats, R., M. Gomez, J. Pla, B. Blasco, and J.A. Ayala. 1989. A new β-lactam-binding protein derived from penicillin-binding protein 3 of *Escherichia coli*. J. Bacteriol. **171:**5194–5198.

235. Pratt, R.F. 1989. β-Lactamase inhibitors, p. 178–205. *In* M. Sandler and H.J. Smith (ed.), Design of enzyme inhibitors as drugs. Oxford University Press, New York.

236. Price, D.A. 1977. Structure-activity relationships of semisynthetic penicillins. Adv. Appl. Microbiol. **11:**17–75.

237. Proctor, P., N.P. Gensmantel, and M.I. Page. 1982. The chemical reactivity of penicillins and other β-lactam antibiotics. J. Chem. Soc. Perkin Trans. **II:**1185–1192.

238. Pullman, B. 1974. Conformational studies in quantum biochemistry, p. 61–89. *In* R. Daudel and B. Pullman (ed.), The world of quantum chemistry. D. Reidel Publ., Dordrecht, Holland.

239. Quinn, J.P., E.J. Dudek, C.A. diVincenzo, D.A. Lucks, and S.A. Lerner. 1986. Emergence of resistance to imipenem during therapy for *Pseudomonas aeruginosa* infections. J. Infect. Dis. **154:**289–294.

240. Rando, R.R. 1975. On the mechanism of action of antibiotics which act as irreversible enzyme inhibitors. Biochem. Pharmacol. **24:**1153–1160.

241. Rao, S.N., and R.A.M. O'Ferrall. 1990. A structure-reactivity relationship for base-promoted hydrolysis and methanolysis of monocyclic β-lactams. J. Am. Chem. Soc. **112:**2729–2735.

242. Rao, V.S.R., and T.K. Vasudevan. 1979. Conformation and activity of β-lactam antibiotics. CRC Crit. Rev. Biochem. **14:**172–206.

243. Rasmussen, B.A., Y. Gluzman, and F.P. Tally. 1990. Cloning and sequencing of the class B β-lactamase gene (*ccrA*) from *Bacteroides fragilis* TAL3636. Antimicrob. Agents Chemother. **34:**1590–1592.

244. Ratcliffe, R.W., and G. Alberts-Schönberg. 1982. The chemistry of thienamycin and other carbapenam antibiotics. p. 227–313. *In* R.B. Morin and M. Gorman (ed.), Chemistry and biology of β-lactam antibiotics. Vol. 2. Nontraditional β-lactam antibiotics. Academic Press, New York.

245. Raviglione, M.C., J.F. Boyle, P. Mariuz, A. Pablos-Mendez, H. Cortes, and A. Merlo. 1990. Ciprofloxacin-resistant methicillin-resistant *Staphylococcus aureus* in an acute-care hospital. Antimicrob. Agents Chemother. **34:**2050–2054.

246. Reading, C., and M. Cole. 1977. Clavulanic acid: a β-lactamase-inhibiting β-lactam from *Streptomyces clavuligerus*. Antimicrob. Agents Chemother. **11:**852–857.

247. Reguera, J.A., F. Baquero, J. Berenguer, M. Martinez-Ferrer, and J.L. Martinez. 1990. β-Lactam-fosfomycin antagonism involving modification of penicillin-binding protein 3 in *Pseudomonas aeruginosa*. Antimicrob. Agents Chemother. **34:**2093–2096.

248. Reinicke, B., P. Blümel, H. Labischinski, and P. Giesbrecht. 1985. Neither an enhancement of autolytic wall degradation nor an inhibition of the incorporation of cell wall material are pre-requisites for penicillin-induced bacteriolysis in *Staphylococci*. Arch. Microbiol. **141:**309–314.

249. Reusch, V. 1984. Lipopolymers, isoprenoids, and the assembly of the gram-positive cell wall. Crit. Rev. Microbiol. **11:**129–155.

250. Reynolds, P.E. 1988. The essential nature of staphylococcal penicillin-binding proteins, p. 343–351. *In* P. Actor, L. Daneo-Moore, M.L. Higgins, M.R.J. Salton, and G.D. Shockman (ed.), Antibiotic inhibition of bacterial cell surface assembly and function. American Society for Microbiology, Washington D.C.

251. Reynolds, P.E., and D.F.J. Brown. 1985. Penicillin-binding proteins of β-lactam-resistant strains of *Staphylococcus aureus*. FEBS Lett. **192:**28–32.

252. Reynolds, P.E., and H. Chase. 1981. β-Lactam-binding proteins: identification as lethal targets and probes of β-lactam accessibility, p. 153–168. *In* M.R.J. Salton and G.D. Shockman (ed.), β-Lactam antibiotics: mode of action, new developments, and future prospects. Academic Press, New York.

253. Richmond, M.H. 1965. Wild-type variants of exopenicillinase from *Staphylococcus aureus*. Biochem. J. **94:**584–593.

253a. Richmond, M.H., and R.B. Sykes. 1973. The β-lactamases of gram-negative bacteria and their possible physiological role. Adv. Microb. Physiol. **9:**31–88.

254. Rogers, H.J., H.R. Perkins, and J.B. Ward. 1980. Microbial cell walls and membranes, p. 1–564. Chapman and Hall, London.

255. Rolinson, G.N. 1986. β-Lactam antibiotics. J. Antimicrob. Chemother. **17:**5–36.

255a. Rolinson, G.N. 1989. β-Lactamase induction and resistance to β-lactam antibiotics. J. Antimicrob. Chemother. **23:**1–2.

256. Rosdahl, V.T. 1973. Naturally occurring constitutive β-lactamase of novel serotype in *Staphylococcus aureus*. J. Gen. Microbiol. **77:**229–231.

257. Rossi, L., E. Tonin, Y.R. Cheng, and R. Fontana. 1985. Regulation of penicillin-binding protein activity: description of a methicillin-inducible penicillin-binding protein in *Staphylococcus aureus*. Antimicrob. Agents Chemother. **27:**828–831.

258. Rubin, L.G., A.A. Medeiros, R.H. Yolken, and E.R. Moxon. 1981. Ampicillin treatment failure of apparently β-lactamase-negative *Haemophilus influenzae* type b meningitis due to novel β-lactamase. Lancet **ii:**1008–1010.

259. Sabath, L.D., M. Jago, and E.P. Abraham. 1965. Cephalosporinase and penicillinase activities of a β-lactamase from *Pseudomonas pyocyanea*. Biochem. J. **96:**739–752.

260. Saino, Y., F. Kobayashi, M. Inoue, and S. Mitsushashi. 1982. Purification and properties of inducible penicillin beta-lactamase isolated from *Pseudomonas maltophilia*. Antimicrob. Agents Chemother. **22:**564–570.

261. Salton, M.R.J., and G.D. Shockman. 1981. β-Lactam antibiotics: mode of action, new developments, and future prospects, p. 1–604. Academic Press, New York.

262. Samraoui, B., B.J. Sutton, R.J. Todd, P.J. Artymiuk, S.G. Waley, and D.C. Phillips. 1986. Tertiary structural similarity between a class A β-lactamase and a penicillin-sensitive D-alanyl carboxypeptidase-transpeptidase. Nature (London) **320:**378–380.

263. Samuni, A. 1975. A direct spectrophotometric assay and determination of Michaelis constants for the beta-lactamase reaction. Anal. Biochem. **63:**17–26.

264. Sargent, M.G. 1968. Rapid fixed-time assay for penicillinase. J. Bacteriol. **95:**1493–1494.

265. Sassiver, M.L., and A. Lewis. 1977. Structure-activity relationships among semisynthetic cephalosporins. 1. The first generation compounds, p. 87. *In* D. Perlman (ed.), Structure-activity relationships among the semisynthetic antibiotics. Academic Press, New York.

266. Sawai, T., S. Hirano, and A. Yamaguchi. 1987. Repression of porin synthesis by salicylate in *Escherichia coli, Klebsiella pneumoniae,* and *Serratia marcescens.* FEMS Microbiol. Lett. **40:**233–237.

267. Sawai, T., K. Matsuba, and S. Yamagishi. 1977. A method for measuring the outer membrane-permeability of β-lactam antibiotics in gram-negative bacteria. J. Antibiot. **30:**1134–1136.

268. Sawai, T., I. Takahashi, and S. Yamagishi. 1978. Iodometric assay method for beta-lactamase with various beta-lactam antibiotics as substrates. Antimicrob. Agents Chemother. **13:**910–913.

269. Schwarz, U., K. Seeger, F. Wengenmayer, and H. Strecker. 1981. Penicillin-binding proteins of *Escherichia coli* identified with a ^{125}I-derivative of ampicillin. FEMS Microbiol. Lett. **10:**107–109.

270. Serfass, D.A., P.M. Mendelman, D.O. Chaffin, and C.A. Needham. 1986. Ampicillin-resistance and penicillin-binding proteins of *Haemophilus influenzae*. J. Gen. Microbiol. **132:**2855–2861.

271. Shaw, E. 1970. Chemical modification by active-site-directed reagents, p. 91–146. *In* P.D. Boyer (ed.), The enzymes: structure and control, vol. 1. 3rd ed. Academic Press, New York.

272. Shepherd, S.T., H.A. Chase, and P.E. Reynolds. 1977. The separation and properties of two penicillin-binding proteins from *Salmonella typhimurium*. Eur. J. Biochem. **78:**521–532.

273. Shlaes, D.M., A. Bouvet, C. Devine, J.H. Shlaes, S. Al-Obeid, and R. Williamson. 1989. Inducible, transferable resistance to vancomycin on *Enterococcus faecalis*. A256. Antimicrob. Agents Chemother. **33:**198–203.

274. Shlaes, D.M., and L.M. Etter. 1990. Synergistic killing of vancomycin-resistant *Enterococci* of classes A, B, and C by vancomycin, penicillin, gentamicin combinations. 30th Intersci. Conf. Antimicrob. Agents Chemother., abstr. 692.

275. Shockman, G.D., and J.F. Barrett. 1983. Structure, function and assembly of cell walls of gram-positive bacteria. Ann. Rev. Microbiol. **37:**501–527.

276. Shute, R.E., D.E. Jackson, and B.W. Bycroft. 1989. Highly conformationally constrained halogenated 6-spiroepoxypenicillins as probes for the bioactive side-chain conformation of benzylpenicillin. J. Computer-Aided Molecular Design **3:**149–164.

277. Simpson, C.N., J.P. Maskell, and J.D. Williams. 1984. The effect of clavulanic acid on the susceptibility of *Bacteroides fragilis* to three acyl-ureidopenicillins, ampicillin, and carbenicillin. J. Antimicrob. Chemother. **14:**133–138.

278. Slocombe, B., M.J. Basker, P.H. Bentley, J.P. Clayton, M. Cole, K.R. Comber, R.A. Dixon, R.A. Edmondson, D. Jackson, D.J. Merrikin, and R. Sutherland. 1981. BRL 17421, a novel β-lactam antibiotic, highly resistant to β-lactamases, giving high and prolonged serum levels in humans. Antimicrob. Agents Chemother. **20:**38–46.

279. Smith, S.M., R.H.K. Eng, P. Bais, P. Fan-Havard, and F. Tecson-Tumang. 1990. Epidemiology of ciprofloxacin resistance among patients with methicillin-resistant *Staphylococcus aureus*. J. Antimicrob. Chemother. **26:**567–572.

280. Song, M.D., S. Maesaki, M. Wachi, T. Takahashi, M. Doi, F. Ishino, Y. Maeda, K. Okonogi, A. Imada, and M. Matsuhashi. 1988. Primary structure and origin of the gene encoding the β-lactam-inducible penicillin-binding protein responsible for methicillin resistance in *Staphylococcus aureus,* p. 352–359. *In* P. Actor, L. Daneo-Moore, M.L. Higgins, M.R.J. Salton, and G.D. Shockman (ed.), Antibiotic inhibition of bacterial cell surface assembly and function. American Society for Microbiology, Washington, D.C.

281. Spratt, B.G. 1975. Distinct penicillin binding proteins involved in the division, elongation, and shape of *Escherichia coli* K12. Proc. Natl. Acad. Sci. USA **72:**2999–3003.

282. Spratt, B.G. 1977. Properties of the penicillin-binding proteins of *Escherichia coli* K12. Eur. J. Biochem. **72:**341–352.

283. Spratt, B.G. 1980. Biochemical and genetical approaches to the mechanism of action of penicillin. Phil. Trans. Royal Soc. **B289:**273–283.

284. Spratt, B.G. 1983. Penicillin-binding proteins and the future of β-lactam antibiotics. J. Gen. Microbiol. **129:**1247–1260.

285. Spratt, B.G., L.D. Bowler, A. Edelman, and J.K. Broome-Smith. 1988. Membrane topology of penicillin-binding proteins 1b and 3 of *Escherichia coli* and the production of water-soluble forms of high-molecular-weight penicillin binding proteins, p. 292–305. *In* P. Actor, L. Daneo-Moore, M.L. Higgins, M.R.J. Salton, and G.D. Shockman (ed.), Antibiotic inhibition of bacterial cell surface assembly and function. American Society for Microbiology, Washington, D.C.

286. Spratt, B.G., and K.D. Cromie. 1988. Penicillin-binding proteins of gram-negative bacteria. Rev. Infect. Dis. **10:**699–711.

287. Suginaka, H., P.M. Blumberg, and J.L. Strominger. 1972. Multiple penicillin-binding components in *Bacillus subtilis, Bacillus cereus, Staphylococcus aureus,* and *Escherichia coli*. J. Biol. Chem. **247:**5279–5288.

288. Suzuki, H., Y. Nishimura, and Y. Hirota. 1978. On the process of cellular division in *Escherichia coli:* a series of mutants of *E. coli* altered in the penicillin-binding proteins. Proc. Natl. Acad. Sci. USA **75:**664–668.

289. Sweet, R.M., and L.F. Dahl. 1970. Molecular architecture of the cephalosporins: insights into biological activity based on structural investigations. J. Am. Chem. Soc. **92:**5489–5507.

290. Sykes, R.B., D.P. Bonner, K. Bush, and N.H. Georgopapadakou. 1982. Azthreonam (SQ 26,776), a synthetic monobactam specifically active against aerobic gram-negative bacteria. Antimicrob. Agents Chemother. **21:**85–92.

291. Sykes, R.B., D.P. Bonner, and E.A. Swabb. 1987. Modern β-lactam antibiotics, p. 171–202. *In* D.J. Tipper (ed.), Antibiotic inhibitors of bacterial cell wall biosynthesis. International Encyclopedia of Pharmacology and Therapeutics, Section 127. Pergamon Press, Oxford.

292. Sykes, R.B., C.M. Cimarusti, D.P. Bonner, K. Bush, D.M. Floyd, N.H. Georgopapadakou, W.H. Koster, W.C. Liu, W.L. Parker, P.A. Principe, M.L. Rathnum, W.A. Slusarchyk, W.H. Trejo, and J.S. Wells. 1981. Monocyclic β-lactam antibiotics produced by bacteria. Nature (London) **291:**489–491.

293. Sykes, R.B., and N.H. Georgopapadakou. 1981. Bacterial resistance to β-lactam antibiotics: an overview, p. 199–214. *In* M.R.J. Salton and G.D. Shockman (ed.), β-Lactam antibiotics: mode of action, new developments and future prospects. Academic Press, New York.

293a. Sykes, R.B., and M. Matthew. 1976. The β-lactamases of gram-negative bacteria and their role in resistance to β-lactam antibiotics. J. Antimicrob. Chemother. **2:**115–157.

294. Ternansky, R.J. and S.E. Draheim. 1989. The synthesis and biological evaluation of pyrazolidinone antibacterial agents, p. 139–156. *In* P.H. Bentley and R. Southgate (ed.), Recent advances in the chemistry of β-lactam antibiotics. Proceedings of the fourth international symposium. Special publication No. 70. Royal Society of Chemistry, London.

295. Then, R.L., and P. Angehrn. 1982. Trapping of nonhydrolyzable cephalosporins by cephalosporinases in *Enterobacter cloacae* and *Pseudomonas aeruginosa* as a possible resistance mechanism. Antimicrob. Agents Chemother. **21:**711–717.

296. Timewell, R., E. Taylor, and I. Phillips. 1981. The β-lactamases of *Bacteroides* species. J. Antimicrob. Chemother. **7:**137–146.

297. Tipper, D.J. 1987. Mode of action of β-lactam antibiotics, p. 133–170. *In* D.J. Tipper (ed.), Antibiotic inhibitors of bacterial cell wall biosynthesis. International Encyclopedia of Pharmacology and Therapeutics, Section 127. Pergamon Press, England.

298. Tipper, D.J., and J.L. Strominger. 1965. Mechanism of action of penicillins: a proposal based on their structural similarity to acyl-D-alanyl-D-alanine. Proc. Natl. Acad. Sci. USA **54:**1133–1141.

299. Deleted.

300. Tipper, D.J., and A. Wright. 1979. The structure and biosynthesis of bacterial cell walls, p. 291–426. *In* I.C. Gunsalus, J.R. Sokatch, and L.N. Ornston (ed.), The bacteria, vol. VII. Academic Press, New York.

301. Tomasz, A. 1979. The mechanism of the irreversible antimicrobial effects of penicillin: how the beta-lactam antibiotics kill and lyse bacteria. Annu. Rev. Microbiol. **33:**113–137.

302. Tomasz, A. 1983. Mode of action of β-lactam antibiotics—a microbiologist's view, p. 15–96. *In* A.L. Demain and N.A. Solomon (ed.), Antibiotics containing the beta-lactam structure I. Handbook of experimental pharmacology, vol. 67/I. Springer-Verlag, Berlin/Heidelberg.

303. Tomasz, A. 1986. Penicillin-binding proteins and the antibacterial effectiveness of beta-lactam antibiotics. Rev. Infect. Dis. **8**(Suppl. 3):260–278.

304. Trias, J., and H. Nikaido. 1990. Outer membrane protein D2 catalyzes facilitated diffusion of carbapenems and penems through the outer membrane of *Pseudomonas aeruginosa*. Antimicrob. Agents Chemother. **34:**52–57.

305. Trias, J., and H. Nikaido. 1990. Protein D2 channel of the *Pseudomonas aeruginosa* outer membrane has a binding site for basic amino acids and peptides. J. Biol. Chem. **265:**15680–15684.

306. Tuomanen, E., and J. Schwartz. 1987. Penicillin-binding protein 7 and its relationship to lysis of nongrowing *Escherichia coli*. J. Bacteriol. **169:**4912–4915.

307. Ubukata, K., N. Yamashita, and M. Konno. 1985. Occurrence of a β-lactam-inducible penicillin-binding protein in methicillin-resistant *Staphylococci*. Antimicrob. Agents Chemother. **27:**851–857.

307a. Ubukata, K., R. Nonoguchi, M. Matsuhashi, and M. Konno. 1989. Expression and inducibility in *Staphylococcus aureus* of the *mec*A gene, which encodes a methicillin-resistant *S. aureus*-specific penicillin-binding protein. J. Bacteriol. 171:2882–2885.

308. Umezawa, H. 1987. Frontiers of antibiotic research, p. 1–363. Proceedings of the 4th Takeda Science Foundation Symposium on Bioscience. Academic Press, Inc., Orlando, FL.

309. Utsui, Y., S. Ohya, T. Magaribuchi, M. Tajima, and T. Yokota. 1986. Antibacterial activity of cefmetazole alone and in combination with fosfomycin against methicillin- and cephem-resistant *Staphylococcus aureus*. Antimicrob. Agents Chemother. **30:**917–922.

310. Vachon, V., D.J. Lyew, and J.W. Coulton. 1985. Transmembrane permeability channels across the outer membrane of *Haemophilus influenzae* type b. J. Bacteriol. **162:**918–924.

310a. Varetto, L., J.-M. Frère, M. Nguyen-Distèche, J.-M. Ghuysen, and C. Houssier. 1987. The pH dependence of the active-site serine DD-peptidase of *Streptomyces* R61. Eur. J. Biochem. **162:**525–531.

311. Vu, H., and H. Nikaido. 1985. Role of β-lactam hydrolysis in the mechanism of resistance of a β-lactamase-constitutive *Enterobacter cloacae* strain to expanded-spectrum β-lactams. Antimicrob. Agents Chemother. **27:**393–398.

312. Waley, S.G. 1987. An explicit model for bacterial resistance: application to β-lactam antibiotics. Microbiol. Sci. **4:**143–146.

313. Ward, J.B. 1984. Biosynthesis of peptidoglycan: points of attack by wall inhibitors. Pharmac. Ther. **25:**327–369.

314. Watanabe, N., T. Nagasu, K. Katsu, and K. Kitoh. 1987. E–0702, a new cephalosporin, is incorporated into *Escherichia coli* cells via the *tonB*-dependent iron transport system. Antimicrob. Agents Chemother. **31:**497–504.

315. Waxman, D.J., and J.L. Strominger. 1982. β-Lactam antibiotics; biochemical modes of action, p. 210–285. *In* R.B. Morin and M. Gorman (ed.), Chemistry and biology of β-lactam antibiotics, vol. 3. The biology of β-lactam antibiotics. Academic Press, New York.

316. Waxman, D.J., and J.L. Strominger. 1983. Penicillin-binding proteins and the mechanism of action of β-lactam antibiotics. Ann. Rev. Biochem. **52:**825–869.

317. Waxman, D.J., R.R. Yocum, and J.L. Strominger. 1980. Penicillins and cephalosporins are active site-directed acylating agents: evidence in support of the substrate analogue hypothesis. Phil. Trans. R. Soc. Lond. B **289:**257–271.

318. Webber, J.A., and J.L. Ott. 1977. Structure-activity relationships in the cephalosporins. II. Recent developments, p. 161–237. *In* D. Perlman (ed.), Structure-activity relationships among the semisynthetic antibiotics. Academic Press, New York.

319. Wiedemann, B., C. Kliebe, and M. Kresken. 1989. The epidemiology of β-lactamases. J. Antimicrob. Chemother. **24(Suppl. B):**1–22.

320. Williamson, R., S.B. Calderwood, R.C. Moellering, and A. Tomasz. 1983. Studies on the mechanism of intrinsic resistance to β-lactam antibiotics in group D streptococci. J. Gen. Microbiol. **129:**813–822.

321. Williamson, R., R. Hakenbeck, and A. Tomasz. 1980. The penicillin-binding proteins of *Streptococcus pneumoniae* grown under lysis-permissive and lysis-protective (tolerant) conditions. FEMS Microbiol. Lett. **7**:127–131.

322. Williamson, R., C. Le Bouguénec, L. Gutmann, and T. Horaud. 1985. One or two low affinity penicillin-binding proteins may be responsible for the range of susceptibility of *Enterococcus faecium* to benzylpenicillin. J. Gen. Microbiol. **131**:1933–1940.

323. Wise, E.M., Jr., and J.T. Park. 1965. Penicillin: its basic site of action as an inhibitor of a peptide cross-linking reaction in cell wall muropeptide synthesis. Proc. Natl. Acad. Sci. USA **54**:75–81.

325. Wolfe, S., M. Khalil, and D.F. Weaver. 1988a. MMPEN: development and evaluation of penicillin parameters for Allinger's MMP2(85) programme. Can. J. Chem. **66**:2715–2732.

327. Wolfe, S., K. Yang, and M. Khalil. 1988b. Conformation-activity relationships and the mechanism of action of penicillin. Can. J. Chem. **66**:2733–2750.

328. Woodruff, W.A., and R.E.W. Hancock. 1989. *Pseudomonas aeruginosa* outer membrane protein F: structural role and relationship to the *Escherichia coli* OmpA protein. J. Bacteriol. **171**:3304–3309.

329. Woodward, R.B. 1980. Penems and related substances. Phil. Trans. R. Soc. Lond. B **289**:239–250.

330. Wrezel, P.W., L.F. Ellis, and F.C. Neuhaus. 1986. In vivo target of benzylpenicillin in *Gaffkya homari*. Antimicrob. Agents Chemother. **29**:432–439.

331. Wyke, A.W., J.B. Ward, M.V. Hayes, and N.A.C. Curtis. 1981. A role in vivo for penicillin-binding protein-4 of *Staphylococcus aureus*. Eur. J. Biochem. **119**:389–393.

332. Yoshihara, E., and T. Nakae. 1989. Identification of porins in the outer membrane of *Pseudomonas aeruginosa* that form small diffusion pores. J. Biol. Chem. **264**:6297–6301.

333. Yoshimura, F., and H. Nikaido. 1985. Diffusion of β-lactam antibiotics through the porin channels of *Escherichia coli* K-12. Antimicrob. Agents Chemother. **27**:84–92.

334. Yotsuji, A., S. Minami, M. Inoue, and S. Mitsuhashi. 1983. Properties of a novel beta-lactamase produced by *Bacteroides fragilis*. Antimicrob. Agents Chemother. **24**:925–929.

335. Yotsuji, A., J. Mitsuyama, R. Hori, T. Yasuda, I. Saikawa, M. Inoue, and S. Mitsuhashi. 1988. Outer membrane permeation of *Bacteroides fragilis* by cephalosporins. Antimicrob. Agents Chemother. **32**:1097–1099.

336. Young, J.D.E., M. Blake, A. Mauro, and Z.A. Cohn. 1983. Properties of the major outer membrane protein from *Neisseria gonorrhoeae* incorporated into model lipid membranes. Proc. Natl. Acad. Sci. USA **80**:3831–3835.

337. Zighelboim, S., and A. Tomasz. 1980. Penicillin-binding proteins of multiply antibiotic-resistant South African strains of *Streptococcus pneumoniae*. Antimicrob. Agents Chemother. **17**:434–442.

338. Zimmerman, W., and A. Rosselet. 1977. Function of the outer membrane of *Escherichia coli* as a permeability barrier to β-lactam antibiotics. Antimicrob. Agents Chemother. **12**:368–372.

10

Lipid A Biosynthesis

Jack Coleman

Lipid A, a constituent of the outer membrane of gram-negative bacteria, is a structurally unique molecule that does not exist in animal cells. The lipid A of most gram-negative bacteria is structurally similar, with analogous biosynthetic steps. These facts, combined with the fact that lipid A is required for the survival of most, if not all gram-negative bacteria, make lipid A biosynthesis an excellent target for antibiotics.

Gram-negative bacteria are surrounded by two distinct lipid bilayers, the inner or cytoplasmic membrane and the outer membrane. The lipid component of the cytoplasmic membrane and of the inner leaflet of the outer membrane is composed of glycerophospholipids. The lipid component of the outer leaflet of the outer membrane, however, contains lipopolysaccharide (LPS), a glucosamine-derived lipid (42). Lipopolysaccharide is composed of three regions (see Fig. 10.1): lipid A, the proximal, hydrophobic region, also called gram-negative endotoxin; the core polysaccharide region containing unique sugars such as heptose phosphates and the 8-carbon sugar 3-deoxy-D-*manno*-octulosonic acid (KDO); and the distal hydrophilic O-antigen polysaccharide, a repeating oligosaccharide that projects out into the medium.

LPS serves several roles for the bacterium: it confers stability to the cell, it acts as a permeability barrier to hydrophobic molecules, it protects against antibiotics and detergents, it renders cells resistant to external phospholipases, it plays a significant role in the interaction with the environment, and it protects the bacterium from host defenses during an infection (17). The permeability-barrier properties of LPS are primarily conferred by the O-antigen region and the outer portion of the core region. The O-antigen region of LPS is also responsible for the attachment of the bacteria to the host cell and for protecting the bacteria from certain host defense mechanisms.

Lipid A, in addition to being the membrane anchor for LPS, is very significant in the pathology of a gram-negative infection. It is responsible for the endotoxic reactions expressed during infection, including lethal toxicity, complement inacti-

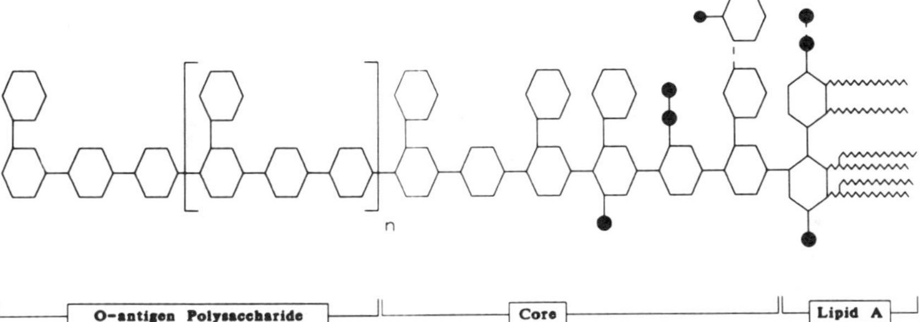

Figure 10.1 Structure of lipopolysaccharide from *E. coli*. Hexagons represent sugar residues, solid black circles phosphate moieties, including phosphoethanolamine, and long wavy lines fatty acids. Dashed lines represent partial substitutions.

vation, pyrogenicity, and activation of macrophages (17,39). The toxic effects of lipid A during gram-negative infections represent a major cause of death among debilitated patients (72). A prevalent drawback of the present selection of antibiotics used against gram-negative bacteria is that many cause the release of toxic LPS at, or just prior to, the time of cell death (27). Antibiotics targeted to lipid A biosynthesis, used alone or in combination with other antibiotics, should prevent the problem of release of toxic materials.

This chapter will discuss the biosynthetic pathway for lipid A, with emphasis on these steps that are likely to be the most effective targets for gram-negative-specific antibiotics. The review will focus on the well studied enteric bacteria *Escherichia coli* and *Salmonella typhimurium*, comparing and contrasting their biosynthetic pathways with other species of gram-negative bacteria.

I. Universal Structure of Lipid A

Unlike the structure of the O-antigen and the core region, the structure of the lipid A portion of LPS remains comparatively constant among the different species of gram-negative bacteria. The structure of the enterobacterial lipid A consists of a disaccharide of glucosamine residues, linked β,1→6, and with phosphate moieties substituted at positions 1 and 4′ (17,59,60,64). It bears five, six, or seven fatty acid moieties depending on the organism (47,48,70), four of which are 3-OH-14:0, which is not found in the glycerophospholipids (see Fig. 10.2). The additional fatty acids are found in acyl-oxy-acyl linkages to the 3-OH-14:0. The identity and number of acyl-oxy-acyl fatty acids is variable between strains. The phosphate moiety at position 1 is occasionally a pyrophosphate (51).

The lipid A structure shown in Figure 10.2 is common in the enteric bacteria and many other gram-negative bacteria. Certain bacteria, however, have modified the structure of their lipid A. For example, *Neisseria gonorrhoeae* lipid A has the shorter fatty acid 3-OH-12:0 instead of 3-OH-14:0 (14,63). This change in

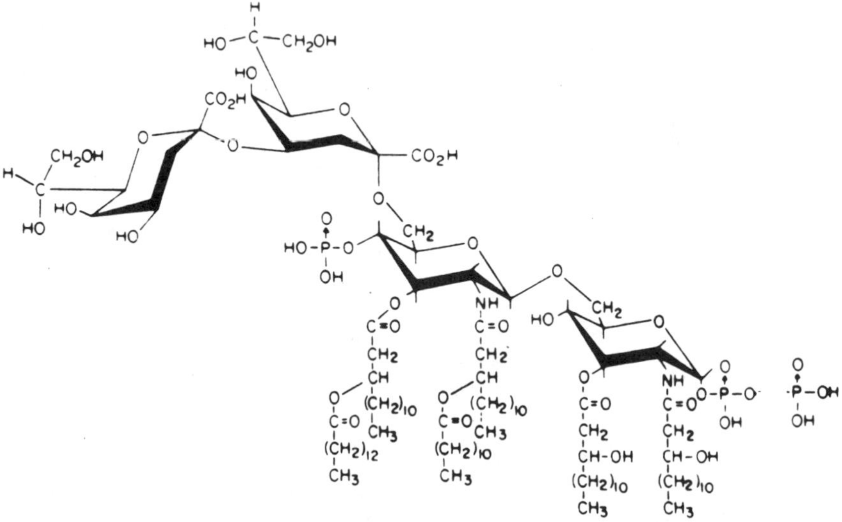

Figure 10.2 Detailed structure of *E. coli* KDO_2-lipid A. Dashed line indicates partial substitution of the phosphate at the 1 position with pyrophosphate. The structure shown is the minimal structure required for growth and assembly of the outer membrane of *Escherichia coli* and *Salmonella typhimurium*.

length of fatty acid appears to have little effect on the toxicity of lipid A in animals.

Certain of the distantly related phototrophic bacteria such as *Rhodopseudomonas capsulatus* and *Rhodopseudomonas sphaeroides* contain 3-OH-10:0 fatty acids instead of the O-linked 3-OH-14:0 (14,32,46). These strains, in addition, have the rare amide-linked 3-keto-14:0 fatty acid linked at the C-2 position. Although most of the lipid A from these strains contains 3-keto-14:0 at the C-2 position, a certain percent of the lipid A contains 3-OH-14:0 instead (46). It is unclear whether a single acyltransferase is able to add either fatty acid, or if there are two separate acyltransferases. These strains have a reduced number of "piggy-backed" fatty acids, some of which are unsaturated (46). The lipid A isolated from these strains is nontoxic, presumably due to the altered acyl-oxy-acyl fatty acid content or the presence of the 3-keto fatty acid (46). Lipid A from the related nonphototrophic bacteria *Thiobacillus versutus* and *Paracoccus dinitrificans* has a structure similar to that from *Rhodopseudomonas capsulatus* and *Rhodopseudomonas sphaeroides*.

Pseudomonas diminuta, a bacterium related to the *Rhodopseudomonas* mentioned above, contains 3-hydroxyl fatty acids, primarily 3-OH-12:0, attached to the 2,2′,3, and 3′ positions of the sugar backbone, with no 3-keto fatty acid as found in some of the *Rhodopseudomonas*. Although the 3-OH fatty acids are characteristic of the enteric bacteria, the sugar backbone of *Pseudomonas diminuta* conspicuously differs from the *Enterobacteriaceae*. The *Pseudomonas dimi-*

nuta lipid A does not possess a phosphate at the 1 position, as most other lipid A molecules do (28). More striking, the disaccharide backbone is not glucosamine, but 2,3-diamino-2,3-dideoxy-D-glucose (28). The fatty acids at the 3 and 3′ positions are N-linked, not O-linked. Roppel and co-workers (61) hypothesize that a known bacterial enzyme capable of adding an amino group to C-3 of sugar molecules in antibiotics and O-antigens may also be responsible for the synthesis of 2,3-diamino-2,3-dideoxy-D-glucose in these bacteria. The lipid A from *Pseudomonas diminuta*, like lipid A from the enteric bacteria, is toxic (28).

The lipid A of *Rhodopseudomonas viridis* and *Rhodopseudomonas palustris* also contains 2,3-diamino-2,3-dideoxy-D-glucose. The sole fatty acid found in the non-toxic lipid A from these organisms is amide-bound 3-OH-14:0, and the lipid A contains little or no phosphate (61). Fatty acids are attached to both the nitrogens, at C-2 and C-3.

The bacterium *Rhodomicrobium vannielii* contains β,1→6-D-glucosamine, as does lipid A from the enteric bacteria. The *Rhodomicrobium vannielii* lipid A, however, lacks phosphate, and C-4′ is partially substituted with a D-*manno*-pyranosyl residue (69).

Lipid A containing 2,3-diamino-2,3-dideoxy-D-glucose represents the most exceptional structural deviation from the enterobacterial lipid A known. Some species of bacteria, such as *Thiobacillus ferrooxidans* and *Brucella abortus*, have a mixed lipid A containing 2,3-diamino-2,3-dideoxy-D-glucose and glucosamine. Indeed, it has been shown that the *E. coli* enzyme that catalyzes the formation of the disaccharide does not distinguish between diacyl-glucosamine and its diamino derivative (69).

Some common structural features can be drawn from the comparison of all known lipid A structures. The sugar backbone always contains at least one, and usually two, glucosamine residues, sometimes substituted with an amino group at C-3. All known lipid A structures contain at least one amino-linked 3-OH-fatty acid. In addition, most lipid A molecules contain a phosphate at C-4 on the nonreducing sugar, and less frequently at position C-1 on the reducing end sugar. Most also have the unique acyl-oxy-acyl moiety.

II. Lipid A Biosynthesis

The biosynthetic pathway of lipid A has only recently been elucidated (4,7,50) and is shown in Figure 10.3. The elucidation of the lipid A biosynthetic pathway was aided by the discovery of a unique fatty acylated monosaccharide, lipid X, in a mutant strain of *E. coli* (44,65). The elucidation of the structure of lipid X (2,3-diacylglucosamine-1-phosphate, Fig. 10.3) revealed that it is a likely precursor of lipid A. Buląwa and Raetz (9) postulated that an activated form of lipid X, UDP-2,3-diacylglucosamine, must be required to form the disaccharide, and demonstrated the existence of this lipid in wild-type *E. coli* cells.

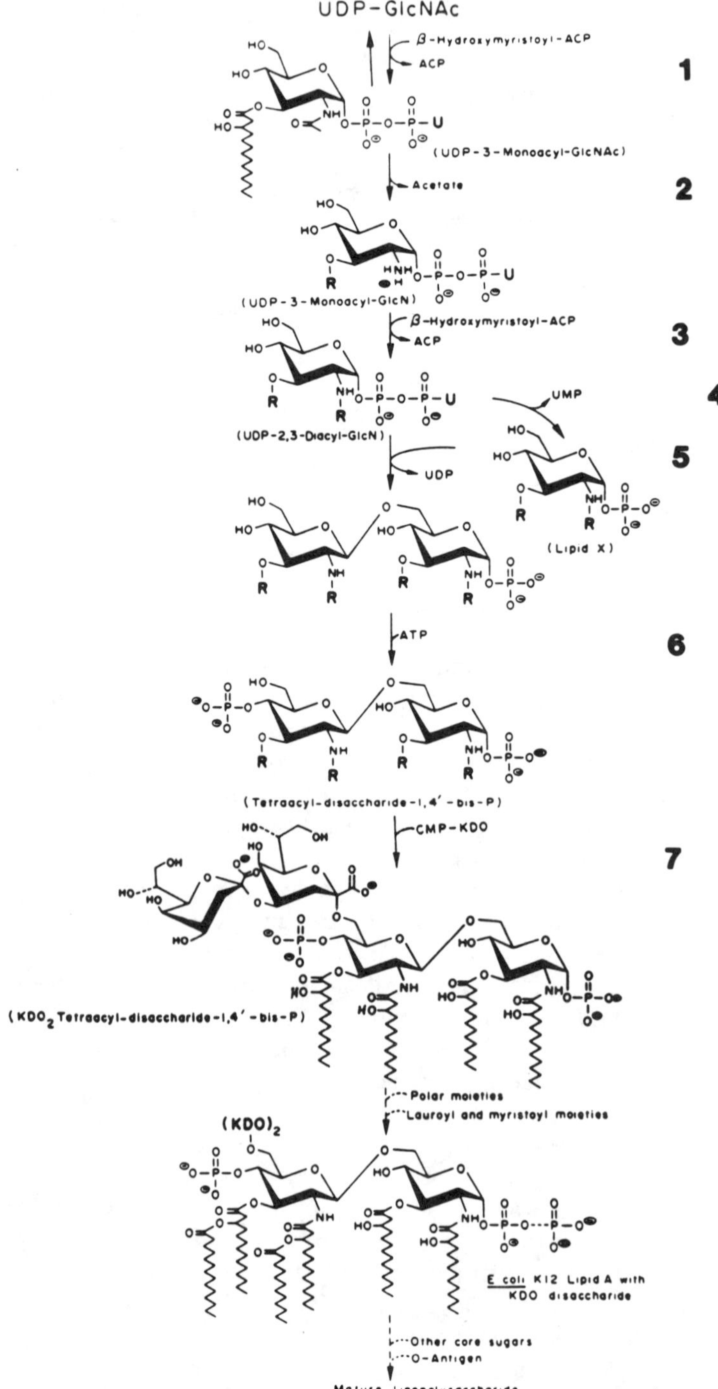

Figure 10.3 Pathway for the enzymatic biosynthesis of lipid A in *E. coli*. R designates an (*R*)-3-OH-14:0 moiety and U designates uridine. Dashed line indicates steps not determined with certainty. Numbers refer to each of the steps in the pathway. 1, UDP-GlcNac acyltransferase (encoded by *lpxA*). 2, UDP-3-*O*-acyl-GlcNAc deacetylase. 3, UDP-3-*O*-acyl-glucosamine acyltransferase. 4, UDP-diacyl-glucosamine hydrolase. 5, disaccharide synthase (encoded by *lpxB*). 6, lipid A 4′-kinase. 7, KDO transferase (encoded by *kdt*).

Direct evidence for the proposed biosynthetic scheme was first obtained by Ray and co-workers (54), who demonstrated the existence of an enzyme, in extracts of *E. coli* cells, that catalyzes the condensation of UDP-2,3-diacylglucosamine with lipid X to form tetraacyldisaccharide-1-phosphate, linked β,1 → 6. This activity was undetectable in extracts prepared from the mutant cells that accumulated lipid X *in vivo*. The gene encoding this disaccharide synthase, *lpxB*, was identified, and its location was mapped to the minute 4 region of the *E. coli* chromosome (43). The gene for *lpxB* has been cloned (16), and the protein has been overproduced (15,16) and purified to near homogeneity (49).

Anderson *et al.* (2) were able to demonstrate a precursor-product relationship between UDP-diacylglucosamine and lipid X, by pulse-labeling *E. coli* cells with $^{32}P_i$. Anderson and Raetz (3) demonstrated the existence of a set of enzymes in *E. coli* that utilizes 3-OH-14:0-acyl carrier protein as an acyl donor to convert UDP-*N*-acetyl-glucosamine to UDP-2,3-diacyl-glucosamine. The first enzyme in the biosynthetic pathway generates UDP-3-*O*-acyl-GlcNAc, and it is encoded by the *lpxA* gene. This gene is part of a long operon including *lpxB* (the disaccharide synthase) and *dnaE* (12,16,24,67; see Fig. 10.4). The biological significance of this operon is not yet known, but it may reflect a requirement of the cell to coordinately control the synthesis of the enzymes encoded within the operon.

Each of the steps in the biosynthetic pathway indicated by solid arrows (Fig. 10.3) have been shown to be catalyzed by extracts of *E. coli*, and the structure of the products has been verified using mass spectrometry and nuclear magnetic resonance (2,3,4,7,8,9,53,54,55,62,65). 3-Deoxy-D-*manno*-octulosonic acid (KDO) is the first core sugar added to lipid A; it is essential for the growth of *E. coli*. Each of the enzymes involved in the synthesis of KDO_2-tetraacyl-disaccharide-1,4′-*bis*-phosphate (Fig. 10.3), with the possible exception of the 4′ kinase are soluble or peripherally membrane bound enzymes. Although the 4′ kinase predominantly fractionates with the membrane, there is significant activity in the soluble fraction (55). For this reason, the 4′ kinase is most likely also a peripheral membrane protein.

The substrates for the formation of KDO_2-tetraacyl-disaccharide-1,4′-*bis*-phosphate are CMP-KDO and tetraacyl-disaccharide-1,4′-*bis*-phosphate. The enzyme(s) responsible for this addition appears to be a peripheral membrane protein (7,40). The presence of a KDO_1-tetraacyl-disaccharide-1,4′-*bis*-phosphate intermediate in the synthesis has not yet been established. The gene for the KDO transferase (*kdt*) has been isolated, and it maps to minute 81 on the *E. coli* chromosome (11). The partially purified *kdt* gene product catalyzes the addition of two residues of KDO onto the disaccharide backbone.

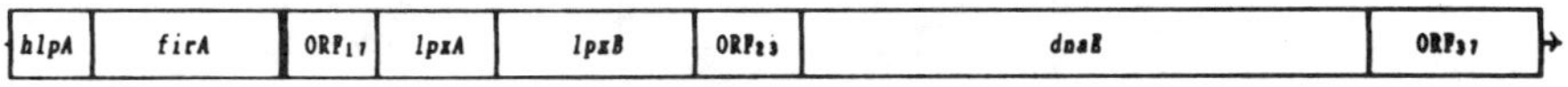

hlpA	*firA*	ORF17	*lpxA*	*lpxB*	ORF23	*dnaE*	ORF37

Figure 10.4 Organization of the *lpx* operon in *E. coli* and *Salmonella typhimurium* (1,12,16,24,33,67). ORF is an unidentified open reading frame. *hlpA*, *firA*, *lpxA*, *lpxB*, and *dnaE* are described in the text. All the genes shown are transcribed, to a certain extent, as one transcript.

Recent work by Brozek and Raetz (8) indicates that the "piggy-backed" fatty acids are added only after the addition of the core KDO sugars. The fatty acyl donor for these fatty acids appears to be acyl acyl-carrier-protein, as is the acyl donor for the first four fatty acids.

Temperature-sensitive mutations have been created in the *lpxA* and *lpxB* genes that are lethal to the bacterium (12,18,38,44, J. Coleman, unpublished results). Mutations at the later stages of lipid A biosynthesis involved in the synthesis of KDO appear to be bacteriostatic (36,57).

Adjacent to the *kdt* gene at minute 81 in the *E. coli* chromosome are most of the genes involved in the addition of the core saccharides to lipid A, in the *rfa* locus. This clustering of genes is analogous to the genes involved in lipid A biosynthesis at the minute 4 region. Upstream from and in the same operon as *lpxA* and *lpxB* are two genes, *hlpA* and *firA*, that may be involved in LPS biosynthesis and assembly. The product of *hlpA*, HLP-I, is an outer membrane protein that binds tightly to LPS (24,31). Conditions that release LPS, such as treatment with the chelator EDTA, also release HLP-I from the cell. It has been postulated that HLP-I is involved in LPS assembly (31). Mutations in *firA* render the cells hypersensitive to hydrophobic antibiotics such as rifampicin (35), a phenotype associated with LPS biosynthesis mutations (37,41,66). Indeed, recent work has shown that some cells deficient in the *firA* gene product no longer synthesize lipid A properly (J. Coleman, unpublished results).

III. History of LPS Antibiotics

Three classes of compounds have been reported to inhibit the synthesis of LPS; diazaborine derivatives, KDO analogues, and CV-1. Each of these antibiotics is able to prevent growth of most gram-negative bacteria tested (19,22,25,26,71). In addition, when these antibiotics are used at subinhibitory concentrations, they are highly synergistic with other hydrophobic antibiotics and detergents (19,22,26).

A convenient assay for LPS biosynthesis is to measure the incorporation of [^{3}H]-galactose into acid-insoluble material. *E. coli* and *Salmonella typhimurium* strains deficient in the galactose epimerase utilize exogenously added galactose primarily for incorporation into LPS, in the core portion. Each of the above LPS inhibitors blocks the incorporation of [^{3}H]-galactose into acid precipitable material (LPS), and they have also been shown to prevent addition of KDO sugars into the LPS. The inhibition of KDO addition is caused by either inhibition of the activation of KDO or inhibition of synthesis of the acceptor molecule for KDO—lipid A.

The first antibiotic discovered to be inhibitory to LPS biosynthesis was 1,2-dihydro-1-hydroxy-6-methyl-2- (propanesulphonyl) -thieno (3–2-D) (1,2,3)-diazaborine (code no. 84474, Fig. 10.5A), a diazaborine derivative (25). Diazaborine derivatives are thought to inhibit the synthesis of fatty acids, causing a

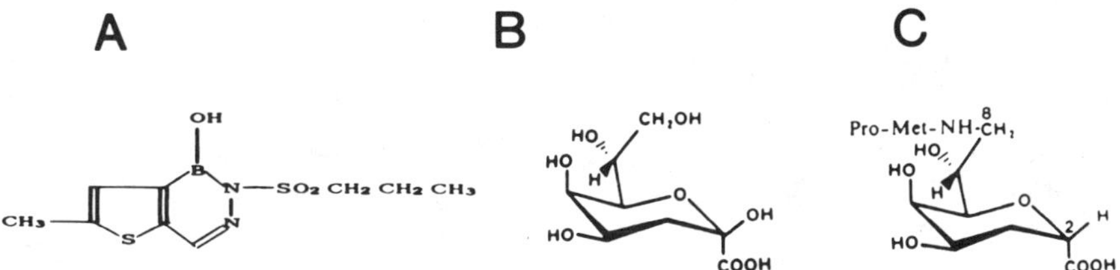

Figure 10.5 Structures of diazaborine code number 84474 (A), β-KDO (B), and Pro-Met-NHdKDO (C) (25,22).

block in both phospholipid and LPS biosyntheses (19). Diazaborine has proven to be too toxic to be used as a therapeutic agent (22).

One of the most studied reactions in the biosynthesis of LPS is the synthesis of 3-deoxy-D-*manno*-octulosonic acid (KDO), the first core sugar added to lipid A (36,56,58). Temperature-sensitive mutations have been isolated in genes involved in the synthesis, activation, and transfer of KDO to lipid A (11,56,57). Strains of bacteria harboring the more severe of these mutations stop LPS biosynthesis and accumulate lipid A precursors at the restrictive temperature (42°C), leading to bacteriostasis (56,57). In the presence of one mutant allele, *kdsB*, involved in KDO activation, virulent strains of *S. typhimurium* become avirulent in mice by rendering the bacteria susceptible to clearance by cells of the reticuloendothelial system (19). These bacteria also become susceptible to complement-mediated killing following incubation at 42°C. The gene encoding CMP-KDO synthetase has been cloned and the product has been purified (20,29,30). Using the purified enzyme Goldman and co-workers (19) designed competitive inhibitors of CMP-KDO synthetase. They found that α-C-(1,5-anhydro-7-amino-2,7-dideoxy-D-*manno*-heptopyranosyl)-carboxylate, a 2-deoxy derivative of KDO, is a good inhibitor of CMP-KDO synthetase *in vitro*. At the same time, Claesson and co-workers (10) obtained similar results. Although both groups found that the compound was effective against the purified enzyme, neither group found activity against intact cells. This lack of activity *in vivo* is due to the inability of the compound to enter the cell. The two groups (19,22) overcame this problem by coupling the inhibitor covalently to a dipeptide, allowing the bacteria to transport the inhibitor via the oligopeptide permease transport system (Fig. 10.5C). Like the mutations in the genes for KDO biosynthesis, these KDO analogues are only bacteriostatic. The inhibition of KDO addition, however, renders the bacteria hypersensitive to host defense mechanisms. In addition, the cells are much more sensitive to hydrophobic antibiotics such as erythromycin (19). The dipeptidyl-2-deoxy-KDO proved to be a potent inhibitor of most species of gram-negative bacteria, with no effect on gram-positive bacteria (10,21). The high concentration of peptidases and peptides in the blood, however, renders these antibiotics ineffective in mammals (10).

Analogues of KDO may be difficult for the bacteria to transport because the structure of KDO is quite unlike anything normally taken up by the cell. Competitive inhibitors of the early steps of LPS biosynthesis are likely to be taken up by one or more of the phosphotransferase systems normally used to transport *N*-acetylglucosamine or glucosamine. The LPS antibiotic, CV-1, is a structural analogue of *N*-acetylglucosamine (see Fig. 10.6). CV-1, naturally produced by the bacterium *Streptomyces* strain CO-1, is a spontaneous structural rearrangement of *N*-carbamoyl-D-1-glucosamine (71). By itself, CV-1 has little activity as an antibiotic, with minimum inhibitory concentrations (MICs) of greater than 100 μg/ml (26). The antibacterial activity of CV-1 against *E. coli* increased significantly, however, when spiramycin was also present (26). Spiramycin is a macrolide antibiotic that is not effective against gram-negative bacteria, presumably due to its inability to cross the outer membrane. CV-1 inhibits the incorporation of galactose into LPS, while it has no effect on the incorporation of adenine into nucleic acids, or threonine into proteins (26). It is interesting to note that CV-1 does inhibit the incorporation of uracil into macromolecular material, possibly reflecting the involvement of uracil in the early stages of lipid A biosynthesis (Fusao Tomita, personal communication). A related observation was found with a mutation in the disaccharide synthase gene (*lpxB*) involved in lipid A biosynthesis, which causes a dramatic decrease in the intracellular concentration of UTP (Jack Coleman, unpublished observations). In addition, mutations in *firA*, another gene in the *E. coli* lipid A biosynthetic operon at minute 4, also prevents incorporation of uracil into RNA (34). CV-1 prevents the incorporation of KDO into LPS, but it does not inhibit the KDO transferase, or CMP-KDO synthetase (Stephen Hammond, personal communication). The inhibition may be at the early stages of KDO biosynthesis. Due to the structural similarity between CV-1 and the glucosamine backbone of the lipid A precursor, however, it is more likely that CV-1 action prevents the formation of a lipid A precursor that is a substrate for the KDO transferase.

CV-1, alone, is not a very potent antibiotic. In combination with certain other antibiotics or with serum, the bactericidal activity is greatly enhanced (26) (Stephen Hammond, personal communication.) At the present time, little is

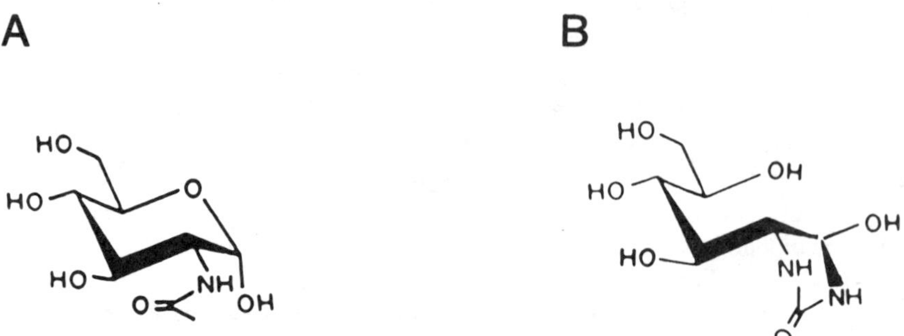

Figure 10.6 Comparative structure of *N*-acetylglucosamine (A) and CV-1 (B) (26).

known about the toxicity or the antibiotic effectiveness of CV-1 in man. Although CV-1 itself may not prove to be a clinically useful antibiotic, it is a good starting point for the design of more effective antibiotics specific for gram-negative bacteria.

IV. The First Committed Step as a Target of Antibiotics: UDP-GlcNAc Acyltransferase

The optimal target for an antibiotic would be the rate-limiting step of a reaction leading to an essential component of the cell. Most often in a biosynthetic pathway, the rate-limiting step is the first committed step. In lipid A biosynthesis the first step unique to lipid A biosynthesis is the acylation of UDP-GlcNAc to form UDP-3-*O*-acyl-GlcNAc (Fig. 10.3). This step is a branch point for both substrates: UDP-GlcNAc is used in the synthesis of LPS and peptidoglycan, and 3-OH-14:0-ACP is used in the synthesis of LPS and glycerophospholipids.

Several mutations have been obtained in the gene encoding UDP-GlcNAc acyltransferase, *lpxA* (12,18,38). Unlike mutations in KDO biosynthesis, these mutations are lethal to the bacterium at the nonpermissive temperature (42°C). Galloway and Raetz (18) have demonstrated that one *lpxA*-defective strain becomes nonviable within 1 hour after shifting to the non-permissive temperature (42°C). Strains containing mutations in *lpxA* are hypersensitive to hydrophobic antibiotics and detergents such as rifampicin, erythromycin, and deoxycholate. The lethality of these *lpxA* mutations indicate that the *lpxA* gene product would make an excellent target for gram-negative specific antibiotics.

An indication that the UDP-GlcNAc acyltransferase is regulated *in vivo* was noted while analyzing a mutation in *lpxB* (the disaccharide synthase). An insertion mutation in the chromosomal copy of *lpxB* has been created, complemented by a plasmid containing a good copy of *lpxB* and a temperature-sensitive replicon. At the nonpermissive temperature for plasmid replication, the cells are nonviable (Jack Coleman, unpublished results). As expected, disaccharide synthase activity decreases exponentially with time after shift to 42°C. Although protein synthesis continues for 2 additional hours, LPS biosynthesis drops precipitously to near zero when the concentration of the disaccharide synthase drops to one-tenth of wild-type levels. Prior to cell death, the bacteria do not accumulate the expected lipid A precursors, lipid X and UDP-diacylglucosamine. The lack of these lipid A precursors under these conditions suggests that lipid A synthesis is stopped at a very early step, at or prior to the first acylation of UDP-GlcNAc. Anderson and co-workers (4) demonstrated that enzymes in extracts of *E. coli* cells are capable of removing the fatty acid from UDP-3-*O*-acyl-GlcNAc. Recent work has shown that the *lpxA* gene product as well as a second soluble protein, not ACP, is required for this reaction (Jack Coleman, unpublished data). Perhaps this second activity of the *lpxA* gene product is a mechanism used by the cell to regulate the synthesis of LPS rapidly at an early step, as seen in the *lpxB* mutation mentioned above.

Because UDP-GlcNAc acyltransferase is an essential enzyme that is likely to be regulated *in vivo*, it should make an excellent target for gram-negative specific antibiotics. An analogue of *N*-acetylglucosamine should have little difficulty in entering the cell. An antibiotic targeted to UDP-GlcNAc acyltransferase should be effective against most species of gram-negative bacteria. Although such an antibiotic would be ineffective against species of bacteria containing 2,3-diamino-2,3-dideoxy-D-glucose instead of *N*-acetylglucosamine in the lipid A backbone, the former derivative is not a common constituent of human pathogens. Another potentially more serious problem is that the precursor UDP-GlcNAc is common to both bacterial and mammalian metabolism; therefore an inhibitor of UDP-GlcNAc acyltransferase may potentially be toxic to the mammalian host.

V. The Lipid A Deacetylase as a Target

The second step of lipid A biosynthesis is the removal of the acetate from the monoacyl precursor by the lipid A deacetylase. This step appears to be common to all species of gram-negative bacteria. In addition, metabolism of 3-acyl-*N*-acetylglucosamine is likely to be unique to gram-negative bacteria, making an inhibitor of this enzyme potentially harmless to the mammalian host.

Anderson and co-workers (4) have noted that the deacetylase is one of the more sensitive enzymes in the lipid A biosynthetic pathway. 1 mM *N*-ethylmaleimide and 1% octylglucoside inhibit the lipid A deacetylase, with little or no effect on the other enzymes in the pathway that were assayed. Inhibition of the lipid A deacetylase is likely to cause a buildup of the detergent-like molecule UDP-3-acyl-*N*-acetylglucosamine within the cell. At the present time, mutations have not been found in the gene(s) encoding the lipid A deacetylase; therefore the lethality of such inhibition is unknown. Since mutations in the first acyltransferase and fifth step, the disaccharide synthase, are lethal (12,13,18), it is expected that mutations in the deacetylase would also be lethal to the cell. The lethality of an antibiotic targeted to the deacetylase could result from inhibition of lipid A biosynthesis, as well as accumulation of a detergent-like molecule within the cell.

The antibiotic CV-1, mentioned earlier, inhibits lipid A biosynthesis at an early step, causing the accumulation of a lipid A precursor within the cell (Stephen Hammond, personal communication). No precursors unique to lipid A biosynthesis would accumulate if the inhibition was targeted to the first acyltransferase. Due to the structural similarity between CV-1 and *N*-acetylglucosamine, it is likely that the target of CV-1 action is the lipid A deacetylase.

Many features distinguish the lipid A deacetylase to make it an ideal target for an antibiotic. Analogues of the substrate are likely to be taken up by the cell by the same pathway as for the uptake of glucosamine or *N*-acetylglucosamine. The substrate for this reaction is unique for gram-negative bacteria. The reaction is essential to the formation of lipid A, required for the survival of most if not all gram-negative bacteria. In addition, the substrate, UDP-3-acyl-GlcNAc, is likely to be deleterious to the cell by acting as a detergent. Because this soluble enzyme is sensitive to many enzyme inhibitors *in vitro*, and possibly *in vivo*, it is likely to be an easily perturbed target *in vivo*.

VI. Conclusions

An important consideration in the design of a new antibiotic, in addition to the target being in an essential biosynthetic pathway, is a reliable, rapid method to screen potential antibiotics. Many screening systems have been developed to identify inhibitors of lipid A biosynthesis.

The first method measures the synergistic action of the putative inhibitor with a hydrophobic antibiotic having little or no activity against gram-negative bacteria. Ichimura and co-workers (26) used the antibiotic spiramycin as the indicator antibiotic to discover CV-1. However, subinhibitory concentrations of rifampicin or novobiocin should work equally well. Gram-negative bacteria are spread on an agar plate containing a non inhibitory concentration of an indicator antibiotic as well as a control plate without antibiotic. A paper disc containing the potential lipid A inhibitor is then placed on top of the agar on each of the plates. A zone of growth inhibition on the plate containing the indicator antibiotic, with less or no inhibition on the control plate, indicates a potential lipid A biosynthesis inhibitor. It should be noted that antibiotics affecting the integrity of the outer membrane by other means will also give positive results with this screening method (5,23,52).

To identify lipid A biosynthesis as the target of the antibiotic, the rate of lipid A biosynthesis can be assayed directly by measuring the rate of incorporation of $^{32}P_i$ into purified lipid A in the presence and absence of antibiotic as described by Galloway and Raetz (18). Alternatively, LPS biosynthesis can be measured as described by Ichimura and coworkers (26) by measuring the incorporation of radioactive galactose into the core region of LPS, an acid insoluble material, in a strain deficient in galactose epimerase. An effectively targeted antibiotic should inhibit lipid A biosynthesis without affecting other macromolecular biosynthesis at early time points (26). This method is capable of testing many potential antibiotics in a short amount of time.

A second method to screen for lipid A biosynthesis inhibitors relies on the fact that lipid A precursors are more soluble than full length LPS. Cells unable to metabolize glucosamine to other sugars (*glmS* and *nagB*) (68) are labelled with radioactive glucosamine in the presence or absence of potential inhibitors. Lipid A precursors isolated from the cells with the use of an acidic Bligh and Dyer (6) two phase partition method, as modified by Nishijima and Raetz (45), can be separated on silica gel thin layer chromatography (TLC) plates as described by Anderson and Raetz (3). If lipid A biosynthesis is inhibited in a step after the first acylation of UDP-GlcNAc, the accumulated lipid A precursor will be separated and its identity inferred by its relative migration on the TLC plate. Complete LPS and other metabolites of glucosamine such as peptidoglycan do not fractionate with the lipid A precursors in the acidic Bligh-Dyer (6) extraction. This screening method, although more labor intensive, is also able to screen large numbers of potential antibiotics in a short amount of time.

In vitro assays have been designed for each of the reactions in lipid A biosynthesis, either individually or combined (3,4,54,55). These assays may be used to positively identify the target of the antibiotic.

Because there is a common structure to the lipid A of all gram-negative bacteria studied to date, an inhibitor of lipid A biosynthesis would be effective against all strains of gram-negative bacteria. In addition, because lipid A is the causative agent of gram-negative sepsis, a lipid A-targeted antibiotic would reduce or eliminate the release of the toxin commonly found with other antibiotics (27). Even at sublethal concentrations, a lipid A antibiotic would render the cells more sensitive to the host defense mechanisms and to hydrophobic antibiotics and detergents. These qualities, along with the unique structure of lipid A, make it an excellent target for a specific antibiotic with little or no effect on the metabolism of the host.

References

1. Aasland, R., J. Coleman, A.L. Holck, C. Smith, C.R.H. Raetz, and K. Kleppe. 1988. The 17K protein, a DNA-binding protein from *Escherichia coli*, is identical to the *firA* gene product. J. Bacteriol. **170**:5916–5918.

2. Anderson, M.S., C.E. Bulawa, and C.R.H. Raetz. 1985. The biosynthesis of gram-negative endotoxin: formation of lipid A precursors from UDP-GlcNAc in extracts of *Escherichia coli*. J. Biol. Chem. **260**:15536–15541.

3. Anderson, M.S., and C.R.H. Raetz. 1987. Biosynthesis of lipid A precursors in *Escherichia coli:* A cytoplasmic acyltransferase that converts UDP-*N*-acetylglucosamine to UDP-3-O-(*R*-3-hydroxymyristoyl)-*N*-acetylglucosamine. J. Biol. Chem. **262**:5159–5169.

4. Anderson, M.S., A.D. Robertson, I. Macher, and C.R.H. Raetz. 1988. Biosynthesis of lipid A in *Escherichia coli:* Identification of UDP-3-O-[(*R*)-3-hydroxymyristoyl]-α-D-glucosamine as a precursor of UDP-N^2,O^3-*bis*[(*R*)-3-hydroxymyristoyl]-α-D-glucosamine. Biochemistry. **27**:1908–1917.

5. Benson, S.A., J.L. Occi, and B.A. Sampson. 1988. Mutations that alter the pore function of the OmpF porin of *Escherichia coli K12*. J. Mol. Biol. **203**:961–970.

6. Bligh, E., and W.J. Dyer. 1959. A rapid method of total lipid extraction and purification. Can. J. Biochem. Physiol. **37**:911–917.

7. Brozek, K.A., K. Hosaka, A.D. Robertson, and C.R. Raetz. 1989. Biosynthesis of lipopolysaccharide in *Escherichia coli:* Cytoplasmic enzymes that attach 3-deoxy-D-*manno*-octulosonic acid to lipid A. J. Biol. Chem. **264**:6956–6966.

8. Brozek, K.A., and C.R.H. Raetz. 1990. Biosynthesis of lipid A in *Escherichia coli:* Acyl carrier protein-dependent incorporation of laurate and myristate. J. Biol. Chem. **265**:15410–15417.

9. Bulawa, C.E., and C.R.H. Raetz. 1984. The biosynthesis of gram-negative endotoxin: identification and function of UDP-diacylglucosamine in *Escherichia coli*. J. Biol. Chem. **259**:4846–4851.

10. Claesson, A., A.M. Jansson, B.G. Pring, S.M. Hammond, and B. Ekström. 1987. Design and synthesis of peptide derivatives of a 3-deoxy-D-*manno*-2-octulosonic acid (KDO) analogue as novel antibacterial agents acting upon lipopolysaccharide biosynthesis. J. Med. Chem. **30**:2309–2313.

11. Clementz, T., and C.R.H. Raetz. 1991. A gene coding for 3-deoxy-D-*manno*-octulosonic-acid transferase in *Escherichia coli*: identification, mapping, cloning, and sequencing. J. Biol. Chem. **266**:9687–9696

12. Coleman, J., and C.R.H. Raetz. 1988. The first committed step of lipid A biosynthesis in *Escherichia coli:* Sequence of the *lpxA* gene. J. Bacteriol. **170**:1268–1274.

13. Coleman, J., and C.R.H. Raetz. 1988. Transcriptional analysis of the *lpx* operon of *Escherichia coli:* Evidence for *in vivo* coupling with *dnaE*. The FASEB J. **2**:a1371-a1371.

14. Cotter, R.J., J. Honovich, N. Qureshi, and K. Takayama. 1987. Structural determination of lipid A from gram negative bacteria using laser desorption mass spectrometry. Biomed. Environ. Mass. Spectrom. **14**:591–598.

15. Crowell, D.N., M.S. Anderson, and C.R.H. Raetz. 1986. Molecular cloning of the genes for lipid A disaccharide synthase and UDP-*N*-acetylglucosamine acyltransferase in *Escherichia coli*. J. Bacteriol. **168**:152–159.

16. Crowell, D.N., W.S. Reznikoff, and C.R.H. Raetz. 1987. Nucleotide sequence of the *Escherichia coli* gene for lipid A disaccharide synthase. J. Bacteriol. **169**:5727–5734.

17. Galanos, C., O. Lüderitz, E.T. Rietschel, and O. Westphal. 1977. Newer aspects of the chemistry and biology of bacterial lipopolysaccharides with special reference to their lipid A component, p. 239–335. In T.W. Goodwin (ed.), International review of biochemistry: biochemistry of lipids II, vol. 14. University Park Press, Baltimore.

18. Galloway, S.M., and C.R.H. Raetz. 1990. A mutant of *Escherichia coli* defective in the first step of endotoxin biosynthesis. J. Biol. Chem. **265**:6394–6402.

19. Goldman, R., W. Kohlbrenner, P. Lartey, and A. Pernet. 1987. Antibacterial agents specifically inhibiting lipopolysaccharide synthesis. Nature **329**:162–164.

20. Goldman, R.C., T.J. Bolling, W.E. Kohlbrenner, Y. Kim, and J.L. Fox. 1986. Primary structure of CTP:CMP-3-deoxy-D-*manno*-octulosonate cytidyly transferase (CMP-KDO synthetase) from *Escherichia coli*. J. Biol. Chem. **261**:15831–15835.

21. Goldman, R.C., C.C. Doran, S.K. Kadam, and J.O. Capobianco. 1988. Lipid A precursor from *Pseudomonas aeruginosa* is completely acylated prior to addition of 3-deoxy-D-*manno*-octulosonate. J. Biol. Chem. **263**:5217–5223.

22. Hammond, S.M., A. Claesson, A.M. Jansson, L.-G. Larsson, B.G. Pring, C.M. Town, and B. Ekström. 1987. A new class of synthetic antibacterials acting on lipopolysaccharide biosynthesis. Nature **327**:730–732.

23. Hirota, Y., H. Suzuki, Y. Nishimura, and S. Yasuda. 1977. On the process of cellular division in *Escherichia coli:* a mutant of *E. coli* lacking a murein-lipoprotein. Proc. Natl. Acad. Sci. U.S.A. **74**:1417–1420.

24. Hirvas, L., J. Coleman, P. Koski, and M. Vaara. 1990. Bacterial "histone-like protein I" (HLP-I) is an outer membrane constituent? FEBS. Lett. **262**:122–126.

25. Högenauer, G., and M. Woisetschläger. 1981. A diazaborine derivative inhibits lipopolysaccharide biosynthesis. Nature **293**:662–664.

26. Ichimura, M., T. Koguchi, T. Yasuzawa, and F. Tomita. 1987. CV-1, a new antibiotic produced by a strain of *Streptomyces* sp.: I. Fermentation, isolation and biological properties of the antibiotic. J. Antibiotics **40**:723–726.

27. Ishiguro, E.E., D. Vanderwel, and W. Kusser. 1986. Control of lipopolysaccharide biosynthesis and release by *Escherichia coli* and *Salmonella typhimurium*. J. Bacteriol. **168**:328–333.

28. Kasai, N., S. Arata, J. Mashimo, Y. Akiyama, C. Tanaka, K. Egawa, and S. Tanaka. 1987. *Pseudomonas diminuta* LPS with a new endotoxic lipid A structure. Biochem. Biophys. Res. Commun. **142**:972–978.

29. Kohlbrenner, W.E., and S.W. Fesik. 1985. Determination of the anomeric specificity of the *Escherichia coli* CTP:CMP-3-deoxy-D-*manno*-octulosonate cytidylyltransferase by ^{13}C NMR spectroscopy. J. Biol. Chem. **260**:14695–14700.

30. Kohlbrenner, W.E., M.M. Nuss, and S.W. Fesik. 1987. ^{31}P and ^{13}C NMR studies of oxygen transfer during catalysis by 3-deoxy-D-*manno*-octulosonate cytidylyltransferase from *Escherichia coli*. J. Biol. Chem. **262**:4534–4537.

31. Koski, P., M. Rhen, J. Kantele, and M. Vaara. 1989. Isolation, cloning, and primary structure of a cationic 16-kDa outer membrane protein of *Salmonella typhimurium*. J. Biol. Chem. **264**:18973–18980.

32. Krauss, J.H., U. Seydel, J. Weckesser, and H. Mayer. 1989. Structural analysis of the nontoxic lipid A of *Rhodobacter capsulatus 37b4*. Eur. J. Biochem. **180**:519–526.

33. Lancy, E.D., M.R. Lifsics, P. Munson, and R. Maurer. 1989. Nucleotide sequences of *dnaE*, the gene for the polymerase subunit of DNA polymerase III in *Salmonella typhimurium*, and a variant that facilitates growth in the absence of another polymerase subunit. J. Bacteriol. **171**:5581–5586.

34. Lathe, R., H. Buc, J.-P. Lecocq, and E.K.F. Bautz. 1980. Prokaryotic histone-like protein interacting with RNA polymerase. Proc. Natl. Acad. Sci. U.S.A. **77**:3548–3552.

35. Lathe, R., and J.-P. Lecocq. 1977. The *firA* gene, a locus involved in the expression of rifampicin resistance in *Escherichia coli* I. Characterization of lambda-*firA* transducing phages constructed *in vitro*. Molec. Gen. Genet. **154**:43–51.

36. Lehmann, V., E. Rupprecht, and M.J. Osborn. 1977. Isolation of mutants conditionally blocked in the biosynthesis of the 3-deoxy-D-*manno*-octulosonic acid—lipid A

part of lipopolysaccharides derived from *Salmonella typhimurium*. Eur. J. Biochem. **76**:41–49.

37. Lindsay, S.S., B. Wheeler, K.E. Sanderson, J.W. Costerton, and K.J. Cheng. 1973. The release of alkaline phosphatase and of lipopolysaccharide during the growth of rough and smooth strains of *Salmonella typhimurium*. Can. J. Microbiol. **19**:335–343.
38. Maurer, R., B.C. Osmond, E. Shekhtman, A. Wong, and D. Botstein. 1984. Functional interchangeability of DNA replication genes in *Salmonella typhimurium* and *Escherichia coli* demonstrated by a general complementation procedure. Genetics **108**:1–23.
39. Morrison, D.C., and J.L. Ryan. 1987. Endotoxins and disease mechanisms. Ann. Rev. Med. **38**:417–432.
40. Munson, R.S., Jr., N.S. Rasmussen, and M.J. Osborn. 1978. Biosynthesis of lipid A: Enzymatic incorporation of 3-deoxy-D-*manno*-octulosonate into a precursor of lipid A in *Salmonella typhimurium*. J. Biol. Chem. **253**:1503–1511.
41. Nikaido, H. 1979. Nonspecific transport through the outer membrane, p. 361–401. In M. Inouye (ed.), Bacterial outer membranes. John Wiley and Sons, Inc., New York.
42. Nikaido, H., and M. Vaara. 1987. Outer membrane, p. 7–22. In F.C. Neidhardt (ed.), *Escherichia coli* and *Salmonella typhimurium*: cellular and molecular biology, vol. 1. American Society for Microbiology, Washington, D.C.
43. Nishijima, M., C.E. Bulawa, and C.R.H. Raetz. 1981. Two interacting mutations causing temperature-sensitive phosphatidylglycerol synthesis in *Escherichia coli* membranes. J. Bacteriol. **145**:113–121.
44. Nishijima, M., and C.R.H. Raetz. 1979. Membrane lipid biogenesis in *Escherichia coli:* Identification of genes for phosphatidylglycerophosphate synthetase and construction of mutants lacking phosphatidylglycerol. J. Biol. Chem. **254**:7837–7844.
45. Nishijima, M., and C.R.H. Raetz. 1981. Characterization of two membrane-associated glycolipids from an *Escherichia coli* mutant deficient in phosphatidylglycerol. J. Biol. Chem. **256**:10690–10696.
46. Qureshi, N., J.P. Honovich, H. Hara, R.J. Cotter, and K. Takayama. 1988. Location of fatty acids in lipid A obtained from lipopolysaccharide of *Rhodopseudomonas sphaeroides* ATCC 17023. J. Biol. Chem. **263**:5502–5504.
47. Qureshi, N., K. Takayama, D. Heller, and C. Fenselau. 1983. Position of ester groups in the lipid A backbone of lipopolysaccharides obtained from *Salmonella typhimurium*. J. Biol. Chem. **258**:12947–12951.
48. Qureshi, N., K. Takayama, and E. Ribi. 1982. Purification and structural determination of non-toxic lipid A obtained from lipopolysaccharide of *Salmonella typhimurium*. J. Biol. Chem. **257**:11808–11815.
49. Radika, K., and C.R. Raetz. 1988. Purification and properties of lipid A disaccharide synthase of *Escherichia coli*. J. Biol. Chem. **263**:14859–14867.
50. Raetz, C.R.H. 1986. Molecular genetics of membrane phospholipid synthesis. Ann. Rev. Genet. **20**:253–295.

51. Raetz, C.R.H. 1987. Structure and biosynthesis of lipid A in *Escherichia coli*, p. 498–503. In F.C. Neidhardt (ed.), *Escherichia coli* and *Salmonella typhimurium* cellular and molecular biology, vol. 1. American Society for Microbiology, Washington, D.C.

52. Raetz, C.R.H., and J. Foulds. 1977. Envelope composition and antibiotic hypersensitivity of *Escherichia coli* mutants defective in phosphatidylserine synthetase. J. Biol. Chem. **252**:5911–5915.

53. Raetz, C.R.H., S. Purcell, M.V. Meyer, N. Qureshi, and K. Takayama. 1985. Isolation and characterization of eight lipid A precursors from a 3-deoxy-D-*manno*-octulosonic acid-deficient mutant of *Salmonella typhimurium*. J. Biol. Chem. **260**:16080–16088.

54. Ray, B.L., G. Painter, and C.R.H. Raetz. 1984. The biosynthesis of gram-negative endotoxin: formation of lipid A disaccharides from monosaccharide precursors in extracts of *Escherichia coli*. J. Biol. Chem. **259**:4852–4859.

55. Ray. B.L., and C.R.H. Raetz. 1987. The biosynthesis of gram-negative endotoxin: A novel kinase in *Escherichia coli* membranes that incorporates the 4′-phosphate of lipid A.J. Biol. Chem. **262**:1122–1128.

56. Rich, P.D., L.W. Fung, C. Ho, and M.J. Osborn. 1977. Lipid A mutants of *Salmonella typhimurium*. Purification and characterization of a lipid A precursor produced by a mutant in 3-deoxy-D-*manno*octulosonate-8-phosphate synthetase. J. Biol. Chem. **252**:4904–4912.

57. Rick, P.D., and M.J. Osborn. 1977. Lipid A mutants of *Salmonella typhimurium*. Characterization of a conditional lethal mutant in 3-deoxy-D-*manno*octulosonate-8-phosphate synthetase. J. Biol. Chem. **252**:4895–4903.

58. Rick, P.D., and D.A. Young. 1982. Relationship between cell death and altered lipid A synthesis in a temperature-sensitive lethal mutant of *Salmonella typhimurium* that is conditionally defective in 3-deoxy-D-*manno*-octulosonate-8-phosphate synthesis. J. Bacteriol. **150**:456–464.

59. Rietschel, E.T. 1984. Handbook of endotoxin, vol 1: Chemistry of endotoxin, p. 1–256. Elsevier Biomedical Press, Amsterdam.

60. Rietschel, E.T., H.W. Wollenweber, U. Zähringer, and O. Lüderitz. 1982. Lipid A, the lipid component of bacterial lipopolysaccharides: relation of chemical structure to biological activity. Klin. Wochenschr. **60**:705–709.

61. Roppel, J., H. Mayer, and J. Weckesser. 1975. Identification of a 2,3-diamino-2,3-dideoxyhexose in the lipid A component of lipopolysaccharides of *Rhodopseudomonas viridis* and *Rhodopseudomonas palustris*. Carbohydr. Res. **40**:31–40.

62. Strain, S.M., I.M. Armitage, L. Anderson, K. Takayama, N. Qureshi, and C.R.H. Raetz. 1985. Location of polar substituents and fatty acyl chains on lipid A precursors from a KDO-deficient mutant of *Salmonella typhimurium*: studies by ^{1}H, ^{13}C, and ^{31}P nuclear magnetic resonance. J. Biol. Chem. **260**:16089–16098.

63. Takayama, K., N. Qureshi, K. Hyver, J. Honovich, R.J. Cotter, P. Mascagni, and H. Schneider. 1986. Characterization of a structural series of lipid A obtained from the lipopolysaccharides of *Neisseria gonorrhoeae*. Combined laser desorption and

fast atom bombardment mass spectral analysis of high performance liquid chromatography–purified dimethyl derivatives. J. Biol. Chem. **261**:10624–10631.

64. Takayama, K., N. Qureshi, and P. Mascagni. 1983. Complete structure of lipid A obtained from the lipopolysaccharides of the heptoseless mutant of *Salmonella typhimurium*. J. Biol. Chem. **258**:12801–12803.

65. Takayama, K., N. Qureshi, P. Mascagni, M.A. Nashed, and L. Anderson. 1983. Fatty acyl derivatives of glucosamine 1-phosphate in *Escherichia coli* and their relation to lipid A. Complete structure of a diacyl GlcN-1-P found in a phosphatidylglycerol-deficient mutant. J. Biol. Chem. **258**:7379–7385.

66. Tamaki, S., T. Sato, and M. Matsuhashi. 1971. Role of lipopolysaccharides in antibiotic resistance and bacteriophase adsorption of *Escherichia coli K-12*. J. Bacteriol. **105**.968–975.

67. Tomasiewicz, H.G., and C.S. McHenry. 1987. Sequence analysis of the *Escherichia coli dnaE* gene. J. Bacteriol. **169**:5735–5744.

68. Vogler, A.P., S. Trentmann, and J.W. Lengeler. 1989. Alternative route for biosynthesis of amino sugars in *Escherichia coli K-12* mutants by means of a catabolic isomerase. J. Bacteriol. **171**:6586–6592.

69. Weckesser, J., and H. Mayer. 1988. Different lipid A types in lipopolysaccharides of phototrophic and related non-phototrophic bacteria. FEMS. Microbiol. Rev. **54**:143–154.

70. Wollenweber, H.W., K.W. Broady, O. Lüderitz, and E.T. Rietschel. 1982. The chemical structure of lipid A. Demonstration of amide-linked 3-acyloxyacyl residues in *Salmonella minnesota* Re lipopolysaccharide. Eur. J. Biochem. **124**:191–198.

71. Yasuzawa, T., M. Yoshida, M. Ichimura, K. Shirahata, and H. Sano. 1987. CV-1, a new antibiotic produced by a strain of *Streptomyces* sp;: II. Structure determination. J. Antibiotics **40**:727–731.

72. Young, L.S. 1985. Gram-negative sepsis, p. 454–474. In G.L. Mandell (ed.), Principles and practice of infectious diseases. John Wiley and Sons, New York.

11

Compromising the Protective Barrier of the Gram-Negative Bacterial Cell Surface

Catherine P. Reese and *Joanna Clancy*

I. Introduction

Antibiotics or synthetic antibacterial agents work on many targets within the bacterial cell, but their net effect is to kill the bacterium (bactericidal agents) or inhibit its growth (bacteriostatic agents). Consequently, screens to detect new drugs often utilize measures of bacterial growth in synthetic media as endpoints. However, there may exist other agents, either in nature or among the myriad collections of synthetically derived compounds, which by themselves are neither static nor cidal and which cannot be detected in standard tests for antibacterial agents. In this chapter we describe methods to detect classes of drugs of this type that may control disease by altering the surface properties of gram-negative bacterial pathogens, thereby enhancing the penetration of antibacterial drugs and/or increasing the sensitivity of the bacteria to normal host defenses.

Advances in our understanding of the physiology and biochemistry of the gram-negative bacterial cell surface suggest that this surface displays characteristics crucial to the survival of a pathogenic bacterium in a vertebrate host (31,128). In particular, this review will focus on two such aspects: the hydrophobic barrier presented by the gram-negative bacterial outer membrane and the surface properties which render an invasive bacterial pathogen resistant to the bactericidal properties of serum. Screens can be designed to detect drugs which increase membrane permeability or which decrease serum resistance. Drugs with such unconventional modes of action are likely to have some advantages over conventional antimicrobial agents. Because they affect specific procaryotic phenotypes, they are less likely to cause host toxicity. Further, they may be effective against bacteria resistant to conventional chemotherapeutic agents.

II. Structure and Function of the Gram-negative Bacterial Cell Surface

A representative structure of the gram-negative cell surface is depicted in Figure 11.1. [Aspects of the bacterial surface not relevant to the subjects of this

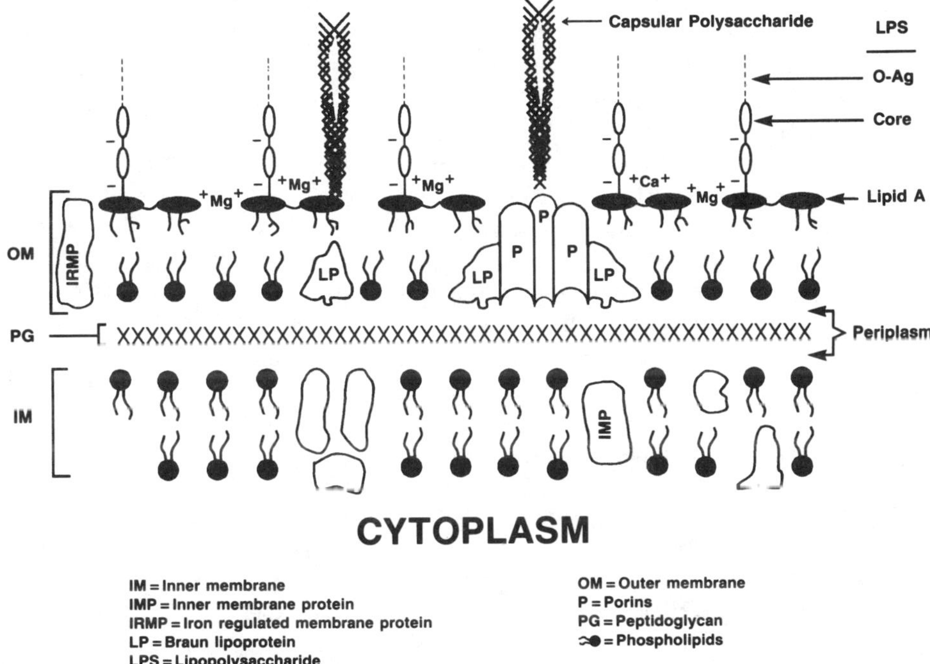

Figure 11.1 Gram-negative bacterial cell envelope. "Adapted, with permission, from the *Annual Review of Biochemistry*, Volume 59, c 1990 by Annual Reviews, Inc."

chapter have been omitted; the interested reader can find further detail in a number of recent reviews (78,86,93,101,102,115,116).] Some strains of gram-negative bacteria manufacture an external layer of polysaccharides termed capsule that is chemically distinct from the O-antigen of their lipopolysaccharide (LPS; see below), but which may be anchored by hydrophobic interactions to lipid A (78,107). Capsules may sometimes contain high-molecular-weight proteins as well; the mechanism by which such proteins adhere to the surface of the bacterium is not well understood (9). Capsules of some invasive gram-negative strains have been shown to be critical to their ability to resist the bactericidal action of serum (see section V.A).

The outer membrane (OM) lies internal to the capsule in encapsulated strains; in nonencapsulated strains it is the surface directly interacting with the bacterium's environment. It is unique among membranes in being an asymmetric bilayer composed of inner and outer leaflets of differing chemical composition (86,93,101,102). The primary component of the outermost leaflet is LPS, a complex molecule composed of lipid A, a core oligosaccharide, and an O-antigen polysaccharide component (15,115,121). Lipid A, which anchors the LPS to the cell surface, is a hydrophobic moiety generally consisting of a number of fatty acid residues linked to a disaccharide which is phosphorylated in *E. coli* at the 1 and 4′ positions (15,115,116). The core oligosaccharide is adjacent to this moiety with 3-deoxy-D-*manno*-octulosonic acid (KDO) residues attached directly to the lipid A and other sugars branching from the KDO residues. Finally, at the distal end of the molecule, the O (somatic)-antigen protrudes from the cell surface.

An important characteristic of the core oligosaccharide region of LPS is the presence of a number of negative charges due to phosphorylation of sugar residues. The negatively charged phosphates on these polysaccharide moieties and on lipid A bind divalent cations, particularly Ca^{2+} and Mg^{2+}; these interactions cross-bridge adjacent molecules and stabilize the membrane. Lack of permeability of the smooth, wild-type membrane to many hydrophobic molecules, including antibiotics, dyes, and detergents, is apparently attributable to the hydrophilic nature of the surface and to the strong LPS-LPS interactions (82,86,93,101,102). Mutants which produce defective LPS lacking portions of the O-antigen chain (rough or deep rough strains) show varying degrees of increased susceptibility to these types of molecules (see 101). Variations in LPS structure also affect the degree of sensitivity of gram-negative bacteria to immunoglobulins and specific components of the complement cascade (discussed in more detail in section V).

In addition to the unique LPS structures on the outer surface, the outer membrane also contains phospholipids, primarily confined to the inner leaflet of the enterobacterial outer membrane, and numerous proteins (93,96,101,102,108). Some important outer-membrane proteins are those that form nonspecific channels (porins) to allow entry of small hydrophilic agents into the cell and others which are involved in specific transport processes; in addition, a murein lipoprotein which anchors the membrane to the underlying peptidoglycan layer has been described (11). While all gram-negative bacteria probably possess these common outer-membrane proteins, pathogenic strains may express some novel proteins which are presumed to convey selective advantages in environments within their hosts. Various iron-regulated membrane proteins (IRMPs; Figure 11.1), for example, are important to the iron economy of bacteria within the relatively iron-poor environment of the interstitial tissues and bloodstream, while other membrane proteins are important for resistance to immunoglobulins or complement-mediated killing (see section V). The peptidoglycan and the inner membrane, a more typical phospholipid bilayer membrane whose structure and functions will not be reviewed here, are shielded from the potentially detrimental effects of the bacterium's environment by the outer membrane. A drug which affects the synthesis, assembly, or stability of the outer membrane might vitiate the vital function of this membrane as a protective, semipermeable barrier. This would allow microbicidal components of the host (*e.g.*, complement proteins) to reach the inner membrane and other host defense components (*e.g*, lysozyme) or exogenous antibacterial agents to reach previously inaccessible targets in the periplasm and cytoplasm.

The discussion which follows has been divided into two sections. The first will address structural changes to the outer membrane that alter cell permeability and will describe screening methods that may uncover compounds which would induce such changes. The subsequent section will focus on various aspects of resistance to serum cidal mechanisms and methods to find agents which can inhibit such serum resistance phenotypes. These two approaches may offer some unique opportunities to find drugs which act by novel mechanisms to alter the ability of some pathogens to cause disease.

III. Agents Affecting Outer Membrane Permeability

An agent which impairs the defensive barrier of the gram-negative outer membrane could act by a variety of mechanisms. Damage to the intact membrane surface can occur upon chelation of divalent cations, as is observed following treatment of enteric bacteria with EDTA. Under appropriate conditions, EDTA treatment leads to sloughing off of LPS and other membrane components and probably to a redistribution of membrane lipids, with "hydrophobic patches," regions of phospholipid bilayer, appearing in the outer membrane (89,101). These areas, and associated changes in membrane fluidity, could then permit entry of previously excluded hydrophobic agents, including antibiotics, dyes, fatty acids, and detergents (82). Other chelating agents shown to have similar effects on outer membrane permeability include nitrilotriacetate (42) and sodium hexametaphosphate (151). Although showing interesting activity *in vitro*, chelating agents such as these are likely to have limited chemotherapeutic application due to host toxicity.

In addition to disruption by direct chelation of divalent cations from the OM, alterations to the membrane barrier can occur by displacement of the bridging cations. This mechanism has been most thoroughly characterized for the polymyxins, and particularly for the deacylated derivative of polymyxin B, polymyxin B nonapeptide (PMBN). The polymyxins are cyclic, polycationic peptides with long fatty acyl tails and antibacterial activity against many gram-negative organisms; they bind to and permeate through the outer membrane, ultimately targeting the cytoplasmic membrane (101,137). Although potent antibacterials, these agents are also quite toxic and their use has therefore been limited. Removal of the fatty acyl tail from colistin (polymyxin E) or from polymyxin B significantly reduces toxicity and antibacterial activity of the compounds, but allows retention of the outer-membrane perturbing effects against many bacteria (54,101,154,161). These effects, most thoroughly profiled with PMBN for *E. coli* and *S. typhimurium*, include increased sensitivity to many hydrophobic antibiotics (*e.g.*, fusidic acid, rifampin, novobiocin, actinomycin D, erythromycin, and clindamycin) as well as enhanced sensitivity of serum-resistant strains to the cidal action of serum (101; see section V). In addition, physical changes in outer-membrane organization are discernible by transmission electron microscopy, which reveals protrusions of outer-membrane material from PMBN-treated cells (152). These alterations in outer-membrane structure and functions apparently result from the insertion of PMBN into the outer leaflet of the membrane by displacement of divalent cations; the binding of the large, polycationic PMBN molecules in positions which formerly held small Ca^{2+} or Mg^{2+} ions changes the overall conformation of the membrane and weakens its barrier properties. There is no apparent sloughing of LPS from PMBN-treated cells. The interested reader is referred to several recent articles that detail the physical interactions of polymyxins, including PMBN, with LPS (112,129,159).

In addition to polymyxins and EDTA, numerous other agents have been shown to alter outer-membrane structure or barrier functions by acting on the preformed membrane. These include members of the aminoglycoside (particularly for *Pseudomonas aeruginosa*; 42,85) and quinolone (12) classes of antibiotics which may gain entry to submembrane target sites by displacing divalent cations; poly-L-lysine, with greater than 20 lysine residues, that appears to act in a manner similar to EDTA (42,152,153); lactoferrin and transferrin, iron-chelating compounds which reportedly contribute to host defenses by sequestering iron from invading pathogens (see section V.A) and by damaging their protective outer-membrane armor (22,23); oligomers of compound 48/80, a polycationic histamine-releasing agent, that show membrane permeability-increasing effects (70,71); and a variety of arylamines, including phenothiazine and thioxanthine structures, used therapeutically as local anesthetics, antipsychotics, and antihistamines (13,73,75). In addition, some host factors, including protamine (152,153), defensins (35,79,81), and bactericidal/permeability-increasing factors (24), may aid in protecting the host by altering cell permeability and surface structures of invading pathogens.

The extent of sensitization of bacteria to hydrophobic or large hydrophilic molecules by agents such as those referred to above can be significant. As shown in Table 11.1, as much as 10- to >100-fold decreases in minimum inhibitory concentrations (MICs) of hydrophobic agents toward *E. coli, S. typhimurium*, and *P. aeruginosa* have been observed in *in vitro* test systems. A number of other gram-negative bacteria, particularly enteric organisms, show similar effects upon treatment with these agents (154,158). In some cases, smooth *S.typhimurium* strains which have a complete, intact LPS have been shown to be made as sensitive as deep rough (Re) mutants by treatment with PMBN (153). Similar patterns of susceptibility to serum cidal activity in the presence of PMBN have also been reported (157; see section V). There are, however, some gram-negative bacteria which are refractory to treatment with this and other polycations, including *Proteus mirabilis, P. vulgaris, Serratia marcescens*, and *Morganella morganii* (158).

In addition to agents that alter outer-membrane permeability by acting on the performed membrane, it should be feasible to find agents which affect the biosynthesis or assembly of the membrane, leading to similar defects in membrane integrity. Lipid A biosynthesis, described elsewhere in this volume (15), offers a very specific target for gram-negative-directed antibacterials. Inhibitors of the

Table 11.1. Sensitization of gram-negative bacteria to hydophobic antimicrobials by various compounds

Sensitizing compound	Hydrophobic antimicrobial[a]	Organism	Sensitization index[b]	References
PMBN	FA	*E. coli*	100	154
	FA	*S. typhimurium*	30	153
	RF	*P. aeruginosa*	100	151
Colistin nonapeptide	ER	*E. coli*	32	54
	FA	*E. coli*	16	54
Poly-L-lysine$_{20}$	FA	*E. coli*	10	153
	FA	*S. typhimurium*	10	153
Protamine	NO	*E. coli*	⩾30	153
EDTA	RF	*P. aeruginosa*	10	151
Hexameta phosphate	RF	*E. coli*	33	151
	RF	*P. aeruginosa*	10	151
Tetracaine	ER	*E. coli*	10	75
Compound 48/80	FA	*E. coli*	>100	Reese[c]

[a] FA, fusidic acid; RF, rifampin; ER, erythromycin; NO, novobiocin.

[b] Sensitization index = (MIC of hydrophobic antimicrobial alone)/(MIC of hydrophobic agent in presence of subinhibitory concentration of sensitizing compound).

[c] Unpublished observation

various steps involved in this aspect of LPS synthesis may, at subinhibitory concentrations, give rise to altered membrane structure and barrier defects (15). Likewise, the incorporation of the eight-carbon sugar KDO into LPS, a step mediated by 3-deoxy-D-*manno*-octulosonate cytidylyltransferase (CMP-KDO synthetase), offers an attractive target for altering membrane function (15,38,41). This has been clearly demonstrated for synthetic antibacterials which specifically inhibit this enzyme and result in enhanced susceptibility of cells to hydrophobic antibacterials and to serum killing (15,38). A compound known as CV-1, a fermentation-derived, weakly active antibacterial that inhibits LPS synthesis, acts synergistically with the macrolide spiramycin to inhibit *E. coli in vitro* (52). CV-1 may act at the level of lipid A deacetylase (15).

Inhibitors of fatty acid synthesis are likely to alter cell permeability by affecting membrane structure. We have observed small permeability changes with thiolactomycin, an inhibitor of the initiation reaction of fatty acid biosynthesis (55, 56) and with cerulenin, an inhibitor of fatty acid elongation (16,55) (Table 11.2). Diazaborines, synthetically derived antibacterials which affect LPS biosynthesis (15,47), also demonstrate some membrane-permeabilizing activity at subinhibitory concentrations (Table 11.2) and have been reported to sensitize *E. coli* to host defenses (77). The specific target for diazaborines has not been defined,

Table 11.2. Changes in E. coli *outer-membrane permeability induced by agents affecting fatty acid synthesis and murein lipoprotein content*[a]

Compound	Sensitization index[b]
Thiolactomycin	4
Cerulenin	2
Diazaborine[c]	4
Globomycin	15
Bicyclomycin	8

[a] MIC of fusidic acid measured in the presence of subinhibitory concentrations of test compounds; similar results obtained by using erythromycin as hydrophobic agent (C. P. Reese, unpublished results).

[b] See footnote b, Table 11.1.

[c] ICI 81,839; see ref. 3.

although a recent report (149) suggests that they target the EnvM protein in *E. coli* and *S. typhimurium* and inhibit fatty acid synthesis, thereby affecting lipid A and phospholipid biosynthesis. Finally, agents which affect steps in synthesis or assembly of the core polysaccharide of the LPS distal to KDO could presumably give rise to cells with a membrane structure comparable to that of deep rough mutants, with concomitant defects in barrier properties.

Aside from affecting LPS structure itself, compounds may alter outer-membrane integrity indirectly. For example, globomycin inhibits processing and secretion of the Braun (murein) lipoprotein which links the outer membrane to the underlying peptidoglycan (51,53). Globomycin increases membrane permeability at subinhibitory doses (Table 11.2) and enhances sensitivity to serum bactericidal activity (see section V). Bicyclomycin apparently inhibits the covalent binding of the lipoprotein to the peptidoglycan (164). Bicyclomycin-treated *E. coli* cells show membrane deformation (133,164) and also exhibit a slightly enhanced susceptibility to hydrophobic agents (Table 11.2). Notably, mutations which alter the structure or reduce the amount of this lipoprotein increase susceptibility of *E. coli* and *S. typhimurium* to cationic dyes and detergents (33,46) and to the hydrophobic antimicrobials rifampin (33) and fusidic acid (C.P. Reese, unpublished results).

Alterations in outer-membrane porins and other proteins may also cause changes in outer-membrane integrity. Porin-deficient mutants may compensate for lack of porin protein by increasing the phospholipid content of the outer membrane, thereby providing a pathway for hydrophobic agents (*e.g.*, see 86). Even slight alterations to porin structure may affect LPS-protein interactions, altering permeability (5,92). Likewise, a mutant strain of *E. coli* producing a defective outer-membrane protein, TraT, has been shown to display increased outer-membrane permeability (140,141). It is important to note that outer-membrane integrity, and indeed its proper assembly, is dependent upon interactions between LPS and OM proteins (93,101). Deep rough mutants show a significantly reduced level of OM protein and a corresponding increase in phospholipid in the outer leaflet, defects which likely contribute to the enhanced permeability of their membranes (99,101). It is possible that agents which alter LPS synthesis or structure will also affect porin/protein content of the membrane, resulting in a *decrease* in the sensitivity of an organism to some antibiotics that normally traverse the membrane via porins (101) and alter other outer-membrane-associated virulence properties (99). Finally, agents whose primary target is the inner or cytoplasmic membrane may have secondary effects on outer-membrane permeability. This may be the cause of permeabilization by some membrane-targeting antipsychotic and antihistaminic agents (see above) as well as by agents known to deenergize the inner membrane, some of which may also cause concomitant changes in protein conformation (see 101).

Alterations to the outer membrane can result in effects other than increased susceptibility to hydrophobic antibiotics or detergents, although this change is probably among the easiest to demonstrate *in vitro*. One may observe leakage of periplasmic proteins, including β-lactamase, from *E. coli* (20,46,142), or enhanced accessibility of an exogenous β-lactam to the periplasmic β-lactamase (42,80). Release of LPS by EDTA, as mentioned above, and by poly-L-lysine and protamine, has been documented (152). Sensitization to host factors, such as bile salts and lysozyme, suggests potential for *in vivo* synergistic effects, as does enhanced susceptibility to phagocytosis (136,156) and to serum cidal factors (see section V). Enhanced interaction of cells with fluorescent probes such as dansyl chloride or *N*-phenyl napthylamine is indicative of increased access to hydrophobic regions of the membrane (42,44,85,150,151). Overall, the effects that have been well documented over many years of work with rough and deep rough mutants of enteric bacteria are generally the same sort that can be observed with agents targeting the outer membrane.

IV. *In Vitro* Screening Strategies for Membrane-permeabilizing Agents

Screening for agents which will alter outer-membrane permeability can exploit any of several of the barrier properties of the normal membrane. As mentioned earlier, synergistic interactions of a membrane-disrupting agent with a hydrophobic antibiotic can be readily detected *in vitro*, and screening can be quite efficient with the aid of microtiter-based formats and turbidimetric assessment of growth. Alternatively, chromogenic β-lactams can be employed as indicators of increased accessibility to a periplasmic enzyme, β-lactamase. These two methods, with several additional possible strategies for follow-up evaluation of putative membrane-active agents, will be described in the following section.

A. *Synergy with Hydrophobic Antibiotics*

A number of authors have described turbidimetric microtiter-based methods to show synergistic interactions of a membrane-perturbing agent and a hydrophobic antibiotic, hereafter referred to as the indicator agent (54,75,151,153,158). In addition, agar-based formats have been applied (52,139) and the reader is referred to a general methodology presented by Coleman (15). In general, the methods that call for titrations of both the indicator and the permeabilizing agent to define the extent of the synergistic effect can be streamlined to an efficient screening method for detection of potentially novel permeabilizing agents. For microtiter-based testing, a single subinhibitory concentration of indicator drug can be incorporated into the growth medium and the medium can be distributed into 96-well microtiter plates by using a semiautomated or robotic plate-filling device. Likewise, the same medium without indicator can be delivered to a separate set of plates. Test samples at a single level or in varying concentrations would be added to identical positions in both plates. Following inoculation and incubation, bacterial growth can be measured by using a multichannel microtiter plate spectrophotometer; growth in wells in the presence of the indicator can then be compared to growth in wells in the absence of the indicator. A conventional antibacterial agent should inhibit growth under both conditions whereas an agent synergizing with the indicator, and therefore presumably altering outer-membrane permeability, should inhibit growth only (or to a greater extent) in the indicator-containing plate. Active compounds detected in this manner can then be further evaluated for their minimal synergistic activities (*i.e.*, the lowest concentration of permeabilizer synergizing with the indicator) and for their sensitization index (see Table 11.1), two parameters that may be useful in comparing relative potencies.

Some precautions are required in selecting a screening strain and test conditions, whether agar- or microtiter-based. As mentioned earlier, various genera and species of gram-negative bacteria respond differently to some of the known permeabilizing agents. Pathogens such as *P. aeruginosa, N. gonorrhoeae*, and *H. influenzae* have permeability properties different from *E. coli* and *S. typhimurium,* and a number of organisms (some *Proteus* strains, for example) seem to be quite resistant to the action of an agent like PMBN, presumably due to properties of their LPS which reduce the number of available sites for interaction with polycationic drugs (*e.g.*, see 101). Such strains may be desirable for screening if they are the target organisms of interest or if one wishes to focus the screen to find permeabilizing agents unrelated to the polycationic type of compounds and chelators which act at the sites of divalent cation interaction in preformed membranes.

In addition to strain considerations, medium composition and incubation temperature can influence assay sensitivity significantly (74, 97, 98, 124, 148, 151, 155). The choice of indicator agent and test level must also be carefully made: the level should be subinhibitory and poised at a concentration that offers the greatest sensitivity in detecting slight changes in permeability. Under these conditions, it is possible to pick up drug-drug interactions that are unrelated to membrane permeability changes and simply represent a combined effect of two conventional antibacterials. A simple follow-up test using an alternate indicator, preferably one with a mechanism of action different from that used for the primary screen, might prove helpful in sorting these agents from true membrane permeabilizers.

B. Chromogenic Cephalosporins

Use of a chromogenic cephalosporin to indicate increased access to periplasmic β-lactamase can provide a rapid and sensitive measure of changes in outer-membrane permeability. It has been shown that susceptibility of numerous gram-negative bacteria to β-lactam antibiotics is a combined effect of outer-membrane permeability and β-lactamase activity (100). Under standardized conditions where β-lactamase activity is not a variable, it should be possible to detect changes in outer-membrane integrity by assessing the rate or level of hydrolysis of a β-lactam substrate. Two compounds that have been successfully used to measure permeability of β-lactams are nitrocefin and pyridine-2-azo-*p*-dimethyl-aniline cephalosporin (PADAC) (42,69,80). For rapid assessment of a change in permeability of preformed membrane, suspensions of cells expressing β-lactamase in the periplasm can be distributed to microtiter plates containing test compounds and chromogenic substrate; hydrolysis of the substrate is measured and results compared with those for untreated control cells to determine if the rate or extent of hydrolysis is increased over that observed with the untreated cells.

Agents acting at the level of synthesis or assembly of the membrane might also be detectable by using the chromogenic endpoint assay. For example, cells could be incubated in the presence of test agents for a period of time sufficient to allow some cell growth and chromogenic substrate hydrolysis assessed after that period of time. If cell turbidity interferes with reading the absorbance of the hydrolyzed cephalosporin, cells could be removed by centrifugation or filtration prior to determining the endpoint. In this case, if the β-lactamase is plasmid-encoded, one should be careful to monitor that the test conditions and/or test compounds do not cause loss of the plasmid. Likewise, one must be aware of whether the β-lactamase is constitutive or inducible and ensure that testing conditions are designed to optimize expression. Assay sensitivity will likely be dependent upon the level of enzyme produced by the screening strain.

C. Evaluation of Active Agents

Follow-up testing of agents detected as having membrane-perturbing activity can be used to further define the target. Agents acting on preformed membranes *vs*. those inhibiting a biosynthetic step can be readily sorted by using cell suspensions and an endpoint measuring immediate changes in cell permeability, such as β-lactam hydrolysis (see above), susceptibility to lysozyme (42), or interaction with the hydrophobic fluorescent probe 1-*N*-phenylnaphthylamine (42,85,151).

Agents interfering with lipid A biosynthesis can be singled out by use of methods described by Coleman (15). Agents targeting the inner membrane as well as the outer membrane may be identified and evaluated by use of a dual-wavelength spectrophotometric assay described by Lehrer *et al.* (80). This assay uses a strain of *E. coli*, ML-35p, constitutive for periplasmic β-lactamase and for cytoplasmic β-galactosidase but lacking lactose permease, so substrate [*o*-nitrophenyl-β-D-galactopyranoside (ONPG)] for the latter enzyme is only hydrolyzed if the cytoplasmic membrane is permeablized. Use of conditions described (79,80) allows one to detect outer-membrane permeablity changes by measuring hydrolysis of PADAC, as described above, while simultaneously measuring inner membrane permeability changes by following ONPG hydrolysis. Finally, assessment of affects of agents on sloughing of LPS, competition with Mg^{2+} or Ca^{2+} ions, interactions of treated cells with dansylated or fluorescent hydrophobic probes, and responses of strains resistant to permeabilizing effects of polycations (*e.g.*, *pmrA* mutants of *S. typhimurium*; see 101) can be used to further characterize a permeabilizing agent's target site and possible mechanism for prioritization and follow-up studies.

V. Detection of Drugs that Inhibit Serum Resistance Phenotypes of Gram-negative Bacteria

A. Serum-resistance Phenotypes

Exposure of most strains of gram-negative bacteria to human and animal serum results in loss of viability, and sometimes lysis, of bacterial cells. By contrast, pathogenic bacteria that colonize normally sterile niches of the vertebrate host must spread through the blood and lymph and must therefore be refractory to the bactericidal components of serum. Epidemiological and experimental data suggest that serum resistance may be an important factor in generalized and some localized infections (21,49,104,122,127,144).

The blood and lymph of vertebrates contain many humoral and cellular components which contribute to killing of gram-negative bacteria. The proteins of the complement cascade, for example, constitute one important system. These proteins are arranged in two distinct pathways (C1, C4, C2 of the classical pathway and B, D, P of the alternative pathway) which produce enzymes, the C3 convertases of the classical (C14b2a) or alternative (C3bBb) pathway. Enzymatic cleavage of complement component C3 by either convertase produces the fragment, C3b, containing an activated internal thioester group which can form covalent bonds with biochemical targets on the surface of a bacterium (83,84). The chemical structure of this surface controls subsequent steps in the cascade. On a nonactivating surface, C3b binds to inhibitory factors H and I; cleavage of C5, the next step in the activation sequence, fails to occur. On an activating surface, stimulatory factor B binds to C3b in preference to factor H, initiating the production of the enzyme (C5 convertase) that catalyzes the cleavage of C5 to C5b. C5b is assembled with other terminal components into a C5b-9 membrane attack complex (MAC) which can form a pore in biological membranes. This pore permits the passage of ionized materials, and its insertion into the hydrophobic surfaces of the outer membrane of gram-negative bacteria is the first step in a complex and poorly understood process which can lead to bacterial death (67,145–147). Other bactericidal factors of the blood and lymph include lysozyme, immunoglobulins of several classes (IgM, IgA and IgG), vertebrate iron-binding proteins (transferrin and lactoferrin), and specialized cells of the immune system (monocytes, macrophages, lymphocytes, and polymorphonuclear leucocytes). Other antibacterial substances may also be present (72).

Serum bactericidal components are also present in the deep tissues of vertebrates. The structure of the capillary wall permits serum proteins to leak into the interstitial spaces. Immunoglobulins IgG and IgA, complement components, lysozyme, lactoferrin, and other serum proteins are found in deep tissues and at mucosal surfaces, but in lower concentrations than in blood. Only the largest molecules (*e.g.*, IgM immunoglobulins) are unable to transit the capillary wall. Secretory IgA is found in highest concentrations at mucosal surfaces. Moreover, when tissues are damaged or infected, the permeability of the wall increases and admits more of the soluble bactericidal components into the interstitial spaces. Leucocytes are also present in greater numbers at sites of infection or trauma (91).

Defensins, a class of related cyclic peptides found in mammalian neutrophils and lung macrophages, are presumably released from azurophilic granules of these cells into phagolysosomes or the extracellular spaces. They are active against gram-positive bacteria, gram-negative bacteria, and fungi (34,79,81). In rabbits, mRNA for another defensin, cryptidin, is localized in specialized mucosal cells (Paneth cells) at the bottom of the intestinal crypts (109). A transcriptional regulatory protein, PhoP, has been shown to affect the expression of cistrons in *Salmonella typhimurium* which convey resistance to defensins (27,39). Interestingly, Tn10-insertion mutants deficient in this phenotype were more sensitive to serum kill than their isogenic defensin-resistant parents (28).

Another group of eucaryotic antibacterial agents, the bactericidal/permeability-increasing factors (BPIs), are 50- to 60-kDa cationic proteins that kill gram-negative bacteria, but not gram-positive bacteria or fungi (see above). The BPIs are tightly associated with primary granules of human and rabbit PMNs and their bactericidal effects may be restricted to the intracellular environment (24). Recently, some of the serine proteases of neutrophils and cytotoxic T-lymphocytes have been shown to possess broad-spectrum antibacterial activity (4), but the mechanism of their antibacterial activity is independent of protease activity and still obscure. By contrast, the antibacterial properties of some enzymes, such as lysozyme, appear to be due to their catalytic activities (144).

Invasive gram-negative bacteria must employ one of two basic strategies to avoid cellular and humoral bactericidal components of the host. They may possess specific phenotypes which decrease their susceptibility to lysis or growth inhibition in serum. Alternatively, they may escape serum kill by penetrating into and multiplying within host cells. The modifications which permit the latter adaptation are largely beyond the scope of this analysis (but see 61). Bacteria which cause local infections and are also exposed to bactericidal components of the host may adopt similar strategies. While many serum resistance phenotypes are expressed at the surface of the bacterium, they extend beyond the hydrophobic barrier created by the lipid A core regions of LPS (Figure 11.1) to function largely in an aqueous milieu at the surface of the bacterium.

External capsules have been shown to enhance survival in serum, especially in nonenteric gram-negative pathogens which possess a truncated LPS as compared to enterobacterial strains. The sialic acid capsules of groups B and C *Neisseria meningitidis, E. coli* K1, and *Treponema pallidum* subsp. *pallidum*, for example, fail to activate complement by the alternative pathway in the absence of specific antibody. Although the role of the capsule is unclear, removal of sialic acid from *T. pallidum* significantly increases alternative pathway activation (32). In the erythrocyte model system, membrane-bound sialic acid is known to inhibit alternative pathway activation of C3b by increasing the binding of factor H, a known inhibitor of the complement activation pathway (25,60). The polyribose phosphate capsule of *Haemophilus influenzae* type b strains apparently cannot form a covalent bond with complement component C3 (84) while structures beneath the capsule bind C3 efficiently (135; reviewed in 60). The surface of *Campylobacter fetus* ssp. *fetus* is covered with a capsule composed of high-molecular-weight proteins that impedes the binding of C3 in the absence of specific antibody. This prevents the activation of the alternative pathway C3 convertase. As a result, such strains are resistant to lysis by MAC and to phagocytosis by polymorphonuclear leukocytes (9).

Variations in the composition of the O-antigenic portion of LPS also affect the ability of a pathogenic bacterium to survive in serum. Isogenic *Salmonella* strains which differ only in a single sugar (abequose in the virulent strain or its epimer, tyvulose, in the avirulent one) within the O-antigenic polysaccharide differ profoundly in virulence. Significantly less C3 is deposited on the virulent strain, so that it is inefficiently phagocytized by peritoneal macrophages. This alteration also protects the bacterial cells from complement-mediated lysis. At the molecular level, the O-antigen from the virulent strain binds factor B inefficiently and is therefore a poor substrate for the alternative pathway C3 convertase (57; reviewed in 60). The length of the O-antigen side chain has also been shown to affect serum resistance. For *Salmonella* and *E. coli* strains, those having more than 20% of their LPS molecules in a longer-chain form (with greater than 13 O-antigen subunits) are serum-resistant, while those with fewer than 20% of their LPS molecules in the long-chain form are serum-sensitive. Resistant strains consume as many of the terminal complement components as sensitive ones, but

MAC is not able to insert into hydrophobic domains of the outer membrane. It is hypothesized that long-chain LPS binds MAC far from its hydrophobic core, preventing formation of the pore which causes bacterial death (59,61–64; reviewed in 60).

Agents which affect the hydrophobic barrier of the LPS decrease concomitantly the ability of some species of pathogenic bacteria to survive in serum. Compounds that interact directly with LPS, such as polymyxin B, PMBN, colistin nonapeptide, or colistin heptapeptide, in combination with complement components, are able to inhibit or kill a variety of gram-negative serum-resistant bacterial strains. Strains whose LPS fails to bind PMBN are not killed by PMBN and serum (54,154,157). The cationic BPIs also work by binding to LPS (24; see section II), but they may function only within phagolysosomes of PMNs. Antibiotics or synthetic agents that inhibit the biosynthesis or assembly of the LPS barrier, such as the diazaborines, also work in synergy with complement to inhibit serum-resistant strains (38,77; J. Clancy, unpublished observations). Bicyclomycin (bicozamycin), an agent that appears to prevent the covalent binding of the Braun lipoprotein to the peptidoglycan (164), also inhibits the serum resistance of smooth *E. coli, Salmonella*, and *Pasteurella* species (J. Clancy, unpublished observations).

A number of outer-membrane proteins have been shown to augment serum resistance in gram-negative bacteria. The TraT protein (TraTp), for example, is encoded by genes on several F-like plasmids, including R100, R6–5, and F, and is known to enhance resistance of *E. coli* strains to killing by complement (30,95,117,118). A study of pathogenic *E. coli* revealed that a significantly higher percentage of enteropathogenic or enteroinvasive strains expressed TraTp than did nonpathogenic fecal isolates (8). In addition, *E. coli* strains expressing TraT proteins were shown to be less efficiently phagocytized by mouse peritoneal macrophages *in vitro* than isogenic strains lacking the protein (1). The Iss ("increased survival in serum") protein whose gene is carried on a plasmid that encodes Colicin V enhances the capacity of a smooth *E. coli* strain to survive in serum. Expression of this phenotype also increases the virulence of this strain, causing a 100-fold reduction in the LD_{50} for day-old chickens as compared to an isogenic strain lacking *iss* (6,7). Other Colicin V plasmids also carry genes which augment serum resistance (103,132). Globomycin (see section III) has been shown by us to inhibit serum resistance in *E. coli* strains whose serum resistance phenotype is the result of TraT or Iss protein expression. This inhibition of TraT or Iss phenotypes is observed in Braun lipoprotein-deficient mutants as well (J. Clancy, E.L. McCormick, and A.H. Redborg, unpublished observations). These observations suggest that the Iss and TraT precursor proteins are cleaved to mature forms by the same signal peptidase that acts on the Braun lipoprotein.

Some outer-membrane proteins may function as decoys to subvert the normal complement-mediated bactericidal mechanisms. *N. gonorrhoeae* strains causing disseminated infections express a specific outer-membrane protein (protein III) that provides an apparent selective advantage in serum. It contains antigenic configurations that are the major sites of recognition by blocking antibody. Here the situation is complicated by the fact that protein III is also present in serum-sensitive strains (143; reviewed in 58). Anti–protein III IgG found in normal human sera (120) has been shown to divert assembled MAC into "protected" sites where it fails to cause bacterial death. Such blocking antibody can compete with bactericidal antibody to inhibit serum kill (66,68,119). A second outer-membrane protein (IA) is reported to have a similar function in at least one model system. Variants of protein IA found on serum-resistant strains have been shown to bind an anti–protein I monoclonal antibody leading to insertion of MAC at locations from which it is released by trypsin digestion. It is hypothesized that such nonbactericidal MAC has not inserted effectively into the outer membrane (65). Strains of *N. gonorrhoeae* have additional mechanisms of serum resistance. For example, binding of C5b-8 to the gonococcal surface stimulates the decay of the alternative pathway C3 convertase by causing the release of factor B (19).

Other outer-membrane proteins have been shown to contribute to the ability of invasive bacteria to survive in serum by enhancing their capacity to acquire iron. The concentration of free ferric ions in the body fluids and plasma of vertebrates is extremely low. Instead, these ions are associated with eucaryotic iron-binding proteins such as transferrin, in blood and lymph, and lactoferrin, in secretions. Pathogenic bacteria from many genera must possess mechanisms of their own for assimilating protein-bound iron or acquiring it from liberated heme (162). Many bacteria are known to produce iron-chelating compounds, called siderophores, and some can utilize siderophores produced by other organisms. Bacteria in the genera *Salmonella, Escherichia*, and *Klebsiella*, for example, manufacture and secrete enterochelin (enterobactin), a phenolate siderophore which is a cyclic trimer of 2,3-dihydroxyl benzoyl serine. Its secretion appears to be essential for growth of the bacterial strain *in vivo* (123). Strains of *E. coli* growing under conditions in which iron is limiting produce a new outer-membrane protein which acts as a receptor for ferric enterochelin (90). Other siderophore-mediated systems of iron acquisition have also been described. In each case the iron-siderophore complex is removed from the external environment by an iron-regulated outer-membrane protein receptor (Figure 11.1) and the iron is transported into and utilized within the bacterium. Complex regulatory networks control the induction and expression of the structural genes for siderophore and receptor synthesis (2,17,106,138,163).

The iron-regulated membrane proteins of other genera of pathogenic bacteria are able to remove iron, not from bacterial or fungal siderophores, but from iron-binding proteins of the host. *N. meningitidis*, for example, does not make siderophores. Rather, it expresses outer-membrane receptors for human lactoferrin and transferrin. It is believed that iron is freed from the iron-transferrin-receptor complex and brought into the interior of the cell (48,130,131). To underscore the ecological significance of iron acquisition in iron-poor environments, it has been observed that many pathogenic bacteria possess multiple systems to acquire iron. *Haemophilus influenzae*, for example, makes a hydroxamate-type siderophore but can also obtain iron from transferrin and hemoglobin (45,110,113; reviewed in 87).

B. Screening Strategies for Agents that Attenuate Serum Resistance

1. Selection and Preparation of Bacterial Strains

As discussed above, diverse phenotypes may contribute to the capacity of gram-negative bacterial pathogens to survive in serum. Moreover, detailed knowledge of the serum-resistance properties of a number of gram-negative pathogens is not yet available. It is important, therefore, in designing screening methods for detection of agents which target serum-resistance phenotypes to incorporate strains with well-characterized biochemical mechanisms of resistance as well as strains whose serum-resistance phenotypes are less well understood. For the well-characterized strains, standard genetic techniques can be used to introduce genes for a number of known serum-resistance phenotypes into a common bacterial strain (*e.g.*, genes for capsules, outer-membrane proteins, and LPS). It is necessary to demonstrate that independent alterations in each of these phenotypes will result in an attenuation in serum resistance that can be detected under the conditions of the assay. Such a strain will be most useful for primary screens. For the less-studied strains, it is important to establish in advance that one or several aspects of the strain contribute to its serum resistance, perhaps by isolating single-site mutants which are demonstrably more serum sensitive than the parent yet are not altered in fundamental aspects of metabolism. The procedures used for culture maintenance and inoculum preparation and the conditions employed for screening for changes in serum resistance are known to have qualitative and/or quantitative effects on serum-resistance properties. Such effects have been comprehensively reviewed (144). Because passage of strains carrying defined serum-resistance phenotypes may cause changes in the expression of these traits, large batches of cells must be prepared for storage and their relevant phenotypes must be confirmed. The bacterial cells used in screening should be derived from frozen samples of this validated stock. For inoculum preparation, the organisms must be in the exponential phase of growth and growing with the same generation time from assay to assay (144). Finally, assay conditions must be established which allow optimal expression of serum-resistance phenotypes. A defined medium can be identified which meets this requirement and large batches of it can be prepared commercially to ensure consistency from test to test. Such precautions are necessary to minimize interassay variability.

2. Standardization of Serum Components

Serum is a heterogeneous mixture which varies in its concentration of bactericidal components. Measurements of total hemolytic activity (CH_{50} assays) of sera for sheep erythrocytes can be employed to standardize sera for complement components of the classical pathway and the membrane attack complex (50). This functional assay measures the ability of a serum sample to lyse 50% of a standard suspension of sheep erythrocytes coated with optimal amounts of rabbit antibody. The activity of the alternative pathway in conjunction with the terminal sequence can be measured by the lysis of unsensitized rabbit erythrocytes without addition of antibody. The assay is made more specific for the alternative pathway by the addition of a chelating agent such as ethylene glycol tetraacetate (with excess magnesium), which inhibits the action of complement component C1 while allowing the formation of the magnesium-dependent alternative pathway convertase (126). Immunochemical assays for individual complement proteins or protein fragments are also available (125). The relative contribution of complement-mediated lysis to bacterial kill in these assays can be assessed by preparation of a serum in which complement component C3 has been inactivated by heating the serum to 56°C. Agents which inhibit bacteria in active, but not heat-inactivated, serum would appear to be working in synergy with C3-mediated bactericidal mechanisms.

Initiation of the classical pathway of complement-mediated killing generally requires the deposition of specific antibody at sites on the bacterial surface. In addition, antibody or antibody complexes may activate serum kill *via* the alternative pathway (61). Sufficient "natural" antibody may be present to inhibit some bacterial species and strains, but hyperimmune antibody preparations may be required to inhibit others. Further, since it is well known that antibody isotypes and subtypes differ in their ability to cause complement-mediated killing (76,144), turbidimetric immunoassay and radial immunodiffusion techniques can be used to measure the concentration of each antibody class (14). Alternatively, the reaction of serum-resistant strains with known positive control substances can be used to estimate antibody concentrations. Standardization of assays can be achieved with the preparation and storage of large batches of bovine or equine sera predetermined to produce the requisite degree of serum kill in combination with known positive control substances such as PMBN.

With regard to iron sequestration phenotypes, many are expressed only under conditions in which iron is limiting. Properly prepared, serum is usually very low in free iron; total protein-bound iron can be measured by combustion and elemental analysis.

3. *Assay Methods*

Most serum resistance assays have used estimates of the number of viable bacteria as an endpoint (reviewed in 144). Estimation of viable cells by dilution and plating is a useful technique if sufficient numbers of replicates, and a large enough sample, are counted to assure statistical accuracy. In practice, viable count protocols are probably impractical for the screening of large numbers of samples, but may be useful for secondary or tertiary assays. A variation of such protocols, in which a small volume of the assay mixture is spread on agar, and significant decreases in viable cells are noted, can be used with good results (J. Clancy, unpublished observations). An additional modification, in which a Steers-Foltz replicator is used to spot small volumes of assay suspensions onto agar, has been employed (114). Assays have been described in which turbidity in liquid media is related to viable cells per unit volume (105). For such assays, it is important that the correlation between turbidity and viable cells under the conditions of the assay be established and that changes in bacterial morphology or lysis of bacteria do not obscure this correlation. This is especially important if interstrain or interspecies comparisons are being made. Alternatively, assays have been developed for some strains in which cell lysis, as measured by a decrease in the turbidity of a liquid culture, is utilized as an endpoint (37,111,160). Other assays have measured morphological changes as an index of the inhibitory effects of serum (18,36,134). Radiometric assays which measure release of ^{32}P (40), ^{51}Cr (29), or ^{14}C (165) have also been described. A rapid colorimetric assay in which bacteria are incubated with serum, medium, and a pH indicator, bromothymol blue, was utilized for the cloning of serum-resistance determinants (94). This assay could be adapted as a screen for inhibitors of serum-resistance traits only for bacterial strains that produce sufficient acid to cause a blue to yellow color change in the indicator dye. It might be difficult to adapt this assay to the screening of fermentation broths because turbid components of the broths can obscure the color change in the medium and ions in the broths may affect the pH of the test medium. Further, assays employing pH indicator dyes may not be effective for isolated fractions of fermentation broths that contain exogenous ions. Finally, for *E. coli* strains that in their viable state fail to transport β-galactosides yet express a β-galactosidase enzyme able to hydrolyze that substrate, bacterial death due to serum components can be measured by the rate of permeation and hydrolysis of the chromogenic substrate, ONPG (26,88,166). This method has also been used to detect agents that increase the permeability of the inner membrane (see section IV.B).

4. *Assay Design*

An effective screening configuration should permit the analysis of large numbers of samples (synthetic agents or fermentation broths); some methods and endpoints are more amenable to this than others (see above). An effective screen should also yield the largest possible ratio of true/false positives and display an acceptable rate of false negatives. The actual rate of false negatives cannot be measured, but the ability of such assays to detect low levels of positive control substances should support the contention that a high rate of false-negative assays will not occur. To decrease the rate of false positives, it may be useful to define assay conditions in which the action of standard antibiotics can be distinguished from that of putative serum-resistance inhibitors. This can be achieved with differential assays. For phenotypes which function solely as serum-resistance determinants, positives can be defined as agents which inhibit in the presence of serum bactericidal components, but not in their absence. For example, a primary screen in which one class of positives is defined as agents that inhibit in active, but not C3-inactivated, serum can be employed. Advance studies should establish that "standard" antibiotics from each of the representative classes are not detected as positives under assay conditions. Some bacterial strains (*e.g.*, smooth *vs.* rough strains of *E. coli*) produce fewer false positives of this kind than others (J. Clancy, unpublished observations).

Effective assay configurations will also provide some indications of the mechanism of action of a given novel antibacterial substance, perhaps even before its structure is elucidated. For example, inhibition of some serum-resistance phenotypes might be expected to affect bacterial growth in serum-free media as well. For such traits, a secondary assay could be employed in which compounds active in serum against the serum-resistant strain are determined to be active in serum-free media against the serum-resistant strain, yet inactive against its isogenic serum-sensitive variant (see above). A similar pattern might be seen for inhibitors of iron-sequestration phenotypes. Here, the compound would inhibit in active and C3-depleted serum, but would fail to inhibit under iron-sufficient conditions. A differential assay in which the bactericidal action of a test agent is measured with and without exogenous ferrous sulfate in the test medium can be used as a primary screen.

If the primary screening strain is expressing multiple serum-resistance phenotypes, secondary assays could use isogenic strains which differ only in the presence or absence of specific serum-resistance genes. The pattern of efficacy of a drug on the bacteria in these secondary assays could provide preliminary evidence of its mechanism of action. Additional assays might employ bacterial strains with related, but nonhomologous, serum-resistance determinants, such as various protein and polysaccharide capsules or unrelated iron sequestration traits, so as to assess the spectrum of activity of a possible anticapsular or anti-iron sequestration agent.

Additional aspects of the screening configuration could include assays in which the activities of agents are assessed in sera deficient in specific bactericidal components. Agents which fail to inhibit in C1-depleted (see above) sera would appear to require the efficient functioning of the classical complement pathway. Antibody-depleted (43), C8-depleted (10), and bentonite-absorbed (144) sera could similarly be used in parallel assays to identify agents which work in synergy with antibody, the terminal MAC, and lysozyme, respectively.

VI. Conclusions

Methods such as those described in this paper could lead to the discovery of novel therapeutic agents that decrease the virulence of gram-negative bacterial pathogens. Specialized surface properties of these microbes have allowed them to subvert host defenses and, in many cases, the potent antimicrobial effects of numerous classes of therapeutic antimicrobials. Agents that could interact synergistically with serum cidal factors or with exogenous antimicrobials to assist their reaching target sites interior to the outer membrane could offer powerful new therapies to combat many gram-negative bacterial infections of humans and domestic animals. As yet, the *in vivo* efficacy of such agents is a matter of speculation; there are scant reports in the literature of *in vivo* testing of agents which specifically and uniquely alter the surface properties of gram-negative bacteria. It is difficult to believe, however, that such agents would not offer significant therapeutic benefits. The vertebrate host displays an elaborate, coordinated cascade of defenses against an invading pathogen; an attenuation of the virulence of the pathogen could increase its vulnerability just enough to permit host defenses to eliminate the pathogen from its parasitic niche. Changes in membrane hydrophobicity and exposure of previously shielded antigenic sites could increase susceptibility to serum killing and to phagocytosis. Alterations of outer-membrane proteins, including porins, could disrupt the flow of nutrients into the cell, compromising growth. Loss of essential periplasmic enzymes following membrane perturbation could likewise affect the dynamics of transport of molecules into and out of the cell and alter a number of aspects of cellular growth and metabolism. Moreover, the possibility of circumventing some pathogens' resistance to commonly used antimicrobial agents by permitting a veritable flood of the compound to reach intracellular targets offers hope for expanded use of drugs already proven safe for therapeutic application. The challenge is to exploit new approaches to drug discovery to find agents that will "kill" *in vivo* in these novel ways.

References

1. Aguero, M.E., L. Aron, A.G. DeLuca, K.N. Timmis, and F.C. Cabello. 1984. A plasmid-encoded outer membrane protein, TraT, enhances resistance of *Escherichia coli* to phagocytosis. Infect. Immun. **46**:740–746.

2. Bagg, A., and J.B. Neilands. 1987. Molecular mechanism of regulation of siderophore-mediated iron assimilation. Microbiol. Rev. **51**:509–518.

3. Bailey, P.J., G. Cousins, G.A. Snow, and A.J. White. 1980. Boron-containing antibacterial agents: effects on growth and morphology of bacteria under various culture conditions. Antimicrob. Agents Chemother. **17**:549–553.

4. Bangalore, N., J. Travis, V.C. Onunka, J. Pohl, and W.M. Shafer. 1990. Identification of the primary antimicrobial domains in human neutrophil cathepsin G. J. Biol. Chem. **265**:13584–13588.

5. Benson, S.A., J.L.L. Occi, and B.A. Sampson. 1988. Mutations that alter the pore function of the OmpF porin of *Escherichia coli* K-12. J. Mol. Biol. **203**:961–970.

6. Binns, M.M., D.L. Davies, and K.G. Hardy. 1979. Cloned fragments of the plasmid ColV,I-K94 specifying virulence and serum resistance. Nature (London) **279**:778–781.

7. Binns, M.M., J. Mayden, and R.P. Levine. 1982. Further characterization of complement resistance conferred on *Escherichia coli* by the plasmid genes *traT* of R100 and *iss* of ColV,I-K94. Infect. Immun. **35**:654–659.

8. Bitter-Sauermann, D., H. Peters, M. Jurs, R. Nehrbass, M. Montenegro, and K.N. Timmis. 1984. Monoclonal antibody detection of IncF group plasmid-encoded TraT protein in clinical isolates of *Escherichia coli*. Infect. Immun. **46**:308–313.

9. Blaser, M.J., P.F. Smith, J.E. Repine, and K.A. Joiner. 1988. Pathogenesis of *Campylobacter fetus* infections: failure of encapsulated *Campylobacter fetus* to bind C3b explains serum and phagocytosis resistance. J. Clin. Invest. **81**:1434–1444.

10. Bloch, E.F., M.A. Schmetz, J. Foulds, C.H. Hammer, M.M. Frank, and K.A. Joiner. 1987. Multimeric C9 within C5b-9 is required for inner membrane damage to *Escherichia coli* J5 during complement killing. J. Immunol. **138**:842–848.

11. Braun, V. 1975. Covalent lipoprotein from the outer membrane of *Escherichia coli*. Biochim. Biophys. Acta **415**:335–377.

12. Chapman, J.S., and N.H. Georgopapadakou. 1988. Routes of quinolone permeation in *Escherichia coli*. Antimicrob. Agents Chemother. **32**:438–442.

13. Chattopadhyay, D., S.G. Dastidar, and A.N. Chakrabarty. 1988. Antimicrobial properties of methdilazine and its synergism with antibiotics and some chemotherapeutic agents. Arzneim.-Forsch./Drug Res. **38 (II)**:869–872.

14. Check, I.J., and M. Piper. 1986. Quantitation of immunoglobulins, p. 138–151. *In* N.R. Rose, H. Friedman, and J.L. Fahey (ed.), Manual of clinical laboratory immunology. American Society for Microbiology, Washington, D.C.

15. Coleman, J. 1991. Lipid A biosynthesis in gram-negative bacteria, p. 274–291. *In* J.A. Sutcliffe and N.H. Georgopapadakou (ed.), Emerging targets in antibacterial and antifungal chemotherapy. Chapman and Hall, New York.

16. Cronan, J.E., Jr., and C.O. Rock. 1987. Biosynthesis of membrane lipids, p. 474–497. *In* F.C. Neidhardt (ed.), *Escherichia coli* and *Salmonella typhimurium*. Cellular and molecular biology, vol. 1. American Society for Microbiology, Washington, D.C.

17. Crosa, J.H. 1989. Genetics and molecular biology of siderophore-mediated iron transport in bacteria. Microbiol. Rev. **53**:517–530.

18. Davis, S.D., E.S. Boatman, D. Gemsa, A. Iannetta, and R.J. Wedgwood. 1969. Biochemical and fine structure changes induced in *Escherichia coli* by human serum. Microbios **1B**:69–86.

19. Denson, P., and C.M. McRill. 1987. C5b-8 promotes alternative pathway C3 convertase decay on *Neisseria*. Fed. Proc. **46**:1196.

20. Dixon, R.A., and I. Chopra. 1986. Leakage of periplasmic proteins from *Escherichia coli* mediated by polymyxin B nonapeptide. Antimicrob. Agents Chemother. **29**:781–788.

21. Durack, D.T., and P.B. Beeson. 1977. Protective role of complement in experimental *Escherichia coli* endocarditis. Infect. Immun. **16**:213–217.

22. Ellison, R.T., III, T.J. Giehl, and F.M. LaForce. 1988. Damage of the outer membrane of enteric gram-negative bacteria by lactoferrin and transferrin. Infect. Immun.**56**:2774–2781.

23. Ellison, R.T., III, F.M. LaForce, T.J. Giehl, D.S. Boose, and B.E. Dunn. 1990. Lactoferrin and transferrin damage of the gram-negative outer membrane is modulated by Ca^{2+} and Mg^{2+}. J. Gen. Microbiol. **136**:1437–1446.

24. Elsbach, P., and J. Weiss. 1988. Phagocytic cells: oxygen-independent antimicrobial systems, p. 445–470. *In* J.I. Gallin, I.M. Goldstein and R. Snyderman (ed.), Inflammation: basic principles and clinical correlates. Raven Press, Ltd., New York.

25. Fearon, D.T. 1978. Regulation by membrane sialic acid of B1H-dependent decay-dissolution of amplification C3 convertase of the alternative complement pathway. Proc. Natl. Acad. Sci. USA **75**:1971–1975.

26. Feingold, D.S., J.N. Goldman, and H.M. Kuritz. 1968. Locus of the lethal event in the serum bactericidal reaction. J. Bacteriol. **96**:2127–2131.

27. Fields, P.I., E.A. Groisman, and F. Heffron. 1989. A *Salmonella* locus that controls resistance to microbicidal proteins from phagocytic cells. Science **243**:1059–1062.

28. Fields, P.I., R.V. Swanson, C.G. Haidaris, and F. Heffron. 1986. Mutants of *Salmonella typhimurium* that cannot survive within the macrophage are avirulent. Proc. Natl. Acad. Sci. USA **83**:5189–5193.

29. Fierer, J., F. Finley, and A.I. Braude. 1974. Release of ^{51}Cr-endotoxin from bacteria as an assay of serum bactericidal activity. J. Immunol. **112**:2184–2192.

30. Fietta, A., E. Romero, and A.G. Siccardi. 1977. Effect of some R factors on the sensitivity of rough *Enterobacteriaceae* to human serum. Infect. Immun. **18**:278–282.

31. Finlay, B.B., and S. Falkow. 1989. Common themes in microbial pathogenicity. Microbiol. Rev. **53**:210–230.

32. Fitzgerald, T.J. 1987. Activation of the classical and alternative pathways of complement by *Treponema pallidum* subsp. *pallidum* and *Treponema vincentii*. Infect. Immun. **55**:2066–2073.

33. Fung, J., T.J. MacAlister, and L.I. Rothfield. 1978. Role of murein lipoprotein in morphogenesis of the bacterial division septum: phenotypic similarity of *lkyD* and *lpo* mutants. J. Bacteriol. **133**:1467–1471.

34. Gabay, J.E., R.W. Scott, D. Campanelli, J. Griffith, C. Wilde, M.N. Marra, M. Seeger, and C.F. Nathan. 1989. Antibiotic proteins of human polymorphonuclear leukocytes. Proc. Natl. Acad. Sci. USA **86**:5610–5615.

35. Ganz, T., M.E. Selsted, and R.I. Lehrer. 1990. Defensins. Eur. J. Haematol. **44**:1–8.

36. Gemsa, D., S.D. Davis, and R.J. Wedgwood. 1966. Lysozyme and serum bactericidal action. Nature (London) **210**:950–951.

37. Glynn, A.A., and C.M. Milne. 1967. A kinetic study of the bacteriolytic and bactericidal action of human serum. Immunology **12**:639–653.

38. Goldman, R., W. Kohlbrenner, P. Lartey, and A. Pernet. 1987. Antibacterial agents specifically inhibiting lipopolysaccharide synthesis. Nature (London) **329**:162–164.

39. Groisman, E.A., and M.H. Saier, Jr. 1990. *Salmonella* virulence: new clues to intramacrophage survival. TIBS **15**:30–33.

40. Hall, W.H., R.E. Manion, and H.H. Zinneman. 1971. Blocking serum lysis of *Brucella abortus* by hyperimmune rabbit immunoglobulin A. J. Immunol. **107**:41–46.

41. Hammond, S.M., A. Claesson, A.M. Jansson, L.-G. Larsson, B.G. Pring, C.M. Town, and B. Ekstrom. 1987. A new class of synthetic antibacterials acting on lipopolysaccharide biosynthesis. Nature (London) **327**:730–732.

42. Hancock, R.E.W., and P.G.W. Wong. 1984. Compounds which increase the permeability of the *Pseudomonas aeruginosa* outer membrane. Antimicrob. Agents Chemother. **26**:48–52.

43. Harlow, E., and D. Lane. 1988. Antibodies: a laboratory manual. Cold Spring Harbor Laboratory, Cold Spring Harbor, NY.

44. Helgerson, S.L., and W.A. Cramer. 1977. Changes in *Escherichia coli* cell envelope structure and the sites of fluorescence probe binding caused by carbonyl cyanide *p*-trifluoromethoxyphenylhydrazone. Biochemistry **16**:4109–4117.

45. Herrington, D.A., and P.F. Sparling. 1985. *Haemophilus influenzae* can use human transferrin as a sole source for required iron. Infect. Immun. **48**:248–251.

46. Hirota, Y., H. Suzuki, Y. Nishimura, and S. Yasuda. 1977. On the process of cellular division in *Escherichia coli*: a mutant of *E. coli* lacking a murein-lipoprotein. Proc. Natl. Acad. Sci. USA **74**:1417–1420.

47. Hogenauer, G., and M. Woisetschlager. 1981. A diazaborine derivative inhibits lipopolysaccharide biosynthesis. Nature (London) **293**:662–664.

48. Holbein, B.E. 1981. Enhancement of *Neisseria meningitidis* infection in mice by addition of iron bound to transferrin. Infect. Immun. **33**:120–125.

49. Howard, C.J., and A.A. Glynn. 1971. The virulence for mice of strains of *Escherichia coli* related to the effects of K antigens on their resistance to phagocytosis and killing by complement. Immunology **20**:767–777.

50. Humphrey, J.H., R.R. Dourmashkin, and S.N. Payne. 1967. The nature of lesions in cell membranes produced by action of C′ and antibody, p. 209–220. *In* P.A. Miescher and P. Grabar (ed.), Mechanisms of inflammation induced by immune reactions. Schwabe, Basel.

51. Hussain, M., S. Ichihara, and S. Mizushima. 1980. Accumulation of glyceride-containing precursor of the outer membrane lipoprotein in the cytoplasmic membrane of *Escherichia coli* treated with globomycin. J. Biol. Chem. **255**:3707—3712.

52. Ichimura, M., T. Koguchi, T. Yasuzawa, and F. Tomita. 1987. CV-1, a new antibiotic produced by a strain of *Streptomyces* sp. I. Fermentation, isolation and biological properties of the antibiotic. J. Antibiotics **40**:723–726.

53. Inukai, M., M. Takeuchi, K. Shimizu, and M. Arai. 1978. Mechanism of action of globomycin. J. Antibiotics **31**:1203–1205.

54. Ito-Kagawa, M., and Y. Koyama. 1984. Studies on the selectivity of action of colistin, colistin nonapeptide and colistin heptapeptide on the cell envelope of *Escherichia coli*. J. Antibiotics **37**:926–928.

55. Jackowski, S. 1991. Fatty acid biosynthesis, p. 152–163. *In* J.A. Sutcliffe and N.H. Georgopapadakou (ed.), Emerging targets in antibacterial and antifungal chemotherapy. Chapman and Hall, New York.

56. Jackowski, S., C.M. Murphy, J.E. Cronan, Jr., and C.O. Rock. 1989. Acetoacetyl-acyl carrier protein synthase. A target for the antibiotic thiolactomycin. J. Biol. Chem. **264**:7624–7629.

57. Jimenez-Lucho, V.E., K.A. Joiner, J. Foulds, M.M. Frank, and L. Leive. 1987. C3b generation is affected by the structure of the O-antigen polysaccharide in lipopolysaccharide from *Salmonellae*. J. Immunol. **139**:1253–1259.

58. Joiner, K.A. 1985. Studies on the mechanism of bacterial resistance to complement-mediated killing and on the mechanism of action of bactericidal antibody. Curr. Top. Microbiol. Immunol. **121**:99–133.

59. Joiner, K.A. 1988. Mechanisms of bacterial resistance to complement-mediated killing, p. 15–30. *In* M.A. Horwitz (ed.), Bacteria-host cell interaction. Alan R. Liss, Inc., New York.

60. Joiner, K.A. 1988. Complement evasion by bacteria and parasites. Annu. Rev. Microbiol. **42**:201–230.

61. Joiner, K.A., E.J. Brown, and M.M. Frank. 1984. Complement and bacteria: chemistry and biology in host defense. Annu. Rev. Immunol. **2**:461–491.

62. Joiner, K.A., N. Grossman, M. Schmetz, and L. Leive. 1986. C3 binds preferentially to long-chain lipopolysaccharide during alternative pathway activation by *Salmonella montevideo*. J. Immunol. **136**:710–715.

63. Joiner, K.A., C.H. Hammer, E.J. Brown, R.J. Cole, and M.M. Frank. 1982. Studies on the mechanism of bacterial resistance to complement-mediated killing. I. Terminal complement components are deposited and released from *Salmonella minnesota* S218 without causing bacterial death. J. Exp. Med. **155**:797–804.

64. Joiner, K.A., C.H. Hammer, E.J. Brown, and M.M. Frank. 1982. Studies on the mechanism of bacterial resistance to complement-mediated killing. II. C8 and C9 release C5b67 from the surface of *Salmonella minnesota* S218 because the terminal complex does not insert into the bacterial outer membrane. J. Exp. Med. **155**:809–819.

65. Joiner, K.A., S.M. Puentes, K.A. Warren, R.A. Scales, and R.C. Judd. 1989. Complement binding of serum-sensitive and serum-resistant transformants of *Neisseria gonorrhoeae*: effect on presensitization with a non-bactericidal monoclonal antibody. Microb. Pathog. **6**:343–350.

66. Joiner, K.A., R. Scales, K.A. Warren, M.M. Frank, and P.A. Rice. 1985. Mechanism of action of blocking immunoglobulin G for *Neisseria gonorrhoeae*. J. Clin. Invest. **76**:1765–1772.

67. Joiner, K.A., M.A. Schmetz, M.E. Sanders, T.G. Murray, C.H. Hammer, R. Dourmashkin, and M.M. Frank. 1985. Multimeric complement component C9 is necessary for killing of *Escherichia coli* J5 by terminal attack complex C5b-9. Proc. Natl. Acad. Sci. USA **82**:4808–4812.

68. Joiner, K.A., K.A. Warren, C. Hammer, and M.M. Frank. 1985. Bactericidal but not nonbactericidal C5b-9 is associated with distinctive outer membrane proteins in *Neisseria gonorrhoeae*. J. Immunol. **134:**1920–1925.

69. Jones, R.N., H.W. Wilson, and W.J. Novick, Jr. 1982. In vitro evaluation of pyridine-2-azo-*p*-dimethylaniline cephalosporin, a new diagnostic chromogenic reagent, and comparison with nitrocefin, cephacetrile, and other beta-lactam compounds. J. Clin. Microbiol. **15**:677–683.

70. Kastu, T., M. Shibata, and Y. Fujita. 1985. Dication and trication which can increase the permeability of *Escherichia coli* outer membrane. Biochim. Biophys. Acta **818**:61–66.

71. Katsu, T., M. Shibata, and Y. Fujita. 1985. Increased permeability of the outer membrane of *Escherichia coli* induced by the dimer in compound 48/80. Chem. Pharm. Bull. **33**:893–895.

72. Kawakami, M., I. Ihara, A. Suzuki, and K. Fukui. 1984. A group of bactericidal factors conserved by vertebrates for more than 300 million years. J. Immunol. **132**:2578–2581.

73. Kristiansen, J.E. 1990. The antimicrobial activity of psychotherapeutic drugs and stereo-isomeric analogues. Danish Medical Bulletin **37**:165–182.

74. Kropinski, A.M.B., V. Lewis, and D. Berry. 1987. Effect of growth temperature on the lipids, outer membrane proteins, and lipopolysaccharides of *Pseudomonas aeruginosa* PAO. J. Bacteriol. **169**:1960–1966.

75. Labedan, B. 1988. Increase in permeability of *Escherichia coli* outer membrane by local anesthetics and penetration of antibiotics. Antimicrob. Agents Chemother. **32**:153–155.

76. Lachmann, P.J. 1979. Complement, p. 283–335. *In* M. Sela (ed.), The antigens, vol 5. Academic Press, Inc., New York.

77. Lam, C., F. Turnowsky, G. Hogenauer, and E. Schutze. 1987. Effect of a diazaborine derivative (Sa 84.474) on the virulence of *Escherichia coli*. J. Antimicrob. Chemother. **20**:37–45.

78. Lambert, P.A. 1988. *Enterobacteriaceae*: composition, structure and function of the cell envelope. J. Appl. Bacteriol. Symposium Supplement. 21S–34S.

79. Lehrer, R.I., A. Barton, K.A. Daher, S.S.L. Harwig, T. Ganz, and M.E. Selsted. 1989. Interaction of human defensins with *Escherichia coli*. Mechanism of bactericidal activity. J. Clin. Invest. **84**:553–561.

80. Lehrer, R.I., A. Barton, and T. Ganz. 1988. Concurrent assessment of inner and outer membrane permeabilization and bacteriolysis in *E. coli* by multiple-wavelength spectrophotometry. J. Immunol. Meth. **108**:153–158.

81. Lehrer, R.I., T. Ganz, and M.E. Selsted. 1990. Defensins: natural peptide antibiotics from neutrophils. ASM News **56**:315–318.

82. Leive, L. 1974. The barrier function of the gram-negative envelope. Ann. N.Y. Acad. Sci. **235**:109–127.

83. Levine, R.P. 1985. Molecular interaction between the third complement protein and the bacterial cell-surface macromolecules, p. 102–120. *In* Bayer-Symposium VIII. The pathogenesis of bacterial infections. Springer-Verlag, Berlin and Heidelberg.

84. Levine, R.P., R. Finn, and R. Gross. 1983. Interactions between C3b and cell-surface macromolecules. Ann. N.Y. Acad. Sci. **421**:235–245.

85. Loh, B., C. Grant, and R.E.W. Hancock. 1984. Use of the fluorescent probe 1-*N*-phenylnaphthylamine to study the interactions of aminoglycoside antibiotics with the outer membrane of *Pseudomonas aeruginosa*. Antimicrob. Agents Chemother. **26**:546–551.

86. Lugtenberg, B., and L. Van Alphen. 1983. Molecular architecture and functioning of the outer membrane of *Escherichia coli* and other gram-negative bacteria. Biochim. Biophys. Acta **737**:51–115.

87. Martinez, J.L., A. Delgado-Iribarren, and F. Baquero. 1990. Mechanisms of iron acquisition and bacterial virulence. FEMS Microbiol. Rev. **75**:45–56.

88. Martinez, R.J., and S.F. Carroll. 1980. Sequential metabolic expressions of the lethal process in human serum-treated *Escherichia coli*: role of lysozyme. Infect. Immun. **28**:735–745.

89. Marvin, H.J.P., M.B.A. ter Beest, D. Hoekstra, and B. Witholt. 1989. Fusion of small unilamellar vesicles with viable EDTA-treated *Escherichia coli* cells. J. Bacteriol. **171**:5268–5275.

90. McIntosh, M.A., and C.F. Earhart. 1977. Coordinate regulation by iron of the synthesis of phenolate compounds and three outer membrane proteins in *Escherichia coli*. J. Bacteriol. **131**:331–339.

91. Mims, C.A. 1987. The pathogenesis of infectious disease. Academic Press, London.

92. Misra, R., and S.A. Benson. 1988. Genetic identification of the pore domain of the OmpC porin of *Escherichia coli* K-12. J. Bacteriol. **170**:3611–3617.

93. Mizushima, S. 1985. Structure, assembly, and biogenesis of the outer membrane, p. 39–75. *In* N. Nanninga (ed.), Molecular cytology of *Escherichia coli*. Academic Press, London.

94. Moll, A., F. Cabello, and K.N. Timmis. 1979. Rapid assay for the determination of bacterial resistance to the lethal activity of serum. FEMS Microbiol. Lett. **6**:273–276.

95. Moll, A., P.A. Manning, and K.N. Timmis. 1980. Plasmid-determined resistance to serum bactericidal activity: a major outer membrane protein, the *traT* gene product, is responsible for plasmid-specified serum resistance in *Escherichia coli*. Infect. Immun. **28**:359–367.

96. Nakae, T. 1986. Outer-membrane permeability of bacteria. CRC Crit. Rev. Microbiol. **13**:1–62.

97. Nicas, T.I., and R.E.W. Hancock. 1983. Alteration of susceptibility to EDTA, polymyxin B and gentamicin in *Pseudomonas aeruginosa* by divalent cation regulation of outer membrane protein H1. J. Gen. Microbiol. **129**:509–517.

98. Nikaido, H. 1976. Outer membrane of *Salmonella typhimurium*. Transmembrane diffusion of some hydrophobic substances. Biochim. Biophys. Acta **433**:118–132.

99. Nikaido, H., and T. Nakae. 1979. The outer membrane of gram-negative bacteria. Adv. Microb. Physiol. **21**:163–250.

100. Nikaido, H., and S. Normark. 1987. Sensitivity of *Escherichia coli* to various β-lactams is determined by the interplay of outer membrane permeability and degradation by periplasmic β-lactamases: a quantitative predictive treatment. Mol. Microbiol. **1**:29–36.

101. Nikaido, H., and M. Vaara. 1985. Molecular basis of bacterial outer membrane permeability. Microbiol. Rev. **49**:1–32.

102. Nikaido, H., and M. Vaara. 1987. Outer membrane, p. 7–22. *In* F.C. Neidhardt (ed.), *Escherichia coli* and *Salmonella typhimurium*. Cellular and molecular biology, vol. 1. American Society for Microbiology, Washington, D.C.

103. Nilius, A., and D.C. Savage. 1984. Serum resistance encoded by Colicin V plasmids in *Escherichia coli* and its relationship to the plasmid transfer system. Infect. Immun. **43**:947–953.

104. Ogata, R.T., C. Winters, and R.P. Levin. 1982. Nucleotide sequence analysis of the complement resistance gene from plasmid R100. J. Bacteriol. **151**:819–827.

105. Olling, S. 1977. Sensitivity of gram-negative bacilli to the serum bactericidal activity: a marker of the host-parasite relationship in acute and persisting infections. Scand. J. Infect. Dis. Suppl. **10**:1–40.

106. Opal, S.M., A.S. Cross, P. Gemski, and L.W. Lyhte. 1990. Aerobactin and α-hemolysin as virulence determinants in *Escherichia coli* isolated from human blood, urine, and stool. J. Infect. Dis. **161**:794–796.

107. Orskov, F., and I. Orskov. 1990. The serology of capsular antigens. Curr. Top. Microbiol. Immunol. **150**:43–63.

108. Osborn, M.J., and H.C.P. Wu. 1980. Proteins of the outer membrane of gram-negative bacteria. Annu. Rev. Microbiol. **34**:369–422.

109. Ouellette, A.J., R.M. Greco, M. James, D. Frederick. J. Naftilan, and J.T. Fallon. 1989. Developmental regulation of cryptdin, a corticostatin/defensin precursor mRNA in mouse small intestinal crypt epithelium. J. Cell Biol. **108**:1687–1695.

110. Payne, S.M., and R.A. Finkelstein. 1978. The critical role of iron in host-bacterial interactions. J. Clin. Invest. **61**:1428–1440.

111. Pelkonen, S., and J. Finne. 1987. A rapid turbidimetric assay for the study of serum sensitivity of *Escherichia coli*. FEMS Microbiol. Lett. **42**:53–57.

112. Peterson, A.A., S.W. Fesik, and E.J. McGroarty. 1987. Decreased binding of antibiotics to lipopolysaccharides from polymyxin-resistant strains of *Escherichia coli* and *Salmonella typhimurium*. Antimicrob. Agents Chemother. **31**:230–237.

113. Pidcock, K.A., J.A. Wooten, B.A. Daley, and T.L. Stull. 1988. Iron acquisition by *Haemophilus influenzae*. Infect. Immun. **56**:721–725.

114. Provonchee, R.B., and S.H. Zinner. 1974. Rapid method for determining serum bactericidal activity. Appl. Microbiol. **27**:185–186.

115. Raetz, C.R.H. 1987. Structure and biosynthesis of lipid A in *Escherichia coli,* p. 498–503. *In* F.C. Neidhardt (ed.), *Escherichia coli* and *Salmonella typhimurium*. Cellular and molecular biology, vol. 1. American Society for Microbiology, Washington, D.C.

116. Raetz, C.R.H. 1990. Biochemistry of endotoxins. Annu. Rev. Biochem. **59**:129–170.

117. Reynard, A.M., and M.E. Beck. 1976. Plasmid-mediated resistance to the bactericidal effects of normal rabbit serum. Infect. Immun. **14**:848–850.

118. Reynard, A.M., M.E. Beck, and R.K. Cunningham. 1978. Effects of antibiotic resistance plasmids on the bactericidal activity of normal rabbit serum. Infect. Immun. **19**:861–866.

119. Rice, P.A., M.S. Blake, and K.A. Joiner. 1987. Mechanisms of stable serum resistance of *Neisseria gonorrhoeae*. Antonie van Leeuwenhoek **53**:565–574.

120. Rice, P.A., and D.L. Kasper. 1982. Characterization of serum resistance of *Neisseria gonorrhoeae* that disseminate. J. Clin. Invest. **70**:157–167.

121. Rick, P.D. 1987. Lipopolysaccharide biosynthesis, p. 648–662. *In* F.C. Neidhardt (ed.), *Escherichia coli* and *Salmonella typhimurium*. Cellular and molecular biology, vol. 1. American Society for Microbiology, Washington, D.C.

122. Roantree, R.J., and L.A. Rantz. 1960. A study of the relationship of the normal bactericidal activity of human serum to bacterial infection. J. Clin. Invest. **39**:72–81.

123. Rogers, H.J. 1973. Iron-binding catechols and virulence in *Escherichia coli*. Infect. Immun. **7**:445–456.

124. Rottem, S., O. Markowitz, and S. Razin. 1978. Thermal regulation of the fatty acid composition of lipopolysaccharides and phospholipids of *Proteus mirabilis*. Eur. J. Biochem. **85**:445–450.

125. Ruddy, S. 1985. Complement, p. 137–162. *In* A.S. Cohen (ed.), Laboratory diagnostic procedures in the rheumatic diseases. Grune & Stratton, Orlando, FL.

126. Ruddy, S. 1986. Complement, p. 175–184. *In* N.R. Rose, H. Friedman, and J.L. Fahey (ed.), Manual of clinical laboratory immunology. American Society for Microbiology, Washington, D.C.

127. Rycroft, A.N., G.L. Thompson, and S.M. Hammond. 1983. The role of cell surface polysaccharide antigens in the pathogenicity of *Escherichia coli*. FEMS Microbiol. Lett. **18**:49–53.

128. Savage, D.C. 1980. Colonization by and survival of pathogenic bacteria on intestinal mucosal surfaces, p. 175–206. *In* G. Bitton, and K.C. Marshall (ed.), Adsorption of microorganisms to surfaces. John Wiley & Sons, Inc., New York.

129. Schindler, M., and M.J. Osborn. 1979. Interaction of divalent cations and polymyxin B with lipopolysaccharide. Biochemistry. **18**:4425–4430.

130. Schryvers, A.B., and L.J. Morris. 1988. Identification and characterization of the human lactoferrin-binding protein from *Neisseria meningitidis*. Infect. Immun. **56**:1144–1149.

131. Schryvers, A.B., and L.J. Morris. 1988. Identification and characterization of the transferrin receptor from *Neisseria meningitidis*. Mol. Microbiol. **2**:281–288.

132. Smith, H.W. 1974. A search for transmissible pathogenic characters in invasive strains of *Escherichia coli*: the discovery of a plasmid-controlled toxin and a plasmid-controlled lethal character closely associated, or identical, with Colicin V. J. Gen. Microbiol. **83**:95–111.

133. Someya, A., K. Tanaka, and N. Tanaka. 1979. Morphological changes of *Escherichia coli* induced by bicyclomysin. Antimicrob. Agents Chemother. **16**:84–88.

134. Spitznagel, J.K., and L.A. Wilson. 1966. Normal serum cytotoxicity for P^{32}-labeled smooth *Enterobacteriaceae*. I. Loss of label, death, and ultrastructural damage. J. Bacteriol. **91**:393–400.

135. Steele, N.P., R.S. Munson, Jr., D.M. Granoff, J.E. Cummins, and R.P. Levine. 1984. Antibody-dependent alternative pathway killing of *Haemophilus influenzae* type b. Infect. Immun. **44**:452–458.

136. Stendahl, O. 1983. The physicochemical basis of surface interaction between bacteria and phagocytic cells, p. 137–152. *In* C.S.F. Easmon, J. Jeljaszewicz, M.R.W. Brown, and P.A. Lambert (ed.), Medical microbiology, vol. 3. Academic Press, London.

137. Storm, D.R., K.S. Rosenthal, and P.E. Swanson. 1977. Polymyxin and related peptide antibiotics. Annu. Rev. Biochem. **46**:723–763.

138. Stuart, S.J., K.T. Greenwood, and R.K.J. Luke. 1980. Hydroxamate-mediated transport of iron controlled by pColV plasmids. J. Bacteriol. **143**:35–42.

139. Sud, I.J., and D.S. Feingold. 1975. Detection of agents that alter the bacterial cell surface. Antimicrob. Agents Chemother. **8**:34–37.

140. Sukupolvi, S., and D. O'Connor. 1987. Amino acid alterations in a hydrophobic region of the TraT protein of R6–5 increase the outer membrane permeability of enteric bacteria. Mol. Gen. Genet. **210**:178–180.

141. Sukupolvi, S., R. Vuorio, S.-Y. Qi, D. O'Connor, and M. Rhen. 1990. Characterization of the *traT* gene and mutants that increase outer membrane permeability from the *Salmonella typhimurium* virulence plasmid. Mol. Microbiol. **4**:49–57.

142. Suzuki, H., Y. Nishimura, S. Yasuda, A. Nishimura, M. Yamada, and Y. Hirota. 1978. Murein-lipoprotein of *Escherichia coli*: a protein involved in the stabilization of bacterial cell envelope. Mol. Gen. Genet. **167**:1–9.

143. Swanson, J. 1981. Surface-exposed protein antigens of the gonococcal outer membrane. Infect. Immun. **34**:804–816.

144. Taylor, P.W. 1983. Bactericidal and bacteriolytic activity of serum against gram-negative bacteria. Microbiol. Rev. **47**:46–83.

145. Taylor, P.W., and H.P. Kroll. 1985. Effect of lethal doses of complement on the functional integrity of target enterobacteria. Curr. Top. Microbiol. Immunol. **121**:135–158.

146. Tomlinson, S., P.W. Taylor, and J.P. Luzio. 1990. Transfer of preformed terminal C5b-9 complement complexes into the outer membrane of viable gram-negative bacteria: effect on viability and integrity. Biochemistry **29**:1852–1860.

147. Tomlinson, S., P.W. Taylor, B.P. Morgan, and J.P. Luzio. 1989. Killing of gram-negative bacteria by complement: fractionation of cell membranes after complement C5b-9 deposition on to the surface of *Salmonella minnesota* Re595. Biochem. J. **263**:505–511.

148. Tsuchido, T., and M. Takano. 1988. Sensitization by heat treatment of *Escherichia coli* K-12 cells to hydrophobic antibacterial compounds. Antimicrob. Agents Chemother. **32**:1680–1683.

149. Turnowsky, F., K. Fuchs, C. Jeschek, and G. Hogenauer. 1989. *envM* genes of *Salmonella typhimurium* and *Escherichia coli*. J. Bacteriol. **171**:6555–6565.

150. Uratani, Y. 1982. Dansyl chloride labeling of *Pseudomonas aeruginosa* treated with pyocin R1: change in permeability of the cell envelope. J. Bacteriol. **149**:523–528.

151. Vaara, M., and J. Jaakkola. 1989. Sodium hexametaphosphate sensitizes *Pseudomonas aeruginosa*, several other species of *Pseudomonas*, and *Escherichia coli* to hydrophobic drugs. Antimicrob. Agents Chemother. **33**:1741–1747.

152. Vaara, M., and T. Vaara. 1983. Polycations as outer membrane disorganizing agents. Antimicrob. Agents Chemother. **24**:114–122.

153. Vaara, M., and T. Vaara. 1983. Polycations sensitize enteric bacteria to antibiotics. Antimicrob. Agents Chemother. **24**:107–113.

154. Vaara, M., and T. Vaara. 1983. Sensitization of gram-negative bacteria to antibiotics and complement by a nontoxic oligopeptide. Nature (London) **303**:526–528.

155. Van Alphen, L., B. Lugtenberg, E.T. Rietschel, and C. Mombers. 1979. Architecture of the outer membrane of *Escherichia coli* K12. Phase transitions of the bacteriophage K3 receptor complex. Eur. J. Biochem. **101**:571–579.

156. van den Brock, P.J. 1989. Antimicrobial drugs, microorganisms, and phagocytes. Rev. Infect. Dis. **11**:213–245.

157. Viljanen, P., H. Kayhty, M. Vaara, and T. Vaara. 1986. Susceptibility of gram-negative bacteria to the synergistic bactericidal action of serum and polymyxin B nonapeptide. Can. J. Microbiol. **32**:66–69.

158. Viljanen, P., and M. Vaara. 1984. Susceptibility of gram-negative bacteria to polymyxin B nonapeptide. Antimicrob. Agents Chemother. **25**:701–705.

159. Wang, H.-Y., B. Tummler, and J.M. Boggs. 1989. Use of spin labels to determine the percentage of interdigitated lipid in complexes with polymyxin B and polymyxin B nonapeptide. Biochim. Biophys. Acta **985**:182–198.

160. Wardlaw, A.C. 1962. The complement-dependent bacteriolytic activity of normal human serum. I. The effect of pH and ionic strength and the role of lysozyme. J. Exp. Med. **115**:1231–1248.

161. Warren, H.S., S.A. Kania, and G.R. Siber. 1985. Binding and neutralization of bacterial lipopolysaccharide by colistin nonapeptide. Antimicrob. Agents Chemother. **28**:107–112.

162. Weinberg, E.D. 1978. Iron and infection. Microbiol. Rev. **42**:45–66.

163. Williams, P.H., and P.J. Warner. 1980. ColV plasmid-mediated, colicin V–independent iron uptake system for invasive strains of *Escherichia coli*. Infect. Immun. **29**:411–416.

164. Williams, R.M., and C.A. Durham. 1988. Bicyclomycin: synthetic, mechanistic, and biological studies. Chem. Rev. **88**:511–540.

165. Wilson, B.M., and A.A. Glynn. 1975. Release of ^{14}C label and complement killing of *Escherichia coli*. J. Immunol. **28**:391–400.

166. Wright, S.D., and R.P. Levine. 1981. How complement kills *E. coli*. I. Location of the lethal lesion. J. Immunol. **127**:1146–1151.

12

Bacterial Virulence Factors as Targets for Chemotherapy

Sarah K. Highlander and *George M. Weinstock*

Bacteria adapt to specific environments where they can grow and reproduce. When the niche is within a second organism, a bacterium may establish a commensal or even symbiotic relationship with the host. In some situations, however, a bacterium may cause damage or otherwise perturb the host, resulting in disease. Such bacteria are pathogens.

To inhabit a host organism, bacteria have developed specific functions that allow them to adapt to the special circumstances and surroundings. For example, bacteria that infect animals must enter the host, move through tissues, attach to target cells, and avoid the animal's immune system. The accessory gene products that permit this lifestyle are known as virulence factors. Loss of one or more of these factors reduces the pathogenicity of the organism.

Because virulence factors are specific to pathogens, they are attractive targets for therapeutics. The ideal inhibitor of such a specialized virulence factor would reduce the survival of the pathogen in the host by allowing the host's defenses to eliminate the microorganism. The therapeutic agent would not be expected to affect the other nonpathogenic bacteria in the host or to affect the host itself.

In this chapter we first present an overview of some of the different types of virulence factors that have been described. Then we present a discussion of some issues and approaches involved in choosing a virulence factor as a target for disease therapy and for the development of a screen for virulence inhibitors. We illustrate these points by focusing on a particular pathogen, *Pasteurella haemolytica*, and describe how this thinking is being applied to the development of a vaccine and therapeutics for the treatment of bovine respiratory disease. For additional information, the reader is referred to a number of reviews discussing virulence factors in general (4,6,9,17,21,29,45,49,69,75,91,99) as well as specific pathogenic microorganisms (38,47,53,66,80,88,103).

I. Virulence Factors

Virulence factors are a heterogeneous group of molecules (Table 12.1) with diverse functions. Many are toxins, which damage or destroy host cells by many

Table 12.1. A sampling of virulence factors

Bacterium	Factor	Regulation {gene}
Bordetella pertussis	Pertussis toxin	Positive {*vir* (*bvg*)}
	Adenylate cyclase	Positive {*vir* (*bvg*)}
	Hemolysin	Positive {*vir* (*bvg*)}
	Filamentous hemaglutinin	Positive {*vir* (*bvg*)}
	Fimbriae	Positive {*vir* (*bvg*)}
		Phase variation
Borrelia hermsii	VMP	Antigenic variation
C. diphtheriae	Toxin	Negative {*DtoxR*}
Escherichia coli	Shiga-like toxin	Negative {*fur*}
	LT enterotoxin	?
	ST enterotoxin	Positive {*crp*}
	Type I pili	Antigen variation
	Pilus tip protein (papG)	?
	inv	?
	hemolysin	Positive {*hlyR*}
Listeria monocytogenes	Listeriolysin	Positive {*prfA*}
	Invasin	?
Neisseria gonorrhoeae	Pilin	Antigenic variation
	Opacity protein	Antigenic variation
	Outer membrane protein	?
	IgA protease	?
Pasteurella haemolytica	Leukotoxin	Negative {*lktR*}
Pseudomonas aeruginosa	Exotoxin A	Positive {*regB*}
Salmonella	Survival in macrophages	Positive {*phoP*}
Shigella flexneri	Invasion proteins	Positive {*virF*}
	Hemolysin	?
Staphylococcus aureus	Enterotoxin A	?
	Enterotoxin B	Positive {*agr*}
	Toxic-shock-syndrome	Positive {*agr*}
	toxin	Positive {*agr*}
	Hemolysin	Positive {*agr*}
	Staphylokinase	
Streptococci (group A)	Morphological changes	Positive {*virR*}
	M protein	?
Treponema pallidum	Fibronectin-binding proteins	?
Vibrio cholerae	Toxin	Positive {*toxR*}
	Pilus	Positive {*toxR*}
	Pilus-colonizing factor	Positive {*toxR*}
	Shiga-like toxin	Negative {*fur*}
Yersinia pestis	V antigen	Negative {*lcrR*}
	OmpS	Positive {*lcrF*}
Yersinia spp.	*ail*	?
	inv	?

(Note): Some of the virulence-related functions associated with a variety of organisms are listed along with the mode of regulation. Positive regulation refers to a regulator that stimulates expression; negative regulation refers to a regulator that inhibits expression. Known regulators are given in braces. In some cases these regulatory loci are clusters of genes, rather than a single regulatory element. The list is not meant to be comprehensive but to convey a sense of the diversity and complexity of virulence functions.

means including inhibiting protein synthesis (44,80), interfering with cellular ion balance through adenylate cyclase activation (12,52,76), or destabilizing host cell membranes (7,92,102). Other virulence factors disrupt cell-mediated and humoral immune responses (4,84) or are surface molecules that have roles in adherence (3,17,103), invasion (46), and resistance to phagocytosis (45,75). Most pathogens have multiple factors. *Staphylococcus aureus*, for example, produces at least 14 extracellular toxins and enzymes (48) while *Bordetella pertussis* produces a toxin, an adenylate cyclase, a filamentous hemagglutinin, and various fimbriae (103).

The role of some virulence factors in disease is clear—such as for cytolytic toxins that destroy cells of the immune system—but others are not. Sometimes the presence of a toxin alone is not sufficient to cause disease, although it is involved in infection or colonization. An essential role of a virulence factor is demonstrated by specifically neutralizing it by mutation or immunization, and then measuring pathogenicity. This test has shown that loss of a single virulence factor can render the bacterium avirulent. For example, a mutant of *Vibrio cholerae* that lacks the toxin-coregulated pilus protein does not cause disease because it cannot attach to host cells (98). This test also shows that loss of other virulence factors, such as the cholera toxin, only reduces the pathogenicity of the organism (39) while not being essential for pathogenicity. Thus, not all virulence factors are of equal importance in disease. Nonessential factors may offer some increase in survival to the bacterium, perhaps by allowing it to deal with special situations in the host (for example, nutritional deficiencies or competition with other bacteria) or by broadening the range of hosts or tissues available to it. In any case, the fact that not all factors have equal effects on pathogenicity complicates the choice of a therapeutic target.

The variety of functions of virulence factors, and the multiplicity of these factors carried by each pathogen, reflect the complexity of the process of infection and survival in the host. Virulence factors allow the bacterium to survive in a specialized niche, *i.e.*, an animal host. There may be one or more adaptations that are required to efficiently succeed at each stage of the process.

To initiate infection, the bacterium must enter the host. This penetration can come through the skin (often facilitated by insect bite, injection, or weakening of the normal defenses by trauma or other damage) or through other host surfaces in contact with the environment (*e.g.*, nasal and oral passages, urogenital and digestive tracts, conjunctiva). Some virulence factors may be required for entry and survival in these initial sites of infection. These factors may not be the best targets for therapeutic agents designed to treat rather than prevent infections. It should also be noted that some infections may be caused by bacteria that already reside in the host as commensals. In this case, some change in the host environment presumably leads to activation of virulence factors and the disease process. This may involve induction of expression of genes in the bacterium that are not active in the commensal state. The regulation of virulence factor expression is discussed below.

A. Adhesion

Following entry, the bacterium must bind to a cell surface to remain in the host. Adherence is usually accomplished by the binding of specific proteins (adhesins) on the bacterial cell surface to carbohydrate receptors on the eukaryotic cells. Most bacteria express several different types of adhesins, allowing multiple means of cell attachment. The process of finding the target cell surface for adherence need not be a passive one. Many pathogens are chemotactic and, although there is little information available, directed movement toward the site of adherence could occur.

A major class of adhesins is the fimbriae (pili) of bacteria, the filamentous structures present on many bacterial surfaces (17). Fimbriae contain a major structural protein and other minor proteins which are often the adhesins responsible for interacting with eukaryotic receptors. The expression of multiple fimbrial adhesins can alternate, a process called phase variation, or be simultaneous, and can result from expression of multiple genes or alteration of a particular coding sequence (91). Nonfimbrial adhesins such as the filamentous hemagglutinin from *Bordetella pertussis* (103) have also been described. Several bacteria bind to fibronectin, an extracellular matrix protein. *Staphylococcus aureus* (85), group A streptococci (3), and *Treponema pallidum* all can adhere to this molecule, although each bacterium binds to a different site with a different adhesin. In the case of *T. pallidum*, three related adhesins each recognize the same tetrapeptide sequence in fibronectin (101). Adhesins may determine the initial site of infection; the availability of different adhesins to a pathogen may give it a wider range of environments in which to survive. Moreover, while some bacteria remain at the site of attachment, others penetrate into other tissues, though little is known about the factors involved in this migration. As mentioned above, the toxin-coregulated pilus protein of *Vibrio cholerae*, required for attachment to host cells, is essential for pathogenesis (98). Thus, attachment proteins are good targets for therapeutic agents. One drawback, however, is the presence of multiple adhesins on some pathogens, which makes it difficult to prevent attachment by interfering with a single adhesin.

B. Invasion

Adherence can also lead to entry into the eukaryotic target cell. A number of pathogens enter and live within the cells to which they bind, a process called invasion (29,45,66,77). Such a strategy can protect the bacterium from the immune system and free it from having to compete with other microorganisms for nutrients. These invasive pathogens include *Shigella* species, *Escherichia coli*,

Yersinia species, *Listeria monocytogenes, Salmonella typhimurium, Legionella pneumophila*, and *Mycobacterium tuberculosis*. In cases where the adherence proteins (invasins) have been identified and studied, they bind to specific receptors that are involved in the uptake process and allow the bacterium to enter the host cell by a normal internalization pathway. For example, the receptor for the invasin of *Yersinia pseudotuberculosis* is a member of the integrin family of proteins (29). The integrins perform functions important in extracellular interactions and phagocytosis and also interact with talin, a part of the intracellular cytoskeleton (90). The target cell type for this phagocytosis-mediated invasion can be quite nonspecific. Some human pathogens can invade insect cells (29), perhaps indicating the conservation through evolution of the internalization mechanism.

The invasion systems used by the various pathogens mentioned above are not all the same (29). For example, some bacteria end up in vacuoles while others are located in the cytoplasm. In many cases the virulence factors for invasion are encoded by clustered chromosomal or plasmid-borne genes (72). It has been possible to move these genes into other bacteria, thereby converting them into invasive microorganisms (46,71). This is usually demonstrable by using tissue culture cells as the eukaryotic target. As with other virulence factors, there can be more than one invasion system carried by a pathogen (71).

Following entry into the eukaryotic host cell, the process of intracellular survival appears to be complex, involving a range of activities affecting both intermediary metabolism as well as more specialized functions (26,27). The process of invasion is obligatory for *Salmonella* pathogenesis since noninvasive strains are avirulent (36), as are strains that cannot survive within the host cell (27). Thus the virulence factors involved in invasion would appear to be good candidates as targets for therapeutics.

C. *Avoidance of the Immune System*

Once a pathogen has entered the host organism, in order to persist and multiply, it must deal with the immune system and other threats from the host environment. The multifaceted nature of the immune system poses numerous challenges to pathogens and, consequently, there are many virulence factors that interfere with different parts of the immune response. As will be described below in detail for *Pasteurella haemolytica*, one mode of defense elaborated by bacteria is the production of leukotoxins. These proteins destroy the leukocytes that provide the cellular immune response at the site of infection. In the case of *P. haemolytica*, this appears to be an essential component of pathogenesis. It was also previously noted that the intracellular localization of invasive microorganisms shields them from the action of the immune system. A common evasion mechanism is the elaboration of a capsule to coat the bacterium (47,49,75). This protects against binding of antibody molecules and phagocytosis. Frequently nonencapsulated mutants of pathogens are avirulent. The observations that capsule formation, invasion, and production of leukotoxins can each be essential for virulence emphasizes the importance in avoiding the immune response for pathogenic microbes.

Other virulence factors directly interact with molecules of the immune response. These include proteases that cleave the secretory immunoglobulin A (IgA proteases) (83), and other mechanisms, collectively called serum resistance, that have evolved to inhibit the complement system [see chapter 11] (49). Usually surface molecules inhibit complement activation or the binding and activity of complement molecules. For example, the surface M protein of group A streptococci binds fibrinogen and fibrin and interferes with opsonization of the bacterium by the complement system (107).

Many bacteria show variation of their surface antigens (91). By the time an immune response is mounted against the infection, the bacteria no longer display the same antigens and thus escape elimination. In addition, a bacterium infecting a previously infected host would show new antigens and escape any immunizing effects of the prior infection. For example, the pilin genes in *Neisseria gonorrhoeae* undergo antigenic variation by a recombinational process (91). Only one pilin gene is expressed, although there are a number of partial genes representing alternative pilin sequences present elsewhere in the chromosome. Either through gene conversion or transformation with DNA from lysed cells, new genes that produce proteins that differ antigenically are assembled and expressed. Another example is the PII opacity protein on the surface of *Neisseria gonorrhoeae* (11). As many as seven different versions of this protein can be made by a strain. These come from different genes and expression is controlled by changing the reading frame of each coding sequence. The streptococcal M protein also shows extensive antigenic variation in a recombination-dependent process (30). In this case, as many as 75 different M protein serotypes are believed to be generated by recombination between the extensive tandem repeats present in the coding sequence for the M protein.

D. Other Virulence Functions

While the advantage of circumventing the immune response is clear, the function of many of the toxins produced by pathogens is less obvious. It has been proposed that the function of diphtheria toxin is to create a layer of dead cells, thus changing the local environment so that *Corynebacterium diphtheriae* can outgrow its competitors, such as the streptococci (29). Enterotoxins are another class of virulence factors whose function in pathogenesis is somewhat obscure. Competition with other bacteria could also provide a role for enterotoxins, causing competitors to be washed out of the environment. Development of therapeutic agents against toxins such as these could help to alleviate the symptoms of a severe infection, but would not necessarily be the most effective means to cure it.

This brief summary of virulence factors should serve to illustrate some major features. One expects a pathogen to produce multiple virulence factors, and these will most likely involve the process of entry into the host, avoiding the immune system, competing with other bacteria, as well as other processes that are currently less well defined. Some of these factors will be essential for survival of the bacterium in the host while others may only contribute subtly to the infective process. Moreover, the importance of the contribution to pathogenicity may not be related to the severity of the symptoms caused or the threat to survival of the host. As mentioned before, the cholera toxin is not as important for protection by immunization as is the adherence protein, yet the toxin causes severe symptoms that can lead to death. The issue then is to identify the key components for survival of the pathogen and target these for therapeutics.

II. Expression of Virulence Factors

Before leaving the subject of virulence factors in general, it is important to comment on the regulation of their expression (21,69). A protein that is necessary for the synthesis of a virulence factor is itself a virulence factor since in its absence the strain cannot produce molecules needed for pathogenesis. In some cases these regulators may control the synthesis of more than one virulence protein. Loss of such a regulator can lead to loss of multiple virulence functions, greatly reducing pathogenicity. Such regulators are potentially more important for pathogenesis than any of the individual genes they modulate.

In many organisms, virulence factors are coordinately regulated by global regulatory systems. Mutation of a positive control system can eliminate expression of some of these and lead to avirulence (105). Most regulation occurs through signal transduction systems that sense changes in the environment of the bacterium and respond by inducing transcription of a wide range of genes. In *Vibrio cholerae*, *Bordetella pertussis*, and *Staphylococcus aureus*, two-component regulatory systems, composed of membrane-sensing and DNA-binding domains, are involved in regulation of their virulence factors. These two-component systems are similar to those involved in chemotaxis and osmotic control (69). For *B. pertussis*, a locus, called *vir* (104), or *bvg* (1), is a positive regulator of the genes adenylate cyclase, hemolysin, and filamentous hemagglutinin. This locus encodes two or three proteins: the BvgA protein is the transcriptional activator (89), while the BvgB/BvgC protein(s) is believed to be involved in transmembrane signalling in response to temperature, magnesium, or nicotinic acid concentration (1). In addition, some of the *B. pertussis* virulence factors, such as fimbriae, are regulated by a phase variation that involves small deletions or insertions in GC-rich DNA sequences within open reading frames (94) or upstream regulatory sequences (109). Expression of virulence factors in *V. cholera* is positively regulated by the *toxR* locus, which encodes a single transmembrane protein that also contains a DNA-binding domain. This protein responds to changes in pH, osmolarity, and temperature and influences transcription of genes encoding cholera toxin (73), toxin-corregulated pilus (100), and accessory colonization factor (82). The accessory gene regulator, or *agr* locus, of *Staphylococcus aureus* also encodes a two-component regulatory system involved in transcriptional control. *agr* controls expression of exoproteins such as enterotoxin B, toxic shock syndrome toxin, hemolysin, and staphylokinase (81,86). Regulation by *agr* may also involve an RNA molecule that modulates expression of virulence factor mRNA molecules (48).

Negative regulation of virulence gene expression has also been described. *Yersinia pestis* virulence is repressed by calcium and low temperature. This low calcium response is modulated, in part, by a regulatory protein encoded by the *lcrR* locus (2). Mutant strains that lack the LcrR protein overproduce the virulence-associated V antigen, but are avirulent in mice. A thermoregulating activator of *Yersinia* outer membrane proteins (Yops), encoded by the *lcrF* locus has also been identified (111). Diptheria toxin regulation in *Corynebacterium diphtheriae* is controlled by an iron-dependent repressor, *DtoxR*, that binds to the *tox* promoter region (33). Similarly, expression of the *Escherichia coli* Shiga-like toxin is negatively regulated by the iron-dependent transcriptional repressor, Fur (10). Recent studies have shown the leukotoxin of *P. haemolytica* is also subject to transcriptional repression (S.K. Highlander, E.A. Wickersham, and G.M. Weinstock, submitted). Leukotoxin expression is described in detail below.

III. The Leukotoxin of *Pasteurella Haemolytica*

To illustrate approaches and issues in developing vaccines and therapeutics, we now describe studies of *Pasteurella haemolytica*, a gram-negative pathogen that causes a severe pneumonia in cattle. The bacterium is commonly found in the upper respiratory tract of cattle, where it is asymptomatic. Induction of the disease is typically a consequence of stress to the animal, caused by transport of animals, extremes in climate, and viral infections (28). These factors cause moderate respiratory distress and can reduce the effectiveness of the normal immune-clearing mechanisms in the animal. Following stress, many *P. haemolytica* organisms are found in the lower respiratory tract, accompanied by a fibrinous pneumonia, though many animals do recover (34). Thus, pasteurellosis reduces an animal's performance in the feedyard and significantly contributes to morbidity and mortality in the feedlot. This bovine respiratory syndrome is known as "shipping fever" and is the major veterinary health problem in the North American beef cattle industry (64,68).

Many attempts to immunize cattle against *P. haemolytica* have been made, but none have been effective (63). Most vaccines were bacterins or preparations of cell extracts (18,20,110). In each case, a significant antibody response to *P. haemolytica* antigens was observed, but protection against respiratory infection could not be shown (19,65). The ineffectiveness of these vaccines was later explained by the discovery and characterization of a leukotoxin secreted by pathogenic *P. haemolytica* strains (92,93). The leukotoxin causes lysis of white blood cells, including macrophages and neutrophils, which are important to the bovine immune response to shipping fever (5,51,62). Because the leukotoxin is a secreted protein, it is primarily found in the medium of *P. haemolytica* cultures. As a result, early vaccine formulations lacked the leukotoxin antigen, and were not protective against its activity, despite the good anti–*P. haemolytica* antibody response that they elicited. These vaccines were unable to provide protection at the infection site, where the leukotoxin destroyed lung macrophages and had inhibited the antibacterial immune response mounted by the host. Recent experiments, using vaccines that contain leukotoxin antigen, effectively protected animals against pasteurellosis challenge (M. Engler and G. Weinstock, unpublished results). This confirmed the cause of previous vaccine failures and established that the leukotoxin is a virulence factor, since inhibition of its activity by prior immunization prevents disease.

IV. Genes and Proteins of the Leukotoxin System

The genetic locus encoding the *P. haemolytica* leukotoxin has at least five contiguous genes contained within a *ca.* 11-kilobase DNA segment (Figure 12.1) (42,60,96) (S.K. Highlander, E.A. Wickersham, and G.M. Weinstock, submitted). The 102-kD leukotoxin, encoded by the *lktA* gene, is a fast-acting cytolysin that causes lysis of cells by disrupting the target cell membrane (15). Its leukotoxic activity is species-specific and it is most active against bovine leukocytes (51,92). The *P. haemolytica* leukotoxin is also weakly hemolytic against most red blood cells (43). This activity is presumably responsible for the organism's name.

The leukotoxin is secreted from the bacterial cell in a process that involves two other products of this locus, encoded by the *lktB* and *lktD* genes. The LktB (80 kD) and LktD (55 kD) proteins are believed to be membrane proteins and function

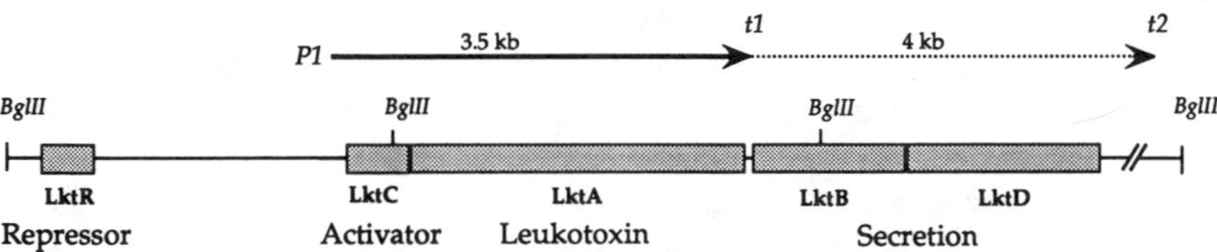

Figure 12.1. Genetic map of the five-gene *P. haemolytica* leukotoxin locus. The leukotoxin, encoded by the *lktA* gene, is flanked by genes that encode a repressor of leukotoxin expression, LktR, an activator, LktC, and proteins required for its secretion (LktB and LktD). The entire locus is contained within three contiguous *Bgl*II fragments that have lengths of 3.4, 3.6, and 6.4 kb, respectively. Leukotoxin transcription is rightward, as drawn, with the *lktCA* message beginning 31 nucleotides upstream of the LktC start codon. Transcription terminator structures are observed within the *lktA-lktB* intergenic region (*t1*) and following the *lktD* open reading frame (*t2*).

to export the leukotoxin from the cell (42,96). A fourth polypeptide, of molecular weight 20 kDa, is encoded by the *lktC* gene, and is required to activate the leukotoxin (42,43,61,96). In the absence of the LktC protein, leukotoxin is synthesized and secreted but it is not active (43). It is believed that the C protein posttranslationally modifies the leukotoxin polypeptide prior to its secretion, though the nature of the modification is unknown.

The transcriptional organization of the leukotoxin locus has not been unequivocally determined. LktC and LktA are produced from a single 3.5-kb mRNA that is transcribed at high levels in *P. haemolytica* (43,97). It is not clear if the downstream LktB and LktD proteins are synthesized from a separate mRNA or from a single read-through transcript that spans all four genes. It is apparent, however, that the mRNA for the B and D proteins is present at much lower amounts than is the transcript for A and C (43). The cause of the differential regulation of transcription of these segments of the locus is unknown, though the B and D genes may be regulated by transcription antitermination in the *lktA-lktB* intergenic region (43,97).

A fifth gene, *lktR*, that is involved in the regulation of leukotoxin expression has been found upstream of the *lktC* and *lktA* genes (S. Highlander, E. Wickersham, and G. Weinstock, submitted). *LktR* encodes a putative protein of 43 kDa that negatively regulates the expression of these genes *in trans*. This regulation is thought to occur at the transcriptional level. The LktR protein may reduce leukotoxin expression in unstressed animals, allowing *P. haemolytica* to be carried commensally, and then respond to a stress signal, leading to increased leukotoxin production and pneumonia. It is not known if LktR controls expression of other genes in *P. haemolytica*. A model of interactions of the leukotoxin proteins is shown in Figure 12.2.

V. Similarities and Distribution of Leukotoxins

The leukotoxin locus is a member of the Repeat Toxin family, which is widely distributed among gram-negative bacteria (Table 12.2) (96). The family derives its name from a short, glycine-rich sequence that is repeated near the carboxy terminus of the leukotoxin protein. While the diseases caused by these organisms are diverse, the toxins expressed are very similar in their genetic organization and activity. Several of these toxins are more active against red blood cells than is the *P. haemolytica* leukotoxin; these are called hemolysins. Many of these hemolysins are also active against leukocytes, and have a broader species-specificity than the *P. haemolytica* leukotoxin (14). Related proteins have also been found in organisms as distantly related as a plant symbiont (22).

A mechanistic role for hemolysin in virulence has not been established, though it has been proposed that hemolysis can provide a source of iron *in vivo* (58). It is likely that the leukotoxic activity of a Repeat Toxin is of greater significance to virulence than is its hemolytic activity. Therefore, the disease model proposed

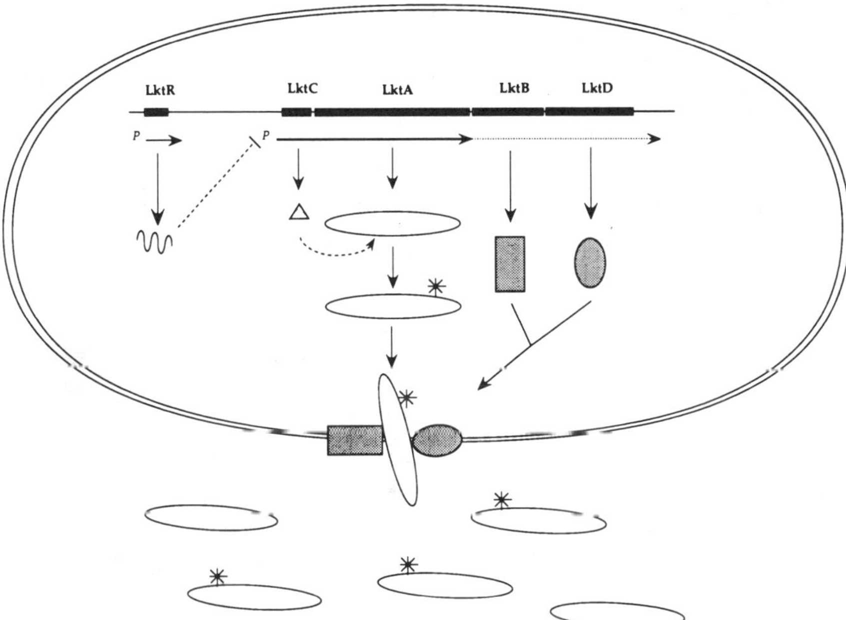

Figure 12.2. Model for the interactions of the proteins of the leukotoxin gene cluster. Transcription of the leukotoxin gene cluster is presumed to be negatively regulated by the product of the *lktR* locus (wavy line) that interacts with DNA sequences upstream of the *lktC* promoter. The LktC protein (triangle) interacts with the leukotoxin to activate it (asterisk). Both active and inactive leukotoxin proteins are secreted from the *P. haemolytica* cell via a mechanism that requires the LktB (rectangle) and LktD (oval) proteins. Though the secretion proteins are shown as spanning the envelope of the cell, no specific secretion mechanism is implied. Each of these interactions is a potential target for therapeutic agents.

for *Pasteurella* also may apply to other organisms that carry Repeat Toxin loci. Typically, an organism resides in its niche within the animal host and does not cause disease. If it moves to other locations within the host, the organism is usually eliminated by the normal clearing mechanism of the immune system. But, if the bacterium expresses a leukotoxic protein, it has the ability to destroy leukocytes and interfere with the host immune defense. Consequently, the bacterium can cause infection of tissues other than its normal site. This is exemplified by the observation that *E. coli* isolated from extraintestinal infections frequently carries a hemolysin locus whereas nonpathogenic strains from the intestine do not (13).

Besides the conservation of the toxin protein sequences and activities within the Repeat Toxin family, the C, B, and D proteins of these gene clusters are also highly conserved (42,95). The B and D secretion proteins are the most similar. In *E. coli*, it has been shown that the hemolysin secretory functions can export

Table 12.2. Repeat Toxin family

Organism (properties)	Associated disease (references)	Toxin
Pasteurella haemolytica (ruminant-specific)	Bovine shipping fever (43, 60, 61)	Leukotoxic
	(small zone)	Hemolytic
Actinobacillus pleuropneumoniae (porcine macrophages)	Swine pleuropneumonia (16)	Leukotoxic
Actinobacillus acetomycetemcomitans (human PMNs)	Jevenile periodontitis (54, 56)	Leukotoxic Hemolytic
Escherichia coli (human leukocytes)	Extraintestinal infection (24)	Leukotoxic
	(large zone)	Hemolytic
Proteus mirabilis (large zone)	Extraintestinal infection (55, 106)	Hemolytic
Proteus vulgaris (large zone)	Extraintestinal infection (55, 106)	Hemolytic
Morganella morganii (large zone)	Extraintestinal infection (55)	Hemolytic
Bordetella pertussis	Whooping cough (37)	Adenylate cyclase

the *P. haemolytica* leukotoxin protein from the cell (43). In addition, the LktC and HlyC proteins are analogous based on their ability to activate their cognate toxins in *E. coli* (32,43). Thus, the proteins are similar both in sequence and in function. Despite the significant homologies observed for the Repeat Toxin proteins, the regions presumed to be involved in regulation of expression are unique to each locus. Thus, the sequences preceding the structural genes for the *E. coli* hemolysin, *P. haemolytica* leukotoxin, *A. pleuropneumoniae* hemolysin, and *A. actinomycetemcomitans* leukotoxin are not conserved (16, 24, 40, 42, 56, 60). In fact, even among the several *E. coli* hemolysin isolates, there are different classes of regulatory sequences (40). Since each of these toxins must be expressed in a different bacterium and tissue, and in response to various signals, it is not surprising that the mechanisms of toxin regulation may be more varied than are the toxin proteins themselves.

The B protein of the Repeat Toxin family has a much broader structural similarity (Table 12.3). It is a member of a family of proteins that function as pumps to export various molecules from cells (8,50). A mammalian version of the B protein is known as the P-glycoprotein or multidrug resistance (MDR) protein (41). This inducible protein removes toxic compounds from the cell, such as antitumor drugs used in cancer chemotherapy. Similar proteins deposit pigments in the eye of insects (79), secrete a mating hormone in yeast (67), and export a protease from *Erwinia chrysanthemum* (57). Recently, a human P-glycoprotein analog has also been associated with the sequence of the cystic fibrosis gene, CF (87). It is striking how the sequence has been conserved.

Table 12.3. LktB homologues

Organism	Protein	Substrate transported	References
Pasteurella haemolytica	LktB	Leukotoxin	(42)
Escherichia coli	HlyB	Hemolysin	(24)
Bordetella pertussis	CyaB	Adenylate cyclase/hemolysin	(37)
Erwinia chrysanthemi	PrtE	Protease	(57)
Saccharomyces cerevisiae	Ste6	α-Factor pheromone	(67)
Drosophila melanogaster	White	Pigment precursor	(79)
Human (tumor cells)	MDR	Antitumor drugs	(35)
Human (cystic fibrosis)	CFTR	Chloride ions	(87)

VI. Targets for Therapeutics Agents

Immunization with a leukotoxin preparation can protect cattle against *P. haemolytica* infection in experimental challenges, but control of shipping fever in the feedlot is much more complex. Many animals that arrive at the feedlot are already experiencing respiratory distress, while others develop pneumonia soon after arrival. As a result, immunizing these animals on arrival is probably less effective, and may even be detrimental, since the immunogen may bind to circulating antileukotoxin antibodies and inhibit the natural immune response (108). For this reason, vaccination prior to transport to the feedlot, or even before arrival at a sale barn, is likely to be most effective. Early inoculation poses other problems, since it is difficult for the feedlot owner to control vaccination procedures of cattle suppliers. Factors such as these tend to reduce the usefulness of a vaccine against shipping fever. To the feedlot operator, a therapy for the disease that does not involve antibiotics is important.

By using the leukotoxin system described above as a model, targets to block the production or activity of the leukotoxin can be identified and examined (Figure 12.2). First, it may be possible to block leukotoxin synthesis. This strategy is based on the discovery of a repressor of leukotoxin expression. There are numerous repressor systems that have been studied in bacterial systems. For example, the genes for catabolic processes are usually regulated by a repressor. In *E. coli*, studies of lactose utilization show that the repressor is inactivated when an inducer is generated by the presence of lactose (70). The inducer binds to the repressor, causing an allosteric change and loss of DNA-binding activity. In another example, the repressor of DNA repair genes in *E. coli*, the LexA protein, is inactivated when the cell's DNA becomes damaged (59). In this case, the inducer is single-stranded DNA and its presence leads to proteolytic inactivation of the repressor. In many other well-characterized repressor systems, it is usually found that an inducer molecule that is generated by some signal leads to inhibition of the repressor's DNA-binding activity.

It is possible that leukotoxin synthesis is repressed in healthy animals that are *P. haemolytica* carriers and that induction of leukotoxin synthesis is important to disease onset. The signal that inactivates the leukotoxin repressor has not been identified, but global regulation of virulence factors by external signals has been identified as a common theme in pathogenesis. It is thus plausible that an inducer leads to loss of DNA-binding activity of the LktR repressor. A therapeutic agent could act by preventing inactivation of the repressor by an inducer. This might occur by preventing inducer binding, by complexing with the inducer itself, or by preventing synthesis of the inducer (if it is an endogenous molecule) or its uptake (if it is exogenous). Despite the mechanism, the result would be continual repression of the *lkt* genes.

A second target for pasteurellosis therapy is activation of the leukotoxin by the C protein. As stated above, in the absence of the C protein the leukotoxin is made and secreted, but it is inactive (43,61). The unmodified leukotoxin protein is also antigenic (G.M. Weinstock and M.J. Engler, unpublished observation). Though the nature of the modification is unknown, it is known that the analogous C protein from the *E. coli* hemolysin system is cytoplasmic, which suggests that the leukotoxin is modified intracellularly, before secretion (78). An agent that blocks the activation of the leukotoxin would not prevent its synthesis or secretion. Such an inhibitor could be particularly useful since the *P. haemolytica* leukotoxin would still be presented to the host as an immunogen, in the absence of its cytotoxic effects.

Toxin secretion, by the LktB and LktD proteins, is another potential target for the development of therapeutic agents. Blocking secretion would cause accumulation of the active leukotoxin inside *P. haemolytica* cells. It would be prevented from acting, by virtue of its intracellular location, though it still could be released by cell lysis or leakage from the bacterium. One would expect, however, that the concentration of leukotoxin in the surrounding medium would be reduced. There is an established therapeutic precedent for this type of target: antagonists of the LktB-homologous P-glycoprotein from eukaryotic cells have been identified in chemical screens (23). Thus, compounds have been identified that inhibit the export activity of the P-glycoprotein, which suggests that similar agents could be found that block secretion of the leukotoxin from *P. haemolytica*.

Finally, the leukotoxin protein itself can be examined as the target for a therapeutic agent. Inhibitors could act by blocking its binding to target cells or by interfering with its activity. Little is known about the receptors for the Repeat Toxin proteins, though they act by forming pores in the plasma membrane of target cells (7). Knowledge of the enzymatic steps involved in target cell lysis could suggest other therapeutic targets.

While this discussion has focused solely on the leukotoxin, it is likely that other virulence factors are involved in pasteurellosis. Simply blocking the action or secretion of the leukotoxin may not be sufficient to control the disease. Other pathogenic organisms that carry Repeat Toxin loci express additional virulence factors such as adhesins, invasins, and additional toxins (29). Similar factors may be expressed by pathogenic *P. haemolytica* and could serve as potential targets that could be exploited to inhibit infection. For example, a drug that blocks leukotoxin expression by preventing repressor inactivation might also be used to maintain repression of the synthesis of other proteins that are virulence factors, such as those involved in invasion or attachment. This is common in many pathogenic organisms where virulence factors are coordinately controlled by a single regulatory locus (74,104). Similarly, agents that prevent the secretion of the leukotoxin could have an additional benefit of preventing export of other virulence proteins by the LktB-LktD secretion system.

VII. Approaches to Screening for Therapeutic Agents

The previous discussion of the *P. haemolytica* leukotoxin system illustrates the role of virulence factors in infection and identifies the targets for chemotherapy. We now consider each target (toxin expression, secretion, and activity) for its potential to be used in screens for compounds that inhibit or interfere with infection and pathogenesis.

As noted earlier, regulatory proteins can control the synthesis of more than one virulence factor. These global regulators provide a single target that can block multiple virulence factors. A regulator can be a repressor, such as the LktR protein, which requires an inducer (usually a small molecule) for toxin expression or virulence. Repressors bind to specific DNA sequences within virulence factor genes and block their expression. When an inducer is present, it interacts with the repressor, prevents repressor-DNA binding, and permits gene expression. An inhibitor of this type of regulation would block inducer binding to the repressor protein.

Another type of virulence regulation occurs by activation. Examples are the positive regulation of virulence imparted by the ToxR protein of *V. cholera* (74) and the *B. pertussis* global regulatory locus, *vir* (104). Activators also bind to specific DNA sequences in virulence genes, but when an activator binds, it turns on expression of the gene. Typically an activator binds only after receiving a signal, such as a small molecule, like cAMP, or a change in some environmental condition, such as osmotic strength. An inhibitor of a positive regulatory protein could interfere either with binding of the activator to the DNA or with the initial signalling step.

The interaction between a protein and a small molecule can be blocked *in vivo*, and there is precedent for interference of sequence-specific protein-DNA interactions. An example of the latter was illustrated by using the resolvase enzyme of transposon *Tn*3 (25,31). Resolvase binds to specific DNA sequences and catalyzes strand cutting and rejoining that results in site-specific recombination. The enzyme also represses its own gene expression when it binds to its recognition site. In these studies, a panel of over 6,000 chemicals was screened for the ability to block the activity of the *Tn*3 resolvase *in vivo*. An antibiotic resistance gene was inserted between two resolvase recognition sites; when resolvase performed its site-specific recombination the resistance marker was lost. Chemicals that prevented site-specific recombination allowed the antibiotic resistance to be maintained and cells carrying the marker were able to grow under antibiotic selection (25). Twenty-six compounds were identified as inhibitors of the resolvase reaction and four were analyzed in detail. Two compounds inhibited the interaction of the transposase protein with its target sequence while two others affected steps in the recombination reaction following DNA binding (31). Thus, this study demonstrates the feasibility of blocking sequence-specific DNA binding.

The *Tn*3 resolvase system involved an *in vivo* screen based on colony formation and selection for the presence of an antibiotic resistance gene. Here, inhibitors that interfered with DNA binding led to gene expression (antibiotic resistance). For virulence factors, compounds that prevent gene expression are desirable. Thus, the selection scheme would require modifications in order to select inhibitors of regulatory proteins. This could be accomplished, but another approach may be more convenient. Using gene fusions of virulence factor genes to *lacZ*, the gene encoding β-galactosidase, it is possible to provide a colorimetric assay for gene expression that could be employed in large-scale screens.

A fusion between *lacZ* and a virulence factor gene tags the gene with a reporter protein, β-galactosidase, that can be easily monitored. In the *P. haemolytica* leukotoxin system, a gene fusion between *lacZ* and the *lktC* gene places β-galactosidase synthesis under the control of the leukotoxin system's regulatory protein, LktR. When the leukotoxin promoter is expressed, it drives β-galactosidase synthesis, and a blue pigment is detected by using the indicator 5-bromo-4-chloro-3-indolyl-β-galactoside (X-Gal). Alternatively, a yellow color is produced by using o-nitro-phenyl galactoside (ONPG) as an indicator. When the repressor gene is present, β-galactosidase synthesis is inhibited and colony color is white (S.K. Highlander, E.A. Wickersham, and G.M. Weinstock, submitted). By using this type of assay, agents could be sought that prevented blue (or yellow) color production under growth conditions that induced leukotoxin expression. Therapeutic agents selected by this method could then be tested for their ability to maintain repression of other *P. haemolytica* virulence factors. A similar approach could be applied to systems under the control of positive regulators. Cells would be white in the absence of the regulator because the gene fusion between *lacZ* and the virulence factor would not be expressed. These cells would be distinguished from uninhibited cells, which would be blue (or yellow) under activating growth conditions. Such a screening system could be set up in *E. coli* and used to test a large library of compounds without being limited by growth eccentricities of the pathogen. Candidates would be rescreened with the pathogen, but this would be limited to the smaller set of compounds that gave a positive result.

Besides inhibition of gene expression, therapeutic agents also can be targeted to toxin secretion and activity. For hemolysins, hemolytic activity provides a simple assay for inhibitors of both steps. Various plate assays for hemolysins exist, the simplest being the formation of a large, clear zone of hemolysis on blood agar plates. In addition, hemolytic activity against red blood cells can be assayed spectrophotometrically.

Though many leukotoxins are also hemolysins, some, like the *P. haemolytica* leukotoxin, have only a weak hemolytic activity. Still, it is possible to use a hemolysin assay to screen for compounds that could inhibit the *P. haemolytica* system. This is possible because there are proteins like the *E. coli* hemolysin that have a strong hemolytic activity and are closely related to the *P. haemolytica* system, as described earlier. Since the secretion proteins of the Repeat Toxin family are highly conserved, it is likely that a compound that inhibits *E. coli* hemolysin secretion also would inhibit secretion of the other toxin proteins. Thus, a protein like the *E. coli* hemolysin, which is closely related to the *P. haemolytica* protein but which has a much stronger hemolytic activity, could be used in a screen. For other leukotoxins, there may not be a convenient cognate hemolytic activity. Here, screens using assays for lysis of leukocytes, such as macrophages or neutrophils, rather than red blood cells, can be devised. These screens can employ colorimetric or fluorescent detection methods.

Using these assays, compounds can be tested for their ability to inhibit lysis of cells, either in a plate test (prevent formation of a hemolytic zone) or by using one of the spectrophotometric methods. These screens would use live bacteria. Thus, agents identified in this *in vivo* screen would be expected to fall into three classes: agents that inhibit hemolysin activity, agents that inhibit its expression, and those that inhibit its secretion. The classes can be distinguished by assaying and comparing intracellular and extracellular hemolysin protein and by monitoring hemolysin specific activity.

VIII. Conclusion

The field of microbial pathogenesis is undergoing a rapid expansion of knowledge due to the application of the techniques of molecular genetics to its study. Much new information about virulence factors has emerged. This information can be used to develop screens for therapeutic compounds that inhibit virulence factors of these bacteria. In this review we have concentrated on the leukotoxin system of *Pasteurella haemolytica*, a pathogen involved in shipping fever. This system illustrates the role of a virulence factor in infection, the conservation of a factor among other bacteria causing apparently unrelated diseases, and targets for chemotherapeutics that affect virulence factor expression, function, and localization. Ultimately, these approaches should provide nonantibiotic therapies for bacterial diseases such as shipping fever. Although the discussion has focused on *P. haemolytica* and leukotoxin and hemolysin proteins, the principles are universally applicable.

References

1. Arico, B., J.F. Miller, C. Roy, S. Stibitz, D. Monack, S. Falkow, R. Gross, and R. Rappuoli. 1989. Sequences required for expression of *Bordetella pertussis* virulence factors share homology with prokaryotic signal transduction proteins. Proc. Natl. Acad. Sci. USA **86**:6671–6675.

2. Barve, S.S., and S.C. Starley. 1990. *lcrR*, a low-Ca2(+)-response locus with dual Ca^{2+}-dependent functions in *Yersinia pestis*. J. Bacteriol. **172**:4661–71.

3. Beachey, E.H., and H.S. Courtney. 1987. Bacterial adherence: the attachment of group A streptoccoci to mucosal surfaces. Rev. Infect. Dis. **9 (Suppl. 5)**:475–481.

4. Bergdoll, M.S. 1970. Enterotoxins, p. 265–326. *In* T. Montie, S. Kadis, and S.J. Ajl (eds.), Microbial Toxins. Academic Press, New York.

5. Berggren, K.A., C.S. Baluyut, R.R. Simonson, W.J. Bemrick, and S.K. Maheswarran. 1981. Cytotoxic effects of *Pasteurella haemolytica* on bovine neutrophils. Am. J. Vet. Res. **42**:1383–1388.

6. Betley, M.J., V.L. Miller, and J.J. Mekalanos. 1986. Genetics of bacterial enterotoxins. Ann. Rev. Microbiol. **40**:577–605.

7. Bhakdi, S., N. Mackman, J.M. Nicaud, and I.B. Holland. 1986. *Escherichia coli* hemolysin may damage target cell membranes by generating transmembrane pores. Infect. Immun. *52*:63–69.

8. Blight, M.A., and I.B. Holland. 1990. Structure and function of haemolysin B, P-glycoprotein and other members of a novel family of membrane translocators. Mol. Microbiol. **4**:873–880.

9. Brubaker, R.R. 1985. Mechanisms of bacterial virulence. Ann. Rev. Microbiol. **39**:21–50.

10. Calderwood, S.B., and J.J. Mekalanos. 1987. Iron regulation of Shiga-like toxin expression in *Escherichia coli* is mediated by the *fur* locus. J. Bacteriol. **169**:4759–4764.

11. Cannon, J.G. 1988. Genetics of Protein II of *Neisseria gonorrhoeae*, p. 75–83. *In* M.A. Horowitz (ed.), Bacterial–host cell interaction. Alan R. Liss, Inc., New York.

12. Cassel, D., and T. Pfeuffer. 1978. Mechanism of cholera toxin action: covalent modification of the guanyl-binding protein of the adenylate cyclase system. Proc. Natl. Acad. Sci. USA **75**:2669–2673.

13. Cavalieri, S.J., G.A. Bohach, and I.S. Snyder. 1984. *Escherichia coli* alpha-hemolysin: characteristics and probable role in pathogenicity. Microbiol. Rev. **48**:326–343.

14. Cavalieri, S.J., and I.S. Snyder. 1982. Cytotoxic activity of partially purified *Escherichia coli* alpha haemolysin. J. Med. Microbiol. **15**:11–21.

15. Chang, Y.F., and H.W. Renshaw. 1986. *Pasteurella haemolytica* leukotoxin: comparison of ^{51}chromium-release, trypan blue dye exclusion, and luminol-dependent chemiluminescence-inhibition assays for sensitivity in detecting leukotoxin activity. Am. J. Vet. Res. **47**:134–138.

16. Chang, Y.F., R. Young, and D.K. Struck. 1989. Cloning and characterization of a hemolysin gene from *Actinobacillus (Haemophilus) pleuropneumoniae*. DNA **8**:635–647.

17. Clegg, S., and G.F. Gerlach. 1987. Enterobacterial fimbriae. J. Bacteriol. **169**:934–938.

18. Confer, A.W., B.A. Lessley, R.J. Panciera, R.W. Fulton, and J.A. Kreps. 1985. Serum antibodies to antigens derived from a saline extract of *Pasteurella haemolytica*: correlation with resistance to experimental bovine pneumonic pasteurellosis. Vet. Immunol. Immunopathol. **10**:265–278.

19. Confer, A.W., R.J. Panciera, R.W. Fulton, M.J. Gentry, and J.A. Rummage. 1985. Effect of vaccination with live or killed *Pasteurella haemolytica* on resistance to experimental bovine pneumonic pasteurellosis. Am. J. Vet. Res. **46**:342–347.

20. Confer, A.W., R.J. Panciera, M.J. Gentry, and R.W. Fulton. 1987. Immunologic response to *Pasteurella haemolytica* and resistance against experimental bovine pneumonic pasteurellosis, induced by bacterins in oil adjuvants. Am. J. Vet. Res. **48**:163–168.

21. DiRita, V.J., and J.J. Mekalanos. 1989. Genetic regulation of bacterial virulence. Ann. Rev. Genet. **23**:455–482.

22. Economou, A., W.D. Hamilton, A.W. Johnston, and J.A. Downie. 1990. The *Rhizobium* nodulation gene *nodO* encodes a Ca^{2+} binding protein that is exported without N-terminal cleavage and is homologous to haemolysin and related proteins. EMBO J. **9**:349–354.

23. Endicott, J.A., and V. Ling. 1989. The biochemistry of P-glycoprotein-mediated multidrug resistance. Ann. Rev. Biochem. **58**:137–171.

24. Felmlee, T., S. Pellett, and R.A. Welch. 1985. Nucleotide sequence of an *Escherichia coli* chromosomal hemolysin. J. Bacteriol. **163**:94–105.

25. Fennewald, M.A., and J. Capobianco. 1984. Isolation and analysis of inhibitors of transposon *Tn*3 site-specific recombination. J. Bacteriol. **159**:404–406.

26. Fields, P.I., E.A. Groisman, and F. Heffron. 1989. A *Salmonella* locus that controls resistance to microbicidal proteins from phagocytic cells. Science **243**:1059–1062.

27. Fields, P.I., R.V. Swanson, C.G. Haidaris, and F. Heffron. 1986. Mutants of *Salmonella typhimurium* that cannot survive within the macrophage are avirulent. Proc. Natl. Acad. Sci. USA **83**:5189–5193.

28. Filion, L.G., P.J. Willson, H. Bielefeldt-Ohmann, L.A. Babiuk, and R.G. Thompson. 1984. The possible role of stress in the induction of pneumonic pasteurellois. Can. J. Comp. Med. **48**:268–274.

29. Finlay, B.B., and S. Falkow. 1989. Common themes in microbial pathogenicity. Microbiol. Rev. **53**:210–230.

30. Fischetti, V.A., K.F. Jones, S.K. Hollingshead, and J.R. Scott. 1988. Molecular approaches to understanding the structure and function of streptococcal M protein, a unique virulence molecule, p. 99–110. *In* M.A. Horowitz (ed.), Bacterial–host cell interaction. Alan R. Liss, Inc., New York.

31. Flanagan, P.M., and M.A. Fennewald. 1989. Analysis of inhibitors of site-specific recombination reaction mediated by *Tn*3 resolvase. J. Molec. Biol. **206**:295–304.

32. Forestier, C., and R.A. Welch. 1990. Nonreciprocal complementation of the *hlyC* and *lktC* genes of the *Escherichia coli* hemolysin and *Pasteurella haemolytica* leukotoxin determinants. Infect. Immun. **58**:828–832.

33. Fourel, G., A. Phalipon, and M. Kaczorek. 1989. Evidence for direct regulation of diphtheria toxin gene transcription by an Fe^{2+}-dependent DNA-binding repressor, DtoxR, in *Corynebacterium diphtheriae*. Infect. Immun. **57**:3221–3225.

34. Frank, G.H., and P.C. Smith. 1983. Prevalence of *Pasteurella haemolytica* in transported calves. Am. J. Vet. Res. **44**:981–985.

35. Gerlach, J.H., J.A. Endicott, P.F. Juranka, G. Henderson, F. Sarangi, K.L. Deuchars, and V. Ling. 1986. Homology between P-glycoprotein and a bacterial haemolysin transport protein suggests a model for multidrug resistance. Nature **324**:485–489.

36. Giannella, R.A., O. Washington, P. Gemski, and S.B. Formal. 1973. Invasion of HeLa cells by *Salmonella typhimurium*: a model for study of invasiveness of *Salmonella*. J. Infect. Dis. **128**:69–75.

37. Galser, P., D. Ladant, O. Sezer, F. Pichot, A. Ullmann, and A. Danchin. 1988. The calmodulin-sensitive adenylate cyclase of *Bordetella pertussis*: cloning and expression in *Escherichia coli*. Mol. Microbiol. **2**:19–30.

38. Gross, R., B. Aricò, and R. Rappuoli. 1989. Genetics of pertussis toxin. Molec. Microbiol. **3**:119–124.

39. Herrington, D.A., R.H. Hall, G. Losonsky, J.J. Mekalanos, R.K. Taylor, and M.M. Levine. 1988. Toxin, toxin-coregulated pili, and the *toxR* regulon are essential for *Vibrio cholerae* pathogenesis in humans. J. Exp. Med. **168**:1487–92.

40. Hess, J., W. Wels, M. Vogel, and W. Goebel. 1986. Nucleotide sequence of a plasmid-encoded hemolysin determinant and its comparison with a corresponding chromosomal hemolysin sequence. FEMS Lett. **34**:1–11.

41. Higgins, C.F., I.D. Hiles, G.P.C. Salmond, D.R. Gill, J.A. Downie, I.J. Evans, I.B. Holland, L. Gray, S.D. Buckel, A.W. Bell, and M.A. Hermodson. 1986. A family of related ATP-binding subunits coupled to many distinct biological processes in bacteria. Nature **323**:448–450.

42. Highlander, S.K., M. Chidambaram, M.J. Engler, and G.M. Weinstock. 1989. DNA sequence of the *Pasteurella haemolytica* leukotoxin gene cluster. DNA **8**:15–28.

43. Highlander, S.K., M.J. Engler, and G.M. Weinstock. 1990. Secretion and expression of the *Pasteurella haemolytica* leukotoxin. J. Bacteriol. **172**:2343–2350.

44. Iglewski, B.H., and D. Kabat. 1975. NAD-dependent inhibition of protein synthesis by *Pseudomonas aeruginosa* toxin. Proc. Natl. Acad. Sci. USA **72**:2284–2288.

45. Isberg, R.R. 1989. Mammalian cell adhesion functions and cellular penetration of enteropathogenic *Yersinia* species. Molec. Microbiol. **3**:1449–1453.

46. Isberg, R.R., and S. Falkow. 1985. A single genetic locus encoded by *Yersinia pseudotuberculosis* permits invasion of cultured animal cells by *Escherichia coli* K-12. Nature (London) **317**:262–264.

47. Jann, K., and B. Jann. 1987. Polysaccharide antigens of *Escherichia coli*. Rev. Infect. Dis. (Suppl. 5) **9**:517–526.

48. Janzon, L., and S. Arvidson. 1990. The role of the delta-lysin gene (*hld*) in the regulation of virulence genes by the accessory gene regulator (*agr*) in *Staphylococcus aureus*. EMBO J. **9**:1391–1399.

49. Joiner, K.A. 1988. Complement evasion by bacteria and parasites. Ann. Rev. Microbiol. **42**:201–230.

50. Juranka, P.F., R.L. Zastawny, and V. Ling. 1989. P-glycoprotein: multidrug-resistance and a superfamily of membrane-associated transport proteins. FASEB J. **3**:2583–2592.

51. Kaehler, K.L., R.J.F. Markham, C.C. Muscoplat, and D.W. Johnson. 1980. Evidence for cytocidal effects of *Pasteurella haemolytica* on bovine peripheral blood mononuclear leukocytes. Am. J. Vet. Res. **41**:1690–1693.

52. Katada, T., and M. Ui. 1982. Direct modification of the membrane adenylate cyclase system by islet-activating protein due to ADP-ribosylation of a membrane protein. Proc. Natl. Acad. Sci. USA **79**:3129–3133.

53. Khardori, N., and V. Fainstein. 1988. *Aeromonas* and *Plesiomonas* as etiological agents. Ann. Rev. Microbiol. **42**:395–419.

54. Kolodrubetz, D., T. Dailey, J. Ebersole, and E. Kraig. 1989. Cloning and expression of the leukotoxin gene from *Actinobacillus actinomycetemocomitans*. Infect. Immun. **57**:1465–1469.

55. Koronakis, V., M. Cross, B. Senior, E. Koronakis, and C. Hughes. 1987. The secreted hemolysins of *Proteus mirabilis*, *Proteus vulgaris*, and *Morganella morganii* are genetically related to each other and to the alpha-hemolysin of *Escherichia coli*. J. Bacteriol. **169**:1509–1515.

56. Kraig, E., T. Dailey, and D. Kolodrubetz. 1990. Nucleotide sequence of the leukotoxin gene from *Actinobacillus actinomycetemocomitans*: homology to the alpha-hemolysin/leukotoxin gene family. Infect. Immun. **58**:920–929.

57. Letoffe, S., P. Delepelaire, and C. Wandersman. 1990. Protease secretion by *Erwinia chrysanthemi*: the specific secretion functions are analogous to those of *Escherichia coli* α-hemolysin. EMBO J. **9**:1375–1382.

58. Linggood, M.A., and P.L. Ingram. 1982. The role of alpha haemolysin in the virulence of *Escherichia coli* for mice. J. Med. Microbiol. **15**:23–30.

59. Little, J.W., and D.W. Mount. 1980. The SOS regulatory system of *Escherichia coli*. Cell **29**:11–22.

60. Lo, R., C.A. Strathdee, and P.E. Shewen. 1987. Nucleotide sequence of the leukotoxin genes of *Pasteurella haemolytica* A1. Infect. Immun. **55**:1987–1996.

61. Lo, R.Y.C., P.E. Shewen, C.A. Strathdee, and C.N. Greer. 1985. Cloning and expression of the leukotoxin gene of *Pasteurella haemolytica* A1 in *Escherichia coli* K-12. Infect. Immun. **50**:667–671.

62. Markham, R., and B.N. Wilkie. 1980. Interaction between *Pasteurella haemolytica* and bovine alveolar macrophages: cytotoxic effect on macrophages and impaired phagocytosis. Am. J. Vet. Res. **41**:18–22.

63. Martin, S.W. 1983. Vaccination: is it effective in preventing respiratory disease or influencing weight gains in feedlot calves? Can. Vet. J. **24**:10–19.

64. Martin, S.W., A.H. Meek, D.G. Davis, J.A. Johnson, and R.A. Curtis. 1981. Factors associated with morbidity and mortality in feedlot calves: The Bruce County beef project, year two. Can. J. Comp. Med. **45**:102–112.

65. Martin, W., S. Acres, E. Janzen, P. Willson, and B. Allen. 1984. A field trial of preshipment vaccination of calves. Can. Vet. J. **25**:145–147.

66. Maurelli, A.T., and P.J. Sansonetti. 1988. Genetic determinants of *Shigella* pathogenicity. Ann. Rev. Microbiol. **42**:127–150.

67. McGrath, J.P., and A. Varshavsky. 1989. The yeast STE6 gene encodes a homologue of the mammalian multidrug resistance P-glycoprotein. Nature **340**:400–404.

68. McMillan, C.W. 1984. Working together, sharing knowledge. In Bovine respiratory disease: a symposium. University Press, College Station, TX.

69. Miller, J.F., J.J. Mekalanos, and S. Falkow. 1989. Coordinate regulation and sensory transduction in the control of bacterial virulence. Science **243**:916–922.

70. Miller, J.H., and W.S. Reznikoff. 1980. The operon. Cold Spring Harbor Laboratory, Cold Spring Harbor, NY.

71. Miller, V.L., and S. Falkow. 1988. Evidence for two genetic loci in *Yersinia enterocolitica* that can promote invasion of epithelial cells. Infect. Immun. **56**:1242–1248.

72. Miller, V.L., B.B. Finlay, and S. Falkow. 1988. Factors essential for the penetration of mammalian cells by *Yersinia*. Curr. Top. Microbiol. Immunol. **138**:15–39.

73. Miller, V.L., and J.J. Mekalanos. 1985. Genetic analysis of the cholera toxin-positive regulatory gene *toxR*. J. Bacteriol. **163**:580–585.

74. Miller, V.L., R.K. Taylor, and J.J. Mekalanos. 1987. Cholera toxin transcriptional activator *toxR* is a transmembrane DNA binding protein. Cell **48**:271–279.

75. Mims, C.A. 1987. The pathogenesis of infectious disease. Academic Press, Inc., London.

76. Moss, J., and S.H. Richardson. 1978. Activation of adenylate cyclase by heat-labile enterotoxin. Evidence of ADP-ribosyl transferase activity similar to that of choleragen. J. Clin. Invest. **62**:281–285.

77. Moulder, J.W. 1985. Comparative biology of intracellular parasitism. Microbiol. Rev. **49**:298–337.

78. Nicaud, J.M., N. Mackman, L.Gray, and I.B. Holland. 1986. The C-terminal, 23 kDa peptide of *E. coli* haemolysin 2001 contains all the information necessary for its secretion by the haemolysin (Hly) export machinery. FEBS Lett. **203**:331–335.

79. O'Hare, K., C. Murphy, R. Levis, and G.M. Rubin. 1984. DNA sequence of the white locus of *Drosophila melanogaster*. J. Mol. Biol. **180**:437–455.

80. Pappenheimer, A.M. 1975. Diphtheria toxin. Ann. Rev. Biochem. **46**:69–94.

81. Peng, H.L., R.P. Novick, B. Kreiswirth, J. Kornblum, and P. Schlievert. 1988. Cloning, characterization, and sequencing of an accessory gene regulator (*agr*) in *Staphylococcus aureus*. J. Bacteriol. **170**:4365–4372.

82. Peterson, K.M., and J.J. Mekalanos. 1988. Characterization of the *Vibrio cholerae* ToxR regulon: identification of novel genes involved in intestinal colonization. Infect. Immun. **56**:2822–2829.

83. Plaut, A.G. 1983. The IgA1 proteases of pathogenic bacteria. Ann. Rev. Microbiol. **37**:603–622.

84. Poindexter, N.J., and P.M. Schlievert. 1985. The biochemical and immunological properties of toxic-shock syndrome toxin-1 (TSST–1) and association with TSS. J. Toxicol. Toxin Rev. **4**:1–39.

85. Proctor, R.A. 1987. The staphylococcal fibronectin receptor: evidence for its importance in invasive infections. Rev. Infect. Dis. **9(Suppl. 4)**:317–321.

86. Recsei, P., B. Kreiswirth, M. ORcilly, P. Schlievert, A. Gruss, and R.P. Novick. 1986. Regulation of exoprotein gene expression in *Staphylococcus aureus* by *agr*. Mol. Gcn. Genet. **202**:58–61.

87. Riordan, J.R., J.M. Rommens, B. Kerem, N. Alon, R. Rozmahel, Z. Grzelczak, J. Zielenski, S. Lok, N. Plavsic, J.-L. Chou, M.L. Drumm, M.C. Iannuzzi, F.S. Collins, and L.-C. Tsui. 1989. Identification of the cystic fibrosis gene: cloning and characterization of complementary DNA. Science **245**:1066–1073.

88. Roth, R.R., and W.D. James. 1988. Microbial ecology of the skin. Ann. Rev. Microbiol. **42**:441–464.

89. Roy, C.R., J.F. Miller, and S. Falkow. 1989. The *bvgA* gene of *Bordetella pertussis* encodes a transcriptional activator required for coordinate regulation of several virulence genes. J. Bacteriol. **171**:6338–6344.

90. Ruoslahti, E., and M.D. Pierschbacker. 1987. New perspectives in cell adhesion: RGD and integrins. Science **238**:491–497.

91. Seifert, H.S., and M. So. 1988. Genetic mechanisms of bacterial antigenic variation. Microbiol. Rev. **52**:327–336.

92. Shewen, P.E., and B.N. Wilkie. 1982. Cytotoxin of *Pasteurella haemolytica* acting on bovine leukocytes. Infect. Immun. **35**:91–94.

93. Shewen, P.E., and B.N. Wilkie. 1985. Evidence for the *Pasteurella haemolytica* cytotoxin as a product of actively growing bacteria. Am. J. Vet. Res. **46**:*1212–1214*.

94. Stibitz, S., W. Aaronson, D. Monack, and S. Falkow. 1989. Phase variation in *Bordetella pertussis* by frameshift mutation in a gene for a novel two-component system. Nature **338**:266–9.

95. Strathdee, C.A., and R.Y.C. Lo. 1987. Extensive homology between the leukotoxin of *Pasteurella haemolytica* A1 and the alpha-hemolysin of *Escherichia coli*. Infect. Immun. **55**:3233–3236.

96. Strathdee, C.A., and R.Y.C. Lo. 1989. Cloning, nucleotide sequence, and characterization of genes encoding the secretion function of the *Pasteurella haemolytica* leukotoxin determinant. J. Bacteriol. **171**:916–928.

97. Strathdee, C.A., and R.Y.C. Lo. 1989. Regulation of expression of the *Pasteurella haemolytica* leukotoxin determinant. J. Bacteriol. **171**:5955–5962.

98. Sun, D.X., J.J. Mekalanos, and R.K. Taylor. 1990. Antibodies directed against the toxin-coregulated pilus isolated from *Vibrio cholerae* provide protection in the infant mouse experimental cholera model. J. Infect. Dis. **161**:1231–1236.

99. Taylor, P.W. 1983. Bactericidal and bacteriolytic activity of serum against gram-negative bacteria. Microbiol. Rev. **47**:46–83.

100. Taylor, R.K., V.L. Miller, D.B. Furlong, and J.J. Mekalanos. 1987. Use of *phoA* gene fusions to identify a pilus colonization factor coordinately regulated with cholera toxin. Proc. Natl. Acad. Sci. USA **84**:2833–2837.

101. Thomas, D.D., J.B. Baseman, and J.F. Alderete. 1985. Putative *Treponema pallidum* cytadhesins share a common functional domain. Infect. Immun. **49**:833–835.

102. Vasil, M.L., R.M. Berka, G.L. Gray, and H. Nakai. 1982. Cloning of a phosphate-regulated hemolysin gene (phospholipase C) from *Pseudomonas aeruginosa*. J. Bacteriol. **152**:431–440.

103. Weiss, A.A., and E.L. Hewlett. 1986. Virulence factors of *Bordetella pertussis*. Ann. Rev. Microbiol. **40**:661–686.

104. Weiss, A.A., E.L. Hewlett, G.A. Myers, and S. Falkow. 1983. *Tn*5-induced mutations affecting virulence factors of *Bordetella pertussis*. Infect. Immun. **42**:33–41.

105. Weiss, A.A., E.L. Hewlett, G.A. Myers, and S. Falkow. 1984. Pertussis toxin and extracytoplasmic adenylate cyclase as virulence factors of *Bordetella pertussis*. **150**:219–222.

106. Welch, R.A. 1987. Identification of two different hemolysin determinants in uropathogenic *Proteus* isolates. Infect. Immun. **55**:2183–2190.

107. Whitnack, E., T.P. Poirier, and E.H. Beachey. 1988. Complement-mediated opsonization of group A streptococci inhibited by the binding of fibrinogen to the surface M protein fibrillae, p. 111–119. *In* M.A. Horowitz (ed.), Bacterial–host cell interaction. Alan R. Liss, Inc., New York.

108. Wilkie, B.N., R.J.F. Markham, and P.E. Shewen. 1980. Response of calves to lung challenge exposure with *Pasteurella haemolytica* after parenteral or pulmonary immunization. Amer. J. Vet. Res. **41**:1773–1778.

109. Willems, R., A. Paul, H.G. van der Heide, A.R. ter Avest and F.R. Mooi. 1990. Fimbrial phase variation in *Bordetella pertussis*: a novel mechanism for transcriptional regulation. EMBO J. **9**:2803–2809.

110. Yates, W.D.G., P.H.G. Stockdale, L.A. Babiuk, and R.J. Smith. 1983. Prevention of bovine pneumonic pasteurellosis with an extract of *Pasteurella haemolytica*. Can. J. Comp. Med. **47**:250–256.

111. Yother, J., T.W. Chamness, and J.D. Goguen. 1986. Temperature-controlled plasmid regulon associated with low calcium response in *Yersinia pestis*. **165**:443–447.

Antifungal Chemotherapy

13

Invasive Fungal Infections: Problems and Challenges for Developing New Antifungal Compounds

Thomas J. Walsh

I. Introduction

Invasive fungal infections have emerged during the past two decades as important pathogens causing formidable morbidity and mortality in an increasingly diverse and progressively expanding population of immunocompromised patients. Those with the acquired immune deficiency syndrome (AIDS) constitute the most rapidly growing group of patients at risk for life-threatening mycoses, especially cryptococcal meningitis [61, 102], disseminated histoplasmosis [125, 125a], and coccidioidomycosis [13, 50a]. These infections are generally community acquired and often develop as a consequence of reactivation of latent infection. Oropharyngeal and esophageal candidiasis are common infections in approximately one-half of all children and adults with AIDS [112, 118]. Selik *et al.* described features of the first 30,632 AIDS patients in the United States reported to the Centers for Disease Control [94]. Esophageal candidiasis occurred with similar frequency in both populations, being found in 15.4% of 350 children (age <13 years) with AIDS in comparison to 10.6% of all AIDS patients and 8.5–16.8% of various subpopulations of adults with AIDS.

Nosocomial fungal infections also have increased in frequency in several populations of susceptible hosts, including very-low-birth-weight infants, cancer patients receiving cytotoxic chemotherapy, organ transplant recipients, burn patients, and surgical patients with complications [4, 7, 98, 113]. According to findings from the National Nosocomial Infection Surveillance System (NNIS), nosocomial fungal infections increased from 1.95 to 4.3 per 1,000 discharges during the period from 1980 to 1990 [9]. Fungemia increased as a cause of nosocomial infections from 5.4% to 10.5%. The rate of fungemia per 1,000 discharges increased from 0.1 to 0.6. During this study period, the proportion of pathogens increased from 6.0 to 9.4%. Among the 24,713 fungal infections reported, 15,032 (61%) were due to *Candida albicans*. Other *Candida* spp.,

Aspergillus spp., *Torulopsis glabrata*, and other fungi constituted 18.9%, 1.3%, 7.6%, and 13.0%, respectively.

While invasive candidiasis is one of the most common nosocomial mycoses [18] and constitutes one of the leading causes of positive blood cultures in many medical centers, *Aspergillus* spp., *Zygomycetes, Fusarium* spp., *Trichosporon* spp., and dematiaceous fungi also have emerged as important nosocomial pathogens refractory to systemic antifungal therapy [7, 98, 115, 57, 83, 5, 35]. Despite the increasing magnitude of these infectious problems, current approaches to prevention and treatment are clearly deficient [114].

Thus invasive fungal infections constitute a major cause of morbidity and mortality among immunocompromised patients. Although recent advances in antifungal chemotherapy have had an impact on these mycoses, expanding populations of immunocompromised patients will require newer approaches to antifungal therapy. These new approaches will include early diagnostic techniques, novel antifungal compounds, and immunomodulation. Discovery, preclinical development, and clinical evaluation of novel antifungal agents are essential elements of this strategy to find antifungal therapy.

The purpose of this chapter is to review the current problems and future challenges of antifungal therapy for the 1990s. This discussion will be organized according to the fungal pathogens, their infections, and the host populations susceptible to these infections.

II. Current Status of Antifungal Chemotherapy for Invasive Mycoses

Amphotericin B has been quite literally the "gold standard" and the drug of choice for systemic fungal infections. The usage of amphotericin B desoxycholate has increased in relation to increased awareness of invasive fungal infections and with improved understanding of amelioration of its toxicity [121]. Nevertheless, the diversity of pathogens and resistance of the mycoses to amphotericin B therapy have posed new challenges for the development of alternative antifungal agents or new formulations of amphotericin B desoxycholate.

Since the development of amphotericin B in the early 1960s, other compounds have become available for treatment of systemic mycoses. These compounds include flucytosine, the imidazoles—miconazole and ketoconazole—and the triazoles—fluconazole and itraconazole. The chemical class, *in vivo* activity, advantages, and disadvantages of these compounds are summarized in Table 13.1. The biochemistry, mechanisms of action, pharmacology, antifungal activity, and toxicity of these compounds has been discussed in greater length in several comprehensive reviews [114, 121, 88, 44, 47, 104, 37].

Amphotericin B is a polyene antifungal agent interacting with ergosterol within the fungal cell membrane. It has broad-spectrum activity and is the drug of choice for the initial treatment of most life-threatening mycoses. Amphotericin B is administered via the intravenous (IV) route in a micellar dispersion with desoxy-

Table 13.1. Antifungal compounds currently available for treatment of systemic mycoses

Antifungal compound	Chemical class	In vivo activity	Advantages	Disadvantages
Amphotericin B	Polyene	*Candida* spp. *C. neoformans* *H. capsulatum* *B. dermatitidis* *C. immitis* *Aspergillus* spp. *Trichosporon* spp. *Fusarium* spp. dematiaceous fungi	Fungicidal, long plasma t1/2, broad spectrum	Acute toxicity, nephrotoxicity, thrombophlebitis, erythroid suppression
Flucytosine	Fluorinated Pyrimidine	Combination with amphotericin B against deep mycoses due to *C. neoformans* and *Candida* spp.	Good CSF penetration	Myelosuppression, diarrhea, hepatotoxicity
Miconazole	Imidazole	Now used only in selected cases of infection due to *Pseydallescheria boydii*	None	Thrombophlebitis, hepatitis, pruritis, GI symptoms, cardiotoxicity
Ketoconazole	Imidazole	*Candida* spp. *C. neoformans* *B. dermatitidis* *H. capculatum* *C. immitis*	Oral formulation	Fungistatic, GI symptoms, hepatitis, endocrine effects, no parenteral formulation
Fluconazole	Triazole	*Candida* spp. *C. neoformans* *C. immitis* *H. capsulatum*[a] *B. dermatitidis*[a] *Aspergillus* spp.[a] dematiaceous fungi	Oral & parenteral, long t1/2, penetrates CSF	Fungistatic, hepatitis
Itraconazole	Triazole	*Candida* spp. *C. neoformans* *C. immitis* *H. capsulatum*[a] *B. dermatitidis*[a] *Aspergillus* spp.[a] dematiaceous fungi	Broad spectrum, oral formulation	GI absorption is pH-dependent,[b] no parenteral formulation, hepatitis

[a] Activity in high doses in animal models only; clinical trials against these pathogens are underway.

[b] Also pertains to ketoconazole.

cholate. Its acute toxic side effects include chills, rigor, and fever, while its more problematic chronic side effects include azotemia, hypokalemia, and renal tubular acidosis, all consequences of its nephrotoxicity.

Flucytosine is a fluorinated pyrimidine that inhibits thymidylate synthetase and DNA synthesis. Due to the potential emergence of resistance, flucytosine is used in combination with amphotericin B. The combination is employed in treatment of cryptococcal meningitis, deeply invasive candidiasis (particularly in disseminated infection), renal candidiasis, and *Candida* endophthalmitis. It is administered orally (PO) or IV and causes side effects of bone marrow suppression, diarrhea, and hepatitis.

Miconazole and ketoconazole belong to the imidazole class of compounds. All antifungal azoles inhibit fungal cytochrome P450 (see chapter 16). Miconazole was the first parenterally administered antifungal azole. However, adverse side effects of hepatitis, pruritis, phlebitis, and cardiotoxicity, as well as the availability of less toxic systemic antifungal azoles, have precluded the use of this agent against most invasive mycoses. Ketoconazole, by comparison, was the first orally administered and systemically absorbed antifungal azole to be implemented. Ketoconazole is used in the treatment of mucosal candidiasis, blastomycosis, chronic cavitary histoplasmosis, coccidioidomycosis, and paracoccidioidomycosis. It does not penetrate effectively into the CNS and is therefore not used to treat cryptococcal infections. Its side effects include hepatitis; suppression of sterol biosynthesis; and dose-limiting anorexia, nausea, and vomiting. The bioavailability of ketoconazole may be impaired by elevated gastric pH, such as that due to oral antacids, H2 receptor-blocking agents, and intrinsic achlorhydria.

Fluconazole and itraconazole are recently developed antifungal triazoles. Fluconazole is utilized in treating cryptococcal meningitis and mucocutaneous candidiasis. Although technically FDA approved for use in "systemic" candidiasis, prospective clinical trials only now are currently investigating fluconazole's role for prevention and treatment of fungemia and deeply invasive or systemic candidiasis. It is available in oral and parenteral formulation. Fluconazole has a long plasma half-life (approximately 24 hours in adults) and is cleared predominantly by the kidney. The bioavailability of orally administered fluconazole is not affected by gastric pH. Side effects consist of hepatitis, nausea, and vomiting. A rare toxic epidermolysis has been reported with fluconazole in patients receiving other possibly causative medications.

Itraconazole is currently available only in an orally administered preparation, but an investigational parenteral formulation may be forthcoming. Itraconazole is active in cryptococcosis, histoplasmosis, sporotrichosis, paracoccidioidomycosis, coccidioidomycosis, and selected cases of aspergillosis. Bioavailability is compromised in conditions of elevated gastric pH. Side effects include hepatitis and an unusual mineralocorticoid-induced syndrome of hypokalemia with peripheral edema.

A. *Opportunistic Yeasts*

1. Candida

Invasive candidiasis is the most common nosocomial mycosis and the most common fungal infection in patients with HIV infection, perhaps reflecting *Candida*'s niche as a component of the endogenous flora of the human alimentary tract. Conditions due to *Candida* spp. include, but are not limited to, oropharyngeal candidiasis [70, 82, 90], esophageal candidiasis [106, 112], gastrointestinal candidiasis [34, 124], hematogenous *Candida* endophthalmitis [14, 31], hepatosplenic candidiasis [107], renal candidiasis [30, 87], *Candida* cystitis [30], epiglottitis [111], peritonitis [32], cholecystitis [2], prosthetic valve endocarditis [48, 66], vascular catheter-associated fungemia, and thrombophlebitis [99, 117]. The National Nosocomial Infections Survey from the Centers for Disease Control estimates that the frequency of deeply invasive candidiasis has increased nearly tenfold during the past decade. This observation, as well as estimates of the other nosocomial mycoses, may underestimate the true magnitude of the problem of nosocomial mycoses due to the marked decline in postmortem rate, the lack of sensitivity of blood culture detection systems for invasive candidiasis, and the lack of clinically available diagnostic markers for invasive candidiasis.

Many infections due to *Candida* spp. are refractory to antifungal therapy. The patterns of resistance to amphotericin B may be classified as *microbiological resistance* or *clinical resistance* [114]. Microbiological resistance is defined as an infection due to a fungus with a minimal inhibitory concentration (MIC) >2.0 μg/ml to amphotericin B. Clinical resistance is defined as the absence of clinical response to amphotericin B by an infection due to a fungus susceptible to amphotericin B. Isolates of *Candida albicans* and *Candida tropicalis*, the two most common causes of invasive candidiasis, were found to be significantly associated with increased mortality in bone marrow transplant recipients when their MICs were elevated in comparison to strains isolated from patients who survived [44]. These findings may be particularly ominous for *C. tropicalis*, due to its increased virulence in granulocytopenic hosts in comparison to *C. albicans* [89, 126].

Candida parapsilosis has been associated with intravascular catheter infections [96] and may exhibit tolerance to amphotericin B; *i.e.*, the minimum lethal concentration (MLC) greatly exceeds the MIC. Such a pattern of resistance may be particularly important in granulocytopenic patients where killing of fungi is dependent almost exclusively upon the antifungal agent. High-level resistance to polyenes has been observed in *Candida lusitaniae* [43, 75] and *Candida guilliermondii* [26]. Isolates of these polyene-resistant fungi have caused fatal refractory disseminated infection. While these infrequently encountered fungi are less virulent than *C. albicans* and *C. tropicalis*, they can cause refractory fatal infections in profoundly compromised hosts [26, 43]. *Torulopsis (Candida) gla-*

brata may emerge as a cause of infection during therapy with antifungal azoles, such as ketoconazole or fluconazole. As such patterns of microbiological resistance in *Candida* species continue to develop in granulocytopenic patients, the need for more effective antifungal therapy becomes ever more critical.

Clinical resistance to amphotcricin B also has been described with increasing frequency. Clinically resistant infections may develop more readily in patients with foreign bodies, profound immunosuppression, or advanced fungal lesions of deep viscera, such as hepatosplenic candidiasis. Among the foreign bodies infected with *Candida* and other yeast-like fungi are vascular catheters (including umbilical vein catheters in neonates and chronically indwelling central silastic venous catheters in cancer patients), prosthetic cardiac valves, and peritoneal dialysis catheters.

2. Cryptococcus Neoformans

Cryptococcus neoformans, which is naturally found in soil, fruits, vegetables, and pigeon excreta, has emerged as a common life-threatening pathogen in patients with HIV infection. *C. neoformans* causes meningoencephalitis in approximately 6–13% of adult patients with AIDS [27, 33, 51]. By contrast, cryptococcal meningoencephalitis, as well as other deeply invasive mycoses, is unusual in children with HIV infection. Fever, headache, and altered mental status in cryptococcal meningoencephalitis are usually indolent, often evolving over the course of weeks to months. A subset of HIV-infected patients with cryptococcal meningoencephalitis may present with altered mental status, papilledema, and other signs of increased intracranial pressure. Such patients may undergo a rapidly fatal deterioration, despite antifungal therapy [27a, 52].

Patients with AIDS-associated cryptococcal meningitis have a high propensity for relapse [128]. Initial therapy of cryptococcal meningitis in HIV-infected patients is often not successful, indicating the need for more effective therapy. Perhaps this lack of initial efficacy is reflected in the late recurrences of cryptococcal meningitis. Chronic suppressive therapy of cryptococcal meningitis in AIDS is essential in order to prevent recurrence of infection. Amphotericin B was initially found to reduce recurrences in comparison to those receiving no suppressive therapy; fluconazole also has been utilized successfully for suppression of recurrences [102]. Nevertheless, despite suppressive therapy, recurrences of cryptococcal meningitis are now being observed in patients with declining CD4 cell counts. Finally, other recent findings indicate that the prostate may be a focus for persistent cryptococcal infection [57].

3. Trichosporon *spp.*

Trichosporonosis is an uncommon but increasingly recognized cause of refractory fatal invasive fungal infections in granulocytopenic and corticosteroid-treated patients [115]. *Trichosporon beigelii* may be inhibited but not killed by safely achievable levels of amphotericin B [123], resulting in intractable fungemia and disseminated tissue infection. Newer strategies for antifungal therapy are clearly needed for disseminated infections caused by this organism.

B. Opportunistic Filamentous Fungi

1. Aspergillus

Invasive pulmonary aspergillosis (IPA) is a particularly perplexing problem in organ transplant recipients [15] and those with granulocytopenia due to intensive cytotoxic chemotherapy [15, 110, 116]. *Aspergillus fumigatus, Aspergillus flavus*, and other less common species that are naturally isolated from soil and air samples generally enter immunocompromised hosts via the respiratory tract. Pathogenic *Aspergillus* spp. have a strong propensity for invasion of blood vessels. This predilection for blood vessel invasion permits the organism to cause thrombosis and hemorrhagic infarction of infected tissue [74]. Although amphotericin B may achieve high pulmonary concentrations, *Aspergillus* may be recovered from pulmonary tissue [19]. These findings suggest that amphotericin B and perhaps other antifungal compounds may not be able to penetrate these infarcted tissue sites.

Early diagnosis and treatment have improved survival in invasive pulmonary aspergillosis [15]. Nevertheless, invasive pulmonary aspergillosis may develop during empirical antifungal therapy [15, 42, 79], progressing relentlessly to multilobar infection of disseminating to cause infection of the central nervous system [120]. Newer antifungal azole compounds, such as itraconazole, appear promising [25]; however, such compounds are fungistatic and may have better activity in non-neutropenic patients where neutrophils may also contribute to further destruction of these invasive fungi. Although recovery from neutropenia and reversal of other immunosuppressive factors is essential for successful treatment of invasive aspergillosis as well as many other deep mycoses, antifungal therapy remains essential for optimal clearance of infection, particularly in patients who receive repeated cycles of immunosuppressive therapy following their initial episode of aspergillosis [50].

2. Zygomycetes

The most common agents of zygomycosis are *Rhizopus, Mucor, Rhizomucor, Absidia,* and *Cunninghamella*. These fungi naturally occur in soil, dust, and decaying foods. The pathogenic *Zygomycetes* also invade blood vessels, resulting in extensive tissue infarction in compromised hosts [83]. The most common syndromes are those of invasive pulmonary zygomycosis, which occurs predominantly in neutropenic patients and organ transplant recipients, and the rhinocerebral syndrome, which most frequently develops in patients with diabetic metabolic acidosis. These organisms may be completely resistant to antifungal therapy *in vitro* and *in vivo*. The extensive surgical debridement of infected sinuses is an essential element in managing these infections, but also underscores the unfortunate paucity of effective agents against Zygomycetes. Another patient population, those receiving deferoxamine, are a newly recognized group at risk for severe pulmonary and disseminated zygomycotic infections. The siderophores of Zygomycetes apparently are capable of successfully competing for iron mobilized by chelation therapy and proliferating in the iron-rich tissue milieu.

3. Pseudallescheria boydii (Scedosporium apiospermium)

This filamentous fungus not only has the same propensity for invading blood vessels as does Aspergillus and Zygomycetes but may also be completely resistant to amphotericin B [85]. The clinical manifestations of *Pseudallescheria* pneumonia, pulmonary hemorrhagic infarction, and CNS infection are similar to those of aspergillosis.

4. Fusarium

This emerging fungal pathogen often does not respond to conventional doses of amphotericin B and may require substantially higher doses (1.0–1.5 mg/kg/day) for successful outcome [67]. Some cases of invasive *Fusarium* infection may be completely refractory to amphotericin B [5].

5. Dematiaceous Fungi

These uncommon causes of deep mycoses are being recognized with increasing frequency. Such pathogens as *Dreschlera, Bipolaris, Cladosporium (Xylohypha)*, and *Dactylaria* have a predilection for invasion of the CNS and elaborate a melanin compound that may be a virulence factor [29, 35].

C. Fungi Causing Endemic Mycoses

1. Histoplasma capsulatum

Histoplasma capsulatum is a dimorphic fungus found in soil and excreta of birds. The conidia of *H. capsulatum* are small enough to be inhaled to the level of the alveolus, where they are confronted by alveolar macrophages and other host defenses. The conidia convert to small, budding yeast cells, which are readily ingested by macrophages. This interaction induces a granulomatous, inflammatory response. Within the macrophages and monocytes, the yeast cells may be disseminated hematogenously to other reticuloendothelial tissues, where granulomas may form. An immune response is elicited and the asymptomatic infection is resolved. In 95% of individuals, the only evidence of this self-limited process is the acquisition of a delayed skin test reaction to histoplasmin and residual calcifications in the lung and, occasionally, the spleen. The infection may not resolve, however, but proceed to progressive primary pulmonary or disseminated histoplasmosis. Quiescent foci of histoplasmosis in the lung or elsewhere have the potential subsequently to become reactivated [125].

Disseminated histoplasmosis usually develops in immunocompromised patients, including infants and children and adults with cellular immunodeficiencies. Patients with AIDS and histoplasmosis generally have represented with a refractory pneumonia or disseminated infection. Fever and weight loss are common but nonspecific symptoms. However, an acute septic shock-like syndrome of disseminated histoplasmosis and AIDS has been reported [125, 125a]. Patients with AIDS and histoplasmosis also appear to require lifelong antifungal suppression of their infection [125, 125a].

2. Coccidioides Immitis

Coccidioides immitis is another endemic fungal pathogen of emerging importance in patients with AIDS in the southwest United States, where it is a common soil fungus [1, 13, 86]. Certain patient populations are apparently more susceptible to progressive pneumonia, complicated pneumonia, and dissemination. Such high-risk populations include those with AIDS, those receiving immunosuppressive chemotherapy, certain racial groups (African Americans, Filipinos, and Native Americans), and women in the third trimester of pregnancy. Primary pulmonary infection may evolve into one of several processes: pulmonary nodules, thin-walled cavities, progressive pneumonia, and pyopneumothorax. Dissemination to extrapulmonary sites may result in cutaneous and soft tissue infection, osteomyelitis, arthritis, and meningitis. The latter is particularly refractory to antifungal therapy and generally requires lifelong suppression.

Virtually all cases of coccidioidomycosis reported thus far in AIDS patients have been observed in adults and virtually all have been disseminated or progressive pulmonary infections. More effective antifungal compounds certainly are warranted for treatment of coccidioidomycosis. Mortality is due to this infection in HIV-infected patients is high. Whether the patient has AIDS or an apparently intact immune system, coccidioidal meningitis is a formidable therapeutic challenge that generally requires lifelong treatment.

Table 13.2 summarizes the relationship between invasive fungi and sites of infection.

Table 13.2. Common opportunistic and pathogenic fungi and site of infection

Organism	Common primary foci of infection and portals of entry	Common secondary foci
Opportunistic yeasts		
Candida spp.	Gastrointestinal tract,	Kidney, liver, spleen, heart,
C. albicans	vascular catheters	eyes, skin (dermis), brain
C. tropicalis		
C. parapsilosis		
C. krusei		
C. lusitaniae		
C. guilliermondii		
C. glabrata		
Trichosporon spp.	Gastrointestinal tract,	Kidney, eyes, skin, lungs
T. beigelii	vascular catheters	
T. capitatum[a]		
Cryptococcus neoformans	Lower respiratory tract	Brain (meningoencephalitis), bone, prostate
Opportunistic hyphomycetes		
Aspergillus spp.	Respiratory tract	Brain (infarction)
A. fumigatus	(pneumonia and sinusitis)	
A. flavus		
A. terreus		
Zygomycetes	Respiratory tract	Brain (infarction)
Rhizopus arrhizus	(pneumonia and sinusitis)	
Cunninghamella bertholletiae		
Absidia corymbifera		
Mucor circinelloides		
Pseudallescheria boydii[b]	Respiratory tract (pneumonia and sinusitis)	Brain (infarction)
Fusarium spp.	Respiratory tract	Skin
F. solani	(pneumonia and sinusitis)	
F. oxysporum		
F. moniliforme		
Dematiaceous hyphomycetes[c]	Respiratory tract	Brain
Pathogenic fungi		
Histoplasma capsulatum	Lower respiratory tract	Spleen, liver, bone marrow
Blastomyces dermatitidis	Lower respiratory tract	Skin, bone, prostate
Coccidioides immitis	Lower respiratory tract	Bone, soft tissue, brain

[a] *Trichosporon captitatum* has been reclassified has *Blastoschizomyces capitatus*.

[b] The synanamorph of *Pseudallescheria boydii* is *Scedosporium apiospermium*, which also may cause infection.

[c] Dematiaceous hyphomycetes causing phaeohyphomycosis include *Cladosporium bantianum (Xylohypha bantiana), Dreschlera* spp., *Bipolaris* spp., and *Dactylaria gallopava*. Black yeasts, such as *Wangiella dermatitidis*, may also cause a refractory deep visceral infection.

III. Increasing Populations of Susceptible Hosts

A. HIV Infection

The deficits of cell-mediated immunity of AIDS predispose the patients to persistent CNS cryptococcosis, disseminated histoplasmosis, systemic coccidioidomycosis, and recurrent mucosal candidiasis. Cryptococcal meningoencephalitis in most patients without AIDS may be treated successfully by an 8-week to 10-week course of amphotericin B with or without flucytosine. Cryptococcal meningoencephalitis in patients with AIDS, however, has a high rate of recurrence, and is treated with an initial course of antifungal therapy followed by a maintenance course of fluconazole for prevention of recurrence of CNS cryptococcosis. This pattern of recurrence following treatment of deep mycoses in AIDS patients also is a feature of histoplasmosis, coccidioidomycosis, and mucosal candidiasis. Thus, this expanding population of patients with AIDS offers new challenges for maintaining remissions of fungal infections.

B. Granulocytopenia

Among the newer problems of invasive candidiasis in granulocytopenic cancer patients is the entity of hepatosplenic candidiasis. This infection develops during granulocytopenia but does not clear following recovery from granulocytopenia. Instead, the infection persists and causes fever, abdominal pain, and expanding lesions in tissue. Since hepatosplenic candidiasis is really a form of disseminated candidiasis, other organs (*e.g.*, kidney and lung) also may be infected. Treatment with amphotericin B may extend from 6 to 12 months. Alternative agents under investigation include liposomal amphotericin B and fluconazole. However, newer strategies to treat this infection are clearly needed.

When pulmonary aspergillosis develops in granulocytopenic patients, amphotericin B may not clear the infection, despite recovery from granulocytopenia. The result is a chronic, necrotizing infection requiring extended courses of therapy. Granulocytopenic patients and other profoundly immunocompromised hosts may benefit particularly from fungicidal agents which may kill the organism in the absence of neutrophils or other effector cells. These patients usually require hospitalization; thus, fungicidal agents do not need to be necessarily given by the oral route. Indeed, many of these patients are unable to tolerate oral medications and often require a parenterally administered compound.

C. Other Populations

1. Surgery Patients

Patients undergoing surgical procedures, particularly those with abdominal surgery and insertion of vascular prostheses, have an increased risk for invasive candidiasis [62, 95]. While broad-spectrum antibiotics and the use of vascular catheters are often necessary in these patients, these factors increase the risk of disseminated candidiasis in susceptible hosts. Patients suffering from severe burns have similar risk factors and often develop severe complications due to invasive candidiasis.

2. *Newborn Infants*

Very-low-birth-weight infants have a particularly high risk for invasive candidiasis [16]. Factors such as broad-spectrum antibiotics, umbilical vein catheters, and quantitative and qualitative neutrophil deficiencies, as well as cellular immune responses, predispose these children to invasive candidiasis. Amphotericin B is the drug of choice for invasive candidiasis; however, the desoxycholate formulation of amphotericin B creates renal toxicity, especially hypokalemia, which may be especially refractory. Newer agents in this population are clearly needed.

Table 13.3 summarizes the relationships among patient populations at risk, the underlying host defense deficit, and commonly associated fungi.

IV. Future Directions of Antifungal Therapy and Prevention

A. *Investigational Antifungal Compounds*

Candidiasis, aspergillosis, cryptococcosis, the infections due to emerging fungal pathogens are increasing challenges to managing immunocompromised patients. The emerging trends in the 1990s for treatment and prevention of invasive fungal infections indicate the need for antifungal agents with specific properties. An antifungal agent used for *treatment* of deeply invasive mycoses *ideally* should be fungicidal and capable of penetrating all tissue sites. Desirable, but not essential, features of an antifungal agent used for *treatment* of deeply invasive fungal infections include oral and parenteral flexibility, long plasma half-life, and broad spectrum. However, when an agent is utilized for *prophylaxis*, these latter attributes also would be beneficial. Minimal toxicity should be a *sine qua non* of new antifungals.

Newer antifungal triazoles continue to be developed. The isobutyl triazole, SCH–39304, has been found to be active in a wide spectrum of experimental mycoses, including disseminated candidiasis [122], invasive aspergillosis [22, 58], cryptococcal meningitis [76], coccidioidomycosis [20, 24], pulmonary blastomycosis [103], disseminated fusariosis [3], and chromoblastomycosis [23]. SCH–39304 has a long half-life in humans (approximately 67 hours) and excellent penetration into tissue [88], including the CSF [122a]. The compound SCH–39304 is a racemic mixture, consisting of one active enantiomer, SCH–42427. This novel triazole was undergoing clinical trials in patients with AIDS and with neoplastic diseases. Unfortunately, recent chronic toxicological studies demon-

Table 13.3. Patient populations frequently at risk for systemic fungal infections

Deficit in host defense	Patient populations	Invasive fungi
Neutrophils		
Granulocytopenia	Recipients of cytotoxic chemotherapy, neoplastic diseases, bone marrow transplantation, organ transplantation	*Candida* spp. (deep infections), *Aspergillus* spp., *Zygomycetes*, *Fusarium* spp. *Trichosporon* spp.
Impaired neutrophil fungicidal activity	Chronic granulomatous disease (CGD), corticosteroid therapy	*Aspergillus* spp. (esp., CGD)
Neutrophils & CMI		
Corticosteroid therapy	Patients receiving treatment for neoplastic diseases, bone marrow transplantation (graft-vs.-host disease), organ transplantation (kidney, liver, heart, lung), autoimmune diseases, neurosurgical procedures, pulmonary disorders	*Candida* spp., *Aspergillus* spp., *Trichosporon* spp., *Fusarium* spp., *Pseudallescheria boydii*, *Zygomycetes*
CMI[a]		
Acquired immunodeficiency syndrome	Potentially any host having close contact with infected blood or sexual secretions	*Candida* spp. (mucosal infections), *Cryptococcus neoformans*, *Histoplasma capsulatum*, *Coccidioides immitis*
Cyclosporin A	Patients receiving treatment for bone marrow transplantation (graft-vs.-host disease), organ transplantation	*Cryptococcus neoformans*, *Candida* spp.
Disruption of mechanical barriers	Vascular catheters; surgical patients; very-low-birth-weight infants; recipients of cytotoxic chemotherapy: neoplastic diseases, bone marrow transplantation, organ transplantation; patients receiving total parenteral nutrition; illicit intravenous drug users; prosthetic valve surgery; disrupted mucosal integrity; recipients of cytotoxic chemotherapy: neoplastic diseases, bone marrow transplantation organ transplantation; gastrointestinal surgery	*Candida* spp.
Disruption of microbiological barriers	Broad-spectrum antibiotics recipients of cytotoxic chemotherapy: neoplastic diseases, bone marrow transplantation, organ transplantation; gastrointestinal surgery; very-low-birth-weigth infants	*Candida* spp.

[a] CMI: cell-mediated immunity; immunity mediated by T-lymphocytes, NK cells, monocytes, and macrophages.

strating hepatic tumors in rodents may alter development of this potent compound, perhaps limiting this promising agent to refractory mycoses.

Several antifungal compounds with different mechanisms of action are being developed. Some of these agents are active against enzymes involved in fungal cell wall biosynthesis [17]. Glucan synthase and chitin synthase are two enzymes involved in the biosynthesis of two fungal cell wall polymers—glucan and chitin (see chapters 18 and 19). Echinocandin and aculeacin compounds, which inhibit glucan synthetase, have been found to be effective agents in experimental disseminated candidiasis. Cilofungin (LY–121019), a semisynthetic lipopeptide of the echinocandin class [21, 40], and other echinocandins are fungicidal agents that disrupt fungal cell wall biosynthesis [78, 104]. Cilofungin has potent activity against *Candida* species, particularly *Candida albicans* and *Candida tropicalis* [46, 53, 67, 72], and was found to be as rapidly fungicidal as amphotericin B in *in vitro* timed-kill assays. Several of the echinocandins also exhibit activity against *Pneumocystis carinii* comparable or exceeding that of trimethoprim-sulfamethoxazole [91A]; *P. carinii* appears to be phylogenetically more closely related to the kingdom Fungi than has been previously appreciated. Cilofungin does not penetrate well into the CNS when administered by intermittent infusion twice daily. However, recent pharmacokinetic studies indicate that cilofungin follows nonlinear saturation kinetics commensurate with continuous infusion, permitting higher tissue concentrations in the CNS [54]. The effect of continuous infusion of this and similar short-half-life antifungal compounds is currently being studied as a possible means of increasing antifungal efficacy. Another glucan synthetase inhibitor, L–671,329, has been found to have properties similar to that of cilofungin but with a broader antifungal spectrum [38]. These compounds are peptide nucleosides that are transported into the fungal cell via peptide permeases. While polyoxins appear to be devoid of useful antifungal activity *in vivo*, the nikkomycins demonstrate activity against several pathogens, including *Candida albicans* [9] and *Coccidioides immitis* [45].

The allylamines are another group of investigational antifungal agents. Terbinafine and naftifine have been among the most extensively investigated of these compounds [77]. The apparent mechanism of action of terbinafine and other allylamines is inhibition of squalene epoxidase in the ergosterol biosynthetic pathway (see chapter 16). Terbinafine has potent *in vitro* activity against *Aspergillus* spp., *Sporothrix schenckii*, and dermatophytes [93]. Although terbinafine was found to be highly active against dermatophytes *in vivo* [109], other animal studies of experimental invasive aspergillosis, candidiasis, or phaeohyphomycosis have revealed little or no activity. For example, Dixon and Polak found good *in vitro* activity but poor *in vivo* antifungal effect of terbinafine against *Cladosporium bantianum (Xylohypha bantiana), Wangiella dermatitidis*, and *Dactylaria constricta* [28]. This disparity between *in vitro* and *in vivo* activity may be due to a short plasma half-life, rapid redistribution into cutaneous tissues, and high protein or lipid binding.

The dimethylmorpholines are another class of antifungal compounds, which include amorolfine and fenpromimorph. These agents inhibit ergosterol biosynthesis, but also appear to have an effect on NADH oxidase and succinate cytochrome C reductase. (See chapter 16.) The inhibition of synthesis of C-14 methyl sterols and 8,14 sterols was associated with irregular deposition of chitin. Amorolfine exhibits broad-spectrum *in vitro* activity, good *in vivo* activity against experimental dermatophytosis and vaginal candidiasis, but little effect in experimental, deeply invasive mycoses [80]. The paucity of activity of amorolfine against deeply invasive mycoses has limited its utility in patients.

Ambruticin is a cyclopropyl-polyene-pyran acid with *in vitro* activity against *C. immitis, H. capsulatum*, and *B. dermatitidis*. Ambruticin was highly effective in achieving microbiological cures and survival in mice with disseminated coccidioidomycosis [84]. Unfortunately, the compound has only limited *in vivo* activity against *C. albicans*.

Pradimicins are a novel group of fungicidal compounds derived from fermentation [73, 91]. Initially, purified compounds were relatively insoluble in aqueous solution; however, water-soluble analogs, especially BMY–28864, were derived for intravenous administration. BMY–28864 was found to be active in murine models of candidiasis, aspergillosis, and cryptococcosis at apparently well-tolerated doses. Benanomicins are another group of fungicidal compounds with low toxicity that are active *in vivo* against candidiasis, cryptococcosis, and aspergillosis [105]. Benanomicins compounds consist of a benzonapthacene quinone chromophore to which D-alanine and a disaccharide are covalently bound. Both pradimicins and benanomicins have broad spectra of activity that include fungi, bacteria, and viruses.

Novel formulations of existing compounds constitute another strategy for administration of antifungal agents. Liposomal amphotericin B is the paradigm of this strategy. Amphotericin B, when formulated into a liposomal suspension, can be administered in doses substantially higher than those achieved with amphotericin B desoxycholate alone [59]. This agent has been found to be particularly effective in management of refractory hepatosplenic candidiasis [60]. The formulation developed by Lopez-Bernstein and colleagues has been modified to yield an amphotericin B lipid complex (ABLC), which will constitute a stable formulation to be administered in multicenter clinical studies. A unilamellar formulation of liposomal amphotericin B is being studied in clinical trials in Western Europe [69, 108]. This formulation has a smaller volume of distribution but achieves higher peak plasma concentrations than does ABLC or desoxycholate amphotericin B. The pharmacodynamic significance of these and other lipid formulations is currently being explored.

Another strategy for administration of amphotericin B is an intranasal amphotericin B spray for the prevention of pulmonary aspergillosis [68]. Patients receiving intranasal amphotericin B spray had fewer episodes of invasive aspergillosis in comparison to untreated control patients. Other studies currently in progress may help to elucidate the role of this prophylactic regimen. Inhaled aerosolized amphotericin B may be another approach for achieving high intrapulmonary concentrations of amphotericin B with potentially minimal systemic toxicity [93].

B. Diagnostic Methodologies

Early diagnosis is an essential element for the timely initiation and optimal outcome of antifungal therapy. Newer methods for detection of cell wall antigens, cytoplasmic antigens, metabolites, and nucleotide fragments may permit a panel of diagnostic tools to guide the initiation and duration of antifungal therapy in the future [10, 36, 39, 49, 63–65, 100, 101, 119, 127].

C. Immunomodulators

Several studies have demonstrated that recombinant human hematopoietic colony-stimulating factors, particularly granulocyte-macrophage colony-stimulating factor (GM-CSF) or granulocyte colony-stimulating factor (G-CSF), shortened the duration and decreased the depth of granulocytopenia in patients treated with cytotoxic chemotherapy. For example, patients receiving melphalan or combination chemotherapy for small-cell lung cancer, soft tissue sarcoma, or autologous bone marrow transplantation had reduced granulocytopenia when administered G-CSF or GM-CSF [6, 11, 12, 71]. Other agents, such as gamma interferon and several interleukins, are being investigated for their immunomodulatory properties against invasive fungal infections. The appropriate use of these promising agents will, it is hoped, translate into reduced frequency and severity of fungal infections in patients receiving cytotoxic chemotherapy.

References

1. Abrams, D., H. Luber, H. Glueck, and *et al.* 1985. Disseminated coccidioidomycosis in AIDS.N. Eng. J. Med. **310**:986–987.
2. Adamson, P.C., M.G. Rinaldi, P.A. Pizzo, and T.J. Walsh. 1989. Amphotericin B in treatment of *Candida* cholecystitis. Ped. Infect. Dis. J. **8**:408–411.
3. Anaissie, E., D.P. Kontoyiannis and G. Bodey. 1990. Fusarial hyalohyphomycosis: Clinical spectrum, prognosis, and treatment. 30th Intersci. Conf. Antimicrob. Agents Chemother., abstr. 1128.
4. Anaissie, E., and G.P. Bodey. 1989. Nosocomial fungal infections. Old problems and new challenges. Infect. Dis. Clin. N. Amer. **3:**867–882.
5. Anaissie, E., H. Kantarjian, J. Ro, and G. Bodey. 1988. The emerging role of *Fusarium* infections in patients with cancer. Medicine **67**:77–83.
6. Antman, K., J. Griffin, A. Elias, L. Ryan, S. Cannistra, D. Oette, M. Whitley, E. Frei 3rd, and L. Schnipper. 1988. Effect of recombinant human granulocyte-

macrophage colony-stimulating factor on chemotherapy-induced myelosuppression. N. Eng. J. Med. **319**:593–598.

7. Armstrong, D., 1989. Problems in management of opportunistic fungal infections. Rev. Infect. Dis. 11 (Suppl. 7):S1591.
8. Becker, J.M., S. Marcus, J. Tallock, D. Miller, E. Krainer, R. Khare, and F. Naider. 1988. Use of the chitin-synthesis inhibitor nikkomycin to treat disseminated candidiasis in mice. J. Infect. Dis. **147**:212–214.
9. Beck-Sague, G., W. Jarvis, S. Banerjee, D. Culver, R. Gaynes, and T.N.N.I.S. System. 1990. Nosocomial fungal infections in U.S. hospitals, 1980–1990. 30th Intersci. Conf. Antimicrob. Agents Chemother., abstr. 1129.
10. Bousgnoux, M.E., C. Hill, D. Moissenet, M.F. DeChauvin, M. Bonnay, I. Vicens-Sprauel, F. Pietri, M. McNeil, L. Kaufman, J. Dupouy-Camet, C. Bohuon, and A. Andremont. 1990. Comparison of antibody, antigen, and metabolite assays for hospitalized patients with disseminated or peripheral candidiasis. J. Clin. Microbiol. **28**:905 – 909.
11. Brandt, S., W.P. Peters, K. Waters, S.K. Atwater, J. Kurtzberg, M.J. Borowitz, R.B. Jones, E.J. Shpall, R.C. Bast, Jr., C.J. Gilbert, and D.H. Oette. 1988. Effect of recombinant human granulocyte-macrophage colony stimulating factor on hematopoietic reconstitution after high dose of chemotherapy and autologous bone marrow transplantation. N. Eng. J. Med. **318**:869–876.
12. Bronchud, M., J. Scarffe, N. Thatcher, D. Crowther, L. Souza, N. Alton, N. Testa, and T. Dexter. 1987. Phase I/II study of recombinant human granulocyte colony-stimulating factor in patients receiving intensive chemotherapy for small-cell lung cancer. Br. J. Cancer. **56**:809–813.
13. Bronniman, D., R. Adam, J. Galgiani, M. Habib, E. Petersen, B. Porter, and J. Bloom. 1987. Coccidioidomycosis in the acquired immunodeficiency syndrome. Ann. Int. Med. **106**:372–379.
14. Brooks, R.G. 1989. Prospective study of *Candida* endophthalmitis in hospitalized patients with candidemia. Arch. Int. Med. **149**:2226–2228
15. Burch, P.A., J.E. Karp, W.G. Merz, J.E. Kuhlman, and E.K. Fishman. 1987. Favorable outcome of invasive aspergillosis in patients with acute leukemia. J. Clin. Oncol. **5**:1985–1993.
16. Butler, K., and C. Baker. 1988. Candida: An increasingly important pathogen in the nursery. Pediatr. Clin. N. Amer. **35**:543–563.
17. Cassone, A. 1986. Cell wall of pathogenic yeasts and implications for antimycotic therapy. Drugs Exp. Clin. Res. **12**:635–643.
18. Centers for Disease Control. 1984. Nosocomial infection surveillance. CDC Surveillance Summary 1983. **33**:9–22.
19. Christiansen, K., E. Bernard, J. Gold, and D. Armstrong. 1985. Distribution and activity of amphotericin B in humans. J. Infect. Dis. **152**:1037–1043.
20. Clemons, K., L. Hanson, A. Perlman, and D. Stevens. 1989. Efficacy of SCH39304 in a systemic model of murine coccidioidomycosis. 31st Intersci. Conf. Antimicrob. Agents Chemother., abstr. 823.

21. DeBono, M., B.J., Abbott, J. Turner, L. Howard, R. Gordee, A. Hunt, M. Barnhart, R. Molloy, K. Willard, D. Fukuda, T. Butler, and D.J. Zeckner. 1988. Synthesis and evaluation of LY121019, a member of a series of semisynthetic analogues of the antifungal lipopeptide echinocandin B. Ann. N.Y. Acad. Sci. **544**:152–167.

22. Defaveri, J., and J. Graybill. 1989. SCH–39304 in treatment of murine pulmonary aspergillosis. 29th Intersci. Conf. Antimicrob. Agents Chemother., abstr. 822.

23. Defaveri, J., and J. Graybill. 1989. Treatment of chromoblastomycosis with antifungal azoles. 29th Intersci. Conf. Antimicrob. Agents Chemother., abstr. 820.

24. Defaveri, J., S. Sun, and J. Graybill. 1989. Treatment of murine coccidioidal meningitis with SCH39304. 29th Intersci. Conf. Antimicrob. Agents Chemother., abstr. 824.

25. Denning, D., Tucker, R.M., Hanson, L.H., and D. Stevens. 1989. Treatment of invasive aspergillosis with itraconazole. Am. J. Med. **86**:791–800.

26. Dick, J.D., B.R. Rosengard, W.G. Merz, R.K. Stuart, G.M. Hutchins, and R. Saral. 1985. Fatal disseminated candidiasis due to amphotericin B-resistant *Candida guilliermondii*. Ann. Int. Med. **102**:67–68.

27. Dismukes, W. 1988. Cryptococcal meningitis in patients with AIDS.J. Infect. Dis. **157**:624–628.

27a. Dismukes W., G. Cloud, S. Thompson, A. Sugar, C. Tuazon, D. Kaufman, M. Grieco, J. Jacobson, W. Powderly, P. Robinson and A.L. B'ham. 1989. Fluconazole versus amphotericin B therapy of acute cryptococcal meningitis. 29th Intersci. Conf. Antimicrob. Agents Chemother., abstr. 1065.

28. Dixon, D., and A. Polak. 1987. In vitro and in vivo studies with three agents of central nervous system phaeohyphomycosis. Chemotherapy **33**:129–140.

29. Dixon, D.M., T.J. Walsh, W.G. Merz, and M.R. McGinnis. 1989. Human central nervous system infections due to *Xylohypha bantiana (Cladosporium trichoides)*. Rev. Infect. Dis. **11**:515–525.

30. Drutz, D.J., and R. Fetchick. 1988. Fungal infections of the kidney and urinary tract. p. 1015–1047. *In* R.W. Schrier, C.W. Gottschalle (eds.), Diseases of the kidney, 4th ed. Little, Brown, & Co., Boston.

31. Edwards, J.E. 1985. Candida endophthalmitis, p. 211–225. *In*: G.P. Bodey, V. Fainstein (eds.), Candidiasis. Raven Press, New York.

32. Elsenberg, E.S., I. Leviton, and R. Soeiro. 1986. Fungal peritonitis in patients receiving peritoneal dialysis: Experience with 11 patients and review of the literature. Rev. Infect. Dis. **8**:309–321.

33. Eng. R., E. Bishburg, S. Smith, and R. Kapila. 1986. Cryptoccocal infections in patients with acquired immune deficiency syndrome. Am. J. Med. **81**:19–23.

34. Eras, P., M.J. Goldstein, and P. Sherlock. 1972. *Candida* infection of the gastrointestinal tract. Medicine **51**:367–379.

35. Fader, R.C., and M.R. McGinnis. 1988. Infections caused by dematiaceous fungi: chromoblastomycosis and phaeohyphomycosis. Infect. Dis. Clin. N. Amer. **2**:925–938.

36. Franklyn, K.M., J.R. Warmington, A.K. Ott, and R.B. Ashman. 1990. An immunodominant antigen of *Candida albicans* shows homology to the enzyme enolase. Immunol. Cell. Biol. **68**:173–178.

37. Fromtling, R. 1988. Overview of medically important antifungal azole derivatives. Clin. Microb. Rev. **1**:187–217.

38. Fromtling, R.A., and G.K. Abruzzo. 1989. L–671, 329, a new antifungal agent. In vitro activity, toxicity, and efficacy in comparison to aculeacin. J. Antibiot. **42**:174–178.

39. Gold, J.W.M., B. Wong, E.M. Bernard, T.E. Kiehn, and D. Armstrong. 1983. Serum arabinitol concentrations and arabinitol/creatinine ratios in invasive candidiasis. J. Infect. Dis. **147**:504–513.

40. Gordee, R.S., D.J. Zeckner, L.F. Ellis, A.L. Thakker, and L.C. Howard. 1984. In vitro and in vivo anti-candida activity and toxicology of LY121019. Ann. N.Y. Acad. Sci. **37**:1054–1065.

41. Graybill, J. 1988. Systemic fungal infections diagnosis and treatment I: Therapeutic agents. Infect. Dis. Clin. N. Amer. **2**:805 825.

42. Group, Internat. Antimicrob. Ther. E.O.R.T.C. 1989. Empiric antifungal therapy in febrile granulocytopenic patients. Am. J. Med. **86**:668–672.

43. Guinet, R.J., A. Chanas, G. Goullier, P. Bonnefoy, and Ambroise-Thomas. 1983. Fatal septicemia due to amphotericin B-resistant *Candida lusitaniae*. J. Clin. Microbiol. **18**:443–444.

44. Hay, R., B. DuPont, and J. Graybill. 1987. First international symposium on itraconazole. Rev. Infect. Dis. **9** (suppl.):1–152.

45. Hector, R. 1990. Evaluation of nikkomycins and Z in murine models of *Coccidioidomycosis, Histoplasmosis*, and *Blastomycosis*. Antimicrob. Agents Chemother. **34**:587–593.

46. Hobbs, M., J.R. Perfect, and D. Durack. 1988. Evaluation of in vitro antifungal activity of LY121019. Euro. J. Microbiol. Infect. Dis. **7**:77–80.

47. Hoeprich, P. 1989. Chemotherapy for systemic mycoses. p. 317–351. *In* Tucker (ed.), Progress in drug research, vol. 33. Billehauser, Basel.

48. Johnston, P., J. Lee, M. Demanski, F. Dressler, E. Tucker, M. Rothenberg, P. Pizzo, and T. Walsh. 1991. Late recurrent *Candida* endocarditis. Chest, in press.

49. Jones, J.M. 1980. Kinetics of antibody responses to cell wall mannan and a major cytoplasmic antigen of *Candida albicans* in rabbits and humans. J. Lab. Clin. Med. **96**:845–860.

50. Karp, J., P. Burch, and W.G. Merz. 1988. An approach to intensive antileukemia therapy in patients with previous invasive aspergillosis. Am. J. Med. **85**:203–206.

50a Knoper, S.R., and J.N. Galgiani. 1988. Coccidioidomycosis. Infect. Dis. Clin. N. Amer. **2**:861–875.

51. Kovacs, J., A. Kovacs, M. Polis, W. Wright, V.J. Gill, C. Tuazon, E. Gelmann, H. Lane, R. Lonfield, G. Overturf, A. Macher, A. Fauci, J. Parrillo, J. Bennett, and

H. Masur. 1985. Cryptococcosis in the acquired immunodeficiency syndrome. Ann. Int. Med. **103**:533–538.

52. Larsen, R., and M. Leal. 1989. Fluconazole compared to amphotericin B as treatment of cryptococcal meningitis. 29th Intersci. Conf. Antimicrob. Agents Chemother., abstr. 1062.

53. Lecciones, J., P. Kelly, J. Lee, R. Schaufele, P. Pizzo, and T. Walsh. 1989. In vitro activity of LY 121019 alone and in combination with other antifungal agents against *C. albicans* and *T. glabrata*, F3, p. 458. Abstr. 89th Annual Meeting of the Amer. Soc. Microbiol.

54. Lee, J., P. Kelly, J. Lecciones, D. Coleman, R. Gordee, P. Pizzo, and T. Walsh. 1990. Cilofungin (LY121019) shows non-linear plasma pharmacokinetics and tissue penetration in rabbits. Antimicrob. Agents Chemother. **34**:2240–2245.

55. Lee, J.W., C. Lin, D. Loebenberg, M. Rubin, P.A. Pizzo, and T.J. Walsh. 1989. Pharmacokinetics and tissue penetration of Sch 39304 in granulocytopenic and non-granulocytopenic rabbits. Antimicrob. Agents Chemother. **33**:1932–1935.

56. Lemieux, C., G. St. Germain, J. Vincelette, L. Kaufman, and L. DeRepentigny. 1990. Collaborative evaluation of antigen detection by a commercial latex agglutination test and enzyme immunoassay in the diagnosis of invasive candidiasis. J. Clin. Microbiol. **28**:249–253.

57. Levitz, S. 1989. Aspergillosis. Infect. Dis. Clin. N. Amer. **3**:1–18.

57a. Lief, M., and F. Sarfarazi. 1986. Prostatic cryptococcosis in acquired immune deficiency syndrome. Urology **28**:318–319.

58. Loebenberg, D., R. Parmegiani, E. Moss, D. Frank, M. McHugh, T. Yarosh-Tomaine, B. Antonacci, F. Menzel, R.S. Hare, and G.H. Miller. 1989. SCH39304, a new antifungal agent. Oral activity in a murine model of pulmonary aspergillosis. 29th Intersci. Conf. Antimicrob. Agents Chemother., abstr. 821.

59. Lopez-Bernstein, G. 1986. Liposomal amphotericin B in the treatment of fungal infections. Ann. Int. Med. **105**:130–131.

60. Lopez-Bernstein, G., G. Bodey, L. Frankel, and K. Mehta. 1987. Treatment of hepatosplenic candidiaisis with liposomal amphotericin B.J. Clin. Oncol. **5**:310–317.

61. Macher, A.M., M.L. De Vinatea, S.M. Tuur, and P. Angritt. 1988. AIDS and the mycoses. Infect. Dis. Clin. N. Amer. **2**:827–839.

62. Marsh, P.K., F.P. Tally, J. Kellum, A. Callow, and S.L. Gorbach. 1982. *Candida* infections in surgical patients. Ann. Surg. **198**:42–47.

63. Mason, A.B., M.E. Brandt and H.R. Buckley. 1988. Enolase activity associated with a *C. albicans* cytoplasmic antigen. Yeast **5**:S231–S240.

64. Matthews, R., and J. Burnie. 1988. Diagnosis of systemic candidiasis by an enzyme-linked dot immunobinding assay for a circulating immunodominant 47-kilodalton antigen. J. Clin. Microbiol. **26**:459–463.

65. Matthews, R.C., J.P. Burnie, and S. Tabaqchali. 1987. Isolation of immunodominant antigens from sera of patients with systemic candidiasis and characterization of serological response to *Candida albicans*. J. Clin. Microbiol. **25**:230–237.

66. McLeod, R., and J.S. Remington. 1977. Postoperative fungal endocarditis. p. 163–236. *In* R.J. Duma (ed.), Infections of prosthetic heart valves and vascular grafts. University Park Press, Baltimore, MD.

67. Melchinger, W., and J. Mueller. 1987. Studies of the in vitro sensitivity of yeast strains isolated from clinical specimens to LY121019, a new antifungal agent. Mykosen **30**:605–608.

67a. Merz, W.G., J. Karp, and M. Hoagland. 1988. Diagnosis and successful treatment of fusariosis in the compromised host. J. Infect. Dis. **158**:1046–1055.

68. Meunier, F., L. Leleux, J. Gerain, D. Ninove, R. Snoeck, and J. Klastersky. 1987. Prophylaxis of aspergillosis in neutropenic cancer patients with nasal spray of amphotericin B: a prospective randomized study. 27th Intersci. Conf. Antimicrob. Agents Chemother., abstr. 1346.

69. Meunier, F., L. Sculier, P. Van Der Auwera, and M. Aoun. 1989. Therapy with amphotericin B (AmB) in phospholipid vesicles (VS104) for proven or suspected aspergillosis in patients with hematological malignancies. 29th Intersci. Conf. Antimicrob. Agents Chemother., abstr. 114.

70. Meunier-Carpentier, F. 1984. Chemoprophylaxis of fungal infections. Am. J. Med. **76**:652–656.

71. Morstyn, G., L. Campbell, L. Souza, M. Alton, N. Keech, J. Keech, M. Green, W. Scheridan, D. Metcalf, and R. Fox. 1988. Effect of granulocyte colony stimulating factor on neutropenia induced by cytotoxic chemotherapy. Lancet **2**:667–672.

72. Odds, F.C. 1988. Activity of cilofungin (LY121019) against *Candida* species in vitro. J. Antimicrob. Chemother. **22**:891–897.

73. Oki, T., M. Kakushima, M. Nishio, H. Kamei, S.M. Hirano, Y. Sawada, and M. Konishi. 1990. BMY–28864, a water soluble pradimicin derivative. 30th Intersci. Conf. Antimicrob. Agents Chemother., abstr. 592.

74. Panos, R.J., L.F. Barr, T.J. Walsh, and H.J. Silverman. 1988. Factors associated with fatal hemoptysis in cancer patients. Chest **94**:1008–1013.

75. Pappagianis, D., M.S. Collins, R. Hector, and J. Remington. 1979. Development of resistance to in *Candida lusitaniae* infection in a human. Antimicrob. Agents Chemother. **16**:123–126.

76. Perfect, J.R., K.A. Wright, M.M. Hobbs, and D.T. Durack. 1989. Treatment of experimental cryptococcal meningitis and disseminated candidiasis with SCH 39304. Antimicrob. Agents Chemother. **33**:1735–1740.

77. Petranyi, G., A. Stutz, N. Ryder, J. Meingasner, and H. Mieth. 1987. Experimental antimycotic activity of naftifine and terbinafine, p. 441–459. *In* R.A. Fromtling (ed.), Recent trends in the discovery, development, and evaluation of antifungal agents. J.R. Prous Science Publishers, Barcelona, Spain.

78. Pfaller, M.A., S.S. Wey, T. Gerarden, A. Houston, and R.P. Wenzel. 1989. Susceptibility of nosocomial isolates of *Candida* species to LY121019 and other antifungal agents. Diagn. Microbiol. Infect. Dis. **12**:1–14.

79. Pizzo, P.A., K. Robichaud, F.A. Gill, and F.G. Witebsky. 1982. Empiric antibiotic and antifungal therapy for cancer patients with prolonged fever and granulocytopenia. Am. J. Med. **72**:101–110.

80. Polak, A., and D. Dixon. 1987. Antifungal activity of amorolfine (Ro 14–4767/002) in vitro and in vivo, p. 555–573. *In* R.A. Fromtling (ed.), Recent trends in the discovery, development, and evaluation of antifungal agents. J.R. Prous Science Publishers, Barcelona, Spain.

81. Powderly, W.G., G.S.Kobayashi, G.P. Herzig, and G. Medoff. 1988. Amphotericin B-resistant yeast infection in severely immunocompromised patients. Am. J. Med. **84**:826–832.

82. Quintiliani, R., N. Owens, R. Quercia, J. Klimek, and C. Nightingale. 1984. Treatment and prevention of oropharyngeal candidiasis. Am. J. Med. **77**:44–48.

83. Rinaldi, M.G. 1989. Zygomycosis. Infect. Dis. Clin. N. Amer. **3**:19–41.

84. Ringel, S. 1987. Ambruticin: a unique oral antibiotic, p. 585–593. *In* R.A. Fromtling (ed.), Recent trends in the discovery, development, and evaluation of antifungal agents. J.R. Prous Science Publishers, Barcelona, Spain.

85. Rippon, J.W. 1988. Pseudallescheriasis, p. 651–680. *In* J.W. Rippon (ed.), Medical mycology: the pathogenic fungi and pathogenic *Actinomycetes*. W.B. Saunders, Philadelphia, PA.

86. Roberts, C. 1984. Coccidioidomycosis in acquired immune deficiency syndrome. Depressed humoral as well as cellular immunity. Am. J. Med. **76**:734.

87. Roy, J.B., J.R. Gejer, and J.A. Ohr. 1984. Urinary tract candidiasis. An Update. Urology. **23**:533–537.

88. Saag, M. and W. Dismukes. 1988. Azole antifungal agents: emphasis on new triazoles. Antimicrob. Agents Chemother. **32**:1–8.

89. Sanford, G., W. Merz, J. Wingard, P. Charache, and R. Saral. 1980. The value of fungal surveillance cultures as predictors of systemic fungal infections. J. Infect. Dis. **142**:503–509.

90. Schechtman, S., L. Fumaro, T. Robin, E. Bottone, and J. Cuttner. 1984. Clotrimazole treatment of oral candidiasis in patients with neoplastic disease. Am. J. Med. **76**:91–94.

91. Sawada, Y., M. Nishio, M. Hatori, H. Yamamoto, M. Kakushima, M. Hirano, T. Miyaki, M. Konishi, and T. Oki. Pradimicins D, E, FA-1, and FA-2, novel antifungal antibiotics. Production, chemistry, and biological activities. 30th Intersci. Conf. Antimicrob. Agents Chemother., abstr. 591.

91a. Schmatz, D.M., M.A. Romancheck, L.A. Pittarelli, R.E. Schwartz, R.A. Fromtling, K.H. Nollstadt, F.L. Vanmiddlesworth, K.E. Wilson, and M.J. Turner. 1990. Treatment of *Pneumocystis carinii* pneumonia with 1,3-β-glucan synthesis inhibitors. Proc. Natl. Acad. Sci. USA **87**:5950–5954.

92. Schmitt, H., E. Bernard, J. Adrade, and *et al.* 1987. Inhibitory and cidal activity of terbinafine versus *Aspergillus*. Antimicrob. Agents Chemother. **32**:780–781.

93. Schmitt, H., E. Bernard, M. Hauser, and D. Armstrong. 1988. Aerosol amphotericin B for prophylaxis and therapy in a rat model of pulmonary aspergillosis. Antimicrob. Agents Chemother. **32**:1676–1679.

94. Selik, R., E. Starcher, and J. Curran. 1987. Opportunistic diseases reported in AIDS patients: frequencies, associations, and trends. AIDS. **1**:175–182.

95. Solomkin, J., A. Flohr, and R. Simmons. 1982. Indications for therapy for fungemia in postoperative patients. Arch. Surg. **117**:1272–1275.

96. Solomon, S.L., R.F. Khabbaazz, R.H. Parker, R.L. Anderson, M.A. Geraghty, R.M. Furman, and W.J. Martone. 1984. An outbreak of *Candida* parapsilosis bloodstream infections in patients receiving parenteral nutrition. J. Infect. Dis. **149**:98–107.

97. Spitzer, E.D., S.J. Travis, and G.S. Kobayashi. 1988. Comparative in vitro activity of LY121019 and amphotericin B against clinical isolates of *Candida* species. Europ. J. Clin. Microbiol. Infect. Dis. **7**:80–81.

98. Stein, D., and A. Sugar. 1989. Fungal infections in the immunocompromised host. Diagn. Microbiol. Infect. Dis. **12** (Suppl.):221–228.

99. Strinden, W.D., R.B. Helgerson, and D.B. Maki, 1985. *Candida* septic thrombosis of the great vein associated with central catheters. Ann. Surg. **202**:653–658.

100. Strockbine, N.A., M.T. Largen, and H.R. Buckley. 1984. Production and charactrization of three monoclonal antibodies to *Candida albicans* proteins. Infect. Immun. **43**:1012–1018.

101. Strockbine, N.A., M.T. Largen, S.M. Zweibel, and H.R. Buckley. 1984. Identification and molecular weight characterization of antigens from *Candida albicans* that are recognized by human sera. Infect. Immun. **43**:715–721.

102. Sugar, A., and C. Saunders. 1989. Oral fluconazole as suppressive therapy of disseminated cryptococcosis in patients with acquire immune deficiency syndrome. Am. J. Med. **85**:481–489.

103. Sugar, A., M. Picard, and L. Noble. 1989. Comparison of SCH 39304, fluconazole, and amphotericin B in the treatment of murine pulmonary blastomycosis. 29th Intersci. Conf. Antimicrob. Agents Chemother., abstr. 825.

104. Symposium. 1990. Fluconazole: a novel advance in therapy for systemic fungal infections. Rev. Infect. Dis. **12**:S263–S389.

105. Taft, C.S., and C.P. Selitrenikoff. 1988. LY121019 inhibits *Neurospora crassa* growth and (1-3)-β-D-glucan synthetase. J. Antibiot. **41**:697–701.

105a. Takeuchi, T., T. Hara, M. Hamada, H. Yamamoto, S. Gomi, Y. Orikasa, M. Sezaki, S. Kondo, and H. Yamaguchi. 1988. Benanomicins A and B, novel antifungal antibiotics. 28th Intersci. Conf. Antimicrob. Agents Chemother., abstr. 1007.

106. Tavitian, A., J. Raufman, L. Rosenthal, J. Weber, C. Webber, and H. Dincsoy. 1986. Ketoconazole-resistant *Candida* esophagitis in patients with acquired immunodeficiency syndrome. Gastroenterol. **90**:443–445.

107. Thaler, M., B. Pastakia, and T. Shawker. 1988. Hepatic candidiasis in cancer patients: the evolving picture of the syndrome. Ann. Int. Med. **108**:88–100.

108. Tollemar, J., O. Olle Ringden, and G. Tyden. 1990. Liposomal amphotericin B (AmBisome) treatment in solid organ and bone marrow transplant recipients. Efficacy and safety evaluation. Clin. Transplantation. **4**:167–175.

109. Vilars, V., and J. TC. 1988. The clinical profile of terbinafine: a new topical and systemic fungicidal drug for the treatment of dermatomycoses, p. 231–234. *In* J.M. Torres-Rodriguez (ed.), 10th Congress of the International Society for Human and Animal Mycology-ISHAM.J.R. Prous Science Publishers, Barcelona, Spain.

110. Walsh, T., and D. Dixon. 1989. Nosocomial aspergillosis: environmental microbiology, hospital epidemiology, diagnosis and treatment. Eur. J. Epidemiol. **5**:131–142.

111. Walsh, T., and W. Gray. 1987. *Candida* epiglottitis in immunocompromised patients. Chest **91**:482–485.

112. Walsh, T., S. Hamilton, and N. Belitsos. 1988. Esophageal candidiasis. Diagnosis and treatment of an increasingly recognized fungal infection. Postgrad. Med. **84**:193–205.

113. Walsh, T., and P. Pizzo. 1988. Nosocomial fungal infections: a classification for hospital-acquired fungal infections and mycoses arising from endogenous flora or reactivation. Ann. Rev. Microbiol. **42**:517–545.

114. Walsh, T., and P. Pizzo. 1988. Treatment of systemic fungal infections: recent advances and current problems. Eur. J. Clin. Microbiol. **7**:460–475.

115. Walsh, T.J. 1989. Trichosporonosis. Medicine **3**:43–65.

116. Walsh, T.J. 1990. Invasive pulmonary aspergillosis in patients with neoplastic diseases. Seminars Resp. Infect. **5**:111–122.

117. Walsh, T.J., C. Bustamente, D. Vlahov, and H.C. Standiford. 1986. *Candida* suppurative peripheral thrombophlebitis: prevention, recognition and management. Infect. Control. **7**:16–22.

118. Walsh, T.J., and K. Butler. 1990. Fungal infections complicating pediatric AIDS, p. 225–244. *In* P.A. Pizzo, and C. Wilfert (eds.), Pediatric AIDS. Williams and Wilkins, Baltimore.

119. Walsh, T.J., H. Buckley, M. Hom-Eng, D.E. Johnson, M. Maret, and P. Gary. 1988. Analysis of a 48-kilodalton cytoplasmic antigen of *Candida albicans* in experimental candidiasis, 0–111, p. 119. Abstr. 10th Congress of the International Society for Human and Animal Mycology. Barcelona, 1988.

120. Walsh, T.J., L.R. Caplan, and D.B. Hier. 1985. *Aspergillus* infections of the central nervous system: a clinicopathological analysis. Ann. Neurology **18**:574–582.

121. Walsh, T.J., P.F. Jarosinski, and R.A. Fromtling. 1990. Increasing usage of systemic antifungal agents. Diagn. Microbiol. Infect. Dis. **13**:37–40.

122. Walsh, T.J., J. Lee, J. Lecciones, P. Kelly, J. Peter, V. Thomas, J. Bacher, and P.A. Pizzo. 1990. Sch 39304 in prevention and treatment of disseminated candidiasis in persistently granulocytopenic rabbits. Antimicrob. Agents Chemother. **34**:1560–11564.

122a. Walsh, T.J., C.L. Lester-McCully, M.G. Rinaldi, J.E. Wallace, F.M. Balis, J.W. Lee, P.A. Pizzo, and D.G. Poplack. 1990. Penetration of SCH–39304, a new antifungal triazole, into cerebrospinal fluid of primates. Antimicrob. Agents Chemother. **34**:1281–1284.

123. Walsh, T.J., G. Melcher, M. Rinaldi, J. Lecciones, D. McGough, J. Lee, D. Callender, M. Rubin, and P.A. Pizzo. 1990. *Trichosporon beigelii*: an emerging pathogen resistant to amphotericin B.J. Clin. Microbiol. **28**:1616–1622.

124. Walsh, T.J., and W.G. Merz. 1986. Pathologic features in the human alimentary tract associated with invasiveness of *Candida tropicalis*. Am. J. Clin. Pathol. **85**:498–502.

125. Wheat, L. 1988. Histoplasmosis. Infect. Dis. Clin. N. Amer. **2**:841–860.

125a. Wheat, L., T. Slama, and M. Zeckel. 1985. Histoplasmosis in the acquired immune deficiency syndrome. Am. J. Med. **78**:203–210.

126. Wingard, J.R., W.G. Merz, and R. Saral. 1979. *Candida tropicalis*: a major pathogen in immunocompromised patients. Ann. Int. Med. **91**:539–543.

127. Wong, B., and K.L. Brauer. 1988. Enantioselective measurement of fungal D-arabinitol in the sera of normal adults and patients with candidiasis. J. Clin. Microbiol. **26**:1670–1674.

128. Zuger, A., E. Louis, R. Holzman, M. Simberkoff, and J. Rahal. 1986. Cryptococcal disease in patients with the acquired immunodeficiency syndrome. Diagnostic features and outcome of treatment. Ann. Int. Med. **104**:234–240.

14

Topoisomerase II Inhibitors: Prospects for New Antifungal Agents

David E. Jackson, D. P. Figgitt, and *Stephen P. Denyer*

I. Introduction

A. Biological Role of Topoisomerases

The DNA of both prokaryotic and eukaryotic cells exhibits several common tertiary structures including supercoils, knots, and catenanes, each of which is involved in various cellular processes (101, 80, 99). These topological forms of DNA may arise as a result of it being in a closed, circular form (rings or loops) such as in plasmid, viral, phage, or bacterial genomic DNA. Alternatively, the genomic DNA of eukaryotes may be complexed with proteins and other cellular molecules, at two points or more, in such a way that the ends are restricted and not free to rotate. Such constrained DNA domains (80, 103) pose certain topological problems that must be resolved by the cellular apparatus for successful propagation of the genetic material. This is achieved in both prokaryotic and eukaryotic organisms by a family of enzymes referred to as DNA topoisomerases. The enzymes are subdivided into type I and type II based on their ability to catalyse single- or double-strand cleavages, respectively; in addition type II enzymes have an energy requirement satisfied by ATP. In bacteria, topoisomerase II is often referred to as bacterial gyrase. *E. coli* contains two type I topoisomerases, the TopA and TopB proteins (84, 87).

The essential role of topoisomerases is reflected in the wide variety of vital cellular functions in which they participate: DNA replication, DNA repair, chromosome segregation, ribosomal RNA synthesis, transcription, and sister chromatid exchange (99, 28, 56, 20). Their involvement in these functions may be indirect, through their role in modulating the superhelical state of DNA and its state of catenation and knotting, or direct, through participation in processes requiring the coupled breakage and rejoining of DNA strands as seen in recombination processes (28).

In eukaryotes, the bulk of topoisomerase activity is found in the nucleus;

topoisomerases have also been isolated from mitochondria and chloroplasts (99, 63). The expression of DNA topoisomerase II may vary throughout the cell cycle (*e.g.*, 24) while topoisomerase I activity remains relatively stable (39). In yeasts, topoisomerase II activity has been shown to be essential for cell survival (10), whereas topoisomerase I is dispensable, particularly in the presence of high levels of type II enzyme (87). A role for topoisomerases in viral replication has also been postulated (81, 53, 4), with both host and virally encoded enzymes being implicated.

In addition to their enzymatic roles, DNA topoisomerases appear to have a structural function in mitotic chromosomes; DNA topoisomerase II is a major polypeptide of the chromosome scaffold (5) and is required for *in vitro* chromatin assembly (11).

B. Structural Aspects of DNA Topoisomerases

The ubiquitous nature of DNA topoisomerases is reflected in the variety of species in which these enzymes have been studied (Table 14.1).

The evolutionary relatedness can be discerned by comparing the DNA sequences of the genes encoding the topoisomerase type I and type II enzymes (Table 14.2). Type II DNA topoisomerases (bacterial gyrase, phage T4 DNA topoisomerase and eukaryotic DNA topoisomerase II) can be aligned with little difficulty (100). The bacterial and phage T4 type II topoisomerases have separate genes encoding different subunits of the enzymes, with bacterial gyrase having an A_2B_2 quaternary structure (51) and the T4 enzyme a hexameric structure containing a pair of each of the three subunits encoded by genes 52, 39, and 60 (58). Eukaryotic type II topoisomerase is a dimer of a single subunit (61, 77, 34). The amino- and carboxyl-terminal halves of the eukaryotic enzyme share respective homologies with the B and A subunits of bacterial gyrase at corresponding positions along the polypeptide chains (36, 37, 96, 107). In a similar manner, the amino acid sequences of the gene 39 and 60 subunits of phage T4 topoisomerase align with the bacterial *gyrB* subunit and the N-terminal part of the eukaryotic topoisomerase II sequences. Similarly, the amino acid sequence of the gene 52 subunit of the phage T4 enzyme aligns with the bacterial *gyr*A subunit and the sequences from the C-terminal part of the eukaryotic type II enzymes (100, 44). Such sequence similarities suggest that different domains in the prokaryotic type II topoisomerases have their counterparts in the eukaryotic enzyme.

By contrast, it appears that the bacterial and eukaryotic type I DNA topoisomerases share no significant sequence similarities (95, 91).

C. Assays for Topoisomerase Activity

Several *in vitro* enzyme assay procedures can be used to measure the activity of DNA topoisomerases. These can be broadly placed into two categories:

Table 14.1. Some sources of DNA topoisomerases

Source	Type	References
HeLa cells	I	57
	II	61
Mouse cells	I	14
Drosophila melanogaster	I	57
	II	77
Wheat germ	I	99
Ustilago maydis	I	75
Plasmodium berghei	I	72
	II	72
Xenopus laevis	I	70
Calf thymus:		
mitochondria	I	55
nucleus	II	79
Saccharomyces cerevisiae	I	2,34
	II	35
Trypanosoma cruzi	I	71
Micrococcus luteus	II	59
Avian erythrocytes	I	94
Human KB cells	I	50
Lentinus edodes	I	52
Carrot cells (*Daucus carota*)	I	12
Pea chloroplast	I	63
E. coli	I	99,18
	II	29
T4 infected *E. coli*	II	58
Bacillus subtilis	II	85
Wheat chloroplasts	II	69

a. Those that determine catalytic activity as measured by relaxation/supercoiling, decatenation/catenation, and unknotting/knotting reactions.

b. Those measuring the inhibitor-enhanced, noncatalytic DNA-cleaving activity of topoisomerases.

In each assay procedure, specific cofactor requirements and assay conditions must be optimised. A comparative review of these assay procedures has recently appeared (3).

II. Topoisomerase Enzymes as Biochemical Targets

A. Inhibitors of DNA Topoisomerases

As a consequence of the fundamental role of DNA topoisomerases in both prokaryotic and eukaryotic cells, the enzymes are emerging as important targets in antibacterial and anticancer therapy (83). In particular, the 4-quinolone antibac-

Table 14.2. Some genes for DNA topoisomerases which have been cloned and sequenced

Enzyme	Source	Type	Gene(s)	References
Topoisomerase I				
	E. coli	I	*Top A*	95
	E. coli	I	*Top B*[a]	19
	S. cerevisiae	I	*Top 1*	37, 91
	Schizosaccharomyces pombe	I	*Top 1*	97
	Rat liver cells	I	*Top 1*	23
	HeLa cells	I	*Top 1*	16,48
Topoisomerase II				
	S. cerevisiae	II	*Top 2*	36, 37
	S. pombe	II	*Top 2*	96, 97
	Drosophila melanogaster	II	*Top 2*	106, 107
Gyrase (topoisomerase II)				
	E. coli	II	*gyrA, gyrB*	1
	Bacillus subtilis	II	*gyrA, gyrB*	62
T4 topoisomerase II	Phage T4	II	*52, 39, 60*	44

[a] Previously referred to as topoisomerase III.

terials and the cytotoxic epipodophyllotoxins have made a major contribution in their respective therapeutic areas. These compounds and other inhibitors of prokaryotic and eukaryotic topoisomerases are summarised in Table 14.3. The precise mechanism(s) by which cell death occurs is not clearly understood, but may include inhibition of DNA synthesis (and induction of SOS repair in bacteria), inhibition of cell division, and DNA degradation (22, 56). In some cases, responsibility for these effects may rest with the formation of drug-stabilised DNA-enzyme ternary complexes (the cleavage complex); antitumour activity may be increased by high levels of topoisomerase II in some tumour cells (8, 42). More recent studies suggest that stimulation of topoisomerase-mediated DNA strand breaks is not in itself not sufficient for cell death. Instead, a collision between a moving replication fork and the drug-stabilised DNA-enzyme ternary complex (the cleavage complex) appears to be critical, particularly for the topoisomerase I inhibitor camptothecin (41, 109). At high levels of topoisomerase II inhibitors, an interaction with other cellular factors unrelated to the replication apparatus has been suggested (109).

B. Mode of Action

Camptothecin is the classical topoisomerase I inhibitor and is believed to act by stabilising the DNA-enzyme cleavage complex (40). In contrast, several mechanisms of action can be attributed to the wide variety of topoisomerase II inhibitors. For a more detailed discussion the reader is referred to recent review articles (22, 56). In brief, there are three classes of inhibitors:

Table 14.3. Examples of DNA Topoisomerase Inhibitors

Class of drug	Examples	Enzyme inhibited	Reference
4-Quinolones	Nalidixic acid [1]	Bacterial gyrase,	33, 89
	Oxolinic acid [2]	Phage T4 topoisomerase	
	Norfloxacin [3]		
Coumarins	Novobiocin [4]	Bacterial gyrase,	30, 43
	Coumermycin A_1 [5]	Eukaryotic topoisomerase II	
Ellipticines	Ellipticine [6]	Eukaryotic topoisomerase II	90
	2-Methyl-9-hydroxyellipticine [7]		
Acridines	*m*-AMSA [8]	Eukaryotic topoisomerase II	68, 90
Anthracyclines	5-Iminodaunorubicin [9]	Eukaryotic topoisomerase II	90
	Doxorubicin [10]	Eukaryotic topoisomerase II	74
Epipodophyllo-toxins	Etoposide [11]	Eukaryotic topoisomerase II	98
	Teniposide [12]		
Alkaloids	Camptothecin [13]	Eukaryotic topoisomerase I	40
Miscellaneous:	Amiloride [14]	Eukaryotic topoisomerase II, bacterial gyrase	6, 7
	Merbarone [15]		21
	Fostriecin [16]		9
	Withangulatin A [17]		49
	Caffeine [18]	Eukaryotic topoisomerase II (Merbarone [15] through Streptonigrin [22])	76
	Fredericamycin A [19]		54
	Distamycin [20]		105
	4′,6-Diamidine-2-phenylindole [21]		105
	Streptonigrin [22]		108
	Genistein [23]	Eukaryotic topoisomerase I and II	66
	Topostin	Eukaryotic topoisomerase I	45, 88

a. DNA-intercalating agents, some of which stabilise the cleavage complex (*e.g.*, *m*-AMSA), while others do not (*e.g.*, amiloride, ethidium bromide).

b. Nonintercalating inhibitors which stabilise a cleavage complex (*e.g.*, etoposide).

c. Competitive inhibitors of the ATP cofactor (*e.g.*, novobiocin).

Although drugs which stabilise the cleavage complex may be subdivided on the basis of their ability to bind DNA (*e.g.*, etoposide and *m*-AMSA), both classes appear to stabilise the complex by slowing the rate of religation of the DNA (67, 73).

The eukaryotic topoisomerase II inhibitor, merbarone, is purported to represent a novel inhibitor class (21). Unlike the other cytotoxic compounds, merbarone neither appears to stabilise the formation of the cleavage complex nor intercalates into DNA. The antileishmanial antimonial drugs stibogluconate and ureastibamine, specific inhibitors of DNA topoisomerase I from *Leishmania donovani* (13), and the newly identified topoisomerase I inhibitors, the topostins (45, 88), await full mechanism of action studies, but may also belong to this category. In addition, compounds such as distamycin, which bind to the minor groove of DNA, inhibit the actions of both topoisomerase I and II (105, 60).

C. Selective Activity of Topoisomerase II Inhibitors

The discovery of DNA gyrase as the cellular target of the quinolone antibacterial, nalidixic acid, highlighted the differences between eukaryotic and prokaryotic topoisomerases and suggested an important avenue for exploitation in the design of antibacterial agents. This selective toxicity can be retained (33) through careful structural modification (15), and a clinically useful range of 4-quinolones has emerged (104). In contrast, the coumarins, as exemplified by novobiocin, while also showing antibacterial activity, do not exhibit the same level of selectivity and thus are more limited in their applications. This difference in selectivity may be attributed to their different mechanism of action.

The halophilic archaebacteria, considered as evolutionary ancestors of both eubacteria and eukaryotes (27), appear to possess a topoisomerase system sensitive to inhibitors of the eukaryotic type II enzyme (82). This sensitivity has led to the recommendation of archaebacteria as a model for *in vitro* screening of potential antitumour agents. Since the homology between yeast and human type II topoisomerase is substantial, it might be anticipated that selective toxicity may be more difficult to achieve with eukaryotic fungi. Indeed, recent studies by Nitiss and Wang have suggested that permeabilised mutant strains of *Saccharomyces cerevisiae* are a useful model for examining antitumour drugs that act on DNA topoisomerases (64). These studies have clearly demonstrated the sensitivity of fungal topoisomerase II to a variety of known inhibitors of the mammalian enzyme system.

III. Antifungal Activity and Topoisomerase II Inhibition

A. *Epipodophyllotoxins*

Preliminary studies by the authors demonstrated antifungal activity in a series of podophyllotoxin analogs (46, 47, 17); this activity is dependent upon appropriate stereochemical and/or substitution patterns at the 4 and 4′ positions, with an obligate requirement for a hydroxyl group at the latter position. Other reports indicating the inhibitory effects of similar podophyllotoxin derivatives against mammalian topoisomerase II have appeared (92, 102).

We sought to correlate the structural parameters necessary for antifungal action with those required for the inhibition of topoisomerase II from both susceptible and resistant fungi. A report of our initial studies has recently appeared (25); a selection of these studies and subsequent work appears in Table 14.4. Employing relaxation assays, in the presence or absence of ATP (25, 26, 38), compounds can be shown to inhibit the type II enzymes selectively. Compounds with a 4′-hydroxyl group and either methylene or carbonyl at the 4 position are the most potent inhibitors of the fungal enzyme in this series and are the ones which display an antifungal action in susceptible organisms.

In addition to podophyllotoxins, other known inhibitors of the mammalian type II enzyme have been evaluated in this assay system. These results have led us to suggest that some agents (*e.g.*, amiloride) may show a degree of selective toxicity between the mammalian and fungal enzyme systems.

B. *Anthracyclines*

Recently a series of closely related benanomicins (89, 31, 32) and pradimicins (93, 65, 78), anthracycline derivatives isolated from actinomycetes, have been shown to exhibit good *in vitro* and *in vivo* antifungal activity. While no clear mechanism of action has been advanced, topoisomerase II inhibition has been shown to occur amongst the structurally related anthracycline antitumour agents (93, 74). These novel antibiotics with low murine toxicity may represent chemical agents which mediate antifungal effects through a selective action on the fungal topoisomerase II enzyme system.

IV. Concluding Remarks

Topoisomerase II is recognised as the primary target for cytotoxic agents in bacterial and mammalian systems and this recognition has led to the development of significantly improved chemotherapeutic agents. In the case of fungi, evidence

Table 14.4. Summary of structural features and in vitro activity of podophyllotoxin derivatives

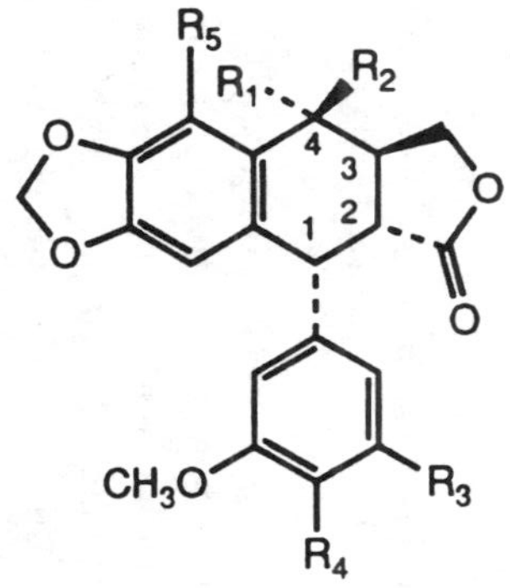

Compound	R_1	R_2	R_3	R_4	R_5	Antifungal activity[a]	Topoisomerase II inhibition[b]
Podophyllotoxin	OH	H	OCH_3	OCH_3	H	–	–
Picropodophyllin[c]	OH	H	OCH_3	OCH_3	H	–	–
Epipodophyllotoxin	H	OH	OCH_3	OCH_3	H	–	–
β-Peltatin	H	H	OCH_3	OCH_3	OH	–	–
4′-Demethylpodophyllotoxin	OH	H	OCH_3	OH	H	–	–
α-Peltatin	H	H	OCH_3	OH	OH	–	–
Podophyllotoxone	=O		OCH_3	OCH_3	H	–	–
Desoxypodophyllotoxin	H	H	OCH_3	OCH_3	H	–	–
4′-Demethyldesoxypodophyllotoxin	H	H	OCH_3	OH	H	+	+
4′-Demethylpodophyllotoxone	=O		OCH_3	OH	H	+	+
3′,4′-Didemethyldesoxypodophyllotoxin	H	H	OH	OH	H	+	+

[a] Whole cells of *Aspergillus niger*. MIC < 80 ug/ml (+).

[b] Enzyme isolated from *Aspergillus niger*. pBR322 relaxation time > 45 min. at 100 ng/ml drug concentration (+).

[c] C-2 epimer of podophyllotoxin.

has emerged to suggest that the fungal topoisomerase II enzyme may serve as a target for antifungal podophyllotoxins and anthracyclines. Amongst the former compounds, activity appears restricted to the filamentous fungi, probably due to impenetrability of many of these agents. This notion is supported by the observation that isolated *Saccharomyces cerevisiae* topoisomerases are sensitive to topoisomerase inhibitors *in vitro*, but little activity is demonstrated in whole cells.

The demonstration that fungal topoisomerases can serve as useful models of the mammalian system poses a significant hurdle for the development of therapeutically useful agents. Differential sensitivity does not yet exist to a degree to be exploited. Nevertheless, the opportunities arising through recognition of this new target cannot be ignored. In this respect, careful attention should focus on the newly characterised benanomycins and pradimicins which appear to exhibit good antifungal activity and selective toxicity; the close structural similarity between these agents and the known topoisomerase-inhibiting anthracycline antitumour agents may be presumptive evidence of a similar target.

Topoisomerase activity can be followed by using several *in vitro* assay techniques. Clearly, structure-activity relationships derived by different methods must be considered with caution; evidence is now emerging to suggest that the drug-induced stimulation of DNA cleavage correlates best with whole cell cytotoxicity, and this may be the most appropriate assay method.

Either of the two strategies available, the use of an isolated fungal enzyme or a permeabilised cell system, can be employed to screen potential topoisomerase active agents. In the final analysis, however, the significance of permeability barriers must be addressed, and clearly a structure-activity relationship must evolve which includes this consideration. In this regard, whole cell inhibition studies should be used to complement a topoisomerase inhibitor screening programme.

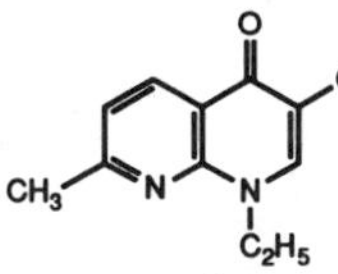

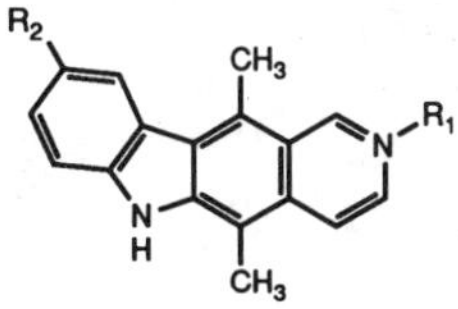

R_1	R_2	
H	H	**6**
CH_3	OH	**7**

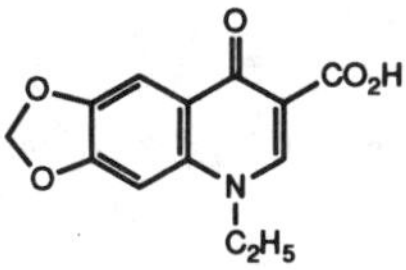

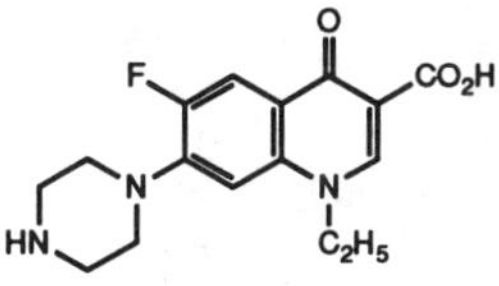

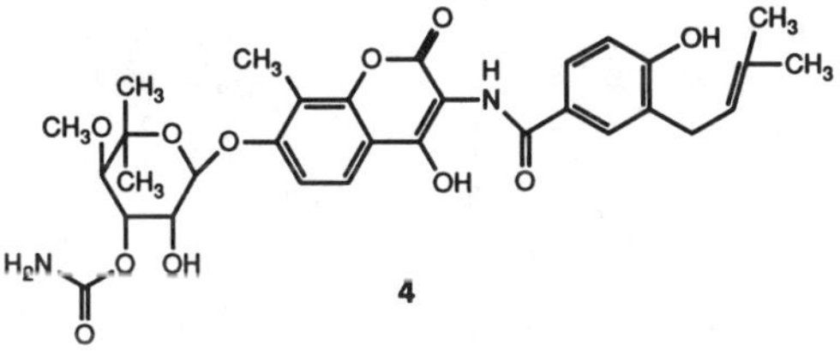

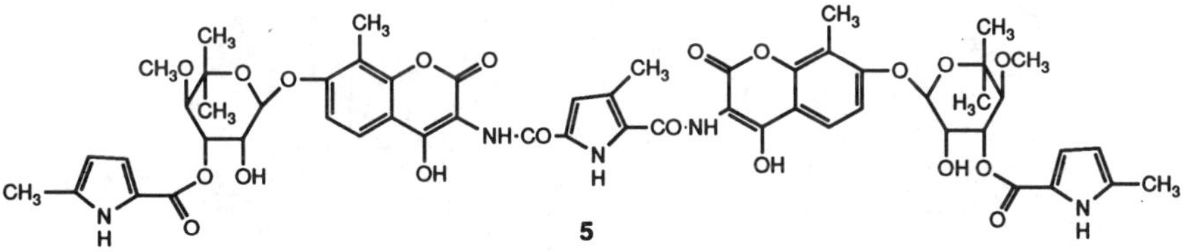

Figure 14.1. Structures of topoisomerase inhibitors.

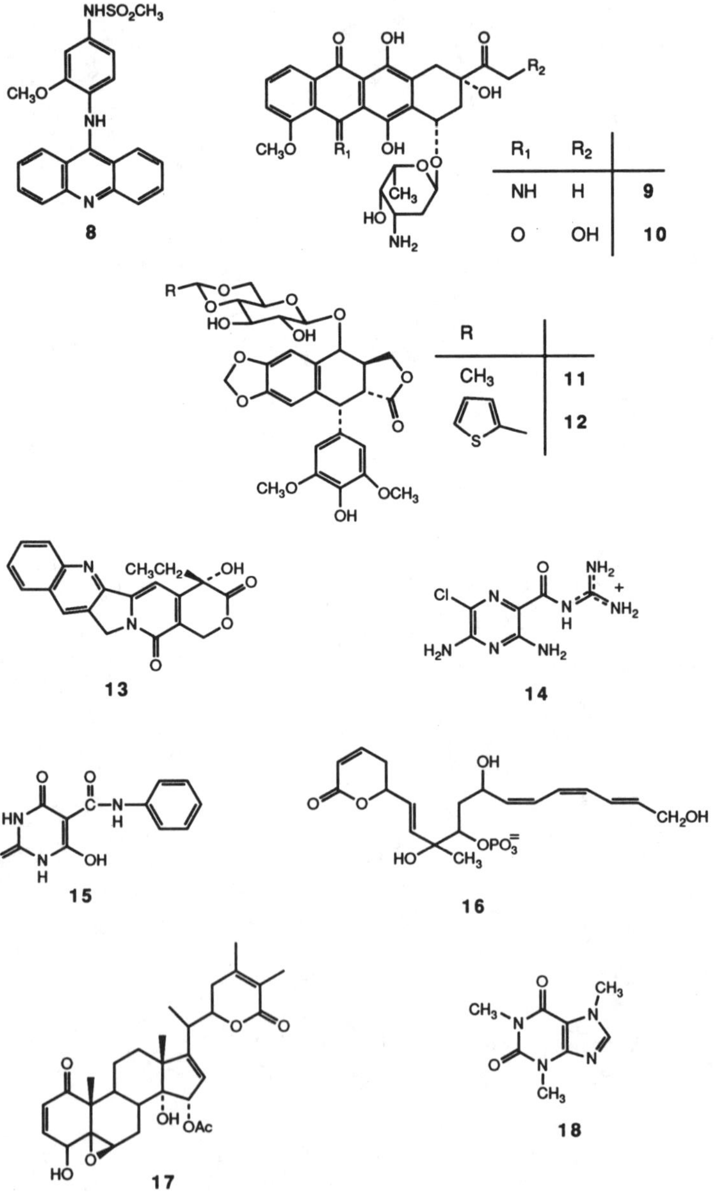

R_1	R_2	
NH	H	**9**
O	OH	**10**

R	
CH_3	**11**
2-thienyl	**12**

Figure 14.2. Structures of topoisomerase inhibitors.

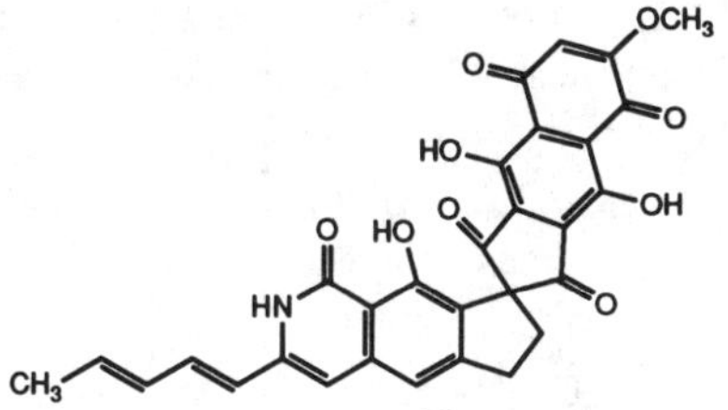

Figure 14.3. Structures of topoisomerase inhibitors.

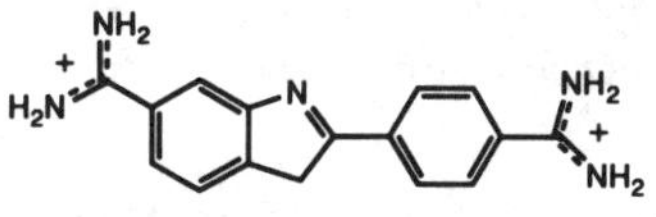

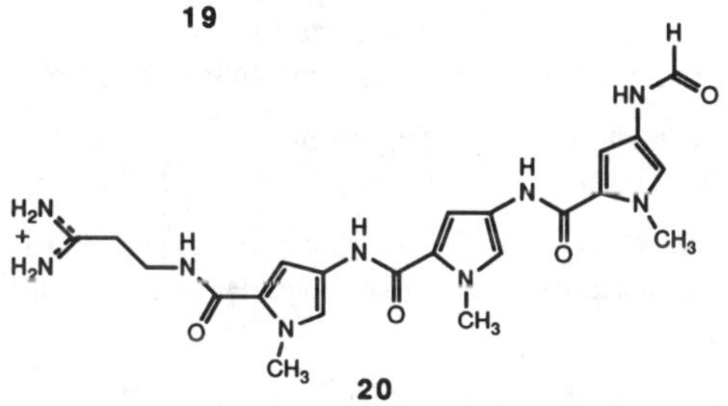

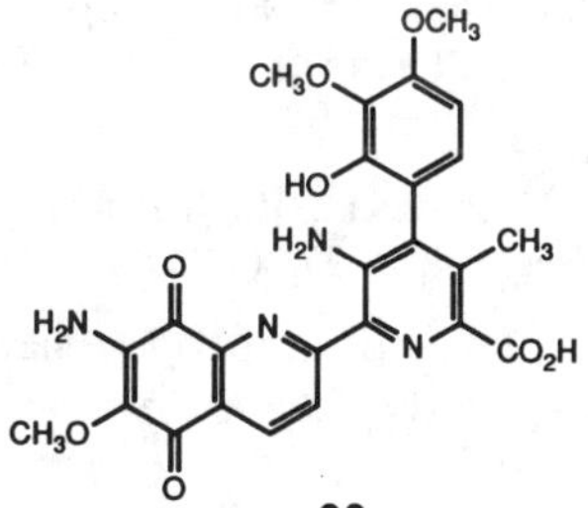

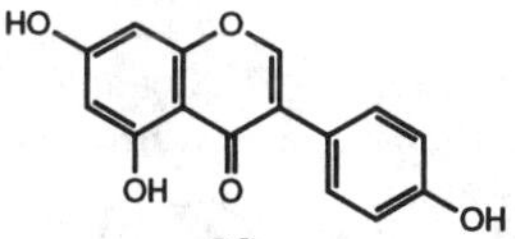

References

1. Adachi, T., M. Mizuuchi, E.A. Robinson, E. Appella, M.H. O'Dea, M. Gellert, and K. Mizuuchi. 1987. DNA sequence of the *E. coli gyr*B gene: application of a new sequencing strategy. Nucleic Acids Res. **15**:771–784.

2. Badaracco, G., P. Plevani, W.T. Ruyechan, and L.M.S. Chang. 1983. Purification and characterization of yeast topoisomerase I.J. Biol. Chem. **258**:2022–2026.

3. Barrett, J.F., J.A. Sutcliffe, and T.D. Gootz. 1990. *In vitro* assays used to measure the activity of topoisomerases. Antimicrob. Agents Chemother. **34**:1–7.

4. Benson, J.D., and E.S. Huang. 1988. Two specific topoisomerase II inhibitors prevent replication of human cytomegalovirus DNA: an implied role in replication of the viral genome. J. Virology **62**:4797–4800.

5. Berrios, M., N. Osheroff, and P.A. Fisher. 1985. In situ localization of DNA topoisomerase II, a major polypeptide component of the *Drosophila melanogaster* nuclear matrix fraction. Proc. Natl. Acad. Sci. USA **82**:4142–4146.
6. Besterman, J.M., L.P. Elwell, S.G. Blanchard, and M. Cory. 1986. Amiloride intercalates into DNA and inhibits DNA topoisomerase II. J. Biol. Chem. **262**:13352–13358.
7. Besterman, J.M., L.P. Elwell, E.J. Cragoe Jr., C.W. Andrews, and M. Corey. 1989. DNA intercalation and inhibition of topoisomerase II. Structure-activity relationships for a series of amiloride analogs. J. Biol. Chem. **265**:2324–2330.
8. Bodley, A.L., and L.F. Liu. 1988. Topoisomerases as novel targets for cancer chemotherapy. Bio/Technology 1315–1319.
9. Boritzki, T.J., T.S. Wolfard, J.A. Besserer, R.C. Jackson, and D.W. Fry. 1988. Inhibition of type II topoisomerase by fostriecin. Biochem. Pharmacol. **37**:4063–4068.
10. Brill, S.J., S. DiNardo, K. Voelkel-Meiman, and R. Sternglanz. 1987. Need for DNA topoisomerase activity as a swivel for DNA replication for transcription of ribosomal RNA. Nature **326**:414–416.
11. Caceres, J.F., D. Blangy, and G.C. Glikin. 1989. Requirement of DNA topoisomerases for *in vitro* chromatin assembly by 3T6 mouse cell extracts. Eur. J. Biochem. **181**:531–537.
12. Carbonera, D., R. Cella, A. Montecucco, and G. Ciarrocchi. 1988. Isolation of a type I topoisomerase from carrot cells. J. Experiment. Bot. **39**:70–78.
13. Chakraborty, A.K., and H.K. Majumder. 1988. Mode of action of pentavalant antimonials: Specific inhibition of type I DNA topoisomerase of *Leishmania donovani*. Biochem. Biophys. Res. Commun. **152**:605–611.
14. Champoux, J.J., and R. Dulbecco. 1972. An activity from mammalian cells that untwists superhelical DNA—a possible swivel for DNA replication. Proc. Natl. Acad. Sci. USA **69**:143–146.
15. Chu, D.T.W., and P.B. Fernandes. 1989. Structure-activity relationships of the fluoroquinolones. Antimicrob. Agents Chemother. **33**:131–135.
16. D'Arpa, P., P.S. Machlin, H. Ratrie III, N.F. Rothfield, D.W. Cleveland, and W.C. Earnshaw. 1988. cDNA cloning of human DNA topoisomerase I: catalytic activity of a 67.7-kDa carboxyl-terminal fragment. Proc. Natl. Acad. Sci. USA **85**:2543–2547.
17. Dewick, P.M., and D.E. Jackson. 1981. Cytotoxic lignans from *Podophyllum*, and the nomenclature of aryltetralin lignans. Phytochemistry **20**:2277–2280.
18. DiGate, R.J., and K.J. Marians. 1988. Identification of a potent decatenating enzyme from *Escherichia coli*. J. Biol. Chem. **263**:13366–13373.
19. DiGate, R.J., and K.J. Marians. 1989. Molecular cloning and DNA sequence analysis of *Escherichia coli topB*, the gene encoding topoisomerase III. J. Biol. Chem. **264**:17924–17930.

20. Downes, C.S., and R.T. Johnson. 1988. DNA topoisomerases and DNA repair. BioEssays **8**:179–184.

21. Drake F.H., G.A. Hofmann, S.-M. Mong, J. Bartus, R.P. Hertzberg, R.K. Johnson, M.R. Mattern, and C.K. Mirabelli. 1989. *In vitro* and intracellular inhibition of topoisomerase II by the antitumor agent merbarone. Cancer Res. **49**:2578–2583.

22. Drlica, K., and R.J. Franco. 1988. Inhibitors of DNA topoisomerases. Biochemistry **27**:2253–2259.

23. Durban, E., M. Bramucci, and R. Cook. 1988. Partial amino acid sequence of rat topoisomerase I: comparison with the predicted sequences for the human and yeast enzymes. Biochem. Biophys. Res. Commun. **154**:358–364.

24. Fairman, R., and D.L. Brutlag. 1988. Expression of the *Drosophila* Type II topoisomerase is developmentally regulated. Biochemistry **27**:560–565.

25. Figgitt, D.P., S.P. Denyer, P.M. Dewick, D.E. Jackson, and P. Williams. 1989. Topoisomerase II: A potential target for novel antifungal agents. Biochem. Biophys. Res. Commun. **160**:257–262.

26. Figgitt, D.P., A.J. Caston, S.P. Denyer, P. Williams, D.E. Jackson, C. Washington, and N. Stevenson. 1988. Disruption of microbial cells using a laboratory microfluidiser. J. Pharm. Pharmacol. **40**(Suppl.):138P.

27. Fox, G.E., E. Stackebrandt, R.B. Hespell, J. Gibson, J. Maniloff, T.A. Dyer, R.S. Wolfe, W.E. Balch, R.S. Tanner, L.J. Magrum, L.B. Zablen, R. Blackmore, R. Gupta, L. Bonen, B.J. Lewis, D.A. Stahl, K.R. Luehrsen, K.N. Chen, and C.R. Woese. 1980. The phylogeny of prokaryotes. Science **209**:457–465.

28. Gellert, M. 1981. DNA topoisomerases. Annu. Rev. Biochem. **50**:879–910.

29. Gellert, M., K. Mizuuchi, M.H. O'Dea, and H.A. Nash. 1976. DNA gyrase: an enzyme that introduces superhelical turns into DNA. Proc. Natl. Acad. Sci. USA **73**:3872–3876.

30. Gellert, M., M.H. O'Dea, T. Itoh, and J.-I. Tomizawa. 1976. Novobiocin and coumermycin inhibit DNA supercoiling catalyzed by DNA gyrase. Proc. Natl. Acad. Sci. USA **73**:4474–4478.

31. Gomi, S., M. Sezaki, S. Kondo, T. Hara, H. Naganawa, and T. Takeuchi. 1988. The structure of new antifungal antibiotics, benanomicins A and B. J. Antibiotics **42**:1019–1028.

32. Gomi, S., M. Sezaki, M. Hamada, S. Kondo, and T. Takeuchi. 1989. Biosynthesis of benanomicins. J. Antibiotics **42**:1145–1150.

33. Gootz, T.D., J.F. Barrett, and J.A. Sutcliffe. 1990. Inhibitory effects of quinolone antibacterial agents on eukaryotic topoisomerases and related test systems. Antimicrob. Agents Chemother. **34**:8–12.

34. Goto, T., P. Laipis, and J.C. Wang. 1984. The purification and characterization of DNA topoisomerases I and II of the yeast *Saccharomyces cerevisiae*. J. Biol. Chem. **259**:10422–10429.

35. Goto, T., and J.C. Wang. 1982. Yeast DNA topoisomerase II. J. Biol. Chem. **257**:5866–5872.

36. Goto, T., and J.C. Wang. 1984. Yeast DNA topoisomerase II is encoded by a single-copy, essential gene. Cell **36**:1073–1080.

37. Goto, T., and J.C. Wang. 1985. Cloning of yeast *TOP1*, the gene encoding DNA topoisomerase I, and construction of mutants defective in both DNA topoisomerase I and DNA topoisomerase II. Proc. Natl. Acad. Sci. USA **82**:7178–7182.

38. Hamelin, C., J. Yelle, and M. Dion. 1986. Fast electrophoresis in superposed agarose minigels for DNA topoisomerase assays. Electrophoresis **7**:323–326.

39. Heck, M.M.S., W.M. Hittelman, and W.C. Earnshaw. 1988. Differential expression of DNA topoisomerases I and II during the eukaryotic cell cycle. Proc. Natl. Acad. Sci. USA **85**:1086–1090.

40. Hertzberg, R.P., M.J. Caranfa, and S.M. Hecht. 1989. On the mechanism of topoisomerase I inhibition by camptothecin: evidence for binding to an enzyme-DNA complex. Biochemistry **28**:4629–4638.

41. Holm, C., J.M. Covey, D. Kerrigan, and Y. Pommier. 1989. Differential requirement of DNA replication for the cytotoxicity of DNA topoisomerase I and II inhibitors in chinese hamster DC3F cells. Cancer Res. **49**:6365–6368.

42. Hong, J.H., K. Okada, T. Kusano, Y. Komazawa, M. Kobayashi, A. Mizutani, N. Kamada, and A. Kuramoto. 1990. Reduced DNA topoisomerase II in VP-16 resistant mouse breast cancer cell line. Biomed. Pharmacother. **44**:41–46.

43. Hsieh, T., and D. Brutlag. 1980. ATP-Dependent DNA topoisomerase from *Drosophila melanogaster* reversibly catenates duplex DNA rings. Cell **21**:115–125.

44. Huang, W.M., S.-Z. Ao, S. Casjens, R. Orlandi, R. Zeikus, R. Weiss, D. Winge, and M. Fang. 1988. A persistent untranslated sequence within bacteriophage T4 DNA topoisomerase gene 60. Science **239**:1005–1012.

45. Ikegami, Y., N. Takeuchi, M. Hanada, Y. Hasegawa, K. Ishii, T. Andoh, T. Sato, K. Suzuki, H. Yamaguchi, S. Miyazaki, K. Nagai, S.-I. Watanabe, and T. Saito. 1990. Topostin, a novel inhibitor of mammalian DNA topoisomerase I from *Flexibacter topostinus* sp. Nov. J. Antibiotics **43**:158–162.

46. Jackson, D.E., and P.M. Dewick. 1984. Aryltetralin lignans from *Podophyllum hexandrum* and *Podophyllum peltatum*. Phytochemistry **23**:1147–1152.

47. Jackson, D.E., and P.M. Dewick. 1984. Biosynthesis of *Podophyllum* lignans—II. Interconversions of aryltetralin lignans in *Podophyllum hexandrum*. Phytochemistry **23**:1037–1042.

48. Juan, C.-C., J. Hwang, A.A. Liu, J. Whang-Peng, T. Knutsen, K. Huebner, C.M. Croce, H. Zhang, J.C. Wang, and L.F. Liu. 1988. Human DNA topoisomerase I is encoded by a single-copy gene that maps to chromosome region 20q12–13.2. Proc. Natl. Acad. Sci. USA **85**:8910–8913.

49. Juang, J.-K., H.W. Huang, C.-M. Chen, and H.J. Liu. 1989. A new compound, withangulatin A, promotes type II DNA topoisomerase-mediated DNA damage. Biochem. Biophys. Res. Commun. **159**:1128–1134.

50. Keller, W. 1975. Characterisation of purified DNA-relaxing enzyme from human tissue culture cells. Proc. Natl. Acad. Sci. USA **72**:2550–2554.

51. Klevan, L., and J.C. Wang. 1980. Deoxyribonucleic acid gyrase–deoxyribonucleic acid complex containing 140 base pairs of deoxyribonucleic acid and an $\alpha_2\beta_2$ protein core. Biochemistry **19**:5229–5234.

52. Kono, H., N. Habuka, and K. Shishido. 1986. Purification and characterization of type I DNA topoisomerase from *Lentinus edodes*. FEMS Microbiol. Lett. **37**:169–172.

53. Kreuzer, K.N. 1989. DNA topoisomerases as potential targets of antiviral action. Pharmac. Ther. **43**:377 – 395.

54. Latham, M.D., C.K. King, P. Gorycki, T.L. Macdonald, and W.E. Ross. 1989. Inhibition of topoisomerases by fredericamycin A. Cancer Chemother. Pharmacol. **24**:167–171.

55. Lazarus, G.M., J.P. Henrich, W.G. Kelly, S.A. Schmitz, and F.J. Castora. 1987. Purification and characterization of a type I DNA topoisomerase from calf thymus mitochondria. Biochemistry **26**:6195–6203.

56. Liu, L.F. 1989. DNA topoisomerase poisons as antitumour drugs. Annu. Rev. Biochem. **58**:351–375.

57. Liu, L.F., and K. Miller. 1981. Eukaryotic DNA topoisomerases: two forms of type I DNA topoisomerases from HeLa cell nuclei. Proc. Natl. Acad. Sci. USA **78**:3487–3491.

58. Liu, L.F., C.-C. Liu, and B.M. Alberts. 1979. T4 DNA topoisomerase: a new ATP-dependent enzyme essential for initiation of T4 bacteriophage DNA replication. Nature **281**:456–461.

59. Liu, L.F., and J.C. Wang. 1978. *Micrococcus luteus* DNA gyrase: active components and a model for its supercoiling of DNA. Proc. Natl. Acad. Sci. USA **75**:2098–2102.

60. McHugh, M.M., J.M. Woynarowski, R.D. Sigmund, and T.A. Beerman. 1989. Effect of minor groove binding drugs on mammalian topoisomerase I activity. Biochem. Pharmacol. **38**:2323–2328.

61. Miller, K.G., L.F. Liu, and P.T. Englund. 1981. A homogenous type II topoisomerase from HeLa cell nuclei. J. Biol. Chem. **256**:9334–9339.

62. Moriya, S., N. Ogasawara, and H. Yoshikawa. 1985. Structure and function of the region of the replication origin of the *Bacillus subtilis* chromosome. III. Nucleotide sequence of some 10,000 base pairs in the origin region. Nucleic Acids Res. **13**:2251–2265.

63. Nielsen, B.L., and K.K. Tewari. 1988. Pea chloroplast topoisomerase I: purification, characterization, and role in replication. Plant Mol. Biol. **11**:3–14.

64. Nitiss, J., and J.C. Wang. 1988. DNA topoisomerase-targeting antitumour drugs can be studied in yeast. Proc. Natl. Acad. Sci. USA **85**:7501–7505.

65. Oki, T., O. Tenmyo, M. Hirano, K. Tomatsu, and H. Kamei. 1990. Pradimicins A, B and C: New antifungal antibiotics. II. In vitro and in vivo biological activities. J. Antibiotics **43**:763–770.

66. Okura, A., H. Arakawa, H. Ok, T. Yoshinari, and Y. Monden. 1988. Effect of genistein on topoisomerase activity and on the growth of [VAL 12] Ha-*ras*-transformed NIH 3T3 cells. Biochem. Biophys. Res. Commun. **157**:183–189.

67. Osheroff, N. 1989. Effect of antineoplastic agents on the DNA cleavage/religation reaction of eukaryotic topoisomerase II: inhibition of DNA religation by etoposide. Biochem. **28**:6157–6160.

68. Pommier, Y., M.R. Mattern, R.E. Schwartz, L.A. Zwelling, and K.W. Kohn. 1984. Changes in deoxyribonucleic acid linking number due to treatment of mammalian cells with the intercalating agent 4′-(9-acridinylamino)methanesulfon-*m*-anisidide. Biochemistry **23**:2927–2932.

69. Pyke, K.A., J. Marrison, and R.M. Leach. 1989. Evidence for a type II topoisomerase in wheat chloroplasts. FEBS Lett. **242**:305–308.

70. Richard, R.E., and D.F. Bogenhagen. 1989. A high molecular weight topoisomerase I from *Xenopus laevis* ovaries. J. Biol. Chem. **264**:4704–4709.

71. Riou, G.F., M. Gabillot, S. Douc-Rasy, A. Kayser, and M. Barrois. 1983. A. type I DNA topoisomerase from *Trypanosoma cruzi*. Eur. J. Biochem. **134**:479–484.

72. Riou, J.-F., M. Gabillot, M. Philippe, J. Schrevel, and G. Riou. 1986. Purification and characterisation of *Plasmodium berghei* DNA topoisomerases I and II: drug action, inhibition of decatenation and relaxation, and stimulation of DNA cleavage. Biochemistry **25**:1471–1479.

73. Robinson, M.J., and N. Osheroff. 1990. Stabilization of the topoisomerase II–DNA cleavage complex by antineoplastic drugs: inhibition of enzyme-mediated DNA religation by 4′-(9-acridinylamino)methanesulfon-*m*-anisidide. Biochem. **29**:2511–2515.

74. Rose, K.M. 1988. DNA topoisomerases as targets for chemotherapy. FASEB J. **2**:2474–2478.

75. Rowe, T.C., J.R. Rusche, M.J. Brougham, and W.K. Holloman. 1981. Purification and properties of a topoisomerase from *Ustilago maydis*. J. Biol. Chem. **256**:10354–10361.

76. Russo, P., M. Peluso, S. Parodi, S. Tornalet, A.M. Pedrini, F. Palitti, and B.A. Kihlman. 1986. DNA-topoisomerase II in mammalian cells could be a target of caffeine (C) and 8-methoxycaffeine (8MC). Eur. J. Cell Biol. **42**:6a.

77. Sander, M., and T.-S. Hsieh. 1983. Double strand DNA cleavage by type II DNA topoisomerase from *Drosophila melanogaster*. J. Biol. Chem. **258**:8421–8428.

78. Sawada, Y., M. Nishio, H. Yamamoto, M. Hatori, T. Miyaki, M. Konishi, and T. Oki. 1990. New antifungal antibiotics, pradimicins D and E. Glycine analogs of pradimicins A and C. J. Antibiotics **43**:771–777.

79. Schomburg, V., and F. Grosse. 1986. Purification and characterisation of DNA topoisomerase II from calf thymus associated with polypeptides of 175 and 150 kDa. Eur. J. Biochem. **160**:451–457.

80. Scovell, W.M. 1986. Supercoiled DNA. J. Chem. Educ. **63**:562–565.

81. Shuman, S., M. Golder, and B. Moss. 1989. Insertional mutagenesis of the vaccinia virus gene encoding a type I DNA topoisomerase: evidence that the gene is essential for virus growth. Virology **170**:302–306.

82. Sioud, M., G. Baldacci, P. Forterre, and A.-M. de Recondo. 1987. Antitumor drugs inhibit the growth of halophilic archaebacteria. Eur. J. Biochem. **169**:231–236.

83. Smith, P.J. 1990. DNA topoisomerase dysfunction a new goal for antitumour chemotherapy. BioEssays **12**:167–172.

84. Srivenugopal, K.S., D. Lockshon, and D.R. Morris. 1984. *Escherichia coli* DNA topoisomerase III: purification and characterisation of a new type I enzyme. Biochemistry **23**:1899–1906.

85. Sugino, A., and K.F. Bott. 1980. *Bacillus subtilis* deoxyribonucleic acid gyrase. J. Bacteriol. **141**:1331–1339.

86. Sugino, A., C.L. Peebles, K.N. Kreuzer, and N.R. Cozzarelli. 1977. Mechanism of action of nalidixic acid: purification of *Escherichia coli nalA* gene product and its relationship to DNA gyrase and a novel nicking-closing enzyme. Proc. Natl. Acad. Sci. USA **74**:4767–4771.

87. Sutcliffe, J.A., T.D. Gootz, and J.F. Barrett. 1989. Biochemical characteristics and physiological significance of major DNA topoisomerases. Antimicrob. Agents Chemother. **33**:2027–2033.

88. Suzuki, K., H. Yamaguchi, S. Miyazaki, K. Nagai, S.-I. Watanabe, T. Saito, K. Ishii, M. Hanada, T. Sekine, Y. Ikegami, and T. Andoh. 1990. Topostin, a novel inhibitor of mammalian DNA topoisomerase I from *Flexibacter topostinus* sp. Nov. J. Antibiotics **43**:154–157.

89. Takeuchi, T., T. Hara, N. Naganawa, M. Okada, M. Hamada, H. Umezawa, S. Gomi, M. Sezaki, and S. Kondo. 1988. New antifungal antibiotics, benanomicins A and B from an *Actinomycete*. J. Antibiotics **41**:807–811.

90. Tewey, K.M., G.L. Chen, E.M. Nelson, and L.F. Liu. 1984. Intercalative antitumor drugs interfere with the breakage-reunion reactions of mammalian DNA topoisomerase II. J. Biol. Chem. **259**:9182–9187.

91. Thrash, C., A.T. Bankier, B.G. Barrell, and R. Sternglanz. 1985. Cloning, characterization and sequence of the yeast DNA topoisomerase I gene. Proc. Natl. Acad. Sci. USA **82**:4374–4378.

92. Thurston, L.S., H. Irie, and S. Tani. 1986. Antitunor agents. 78. Inhibition of human DNA topoisomerase II by podophyllotoxin and α-peltatin analogues. J. Med. Chem. **29**:1547–1550.

93. Tomita, K., N. Nishio, K. Saitoh, H. Yamamoto, Y. Hoshino, H. Ohkuma, M. Konishi, T. Miyaki, and T. Oki. 1990. Pradimicins A, B and C: New antifungal antibiotics. I. Taxonomy, production, isolation and physico-chemical properties. J. Antibiotics **43**:755–762.

94. Trask, D.K., and M.T. Muller. 1983. Biochemical characterization of topoisomerase I purified from avian erythrocytes. Nucleic Acids Res. **11**:2779–2800.

95. Tse-Ding, Y.-C., and J.C. Wang. 1986. Complete nucleotide sequence of the *topA* gene encoding *Escherichia coli* DNA topoisomerase I. J. Mol. Biol. **191**:321–331.

96. Uemura, T., K. Morikawa, and M. Yanagida. 1986. The nucleotide sequence of the fission yeast DNA topoisomerase II gene: structural and functional relationships to other DNA topoisomerases. EMBO J. **5**:2355–2361.

97. Uemura, T., K. Morino, S. Uzawa, K. Shiozaki, and M. Yanagida. 1987. Cloning and sequencing of *Schizosaccharomyces pombe* DNA topoisomerase I gene, and the effect of gene disruption. Nucleic Acids Res. **23**:9727–9739.

98. van Maanen, J.M.S., J. Retel, J. de Vries, and H.M. Pinedo. 1988. Mechanism of action of antitumour drug etoposide: A review. J. Nat. Cancer Inst. **80**:1526–1533.

99. Wang, J.C. 1985. DNA topoisomerases. Annu. Rev. Biochem. **54**:665–697.

100. Wang. J.C. 1987. Recent studies of DNA topoisomerases. Biochim. Biophys. Acta. **909**:1–9.

101. Wang, J.C. 1987. DNA topoisomerases: From a laboratory curiosity to a subject in cancer chemotherapy. NCI Monographs **4**:3–6.

102. Wang, Z.-Q., Y.-H. Kuo, D. Schnur, J.P. Bowen, S.-Y. Liu, F.-S. Han, J.-Y. Chang, Y.-C. Cheng, and K.-H. Lee. 1990. Antitumor agents 113. New 4β-arylamino derivatives of 4′-O-demethylepipodophyllotoxin and related compounds as potent inhibitors of human DNA topoisomerase II. J. Med. Chem. **33**:2660–2666.

103. Wasserman, S.A., and N.R. Cozzarelli. 1986. Biochemical topology: applications to DNA recombination and replication. Science **232**:951–960.

104. Wolfson, J.S., and D.C. Hooper (eds.) 1989. Quinolone Antimicrobial Agents. ASM Washington.

105. Woynarowski, J.M., M. McHugh, R.D. Sigmund, and T.A. Beerman. 1988. Modulation of topoisomerase II catalytic activity by DNA minor groove binding agents distamycin, Hoechst 33258, and 4′,6-diamidine-2-phenylindole. Mol. Pharmacol. **35**:177–182.

106. Wyckoff, W., and T.-S. Hsieh. 1988. Functional expression of a *Drosophila* gene in yeast: genetic complementation of DNA topoisomerase II. Proc. Natl. Acad. Sci. USA **85**:6272–6276.

107. Wyckoff, E., D. Natalie, J.M. Nolan, M. Lee, and T.-S. Hsieh. 1989. Structure of the *Drosophila* DNA topoisomerase II gene. Nucleotide sequence and homology among topoisomerase II. J. Mol. Biol. **205**:1–13.

108. Yamashita, Y., S.-Z. Kawada, N. Fujii, and H. Nakano. 1990. Induction of mammalian DNA topoisomerase II dependent DNA cleavage by antitumor antibiotic streptonigrin. Cancer Res. **50**:5841–5844.

109. Zhang, H., P. D'Arpa, and L.F. Liu. 1990. A model for tumor cell killing by topoisomerase poisons. Cancer cells **2**:23–27.

15

Functions for Sterols in Yeast Membranes

Leo W. Parks, R. Todd Lorenz, and *Warren M. Casey*

I. Introduction

Most of the clinically useful antifungal agents either interact directly with the sterols in the fungal membrane or interfere with the synthesis of fungal sterols. The characteristics of these agents have been studied extensively in recent years, but their precise mechanisms in altering fungal growth are only poorly understood. Our research on *Saccharomyces cerevisiae* has produced evidence that there are multiple functions for sterols in this organism. It seems logical that such is the case for the large number of fungi that are pathogenic to various plants or animals. The challenge, therefore, is to define the different functions of the principal fungal sterols and to assess the specific features of the sterol that are of greatest significance in those functions. Once these functions and physiological processes are identified, they may be of value in the biorational design of specific inhibitors and agents that are targeted to essential functions in fungi. Such agents might be effective in smaller amounts and have less deleterious effects on the host than those presently in use.

The principal sterol of yeast and many other fungi is ergosterol. Over a century ago it was isolated from ergot of rye (60), and a quarter century later it was purified from *S. cerevisiae* (57). Ergosterol is a 28-carbon alcohol with a conjugated diene in the B ring. Like most plant and fungal sterols, it is transalkylated at position 24, the donor being S-adenosylmethionine (42).

The term "function" will be used here to mean that the phenomenon being discussed is significant to the growth and survival of the cell. While the concept of function would appear to be rather simple to define, in fact, it is somewhat ambiguous. Activities of specific sterols or features of the sterol molecule can be demonstrated *in vitro*, but their importance in cell physiology is more difficult to assess. Sterols are often described simply as hydrophobic, passive structural components of membranes. Our experiments are based on the assumption that whatever roles sterols play in the cell, they have evolved, been perfected, and

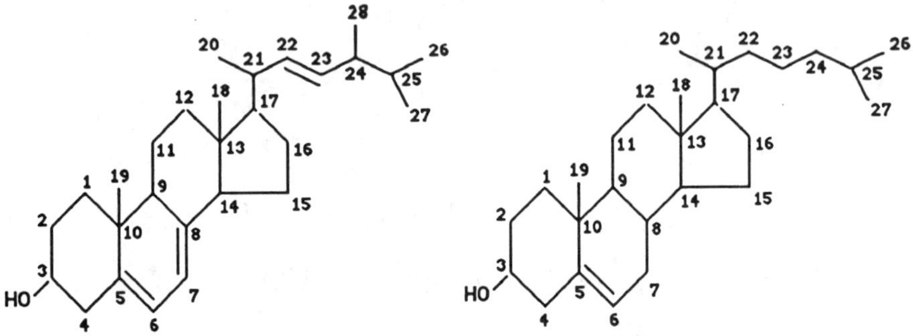

ERGOSTEROL

CHOLESTEROL

been maintained by the species for survival advantage; otherwise they would have been eliminated by natural selection. It follows, then, that interference in any of the sterol functions would be deleterious to the organism to some degree.

Baker's yeast, *S. cerevisiae*, is a convenient model with which to study the functions of sterols in fungi. It is easy to culture in large quantity, has a well-defined genetic mechanism, is generally regarded as nonpathogenic, has been well studied and a wealth of biochemical information exists, is readily amenable to molecular biological manipulation, and has very simple nutritional requirements for the wild type. With its stable haploid and diploid phases, mutant isolation and characterization can be performed with relative ease.

II. Defining Multiple Functions for Ergosterol

Sterols appear to be a component of all fungal cells (65). Those organisms that are unable to synthesize sterols must obtain them in the diet (66) for at least some phases of their life cycle. It has been assumed that the primary role of sterols is to modulate membrane fluidity and thus prevent dramatic changes with fluctuating environmental conditions. While widely presented, this notion is not supportable in warm-blooded animals where the temperature is well controlled. Over two decades ago the first evidence emerged suggesting that there might be at least dual roles for sterols in yeast (1). Subsequent research by two different groups authenticated that prediction (49, 53). However, a single, unifying model for the functions of sterols did not emerge. Recently, we decided to investigate the possibility that there are multiple roles for sterols in yeast. In designing our experiments we made three simple assumptions: (1) if there are multiple roles for sterols, it seems highly unlikely that each function would have the same quantitative requirement for sterols; (2) it is unlikely that the different roles would have identical structural requirements for the different features of the sterol molecule; and (3) ergosterol is a consensus sterol that is able to fulfill all of the quantitative and qualitative requirements for sterols in yeast.

We constructed the sterol auxotrophic yeast, RD5R (*hem1, erg7, erg3*), to conduct a series of structure-function experiments. We selected the saturated

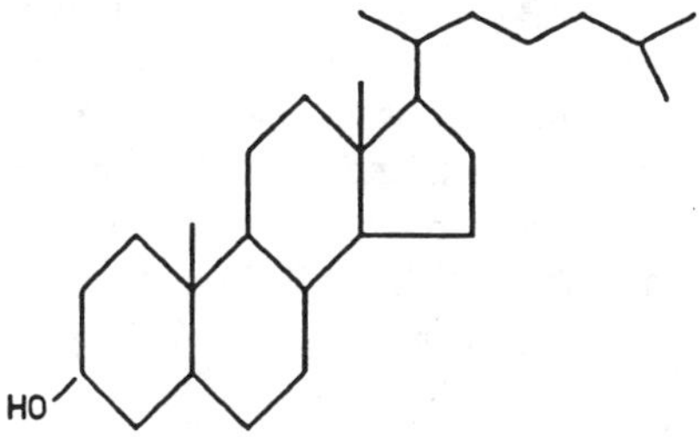

HO

CHOLESTANOL

stanol, cholestanol, as the simplest structure for determining the structural requirements of the different functions. By growing yeast cultures in varying amounts of sterols with different structural characteristics, we obtained evidence for at least four different sterol functions (51, 54). We have called these functions sparking, critical domain, domain, and bulk. These represent the minimal number of functions, and all can be accommodated with free ergosterol.

In subsequent experiments, we have focused on a study of two of those functions, sparking and bulk (53). These are the most and least structurally demanding of the various functions, respectively. Sparking function represents a microrequirement for sterol (10 ng/ml) that has an unsaturation between carbons 5 and 6 (C5=6). Bulk function involves a quantitatively variable pool of free sterol and requires only the basic cholestanol structure with the 3β-hydroxy group. It should be noted that our assignment of features for a function represents only the minimal structural requirements to satisfy that function, and *in vivo* the natural sterol, ergosterol, may be more effective at lower concentrations.

Numerous sterols and stanols have been examined for their ability to satisfy the bulk and sparking functions in several yeast sterol auxotrophs. Most of the tested compounds satisfied the bulk function, but only compounds having a C5=6 unsaturation or that were capable of being desaturated at C5–6 could fulfill the sparking function. Unsaturation in the A ring or β-saturation of the C5=6 double bond made the compound unsuitable for either function. The methyl group of ergosterol at C24 is not required for the sparking function, but those stanols possessing the side chain aklylation were more readily desaturated at C5–6. Thus, some of the sterols and stanols that lacked the C28 methyl group were incapable of satisfying the sparking function while analogous compounds that had the C28 methyl were more facile to desaturation at C5–6 (53).

A sterol auxotroph, SPK14 (*MATα, hem1, erg6, erg7, ura*) was constructed to define further the ability of growth-limiting amounts of C5=6 unsaturated sterol to spark growth on bulk levels of cholestanol (28). Although the C5=6 is required in sparking sterols, the additional presence of the C22 unsaturation was found to enhance sparking better than either the C7=8 unsaturation or the C24 methyl group. Ergosterol initiated growth faster and elicited greater cell yield than did any other sterol. It appears that, although there is an absolute requirement only for the C5=6 in sparking, additional features of ergosterol facilitate its overall effectiveness (28). These results support previous findings that the C5=6 configuration plays a critical role in the physiology of the cells (53).

Mutant GL7 of *Saccharomyces* requires a sterol and an unsaturated fatty acid for growth (20). T. Buttke showed that the organism cultured with oleic acid or petroselenic acid grew much more slowly with 7-dehydrocholesterol than with

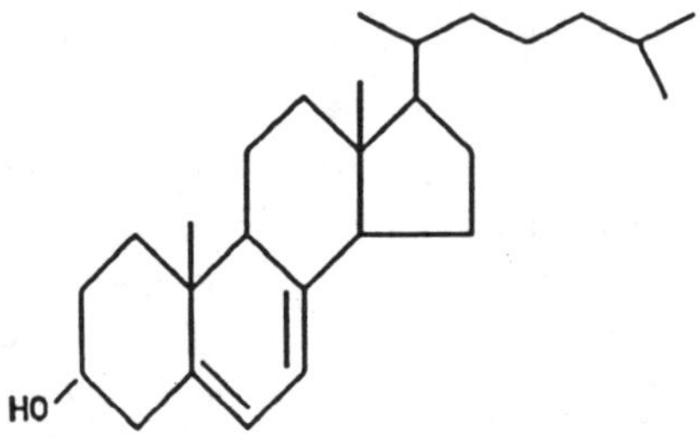

HO

7-DEHYDROCHOLESTEROL

ergosterol (10), demonstrating the importance of the features of the side chain of ergosterol. Recent experiments have shed some light on other possible functions for the side-chain features of ergosterol (32). The structural gene *ERG6* encodes sterol methyl transferase, the enzyme that transfers the methyl group from S-adenosylmethionine to zymosterol in *Saccharomyces*. We have shown that some mutants defective in that gene, *ERG6*, do not produce detectable levels of side-chain alkylated sterols (32). One interesting feature of these mutants is that they produce eight-times-higher levels of steryl esters than the homologous wild-type strains. In addition, one *erg6* mutant had a 40% increase in the oleic acid content of phosphatidylethanolamine.

While the *ERG6* gene product is not required for the sparking function, it is needed for some normal membrane functions (19). A null mutant of *ERG6* has been isolated. This strain, like the point mutant described above, produces no measurable ergosterol. A detailed analysis of that organism revealed that the cell was defective in conjugation, hypersensitive to cycloheximide, resistant to nystatin, defective in genetic transformation, and took up tryptophan only poorly. Thus, while the absence of the C28 methyl group does not decrease the growth rate of the cells under usual laboratory conditions, it does play a critical role in many of the membrane activities of the organism (19). Cells grown anaerobically in the presence of iminosqualene appear to require sterols having the C28 methyl group (46, 47). The significance of the different features of ergosterol in sustaining anaerobic growth of yeast has been reported (40).

Very recently we have obtained evidence that the side-chain features of ergosterol also play an important role in the regulation of ergosterol synthesis (Casey and Parks, unpublished experiments). We have prepared some interesting analogs of ergosterol that possess an endoperoxide structure in the B ring.

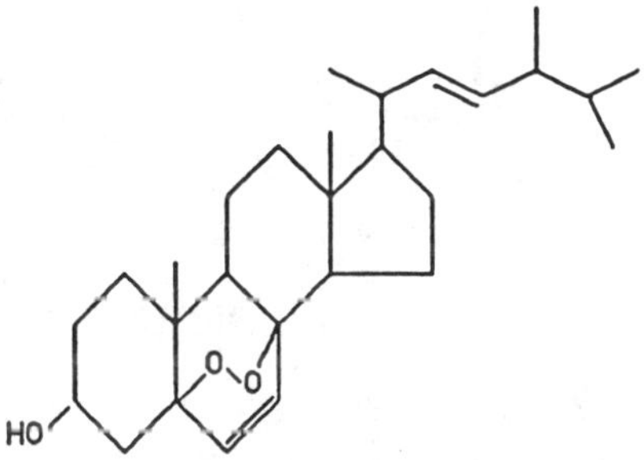

HO O O

ERGOSTEROL PEROXIDE

Using these compounds in addition to a number of other ergosterol and cholesterol analogs, we were able to show that effective feedback inhibition of ergosterol synthesis required both the C22=23 unsaturation and the C28 methyl group. This result is consistent with our observation that *erg6* mutants produce high levels of steryl esters. Esterification is an effective sink for excess sterol in yeast (27). The site of regulation by side-chain features is under further study, but it appears to precede the sterol cyclization event.

As will be discussed below, sterol modifications cause changes in the phospholipid composition of the cell. The fatty acid composition of membrane phospholipids also modulates sterol synthesis. Depriving sterol mutant GL7 of unsaturated fatty acids caused the organism to reduce markedly the activity of squalene epoxidase (11). Restoration of the activity required *de novo* protein synthesis.

III. Effect of Structural Features of Sterols on Identified Biochemical Activities

Many physical measurements conducted in the Sixties and Seventies indicated that sterols had a condensing effect on phospholipid bilayers. The addition of cholesterol to phospholipid bilayers tended to buffer the bilayer from abrupt phase transitions. Most of that work was done with synthetic mixtures of phospholipids and sterols, and it is difficult to extrapolate to natural membranes. It is clear that membrane sterols are involved in maintaining the proper membrane fluidity and integrity (7, 13, 18, 21, 25). Model systems show that in order for a sterol to fulfill bulk membrane requirement, it must have a β-hydroxyl group, a planar nucleus, and an alkyl side chain (13, 18, 25, 27, 38). Even with these restrictions a large number of sterol structures are possible, and many are found even within the same organism.

The assumption is often made that most sterols are as effective as cholesterol in controlling membrane properties. Ergosterol and stigmasterol, however, are

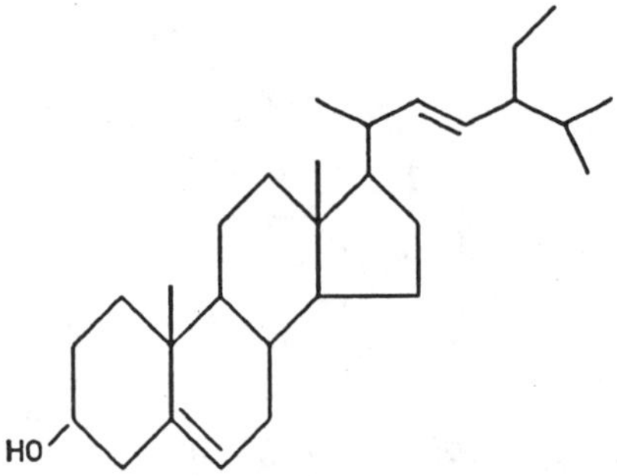

STIGMASTEROL

less efficient than cholesterol in reducing the permeability of egg phosphatidylcholine liposomes to glycerol and erythritol (18). At concentrations greater than 10 mole%, cholesterol still aligns egg phosphatidylcholine bilayers, but ergosterol disrupts the bilayer (55). The alkylated side-chain sterols, stigmasterol and sitosterol, are less effective than cholesterol in liquefying dipalmitoylphosphatidylcholine monolayers (21).

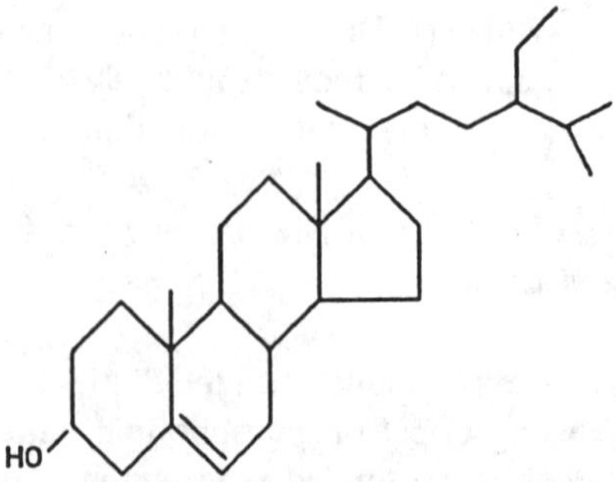

SITOSTEROL

We modified the sterols in natural membranes and determined some physical consequences of the changes (9). Using the fluorescence anisotropy of a probe,

1,6-diphenyl-1,3,5-hexatriene (DPH), we compared the properties of yeast membranes that had altered sterol compositions, either by mutation or supplementation of the growth medium. The sterol mutants that were used have defects late in the ergosterol biosynthetic pathway; they synthesize precursor sterols other than ergosterol and grow in the absence of ergosterol. Mutants were selected on the basis of nystatin resistance (5, 33). Detailed descriptions of the sterol mutants and the associated biochemical defects have been published (23, 42). The sterol auxotroph that was used is defective in heme biosynthesis (*hem1*) and in the early stages of ergosterol formation (*erg7*), and hence it requires an exogenous source of sterol, unsaturated fatty acid, and methionine.

The fluorescence anisotropy of the probe decreased with increasing temperatures, indicating an increase in mobility of DPH in the membrane. Using the mutant organisms, we observed that the rate of decrease in anisotropy with respect to temperature varied with the mutants. Whereas the wild type gave a fairly straight linear response to temperature, we observed pronounced discontinuities in plots of the data obtained from the cells (*erg3*) producing ergosta-7,22-dienol or *erg6* which synthesizes nonalkylated cholesta-derivatives (zymosterol, cholesta-

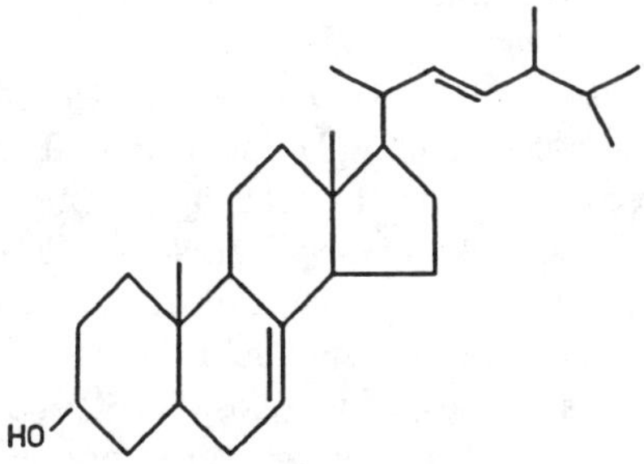

HO

ERGOSTA-7,22-DIENOL

5,7,22,24-tetraenol). Using lipids isolated from the mutants to prepare liposomes, we verified that the physical changes observed in the sterol mutants are dependent on the sterol present. We concluded that the presence of ergosterol in the wild-type plasma membranes prevents the changes in the physical properties of the membranes seen in the mutants that synthesized the aberrant sterols.

When the auxotrophs were fed altered sterols that are not obtainable in mutants, substantial accommodation by the cells was observed (31). Plasma membranes were isolated from a sterol auxotroph RD5R that had been fed different sterols. These preparations showed no discontinuity in plots of the steady-state fluorescence anisotropy of DPH. Liposomes prepared from the phospholipids that were produced by RD5R and the sterol that the cells had been fed showed similar patterns of anisotropy with no discontinuities. However, when liposomes were prepared with phospholipids extracted from a culture grown on a specific sterol and a sterol different from the one that was used for the growth of that culture, a discontinuity was found. Thus, compensatory changes were made in the phospholipids as a function of the sterol in the growth medium, and this appeared to be specific for each exogenously supplied sterol. These changes in plasma membrane phospholipid in response to exogenous sterol were dependent on the sterol that had been supplied in the growth medium. Only the sterol on which the culture was grown was effective in eliminating the discontinuity in the anisotropy plots of the reconstituted liposomes.

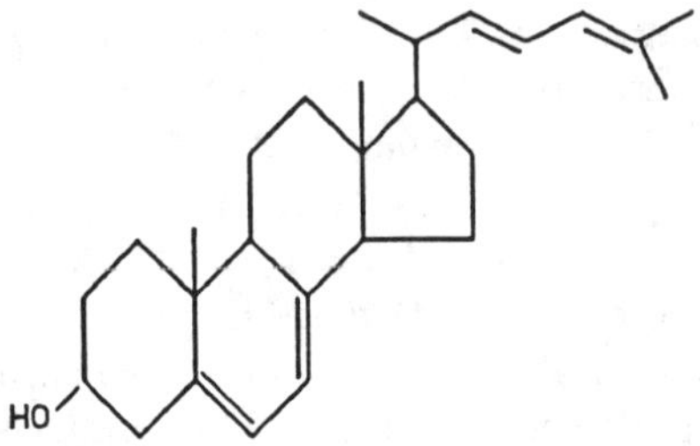

The changes in the physical behavior of the membranes described above were accompanied by alterations in the phospholipid composition (31). Both mutants and auxotrophs showed changes in the total lipid composition of the plasma membrane when different sterols were available to the cells. Quantitation of the lipids extracted from the auxotroph grown on different sterols showed no major differences in the total amount of phospholipid present. There were significant changes in the percentages of phosphatidylinositol (PI) and phosphatidylethanoline (PE) in cells grown on stigmasterol, sitosterol, or ergosterol in comparison to cholesterol. These differences can be interpreted as a cellular response to growth on cholesterol which results in a plasma membrane having fluid-state properties similar to those of cells grown on stigmasterol, sitosterol, or ergosterol. The cells compensated for growth on cholesterol by increasing the amount of PI and decreasing the amount of PE. Analyses of total fatty acids showed changes in palmitic (C16:0), palmitoleic (C16:1), and oleic (C18:1) acid percentages in cells adjusted to growth on cholesterol compared to cells grown on the alkylated sterols.

These results indicate that membrane phospholipid composition can be modulated in response to the sterol that is provided. This requires that the cell has the ability to discriminate and to respond to the distinctive features of the sterol molecule and produce an appropriate head-group and fatty acid composition that accommodates the sterol structure.

It has been shown that ergosterol enhances the synthesis of PI and promotes a rapid turnover of phosphoinositides in cells previously grown on cholesterol (17).

Ergosterol also appears to increase the transmethylation rate of intermediates in the synthesis of phosphatidylcholine (27).

If cells were forced by inhibitors to synthesize sterols to which a physiologically acceptable accommodation was not possible, they would not grow. Lanosterol is one such sterol that is not accommodated well by the membrane. It will be interesting to study cells that are resistant to such growth inhibition.

Cultures of the sterol auxotroph, GL7, that were starved for sterol were shown to arrest growth in the unbudded or G_1 phase (14). The addition of ergosterol apparently stimulates a protein kinase concurrently with the reinitiation of growth. Ergosterol also stimulated PI kinase activity under similar conditions. Cholesterol was a much less effective activator than ergosterol. The stimulatory effect of ergosterol occurred in the absence of protein synthesis, as the response to the sterol was unaffected by the protein synthetic inhibitor, cycloheximide.

IV. Anticipated Targets for Sterol Biosynthesis Inhibitors

The specific causes of growth inhibition by interfering with sterol biosynthesis are not clearly understood. Aberrant sterols may interfere with normal sterol functions. Alternatively, if ergosterol is absent, critical physiological functions dependent on the distinctive structural features of ergosterol may not be accomplished. While the ultimate effect from either of these causes is essentially the same, the biochemical mechanisms are different.

The limited effectiveness of 14-methyl sterols in model sterol-phospholipid interactions led to the prediction that these compounds should be unsatisfactory for biological activity (15, 16, 26, 67). Removal of the C14 methyl group from lanosterol would appear to be essential in order to establish the planar alpha face for biological activity of the sterols (6, 24, 39). Sterol mutants (*erg11*) have been isolated, however, which are defective in the removal of the C14 methyl group (44, 63). Genetic analysis of these mutants revealed the presence of a second mutation that was in the *ERG3* gene (62). The *erg3* defect impairs the ability of the organism to introduce the C5=6 unsaturation in the B ring. Segregational analysis revealed that the spores singly defective in *erg11* did not germinate. It is essential that in the double mutant the *erg3* mutation be leaky; that is, there must be a small amount of C5–6 desaturase activity. The major sterol in the double mutant, *erg3, erg11* is 14-methyl-ergosta-8,24(28)-dienol. This sterol will satisfy the bulk sterol function, but it is incapable of serving as a sparking sterol (54). A careful analysis of the double mutant revealed the presence of a small amount of ergosterol, presumably to provide the essential sparking function. Thus, the C14 methyl group in itself is not toxic to the cell, but is incapable of providing one of the needed sterol functions. It is interesting that numerous *erg3* mutants have been isolated, but every one has been shown to be leaky. This is in contrast to the *erg6* mutants, which lack detectable transmethylase activity and are totally devoid of C28 sterols.

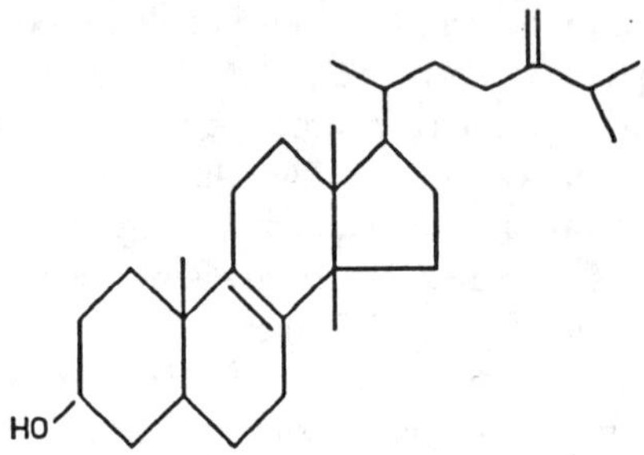

14-METHYL-ERGOSTA-8,24(28)-DIENOL

One of the most potent and naturally occurring antifungal agents is 15-aza-24-methylene-8,14-cholestadienol (52). It is produced by the fungus *Scytalidium flavobrunneum* and exerts multiple effects of the biosynthesis of ergosterol. The 15-azasterol has been shown to inhibit the sterol methyltransferase (4), the 24-methylene sterol reductase (34), and C14=15 reductase (8). The most sensitive enzyme is the latter reductase with a noncompetitive K_i of 2 nM. The other two enzymes have competitive K_is of 43 μM and 17μM. Other azasterols have been shown to affect the methyl transferase and methylene reductase enzymes (45).

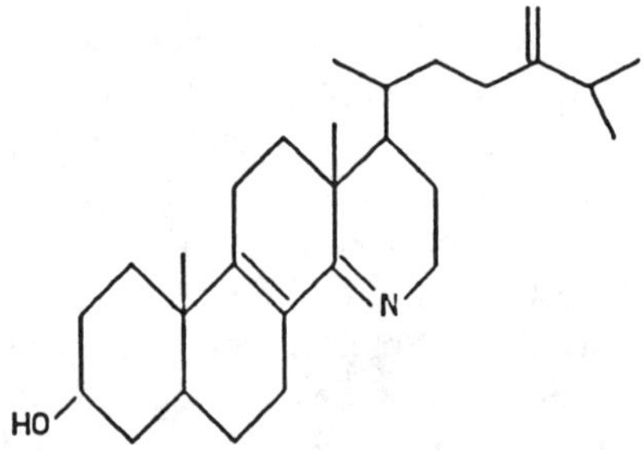

15-AZASTEROL

When treated with very low, but growth-inhibiting, levels of 15-azasterol, yeast cells produce ergosta-8,14-dienol (ignosterol) as their principal sterol. If, however, a small amount of ergosterol is available from exogenous sources, cell growth is unaffected by the 15-azasterol (61). Here also it would appear that it is the absence of ergosterol and the inability to satisfy one or more of the essential sterol functions that results in growth inhibition.

Heme-deficient mutants are unable to grow unless they are provided a suitable C5=6 sterol, an unsaturated fatty acid, and methionine. Indeed, the first sterol mutants were heme-deficient mutants. Cytochrome P–450 hemoproteins are involved in sterol synthesis for the oxidative demethylation of lanosterol (2, 3) and

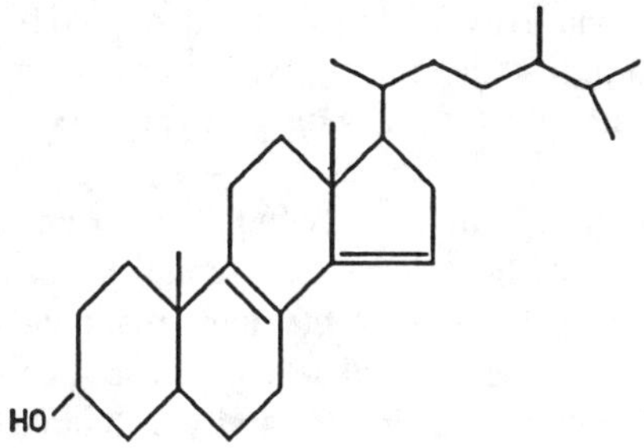

IGNOSTEROL

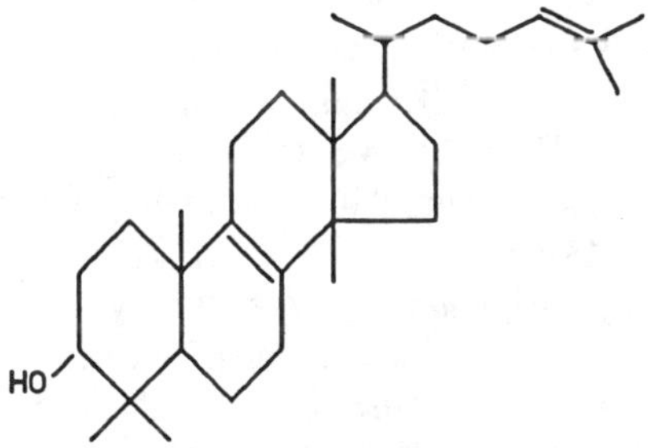

LANOSTEROL

the desaturation of the C22 position of the side chain (22). There appears to be a requirement for higher amounts of hemoprotein for the latter function (56). Cytochrome b_5 hemoproteins are necessary for the desaturation of the C5–6 position of the B ring of the sterol molecule (41, 50). In addition to the catalytic roles that hemes play, there is also a regulatory role. Heme competency results in a fivefold stimulation in hydroxymethylglutaryl–coenzyme A (HMG-CoA) reductase activity and a concomitant enhancement of sterol precursor synthesis (29).

While the heme-dependent reactions may be suitable targets for antifungal activity, sterol biosynthesis is not the only vulnerable pathway. We have shown that although some azoles do inhibit C14 demethylation, another sensitive physiological site may be effecting growth inhibition (61).

V. Conclusions and Future Directions

Sterols have multiple functions in the physiology of the yeast cell. As the predominant sterol in yeast, ergosterol appears best to fulfill the multiple roles. Preferences for different features of the sterol for optimal function in different physiological roles have been demonstrated. Ergosterol represents a consensus structure that satisfies the specific and cumulative needs of the sterol functions for the survival of the cells. Any activity that is disrupted by changes in the sterol structure is a potential antifungal target. Mutants resistant to known fungicides provide valuable insight into the physiological processes that are subjects for further study.

The ubiquity of the sterols in various fungi has been established (64). It is tempting to speculate that the diverse functions that have been described in *Saccharomyces* for sterols also exist in the many plant and animal fungal pathogens. However, a direct experimental test of the possibilities must await the isolation and characterization of a similar family of mutants and auxotrophs that were used in the studies with *Saccharomyces*. It is known that there are differences in the synthesis and composition of sterols in different fungi. Some do not produce ergosterol as their major sterol. Some insect pathogens (Uredinales) have desmosterol as the principal sterol (58). The rust fungi have mainly C29 sterols (65), while fucosterol is the predominant sterol of many Oomycetes. Although the pythiaceous fungi (*Phytophthora* and *Pythium*) do not synthesize sterols, they are required for reproduction (66). We have shown that even under vegetative conditions, sterols impart a substantial increase in the efficiency of energy metabolism in *Phytophthora* (43), and it has been speculated that during vegetative growth without sterols, long-chain fatty alcohols may serve as functional replacements for the sterols (35). The stereochemistry of the side chain of the sterol molecule is important in *Phytophthora* in the synthesis of functional spores and in the growth response to the sterols, but the induction of the sexual structures is less discriminatory (37). Some fungi produce mixtures of sterols (66). These are not simply precursors of a predominant sterolic endproduct. Evidence has been obtained for the different sterols having different physiological functions (36). Again, it must be assumed that the complex of sterols found in different organisms must impart a physiological advantage to the cell; otherwise a simpler mixture of sterols would have been selected in the evolution of the organism.

Considerable biochemical homology has been found amongst different fungi, however. Most are susceptible to varying degrees to the usual antifungal agents. Similar mechanisms for biochemical activities exist. Indeed, the genes for sterol biosynthetic enzymes can be transferred and shown to be expressed in quite phylogenically distant organisms (D. Hollomon, personal communication). While a case cannot be made for identical physiological roles for sterols in the many different fungi, it is clear that a model system such as is found in *Saccharomyces* has substantial heuristic value.

Finally, based on some recent experiments in *Saccharomyces*, a new direction may be imminent in antifungal development. Most antifungal agents are toxic to the host. Even in plant systems, antifungal agents inhibiting ergosterol synthesis alter the sterol composition of the host (48), and phytophagous insects that were allowed to consume those plants had severe developmental aberrations (12). It is reasonable to suggest that, if it were possible to reduce the minimal inhibitory concentration (MIC) of antifungal agents, toxicity effects could be reduced, and agents heretofore deemed unacceptable may prove clinically useful. It was found recently that an inhibitor of HMG-CoA reductase in combination with azoles reduced the MIC of azoles by as much as 30-fold in *Saccharomyces* (30, 59). The mechanism of this effect is being studied further, but it appears that the enhanced sensitivity of the yeast to azoles is not simply due to sterol depletion.

Acknowledgments

Research conducted in the authors' laboratory was supported by the National Science Foundation (DCB8814387), the National Institutes of Health (DK-37222), and the North Carolina Agricultural Research Service.

References

1. Adams, B.G., and L.W. Parks. 1967. Evidence for dual physiological forms of ergosterol in *Saccharomyces cerevisiae*. J. Cell. Physiol. **70,** 161–168.
2. Aoyama, Y., and Yo Yoshida. 1978. The 14α-demethylation of lanosterol by a reconstituted cytochrome P–450 system from yeast. Biochem. Biophys. Res. Commun. **85,** 28–34.
3. Aoyama, Y., and Yo Yoshida. 1978. Interaction of lanosterol to cytochrome P–450 purified from yeast microsomes: evidence for contribution of cytochrome P–450 to lanosterol metabolism. Biochem. Biophys. Res. Commun. **82,** 33–38.
4. Bailey, R.B., P.R. Hays, and L.W. Parks. 1976. Homoazasterol-mediated inhibition of yeast sterol biosynthesis. J. Bacteriol. **128,** 730–734.
5. Bard, M. 1972. Biochemical and genetic aspects of nystatin resistance in *Saccharomyces cerevisiae*. J. Bacteriol. **111,** 649–657.
6. Bloch, K. 1981. Sterol structure and membrane function. Curr. Top. Cell. Regul. **18,** 289–299.
7. Bloch, K. 1983. Sterol structure and membrane function. CRC Crit. Rev. Biochem. **14,** 47–92.
8. Bottema, C.D.K., and L.W. Parks. 1978. 14-Sterol reductase in *Saccharomyces cerevisiae*. Biochim. Biophys. Acta **531,** 301–307.
9. Bottema, C.D.K., R.J. Rodriguez, and L.W. Parks. 1985. Influence of sterol structure on yeast plasma membrane. Biochim. Biophys. Acta **813,** 313–320.
10. Buttke, T.M., S.D. Jones, and K. Bloch. 1980. Effect of sterol side chains on growth and membrane fatty acid composition of *Saccharomyces cerevisiae*. J. Bacteriol. **144,** 124–130.
11. Buttke, T.M., S.L. Brint, and M.R. Lowe. 1988. Regulation of squalene epoxidase activity by membrane fatty acid composition. Lipids **23,** 68–71.
12. Costet, M.F., M. El Achouri, M. Charlet, R. Lanot, P. Benveniste, and J.A. Hoffman. 1987. Ecdysteroid biosynthesis and embryonic development are disturbed in insects (*Locusta migratoria*) reared on a plant diet (*Triticum sativum*) with a selectively modified sterol profile. Proc. Natl. Acad. Sci. USA **84,** 643–647.

13. Cullis, P.R., and B. DeKruyff. 1979. Lipid polymorphism and the functional roles of lipids in biological membranes. Biochim. Biophys. Acta **559,** 399–420.

14. Dahl, C., H.-P. Biemann, and J. Dahl. 1987. A protein kinase antigenically related to pp 60v-src possibly involved in yeast cell cycle control: positive *in vivo* regulation by sterol. Proc. Natl. Acad. Sci. USA **84,** 4012–4016.

15. Dahl, C.E., J.S. Dahl, and K. Bloch. 1980. Effect of cycloartenol and lanosterol on artificial and natural membranes. Biochem. Biophys. Res. Commun. **92,** 221–228.

16. Dahl, C.E., J.S. Dahl, and K. Bloch. 1980. Effect of alkyl-substituted precursors of cholesterol on artificial and natural membranes and on the viability of *Mycoplasma capriocolum*. Biochemistry **19,** 1462–1467.

17. Dahl, J.S., and C.E. Dahl. 1985. Stimulation of cell proliferation and polyphosphoinositide metabolism in *Saccharomyces cerevisiae* GL7 by ergosterol. Biochem. Biophys. Res. Commun. **133,** 844–850.

18. Demel, R.A., and B. DeKruyff. 1976. The function of sterols in membranes. Biochim. Biophys. Acta **457,** 109–132.

19. Gaber, R.F., D.M. Copple, B.K. Kennedy, M. Vidal, and M. Bard. 1989. The yeast gene *ERG6* is required for normal membrane function but is not essential for the cell-cycle-sparking sterol. Mol. Cell. Biol. **9,** 3447–3456.

20. Gollub, E.G., K. Lin, J. Dogan, M. Aldersberg, and D. Sprinson. 1977. Yeast mutants deficient in heme biosynthesis and a heme mutant additionally blocked in cyclization of 2,3-oxidosqualene. J. Biol. Chem. **252,**2846–2854.

21. Green, C. 1977. Sterols in cell membranes and model membrane systems. Internat. Rev. Biochem. **14,** 101–152.

22. Hata, S., T. Nashino, M. Komori, and H. Katsuki. 1981. Involvement of cytochrome P–450 in Δ^{22}-desaturation in ergosterol biosynthesis in yeast. Biochem. Biophys. Res. Commun. **103,** 272–277.

23. Henry, S.A. 1982. "Membrane lipids of yeast: Biochemical and genetic studies." In: The molecular biology of the yeast *Saccharomyces*: Metabolism and gene expression. Cold Spring Harbor Monograph Series. pp. 101–158.

24. Huang, C. 1977. A structural model of the cholesterol-phosphatidylcholine complexes in bilayer membranes. Lipids **12,** 348–355.

25. Ladbrooke, B.D., and D. Chapman. 1969. Thermal analysis of lipids, proteins and biological membranes: a review and summary of recent studies. Chem. Phys. Lipids **3,** 304–367.

26. Lala, A.K., H.K. Lin, and K. Bloch. 1978. The effect of some alkyl derivatives of cholesterol on the permeability properties and microviscosities of model membrane. Bioorg. Chem. **7,** 437–445.

27. Lewis, T.A., R.J. Rodriguez, and L.W. Parks. 1987. Relationship between intracellular sterol content and sterol esterification and hydrolysis in *Saccharomyces cerevisiae*. Biochim. Biophys. Acta **921,** 205–212.

28. Lorenz, R.T., W.M. Casey, and L.W. Parks. 1989. Structural discrimination in the sparking function of sterols in the yeast *Saccharomyces cerevisiae*. J. Bacteriol. **171,** 6169–6173.

29. Lorenz, R.T., and L.W. Parks. 1987. Regulation of ergosterol biosynthesis and sterol uptake in a sterol-auxotrophic yeast. J. Bacteriol. **169,** 3707–3711.

30. Lorenz, R.T., and L.W. Parks. 1990. Effects of lovastatin (Mevinolin) on sterol levels and on activity of azoles in *Saccharomyces cerevisiae*. Antimicrob. Agents Chemother. **34,** 1660–1665.

31. Low, C., R.J. Rodriguez, and L.W. Parks. 1985. Modulation of yeast plasma membrane composition of a yeast sterol auxotroph as a function of exogenous sterol. Arch. Biochem. Biophys. **240,** 530–538.

32. McCammon, M.T., M-A. Hartmann, C.D.K. Bottema, and L.W. Parks. 1984. Sterol methylation in *Saccharomyces cerevisiae*. J. Bacteriol. **157,** 475–483.

33. Molzahn, S.W., and R.A. Woods. 1972. Polyene resistance and the isolation of sterol mutants in *Saccharomyces cerevisiae*. J. Gen. Microbiol. **72,** 339–348.

34. Neal, W.D., and L.W. Parks. 1977. Sterol 24(28)-methylene reductase in *Saccharomyces cerevisiae*. J. Bacteriol. **129,** 1375–1378.

35. Nes, W.D. 1988. Phytophthorols—novel lipids produced by *Phytophthora cactorum*. Lipids **23,** 9–16.

36. Nes, W.D., and R.C. Heupel. 1986. Physiological requirement for biosynthesis of multiple 24-beta methyl sterols in *Gibberella fujikuroi*. Arch. Biochem. Biophys. **244,** 211–217.

37. Nes, W.D., and A.E. Stafford. 1984. Side-chain structural requirements for sterol-induced regulation of *Phytophthora cactorum* physiology. Lipids **19,** 544–549.

38. Nes, W.R. 1974. Role of sterols in membranes. Lipids **9,** 596–612.

39. Nes, W.R., and M.L. McKean. 1977. Biochemistry of steroids and other isopentanoids. University Park Press, Baltimore, MD, pp. 536–557.

40. Nes, W.R., B.C. Sekula, W.D. Nes, and J.H. Adler. 1978. The functional importance of structural features of ergosterol in yeast. J. Biol. Chem. **253,** 6218–6225.

41. Osumi, T., T. Nishino, and H. Katsuki. 1979. Studies of the Δ^5-desaturation in ergosterol biosynthesis in yeast. J. Biochem. **85,** 819–826.

42. Parks, L.W. 1978. Metabolism of sterols in yeast. CRC Crit. Rev. Microbiol. **6,** 301–341.

43. Parks, L.W., C. McLean-Bowen, C.K. Bottema, F.R. Taylor, R. Gonzales, B.W. Jensen, and J.L. Ramp. 1982. Aspects of sterol metabolism in yeast and *Phytophthora*. Lipids **17,** 187–196.

44. Pierce, A.M., R.B. Mueller, A.M. Unrau, and A.C. Oehlschlager. 1978. Metabolism of Δ^{24}-sterols by yeast mutants blocked in removal of the C-14 methyl group. Canad. J. Biochem. **56,** 794–800.

45. Pierce, A.M., A.M. Unrau, A.C., Oehlschlager, and R.A. Woods. 1979. Azasterol inhibitors in yeast. Inhibition of the Δ^{24}-sterol methyltransferase and the 24-methylene sterol $\Delta^{24(28)}$-reductase in sterol mutants of *Saccharomyces cerevisiae*. Canad. J. Biochem. **57,** 201–208.

46. Pinto, W.J., and W.R. Nes. 1983. Stereochemical specificity for sterols in *Saccharomyces cerevisiae*. J. Biol. Chem. **258,** 4472–4476.

47. Pinto, W.J., R. Lozano, B.C. Sekula, and W.R. Nes. 1983. Stereochemically distinct roles for sterol in *Saccharomyces cerevisiae*. Biochem. Biophys. Res. Commun. **111,** 47–54.

48. Rahier, A., P. Schmitt, B. Huss, P. Benveniste, and E.H. Pommer. 1986. Chemical structure-activity relationships of the inhibition of sterol biosynthesis by N-substituted morpholines in higher plants. Pestic. Biochem. Physiol. **25,** 112–124.

49. Ramgopal, M., and K. Bloch. 1983. Sterol synergism in yeast. Proc. Natl. Acad. Sci. USA **80,** 712–715.

50. Reddy, V.V.R., D. Kupfer, and E. Caspi. 1977. Mechanism of C-5 double bond introduction in the biosynthesis of cholesterol in rat liver microsomes. J. Biol. Chem. **252,** 2797–2801.

51. Rodriguez, R.J., C. Low, C.D.K. Bottema, and L.W. Parks, 1985. Multiple functions for sterols in *Saccharomyces cerevisiae*. Biochim. Biophys. Acta **837,** 336–343.

52. Rodriguez, R.J., and L.W. Parks. 1980. Growth and antifungal homoazasterol production in *Geotrichum flavobrunneum*. Antimicrob. Agents Chemother. **18,** 822–828.

53. Rodriguez, R.J., and L.W. Parks. 1983. Structural and physiological features of sterols necessary to satisfy bulk membrane and sparking requirements in yeast sterol auxotrophs. Arch. Biochem. Biophys. **225,** 861–871.

54. Rodriguez, R.J., F.R. Taylor, and L.W. Parks. 1982. A requirement for ergosterol to permit growth of yeast sterol auxotrophs on cholesterol. Biochem. Biophys. Res. Commun. **106,** 435–441.

55. Semer, R., and E. Gelerinter. 1979. A spin label study of the effect of sterols on egg lecithin bilayers. Chem. Phys. Lipids. **23,** 201–211.

56. Shinabarger, D.L., G.A. Keesler, and L.W. Parks. 1989. Regulation by heme of sterol uptake in *Saccharomyces cerevisiae*. Steroids **53,** 607–623.

57. Smedley-Maclean, I., and E.M. Thomas. 1920. The nature of yeast fat. Biochem. J. **14,** 483–493.

58. Starr, A.M., R.W. Lichtwardt, J.D. McChesney, and T.A. Baer. 1979. Sterols synthesized by cultured Trichomycetes. Arch. Microbiol. **120,** 185–189.

59. Sud, I.J., and D.S. Feingold. 1985. Effect of ketoconazole in combination with other inhibitors of sterol synthesis on fungal growth. Antimicrob. Agents Chemother. **28,** 532–534.

60. Tanret, C. 1889. Sur un nonveau principe immédiat de l'ergot de seigle, l'ergostérine. C.R. Seances Acad. Sci. **108,** 98–100.

61. Taylor, F.R., R.J. Rodriguez, and L.W. Parks. 1983. Relationship between antifungal activity and exhibition of sterol biosynthesis by miconazole, clotrimazole, and 15-azasterol. Antimicrob. Agents Chemother. **23,** 515–521.

62. Taylor, F.R., R.J. Rodriguez, and L.W. Parks. 1983. Requirement for a second sterol biosynthetic mutation for viability of a sterol C-14 demethylation defect in *Saccharomyces cerevisiae*. J. Bacteriol. **155,** 64–68.

63. Trocha, P.J., S.J. Jasne, and D.B. Sprinson. 1977. Yeast mutants blocked in removing the methyl group of lanosterol at C-14. Separation of sterols by high-pressure liquid chromatography. Biochemistry **16,** 4721–4726.

64. Weete, J.D. 1980. Lipid biochemistry of fungi and other organisms. Plenum Press, New York.

65. Weete, J.D. 1989. Structure and function of sterols in fungi. Adv. Lipid Res. **23,** 115–167.

66. Weete, J.D., M. Fuller, and M.Q. Huang. 1989. Fatty acids and sterols of selected hypochytriomycetes and chytridiomycetes. Expl. Mycol. **13,** 183–195.

67. Yeagle, P.L., R.B. Martin, A.K. Lala, H.K. Lin, and K. Bloch. 1977. Differential effects of cholesterol and lanosterol on artificial membranes. Proc. Natl. Acad. Sci. USA **74,** 4924–4926.

16

Biochemical Aspects of Ergosterol Biosynthesis Inhibition

Keith Barrett-Bee and *Neil Ryder*

I. Introduction

The major product of sterol biosynthesis in fungi and some trypanosomes is ergosterol, unlike in mammalian systems which synthesize cholesterol as the major membrane lipid (12, 13). The two sterols differ in a few minor ways: cholesterol has a second double bond ($\Delta^{5(6)}$) in the B ring (Figure 16.1) and has a fully saturated side chain without a methyl group at C24. These small differences are clearly very important as ergosterol has been shown to be essential for the aerobic growth of most fungi. This requirement is demonstrated by the sparking phenomenon discussed by Nes *et al.* (55), who described the essential structural parts of the sterol molecule needed for growth. Some fungi, however (*Pythium* and *Phytophthora*), use an alternative terpene-like compound instead of sterols (33).

Many of the most successful antifungal agents available today in the medical and agrochemical fields interfere in some way with sterol biosynthesis or function. The polyenes, which are thought to act by complexing directly with membrane ergosterol, are beyond the scope of this chapter but have been reviewed by Hamilton-Miller (36). The most common group of compounds used as antifungal agents are the azoles, which interfere at the stage of demethylation of the sterol nucleus (82). The allylamine and thiocarbamate types of compounds interfere at the stage of squalene epoxidation (8). However, there are potentially more antifungal targets within the sterol biosynthesis pathway.

In a series of experiments, Nes *et al.* supplied anaerobically grown *Saccharomyces cerevisiae* with different types of sterols to "spark" growth (55). The degree of growth was assessed and the structures of the sterols were compared. The following parts of the sterol molecule were found to be important (Figure 16.1). In the tetracyclic nucleus, the 3β-OH was obligatory for growth, whereas the presence of methyl groups at C14 or C4 did not allow growth. There appeared

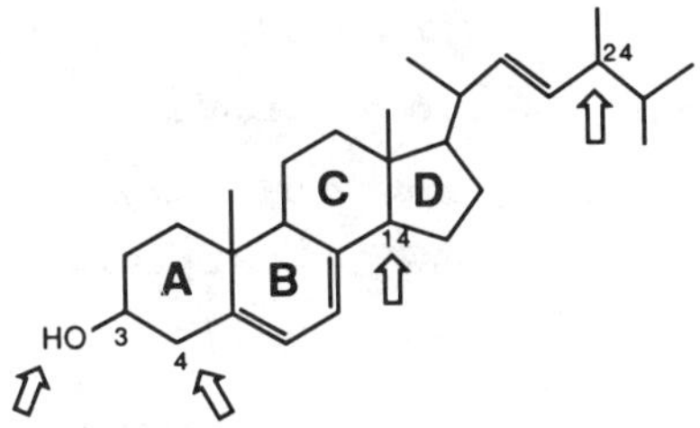

Figure 16.1. The structure of ergosterol. The arrows indicate the parts of the molecule which have been shown to be essential for the growth of fungi.

to be little advantage in the Δ^{5-7} conjugated diene over the Δ^{5} olefin or even fully saturated ring sterols. In the side chain, the β methyl at position 24 was obligatory for growth, with better growth observed with 24 methyl Δ^{22} side chains. A *trans* double bond at 22 was essential since *cis* isomers did not support growth. A more comprehensive report of the functions of sterols in yeast membranes is given in this book in chapter 15. Clearly, if those features that are essential for growth could be exploited as antifungal drug targets, then it should be possible to obtain an effective antifungal agent.

Figure 16.2 shows the biosynthetic pathway from acetate to lanosterol. This is well established and is essentially similar in fungi and mammalian systems. Figure 16.3 shows the synthetic scheme of sterols with the possible interrelationships of the metabolic steps from lanosterol to ergosterol, together with the preferred route in *C. albicans*. The sequence of these interconversions varies among fungi and the biosynthetic route differences may account for the claimed different modes of inhibition reported for some antifungal agents. The bars on these charts show the points of inhibition of known antifungal agents, which we will discuss in relation to our own and published work.

II. Pre-lanosterol Steps

A. Acetate to Mevalonate

The entire carbon skeleton of the sterol molecule is derived from acetyl-CoA, with the exception of the C24 methyl group in the ergosterol side chain. The initial biosynthetic sequence involves condensation of two acetyl-CoA units to form acetoacetyl-CoA, and then addition of a third unit to form 3-hydroxy-3-methylglutaryl-CoA (HMG-CoA), which is reduced by NADPH to give mevalonic acid. These steps are catalysed by the cytosolic enzymes acetoacetyl-CoA thiolase and HMG-CoA synthase, and HMG-CoA reductase, a mitochondrial enzyme in yeasts (90). In mammalian cholesterol biosynthesis, HMG-CoA reduc-

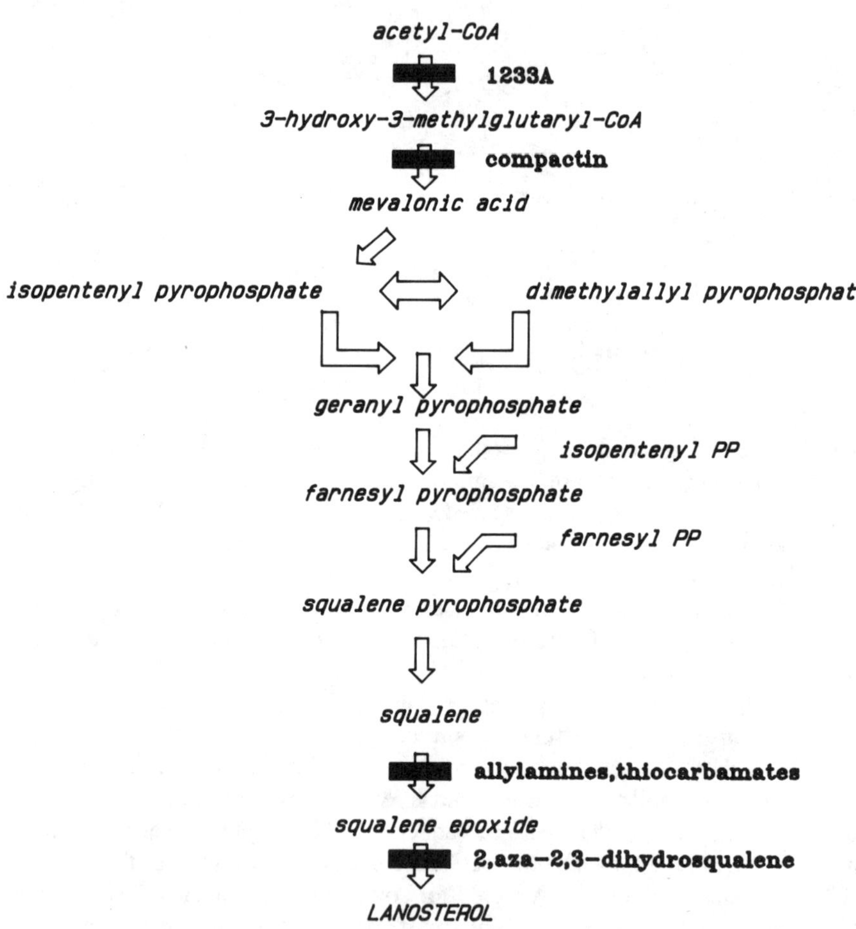

Figure 16.2. The biosynthesis of lanosterol from acetyl-CoA. The reactions for which antifungal inhibitors are known are indicated by a dark bar. For details see text.

tase is the rate-limiting step and its complex regulatory mechanisms have been extensively investigated. However, in yeast, ergosterol appears to regulate its synthesis by repression of the thiolase and synthase rather than the reductase (81, 90).

The fungal metabolite, citrinin, has been reported to inhibit rat liver and yeast acetoacetyl-CoA thiolases at the relatively high concentration of 0.2 mM (45). Cerulenin, an antifungal antibiotic which primarily inhibits fatty acid biosynthesis, also inhibits ergosterol biosynthesis at the point of HMG-CoA synthase (58). More recently, the β-lactone antibiotic 1233A has been identified as a specific inhibitor of the synthase (59), with an IC_{50} of about 0.1 μM. This compound, also isolated as L–659,699 (35) and F–244 (89), has antifungal activity *in vitro*

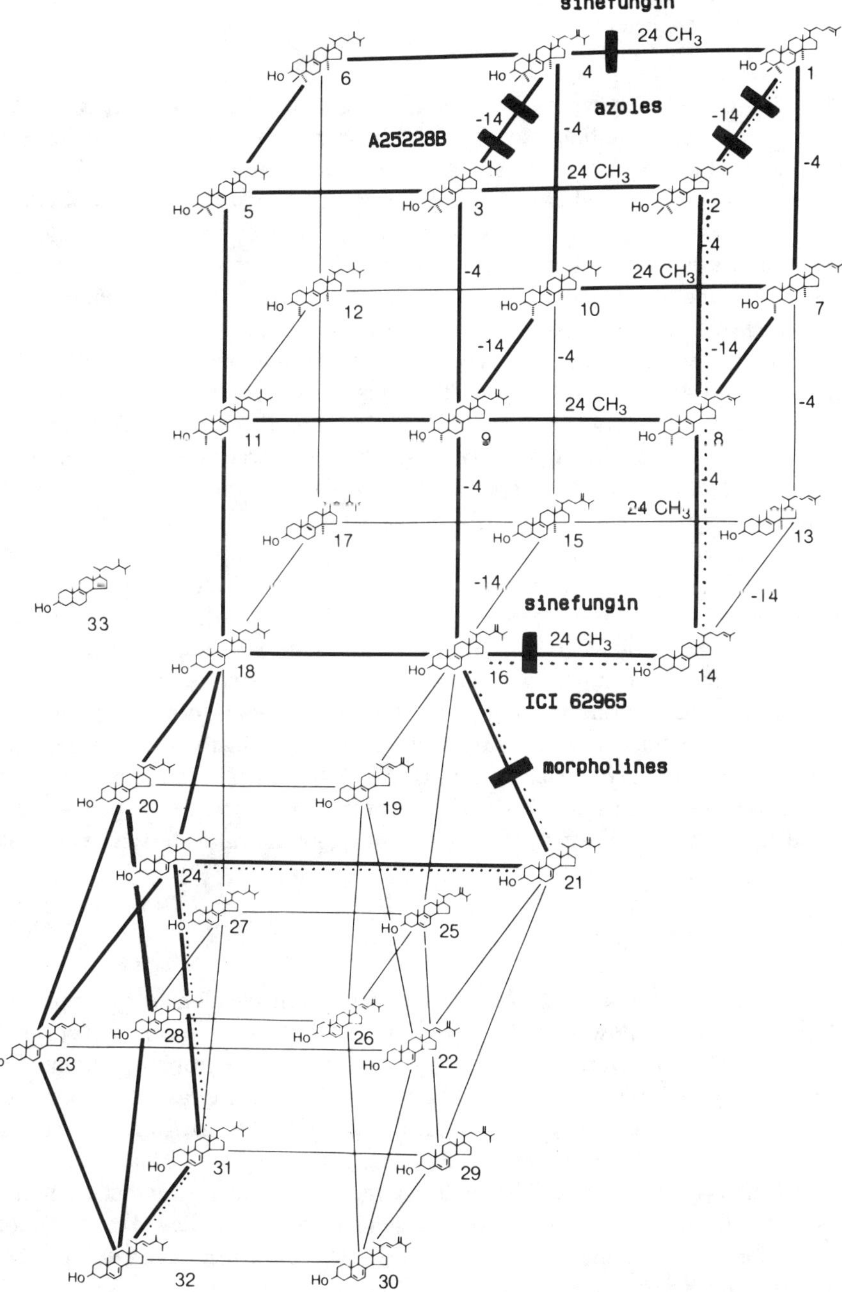

Figure 16.3. A generalized biosynthetic pathway from lanosterol to ergosterol. The figure shows the alternative pathways that exist. The darker lines represent the more preferred transformations; the dotted line is the main route followed in *C. albicans*. Transformations where antifungal agents are known to inhibit are shown by bars; these are explained in the text. −4, −14, 24, etc., are the chemical reactions, *i.e.*, C4 demethylation, C14 demethylation, and 24 methenylation. The numbers represent individual sterols: 1, lanosterol; 4, 24-methylene lanosterol; 14, zymosterol, 16; fecosterol, 32; ergosterol; 33, the product of delta 14 reductase inhibition ignosterol (after Pierce *et al*., *Can. J. Biochem*. 56:135.)

(61), but is not selective as it also inhibits growth of mammalian cells. The structures of these inhibitors and others which interfere with biosynthetic steps between acetate and lanosterol are shown in Figure 16.4.

Several microbial products have been identified as potent inhibitors of HMG-CoA reductase. A large number of synthetic derivatives are also available as a result of efforts to find clinical agents for reduction of serum cholesterol levels. Compactin, the first of these agents to be found, is active in the nanomolar range against HMG-CoA reductase from rat liver (28). It is also active in insect tissue (50) and higher plant cells (73). The related compound mevinolin was found to inhibit yeast HMG-CoA reductase with a Ki of 3.5 nM (4). Compactin is a potent inhibitor of cell-free ergosterol biosynthesis in *Candida albicans* (40). However, these and similar compounds have little effect on ergosterol biosynthesis in intact *Candida* cells (Ryder, unpublished data) and have only very weak growth inhibitory activity against clinically relevant fungi. It would appear that these compounds, like mevalonate, to which they are structurally related, do not easily penetrate the fungal cell membrane. Another compound inhibiting mevalonate biosynthesis is L–660,631, an unstable acetylenic fatty acid derivative with broad-spectrum antimicrobial activity (60). Its precise mode of action was not reported. Compactin has been reported to inhibit growth of some yeasts with minimal inhibitory concentration (MIC) values as low as 0.1 μg/ml (40), suggesting that HMG-CoA reductase may be a suitable target for antifungals, if cell penetration is not a problem, since clinically used HMG-CoA reductase inhibitors appear to be well tolerated in patients. HMG-CoA synthase might be an even better target, as indicated above. Thus potent inhibitors of fungal mevalonate synthesis could have therapeutic potential.

B. Mevalonate to Squalene

Isopentenyl pyrophosphate (IPP) is formed from mevalonate by sequential phosphorylations followed by decarboxylation. The latter step can be blocked in liver extracts by 6-fluoromevalonate (54, 64) after it is pyrophosphorylated to form a potent inhibitor of pyrophosphomevalonate decarboxylase. The formation of squalene involves the condensation of six IPP units to form presqualene pyrophosphate followed by removal of the pyrophosphate. This sequence is carried out by the enzymes IPP isomerase, farnesyl diphosphate synthetase (prenyl transferase), and squalene synthetase. Poulter has recently reviewed the enzymology of this biosynthetic sequence (63). All three enzymes have been purified from fungal sources.

A number of mechanism-based inhibitors structurally related to intermediates of the biosynthetic pathway have been described (63), but none of these is likely to have activity against whole fungal cells because of their inability to penetrate intact cells. Assessment of this area as a target for antifungal chemotherapy therefore awaits the discovery of novel inhibitors. Squalene synthetase appears

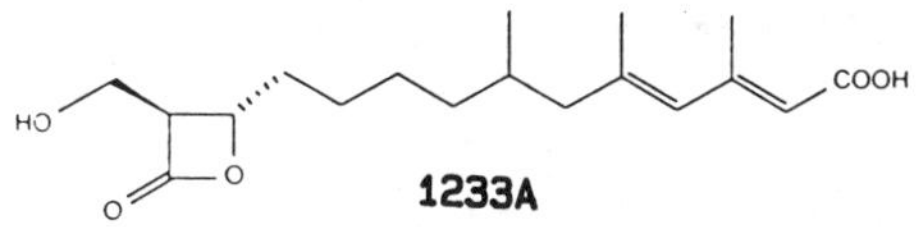

1233A

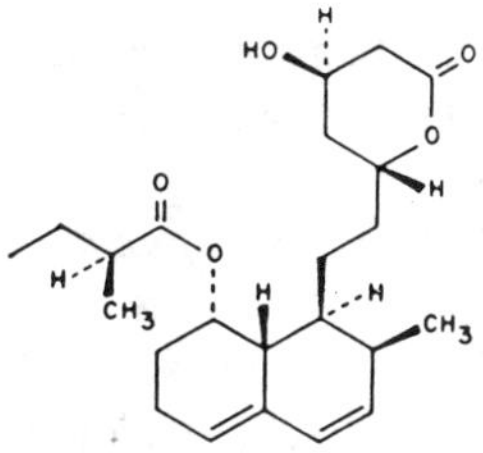

compactin

ALLYLAMINES

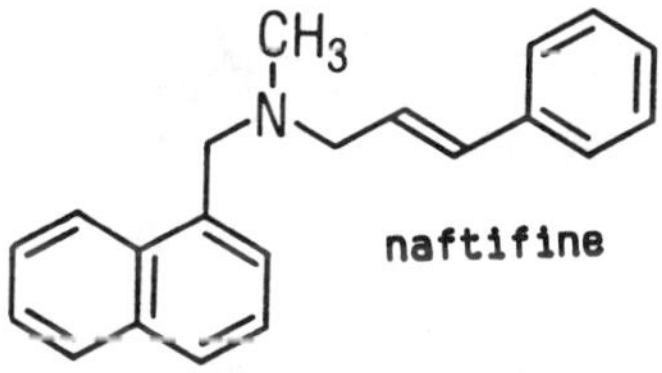

naftifine

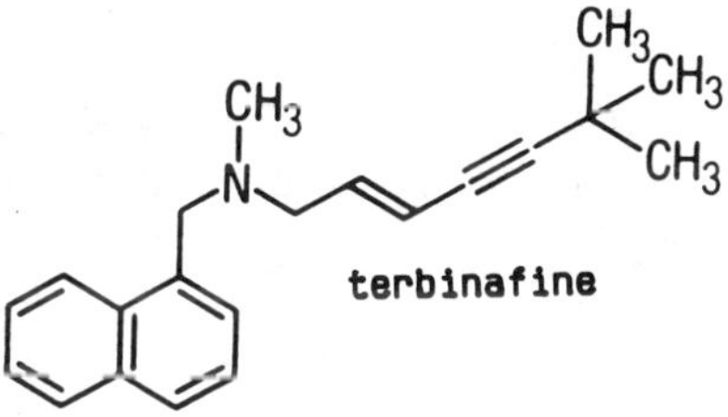

terbinafine

THIOCARBAMATES

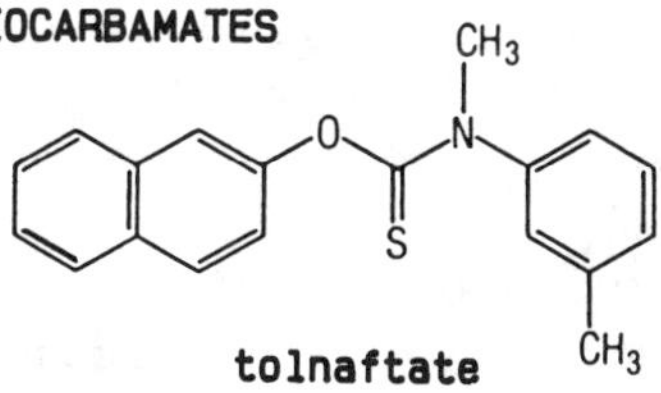

tolnaftate

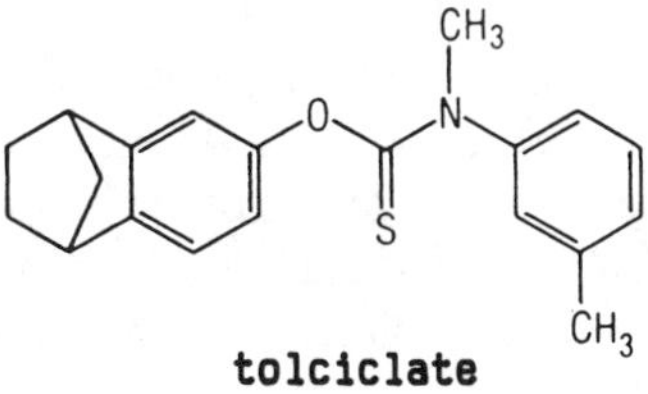

tolciclate

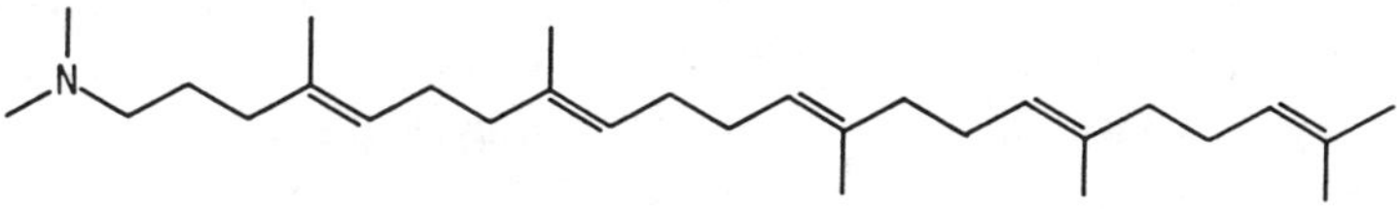

2-aza-2, 3-dihydrosqualene

Figure 16.4. Structures of inhibitors of the ergosterol biosynthetic pathway from acetate to lanosterol.

to be the most interesting as a potential target, since it forms the first intermediate in the pathway that is fully committed to sterol production.

C. Squalene Epoxidase

Epoxidation at the 2,3-position is the first step in the conversion of the 30-carbon-chain squalene to the tetracyclic sterol skeleton. This reaction is performed by a microsomal enzyme complex consisting of a flavoprotein with NAD(P)H cytochrome C reductase activity, and a terminal oxidase which is not of the cytochrome P–450 type. Squalene epoxidase is the first enzyme in the pathway to require molecular oxygen and is of particular interest in fungi as the step at which ergosterol biosynthesis is modulated by availability of oxygen. The enzymology of the fungal and mammalian epoxidases has recently been reviewed (69).

1. Allylamines

With the development of the allylamine antifungals over the last decade, squalene epoxidase has emerged as a major target enzyme for antifungals. Naftifine, the first of these synthetic compounds to be found, has activity against a wide range of pathogenic fungi (32) and is used clinically as a topical antifungal. Terbinafine is a more potent derivative of this class and shows both oral and topical efficacy in the clinic. Readers are referred to a recent comprehensive review (74) for details of the biological and clinical properties of these drugs. Both naftifine and terbinafine are potent inhibitors of fungal ergosterol biosynthesis (75, 65), acting by direct specific inhibition of squalene epoxidase (70). This inhibition is reversible and noncompetitive with respect to squalene, NAD(P)H, and FAD (70). Although fungal and mammalian squalene epoxidases appear to be quite similar in properties (69), the rat liver enzyme is several orders of magnitude less sensitive than the *C. albicans* enzyme to inhibition by allylamines (66, 70). This has important implications for chemotherapeutic target enzymes in general, because it demonstrates the possibility of obtaining a highly selective inhibitor for a fungal enzyme without adverse effects on the equivalent mammalian enzyme.

The exact molecular mechanism of inhibition and selectivity by the allylamines is not yet known. One possible model (68) involves the inhibitor binding to two separate sites on the epoxidase, resulting in the observed potent inhibition by the principle of entropic binding as has been postulated for the HMG-CoA reductase inhibitors (53). In this speculative model, the allylamine side chain would bind to a lipophilic site, positioning the naphthalene ring in the adjacent squalene-binding active site. The model is compatible with the known experimental evidence and structure-activity relationships of the allylamines. More than one thousand compounds of this class have been synthesized (8, 52, 72). In all cases examined, antifungal activity is associated with epoxidase inhibition; active compounds retain a lipophilic side chain, although considerable variation in both side-chain and ring structure is permitted. Neither the allylamine nitrogen nor the double bond is essential for epoxidase inhibition (56). It seems clear that squalene epoxidase inhibition provides the biochemical basis for the antifungal activity of the allylamines, since these two parameters correlate quite well with each other in different fungi (67, 68). The primary fungicidal action and an important characteristic for the clinical efficacy of these compounds (75), is the intracellular accumulation of squalene resulting from epoxidase inhibition (67).

2. *Thiocarbamates*

Subsequent to the discovery of the mechanism of action of the allylamines, a second class of antifungals, the thiocarbamates, was also found to act by inhibition of squalene epoxidase. This mechanism has now been demonstrated for the topical compounds tolnaftate (8, 52, 72), tolciclate (72), and piritetrate (51). These compounds are primarily active against dermatophytes and have little effect on *C. albicans*, apparently due to poor penetration of the *Candida* cell envelope (8, 72). As with the allylamines, the thiocarbamates are selective for the fungal rather than rat liver epoxidase (72). Currently it is not known whether the mechanism of inhibition of squalene epoxidase by thiocarbamates is identical to that of the allylamines. These two classes of compounds are chemically quite distinct, but have some degree of structural similarity.

3. *Squalene Derivatives*

A further approach to inhibition of squalene epoxidase is provided by modification of the substrate, squalene. One such compound, 2-aza-2,3-dihydrosqualene, was originally designed as a cyclase inhibitor (26), but found to inhibit squalene epoxidase also (71). It is, however, more effective against rat liver epoxidase (IC_{50} = 2 μM) than the fungal enzyme. A similar degree of activity against mammalian epoxidase has been reported for trisnorsqualene cyclopropylamine (79), and a number of other squalene derivatives with weaker epoxidase inhibitory activity have recently been described (19, 78, 80). This approach seems unlikely to deliver compounds with a high degree of selectivity and therefore appears unpromising as a potential source of antifungal drugs.

Squalene epoxidase has several attractive features as a target for antifungal drugs; it is not a member of the cytochrome P–450 superfamily, it can be inhibited selectively, and inhibition can result in fungicidal activity. It thus appears to be one of the most promising known targets for antifungals; further classes of inhibitors may be awaiting discovery.

D. 2,3-Oxidosqualene Cyclase

2,3-Oxidosqualene is cyclised to lanosterol, the initial precursor of the vast array of steroid structures formed by fungi and mammals. This cyclisation, which has been described as the most complex enzyme-catalysed reaction known, is carried out by a single protein without requirement for any organic cofactors. The key position of the cyclase and the complexity of its action have encouraged a large amount of work on the mechanism of cyclisation and its inhibition. The design of cyclase inhibitors is described in chapter 17 of this book, so only an overview will be given here.

The structure and activity of various cyclase inhibitors have been reviewed by Cattel *et al.* (19). Many of these are derived from structures related to squalene, such as 2,3-iminosqualene (23), reported to inhibit liver microsomal cyclase by 90% at 1.4 μM. Various azasqualene derivatives have been synthesized as mimics of a carbocationic intermediate of cyclisation (20). These compounds inhibit cyclases from animals and higher plants (28) and also from fungi (21, 71). Related structures such as N,N-diethylazasqualene were reported to inhibit ergosterol biosynthesis in yeast and to have antifungal activity *in vitro* (5), although at relatively high concentrations (10–100 μM). Further cyclase inhibitors described in the literature include 4,4,10β-trimethyl-trans-decal-3β-ol (22) and azadecaline derivatives which are postulated to act as analoges of a high-energy intermediate in the reaction (28). The first cyclase inhibitor to be identified was 1-dodecylimidazole, which inhibited sterol biosynthesis in rat liver (3) and in the fungus, *Ustilago maydis* (37). In contrast, Mercer *et al.* (48) reported that this compound and related heterocycles inhibited only the rat liver cyclase but had no effect on the enzyme from yeast. These authors suggest that the heterocycle blocks the active site, occupied by the incipient sterol ring A, while the side chain binds to a nearby lipophilic site, which could vary between the fungal and mammalian enzymes, thus explaining the selective inhibition. Evidence obtained with the currently available inhibitors suggests that the cyclase may be a very promising target enzyme for development of novel antifungals. As in the case of squalene epoxidase, there appears to be potential for highly selective inhibition of the fungal enzyme. More potent cyclase inhibitors are required to test the suitability of this target as a basis for antifungal action.

III. Postlanosterol Steps

Once lanosterol has been formed by the cyclisation of squalene epoxide, it undergoes several sequential transformations to form ergosterol. The exact route that the sterol molecule follows depends on the availability of the substrate and the specific interaction between the individual enzyme and that substrate in a given fungus (Figure 16.3). This is true for many of the steps, but some must always preceed others; *e.g.*, demethylation always occurs before the isomerisation of double bonds in the B ring of the nucleus. Thus different genera of fungi may follow different biosynthetic routes.

A. C14 Demethylation

One of the earliest steps in the lanosterol pathway is the demethylation of the ring system at the C14 position. The demethylation step utilises a cytochrome P–450-containing monooxygenase enzyme (2). Removal of the methyl group is a two-stage oxidative reaction. The first stage is a hydroxylation to the hydromethyl derivative followed by a second hydroxylation, leading to loss of a water molecule and the methyl group as formaldehyde which is then converted to formic acid. The loss of water produces a double bond by withdrawing a proton from the C15 carbon atom. The Δ^{14} double bond produced is subsequently reduced to give the demethylated sterol in a NAPDH-dependent reaction (2). The Δ^{14} sterol is a transient intermediate since it is rapidly reduced to the next intermediate in the pathway. The mechanism is shown in Figure 17.6 of chapter 17 of this book.

The major antifungal class, the azoles (Figure 16.5), interferes at the demethylation stage of the pathway. Some of these compounds are important as clinical agents (miconazole, ketoconazole, fluconazole) and others as agrochemicals (dichlorbutrazole, fenarimol) (82).

The lone-pair electrons which exist on the azole nitrogen atoms of the antifungal azoles interact with the haem group of cytochrome P–450 in a tight-binding type II manner (96). This interaction is stabilized at the binding site by the hydrophobic parts of the molecule, preventing the oxidation of the methyl group and its subsequent removal. The inhibition has been shown to be noncompetitive with regard to the substrate (9). This type of inhibition is important because it leads to a greater reduction in flow through a metabolic pathway than does a competitive inhibitor. The reason for this is that the accumulation of intermediates will not compete for the binding site of the drug and reduce its inhibition of the reaction.

The interaction with cytochrome P–450 also occurs with mammalian systems, particularly with steroid-synthesising enzymes, resulting in some toxicity (77). However, inhibition is competitive with regard to substrate (9) and this may explain the relatively low level of toxicity seen in the clinic with compounds such as ketoconazole.

If fungi are grown for some time in the presence of azoles and the sterols are extracted and examined by gas chromatography–mass spectrometry (GC-MS), there is an accumulation of trimethylated intermediates. In *Ustilago maydis* the precursor is 24-methylene lanosterol whilst in *C. albicans* it is lanosterol (10). (See also Figure. 16.3.) The different intermediates that accumulate also demonstrate that different sterol biosynthetic pathways exist among fungi. Broken-cell systems utilising radiolabeled precursors have been used to demonstrate inhibition of demethylation. Thin-layer chromatography is used to separate the intermediate

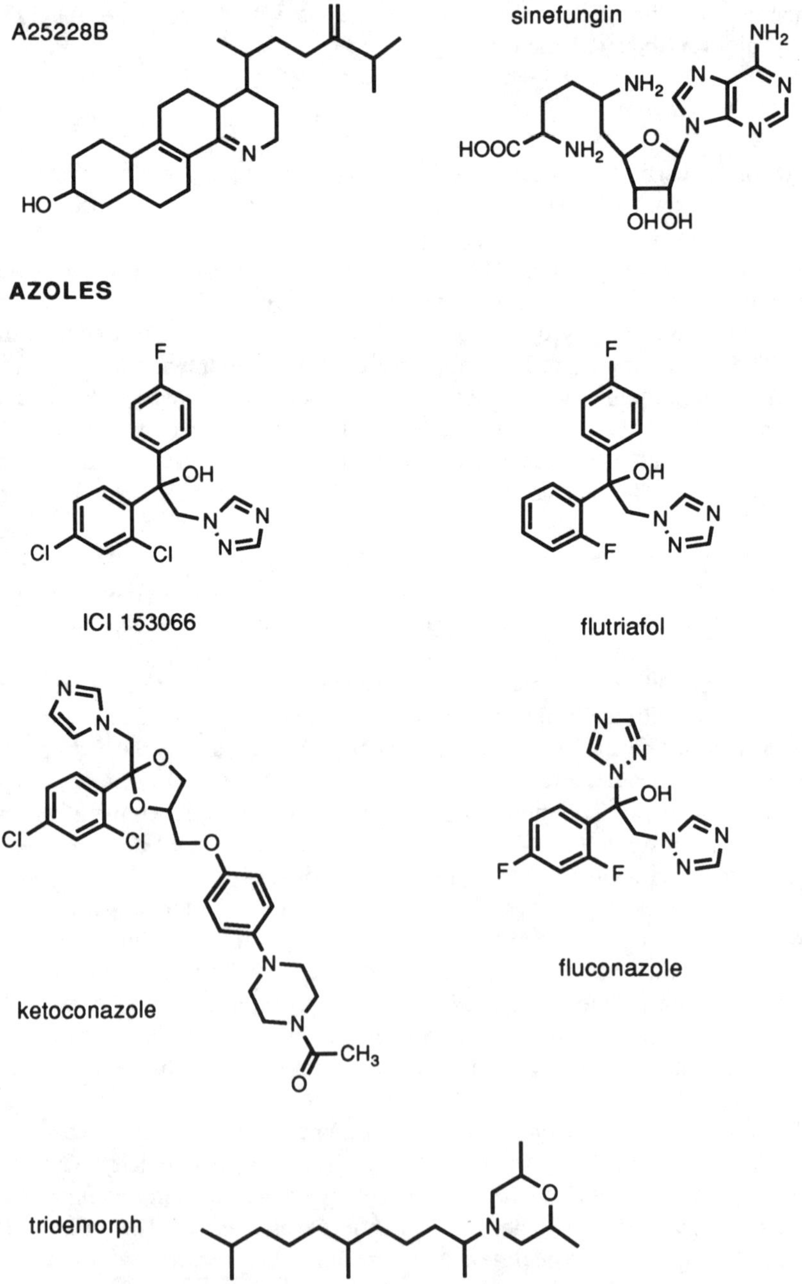

A25228B

sinefungin

AZOLES

ICI 153066

flutriafol

ketoconazole

fluconazole

tridemorph

Figure 16.5. Structures of antifungal agents which interfere in the biosynthetic pathway from lanosterol to ergosterol.

sterols produced from [^{14}C]mevalonate, which build up during inhibition (41, 47). The fact that the fungus is unable to remove the methyl groups at the C4 position and to further metabolise lanosterol indicates the strict structural requirements of the C4 demethylase enzyme.

The trimethylated sterols that accumulate after inhibition by the azole antifungals are bulkier than ergosterol, and it is believed that they cannot pack into the membranes as well as ergosterol. These membranes function less efficiently (95), and there is an interference with the function of membrane-bound enzymes. Vanden Bossche has suggested that one effect is the alteration of the activity of membrane-bound enzymes such as chitin synthase and the inappropriate synthesis of chitin (92). This has been demonstrated in *C. albicans* incubated with ketoconazole and also in *U. maydis* incubated with imazalil (43).

It has also been shown that when yeasts are grown in the presence of azole antifungals there is an inhibition of the uptake of compounds such as amino acids and nucleotides (91). This inhibition has been shown to be noncompetitive with respect to the substrate; *i.e.*, there is a reduction in flow through the permeation system rather than an interference with the affinity of the substrate for the receptor (10).

The mycelial form of *C. albicans* is very important in the pathogenic process (57), and this form is particularly susceptible to the azoles (15). This increased susceptibility may be due to the difference in the cellular content of ergosterol; the mycelial form contains ten times more than the yeast form (39). Possibly, if there is a greater requirement for ergosterol biosynthesis, then a smaller perturbation in its rate of synthesis would have a greater impact on the cell growth by disturbing the steady state. Also, the greater amounts of chitin and chitin synthase required in mycelia may be important if perturbation of enzymes such as chitin synthase are important. Germ-tube formation, the initial step in mycelial growth, is particularly susceptible to azoles; this stage of hyphal growth requires a highly coordinated biosynthetic switching and synthesis of chitin which may be very sensitive to membrane disturbance (16). In addition, mycelial forms have also been shown to take up more radiolabeled azole than yeasts, and this may be a factor in the increased susceptibility to azoles (39).

B. Δ^{14} Reductase

In the C14 demethylation step described above, the sterol goes through an intermediate state with a double bond at the C14 position; this olefin is then reduced to the C4 dimethyl derivative (46). Antifungal compounds have been described which inhibit this reductase; *i.e.*, inhibition of the growth of fungi will induce an accumulation of sterols possessing a Δ^{14} bond. Subsequent enzymes in the pathway, however, do not metabolise this unnatural sterol normally. It is metabolised in *C. albicans* by the C4 demethylase enzymes and Δ^{24} methenylase to form Δ^{14}-fecosterol. Under normal conditions fecosterol would undergo ring isomerisation at the Δ^7 position. However, this unnatural sterol is not able to be further metabolised by the Δ^7-Δ^8 isomerase. The consequence is that the sterol acts as a substrate for the Δ^{22} reductase and is reduced to the unnatural saturated side-chain sterol, ignosterol. This sterol (ignosterol, shown as No. 33 in Figure 16.3) is a Δ^{14} derivative of sterol 18. Ignosterol is not able to be further metabolised and accumulates in the organism to become the predominant sterol. All fungi normally follow a sterol biosynthetic route that passes through fecosterol; hence, this type of inhibitor induces the accumulation of ignosterol in all species of fungi (42).

At higher concentrations of this type of inhibitor, the Δ^{22} reductase enzyme is also inhibited, and the triene, Δ^{14} fecosterol, is accumulated as well. An example of this type of inhibitor is the nitrogen-containing sterol A25822B (93), which is active against many species of fungi. Fenpropidine also acts exclusively at this stage in *C. albicans* (17). In many species of fungi, the morpholine antifungal compounds such as fenpropimorph act at this stage, but many morpholines have multiple points of inhibition (42).

C. Δ^7-Δ^8 *Isomerase*

Following the removal of the C4 and C14 methyl groups and the methenylation of the side chain, the next reaction in the sequence is the isomerisation of the B-ring double bonds in fecosterol (16 in Figure 16.3). Δ^7-Δ^8 isomerisation has been shown to be a reversible reaction by tritium-transfer experiments and does not require any cofactors such as NADH. The morpholine antifungal agents are able to inhibit this enzyme in plant pathogens and are used as crop protectants (14). Many of the compounds have a mixed mode of inhibition, also inhibiting the Δ^{14} reductase (42).

Tridemorph is claimed to be a selective Δ^7-Δ^8 isomerase inhibitor in *U. maydis*, whereas fenpropimorph inhibits both Δ^{14} and Δ^7-Δ^8 isomerases (87). The CNS drug trifluperidol has been shown to inhibit production of Δ^7 sterols in cultures of *S. cerevisiae* and to reduce fungal growth (87). When *C. albicans* was grown in the presence of this agent there was no inhibition of growth despite an accumulation of Δ^7 sterols shown by GC-MS (Barrett-Bee, unpublished results). This must mean that in this species of fungus there is not as strict a structural requirement for the sterols in the cell membranes as in *S. cerevisiae*. GCMS analysis of sterols from *C. albicans* grown in the presence of tridemorph indicated that the drug was a mixed inhibitor. Since A25822B has been shown to be fungitoxic, then either ignosterol (the product of inhibition of Δ^{14} reduction) is toxic *per se* or it is incapable of being metabolised by the Δ^7-Δ^8 isomerase. If the 5,7 diene is not necessary for fungal growth as suggested by Ness *et al.* (55), then the Δ^{14} inhibition is the key step in fungitoxicity. Fenpropidine was shown to be a pure reductase inhibitor in *C. albicans* (Table 16.3).

D. 24 Methenylation

One stage in the biosynthesis of ergosterol that does not exist in the biosynthesis of cholesterol is the addition of a methyl group at the C24 position in the sterol side chain. This occurs early in many fungi and before the C4 demethylation. However, in *C. albicans*, it occurs at the level of zymosterol (14 in Figure 16.3) (10). Inhibitors of this step would be selective and not expected to inhibit any mammalian enzymes and so there would be less potential for toxicity.

The cellular localisation and kinetic properties of the sterol Δ^{24} methyl transferase have been studied (62). Mechanistically, the methyl donor for the 24 methenylation is S-adenosyl methionine (SAM). A positively charged sulphonium ion in SAM renders the methyl group electrophilic and susceptible to the nucleophilic Δ^{24} double bond. The ensuing reaction results in S-CH_3 cleavage, formation of a C24-CH_3 bond, and generation of a transient carbonium ion at C25. A hydride shift followed by proton loss from the newly introduced methyl group gives rise to the new sterol (88) (fecosterol in *C. albicans*).

High concentrations of S-adenosyl homocysteine, the product of the reaction, inhibit 24 methenylation. Analogs of this compound have been sought and sinefungin was discovered; this natural product had been known to have antifungal properties for some time (34). When its potency against the enzyme and growing cells is compared, however, it is a more potent antifungal than a 24-methenylation inhibitor. This result indicates that other fungal pathways are inhibited at lower concentrations of the agent. This compound also inhibits methylation reactions in mammalian systems that depend on SAM, leading to toxicity.

Some 25 years ago it was shown that a series of nitrogen-containing side-chain analogs of cholesterol were able to inhibit the production of cholesterol leading to rises in desmosterol, consistent with inhibition of $\Delta^{24(28)}$ reductase (24). Later it was shown that the mechanism of 28 reduction and 24 transmethylation were very similar and involved attack of an electrophile on the electron-rich centre at C24 (1). This result suggested to us that an alternative approach to analogs of SAM is to look for nitrogen-containing substrate-derived competitive inhibitors. We have studied the natural product tomatidine and natural and synthetic analogs as potential inhibitors of this enzyme reaction (Table 16.1).

When *C. albicans* was grown in the presence of tomatidine and the sterols were extracted and examined by GCMS there was an accumulation of zymosterol (14 in Figure 16.3). When a similar experiment was performed by using the dermatophyte *Trichophyton quinckeanum* as the fungus, the product was lanosterol (1 in Figure 16.3). This result indicates that the dermatophyte organism will not demethylate lanosterol and the natural substrate for the C14 demethylase stage is 24 methyl lanosterol (4 in Figure 16.3). This gives us information on the biosynthetic pathway in this organism. This experiment indicated that in *C. albicans*, the natural route for sterol biosynthesis is methenylation at the stage of zymosterol.

Table 16.1. The inhibition of methenylation of sterols in whole and broken cell preparations of C. albicans.

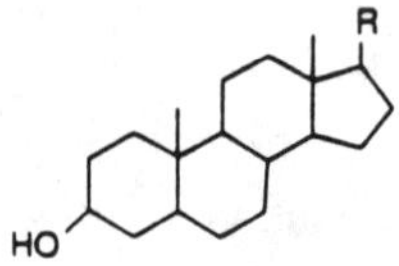

Compound No.	R	IC_{50} (ug/ml)	
		Whole Cells	Broken Cells
1	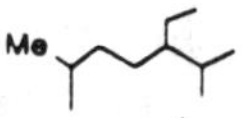		a
2	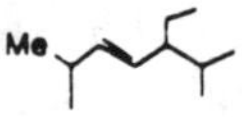		a
3	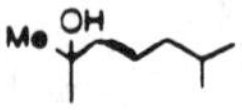		a
4	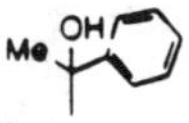		1
5	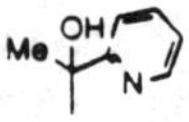	>40	5.
6	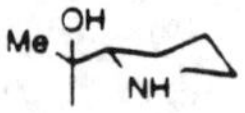	0.032	0.007
7	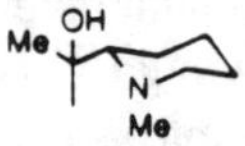	0.023	0.004
8	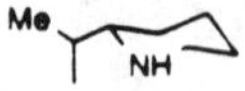		0.022
9	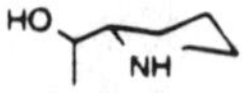	10	2.3
10	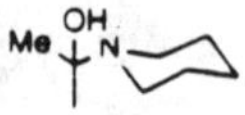	0.065	0.085
11	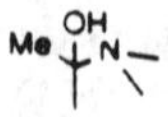		0.023
12	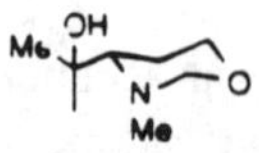		0.34
13	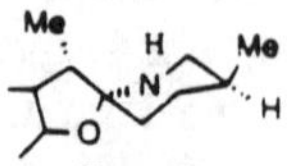 Tomatidine	0.045	0.014
14	PD.UK50-98	28	.015
	Lanosterol Cholesterol		a

[a]Less than 20% inhibition at 1 μg/ml

When a broken-cell preparation of *C. albicans* was incubated with radiolabeled SAM, the methenylation of sterol pathway intermediates could be measured. Compounds were added to these systems and structure-activity relationships could be determined from relative inhibition potencies of the compounds. Since SAM is able to enter intact cells, a similar study could be performed and the comparative IC_{50} values give a measure of penetrability of the compounds. Table 16.1 gives a summary of selected compounds where structure-activity relationships can be obtained. It is clear that tomatidine (13 in Table 16.1) is able to inhibit the enzyme both in intact and broken cells whereas the Phillips-Duphar UK–50–98 compound penetrates cells only poorly, as reflected by its poor antifungal activity.

The broken-cell assay showed good endogenous substrate specificity; the addition of a range of exogenous sterols (*e.g.*, lanosterol or cholesterol) had little or no effect. Suboptimal potential suicide substrates, aromatic (**4**), olefinic (**2**), or acetylenic (**3**) C22–23 side-chain analogs all gave poor inhibition. Secondary (**6, 8, 9**) and tertiary (**7**) C23 amino derivatives in general were the best inhibitors of the enzyme. The analog of tomatidine where the nitrogen was replaced with an oxygen atom, solacidine, had no activity.

The following structure-activity patterns were observed. Linkage of the secondary or tertiary amine alpha to nitrogen (**6, 7**) generally gave better inhibition than direct N-substitution (**10**) (IC_{50} = 10 nM compared to 250 nM). Acyclic N-linked systems (**11**) were better than cyclic piperidine or morpholine groups (**10, 12**). The C20 hydroxyl was not essential for activity; it may, however play a secondary conformational role, based on the interesting result with tomatidine (**13**). The C21 methyl group was essential for good activity (**9**), again confirming its structural requirement in the sterol side chain.

To further explore this chemical series ICI 62,965 (**7**) was chosen as an example. The degree of inhibition of ergosterol biosynthesis and zymosterol accumulation in *C. albicans* grown in various concentrations was measured. Table 16.2 shows that for significant inhibition of growth over a 16-hour period there must be around 90% inhibition of ergosterol biosynthesis, with subsequent accumulation of zymosterol.

E. C4 Demethylation

There are no known inhibitors of the mixed-function oxidase-dependent enzyme complex that catalyses the C4 demethylation. This enzyme removes both of the C4 methyl groups in two steps (49). The initial β-methyl group is epimerised during this process and the cycle repeats. Each demethylation step occurs by an oxidation of the methyl group to the corresponding keto group via the hydroxyl. The keto group is further oxidised to a carboxyl group. All these steps take place in the presence of molecular oxygen and pyridine nucleotides via the mixed-function oxidase. The subsequent decarboxylation is an anaerobic process, with CO_2 released and a 3-keto group formed. The latter group is reduced by 3-keto

Table 16.2. The effect of ICI 62,965 on growth and sterol accumulation by C. albicans.

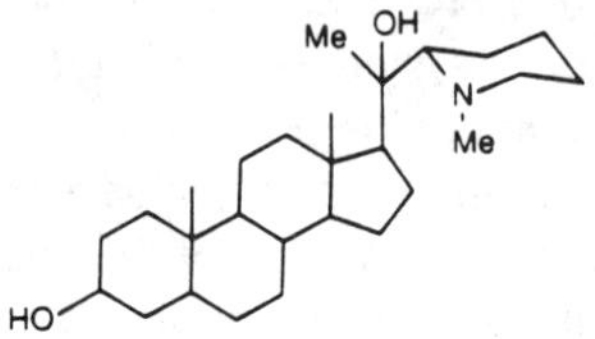

ICI 62,965

Concentration	% Growth Inhib.	% Ergosterol	% Zymosterol
0	0	100	0
10 nM	10	94	7
100 nM	12	83	17
1 uM	12	84	16
10 uM	50	5	95
100 uM	80	1	99

C. albicans yeasts were grown in a shake flask for 16 h at 37°C and the OD_{650} measured. Sterols were extracted and quantitated by Gas Chromatography; identity was confirmed by GC-MS.

reductase which inverts the existing methyl group stereochemistry so that the cycle can repeat in an exactly analogous way. The overall reaction is sensitive to KCN but not to CO. Inhibitors of this pathway would be expected not to suffer from the potential toxicity seen with the C14 demethylase inhibitors which have the potential to inhibit mammalian cytochrome P–450-containing enzymes.

F. Late Stages

There are no known inhibitors of the later stages of the pathway that are antifungals. This may be because the organisms are more tolerant to the small changes in the structure of the ring, as exemplified by our results with trifluperidol in *C. albicans*. These observations are in agreement with the work of Nes *et al.* with his sparking experiments (55).

IV. Ergosterol Synthesis Inhibitors as Antifungals

A. Synergy

Mixtures of sulphonamides and dihydrofolate reductase inhibitors have been shown to be synergistic in the treatment of bacterial infections. These two agents inhibit different steps in the folic acid biosynthetic pathway in bacteria and are clinically synergistic (*e.g.*, Septrin). In the same way, it would be expected that different types of inhibitors of the sterol biosynthetic pathway may act in concert. Combinations of naftifine, 15 azasterol, triarimol, and mevinolin with ketocona-

Table 16.3. The accumulation of sterols extracted from C. albicans *when grown in the presence of mixtures of ergosterol biosynthesis inhibitors*

	Major sterols accumulated			
Compounds	Terbinafine	62,965	153,066	Fenpropidine
Terbinafine	squal^{++}			
ICI 62,965	squal^{++} zymo^{--}	zymo^{++}		
ICI 153,066	squal^{++} lano^{--}	zymo^{--} lano^{--}	lano^{++}	
Fenpropidine	squal^{++} igno^{--}	zymo^{--} lano^{--} Δ 14 zymo^{++}	lano^{++} igno^{--}	igno^{++}

(Note): *C. albicans* B2630 was grown in shake flasks at 37°C in the presence of the ergosterol biosynthesis inhibitors above. Cells were harvested; the sterols were extracted and analysed by GC-MS. The predominant sterols are: lanosterol (lano), zymosterol (zymo), squalene (squal), ignosterol (igno), ergosterol (ergo), ergosta-8:14:24-triene (Δ 14 zymo). $^{++}$represents an increase, $^{--}$ represents a decrease.

zole have been reported to reduce the MIC for *C. albicans* by a factor of four (88). We have extended these studies in *C. albicans* to combinations of various types of inhibitors. Both growth of the organism and the sterols produced were measured.

Table 16.3 shows a summary of the changes in sterols in the presence of different ergosterol biosynthesis inhibitors. When there were two points of inhibition in the pathway, the balance of sterols was changed with an accumulation of the first blockage product in the pathway. In the case of inhibitors of 24 methenylation and Δ 14 reduction, however, the product was a novel sterol, Δ 14 zymosterol.

Combinations of sterol biosynthesis inhibitors were examined for effects on the growth *C. albicans* (Table 16.4). After 24 hours the cultures were sampled and diluted into fresh broth diluting the drugs at least 100-fold. The culture dishes were incubated for another 24 hours for growth assessment. When no growth was observed, it was interpreted to mean that the original treatment had been fungicidal. It can be seen from Table 16.5 that fungicidal effects are seen with all combinations of inhibitors except Δ 24 methenylase inhibitors and C14 demethylase inhibitors. This observation indicates that whilst most ergosterol biosynthesis inhibitors are fungistatic alone, combinations can be fungicidal. The implication of this type of treatment for immunocompromised patients is considerable.

B. Resistance

The development of resistance in fungi to fungicides with specific mechanisms of action is quite common in the laboratory. Whilst resistance has proved trouble-

Table 16.4. The effect of mixtures of ergosterol biosynthesis inhibitors on the growth of C. albicans

	Minimum inhibitory concentration (μg/ml)[a]				
Compounds	ICI 59,620	Naftifine	ICI 153,066	Trifluperidol	Fenpropidine
ICI 59,620	25				
Naftifine	2.5 + 1.6	200			
ICI 153,066	NS	0.1 + 6.2	25		
Trifluperidol	6.2 + 6.2	25 + 6.2	100 + 0.006	>400	
Fenpropidine	3.1 + 3.1	6.2 + 6.2	NS	6.2 + 100	100

[a] *C. albicans* B2630 (2 × 10^4 yeasts /ml) was incubated in yeast nitrogen broth for 24 hours at 37°C in microtitre plates. Compounds were serially dilute across the plates for each of a pair of inhibitors. The growth was assessed and the minimum inhibitory concentration determined. The values quoted are the minimum synergistic concentrations. The first value is for the compound in the vertical column and the second value is for the compound in the horizontal row. NS, no synergy.

some among plant pathogens in the field (25), few clinical isolates have been found. Clinical isolates of pathogenic organisms less susceptible to the azoles have, however, appeared during treatment with ketoconazole (76, 94). This loss of susceptibility has been shown to result from changes in the cytochrome P–450 (38), but, in most cases the loss of susceptibility has resulted from an impairment of accumulation of azoles. This has been observed to occur in *C. albicans* (76), and *Drechslera halodes* and in several phytopathogenic organisms (7, 27).

Site-specific inhibitors such as dimethylmorpholines should be prone to resistance development but long usage in crop protection has yielded few instances of failure (18). By contrast, fungal strains resistant to tridemorph and fenpropimorph have been readily produced in the laboratory. This apparent discrepancy probably is explained by mutations conferring resistance yielding a less fit patho-

Table 16.5. The candicidal effects of mixtures of ergosterol biosynthesis inhibitors

	Minimum fungicidal concentration (μg/ml)				
Compounds	59,620	Terbinafine	153,066	Trifluperidol	Fenpropidine
ICI 59,620	100				
Terbinafine	25 + 12.5	100			
ICI 153,066	12.5 + 50	3.1 + 3.1	NC		
Trifluperidol	400 + 25	50 + 6.2	400 + .4	NC	
Fenpropidine	6.2 + 3.1	6.2 + 6.2	NS	25 + 400	NC

(Note): *C. albicans* was grown as described in Table 16.4. After a 24-hour growth period, 0.010 ml of incubation fluid was transferred to a microtitre plate of fresh medium and growth continued for 48 hours. The plates were then visually scored for regrowth of the organism. The values are for the minimum concentration of the mixture which in the initial plate prevented regrowth in the second incubation. The values given are for the compound in the vertical column plus the compound in the horizontal row. NC, no killing; NS, no synergy.

gen. This also probably explains the lack of significant resistance to C14 demethylase inhibitors in the clinic.

C. Toxicity

Since the biosynthetic pathway to ergosterol is very similar to the mammalian pathway to cholesterol there is clearly potential for mammalian toxicity. The major toxicity of existing antifungal agents is, however, not observed in the cholesterol biosynthesis pathway but rather on enzyme pathways that metabolise sterols to form hormones, particularly the sex hormones. This may be because cholesterol is mostly obtained from the diet. Azole antifungals inhibit cytochrome P–450-containing enzymes, thus disrupting biosynthesis of sex hormones. The major clinical manifestation of ketoconazole toxicity is inhibition of testosterone biosynthesis (77) and rare liver toxicity of unknown etiology.

V. The Future

The inhibition of ergosterol biosynthesis is universally a way to prevent the growth of fungi (apart from the very few plant pathogens mentioned earlier). This is an important concept when dealing with unknown pathogens. Inhibition of biosynthesis of cell wall macromolecules appears to be more specific and limited to certain fungi; glucan synthesis appears to be specific to organisms such as *Candida* (see chapter 19 of this book), and inhibition of chitin synthesis is more important in retarding growth of filamentous fungi.

The use of inhibitors of ergosterol biosynthesis in the treatment of fungal infections is still in its infancy. At present, the C14 demethylase inhibitors, the azoles, are the major therapeutic agents in clinical use. These compounds are predominantly fungistatic and so provide the fungus with an escape route, particularly in the immunocompromised patient when therapy is withdrawn. One of the most promising areas of antifungal research may be the inhibition of 24 methenylation. This part of the sterol molecule is essential for growth and, importantly, does not occur in mammalian systems. This should provide a means of avoiding potential mechanism-based toxicity. Additionally the process is not cytochrome P–450-based, avoiding the potential problems seen with the azoles. Inhibitors of the pathway have been identified and there is evidence that these can be nonsteroidal, making chemical syntheses simpler (17). Strategies for this class of inhibitors include a mechanism-based assay followed by antifungal testing and appropriate animal models. (See chapter 21 of this book.)

In vitro experiments indicate that it is possible to produce fungicidal combinations of ergosterol biosynthesis inhibitors. In addition to the use of synergistic pairs of compounds, hybrid molecules that possess two toxic moieties, such as a nitrogen-containing side chain and a nitrogen situated at the 14 or 15 position, may well produce potent and fungicidal compounds. Such compounds would be invaluable in the difficult-to-treat infections such as aspergillosis and deep-seated infections. These diseases are presently treated with azoles such as fluconazole, with the risk of relapse when treatment is stopped, or the fungicidal compound, amphotericin, which is overtly toxic.

References

1. Akhtar, M., K.A. Munday, A.D. Rahimtula, I.A. Watkinson, and D.C. Wilton. 1969. Mechanism of reduction of double bonds in biological systems: conversion of desmosterol into cholesterol. Chem. Comm. **1969:**1287–1288.
2. Aoyama, Y., Y. Yoshida, and R. Sato. 1984. Yeast cytochrome p–450 catalysing lanosterol 14α-demethylation. J. Biol. Chem. **259:**1661–1666.
3. Atkin, S.D., B. Morgan, K.H. Baggaley, and J. Green 1972. The isolation of 2,3-oxidosqualene from the liver of rats treated with 1-dodecylimidazole, a novel hypocholesterolaemic agent. Biochem. J. **130:**153 – 157.
4. Bach, T.J., and H.K. Lichtenthaler. 1982. Mechanism of inhibition by mevinolin (MK 803) of microsome-bound radish and of partially purified yeast HMG-CoA reductase (EC.1.1.1.34.) Z. Naturforsch. **38c:**212–219.
5. Balliano, G., F. Viola, M. Ceruti, and L. Cattel. 1988. Inhibition of sterol biosynthesis in *Saccharomyces cerevisiae* by *N,N*-diethylazasqualene and derivatives. Biochim. Biophys. Acta. **959:**9–19.
6. Baloch, R.I., E.I. Mercer, T.E. Wiggins, and B.C. Baldwin. 1984. Where do morpholines inhibit sterol biosynthesis? Brit. Crop Protect. Conf. 1984. **3:**893–898.
7. Barrett-Bee, K. 1991. Resistance to azole antifungal agents in several species. J. Med. Vet. Mycol., in press.
8. Barrett-Bee, K., A.C. Lane, and R.W. Turner. 1986. The mode of antifungal action of tolnaftate. J. Med. Vet. Mycol. **24:**155–160.
9. Barrett-Bee, K., J. Lees, P.E. Pinder, J. Campbell, and L. Newboult. 1988. Biochemical studies with a novel antifungal agent ICI 195,739. Ann. N.Y. Acad. Sci. **544:**231–244.
10. Barrett-Bee, K., L. Newboult, and P.E. Pinder. 1991. Biochemical changes associated with the antifungal action of the triazole ICI 153,066 on *C. albicans* and *T. quickeanum*. FEMS, Lett. **79:**127–132.
11. Barrett-Bee, K., and P.E. Pinder. 1984. Resistance to ergosterol biosynthesis inhibitors observed in several fungal species. p. 139. *In* C. Nombela (ed.) Proceedings of FEMS Symposium, The development of antifungal agents. S.E.M., Madrid.
12. Block, K. 1979. Speculations on the evolution of sterol structure and function. CRC Crit. Rev. Biochem. **91:**1–5.
13. Block, K. 1981. Sterol structure and membrane function. Curr. Top. Cell Regul. **18:**289–299.
14. Bohnen, K. and A. Pfinner. 1979. Fenpropimorph, ein neus systemisches Fungizid zur Bekaempfung von echten Meltau und Rostkrankheiten im Getreidebau, Meded Rijksfac Landbouwwetensch Gent **44:**487–497.

15. Borgers, M. 1985. Antifungal azole derivatives, p. 133–153. *In* D. Greenwood and F. O'Grady (ed.), Scientific basis of antimicrobial chemotherapy. Cambridge University Press, Cambridge.

16. Borgers, M., M. De Brabander, H. Vanden Bossche, and J. Van Cutsem. 1979. Promotion of pseudomycelium formation of *Candida albicans* in culture: A morphological study of the effects of miconazole and ketoconazole. Postgrad. Med. J. **55**:687–691.

17. Boyle, F.T. 1990. Drug discovery: a chemists approach, p. 3–30. *In* J.F. Ryley (ed.), Chemotherapy of fungal diseases, handbook of experimental pharmacology, vol. 96. Springer Verlag, Berlin.

18. Butters, J., J. Clark, and D.W. Hollman. 1984. Resistance to inhibitors of sterol biosynthesis in barley powdery mildew. Meded. Rijksfac. Landbouwwetensch Gent. **49**:143 – 151.

19. Cattel, L., M. Ceruti, G. Balliano, F. Viola, G. Grosa, and F. Schuber. 1989. Drug design based on biosynthetic studies: synthesis, biological activity, and kinetics of new inhibitors of 2,3-oxidosqualene cyclase and squalene epoxidase. Steroids **53**:363–391.

20. Cattel, L., M. Ceruti, F. Viola, L. Delprino, G. Balliano, A. Duriatti, and P. Bouvier-Nave. 1986. The squalene-2,3-epoxide cyclase as a model for the development of new drugs. Lipids **21**:31–38.

21. Ceruti, M., F. Viola, G. Balliano, G. Grosa, P. Caputo, and N. Gerst. 1988. Synthesis of a squalenoid oxaziridine and other new classes of squalene derivatives, as inhibitors of sterol biosynthesis. Eur. J. Med. Chem. **23**:533–537.

22. Chang, T.-Y., E.S. Schiavoni, Jr., K.R. McCrae, J.A. Nelson, and T.A. Spencer. 1979. Inhibition of cholesterol biosynthesis in chinese hamster ovary cells by 4,4-10β-trimethyl-trans-decal-3β-ol. J. Biol. Chem. **254**:11258–11263.

23. Corey, E.J., P.R. Ortiz de Montellano, K. Lin, and P.D.G. Dean. 1967. 2,3-iminosqualene, a potent inhibitor of the enzymatic cyclization of 2,3-oxidosqualene to sterols. J. Am. Chem. Soc. **89**:2797–2798.

24. Counsell, R.E., P.D. Klimstra, R.E. Ranney, and D.L. Cook. 1962. Hypocholesterolemic agents. 1. 20α-(2-dialkylaminoethyl)aminopregn-5-en-3α-ol derivatives. J. Med. Pharm. Chem. **5**:720–729.

25. Dekker, J. 1984. Development of resistance to antifungal agents, p. 89–112. *In* A.P.J. Trinci and J.F. Ryley (ed.), Mode of action of antifungal agents. Cambridge University Press, Cambridge.

26. Delprino, L., G. Balliano, L. Cattel, P. Benveniste, and P. Bouvier. 1983. Inhibition of higher plant, 2,3-oxidosqualene cyclase by 2-aza-2,3-dihydrosqualene and its derivatives. J. Chem. Soc. Chem. Commun. (1983):381–382.

27. DeWaard, M.A., and J.G. van Nistleroy. 1980. An energy-dependent efflux mechanism for fenarimol in a wild-type strain and fenarimol-resistant mutant of *Aspergillus nidulans*. Pestic. Biochem. Biophys. **13**:255–266.

28. Duriatti, A., P. Bouvier-Nave, P. Benveniste, F. Schuber, L. Delprino, G. Balliano, and L. Cattel. 1985. In vitro inhibition of animal and higher plant 2,3-oxidosqualene-

sterol cyclases by 2-aza-2,3-dihydrosqualene and derivatives, and by other ammonium-containing molecules. Biochem. Pharmacol. **34:**2765–2777.

29. Endo, A., M. Kuroda, and K. Tanzawa. 1976. Competitive inhibition of 3-hydroxy-3-methylglutaryl coenzyme A reductase by ML–236A and ML–236B fungal metabolites, having hypocholesterolcmic activity. FEBS Lett. **72:**323–326.
30. Floss, H.G., L. Mascaro, M.D. Tsai, and R.W. Woodard. 1979. Stereochemistry of enzymatic transmethylation, p. 135–141. *In* E. Usdin, R.T. Borchardt, C.R. Creveling (ed.), Transmethylation: developments in neuroscience, vol. 5. Elsevier/North Holland, New York.
31. Fryberg, M., and A.C. Oehlschlager. 1976. Sterol biosynthesis in antibiotic sensitive and resistant *Candida*. Arch. Biochem. Biophys. **173:**171–177.
32. Georgopoulos, A., G. Petranyi, H. Mieth, and J. Drews. 1981. In vitro activity of naftifine, a new antifungal agent. Antimicrob. Agents Chemother. **19:**386–389.
33. Goodwin, T.W. 1973. Comparative biochemistry of sterols in eukaryotic microorganisms, p. 1–41. *In* J.A. Erwin (ed.), Lipids and biomembranes of eukaryotic microorganisms. Academic Press, New York.
34. Gordee, R.S., and T.F. Butler. 1973. A9145 a new adenine-containing antifungal antibiotic. II. Biological activity. J. Antibiot. **26:**466–470.
35. Greenspan, M.D., J.B. Judkovitz, C.H.L. Lo, J.W. Chen, A.W. Alberts, V.M. Hunt, M.N. Chang, S.S. Yang, K.L. Thompson, Y.P. Chiang, J.C. Chabala, R.L. Monaghan, and R.L. Schwart. 1987. Inhibition of hydroxymethylglutaryl-coenzyme A synthase by L–659–699. Proc. Natl. Acad. Sci. USA **84:**7688–7492.
36. Hamilton-Miller, J.M.T. 1973. Chemistry and biology of the polyene macrolide antibiotics. Bacteriol. Rev. **37:**166–196.
37. Henry, M.J., and H.D. Sisler. 1979. Effects of miconazole and dodecylimidazole on sterol biosynthesis in *Ustilago maydis*. Antimicrob. Agents Chemother. **15:**603–607.
38. Hitchcock, C.A., K. Barrett-Bee, and N.J. Russell. 1987. Inhibition of 14α-sterol demethylase activity in *Candida albicans*–Darlington does not correlate with resistance to azole. J. Med. Vet. Mycol. **25:**329–333.
39. Hitchcock, C.A.,N.J. Russell, and K. Barrett-Bee. 1989. The lipid composition and permeability to the triazole antifungal antibiotic ICI 153,066 of serum grown mycelial cultures of *C. albicans*. J. Gen. Micro. **135:**1949–1955.
40. Ikeura, R., S. Murakawa, and A. Endo. 1988. Growth inhibition of yeast by compactin (ML–236B) analogues. J. Antibiot. **41:**1148–1150.
41. Kato, T., and Y. Kawase. 1976. Selective inhibition of the demethylation at C-14 in ergosterol biosynthesis by the fungicide Denmert. Agri. Biol. Chem. **40:**2379–2388.
42. Kerkanaar, A. 1987. The mode of action of dimethylmorpholines, p. 523–542. *In* R.A. Fromtling (ed.), Recent trends in the discovery, development and evaluation of antifungal agents. J.R. Prous Science, Barcelona.
43. Kerkenaar, A., and D. Barug. 1984. Fluorescence microscope studies of *Ustilago maydis* and *Penicillium italicum* after treatment with imazalil and fenpropimorph. Pestic. Sci. **15:**199–205.

44. Kerkenaar, A., M. Uchiyama, and G.G. Versluis. 1981. Specific effects of tridemorph on sterol biosynthesis in *Ustilago maydis*. Pestic. Biochem. Physiol. **16:**97–104.

45. Kuroda, M., Y. Hazama-Shimada, and A. Endo. 1977. Inhibition of sterol synthesis by citrinin in a cell-free system from rat liver and yeast. Biochim. Biophys. Acta. **486:**254–259.

46. Lee, W.H., B.N., Lutsky, and G.S. Schroepfer. 1969. 5α cholest-8(14)-en-3β-ol, a possible intermediate in the biosynthesis of cholesterol. J. Biol. Chem. **244:**5440–5448.

47. Marriott, M.S. 1980. Inhibition of sterol biosynthesis in *Candida albicans* by imidazole containing antifungals. J. Gen. Micro. **117**:265–275.

48. Mercer, E.I., P.K. Morris, and B.C. Baldwin. 1985. Differences in the inhibitory effects of *N*-(1-n-dodecyl)-heterocycles on the 2,3-oxidosqualene lanosterol-cyclase of rat liver and yeast. Comp. Biochem. Physiol. **80B:**341–346.

49. Miller, W.L., and J.L. Gaylor. 1970. Investigation of the component reactions of oxidative sterol demethylation, oxidation of a 4α methyl to a 4α carboxylic-acid during cholesterol biosynthesis. J. Biol. Chem. **245:**5369 5381.

50. Monger, D.J., W.A. Lim, F.J. Kezdy, and J.H. Law. 1982. Compactin inhibits insect HMG-CoA reductase and juvenile hormone biosynthesis. Biochem. Biophys. Res. Commun. **105:**1374–1380.

51. Morita, T., K. Iwata, and Y. Nozawa. 1989. Inhibitory effect of a new antimycotic agent, piritetrate, on ergosterol biosynthesis in pathogenic fungi. J. Med. Vet. Mycol. **27:**17–25.

52. Morita, T., and Y. Nozawa. 1985. Effects of antifungal agents on ergosterol biosynthesis in *Candida albicans* and *Trichophyton mentagrophytes*: differential inhibitory sites of naphthiomate and miconazole. J. Invest. Dermatol. **85:**434–437.

53. Nakamura, C.E., and R.H. Abeles. 1985. Mode of interaction of β-hydroxy-β-methylglutaryl coenzyme A reductase with strong binding inhibitors: compactin and related compounds. Biochemistry **24:**1364–1376.

54. Nave, J.-F., H. d'Orchymont, J.-B. Ducep, F. Piriou, and M.J. Jung. 1985. Mechanism of the inhibition of cholesterol biosynthesis by 6-fluoromevalonate. Biochem. J. **227:**247–254.

55. Nes, W.R., B.C. Sekula, W.D. Nes, and J.H. Adler. 1978. The functional importance of structural features of ergosterol in yeast. J. Biol. Chem. **253:**6218–6225.

56. Nussbaumer, P., N.S. Ryder, and A. Stutz. 1991. Allylamine antimycotics: recent trends in structure-activity relationships and syntheses. Pestic. Sci., in press.

57. Odds, F. 1988, *Candida* and candidosis, 2nd ed. Bailliere Tindall, London.

58. Ohno, T., T. Kesado, J. Awaya, and S. Omura. 1974. Target of inhibition by the anti-lipogenic antibiotic cerulenin of sterol synthesis in yeast. Biochem. Biophys. Res. Comm. **57:**1119–1124.

59. Omura, S., H. Tomoda, H. Kumagai, M.D. Greenspan, J.B. Yodkovitz, J.S. Chen, A.W. Alberts, I. Martin, S. Mochales, R.L. Monaghan, J.C. Chabal, R.E. Schwartz, and A.A. Patchett. 1987. Potent inhibitory effect of antibiotic 1233A on cholesterol

biosynthesis which specifically blocks 3-hydroxy-3-methylglutaryl coenzyme A synthase. J. Antibiot. **11:**1356–1357.

60. Onishi, J.C., G.K. Abruzzo, R.A. Fromtling, G.M. Garrity, J.A. Milligan, B.A. Pelak, W. Rozdilsky, and B. Weissberger. 1988. Mode of action of L–660,631 in *Candida albicans*. Ann. N.Y. Acad. Sci. **544:**229.

61. Onishi, J.C., G.K. Abruzzo, R.A. Fromtling, G.M. Garrity, J.A. Milligan, B.A. Pelak, W. Rozdilsky, and B. Weissberger. 1988. Mode of action of β-lactone 1233A in *Candida albicans*. Ann. N.Y. Acad. Sci. **544:**230.

62. Parks, L.W. 1958. S-Adenosylmethionine and ergosterol synthesis. J. Am. Chem. Soc. **80:**2023–2024.

63. Poulter, C.D. 1990. Isopentenyl diphosphate to squalene-enzymology and inhibition, p. 169–188. *In* P.J. Kuhn, A.P.J. Trinci, M.J. Jung, M.W. Goosey, and L.G. Copping (eds.), Biochemistry of cell walls and membranes of fungi. Springer-Verlag, Berlin.

64. Reardon, J.E., and R.H. Abeles. 1987. Inhibition of cholesterol biosynthesis by fluorinated mevalonate analogues. Biochemistry **26:**4717–4722.

65. Ryder, N.S. 1985. Specific inhibition of fungal sterol biosynthesis by SF 86–327, a new allylamine antimycotic agent. Antimicrob. Agents Chemother. **27:**252–256.

66. Ryder, N.S. 1987. Squalene epoxidase as the target of antifungal allylamines. Pestic. Sci. **21:**281–288.

67. Ryder, N.S. 1988. Mode of action of allylamines, p. 151–167. *In* D. Berg and M. Plempel (eds.), Sterol biosynthesis inhibitors: pharmaceutical and agrochemical aspects. Ellis Horwood, Chichester, U.K.

68. Ryder, N.S. 1990. Inhibition of squalene epoxidase and sterol side-chain methylation of allylamines. Biochem. Soc. Trans. **18:**45–46.

69. Ryder, N.S. 1990. Squalene epoxidase—enzymology and inhibition, p. 189–203. *In* P.J. Kuhn, A.P.J. Trinci, M.J. Jung, M.W. Goosey, and L.G. Copping (eds.), Biochemistry of cell walls and membranes of fungi. Springer-Verlag, Berlin.

70. Ryder, N.S., and M.C. Dupont. 1985. Inhibition of squalene epoxidase by allylamine antimycotic compounds: a comparative study of the fungal and mammalian enzymes. Biochem. J. **230:**765–770.

71. Ryder, N.S., M.C. Dupont, and I. Frank. 1986. Inhibition of fungal and mammalian sterol biosynthesis by 2-aza-2,3-dihydrosqualene. FEBS Lett. **204:**239–242.

72. Ryder, N.S., I. Frank, and M.C. Dupont. 1986. Ergosterol biosynthesis inhibition by the thiocarbamate antifungel agents tolnaftate and tolciclate. Antimicrob. Agents Chemother. **29:**858–860.

73. Ryder, N.S., and L.J. Goad. 1980. The effect of the 3-hydroxy-3-methylglutaryl CoA reductase inhibitor ML–236B on phytosterol synthesis in *Acer pseudoplatanus* tissue culture. Biochim. Biophys. Acta. **619:**424–427.

74. Ryder, N.S., and H. Mieth. 1990. Allylamine antifungal drugs. Curr. Top. Med. Mycol., in press.

75. Ryder, N.S., G. Seidl, and P.F. Troke. 1984. Effect of the antimycotic drug naftifineon growth of and sterol biosynthesis in *Candida albicans*. Antimicrob. Agents Chemother. **25**:483–487.

76. Ryley, J.F., R.G. Wilson, and K. Barrett-Bee. 1984. Azole resistance in *Candida albicans*. Sabouraudia **22**:53–63.

77. Santen, R.J., H. Vanden Bossche, J. Symoens, J. Brugmans, and R. Decoster. 1983. Site of action of low-dose ketoconazole on androgen biosynthesis in men. J. Clin. Endocrinol. Metab. **57**:732–736.

78. Sen, S.E., and G.D. Prestwich. 1989. Trisnorsqualene alcohol, a potent inhibitor of vertebrate squalene epoxidase. J. Am. Chem. Soc. **111**:1508–1510.

79. Sen, S.E. and G.D. Prestwich. 1989. Trisnorsqualene cyclopropylamine: a reversible, tight-binding inhibitor of squalene epoxidase. J. Am. Chem. Soc. **111**:8761–8763.

80. Sen, S.E., C. Wawrzenczyk, and G.D. Prestwich. 1990. Inhibition of vertebrate squalene epoxidase by extended and truncated analogues of trisnorsqualene alcohol. J. Med. Chem. **33**:1698–1701.

81. Servouse, M., and F. Karst. 1986. Regulation of early enzymes of ergosterol biosynthesis in *Saccharomyces cerevisiae*. Biochem. J. **240**:541–547.

82. Sisler, H.D., R.C. Walsh, and B.N. Ziogas. 1983. Ergosterol biosynthesis a target for fungitoxic action, p. 6218–6225. *In* J. Miyamo and P.C. Kearney (ed.), Proceedings of the Fifth International Congress of Pesticide Chemistry, vol. 3. Pergamon Press, Elmsford.

83. Sobus, M.T., C.E. Holmlund, and N.F. Whittaker. 1977. Effects of the hypocholesterolemic agent trifluperidol on the sterol, steryl ester and fatty acid metabolism of *S. cerevisiae*. J. Bacteriol. **130**:1310–1316.

84. Steel, C.C., R.I. Baloch, E.I. Mercer, and B.C. Baldwin. 1989. The intracellular location and physiological effects of abnormal sterols in fungi grown in the presence of morpholine and functionally related fungicides. Pestic. Biochem. Physiol. **33**:101–111.

85. Stutz, A. 1988. Synthesis and structure-activity correlations within allylamine antimycotics. Ann. N.Y. Acad. Sci. **544**:46–62.

86. Stutz, A., A. Georgopoulos, W. Granitzer, G. Petranyi, and D. Berney. 1986. Synthesis and structure-activity relationships of naftifine-related allylamine antimycotics. J. Med. Chem. **29**:112–125.

87. Stutz, A., and G. Petranyi. 1984. Synthesis and antifungal activity of (E)-*N*-(6,6 dimethyl-2-hepten-4-ynyl)-*N*-methyl-1-naphthalene-methanimine (SF 86–327) and related allylamine derivatives with enhanced oral activity. J. Med. Chem. **27**:1539–1543.

88. Sud. I.J., and D.S. Feingold. 1985. Effect of ketoconazole in combination with other inhibitors of sterol synthesis on fungal growth. Antimicrob. Agents. Chemother. **28**:532–534.

89. Tomoda, H., H. Kumagai, H. Tanaka, and S. Omura. 1987. F–244 specifically inhibits 3-hydroxy-3-methylglutaryl coenzyme A synthase. Biochim. Biophys. Acta. **922**:351–356.

90. Trocha, P.J., and D.B. Sprinson. 1976. Location and regulation of early enzymes of sterol biosynthesis in yeast. Arch. Biochem. Biophys. **174:**45–51.

91. Vanden Bossche, H. 1974. Biochemical effects of miconazole on fungi. 1. Effects on the uptake and/or utilisation of purines, pyrimidines, amino acids and glucose by *Candida albicans*. Biochem. Pharmacol. **26:**887–899.

92. Vanden Bossche, H., G. Willemsens, P. Marichal, W. Cools, and W. Lauwers. 1985. The molecular basis for the antifungal activation of *N*-substituted azole derivatives, p. 321–341. *In* A.P.J. Trinci and J.F. Ryley (ed.), Mode of action of antifungal agents. Cambridge University Press, Cambridge.

93. Walsh, R.C., and H.D. Sisler. 1982. A mutant of *Ustilago maydis* deficient in sterol carbon-14 demethylation characteristics and sensitivity to inhibitors of ergosterol biosynthesis. Pestic. Biochem. Physiol. **18:**122–131.

94. Warnock, D.W., G.M. Johnson, and M.D. Richardson. 1983. Modified response to ketoconazole of *Candida albicans* from a treatment failure. Lancet **i:**642–643.

95. Yeagle, P.L., R.B. Martin, A.K. Lala, H.K. Lin, and K. Bloch. 1977. Differential effects of cholesterol and lanosterol on artificial membranes. Proc. Natl. Acad. Sci. USA **74:**4924–4926.

96. Yoshida, Y. 1988. Cytochrome P450 of fungi: primary target for azole antifungal agents, p. 388–418. *In* M.R. McGinnis (ed.), Current topics in medical mycology, vol. 2. Springer Verlag, Berlin.

17

Rationally Designed Inhibitors of Sterol Biosynthesis

A. C. Oehlschlager and *Eva Czyzewska*

Dedicated to the late Professor A. M. Unrau, a colleague and friend.

I. Introduction

Sterols play vital hormonal, regulatory, and architectural roles in all living organisms. As knowledge of their biosynthetic pathways has grown, so has the possibility for manipulation. One of the most exciting and promising strategies to develop over the past 15 years has been the interference of specific sterol biosynthetic enzymes by mechanism-based inhibitors. Because sterol biosynthesis and its regulation vary among animals, plants, and fungi, this approach offers significant prospects for the rational design of chemotherapeutic agents aimed at the control of disease states ranging from arteriosclerosis to fungal infections.

This chapter reviews current research aimed at the development of mechanism-based inhibitors of sterol biosynthesis. It is intended to critically assess the relative activities of rationally designed substrate and intermediate mimics and to uncover trends that could lead to optimized structures.

We initially present an overview of sterol biosynthesis in plants, animals, and fungi highlighting the key differences in the pathways. This is followed by a description of the mechanism of each key enzyme and an analysis of the activity of known rationally designed inhibitors. There are numerous reviews available allowing readers to broaden and deepen their knowledge on this and similar matters [13–15, 21, 51, 58, 73].

Since the biosynthetic steps prior to squalene oxide (**1**) cyclization are essentially the same in plants, animals, and fungi, we begin our treatment with squalene synthase, the enzyme catalyzing the first committed step in sterol biosynthesis (Figure 17.1). Diversity is first apparent in the products of squalene epoxide cyclization. In vertebrates, lanosterol (**2**) is the immediate cyclization product. Following oxidative removal of the 14α-, 4α-, and 4β-methyl groups (**2** → **3** → **4** → **5** → **6**), isomerization of nuclear unsaturation from Δ^8 to Δ^7 (**6** → **7**), introduction of Δ^5 unsaturation (**7** → **8**), and reduction of Δ^7 and Δ^{24} (**8** → **9** → **10**) produce cholesterol (**10**). This is perhaps the most widely distributed sterol in nature. It plays key roles

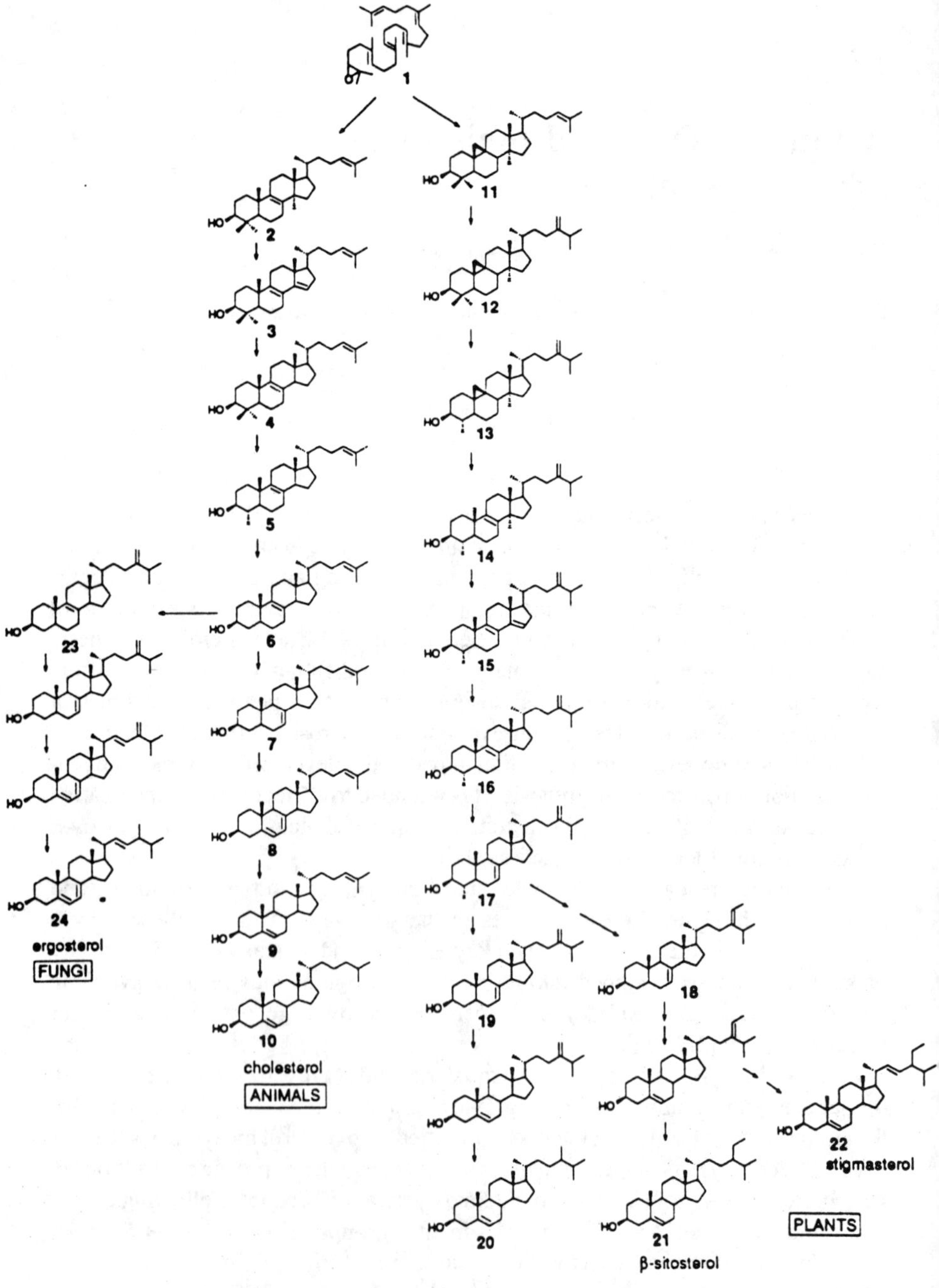

Figure 17.1. Pathways of sterol biosynthesis in fungi, animals, and plants.

as a structural component of animal cell membranes and as a precursor of essential animal hormones. It is also found in small quantities in plants and fungi, where its regulatory role is just beginning to be understood [21].

In photosynthetic plants and algae, cyclication of squalene oxide (**1**) yields cycloartenol (**11**). Although the sequence of events that converts cycloartenol to major plant sterols varies with species, the predominant next step is alkylation at C-24 to produce the $\Delta^{24(28)}$-methylene derivative. (**12**). Removal of one of the C-4 methyl groups (**12** → **13**) usually precedes opening of the cyclopropyl ring (**13** → **14**) and removal of the 14α-methyl (**14** → **15**). Isomerization of the nuclear unsaturation from Δ^8 to Δ^7 (**16** → **17**) provides an intermediate (**17**) that usually undergoes further side-chain alkylation and is demethylated at C-4 (**17** → **18**) or can simply be demethylated (**17** → **19**) to eventually yield campesterol (24α-Me) and dihydrobrassicasterol (24β-Me) (**20**). The major plant sterols β-sitosterol (**21**) and stigmasterol (**22**) are derived from **18**.

Fungal sterol biosynthesis follows the vertebrate pathway to the point of zymosterol (**6**), which in fungi is methylated at C-24 (**6** → **23**). Conversion of the methylated product (**23**) to the major fungal sterol ergosterol (**24**), follows the same pattern of changes in nuclear unsaturation ($\Delta^8 \rightarrow \Delta^7 \rightarrow \Delta^{5,7}$) as for cholesterol biosynthesis, while side-chain modifications require only introduction of Δ^{22} unsaturation and reduction of $\Delta^{24(28)}$.

Thus, vertebrates, plants, and fungi use different post-squalene-oxide pathways to produce different bulk sterols. In vertebrates and fungi, the initial product of cyclization is lanosterol (**2**), whereas in photosynthetic organisms (algae and plants) cycloartenol (**11**) is formed. The latter possess a unique enzyme, cycloeucalenol (**13**)-obtusifoliol (**14**) isomerase, that cleaves the cyclopropyl ring. Selective inhibition of this enzyme might offer prospects for the development of herbicides.

In higher plants and fungi, but not in vertebrates, C-24 alkylation occurs [65]. Inhibition of the enzyme mediating this process could lead to a decrease in sterol biosynthesis in fungi compared to vertebrates. Likewise, inhibition of fungal 24-sterol methyltransferases (24-SMTs), in preference to plant 24-SMTs, would be a logical approach to the development of a control for phytopathogenic fungi. The prospects for this approach seem reasonably good since the substrate specificities for the 24-SMTs of higher plants and fungi differ appreciably. Plant 24-SMTs prefer Δ^{24} sterols containing 14α- and 4-methyl substitution, while the absence of these appendages is crucial to substrate acceptance by fungal 24-SMTs [21].

Selective inhibition of fungal 14α-demethylase in the presence of plant demethylases might also offer opportunities for the development of agricultural fungicides since the fungal enzyme prefers substrates with methyl groups at C-4 and unsaturation at C-24 in the side chain, while plant systems favor C-4 monomethyl derivatives containing $\Delta^{24(28)}$ unsaturation [21].

Research on the development of inhibitors of sterol biosynthesis as control agents has focused on establishing the detailed mechanisms by which enzymes in this pathway operate. Emerging from these studies are mimics of presumptive intermediates of key enzymes which are often potent inhibitors. When administered, these intermediate analogs often cause accumulation of sterols that are the substrates for the inhibited enzymes. Most inhibitors interfere with sterol biosynthesis in the invading and host organisms, affecting the bulk sterol composition in both. For example, treatment of fungus-infected plants with allylamines that inhibit fungal squalene epoxidase also results in modification of the sterol composition in the host plant [71]. Plant growth is not decreased, however, suggesting that a variety of sterols can serve as membrane components and that the necessary regulatory sterols are still present.

An area requiring further work is the definition of the structural requirements of sterols that perform regulatory functions in cell division and growth (chapter 15). For example, normal growth can be restored to celery cell cultures treated with a 14α-demethylase inhibitor by addition of bulk cholesterol (50 μM) and a trace amount of stigmasterol (0.05 μM) or bulk stigmasterol (50 μM) alone [47]. This suggests that at least some 24-ethyl sterol is essential for growth. Cell division can be halted by addition (0.05 μM) of the 14α-methyl sterol, obtusifoliol (**13**), but not lanosterol (**2**) or cycloartenol (**11**) [47]. Until the regulatory functions of sterols in fungal and plant cells are well understood it will be difficult to rationalize the differential effects of sterol biosynthesis inhibitors on target and non-target organisms.

II. Squalene Synthase

Squalene synthase (EC 2.5.1.21; SS) plays an important role in sterol biosynthesis by catalyzing the head to head condensation of two molecules of farnesyl pyrophosphate (FPP, **25**, Figure 17.2). The enzyme is bound to the subcellular membranes of the endoplasmic reticulum in yeast and mammalian liver [72]. Preparations derived from yeast microsomes contain lower levels of membrane-bound phosphatases and are more active than those from liver microsomes [2, 52, 72]. Not until 1988 did Sasiak and Rilling [85] develop methods to obtain stable solubilized preparations of SS. Using non-ionic detergents, these workers determined that a single 47,000-dalton polypeptide is responsible for both presqualene pyrophosphate (PPP) and squalene synthesis [52, 85]. The overall reaction occurs through two distinct steps and involves an isolated chiral intermediate, presqualene pyrophosphate (PPP, **26**) [72].

$$2\ \mathrm{FPP} \rightarrow \mathrm{PPP} + \mathrm{H}^+ + \mathrm{PP_i}$$
$$\mathrm{PPP} + \mathrm{NAD(P)H} \rightarrow \mathrm{Squalene} + \mathrm{PP_i} + \mathrm{NAD(P)}^+$$

$$2\ \mathrm{FPP} + \mathrm{NAD(P)H} \rightarrow \mathrm{Squalene} + 2\ \mathrm{PP_i} + \mathrm{H}^+ + \mathrm{NAD(P)}^+$$

The first step requires Mg^{2+} or Mn^{2+} [2, 72], and proceeds by the loss of inorganic pyrophosphate and a hydrogen from one of the FPPs to yield **26**. Rilling first isolated PPP from washed yeast microsomes incubated with radiolabelled

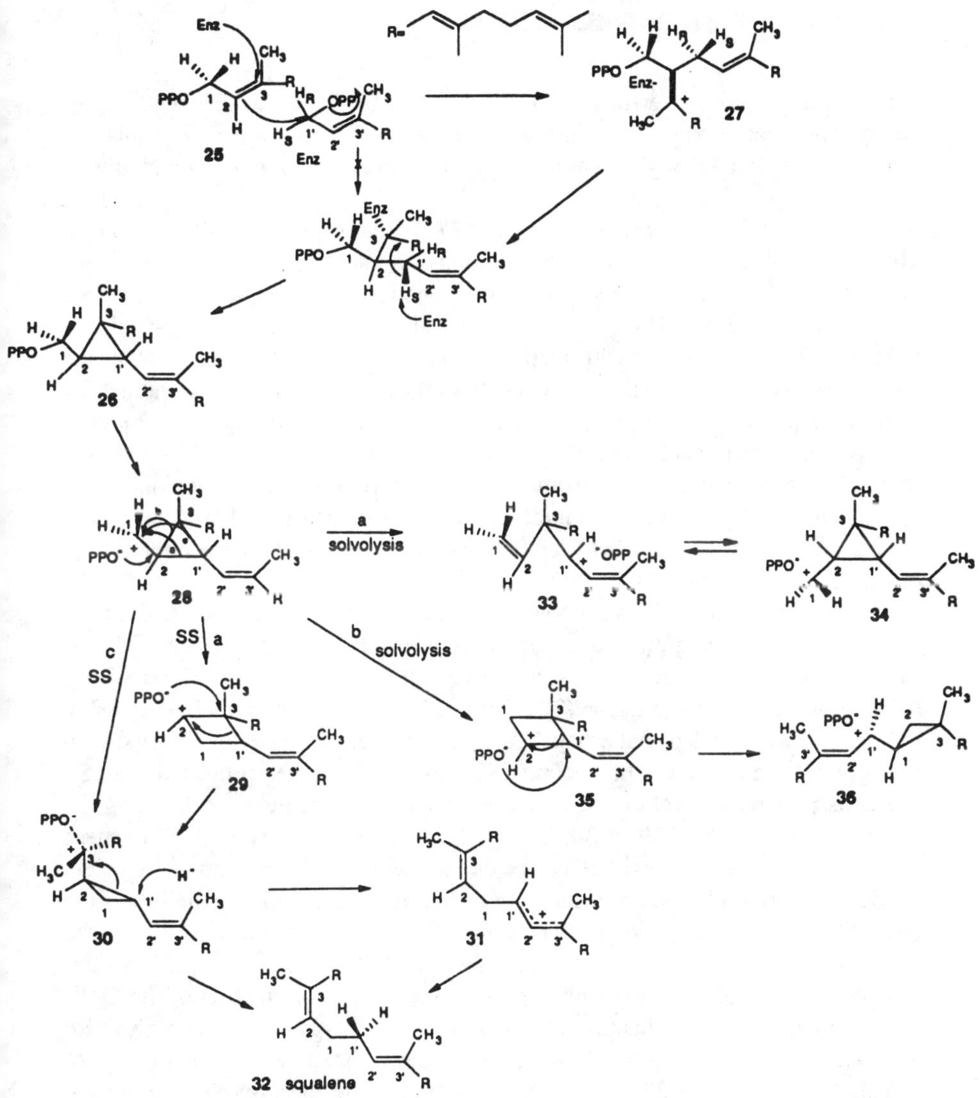

Figure 17.2. Hypothetical mechanism for squalene synthase.

[1-$^3H_2$-^{14}C]-FPP, but without NADPH—the cofactor required to drive the conversion of PPP to squalene in the second step. Comparison of the $^3H/^{14}C$ ratio in PPP (0.67) and FPP (1.0) demonstrated stoichiometric H-atom loss from C-1 of one FPP during the synthesis of PPP [103]. Thus, the two FPPs are not equivalent in the condensation process. Confirmation of two binding sites came from Ortiz de Montellano *et al.* [70], who showed that they have different affinities for FPP. They also observed that 2-methyl-FPP and 3-desmethyl-FPP were acceptable to one site but not to the other. These substrate analogs did not alter the binding of FPP at one site and indeed reacted with it to form 11-methyl squalene and 10-desmethyl squalene, respectively.

FPP is also a substrate for prenyl transferases, mediating the biosynthesis of ubiquinones, carotenoids, and other polyisoprenoids, which suggests that analogs of FPP that inhibit SS might have potential to repress polyisoprenoid biosynthesis [16, 52, 85].

Altman *et al.* [4] have proposed (Figure 17.2) that formation of PPP from FPP involves initial attack by a nucleophilic group of SS at C-3 of the C-2, C-3 double bond in one FPP; polarizes this double bond; and promotes its nucleophilic attack at C-1′ of the second. This results in formation of a bond between C-2 of one FPP and C-1′ of the other with displacement of pyrophosphate anion from the second. If ionization of the pyrophosphate in the second FPP precedes attack by the double bond, then it is easy to envision that cationic intermediate **27** would be involved in the generation of PPP. How the latter is transferred from its site of formation to that of its utilization is not clear. Since no activation of SS accompanies serial washing of soluble proteins from microsomal preparations of the enzyme, direct coupling between these two processes without the intervention of a soluble carrier protein is likely [1].

In the second step, PPP is proposed to ionize to **28** and thence *via* C-1′, C-2, or C-1′, C-3 bond migration to a cyclobutyl (**29**) or a cyclopropylcarbinyl (**30**) cation, respectively. The latter or a subsequent squalene cation (**31**) is finally reduced by NADPH to squalene (**32**, Figure 17.2). Recent work by Rilling using solubilized SS revealed that when NADPH is present activity for the first step was greater than that of the second [85]. Poulter [74] has proposed that SS exerts its primary control on the conversion of PPP to squalene by directing the rearrangement of **28** to **29** or **30**. The preference for conversion of **28** to these intermediates by SS is considered to be due to the ability of the enzyme to orient the departing pyrophosphate in such a way that it is antiparallel to the migrating C-1′, C-2, or C-1′, C-3 bond. According to Coates and Robinson [30], in the case of C-1′, C-2 bond migration to give the cyclobutylcationic intermediate, **29**, this conformational preference must be accompanied by orientation of the C-2′, C-3′ π-bond so that stabilization of any developing charge at C-1′ is minimized; otherwise, cleavage of **28** to **33** would be expected. Solvolysis of analogs of **26** reveals that in solution **28** preferentially (>98%) rearranges by cleavage of the C-1′, C-2 bond to give the ring-opened allylic cation, **33** (and thence **34**), or migration of the C-2, C-3 bond to give **35** (and thence **36**) [74].

As judged from the solution chemistry of presumptive intermediates, **29** and **30**, little enzymatic control is necessary to direct the formation of squalene. Thus, generation of an analog of **29** (R=Me) by solvolysis of the corresponding tosylate gave mixtures rich (>99%) in products arising from hydroxide neutralization of **31** (R=Me) and minor (0.3%) in products resulting from neutralization of **30** (R=Me) [74]. Solvolysis of the p-nitrobenzoate analog of **30** (R=Me) gave an identical product distribution [74].

The absolute sterochemistry of the conversion of **25** to **26** and of the latter to squalene is known. As predicted by the processes envisioned in Figure 17.2, the configuration at C-1 in squalene is opposite to that at C-1 in **25**. The configuration at C-1′ in squalene is determined by the relative stereochemistries of loss and replacement of the C-1′ hydrogen. The pro **S** hydrogen has been shown to be the one lost in the first step and it is replaced by the B-side (pro **S**) hydrogen of NADPH in the second. Although it is unclear whether intermediate **30** or **31** is the species undergoing reduction, the overall sterochemistry of conversion of **26** to squalene suggests that if **30** undergoes reduction, then inversion of configuration at C-1 is involved. This is equivalent to reduction of **31** from the α-face.

Poulter [74] has pointed out that rearrangement of **28** to both **29** and **30** requires much less charge separation between the pyrophosphate anion and the paired carbocation than rearrangement to **33**. Furthermore, the process involving conversion of **26** to squalene requires a maximum leeway in any direction of only 1.5 Å to accommodate the entire reaction sequence.

Inhibition of SS has been achieved by administration of substrate, substrate analogs, and intermediate mimics. High concentrations of **25** ($\leq$25 mM) [2, 72] inhibit the formation of squalene, but, surprisingly, have no effect on PPP synthesis. Agnew suggested that this result may be the consequence of the similarity of **25** to **26** representing true competitive inhibition. Alternatively, the inhibition could simply be due to a detergent effect [1, 72]. Examination of the relative inhibitory potency of analogs of **25** for partially purified SS revealed that both hydrophobic interactions of the chain and polar interactions of the oxygens of the pyrophosphate are required for maximum affinity [17]. Thus, terpenoid (phosphinylmethyl)phosphonates (e.g., **37**, I_{50} 12.5μM) were efficient inhibitors of SS (I_{50}s 12 $\rightarrow$ 67 μM) as long as the hydrocarbon chain had a length comparable to farnesol. Replacement of the allylic or pyrophosphate ether oxygen by a methylene or difluoromethylene had little effect on the binding efficiency. The requirement for both anionic phosphorus moieties is apparent from the lower inhibitory activity of terpenoid phosphates compared to (phosphinylmethyl)phosphonates. Similar studies by Corey and Volante [33] have shown that farnesylpyrophosphate mimics such as **38** (I_{50} 0.2 μM) inhibit the conversion of mevalonic acid to squalene by liver homogenates.

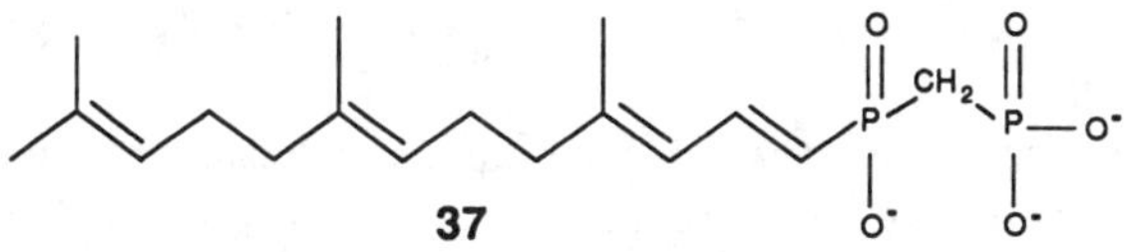

37

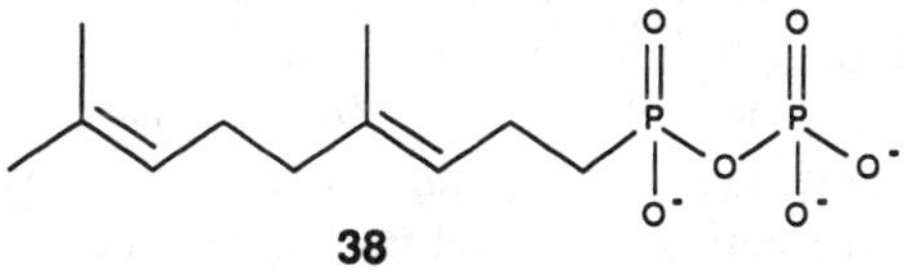

38

Mimics of reaction intermediates have been used to probe the involvement of the carbocationic intermediates proposed in Figure 17.2, as well as the number of active sites in SS. Since the overall reaction is composed of two steps, it is logical that one active site is involved in synthesis of **26** and another in its conversion to squalene. Alternatively, one active site could catalyze the formation of both **26** and squalene. The approach taken has been to administer inhibitors that are mimics of the presumptive intermediates and determine if one or both of the steps are blocked.

The ammonium ion analogs of intermediates **28** and **30** were shown to be efficient inhibitors of SS only if inorganic pyrophosphate was added. Likewise, sulfonium ion mimics of intermediates **27** and **31** are efficient inhibitors of SS only in the presence of added inorganic pyrophosphate [Oehlschlager, A. C. *et al.*, unpublished results]. This is good evidence for both the involvement of these cationic intermediates and the existence of a tight ion pair in which pyrophosphate is the anion. With a phosphonophosphate tethered by a 3-carbon fragment (**39**) to an ammonium ion analog of intermediate **28**, efficient inhibition (I_{50} 3 μM) is achieved with very low levels or no added inorganic pyrophosphate. Poulter estimates that the tethered phosphonophosphate effectively increased phosphate concentration in the active site by 300-fold [75].

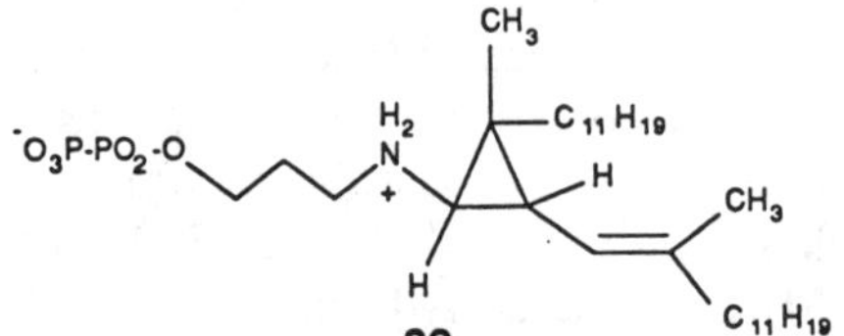

$^{-}O_3P\text{-}PO_2\text{-}O$ ... $\overset{+}{N}H_2$ ... CH_3, $C_{11}H_{19}$, H, H, CH_3, $C_{11}H_{19}$

39

While each of these intermediate mimics should block only one partial reaction of SS, each efficiently blocks both hydrogen loss from C-1 of **25** which occurs in the first step and the production of squalene in the second step. Since the inhibition of both partial reactions of SS by each of these inhibitors occurs approximately to the same extent, the weight of evidence suggests both steps occur in only one site. As was the case with analogs of **25**, when the alkyl residues on these inhibitors are shortened, activity is reduced [Oehlschlager, A. C. *et al.*, unpublished results].

The most efficient inhibitors of SS prepared to date have been mimics of intermediates presumed to be involved in the catalytic process. The construction of these molecules is a formidable task and future efforts in this area should probably be centered on attachment of simpler hydrocarbon residues to the region containing the polar groups. The combination of a cation and teathered anion leads to a dramatic increase in activity, suggesting that zwitterionic species would be attractive inhibitor candidates. In addition more work is needed to define the differences (if any) between fungal and plant as well as between fungal and mammalian SS. To date, there do not appear to be significant differences which would lead to the development of selective inhibitors.

III. Squalene Epoxidase

Squalene 2,3-monooxygenase (EC 1.14.99.7, SE) mediates the epoxidation of squalene (**32**) to 3**S**-squalene-2,3-epoxide (**1**). This is the first step during sterol biosynthesis that requires oxygen. Although in most cells the primary site of regulation of sterol biosynthesis is hydroxyglutaryl CoA reductase, in Chinese hamster ovary cells squalene epoxidase also plays an important regulatory role [36, 37]. Squalene epoxidase from rat liver and human pathogenic fungi has received the most attention while relatively little information is available for SE from plant sources [12]. Like squalene synthase and squalene oxide-lanosterol cyclase, SE is a microsomal membrane-bound protein [54]. It is moderately stable, although activity decreases when it is stored at −25°C and is rapidly lost at 37°C [54].

The enzyme is a flavoprotein monooxygenase requiring flavin adenine dinucleotide (FAD), cytochrome P-450 reductase, NADPH (NADH in the case of fungal SE), and supernatant protein factor (SPF) for activity [63]. The purified enzyme does not contain a heme residue and no metals have been detected in the active site [82]. As expected for an oxygen requirement, SE activity is depressed in anaerobically grown *Saccharomyces cerevisiae* and restored after aeration [54]. The most notable difference between rat liver and candidal (although not of *S. cerevisiae* [54]) SE is in the cofactor requirement. This suggests a difference in electron transport systems [82].

Squalene is the only natural substrate for SE. The enzyme prefers endogenous membrane-bound substrate, but will accept exogenous squalene if anionic phospholipids such as phosphatidylinositol (PI) [63], phosphatidylglycerol (PG), or phosphatidylserine (PS) [29] are provided. The SPF facilitates access of squalene to the enzyme by acting as a carrier [94] or by direct action within the membrane [39]. A more recent study indicates that both SPF and anionic phospholipids stimulate rat liver SE by controlling microsomal uptake and membrane localization of the substrate [29].

It has also been shown in *S. cerevisiae* that membrane-associated unsaturated fatty acids are required for SE activity. A minimum of 20% and a maximum of 70% free fatty acids in the phospholipid is optimal [22]. The requirement can be satisfied by unsaturated fatty acids of various chain lengths and number of unsaturations. However, activity was maximal when C-10 and C-12 saturated fatty acids were absent. The enzyme was totally inactive when the concentration of saturated C-10 and C-12 fatty acids approached 37% of total phospholipids. *De novo* protein synthesis was required for restoration of SE activity in unsaturated fatty-acid-deficient cells [22].

Both mammalian and fungal SE are inhibited by Cu^{2+}, thiol reagents [39], deoxycholate [69], and digitonin [38]. The detergent, Triton X–100, stimulates rat liver preparations, but strongly inhibits SE from *Candida* [82]; however, both are active when Tween-80 is used. Free squalene is only a poor substrate for SE in *S. cerevisiae* despite use of Tween-80. This can be overcome by the use of endogenously synthesized squalene. Either sodium [^{14}C]-acetate [48] or farnesyl-pyrophosphate (**25**) [54] can act as a precursor. Under optimal conditions the specific activity of freshly prepared SE was 0.1 nmol mg^{-1} min^{-1}.

Few antimycotic agents have been shown to interfere with fungal sterol biosynthesis by inhibition of squalene epoxidase [96]. Two compounds proven to be most active are the allylamines: naftifine and terbinafine (chapter 16). Both exhibit activity toward a broad range of fungi but are very poor inhibitors of mammalian and plant SE. It is unlikely that these act as substrate analogs and their high selectivity may imply inhibition of a component that is unique to the fungal epoxidase [83].

The design of mechanism-based inhibitors for squalene epoxidase is difficult because the mechanism of flavoprotein-mediated epoxidation is poorly understood. Oxidases use either transition metals or π systems to increase the rates of the otherwise sluggish reactions of oxygen with organic substrates. The reduced forms of flavoproteins lacking metals (**40**, Figure 17.3) are considered to initially react with molecular oxygen in a single electron transfer to give the superoxide anion (**41**) and the flavin radical (**42**) [108]. Superoxide anion has insufficient reactivity in a chemical sense to be the oxidant of most organic substrates oxidized by this class of enzymes. Further electron transfer and protonation of **41** give rise to hydrogen peroxide and **43** which can combine to produce flavin 4α-hydroperoxides (**44**). The latter or a tautomer (**45**) is likely to be the active species in flavoprotein-mediated oxidations involving both nucleophilic (Baeyer-Villiger) and electrophilic (amine and sulfide oxidation) oxygenations [108].

As is the case for the chemical oxidation of organic sulfur and nitrogen compounds, epoxidation generally proceeds well with electrophilic oxygen sources. As a working hypothesis, it is reasonable to assume that the mechanism of squalene epoxidation by a nonmetallic flavoprotein monooxygenase is similar to that of oxidation of sulfides and amines which proceeds *via* the 4α-hydroperoxide (**44**). Such a process could involve stepwise formation of the two new carbon-oxygen bonds and proceed through intermediates **46** and **47**. Alternatively, squalene epoxide (**32**) and **48** could be formed directly. In the proposed mechanism, the oxidized flavin (**43**) is derived from the reduced flavin (**42**) as well as from **48**. Thus, its formation is decoupled from the oxygen transfer step. This intermediate is common to many flavoprotein oxidases and has been a logical target for interception in related systems [93].

If these scenarios reflect reality, one could imagine interruption at several points. The amine **49** and its quaternary ammonium analog inhibit SE from rat liver, perhaps by interaction with **43** to form adducts (**50**) (Figure 17.4a). These compounds also inhibited *C. albicans* SE [84, 90]. Prestwich [91] has shown that squalene epoxidase from pig liver was very effectively inhibited by trisnorsqualene alcohol, trisnorsqualene thiol, and trisnorsqualene hydroperoxide.

The most potent inhibitor for rat liver SE, **49**, exhibits an I_{50} of 2.4 μM but

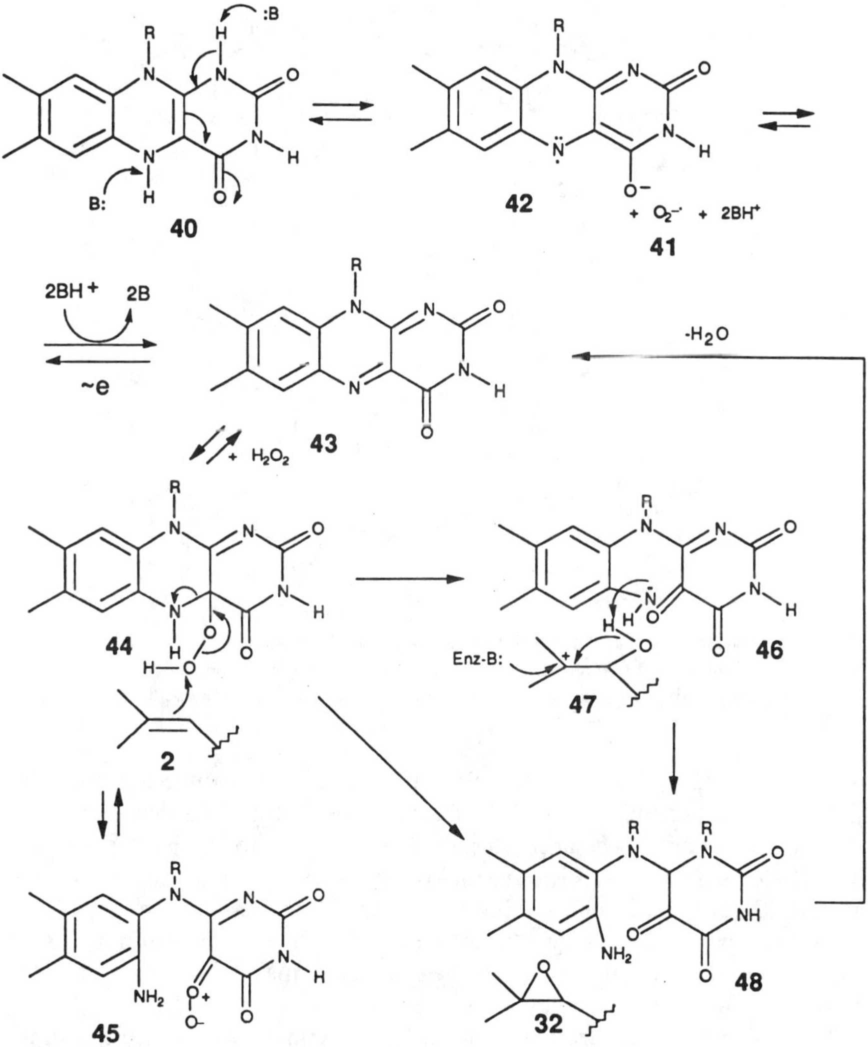

Figure 17.3. Hypothetical mechanism for squalene oxidase.

gives only an I_{50} of 20 μM for pig liver SE [82]. A related secondary amine carrying a cyclopropane in the place of the nitrogen methyls is the most potent inhibitor of pig liver SE (I_{50}=2 μM). It is attractive to propose that this could be due to a single electron transfer from the nitrogen center to produce a cyclopropylamine radical cation, which rearranges to give a homoallyl imminium radical cation, which is then captured by the enzyme. This is the case with related monoamine oxidases [93], but appears to be ruled out for SE because the time-dependent loss in activity that occurs with these compounds can be completely

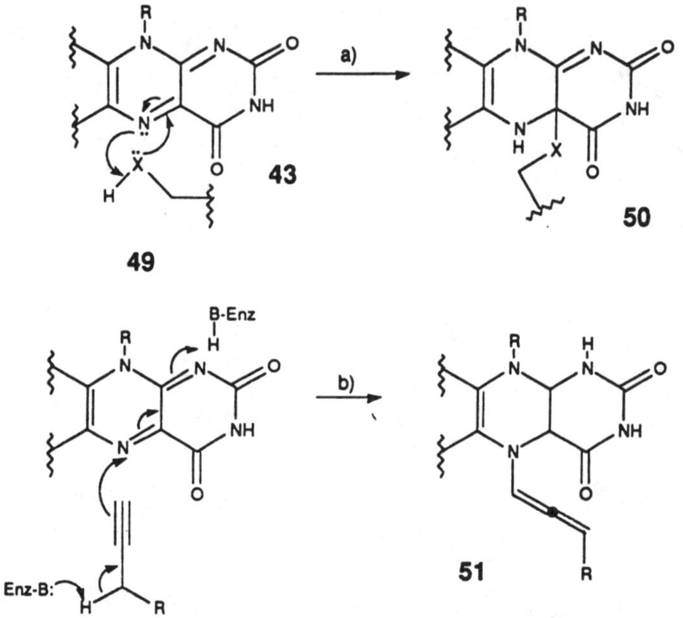

Figure 17.4. a) Potential interaction of known SE heteroatom-containing inhibitors with oxidized flavin intermediate: b) Potential interaction of known alkyne-containing SE inhibitors with oxidized flavin intermediate.

reversed by addition of excess squalene [92]. It is also possible that amine **49** and related compounds are active in their protonated form and thus mimic **47**.

With the aid of molecular mechanics, Prestwich has found that two or more of the following characteristics are required for pig liver SE inhibition by squalene analogs [90]: (1) a hydrophobic region that mimics the terminal isopropylidene methyl groups; (2) an overall size and geometry similar to the substrate; (3) an unpolarized, reactive unsaturation in the position normally epoxidized; and (4) a hydroxyl group in the isopropylidene region.

The requirement for electron donation by the terminal squalene olefin in the proposed mechanism readily explains the weak inhibitory activity of squalene analogs in which the terminal double bond is substituted with electron-withdrawing groups.

Several inhibitors of rat liver SE examined by Ceruti *et al.* and Prestwich have contained allenes, alkynes, and cyclopropanes as mimics of the terminal double bond of squalene which, if oxidized, would yield unstable cyclopropenes or epoxyallenes which could irreversibly react with the enzyme [26, 90]. It is also possible that these substrate analogs, especialy the acetylenes, react with **43** to form covalent adducts (**51**) (Figure 17.4b). This type of irreversible inhibition is analogous to that proposed for the inactivation of flavoprotein monoamine oxidases by propargyl amines [93]. The nature of the inhibition by these compounds has not yet been reported.

The symmetry of squalene has allowed the preparation of bifunctional analogs. Within the series tested, a bifunctional allene was the most potent inhibitor of rat liver SE [26, 90].

Based on Prestwich's observations that heteroatom-containing squalene derivatives are the best rationally designed inhibitors of pig liver SE, it seems logical that combination of these functionalities with alkynes and allenes might yield potent inhibitors of this enzyme. It is noteworthy that the very powerful SE inhibitors, naftifine and terbinafine, although not resembling **2** in structure, possess large hydrophobic groups that could mimic a coiled chain of squalene.

IV. 2,3-Squalene Oxide Cyclase

The formation of the sterol nucleus is mediated by a group of squalene-2,3-oxide cyclases (EC 5.4.99.7, SOCs, Figure 17.5). In photosynthetic plants thc

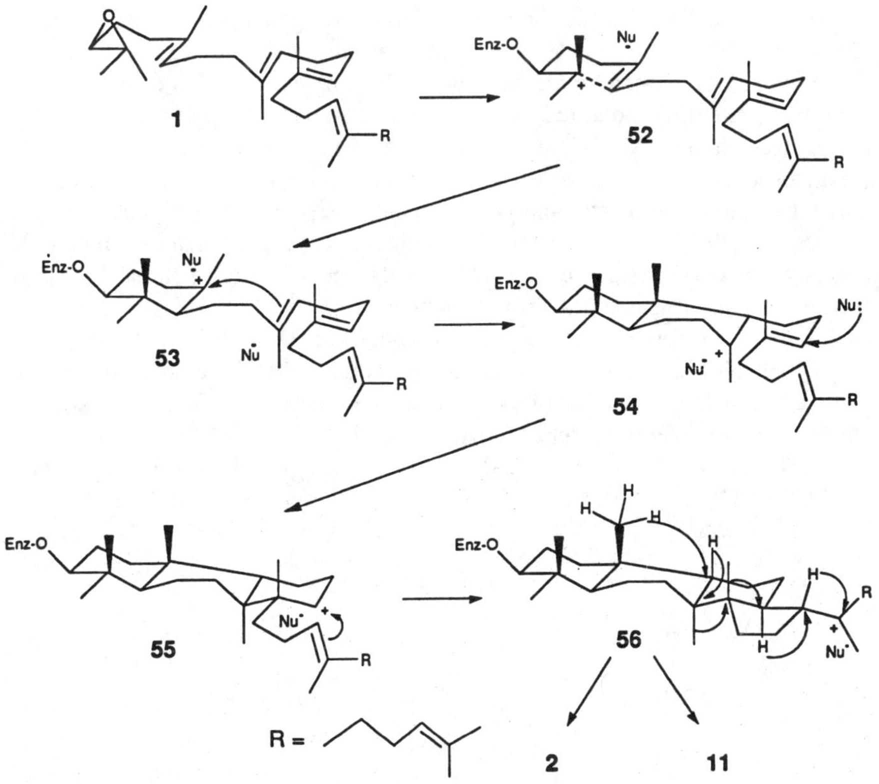

Figure 17.5. Proposed mechanism for 2,3-oxidosqualene cyclase.

usual product of this cyclization is cycloartenol (**11**) but ring systems such as β-amyrin can also be produced [23]. In nonphotosynthetic plants (fungi) and mammalian systems, lanosterol (**2**) is the product. The enzyme has been obtained in partially purified form from animal liver [18, 24, 106], yeast [20], and plant cells [25]. There appear to be no cofactor requirements [90], although mammalian squalene-2,3-epoxide-lanosterol cyclase is stimulated by a soluble protein factor and anionic phospholipids [24]. This enzyme catalyzes the cyclization of squalene epoxide to a fusidane which then undergoes a backbone rearrangement to **2** or **11**. The solution analog of the backbone rearrangment of **56** to **2** is essentially quantitative, indicating little enzymatic control is necessary in this process [Arigoni, D., personal communication]. The initial cyclization could proceed *via* a completely concerted process not involving generation of intermediates. Alternatively, and more likely, the formation of each ring could be accompanied by formation of a stabilized carbocation intermediate, **52–56**. Sulfhydryl groups have been implicated as essential for retaining activity [35].

Stereoelectronic analysis of the cyclization suggests that the two main points of control in the initial cyclization are the folding of the B ring in a boat conformation (**53** → **54**) and the anti-Markovnikov cyclization leading to ring C (**54** → **55**) [104–107]. Model studies indicate that the solution analog of the initial epoxide ring opening proceeds with a high degree of π-bond participation from the first double bond (**52** → **53**), but that additional remote double bonds do not accelerate epoxide cleavage [105]. This has been interpreted as evidence for a stepwise mechanism, at least for related solution cyclizations. Strong evidence for the stepwise hypothesis has recently been reported by Johnson and co-workers [49]. They found that the yields in polyene cyclizations increased dramatically when groups were added to the squalene chain at positions that allow them to stabilize the cationic charges generated.

The substrate requirements for SOCs are not rigid. They will cyclize unnatural analogs of squalene epoxide in which methyls attached to the epoxide have been replaced by hydrogen or ethyl [32] as well as diepoxysqualene [31]. The minimum structural requirements for tetracyclization are the 2,3-epoxide and tetra π-bond sequence. The methyls at C-6, C-10, and C-15 are not individually necessary [32, 101, 102].

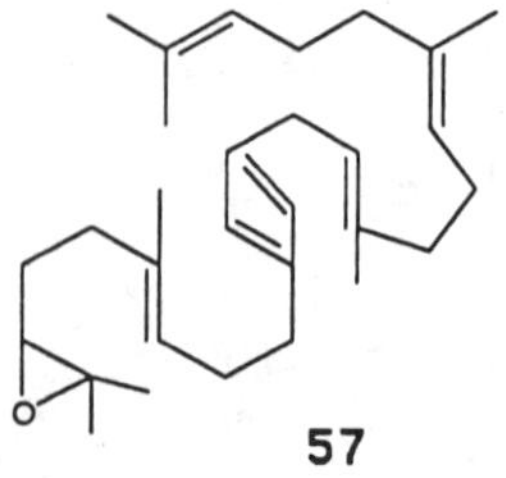

O

57

Table 17.1. Inhibition of SOC by analogs of intermediate **52:** *data from reference 25 except where noted*

	I_{50} (μM)		
Analog	Rat liver	*S. cerevisiae*	3T3 Fibroblasts
Tertiary amines			
58a	7.5	10	0.3 [44]
58b	5.1		
58c	3.2	14	
58d	3.5		
58e	2.0	40	
58f	1.9	0.1	
58g	3.0	6.0	
58h	6.5	4.0	
58i	18	1.0	
Amine oxides			
59a	3.7	16	
59b	1.5	14	
Related nitrogen derivatives			
60	0.4	10	
61	3.9		

The most dramatic structural changes to be introduced into squalene analogs and be accepted by SOC have been reported by Kyler [56, 57]. This group found that compounds related to **57** were cyclized with good efficiency by SOC derived from bakers' yeast. The observation that analogs substituted on the β-face did not cyclize suggests that structural features perturbing this face, but not the α-face of the substrate folded in its chair-boat-chair conformation, interfere with the enzymatic cyclization [56, 57].

Inhibition of SOC has been achieved by both substrate analogs and mimics of the presumptive carbocationic intermediates (Figure 17.5). The extensive studies of Ceruti's group have shown that the most effective inhibitors are those that possess a positive charge at C-2 of the squalene skeleton. These apparently mimic intermediate **52** [25, 89]. There are significant differences in the activity of a given inhibitor for SOCs derived from plant, animal, and fungal sources (Table 17.1). Unfortunately, few of the mimics of **52** examined are more active for fungal SOC than for the plant or animal enzyme.

The 2-aza-dihydrosqualenes tested fall into four structural categories: amines (**58**), amine oxides (**59**), epoxide analogs (**60**), and nitrogen derivatives (**61**). Within the first group only **58f, 58h,** and **58i** show higher activity toward yeast than liver SOC. The distinguishing structural feature of these is the location of ethyl groups on the nitrogen presumed to bear a positive charge. Systematic analysis of the inhibitory activity of amine mimics of **52** reveals that, in the systems examined, methyl (**58a** and **58b**), ethyl (**58c**), as well as cyclic (**58d**) nitrogen substituents are tolerated. The structure of the major hydrocarbon tail attached to nitrogen is also important. A cyclized chain as in **58f** is very effective as are nonbranched hydrocarbon (**58e**), squalene (**58a, 58b** and **58c**), and squalene chains.

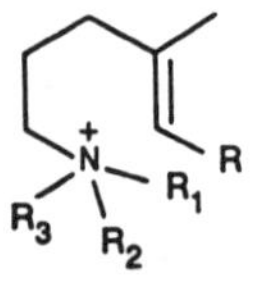

58a, $R_1 = H; R_2 = R_3 = CH_3$

58b, $R_1 = R_2 = R_3 = CH_3$

58c, $R_1 = H; R_2 = R_3 = Et$

58d, $R_1 = H; R_2 = R_3 = (CH_2)_5$

59a, $R_1 = R_2 = CH_3; R_3 = O^-$

59b, $R_1 = R_2 = C_2H_5; R_3 = O^-$

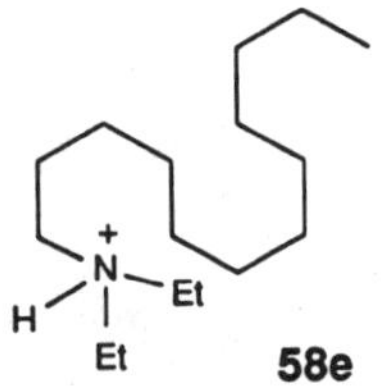

58e

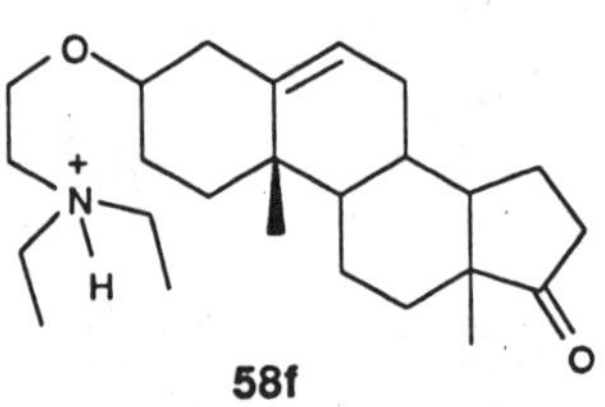

58f

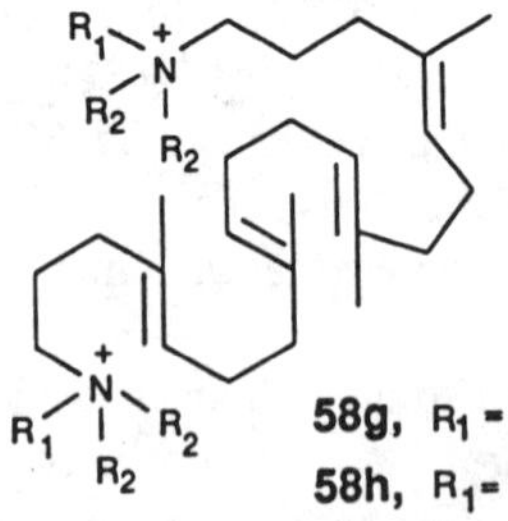

58g, $R_1 = H; R_2 = Et$

58h, $R_1 = Me; R_2 = Et$

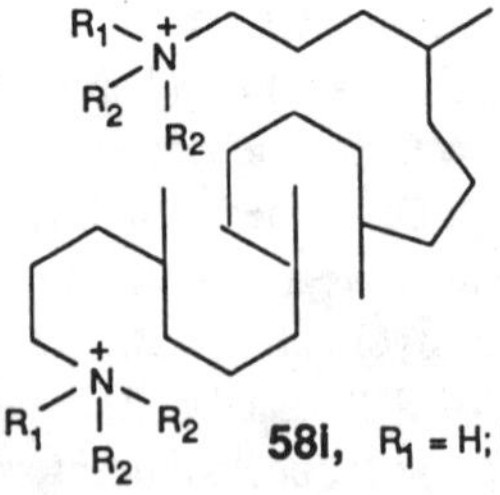

58i, $R_1 = H; R_2 = Et$

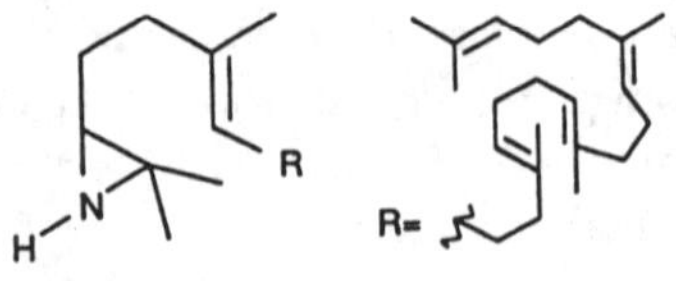

60

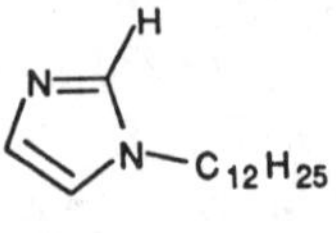

61

Mimicking the dipoles generated during the enzymatic cleavage of the epoxide by incorporation of an amine oxide into the squalene framework yielded inhibitors which were more powerful than the parent tertiary amines (**58a** *vs*. **59a, 58c** *vs*. **59b**). Removal of the double bond in the squalene moiety or its replacement by a shorter chain results in a decrease in activity [25]. Unfortunately, although all

Table 17.2. Inhibition of SOCs by mimics of intermediates **52** *and* **54**

Compound	Mimic of	Rat liver	Bramle cells	Maize seedlings: Cycloartenol	Maize seedlings: β-Amyrin	3T3 Fibroblasts
		I_{50} (μM)				
62	52	9 [34]	>100 [34] (peas)			
		2 [99]	65 [81]			
63	54		10 [99]	1.0 [99]	NI[a]	0.02 [45]

[a]NI, no inhibition

amine oxides examined were significantly active, none was more active in yeast than in liver systems (Table 17.1).

Nitrogen-containing analogs of **52** containing imine and 3-hydroperoxy functions were not significantly active for liver or fungal SOCs. The ethyleneimine analog of **2** (**60**) was active against both liver and fungal SOC. Dipolar analogs of **52** such as amides were not significantly active, although azole **61** was [25].

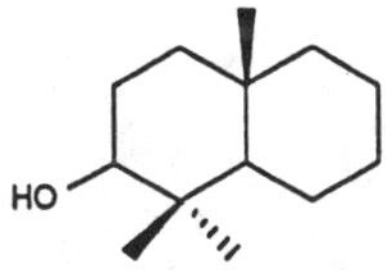

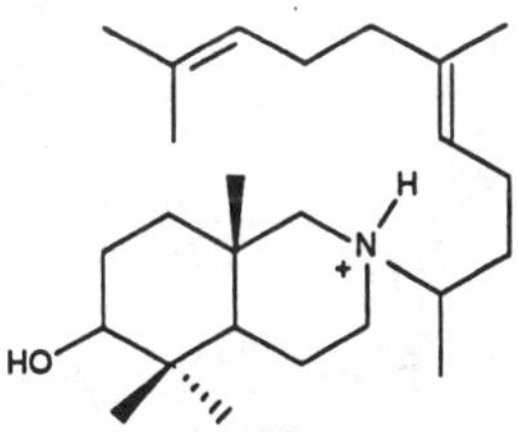

Several decalin mimics of intermediate **52** have been investigated as inhibitors of SOCs (Table 17.2) [81]. The only derivative in this class to be active was **62**, which presumably binds to the active site *via* its 3β-hydroxyl. Benveniste has studied the inhibitory power of 8-aza-decalins that should be mimics of intermediate **54**. Only those with squalene (**63**) and squalene side chains were inhibitory [45, 99]. Activity is lowered when the branch methyls are removed from the side chain or when the 3β-hydroxyl is removed [99].

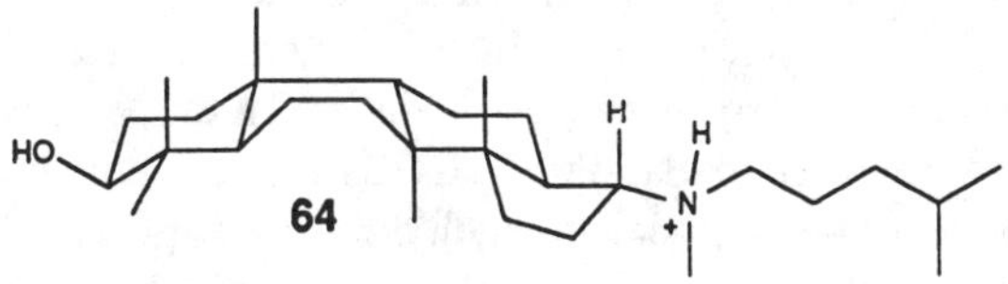

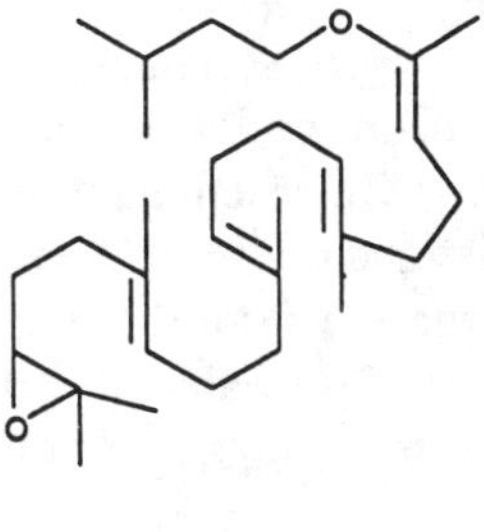

There have been a few attempts to inhibit SOCs at the stage of the fusidane cation (**56**). 20-Azadammaranol (**64**) was found not to inhibit squalene-2,3-epoxide-β-amyrin synthetase [25]. An alternative strategy involving **65** was successful. The isomeric vinyl ethers inhibited SOC derived from pea seedling (I_{50}~300 μM, K_m for **1**=250 μM) [28].

The point of divergence of fungal and plant SOCs is in the final generation of lanosterol *vs* cycloartenol, respectively. It is reasonable to expect the plant SOCs to accommodate skeletons resembling cycloartenol while the fungal SOCs should accommodate lanosterol skeletons more readily. Since the best inhibitors of all SOCs are those that mimic preesumptive intermediate **52**, a promising approach to the design of fungal-selective SOC inhibitors would be to incorporate a polar functional group into a lanosterol skeleton as in **66**. Groups of choice would be tertiary amino, amine oxide, aziridine, oxaziridine, amidine, and azole. The rigidity of the hydrocarbon chain would reduce the requirement for SOC orientation and should favor binding. A hint that this approach would be fruitful comes from the observation that the polycyclic tertiary amine, **58f**, is both highly active and selective for yeast SOC [25].

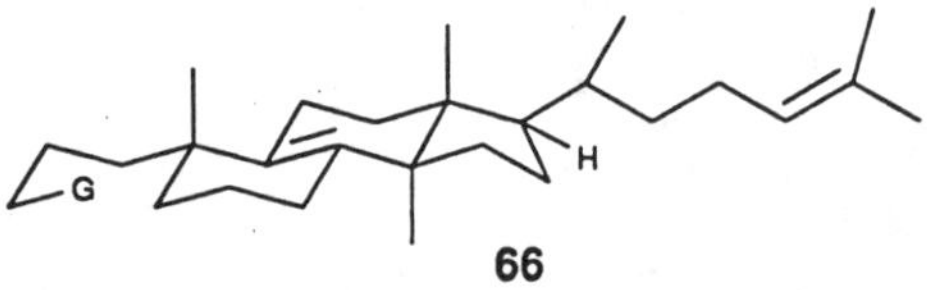

G H

66

V. Sterol 14α-Demethylase

In fungi and mammals, removal of the 14α-methyl (C–32) of lanosterol (**2**) is the first transformation after cyclization of squalene oxide (**1**). This step is an important regulatory process because enzymes that convert lanosterol and its metabolites to the predominant sterols of fungi and mammals generally do not accept substrates containing a 14α-methyl [73]. The demethylation is catalyzed in *S. cerevisiae* by lanosterol-14α-methyl demethylase (14-DM), a 58,000-dalton, membrane-bound cytochrome P–450 oxidase requiring oxygen and NADPH cytochrome P–450 reductase [111–113]. Although lanosterol is the preferred substrate, 24,25-dihydrolanosterol is also demethylated by the rat liver and *S. cerevisiae* systems [5, 73, 113]. Semianaerobic cultures of yeast are rich in cytochrome P–450, but anaerobic and aerobic ones are deficient in this oxidase [73]. The reaction proceeds by three monooxidations *via* C-32 hydroxymethylene (**67**) and C-32 oxo (**68**) intermediates with direct formation of formate and a $\Delta^{8,14}$-diene (**3**) from the third oxidation (Figure 17.6) [6, 8]. The intermediates are more tightly bound to the enzyme and are metabolized more rapidly than the substrate, suggesting that the initial oxidation is the rate-determining step [6, 8].

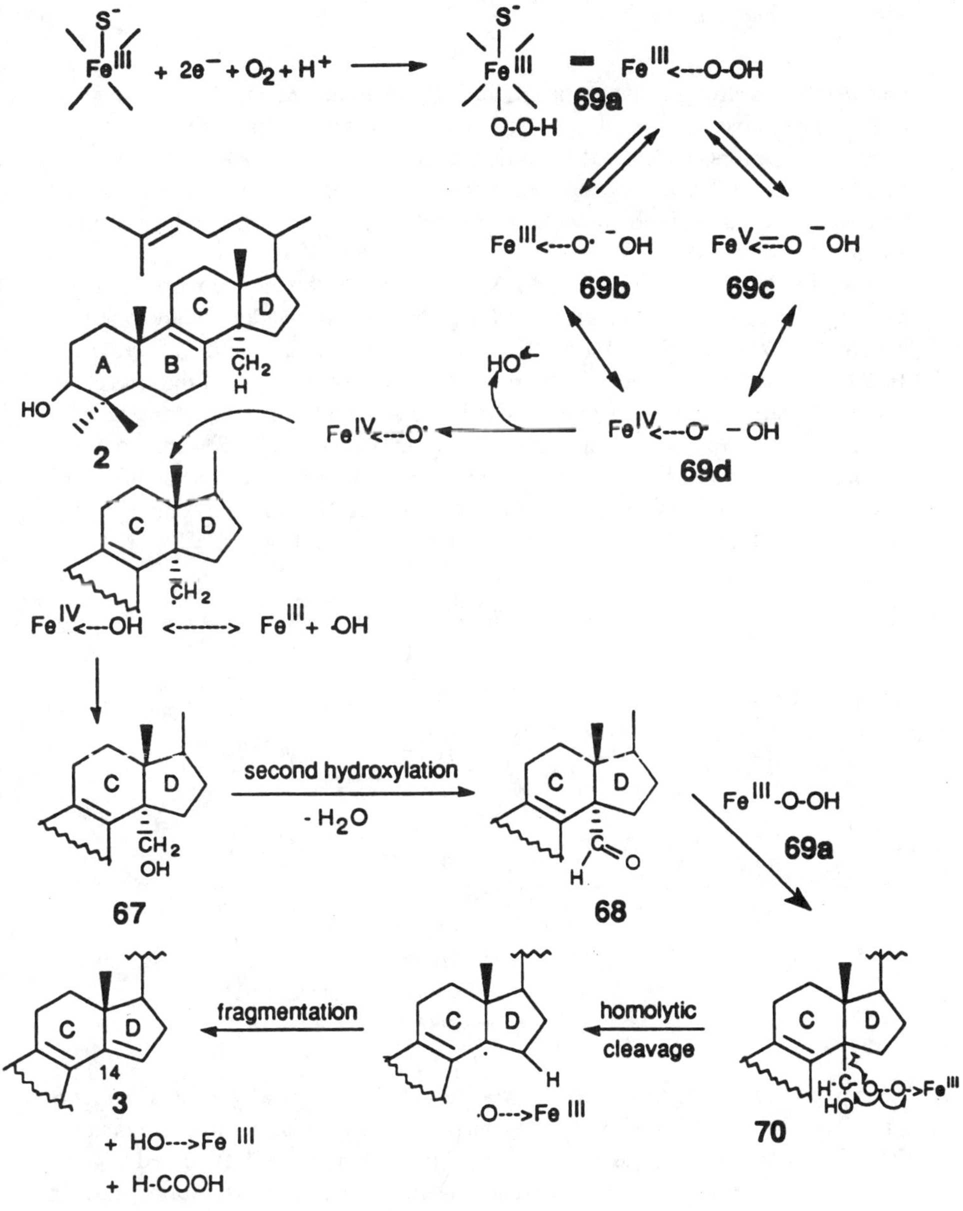

Figure 17.6. Proposed mechanism for lanosterol 14α-methyl demethylase.

Reasoned descriptions of the alternate mechanisms by which 14-DM could operate have been advanced [3, 46, 95]. The first two hydroxylations are envisioned to occur by initial formation of peroxy iron(III) complex (**69a**) which fragments to a mono-oxygen species (**69b**) that can have several formulations. The observation that cytochrome P–450 hydroxylations sometimes occur with racemization is strong evidence that radical intermediates are involved and support **69d** as a reactive species [43, 109]. Peroxy iron intermediate, **69a,** would be expected to possess strong nucleophilic character and should add to the carbonyl of **68** to produce **70**. Homolytic cleavage of the peroxy bond of the latter would give an alkoxy radical which would be expected to fragment to produce formate and a $\Delta^{8,14}$-diene (**3**).

Selective inhibition of fungal 14-DM is the basis of a viable control strategy. Most agricultural fungicides as well as most antifungals used in the treatment of human mycosis act by inhibiting this step [42, 50]. Inhibition of *S. cerevisiae* 14-DM by antifungal azoles has been shown to involve binding of the side chain to the substrate binding site of both the reduced and oxidized forms of the enzyme in such a way as to place the triazole nitrogen into the coordination sphere of the cytochrome P–450 heme iron. Spectral changes accompanying binding of azoles to the reduced form of yeast 14-DM were consistent with replacement of the normal sixth ligand of the ferrous ion by a σ-bonded nitrogen [7, 86, 114].

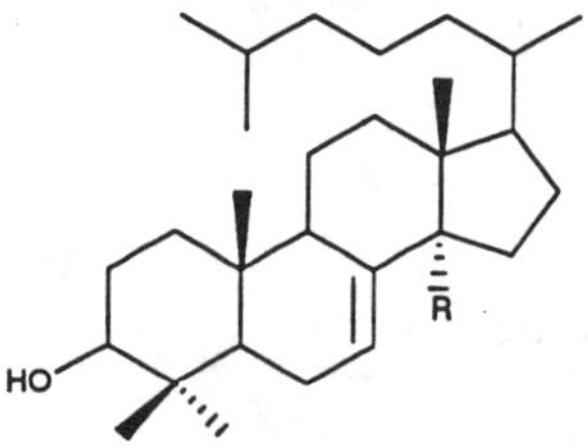

HO

R

71a, R= CHF_2
71b, R= CH_2-CHF_2
71c, R= CH_2-C≡CH
71d, R= CHOH-C≡CH
71e, R= CH=CH_2
71f, R= C≡CH

Frye and Robinson have recently prepared the first rationally designed inhibitors of 14-DM [41]. The C-32 functionalized substrate and intermediate analogs (**71a–71f**) were designed as irreversible inhibitors and indeed exhibited affinity for yeast 14-DM that was stronger than lanosterol (Table 17.3). Although their I_{50}s were considerably higher than those of azoles, substitution at C-32 may prove to be a viable approach to the design of 14-DM inhibitors. Candidates not yet examined but chemically accessible are lanosterols containing ether, oxygen, nitrogen, and sulfur at C-32.

VI. $\Delta^{8(14)}$-Reductase

Removal of a 14α-methyl generates a $\Delta^{8(14)}$ unsaturated sterol, **3** (Figure 17.7). This unsaturation is the first to be removed during the subsequent transformations

Table 17.3 Inhibition of S. cerevisiae *14 DM by 32-substituted lanost-7-en-3β-ols*

Compound	I_{50} (μM)
lanosterol	(23)[a]
71a	7.3
71b	7.6
71c	1.2
71d	0.005/0.57[b]
71e	22.3
71f	1.2

[a] K_m.

[b] Diastereomers.

to Δ^5 sterols. It is important that this process occurs because $\Delta^{8(14)}$ sterols do not readily fulfill the architectural requirement normally met by $\Delta^{5,7}$ and Δ^5 sterols [21].

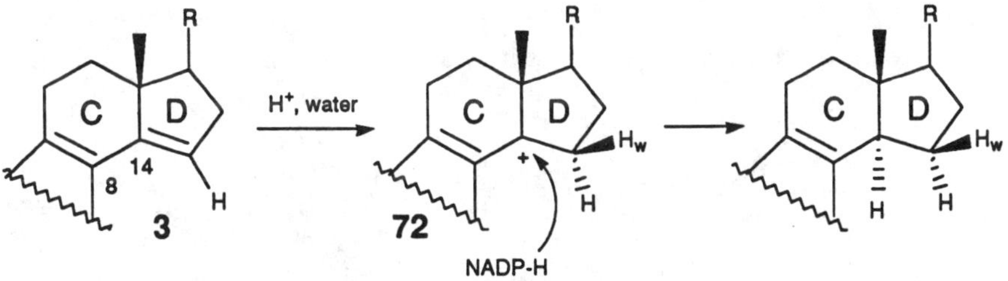

Figure 17.7. Proposed mechanism for $\Delta^{8(14)}$ sterol reductase.

Reduction is mediated by an enzyme that operates by initial addition of a proton from the medium to the β-face of C-15 (Figure 17.7). This is presumed to give an intermediate C-14 cation (**72**) that is neutralized by hydride (NADPH) attack from the α-face [3].

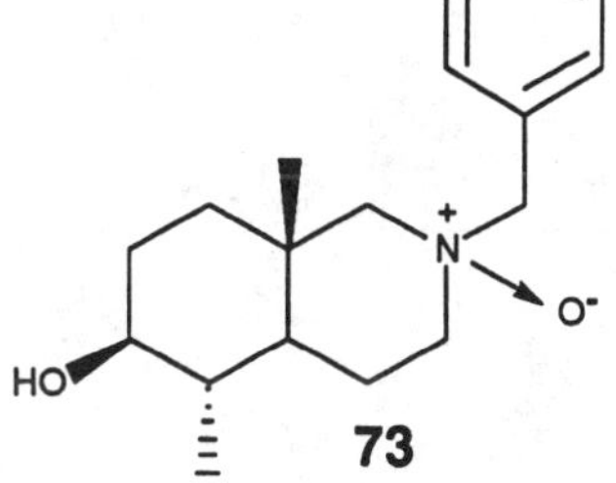

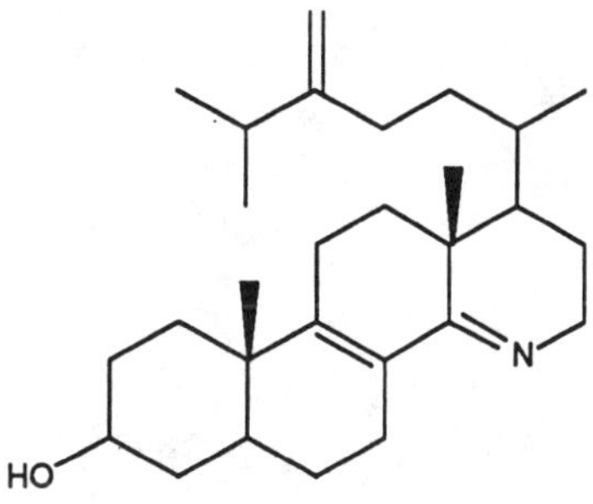

Blockage of this enzyme confers antifungal activity because fungal cells are more sensitive to changes in bulk sterol composition than plant cells [21]. Benveniste has found that amine oxide, **73,** is a potent inhibitor of $\Delta^{8(14)}$-reductase in bramble cells [100]. The naturally occurring 15-azasterol, **74**, strongly inhibits yeast (K_i=2 nM) and plant 14-DM [19, 87]. Both of these inhibitors presumably act by mimicking the intermediate **72**. Based on the potent inhibitory activity of **74** in the yeast system, one would expect that the placement of heteroatoms at C-14 and C-15 in sterol skeletons would yield strong inhibitors of this enzyme.

VII. Δ^{24} Sterol Methyltransferase

S-Adenosyl-L-methionine:Δ^{24}-sterol methyltransferase (EC 2.1.1.43; 24-SMT) is an enzyme that catalyzes transfer of the methyl of S-adenosyl-L-methionine (SAM) to C-24 of Δ^{24} sterols to produce $\Delta^{24(28)}$ sterols (Figure 17.8) [53]. It is a 150,000-dalton, membrane-bound protein that has been partially purified from *S. cerevisiae* [59] and *Candida spp.* [10]. It is relatively unstable in its solubilized form, losing 60% of its activity at 4°C over 1 day [10]. It is present

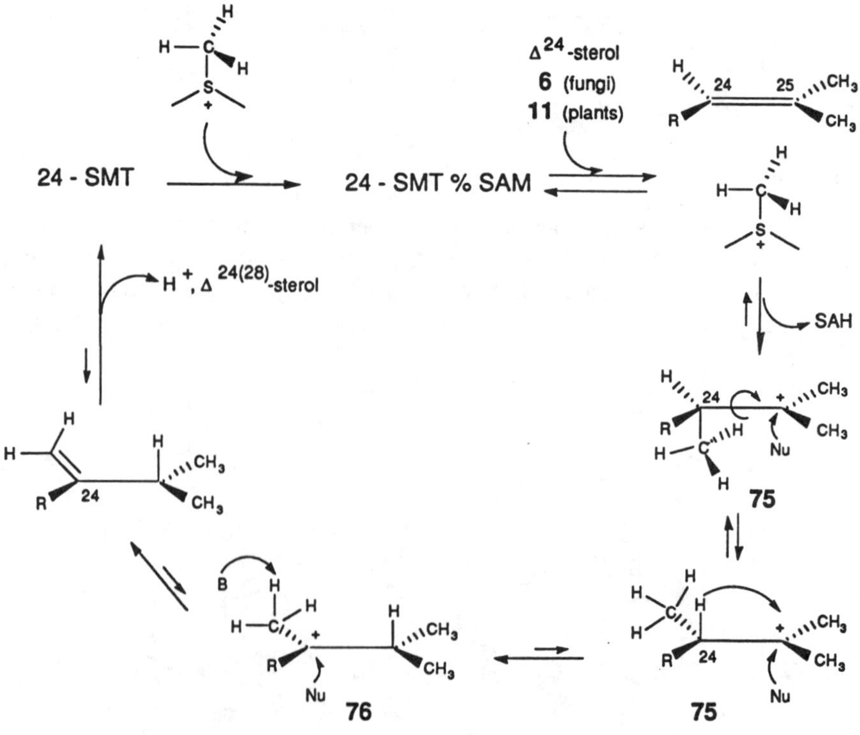

Figure 17.8. Mechanism of yeast 24–SMT.

in plants and fungi, but not in mammalian sterol biosynthetic systems, making it a logical target for development of antifungal agents.

The preferred substrate in fungi is zymosterol (**6**, Figure 17.1) although 4α-methyl (**5**) and 4,4-dimethyl (**4**) zymosterol are also metabolized as poor substrates. Lanosterol (**2**) is not a substrate [40, 60]. In photosynthetic organisms, double methylation at C-24 is required for production of the principal plant sterols. The first of these is catalyzed by 24-SMTs that prefer cycloartenol, **11**, as a substrate [73].

Gaylor has shown that Δ^7, Δ^5, and $\Delta^{5,7}$ sterols lacking C-4 and C-14 methyls are moderately good substrates for the 24-SMT of *S. cerevisiae* [60]. Ator *et al.* has recently observed that the substrate specificity of the transferase of *Candida spp.* parallels that from *S. cerevisiae* in that the presence of C-4 or C-14 methyls causes a complete loss of substrate binding [10]. This study revealed that orbital hybridization at C-9 and C-14 of substrates substantially influences the values of K_m and V_{max}. An increase in K_m attends a change in C-9 from sp^2 to sp^3 which alters the conformations of rings B and C. Decreases in V_{max} are observed when C-14 is changed from sp^3 to sp^2, a modification that alters the conformation of ring D and the position of the side chain [10].

Mechanistically, these methylations are proposed to involve nucleophilic attack by the π-electrons of the Δ^{24} double bond on the S-methyl group of SAM [9]. The resulting carbonium ion intermediate (**75**) undergoes a C-24 $\rightarrow$ C-25 hydride shift producing a C-24 carbonium ion (**76**). Loss of a hydrogen from the methyl generates the $\Delta^{24(28)}$ product.

Sterols in which C-24 or C-25 has been replaced by positively charged groups, such as sulfonium, protonated tertiary amine, or quaternary ammonium, are potent inhibitors of yeast and plant 24-SMTs (Table 17.4). Structure-activity relationships obtained by using the *S. cerevisiae* 24-SMT and a series of tertiary amines designed to mimic the C-25 carbonium ion intermediate (**75**) reveal that inhibition is significantly lowered when the nuclear structure is altered from the Δ^8 unsaturation in the preferred substrate, **6**. For example, 25-azasterols containing a Δ^8 unsaturation (**77c**) are more inhibitory than those with a saturated nucleus (**77e**) or Δ^5 analogs (**77b**) [67, 77]. Placement of C-4 and C-14 methyls on the nucleus (**77d**) of 25-azasterols predictably results in a significant decrease in activity. While 25-azasterols function quite well, presumably in their protonated form, as mimics of **75**, the corresponding quaternary ammonium (**77a**) and sulfonium (**77i**) ions are clearly superior in yeast systems [10, 11, 68]. Likewise the bridged structure 24-epiiminolanosterol (**77h**) has been shown to strongly inhibit 24β-methylsterol production in the fungus *Gibberella fujikuori* [65] and higher plants [97].

Ator *et al.* [10] has recently shown that when the charged atom is part of a heterocyclic ring (**77q**) or an amidine (**77r**) there is sufficient interaction with the *Candida* 24-SMT to cause binding even when attached to a nucleus containing C-4 and C-14 methyl groups. Comparison of the inhibition by **77b** with that by

Table 17.4. Inhibition of 24-SMTs by side-chain heteroatom analogs of presumptive intermediates

Compound	Structure	*S. cerevisiae* I_{50} (μM)[a] [67, Oehlschlager, *et al.*, unpublished]	*Candida* spp. K_i (μM)[b] [10]
77a	Δ^5	K_i=0.0028	
77b	Δ^5	0.14, K_i=0.0054	(0.063)
77c	Δ^8	0.08, K_i=0.0051	
77d	Δ^8, 4,4,14 - trimethyl	1.14	
77e	Δ^0	0.14	
77f	Δ^5	0.22	
77g	Δ^5	4.8, K_i=0.0072	
77h	Δ^8, 4,4,14 - trimethyl	<1[c]	
77i	Δ^5	0.07	
77j	Δ^5	0.25	
77k	Δ^5	0.14	

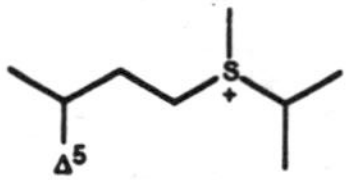

Compound	Structure	*S. cerevisiae* I_{50} (μM)[a] [67, Oehlschlager, *et al.*, unpublished]	*Candida* spp. K_i (μM)[b] [92]
77l	Δ^5		(0.030)
77m	Δ^5		(1.1)
77n	Δ^8, 4,4,14 - trimethyl		(198)
77o	Δ^0	0.8	
77p	Δ^0	−8	
77q	Δ^8, 4,4,14 - trimethyl		(0.020)
77r	Δ^8, 4,4,14 - trimethyl		(0.040)

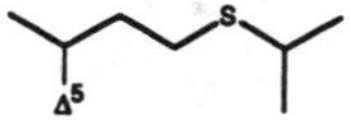

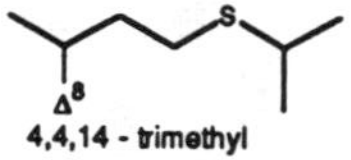

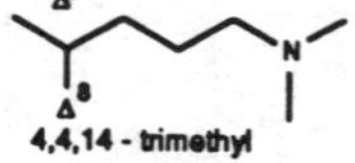

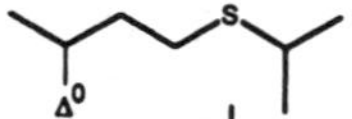

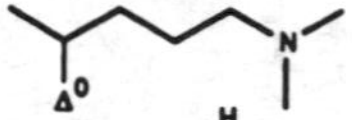

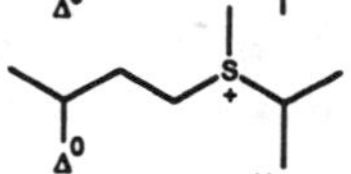

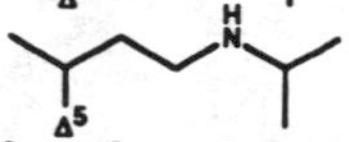

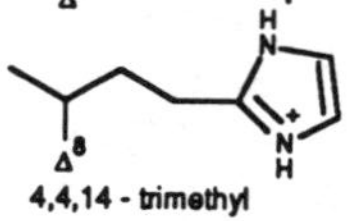

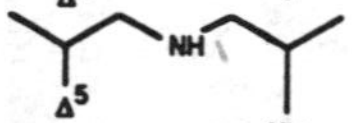

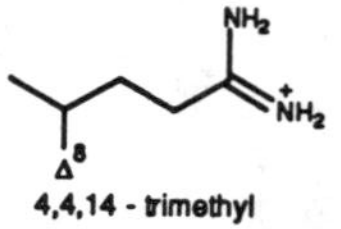

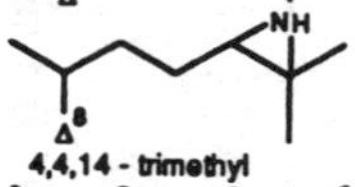

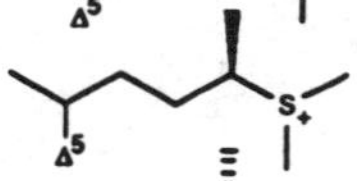

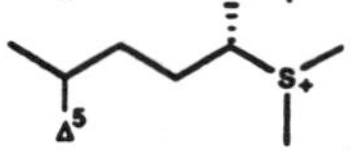

[a] In whole cells. [b] Partially purified enzyme *vs.* sterol substrate [c] Whole cells of *Gibberella fujikuori*.

77d in the *Saccharomyces* 24-SMT suggests that the 4,14-desmethyl analogs of **77q** and **77r** should be at least an order of magnitude more potent.

Sterols containing heteroatoms at C-24 are also excellent inhibitors of yeast 24-SMTs. Comparison of the relative inhibitory power of the uncharged thioether (**77m**) with that of its sulfonium analog, **77l**, reveals that these sterols interact with the enzyme through their resemblance to intermediate **76**. It is clear from the lower inhibitory activity of the 4,4,14-trimethyl-substituted thioether **77n** compared to **77m** that if no charge is present in the side-chain, binding is dramatically decreased by nuclear methylation [10].

Benveniste has conducted extensive studies of inhibitors of the 24-SMTs of maize and bramble [64, 77, 78, 88]. These mimics of **75** contained the cycloartenol (**11**) skeleton because this is the preferred nucleus for plant 24-SMTs. Cationic charges carried at C-25 by nitrogen, sulfur, or arsenic were equally effective in promoting binding while the presence of an amine oxide at C-25 was the most effective. Although the C-25 cationic intermediate being mimicked possesses an alkyl group at C-24, addition of this appendage did not enhance activity in either the yeast or plant systems.

Detailed stereochemical and kinetic analysis of yeast 24-SMTs has been carried out with the aid of the mechanism-based inhibitors described above (Figure 17.8). That the methyl transfer involves inversion of configuration is deduced from the observation that in both *S. cerevisiae* [59] and *Candida spp*. [10] kinetic analysis is consistent only with direct transfer and not the intermediacy of an alkylated enzyme. That the alkylation probably occurs on the *si*-face of Δ^{24} in yeast is deduced from the slightly better inhibitory power of **77k** compared to **77j** for *S. cerevisiae* 24-SMT [Oehlschlager, A. C. *et al.*, unpublished results]. Similar experiments in plants suggest that alkylation takes place from the *re*-face in these systems [78]. The C-24 → C-25 hydrogen migration has been shown in *S. cerevisiae* to occur across the *re*-face of the Δ^{24} unsaturation [9]. The stereochemistry of C-28 hydrogen loss in yeast is deduced from the above and two additional experiments. The first involves the observation that methionine possessing a chiral methyl group ($C^1H^2H^3H$) was transformed by the yeast *Claviceps paspali* and *S. cerevisiae* to ergosterol with a C-24 methyl of the same chirality [9]. This defines the overall stereochemistry at this center and, when coupled with the observation that addition of hydrogen to $\Delta^{24(28)}$ sterols in *C. paspali* occurs in a *trans* manner [Arigoni, D., personal communication], defines the loss of this hydrogen as occurring on the face opposite to the original alkylation.

The stereochemically refined mechanism that emerges for yeast from the composite of these experiments (Figure 17.8) involves initial transfer of the methyl to the *si*-face of a Δ^{24} sterol. This is followed by migration of the C-24 hydrogen across the opposite (*re*) face to C-25. A slight rotation (60°) to align the C-24 hydrogen with the empty *p* orbital at C-25 is required. The loss of hydrogen at C-28 then occurs from the *re*-face [9]. As Arigoni has pointed out, this process involves passage of the C-24 hydrogen near to where one would expect the base

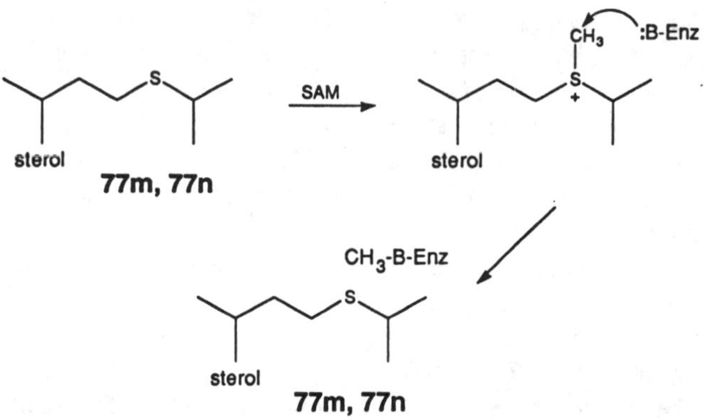

SAM

sterol 77m, 77n → sterol (CH_3 :B-Enz) → CH_3-B-Enz, sterol 77m, 77n

Figure 17.9. Presumed mechanism of 24-SMT inhibition by 24-thiasterols.

abstracting the hydrogen from C-28 to be located [9]. This mechanism requires some rotation of the isopropyl during the C-24 to C-25 hydrogen migration and suggests that this region could accommodate larger structures. It is interesting to speculate on the location of the carbonium-stabilizing groups in this region. It appears that the same nucleophilic site could stabilize both C-24 and C-25 carbonium intermediates. This would lead one to expect that cationic inhibitors should be effective if their positive charge is located within the region normally spanned by these positions.

The kinetics of interaction of the substrate analogs in Table 17.4 with yeast and maize 24-SMTs have revealed distinctly different patterns of inhibition. Thioethers **77m** and **77n** are competitive inhibitors with respect to sterol and uncompetitive with respect to SAM in *Candida* transferase [10]. These thioethers irreversibly inactivate this system by alkylation of the enzyme (Figure 17.9) [10].

In the same system **77b** and **77l** were parabolic competitive inhibitors *vs* sterol and parabolic noncompetitive inhibitors *vs* SAM. This is consistent with the binding of two equivalents of these substrate analogs to the enzyme interrupting the ordered binding of sterol and SAM. It was envisioned that one equivalent of inhibitor binds to the sterol binding site while the second binds to the SAM site [10].

Azasterol **77b** exhibits a different pattern of inhibition with *S. cerevisiae* methyltransferase. In this system, it is uncompetitive with respect to sterol and competitive with respect to SAM [68]. This is most consistent with sequential binding of the substrates with SAM being the first to bind. Rahier *et al.* [78] observed the related 25-azacycloartenol to be uncompetitive with respect to both substrates in the maize methyltransferase. A related pattern of inhibition was found for **77q** and **77r** in the *Candida* methyltransferase. Both of these charged inhibitors were noncompetitive with respect to both substrates [10]. When their ability to bind to the enzyme while it was complexed with the product of methyl transfer, SAH, was examined, it was found that the binding of the latter and these inhibitors to the enzyme is synergistic. Parallel experiments with **77m** revealed mutually exclusive binding with SAH [10]. Thus, the positively charged inhibitors form a ternary complex with SAH while the uncharged analogs do not.

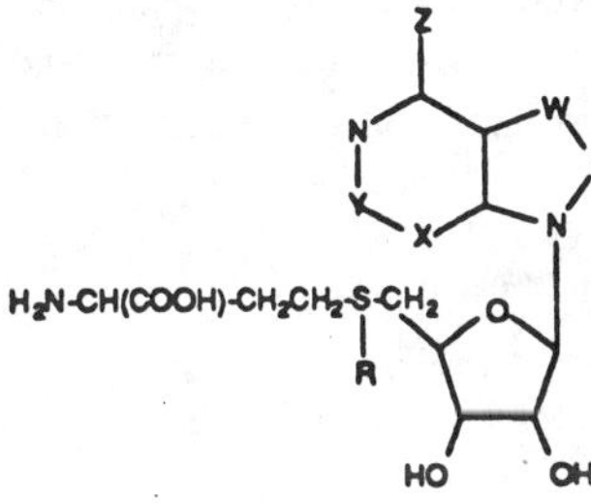

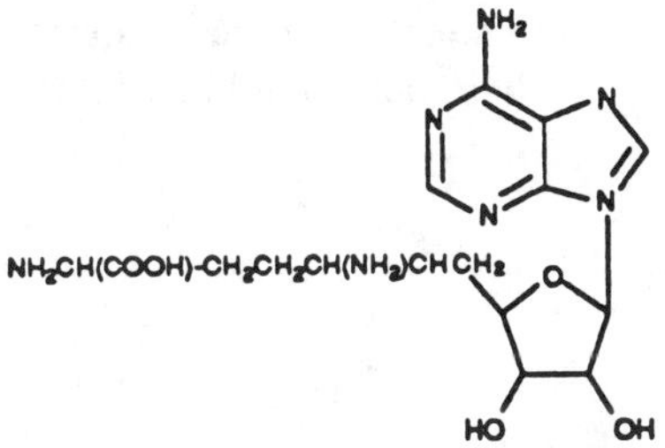

		R	V	W	X	Y	Z
78a	SAM	CH_3	CH	N	N	CH	NH_2
78b	SAH		CH	N	N	CH	NH_2
78c	8-Aza-SAH		N				
78d	N6-Methyl-SAH						$NHCH_3$
78e	2-Aza-SAH					N	
78f	SUH				A		

78g

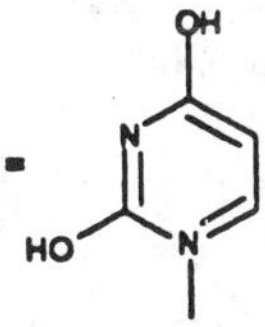

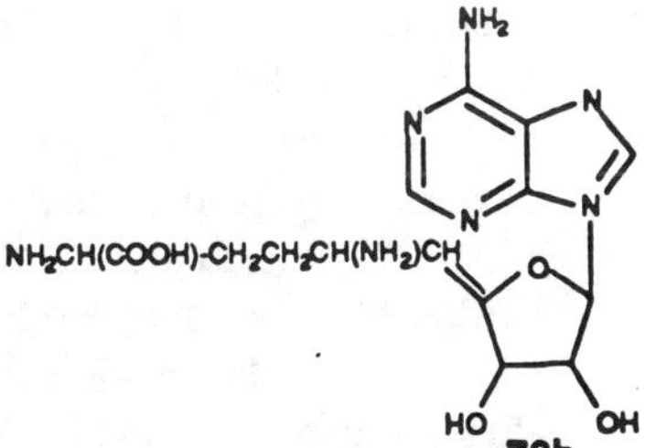

78h

Transfer of the S-methyl of SAM (**78a**) generates S-adenosylhomocysteine (SAH, **78b**), which shows product inhibition to 24-SMT in yeast [55]. Parks has examined the inhibitory activity of a variety of analogs of SAH varied either in the base residue or in the amino acid residue [55]. The most potent inhibitors among the base-modified analogs were **78b, 78c, 78d, 78e,** and **78f**. All these analogs exhibited I_{50}s less than 50 μM. Within the amino-acid-modified series examined, only antifungal metabolites isolated from *Streptomyces griseolus* [55], **78g** and **78h**, stand out as clearly potent ($K_i \approx 1.6$ μM). These studies reveal the most promising approach to inhibitors of 24-SMTs by using SAM analogs will be the design of structures containing the adenosyl group with amino acid modifications that allow normal binding of the amine and carboxyl. The key to efficient binding of **78g** and **78h** seems to be the location of an amino function near the sight normally occupied by the sulfonium moiety of SAM. This suggests that amino acid modifications containing positively charged groups in this region would be excellent candidates for study.

A final clue to the design of new 24-SMT inhibitors lies in the kinetics of

Saccharomyces [59, 68] and *Candida* [10] methyltransferases that reveal that methyl transfer does not involve the intermediacy of a methylated enzyme. Since direct methyl transfer occurs in these systems, a ternary complex between the methyltransferase, SAM, and sterol must be formed. This deduction, coupled with the indication that a positive charge is required in one substrate partner for formation of a ternary complex, suggests that a rational inhibitor design should include both sterol and SAM components with a charge on one such as in **78i**.

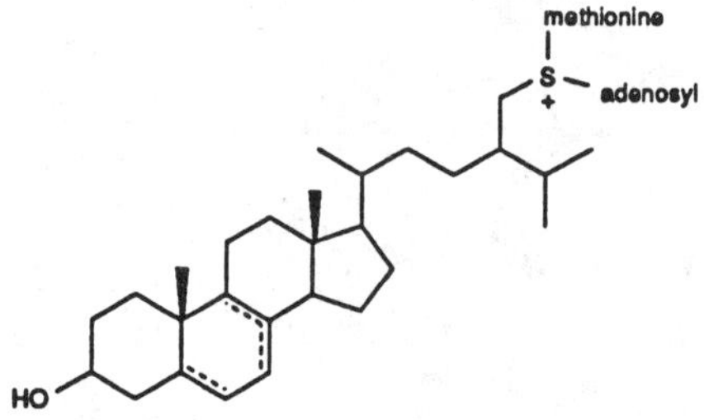

VIII. Cycloeucalenol—Obtusifoliol Isomerase

Cyclization of squalene epoxide in nonphotosynthetic eukaryotic organisms such as fungi and animals produces lanosterol (**2**) whereas in photosynthetic organisms such as algae and higher plants this process leads to cycloartenol (**11**). Further elaboration of the latter to the major plant sterols, stigmasterol (**22**) and β-sitosterol (**21**), requires cleavage of the cyclopropane ring of **11**. This occurs after removal of a methyl group from C-4 and alkylation of C-24 [73]. The enzyme converts **13** to its Δ^8 isomer, obtusifoliol (**14**), by a *cis* regiospecific elimination of the 9β,19-cyclopropane bond and the 8β-hydrogen. It is unlikely that these processes are simultaneous and generation of a C-8 carbonium ion intermediate is presumed.

Benveniste has shown that the 8-aza-decalols (**63**, I_{50} = 0.025 μM) are potent inhibitors of this enzyme in maize, even though the nitrogen is located at a skeletal position adjacent to that at which the positive charge is expected to be generated [99]. The carbocyclic analog of **63** was not inhibitory [79, 99, 100]. Comparison of the inhibitory strength of **63** with derivatives not containing an alkyl group on the nitrogen reveals that increasing hydrophobicity increases inhibitory activity [99].

IX. $\Delta^8 \rightarrow \Delta^7$ Sterol Isomerase

This enzyme moves the Δ^8 double bond, formed during the cyclization and rearrangement of squalene epoxide to lanosterol (**2**), to the Δ^7 position (Figure 17.10). This migration is a preferred but not obligatory prerequisite to introduction of Δ^5 unsaturation in fungi, mammals, and plants [73]. The isomerization is not reversible in the case of rat liver and higher plants but was demonstrated to be reversible in the yeast *S. cerevisiae* [66, 73].

In plants, the isomerase prefers substrates with only $\Delta^{8(9)}$ unsaturation (**17**). Those with $\Delta^{8(14)}$ and $\Delta^{5,8(9)}$ unsaturation are not substrates, even though isomerization of the latter would yield a thermodynamically stable $\Delta^{5,7}$-conjugated diene [98]. The preferred isomerase substrates in yeast and plants contain 24(28)-methylene groups. If this remote function is missing, substrate conversion by the yeast system is appreciably slower [110]. A point of difference between the yeast and plant isomerases is in the effect of 4,14-alkylation on substrate acceptability. While isomerases from both sources will not operate on sterols containing 14α-methyls, the plant enzyme prefers substrates containing a 4α-methyl [98]. This substituent dramatically lowers substrate acceptance for the yeast system [66].

In all systems, this isomerase is presumed to operate by a mechanism (Figure 17.10) similar to that described for isopentenyl pyrophosphate:dimethylallyl pyrophosphate isomerase. It is postulated that initial addition of a proton to the α-face of C-9 in a Δ^8 sterol (**6, 13,** or **17**) gives a stabilized carbonium ion (**79**) [99] which is converted to the Δ^7 sterol by removal of the 7β (rat liver and higher plants) or the 7α (yeast)-hydrogen [73].

Two systemic morpholine-based fungicides, tridemorph, fenpropimorph, widely used on crops, are known inhibitors of this enzyme [98]. They are postulated to be mechanism-based inhibitors which interact with the enzyme in their protonated form as mimics of the presumptive carbonium ion intermediate, **79**. This view is supported by the observation that the inhibitory activity of fenpropimorph (pKa = 7.5) increased by a factor of 20 as the pH was increased from 6 to 8.5 [98]. The inhibitory power of its N-methyl derivative (a quaternary ammonium ion) changed only slightly in this pH range.

Benveniste has shown that 8-aza-decalols such as **63** are excellent inhibitors of $\Delta^8 \rightarrow \Delta^7$ sterol isomerase in maize (I_{50} = 0.2μM) [79, 99, 100]. These compounds are also presumed to act due to their resemblance of the C-8 carbonium ion (**79**) produced during the isomerization. If the nitrogen at position 8 is omitted from the structure, no inhibitory activity is observed [79].

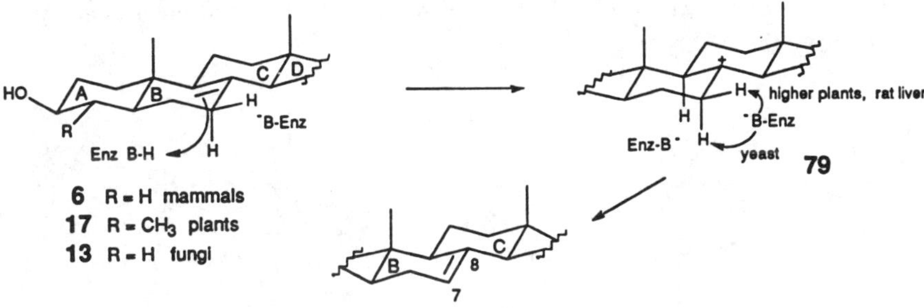

Figure 17.10. Proposed mechanism for $\Delta^8 \rightarrow \Delta^7$ sterol isomerase.

Since these inhibitors were specifically constructed with structures to match the A and B ring of the preferred substrate of plant $\Delta^8 \rightarrow \Delta^7$ isomerases, which contain a 4α-methyl, the analogs lacking this alkyl appendage would be worthy of investigation as selective inhibitors of the corresponding fungal isomerase.

Reversible inhibitors of $\Delta^8 \rightarrow \Delta^7$ sterol isomerase based on the concept of internal stabilization or ion collapse have not been investigated. The first strategy involves placement of heteroatoms capable of lone-pair donation adjacent to the position acquiring the positive charge as in **80** or **81**. Alternatively, a strategically placed silicon might be expected to lead to fragmentation of the intermediate as in **82**.

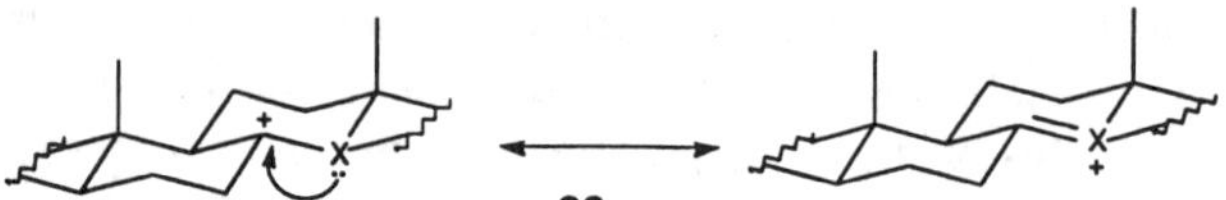

80

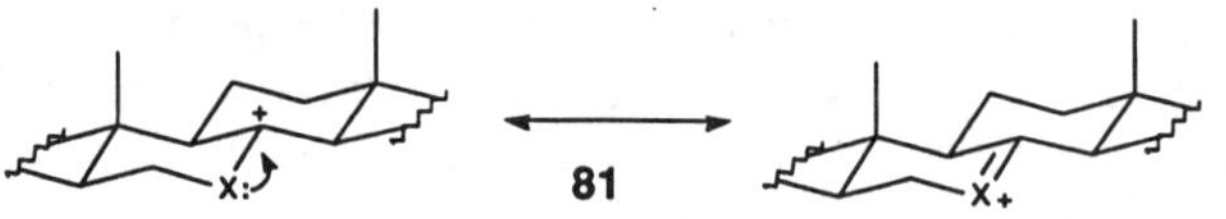

81

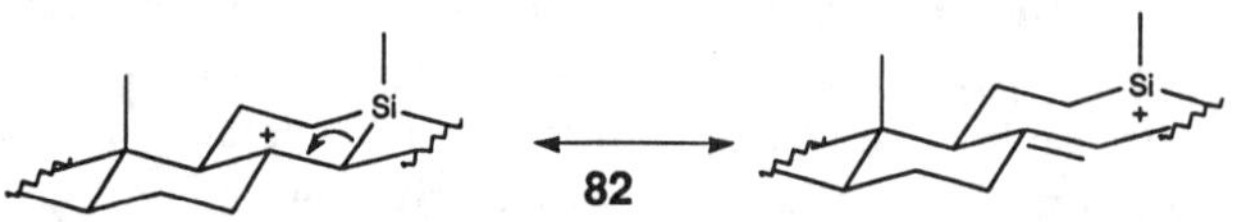

82

The success of epoxides and allylic fluorides as irreversible inhibitors of isopentenyl pyrophosphate:dimethylallyl pyrophosphate isomerase [61, 62, 76, 80], coupled with the different stereochemistry for removal of the hydrogen at C-7 for yeast $\Delta^8 \rightarrow \Delta^7$ sterol isomerase (7α) compared to mammalian and plant systems (7β) [73], suggests design of specific inhibitors of yeast isomerase might be possible. Thus, one could envision that **83** or **84** might interact with the base responsible for removal of the 7α hydrogen in yeast, while the epimers of these potential inhibitors would be selective for the mammalian and plant isomerases that remove the 7β hydrogen.

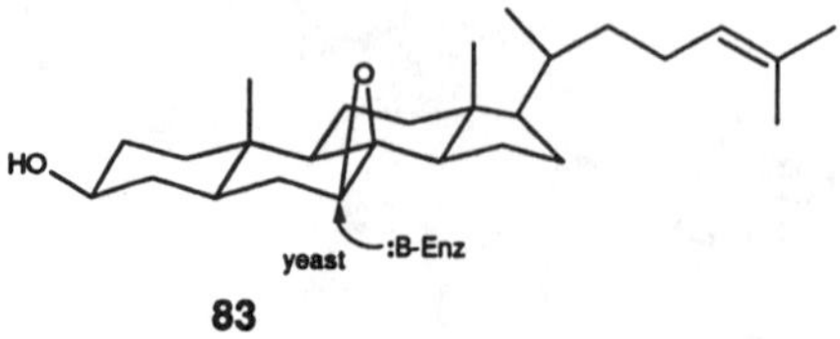

83

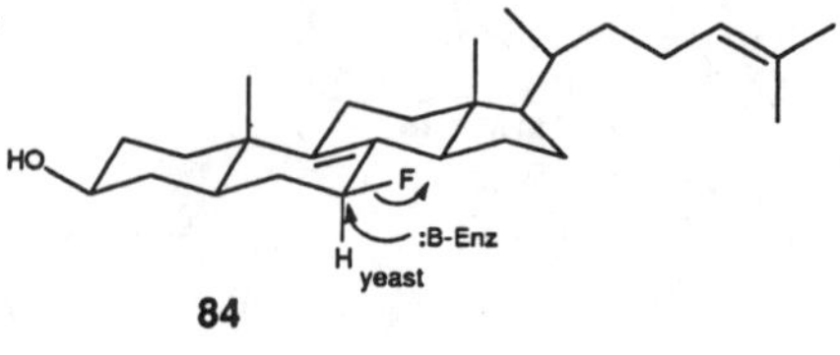

84

X. Later Transformations

The foregoing enzymatic processes yield Δ^7 sterols carrying nuclear and side-chain methylene appendages that vary with the species. All sterol-producing organisms convert Δ^7 unsaturation to Δ^5 or $\Delta^{5,7}$ unsaturation. Because of the late involvement of these transformations, as well as the reduction of $\Delta^{24(28)}$ and Δ^{22} unsaturation, these enzymes have not been regarded as prime targets for inhibition. Sterols containing nitrogen at positions C-24 (**77f**) and C-23 (**77g**) in the side chain are inhibitors of $\Delta^{24(28)}$-reductase (I_{50}= 0.12 μM and I_{50}=0.7 μM, respectively) in yeast [67]. These could act in their unprotonated form as substrate analogs or in their protonated form as intermediate mimics.

Removal of the methyl groups at C-4 is also an important step in the generation of sterols that fulfill architectural and hormonal roles. These enzymes require molecular oxygen but are not cytochrome P–450-dependent [58]. Inhibition of this enzyme by rationally designed mimics of substrates or intermediates has not been reported. By analogy with the inhibitors of SE and 14-DM one would suspect that sterols containing C-4 methylene groups carrying heteroatoms or those in which C-4 has been replaced by a heteroatom would be logical candidates.

XI. Conclusions

There has been a rapid expansion in the search for rationally designed inhibitors of most enzymes of therapeutic interest and those mediating sterol biosynthesis are no exception. The best chances for development of clinically and agriculturally useful sterol biosynthesis inhibitors appears to lie in the inhibition of enzymes biosynthetically near to the initial sterol-forming cyclization. The two enzymes that seem to offer the most promise for highly selective inhibition are 14-DM and 24-SMT. Of these, the latter is clearly a preferable target for the development of antifungal agents affecting mammals since they lack this enzyme. The work to date shows that very efficient ($K_i \approx 2$ nM) inhibition of yeast 24-SMT is possible through mimics of both the sterol [Oehlschlager, A. C. *et al.*, unpublished results] and S-adenosylmethionine substrates [55].

Many commercial azole fungicides are inhibitors of 14-DM. While these are probably mechanism based, they are not rationally designed. The difficulty of preparing C-32 and C–15-substituted sterols that would be expected to be good 14-DM inhibitors has slowed work on this system, and efficient, rationally designed inhibitors have only recently been reported.

Acknowledgments

We thank the Natural Sciences and Research Council of Canada and the British Columbia Health Care Research Foundation for financial support.

References

1. Agnew, W.S. and Popjak, G. 1978. Squalene synthetase. Stoichiometry and kinetics of presqualene pyrophosphate and squalene synthesis by yeast microsomes. J. Biol. Chem. **253**:4566–4573.

2. Agnew, W.S. 1985. Squalene synthetase. Methods Enzymol., **110**:359–373.

3. Akhtar, M., Alexander, K., Boar, R.B., McGhie, J.F. and Barton, D.H.R. 1978. Chemical and enzymatic studies on the characterization of intermediates during the removal of the 14α-methyl group in cholesterol biosynthesis. Use of 32-functionalized lanosterol derivatives. Biochem. J. **169**:449–463.

4. Altman, L.J., Kowerski, R.C. and Laungani, F.R. 1978. Studies in terpene biosynthesis. Synthesis and resolution of presqualene and prephytoene alcohols. J. Am. Chem. Soc. **100**:6174–6182.

5. Aoyama, Y. and Yoshida, Y. 1978. Interaction of lanosterol to cytochrome P–450 purified from yeast microsomes: evidence for contribution of cytochrome P–450 to lanosterol metabolism. Biochem. Biophys. Res. Commun. **82**:33–38.

6. Aoyama, Y., Yoshida, Y., Sonoda, Y. and Sato, Y. 1987. Metabolism of 32-hydroxyl-24,25-didydrolanosterol by purified cytochrome P–$450_{14\text{-DM}}$ from yeast. Evidence for contribution of the cytochrome to whole process of lanosterol 14α-demethylation. J. Biol. Chem. **262**:1239–1243.

7. Aoyama, Y., Yoshida, Y., Nishino, T., Katsuki, H., Maitra, U.S., Mohan, V.P. and Sprinson, D.B. 1987. Isolation and characterization of an altered cytochrome P–450 from a yeast mutant defective in lanosterol 14α-demethylation. J. Biol. Chem. **262**:14260–14264.

8. Aoyama, Y., Yoshida, Y., Sonoda, Y. and Sato, Y. 1989. Deformylation of 32-oxo-24,25-dihydrolanosterol by the purified cytochrome P–$450_{14\text{-DM}}$ (lanosterol 14α-demethylase) from yeast. Evidence confirming the intermediate step of lanosterol 14α-demethylation. J. Biol. Chem. **264**:18502–18505.

9. Arigoni, D. 1978. Stereochemical studies of enzymic C-methylations. Ciba Found. Symp. **60**:243–261.

10. Ator, M.A., Schmidt, J., Adams, J.L. and Dolle, R.E. 1989. Mechanism and inhibition of Δ^{24}-sterol methyltransferase from *Candida albicans* and *Candida tropicalis*. Biochemistry **28**:9633–9640.

11. Avruch, L., Fisher, S., Pierce, H.D., Jr. and Oehlschlager, A.C. 1976. The induced biosynthesis of 7-dehydrocholesterols in yeast: potential sources of new provitamin D_3 analogs. Can. J. Biochem. **54**:657–665.

12. Bach, T.J. 1985. Selected natural and synthetic enzyme inhibitors of sterol biosynthesis as molecular probes for *in vivo* studies concerning the regulation of plant growth. Plant Sci., **39**:183–187.

13. Bach, T.J. 1988. Current trends in the development of sterol biosynthesis inhibitors: early aspects of the pathway. J. Am. Oil. Chem. Soc. **65**:591–595.

14. Baldwin, B.C. 1983. Fungicidal inhibitors of ergosterol biosynthesis. Biochem. Soc. Trans. **11**:659–663.

15. Benveniste, P. 1986. Sterol biosynthesis. Ann. Rev. Plant Physiol. **37**:275–308.

16. Bertolino, A., Altman, L.J., Vasak, J. and Rilling, H.C., 1978. Polyisoprenoid amphiphilic compounds as inhibitors of squalene synthesis and other microsomal enzymes. Biochim. Biophys. Acta **530**:17–23.

17. Billes, S.A., Foster, C., Gordon, E.M., Harrity, T., Scott, W.A. and Ciosek, C.P., 1988. Isoprenoid (phosphinylmethyl)phosphonates as inhibitors of squalene synthetase. J. Med. Chem. **31**:1869–1871.

18. Bloch, K., Corey, E.J., Dean, P.D.G. and Ortiz de Montellano, P.R. 1967. A soluble 2,3-oxidosqualene cyclase. J. Biol. Chem. **242**:3014–3019.

19. Bottema, C.K. and Parks, L.W. 1978. Δ^{14}-Sterol reductase in *Saccharomyces cerevisiae*. Biochem. Biophys. Acta **531**:301–307.

20. Bujons, J., Guajards, R. and Kyle, K.S. 1988. Enantioselective enzymatic sterol synthesis by using ultrasonically stimulated bakers' yeast. J. Am. Chem. Soc. **110**:604–606 for leading references.

21. Burden, R.S., Cooke, D.T. and Carter, G.A. 1989. Inhibitors of sterol biosynthesis and growth in plants and fungi. Phytochem. **28**:1791–1804.

22. Buttke, T.M., Brint, S.L. and Lowe, M.R. 1988. Regulation of squalene epoxidase activity by membrane fatty acid composition in yeast. Lipids **23**:68–71.

23. Capstack, E., Jr., Rosin, N., Blondin, G.A. and Nes, W.R. 1965. Squalene in *Pisum sativum*. Its cyclization to β-amyrin and labelling pattern. J. Biol. Chem. **240**:3258–3263.

24. Caras, I.W. and Bloch, K. 1979. Effects of a supernatant protein activator on microsomal squalene-2,3-oxide-lanosterol cyclase. J. Biol. Chem. **254**:11816–11821.

25. Cattel, L., Ceruti, M., Viola, F., Delprino, L., Balliano, G., Duriatti, A. and Bouvier-Nave, P. 1986. The squalene-2,3-epoxide cyclase as a model for the development of new drugs. Lipids **21**:31–38.

26. Ceruti, M., Viola, F., Grosa, G., Balliano, G., Delprino, L. and Cattel, L. 1988. Synthesis of squalenoid acetylenes and allenes, as inhibitors of squalene epoxidase. J. Chem. Res. (S):18–19.

27. Ceruti, M., Viola, F., Balliano, G., Grosa, G. Caputo, O., Gerst, N., Schuber, F. and Cattel, L. 1988. Synthesis of a squalenoid oxaziridine and other new classes of squalene derivatives as inhibitors of sterol biosynthesis. Eur. J. Med. Chem. **23**:533–537.

28. Ceruti, M., Viola, F., Dosis, F., Cattel, L., Bouvier-Nave, P. and Ugliengo, P. 1988. Stereospecific synthesis of squalenoid epoxide vinyl ethers as inhibitors of 2,3-oxidosqualene cyclase. J. Chem. Soc. Perkin Trans. I:461–469.

29. Chin, J. and Bloch, K. 1984. Role of supernatant protein factor and anionic phospholipid in squalene uptake and conversion by microsomes. J. Biol. Chem. **259**:11735–11738.

30. Coates, R.M. and Robinson, W.H., 1971. Stereoselective total synthesis of (±)-presqualene alcohol. J. Am. Chem. Soc. **93**:1785–1786.

31. Corey, E.J. and Gross, S.K. 1967. Formation of sterols by the action of 2,3-oxidosqualene-sterol cyclase on the facticious substrates 2,3:22,23-dioxidosqualene and 2,3-oxido-22,23-dihydrosqualene. J. Am. Chem. Soc. **89**:4561–4562.
32. Corey, E.J., Lin, K. and Jautelat, .M. 1968. Studies on the action of 2,3-oxidosqualene-sterol cyclase on unnatural substrates produced by alkylidene transfer from sulfonium alkylides to 4,8,13,17,21-pentamethyldocosa-4,8,12,16,20-pentaenal. J. Am. Chem. Soc. **90**:2724–2726.
33. Corey, E.J. and Volante, R.P., 1976. Application of unreactive analogs of terpenoid pyrophosphates to studies of multistep biosynthesis. Demonstration that "presqualene pyrophosphate" is an essential intermediate on the path to squalene. J. Am. Chem. Soc. **98**:1291–1293.
34. Duriatti, A., Bouvier-Nave, P., Benveniste, P., Schuber, F., Delprins, L., Balliano, G. and Cattel, L. 1985. *In vitro* inhibition of animal and higher plant 2,3-oxidosqualene-sterol cyclases by 2-aza-2,3-dehydrosqualene and derivatives, and by other ammonium-containing molecules. Biochem. Pharmacol. **34**:2765–2777.
35. Duriatti, A. and Schuber, F. 1988. Partial purification of 2,3-oxidosqualene-lanosterol cyclase from hog liver. Evidence for a functional thiol residue. Biochem. Biophys. Res. Commun. **151**:1378–1385.
36. Eilenberg, H. and Shechter, I. 1984. A possible regulatory role of squalene epoxidase in chinese hamster ovary cells. Lipids. **19**:539–543.
37. Eilenberg, H. and Shechter, I. 1987. Regulation of squalene epoxidase activity and comparison of catalytic properties of rat liver and chinese hamster ovary cell-derived enzymes. J. Lipid Res. **28**:1398–1404.
38. Eilenberg, H., Kliner, E., Przedecki, F. and Shechter, I. 1989. Inactivation and activation of various membranal enzymes of the cholesterol biosynthetic pathway of digitonin. J. Lipid. Res. **30**:1127–1135.
39. Friedlander, E.J., Caras, I.W., Leu Fen Hou Lin and Bloch, K. 1980. Supernatant protein factor facilitates intermembrane transfer of squalene. J. Biol. Chem. **255**:8042–8045.
40. Fryberg. M., Avruch, L., Oehlschlager, A.C. and Unrau, A.M. 1975. Nuclear demethylation and C-24 alkylation during ergosterol biosynthesis in *Saccharomyces cerevisiae*. Can. J. Biochem. **53**:881–889.
41. Frye, L.L. and Robinson, C.H. 1988. Novel inhibitors of lanosterol 14α-methyl demethylase, a critical enzyme in cholesterol biosynthesis. J. Chem. Soc. Chem. Commun.:129–131.
42. Gadher, P. Mercer, E.I., Baldwin, B.C. and Wiggins, T.C. 1983. A comparison of the potency of some fungicides as inhibitors of sterol 14α-demethylation. Pestic. Biochem. Physiol. **19**:1–10.
43. Gelb, M.H., Heimbrook, D.C., Malkonen, P. and Sligar, S.G. 1982. Stereochemistry and deuterium isotope effects in camphor hydroxylation by the cytochrome P–450 monoxygenase system. Biochem. **21**:370–377.
44. Gerst, N., Schuber, F., Viola, F. and Cattel, L. 1986. Inhibition of cholesterol biosynthesis in 3T3 fibroblasts by 2-aza-2,3-dihydrosqualene, a rationally designed 2,3-oxidosqualene cyclase inhibitor. Biochem. Pharmacol. **35**:4243–4250.

45. Gerst, N., Duriatti, A., Schuber, F., Taton, M., Benveniste, P. and Rahier, A. 1988. Potent inhibition of cholesterol biosynthesis in 3T3 fibroblasts by N-[(1,5,9)-trimethyldecyl]-4α, 10-dimethyl-8-aza-*trans*-decal-3β-ol, a new 2,3-oxidosqualene cyclase inhibitor. Biochem. Pharmacol. **37**:1955–1964.
46. Gibbons, G.F., Pullinger, C.R. and Mikropaulos, K.A. 1979. Studies on the mechanism of 14α-demethylation; a requirement for two distinct types of mixed function oxidases. Biochem. J. **183**:309–315.
47. Haughan, P.A., Lenton, J.R. and Goad, L.J. 1988. Sterol requirements and paclobutrazol inhibition of a celery cell culture. Phytochem. **27**:2491–2500.
48. Jahnke, L. and Kein, H.P. 1983. Oxygen requirements for formation and activity of the squalene epoxidase in *Saccharomyces cerevisiae*. J. Bacteriol. **155**:488–492.
49. Johnson, W.S., Telfer, S.J., Cheng, S. and Schubert, U. 1987. Cation-stabilizing auxiliaries: A new concept in biomimetic polyene cyclizations. J. Am. Chem. Soc. **109**:2517–2518.
50. Koller, W. (1987). Isomers of sterol synthesis inhibitors: fungicidal effects and plant growth regulator activities. Pestic. Sci., **18**:129–147.
51. Koller, W. and Scheinpflung, H. 1987. Fungal resistance to sterol biosynthesis inhibitors: a new challenge. Plant Disease **71**:1066–1074.
52. Kuswik-Rabiega, G. and Rilling, H.C. 1987. Squalene synthetase—Solubilization and partial purification of squalene synthetase—Copurification of presqualene pyrophosphate and squalene synthetase activity. J. Biol. Chem. **262**:1505–1509.
53. Lederer, E. 1977. Biological C-alkylation reactions, p. 89–126. In Salvatore, F., Borek, E., Zappia, V., Williams-Ashman, H.G. and Schlenk F., Eds. Biochemistry of S-adenosylmethinine. Columbia University Press, New York.
54. M'Baya, B. and Karst, F. 1987. *In vitro* assay of squalene epoxidase of *Saccharomyces cerevisiae*. Biochem Biophys. Res. Comm. **147**:555–564.
55. McCammon, M. and Parks, L.W. 1981. Inhibition of sterol transmethylation by S-adenosylhomocysteine analogs. J. Bacteriol. **145**:106–112.
56. Medina, J.C. and Kyler, K.S. 1988. Enzymatic cyclization of hydroxylated surrogate squalenoids with bakers' yeast. J. Am. Chem. Soc., **110**:4818–4821.
57. Medina, J.C., Guajardo, R. and Kyler, K.S. 1989. Vinyl group rearrangement in the enzymatic cyclization of squalenoids: Synthesis of 30-oxysterols. J. Am. Chem. Soc. **111**:2310–2311.
58. Mercer, E.J. 1984. The biosynthesis of ergosterol. Pestic. Sci.**15**:133–155.
59. Moore, J.T., Jr. and Gaylor, J.L. 1969. Isolation and purification of an S-adenosylmethionine:Δ^{24}-sterolmethyltransferase from yeast. J. Biol. Chem. **244**:6334–6340.
60. Moore, J.T., Jr. and Gaylor, J.L. 1970. Investigation of an S-adenosylmethionine: Δ^{24}-sterolmethyltransferase in ergosterol biosynthesis in yeast. Specificity of sterol substrates and inhibitors. J. Biol. Chem. **245**:4684–4688.
61. Muehlbacher, M. and Poulter, C.D. 1985. Isopentenyl diphosphate:dimethylallyl diphosphate isomerase. Irreversible inhibition of the enzyme by active-site-directed covalent attachment. J. Am. Chem. Soc. **107**:8307–8308.

62. Muehlbacher, M. and Poulter, C.D. 1988. Isopentenyl diphosphate isomerase: inactivation of the enzyme with active-site-directed irreversible inhibitors and transition-state analogues. Biochem. **27**:7315–7328.

63. Nakamura, M. and Sato, R. 1979. The roles of soluble factors in squalene epoxidation by rat liver microsomes. Biochem. Biophys. Res. Commun. **89**:900–906.

64. Narula, A.S. Rahier, A., Benveniste, P. and Schuber, F. 1981. 24-Methyl-25-azacycloartanol, an analogue of a carbonium ion high-energy intermediate, is a potent inhibitor of (S)-adenosyl-L-methionine:sterol C–24-methyltransferase in higher plant cells. J. Am. Chem. Soc. **103**:2408–2409.

65. Nes, W.D., Hanners, P.K. and Parish, E.J. 1986. Control of fungal sterol C-24 transmethylation: Importance to developmental regulation. Biochem. Biophys. Res. Commun. **139**:410–415.

66. Oehlschlager, A.C., Fryberg, M. and Unrau, A.M. 1973. Biosynthesis of ergosterol in yeast. Evidence for multiple pathways. J. Am. Chem. Soc. **95**:5747–5757.

67. Oehlschlager, A.C. Pierce, H.D., Jr., Pierce, A.M., Angus, R.H., Quantin-Martenot, E., Unrau, A.M. and Srinivasan, R. 1980. The use of mutants and azasterols in studies of yeast steroid biosynthesis, p. 395–405. In Mazliak, P., Benveniste, P., Costes, C. and Douce, R., Eds. Biogenesis and function of plant lipids. Elsevier/Holland Biomedical Press, New York.

68. Oehlschlager, A.C., Anugs, R.H., Pierce, A.M., Pierce, H.D., Jr. and Srinivasan, R. 1984. Azasterol inhibition of Δ^{24}-sterol methyltransferase in *Saccharomyces cerevisiae*. Biochemistry. **23**:3582–3589.

69. Ono, T. and Bloch, K. 1975. Solubilization and partial characterization of rat liver squalene epoxidase. J. Biol. Chem. **250**:1571–1579.

70. Ortiz de Montellano, P.R., Castillo, R., Vinson, W. and Wei, J.S. 1976. Squalene biosynthesis. Role of the 3-methyl group in farnesyl pyrophosphate. J. Am. Chem. Soc. **98**:3020–3021.

71. Petranyl, G., Ryder, N.S. and Stütz, A. 1984. Allylamine derivatives: new class of synthetic antifungal agents inhibiting fungal squalene epoxidase. Science **224**:1239–1241.

72. Popjak, G. and Agnew, W.S., 1978, Squalene synthetase. Mol. and Cell. Biochem. **27**:97–116.

73. Porter, J.W. and Spurgeon, S.L., Eds. 1981. Biosynthesis of isoprenoid compounds. Vol. 1., John Wiley & Sons, New York.

74. Poulter, C.D., Musico, O.J. and Goodfellow, R.J. 1974. Biosynthesis of head-to-head terpenes. Carbonium ion rearrangements which lead to head-to-head terpenes. Biochemistry **13**:1530–1537.

75. Poulter, C.D., Capson, T.L., Thompson, M.D. and Bard, R.S., 1989. Squalene synthetase. Inhibition by ammonium analogs of carbocationic intermediates in the conversion of presqualene diphosphate to squalene, J. Am. Chem. Soc. **111**:3734–3739.

76. Poulter, C.D., Muehlbacher, M. and Davis, D.R. 1989. Mechanism of active-site-directed irreversible inhibition by 3-(fluoromethyl)-3-butenyl diphosphate. J. Am. Chem. Soc. **111**:3740–3742.

77. Rahier, A., Genot, J.C., Narula, A.S., Beneveniste, P. and Schmitt, P. 1980. 25-Azacycloartanol, a potent inhibitor of S-adenosyl-L-methionine-sterol C-24 and C-28 methyltransferases in higher plant cells. Biochem. Biophys. Res. Commun. **92**:20–25.

78. Rahier, A., Genot, J.C., Schuber, F., Benveniste, P. and Narula, A.S. 1984. Inhibition of S-adenosyl-L-methionine sterol-C–24-methyltransferase by analogues of a carbocationic ion high-energy intermediate. Structure activity relationships for C-25 heteroatom (N, As, S) substituted triterpenoid derivatives. J. Biol. Chem. **259**:15215–15223.

79. Rahier, A., Taton, M., Schmitt, P., Benveniste, P., Place, P. and Anding, C. 1985. Inhibition of $\Delta^8 \rightarrow \Delta^7$-sterol isomerase and of cycloeucalenol-obtusifoliol isomerase by N-benzyl-8-aza-4α, 10-dimethyl-*trans*-decal–3β-ol, an analogue of carbocationic high energy intermediate. Phytochem. **24**:1223–1232.

80. Reardon, J.E. and Abeles, R.H. 1986. Mechanism of action of isopentenyl pyrophosphate isomerase: evidence for a carbonium ion intermediate. Biochem. **25**:5609–5616.

81. Ruhl, K.K., Anzalone, L., Arguropoulos, E.D., Gayen, A.K. and Spencer, T.A. 1989. Azadecalin analogs of 4,4,10β-trimethyl-*trans*-decal-3β-ol. Synthesis and assay as inhibitors of oxidosqualene cyclase. Bioorg. Chem. **17**:108–120.

82. Ryder, N.S. and Dupont, M-C. 1984. Properties of a particulate squalene epoxidase from *Candida albicans*. Biochem. Biophys. Acta **794**:466–471.

83. Ryder, N.S. and Dupont, M.-C. 1985. Inhibition of squalene epoxidase by allylamine antimycotic compounds. A comparative study of the fungal and mammalian enzymes. Biochem. J. **230**:765–770.

84. Ryder, N.S., Dupont, M.-C. and Frank, I. 1986. Inhibition of fungal and mammalian sterol biosynthesis by 2-aza-2,3-dihydrosqualene. FEBS Letters **204**:239–242.

85. Sasiak, K. and Rilling, H.C., 1988. Purification to homogeneity and some properties of squalene synthetase. Arch. Biochem. Biophys. **260**:622–627.

86. Schenkman, J.B., Remmer, H. and Estabrook, R.W. 1967. Spectral studies of drug interaction with hepatic microsomal cytochrome (rat). Mol. Pharmacol. **3**:113–123.

87. Schmitt, P., Scheid, F. and Benveniste, P. 1980. Accumulation of $\Delta^{8,14}$ sterols in suspension cultures of bramble cells cultured with an azasterol antimycotic agent (A25822B). Phytochem. **19**:525–530.

88. Schmitt, P., Narula, A.S., Benveniste, P. and Rahier, A. 1981. Manipulation by 25-azacycloartanol of the relative percentage of C-10, C-9 and C-8 side-chain sterols in suspension cultures of bramble cells. Phytochem. **20**:197–201.

89. Schmitt, P., Gonzales, R., Benveniste, P., Ceruti, M. and Cattel, L. 1987. Inhibition of sterol biosynthesis and accumulation of 2,3-oxidosqualene in bramble cell suspension cultures treated with 2-aza-2,3-dihydrosqualene and 2-aza-2,3-dehydrosqualene-N-oxide. Phytochem. **26**:2709–2714.

90. Sen, S.E. and Prestwich, G.D. 1989. Squalene analogs containing isopropylidene mimic as potential inhibitors of pig liver squalene epoxidase and oxidosqualene cyclase. J. Med. Chem. **32**:2152–2158.

91. Sen, S.E. and Prestwich, G.D. 1989. Trisnorsqualene alcohol, a potent inhibitor of vertebrate squalene epoxidase. J. Am. Chem. Soc. **111**:1508–1510.

92. Sen, S.E. and Prestwich, G.D. 1989. Trisnorsqualene cyclopropylamine: a reversible tight-binding inhibitor of squalene epoxidase. J. Am. Chem. Soc. **111**:8761–8762.

93. Silverman, R.B. 1988, Mechanism-based enzyme inactivation: chemistry and enzymology. Vol. 2. p. 89–190, CRC Press, Boca Raton, Florida.

94. Srikantalah, M.V., Hansbury, E., Loughran, E.D. and Scallen, T.J. 1976. Purification and properties of sterol carrier protein. J. Biol. Chem. **251**:5496–5504.

95. Stevenson, D.E., Wright, J.N. and Akhtar, M. 1988. Mechanistic consideration of P–450 dependent enzymic reactions: studies on oestriol biosynthesis. J. Chem. Soc. Perkin Trans. I:2043–2052.

96. Stutz, A. 1987. Allylamine derivatives—a new class of active substances in antifungal chemotherapy. Angew. Chem. Int'l. Ed. Engl. **26**:320–328.

97. Tal, B. and Nes, W.D. 1987. Regulation of sterol biosynthesis: Importance of the C-24 alkyl group to growth of sunflower suspension cultures. Plant Physiol. Suppl. **83**:969.

98. Taton, M. Benveniste, P. and Rahier, A. 1987. Mechanism of inhibition of sterol biosynthesis enzymes by N-substituted morpholines. Pestic. Sci. **21**:269–280.

99. Taton, M., Benveniste, P. and Rahier, A. 1987. Use of rationally designed inhibitors to study sterol and triterpenoid biosynthesis. Pure and Appl. Chem. **59**:287–294.

100. Taton, M., Benveniste, P. and Rahier, A. 1987. Comparative study of the inhibition of sterol biosynthesis in *Rubus fruticosus* suspension cultures and *Zea mays* seedlings by N-(1,5,9-trimethyldecyl)-4α,10-dimethyl-8aza-*trans*-decal-3β-ol and derivatives. Phytochem. **26**:385–392.

101. van Tamelen, E.E., Sharpless, K.B., Willett, J.D., Clayton, R.B. and Burlingame, A.L. 1967. Biological activities of some terminally modified squalene and squalene-2,3-oxide analogs. J. Am. Chem. Soc. **89**, 3920–3922.

102. van Tamelen, E.E., Sharpless, K.B., Hanzlik, R., Clayton, R.B., Burlingame, A.L. and Wszolek, P.C. 1967. Enzymatic cyclization of *trans, trans, trans*-18,19-dihydrosqualene-2,3-oxide. J. Am. Chem. Soc. **89**:7150–7151.

103. van Tamelen, E.E. and Schwartz, M.A., 1971. Mechanism of presqualene pyrophosphate–squalene biosynthesis. J. Am. Chem. Soc. **93**:1780–1782.

104. van Tamelen, E.E., Lees, R.G. and Grieder, A. 1974. Cyclization of a terpenoid diene with preformed A-B-D rings and its significance for the mechanism of terpenoid terminal epoxide cyclizations. J. Am. Chem. Soc. **96**:2255–2256.

105. van Tamelen, E.E. and James, D.R. 1977. Overall mechanism of terpenoid terminal epoxide polycyclization. J. Am. Chem. Soc. **99**:950–952.

106. van Tamelen, E.E. and Hopla, R.E. 1979. Generation of the onocerin system by lanosterol 2,3-oxidosqualene cyclase. Implications for the cyclization process. J. Am. Chem. Soc. **101**:6112–6114.

107. van Tamelen, E.E., Leopold, E.J., Marson, S.A. and Walspe, H.R. 1982. Action of 2,3-oxidosqualene lanosterol cyclase on 15′-nor-18,19-dihydro-2,3-oxidosqualene. J. Am. Chem. Soc. **104**:6479–6480.

108. Walsh, C. 1979. Enzymatic reaction mechanisms, p. 358–431, W.H. Freeman and Co., San Francisco.

109. White, R.E., Miller, J.P., Favreau L.V. and Battacharyya, A. 1986. Stereochemical dynamics of aliphatic hydroxylation by cytochrome P–450. J. Am. Chem. Soc. **108**:6024–6031.

110. Yabusaki, Y., Nishino, T., Ariga, N. and Katsuki, H. 1979. Studies on $\Delta^8 \rightarrow \Delta^7$ isomerization and methyl transfer of sterols in ergosterol biosynthesis in yeast. J. Biochem. **85**:1531–1537.

111. Yoshida, Y., Kumaoka, H. and Sato, R. 1974. Studies on the microsomal electron-transport system of anaerobically grown yeast. I. Intracellular localization and characterization. J. Biochem. (Tokyo) **75**:1201–1210.

112. Yoshida, Y., Aoyama, Y., Kumaoka, H. and Kubota, S. 1977. A highly purified preparation of cytochrome P–450 from microsomes of anaerobically grown yeast. Biochem. Biophys. Res. Commun. **78**:1005–1010.

113. Yoshida, Y. and Aoyama, Y. 1984. Yeast cytochrome P–450 catalyzing lanosterol 14α-demetylation. I. Purification and spectral properties. J. Biol. Chem. **259**:1655–1660.

114. Yoshida, Y.M., Aoyama, Y., Takano, H. and Kato, T. 1986. Stereoselective interaction of enantiomers of diniconazole, a fungicide, with purified P–450/14-DM from yeast. Biochem. Biophys. Res. Commun. **137**:513–519.

18

Chitin Synthase as a Chemotherapeutic Target

N.H. Georgopapadakou

I. Introduction

The fungal cell wall is a major and essential fungal structure, entirely lacking in animal cells, and thus an attractive chemotherapeutic target. Its biosynthesis is central to cell division and morphogenesis. Like its bacterial counterpart, the fungal cell wall determines cell shape, provides rigidity and osmotic support, and plays an important role in adherence and pathogenicity (73, 86). It may, in addition, constitute a permeability barrier (113).

The fungal cell wall is a multilayer structure composed mostly of carbohydrates, either free (unattached) or covalently linked to proteins. Its composition varies from species to species and, in some cases (like in *Histoplasma capsulatum* and *Cryptococcus neoformans*), it extends to form a capsule. Nonetheless, it consists of specific components: carbohydrate polymers of N-acetylglucosamine (chitin, chitosan) and glucose (cellulose, glucans), and mannoproteins. The former two provide the structural scaffolding for mannoproteins. Fungal cell walls are covered in several excellent reviews (19, 25, 27, 45), while glucan synthesis is specifically covered in chapter 19 of this volume.

The present review focuses on chitin synthesis in two fungi: the innocuous but extensively studied baker's yeast, *Saccharomyces cerevisiae*, and the pathogenic but less well-studied *Candida albicans*. The two organisms vary greatly in their DNA relatedness and ploidy; *S. cerevisiae* is haploid while *C. albicans* is diploid (94, 98). Nevertheless, studies of *S. cerevisiae* have provided much of our knowledge of the molecular mechanisms involved in chitin synthesis, and sufficient similarities exist between the two organisms for the *S. cerevisiae* system to be a model for *C. albicans*.

A. Structure and Distribution of Chitin

Chitin is a linear homopolymer of β–1,4-linked N-acetyl-D-glucosamine (GlcNAc) (Figure 18.1). It was the first cell wall component to be discovered in

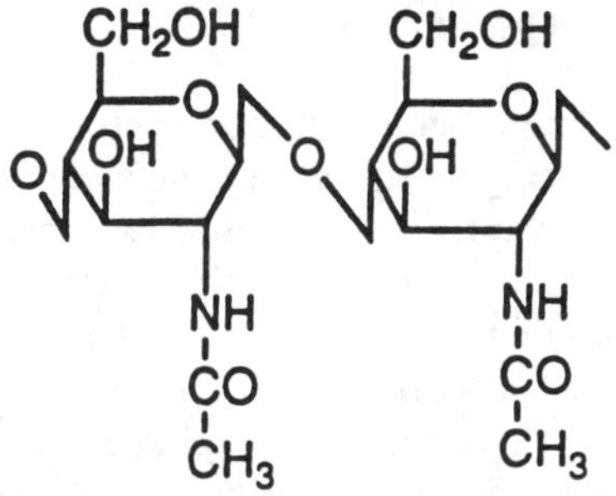

Figure 18.1. The structure of GlcNAc dimer, the stereochemical repeating unit in chitin.

fungi (51) and represents the most abundant aminopolysaccharide in nature, being also present in the exoskeleton of insects and crustaceans (20). There are three distinct forms of chitin (α, β, γ) by X-ray crystallography, based on differences in alignment and packing of the poly-GlcNAc chains. In α-chitin, the most stable of the three and the form found in fungi and arthropods, the chains are antiparallel. They are stabilized by intra- and interchain hydrogen bonds; the former between the C-3 hydroxyl group of each sugar and the pyranose ring oxygen of the next, and the latter between the keto and amino groups of the acetamido moiety in adjacent chains (74, 85). Thus, the *stereochemical* repeating unit in α-chitin is the GlcNAc dimer (diacetylchitobiose).

Chitosan is an extensively deacetylated form of chitin found in some fungal cell walls (7); deacetylation probably occurs at the polymer stage. Partial deacetylation ($\leq$15% of the GlcNAc monomers) also occurs in chitin (11), although some could be the result of extraction procedures.

In *S. cerevisiae* and the yeast form of *C. albicans*, chitin is a minor ($<$3%) component, being localized primarily in the septal region and the bud scars (23, 111). Chitin is more abundant in the mycelial form of *C. albicans*, where it is localized in the apical region in the hyphae (13). There is also wall chitin, associated with glucan (76, 99, 104).

Chitin is synthesized on the cytoplasmic surface of the plasma membrane and is extruded perpendicularly to the cell surface, through layers of other cell wall components (mannoproteins, glucan) (24), as microfibrils. The chitin microfibrils ($\sim$100-mers in *S. cerevisiae* (64)) subsequently form crystalline chitin outside the cell through extensive hydrogen bonding (108).

Chitin (noncrystalline) is hydrolyzed to GlcNAc dimers through the action of chitinases, 30- to 50-kDa enzymes analogous to autolysins in bacteria, involved in cell separation (28). The enzyme from *S. cerevisiae* is periplasmic (40), as is probably the enzyme from *C. albicans* (5). Although these enzymes are beyond the scope of this review, it is noted that they are strongly inhibited (IC_{50}, 0.3 μM for *C. albicans* chitinase) by the aza-oligosaccharide antibiotic allosamidin (35).

B. Chitin Synthesis and Fungal Physiology

Chitin synthesis is an important event in the cell cycle of *S. cerevisiae*, and presumably other fungi, being under both spatial and temporal control. A burst of chitin synthesis occurs shortly before bud emergence; a ring of chitin is laid down at the presumptive bud emergence site (55). However, bud emergence can occur in the absence of chitin ring formation (23) and aberrant septa can be made away from the mother-bud junction (101). A second burst of chitin synthesis, occurring at the end of mitosis, deposits the primary septum in the mother-cell side, in the area defined by the chitin ring. After cell separation, the chitinous septum remains with the mother cell and becomes a bud scar. Chitin can be easily visualized in cells by the intense blue fluorescence produced after staining with Calcofluor White M2R (λ_{ex}=365 nm, λ_{em}=420 nm; 39). The dye interacts specifically with polysaccharides having contiguous (1 $\rightarrow$ 4) β-linked D-glucopyranosyl residues such as chitin and cellulose (β(1 $\rightarrow$ 4) glucan) (110). If added during cell growth, it affects the assembly of chitin microfibrils (41) and causes overproduction of chitin (90).

Several cell-cycle mutants of *S. cerevisiae* (*cdc*28, *cdc*24, *cdc*7, *cdc*4, *cdc*3) arrested in the G1 phase (the gap before DNA synthesis, critical for controlling cell division) overproduce chitin, inserting it over the entire cell surface (87). Chitin overproduction has been suggested to be due to chitin synthase activation rather than induction, at least in the *cdc*24 mutants (32). α-Factor binding to MATa cells of *S. cerevisiae*, which arrests cells in the G1 phase (103), causes localized chitin over-production (93). Interestingly, some ergosterol synthesis inhibitors cause both G1 arrest (70) and delocalized chitin overproduction (49, 107) (Georgopapadakou and Bertasso, manuscript in preparation), consistent with the suggestion (87) that an increase in chitin synthesis may be a nonspecific response to cell-cycle arrest. Nevertheless, the reverse has also been postulated, on the basis of membrane fluidity changes brought about by ergosterol depletion (107).

II. Chitin Synthase

Yeast chitin synthases (EC 2.4.1.16) are membrane-bound enzymes, with their catalytic site facing the cytoplasm (37), usually isolated largely in an inactive (zymogenic) form. The major (*in vitro*) chitin synthase (Chs1) from *S. cerevisiae* is activated by an enzyme preparation from yeast (activating factor) (26, 106) or exogenously by trypsin (38). The action of the endogenous activating factor can be blocked with a heat-stable protein from yeast, while that of trypsin can be blocked with soybean inhibitor. Chitin synthase activity occurring in the absence of exogenous proteolytic activation is referred to as basal activity.

For purification and assays, chitin synthase is usually solubilized with digitonin (14, 38).

Digitonin has also been used to permeabilize fungal cells and assay chitin synthase *in situ* (46, 50, 52). Organic solvents, such as dimethyl sulfoxide (50), toluene (36), toluene/ethanol (72, 97), and toluene/ethanol/Triton X–100 (84), have also been used.

A. *Preparation and Assays*

1. S. Cerevisiae *Enzyme*

For assays, a particulate fraction from spheroplasts of log-phase cells, prepared with zymolyase or β-glucuronidase, is used (18, 38, 43) (Figure 18.2). Alternatively, a particulate fraction may be obtained from cells disrupted mechanically with glass beads (46). After treatment with trypsin and trypsin inhibitor (excessive protease treatment reduces activity), the chitin synthase preparation (primarily, Chs1) is incubated with UDP-[^{14}C]GlcNAc. The reaction is stopped with aqueous trichloroacetic acid (TCA) and the resulting TCA-insoluble chitin is separated from the TCA-soluble substrate and counted. Activity is expressed as incorporation of GlcNAc into chitin either in nmoles/min/10^9 protoplasts or μmoles/min/mg protein.

In the particulate form, chitin synthase is stable in 33% glycerol at −20°C for at least 2 months. It has a pH optimum of 6.2 and a K_m for UDP-GlcNAc of 0.9

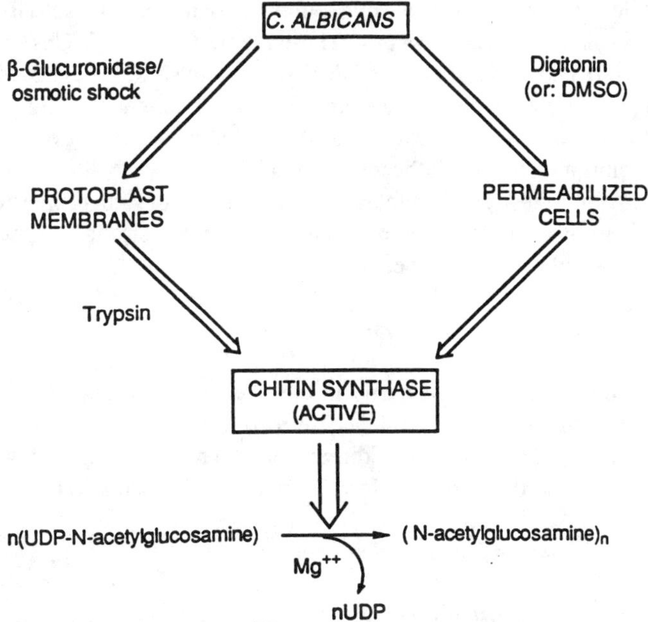

Figure 18.2. Chitin synthase assay.

mM. The enzyme requires divalent cations (1–20 mM Mg^{2+} or 1 mM Mn^{2+}) for activity and is stimulated three- to five fold by GlcNAc (K_a, 4.7 mM). Purified preparations of Chs1 require, in addition, phosphatidylserine and digitonin for activity (64).

Digitonin-solubized chitin synthase from particulate fractions (protoplasts or broken cells) requires exogenous (trypsin) activation, while chitin synthase in digitonin-permeabilized cells does not (46). In the latter case, proteolytic activation probably occurs endogenously, as protease inhibitors decrease it.

Chs1, the major *in vitro* but dispensable *in vivo* chitin synthase, is the only enzyme purified so far (64). It is a ~60-kDa protein by SDS gel analysis (130 kDa based on nucleic acid sequence) but a 570-kDa protein in the undenatured form, indicating an oligomer. It is attractive to speculate that the putative subunits form a pore through which the linear chitin polymer is extruded during synthesis. The enzyme does not require primer for GlcNAc polymerization (64) and acts processively rather than randomly; that is, it completes a chitin microfibril before starting a new one.

2. C. Albicans *Enzyme*

For assays, a particulate form of the enzyme, prepared from spheroplasts of log-phase yeast cells and probably representing the major form, Chs1, is used (13, 50, 54). Activity varied with growth phase (6) and strain in eight strains examined (Georgopapadakou and Smith, unpublished results). Activity is stable in 33% glycerol for at least 2 weeks. The enzyme (primarily, Chs1) has a pH optimum of 8.0 and a K_m for UDP-GlcNAc of 1 mM. Like the *S. cerevisiae* enzyme, it is stimulated by magnesium (5 mM for maximal stimulation) and GlcNAc (30 mM for maximal stimulation). A molecular weight of 400 kDa by Sepharose column analysis has been reported (14).

Digitonin-solubilized chitin synthase retains most of the properties of the particulate form of the enzyme, except that the former is strongly stimulated (up to 20-fold) by some organic solvents (50).

B. Cellular Location: the Chitosome Controversy

Chitin synthase is located primarily in the plasma membrane (37, 63). However, chitin synthase-enriched vesicles, termed "chitosomes," have also been isolated (8, 68, 71). Chitosomes could represent the transport form of the enzyme or plasma membrane domains that form vesicles during isolation.

C. The Nature of the Endogenous Activation

Early on, it was suggested (106) that the endogenous activating factor might be proteinase B, a serine protease. This opened an obvious chemotherapeutic opportunity, as serine proteases are probably the best-characterized enzymes and serine protease inhibitors are readily available from other programs. Unfortunately, further studies have indicated that proteinase B is not responsible for the endogenous activation. Mutants lacking proteinase B made normal amounts of chitin (114), as did mutants lacking vacuoles. (Proteases are thought to be mostly, though not exclusively (42), sequestered in vacuoles.) Later on, KEX2 protease, a calcium-requiring neutral protease (105), appeared to be an attractive candidate for activation, but again experiments showed this not to be the case (E. Cabib, personal communication).

Recently, the demonstration of three chitin synthases in *S. cerevisiae* and *C. albicans*, with two components (Chs2 and Chs3, the latter nonzymogenic) being essential, has complicated matters. The nonessential Chs1, relegated to the role of repair enzyme, is not activated by proteinase B, as its deletion does not cause lysis of daughter cells in unbuffered media (Cabib, personal communication). Less attention has been given to related regulatory issue, the termination of septal chitin synthesis. It could involve deactivation of chitin synthase, triggering of a chitinase, or some other process. At any rate, the prospects of the endogenous proteolytic activation as a chemotherapeutic target have dimmed considerably after these developments.

D. Enzyme Multiplicity and other Properties

The presence of multiple, genetically distinct forms of chitin synthase in *S. cerevisiae* was revealed when the structural gene of Chs1 was cloned and disrupted with no effect on chitin content or cell viability (17). Chs1 has repair function *in vivo*; daughter cells of *chs1* mutants lyse in unbuffered media (28). Further studies led to the identification of the structural gene for a second chitin synthase, Chs2 (81, 92). Cloning and disruption of this structural gene, CHS2, indicated that Chs2 is essential under certain conditions (100), while nonessential under others (16). Some of the results obtained are reminiscent of earlier findings with glucosamine auxotrophs of *S. cerevisiae* (4). A third chitin synthase activity, Chs3, was detected in a double-disruption mutant lacking both Chs1 and Chs2 (9, 16). The three chitin synthases may function in cell wall synthesis (Chs3), septum synthesis (Chs2), and repair (Chs1), with Chs3 and Chs2 being able to compensate for each other.

Three chitin synthases have also been found in *C. albicans*. The structural gene for Chs1 of *C. albicans* was isolated by heterologous expression in *S. cerevisiae* (3). Two additional chitin synthase genes were detected by the polymerase chain reaction by using as primers small, conserved regions of *S. cerevisiae* CHS1 and CHS2 and *C. albicans* CHS1 genes (30). In another study with wild-type *C. albicans*, two types of chitin synthase activity were detected: a major one, stimulated by trypsin and inhibited by cobalt, and a minor one, unaffected by trypsin or cobalt (Georgopapadakou and Bertasso, unpublished results).

E. Effects of Antifungals

Amphotericin, nystatin, 5-fluorocytosine, haloprogin, and aculeacin had no effect in particulate chitin synthase preparations (predominantly, Chs1) from *C. albicans* spheroplasts up to 1 mM (Georgopapadakou and Smith, unpublished results). Miconazole inhibited synthesis by ~50% at 1 mM.

Particulate chitin synthase preparations (most likely, Chs1) from *C. albicans* grown in the presence of ergosterol-inhibiting concentrations of azoles, allylamines, morpholines, or 15-azasterol showed increased activity (Georgopapadakou and Bertasso, manuscript in preparation). This chitin synthase stimulation may be secondary to membrane fluidity changes resulting from bulk ergosterol depletion (see also ref. 82). The compounds also caused deposition of chitin over the entire cell surface of growing *C. albicans*, which was detected as bright fluorescent patches after staining with Calcofluor. Increased chitin deposition was also observed with ketoconazole in *S. cerevisiae* lacking Chs1, suggesting stimulation of the physiologically relevant Chs2 or Chs3. Since increased chitin deposition can arise in *S. cerevisiae* from cell-cycle arrest (87), the effect of ketoconazole and other ergosterol synthesis inhibitors on chitin synthesis *in vivo* is probably secondary to cell-cycle arrest rather than the reverse, as had been previously proposed (107).

III. Chitin Synthase Inhibitors

A. Polyoxins, Nikkomycins, and Related Compounds

Polyoxins and nikkomycins are closely related nucleoside-peptide antibiotics produced by *Streptomyces cacaoi* (61, 62) and *Streptomyces tendae* (31), respectively. They have been used exclusively as agricultural fungicides, as they are virtually inactive against human pathogens such as *C. albicans* and *Aspergillus fumigatus*. Nevertheless, in millimolar concentrations, they produce morphological changes in *C. albicans* (swelling, cell chains) reminiscent of those produced in bacteria by β-lactams (10). Nikkomycins X and Z have been recently found to be active against the dimorphic fungi *Coccidioides immitis* and *Blastomyces dermatidis* in a mouse infection model, with nikkomycin Z being the most active compound (57).

Both polyoxins and nikkomycins are thought to be analogs of the UDP-GlcNAc substrate (Figure 18.3), although their binding is stronger, with K_is in the 0.1–1 μM range. Attempts at establishing structure-activity relationships have been met with limited success (58, 59, 78). The essential structural features for inhibition are shown in Figure 18.4.

The related 10381A1 (2) and valclavam (89) do not inhibit chitin synthase in permeabilized *C. albicans* up to 1 mM. With the latter compound, which contains a β-lactam nucleus found to inhibit some serine proteases (47, 80), no inhibition

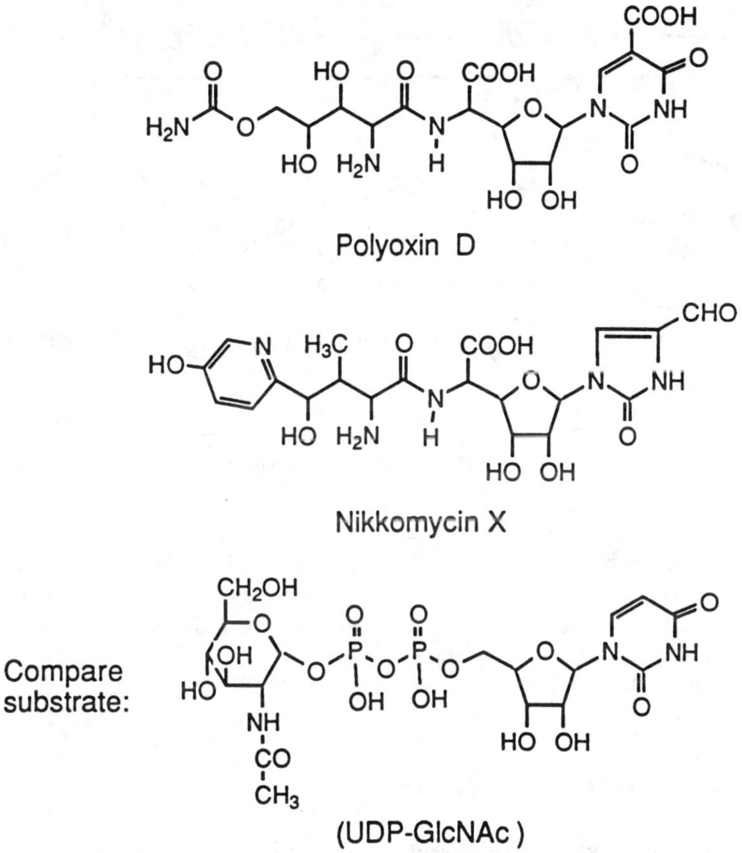

Polyoxin D

Nikkomycin X

Compare substrate:

(UDP-GlcNAc)

Figure 18.3. Structures of chitin synthase inhibitors and the UDP-GlcNAc substrate.

of the endogenous proteolytic activation was observed in permeabilized cells up 1 mM (Georgopapadakou and Bertasso, unpublished results). Arginosuccinate, which also resembles part of the basic polyoxin structure, did not inhibit chitin synthase in permeabilized *C. albicans* up to 4 mM (Georgopapadakou and Smith, unpublished results).

The transport and thus antifungal activity of polyoxins and nikkomycins is antagonized by peptides (48, 75, 112). Both *S. cerevisiae* and *C. albicans* are resistant to polyoxins, though chitin synthesis is inhibited *in vitro*. In the past 10 years, several efforts to attach amino acids to the basic polyoxin structure and thereby increase transport in *C. albicans* have been made. Of the resulting synthetic derivatives, tripeptidyl polyoxins had modest activity against *C. albicans* (50% growth inhibition at 250 μM), but substantially reduced activity against chitin synthase (IC_{50}s $\geqslant$ 100 μM) (79). Dipeptidyl polyoxins containing α-amino fatty acids were more promising, inhibiting growth of *C. albicans* at

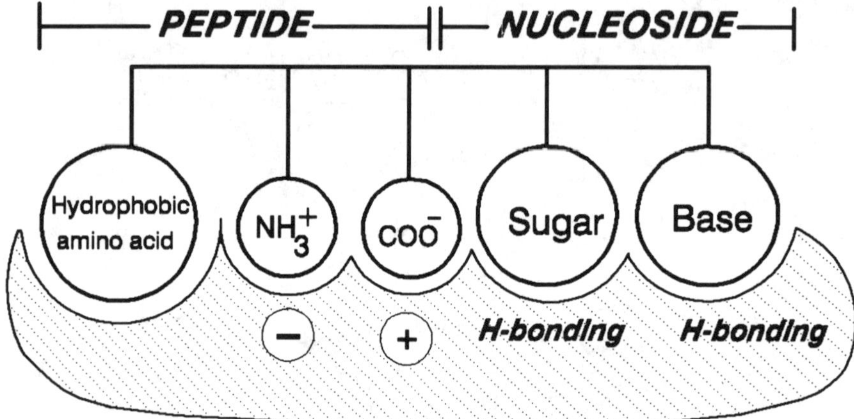

Figure 18.4. Minimal scheme of the interaction of polyoxins and nikkomycins with the chitin synthase active site.

~50 μM and chitin synthase (IC_{50} of best compound, 0.5 μM) (65). Nikkomycin derivatives produced by mutasynthesis were active against the chitin synthase of the toadstool, *Coprinus cinereus* (34), and had some antifungal activity (12), though not against medically important fungi (33).

B. Other Inhibitors

1. Lipids and Related Compounds

As noted earlier, phosphatidylserine stimulates activity in isolated *S. cerevisiae* Chs1. However, it weakly inhibits activity in digitonin-treated *C. albicans* cells (IC_{50}=1 mM). None of the other phospholipids tested (phosphatidylethanolamine, -glycerol, -inositol, cardiolipin) had any effect up to 1 mM (Georgopapadakou and Smith, unpublished results).

Of the fatty acids, *cis* unsaturated fatty acids (linoleic, oleic, palmitic) inhibited *C. albicans* chitin synthase at ~1 mM, while *trans* unsaturated (*t*-vaccenic, elaidic) and saturated (palmitic, stearic) fatty acids were without effect.

The antibiotic citrinin inhibited chitin synthase activity in permeabilized *C. albicans* also with an IC_{50} of 1 mM, as did nalidixic acid and some other amphiphilic antibiotics (Georgopapadakou and Bertasso, unpublished results).

Avermectin, a lipophilic macrocyclic lactone widely used as an antiparasitic agent and active against the fungus *Mucor miehei*, inhibited chitin synthase from that organism (IC_{50}, 5–50 μM) (29). However, its mechanism of inhibition of chitin synthase is obscure (it may act on the membrane) and is inactive against fungal pathogens.

The above results are not surprising, considering the fact that chitin synthase is membrane bound. None of the compounds appears to be a promising lead for nonpolyoxin inhibitors, despite the specificity for *cis* fatty acids.

2. *Halogenated Aromatic Compounds*

Pentachloronitrobenzene inhibited *Agaricus bisporus* chitin synthase at 1 μM (88), but was inactive toward *C. albicans* preparations up to 1 mM (Georgopapadakou and Bertasso, unpublished results). This result highlights the danger of screening compounds with chitin synthase preparations from nonpathogens. Hexachlorophene inhibited chitin synthase from *C. cinereus* with a K_i of 8 μM, but inhibition was nonspecific, as it was reversed by phosphatidylcholine (83). Other halogenated aromatic compounds inhibited the activation of chitin synthase (15).

3. *Inhibitors of the Insect Enzyme*

Several fungicides and insecticides were screened for inhibition of chitin synthesis in *Phycomyces* and cockroach (69). Generally, they had higher activity in the cockroach system, although little is known about either system to provide a conceptual framework for the findings. Fungal and insect chitin synthases may be structurally and mechanistically very different enzymes. Further, polyoxin D was less active than halogenated aromatic compounds in the *Phycomyces* system, casting doubt as to the relevance of the findings to fungal pathogens.

C. *Screening for Inhibitors*

A primary screen for chitin synthase inhibitors involving *Neurospora crassa* spheroplasts (67, 95, 96) is described in chapter 21. A concern when using this system is its relevance to human pathogens. Also, chitin synthesis in protoplasts is independent of the cell cycle (45), and may thus represent activity of a nonphysiological chitin synthase. Another screen, involving incorporation of UDP-[^{14}C]-GlcNAc in DMSO-permeabilized cells of *C. albicans*, has been used in the author's laboratory. A drawback of this assay is that it may measure a nonphysiological chitin synthase, and thus overlook specific inhibitors of the physiologically relevant form. Mutants producing exclusively Chs2 or Chs3 are obviously valuable for screening for compounds with therapeutic potential.

IV. Conclusions and Future Direction

While chitin is one of the most abundant polymers on earth, its importance in the physiology of major pathogenic fungi has been recognized relatively recently (1). The relation between chitin synthesis and cell morphology and growth has not been analyzed sufficiently for distinct chemotherapeutic opportunities to emerge. At the biochemical level, the coupling of chitin synthesis to extrusion through the cell envelope *in vivo* may not be mimicked in enzyme studies, and conventional isolation may result in an altered enzyme. Similarly, the presence of polyoxin-inhibitable activity in permeabilized cells, though useful in screening, is not equivalent to an isolated enzyme.

Complicating the situation further is accumulating evidence that there are multiple chitin synthases in fungi, the essential component(s) being only a minor *in vitro* activity. This enzyme multiplicity is reminiscent of the penicillin-sensitive enzymes involved in the biosynthesis of bacterial peptidoglycan, detected in the early 1970s as penicillin-binding proteins. It is recalled that the major and best understood penicillin-sensitive enzymes, the DD-carboxypeptidases, are generally dispensable. A reactive polyoxin derivative, such as azido-nikkomycin Z, could be valuable in detecting polyoxin-binding proteins in fungal pathogens, and eventually assigning physiological functions to individual proteins.

The chitin synthase multiplicity has several practical implications. First, conventional activity assays measure the predominant, though nonessential, chitin synthase (Chs1). This might become an issue when screening for inhibitors of Chs2 and Chs3. Even with substrate analogs, such as polyoxins and nikkomycins, differential activity toward the different chitin synthase forms apparently exists (21). Second, in dimorphic pathogens, such as *C. albicans*, the distribution and cell-cycle fluctuation of the minor but essential chitin synthase activities in the different forms (yeast, hyphal) is unknown. Third, the effect of compounds that stimulate chitin synthase (e.g., azoles) on the minor enzyme form(s) is unclear, although the availability of *S. cerevisiae* mutants expressing only those enzymes makes experimentation straightforward. Clearly much work needs to be done to untangle the puzzle.

In the area of polyoxins/nikkomycins, improved transport is an obvious aim. Since *C. albicans* excretes proteases, nonpeptidic linkages of the side chain should be explored. It is noted that even if improved transport is achieved, the potential for resistance is all too real once antifungal pressure is applied. The potential for transport-associated resistance should be therefore assessed at an early stage.

Difficult, but more rewarding in the long run, may be to discover novel chitin synthase inhibitors not subject to peptide transport constraints. Assaying specifically for the essential enzymes is of course a prerequisite.

Synergy between chitin synthase and glucan synthase inhibitors is a welcome development (56). Unfortunately, there is no synergy with ergosterol synthesis inhibitors or compounds affecting membrane integrity directly, such as amphotericin B. Synergy with established antifungals is obviously desirable.

The importance of chitinase as a chemotherapeutic target has not yet been fully evaluated. Intuitively, it may function as a counterweight to chitin synthase, maintaining cell wall plasticity during growth and cell separation after budding. Consistent with its morphogenetic role, chitinase may be a zymogen, subject to endogenous activation (60). The effect of CHS disruption on chitinase activity will therefore be most interesting. On the other hand, inhibition of chitinase could result in cell-cycle arrest and chitin overproduction. Although cell-cycle arrest has been shown to cause chitin overproduction, the reverse has not yet been unequivocally demonstrated.

Finally, an interesting recent development is the discovery of chitin in *P. carinii* by using Calcofluor staining (66, 109). This opens the possibility of intervention at the level of chitin synthase in this born-again fungus. (See chapter 22.) The obvious next step is to test polyoxins and nikkomycins for anti-*pneumocystis* activity.

Acknowledgments

I wish to thank Sandra Smith and Anne Bertasso for carrying out some of the experiments described, and Joyce Sutcliffe and Enrico Cabib for critically reviewing the manuscript.

References

1. Adams D.J., and G.W. Gooday. 1983. Chitin synthesis as a target—current progress. Abh. Akad. Wiss. DDR Abt. Math. Naturwiss. Tech. (Syst. Fungiz. Antifungale Verbind.) **1983**:39–45.
2. Argoudelis, A., A.L. Laborde, O. Sebek, and S.E. Truesdell. October 1986. World patent 8,605,785.
3. Au-Young, J., and P.W. Robbins. 1990. Isolation of a chitin synthase gene (CHS1) from *Candida albicans* by expression in *Saccharomyces cerevisiae*. Mol. Microbiol. **4**:197–207.
4. Ballou, C.E., S.K. Maitra, J.S. Walker, and W.L. Whelan. 1977. Developmental defects associated with glucosamine auxotrophy in *Saccharomyces cerevisiae*. Proc. Natl. Acad. Sci. USA **74**:4351–4355.
5. Barrett-Bee, K., and M. Hamilton. 1984. The detection and analysis of chitinase activity from the yeast form of *Candida albicans*. J. Gen. Microbiol. **130**:1857–1861.
6. Barrett-Bee, K.J., J. Lees, and W. Henderson. 1982. Variation in the activities of enzymes associated with cell wall metabolism during a growth cycle of *Candida albicans*. FEMS Microbiol. Lett. **15**:275–278.
7. Bartnicki-Garcia, S. 1968. Cell wall chemistry, morphogenesis and taxonomy of fungi. Annu. Rev. Biochem. **22**:87–108.
8. Bartnicki-Garcia, S., C.E. Bracker, E. Reyes, and J. Ruiz-Herrera. 1978. Isolation of chitosomes from taxonomically diverse fungi and synthesis of chitin microfibrils in vitro. Exp. Mycol. **2**:173–192.
9. Baymiller, J., and J. McCullough. 1990. Analysis of a *Saccharomyces cerevisiae* strain which grows without a functional CHS2 gene, abstr. H–133, p. 176. Abstr. 90th Annu. Meet. Am. Soc. Microbiol. 1990.

10. Becker, J.M., N.L. Covert, P. Shenbagamurthi, A.S. Steinfeld, and F. Naider. 1983. Polyoxin D inhibits growth of zoopathogenic fungi. Antimicrob. Agents Chemother. **23**:926–929.

11. Blackwell, J. 1988. Physical methods for determination of chitin structure and conformation. Methods Enzymol. **161**:435–442.

12. Bormann, C., S. Mattern, H. Schrempf, H.-P. Fiedler, and H. Zahner. 1989. Isolation of *Streptomyces tendae* mutants with an altered nikkomycin spectrum. J. Antibiot. **42**:913–918.

13. Braun, P.C., and R.A. Calderone. 1978. Chitin synthesis in *Candida albicans*: comparison of yeast and hyphal forms. *J. Bacteriol.* **133**:1472–1477.

14. Braun, P.C., and R.A. Calderone. 1979. Regulation and solubilization of *Candida albicans* chitin synthetase. J. Bacteriol. **140**:666–670.

15. Brillinger, G.U. 1979. Metabolic products of microorganisms. 181. Chitin synthase from fungi, a test model for substances with insecticidal properties. Arch. Microbiol. **121**:71–74.

16. Bulawa, C.E., and B.C. Osmond. 1990. Chitin synthase I and chitin synthase II are not required for chitin synthesis *in vivo* in *Saccharomyces cerevisiae*. Proc. Natl. Acad. Sci. USA **87**:7424–7428.

17. Bulawa, C.E., M. Slater, E. Cabib, J. Au-Young, A. Sburlati, W. Lee Adair, Jr., and P. Robbins. 1986. The *S. cerevisiae* structural gene for chitin synthase is not required for chitin synthesis in vivo. Cell **46**:213–225.

18. Cabib, E. 1972. Chitin synthetase system from yeast. Methods Enzymol. **28**:572–580.

19. Cabib, E. 1975. Molecular aspects of yeast morphogenesis. Ann. Rev. Microbiol. **29**:191–214.

20. Cabib, E. 1987. The syntheis and degradation of chitin. Adv. Enzymol. **59**:59–101.

21. Cabib, E. 1991. Differential inhibition of chitin synthases 1 and 2 from *Saccharomyces cerevisiae* by polyoxin D and nikkomycins. Antimicrob. Agents Chemother. **35**:170–173.

22. Cabib, E., and B. Bowers. 1971. Chitin and yeast budding. Localization of chitin in yeast bud scars. J. Biol. Chem. **246**:152–159.

23. Cabib, E., and B. Bowers. 1975. Timing and function of chitin synthesis in yeast. J. Bacteriol. **124**:1586–1593.

24. Cabib, E., B. Bowers, and R.L. Roberts. 1983. Vectorial synthesis of a polysaccharide by isolated plasma membranes. Proc. Natl. Acad. Sci. USA **80**:3318–3321.

25. Cabib, E., B. Bowers, A. Sburlati, and S.J. Silverman. 1988. Fungal cell wall synthesis: the construction of a biological structure. Microbiol. Sci. **5**:370–375.

26. Cabib, E., and V. Farkas. 1971. The control of morphogenesis: an enzymatic mechanism for the initiation of septum formation in yeast. Proc. Natl. Acad. Sci. USA **68**:2052–2056.

27. Cabib, E., and R. Roberts. 1982. Synthesis of the yeast cell wall and its regulation. Ann. Rev. Biochem. **51**:763–793.

28. Cabib, E., A. Sburlati, B. Bowers, and S.J. Silverman. 1989. Chitin synthase 1, an auxiliary enzyme for chitin synthesis in *Saccharoymces cerevisiae*. J. Cell. Biol. **108**:1665–1672.
29. Calcott, P.H., and P.O. Fatig. 1984. Inhibition of chitin metabolism by avermectin in susceptible organisms. J. Antibiot. **37**:253–259.
30. Chen-Wu, J., and P.W. Robbins. 1990. Chitin synthase gene expression in *C. albicans*, abstr, S14, p. 3. Second Conference on Candida and Candidiasis: Biology, Pathogenesis, and Management, April 1–4, 1990, Philadelphia.
31. Dahn, U., H. Hagermeier, H. Hohne, W.A. Konig, G. Wolf, and H. Zähner, 1976. Stoffwechselprodukte von Mikroorganismen. 154. Mitteilung. Nikkomycin, ein neuer Hemmstoff der Chitinisynthese bei Pilzen. Arch. Microbiol. **107**:143–160.
32. Dankova, R., and V. Farkas. 1987. Chitin synthase activity and the rate of chitin formation in cell-division cycle mutant *Saccharomyces cerevisiae cdc*24. J. Basic Microbiol. **27**:185–190.
33. Dĕcker, H., F. Walz, C. Bormann, H. Zähner, and H.-P. Fiedler. 1990. Metabolic products of microorganisms. 255. Nikkomycins W^z and W^x, new chitin synthetase inhibitors from *Streptomyces tendae*. J. Antibiot. **43**:43–48.
34. Delzer, J., H.-P. Fiedler, H. Muller, H. Zahner, R. Rathman, K. Ernst, and W.A. Konig. 1984. New nikkomycins by mutasynthesis and directed fermentation. J. Antibiot. **37**:80–82.
35. Dickinson, K., V. Keer, C.A. Hitchcock, and D.J. Adams. 1989. Chitinase activity from *Candida albicans* and its inhibition by allosamidin. J. Gen. Microbiol. **135**:1417–1421.
36. Dominguez, A., M.V. Elorza, J.R. Villanueva, and R. Sentadreu. 1980. Chitin synthase activity in *Saccharomyces cerevisiae*—effect of inhibition of cell-division and of the synthesis of RNA and protein. Curr. Microbiol. **3**:263–266.
37. Duran, A., B. Bowers, and E. Cabib. 1975. Chitin synthetase zymogen is attached to yeast plasma membrane. Proc. Natl. Acad. Sci. USA **72**:3952–3955.
38. Duran, A., and E. Cabib. 1978. Solubilization and partial purification of yeast chitin synthetase. Confirmation of the zymogenic nature of the enzyme. J. Biol. Chem. **253**:4419–4425.
39. Duran, A., E. Cabib, and B. Bowers. 1979. Chitin synthetase distribution on the yeast plasma membrane. Science **203**:363–365.
40. Elango, N., J.U. Correa, and E. Cabib. 1982. Secretory character of a yeast chitinase. J. Biol. Chem. **257**:1398–1400.
41. Elorza, M.V., H. Rico, and R. Sentandreu. 1983. Calcofluor white alters the assembly of chitin fibrils in *Saccharomyces cerevisiae* and *Candida albicans* cells. J. Gen. Microbiol. **129**: 1577–1582.
42. Emter, O., and D.H. Wolf. 1984. Vacuoles are not the sole compartments of proteolytic enzymes in yeast. FEBS Lett. **166**:321–325.
43. Fähnrich, M., and J. Ahlers. 1981. Improved assay and mechanism of the reaction catalyzed by the chitin synthase from *Saccharomyces cerevisiae*. Eur. J. Biochem. **121**:113–118.

44. Farkas, V. 1979. Biosynthesis of cell walls of fungi. Microbiol. Rev. **43**:117–144.

45. Farkas, V., and A. Svodoba. 1980. Kinetics of β-glucan and chitin formation by cells and protoplasts of the yeast *Saccharomyces cerevisiae*. Curr. Microbiol. **4**:99–103.

46. Fernandez, M.P., J.U. Correa, and E. Cabib. 1982. Activation of chitin synthetase in permeabilized cells of a *Saccharomyces cerevisiae* mutant lacking proteinase B. J. Bacteriol. **152**:1255–1264.

47. Frere, J.-M., B. Joris, O. Dideberg, P. Charlier, and J.-M. Ghuysen. 1988. Penicillin-recognizing enzymes. Biochem. Soc. Trans. **16**:934–938.

48. Furter, R., and D.M. Rast. 1985. A comparison of the chitin synthase-inhibitory and antifungal efficacy of nucleoside-peptide antibiotics: structure-activity relationships. FEMS Microbiol. Lett. **28**:205–211.

49. Georgopapadakou, N.H. 1988. Effects of drugs on lipids and membrane integrity of fungi, p. 100–119. *In* G.G. Jackson, H.D. Schlumberger, and H.-J. Zeiler (ed.), Perspectives in antiinfective therapy. Springer-Verlag, Berlin.

50. Georgopapadakou, N.H., and S.A. Smith. 1985. Chitin synthase in *Candida albicans*: comparison of digitonin-permeabilized cells and spheroplast membranes. J. Bacteriol. **162**:826–829.

51. Glaser, L., and D.H. Brown. 1957. The synthesis of chitin in cell free extracts of *Neurospora crassa*. J. Biol. Chem. **228**:729–742.

52. Gooday, G.W., and A. de Rousset-Hall. 1975. Properties of chitin synthetase from *Coprinus cinereus*. J. Gen. Microbiol. **89**:137–145.

53. Gow, N.A.R., G.W. Gooday, R.J. Newsam, and K. Gull. 1980. Ultrastructure of the septum in *Candida albicans*. Curr. Microbiol. **4**:357–359.

54. Hardy, J.C., and G.W. Gooday. 1983. Stability and zymogenic nature of chitin synthase from *Candida albicans*. Curr. Microbiol. **9**:51–54.

55. Hayashibe, M., and S. Katohda. 1973. Initiation of budding and chitin ring. J. Gen. Appl. Microbiol. **19**:23–39.

56. Hector, R.F., and P.C. Braun. 1986. Synergistic action of nikkomycins X and Z with papulacandin B on whole cells and regenerating protoplasts of *Candida albicans*. Antimicrob. Agents Chemother. **29**:389–394.

57. Hector, R.F., B.L. Zimmer, and D. Pappagianis. 1990. Evaluation of Nikkomycins X and Z in murine models of coccidioidomycosis, histoplasmosis, and blastomycosis. Antimicrob. Agents Chemother. **34**:587–593.

58. Hori, M., K. Kakiki, and T. Misato. 1974. Further study on the relation of polyoxin structure to chitin synthetase inhibition. Agric. Biol. Chem. **38**:691–698.

59. Hori, M., K. Kakiki, S. Suzuki, and T. Misato. 1971. Studies on the mode of action of polyoxins. Part III. Relation of polyoxin structure to chitin synthetase inhibition. Agric. Biol. Chem. **35**:1280–1291.

60. Humphreys, A.M., and G.W. Gooday. 1984. Properties of chitinase activities from *Mucor mucedo*: evidence for a membrane-bound zymogenic form. J. Gen. Microbiol. **130**:1359–1366.

61. Isono, K., J. Nagatsu, Y. Kawashima, and S. Suzuki. 1965. Studies on polyoxins, antifungal antibiotics. I. Isolation and characterization of polyoxins A and B. Agric. Biol. Chem. **29**:848–854.

62. Isono, K., K. Asahi, and S. Suzuki. 1969. Studies on polyoxins, antifungal antibiotics. XIII. The structure of polyoxins. J. Amer. Chem. Soc. **91**:7490–7505.

63. Kang, M.S., J. Au-Young, and E. Cabib. 1985. Modification of yeast plasma membrane density by concanavalin A attachment. J. Biol. Chem. **260**:12680–12685.

64. Kang, M.S., N. Elango, E. Mattia, J. Au-Young, P.W. Robbins, and E. Cabib. 1984. Isolation of chitin synthetase from *Saccharomyces cerevisiae*. J. Biol. Chem. **259**:14966–14972.

65. Khare, R.K., J.M. Becker, and F.R. Naider. 1988. Synthesis and anticandidal properties of polyoxin L analogues containing α-amino fatty acids. J. Med. Chem. **31**:650–656.

66. Kim, Y.K., S. Parulekar, P.K.W. Yu, R.J. Pisani, T.F. Smith, and J.P. Anhalt. 1990. Evaluation of calcofluor white stain for detection of *Pneumocystis carinii*. Diagn. Microbiol. Infect. Dis. **13**:307–310.

67. Kirsch, D.R., and Lai, M.H. 1986. A modified screen for the detection of cell wall-acting antifungal compounds. J. Antibiot. **39**:1620–1622.

68. Leal-Morales, C.A., C.E. Bracker, and S. Bartnicki-Garcia. 1988. Localization of chitin synthetase in cell-free homogenates of *Saccharomyces cerevisiae*: chitosomes and plasma membrane. Proc. Natl. Acad. Sci. USA **85**:8516–8520.

69. Leighton, T., E. Marks, and F. Leighton. 1981. Pesticides: insecticides and fungicides are chitin synthesis inhibitors. Science **213**:905–907.

70. Marcireau, C., M. Guilloton, and F. Karst. 1990. In vivo effects of fenpropimorph on the yeast *Saccharomyces cerevisiae* and determination of the molecular basis of the antifungal property. Antimicrob. Agents Chemother. **34**:989–993.

71. Martinez, A.F., and J. Schwencke. 1988. Chitin synthetase activity is bound to chitosomes and to the plasma membrane in protoplasts of *Saccharomyces cerevisiae*. Biochim. Biophys. Acta **946**:328–336.

72. Masson, J.-M.V. Guillermet, A. Marinos, and F. Le Goffic. 1987. Chitin synthase as a target for the design of anti-fungal agents. Eur. J. Med. Chem. **22**:377–381.

73. McCourtie, J., and L.J. Douglas. 1985. Extracellular polymer of *Candida albicans*: isolation, analysis and role in adhesion. J. Gen. Microbiol. **131**:495–503.

74. Minke, R., and J. Blackwell. 1979. The structure of α-chitin. J. Mol. Biol. **120**:167–182.

75. Mitani, M., and Y. Inoue. 1968. Antagonists of antifungal substance polyoxin. J. Antibiot. **21**:492–496.

76. Molano, J., B. Bowers, and E. Cabib. 1980. Distribution of chitin in the yeast cell wall. An ultrastructural and chemical study. J. Cell. Biol. **85**:199–212.

77. Montgomery, G.W.G., D.J. Adams, and G.W. Gooday. 1984. Studies on the purification of chitin synthase from *Coprinus cinereus*. J. Gen. Microbiol. **130**:291–297.

78. Müller, H., R. Furter, H. Zähner, and D. Rast. 1981. Metabolic products of microorganisms. 203. Inhibition of chitosomal chitin synthetase and growth of *Mucor rouxii* by nikkomycin Z, nikkomycin X, and polyoxin A: a comparison. Arch. Microbiol. **130**:195–197.

79. Naider, F., P. Shenbagamurthi, A.S. Steinfeld, H.A. Smith, C. Boney, and J.M. Becker. 1983. Synthesis and biological activity of tripeptidyl polyoxins as antifungal agents. Antimicrob. Agents Chemother. **24**:787–796.

80. Navia, M.A., J.P. Springer, T. -Y. Lin, H. Williams, R.A. Firestone, J.M. Pisano, J.B. Doherty, P.E. Finke, and K. Hoogsteen. 1987. Crystallographic study of a β-lactam-inhibitor complex with elastase at 1.8 A resolution. Nature **327**:79–82.

81. Orlean, P. 1987. Two chitin synthases in *Saccharomyces cerevisiae*. J. Biol. Chem. **262**:5732–5739.

82. Pesti, M., J.M. Campbell, and J.F. Peperdy. 1981. Alteration of ergosterol content and chitin synthase activity in *Candida albicans*. Curr. Microbiol. **5**:187–190.

83. Pfefferle, W., H. Anke, M. Bross, and W. Steglich. 1990. Inhibition of solubilized chitin synthase by chlorinated aromatic compounds isolated from mushroom cultures. Agric. Biol. Chem. **54**:1381–1384.

84. Ram, S.P., P.A. Sullivan, and M.G. Shepherd. 1983. The *in situ* assay of *Candida albicans* enzymes during yeast growth and germ-tube formation. J. Gen. Microbiol. **129**:2367–2378.

85. Ramakrishnan, C., and N. Prasad. 1972. Rigid-body refinement and conformation of α-chitin. Biochim. Biophys. Acta **261**:123–135.

86. Rhodes, J.C. 1988. Virulence factors in fungal pathogens. Microbiol. Sciences **5**:252–254.

87. Roberts, R.L., B. Bowers, M.L. Slater, and E. Cabib. 1983. Chitin synthesis and localization in cell division cycle mutants of *Saccharomyces cerevisiae*. Mol. Cell. Biol. **3**:922–930.

88. Rodewald, R., L. Hoesch, and D.M. Rast. 1987. Chitin synthase—the (or a major) target of some fungicides with insufficiently known mode of action, abstr. K–152, p. 228. Abstr. 87th Annu. Meet. Am. Soc. Microbiol. 1987.

89. Roehl, F., J. Rabenhorst, and H. Zaehner. 1987. Metabolic products of microorganisms. 241. Biological properties and mode of action of clavams. Arch. Microbiol. **147**:315–320.

90. Roncero, C., and A. Duran. 1985. Effect of Calcofluor White and Congo Red on fungal cell wall morphogenesis: in vivo activation of chitin polymerization. J. Bacteriol. **163**:1180–1185.

91. Ryder, N.S. and J.F. Peberdy. 1977. Chitin synthetase in *Aspergillus nidulans*: properties and proteolytic activation. J. Gen. Microbiol. **99**:69–76.

92. Sburlati, A., and E. Cabib. 1986. Chitin synthetase 2, a presumptive participant in septum formation in *Saccharomyces cerevisiae*. J. Biol. Chem. **261**:15147–15152.

93. Schekman, R., and V. Brawley. 1979. Localized deposition of chitin on the yeast cell surface in response to mating pheromone. Proc. Natl. Acad. Sci. USA **76**:645–649.

94. Scherer, S., and P.T. Magee. 1990. Genetics of *Candida albicans*. Microbiol. Rev. **54**:226–241.

95. Selitrennikoff, C.P. 1983. Use of a temperature-sensitive, protoplast-forming *Neurospora crassa* strain for the detection of antifungal antibiotics. Antimicrob. Agents Chemother. **23**:757–765.

96. Selitrennikoff, C.P., B.L. Lilley, and R. Zucker. 1981. Formation and regeneration of protoplasts derived from a temperature-sensitive osmotic strain of *Neurospora crassa*. Exp. Mycol. **5**:155–161.

97. Sentadreu, R., A. Martinez-Ramon, and J. Ruiz-Herrera. 1984. Localization of chitin synthase in *Mucor rouxii* by an autoradiographic method. J. Gen. Microbiol. **130**:1193–1199.

98. Shepherd, M.G., R.T.M. Poulter, and P.A. Sullivan. 1985. *Candida albicans*: biology, genetics and pathogenicity. Ann. Rev. Microbiol. **39**:579–614.

99. Sietsma, J.H., and J.G.H. Wessels. 1979. Evidence for covalent linkages between chitin and β-glucan in a fungal cell. J. Gen. Microbiol. **114**:99–108.

100. Silverman, S.J., A. Sburlati, M.L. Slater, and E. Cabib. 1988. Chitin synthase 2 is essential for septum formation and cell division in *Saccharomyces cerevisiae*. Proc. Natl. Acad. Sci. USA **85**:4735–4739.

101. Slater, M.L., B. Bowers, and E. Cabib. 1985. Formation of septum-like structures at locations remote from the budding sites in cytokinesis-defective mutants of *Saccharomyces cerevisiae*. J. Bacteriol. **162**:763–767.

102. Soll, D.R., and L. Mitchell. 1983. Filament ring formation in the dimorphic yeast *Candida albicans*. J. Cell. Biol. **96**:486–493.

103. Sprague, G., Jr., L. Blair, and J. Thorner. 1983. Cell interactions and regulation of cell type in the yeast *Saccharomyces cerevisiae*. Ann. Rev. Microbiol. **37**:623–660.

104. Surarit, R., P.K. Gopal, and M.G. Shepherd. 1988. Evidence for a glycosidic linkage between chitin and glucan in the cell wall of *Candida albicans*. J. Gen. Microbiol. **134**:1723–1730.

105. Thorner, J. 1985. Pheromone-processing protease of the yeast *Saccharomyces cerevisiae*. Nature **314**:384.

106. Ulane, R., and E. Cabib. 1976. The activating system of chitin synthetase from *Saccharomyces cerevisiae*. J. Biol. Chem. **251**:3367–3374.

107. Vanden Bossche, H. 1985. Biochemical targets for antifungal azole derivatives: hypothesis on the mode of action, p. 313–351. *In* M.R. McGinnis (ed.), Current topics in medical mycology. Springer-Verlag, New York.

108. Vermeulen, C.A., and J.G.H. Wessels. 1986. Chitin biosynthesis by a fungal membrane preparation. Evidence for a transient noncrystalline state of chitin. Eur. J. Biochem. **158**:411–415.

109. Walker, A., R.E. Garner, and M.N. Horst. 1990. Immunochemical detection in *Pneumocystis carinii*. Infect. Immun. **58**:412–415.

110. Wood, P.J. 1980. Specificity on the interaction of direct dyes with polysaccharides. Carbohydr. Res. **85**:271–287.

111. Yamaoka, F., Y. Kagei, S. Tomita, Y. Kondo, and S. Hirano. 1989. The structure of chitin-glucan complex isolated from yeast bud scars. Agric. Biol. Chem. **53**:1255–1259.

112. Yadan, J.-C., M. Gonneau, P. Sarthou, and F. Le Goffic. 1984. Sensitivity of nikkomycin Z in *Candida albicans*: role of peptide permeases. J. Bacteriol. **160**:884–888.

113. Zlotnik, H., M.P. Fernandez, B. Bowers, and E. Cabib. 1984. *Saccharomyces cerevisiae* mannoproteins form an external cell wall layer that determines well porosity. J. Bacteriol. **159**:1018–1926.

114. Zubenko, G.S., A.P. Mitchell, and E.W. Jones. 1979. Septum formation, cell division and sporulation in mutants of yeast deficient in proteinase B. Proc. Natl. Acad. Sci. USA **76**:2395–2399.

19

Glucan Biosynthesis in Fungi and its Inhibition

Jan S. Tkacz

I. Introduction

The fungal pathogen and its mammalian host share a eukaryotic heritage, and, viewed from the standpoint of anabolic metabolism, they appear more alike than dissimilar. Therefore, the identification and the exploitation of physiological and biochemical differences are the underlying challenges to be met in the development of new selective antifungal agents for use in human medicine. One prominent area of interest as a target is fungal cell-wall metabolism (19). A fungal cell encases itself in a net of polysaccharides, chiefly β-glucans and chitin, which provide the strenghtening and rigidifying framework for the cell wall. The critical functions of the wall are to counteract the osmotic forces that would otherwise bring about the lysis of the cell and also to exclude larger molecules from gaining access to the plasma membrane. Mammalian cells do not possess a cell wall and lack the capability to produce β-glucans and chitin altogether. Spurred by expectations of selectivity and efficacy as well as the paradigm provided by clinically useful bacterial cell-wall inhibitors (e.g., β-lactams, glycopeptides), efforts have been made during the past two decades to discover antibiotics directed against fungal wall synthesis. The chitin synthesis inhibitors which have been found are discussed in chapter 18 of this volume. Therefore, this chapter will focus on inhibitors of glucan synthesis; these can be grouped structurally as the papulacandin-type and the echinocandin-type compounds. Before discussing these antibiotics and their therapeutic potential, however, the nature of fungal β-glucan will be considered.

II. Fungal Cell Walls and Glucan

Fungi exhibit considerable diversity in cell-wall structure, although a correlation with taxonomic status can be made (7). The common pathogenic fungi which have been studied, especially those classified as ascomycetes or *Fungi imperfecti*, have walls rich in polymers containing β-linked residues of glucose, glucosamine, or N-acetylglucosamine (7, 31). *Cryptococcus* species, for example, have walls that are 5–8% chitin and 30–50% glucan [both α(1,3) and β(1,3)]. Wall composition of *Histoplasma capsulatum* varies with serotype; in one type, the cell wall contains approximately equal amounts of β(1,3)-glucan (25%) and chitin (30%), and, in another, the β(1,3)-glucan and chitin contents are reduced to 12% and 8%, respectively, with α(1,3)-glucan accounting for 30% of the mass (101). In the dimorphic pathogen *Paracoccidioides brasiliensis*, the main glucose polymer is a β(1,3)-glucan in the mycelial form and an α(1,3)-glucan in the yeast form (107).

A. Saccharomyces Cerevisiae *and* Candida Albicans *Glucans*

Most of what is known regarding the structure, organization, and biosynthesis of fungal glucans has been learned by the study of *Saccharomyces cerevisiae*, and comprehensive reviews have been published (16, 18, 26, 31). Structural characterization of fungal β-glucans has been a formidable problem owing to two inherent properties of these polymers (26, 31). First, most of the cell-wall glucan is insoluble even in hot alkali, and, second, the glucose residues which comprise the glucans do not appear to form regular or repeating structures. Consequently, from chemical analysis of complex glucan fractions, it is possible to discern only an "average" structure. For the most part, glucan fractionation procedures have been based on sequential extraction of cells or walls with hot alkali and then hot acid following the scheme in Figure 19.1. More convenient enzymatic methods permitting the analysis of various glucan species are now available (11, 101, 118, 131).

In the case of *S. cerevisiae* (26, 31), the major glucan fraction is the alkali-insoluble/acid-insoluble material which comprises about 60% of the total glucan and is composed almost exclusively of β(1,3)-linked glucose residues. The polymer length averages 1,500 units with 3% of the residues forming β(1,6) branch points. This glucan is considered to be responsible for the strength and rigidity of the wall since it appears as a fibrillar network retaining the shape of the yeast cell by electron microscopy. A second glucan fraction, representing 8% of the total, is also insoluble in alkali but can be extracted into acid. This material has a degree of polymerization (DP) of 140, a high proportion of β(1,6)-linkages (65%), and is highly branched with β(1,3)-bonds present both as interresidue (5%) and as interchain (14%) linkages. The final glucan fraction contains material soluble in alkali. This complex accounts for 33% of the total glucan and appears to consist of a lightly branched β(1,3)-glucan core resembling the alkali-insoluble/acid-insoluble material. Side chains consisting mainly of β(1,3)-linked or β(1,6)-linked residues or alternatively of a mixture of β(1,3)- and β(1,6)-linked moieties are attached to the core. This glucan (DP, 1,500) is covalently associated with mannoprotein. It is not known for any yeast glucan whether the structure is arboreal, laminated, or comb-like (26).

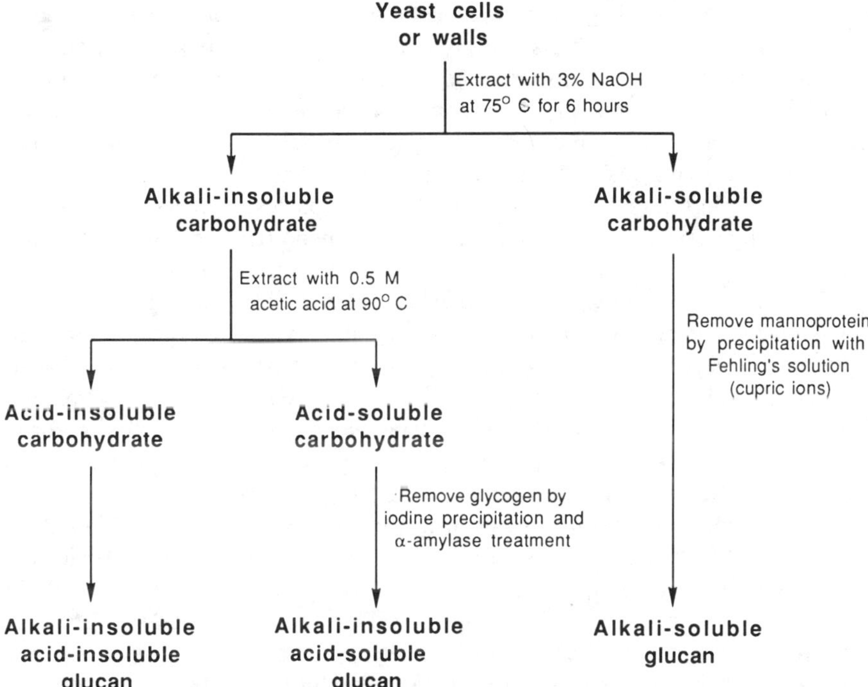

Figure 19.1. Scheme for the fractionation of yeast glucans by extraction with alkali and acid [adapted from Manners and co-workers (26)].

Althought the transition of *Candida albicans* from the yeast form to the mycelial form is accompanied by a three- to four-fold increase in chitin content of the cell wall, glucan remains the major wall polymer in both forms (31, 117). The alkali-insoluble/acid-insoluble glucan of both forms is highly branched and contains nearly equal amounts of β(1,3)-linked (30–40%) and β(1,6)-linked (43–53%) glucose residues (34). As in *S. cerevisiae*, the alkali-insoluble/acid-soluble material from *Candida* is a branched polymer composed chiefly of β(1,6)-linked glucose units (67–77%). During germ-tube induction in yeast cells of this organism, the synthesis of β(1,6)-glucan is apparently suppressed because alkali-insoluble/acid-insoluble glucan from walls of germ-tube-forming cells is markedly enriched in β(1,3)-linkages (67%). Chitin, in addition to being a major component of bud scars in *C. albicans*, is present throughout the innermost region of the cell wall and is apparently covalently associated with β(1,6)-glucan molecules (126).

1. β(1,3)-Glucan Synthesis

Several laboratory groups have investigated the enzymology of β(1,3)-glucan synthesis in *S. cerevisiae*. One (or more) β-glucan synthase is associated with membrane fragments that can be obtained either through mechanical disruption of the cells (62) or osmotic lysis of protoplasts (17, 115). Glycerol and sucrose appear to stabilize the activity as do bovine serum albumin and sodium fluoride (63). UDP-glucose is the only sugar donor used as substrate (K_m, 1.7–3.8 mM); no product is formed from either ADP-glucose or GDP-glucose. Unlike many other glycosyl transferases, the β-glucan synthase has no requirement for divalent cations. Dolichol-linked sugar phosphates or diphosphates, such as those involved as intermediates in the synthesis of cell-wall mannoproteins, are not used in the formation of β(1,3)-glucan in this yeast (115) or other fungi (30, 98). The reaction products are UDP and a water-insoluble, alkali-soluble linear β(1,3)-glucan which under some conditions is visible to the naked eye as turbidity. The glucan appears as a disorganized microfibrillar mass by electron microscopy. Glucan synthases with very similar properties have been demonstrated in membrane preparations of *C. albicans* (3, 4, 85, 86) and other taxonomically diverse fungi (99, 127). In *S. cerevisiae*, the enzyme is thought to be a component of the plasma membrane (115). The activities of most fungal β(1,3)-glucan synthases are stimulated by nucleotide triphosphates, especially ATP and GTP and their analogs (83, 85, 86, 116, 127). Analysis of this stimulation has provided evidence for a dissociable soluble protein which binds to the membrane-bound catalytic portion of the enzyme in the presence of GTP and activates it (47). There is evidence for an additional endogenous regulatory factor, but its identity has not been established (38). The role of primer or acceptor in β(1,3)-glucan synthesis also needs clarification (3, 4). The β(1,3)-glucan synthase from plants (callose synthase), which has been more extensively studied than the fungal enzyme (24, 50, 100, 121), requires Ca^{2+}, uses a β-glucoside as an allosteric effector, and is not regulated by GTP. At less than saturating levels of Ca^{2+}, polyamines or Mg^{2+} stimulate the enzyme, and the detergent-solubilized enzyme requires phospholipid for activity. Photoaffinity labeling with 5-azido-UDP-glucose provides evidence that a 57-kD peptide is a component of the enzyme (33).

2. *β(1,6)-Glucan Synthesis*

Nothing is known regarding the fungal enzyme which forms β(1,6)-glucan or which introduces β(1,6)-branch points into β(1,3)-glucan. However, the synthesis of glucan containing β(1,6)-linkages was reported for a crude envelope (wall and membrane) fraction from *S. cerevisiae* (65). The formation of β(1,6)-glucan GDP-glucose in yeast membrane preparations has also been reported (6) though not confirmed (115). Analysis of K1 killer toxin action has made possible the genetic dissection of this aspect of cell-wall synthesis (15). The interaction of the killer toxin with susceptible yeast cells initially involves binding to β(1,6)-glucan in the cell wall (44), and mutants with defects in the formation of β(1,6)-glucan have been isolated as resistant to killer toxin. Three genes (*KRE*1, *KRE*5, and *KRE*6) have been defined, and two (*KRE*1 and *KRE*5) have been cloned, sequenced, and used to prepare null mutants by gene disruption. From the open reading frame, the *KRE*1 gene product appears to be a 32-kD protein with a high serine and threonine content (35%) and a hydrophobic N-terminal region that could act as a signal sequence (11). Although N-glycosylation sites are absent, the abundance of hydroxy-amino acids suggests that the protein might be extensively O-glycosylated. A *kre*1 null mutant is completely resistant to killer toxin and grows slightly more slowly than a wild-type cell. Its $\beta(1,6)$-glucan is reduced in amount (approximately 40% relative to the wild type) as well as in size. The protein predicted from the *KRE*5 gene sequence is a 160-kD peptide with a hydrophobic N-terminal sequence, 13 potential N-glycosylation sites, and a C-terminal HDEL signal for retention in the endoplasmic reticulum (68). Disruption of the *KRE*5 gene resulted in cells that were able to grow only poorly, even in osmotically protected media. Microscopic examination revealed that the cells were large and multiply budded, presumably as the result of defective cell separation. No $\beta(1,6)$-linkages were found in the alkali-insoluble glucan from the disruptant, and its content of soluble $\beta(1,6)$-glucan was reduced at least 18-fold relative to the wild type. The null phenotype of this strain resembles that of a mutant hypersensitive to $\beta(1,3)$-glucanase (119). Epistasis analysis of these *KRE* genes suggested that they participate in a pathway for $\beta(1,6)$-glucan synthesis with the *KRE*5 and *KRE*6 products in the endoplasmic reticulum mediating the synthesis of a core structure or backbone which is secreted to the cell surface to be further modified by the *KRE*1 product (68). Further elucidation of the biological activities associated with the *KRE*1 and *KRE*5 gene products will provide useful insights into the mechanism of wall formation. Analysis of a mutant with an abnormally high content of $\beta(1,6)$-glucan might also prove instructive (46).

The biogenesis of the cell wall involves more than the polymerization of sugars; it is a complex biochemical process which the cell must regulate with respect to both cellular location and timing in the cell cycle. (See discussion in section VI.) Although much of the system remains a "black box," it is clear that there are several aspects of fungal glucan and cell-wall metabolism which might constitute points of inhibition for antibiotics. We will now turn to a consideration of the two classes of antifungal compounds which act by interfering with the synthesis of wall glucan.

III. Papulacandin-type Antibiotics

A. *Structure*

Papulacandins A and B are acylated spirocyclic disaccharide antibiotics produced by the fungus *Papularia sphaerosperma* (134–136). As shown in Figure 19.2, each has a disaccharide core of galactosyl-glucose; the galactose residue is connected by a $\beta(1,4)$-linkage. A spirocyclic system is attached at the C-1 of

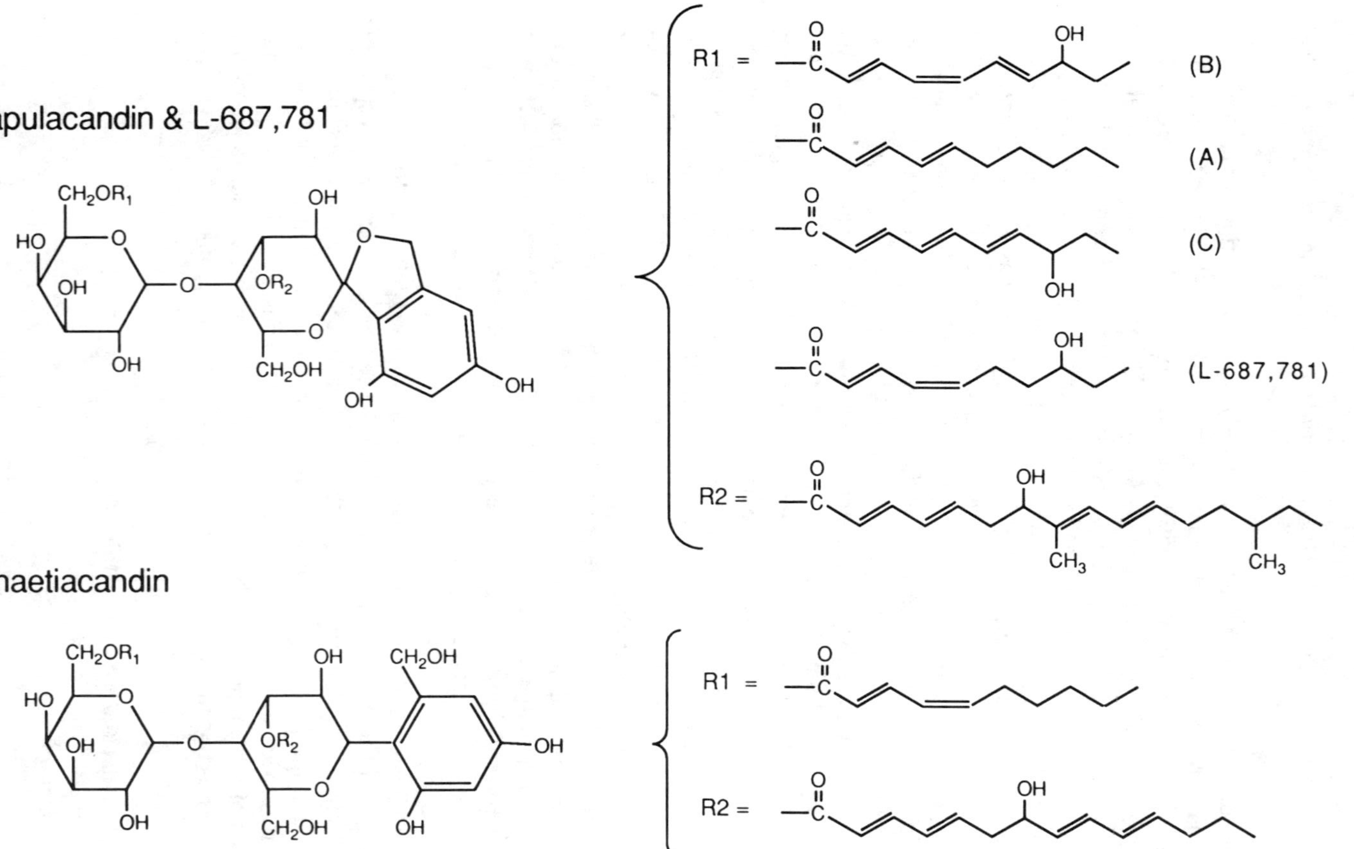

Figure 19.2. Structures of the members of the papulacandin family of antibiotics.

glucose, and the C-3 position of this sugar is esterified with all-*trans* 7-hydroxy-8,14-dimethyl-2,4,8,10-hexadecatetraenoic acid. Papulacandins A and B differ from each other according to the fatty acid that acylates the C-6 position of the galactose moiety. In papulacandin B, it is a C_{10}-carbyoxylic acid with a hydroxyl substitution at C-8 and three conjugated double bonds in the *trans-cis-trans* configuration at C-2, C-4, and C-6. This fatty acid in papulacandin A lacks the hydroxyl group and the third double bond and is thus 2,4-decadienoic acid. During purification of papulacandin B, a rearrangement to form the all-*trans* hydroxy-decatrienoic acid side chain apparently accounts for the formation of papulacandin C. Papulacandin D, on the other hand, is an authentic fermentation product of the fungus. It consists solely of the substituted glucosyl component common to papulacandins A and B.

L–687,781 is a new papulacandin from *Dictyochaeta simplex*. Structurally, it is identical to papulacandin B except that the fatty acid at C-6 of the galactose contains only two conjugated double bonds at C-2 and C-4. As in papulacandin B, these bonds are in the *trans-cis* configuration, respectively (112) (F. Vanmiddlesworth, personal communication).

The fungus *Monochaetia dimorphospora* synthesizes a compound related to papulacandins A and B and known as chaetiacandin (55, 56). Like the former, it is an acylated galactosyl-glucose disaccharide (Figure 19.2). The fatty acid esterified to C-6 of the galactosyl unit is 2,4-decadienoic acid with the conjugated double bonds in the *trans-cis* configuration, respectively. The C-3 esterification of the glucose residue involves all-*trans* 7-hydroxy-2,4,8,10-tetradecatetraenoic acid. In contrast with the papulacandins, the glucose moiety of chaetiacandin does not participate in a spiroketal linkage; rather, the C-1 of this residue is linked directly to C-2 of 3,5-dihydroxybenzyl alcohol.

B. Antifungal Spectrum

Papulacandin B and chaetiacandin were isolated as antifungal antibiotics. They are strongly active against *Candida albicans, C. tropicalis, C. krusei,* and *C. parakeusei* with minimal inhibitory concentration (MIC) values of 0.1 to 0.2 μg/ml (56, 136). *C. parapsilosis, C. utilis, Torulopsis dattila, T. famata, T. glabrata, Saccharomyces cerevisiae*, and *Microsporum canis* are also quite sensitive to papulacandin B. Both antibiotics are ineffective against *Candida guilliermondii, Cryptococcus neoformans, Sporotrichum schenckii, Trichophyton asteroides, T. mentagrophytes, T. rubrum, Aspergillus fumigatus, A. niger, Phoma* sp., bacteria, or protozoa.

C. Mode of Action

Traxler *et al.* noted that papulacandin B caused the lysis of the growing buds in dividing *Candida albicans* cells; in contrast, resting cells were not killed (136).

Essentially the same results were obtained with *Saccharomyces cerevisiae* (5, 57). Since the growing bud is the insertion point for newly made wall polymers in these yeasts, the findings pointed to an inhibition of wall synthesis by the antibiotic. Observations consistent with this view were made by electron microscopy with *C. albicans*; the compound caused distortion of the cell surface, improper septation, and in some cases, cytoplasmic degeneration (12). By following the incorporation of [U-^{14}C]glucose into wall polymers, Baguley *et al.* found that papulacandin B inhibited the synthesis of β-glucan but not mannan in both intact cells and protoplasts of *S. cerevisiae* and in protoplasts of *C. albicans* (5). The effect on β-glucan synthesis in *S. cerevisiae* was more dramatic in protoplasts than in intact cells. For these experiments, labeled cells or protoplasts were fractionated by a scheme that relied on alkali insolubility of the branched β(1,3)-glucan and alkali solubilization/cupric ion precipitation of the mannan fraction. The alkali-soluble/cupric ion nonprecipitable fraction, which was taken to be largely "soluble glucan," was poorly characterized, and might contain other cellular materials derived from labeled glucose. Thus, the data regarding the effect of the antibiotic on glucan synthesis in this fraction cannot be interpreted. In a more detailed radioisotopic study with *C. albicans* protoplasts, regenerated walls were first separated from other cellular constituents prior to fractionation; papulacandin B diminished the synthesis of both alkali-insoluble and alkali-soluble glucans (28). Examining wall regeneration in *S. cerevisiae* protoplasts by electron microscopy, Kopecka concluded that papulacandin B blocks the formation of the fibrillar network that consists of alkali-insoluble β(1,3)-glucan with β(1,6)-branches while allowing the formation of disconnected microfibrils of linear, alkali-soluble β(1,3)-glucan (58). These observations suggested that papulacandin B inhibits the enzyme which introduces β(1,6)-branches into growing chains of β(1,3)-glucan rather than the enzyme that elongates these chains. Nevertheless, papulacandin B inhibition of β(1,3)-glucan synthase in membrane preparations from a number of fungi has been reported by several groups (48, 90, 98, 129, 143). Activities of chitin synthase and mannan "synthase" were not sensitive (90, 143). The inhibition was carefully examined by Kang *et al.*, who reported that it became less severe as the concentration of UDP-glucose substrate decreased, and at very low substrate levels, papulacandin B actually stimulated enzyme activity (48). These results were taken as an indication that the glucan synthase undergoes a substrate-dependent allosteric change in conformation and that papulacandin B binds preferentially to the form that is favored by high substrate concentrations. These workers also found that the ability of the antibiotic to inhibit the growth of a fungus was not correlated with its potency as an inhibitor of the glucan synthase from that organism. The reason for this is not evident, but the inability of cell-free preparations to reflect the constitution of the glucan-forming enzyme system *in vivo* (with respect to a transmembrane electrochemical potential, for example) was suggested as a potential factor. The possible existence of another cellular target for the antibiotic could not be excluded. To elucidate the target of papulacandin B *in vivo*, Perez *et al.* assayed the activity of $\beta(1,3)$-glucan synthase in membrane prepared from cells of *Geotrichum lactis* that had been incubated in the presence of the antibiotic (89). Antibiotic treatment reduced enzyme activity, but no reduction in the activity of chitin synthase was noted. In this organism, the antibiotic does not alter the growth rate, but branching of the mycelia accompanied by decreases in the synthesis of both alkali-insoluble and alkali-soluble glucan occurs. From a microbiological point of view, it seems more than coincidental that *Papularia sphaerosperma* and *Monochaetia dimorphospora*, the two organisms which produce papulacandins A to D and chaetiacandin, are endophytic fungi since plants often respond to mechanical injury or microbial infection by synthesizing callose, a $\beta(1,3)$-glucan (49).

D. Chemical Modification

Initial tests indicated that the acute toxicity of papulacandin-type antibiotics was very low. The LD_{50} in mice for papulacandin B administered subcutaneously was above 1,000 mg/kg, and intraperitoneal administration of chaetiacandin at the same dose produced no symptoms of toxicity. Subcutaneous administration of papulacandin B to mice that had been intravenously challenged with 2×10^6 cells of *C. albicans* cured infection, but the unexpectedly high ED_{50} of 80 mg/kg obtained in this study implied that the compound lacked bioavailability or was metabolized *in vivo* (136). Serum itself lowered the activity of chaetiacandin (56).

Structure-activity relationships for the anti-*Candida* action of the papulacandins have been explored by Römmele *et al.* (104). Although papulacandin B is the most active member of this natural product series, its MIC is only two- to fourfold less than that of papulacandins A and C, indicating that the nature of the fatty acid on the galactose residue does not greatly influence activity. In fact, even papulacandin D which lacks the acylated galactose moiety can inhibit the growth of *Candida* with an MIC only five to 20 times higher than those for the disaccharide forms of the antibiotic. When the galactose moiety is present without its fatty acid, the compound apparently cannot penetrate the yeast and is totally inactive. The fatty acid residue attached to the glucose unit of either papulacandin B or D is essential since its removal or alteration by hydrogenation inactivates the compound. In an attempt to improve *in vivo* efficacy, Traxler *et al.* prepared a large number of papulacandin B derivatives and compared the antifungal MIC spectrum of each with its efficacy in mice infected with *C. albicans* (137). Substitutions at the two phenolic hydroxyl groups, additions to the carbons proximal to these hydroxyls in the aromatic ring, and derivatization at C-6 of the glucose moiety were tried, with mostly disappointing results. Modification of the glucose group could not be tolerated, and only certain alkyl ether and acylamino derivatives of the aromatic ring gave improved activity in the animal model. Even though efficacy could be increased three- to sixfold by these substitutions, the

ED_{50} values remained unacceptably high (14–30 mg/kg). Moreover, derivatization did not extend the antifungal spectrum to include filamentous fungi. In light of these results, it is unlikely that a papulacandin could be developed for clinical use.

IV. Echinocandin-type Antibiotics

A. Structure

Echinocandin was discovered independently at Ciba-Geigy as A32204 (9), at Sandoz as SL7810 (51, 133), and at Lilly as A30912. Several other structurally related antibiotics are now known. These include aculeacin (78, 109, 145), mulundocandin (80, 105), L–671,329 (32, 113, 144), and antibiotic S31794/F-1 (1, 25). A general structure for the members of this group is presented in Figure 19.3. They are cyclic hexapeptides with acyl side chains. In each case, the peptide contains threonine and 4-hydroxyproline together with three unusual amino acids: 3,4-dihydroxyhomotyrosine (9, 52), 4,5-dihydroxyornithine (9), and 3-hydroxy-4-methylproline (9, 59). The sixth amino acid (R_1, Figure 19.3) can be threonine as in echinocandin and aculeacin (9, 109, 133), serine as in mulundocandin (80), or 3-hydroxyglutamine as in L–671,329 and antibiotic S31794/F-1 (1, 25, 144). Incomplete hydroxylation of the homotyrosine and ornithine residues accounts for the occurrence of echinocandin as a series of compounds (echinocandins B, C, and D) (133). Each member of the echinocandin group is acylated with a different fatty acid (R_2, Figure 19.3). Aculeacin contains palmitic acid (109) whereas the echinocandins have linoleic acid (9, 51). The branched-chain fatty acids, 12-methylmyristic acid and 10,12-dimethylmyristic acid, are present in mulundocandin and L–671,329, respectively (80, 144). Antibiotic S31794/F-1 has a myristoyl side chain (1, 25). All members of the echinocandin class of antibiotics are produced by *Aspergillus* species (9, 78, 105, 133) with the exception of L–671,329, which is made by *Zalerion arboricola* (113), and antibiotic S31794/F-1, which is a product of *Acrophialophora limonispora* nov. sp. (25). A *Cryptosporiopsis* isolate synthesizes a structurally related antibiotic, sporiofungin. This compound resembles L–671,329 in that it contains a branched-chain C_{16} fatty acid and residues of hydroxyasparagine and serine in place of hydroxyglutamine and threonine, respectively (23, 138). Also the homotyrosine residue in sporiofungin, like that of echinocandin C, is not fully hydroxylated. The structure of echinocandin D has been verified by chemical synthesis (29, 60, 61, 106). No systematic analysis of the structure-activity relationships in the cyclic peptide portions of these antibiotics has been published.

B. Antifungal Spectrum

The members of the echinocandin group uniformly have high activity against *Candida albicans* with MIC values generally less than 1μg/ml (9, 32, 45, 78,

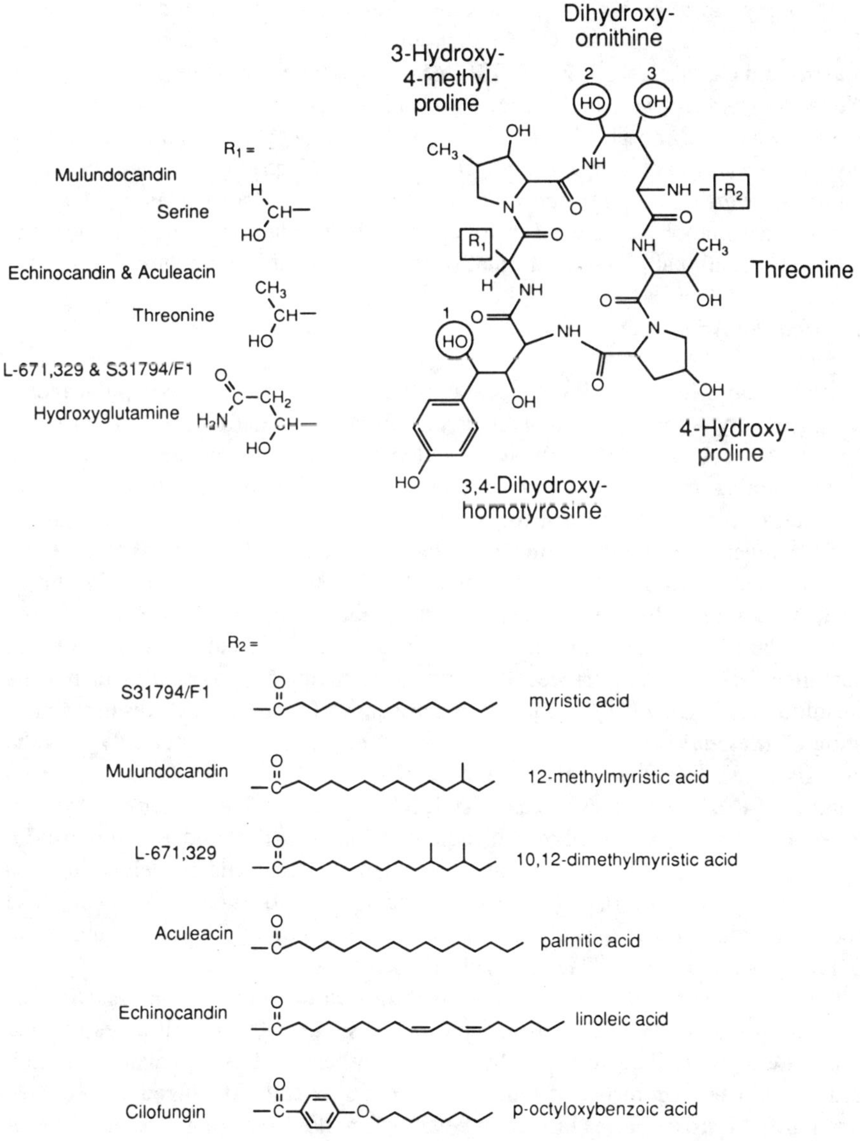

Figure 19.3. General structure of echinocandin-type antibiotics. R_1 indicates the amino acid in the cyclic hexapeptide core which varies from one member of the group to another, R_2 shows the hydrophobic side chains of the naturally occurring and the semisynthetic members of the group. Circles indicate the hydroxyl functions which may be missing in the members of the echinocandin series: echinocandin B, all three present; echinocandin C, hydroxyl No. 1 absent; echinocandin D, all numbered hydroxyls absent.

105). They are quite active against other *Candida* species including *C. krusei, C. parakrusei, C. parapsilosis, C. pseudotropicalis, C. pulcherrima, C. stellatoidea,* and *C. tropicalis* (32, 45, 78). Yeasts such as *Saccharomyces cerevisiae, Hansenula californica,* and various *Torulopsis* species are also sensitive, but filamentous or dimorphic pathogens including *Aspergillus fumigatus, Blastomyces dermatitidis, Cryptococcus neoformans, Geotrichum candidum, Histoplasma capsulatum, Paracoccidioides brasiliensis, Sporotrichum schenckii,* and the dermatophytes are insensitive. Overall, the spectrum of antimicrobial activity is remarkably similar to that of the papulacandin-type antibiotics.

C. *Mode of Action*

Biological effects of the echinocandin-type antibiotics and the papulacandins are also very similar. Mizoguchi *et al.* observed that treatment of growing *Saccharomyces cerevisiae* cells with aculeacin A caused lysis at the tips of the buds resulting in release of intracellular material (77). Nongrowing cells were resistant to the lethal effect of the antibiotic. *C. albicans* cells responded in the same manner when treated with either aculeacin or echinocandin (20, 145, 147). Osmotic stabilization of the medium with 0.8–1.0 M sorbitol afforded significant protection against the lethal effects of these antibiotics on *C. albicans*. At the subcellular level, these antibiotics initially produced abnormalities in the cell wall accompanied or possibly preceded by invagination and convolution of the plasma membrane (12, 20, 77, 145). Electron microscopy of surface structures in regenerating *C. albicans* protoplasts revealed that aculeacin interfered not only with the genesis of alkali-insoluble glucan microfibrils but also with their association into bundles (146). The synthesis of alkali-insoluble glucan in *S. cerevisiae* was shown to be preferentially inhibited by aculeacin in whole cells (77) and by echinocandin in spheroplasts (5). The formation of mannan, protein, and nucleic acids was unaffected. In a strain of *S. cerevisiae* that diverted exogenous [U-^{14}C]galactose specifically into glucans, aculeacin affected the labeling of all glucans, including β(1,6)-glucan (131) (J. Tkacz, unpublished results).

The addition of aculeacin to a mid–log phase culture of the fission yeast *Schizosaccharomyces pombe* resulted in rapid cell lysis at both the cell poles and the cell plate regions, the two locations of wall synthesis in this organism (76). Cell death could be diminished if the cells were osmotically stabilized with 0.8 M mannitol. Microscopic examination revealed marked swelling at either one or both poles in many of the treated cells. Very similar observations have been made with papulacandin B (143). The areas of polar distention in aculeacin-treated cells had an increased ability to bind the fluorescent brightener Calcofluor White. Chemical analysis of the wall polymers indicated that aculeacin caused a decrease in alkali-insoluble β-glucan and mannan accompanied by a significant increase in alkali-soluble glucan, which in this yeast is β(1,3)-glucan (75). Slight stimulation in α(1,3)-glucan synthesis was also reported with papulacandin B (143).

Electron microscopy of cells that had been inhibited with aculeacin before fixation with glutaraldehyde and treatment with glucanases revealed that the cell wall surrounding the distended pole of the cell was sensitive to digestion with Novozym 234, a source of α(1,3)-glucanase, whereas that covering the tubular portion of the cell was attacked by Zymolyase 60,000, a source of β(1,3)-glucanase. Normally in *S. pombe*, the α(1,3)-glucan of the wall is present as a thin layer underlying the β-glucan and adjoining the periplasmic space. The results obtained with this yeast demonstrate that aculeacin does not inhibit the synthesis of α(1,3)-glucan. Similar observations have been made with papulacandin B and the dimorphic pathogen, *Paracoccidioides brasiliensis* (21). The antibiotic did not suppress the growth of the yeast form which has walls of α(1,3)-glucan but did inhibit the mycelial phase which has walls rich in β(1,3)-glucan. Furthermore, the mycelia responded to papulacandin B treatment by shifting synthesis from β(1,3)-glucan to α(1,3)-glucan. The transformation of yeast to mycelium was also inhibited by the antibiotic.

Using membrane preparations from *S. cerevisiae*, Yamaguchi *et al.* demonstrated the inhibition of β(1,3)-glucan synthase activity by aculeacin; no effect upon mannan "synthase" was observed, although a 1.5-fold stimulation of chitin synthase activity occurred with high levels of the antibiotic (145, 147). Aculeacin was also active against the β(1,3)-glucan synthase from *C. albicans* (147) as was a derivative of echinocandin B (111). In the latter case it was additionally shown that inhibition was not dependent upon GTP and did not follow simple competitive or noncompetitive kinetics. Mixed or uncompetitive inhibition of the glucan synthase from *Neurospora crassa* was reported for echinocandin B, with an apparent K_i of 1.1 μM (98). Since attempts to solubilize and purify fungal β(1,3)-glucan synthase have not been successful, it is important to emphasize that all work showing effects of the papulacandin- and echinocandin-type antibiotics upon β(1,3)-glucan synthase activity has been performed with membrane-bound enzyme preparations. For this reason, it is difficult to know whether the observed inhibitions result from the interaction of the antibiotics with the protein at a site distinct from the active site or from the perturbation of a specific lipid microenvironment required for enzyme activity. In the latter connection, it is interesting that four cerebrosides produced by a *Pachybasium* species are synergistic with aculeacin but are themselves devoid of antifungal activity (120). These cerebrosides are also known to stimulate fruiting body formation in certain fungi. The probable existence of more than one form of glucan synthase in a membrane preparation is another complicating issue. There are data implicating both nucleotide triphosphate activation and substrate-dependent allosteric conformational changes in glucan synthase, and it has been noted that membranes prepared from *C. albicans* in buffers lacking sucrose or 2-mercaptoethanol have glucan synthase activity which is insensitive to echinocandin (111). Thus, while it is clear that both the echinocandin- and papulacandin-type antibiotics interfere with the synthesis of various β-glucans in susceptible organisms, the mechanism by which the compounds do so has not been fully resolved. It may be premature to think of them solely as inhibitors of $\beta(1,3)$-glucan synthase.

There is another line of evidence that is consistent with this view. Rine *et al.* developed a genetic strategy to identify the cellular target of a metabolic inhibitor which acts by interacting specifically with a protein (103). The approach relies on overexpressing the target protein by means of a high-copy-number vector. The gene encoding the target protein can be recovered selectively from a DNA library because cells containing a plasmid capable of overexpressing the target protein require more drug to inhibit growth. A plasmid library of *S. cerevisiae* genomic DNA has been screened for clones conferring resistance to levels of aculeacin sufficient to inhibit the host cells (J. Becker, personal communication). Although a specific segment of DNA has been recovered, it apparently does not contain an open reading frame which might constitute the gene being sought, and it is not clear how the segment can mediate resistance.

Ultraviolet irradiation has been used to induce mutations to aculeacin resistance in *C. albicans* (69, 70). These mutants were between 100- and 400-fold more resistant than the parent and showed cross-resistance to papulacandin B but not amphotericin B or 5-fluorocytosine. The total lipid content in the mutants was twice as high as that of the parental cells. Increases in unesterified ergosterol, free fatty acids, and lysophospholipids were found which could have resulted from an elevated lipase (deacylase) activity. Such activity could also have accounted for resistance because deacylation inactivates aculeacin (130). Nevertheless, no evidence that the mutants alter the antibiotic was found, and resistance was ascribed to a change either in the lipid environment of the target enzyme or in the permeability characteristics of the cells. There are no published data on the metabolic alteration of echinocandin-type antibiotics by fungi, and it will be particularly important to determine whether the natural insensitivity of certain organisms reflects a capacity to inactivate the antibiotic.

V. Cilofungin (LY121019), a Semisynthetic Echinocandin

Through toxicity studies in dogs, researchers at Lilly identified erythrocyte lysis as a primary stumbling block for the clinical application of echinocandin B (43). Based on the theory that membrane disruption is a consequence of the amphipathic nature of the antibiotic, derivatives with new hydrophobic side chains replacing the native linoleoyl group were prepared in the hope of eliminating this undesirable characteristic while preserving the antibiotic activity. The approach involved the enzymatic deacylation of echinocandin B and the subsequent chemical reacylation of the peptide nucleus (22, 23). Deacylation was accomplished with live cells of *Actinoplanes utahensis* (10, 22, 23), though the purified deacylating enzyme from this organism (130) or the polymyxin acylase from *Pseudomonas* strain M-6-3 (53, 87) may also be employed. Once deacylated, the peptide is devoid of antibiotic activity. A large number of acyl, alkoxybenzoyl, and aralkanoyl derivatives were prepared and tested for activity against *Candida albicans* both in MIC assays and in infected mice. In general, reacylation restored antibiotic activity, with lengthening of the side chain resulting in a decrease in the ED_{50} but an increase in the hemolytic potential (23). As with papulacandin, derivitization did not extend the spectrum of activity to other pathogens. The hemolytic properties of other members of the echinocandin-group have also been examined. L-671,329, having a branched C_{16} side chain, was found less lytic than aculeacin which has an unbranched C_{16} side chain (32). Several of the semisynthetic echinocandin B derivatives exhibited *in vitro* and *in vivo* activity against *Candida* equal or superior to that of the parent compound (23). Compounds that were also less hemolytic were evaluated for toxicity in dogs. From these studies, LY121019, the p-(octyloxy)benzoyl analog of echinocandin B now known as cilofungin (Figure 19.3), emerged as a clinical candidate.

A. Mode of Action

The mode of action of cilofungin is indistinguishable from that of the parent compound and other members of the echinocandin class. Cilofungin is equally active against the yeast form and the mycelium form of *C. albicans* (36, 125). Like echinocandin, it is fungicidal to growing cells (35) and possibly also to nongrowing cells (93). It strongly inhibits the incorporation of [^{14}C]glucose into alkali-insoluble glucan in this organism with little or no effect on chitin or mannoprotein formation (36, 94). Neither DNA nor RNA nor protein synthesis is affected in short-term experiments (36), but ergosterol production is decreased by 50% on prolonged exposure (94). The significance of the effect on membrane sterol formation is as yet unexplained (and is discussed further in section VI). Noncompetitive inhibition of $\beta(1,3)$-glucan synthase by cilofungin has been observed with cell-free membrane preparations from *C. albicans* (apparent K_i, 2.5 μM) and *Neurospora crassa* (apparent K_i, 16 μM) (128, 129). The MIC of cilofungin for the *C. albicans* strain used as the source of glucan synthase in this experiment was not reported, but typical MIC values for *C. albicans* are below 1 μg/ml (1 μM).

B. Microbiological Evaluation

The susceptibility of cilofungin of clinical isolates of *C. albicans* and *C. tropicalis*, the two most virulent species that account for approximately 70% of all deep-seated candidiasis in humans, has been exhaustively documented with organisms of diverse geographic origins (13, 14, 36, 39, 40, 42, 71, 73, 84, 95–97, 102, 123, 125, 132). In many of these studies where the susceptibilities to cilofungin and to amphotericin B, the principal therapeutic agent for such infections, were directly compared, very similar MIC values were found. These investigations also served as further confirmation of the narrow antifungal spectrum of cilofungin, though the sensitivity of a newly recognized pathogen, *Malassezia pachydermatis*, was uncovered (40). Inoculum size, pH, and medium composition were identified as important variables for cilofungin bioassays (39, 67, 95, 125), and several workers observed a paradoxical dose response or "Eagle effect" with this compound (13, 36, 39, 42, 73, 95, 96, 102, 123, 132) and other members of the echinocandin group (32, 45, 145, 147). The Eagle effect is not observed in Antibiotic Medium 3, and the use of this medium for the determinations of MIC and MFC values of echinocandin-type antibiotics has been recommended by several investigators (36, 39, 73, 102).

In some patients with serious *Candida* infections, the administration of 5-fluorocytosine along with amphotericin B produces a synergy that permits a lowering of the polyene dosage. By broth assay, cilofungin does not appear to be synergistic with either amphotericin B or 5-fluorocytosine (93, 102); in fact, antagonism between cilofungin and these compounds has been reported (14). Similarly, no synergy between aculeacin and amphotericin B has been found (120). The situation with azoles such as ketoconazole and econazole is much the same, but at concentrations of cilofungin where paradoxical growth occurs, synergy with the azoles was observed (93, 102).

C. Preclinical Evaluation

Cilofungin has shown significant *in vivo* activity in several animal models of *C. albicans* infection (35). In mice immunocompromised by X-ray irradiation and infected systemically, intraperitoneal administration afforded protection with an ED_{50} of 7.4 mg/kg. Oral administration cleared *C. albicans* from the gastrointestinal tracts of mice. Local application of the compound cured superficial infections in guinea pigs and reduced vaginal infections in rats. In an independent study with mice, cilofungin and amphotericin B were compared directly (79). Cilofungin (at 6.25 and 62.5 mg/kg/day) prolonged survival without producing the pathological changes in the kidneys seen with amphotericin B (at 0.625 or 6.25 mg/kg/day). The two drugs have also been compared in rabbits. Both produced similar reductions in yeast counts from the renal cortex, renal pelvis, and urine (91). With respect to endophthalmitis, cilofungin reduced the number of positive cultures from the vitreous fluid but was not as effective as amphotericin B in reducing infection of retina or choroid. Only amphotericin B suppressed candidal endocarditis. In another study, cilofungin neither improved survival of the rabbits nor reduced the size of the *Candida* population in the aortic valve (88). In both of the rabbit studies, a relatively short serum half-life for cilofungin was observed. Perfect *et al.* found that the peak level of 160 μg/ml measured immediately after a single intravenous injection (50 mg/kg) decreased to an undetectable level by 90 minutes (91). Padula and Chambers reported a serum half-life of 1.3 hours (88). Toxicological evaluation of cilofungin appears 20-fold less toxic than amphotericin B (35). Cilofungin is not overtly toxic to polymorphonuclear leucocytes (neutrophils), nor does it impede taxis, adherence, or phagocytosis (72, 142).

D. Current Status

With its novel and fungicidal mode of action, high potency, and relatively low toxicity, cilofungin shows promise as an alternative to the standard amphotericin B therapy for disseminated candidiasis in humans, and clinical trials are underway (37). The proportedly low potential for resistance is another favorable factor (35). Nevertheless, cilofungin is not without limitations (124). The fact that it is not orally absorbed restricts the route of administration to injection. It is not water soluble, necessitating the development of a suitable vehicle for formulation. Furthermore, if serum half-life proves to be problematical in humans, administration will be additionally confined to continuous intravenous infusion. Penetration into the cerebrospinal fluid and the vitreous humor of the eye is poor, and the drug's limited antifungal spectrum makes the accurate diagnosis of the pathogen a crucial factor (64). Offsetting these drawbacks is the realization that the development of other echinocandin-type derivatives can be envisioned, especially now that the chemical synthesis of echinocandin D has been achieved (29, 60, 61, 106). The willingness of pharmaceutical companies to embark on new development programs with this group of antibiotics will depend not only on the experience with cilofungin in the initial clinical trials but also on the performance of new competing azole compounds which are also presently under evaluation in the clinic (37, 124).

IV. Outlook for Inhibitors of Fungal Wall Synthesis

During the formation of new wall in a fungal cell, sugars are polymerized, growing unimolecular polymers associate into fibrous bundles, existing wall fabric is rearranged to accommodate new material, and components are cross-linked. The entire process must be regulated both spatially and temporally. At present our knowledge of many aspects of the process is rudimentary or still only at the working-hypothesis stage; clearly, there is much yet to learn (16, 18, 117). Since unbranched β(1,3)-glucans associate to form hydrogen-bonded triple helices with the chains running in parallel (108), intertwining during synthesis is envisioned. Interaction of nascent polymer chains might be facilitated by the presence of multiple active sites in a single enzyme molecule, an arrangement postulated for callose synthase in plants (24). Localized control of polymerase activity through GTP activation in the case of β(1,3)-glucan synthase and proteolysis for activation of chitin synthase zymogen are possibilities suggested from *in vitro* experiments whose psychological relevance needs to be established. Introduction of β(1,6)-linkages in glucan may be part of a maturation process (122). Endoglucanase and transglycosylase reactions are thought to have space-making and remodeling roles, but experimental proof is lacking. Certain mannoproteins may associate with structural polymers by a self-assembly process (81). Mutants of *S. cerevisiae* with defects in β(1,3)-glucan synthesis which would aid analysis of these facets of wall synthesis are currently unavailable. In a null mutant created by disrupting the gene for phosphoglucose isomerase, the synthesis of UDP-glucose is effectively isolated from the remainder of cellular metabolism resulting in glucose auxotrophy (2). The impact of glucose starvation on wall synthesis in this mutant should be explored since the consequences should mimic those of glucan-synthesis inhibition.

Despite our elementary understanding of fungal wall synthesis, we can be optimistic that the process constitutes a selective target for the development of effective antifungal compounds. The clarification of the mechanisms by which papulacandins and echinocandins interfere with glucan synthesis and cause cell death is bound to provide valuable guidelines for future work. Given the diversity of cell-wall structure in fungi, however, restricted antifungal spectrum for cell-wall-active agents will always be a concern. Screening methodologies are required which will allow inhibitors of wall metabolism to be recognized easily. Certainly, the development of an assay for fungal β(1,3)-glucan synthase has made it possible to screen for naturally occurring inhibitors of the enzyme, but the compounds found thus far [the neopeptins (110, 139), phosphazomycin (141), and cystargin (140)] lack potency as enzyme inhibitors and do not have antifungal activity. Furthermore, biochemical assays of this type are prone to interference by trivial compounds such as fatty acids (82) and limit the search to a single target reaction while ignoring other, possibly equally valid, but as-yet-undefined targets. (See chapter 21.) Screens in which groups of enzymatic reactions are targeted modeled after the plate assay for inhibitors of fungal wall synthesis described by Selitrennikoff (54, 114) but employing a relevant pathogen may represent a more efficient use of resources. Further elucidation of the mechanism by which the 10.7-kD killer toxin from *Hansenula mrakii* selectively inhibits glucan synthesis in *Saccharomyces cerevisiae* (148) could yield information of value in the search for other inhibitors of wall formation.

The echinocandin antibiotics, with the possible exception of cilofungin (93), are analogous to β-lactams in that they kill only growing cells. This is likely to be the basis of the antagonism or lack of synergy between antibiotics of the echinocandin type and clincally used compounds such as amphotericin B, 5-fluorocytosine, and ketoconazole (14, 93, 120). However, there are indications that synergy with compounds that suppress chitin synthesis is worth more thorough investigation. Using chemical means to measure the chitin content of *Candida albicans*, Pfaller *et al.* observed a four- to six-fold increase in chitin synthesis following exposure to the cells to cilofungin (94). Similarly, Hector and Braun found that treatment of *C. albicans* with papulacandin B caused a stimulation of chitin synthesis as indicated by Calcofluor binding and that combinations of papulacandin B and nikkomycin X or Z were more effective than either drug alone in inhibiting the growth of the organism (41). Synergy between aculeacin A and polyoxin B has been observed (120). As inhibitors of chitin synthase, the polyoxins and nikkomycins block chitin formation in a very specific fashion. A decrease in chitin synthesis can also result indirectly from suppressing glucosamine production with a compound such as anticapsin, an inhibitor of fungal glucosamine-6-phosphate synthase (EC 2.6.1.16) (74). It is interesting in this context that synergy between anticapsin and cilofungin has also been found (93). Bearing in mind that chitin synthesis can also be stimulated by other metabolic changes such as lowering of the ergosterol content of the membrane (8, 92) and that increase in chitin synthesis in response to cilofungin may be accompanied by suppression of ergosterol formation (94), one might reasonably infer that a synthetic balance exists between the formation of the two structural polymers, chitin and β-glucan, and that a decrease in one is counteracted by an increase in the other. If such a compensatory mechanism involves membrane sterol, however, the lack of synergy between inhibitors of ergosterol synthesis and cilofungin requires experimental exploration. And, as discussed earlier, chitin need not be the only polymer to compensate for a lack of β-glucan. Enhancement of α(1,3)-glucan synthesis accompanies inhibition of β-(1,3)-glucan formation by aculeacin in *Schizosaccharomyces pombe* (75). Currently, no low-molecular-weight compounds capable of blocking α(1,3)-glucan synthesis are known, but synergy between such a compound and echinocandin-type antibiotics would be anticipated.

Finally, the clinical utility of echinocandin-type compounds may not be limited to systemic infections caused by certain species of *Candida* and *Malassezia pachydermatis*. The phylogenetic status of *Pneumocystis carinii* has been the subject of much discussion, and on the basis of homology studies with RNA from the small subunits of ribosomes from various organisms, *Pneumocystis carinii* appears to be more closely related to fungi than to animal cells (27; see also chapter 22). Nevertheless, antifungal drugs are ineffective at clearing infections. The report of Matsumoto *et al.* that the cyst wall of *Pneumocystis carinii* contains a layer which is sensitive to β(1,3)-glucanase (Zymolyase) (66) raised the possibility that inhibitors of glucan synthesis might exhibit anti-*Pneumocystis* activity, and the recent study by Schmatz *et al.* confirms that this is the case for L-671,329 in rats (112). Inasmuch as *Pneumocystis* pneumonia is the most common secondary infection afflicting patients with aquired immunodeficiency syndrome, or AIDS, there is hope that echinocandin-type antibiotics may one day fill an important chemotherapeutic niche.

Acknowledgments

I wish to thank my colleagues G. Bills, M. Hammond, M. Kurtz, J. Onishi, D. Schwartz, R. Schwartz, and F. Vanmiddlesworth for helpful and stimulating discussions on subjects covered in this chapter and for their critical review of the manuscript.

References

1. Abbott, B.J., and D.S. Fukuda. 1981. S31794/F-1 Nucleus. U.S. Patent 4,304,716
2. Aguilera, A. 1986. Deletion of the phosphoglucose isomerase structural gene makes growth and sporulation glucose dependent in *Saccharomyces cerevisiae*. Mol. Gen. Genet. **204**:310–316.
3. Andaluz, E., A. Guillen, P. Caceres, and G. Larriba. 1985. Preliminary characterization of two glucan synthetase preparations and their reaction products from *Candida albicans*. Microbiologia **1**:5–17.
4. Andaluz, E., A. Guillen, and G. Larriba. 1986. Preliminary evidence for a glucan acceptor in the yeast *Candida albicans*. Biochem. J. **240**:495–502.
5. Baguley, B.C., G. Römmele, J. Gruner, and W. Wehrli. 1979. Papulacandin B: an inhibitor of glucan synthesis in yeast spheroplasts. Eur. J. Biochem. **97**:345–351.
6. Balint, S., V. Farkas, and S. Bauer. 1976. Biosynthesis of β-glucans catalyzed by a particulate enzyme preparation from yeast. FEBS Lett. **64**:44–47.
7. Bartnicki-Garcia, S. 1968. Cell wall chemistry, morphogenesis, and taxonomy of fungi. Annu. Rev. Microbiol. **22**:87–108.
8. Barug, D., R.A. Samson, and A. Kerkenaar. 1983. Microscopic studies of *Candida albicans* and *Torulopsis glabrata* after *in vivo* treatment with bifonazole. Arzneim.-Forsch./Drug Res. **33**:528–537.
9. Benz, F., F. Knüsel, J. Nüesch, H. Treichler, W. Voser, R. Nyfeler, and W. Keller-Schierlein. 1974. Stoffwechselprodukte von Mikroorganismen. Echinocandin B, ein neuartiges Polypeptid-Antibioticum aus *Aspergillus nidulans* var. *echinulatus*: Isolierung und Bausteine. Helv. Chim. Acta **57**:2459–2477.
10. Boeck, L.D., D.S. Fukuda, B.J. Abbott, and M. Debono. 1989. Deacylation of echinocandin B by *Actinoplanes utahensis*. J. Antibiot. **42**:382–388.
11. Boone, C., S.S. Sommer, A. Hensel, and H. Bussey. 1990. Yeast *KRE* genes provide evidence for a pathway of cell wall β-glucan assembly. J. Cell Biol. **110**:1833–1843.
12. Bozzola, J.J., R.J. Mehta, L.J. Nisbet, and J.R. Valenta. 1984. The effect of aculeacin A and papulacandin B on morphology and cell wall ultrastructure in *Candida albicans*. Can. J. Microbiol. **30**:857–863.
13. Bulo, A.N., S.F. Bradley, and C.A. Kauffman. 1988. Susceptibility of yeast-like fungi to a new antifungal agent, LY121019. Mycoses **31**:330–333.
14. Bulo, A.N., S.F. Bradley, and C.A. Kauffman. 1989. The effect of cilofungin (LY121019) in combination with amphotericin B or flucytosine against *Candida* species. Mycoses **32**:46–52.
15. Bussey, H., C. Boone, H. Zhu, T. Vernet, M. Whiteway, and D.Y. Thomas. 1990. Genetic and molecular approaches to synthesis and action of the yeast killer toxin. Experientia **46**:193–200.

16. Cabib, E., B. Bowers, A. Sburlati, and S.J. Silverman. 1988. Fungal cell wall synthesis: the construction of a biological structure. Microbiol. Sci. **5**:370–375.

17. Cabib, E., and M.S. Kang. 1987. Fungal 1,3-β-glucan synthase. Methods Enzymol. **138**:637–642.

18. Cabib, E., and E.M. Shematek. 1981. Structural polysaccharides of plants and fungi: comparative and morphologenetic aspects, p. 51–90. *In* V. Ginsburg and P. Robbins (eds.), Biology of Carbohydrates. Vol. 1. J. Wiley & Sons, New York.

19. Cassone, A. 1986. Cell wall of pathogenic yeasts and implications for antimycotic therapy. Drugs Exp. Clin. Res. **12**:635–643.

20. Cassone, A., R.E. Mason, and D. Kerridge. 1981. Lysis of growing yeast-form cells of *Candida albicans* by echinocandin: a cytological study. Sabouraudia **19**:97–110.

21. Davila, T., G. San-Blas, and F. San-Blas. 1986. Effect of papulacandin B on glucan synthesis in *Paracoccidioides brasiliensis*. J. Med. Vet. Mycol. **24**:193–202.

22. Debono, M., B.J. Abbott, D.S. Fukuda, M. Barnhart, K.E. Willard, R.M. Molloy, K.H. Michel, J.R. Turner, T.F. Butler, and A.H. Hunt. 1989. Synthesis of new analogs of echinocandin B by enzymatic deacylation and chemical reacylations of the echinocandin B peptide: synthesis of the antifungal agent cilofungin (LY121019). J. Antibiot. **42**:389–397.

23. Debono, M., B.J. Abbott, J.R. Turner, L.C. Howard, R.S. Gordee, A.S. Hunt, M. Barnhart, R.M. Molloy, K.E. Willard, D. Fukuda, T.F. Butler, and D.J. Zeckner. 1988. Synthesis and evaluation of LY121019, a member of a series of semisynthetic analogues of the antifungal lipopeptide echinocandin B. Ann. N.Y. Acad. Sci. **544**:152–167.

24. Delmer, D.P. 1987. Cellulose biosynthesis. Annu. Rev. Plant Physiol. **38**:259–290.

25. Dreyfuss, M.M., and H. Tscherter. 1979. Antibiotic S31794/F-1. U.S. Patent 4,173,629.

26. Duffus, J.H., C. Levi, and D.J. Manners. 1982. Yeast cell-wall glucans. Adv. Microbiol. Physiol. **23**:151–181.

27. Edman, J.C., J.A. Kovacs, H. Masur, D.V. Santi, H.J. Elwood, and M.L. Sogin. 1988. Ribosomal RNA sequence shows *Pneumocystis carinii* to be a member of the fungi. Nature **334**:519–522.

28. Elorza, M.V., A. Murgui, H. Rico, F. Miragall, and R. Sentandreu. 1987. Formation of a new cell wall by protoplasts of *Candida albicans*: effect of papulacandin B, tunicamycin and nikkomycin. J. Gen. Microbiol. **133**:2315–2325.

29. Evans, D.A., and A.E. Weber. 1987. Synthesis of the cyclic hexapeptide echinocandin D. New approaches to the asymmetric synthesis of β-hydroxy-α-amino acids. J. Am. Chem. Soc. **109**:7151–7157.

30. Fevre, M. 1983. Inhibitors of synthesis of lipid-linked saccharides also inhibit β-glucan synthesis by cell-free extracts of the fungus *Saprolegnia monoica*. J. Gen. Microbiol. **129**:3007–3013.

31. Fleet, G.H. 1985. Composition and structure of yeast cell walls, p. 24–56. *In* M.R. McGinnis (ed.), Current Topics in Medical Mycology. Vol. 1. Springer-Verlag, New York.

32. Fromtling, R.A., and G.K. Abruzzo, 1989. L–671,329, a new antifungal agent. III. *In vitro* activity, toxicity and efficacy in comparison to aculeacin. J. Antibiot. **42**:174–178.

33. Frost, D.J., S.M. Read, R.R. Drake, B.E. Haley, and B.P. Wasserman. 1990. Identification of the UDP-glucose-binding polypeptide of callose synthase from *Beta vulgaris* L. by photoaffinity labeling with 5-azido-UDP-glucose. J. Biol. Chem. **265**:2162–2167.

34. Gopal, P.K., M.G. Shepherd, and P.A. Sullivan. 1984. Analysis of wall glucans from yeast, hyphal and germ-tube forming cells of *Candida albicans*. J. Gen. Microbiol. **130**:3295–3301.

35. Gordee, R.S., D.J. Zeckner, L.F. Ellis, A.L. Thakkar, and L.C. Howard. 1984. *In vitro* and *in vivo* anti-*Candida* activity and toxicology of LY121019. J. Antibiot. **37**:1054–1065.

36. Gordee, R.S., D.J. Zeckner, L.C. Howard, W.E. Alborn Jr., and M. Debono. 1988. Anti-*Candida* activity and toxicology of LY121019, a novel semisynthetic polypeptide antifungal antibiotic. Ann. N.Y. Acad. Sci. **544**:294–309.

37. Graybill, J.R. 1989. New antifungal agents. Eur. J. Clin. Microbiol. Infect. Dis. **8**:402–412.

38. Guillen, A., F. Leal, E. Andaluz, and G. Larriba. 1985. Endogenous factors that modulate yeast glucan synthetase in cell-free extracts. Biochim. Biophys. Acta **842**:151–161.

39. Hall, G.S., C. Myles, K.J. Pratt, and J.A. Washington, 1988. Cilofungin (LY121019), an antifungal agent with specific activity against *Candida albicans* and *Candida tropicalis*. Antimicrob. Agents Chemother. **32**:1331–1335.

40. Hanson, L.H., and D.A. Stevens. 1989. Evaluation of cilofungin, a lipopeptide antifungal agent, *in vitro* against fungi isolated from clinical specimens. Antimicrob. Agents Chemother. **33**:1391–1392.

41. Hector, R.F., and P.C. Braun. 1986. Synergistic action of nikkomycins X and Z with papulacandin B on whole cells and regenerating protoplasts of *Candida albicans*. Antimicrob. Agents Chemother. **29**:389–394.

42. Hobbs, M., J. Perfect, and D. Durack. 1988. Evaluation of *in vitro* antifungal activity of LY121019. Eur. J. Clin. Microbiol. Infect. Dis. **7**:77–80.

43. Howard, L.C., M.D. Gunnoe, M. Debono, B.J. Abbott, and J.R. Turner. 1982. Utilization of *in vitro* erythrocyte fragility to predict toxicology of a group of antifungal echinocandin B analogs. Toxicologist **2**:1984.

44. Hutchins, K., and H. Bussey. 1983. Cell wall receptor for yeast killer toxin: involvement of (1→6)-β-D-glucan. J. Bacteriol. **154**:161–169.

45. Iwata, K., Y. Yamamoto, H. Yamaguchi, and T. Hiratani. 1982. *In vitro* studies of aculeacin A, a new antifungal antibiotic. J. Antibiot. **35**:203–209.

46. Jamas, S., C.-K. Rha, and A.J. Sinskey. 1986. Morphology of yeast cell wall as affected by genetic manipulation of β(1→6) glycosidic linkage. Biotechnol. Bioeng. **28**:769–784.

47. Kang, M.S., and E. Cabib. 1986. Regulation of fungal cell wall growth: a guanine nucleotide-binding, proteinaceous component required for activity of (1→3)-β-D-glucan synthase. Proc. Natl. Acad. Sci. USA **83**:5808–5812.

48. Kang, M.S., P.J. Szaniszlo, V. Notario, and E. Cabib. 1986. The effect of papulacandin B on (1→3)-β-D-glucan synthetases. A possible relationship between inhibition and enzyme conformation. Carbohydr. Res. **149**:13–21.

49. Kauss, H. 1987. Some aspects of calcium-dependent regulation in plant metabolism. Annu. Rev. Plant Physiol. **38**:47–72.

50. Kauss, H., and W. Jeblick. 1987. Solubilization, affinity chromatography and Ca^{2+}/polyamine activation of the plasma membrane-located 1,3-β-D-glucan synthase. Plant Sci. **48**:63–69.

51. Keller-Juslen, C., M. Kuhn, H.R. Loosli, T.J. Petcher, H.P. Weber, and A. von Wartburg. 1976. Struktur des Cyclopeptid-Antibioticums SL7810 (=Echinocandin B). Tetrahedron Lett. **1976**:4147–4150.

52. Keller-Schierlein, W., and J. Widmer. 1976. Stoffwechselprodukte von Mikroorganismen. Uber die aromatische Aminosaure des Echinocandins B: 3,4-Dihydroxyhomotyrosin. Helv. Chim. Acta **59**:2021–2031.

53. Kimura, Y., H. Matsunaga, N. Yasuda, T. Tatsuki, and T. Suzuki. 1987. Polymyxin acylase: a new enzyme for preparing starting materials for semisynthetic polymyxin antibiotics. Agric. Biol. Chem. **51**:1617–1623.

54. Kirsch, D.R., and M.H. Lai. 1986. A modified screen for the detection of cell wall-acting antifungal compounds. J. Antibiot. **39**:1620–1622.

55. Komori, T., and Y. Itoh. 1985. Chaetiacandin, a novel papulacandin. II. Structure determination. J. Antibiot. **38**:544–546.

56. Komori, T., M. Yamashita, Y. Tsurumi, and M. Kohsaka. 1985. Chaetiacandin, a novel papulacandin. I. Fermentation, isolation and characterization. J. Antibiot. **38**:455–459.

57. Kopecka, M. 1984. Lysis of growing cells of *Saccharomyces cerevisiae* induced by papulacandin B. Folia Microbiol. (Praha) **29**:115–119.

58. Kopecka, M. 1984. Papulacandin B: inhibitor of biogenesis of (1→3)-β-D-glucan fibrillar component of the cell wall of *Saccharomyces cerevisiae* protoplasts. Folia Microbiol. (Praha) **29**:441–449.

59. Koyama, G. 1974. Metabolites of microorganisms. The crystal and molecular structure of 3-hydroxy-4-methylproline. Helv. Chim. Acta **57**:2477–2483.

60. Kurokawa, N., and Y.Ohfune. 1986. Total synthesis of echinocandins. I. Stereocontrolled syntheses of the constituent amino acids. J. Am. Chem. Soc. **108**:6041–6043.

61. Kurokawa, N., and Y. Ohfune. 1986. Total synthesis of echinocandins. II. Total synthesis of echinocandin D via efficient peptide coupling reactions. J. Am. Chem. Soc. **108**:6043–6045.

62. Larriba, G., M. Morales, and J. Ruiz-Herrera. 1981. Biosynthesis of β-glucan microfibrils by cell-free extracts from *Saccharomyces cerevisiae*. J. Gen. Microbiol. **124**:375–383.

63. Leal, F., J. Ruiz-Herrera, J.R. Villanueva, and G. Larriba. 1984. An examination of factors affecting the instability of *Saccharomyces cerevisiae* glucan synthetase in cell free extracts. Arch. Microbiol. **137**:209–214.

64. Lew, M.A. 1989. Diagnosis of systemic *Candida* infections. Ann. Rev. Med. **40**:87–97.

65. Lopez-Romero, E., and J. Ruiz-Herrera. 1977. Biosynthesis of β-glucans by cell-free extracts from *Saccharomyces cerevisiae*. Biochim. Biophys. Acta **500**:372–384.

66. Matsumoto, Y., S. Matsuda, and T. Tegoshi. 1989. Yeast glucan in the cyst wall of *Pneumocystis carinii*. J. Protozool. **36**:21S–22S.

67. McIntyre, K.A., and J.N. Galgiani. 1989. pH and other effects on the antifungal activity of cilofungin (LY121019). Antimicrob. Agents Chemother. **33**:731–735.

68. Meaden, P., K. Hill, J. Wagner, D. Slipetz, S.S. Sommer, and H. Bussey. 1990. The yeast *KRE5* gene encodes a probable ER protein required for $(1\rightarrow6)$-β-glucan synthesis and normal cell growth. Mol. Cell. Biol. **10**:3013–3019.

69. Mehta, R.J., J.M. Boyer, and C.H. Nash. 1984. Aculeacin resistant mutants of *Candida albicans*: alterations in cellular lipids. Microbios Lett. **27**:25–29.

70. Mehta, R.J., C.H. Nash, S.F. Grappel, and P. Actor. 1982. Aculeacin A resistant mutants of *Candida albicans*. J. Antibiot. **35**:707–711.

71. Melchinger, W., and J. Muller. 1987. Studies on the *in vitro*-sensitivity of yeast strains isolated from clinical specimens to LY121019, a new antifungal agent. Mykosen **30**:605–608.

72. Meshulam, T., S.M. Levitz, R.D. Diamond, and A.M. Sugar. 1989. Effect of cilofungin (LY121019), a fungal cell wall synthesis inhibitor, on interactions of *Candida albicans* with human neutrophils. J. Antimicrob. Chemother. **24**:741–745.

73. Meunier, F., C. Lambert, and P. Van der Auwera. 1989. *In-vitro* activity of cilofungin (LY121019) in comparison with amphotericin B. J. Antimicrob. Chemother. **24**:325–331.

74. Milewski, S., H. Chmara, and E. Borowski. 1986. Antibiotic tetaine—a selective inhibitor of chitin and mannoprotein biosynthesis in *Candida albicans*. Arch. Microbiol. **145**:234–240.

75. Miyata, M., T. Kanbe, and K. Tanaka. 1985. Morphological alterations of the fission yeast *Schizosaccharomyces pombe* in the presence of aculeacin A: spherical wall formation. J. Gen. Microbiol. **131**:611–621.

76. Miyata, M., J. Kitamura, and H. Miyata. 1980. Lysis of growing fission-yeast cells induced by aculeacin A, a new antifungal antibiotic. Arch. Microbiol. **127**:11–16.

77. Mizoguchi, J., T. Saito, K. Mizuno, and K. Hayano. 1977. On the mode of action of a new antifungal antibiotic, aculeacin A: inhibition of cell wall synthesis in *Saccharomyces cerevisiae*. J. Antibiot. **30**:308–313.

78. Mizuno, K., A. Yagi, S. Satoi, M. Takada, M. Hayashi, K. Asano, and T. Matsuda. 1977. Studies on aculeacin. I. Isolation and characterization of aculeacin A.J. Antibiot. **30**:297–302.

79. Morrison, C.J., and D.A. Stevens. 1990. Comparative effects of cilofungin and amphotericin B on experimental murine candidiasis. Antimicrob. Agents Chemother. **34**:746–750.

80. Mukhopadhyay, T., B.N. Ganguli, H.W. Fehlhaber, H. Kogler, and L. Vertesy. 1987. Mulundocandin, a new lipopeptide antibiotic. II. Structure elucidation. J. Antibiot. **40**:281–289.

81. Murgui, A., M.V. Elorza, and R. Sentandreu. 1986. Tunicamycin and papulacandin B inhibit incorporation of specific mannoproteins into the wall of *Candida albicans* regenerating protoplasts. Biochim. Biophys. Acta **884**:550–558.

82. Niwano, M., E. Mirumachi, M. Uramoto, and K. Isono. 1984. Fatty acids as inhibitors of microbiol cell wall synthesis. Agric. Biol. Chem. **48**:1359–1360.

83. Notario, V., H. Kawai, and E. Cabib. 1982. Interaction between yeast β-(1→3)glucan synthetase and activating phosphorylated compounds. A kinetic study. J. Biol. Chem. **257**:1902–1905.

84. Odds, F.C. 1988. Activity of cilofungin (LY121019) against *Candida* species *in vitro*. J. Antimicrob. Chemother. **22**:891–897.

85. Orlean, P.A.B. 1982. (1,3)-β-D-Glucan synthase from budding and filamentous cultures of the dimorphic fungus *Candida albicans*. Eur. J. Biochem. **127**:397–403.

86. Orlean, P.A.B., and S.M. Ward. 1983. Sodium fluoride stimulates (1,3)-β-glucan synthase from *Candida albicans*. FEMS Microbiol. Lett. **18**:31–35.

87. Pache, W., C. Keller, and M. Kuhn. 1978. Cleavage of echinocandin B with polymoxin acylase liberating the fatty acid and reacylation of the peptide moiety. Experientia **34**:1670–1671.

88. Padula, A., and H.F. Chambers. 1989. Evaluation of cilofungin (LY121019) for treatment of experimental *Candida albicans* endocarditis in rabbits. Antimicrob. Agents Chemother. **33**:1822–1823.

89. Perez, P., I. Garcia-Acha, and A. Duran. 1983. Effect of papulacandin B on the cell wall and growth of *Geotrichum lactis*. J. Gen. Microbiol. **129**:245–250.

90. Perez, P., R. Varona, I. Garcia-Acha, and A. Duran. 1981. Effect of papulacandin B and aculeacin A on β-(1,3)-glucan synthase from *Geotrichum lactis*. FEBS Lett. **129**:249–252.

91. Perfect, J.R., M.M. Hobbs, K.A. Wright, and D.T. Durack. 1989. Treatment of experimental disseminated candidiasis with cilofungin. Antimicrob. Agents Chemother. **33**:1811–1812.

92. Pesti, M., J.M. Campbell, and J.F. Peberdy. 1981. Alteration of ergosterol content and chitin synthase activity in *Candida albicans*. Curr. Microbiol. **5**:187–190.

93. Pfaller, M., R. Gordee, T. Gerarden, M. Yu, and R. Wenzel. 1989. Fungicidal activity of cilofungin (LY121019) alone and in combination with anticapsin or other antifungal agents. Eur. J. Clin. Microbiol. Infect. Dis. **8**:564–567.

94. Pfaller, M., J. Riley, and T. Koerner. 1989. Effects of cilofungin (LY121019) on carbohydrate and sterol composition of *Candida albicans*. Eur. J. Clin. Microbiol. Infect. Dis. **8**:1067–1070.

95. Pfaller, M.A., T. Gerarden, M. Yu, and R.P. Wenzel. 1988. Influence of *in vitro* susceptibility testing conditions on the anti-candidal activity of LY121019. Diagn. Microbiol. Infect. Dis. **11**:1–9.

96. Pfaller, M.A., S. Wey, T. Gerarden, A. Houston, and R.P. Wenzel. 1989. Susceptibility of nosocomial isolates of *Candida* species to LY121019 and other antifungal agents. Diagn. Microbiol. Infect. Dis. **12**:1–4.

97. Picard, M., T. Meshulam, and A.M. Sugar. 1988. *In vitro* activity of LY121019 against *Candida* species and *Torulopsis glabrata*. Eur. J. Clin. Microbiol. Infect. Dis. **7**:432–433.

98. Quigley, D.R., and C.P. Selitrennikoff. 1984. $\beta(1\rightarrow3)$Glucan synthase activity of *Neurospora crassa*: kinetic analysis of negative effectors. Exp. Mycol. **8**:320–333.

99. Quigley, D.R., and C.P. Selitrennikoff. 1984. $\beta(1\rightarrow3)$Glucan synthase activity of *Neurospora crassa*: stabilization and partial characterization. Exp. Mycol. **8**:202–214.

100. Read, S.M., and D.P. Delmer. 1987. Inhibition of mung bean UDP-glucose: $(1\rightarrow3)$-β-glucan synthase by UDP-pyridoxal. Evidence for an active-site amino group. Plant Physiol. **85**:1008–1015.

101. Reiss, E. 1977. Serial enzymatic hydrolysis of cell walls of two serotypes of yeast-form *Histoplasma capsulatum* with $\alpha(1\rightarrow3)$-glucanase, $\beta(1\rightarrow3)$glucanase, Pronase, and chitinase. Infect. Immun. **16**:181–188.

102. Rennie, R.P., and L. Hellman. 1989. Comparative activity *in vitro* of cilofungin (LY121019) with other agents used for treatment of deep-seated *Candida* infections. Mycoses **32**:145–150.

103. Rine, J., W. Hansen, E. Hardeman, and R.W. Davis. 1983. Targeted selection of recombinant clones through gene dosage effects. Proc. Natl. Acad. Sci. USA **80**:6750–6754.

104. Römmele, G., P. Traxler, and W. Wehrli. 1983. Papulacandins—the relationship between chemical structure and effect on glucan synthesis in yeast. J. Antibiot. **36**:1539–1542.

105. Roy, K., T. Mukhopadhyay, G.C.S. Reddy, K.R. Desikan, and B.N. Ganguli. 1987. Mulundocandin, a new lipopeptide antibiotic. I. Taxonomy, fermentation, isolation and characterization. J. Antibiot. **40**:275–280.

106. Sakaitani, M., and Y. Ohfune. 1989. Stereoselective hydroxylation of a peptide side chain. The synthesis of the echinocandin right-half equivalent. Tetrahedron Lett. **30**:2251–2254.

107. San-Blas, G., and F. San-Blas. 1977. *Paracoccidioides brasiliensis*: cell wall structure and virulence. Mycopathologia **62**:77–86.

108. Sarko, A., H.C. Wu, and C.T. Chuah. 1983. Multiple helical glucans. Biochem. Soc. Trans. **11**:139–142.

109. Satoi, S., A. Yagi, K. Asano, K. Mizuno, and T. Watanabe. 1977. Studies on aculeacin. II. Isolation and characterization of aculeacins B, C, D, E, F and G.J. Antibiot. **30**:303–307.

110. Satomi, T., H. Kusakabe, G. Nakamura, T. Nishio, M. Uramoto, and K. Isono. 1982. Neopeptins A and B, new antifungal antibiotics. Agric. Biol. Chem. **46**:2621–2623.

111. Sawistowska-Schröder, E.T., D. Kerridge, and H. Perry. 1984. Echinocandin inhibition of 1,3-β-D-glucan synthase from *Candida albicans*. FEBS Lett. **173**:134–138.

112. Schmatz, D.M., M. Romancheck, L.A. Pittarelli, R.E. Schwartz, R.A. Fromtling, K.H. Nollstadt, F.L. Vanmiddlesworth, K.E. Wilson, and M.J. Turner. 1990. Treatment of *Pneumocystis carinii* pneumonia with β 1,3 glucan synthesis inhibitors. Proc. Natl. Acad. Sci. USA **87**:5950–5954.

113. Schwartz, R.E., R.A. Giacobbe, J.A. Bland, and R.L. Monaghan. 1989. L-671,329, a new antifungal agent. I. Fermentation and isolation. J. Antibiot. **42**:163–167.

114. Selitrennikoff, C.P. 1983. Use of a temperature-sensitive, protoplast-forming *Neurospora crassa* strain for the detection of antifungal antibiotics. Antimicrob. Agents Chemother. **23**:757–765.

115. Shematek, E.M., J.A. Braatz, and E. Cabib. 1980. Biosynthesis of the yeast cell wall. I. Preparation and properties of β-(1→3)glucan synthetase. J. Biol. Chem. **255**:888–894.

116. Shematek, E.M., and E. Cabib. 1980. Biosynthesis of the yeast cell wall. II. Regulation of β-(1→3)glucan synthetase by ATP and GTP.J. Biol. Chem. **255**:895–902.

117. Shepherd, M.G. 1987. Cell envelope of *Candida albicans*. CRC Crit. Rev. Microbiol. **15**:7–25.

118. Shibata, N., K. Mizugami, K. Takano, and S. Suzuki. 1983. Isolation of mannan-protein complexes from viable cells of *Saccharomyces cerevisiae* X2180-1A wild type and *Saccharomyces cerevisiae* X2180-1A-5 mutant strains by the action of Zymolyase-60,000. J. Bacteriol. **156**:552–558.

119. Shiota, M., T. Nakajima, A. Satoh, M. Shida, and K. Matsuda. 1985. Comparison of β-glucan structures in a cell wall mutant of *Saccharomyces cerevisiae* and the wild type. J. Biochem. **98**:1301–1307.

120. Sitrin, R.D., G. Chan, J. Dingerdissen, C. DeBrosse, R. Mehta, G. Roberts, S. Rottschaefer, D. Staiger, J. Valenta, K.M. Snader, R.J. Stedman, and J.R.E. Hoover. 1988. Isolation and structure determination of *Pachybasium* cerebrosides which potentiate the antifungal activity of aculeacin. J. Antibiot. **41**:469–480.

121. Sloan, M.E., P. Rodis, and B.P. Wasserman. 1987. CHAPS Solubilization and functional reconstitution of β-glucan synthase from red beet (*Beta vulgaris* L.) storage tissue. Plant Physiol. **85**:516–522.

122. Sonnenberg, A.S.M., J.H. Sietsma, and J.G.H. Wessels. 1985. Spatial and temporal differences in the synthesis of (1→3)-β and (1→6)-β linkages in a wall glucan of *Schizophyllum commune*. Exp. Mycol. **9**:141–148.

123. Spitzer, E.D., S.J. Travis, and G.S. Kobayashi. 1988. Comparative *in vitro* activity of LY121019 and amphotericin B against clinical isolates of *Candida* species. Eur. J. Clin. Microbiol. Infect. Dis. **7**:80–81.

124. Stevens, D.A. 1988. The new generation of antifungal drugs. Eur. J. Clin. Microbiol. Infect. Dis. **7**:732–735.

125. Strippoli, V., F.D. D'Auria, and N. Simonetti. 1988. A study of the antifungal activity of LY121019, a new echinocandin derivative. Chemioterapia **7**:33–37.

126. Surarit, R., P.K. Gopal, and M.G. Shepherd. 1988. Evidence for a glycosidic linkage between chitin and glucan in the cell wall of *Candida albicans*. J. Gen. Microbiol. **134**:1723–1730.

127. Szaniszlo, P.J., M.S. Kang, and E. Cabib. 1985. Stimulation of $\beta(1\rightarrow3)$glucan synthetase of various fungi by nucleoside triphosphates: generalized regulatory mechanism for cell wall biosynthesis. J. Bacteriol. **161**:1188–1194.

128. Taft, C.S., and C.P. Selitrennikoff. 1988. LY121019 inhibits *Neurospora crassa* growth and (1–3)-β-D-glucan synthase. J. Antibiot. **41**:697–701.

129. Taft, C.S., T. Stark, and C.P. Selitrennikoff. 1988. Cilofungin (LY121019) inhibits *Candida albicans* (1–3)-β-D-glucan synthase activity. Antimicrob. Agents Chemother. **32**:1901–1903.

130. Takeshima, H., J. Inokoshi, Y. Takada, H. Tanaka, and S. Omura. 1989. A deacylation enzyme for aculeacin A, a neutral lipopeptide antibiotic, from *Actinoplanes utahensis*: purification and characterization. J. Biochem. **105**:606–610.

131. Tkacz, J.S. 1984. *In vivo* synthesis of β–1,6-glucan in *Saccharomyces cerevisiae*, p. 287–295. *In* C. Nombela (ed.), Microbial Cell Wall Synthesis and Autolysis. Elsevier Science Publishers, New York.

132. Torres-Rodriguez, J.M., A. Carrillo-Munoz, C. Gallach-Bau, and N. Madrenys. 1989. Susceptibility of *Candida* species to cilofungin (LY–121019). Mycoses **32**:316–318.

133. Traber, R., C. Keller-Juslen, H.-R. Loosli, M. Kuhn, and A. von Wartburg. 1979. Cyclopeptid-Antibiotika aus *Aspergillus* Arten. Struktur der Echinocandine C und D. Helv. Chim. Acta **62**:1252–1267.

134. Traxler, P., H. Fritz, H. Fuhrer, and W.J. Richter. 1980. Papulacandins, a new family of antibiotics with antifungal activity. Structures of papulacandins A, B, C and D. J. Antibiot. **33**:967–978.

135. Traxler, P., H. Fritz, and W.J. Richter. 1977. Zur Struktur von Papulacandin B, einem neuen antifungischen Antibiotikum. Helv. Chim. Acta **60**:578–584.

136. Traxler, P., J. Gruner, and J.A.L. Auden. 1977. Papulacandins, a new family of antibiotics with antifungal activity. I. Fermentation, isolation, chemical and biological characterization of papulacandins A, B, C, D and E.J. Antibiot. **30**:289–296.

137. Traxler, P., W. Tosch, and O. Zak. 1987. Papulacandins—synthesis and biological activity of papulacandin B derivatives. J. Antibiot. **40**:1146–1164.

138. Tscherter, H., and M.M. Dreyfuss. 1982. Antibiotics S41062/F-1, -6, and -7 useful as antimycotic agents. Belgian patent 889,955.

139. Ubakata, M., M. Uramoto, and K. Isono. 1984. The structure of neopeptins, inhibitors of fungal wall biosynthesis. Tetrahedron Lett. **25**:423–426.

140. Uramoto, M., Y. Itoh, R. Sekiguchi, K. Shin-ya, H. Kusakabe, and K. Isono. 1988. A new antifungal antibiotic, cystargin: fermentation, isolation, and characterization. J. Antibiot. **41**:1763–1768.

141. Uramoto, M., Y.-C. Shen, N. Takizawa, H. Kusakabe, and K. Isono. 1985. A new antifungal antibiotic, phosphazomycin A. J. Antibiot. **38**:665–668.

142. Van der Auwera, P., and F. Meunier. 1989. *In-vitro* effects of cilofungin (LY121019), amphotericin B and amphotericin B-deoxycholate on human polymorphonuclear leucocytes. J. Antimicrob. Chemother. **24**:747–763.

143. Varona, R., P. Perez, and A. Duran. 1983. Effect of papulacandin B on β-glucan synthesis in *Schizosaccharomyces pombe*. FEMS Microbiol. Lett. **20**:243–247.

144. Wichmann, C.F., J.M. Liesch, and R.E. Schwartz. 1989. L-671,329, a new antifungal agent. II. Structure determination. J. Antibiot. **42**:168–173.

145. Yamaguchi, H., T. Hiratani, M. Baba, and M. Osumi. 1985. Effect of aculeacin A, a wall-active antibiotic, on synthesis of the yeast cell wall. Microbiol. Immunol. **29**:609–623.

146. Yamaguchi, H., T. Hiratani, M. Baba, and M. Osumi. 1987. Effect of aculeacin A on reverting protoplasts of *Candida albicans*. Microbiol. Immunol. **31**:625–638.

147. Yamaguchi, H., T. Hiratani, K. Iwata, and Y. Yamamoto. 1982. Studies on the mechanism of antifungal action of aculeacin A. J. Antibiot. **35**:210–219.

148. Yamamoto, T., T. Hiratani, H. Hirata, M. Imai, and H. Yamaguchi. 1986. Killer toxin from *Hansenula mrakii* selectively inhibits cell wall synthesis in a sensitive yeast. FEBS Lett. **197**:50–54.

20

Virulence-Associated Mannoproteins of Candida Albicans

Richard A. Calderone and *Makiko Fukayama*

I. Introduction

The cell surface of *Candida albicans* has been the subject of considerable study over the past 10 years (69). What was originally thought to be a relatively static arrangement of three polysaccharides (glucan, chitin, and mannan), lipid, and some protein is now seen as an exceedingly complex mix of numerous components. There is good evidence, for example, that classes of mannoproteins exist which differ in regard to the degree of glycosylation, molecular weight, and function (enzymatic, structural, immunosuppression, ligand/receptors). Furthermore, mannoprotein can be growth-form specific, *i.e.*, found exclusively on germ tubes (15,77), variable in expression, found in high numbers on exponential but not on stationary-phase yeast cells (or vice versa) (8–10), or spatially rearranged so as to appear as a cell-surface component in hyphae (or germ tubes) but buried within the wall in yeast-phase cells (57).

Two complementary approaches are used to dissect the cell surface of *C. albicans*. The first approach catalogs mannoproteins according to molecular mass and association with growth form and expression (variable *vs.* constantly expressed). The outcome of this exhaustive experimentation is a topographical map of these macromolecules. The second approach involves the detailed study of selected mannoproteins. These studies may focus upon function, synthesis, and the questions raised above in regard to the first approach. Both approaches are complementary and essential to the understanding of the cell-surface structure for this important human pathogen.

This review will focus on the function of selected mannoproteins, which is possible in the case of the integrin-like macromolecules, emphasizing their role in adhesion. Since these macromolecules contribute to the antigenicity of the organism and appear to be extremely important in recognition and invasion of host cells, the biosynthetic and regulatory pathways of these cellular mannoproteins potentially may provide targets for antifungal agents.

II. Adhesins of *C. Albicans*

The number of patients with systemic candidiasis is increasing, and means for its prevention and treatment are urgently needed. Deep-seated candidiasis generally results from widespread and hematogenous dissemination of the organism, which in most cases is derived from resident fungi, *i.e.*, endogenous infection (1). Adherence to host tissues is important as a first step in systemic infection, since colonization by *Candida* species is a necessary prerequisite for the establishment of certain infections. The specific attachment of microorganisms is mediated by surface constituents called adhesins. On the other hand, host cells have receptors or adhesive structures on their surfaces. Since pseudohyphal forms of *C. albicans* usually are not observed *in vivo* until conditions are favorable for tissue invasion, blastoconidia must be able to adhere to mucosal epithelial cells. Thus, adherence of blastoconidia is one important factor which enables the organism to persist on the mucosa. Other virulence determinants of *Candida* might include the ability to form hyphae (72), to resist phagocytosis (75), and to produce extracellular hydrolytic enzymes such as proteinase (56) and phospholipase (2).

Adherence to epithelial cells (36), epidermal corneocytes (60), intestinal epithelial cells (52), vascular endothelium (37), and fibrin-platelet matrices (47) *in vitro* correlates with pathogenicity of *Candida* species *in vivo*. Furthermore, more direct evidence has been obtained for the importance of adhesion in the infective process. Lehrer *et al.* (41) showed that *Candida* mutants with reduced ability to adhere *in vitro* are also less virulent in a mouse model of vaginal candidiasis and in the rabbit model of endocarditis (13). Thus, it seems clear that adhesion is an important event in the initiation of an infection (3,17).

A. *Is the* Candida *Adhesin a Mannoprotein?*

Because the cell surface of *Candida albicans* is rich in mannoproteins, it is likely that such macromolecules might mediate cell-cell or cell-matrix adherence for the organism. Indeed, several lines of evidence suggest that the *Candida* adhesin is a mannoprotein. Adhesion of *C. albicans* strains to epithelial cells can be significantly increased by growing the organisms in a medium containing a high concentration of sucrose or galactose as the carbon source (19,50,51). This increase in adhesion correlates with the accelerated production of a fibrillar surface layer which is mannoprotein, as revealed by cytochemical (80) and agglutination studies (14) with concanavalin A (a lectin which binds to α-D-mannosyl residues). Moreover, tunicamycin, an antibiotic which inhibits protein glycosylation, also decreased adhesion to buccal cells as compared with untreated organisms (18). Interestingly, this fibrillar mannoprotein inhibited intracellular killing of *C. albicans* by neutrophils following yeast phagocytosis (29).

Additional evidence for a mannoprotein-mediated adherence is provided by studies showing diminished attachment of *Candida* pretreated with concanavalin A (64,65), proteolytic enzymes (39,73), or reducing agents (39). Further, extracted cell-wall fragments of *Candida* rich in mannan can attach to vaginal epithelial cells; attachment is inhibited by treatment with α-mannosidase and proteases (39). Critchley and Douglas (16) purified a mannoprotein from a crude culture filtrate which blocked adherence of *C. albicans* more effectively than the original material. Purification was accomplished by Con A–affinity chromatography, which yielded two active fractions following elution with α-methyl mannoside. These fractions were pooled and subsequently chromatographed on DEAE-cellulose. The active material from this column was about 30-fold more active in blocking adherence than the original culture filtrate. Treatment of the culture filtrate material with heat, dithiothreitol, HCl, or various proteases (except papain) significantly decreased the inhibitory activity of the filtrate in blocking adherence of the organism. However, periodate or mannosidase treatments had no effect. This observation indicates that the protein component of the mannoprotein may be more important in recognition of host cells than the oligosaccharide; however, more work is needed to verify this.

B. Adhesion-Deficient C. Albicans

A strain of C. *albicans* with reduced ability to adhere to a variety of mammalian cells including buccal (26), vaginal (41), and endothelial cells (Edwards, personal communication) has been described. Accordingly, this strain is avirulent in animal models of endocarditis (13) and murine vaginitis (41). The strain of *C. albicans* used in these studies (designated m–10) was a spontaneous, cerulenin-resistant mutant with unimpaired ability to germinate and a growth rate similar to parental cells. The major antigens of m-10 have been studied and compared with those of parental cells (strain 4918). We found that strain m-10 lacked a mannoprotein of low electromobility when studied by crossed immunoelectrophoresis and compared to its parent, strain 4918 (12). This antigen was identified as a mannoprotein by its reactivity with Con A. Further, antibody to the mannoprotein could be found in all patient sera (10/10) examined by crossed immunoelectrophoresis. Studies are underway to purify this component and determine its functional characteristics as an adhesin. Table 20.1 summarizes the differences in virulence, adherence, and biochemical properties of m-10 *vs.* its parental strain, 4918.

C. Adhesins other than Mannoprotein

Several investigators have shown that solubilized preparations of chitin (CSE) extracted from *C. albicans* cell walls blocked attachment of yeast cells to human vaginal epithelial and intestinal mucosal cells but not to buccal epithelial cells (40,42,67,68). The CSE was without effect when preincubated with *C. albicans* prior to measuring adherence to buccal epithelial cells, but inhibited attachment

Table 20.1. C. albicans *m−10: phenotypic properties compared to parental strain 4918 (wt)*

In vitro *adherence*[a]	*% Of wt*
Fibrin platelets	*20*
Vaginal epithelial cells	*50*
Endothelial cells	*50*
EAiC3b	*47*
Virulence	
Rabbit endocarditis	10^7 I.D.$_{50}$ (wt 10^4)
Murine vaginitis	30% of control
Antigenic/biochemical	
Yeast cells	Lacks mannoprotein of low electromobility
Hyphae	Lacks 47–50-kDa mannoprotein, reduced 68–71, 55 kDa

[a] Fibrin platelet–matrices derived from rabbit, vaginal, and endothelial cells (umbilical cord) from human; EAiC3b = sheep erythrocytes conjugated with antibody (A) and iC3b.

when preincubated with human vaginal epithelial cells. N-acetylglucosamine (NAGA) also inhibited attachment but was not as effective as the CSE or nonextracted chitin. CSE also blocked attachment of *C. albicans in vivo* to vaginal tissue (42,67) and intestinal mucosa (68). Thus far, the active material from the crude CSE has not been identified. Fractionation of CSE by gel chromatography on Sephadex G–150 resulted in the isolation of two fractions, one of which was inhibitory (40). This fraction contained several proteins and some lipid, but no amino sugars.

D. Host Influences on Adherence of C. Albicans

Chief among the factors which influence the extent of colonization by *C. albicans* is the role other organisms, especially bacteria, play in the process. These relationships between *Candida* and bacteria have been studied in both the oral cavity (43) and the gastrointestinal tract (33–35). In the former study, streptococci inhibited the colonization of the oral mucosa by *C. albicans*. Likewise, *C. albicans* adheres in the gut tissue of normal mice to a significantly lower extent than in mice receiving various antibiotics which reduce the level of gastrointestinal bacteria (33–35). Antibiotics which reduce the anaerobic bacterial population were especially able to predispose animals to mucosal colonization by *C. albicans*. Mice treated with clindamycin, vancomycin, or penicillin G had higher levels of *Candida* in the lumenal wall of the intestine and were more susceptible to *Candida* dissemination. On the other hand, animals treated with gentamicin or erythromycin prior to challenge with *C. albicans* had much lower gut populations of *Candida*. No disseminated infection occurred. Following treatment with various antibiotics, colonization by *Candida* increased as the total anaerobic population was reduced, and the numbers of enteric bacilli increased significantly on mucosal surfaces as well as in the cecal contents. Similar observations have also been made by using an *in vitro* adherence assay (35). Intestinal slices from animals variously treated were infected *in vitro* with *C. albicans*. Adherence was greater to intestinal slices from antibiotic-treated animals, although the extent of adherence was reduced when volatile fatty acids or intestinal filtrates (free of bacteria) were incubated with the intestinal slices and *Candida*. This observation may indicate that specific bacteria and/or bacterial components reduce the association of *Candida* with the gut intestinal mucosa.

The nature of the host receptor on different mucosal surfaces which recognizes the *Candida* ligand (adhesin) is not completely understood. Initial studies indicated that fibronectin, a high-molecular-weight glycoprotein found in a variety of mammalian cells, may have receptor function for *C. albicans* (70). *In vitro* studies have shown that the yeast form of *C. albicans* binds to fibronectin-coated glass coverslips. Further, inhibition of attachment to buccal epithelial cells was observed when *Candida* was preincubated with fibronectin. These studies have been extended more recently by Kalo *et al.* (32). They showed that *C. albicans* attached to a greater extent on the intermediate type of human vaginal epithelial cells (I cells) rather than to the superficial type of vaginal mucosal cell (S cells). The same investigators also showed by an immunofluorescence assay that a greater percentage of I cells (53%) was positive for fibronectin than S cells (approximately, 24%). In the presence of soluble fibronectin, adherence of *C. albicans* to I cells was reduced by approximately 66%. Thus, these data point to fibronectin as a receptor for *C. albicans*. As will be described later, *C. albicans* appears to possess a cell-surface component(s) which is similar to the integrin family of cell-surface receptors of mammalian cells. The "integrins" promote cell adhesion (both cell-cell and cell-matrix) and homing mechanisms (among other functions) for mammalian cells. Among the ligands recognized by these receptors are fibronectin, laminin, iC3b (complement C3 product), and vitronectin. Further studies are needed to determine which of these receptors is responsible for adherence to mammalian cells.

III. Complement Receptors

Sheep red blood cells (E) which are sensitized with anti-SRBC antibody (A) can be used as a matrix to generate intermediates in the complement cascade. Of primary importance in this regard are the EAC3b, EAiC3b, and EAC3d complement C3 ligands. It has recently been established that *C. albicans* possesses cell-surface-binding proteins which recognize the iC3b and C3d ligands but not the C3b ligand (11,20,23,25). Membrane complement receptors (CRs) have been demonstrated in various types of mammalian cells. In fact, nine distinct CRs are believed to exist although only five of these have been characterized thoroughly. The best known of these receptors include CR1 (specificity for C3b, C4b, iC3b), CR2 (specificity for iC3b, C3dg, C3d), and CR3 (specificity for iC3b, C3dg, C3d) (61). These receptors are biochemically and antigenically distinct, are functionally different and variable as to cell distribution (61). CR1, for example, has four distinct allotypes ranging from 160 K Mr to 250 K Mr and is found on a variety of cell types including erythrocytes, monocytes/macrophages, neutrophils, eosinophils, and on B and some T cells. CR2 is a 140 K Mr protein found exclusively on B lymphocytes, while CR3 is composed of a 165 K Mr α-chain and a 95 K Mr β-chain and (like CR1) is found on a variety of mammalian cell types such as monocytes/macrophages, neutrophils, NK- and ADCC-effector lymphocytes, and some T lymphocytes. Functionally, CR1 and CR3 are involved in recognition and phagocytosis of microorganisms following their opsonization with the C3 ligands iC3b and C3b. CR2, on the other hand, appears to be associated with proliferation of resting B lymphocytes (22,53). Additionally, the CR3 of neutrophils, monocytes, and macrophages appears to promote leukocyte adhesion and homing to endothelial cells (76).

That some cell-surface proteins of *C. albicans* are functionally similar to the CR of mammalian cells was first demonstrated by Heidenreich and Dierich (25). These investigators demonstrated binding of iC3b and C3d but not C3b. Presumably, both blastoconidia and hyphal forms of the organism exhibited binding activity. Activity was limited to *C. albicans* and *C. stellatoidea;* other *Candida* species, such as *C. tropicalis, C. krusei,* and *C. parapsilosis,* did not exhibit binding activity. D-glucose and D-mannose minimally inhibited the binding of both ligands to hyphae of the organism. Edwards *et al.* (20) obtained results similar to those of Heidenreich and Dierich although minimal binding of ligands was observed with blastoconidia.

Remarkably, while proteins with similar binding specificities exist for mammalian cells and *C. albicans,* a functional assignment for these proteins in *Candida* remains somewhat speculative. The *Candida* CR-like proteins may have invasive functions as postulated by several investigators. Heidenreich and Dierich suggested that the CR-like proteins of *Candida* may promote the formation of complement bridges between those organisms that are opsonized (with iC3b, for example) and other *Candida* cells with free binding sites for iC3b. Such cells would tend to form aggregates which may reduce their accessibility to phagocytes. Edwards *et al.* have discussed other possible invasion functions of the CR-like proteins from *C. albicans,* including: (1) *Candida* CR may function as a cofactor promoting the degradation of bound complement ligands, as has been observed with human CR. If degraded, the complement ligands would not be available for complement-mediated phagocytosis. (2) *Candida* may compete with mammalian cell CR for C3 ligands, thereby blocking the interaction of *Candida*-bound C3 with receptors on phagocytic cells. Or, shedding of CR-like proteins may inhibit the recognition process by phagocytes by competing with receptors on phagocytic cells. We have shown recently, in fact, that the C3d-binding protein of *C. albicans* is found in culture supernatants from growing cells (66). (3) *Candida* CR may function as a ligand (adhesin) and promote the attachment of the organism to host cells. Similar functions have been proposed for the mammalian CR3. (4) Finally, molecular mimicry of mammalian cell-surface proteins may promote nonrecognition of *C. albicans* by host defenses. While this latter possibility cannot be ruled out, patient sera contain antibody(ies) which recognize the purified C3d-binding protein as well as the iC3b-binding protein from semipurified preparations (58).

More recently, Ollert *et al.* (58) have demonstrated that an avirulent strain of *C. albicans* (m–10) derived from a virulent, parental strain (4918) lacked the iC3b-binding protein but not the C3d-binding protein. This observation was made by using direct assays for CR activity from hyphal forms of both strains as well as from the analysis of whole cell extract, culture filtrates, and DEAE-fractionated material prepared from 4918 and m-10 cells. In parental cells, proteins of 68–71 kDa, 55 kDa, and 50 kDa were observed in Western blots with patient sera. For m-10 strain, only weak reactions occurred with the 68–71- and 55-kDa proteins, while the 50-kDa protein was not observed. The patient serum had been shown to block binding of iC3b to *C. albicans*. Adherence of strain m-10 (in comparison to 4918) to fibrin platelets (13) as well as buccal (26) and vaginal epithelial cells (41) is considerably reduced. It is also relatively avirulent in a rabbit model of endocarditis (13) and in murine vaginitis (41). Thus, there is a direct correlation between the absence of the iC3b-binding protein and decreased adherence/virulence for *C. albicans*.

Other investigators have postulated a role for iC3b receptor in allowing *C. albicans* to escape phagocytosis (27). When the organism was grown in 50 mM glucose, iC3b-receptor expression increased approximately 44%, although the percentage of increase was strain dependent; *i.e.*, receptor expression in strain B311 increased only 16% when grown in a high-glucose medium. Correspondingly, cells grown in high glucose and expressing higher levels of iC3b receptor were less readily phagocytized. Thus, the presence of the iC3b-binding protein (receptor) may reduce the extent of phagocytosis of yeast cells. The observation, however, because of its strain specificity, may need to be examined with multiple isolates from different sites of infection.

A. *Characterization of the* Candida *Complement-binding Proteins*

The initial studies of Heidenreich and Dierich (25) on the C3d and iC3b proteins of *C. albicans* established that D-mannose or D-glucose could inhibit binding of EAC3d or EAiC3b to *C. albicans* when either was incorporated in the assay medium. However, Calderone *et al.* (11) were unable to block binding with either sugar or other hexoses. Instead, mannan/mannoprotein, extracted from hyphal cells and purified by Con A–affinity chromatography, effectively inhibited binding of EAC3d to *C. albicans* (50% inhibition at 6 μg/ml). On the other hand, non-mannosylated protein fractions (the wash from a Con A–affinity column) did not block binding of EAC3d to hyphal forms of the organism. This observation indicated that the receptor from *C. albicans* was extractable in a functional form and that the receptor was a mannoprotein. Subsequent studies showed that the purified receptor (described below) could bind Con A. In this regard, the *C. albicans* receptor is like the mammalian cell receptor in that both are glycosylated.

Purification of the C3d receptor was pursued in two ways (11,44). Hyphal extracts were either chromatographed on DEAE-Trisacryl and subsequently affinity-purified by ligand (C3d)-affinity chromatography or were affinity-purified by using a monoclonal antibody (CA-A). The monoclonal CA-A had been shown to block EAC3d binding to hyphae of *C. albicans* (11). By either procedure, two mannoproteins of 60–62- and 68–70-kDa molecular weights were observed. By using high-pressure liquid chromatography (HPLC) subsequent to affinity chromatography with CA-A, the two mannoproteins could be resolved and each tested for its ability to inhibit EAC3d-binding to *C. albicans* (44). The 60–62-kDa glycoprotein was found to be inhibitory, and it therefore probably represents the C3d-binding protein of *C. albicans*. The relationship of the 68–70-kDa glycoprotein to the 60–62-kDa binding protein is unknown at this time. Studies are now directed toward its purification and comparison to the 60–62-kDa glycoprotein by peptide analysis following protease digestion.

As stated previously, fractions of *C. albicans* hyphae rich in mannoprotein contain receptor activity. More direct proof for the glycosylated nature of the *Candida* CR proteins was established by direct staining of both the 60–62-kDa and 68–70-kDa proteins with Con A following SDS-PAGE analysis (44). Additionally, treatment of partially purified fractions containing C3d-binding activity with Con A–Sepharose beads results in loss of C3d-binding activity. Interestingly, *C. albicans* hyphae, when treated with Con A at saturating concentrations, still bind EAC3d, indicating that while the C3d-binding protein is mannosylated, the C3d-binding domain is probably not glycosylated (11). In fact, this observation is quite similar to that described for mammalian CR (82). In this system, the oligosaccharide, while not involved in the binding of C3d, stabilizes the protein (82). Further studies are needed to assess the role of the oligosaccharide for the *C. albicans* CR.

The C3d-binding protein of *C. albicans* is released into the culture medium by growing cells (66). The extracellular form of the C3d-binding protein still binds C3d as it inhibits the interaction of EAC3d with *Candida* hyphae. The C3d-binding protein from culture filtrate (CF) was isolated by preparative isoelectric focusing (IEF) followed by gel filtration by using Protein Pak 300 SW. These procedures resulted in a 32-fold increase in activity and an overall yield of 42%. SDS-PAGE analysis in the presence of mercaptoethanol revealed a doublet with apparent molecular weights of 60 and 55 kDa. When analyzed under nonreducing conditions, the protein instead migrated as a diffuse band in the 50–55-kDa region. In Western blot analysis the protein(s) was detected by using a monospecific, polyclonal hyperimmune rabbit serum prepared against the purified protein. The protein had an apparent pI of 3.9–4.1 and contained approximately 29.8% glutamic acid/glutamine by amino acid analysis. Additionally, the mannoprotein receptor was susceptible to cleavage with the glycosidase Endo F, suggesting that the oligosaccharide is of the high-mannose and biantennary-complex type. Endo H and Endo N glycosidases were relatively ineffective in cleaving the oligosaccharide, as was O-glycanase. However, the Endo F–cleaved receptor was still able to bind Con A indicating that oligosaccharide still remained. The molecular size of the Endo F–treated protein was approximately 45 kDa. The nativc protein was subjected to sizing by sucrose-density gradient centrifugation. The apparent molecular weight of the native protein, as determined by its co-migration with catalase, was approximately 240 kDa.

As previously stated, CR proteins have been described for a variety of mammalian cells (61). These proteins function in various ways according to the specific cell type but include phagocytic recognition of complement-coated microorganisms and cell growth and adhesion of certain types of white cells to tissues. There is good indication that at least the iC3b-binding protein of *C. albicans* shares epitope homology with its mammalian counterpart (CR3). This observation was first made by Edwards *et al.* (20). While antibodies to the human CR1 and CR3 (C3D9, 1B4, C511, 2B6, anti-B2, Mol, and anti-Mac 1) did not block binding of EAiC3b and EAC3d to *C. albicans,* Mol did bind to *Candida* when examined by immunofluorescence. Anti-CR2 antibodies (HB5 and anti-GP140) did block adherence of EAiC3b and EAC3d to *C. albicans* hyphae. Inhibition occurred, however, only at high concentrations of antibody and at low concentrations of C3 ligands.

Similar observations were noted by Gilmore *et al.* (23). These investigators showed binding of the anti-CR3 antibodies OKM-1 and Mol to *C. albicans* yeast and hyphal forms but not of antibodies to CR1 and CR2. The anti-CR3 antibodies which reacted with *C. albicans* were specific for the α-chain and not the β-chain of the CR3.

Eigentler *et al.* (21) used the antihuman CR3 monoclonal antibody OKM-1 to affinity purify a 130-kDa protein from solubilized, ^{125}I-labeled *C. albicans* cells. Following iodination of intact *C. albicans* cells, proteins were extracted by mechanical treatment and separated from nonsolubilized fragments by centrifugation. OKM-1 IgG, coupled to Sepharose, was then incubated with the ^{125}I-labeled protein for 1 hour at room temperature (RT). Bound proteins were eluted and analyzed by SDS-PAGE under reducing conditions. The monoclonal antibody OKM-1 was chosen for purification of the iC3b-binding protein of *C. albicans* since it reacted with pseudohyphae when analyzed by immunofluorescent straining. OKM-1 and M–522 (also slightly reactive with *Candida*) are both monoclonals with specificity for the α-chain of the human CR3. Further, a rabbit anti-*Candida* IgG which blocked iEAC3b adherence inhibited the binding of OKM-1. On other monoclonals tested, none with specificity to the β-chain of CR3 or to the α-chain (Mn-41 or Leu 15) was reactive with *C. albicans* pseudohyphae. The ^{125}I proteins of *Candida* reactive with OKM-1 included proteins of 130 kDa (major band) as well as 50 and 100 kDa (minor bands). In comparison, Hostetter and Kendrick (28) used a monoclonal (Mab-Bu–15) which recognizes CD11c (α-

Table 20.2. Characteristics of the integrin family of cell-surface receptors

Two nonconvalently linked subunits
α-subunit—95–130 kDa
10 classes
β-subunit—130–210 kDa
3 classes (β_1, β_2, β_3)
Ligand specificity—fibronectin, iC3b, laminin, others:
Cell-cell or cell-matrix interactions

subunit of the human receptor p150.95) to isolate proteins of 185 kDa (nonreducing conditions) and 70, 67 and 55 kDa (reducing conditions). These proteins are very similar in size to those found in wild-type hyphal cells described by Ollert *et al.* (58). (See previous discussion.) Thus, the pattern revealed by immunoreactivity with monoclonals to the human receptors (CR3 or p150.95) varies directly with the monoclonal used for purifying the *Candida* proteins.

Additional studies by Hostetter and Kendrick (28) focused on the cloning and sequencing of cDNA encoding the iC3b receptor on *C. albicans*. The Bu-15 monoclonal to human CR4 (p150.95) was used to screen a cDNA library obtained from the hyphal phase of *C. albicans* B311. Three immunoreactive clones of 1.1, 3.8, and 3.9 kbp were isolated. By Western blot analysis these three clones produced proteins reactive with the Mab Mol (antihuman CR3). The amino acid sequence of the *C. albicans* protein possessed regions homologous with the carboxy-terminal region of CD11b and to a lesser extent with the α-subunit of CR4 (CD11c). Both CR3 and CR4 are leukocyte adhesion proteins that belong to the integrin family of cell-surface macromolecules (31). (See Table 20.2 for characteristics of integrins.)

Table 20.3 summarizes the information compiled to date in regard to the complement-binding proteins of *C. albicans*. For simplicity, these macromolecules are treated as unique proteins and are designated as CR2 and CR3. However, there is no direct proof that they are distinct proteins. Thus, a single protein may possess a distinct ligand-binding site for iC3b and C3d or ligand specificity could

Table 20.3. Complement-binding proteins of Candida albicans

	CR2	CR3
Ligand specificity	iC3b, C3d	iC3b, C3d
MW	60–62, 68 kD	130, 185, 70, 67, 55 kD
Glycosylated	(+)	Unknown
Function	Unknown	Adhesin, virulence?
Cross-reactivity with human CR	(±)	(±) (α-chain)
Hyphae	+	+
Yeast	(+)[a]	(+)[a]
pI	3.9–4.1	Unknown

[a] But lower than hyphae.

Table 20.4. Antibodies to mammalian integrins: reactivity with Candida albicans[a]

Monoclonal antibody (reference)	Specificity	*Candida*-reactive macromolecule
Anti-Mol-94 (24)	CR3, α-chain	165 kDa
OKM-1 (21)	CR3, α-chain	130, 100, 50 kDa
Anti-integrin (48)	Integrin, β-chain	95 kDa
BU-15 (28)	pl50.95, α-chain	185 kDa (Non-red)

[a] Anti-integrin was prepared in rabbits against a synthetic peptide from the COOH-terminal end of the vertebrate integrin. All other antibodies are monoclonals prepared against human CR.

be associated with different proteins as in the human CR. Either possibility could be correct; however, it seems unlikely that one protein could have a single binding site for both iC3b and C3d since, as described previously, a mutant (m–10) strain has been characterized which has reduced binding activity to iC3b but not C3d. Molecular weight determinations of CR3 vary, making assessments difficult. Other characteristics include: the CR2 of *C. albicans* is glycosylated while direct determinations of the CR3 protein have not been completed. The CR3 appears to promote *Candida* attachment to endothelium (24) and may also be associated with invasiveness and virulence (58).

Homology of the *Candida* CR3 with human CR3 is apparent based upon the cross-reactivity of monoclonal antibodies (to human CR3) with *Candida* (Table 20.4). When cross-reactivity has been demonstrated, the monoclonals generally (but not always) have a specificity for the α-chain of CR3 but not the β-chain. Antihuman CR2 antibodies are weakly reactive. Likewise, a DNA probe with specificity for the mammalian CR2 does not react with a cDNA library from *C. albicans* (unpublished observation). Both CR2 and CR3 activities are associated with hyphal forms of the organism. These activities are reduced or negligible in yeast cells when assayed by rosetting assays but can be detected by using radiolabeled iC3b; the latter assay has not been tried with C3d. Finally, the CR2-like protein of *C. albicans* has been partially characterized biochemically (66) and a partial amino acid sequence obtained from the N-terminal region of the protein (unpublished data).

IV. Laminin- and Fibrinogen-binding Proteins

C. albicans also possesses cell-surface proteins which bind fibrinogen (5,6,59,79) as well as laminin (7). Binding of both proteins is observed in filamentous forms of the organism while yeast forms (blastoconidia) bind little if any, similar to the observations of others in regard to the complement ligands iC3b and C3d. (See discussion in section III.A.) In fact, binding of fibrinogen relative to that of blastoconidia was 12 times greater for mycelium and 7.7 times greater for germ tubes (5). Binding of fibrinogen by germ tubes and hyphae was observed under electron microscopy as patchy but dense deposits (by ferritin staining) over the cell-wall coat, primarily associated with the flocculent surface layer (79). In thin sections, fibrinogen-coated gold particles were observed as clusters that extended from the surface through the outer positions of the cell wall but not (or weakly) within the inner cell-wall layers or cytoplasm. Immunostaining was reduced if hyphae were aged or if treated with formaldehyde, mercaptoethanol, trypsin, or α-mannosidase. Fibrinogen binding was inhibited by a dialyzed culture filtrate but less so with mannan. Neither chitin nor mannose nor other sugars inhibited fibrinogen binding (5). The nature of the fibrinogen-binding protein is unknown.

More recently, Bouchara *et al.* (7) demonstrated the presence of laminin receptors in the outermost fibrillar layer of *C. albicans* germ tube cell walls and their absence on nongerminating blastoconidia. In this assay, following incubation of hyphae with laminin, cells were reacted with an anti-laminin antiserum and subsequently a fluorescein-conjugated anti-IgG antibody. For immune electron microscopy, protein A–sensitized gold particles were substituted for the fluorescent antibody. As in the case of fibrinogen, labeling was distributed over the outermost fibrillar layer of the germ tube cell wall but not in the mother-cell blastoconidia. Binding was demonstrated to be saturable and specific with approximately 8,000 binding sites per cell, and a dissociation constant (Kd) of 1.3×10 nM. Binding of radiolabeled laminin was inhibited by trypsin, by laminin, or by heat treatment of cells, but not by glucose, mannose, NAGA, fibronectin, bovine serum albumin, or mannan. Cell-wall components, extracted by dithiothreitol and iodoacetamide, were separated by SDS-PAGE and transferred to nitrocellulose sheets. The proteins were reacted with laminin and subsequently anti-laminin and alkaline phosphatase-conjugated anti-IgG. SDS-PAGE analyses of extracts revealed approximately 30 proteins by Coomassie blue staining. By Western blot analysis, however, a protein of 68 kDa and a doublet of 60–62 kDa are observed and are presumed to be the cell-surface laminin-binding proteins.

Proteins of similar molecular size have been described as the C3d-binding proteins (44, see above) as well as the adhesins which promote the attachment of germ tubes to neutral polystyrene plastic petri dishes (78). Further study is needed to determine if these proteins are related.

V. Acidic Proteinase

Strains of *C. albicans* are known to hydrolyze proteins such as serum albumin. Both intra- and extracellular proteinases have been described although more information is available on the latter. A proteinase has been purified from culture filtrates of the yeast form of the organism by DEAE-cellulose chromatography and affinity chromatography (62). SDS-PAGE under nonreducing conditions revealed a single band of Mr 45,000 which stained with periodic acid–Schiff (PAS) suggesting a glycoprotein. The pI of the enzyme was 4.45 and enzyme activity was optimal between pH 2.5 and 3.9 when hemoglobin was used as a substrate. Above pH 8.4 the enzyme underwent denaturation. The enzyme was defined as an acidic proteinase since it was inhibited by pepstatin.

Antibody to the protease (56) has been found in patient sera, and the enzyme has been demonstrated in tissues (4). Therefore, its role in virulence has been studied. Samples of nonkeratinized human oral epithelium were infected with blastoconidia of *C. albicans* (serotype A) *in vitro*. Immune electron microscopy demonstrated the enzyme on the surface of both yeast and germinating cells which had penetrated the epithelial cell surface. In comparison to this strain, a serotype B *C. albicans* had protease activity in the yeast form but not in germ tubes and filamentous forms. Interestingly, other investigators have shown that serotype A *C. albicans* isolates predominate in cases of denture stomatitis (49). Scanning electron microscopy (SEM) was utilized to determine if structural damage occurred in the epithelial surface during penetration by *C. albicans* hyphae, as might be expected if a protease was involved in invasiveness (4). Although no proteolytic damage was directly visualized in this study, other investigators have observed signs of damage to host cells during invasion (30). As penetration occurred, the deep layers of the invading hyphal cell wall were observed in close association with the epithelial surface. Damage to epithelial cells was seen only at the point of entry by the hyphae. Loss of cytoplasmic components in the vicinity of the invading hyphae was observed only occasionally. Tissue damage induced by proteolytic (or other enzymatic) cleavage must thus be local but nevertheless essential to invasion.

Borg and Ruchel (4) observed a correlation between adherence of the organism to the buccal mucosal surface and differential expression of secretory proteinase by different *Candida* species. Thus, *C. parapsilosis* did not adhere to buccal mucosa and did not produce (or carry) proteinase antigens. In contrast, *C. tropicalis* and *C. albicans* strongly adhere to tissue and produce high levels of proteinase. Further, adherence of *C. albicans* to oral mucosa was inhibited by approximately 90% when pepstatin was added, suggesting some role for the proteinase in fungal attachment.

While these studies indirectly support a role for the proteinase in the invasive process of *C. albicans,* other investigators (38,46) have shown that proteinase-negative strains are less virulent in mice compared to the parental, proteinase-positive strains. The interpretation of these experiments has been complicated by the slower growth rate of the proteinase-negative strain. However, Kwon-Chung *et al.* (38) isolated a proteinase-positive revertant with a growth rate similar to its proteinase-negative parent. Proteinase production by the revertant was approximately 50% of the parental cells; however, in a time-course study of proteinase production, the revertant eventually produced enzyme at levels comparable to parental cells. Mortality rates for the parental, proteinase-negative, and revertant strains were (at 20 days post-infection) 90%, 0%, and 80%, respectively. Thus, the differences in growth rates among the strains did not significantly affect the virulence in *C. albicans*.

The nucleotide sequence of the *C. albicans* aspartic proteinase gene has been deduced (45). Hybridization studies using a *Saccharomyces* PrA probe have indicated a 72% nucleotide sequence conservation between the yeast and the *C. albicans* genes and, at the amino acid level, an 85% homology. Some differences exist, however, as in the NH_2-terminal signal peptide, suggesting a different targeting for each protein.

In summary, aspartic proteinase appears to be made during tissue invasion, its production correlates with the degree of virulence among isolates of *Candida*, and antibody to the protease is observed in patient sera. The enzyme has been purified but varies biochemically among isolates. It may have a major role in virulence, since nonproducing strains derived from virulent producers are avirulent while revertant, producing strains are virulent. Transformation and/or gene disruption experiments will facilitate advances in molecular studies of this protein and its role in virulence. Such experiments would represent the cleanest approach to the elucidation of this enzyme in the virulence of *Candida albicans*.

VI. Ligand/Receptor Assays

Adherence of *Candida* to a variety of surfaces has been observed, including epithelial cells, vascular endothelial cells, neutrophils, and plastic. (See previous discussion.) Adherence is measured visually by counting either adhering *Candida* or the number of host cells with attached *Candida* (26) or by using radiolabeled *Candida* to determine the total amount of adhering cells (36,47).

The adherence of *Candida* varies according to the types of mucosal epithelial cells used in the study: exfoliated cells (13), cultured monolayers (63), or tissue outgrowth (74). Although exfoliated epithelial cells are convenient to use, it is often difficult to standardize the conditions of this assay. Host cell viability, contamination with commensal bacteria, and adsorption of immunoglobulins and other components of salivary or vaginal secretions may interfere with assays. Other factors that should be taken into consideration include donor age, phase of the menstrual cycle, use of oral contraceptives or spermicidal creams, and antecedent bacterial or viral infections. Moreover, donor-to-donor or day-to-day variability has been reported. Because of these disadvantages with exfoliated cells, methods for measuring *Candida* adhesion to more uniform cell populations such as tissue culture cells have been devised (63).

Candida adhesion is also influenced by the growth conditions of organisms *in vitro* including growth phase, temperature of growth, growth medium composition, or germ tube formation. Furthermore, a problem inherent in studying *Candida,* or any microorganism whose cells stick together, is the need either to account for or eliminate microbial coadherence (adherence of cells to one another). The following assays have been utilized with exfoliated cells but can be adapted to any cell system.

A. Visual Assay

Buccal or vaginal epithelial cells are collected with sterile cotton swabs and suspended in phosphate-buffered saline (PBS) (26). Following two washings with PBS, the epithelial cells are resuspended in PBS and adjusted to a concentration of 1×10^5 cells/ml. For preparation of *C. albicans,* cells are grown overnight at room temperature in phytone peptone broth containing 1 mg/ml glucose, or cells from a Sabouraud modified agar slant are harvested, washed twice, and resuspended in PBS at a cell concentration of 1×10^7 cells/ml.

Equal volumes (1.0 ml) of epithelial cells and yeast are incubated in plastic tubes in a shaker (200 rpm) at 37°C for 1 hour. The epithelial cell–yeast mixture is passed through polycarbonate filters (10–12 μm pore size) to remove nonadhering yeasts. After washing the filters three times with 5 ml of PBS, filters are placed face down on a microscope slide which are fixed and stained with crystal violet. The percentage of adherence is determined by counting the number of epithelial cells with ten or more adherent yeasts out of a total of 100 cells.

B. Radiometric Adherence Assay

For labeling yeasts before incubation with epithelial cells, *C. albicans* cells are pulse-labeled with ^{3}H-leucine for 2 hours at 37°C (36,47). The radiolabeled yeasts are harvested, washed twice with PBS, and readjusted to 1×10^7 cells/ml. Adherence assays are performed as described above. The filters (containing the adhering *Candida*) are transferred to scintillation vials, scintillation fluid is added, and the radioactivity determined in a liquid scintillation counter. The quantitation of adherent *C. albicans* out of total inocolum is determined by the following calculation: (cpm sample − cpm background)/cpm inoculum × 100.

VII. Conclusions and Future Directions

The virulence factors of *C. albicans* include the acidic proteinase and, most likely, the CR-like cell-surface mannoproteins involved in host-cell recognition. A nonadhering, avirulent strain of *C. albicans* has reduced levels of a CR-like protein specific for the C3 ligand iC3b, indicating that the CR-like protein may have a role in adhesion. The similarity of the *Candida* CR to that of mammalian cells makes finding an agent selective for *C. albicans* more difficult. Infusions of the purified CR-like protein in patients would likely be impractical and expensive since early diagnosis of systemic candidiasis is not routinely made. Recently, Klotz and Smith demonstrated that peptide sequences common to ligands such as iC3b, laminin, and fibronectin blocked the adherence of *C. albicans* to the subendothelial extracellular matrix of endothelial cells (Klotz, S. and R.L. Smith, Abstract, Conference of the Southern AFCR Immunology and Rheumatology, 1990. "*Candida albicans* Adherence to Subendothelial Extracellular Matrix Is Inhibited by Arginine-Glycine-Asparatic Acid Peptides" (p. 13A). These sequences (R-G-D peptides) are recognized by the mammalian CR3 and, presumably, the *Candida* CR3-like protein. These data may argue for the use of commercially available peptides as a therapeutic measure rather than preventing the synthesis of the CR-like proteins. Future studies, however, should be directed toward an understanding of the *Candida* CR since the CR3 of *C. albicans* is probably one of the specific adhesins of the organism.

Several investigators have reported that sublethal concentrations of antifungal drugs may block adherence (54,55,71,81). Amphotericin B, for example, inhibited the attachment of yeast cells to mammalian cells although the degree of inhibition depended upon the age of the yeast culture and serum concentration (55). Significantly, drug concentrations that prevented adhesion were not cidal for *C. albicans* (54). Likewise, ketoconazole (71) and other antifungals (81) have been reported to have similar inhibitory effects. Such studies suggest that new compounds which block adherence may be effective even though they have no antifungal activity *per se*.

Recent evidence by Frey *et al.* ("The Effect of Synthetic Protease Inhibitors on *Candida albicans* Extracellular and Protease Activity and Adherence to Endothelial Cells." Abstract, presented at the Second Conference on *Candida* and Candidiosis, p. 9, 1990) suggests that protease inhibitors significantly reduce growth, adherence, and protease activity of *C. albicans*. To screen for new antifungal compounds which function as protease inhibitors, one needs to develop a plate assay which incorporates the inhibitors as well as the protease substrate.

Similarly, if the synthesis (or regulation) of CR-like proteins from *Candida* is to be inhibited, a specific and sensitive assay needs to be developed to identify potentially useful compounds. Such an assay might utilize anti-mammalian CR3 (or anti-*Candida* CR3) MAb. Cultures of *C. albicans* in microtiter wells are reacted with a putative inhibitor. At a designated time, the organism is incubated with the anti-CR3 antibody and a second, conjugated antibody. Inhibition of MAb binding is then determined in a spectrophotometric assay. As a specificity control, mutant strains of *C. albicans* lacking CR3 (already available) could be used.

Acknowledgments

The authors wish to thank Ms. Stephanie Coleman for typing the manuscript. Some of these observations were supported by grants from the Public Health Service (HLB 21370 and AI 25738 to R.A.C.).

References

1. Ahern, D.G. 1978. Medically important yeasts. Annu. Rev. Microbiol. **32:**59–68.
2. Banno, Y., Y. Yamada and Y. Nozawa. 1985. Secreted phospholipases of the dimorphic fungus, *Candida albicans;* separation of three enzymes and some biological properties. Sabouraudia. J. Med. Vet. Mycol. **23:**47–54.
3. Beachey, E.H. 1981. Bacterial adherence: adhesin-receptor interactions mediating the attachment of bacteria to mucosal surfaces. J. Infect. Dis. **143:**325–345.
4. Borg, M. and R. Ruchel. 1988. Expression of extracellular acid proteinase by proteolytic *Candida* spp. during experimental infection of the oral mucosa. Infect. Immun. **56:**626–631.
5. Bouali, A., R. Robert. G. Tronchin and J.M. Senet. 1987. Characterization of binding of human fibrinogen to the surface of germ tubes and mycelium of *Candida albicans*. J. Gen. Microbiol. **133:**545–551.
6. Bouali, A., R. Robert, G. Tronchin and J.M. Senet. 1986. Binding of human fibrinogen to *Candida albicans in vitro:* a preliminary study. J. Med. Vet. Mycol. **24:**345–348.
7. Bouchara, J., G. Tronchin, V. Annaix, R. Robert and J.M. Senet. 1990. Presence of laminin receptors on *Candida albicans* germ tubes. Infect. Immun. **58:**48–54.
8. Brawner, D.L. and J.E. Cutler. 1986. Ultrastructural and biochemical studies of two dynamically expressed cell surface determinants on *Candida albicans*. Infect. Immun. **51:**327–336.
9. Brawner, D.L. and J.E. Cutler. 1986. Variability in expression of cell surface antigens of *Candida albicans* during morphogenesis. Infect. Immun. **51:**337–343.
10. Brawner, D.L. and J.E. Cutler. 1984. Variability in expression of a cell surface determinant on *Candida albicans* as evidenced by an agglutinating monoclonal antibody. Infect. Immun. **43:**966–972.
11. Calderone, R.A., L. Linehan, E. Wadsworth and A.L. Sandberg. 1988. Identification of C3d receptors on *Candida albicans*. Infect. Immun. **56:**252–258.
12. Calderone, R.A. and E. Wadsworth. 1985. Characterization with crossed immunoelectrophoresis of some antigens differentiating a virulent *Candida albicans* from its derived avirulent strain. Proc. Soc. Exp. Biol. Med. **185:**325–334.
13. Calderone, R.A., R.L. Cihlar, D. Lee, K. Hoberg and W.M. Scheld. 1985. Yeast adhesion in the pathogenesis of endocarditis due to *Candida albicans:* studies with adherence-negative mutants. J. Infect. Dis. **152:**710–715.
14. Cassone, A., E. Mattia and L. Boldrini. 1978. Agglutination of blastospores of *Candida albicans* by concanavalin A and its relationship with the distribution of mannan polymers and the ultrastructural of the cell wall. J. Gen. Microbiol. **105:**263–273.
15. Chaffin, W.L., J. Szkudlarek and K.J. Morrow. 1988. Variable expression of a surface determinant during proliferation of *Candida albicans*. Infect. Immun. **56:**302–309.

16. Critchley, I.A. and L.J. Douglas. 1987. Isolation and partial characterization of an adhesin from *Candida albicans*. J. Gen. Microbiol. **133**:629–636.

17. Douglas, L.J. 1987. Adhesion of *Candida* species to epithelial surfaces. In: CRC Crit. Rev. in Microbiol. **15**:27–43. CRC Press, Boca Raton, FL.

18. Douglas, L.J. and J. McCourtie. 1983. Effect of tunicamycin treatment on the adherence of *Candida albicans* to human buccal epithelial cells. FEMS Microbiol. Lett. **16**:199–202.

19. Douglas, L.J., J.G. Houston and J. McCourtie. 1981. Adherence of *Candida albicans* to human buccal epithelial cells after growth on different carbon sources. FEMS Microbiol. Lett. **12**:241–243.

20. Edwards, J.E., Jr., T.A. Gaither, J.J. O'Shea, D. Rotrosen, T.J. Lawley, S.A. Wright, M.M. Frank and I. Green. 1986. Expression of specific binding sites on *Candida* with functional and antigenic characteristics of human complement receptors. J. Immunol. **137**:3577–3583.

21. Eigentler, A., T.F. Schulz, C. Larcher, E.M. Breitwieser, B. Myones, A.L. Petzer and M.P. Dierich. 1989. C3bi-binding protein on *Candida albicans:* temperature-dependent expression and relationship to human complement receptor type 3. Infect. Immun. **57**:616–622.

22. Fingeroth, J.D., M.E. Heath and D.M. Ambrosino. 1989. Proliferation of resting B cells is modulated by CR2 and CR1. Immunol. Lett. **21**:291–302.

23. Gilmore, B.J., E.M. Resinas, J.S. Lorenz and M.K. Hostetter. 1988. An iC3b receptor on *Candida albicans:* structure, function and correlates for pathogenicity. J. Infect. Dis. **157**:38–46.

24. Gustafson, K.S., G.M. Vercellotti, J.S. Lorenz and M.K. Hostetter. 1989. The iC3b receptor on *Candida albicans* mediates adhesion in a glucose-dependent reaction. Comp. Inflam. **6**:339.

25. Heidenreich, F. and M.P. Dierich. 1985. *Candida albicans* and *Candida stellatoidea,* in contrast to other *Candida* species, bind iC3b and C3d but not C3b. Infect. Immun. **50**:598–600.

26. Hoberg, K., R. Cihlar and R.A. Calderone. 1986. Characterization of cerulenin-resistant mutants of *Candida albicans*. Infect. Immun. **51**:102–109.

27. Hostetter, M.K., J.S. Lorenz, L. Preus and K.E. Kendrick. 1990. The iC3b receptor on *Candida albicans:* subcellular localization and modulation of receptor expression by glucose. J. Infect. Dis. **161**:761–768.

28. Hostetter, M.K. and K.E. Kendrick. 1989. Cloning and sequencing of cDNA encoding the iC3b receptor on *Candida albicans*. Compl. and Inflam. **6**:348.

29. Houston, J.G. and L.J. Douglas. 1989. Interaction of *Candida albicans* with neutrophils: effect of phenotypic changes in yeast cell-surface composition. J. Gen. Microbiol. **135**:1885–1893.

30. Howlett, J.A. and C.A. Squier. 1980. *Candida albicans* ultrastructure: colonization and invasion of oral epithelium. Infect. Immun. **29**:252–260.

31. Hynes, R.O. 1987. Integrins: a family of cell surface receptors. Cell **48**:549–554.

32. Kalo, A., E. Segal, E. Sahar and D. Dayan. 1988. Interaction of *Candida albicans* with genital mucosal surfaces: involvement of fibronectin in adherence. J. Infect. Dis. **157:**1253–1256.

33. Kennedy, M.J., A.L. Rogers and R.J. Yancey, Jr. 1988. An anaerobic, continuous-flow culture model of interactions between intestinal microflora and *Candida albicans*. Mycopathologia **103:**125–134.

34. Kennedy, M.J. and P.A. Volz. 1985. Effect of various antibiotics on gastrointestinal colonization and dissemination by *Candida albicans*. Sabouraudia. J. Med. Vet. Mycol. **23:**265–273.

35. Kennedy, M.J. and P.A. Volz. 1985. Ecology of *Candida albicans* gut colonization: inhibition of *Candida* adhesion, colonization, and dissemination from the gastrointestinal tract by bacterial antagonism. Infect. Immun. **49:**654–663.

36. King, R.D., J.C. Lee and A.L. Morris. 1980. Adherence of *Candida albicans* and other *Candida* species to mucosal epithelial cells. Infect. Immun. **27:**667–674.

37. Klotz, S.A., D.J. Drutz, J.L. Harrison and M. Huppert. 1983. Adherence and penetration of vascular endothelium by *Candida* yeast. Infect. Immun. **42:**374–384.

38. Kwon-Chung, K.J., D. Lehmann, C. Good and P.T. Magee. 1985. Genetic evidence for role of extracellular proteinase in virulence of *Candida albicans*. Infect. Immun. **49:**571–575.

39. Lee, J.C. and R.D. King. 1983. Characterization of *Candida albicans* adherence to human vaginal epithelial cells *in vitro*. Infect. Immun. **41:**1024–1030.

40. Lehrer, N., E. Segal, H. Lis and Y. Gov. 1988. Effect of *Candida albicans* cell wall components on the adhesion of the fungus to human and murine vaginal mucosa. Mycopathologia **102:**115–121.

41. Lehrer, N., E. Segal, R.L. Cihlar and R.A. Calderone. 1986. Pathogenesis of vaginal candidiasis: studies with a mutant which has reduced ability to adhere *in vitro*. J. Med. Vet. Mycol. **24:**127–131.

42. Lehrer, N., E. Segal and L. Barr-Nea. 1983. *In vitro* and *in vivo* adherence of *Candida albicans* to mucosal surfaces. Ann. Microbiol. (Paris) **134B:**293–306.

43. Liljemark, W.F. and R.J. Gibbons. 1973. Suppression of *Candida albicans* by human oral streptococci in gnotobiotic mice. Infect. Immun. **8:**846–849.

44. Linehan, L., E. Wadsworth and R. Calderone. 1988. *Candida albicans* C3d receptor, isolated by using a monoclonal antibody. Infect. Immun. **56:**1981–1986.

45. Lott, T.J., L.S. Page, P. Boiron, J. Benson and E. Reiss. 1989. Nucleotide sequence of the *Candida albicans* aspartyl proteinase gene. Nucleic Acid Res. **17:**1779.

46. MacDonald, F. and F.C. Odds. 1983. Virulence for mice of a proteinase-secreting strain of *Candida albicans* and a proteinase deficient mutant. J. Gen. Microbiol. **129:**431–438.

47. Maisch, P.A. and R.A. Calderone. 1980. Adherence of *Candida albicans* to a fibrin-platelet matrix formed *in vitro*. Infect. Immun. **27:**650–656.

48. Marcantonio, E. and R.O. Hynes. 1988. Antibodies to the conserved cytoplasmic domain of the integrin 1 subunit react with proteins in vertebrates, invertebrates and fungi. J. Cell. Biol. **106:**1765–1772.

49. Martin, M.V. and C.J. Lamb. 1982. Frequency of *Candida albicans* serotypes in patients with denture-induced stomatitis and in normal denture wearers. J. Clin. Path. **35:**888–891.

50. McCourtie, J. and L.J. Douglas. 1984. Relationship between cell surface composition, adherence and virulence of *Candida albicans*. Infect. Immun. **45:**6–12.

51. McCourtie, J. and L.J. Douglas. 1981. Relationship between cell surface composition of *Candida albicans* and adherence to acrylic after growth on different carbon sources. Infect. Immun. **32:**1234–1241.

52. Mchcntcc, J.F. and R.J. Hay. 1989. *In vitro* adherence of *Candida albicans* strains to murine gastrointestinal mucosal cells and explants and the role of environmental pH. J. Gen. Microbiol. **135:**2181–2188.

53. Melchers, F., A. Erdei, T. Schulz, and M.P. Dierich. 1985. Growth control of activated, synchronized murine B cells by the C3d fragment of human complement. Nature **317:**264–267.

54. Merkel, G.J. and C.L. Phelps. 1989. The effects of amphotericin B on the interaction of *Candida albicans* with fibroblast cultures. Can. J. Microbiol. **35:**255–259.

55. Merkel, G.J. and C.L. Phelps. 1989. Conditions affecting the amphotericin B mediated inhibition of *Candida albicans* attachment to cell cultures. Can. J. Microbiol. **35:**260–264.

56. Odds, F.C. 1985. *Candida albicans* proteinase as a virulence factor in the pathogenesis of *Candida* infections. Zentralbl. Bakteriol. Mikrobiol. Hyg. Abt. **260:**539.

57. Ollert, M.K. and R.A. Calderone. 1990. A monoclonal antibody that defines a surface antigen on *Candida albicans* hyphae cross-reacts with yeast cell protoplasts. Infect. Immun. **58:**625–631.

58. Ollert, M.K., E. Wadsworth and R.A. Calderone. 1990. Reduced expression of the functionally active complement receptor for iC3b but not for C3d on an avirulent mutant of *Candida albicans*. Infect. Immun. **58:**909–913.

59. Page, S. and F.C. Odds. 1988. Binding of plasma proteins to *Candida* species *in vitro*. J. Gen. Microbiol. **134:**2693–2702.

60. Ray, T.L., K.B. Digre and C.D. Payne. 1984. Adherence of *Candida* species to human epidermal corneocytes and buccal mucosal cells: correlation with cutaneous pathogenicity. J. Invest. Dermatol. **83:**37–41.

61. Ross, G.D. 1986. Opsonization and membrane complement receptors, p. 87–114. In G.D. Ross (ed.), Immunobiology of the Complement System, Academic Press, Inc., NY.

62. Ruchel, R. 1981. Properties of a purified proteinase from the yeast *Candida albicans*. Biochim. Biophys. Acta **659:**99–113.

63. Samaranayake, L.P. and T.W. MacFarlane. 1981. The adhesions of the yeast *Candida albicans* to epithelial cells of human origin *in vitro*. Arch. Oral Biol. **26:**815–820.

64. Sandin, R.L. and A.L. Rogers. 1982. Inhibition of adherence of *Candida albicans* to human epithelial cells. Mycopathologia **77**:23–26.

65. Sandin, R.L., A.L. Rogers, R.J. Patterson and E.S. Beneke. 1982. Evidence of mannose-related adherence of *Candida albicans* to human buccal cells *in vitro*. Infect. Immun. **35**:79–85.

66. Saxena, A. and R.A. Calderone. 1990. Purification and characterization of the extracellular C3d-binding protein of *Candida albicans*. Infect. Immun. **58**:309–314.

67. Segal, E., A. Soroka and N. Lehrer. 1984. Attachment of *Candida* to mammalian tissues: clinical and experimental studies Zbl. Bakt. Hyg. A. **257**:257–265.

68. Segal, E. and D. Savage. 1986. Adhesion of *Candida albicans* to mouse intestinal mucosa *in vitro:* development of the assay and test of inhibitors. J. Med. Vet. Mycol. **24**:477–479.

69. Shepherd, M.G. 1987. Cell Envelope of *Candida albicans*. In: CRC Crit. Rev. in Microbiol. **15**:7–25. CRC Press, Inc., Boca Raton, FL.

70. Skerl, K.G., R.A. Calderone, E. Segal, T. Sreevalsan and W.M. Scheld. 1983. *In vitro* binding of *Candida albicans* yeast cells to human fibronectin. Can. J. Microbiol. **30**:221–227.

71. Sobel, J.D. and N. Obedeanu. 1983. Effects of subinhibitory concentrations of ketoconazole on *in vitro* adherence of *Candida albicans* to vaginal epithelial cells. Eur. J. Clin. Microbiol. **2**:445–452.

72. Sobel, J.D., G. Muller and H.R. Buckley. 1984. Critical role of germ tube formation in the pathogenesis of candidal vaginitis. Infect. Immun. **44**:576–580.

73. Sobel, J.D., G. Muller, P.G. Myers, D. Kaye and M.E. Levison. 1981. Adherence of *Candida albicans* to human vaginal and buccal epithelial cells. J. Infect. Dis. **143**:76–82.

74. Sobel, J.D., P. Myers, M.E. Levison and D. Kaye. 1982. Comparison of bacterial and fungal adherence to vaginal exfoliated epithelial cells and human vaginal epithelial tissue culture cells. Infect. Immun. **35**:697–701.

75. Stanley, V.C. and R. Hurley. 1969. The growth of *Candida* species in cultures of mouse peritoneal macrophages. J. Pathol. **97**:357–366.

76. Stoolman, L.M. 1989. Adhesion molecules controlling lymphocyte migration. Cell **56**:907–910.

77. Sundstrom, P.M., M.R. Tam, E.J. Nichols and G.E. Kenny. 1988. Antigenic differences in the surface mannoproteins of *Candida albicans* as revealed by monoclonal antibodies. Infect. Immun. **56**:601–606.

78. Tronchin, G., J.P. Bouchara, R. Robert and J.M. Senet. 1988. Adherence of *Candida albicans* germ tubes to plastic: ultra-structural and molecular studies of fibrillar adhesins. Infect. Immun. **56**:1987–1993.

79. Tronchin, G., R. Robert, A. Douali and J.M. Senet. 1987. Immunocytochemical localization of *in vitro* binding of human fibrinogen to *Candida albicans* germ tube and mycelium. Ann. Inst. Pasteur (Microbiol.) **138**:177–187.

80. Tronchin, G., D. Poulain, J. Herbaut and J. Biguet. 1981. Cytochemical and ultrastructural studies of *Candida albicans*. II. Evidence for a cell wall coat using concanavalin A. J. Ultrastruct. Res. **75**:50–59.

81. Vuddhakul, V., J.G. McCormack, W.K. Seow, S.E. Smith and Y.H. Thong. Inhibition of adherence of *Candida albicans* by conventional and experimental antifungal drugs. J. Antimicrob. Chemother. **21**:755–763.

82. Weis, J.T. and D.T. Fearon. 1985. The identification of N-linked oligosaccharides on the human CR2/Epstein-Barr virus receptor and their function in receptor metabolism, plasma membrane expression and ligand binding. J. Biol. Chem. **260**:13824–13830.

21

Screening for Antifungal Activity

John F. Ryley and *Keith Barrett-Bee*

I. Introduction

The purpose of antifungal screening is not to define the therapeutic properties of an individual compound, but rather to indicate which of a large number of compounds or samples are worthy of further study. The purpose is simply to provide a YES/NO answer to the question, "Is there sufficient interaction between this sample and fungi to warrant further investigation?" The attempt is to eliminate those samples which are of no definite interest, rather than to be dogmatic about samples which may be of interest. One wishes to reduce the size of a batch of samples considerably, so that those which remain can be subject to more careful study. Screening is not testing a handful of compounds for activity; it is essentially playing the numbers game. If compounds identified as interesting by the screen subsequently prove to be of no value (false positives), that does not matter; it *is* important that the screen does not reject compounds which really are of interest (false negatives). Once the preliminary process of screening is complete, the few samples which have been identified as possibly of interest can be evaluated in more comprehensive tests to determine their real potential and limitations. It is important to recognize the distinction between *screening* and *evaluation*. Screening is the detection of potentially interesting activity; evaluation is the quantitation and characterization of that activity.

The nature of the screen will depend on a number of factors, but important among these will be the weekly throughput of samples required and the technical effort available. Thus for a small-scale directed synthetic programme, where only a few costly samples are generated, but where the chance of useful activity is high, a more complicated and comprehensive screen may be appropriate. For a random survey of a large collection of synthetic chemicals, where the number of compounds for screening each week is high, but where the chance of useful activity is low, a simpler screen with high throughput is more desirable. For a campaign aimed at identifying natural product activity, where it may be necessary to screen several thousand samples each week, only the simplest of screens is possible.

A useful antifungal drug must be toxic to the parasite, but not to the host; since fungi are eukaryotic organisms, this selective toxicity may be thought particularly difficult to achieve. In the hope of more readily obtaining selective toxicity, screens based on a particular pathway unique to the target fungus have been used; such systems include the processes responsible for the synthesis of cell wall chitin, glucan, and/or mannan, and those involved at specific stages of ergosterol biosynthesis. Figure 21.1 indicates diagrammatically some of the potential specific intracellular targets for an antifungal agent and the types of compounds known to be active in these specific areas. In general such screens can be simplified or mechanized to achieve a very high throughput of samples. In association with a directed synthetic campaign, activity in a specific enzyme screen will encourage the chemist that he is on the right track. Random screening of a chemical collection may identify lead activity of a specific type which may subsequently be improved by chemical modification; in a natural product setting it will eliminate a lot of compounds of other types already known. But it must be emphasized and recognized that by seeking inhibition of a single enzyme or restricted process rather than a nonspecific inhibition of overall growth, unexpected but useful antifungal activity of another type may well be missed.

Activity *in vitro* is no guarantee of activity *in vivo,* nor is the degree of activity *in vitro* necessarily predictive of the extent of activity *in vivo*. Moreover, lack of activity or very poor activity *in vitro* is no guarantee of lack of useful activity *in vivo*. This is particularly a problem with azoles, but it is by no means restricted to this group of compounds (38). A recent example of the latter situation is cispentacin, an antibiotic isolated from the culture filtrate of *Bacillus cereus* L450-B2 by Konishi *et al.* (23). Cispentacin demonstrated only weak activity *in vitro* against fungi, and then only under certain conditions, but showed very marked oral activity in mice with vaginal, lung, or systemic infections with *Candida albicans* or a systemic infection with *Cryptococcus neoformans* (23,26). It is similarly important to realize that a number of clinically useful azoles would never have been identified had reliance been placed on a primary *in vitro* screen. Similarly, activity against an isolated enzyme or in a cell-free preparation is no guarantee that useful activity will be possible with intact cells. Permeability barriers, uptake mechanisms, and metabolism may influence the broken cell/intact cell correlation, while absorption, metabolism, and distribution are among factors germane to the *in vitro*/*in vivo* situation. Unfortunately, only positives from the *in vitro* screens are normally further tested *in vivo*. The ideal screening procedure is undoubtedly one carried out *in vivo;* the use of an *in vitro* or an enzyme screen is a second-best situation enforced through the smallness and/or multiplicity of samples, resource constraints, and a desire to avoid animal experimentation. The ideal screen should allow a high throughput of samples, should be technically simple and economic to operate, should require only small

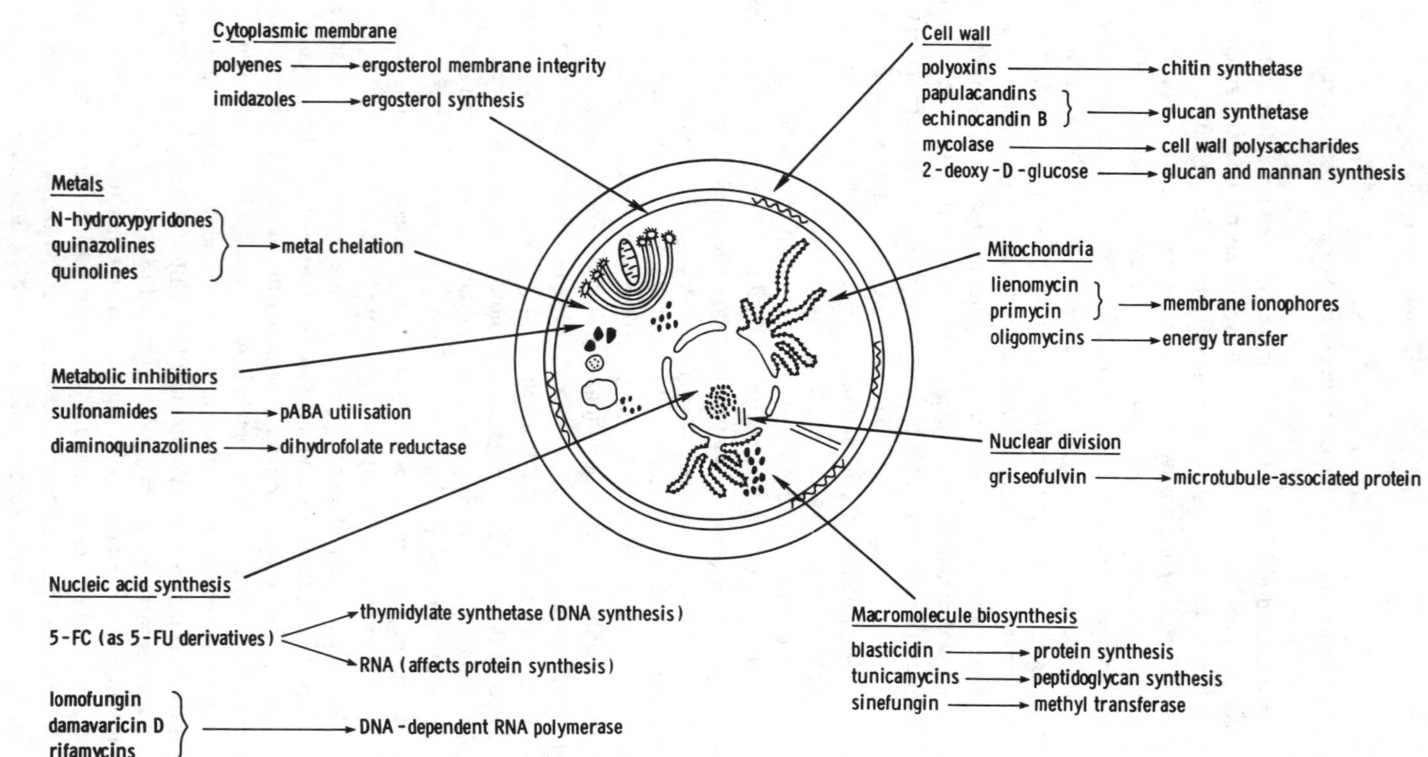

Figure 21.1. Potential targets for antifungal agents.

amounts of compound, should be easy and unambiguous to read, and should be sensitive enough not to miss potential positives.

In Table 21.1 we have listed just a few synthetic chemicals and natural products having antifungal action resulting from their ability to inhibit specific metabolic processes. It needs to be emphasized that not one of these substances was discovered in the first place by using a specific target-oriented selective screen; rather, inhibitory activity has first been found against intact fungi, and the specific nature of that activity has subsequently been identified or characterized in fungal or other systems. A much more comprehensive review of synthetic chemicals having some sort of antifungal activity was given by Ryley *et al.* (33). This review also covered natural products, an area which has been well updated by Debono and Gordee (10). An amazing number of natural products having antifungal activity *in vitro* have been discovered, and in many cases given a name; it is important to realize just how few have any trace of activity *in vivo,* and to note that only amphotericin B, nystatin, griseofulvin, and possibly cilofungin have any useful therapeutic value. The synthetic chemicals listed in Table 21.1 all have therapeutic value; this list does not, however, emphasize the facts that the majority of synthetic chemicals which have been found to have some sort of activity *in vitro* do not have any therapeutic potential *in vivo,* that the majority of marketed synthetic antifungals have utility only for topical medication, and that the azoles far outstrip all other chemical classes combined. In what follows we will consider antifungal screening at the three levels discussed above. A possible summary flow chart for the various screens is given in Figure 21.2.

II. Selective Screens

Under this heading we will consider just a few examples of screens based on specific modes of drug action. Such screens will be most appropriate in association with a programme of directed chemical synthesis, but are used on occasions in the random screening of synthetic chemicals or natural product broths and extracts.

A. *Cell Wall Synthesis in* Candida Albicans

Inhibition of the synthesis of cell wall mannan, glucan, or chitin is theoretically attractive in the search for antifungal agents with selective toxicity; it is particularly attractive following the demonstration of synergism between nikkomycins (chitin synthase inhibitors) and papulacandin B (glucan synthase inhibitor) by Hector and Braun (16). Further work (17) has shown the nikkomycins to have promising activity *in vivo* in murine models of coccidioidomycosis, blastomycosis, and histoplasmosis. The dimorphic fungi involved have walls containing 10–20% chitin, but the amount of chitin in the cell wall does not appear to be the sole determinant of susceptibility. Thus *C. albicans* with a much lower content

Table 21.1. Some synthetic and natural product antifungal agents

Compound	Structural class	Molecular target[a]	Route of administration
Synthetic chemicals			
5-fluorocytosine	Pyrimidine	RNA function/DNA synthesis	Oral/iv
Tolnaftate	Thicarbamate	Squalene epoxidase	Topical
Ciclopiroxolamine	Hydroxypyridone	Transmembrane transport	Topical
Amorolfine	Morpholine	Δ^{14}-reductase/$\Delta^{8}-\Delta^{7}$ isomerase	Topical
Clotrimazole	Azole	C-14 demethylase	Topical
Miconazole	Azole	C-14 demethylase	Topical
Ketoconazole	Azole	C-14 demethylase	Oral/topical
Itraconazole	Azole	C-14 demethylase	Oral
Fluconazole	Azole	C-14 demethylase	Oral/iv
Naftifine	Allylamine	Squalene epoxidase	Topical
Terbinafine	Allylamine	Squalene epoxidase	Oral
Natural products			
Polyoxins	Nucleoside	Chitin synthase	—
Nikkomycins	Nucleoside	Chitin synthase	—
Bacilysin (tetaine)	Epoxypeptide	Glucosamine-6-phosphate synthetase	—
Echinocandin	Cyclic lipopeptide	Glucose synthase	—
Cilofungin (semisynthetic)	Cyclic lipopeptide	Glucose synthase	iv
Papulacandin	Glycolipid	Glucose synthase	—
A 25822B	Azasterol	Δ^{14}-reductase	—
Mevinolin	Sesquiterpene	HMGCoA reductase	—
Griseofulvin	Dimethoxycoumarin derivative	Microtubule-associated protein	Oral
Ambruticin	Cyclopropyl-pyran	Amino acid transport	—
Nystatin	Polyene	Membrane ergosterol	Topical/oral
Amphotericin B	Polyene	Membrane ergosterol	iv/topical

[a] See Polak (28) and Kerridge and Vanden Bossche (21).

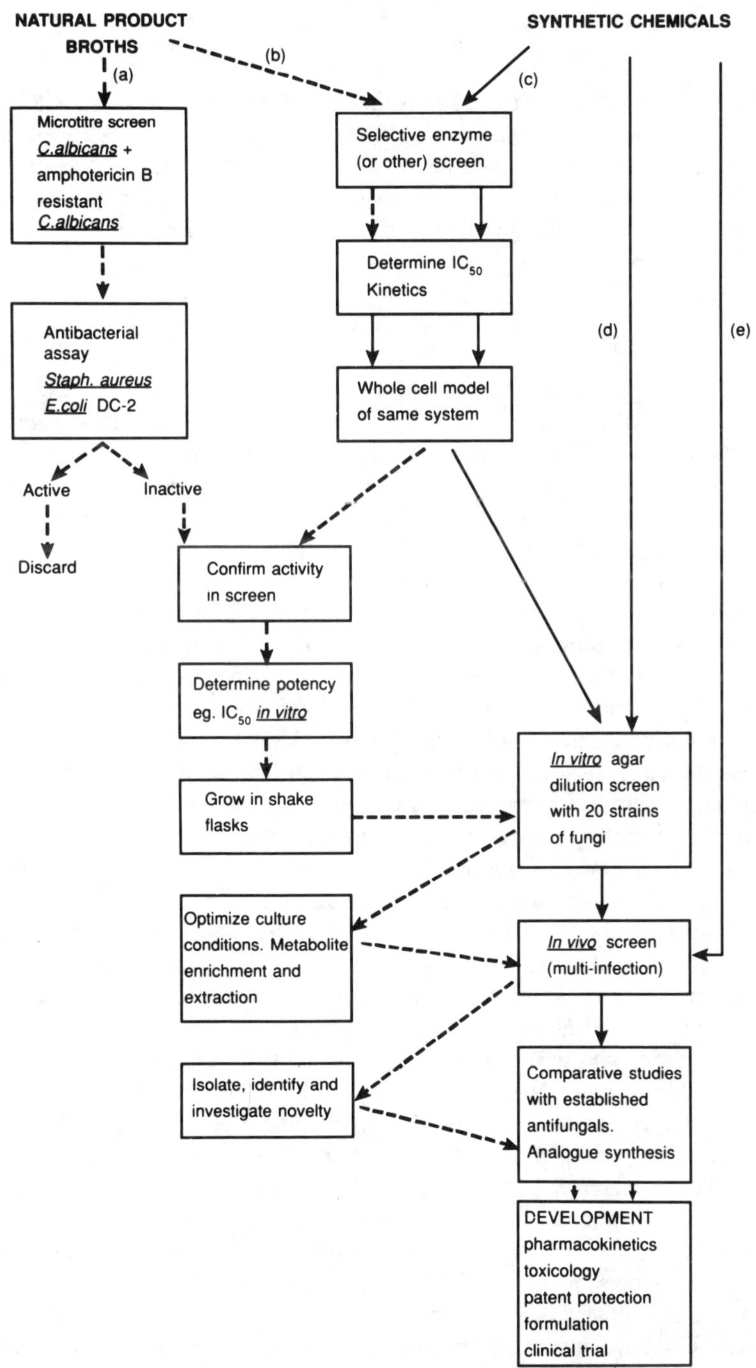

Figure 21.2. Flow chart for antifungal screening of synthetic chemicals and natural product broths. Natural product routing shown by dotted lines, synthetic chemicals by solid line. (a) and (b) Alternative routes. (c) Route for rationally synthesized compounds; preferably run (d) and even (e) in parallel. (d) Route for random chemicals. (e) Preferred direct route.

of chitin is less susceptible to nikkomycin, but *Aspergillus fumigatus* with higher levels of chitin is more resistant still to the nikkomycins. Some cell wall inhibitors—such as the polyoxins and bacilysin—enter cells by means of a dipeptide permease; activity against intact cells depends, therefore, on the presence of a suitable permease and the absence of competing peptides in the environment.

1. Chitin

We have chosen to work with enzyme preparations from *C. albicans*. Over 50% of the cell wall of *C. albicans* consists of mannoproteins—mannose-containing polymers hydroxy-linked to protein via a serine or threonine residue, and polymers hydroxy-linked to protein via chitobiose. The rest of the cell wall is glucan, with a small amount of chitin. Although the chitin content is low, it does appear to be important, particularly in connection with cell budding and hyphal elongation. Chitin synthase is important for the synthesis of chitin for new cells, while chitinase is important in the remodelling of cell walls during growth and division. For our chitin synthase screen, a cell-free preparation of *C. albicans* was obtained by shaking with Ballotini glass beads in a Braun disintegrator and centrifuging as described by Barrett-Bee *et al.* (6). Enzyme activity was assayed by the method of Adams and Gooday (1) using UDP [^{14}C] N-acetyl glucosamine as substrate and measuring the transfer of label to insoluble material. Activity was defined in terms of percentage inhibition relative to untreated control. Where sufficiently interesting inhibition was obtained, its nature (competitive or otherwise) was investigated by varying the concentrations of substrate and inhibitor and constructing Lineweaver-Burk and Dixon plots.

Compounds highlighted by this cell-free screen and considered worthy of further study were then tested in a whole-cell model of chitin synthesis. Midexponential-phase *C. albicans* yeasts were harvested, washed, and incubated with [^{3}H] methyl-N-acetyl-D-glucosamine (2×10^7 cells and 0.05 μCi substrate per ml). The reaction was terminated after 1 h (while still on the linear portion of the curve) by addition of 4N NaOH and heating at 100°C for 1 h; radiolabel associated with macromolecules was collected by filtration and counted. This had previously been shown to reside in the chitin fraction of the cell wall, and not to be redistributed to macromolecules in general. Inhibition was measured in the presence of compounds and potency expressed as an IC_{50}.

In our experience these chitin synthase screens have not been very productive. Nikkomycin shows an IC_{50} of 0.02 μg/ml compared with 30–40 μg/ml for polyoxin in the cell-free screen, and is effective in whole cells at 3 μg/ml where polyoxin is inactive at a concentration of 1 mg/ml. We have screened collections of compounds having ribose or arabinose linked to a heterocycle moiety, compounds related to known herbicides with chitin synthase activity, miscellaneous synthetic chemicals, and a variety of plant extracts. Where activity has been found in the cell-free screen it has seldom transferred—presumably due to permeability barriers—to activity in whole cells, let alone to antifungal activity in whole cells.

2. *Mannan*

We initially studied mannan biosynthesis by following the uptake of radiolabel from GDP [^{14}C] mannose into macromolecules in broken-cell preparations at 37°C for 1 h. The reaction was terminated by the addition of trichloroacetic acid and the non–mannoprotein mannose was isolated by β-elimination of the serine-mannose linkages with NaOH as described by Nakajima and Ballou (24). The resultant macromolecules were reprecipitated with trichloroacetic acid and radioactivity was counted. This gave a measure of radiolabel incorporated into the macromolecular mannan, which has a 1:6 backbone with 1:3 and 1:2 branches (3); statistically most will be in the branched regions. Inhibition was measured relative to untreated controls.

Using this test we screened a group of compounds which were potential analogs of the GDP mannose substrate. Interestingly enough, purine-like compounds tended to *stimulate* mannan synthetase—up to 50%—while pyrimidine-like compounds could give up to 25% inhibition. Similar mild stimulation or inhibition was found when a miscellaneous series of 110 compounds, active *in vitro* against a range of *Candida* species, but with limited activity towards dermatophytes, was screened.

The importance of the mannan 1:6 backbone relative to its branches was shown by Ballou *et al.* (4) by the use of mutants lacking parts of the molecule. Nakajima and Ballou (25), using a *Saccharomyces cerevisiae* preparation, showed that a tetramer containing 1:2 and 1:3 linkages would react with GDP mannose to produce a pentamer containing a 1:6 linkage, thus mimicking backbone elongation. We have demonstrated, using high-resolution NMR that a similar situation exists with *C. albicans*.

We developed a screen based on this tetramer→pentamer conversion, and by use of a multipipetting system and microtitre plates have adapted it to screen 100 compounds a week. The assay system contains 1 mM tetramer (prepared from *S. cerevisiae*), 1 mM GDP mannose, 94 nCi GDP [^{14}C] mannose, 12.5 mM Mg^{2+}, 7.5 mM Mn^{2+} and 5 μl compound at 2 mg/ml in DMSO (approximately 1 mM final concentration) in a total volume of 50 μl; each compound is assayed in triplicate. Following incubation at 37°C for 50 min, the reaction is stopped by adding 150 μl ice-cold ethanol to each well and the plates are left in the cold room overnight. They are then centrifuged at 2,600 rpm in a Beckman centrifuge; 100 μl of supernatant is transferred to Skatron micronic tubes and vortex evaporated for 1 h at 50°C. The residue is taken up into 20 μl water, mixed by vortex for 2 min, allowed to dissolve for 30 min at room temperature, and vortexed for a further 1 min, after which 5 μl of each replicate is spotted on a cellulose TLC plate (nine samples representing three compounds per plate). The samples are run for 8 h in the cold room by using a mixture of butanol:ethyl acetate:acetic acid:water in the ratio 8:6:5:8. The dried plates are autoradiographed for 24 h; the positions of the bands are marked; and the pentamer band is cut out, placed in 4 ml scintillant, and counted. Control counts of around 1,500 cpm are obtained; inhibition is measured relative to controls.

From a collection of 57 guanine derivatives, only GDP glucose and 7-methyl guanosine inhibited polymerization by more than 50%; no activity was found during a screening campaign with general chemicals. GDP glucose was shown to be a competitive inhibitor for GDP mannose with a K_i of 0.7 mM. When inhibition of tetramer to pentamer conversion was compared to inhibition of macromolecular incorporation, there was little correlation between active compounds. This probably reflects the fact that macromolecular incorporation is into the side chains (with α–1:2 and α–1:3 linkages) rather than backbone (α–1:6 links).

3. *Glucan*

We have not had much success in trying to establish a screen for inhibitors of glucan synthesis by using broken-cell preparations of *C. albicans* and UDP [^{14}C] glucose. However, some compounds active in enzyme assays have been retested by using whole cells, which makes possible observations on glucan synthesis. Exponential-phase cells (2×10^6/ml) were incubated in the presence of radiolabelled glucose (1.85 μCi) in a volume of 100 μl for 1 h and the reaction was stopped by the addition of 100 μl 4N NaOH. The cell wall fractions were separated by the method of Baguley *et al.* (2) and radiolabel was measured in the mannan and glucan fractions.

B. *Protoplast Regeneration in* Neurospora Crassa

Selitrennikoff *et al.* (35) described a temperature-sensitive osmotic mutant (*os–1*) of *Neurospora crassa* which forms and grows as protoplasts when cultured under special conditions. Growth as protoplasts can be serially maintained at 37°C, but if the incubation temperature is reduced to 22°C, cell wall regeneration takes place and growth in the hyphal form is resumed. We have already considered inhibition of the enzymes responsible for cell wall polysaccharide synthesis at the cell-free and the intact cell levels. Selitrennikoff (34) described how this mutant could be used in an alternative screen to identify compounds which inhibit cell wall polysaccharide synthesis at the level of the intact cell.

Protoplasts grown at 37°C in Vogel's medium with added sorbitol (7.5% w/v) as an osmotic stabilizer are incorporated at a final density of 5×10^4 per ml into a molten medium containing 1.2% agar, and plates are poured. Filter paper discs loaded with samples for screening are placed on the solidified medium, and incubation is carried out for 6 days at 22°C. Compounds with general antifungal activity produce clear zones of inhibition (e.g., amphotericin B, azoles, cycloheximide, etc.), while compounds with selective action on cell wall synthesis produce hazy zones with a fuzzy border; microscopial examination indicates that within these zones, growth occurs as protoplasts. Nikkomycin and polyoxin were identified in this screen as specific inhibitors of cell wall synthesis, but surprisingly, papulacandin B gave a clear zone of inhibition, presumably due to inhibition of unidentified processes in addition to its known inhibition of β–1:3-glucan synthesis.

Kirsch and Lai (22) were discouraged from using Selitrennikoff's screen because of the need to prepare and maintain large-scale cultures of protoplasted cells; they pointed out, moreover, that zones were difficult to score because protoplasts (in the presence of active compounds) grow poorly at 22°C. They therefore described an alternative screen based on an inversion of Selitrennikoff's approach. A medium containing 3.5% Difco morphology agar, 0.1% yeast extract, 1.5% agar, 2.5% L-sorbose, and 7.5% D-sorbitol was inoculated with spores of *N. crassa os–1* to a final concentration of 5×10^4–10^5 per ml, and plates were poured. Discs loaded with samples for screening were placed on the agar, and plates were incubated at 37°C for 36–48 h. D-sorbitol was included in the medium as an osmotic stabilizer, and L-sorbose as an inhibitor of glucan synthesis. Normal growth under these conditions is in the hyphal form. Discs loaded with active antifungals produced clear zones, while discs containing inhibitors of chitin synthesis produced turbid zones, and in addition an unidentified orange pigment. Microscopical examination of the turbid zones revealed enlarged and ballooned hyphae and protoplasts in addition to some normal hyphae. In order to produce protoplasts, inhibitors of both chitin and glucan synthesis need to be present; since it is unlikely that samples for screening will contain both, Kirsch and Lai included L-sorbose in the medium as a glucan synthesis inhibitor. Their screen therefore is specific for inhibitors of chitin synthesis—either compounds like polyoxins and nikkomycins which inhibit chitin synthesis from N-acetyl glucosamine, or compounds like bacilysin, which inhibits N-acetyl glucosamine synthesis in the first place. Approximately 0.7% of their randomly isolated actinomycetes produced activity. Were adequate supplies of polyoxin available, this could replace L-sorbose in the medium, giving a screen specific for glucan synthesis inhibitors.

It is important to realize that a screen based on *N. crassa* may well be irrelevant for the discovery of agents effective against fungi of medical significance, and that the screen based on the uptake of [^{14}C] glucose by *C. albicans* as previously described is probably more relevant. Moreover, the limitations of a screen based on a specific mode of action should be constantly kept in mind.

C. Morphology Screens

A number of selective antifungal agents produce specific and characteristic effects on fungal cells. For example, griseofulvin induces hyphal curling (9), and validamycins stimulate branch formation (18), while a number of agents which are able to inhibit cell wall synthesis or interact with the cytoplasmic membrane induce characteristic swellings. Gunji *et al.* (15) described a screen to detect novel microbial products capable of inducing morphological changes. Petri dishes were prepared containing 10 ml of a medium consisting of 1% malt extract, 0.1% yeast extract, 0.3% sucrose, and 0.7% agar seeded with 2.5×10^4 cells/ml of *C. albicans* or *Mucor racemosus*. Pulp discs loaded with acetone extracts of cultures of soil microorganisms were placed on the agar and incubated at 26.5°C. Morphological changes in fungi growing round the discs were observed by microscopical examination after 6, 24, and 48 h. Approximately 25% of 2,000 broths tested showed activity requiring further investigation; this is basically too high a hit rate for a primary screen. Numbers were reduced to 81 (4%) by using a similar screen with six other organisms (which included *Trichophyton mentagrophytes* and *A. fumigatus*). Further studies such as extractability with n-butanol and antagonism of activity by osmotic stabilizers or ergosterol allowed classification of the active compounds into five groups. By far the biggest group consisted of polyenes, but three apparently novel compounds were identified: one produced effects like griseofulvin, but with a quite different antifungal spectrum; one produced unbalanced growth of yeast cells; and one produced morphological changes similar to those caused by polyoxin, and was probably related to gougeroxymycin. Again the relevance of using fungi which are not human pathogens in a screen needs to be considered.

D. Ergosterol Biosynthesis

The majority of antifungal agents in use today are inhibitors of ergosterol biosynthesis; preeminent among these are the azoles which inhibit the C-14 demethylation process. Because this is a cytochrome P–450 system, a variety of potential toxicological problems attend the azoles. Allylamines and thiocarbamates inhibit biosynthesis at the earlier stage of squalene epoxidase, while morpholines such as amorolfine affect Δ^{14} reductase, and possibly also $\Delta^7 - \Delta^8$ isomerase (28). Although the therapeutically useful antifungal spectrum of current allylamines and morpholines is not as wide as that of the azoles, problems of P–450-based toxicity are absent.

1. 2,3-Oxidosqualene Lanosterol-cyclase

One of the key enzymes in ergosterol biosynthesis is 2,3-oxidosqualene lanosterol-cyclase, which acts in the biosynthetic pathway immediately after squalene epoxidase. Jolidon *et al.* (19) have chosen this enzyme as a target in their search for new antifungal agents. They prepared protoplasts of *C. albicans* by digestion with zymolyase–100T and subsequently lysed them in 100 mM phosphate buffer. The lysate, centrifuged at 5,000*g* and adjusted to a protein content of 10 μg/

ml, behaves as an essentially pure preparation of the cyclase as judged by the incorporation of [^{14}C] oxidosqualene into lanosterol in the presence of 0.1% Decyl Poe (a non-ionic detergent; n-decylpentaoxyethylene). The detergent prevents further conversion of lanosterol to fungal sterols. Samples have been assayed in this system and in a similar one based on rat liver in order to detect compounds with the potential of differential toxicity. As a result, two enzyme inhibitors, Ro 42–5604 and Ro 43–0688, have been found, which also show high antifungal activity *in vitro* against yeasts, dermatophytes, and dimorphic fungi (Scrip No. 1498, March 21st 1990, p. 25); in no case has activity been transferrable to the *in vivo* situation. This has obviously proved to be a useful screen to use in connection with custom-synthesized compounds, and could well lead to the first example of a rationally derived antifungal drug.

2. Δ^{14} Reductase and $\Delta^{8}-\Delta^{7}$ Isomerase

Compounds inhibiting the two enzymes Δ^{14} reductase and $\Delta^{8}-\Delta^{7}$ isomerase include the morpholines tridemorph, fenpropimorph, and amorolfine, the piperidines fenpropidin and piperalin, and azasteroid A 25822B, and N-substituted 8-azadecalins. Baloch and Mercer (5) describe assays for these two enzymes based on *S. cerevisiae*. The $\Delta^{8}-\Delta^{7}$ isomerase assay uses an acetone powder of the yeast and measures the conversion of fecosterol to episterol as detected by gas chromatography. The Δ^{14} reductase assay uses a freshly prepared cell extract which brings about the conversion of 5α-ergosta-8,14,24(28)-trien–3β-ol to a variety of products identified by gas chromatography/mass spectroscopy (GC/MS). They, and Kerkenaar (20), discuss structure-activity relationships with agricultural fungicides in view. No doubt the assays could be modified to screen for compounds active against human pathogens. These enzymes are further discussed in relation to amorolfine by Polak (28). Her studies, however, have been limited to retrospective observations on the sterol patterns of whole cells of *C. albicans, T. mentagrophytes,* or *Histoplasma capsulatum* exposed to amorolfine, rather than using cell-free extracts from these fungi for prospective screening purposes (27). In the case of *C. albicans* the major sterol accumulating in place of ergosterol was ignosterol; with *H. capsulatum* at low levels of amorolfine, ergosta-8,14,24(28)-trienol and ignosterol were the main products of biosynthesis, while with *T. mentagrophytes,* squalene was an important end-product, particularly at higher amorolfine concentrations. These marked interspecies differences in sterol patterns suggest that caution would need to be exercised in establishing screens for these enzymes and in interpreting data in relation to potential therapeutic utility.

3. Δ^{24} Methenylation

Having considered the various stages of ergosterol biosynthesis as potential targets, we feel Δ^{24} methenylation the most attractive option (7). Moreover, preliminary experiments with combinations of inhibitors of different stages of ergosterol biosynthesis suggest that there is the possibility of synergy, and also fungicidal activity where the individual components of the mixture are only fungistatic. We therefore describe a screen we have used to detect inhibitors of Δ^{24} methenylation; this screen is more appropriate for use in a programme of directed synthesis rather than for random screening. The test measures inhibition of the transfer of the tritiated methyl group from radiolabelled S-adenosyl methionine (SAM) to sterol, and can be done either with a broken-cell preparation of *C. albicans* or with whole cells.

C. albicans yeasts are suspended at 2×10^9/ml in phosphate buffer (100 mM; pH 7.4) containing 30 mM nicotinamide, 5 mM $MgCl_2$, and 5 mM N-acetylcysteine and broken by shaking with Ballotini beads in a Braun disintegrator for a total of 3 min interspersed with periods of cooling. The cellular homogenate is then centrifuged at 8,000g for 20 min at 4°C and the supernatant is used for the assay. The system contains in a final volume of 1 ml: 0.9 ml homogenate, 1 μmole NADP, 1 μmole NAD, 3 μmoles glucose-6-phosphate, 5 μmoles ATP, 3 μmoles reduced glutathione, 0.7 i.u. glucose-6-phosphate dehydrogenase, 2 μmoles $MnCl_2$, 3 μmoles $MgCl_2$, 0.5 μCi [^{3}H] methyl-S-adenosyl methionine, and inhibitor (added in 10 μl DMSO). Following incubation at 37°C for 30 min, the reaction is stopped by the addition of 2 ml 15% KOH in 90% ethanol and the mixture is heated at 80°C for 90 min. Radioactivity is extracted with petroleum ether and counted. Experiments with whole cells utilize a suspension of 2×10^9/ml in 100 mM phosphate buffer pH 6.5 containing 56 mM glucose and 0.05 μCi/ml SAM; incubation and processing is as before.

We have used this screen to test large series of synthetic sterols; particularly good activity was found with derivatives containing an N substitution in the side chain such as ICI 62,965 and tomatidine (7). GC/MS of sterols extracted from *C. albicans* cells incubated with these inhibitors showed accumulation of zymosterol in place of ergosterol.

4. *Synergistic Mixtures*

Having identified inhibitors of Δ^{24} methenylation, we then investigated active compounds for antifungal activity in combination with inhibitors of other stages of the sterol synthetic pathway such as the allylamine terbinafine (squalene epoxidase), trifluperidol ($\Delta^8 - \Delta^7$ isomerase), the azole ICI 153,066 (C-14 demethylase), and the azasterol A 25822B or the similar ICI 173,868 (Δ^{14} reductase). Pronounced synergism and fungicidal activity was found, particularly with combinations containing Δ^{14} reductase inhibitors; fungicidal activity appeared to be due to the production of a novel 8:14:24 triene. Unfortunately the combinations we investigated only demonstrated these effects *in vitro* and showed no enhancement of activity in animal models of infection.

Sud and Feingold (36) studied combinations of ketoconazole with an earlier allylamine, naftifine, 15-azasterol, triarimol (a nonazole inhibitor of C-14 demethylase), and mevinolin (an inhibitor of 3-hydroxy-3-methylglutaryl-coenzyme A reductase). Their studies utilized a range of fungi, but were confined to investigations *in vitro*. At levels below their minimum inhibitory concentrations (MICs), these agents were able to bring about a fourfold or greater decrease in the MIC of ketoconazole. Results were variable from agent to agent and between fungi; in general yeasts were found to be less susceptible to drug combinations than were filamentous fungi. Gadebusch and Valiant (13,14) have sought to patent mixtures of ketoconazole and terbinafine and ketoconazole and mevinolin, but again their claims are supported only by *in vitro* data.

Trypanosoma cruzi, the agent of Chagas' disease, or South American trypanosomiasis, has ergosterol as its principal sterol, and is susceptible to inhibition by ketoconazole. Urbina *et al*. (37) showed synergism between ketoconazole and terbinafine against this parasite in a number of *in vitro* systems. This synergism has now been demonstrated in mice, as has a remarkable synergism between the much more potent azole ICI 195,739 and terbinafine; interestingly enough, this latter combination did not show synergism *in vitro*. Basic *in vitro* screening tests and studies *in vitro* of potential drug combinations may be very misleading; there is no reliable substitute for doing either type of study *in vivo* in the first instance.

5. *Evaluation of Selective Activity*

Detection of activity in any screening system may be followed by quantitation of that activity in the same system. With natural product samples, some extraction and purification procedures will be necessary prior to such quantitation. In the case of interesting activity in cell-free systems, the logical progression is to see whether this specific form of activity is retained with intact cells (Figure 21.2). Ultimately, however, active compounds need to be examined in the sort of systems described in the next two sections to establish the spectrum of useful activity and whether this activity is expressed in *in vivo* models of infection.

III. General *In Vitro* Screens

A. *Preliminary Screening*

Currently useful antifungal drugs include wide-spectrum agents such as amphotericin B and the azoles, and compounds with a more restricted spectrum of activity like griseofulvin (dermatophytes only), allylamines (effectively dermatophytes only), 5-fluorocytosine (yeasts only), and cilofungin (*C. albicans* and *C. tropicalis* only). Although the widest possible spectrum of activity is desirable, more restricted activity is not to be despised. A screen therefore should ideally encompass a wide range of relevant organisms so that narrow-spectrum activity is not missed. With a programme of directed chemical synthesis, this should present little difficulty, since throughput of compounds will be modest. With a concerted empirical attack on a large chemical collection, or a natural product campaign, where a high weekly throughput of samples is essential, a single organism or very restricted group of organisms may need to be used to make screening possible.

Three basic methods of screening are available: agar dilution, agar diffusion, and broth dilution. In an agar dilution system, the test compound is incorporated at a known level (*e.g.*, 50 μg/ml) in a molten agar-based medium such as Kimmig's; a plate is poured and subsequently inoculated. By using a multipoint inoculator a large number (around 20) of organisms can be tested on a single plate. Plates are incubated and growth of each organism is scored. When activity is found, the test is repeated by using plates containing serial dilutions of the compound to determine the MIC against each organism. With many organisms growing on the same plate, time and temperature of incubation will not be optimal for all; nevertheless, it is possible to achieve a compromise which allows detection of activity against each organism, thus fulfilling the purposes of the screen. Each compound—and where repeat testing is required, each dilution of each compound—requires a separate plate, but results generated are comprehensive. We have had no difficulty maintaining a throughput of 200 compounds a week in such a system, and in fact expanded it for a period to accommodate an additional 100 natural product samples weekly.

For an agar diffusion assay each plate is seeded with a single organism. Samples for screening are put in wells cut in the agar or in small stainless-steel cylinders or on filter paper discs placed on the agar. Following incubation the zones of inhibition of growth round each sample are measured; the more active the compound, the larger the zone. Although only a single organism is incorporated in each plate, one plate can be used to test several compounds. Moreover, repeat testing is not necessary, since some idea of the potency can be gained from the zone size. With poorly soluble compounds which diffuse insufficiently in the agar, the zone of inhibition may not reflect the true antifungal potential of the compound; itraconazole is a good example of this problem. Since quantitation in terms of MIC with unknown samples is not possible, and since realistically only a limited number of organisms can be used with such a system (more organisms can be inoculated per unit area in an agar dilution assay than samples can be assayed in the same area in an agar diffusion assay), we have preferred the agar dilution assay. However, Debono and Gordee (10) find an agar diffusion screen based on *C. albicans* and *Aspergillus* spp. very satisfactory in the preliminary screening of natural product samples.

In a broth dilution assay, compound is incorporated in a liquid medium which is then inoculated with the fungus and incubated. After a specified time, growth—or absence of growth—can be detected by visual observation or can be quantified by nephelometry or spectrophotometry. Although each dilution of each compound and each organism requires by a separate tube, the broth dilution system is amenable to miniaturization by using microtitre plates and automation. We have not used the broth dilution system for primary screening of synthetic chemicals, but have found it useful for secondary evaluation of azoles in an attempt to establish *in vitro*/*in vivo* correlations (8). Azoles usually give anomalous end-points in conventional MIC determinations, since growth restriction can occur over a very wide range of drug concentrations. Using a broth dilution assay with quantitation of growth it is possible to calculate an IC_{50} (concentration to inhibit growth by 50%) which is more useful in correlation studies (7,12). Moreover, we used a miniaturized broth dilution screen based on normal and amphotericin B–resistant strains of *C. albicans* for a while in connection with a natural product campaign (Figure 21.2). We have used amphotericin B–resistant *C. albicans* in both our screens to identify polyenes at an early stage and thus eliminate them from further consideration. Etienne *et al.* (11) on the contrary suggest the use of normal and polyene-resistant strains of *S. cerevisiae* in screens to identify new polyene macrolides.

Although they are of considerable therapeutic significance and show apprecia-ble activity *in vivo,* azoles as a class do not show marked activity *in vitro* under normal conditions. The mycelial phase of *C. albicans* is, however, extremely sensitive to active azoles, and when screening this type of compound we have found it advantageous to run a supplementary mycelial screen alongside the standard agar dilution screen—which includes several strains of *C. albicans* growing in the yeast phase. This mycelial screen is performed by using Eagle's minimal essential medium containing 1% foetal calf serum with overnight incuba-tion at 37°C in 5% CO_2:air in multiwell plastic plates; cultures are examined for mycelial growth by means of an inverted microscope.

B. *Follow-up of Activity*

The biggest problem with an *in vitro* screen is interpreting the results and ascribing significance to them. With antifungals in general and with azoles in particular, correlation of activity *in vitro* with activity *in vivo* is poor; the pre-dictive value of *in vitro* screens is limited. Where a defined synthetic chemical has shown activity in an agar dilution screen against a range of organisms it is appropriate to test that compound *in vivo*. Only if interesting *in vivo* activity is found is it reasonable to do further evaluation *in vitro,* investigating such features as fungicidal as opposed to fungistatic activity, synergy or antagonism with other compounds, pH effects, and studies on the development of resistance. On the other hand, when an undefined natural product sample shows activity *in vitro,* problems start! It is necessary to confirm activity and if possible define culture conditions which result in the production of larger amounts of that activity (Figure 21.2). During one natural product campaign, using 192 actinomycetes cultured on four different media each week, we introduced a simple antibacterial screen, reasoning that broths showing activity in both situations were displaying general rather than hoped-for selective toxicity. Methods have to be devised to extract and purify the active principle, and at as early a stage in the extraction and identification procedure as possible, to determine whether the substance is novel or whether an already-known antibiotic has been rediscovered yet again. The technology by which this is achieved involves a variety of up-to-date physico-chemical methods in conjunction with a comprehensive data bank containing information on all known antibiotics. Testing of natural product samples *in vivo* may have to await the isolation of a pure product, although we have had some success leading to the discovery of an antifungal azasteroid (ICI 173,868) by using cultures or extracts of *Calcarisporium thermophilum* freeze-dried onto mouse food which was subsequently fed to animals in the screen.

More comprehensive reviews of the philosophy and practice of screening *in vitro,* along with more detailed descriptions of methodology, may be found in Ryley *et al.* (33), Ryley and Rathmell (32), and Wilson and Ryley (38).

IV. *In Vivo* Screens

The ultimate laboratory test for a potential antifungal agent is its performance in an animal model of infection. Because of the problems of poor *in vitro*/*in vivo* correlation already alluded to, screening *in vivo* as well as evaluation *in vivo* is desirable wherever possible. The limitations of *in vivo* screening involves such factors as quantity of compound required/available, limitations of numbers of samples that can be screened because of the technical effort entailed, and the spectrum of activity which can be investigated at the screening stage; also appropriate for consideration at this time is the ethical issue of the use of animals rather than alternative (albeit unsatisfactory) *in vitro* technology.

A. *Mouse Protection Test*

The simplest type of screen is a mouse protection test such as that used by Richardson *et al.* (29) in the discovery of fluconazole. Mice are inoculated intravenously with an inoculum of *C. albicans* uniformly lethal for untreated animals within 48 h (around 10^7 yeasts). Groups of five mice are treated orally or by injection with test compounds 1, 4, and 24 h after inoculation. Mortality is assessed over 48 h. If several dose levels of compound are used, it is possible to calculate the 50% effective dose (ED_{50}). In a screening situation a single arbitrary dose would be used in the first instance, and those samples showing activity would be retested at a range of doses in order to determine an ED_{50}, which would allow some assessment of potential interest in the compound. This screen could be used with lethal infections of other fungi of potential interest, *e.g.*, *C. neoformans* or *A. fumigatus*. The model is attractive from the point of view of simplicity of operation and the relatively small quantities of compound required; the ED_{50} for fluconazole, for instance, is 0.06–0.08 mg/kg, depending on the route of administration. The system is, however, a very artificial one, the compound just tipping the very delicate balance between death and life in this very acute infection. If a smaller inoculum and a longer period of observation are used, considerably higher doses of compound are required to suppress infection sufficiently to control mortality. Of present concern also is the use of animals in acute experiments where mortality is the end point.

B. Multiple Infection Models

Because at the time we were interested in therapy of the more prevalent though less serious superficial fungal infections, we chose to screen compounds against vaginal infections with *C. albicans* and against dermatophyte infections. In order to conserve compound at the screening stage, we chose to work in mice rather than the more usual rat and guinea pig respectively, and to further conserve compound and animals, we used mice carrying a dual infection (full details given in 32 and 33). Female mice were treated with estradiol to render the vagina susceptible to infection and were inoculated into the vagina with *C. albicans* and on the back with *T. mentagrophytes var quinckaenum*. Mice were used orally at an initial screening level of 250 mg/kg with compound once daily for 5 days; infections were assessed on day 6 or 7. Ringworm lesions were scored visually, and vaginal infections were monitored by plating out a sample obtained with a wire loop onto BiGGY agar (Difco) and scoring growth following incubation. Active samples were retested to determine the minimal dose necessary to give complete control of the two infections. Although more involved than the simple mouse protection test, the screen is technically simple to perform and read; scoring of infections rather than labour-intensive counting procedures is adequate at the screening stage. Compound requirements are greater than for the acute mouse protection test, but the results obtained are more realistic in terms of treatment of chronic infections. Finally, the use of nonlethal infections is ethically more acceptable.

Although ideally we would like a wide-spectrum antifungal agent—which would be detected by any antifungal screen—new compounds with restricted activity could be useful. Although the mouse protection test with *C. albicans* would detect interesting azoles and amphotericin B, it would miss compounds such as griseofulvin and terbinafine. Our dual-infection model would identify activity in azoles, griseofulvin, and terbinafine, but would miss amphotericin B, since this drug does not appear to reach the vagina. For a screening campaign with azoles, either screen would be completely satisfactory; compounds identified in the screen as being of particular interest would then be evaluated in a series of other models to determine more accurately their potency and spectrum of activity. With samples where wide-spectrum activity may not necessarily be expected, the choice is either to concentrate on an infection of particular interest and possibly miss some useful compounds or to modify or enlarge the screen to encompass a wider range of fungi. We have chosen the latter approach and developed a screen using mice carrying both vaginal and nonlethal systemic infections with *C. albicans,* a superficial infection with *T. quinckeanum,* and a respiratory infection with *C. neoformans* all in the same mouse; we are currently attempting to introduce a fifth infection with *A. fumigatus* into the mice in such a way that it does not interfere with the other four distinct infections! We can induce discrete, nondisseminating subcutaneous infections by injection of spores of *A. fumigatus* in sloppy agar or with the simultaneous formation of an air sac; unfortunately these infections do not respond very well to high doses of amphotericin B or the azole ICI 195,739 compounds which have marked activity against systemic infections in mice. The advantage of the model is that activity against all four infections is obtained with one lot of compound and one lot of mice. The technical effort required is of course greater than with the simpler screens, and would not be justified at the *screening* stage with azoles, for instance. When compound for screening is at a premium, or for purposes of further evaluation, the model deserves further consideration; full details are given by Ryley and McGregor (31).

V. Conclusions and Further Prospects

The purpose of screening is to identify compounds which merit further investigation. The screens just considered will not only identify such compounds; they will give some indication of potency and spectrum, and hence suggest a degree of interest. Further evaluation of particularly interesting compounds will involve experiments with vaginal infections in mice and rats, superficial infections with *Candida* and dermatophytes in guinea pigs, intestinal infections with *C. albicans* in mice, and systemic infections with a whole range of fungi in mice, rats, and guinea pigs. Such models for drug evaluation are discussed in more detail by Ryley (30).

Target-oriented selective enzyme screens are the fashion, and we describe a number of screens of this type. We are not, however, aware of any antifungal drug which has been designed and discovered by the use of such selective screens; mode-of-action studies have been retrospective rather than prospective. The rationally designed antifungal drug is something for the future. In the meantime we feel there is much to be said for screens utilizing whole organisms, and preferably *in vivo* rather than *in vitro* screens. If we really want new antifungal agents, by all means let us pursue the rational pathway—but not to the exclusion of the empirical approach, which has served us so well in the past.

References

1. Adams, D.J., & G.W. Gooday. 1980. A rapid chitin synthase preparation for the assay of potential fungicides and insecticides. Biotech. Lett. **2**:75–78.

2. Baguley, B.C., G. Rommelle, J. Gruner, & W. Wehrli. 1979. Papulacandin B: an inhibitor of glucan synthesis in yeast spheroplasts. Eur. J. Biochem. **97:**345–351.
3. Ballou, C. 1976. Structure and biosynthesis of the mannan component of the yeast cell envelope. Adv. Microbial Physiol. **14:**93–158.
4. Ballou, L., R.E. Cohen, & C.E. Ballou. 1980. *Saccharomyces cerevisiae* mutants that make mannoproteins with a truncated carbohydrate outer chain. J. Biol. Chem. **255:**5986–5991.
5. Baloch, R.I., & E.I. Mercer. 1987. Inhibition of sterol $\Delta^8 \rightarrow \Delta^7$-isomerase and Δ^{14}-reductase by fenpropimorph, tridemorph and fenpropidin in cell-free enzyme systems from *Saccharomyces cerevisiae*. Phytochemistry **26:**663–668.
6. Barrett-Bee, K.J., J. Lees, & W. Henderson. 1982. Variation in the activities of enzymes associated with cell wall metabolism during a growth cycle of *Candida albicans*. FEMS Microb. Lett. **15:**275–278.
7. Boyle, F.T. 1990. Drug discovery: a chemist's approach, p. 3–29. *In* J.F. Ryley (ed.), Handbook of Experimental Pharmacology, vol. *96,* Chemotherapy of Fungal Diseases. Springer-Verlag, Heidelberg.
8. Boyle, F.T., J.F. Ryley, & R.G. Wilson. 1987. In vitro–in vivo correlations with azole antifungals, p. S1:31–S1:41. *In* R.A. Fromtling (ed.), Recent Trends in the Discovery, Development and Evaluation of Antifungal Agents. J.R. Prous Science Publishers, Barcelona.
9. Brian, P.W., P.J. Curtis, & H.G. Hemming. 1946. A substance causing abnormal development of fungal hyphae produced by *Penicillium janczewskii* Zal. I. Biological assays, production, and isolation of 'Curling Factor'. Trans. Br. Mycol. Soc. **29:**173–187.
10. Debono, M., & R.S. Gordee. 1990. Drug discovery: Nature's approach, p. 77–109. *In* J.F. Ryley (ed.), Handbook of Experimental Pharmacology, vol. 96, Chemotherapy of Fungal Diseases. Springer-Verlag, Heidelberg.
11. Etienne, G., E. Armau, & G. Tiraby. 1990. A screening method for antifungal substances using *Saccharomyces cerevisiae* strains resistant to polyene macrolides. J. Antibiotics **43:**199–206.
12. Flint, O.P., & F.T. Boyle. Structure-teratogenicity relationships among antifungal triazoles, p. 231–249. *In* J.F. Ryley (ed.), Handbook of Experimental Pharmacology, vol. 96, Chemotherapy of Fungal Diseases. Springer-Verlag, Heidelberg.
13. Gadebusch, H.H., & M.E. Valiant. 1988a. Controlling mycotic infections. U.K. Patent GB2197194.
14. Gadebusch, H.H., & M.E. Valiant, 1988b. Controlling mycotic infections. U.K. Patent GB2200550.
15. Gunji, S., K. Arima, & T. Beppu. 1983. Screening of antifungal antibiotics according to activities inducing morphological abnormalities. Agric. Biol. Chem. **47:**2061–2069.
16. Hector, R.F., & P.C. Braun. 1986. Synergistic action of nikkomycins X and Z with papulacandin B on whole cells and regenerating protoplasts of *Candida albicans*. Antimicrob. Agents Chemother. **29:**389–394.

17. Hector, R.F., B.L. Zimmer, & D. Pappagianis. 1990. Evaluation of nikkomycins X and Z in murine models of coccidioidomycosis, histoplasmosis, and blastomycosis. Antimicrob. Agents Chemother. **34:** 587–593.

18. Iwasa, T., E. Higashide, H. Yamamoto, & M. Shibata. 1971. Studies on validamycins, new antibiotics. II. Production and biological properties of validamycins A and B. J. Antibiot. **24:**107–113.

19. Jolidon, S., A.-M. Polak, P. Guerry, & P.G. Hartman. 1990. Inhibitors of 2,3-oxidosqualene lanosterol-cyclase as potential antifungal agents. Biochem. Soc. Trans. **18:**47–48.

20. Kerkenaar, A. 1990. Inhibition of the sterol Δ^{14}-reductase and $\Delta^{8}\rightarrow\Delta^{7}$ isomerase in fungi. Biochem. Soc. Trans. **18:**59–61.

21. Kerridge, D., & H. Vanden Bossche. 1990. Drug discovery: a biochemist's approach, p. 31–76. *In* J.F. Ryley (ed.), Handbook of Experimental Chemotherapy, vol. 96, Chemotherapy of Fungal Diseases. Springer-Verlag, Heidelberg.

22. Kirsch, D.R., & M.H. Lai. 1986. A modified screen for the detection of cell wall-acting antifungal compounds. J. Antibiot. **39:**1620–1622.

23. Konishi, M., M. Nishio, K. Saitoh, T. Miyaki, T. Oki, & H. Kawaguchi. 1989. Cispentacin, a new antifungal antibiotic. I. Production, isolation, physico-chemical properties and structure. J. Antibiot. **42:**1749–1755.

24. Nakajima, T., & C.E. Ballou. 1974. Characterization of the carbohydrate fragments obtained from *Saccharomyces cerevisiae* mannan by alkaline degradation. J. Biol. Chem. **249:**7679–7684.

25. Nakajima, T., & C.E. Ballou. 1975. Yeast manno-protein biosynthesis: solubilization and selective assay of four mannosyltransferases. Proc. Nat. Acad. Sci. USA **72:**3912–3916.

26. Oki, T., M. Hirano, K. Tomatsu, K. Numata, & H. Kamei. 1989. Cispentacin, a new antifungal antibiotic. II. *In vitro* and *in vivo* antifungal activities. J. Antibiot. **42:**1756–1762.

27. Polak, A. 1988. Mode of action of morpholine derivatives. Ann. N.Y. Acad. Sci. **544:**221–228.

28. Polak, A. 1990. Mode of action studies, p. 153–182. *In* J.F. Ryley (ed.), Handbook of Experimental Pharmacology, vol. 96. Chemotherapy of Fungal Diseases. Springer-Verlag, Heidelberg.

29. Richardson, K., K.W. Brammer, M.S. Marriott, & P.F. Troke. 1985. Activity of UK-49,858, a bis-triazole derivative, against experimental infections with *Candida albicans* and *Trichophyton mentagrophytes*. Antimicrob. Agents Chemother. **27:**832–835.

30. Ryley, J.F. 1990. Screening and evaluation in vivo, p. 129–147. *In* J.F. Ryley (ed.), Handbook of Experimental Pharmacology, vol. 96. Chemotherapy of Fungal Diseases, Springer-Verlag, Heidelberg.

31. Ryley, J.F., & S. McGregor. 1988. A multi-infection model for antifungal screening *in vivo*. J. Antimicrob. Chemother. **22:**353–358.

32. Ryley, J.F., & W.G. Rathmell. 1984. Discovery of antifungal agents: *in vitro* and *in vivo* testing, p. 63–87. *In* A.P.J. Trinci & J.F. Ryley (ed.), Mode of Action of Antifungal Agents, British Mycological Society Symposium No. 9, Cambridge University Press.
33. Ryley, J.F., R.G. Wilson, M.B. Gravestock, & J.P. Poyser. 1981. Experimental approaches to antifungal chemotherapy. Adv. Pharmacol. Chemother. **18:**49–176.
34. Selitrennikoff, C.P. 1983. Use of a temperature-sensitive, protoplast-forming *Neurospora crassa* strain for the detection of antifungal antibiotics. Antimicrob. Agents Chemother. **23:**757–765.
35. Selitrennikoff, C.P., B.L. Lilley, & R. Zucker, 1981. Formation and regeneration of protoplasts derived from a temperature-sensitive *osmotic* strain of *Neurospora crassa*. Exp. Mycol. **5:**155–161.
36. Sud, I.J., & D.S. Feingold. 1985. Effect of ketoconazole in combination with other inhibitors of sterol synthesis on fungal growth. Antimicrob. Agents Chemother. **28:**532–534.
37. Urbina, J.A., K. Lazardi, T. Aguirre, M.M. Piras, & R. Piras. 1988. Antiproliferative synergism of the allylamine SF 86–327 and ketoconazole on epimastigotes and amastigotes of *Trypanosoma* (*Schizotrypanum*) *cruzi*. Antimicrob. Agents Chemother. **32:**1237–1242.
38. Wilson, R.G., & J.F. Ryley. 1990. Screening and evaluation in vitro, p. 111–128. *In* J.F. Ryley (ed.), Handbook of Experimental Pharmacology, vol. 96, Chemotherapy of Fungal Diseases. Springer-Verlag, Heidelberg.

22

Chemotherapeutic Targets in *Pneumocystis carinii*

Gregg Y. Lipschik and *Joseph A. Kovacs*

I. Introduction

A. Pneumocystis Carinii *Pneumonia: Scope of the Problem*

P. carinii pneumonia, the most common life-threatening infection in patients with the acquired immunodeficiency syndrome (AIDS) as well as the most common index diagnosis in AIDS, is caused by an enigmatic pathogen whose metabolism and basic biology are poorly understood (11,12,16). Prior to AIDS, *P. carinii* pneumonia was largely a rare disease of immunosuppressed children, with a yearly incidence of fewer than 100 cases in the early 1970s (96). The AIDS epidemic has made *P. carinii* pneumonia an enormously important public health problem; it has become among the most frequent causes of pneumonia requiring hospitalization and is probably the most common cause of death in AIDS. A recent estimate predicts 150,000 cases of *P. carinii* pneumonia in the next 3–4 years (35). Approximately 80% of patients with AIDS will have at least one episode of *P. carinii* pneumonia (67). The incidence of *P. carinii* pneumonia in patients with cancer may also be rising (32), further highlighting the need for improved diagnosis, therapy, and prophylaxis of this once-obscure illness.

Although *P. carinii* is one of the most important pulmonary pathogens infecting immunosuppressed hosts, standard therapy is limited to two drugs, trimethoprim/sulfamethoxazole and pentamidine, both of which frequently cause serious adverse effects, particularly in AIDS patients, and are ineffective in 10–40% of cases (31,33,51). Although survival in *P. carinii* pneumonia has improved recently, this has been due to earlier recognition and improved diagnosis rather than to the development of more effective therapeutic regimens. Improvements in therapy for *P. carinii* pneumonia have lagged diagnostic advances for several reasons. Until recently, little was known of the intermediary metabolism of *P. carinii;* only a handful of surveys of the organism's metabolic potential had been published (59,74,77). In addition, *in vitro* cultivation of *P. carinii* is unreliable and not applicable to human-derived organisms. Finally, the corticosteroid-treated rat model of *P. carinii* pneumonia (27), the most widely used model of the infection, is cumbersome and unpredictable.

There is also a limited repertoire of prophylactic therapies for the prevention of *P. carinii* pneumonia. Prophylaxis is warranted for patients at high risk of developing *P. carinii* pneumonia, including AIDS patients with a history of *P. carinii* pneumonia or who have CD4 counts below 200 cells/mm^3 (25,57), children undergoing therapy for acute lymphocytic leukemia (37,99), and several other severely immunosuppressed patient groups. Trimethoprim/sulfamethoxazole is the prophylactic agent of choice but, as for therapy, adverse reactions are frequent in AIDS patients and may necessitate discontinuation of the drug (25). Aerosolized (nebulized) pentamidine is emerging as a viable, convenient alternative but has a high breakthrough rate (54). Several other drugs, including pyrimethamine/sulfadoxine (71) and dapsone (64), have been evaluated in uncontrolled trials as prophylactic agents for *P. carinii* pneumonia. Although zidovudine appears to reduce recurrences of *P. carinii* pneumonia, probably contributing to its ability to prolong survival in AIDS patients (26), this drug alone is inadequate prophylaxis for *P. carinii* pneumonia.

In the past few years, elucidation of aspects of the organism's metabolism and the development of methods for short-term *in vitro* cultivation of *P. carinii* have enabled the screening of novel compounds including potent dihydrofolate reductase (DHFR) inhibitors such as trimetrexate, newer sulfonamides and sulfones, clindamycin with primaquine, and several new 8-aminoquinolines (2,8,48,81). The recent application of the powerful methods of molecular biology to the study of *P. carinii* has begun to bear fruit; several genes have been cloned and sequenced (20–22,88,101). Their products not only offer intriguing clues into the taxonomy of the organism but also serve as potential chemotherapeutic targets.

B. Phylogeny and Metabolism of P. Carinii

Although long presumed to be a protozoan based on morphological considerations and the success of therapy with pentamidine, recent studies suggest that *P. carinii* is a fungus related to *Saccharomyces cerevisiae* or to the *Myxomycota* or *Zygomycota* (21,49,58,89,92,97,100). In the strongest evidence supporting this classification, Edman *et al.* (21), Watanabe *et al.* (97) and Stringer *et al.* (89) used ribosomal RNA sequence homologies to show phylogenetic similarity to these fungi. Other investigators have demonstrated ultrastructural similarities of fungal and *P. carinii* cell walls (92) as well as the presence in *P. carinii* of yeast glucan (1,3,-β-glucan) and β-glucan synthase activity (58,100) and chitin (93). Finally, DHFR and thymidylate synthase (TS) activities reside on different proteins in *P. carinii,* a characteristic of fungi rather than protozoans, which have a bifunctional DHFR/TS molecule (20,22,28,49). Interestingly, cholesterol is the major sterol present in *P. carinii* (46); the organism lacks ergosterol and other fungal sterols, consistent with the observation that it is relatively resistant to polyene antifungals such as amphotericin B (19).

Studies of intermediary metabolism in *P. carinii* have been hampered by difficulties in purifying the organism from contaminating rat cells and thus in unambiguously attributing metabolic activities to *P. carinii*. Pesanti and Cox (77) and Pesanti (74) were the first to describe metabolic activities of *P. carinii* organisms. They demonstrated that *P. carinii* was able to metabolize [^{14}C]-glucose to $^{14}CO_2$, synthesize proteins from [^{3}H]-amino acids, synthesize RNA from [^{3}H]-uridine, and utilize molecular oxygen. Pesanti (74) also suggested the presence in the organism of superoxide dismutase, among the several antioxidant enzymes assayed for. Mazer *et al.* (59) used a tetrazolium dye technique to demonstrate and localize lactate, succinate, and glutamate dehydrogenases in *P. carinii,* implying the potential for glycolysis, the Krebs cycle, and protein metabolism. More recently the metabolic repertoire of *P. carinii* has been shown to include DHFR (2,20,49), TS (22), β-glucan synthase (100), catalase and glucose-6-phosphate dehydrogenase (75), *de novo* folate synthesis (48), and polyamine metabolism (54a). Several of these are obvious chemotherapeutic targets.

The therapeutic response of *P. carinii* pneumonia to trimethoprim/sulfamethoxazole in rats and humans (38,40) implied the presence in *P. carinii* of DHFR and the enzymes required for *de novo* folate synthesis. Allegra *et al.* (2) demonstrated DHFR activity in lysates of the organism and showed that *P. carinii* cannot take up available preformed folates. This suggests that *P. carinii* must synthesize folates *de novo* from pteridines and para-aminobenzoic acid (PABA), presumably via a pathway involving the enzyme dihydropteroate synthase (DHPS: Figure 22.1). Kovacs *et al.* (48) subsequently showed that *P. carinii* can assimilate and incorporate [^{3}H]-PABA into reduced folates, principally 10-formyl tetrahydrofolate. DHPS activity has recently been demonstrated in lysates of the organism (62). As will be discussed, this pattern of folate metabolism presents an exploitable chemotherapeutic target, with a fortuitous bonus: mammalian cells cannot synthesize folates *de novo*.

Two enzymes in the folate metaboic pathway of *P. carinii* have now been cloned, sequenced, and expressed heterologously (20,22). The DHFR gene was isolated by its ability to confer trimethoprim resistance to *E. coli* (20); native *P. carinii* DHFR has also been partially purified and characterized (49). The TS gene was cloned by using degenerate oligonucleotide primers based on conserved regions of the TS amino acid sequence to amplify *P. carinii* genomic DNA; amplified fragments were then used to probe a cDNA library (22).

The reported response of *P. carinii* pneumonia to α-difluoromethylornithine (DFMO), a specific inhibitor of ornithine decarboxylase, the rate-limiting enzyme of polyamine synthesis, has prompted the study of polyamine metabolism in *P.*

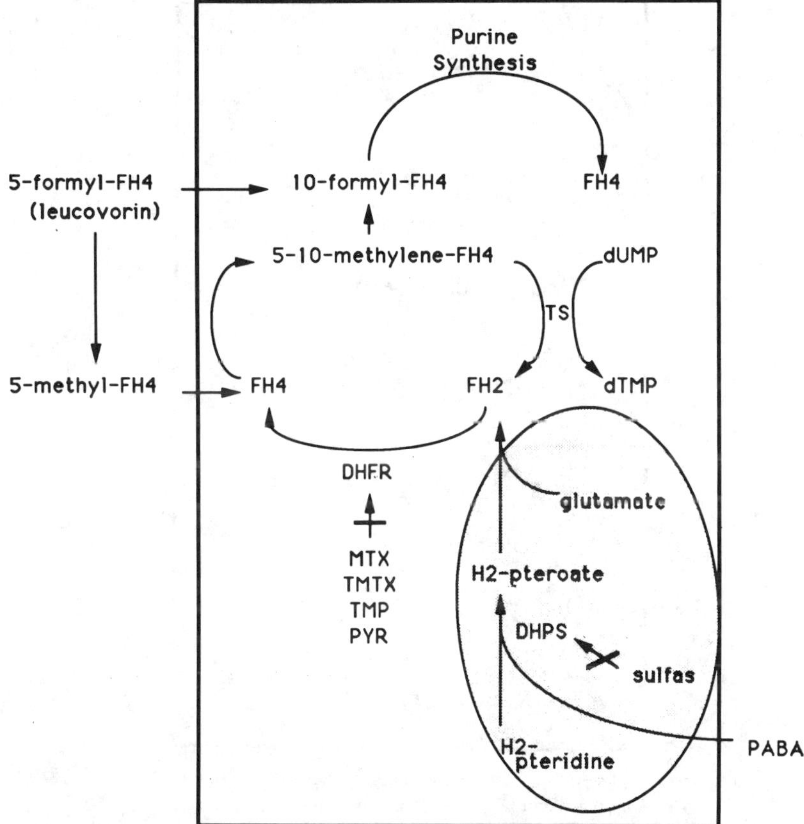

Figure 22.1. Cellular folate metabolism. The rectangle represents the cell membrane; the shaded oval represents *P. carinii*. Pathways within the oval are unique to *P. carinii* while those in the rectangle but outside the oval are common to the pathogen and host. Exogenous PABA is incorporated by *P. carinii* but not by the host. Preformed folates (5-formyl-FH4 and 5-methyl-FH4) are taken up and incorporated by host cells but not by *P. carinii*. Abbreviations: FH4, tetrahydrofolate; FH2, dihydrofolate; dUMP, deoxyuridine monophosphate; dTMP, deoxythymidine monophosphate; TS, thymidilate synthase; DHFR, dihydrofolate reductase; DHPS, dihydropteroate synthase; H2, dihydro; MTX, methotrexate; TMTX, trimetrexate; TMP, trimethoprim; PYR, pyrimethamine; PABA, paraaminobenzoic acid.

carinii (30,60,69). *P. carinii* contain a different profile of polyamines than mammalian cells and are able to incorporate [^{3}H]-ornithine and [^{3}H]-arginine into polyamines (Figure 22.2) (54a).

Thus, over the past several years a number of potential chemotherapeutic targets have been characterized. These enzyme systems provide a rational approach to developmental therapeutics for *P. carinii* pneumonia. The obvious next step is screening of potentially useful compounds in preclinical studies.

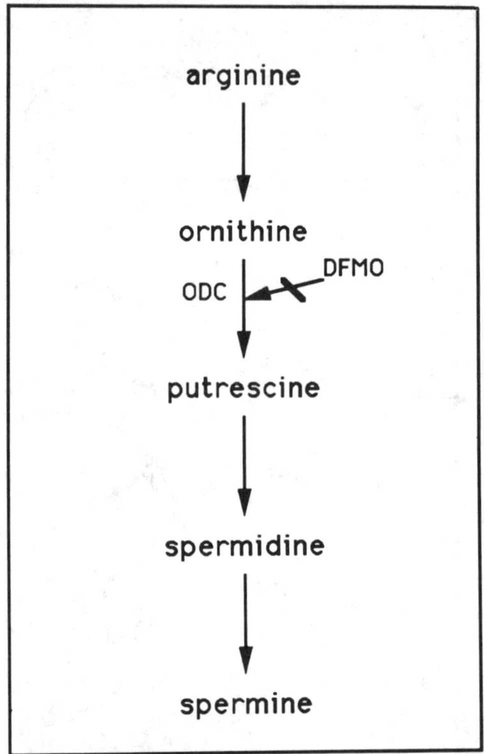

Figure 22.2. Polyamine biosynthetic pathway in eukaryotes. Abbreviations: ODC, ornithine decarboxylase; DFMO, α-difluoromethylornithine.

C. In Vitro *and* In Vivo *Drug Screening*

1. Animal Models

The corticosteroid-treated rat model of *P. carinii* pneumonia has scarcely been improved upon since its description in 1966 (27). Briefly, administration of steroids to rats results in spontaneous *P. carinii* pneumonia of varying severity in most rats within 6–12 weeks. Concomitant administration of tetracycline prevents concurrent bacterial infections. Some investigators believe that a low protein diet enhances the intensity of the induced pneumonia. Using this model, trimethoprim/sulfamethoxazole was first demonstrated to be effective therapy and prophylaxis for *P. carinii* pneumonia (40). More recently, a variety of compounds have been screened in the rat model, with some promising new therapeutic and prophylactic agents, including diaminodiphenylsulfone (dapsone) and aerosolized pentamidine, emerging from this work (29,39,42,45,81,94).

The empiric success of the rat model, however, belies its inherent limitations. Large-scale drug screening using this model is not practical. A minimum of five to ten animals are needed to evaluate a single drug at one dose, and many milligrams of each compound are necessary to complete a single study. Thus screening a given class of drugs, such as DHFR inhibitors, some of which may be available only in microgram quantities, is not possible. Moreover, despite prophylactic tetracycline, steroid-treated rats often develop bacterial or fungal infections which may complicate interpretation of response to a drug. In addition, the development of *P. carinii* pneumonia in immunosuppressed rats is quite variable, and testing rats for the presence of the infection prior to initiating experimental therapy is not practical. Of necessity then, experimental infections are not uniform in severity, clouding the interpretation of a therapeutic response. Alternatively, drugs may be tested as prophylactic agents, initiating therapy at the same time as or soon after immunosuppression is begun. A recent modification of the rat model, involving intratracheal instillation of a measured inoculum, may obviate some of these problems (6). Finally, pharmacokinetics in the rat model may differ markedly from that in humans, rendering dosage and interval data less useful.

2. In Vitro *Drug Screening*

In vitro drug screening methods for *P. carinii* offer, in addition to avoiding some of the unreliability of the rat model, the important advantages of speed, lower cost, and the need for only small quantities of drugs. *In vitro* methods suffer from several problems. *P. carinii* cannot be reliably cultured for more than a few days, and it is not possible to completely purify the organism from host tissues. In addition, methods requiring enumeration of organisms are extremely labor-intensive.

Nevertheless, screening methods based on short-term culture of the organism with various feeder cells and enumeration of cultured organisms have been published (9,18,78), although these techniques have been plagued by lack of reproducibility between laboratories. These methods have been used to demonstrate sensitivity of *P. carinii* to pentamidine and chloroquine (73), trimethoprim/sulfamethoxazole and pentamidine (18), trimetrexate and piritrexim (80), clindamycin/primaquine (81), and several pyrimethamine analogs (82). Recently, Cushion and Ebbets (17) described a method of axenic culture of *P. carinii* which they propose as useful for drug screening.

Identification and characterization of target enzyme systems unique to *P. carinii* would provide the basis of an excellent method of drug screening, unfettered by the clumsiness of the rat model and the unreliability of culture-based methods. The recent cloning and heterologous expression of *P. carinii* DHFR and TS are steps in this direction (20,22), as is the partial purification and characterization of *P. carinii* DHFR (49). The demonstration that *P. carinii* DHFR was susceptible to inhibition by the lipid-soluble antifolate trimetrexate (2) provided the theoretical basis for a successful clinical trial of trimetrexate in *P. carinii* pneumonia (1). The availability of recombinant proteins from *P. carinii* potentially enables the use of a powerful technology in developing new antipneumocystis drugs: molecular modeling of new agents based on x-ray crystallographic considerations of target enzyme tertiary structure.

Several other studies of intermediary metabolism in *P. carinii* have resulted in the development of viable methods of drug screening. Kovacs *et al.* (48) showed that *P. carinii* incorporates [^{3}H]-PABA into reduced folates (Figure 22.1); mammalian cells are unable to synthesize folates and require a source of preformed folates. This demonstration of a metabolic activity unambiguously attributable to *P. carinii* not only provides a rationale for the use of sulfonamides in *P. carinii* pneumonia but also provides an assay system for *in vitro* screening of antipneumocystis drugs which is not confounded by the presence of a feeder cell layer or contaminating host cells. In this system trimethoprim/sulfamethoxazole, sulfonamides, and pentamidine were effective inhibitors of *de novo* folate synthesis in *P. carinii*. In addition, this study showed that under conditions similar to those used by others for *in vitro* culture (9,18,78), *P. carinii* organisms remain metabolically active for only 2–3 days.

Lipschik *et al.* (54a) have shown that *P. carinii* can incorporate [^{3}H]-ornithine into polyamines (Figure 22.2) and proposed this as a method of screening for drug activity. Not only DFMO, a specific inhibitor of polyamine metabolism, but also pentamidine, whose mode of action is not known, were active in this system.

II. Chemotherapeutic Targets in Nucleotide Precursor Synthesis

A. Introduction

Enzymes in the pathway which produces reduced folates for nucleotide synthesis are particularly attractive chemotherapeutic targets in *P. carinii* (Figure 22.1). The established effectiveness of trimethoprim/sulfamethoxazole for *P. carinii* pneumonia, the recent elucidation of aspects of folate metabolism in *P. carinii* (see above), and the availability of many novel inhibitors of this pathway establish this as a primary target for investigation.

B. Dihydropteroate Synthase

The demonstration that *P. carinii* are unable to utilize preformed folates suggested the presence in the organism of the enzymes required to synthesize folates *de novo* from pteridines and PABA (2). DHPS, which catalyzes the formation of dihydropteroate from PABA and a dihydropteridine, 6-hydroxymethyldihydropterin (Figure 22.1), has now been found in crude lysates of *P. carinii* (62);

the unpurified enzyme had relatively low specific activity, 14 units/mg protein compared with 450 units/mg for *Escherichia coli*. The K_ms for PABA and 6-hydroxymethyldihydropterin were 71 μM and 81 μM, respectively.

Sulfonamides and sulfones are competitive inhibitors of bacterial DHPSs, and they appear to be among the most active compounds known against *P. carinii* (Figure 22.3). Several sulfonamides have activity against *P. carinii*. Sulfamethoxazole inhibited PABA incorporation into reduced folates both alone and in combi-

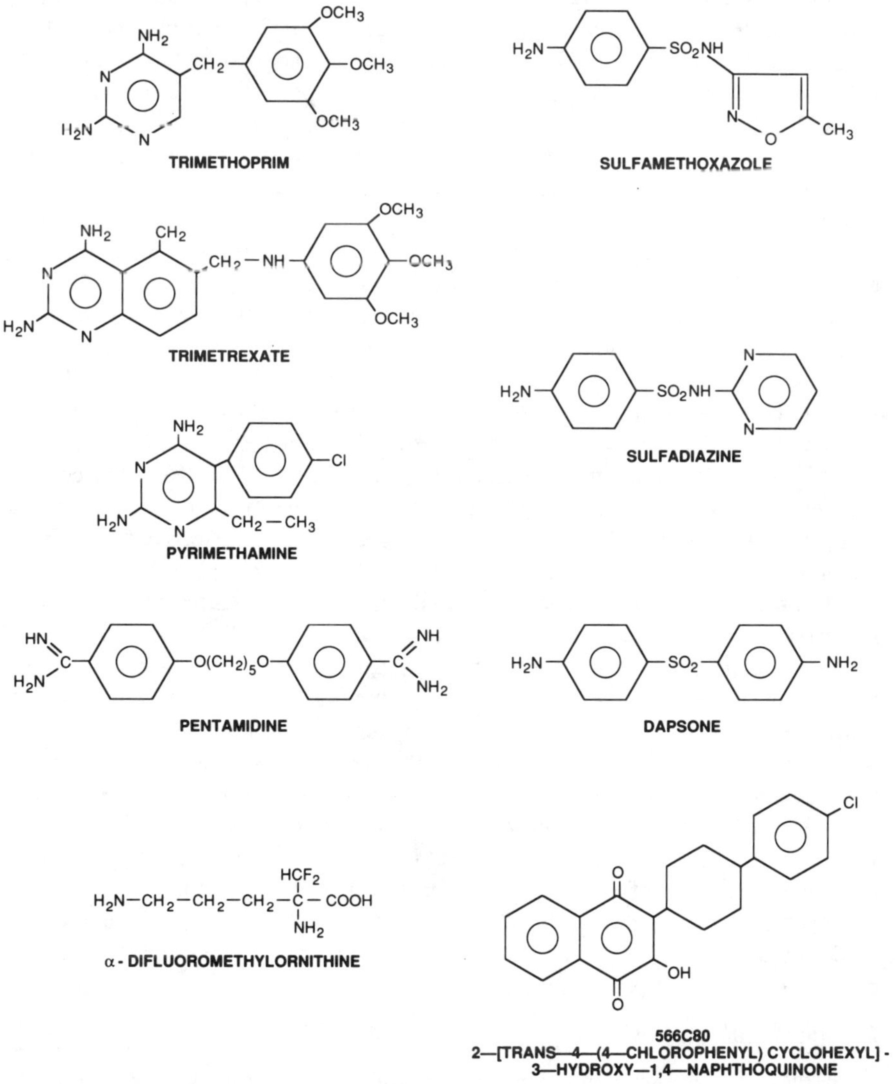

TRIMETHOPRIM

SULFAMETHOXAZOLE

TRIMETREXATE

SULFADIAZINE

PYRIMETHAMINE

PENTAMIDINE

DAPSONE

α - DIFLUOROMETHYLORNITHINE

566C80
2—[TRANS—4—(4—CHLOROPHENYL) CYCLOHEXYL] -
3—HYDROXY—1,4—NAPHTHOQUINONE

Figure 22.3. Structures of several established and investigational anti-*Pneumocystis* agents.

nation with trimethoprim (48) and also inhibited *P. carinii* DHPS, with a K_i of 59 μM (62), but was not active alone in an *in vitro* testing system based on exclusion of vital dyes (73). Sulfamethoxazole in combination with the DHFR inhibitor trimethoprim is the preferred initial therapy for most episodes of *P. carinii* pneumonia (52). Interestingly, sulfamethoxazole alone was very active in the rat model of *P. carinii* pneumonia as judged by histologic examination of lungs (95).

Sulfadiazine was effective in the rat model of *P. carinii* pneumonia whether used alone (95) or in combination with pyrimethamine (27), but the drug was a weak inhibitor of *P. carinii* DHPS (62). The literature abounds with uncontrolled trials and anecdotal reports of the clinical efficacy of pyrimethamine/sulfadiazine (24,47,102), but no rigorous trial has been published. Although sulfadiazine added little therapeutic benefit to the DHFR inhibitor trimetrexate in a recent clinical trial (1), its use was associated with a dramatically decreased incidence of relapse.

Clinical trials of the sulfone, dapsone, were undertaken following demonstration of its efficacy in the rat model of *P. carinii* pneumonia (42); dapsone was the most active of several antifungals, antimalarials, and other agents tested. The drug was also an effective inhibitor of PABA incorporation into reduced folates (48) and was the most potent inhibitor of *P. carinii* DHPS of five sulfa drugs tested (62). Dapsone in combination with trimethoprim appeared to be effective therapy for patients with mild first episodes of *P. carinii* pneumonia (53) but was less effective when administered alone (65). The combination was as effective but better tolerated than trimethoprim/sulfamethoxazole in a preliminary report of a randomized study (61). Dapsone was also shown to be an effective prophylactic agent in one preliminary, uncontrolled study (64).

Several other sulfonamides, sulfones, and related compounds have been studied *in vitro* and *in vivo* for activity against *P. carinii*. Sulfadoxine and sulfaquinoxaline were relatively poor inhibitors of crude *P. carinii* DHPS (62). Sulfadoxine was active in the rat model of *P. carinii* pneumonia as assessed by histologic scoring (95). The diformyl derivative of dapsone, 4,4′-sulfonylbisformanilide, was as effective as trimethoprim/sulfamethoxazole in experimental *P. carinii* pneumonia (43). This compound, which is metabolized to dapsone, appears to be a safe and effective antimalarial (14). Finally, a hypoglycemic sulfonylurea with known antibacterial activity, carbutamide, was effective therapy and prophylaxis in the rat model of *P. carinii* pneumonia (41). The large number of available sulfonamides, sulfones, and related compounds highlights the need for an *in vitro* method practical for large-scale drug screening, preferably employing the target enzyme of these compounds, DHPS.

C. Dihydrofolate Reductase

DHFR catalyzes the reduction of dihydrofolate to tetrahydrofolate (Figure 22.1), ultimately providing a single carbon donor for purine and thymidylate synthesis. DHFR is the target of the classical antifolate methotetrexate as well as the diaminopyrimidines, trimethoprim and pyrimethamine (Figure 22.3). *P. carinii* DHFR has a molecular weight of about 26,000 Da (49); its predicted size from cloning and sequencing of the DHFR gene is 23,868 Da (20). The K_m for dihydrofolate is 17.6 μM and that for NADPH is 40.2. Interestingly, trimethoprim and pyrimethamine are weak inhibitors of the enzyme (IC_{50}s of 39,600 nM and 2800 nM, respectively) (2), compared with methotrexate and its lipid soluble analog, trimetrexate (IC_{50}s of 1.4 nM and 26.1 nM, respectively).

As discussed above, trimethoprim in combination with sulfamethoxazole (83,98) or dapsone (53) has established clinical efficacy against *P. carinii* pneumonia. Numerous *in vitro* and *in vivo* studies support the use of trimethoprim/ sulfamethoxazole (5,18,42,79). Trimethoprim alone, however, appears to be relatively ineffective against *P. carinii* both *in vitro* and in the rat model (48,73,80,95); this is somewhat predictable given its poor inhibition of *P. carinii* DHFR (2).

Antifols with a more classic pteridine-like structure, such as methotrexate, are ineffective antibiotic agents despite being potent DHFR inhibitors; they require a folate-specific carrier for transmembrane transport which is generally not found in microorganisms, including *P. carinii* (Figure 22.1) (2). This suggests that a drug similar in structure to methotrexate but modified to be lipid-soluble in order to readily cross the cell membrane would be an effective drug against *P. carinii*. Trimetrexate, a lipid-soluble quinazoline analog of methotrexate, is such an agent (Figure 22.3). Because trimetrexate is a potent inhibitor of both mammalian and *P. carinii* DHFR, its administration can potentially cause severe host toxicity. An understanding of folate metabolism in *P. carinii* provided a mechanism for obviating this problem. Coadministration of calcium leucovorin, a reduced folate actively taken up by mammalian cells, can prevent host toxicity by bypassing the DHFR blockade. Anti–*P. carinii* activity of trimetrexate is preserved because *P. carinii* lack the folate-specific carrier required for uptake of this preformed folate.

In a preliminary trial trimetrexate produced a 63–71% response with or without sulfadiazine as initial or salvage therapy in AIDS patients with *P. carinii* pneumonia (1). As single-agent initial therapy, trimetrexate was associated with a high relapse rate (60%); no relapses were seen when the drug was coadministered with sulfadiazine. Its apparent safety makes trimetrexate a valuable therapeutic option in patients failing conventional therapy. Piritrexim isethionate, another lipid-soluble methotrexate analog, appears to be a potent inhibitor of *P. carinii* DHFR and was effective against the organism *in vitro,* although it was disappointing as a prophylactic agent in the rat model (50,80). In a recent pilot study, piritrexim was safe and effective therapy for *P. carinii* pneumonia in patients with AIDS, although its use (without a sulfonamide) was associated with a greater-than-expected number of early recurrences (23). Trimetrexate and piritrexim therapy require coadministration of leucovorin, as discussed above.

Rosowsky *et al.* (82) recently reported that several lipid-soluble 2,4-diamino-pyrimidine analogs of pyrimethamine have activity against *P. carinii in vitro* and are potent inhibitors of DHFR. These drugs have not been tried in the rat model of *P. carinii* pneumonia.

These emerging therapeutic strategies utilizing novel DHFR inhibitors take advantage of recently developed insights into folate metabolism and transport in *P. carinii,* as well as some lessons learned from cancer chemotherapy. Again, as new compounds become available, the need for a rapid, efficient system for *in vitro* screening becomes acutely apparent. The availability of recombinant DHFR (20) should enable such screening.

D. Thymidylate Synthase

The TS gene from *P. carinii* encodes a 297-amino-acid protein with a molecular weight of 34,269 Da (22). The deduced amino acid sequence has 65% homology with that of *S. cerevisiae,* further supporting the phylogenetic grouping of *P. carinii* with the fungi. While no selective inhibitors of TS are currently available, the ability to produce catalytically active enzyme provides a powerful tool for the development of such novel agents.

III. Other Targets

A. Diamidines

Pentamidine has numerous effects on protozoan metabolism (63), but the specific mode of action of the diamidines is not known (Figure 22.3). Pentamidine has been a mainstay of therapy for *P. carinii* pneumonia since 1958 when Ivady and Paldy (44) first reported successful treatment of interstitial plasma cell pneumonia in infants with this drug, then an established antitrypanosomal and antileishmanial agent. This represented the first report of effective therapy for confirmed *P. carinii* pneumonia. Pentamidine continues to be regarded as an alternative to trimethoprim/sulfamethoxazole as first-line therapy. The two drugs have been considered equally effective (34,51,86,98), but a recent prospective, randomized, noncrossover study in AIDS patients found that survival without mechanical ventilation was significantly higher in patients receiving trimethoprim/sulfamethoxazole (83). Aerosolization of pentamidine, a particularly convenient mode of delivering the drug, is earning a place in both the therapy and prophylaxis of *P. carinii* pneumonia (3,15,66).

Diminazene, an older diamidine still in use as a veterinary trypanocide, was as effective as pentamidine in the rat model of *P. carinii* pneumonia, as assessed by histologic scoring (94). Hydroxystilbamidine, another diamidine, was actually reported to be effective against *P. carinii* pneumonia in humans in the same study that originally described pentamidine therapy (44), but its potential toxicities have precluded its further development. Hydroxystilbamidine was also effective in treating and preventing pneumocystosis in the rat model (27). These findings, as well as their own preliminary study demonstrating that replacement of the amidine group of diamidines with imidazoline moieties resulted in compounds with enhanced antipneumocystis activity and reduced toxicity (90), prompted Jones *et al.* (45) to test several imidazoline-substituted pentamidine analogs against experimental *P. carinii* pneumonia. One of these compounds, 1,3-di(4-imidazolino-2-methoxyphenoxy)propane (DIMP), appeared to be more potent and no more toxic than pentamidine. These investigators also noted several structure-activity relationships of their substituted analogs which would allow production of a number of potentially useful new compounds; a rapid *in vitro* screening method would be ideal for the initial evaluation of these compounds.

B. Polyamine Metabolism

Ornithine decarboxylase catalyzes the rate-limiting step in the biosynthesis of the polyamines, trivially named putrescine, spermidine, and spermine. Inhibition of polyamine synthesis by the mechanism-based ornithine decarboxylase inhibitor DFMO (Figure 22.3) is effective therapy for West African trypanosomiasis (72). Based on the presumed protozoan phylogeny of *P. carinii,* Golden *et al.* (30) used DFMO in an uncontrolled trial as salvage therapy for *P. carinii* pneumonia in patients with AIDS. This trial and several other preliminary reports (10,60,69,70) have demonstrated some efficacy of the drug, but no large, controlled trial has been published.

DFMO has had equivocal efficacy in experimental *P. carinii* pneumonia (13,42,94) and *in vitro* against the organism (18). Ornithine decarboxylase activity has not been demonstrable in crude lysates of *P. carinii* under a variety of assay conditions (54a, 76). The organism does, however, synthesize polyamines *de novo* from ornithine, and this incorporation is blocked by clinically achievable levels of DFMO and an analog, monofluoromethyldehydroornithine. Several other substrate and product analogs of ornithine decarboxylase are available, but have not been tested *in vitro* or *in vivo*.

C. New Approaches to Chemotherapy

A variety of older as well as novel compounds have recently undergone *in vitro* or *in vivo* testing against *P. carinii,* and some potentially useful agents have emerged. Queener *et al.* (81) demonstrated that the antimalarial primaquine, in combination with the conventional antibiotic clindamycin, is effective therapy and prophylaxis for experimental *P. carinii* pneumonia as well as inhibiting growth of organisms in culture. The use of this combination was suggested by the enhanced antimalarial effect of primaquine when combined with mirincamycin, a lincosamide similar in structure to clindamycin (85); clindamycin had also been used successfully to treat toxoplasmosis in AIDS (68). A preliminary report of this combination as salvage therapy for *P. carinii* pneumonia documented improvement in all six patients enrolled (91).

Walzer *et al.* (94) tested a variety of drugs including the antimalarials quinine and quinacrine; these compounds had little or no activity in the rat model. Interestingly, several cationic trypanocides tested, including diminazene (discussed above), imidocarb, and amicarbalide, as well as quinapyramine and isometamidium, had activity greater than or equal to pentamidine. The former three drugs share structural features, including cationic charge, with the polyamines, and interference with polyamine metabolism has been proposed as the basis of their antiparasitic effect (4).

Based on the broad *in vitro* antiprotozoal effects of the electron-transport-inhibiting naphthoquinones (36), Hughes *et al.* (39) tried a hydroxynaphthoquinone, 566C80 (2-[trans-4-(4-chlorophenyl)cyclohexyl]-3-hydroxy-1,4-naphthoquinone; Figure 22.3), in experimental *P. carinii* pneumonia. The drug was at least as effective as trimethoprim/sulfamethoxazole for both therapy and prophylaxis of the infection. Clinical trials of 566C80 are now underway; the drug may have the unique advantage of having activity against other opportunistic infections, *e.g.*, toxoplasmosis.

9-Deazainosine, an inosine analog with *in vitro* activity against several parasitic protozoa, inhibited growth of cultured *P. carinii* (7) and was also effective in the rat model (87). Although assumed to be a nucleotide antimetabolite, the drug produced profound ultrastructural changes which led the investigators to hypothesize a specific effect on ion-water homeostasis.

Based on the demonstration that the *P. carinii* cyst wall contains 1,3-β-glucan and that the organism may contain β-glucan synthase (58,100), Schmatz *et al.* (84) recently evaluated several 1,3-β-glucan synthesis inhibitors in experimental *P. carinii* pneumonia. Two compounds, an echinocandin (L-671,329) and a papulacandin derivative (L-687,781), produced greater reductions in *P. carinii* cyst counts than trimethoprim/sulfamethoxazole, but the experimental design did not allow a difference in survival to become apparent. Therapy with these compounds or their analogs has the potential advantage of complete specificity for the pathogen and therefore low host toxicity; mammalian cells do not contain 1,3-β-glucan synthase (84).

IV. Summary and Future Prospects

The past 10 years have witnessed an explosion in the incidence of *P. carinii* pneumonia. It is only recently, however, that the elucidation of aspects of the metabolism of *P. carinii* and the application of the powerful methods of molecular biology to the study of the organism have begun to enable the development of novel, rational therapeutic approaches for this disease. The emergence of trimetrexate and dapsone as clinically useful agents attests to this. The availability of functional recombinant proteins such as dihydrofolate reductase should also prove valuable for screening and designing new therapies. One intriguing approach is suggested by the recent mapping of the topoisomerase I gene of *P. carinii* (56). Cloning and expression of this gene would provide a substrate for testing several classes of topoisomerase "poisons," including the camptothecins (55). Conceivably, newly described metabolic functions of *P. carinii,* including the presence of 1,3-β-glucan synthase and polyamine metabolism, will yield rational chemotherapeutic targets; novel inhibitors of these and other pathways are already under investigation. As our knowledge of the biology of this elusive pathogen broadens, we can expect dramatic improvements in our ability to treat and prevent the disease it causes.

References

1. Allegra, C.J., B.A. Chabner, C.U. Tuazon, D. Ogata-Arakaki, B. Baird, J.C. Drake, J.T. Simmons, E.E. Lack, J.H. Shelhamer, F. Balis, R. Walker, J.A. Kovacs, H.C. Lane, and H. Masur. 1987. Trimetrexate for the treatment of *Pneumocystis carinii* pneumonia in patients with the acquired immunodeficiency syndrome. N. Engl. J. Med. **317:**978–985.
2. Allegra, C.J., J.A. Kovacs, J.C. Drake, J.C. Swan, B.A. Chabner, and H. Masur. 1987. Activity of antifolates against *Pneumocystis carinii* dihydrofolate reductase and identification of a potent new agent. J. Exp. Med. **165:**926–931.
3. Armstrong, D., and E. Bernard, 1988. Aerosol pentamidine. Ann. Intern. Med. **109:**852–854.
4. Bacchi, C.J. 1981. Content, synthesis, and function of polyamines in trypanosomatids: relationship to chemotherapy. J. Protozool. **28:**20–27.
5. Bartlett, M.S., R. Eichholz, and J.W. Smith, 1985. Antimicrobial susceptibility of *Pneumocystis carinii* in culture. Diagn. Microbiol. Infect. Dis. **3:**381–387.
6. Bartlett, M.S., J.A. Fishman, S.F. Queener, M.M. Durkin, M.A. Jay, and J.W. Smith. 1988. New rat model of *Pneumocystis carinii* infection. J. Clin. Microbiol. **26:**1100–1102.
7. Bartlett, M.S., J.J. Marr, S.F. Queener, R.S. Klein, and J.W. Smith. 1986. Activity of inosine analogs against *Pneumocystis carinii* in culture. Antimicrob. Agents. Chemother. **30:**181–183.
8. Bartlett, M.S., S.F. Queener, M.A. Jay, M.M. Durkin, and J.W. Smith. 1988. Inhibition of *Pneumocystis* in culture by 8-aminoquinolines. Program Abstr. 28th Intersci. Conf. Antimicrob. Agents Chemother. (abstract 1023).
9. Bartlett, M.S., P.A. Verbanac, and J.W. Smith. 1979. Cultivation of *Pneumocystis carinii* with WI-38 cells. J. Clin. Microbiol. **10:**796–799.
10. Brennessel, D.J., and J.A. Goldstein. 1986. Eflornithine hydrochloride (DFMO) in the therapy (rx) of *Pneumocystis carinii* pneumonia (PCP) in AIDS. Abstracts of the Second International Conference on AIDS (abstract).
11. Centers for Disease Control. 1985. Update: acquired immunodeficiency syndrome—United States. MMWR **34:**245–248.

12. Centers for Disease Control. 1987. AIDS weekly surveillance report—United States. October 12, 1987.

13. Clarkson, A.B., D.E. Williams, and C. Rosenberg. 1988. Efficacy of DL-α-difluoromethylornithine in a rat model of *Pneumocystis carinii* pneumonia. Antimicrob. Agents Chemother. **32:**1158–1163.

14. Clyde, D.F., C.C. Robert, V.C. McCarthy, and R.M. Miller. 1971. Prophylaxis of malaria in man using the sulfones DFD and DDS alone and with chloroquine. Milit. Med. **136:**836–841.

15. Conte, J.E. Jr., H. Hollander, and J.A. Golden. 1987. Inhaled or reduced-dose intravenous pentamidine for *Pneumocystis carinii* pneumonia. A pilot study. Ann. Intern. Med. **107:**495–498.

16. Curran, J.W., W.M. Morgan, A.M. Hardy, H.W. Jaffe, W.W. Darrow, and W.R. Dowdle. 1985. The epidemiology of AIDS: current status and future prospects. Science **229:**1352–1357.

17. Cushion, M.T., and D. Ebbets. 1990. Growth and metabolism of *Pneumocystis carinii* in axenic culture. J. Clin. Microbiol. **28:**1385–1394.

18. Cushion, M.T., D. Stanforth, M.J. Linke, and P.D. Walzer. 1985. Method of testing the susceptibility of *Pneumocystis carinii* to antimicrobial agents in vitro. Antimicrob. Agents Chemother. **28:**796–801.

19. Cushion, M.T., and P.D. Walzer. 1984. Growth and serial passage of *Pneumocystis carinii* in the A549 cell line. Infect. Immun. **44:**245–251.

20. Edman, J.C., U. Edman, M. Cao, B. Lundgren, J.A. Kovacs, and D.V. Santi. 1989. Isolation and expression of the *Pneumocystis carinii* dihydrofolate reductase gene. Proc. Natl. Acad. Sci. USA **86:**8625–8629.

21. Edman, J.C., J.A. Kovacs, H. Masur, D.V. Santi, H.J. Elwood, and M.L. Sogin, 1988. Ribosomal RNA sequence shows *Pneumocystis carinii* to be a member of the fungi. Nature **334:**519–522.

22. Edman, U., J.C. Edman, B. Lundgren, and D.V. Santi. 1989. Isolation and expression of the *Pneumocystis carinii* thymidylate synthase gene. Proc. Natl. Acad. Sci. USA **86:**6503–6507.

23. Falloon, J., J. Kovacs, C. Allegra, D. O'Neill, C. Tuazon, P. Frame, M. Dohn, P. Joseph, I. Feuerstein, S. LaFon, M. Rogers, and H. Masur. 1990. A pilot study of piritrexim (PTX) with leucovorin (LCV) for the treatment of pneumocysis pneumonia. Abstracts of the Sixth International Conference on AIDS (abstract Th.B.399).

24. Farinas, E., and J.A. Quel. 1971. Hypertension and duodenal ulcer associated with *Pneumocystis carinii* pneumonitis and dysgammaglobulinemia. Ill. J. Med. **139:**138.

25. Fischl, M.A., G.M. Dickinson, and L. La Voie. 1988. Safety and efficacy of sulfamethoxazole and trimethoprim chemoprophylaxis for *Pneumocystis carinii* pneumonia in AIDS. JAMA **259:**1185–1189.

26. Fischl, M.A., D.D. Richman, M.H. Grieco, M.S. Gottlieb, P.A. Volberding, O.L. Laskin, J.M. Leedom, J.E. Groopman, D. Mildvan, R.T. Schooley, G.G. Jackson, D.T. Durack, D. King, and the AZT Collaborative Working Group. 1987. The efficacy of azidothymidine (AZT) in the treatment of patients with AIDS and AIDS-

related complex. A double-blind, placebo-controlled trial. N. Engl. J. Med. **317:**185–191.

27. Frenkel, J.K., J.T. Good, and J.A. Shultz. 1966. Latent *Pneumocystis* infection of rats, relapse, and chemotherapy. Lab. Invest. **15:**1559–1577.

28. Garrett, C.E., C.E. Coderre, T.D. Meek, E.P. Garvey, D.M. Claman, S.M. Beverley, and D.V. Santi. 1984. A bifunctional thymidilate synthetase–dihydrofolate reductase in protozoa. Mol. Biochem. Parasitol. **11:**257–265.

29. Girard, P.M., M. Brun Pascaud, R. Farinotti, L. Tamisier, and S. Kernbaum. 1987. Pentamidine aerosol in prophylaxis and treatment of murine *Pneumocystis carinii* pneumonia. Antimicrob. Agents Chemother. **31:**978–981.

30. Golden, J.A., A. Sjoerdsma, and D.V. Santi. 1984. *Pneumocystis carinii* pneumonia treated with alpha-difluoromethylornithine. A prospective study among patients with the acquired immunodeficiency syndrome. West. J. Med. **141:**613–623.

31. Gordin, F.M., G.L. Simon, C.B. Wofsy, and J. Mills. 1984. Adverse reactions to trimethoprim-sulfamethoxazole in patients with the acquired immunodeficiency syndrome. Ann. Intern. Med. **100:**495–499.

32. Haron, E., G.P. Bodey, M.A. Luna, R. Dekmezian, and L. Elting. 1988. Has the incidence of *Pneumocystis carinii* pneumonia in cancer patients increased with the AIDS epidemic? Lancet **2:**904–905.

33. Haverkos, H.W. 1984. Assessment of therapy for *Pneumocystis carinii* pneumonia. PCP Therapy Project Group. Am. J. Med. **76:**501–508.

34. Haverkos, H.W. 1987. Assessment of therapy for toxoplasma encephalitis. Am. J. Med. **82:**907–914.

35. Hopewell, P.C. 1988. *Pneumocystis carinii* pneumonia: diagnosis. J. Infect. Dis. **157:**1115–1119.

36. Hudson, A.T. 1988. Antimalarial hydroxynaphthoquinones, p. 266–283. *In* P.R. Leeming (ed.), Topics in medicinal chemistry. Special publication no. 65. Proceedings of the 4th SCI-RSC Medicinal Chemistry Symposium. Royal Society of Chemistry, London.

37. Hughes, W.T. 1984. Five year absence of *Pneumocystis carinii* in a pediatric oncology center. J. Infect. Dis. **150:**305–306 (letter).

38. Hughes, W.T., S. Feldman, and S.K. Sanyal. 1975. Treatment of *Pneumocystis carinii* pneumonitis with trimethoprim-sulfamethoxazole. Can. Med. Assoc. J. **112:**47–50.

39. Hughes, W.T., V.L. Gray, W.E. Gutteridge, V.S. Latter, and M. Pudney, 1990. Efficacy of a hydroxynaphthoquinose, 566C80, in experimental *Pneumocystis carinii* pneumonitis. Antimicrob. Agents Chemother. **34:**225–228.

40. Hughes, W.T., P.C. McNabb, T.D. Makres, and S. Feldman. 1974. Efficacy of trimethoprim and sulfamethoxazole in the prevention and treatment of *Pneumocystis carinii* pneumonitis. Antimicrob. Agents Chemother. **5:**289–293.

41. Hughes, W.T., and B.L. Smith McCain, 1986. Effects of sulfonylurea compounds on *Pneumocystis carinii*. J. Infect. Dis. **153:**944–497.

42. Hughes, W.T., and B.L. Smith. 1984. Efficacy of diaminodiphenylsulfone and other drugs in murine *Pneumocystis carinii* pneumonitis. Antimicrob. Agents Chemother. **26:**436–440.

43. Hughes, W.T., B.L. Smith, and D.P. Jacobus. 1986. Successful treatment and prevention of murine *Pneumocystis carinii* pneumonitis with 4,4′-sulfonylbisformanilide. Antimicrob. Agents Chemother. **29:**509–510.

44. Ivady, G., and L. Paldy. 1958. Ein neues behandlungsverfahren der interstitiellen plasmazelligen pneumonie fruhgeborener mit funfwertigen stibium und aromatischen diamidien. Moatsschr. Kinderheilkd. **106:**10.

45. Jones, S.K., J.E. Hall, M.A. Allen, S.D. Morrison, K.A. Ohemeng, V.V. Reddy, J.D. Geratz, and R.R. Tidwell. 1990. Novel pentamidine analogs in the treatment of experimental *Pneumocystis carinii* pneumonia. Antimicrob. Agents Chemother. **34:**1026–1030.

46. Kaneshiro, E.S., M.T. Cushion, P.D. Walzer, and K. Jayasimhulu. 1989. Analysis of *pneumocystis* fatty acids. J. Protozool. **36:**69S–70S.

47. Kirby, H.B., B. Kenamore, and J.C. Guckian. 1971. *Pneumocystis carinii* pneumonia treated with pyrimethamine and sulfadiazine. Ann. Intern. Med. **75:**505–509.

48. Kovacs, J.A., C.J. Allegra, J. Beaver, D. Boarman, M. Lewis J.E. Parrillo, B. Chabner, and H. Masur. 1989. Characterization of de novo folate synthesis in *Pneumocystis carinii* and *Toxoplasma gondii:* Potential for screening therapeutic agents. J. Infect. Dis. **160:**312–320.

49. Kovacs, J.A., C.J. Allegra, and H. Masur. 1990. Characterization of dihydrofolate reductase of *Pneumocystis carinii* and *Toxoplasma gondii*. Exp. Parasitol. **71:**60–68.

50. Kovacs, J.A., C.J. Allegra, J.C. Swan, J.C. Drake, J.E. Parrillo, B.A. Chabner, and H. Masur. 1988. Potent antipneumocystis and antitoxoplasma activities of piritrexim, a lipid-soluble antifolate. Antimicrob. Agents Chemother. **32:**430–433.

51. Kovacs, J.A., J.W. Hiemenz, A.M. Macher, D. Stover, H.W. Murray, J. Shelhamer, H.C. Lane, U. Urmacher, C. Honig, D.L. Longo, M.M. Parker, C. Natanson, J.E. Parrillo, A.S. Fauci, P.A. Pizzo, and H. Masur. 1984. *Pneumocystis carinii* pneumonia: A comparison between patients with the acquired immunodeficiency syndrome and patients with other immunodeficiencies. Ann. Intern. Med. **100:**663–671.

52. Kovacs, J.A., and H. Masur. 1988. *Pneumocystis carinii* pneumonia: therapy and prophylaxis. J. Infect. Dis. **158:**254–259.

53. Leoung, G.S., J. Mills, P.C. Hopewell, W. Hughes, and C. Wofsy. 1986. Dapsone-trimethoprim for *Pneumocystis carinii* pneumonia in the acquired immunodeficiency syndrome. Ann. Intern. Med. **105:**45–48.

54. Leoung, G.S., A.B. Montgomery, D.A. Abrams, K. Korkery, L. Wardlaw, and D.W. Feigal. 1988. Aerosol pentamidine for *Pneumocystis carinii* pneumonia: a randomized trial of 439 patients. Abstracts of the Fourth International Conference on AIDS (abstract 7166).

54a. Lipschik, G.Y., H. Masur, and J.A. Kovacs. 1991. Polyamine metabolism in *Pneumocystis carinii*. J. Inf. Dis. **163:**900–905.

55. Liu, L.F. 1989. DNA topoisomerase poisons as anti-tumor drugs. Ann. Rev. Biochem. **58:**351–375.

56. Lundgren, B., R. Cotton, J.D. Lundgren, J.C. Edman, and J.A. Kovacs. 1990. Identification of *Pneumocystis carinii* chromosomes and mapping of five genes. Infect. Immun. **58:**1705–1710.

57. Masur, H., F.P. Ognibene, R. Yarchoan, J.H. Shelhamer, B.F. Baird, W. Travis, A.F. Suffredini, L. Deyton, J.A. Kovacs, J. Falloon, R. Davey, M. Polis, J. Metcalf, M. Baseler, R. Wesley, V.J. Gill, A.S. Fauci, and H.C. Lane. 1989. CD4 counts as predictors of opportunistic pneumonias in human immunodeficiency virus (HIV) infection. Ann. Intern. Med. **111:**223–231.

58. Matsumoto, Y., S. Matsuda, and T. Tegoshi. 1990. Yeast glucan in the cell wall of *Pneumocystis carinii*. J. Protozool. **36:**21S–22S.

59. Mazer, M.A., J.A. Kovacs, J.C. Swan, J.E. Parrillo, and H. Masur. 1987. Histoenzymological study of selected dehydrogenase enzymes in *Pneumocystis carinii*. Infect. Immun. **55:**727–730.

60. McLees, B.D., J.L.R. Barlow, R.J. Kuzma, D.C. Baringtang, P.J. Schecter, and A. Sjoerdsma. 1987. Studies on successful eflornithine treatment of *Pneumocystis carinii* pneumonia in AIDS patients failing conventional therapy. Abstracts of the Third International Conference on AIDS (abstract Th.4.2).

61. Medina, I., G. Leoung, J. Mills, P. Hopewell, D. Feigal, and C. Wofsy. 1987. A randomized double bind trial of trimethoprim-sulfamethoxazole versus dapsone-trimethoprim for first episode *Pneumocystis carinii* pneumonia in AIDS. Program Abstr. 27th Intersci. Conf. Antimicrob. Agents Chemother. (abstract 941).

62. Merali, S., Y. Zhang, D. Sloan, and S. Meshnick. 1990. Inhibition of *Pneumocystis carinii* dihydropteroate synthetase by sulfa drugs. Antimicrob. Agents Chemother. **34:**1075–1078.

63. Meshnick, S.R. 1981. The chemotherapy of African trypanosomiasis, p. 165–199. *In* J.M. Mansfield (ed.), Parasitic diseases. Dekker, New York.

64. Metroka, C.E., N. Braun, H. Josefberg, and D. Jacobus. 1988. Successful chemoprophylaxis for *Pneumocystis carinii* pneumonia with dapsone in patients with AIDS and ARC. Abstracts of the Fourth International Conference on AIDS (abstract 7157).

65. Mills, J., G. Leoung, I. Medina, W. Hughes, P. Hopewell, and C. Wofsy. 1986. Dapsone is ineffective therapy for *Pneumocystis* pneumonia in patients with AIDS. Clin. Res. **34:**101A (abstract).

66. Montgomery, A.B., R.J. Debs, J.M. Luce, K.J. Corkery, J. Turner, E.N. Brunette, E.T. Lin, and P.C. Hopewell. 1987. Aerosolised pentamidine as sole therapy for *Pneumocystis carinii* pneumonia in patients with acquired immunodeficiency syndrome. Lancet **2:**480–483.

67. Murray, J.F., S.M. Garay, P.C. Hopewell, J. Mills, G.L. Snider, and D.G. Stover. 1987. NHLBI workshop summary: pulmonary complication of the acquired immune deficiency syndrome: an update. Am. Rev. Respir. Dis. **135:**504–509.

68. Navia, B.A., C.K. Petito, J.W.M. Gold, E.-S. Cho, B.D. Jordan, and R.W. Price. 1986. Cerebral toxoplasmosis complicating the acquired immune deficiency syn-

drome: clinical and neuropathological findings in 27 patients. Ann. Neurol. **19:**224–238.

69. Neibart, E., H.S. Sacks, G. Hammer, and S.Z. Hirschman. 1986. Difluoromethylornithine in the treatment of *Pneumocystis* pneumonia. Program Abstr. 26th Intersci. Conf. Antimicrob. Agents Chemother. (abstract 693).
70. Paulson, Y.J., T.M. Gilman, C.T. Boylen, O.P. Sharma, and P.N.R. Heseltine. 1986. Eflornithine treatment of *Pneumocystis carinii* pneumonia in patients failing other therapy. Program Abstr. 26th Intersci. Conf. Antimicrob. Agents Chemother. (abstract 694).
71. Pearson, R.D., and E.L. Hewlett. 1987. Use of pyrimethamine-sulfadoxine (Fansidar) in prophylaxis against chloroquine-resistant *Plasmodium falciparum* and *Pneumocystis carinii*. Ann. Intern. Med. **106:**714–718.
72. Pepin, J., F. Milford, C. Guern, and P.J. Schecter. 1987. Difluoromethylornithine for arseno-resistant *Trypanosoma brucei* Gambiensi sleeping sickness. Lancet **2:**1431–1433.
73. Pesanti, E.L. 1980. In vitro effects of antiprotozoan drugs and immune serum on *Pneumocystis carinii*. J. Infect. Dis. **141:**775–780.
74. Pesanti, E.L. 1984. *Pneumocystis carinii:* oxygen uptake, antioxidant enzymes, and susceptibility to oxygen-mediated damage. Infect. Immun. **44:**7–11.
75. Pesanti, E.L. 1989. Enzymes of *Pneumocystis carinii:* electrophoretic mobility on starch gels. J. Protozool. **36:**2S–3S.
76. Pesanti, E.L., M.S. Bartlett, and J.W. Smith. 1988. Lack of detectable activity of ornithine decarboxylase in *Pneumocystis carinii*. J. Infect. Dis. **158:**1137–1138 (letter).
77. Pesanti, E.L., and C. Cox. 1981. Metabolic and synthetic activities of *Pneumocystis carinii* in vitro. Infect. Immun. **34:**908–914.
78. Pifer, L.L., W.T. Hughes, and M.J. Murphy, Jr. 1977. Propagation of *Pneumocystis carinii* in vitro. Pediatr. Res. **11:**305–316.
79. Pifer, L.L., D.D. Pifer, and D.R. Woods. 1983. Biological profile and response to anti-pneumocystis agents of *Pneumocystis carinii* in cell culture. Antimicrob. Agents Chemother. **24:**674–678.
80. Queener, S.F., M.S. Bartlett, M.A. Jay, M.M. Durkin, and J.W. Smith. 1987. Activity of lipid-soluble inhibitors of dihydrofolate reductase against *Pneumocystis carinii* in culture and in a rat model of infection. Antimicrob. Agents Chemother. **31:**1323–1327.
81. Queener, S.F., M.S. Bartlett, J.D. Richardson, M.M. Durkin, M.A. Jay, and J.W. Smith. 1988. Activity of clindamycin with primaquine against *Pneumocystis carinii* in vitro and in vivo. Antimicrob. Agents. Chemother. **32:**807–813.
82. Rosowsky, A., J.H. Freisheim, J.B. Hynes, S.F. Queener, M. Bartlett, J.W. Smith, H. Lazarus, and E.J. Modest. 1989. Tricyclic 2,4-diaminopyrimidines with broad antifolate activity and the ability to inhibit *Pneumocystis carinii* growth in cultured human lung fibroblasts in the presence of leucovorin. Biochem. Pharmacol. **38:**2677–2684.

83. Sattler, F.R., R. Cowan, D.M. Nielsen, and J. Ruskin. 1988. Trimethoprim-sulfamethoxazole compared with pentamidine for treatment of *Pneumocystis carinii* pneumonia in the acquired immunodeficiency syndrome. A prospective, noncrossover study. Ann. Intern. Med. **109:**280–287.

84. Schmatz, D.M., M.A. Romancheck, L.A. Pittarelli, R.E. Schwartz, R.A. Fromtling, K.H. Nollstadt, F.L. Van middlesworth, K.E. Wilson, and M.J. Turner. 1990. Treatment of *Pneumocystis carinii* pneumonia with 1,3-β glucan synthesis inhibitors. Proc. Natl. Acad. Sci. USA **87:**5950–5954.

85. Schmidt, L.H. 1985. Enhancement of the curative activity of primaquine by concomitant adminstration of mirincamycin. Antimicrob. Agents Chemother. **27:**151–157.

86. Siegel, S.E., L.J. Wolff, R.L. Baehner, and D. Hammond. 1984. Treatment of *Pneumocystis carinii* pneumonitis. A comparative trial of sulfamethoxazole-trimethoprim v pentamidine in pediatric patients with cancer: report from the Children's Cancer Study Group. Am. J. Dis. Child. **138:**1051–1054.

87. Smith, J.W., M.S. Bartlett, S.F. Queener, M.M. Durkin, M.A. Jay, M.T. Hull, R.S. Klein, and J.J. Marr. 1987. *Pneumocystis carinii* pneumonia therapy with 9-deazainosine in rats. Diagn. Microbiol. Infect. Dis. **7:**113–118.

88. Sogin, M.L., and J.C. Edman. 1989. A self-splicing intron in the small subunit rRNA gene of *Pneumocystis carinii*. Nucleic Acids Res. **17:**5349–5359.

89. Stringer, S.L., J.R. Stringer, M.A. Blase, P.D. Walzer, and M.T. Cushion. 1989. *Pneumocystis carinii:* sequence from ribosomal RNA implies a close relationship with fungi. Exp. Parasitol. **68:**450–461.

90. Tidwell, R.R., S.K. Jones, J.D. Geratz, K.A. Ohemeng, M. Cory, and J.E. Hall. 1990. Analogues of 1,5-bis(4-amidinophenoxy)pentane (pentamidine) in the treatment of experimental *Pneumocystis carinii* pneumonia. J. Med. Chem. **33:**1252–1257.

91. Toma, E., S. Fournier, M. Poisson, R. Morisset, D. Phaneuf, and C. Vega. 1989. Clindamycin with primaquine for *Pneumocystis carinii* pneumonia. Lancet **1:**1046–1048.

92. ul Haque, A., S.B. Plattner, R.T. Cook, and M.N. Hart. 1987. *Pneumocystis carinii*. Taxonomy as viewed by electron microscopy. Am. J. Clin. Pathol. **87:**504–510.

93. Walker, A.N., R.E. Garner, and M.N. Horst. 1990. Immunocytochemical detection of chitin in *Pneumocystis carinii*. Infect. Immun. **58:**412–415.

94. Walzer, P.D., C.K. Kim, J. Foy, M.J. Linke, and M.T. Cushion. 1988. Cationic antitrypanosomal and other antimicrobial agents in the therapy of experimental *Pneumocystis carinii* pneumonia Antimicrob. Agents Chemother. **32:**896–905.

95. Walzer, P.D., C.K. Kim, J.M. Foy, M.J. Linke, and M.T. Cushion. 1988. Inhibitors of folic acid synthesis in the treatment of experimental *Pneumocystis carinii* pneumonia. Antimicrob. Agents Chemother. **32:**96–103.

96. Walzer, P.D., D.J. Krogstad, P.G. Rawson, and M.G. Schultz. 1974. *Pneumocystis carinii* pneumonia in the United States: epidemiologic, diagnostic, and clinical features. Ann. Intern. Med. **80:**83–93.

97. Watanabe, J., H. Hori, K. Tanabe, and Y. Nakamura. 1989. Phylogenetic association of *Pneumocystis carinii* with the 'Rhizopoda/Myxomycota/Zygomycota group' indi-

cated by comparison of 5S ribosomal RNA sequences. Mol. Biochem. Parasitol. **32:**163–167.

98. Wharton, J.M., D.L. Coleman, C.B. Wofsy, J.M. Luce, W. Blumenfeld, W.K. Hadley, L. Ingram Drake, P.A. Volberding, and P.C. Hopewell. 1986. Trimethoprim-sulfamethoxazole or pentamidine for *Pneumocystis carinii* pneumonia in the acquired immunodeficiency syndrome. A prospective randomized trial. Ann. Intern. Med. **105:**37–44.

99. Wilber, R.B., S. Feldman, W.J. Malone, M. Ryan, R.J. Aur, and W.T. Hughes. 1980. Chemoprophylaxis for *Pneumocystis carinii* pneumonitis: outcome of unstructured delivery. Am. J. Dis. Child. **134:**643–648.

100. Williams, D.J., J.A. Radding, M.Y.K. Armstrong, and F.F. Richards. 1990. *Pneumocystis carinii* contains a beta-glucan synthesizing activity. Clin. Res. **38:**363A (abstract).

101. Yoganathan, T., H. Lin, and G.A. Buck. 1989. An electrophoretic karotype and assignment of ribosomal genes to resolved chromosomes of *Pneumocystis carinii*. Mol. Microbiol. **3:**1473–1480.

102. Young, R.C., and V.T. Devita. Jr. 1976. Treatment of *Pneumocystis carinii* pneumonia: current status of the regimens of pentamidine isethionate and pyrimethamine-sulfadiazine. Natl. Cancer Inst. Monogr. **43:**193–200.

Index

Page references to chemical structures or figures are given in **boldface**

E

M

T

U